Code of Federal Regulations

CODE OF FEDERAL REGULATIONS

T0175184

Title 7
Agriculture

Parts 1600 to 1759

Revised as of January 1, 2022

Containing a codification of documents
of general applicability and future effect

As of January 1, 2022

Published by the Office of the Federal Register
National Archives and Records Administration
as a Special Edition of the Federal Register

Table of Contents

Cite this Code: CFR

To cite the regulations in this volume use title, part and section number. Thus, 7 CFR 1700.1 *refers to title 7, part 1700, section 1.*

Explanation

The Code of Federal Regulations is a codification of the general and permanent rules published in the Federal Register by the Executive departments and agencies of the Federal Government. The Code is divided into 50 titles which represent broad areas subject to Federal regulation. Each title is divided into chapters which usually bear the name of the issuing agency. Each chapter is further subdivided into parts covering specific regulatory areas.

Each volume of the Code is revised at least once each calendar year and issued on a quarterly basis approximately as follows:

Title 1 through Title 16..........as of January 1
Title 17 through Title 27..........as of April 1
Title 28 through Title 41..........as of July 1
Title 42 through Title 50..........as of October 1

The appropriate revision date is printed on the cover of each volume.

LEGAL STATUS

The contents of the Federal Register are required to be judicially noticed (44 U.S.C. 1507). The Code of Federal Regulations is prima facie evidence of the text of the original documents (44 U.S.C. 1510).

HOW TO USE THE CODE OF FEDERAL REGULATIONS

The Code of Federal Regulations is kept up to date by the individual issues of the Federal Register. These two publications must be used together to determine the latest version of any given rule.

To determine whether a Code volume has been amended since its revision date (in this case, January 1, 2022), consult the "List of CFR Sections Affected (LSA)," which is issued monthly, and the "Cumulative List of Parts Affected," which appears in the Reader Aids section of the daily Federal Register. These two lists will identify the Federal Register page number of the latest amendment of any given rule.

EFFECTIVE AND EXPIRATION DATES

Each volume of the Code contains amendments published in the Federal Register since the last revision of that volume of the Code. Source citations for the regulations are referred to by volume number and page number of the Federal Register and date of publication. Publication dates and effective dates are usually not the same and care must be exercised by the user in determining the actual effective date. In instances where the effective date is beyond the cutoff date for the Code a note has been inserted to reflect the future effective date. In those instances where a regulation published in the Federal Register states a date certain for expiration, an appropriate note will be inserted following the text.

OMB CONTROL NUMBERS

The Paperwork Reduction Act of 1980 (Pub. L. 96–511) requires Federal agencies to display an OMB control number with their information collection request.

Many agencies have begun publishing numerous OMB control numbers as amendments to existing regulations in the CFR. These OMB numbers are placed as close as possible to the applicable recordkeeping or reporting requirements.

PAST PROVISIONS OF THE CODE

Provisions of the Code that are no longer in force and effect as of the revision date stated on the cover of each volume are not carried. Code users may find the text of provisions in effect on any given date in the past by using the appropriate List of CFR Sections Affected (LSA). For the convenience of the reader, a "List of CFR Sections Affected" is published at the end of each CFR volume. For changes to the Code prior to the LSA listings at the end of the volume, consult previous annual editions of the LSA. For changes to the Code prior to 2001, consult the List of CFR Sections Affected compilations, published for 1949-1963, 1964-1972, 1973-1985, and 1986-2000.

"[RESERVED]" TERMINOLOGY

The term "[Reserved]" is used as a place holder within the Code of Federal Regulations. An agency may add regulatory information at a "[Reserved]" location at any time. Occasionally "[Reserved]" is used editorially to indicate that a portion of the CFR was left vacant and not dropped in error.

INCORPORATION BY REFERENCE

What is incorporation by reference? Incorporation by reference was established by statute and allows Federal agencies to meet the requirement to publish regulations in the Federal Register by referring to materials already published elsewhere. For an incorporation to be valid, the Director of the Federal Register must approve it. The legal effect of incorporation by reference is that the material is treated as if it were published in full in the Federal Register (5 U.S.C. 552(a)). This material, like any other properly issued regulation, has the force of law.

What is a proper incorporation by reference? The Director of the Federal Register will approve an incorporation by reference only when the requirements of 1 CFR part 51 are met. Some of the elements on which approval is based are:

(a) The incorporation will substantially reduce the volume of material published in the Federal Register.

(b) The matter incorporated is in fact available to the extent necessary to afford fairness and uniformity in the administrative process.

(c) The incorporating document is drafted and submitted for publication in accordance with 1 CFR part 51.

What if the material incorporated by reference cannot be found? If you have any problem locating or obtaining a copy of material listed as an approved incorporation by reference, please contact the agency that issued the regulation containing that incorporation. If, after contacting the agency, you find the material is not available, please notify the Director of the Federal Register, National Archives and Records Administration, 8601 Adelphi Road, College Park, MD 20740-6001, or call 202-741-6010.

CFR INDEXES AND TABULAR GUIDES

A subject index to the Code of Federal Regulations is contained in a separate volume, revised annually as of January 1, entitled CFR INDEX AND FINDING AIDS. This volume contains the Parallel Table of Authorities and Rules. A list of CFR titles, chapters, subchapters, and parts and an alphabetical list of agencies publishing in the CFR are also included in this volume.

An index to the text of "Title 3—The President" is carried within that volume.

The Federal Register Index is issued monthly in cumulative form. This index is based on a consolidation of the "Contents" entries in the daily Federal Register.

A List of CFR Sections Affected (LSA) is published monthly, keyed to the revision dates of the 50 CFR titles.

REPUBLICATION OF MATERIAL

There are no restrictions on the republication of material appearing in the Code of Federal Regulations.

INQUIRIES

For a legal interpretation or explanation of any regulation in this volume, contact the issuing agency. The issuing agency's name appears at the top of odd-numbered pages.

For inquiries concerning CFR reference assistance, call 202–741–6000 or write to the Director, Office of the Federal Register, National Archives and Records Administration, 8601 Adelphi Road, College Park, MD 20740-6001 or e-mail *fedreg.info@nara.gov.*

THIS TITLE

Title 7—AGRICULTURE is composed of fifteen volumes. The parts in these volumes are arranged in the following order: Parts 1–26, 27–52, 53–209, 210–299, 300–399, 400–699, 700–899, 900–999, 1000–1199, 1200–1599, 1600–1759, 1760–1939, 1940–1949, 1950–1999, and part 2000 to end. The contents of these volumes represent all current regulations codified under this title of the CFR as of January 1, 2022.

The Food and Nutrition Service current regulations in the volume containing parts 210–299 include the Child Nutrition Programs and the Food Stamp Program. The regulations of the Federal Crop Insurance Corporation are found in the volume containing parts 400–699.

All marketing agreements and orders for fruits, vegetables and nuts appear in the one volume containing parts 900–999. All marketing agreements and orders for milk appear in the volume containing parts 1000–1199.

For this volume, Gabrielle E. Burns was Chief Editor. The Code of Federal Regulations publication program is under the direction of John Hyrum Martinez, assisted by Stephen J. Frattini.

Title 7—Agriculture

(This book contains parts 1600 to 1759)

SUBTITLE B—REGULATIONS OF THE DEPARTMENT OF AGRICULTURE (CONTINUED)

Subtitle B—Regulations of the Department of Agriculture (Continued)

CHAPTER XVI [RESERVED]

CHAPTER XVII—RURAL UTILITIES SERVICE, DEPARTMENT OF AGRICULTURE

EDITORIAL NOTE: Nomenclature changes to chapter XVII appear at 59 FR 66440, Dec. 27, 1994.

PART 1700—GENERAL INFORMATION

Subpart A—General

Subpart B—Agency Organization and Functions

Subpart C—Loan and Grant Approval Authorities

Subpart D—Substantially Underserved Trust Areas

AUTHORITY: 5 U.S.C. 301, 552; 7 U.S.C. 901 *et seq.*, 1921 *et seq.*, 6941 *et seq.;* 7 CFR 2.7.

SOURCE: 63 FR 16085, Apr. 2, 1998, unless otherwise noted.

Subpart A—General

§ 1700.1 General.

(a) The Rural Electrification Administration (REA) was established by Executive Order No. 7037 on May 11, 1935. Statutory authority was provided by the Rural Electrification Act of 1936 (RE Act) (7 U.S.C. 901). The RE Act established REA as a lending agency with responsibility for developing a program for rural electrification.

(b) The Secretary of Agriculture (Secretary) established the Rural Utilities Service (RUS) on October 20, 1994, pursuant to the Department of Agriculture Reorganization Act of 1994, (7 U.S.C. 6941 *et. seq.*). RUS was assigned responsibility for administering electric and telecommunications loan and loan guarantee programs previously administered by REA, including water and waste loans and grants previously administered by the Rural Development Administration, along with other functions as the Secretary determined appropriate. The rights, interests, obligations, duties, and contracts previously vested in REA were transferred to, and vested in RUS.

[63 FR 16085, Apr. 2, 1998, as amended at 84 FR 59920, Nov. 7, 2019]

§ 1700.2 Availability of information.

(a) The offices of RUS are located in the South Building of the United States Department of Agriculture at 1400 Independence Avenue, SW, Washington, DC 20250–1500. Hours of operation are from 8:15 AM to 4:45 PM, Eastern time on Federal Government business days.

(b) Information about RUS is available for public inspection and copying as required by the Freedom of Information Act, 5 U.S.C. 552 *et seq.* Information about availability and costs of agency publications and other agency materials is available from the Director, Program Development and Regulatory Analysis, Rural Utilities Service, United States Department of Agriculture, Room 5159–S, 1400 Independence Avenue, SW., STOP 1530, Washington, DC 20250–1530. Phone 202–720–9450. FAX 202–720–8435.

(c) RUS issues indexes of publications in conformance with the Freedom of

Information Act and Department of Agriculture regulations at 7 CFR part 1. Many RUS documents, including regulations and delegations of authority for headquarters and field staff are available on the world wide web at *http://www.usda.gov/rus.*

[63 FR 16085, Apr. 2, 1998, as amended at 71 FR 8435, Feb. 17, 2006]

§ 1700.3 Requests under the Freedom of Information Act.

Department of Agriculture procedures for requests for records under the Freedom of Information Act are found at 7 CFR part 1. Requests must be in writing and may be submitted in person or by mail to United States Department of Agriculture, Rural Utilities Service, Room 5159–S, 1400 Independence Avenue, SW., STOP 1530, Washington, DC 20250–1530; or by FAX to 202–401–1977. As set forth in 7 CFR 1.16, fees may be charged for processing of requests for records. An appeal of the agency determination concerning the request for official records shall be made in writing to the Administrator, Rural Utilities Service, United States Department of Agriculture, Room 5135–S, 1400 Independence Avenue, SW., STOP 1510, Washington, DC 20250–1510.

[71 FR 8435, Feb. 17, 2006]

§ 1700.4 Public comments on proposed rules.

RUS requires that all persons submitting comments to a proposed rule or other document published by the agency in the FEDERAL REGISTER, submit comments as specified in the published notice. Copies of comments submitted are available to the public in conformance with 7 CFR part 1.

[71 FR 8435, Feb. 17, 2006]

§§ 1700.5–1700.24 [Reserved]

Subpart B—Agency Organization and Functions

§ 1700.25 Office of the Administrator.

The Administrator of the Rural Utilities Service (RUS) is appointed by the President. The Under Secretary, Rural Development delegated to the Administrator, in 7 CFR part 2, responsibility for administering the programs and activities of RUS. The Administrator is aided directly by Deputy Administrators and by Assistant Administrators for the electric program, telecommunications program, the water and environmental programs, and program accounting and regulatory analysis, and by other staff offices. The work of the agency is carried out as described in this part.

[79 FR 44117, July 30, 2014, as amended at 84 FR 59920, Nov. 7, 2019]

§ 1700.26 Deputy Administrator.

The Deputy Administrator aids and assists the Administrator. The Deputy Administrator provides overall policy direction to all RUS programs. The Deputy Administrator reviews agency policies and, as necessary, implements changes and participates with the Administrator and other officials in planning and formulating the programs and activities of the agency, including the making and servicing of loans and grants.

[79 FR 44117, July 30, 2014]

§ 1700.27 Chief of Staff.

The Chief of Staff aids and assists the Administrator and the Deputy Administrator. The Chief of Staff advises the Administrator regarding policy initiatives and operational issues and assists the Administrator and the Deputy Administrator in developing and planning agency program initiatives. The Chief of Staff is responsible for implementation of overall policy initiatives and provides direction to all RUS programs.

[71 FR 8436, Feb. 17, 2006]

§ 1700.28 Electric Program.

RUS, through the Electric Program, makes loans and loan guarantees for rural electrification and the furnishing of electric service to persons in rural areas.

(a) *The Assistant Administrator, Electric Program,* directs and coordinates the rural electrification programs, participating with the Administrator, and others, in planning and formulating the programs and activities of the agency, and performs other activities as the Administrator may prescribe from time to time.

(b) *Primary point of contact with borrowers.* Two regional divisions, one for the Northern Region and one for the Southern Region, are the primary points of contact between RUS and its electric distribution borrowers. Each office administers the rural electric program for its assigned geographical area through headquarters staff and general field representatives. The Power Supply Division is the primary point of contact between RUS and its electric power supply borrowers.

(c) *Staff office.* The Electric Staff Division is responsible for engineering aspects of RUS' standards, specifications and other requirements for design, construction, and technical operation and maintenance of RUS borrowers' electric systems. The Electric Staff Division oversees the activities of Technical Standards Committees "A" and "B", Electric, which determine whether engineering specifications, drawings, material and equipment are acceptable for use in RUS borrowers' electric systems. The Office of the Assistant Administrator prepares analyses of loan making activities and the business and regulatory environment of RUS borrowers and recommends policies and procedures.

[63 FR 16085, Apr. 2, 1998. Redesignated at 71 FR 8436, Feb. 17, 2006]

§ 1700.29 Telecommunications Program.

RUS, through the Telecommunications Program, make loans and loan guarantees to furnish and improve telecommunications service in rural areas.

(a) *The Assistant Administrator, Telecommunications Program,* directs and coordinates the rural telecommunications programs, including the distance learning and telemedicine program, and in conjunction with the Administrator and Deputy Administrator, and others, the planning and formulating of programs and activities of the agency, and performs other activities as the Administrator may prescribe from time to time.

(b) *Primary point of contact with borrowers.* Area offices are the primary points of contact between RUS and all telecommunications program borrowers. Each office administers the rural telecommunications program for its assigned geographical area with assistance of field representatives located in areas assigned to them.

(c) *Staff offices.* The Telecommunications Staff Division is responsible for engineering aspects of design, construction, and technical operation and maintenance of rural telecommunications systems and facilities, including the activities of Technical Standards Committees "A" and "B", Telecommunications, which determine whether engineering specifications, drawings, material, and equipment are acceptable for use in RUS financed telecommunications systems. The Advanced Telecommunications Services office prepares analyses of loan making activities and the business and regulatory environment of RUS borrowers and recommends policies and procedures.

[63 FR 16085, Apr. 2, 1998; 63 FR 18307, Apr. 15, 1998. Redesignated at 71 FR 8436, Feb. 17, 2006, as amended at 84 FR 59920, Nov. 7, 2019]

§ 1700.30 Water and Environmental Programs.

RUS, through the Water and Environmental Programs, provides loan and grant funds for water and waste disposal projects serving the most financially needy rural communities.

(a) *The Assistant Administrator, Water and Environmental Programs,* develops and institutes plans, procedures, and policies for the effective, efficient, and orderly management of Water and Environmental Programs responsibilities; provides leadership to ensure execution of policies and procedures by the Water and Waste Disposal programs and support functions; and performs other activities as the Administrator or Deputy Administrator may prescribe from time to time.

(b) *Primary point of contact.* The State Rural Development Offices are the primary points of contact between RUS and loan and grant recipients.

(c) *The Engineering and Environmental Staff* is responsible for engineering staff activities at all stages of Water and Waste Disposal programs implementation, including review of preliminary engineering plans and specifications, procurement practices, contract awards, construction monitoring, and system operation and maintenance.

This staff develops agency engineering practices, policies, guidelines, and technical data relating to the construction and operation of water and waste disposal systems, and for implementing the National Environmental Policy Act, and other environmental requirements as they apply to all agency programs and activities.

[63 FR 16085, Apr. 2, 1998. Redesignated at 71 FR 8436, Feb. 17, 2006]

§ 1700.31 Distance Learning and Telemedicine Loan and Grant Program.

RUS, through the Telecommunications Program, makes grants and loans to furnish and improve telemedicine services and distance learning services in rural areas.

(a) *The Assistant Administrator, Telecommunications Program,* directs and coordinates the distance learning and telemedicine program.

(b) *Primary point of contact with borrowers.* The area offices, described in §1700.28(b) support the distance learning and telemedicine program. Each office administers the distance learning and telemedicine program for its assigned geographical area with assistance of field representatives located in areas assigned to them.

[63 FR 16085, Apr. 2, 1998; 63 FR 18307, Apr. 15, 1998. Redesignated at 71 FR 8436, Feb. 17, 2006]

§ 1700.32 Program Accounting and Regulatory Analysis.

RUS, through Program Accounting and Regulatory Analysis, monitors and administers applicable regulations, RUS policy, and accounting requirements. The staffs assist the Assistant Administrator with respect to management, information systems, budgets, and other such matters.

(a) *The Assistant Administrator, Program Accounting and Regulatory Analysis,* directs and coordinates program accounting and financial services with respect to electric and telecommunications borrowers and directs and coordinates the regulatory actions of the agency.

(b) This division monitors borrowers' accounting operations in order to ensure compliance with applicable statutory and regulatory requirements and with the requirements of the Office of Management and Budget.

(c) The two regional branches (the Northern Region and the Southern Region) work directly with borrowers. Each regional office has a staff of headquarters and field accountants. The Technical Accounting and Auditing Staff monitors industry developments, including the standards of the Financial Accounting Standards Board, and recommends Agency policies and procedures.

(d) Program Development and Regulatory Analysis directs and administers the preparation, clearance, processing, and distribution of RUS submissions to the Office of the Federal Register in the form of proposed and final rules and notices and RUS bulletins and staff instructions.

[63 FR 16085, Apr. 2, 1998. Redesignated at 71 FR 8436, Feb. 17, 2006]

§ 1700.33 Financial Services Staff.

The Financial Services Staff evaluates the financial condition of financially troubled borrowers in order to protect the Government's interests.

[63 FR 16085, Apr. 2, 1998. Redesignated at 71 FR 8436, Feb. 17, 2006]

§ 1700.34 Assistance to High Energy Cost Rural Communities.

RUS, through the Electric Program, makes grants and loans to assist high energy cost rural communities. The Assistant Administrator, Electric Program, directs and coordinates the assistance to high energy cost rural communities program and serves as the primary point of contact for applicants, grantees, and borrowers.

[70 FR 5351, Feb. 2, 2005. Redesignated at 71 FR 8436, Feb. 17, 2006]

§§ 1700.35–1700.49 [Reserved]

Subpart C—Loan and Grant Approval Authorities

§§ 1700.50–1700.52 [Reserved]

§ 1700.54 Electric Program.

(a) *Administrator:* The authority to approve the following loans, loan guarantees, and lien accommodations and subordinations of liens is reserved to the Administrator:

(1) All discretionary hardship loans.

(2) All loans, loan guarantees, and lien accommodations and subordinations of liens to finance operating costs.

(3) All loans, loan guarantees, and lien accommodations and subordinations of liens of more than $20,000,000 for distribution borrowers or more than $50,000,000 for power supply borrowers.

(4) All loans, loan guarantees, and lien accommodations and subordinations of liens for distribution borrowers that are members of a power supply borrower that is in default of its obligations to the Government or that is currently assigned to the Financial Services Staff, unless otherwise determined by the Administrator.

(5) All loans, loan guarantees, and lien accommodations and subordinations of liens that require an Environmental Impact Statement.

(6) Certifications and findings required by the RE Act or other applicable laws and regulations, the placing and releasing of conditions precedent to the advance of funds, and all security instruments, loan contracts, and all other necessary documents relating to the authorities reserved in this section.

(7) Execution of all loan contracts, security instruments, and all other documents in connection with loans, loan guarantees, and lien accommodations approved by the Administrator.

(b) *The Assistant Administrator, Electric Program,* has the authority to approve the following loans, loan guarantees, and lien accommodations and subordinations of liens, except for those approvals reserved to the Administrator:

(1) Loans, loan guarantees, and lien accommodations and subordinations of liens for distribution borrowers in amounts not exceeding $20,000,000.

(2) Loans, loan guarantees, and lien accommodations and subordinations of liens for power supply borrowers in amounts not exceeding $50,000,000.

(3) Execution of all loan contracts, security instruments, and all other documents in connection with loans, loan guarantees, and lien accommodations approved by the Assistant Administrator, Electric Program.

(c) *Directors, Regional Divisions,* have the authority to approve, for distribution borrowers:

(1) Loans, loan guarantees, and lien accommodations and subordinations of liens in amounts not exceeding $15,000,000 except for those approvals reserved to the Administrator.

(2) All certifications and findings required by the RE Act or other applicable laws and regulations, the imposing and releasing of conditions precedent to the advance of loan funds, and all security instruments, loan contracts, and all other documents relating to the delegations set forth in paragraph (c)(1) of this section.

(d) *Director, Power Supply Division,* has the authority to approve for power supply borrowers:

(1) Loans, loan guarantees, and lien accommodations and subordinations of liens in amounts not exceeding $30,000,000, except for those approvals reserved to the Administrator.

(2) All certifications and findings required by the RE Act or other applicable laws and regulations, the placing and releasing of conditions precedent to the advance of funds, and all security instruments, loan contracts or all other documents relating to the delegations set forth in paragraph (d)(1) of this section.

§ 1700.55 Telecommunications Program.

(a) *Administrator:* The authority to approve the following loans, loan guarantees, and lien accommodations is reserved to the Administrator:

(1) All loans, loan guarantees, and lien accommodations and subordinations of liens to finance operating costs.

(2) All loans, loan guarantees, or lien accommodations and subordinations of liens of $25,000,000 or more.

(3) Loans and loan guarantees with acquisition costs of $5,000,000 or more.

(4) Loans and loan guarantees containing funds to refinance outstanding debt of more than $5,000,000.

(5) All loan contracts, security instruments, and all other documents to be executed in connection with loans

and loan guarantees approved by the Administrator.

(b) *Assistant Administrator, Telecommunications Program,* has the authority to approve the following loans, loan guarantees, and lien accommodations, except for those approvals reserved to the Administrator:

(1) Loans, loan guarantees, and lien accommodations and subordinations of liens not to exceed $25,000,000 except for those reserved to the Administrator.

(2) Loans and loan guarantees with acquisition costs where the acquisition portion of the loan is less than $5,000,000.

(3) Loans and loan guarantees including refinancing amounts that do not exceed $5,000,000.

(4) Distance learning and telemedicine loans and loan guarantees that do not exceed $5,000,000.

(5) Loan contracts, security instruments, and other documents to be executed in connection with loans and loan guarantees approved by the Assistant Administrator, Telecommunications Program.

(c) *Area Directors* have the authority to approve the following loans, loan guarantees, and lien accommodations, except for those approvals reserved to the Administrator:

(1) Loans, loan guarantees, and lien accommodations and subordinations of liens of less than $10,000,000.

(2) Loans and loan guarantees with acquisition costs of less than $2,000,000.

(3) Loans and loan guarantees including refinancing amounts of less than $2,000,000.

(4) Any modifications in the method of carrying out loan purposes.

§ 1700.56 Water and Environmental Programs.

The State Rural Development Offices have the responsibility for making and servicing water and waste loans and grants.

§ 1700.57 Distance Learning and Telemedicine Loan and Grant Program.

(a) *Administrator:* The authority to approve the following loans and lien accommodations is reserved to the Administrator:

(1) Grants or loan and grant combinations.

(2) The number selected from each state for financial assistance for grant approval and loans or grants approved.

(3) Extension of principal and interest repayments for rural development purposes.

(4) Loan contracts, security instruments, and all other documents to be executed in connection with loans and loan guarantees approved by the Administrator.

(b) *Assistant Administrator, Telecommunications Program,* has the authority to approve the following loans and lien accommodations and subordinations of liens:

(1) Loans, that do not also include requests for grant funds, except for those reserved to the Administrator.

(2) Loan contracts, security instruments, and all other documents to be executed in connection with loans and loan guarantees approved by the Assistant Administrator, Telecommunications Program.

§ 1700.58 Assistance to high energy cost rural communities.

(a) *Administrator:* The authority to approve the following is reserved to the Administrator:

(1) Allocation of appropriated funds among high energy cost community assistance programs;

(2) Awards of grants and loans to extremely high energy cost communities;

(3) Awards of grants and loans to the Denali Commission;

(4) Awards of grants to State entities for State bulk fuel revolving funds; and

(5) Grant agreements, loan contracts, security instruments and all other documents executed in connection with grants and loans agreements approved by the Administrator.

(b) *The Assistant Administrator, Electric Program* has the authority to make any required certifications and to approve all grant and loan servicing actions not specifically reserved to the Administrator.

[70 FR 5351, Feb. 3, 2005]

§§ 1700.59–1700.99 [Reserved]

Subpart D—Substantially Underserved Trust Areas

SOURCE: 77 FR 35250, June 13, 2012, unless otherwise noted.

§ 1700.100 Purpose.

This subpart establishes policies and procedures for the Rural Utilities Service (RUS) implementation of the Substantially Underserved Trust Areas (SUTA) initiative under section 306F of the Rural Electrification Act of 1936, as amended (7 U.S.C. 906f). The purpose of this rule is to identify and improve the availability of eligible programs in communities in substantially underserved trust areas.

§ 1700.101 Definitions.

Administrator means the Administrator of the Rural Utilities Service, or designee or successor.

Applicant means an entity that is eligible for an eligible program under that program's eligibility criteria.

Borrower means any organization that has an outstanding loan or loan guarantee made by RUS for a program purpose.

Completed application means an application that includes the elements specified by the rules for the applicable eligible program in form and substance satisfactory to RUS.

ConAct means the Consolidated Farm and Rural Development Act, as amended (7 U.S.C. 1921 *et seq.*).

Credit support means equity, cash requirements, letters of credit, and other financial commitments provided in support of a loan or loan guarantee.

Eligible community means a community as defined by 7 CFR 1700.103.

Eligible program means a program as defined by 7 CFR 1700.102.

Financial assistance means a grant, combination loan and grant, loan guarantee or loan.

Financial feasibility means the ability of a project or enterprise to meet operating expenses, financial performance metrics, such as debt service coverage requirements and return on investment, and the general ability to repay debt and sustain continued operations at least through the life of the RUS loan or loan guarantee.

Matching fund requirements means the applicant's financial or other required contribution to the project for approved purposes.

Nonduplication generally means a restriction on financing projects for services in a geographic area where reasonably adequate service already exists as defined by the applicable program.

Project means the activity for which financial assistance has been provided.

RE Act means the Rural Electrification Act of 1936, as amended (7 U.S.C. 901 *et seq.*).

RUS means the Rural Utilities Service, an agency of the United States Department of Agriculture, successor to the Rural Electrification Administration.

Substantially underserved trust area means a community in trust land with respect to which the Administrator determines has a high need for the benefits of an eligible program.

Trust land means "trust land" as defined in section 3765 of title 38, United States Code as determined by the Administrator under 7 CFR 1700.104.

Underserved means an area or community lacking an adequate level or quality of service in an eligible program, including areas of duplication of service provided by an existing provider where such provider has not provided or will not provide adequate level or quality of service.

§ 1700.102 Eligible programs.

SUTA does not apply to all RUS programs. SUTA only applies to eligible programs. An eligible program means a program administered by RUS and authorized in paragraph (a) of the RE Act, or paragraphs (b)(1), (2), (14), (22), or (24) of section 306(a) (7 U.S.C. 1926(a)(1), (2), (14), (22), (24)), or sections 306A, 306C, 306D, or 306E of the Con Act (7 U.S.C. 1926a, 1926c, 1926d, 1926e).

§ 1700.103 Eligible communities.

An eligible community is a community that:

(a) Is located on Trust land;

(b) May be served by an RUS administered program; and

(c) Is determined by the Administrator as having a high need for benefits of an eligible program.

§ 1700.104 Financial feasibility.

Pursuant to normal underwriting practices, and such reasonable alternatives within the discretion of RUS that contribute to a financial feasibility determination for a particular eligible program or project, the Administrator will only make grants, loans and loan guarantees that RUS finds to be financially feasible and that provide eligible program benefits to substantially underserved trust areas. All income and assets available to and under the control of the Applicant will be considered as part of the Applicant's financial profile.

§ 1700.105 Determining whether land meets the statutory definition of "trust land."

The Administrator will use one or more of the following resources in determining whether a particular community is located in Trust land:

(a) Official maps of Federal Indian Reservations based on information compiled by the U. S. Department of the Interior, Bureau of Indian Affairs and made available to the public;

(b) Title Status Reports issued by the U. S. Department of the Interior, Bureau of Indian Affairs showing that title to such land is held in trust or is subject to restrictions imposed by the United States;

(c) Trust Asset and Accounting Management System data, maintained by the Department of the Interior, Bureau of Indian Affairs;

(d) Official maps of the Department of Hawaiian Homelands of the State of Hawaii identifying land that has been given the status of Hawaiian home lands under the provisions of section 204 of the Hawaiian Homes Commission Act, 1920;

(e) Official records of the U.S. Department of the Interior, the State of Alaska, or such other documentation of ownership as the Administrator may determine to be satisfactory, showing that title is owned by a Regional Corporation or a Village Corporation as such terms are defined in the Alaska Native Claims Settlement Act (43 U.S.C. 1601 *et seq*);

(f) Evidence that the land is located on Guam, American Samoa or the Commonwealth of the Northern Mariana Islands, and is eligible for use in the Veteran's Administration direct loan program for veterans purchasing or constructing homes on communally-owned land; and

(g) Any other evidence satisfactory to the Administrator to establish that the land is "trust land" within the meaning of 38 U.S.C. 3765(1).

§ 1700.106 Discretionary provisions.

(a) To improve the availability of eligible programs in eligible communities determined to have a high need for the benefits of an eligible program, the Administrator retains the discretion, on a case-by-case basis, to use any of the following SUTA authorities individually or in combination to:

(1) Make available to qualified applicants financing with an interest rate as low as 2 percent;

(2) Extend repayment terms;

(3) Waive (individually or in combination) non-duplication restrictions, matching fund requirements, and credit support requirements from any loan or grant program administered by RUS; and

(4) Give the highest funding priority to designated projects in substantially underserved trust areas.

(b) Requests for waivers of non-duplication restrictions, matching fund requirements, and credit support requirements, and requests for highest funding priority will be reviewed on a case-by-case basis upon written request of the applicant filed pursuant to 7 CFR 1700.108.

(c) Notwithstanding the requirements in paragraph (b) of this section, the Administrator reserves the right to evaluate any application for an eligible program for use of the discretionary provisions of this subpart without a formal, written request from the applicant.

§ 1700.107 Considerations relevant to the exercise of SUTA discretionary provisions.

(a) In considering requests to make available financing with an interest rate as low as 2 percent, and extended repayment terms, the Administrator will evaluate the effect of and need for such terms on the finding of financial feasibility.

(b) In considering a request for a non-duplication waiver, the Administrator will consider the offerings of all existing service providers to determine whether or not granting the non-duplication waiver is warranted. A waiver of non-duplication restrictions will not be given if the Administrator determines as a matter of financial feasibility that, taking into account all existing service providers, an applicant or RUS borrower would not be able to repay a loan or successfully implement a grant agreement. Requests for waivers of non-duplication restrictions will be reviewed by taking the following factors into consideration:

(1) The size, extent and demographics of the duplicative area;

(2) The cost of service from existing service providers;

(3) The quality of available service; and

(4) The ability of the existing service provider to serve the eligible service area.

(c) Requests for waivers of matching fund requirements will be evaluated by taking the following factors into consideration:

(1) Whether waivers or reductions in matching or equity requirements would make an otherwise financially infeasible project financially feasible;

(2) Whether permitting a matching requirement to be met with sources not otherwise permitted in an affected program due to regulatory prohibition may be allowed under a separate statutory authority; and

(3) Whether the application could be ranked and scored as if the matching requirements were fully met.

(d) Requests for waivers of credit support requirements will be evaluated taking the following factors into consideration:

(1) The cost and availability of credit support relative to the loan security derived from such support;

(2) The extent to which the requirement is shown to be a barrier to the applicant's participation in the program; and

(3) The alternatives to waiving the requirements.

(e) The Administrator may adapt the manner of assigning highest funding priority to align with the selection methods used for particular programs or funding opportunities.

(1) Eligible programs which use priority point scoring may, in a notice of funds availability or similar notice, assign extra points for SUTA eligible applicants as a means to exercise a discretionary authority under this subpart.

(2) The Administrator may announce a competitive grant opportunity focused exclusively or primarily on trust lands which incorporates one or more discretionary authorities under this subpart into the rules or scoring for the competition.

§ 1700.108 Application requirements.

(a) To receive consideration under this subpart, the applicant must submit to RUS a completed application that includes all of the information required for an application in accordance with the regulations relating to the program for which financial assistance is being sought. In addition, the applicant must notify the RUS contact for the applicable program in writing that it seeks consideration under this subpart and identify the discretionary authorities of this subpart it seeks to have applied to its application. The required written request memorandum or letter must include the following items:

(1) A description of the applicant, documenting eligibility.

(2) A description of the community to be served, documenting eligibility in accordance with 7 CFR 1700.103.

(3) An explanation and documentation of the high need for the benefits of the eligible program, which may include:

(i) Data documenting a lack of service (i.e. no service or unserved areas) or inadequate service in the affected community;

(ii) Data documenting significant health risks due to the fact that a significant proportion of the community's residents do not have access to, or are not served by, adequate, affordable service.

(iii) Data documenting economic need in the community, which may include:

(A) Per capita income of the residents in the community, as documented by the U.S. Department of Commerce, Bureau of Economic Analysis;

(B) Local area unemployment and not-employed statistics in the community, as documented by the U.S. Department of Labor, Bureau of Labor Statistics and/or the U.S. Department of the Interior, Bureau of Indian Affairs;

(C) Supplemental Nutrition Assistance Program participation and benefit levels in the community, as documented by the U.S. Department of Agriculture, Economic Research Service;

(D) National School Lunch Program participation and benefit levels in the community, as documented by the U.S. Department of Agriculture, Food and Nutrition Service;

(E) Temporary Assistance for Needy Families Program participation and benefit levels in the community, as documented by the U.S. Department of Health and Human Services, Administration for Children and Families;

(F) Lifeline Assistance and Link-Up America Program participation and benefit levels in the community, as documented by the Federal Communications Commission and the Universal Service Administrative Company;

(G) Examples of economic opportunities which have been or may be lost without improved service.

(H) Data maintained and supplied by Indian tribes or other tribal or jurisdictional entities on "trust land" to the Department of Interior, the Department of Health and Human Services and the Department of Housing and Urban Development that illustrates a high need for the benefits of an eligible program.

(4) The impact of the specific authorities sought under this subpart.

(b) The applicant must provide any additional information RUS may consider relevant to the application which is necessary to adequately evaluate the application under this subpart.

(c) RUS may also request modifications or changes, including changes in the amount of funds requested, in any proposal described in an application submitted under this subpart.

(d) The applicant must submit a completed application within the application window and guidelines for an eligible program.

§ 1700.109 RUS review.

(a) RUS will review the application to determine whether the applicant is eligible to receive consideration under this subpart and whether the application is timely, complete, and responsive to the requirements set forth in 7 CFR 1700.107.

(b) If the Administrator determines that the application is eligible to receive consideration under this subpart and one or more SUTA requests are granted, the applicant will be so notified.

(c) If RUS determines that the application is not eligible to receive further consideration under this subpart, RUS will so notify the applicant. The applicant may withdraw its application or request that RUS treat its application as an ordinary application for review, feasibility analysis and service area verification by RUS consistent with the regulations and guidelines normally applicable to the relevant program.

§§ 1700.110–1700.149 [Reserved]

§ 1700.150 OMB Control Number.

The reporting and recordkeeping requirements contained in this part have been approved by the Office of Management and Budget and have been assigned OMB control number 0572–0147.

PART 1703—RURAL DEVELOPMENT

Subparts A–B [Reserved]

Subpart C—Rural Business Incubator Program [Reserved]

Subparts D–G [Reserved]

Subpart H—Deferments of RUS Loan Payments for Rural Development Projects

AUTHORITY: 7 U.S.C. 901 *et seq.*

SOURCE: 54 FR 6870, Feb. 15, 1989, unless otherwise noted. Redesignated at 55 FR 39394, Sept. 27, 1990.

Subparts A–B [Reserved]

Subpart C—Rural Business Incubator Program [Reserved]

§§ 1703.80–1703.99 [Reserved]

Subparts D–G [Reserved]

Subpart H—Deferments of RUS Loan Payments for Rural Development Projects

SOURCE: 58 FR 21639, Apr. 23, 1993, unless otherwise noted. Redesignated at 64 FR 14356, Mar. 25, 1999.

§ 1703.300 Purpose.

This subpart H sets forth RUS' policies and procedures for making loan deferments of principal and interest payments on direct loans or insured loans made for electric or telephone purposes, but not for loans made for rural economic development purposes, in accordance with subsection (b) of section 12 of the RE Act. Loan deferments are provided for the purpose of promoting rural development opportunities.

[82 FR 55925, Nov. 27, 2017]

§ 1703.301 Policy.

It is RUS's policy to encourage borrowers to invest in and promote rural development and rural job creation projects that are based on sound economic and financial analyses. Borrowers are encouraged to use this program to promote economic, business and community development projects that will benefit rural areas.

§ 1703.302 Definitions and rules of construction.

(a) *Definitions*. For the purpose of this subpart, the following terms will have the following meanings:1

Administrator means the Administrator of RUS.

Borrower means any organization which has an outstanding direct loan or insured loan made by RUS for the provision of electric or telephone service.

Cushion of credit payment means a voluntary unscheduled payment on an RUS note made after October 1, 1987, credited to the cushion of credit account of a borrower.

Deferment means a re-amortization of a payment of principal and/or interest on an RUS direct loan or insured loan for over either a 5- or 10 year period, with the first payment beginning on the date of the deferment.

Direct loan means a loan that is made by the Administrator pursuant to section 4 or section 201 of the RE Act (7 U.S.C. 901 *et seq.*) for the provision of electric or telephone service in rural areas and does not include a loan made to promote economic development in rural areas.

Financially distressed borrower means an RUS-financed borrower determined by the Administrator to be either:

(i) In default or near default on interest or principal payments due on loans made or guaranteed under the RE Act;

(ii) A borrower that was in default or near default, but is currently participating in a workout or debt restructuring plan with RUS; or

(iii) Experiencing a financial hardship.

Insured loan means a loan that is made, held, and serviced by the Administrator, and sold and insured by the Administrator, pursuant to Section 305 of the RE Act (7 U.S.C. 901 *et seq.*) for the provision of electric or telephone service in rural areas and does not include a loan made to promote economic development in rural areas.

Job creation means the creation of jobs in rural areas, or in close enough proximity to rural areas so that it is likely that the majority of the jobs created will be held by residents of rural areas.

Project means a rural development project that a borrower proposes and the Administrator approves as qualifying under this subpart.

RE Act means the Rural Electrification Act of 1936, as amended (7 U.S.C. 901 *et seq.*).

REA means the Rural Electrification Administration formerly an agency of the United States Department of Agriculture and predecessor agency to RUS with respect to administering certain electric and telephone loan programs.

RTB means the Rural Telephone Bank (telephone bank), a body corporate and an instrumentality of the United States, that obtains supplemental funds from non-Federal sources and utilizes them in making loans, operating on a self-sustaining basis to the extent practicable (section 401, RE Act).

RUS means the Rural Utilities Service, an agency of the United States Department of Agriculture established pursuant to Section 232 of the Federal Crop Insurance Reform and Department of Agriculture Reorganization Act of 1994 (Pub. L. 103–354, 108 Stat. 3178), successor to REA with respect to administering certain electric and telephone programs. See 7 CFR 1700.1.

Technical assistance means market research, product or service improvement, feasibility studies, environmental studies, and similar activities that benefit rural development or rural job creation projects.

(b) *Rules of construction.* Unless the context otherwise indicates; "includes" and "including" are not limiting, and "or" is not exclusive. The terms defined in § 1703.302(a) include both the plural and the singular.

[58 FR 21639, Apr. 23, 1993, as amended at 59 FR 66440, Dec. 27, 1994]

§ 1703.303 Eligibility criteria for deferment of loan payments.

The deferment of loan payments may be granted to any borrower that is not financially distressed, delinquent on any Federal debt, or in bankruptcy proceedings. However, the deferment of loan payments will not be granted to a borrower during any period in which the Administrator has determined that no additional financial assistance of any nature should be provided to the borrower pursuant to any provision of the RE Act. The determination to suspend eligibility for the deferment of loan payments under this subpart will be based on:

(a) The borrower's demonstrated unwillingness to exercise diligence in repaying loans made by RUS or RTB or guaranteed by RUS that results in the Administrator being unable to find that such loans, would be repaid within the time agreed; or

(b) The borrower's demonstrated unwillingness to meet the requirements in RUS's or RTB's legal documents or regulations.

§ 1703.304 Restrictions on the deferment of loan payments.

(a) The deferment must not impair the security of any loans made RUS or RTB, or guaranteed by RUS, pursuant to the RE Act.

(b) At no point in time may the amount of the debt service payments deferred exceed 50 percent of the total cost of a community, business, or economic development project for which a deferment is provided.

(c) A borrower may defer debt service payments only in an amount equal to the investment made by such borrower in a rural development project. The investment must not be made from:

(1) Proceeds of loans made or guaranteed pursuant to the RE Act, or grants made pursuant to the RE Act or section 2331 through section 2335A of the Rural Economic Development Act of 1990 (7 U.S.C. 950aaa *et seq.*);

(2) Funds necessary to make timely payments of principal and interest on loans made, guaranteed or lien accommodated pursuant to the RE Act;

(3) Insurance proceeds from mortgaged property;

(4) Damage awards and sale proceeds resulting from eminent domain and similar proceedings involving mortgaged property;

(5) Sale proceeds from mortgaged property sales requiring specific Administrator approval; and

(6) Funds which are restricted by RUS or RTB loan instruments to be held in trust for the Government or to be held for any other specific purpose.

(d) Any investment made in a rural development project prior to the date of the application for a deferment based on such project cannot be used to satisfy the requirements of this section.

§ 1703.305 Requirements for deferment of loan payments.

(a) Except as otherwise provided in paragraph (b) of this section, the borrower must make a cushion of credit payment equal to the amount of the payment deferred and subject to the following rules:

(1) Cushion of credit payments made prior to the date that an application for deferral has been approved by RUS cannot be used to satisfy the requirements of this section;

(2) Once a cushion of credit payment has been made to satisfy the requirements of paragraph (a) of this section, it must remain on deposit in the cushion of credit account on the date of the deferral or the deferral will not take place; and

(3) The cushion of credit payment must be received by RUS on the date the payment being deferred is due, or within 30 days prior to this date.

(b) A borrower may elect to consolidate in one application filed pursuant to § 1703.311, all of the related deferrals it wishes to receive in a twelve month period following application approval. In such a case, the requirement contained in paragraph (a)(1) of this section may alternatively be satisfied by depositing an amount equal to the aggregate deferrals covered by such application into the cushion of credit account at the time the first cushion of credit payment is due under paragraph (a)(1) of this section.

§ 1703.306 Limitation on funds derived from the deferment of loan payments.

Funds derived from the deferment of loan payments will not be used:

(a) To fund or assist projects which would, in the judgement of the Administrator, create a conflict of interest or the appearance of a conflict of interest. The borrower must disclose to the Administrator information regarding any potential conflict of interest or appearance of a conflict of interest;

(b) For any purpose not reasonably related to the project as determined by the Administrator;

(c) To transfer existing employment or business activities from one area to another; or

(d) For the borrower's electric or telephone operations, nor for any operations affiliated with the borrower unless the Administrator has specifically informed the borrower in writing that the affiliated operations are part of the approved purposes.

§ 1703.307 Uses of the deferments of loan payments.

The deferment of loan payments will be made to enable the borrower to provide funding and assistance for rural development and job creation projects. This includes, but is not limited to, the borrower providing financing to local businesses, community development assistance, technical assistance to businesses, and other community, business, or economic development projects that will benefit rural areas.

§ 1703.308 Amount of deferment funds available.

(a) The total amount of deferments made available for each fiscal year under this program will not exceed 3 percent of the total payments due during fiscal year 1993 from all borrowers on direct loans and insured loans made under the RE Act. For each subsequent fiscal year after 1993, the total amount of deferments will not exceed 5 percent of the total payments due for the year from all borrowers on direct loans and insured loans.

(b) The total amount of annual deferments are subject to limitations established by appropriations Acts.

§ 1703.309 Terms of repayment of deferred loan payments.

(a) Deferments made to enable the borrower to provide financing to local businesses will be repaid over a period of 60 months, in equal installments, with payments beginning on the date of the deferment, and continuing in such a manner until the total amount of the deferment is repaid. The deferment payments will be made on either a monthly or quarterly basis depending on the existing repayment terms of the direct loan or insured loan being deferred. The deferment will not accrue interest.

(b) In the case of deferments made to enable the borrower to provide community development assistance, technical assistance to businesses, and for other community, business, or economic development projects not included in paragraph (a) of this section, the deferment will be repaid over a period of 120 months, in equal installments, with payments beginning on the date of the deferment and continuing in such a manner until the total amount of the deferment is repaid. The deferment payments will be made on either a monthly or quarterly basis depending on the existing repayment terms of the direct loan or insured loan being deferred. The deferment will not accrue interest.

(c) The maturity date of a loan may not be extended as a result of a deferment.

(d) If the required payment is not made by the borrower or received by the Administrator when due, the Administrator will reduce the borrower's cushion of credit account established under this subpart in an amount equal to the deferment payment required.

(e) The balance in a borrower's cushion of credit account shall not be reduced by the borrower below the level of the unpaid balance of the payment deferred.

§ 1703.310 Environmental considerations.

Prospective recipients of funds received from the deferment of loan payments are encouraged to consider the potential environmental impact of their proposed projects at the earliest planning stage and plan development in a manner that reduces, to the extent practicable, the potential to affect the quality of the human environment adversely.

§ 1703.311 Application procedures for deferment of loan payments.

(a) A borrower applying for a deferment must:

(1) Submit a certified board resolution to the Administrator requesting a deferment of principal and interest. The resolution must:

(i) Be signed by the president or vice president of the borrower;

(ii) Contain information on the total amount of deferment requested for each specific project;

(iii) Contain information on the type of project and the length of deferment requested as defined in § 1703.309; and

(iv) Specify which officer of the borrower has been given the authority to certify to those matters required in this section;

(2) Submit certification by the appropriate officer to the Administrator that the proposed project will not violate the limitations set forth in § 1703.306 and disclose all information regarding any potential conflict of interest or appearance of a conflict of interest that would allow the Administrator to make an informed decision;

(3) Submit certification by the appropriate officer to the Administrator that an investment in the rural development project will be made by the borrower in an amount equal to the deferred debt service payment;

(4) Submit certification by the appropriate officer to the Administrator that the amount of the deferment will not exceed 50 percent of the total cost of the project for which the deferment is provided;

(5) Submit certification by the appropriate officer to the Administrator that it will make a cushion of credit payment necessary to satisfy the requirement of § 1703.305(a);

(6) Submit certification by the appropriate officer to the Administrator that it will comply with § 1703.313 and provide documentation showing that its total investments, including the proposed investment, will not exceed the investment limitations specified in

7 CFR part 1717, Subpart N, Investments, Loans and Guarantees by Electric Borrowers, or 7 CFR Part 1744, Post Loan Policies and Procedures Common to Guaranteed and Insured Loans. The documentation must provide a list of each rural development project the borrower has invested in to date, including the investment amounts;

(7) Submit to the Administrator written identification of the direct loan(s) and/or insured loan(s) for which payments are to be deferred;

(8) Submit to the Administrator a written narrative which contains information regarding the proposed rural development or job creation project such as the manner in which the project will promote community, business, or economic development in rural areas, the nature of the project, its location, the primary beneficiaries, and, if applicable, the number and type of jobs to be created; and

(9) Submit to the Administrator a letter of approval from the state regulatory authority, if applicable, granting its approval for the borrower to defer direct loan payment(s) and/or insured loan payment(s) and invest the amount in a rural development project.

(b) The Administrator reserves the right to determine that special circumstances require additional data from borrowers before acting on a deferment. The Administrator also reserves the right to require, as a condition of approving a loan payment deferment pursuant to this subpart, that the borrower execute and deliver any amendments or supplements to its loan documents that may be necessary or appropriate to achieve the purposes outlined in §1703.300.

(c) The Administrator will decide whether the borrower is eligible for the deferment and will notify the borrower of the decision.

§1703.312 RUS review requirements.

Borrowers shall ensure that funds are invested in the rural development project as approved by RUS. The Administrator reserves the right to review the books and copy records of borrowers receiving loan payment deferments as necessary to ensure that the investments in the rural development project are in accordance with this subpart and the representations and purposes stated in the borrower's completed application. If an audit discloses that the amount deferred was not used for the purposes stated in the completed application, the borrower shall be required to promptly repay the amount deferred and the benefits of the deferment to the borrower will be recaptured by RUS. The borrower is responsible for ensuring that disbursements and expenditures of funds covering the investment in the rural development project are properly supported with certifications, invoices, contracts, bills of sale, cancelled checks, or any other forms of evidence determined appropriate by the Administrator and that such supporting material is available at the borrower's premises for review by the RUS field accountant, borrower's certified public accountant, the Office of Inspector General, the General Accounting Office and any other accountant conducting an audit of the borrower's financial statements for this rural development program.

§1703.313 Compliance with other regulations.

(a) Investments in a rural economic development project made by an electric borrower under this subpart are subject to the provisions of 7 CFR part 1717, Subpart N, Investments, Loans and Guarantees by Electric Borrowers.

(b) Investments in a rural economic development project made by a telephone borrower under this subpart are subject to the provisions of 7 CFR Part 1744, Post Loan Policies and Procedures Common to Guaranteed and Insured Loans.

PART 1709—ASSISTANCE TO HIGH ENERGY COST COMMUNITIES

Subpart A—General Requirements

Sec.

Subpart B—RUS High Cost Energy Grant Program

Subpart C—Bulk Fuel Revolving Fund Grant Program

Subparts D–F [Reserved]

Subpart G—Recovery of Financial Assistance Used for Unauthorized Purposes

AUTHORITY: 5 U.S.C. 301, 7 U.S.C. 901 *et seq.*

SOURCE: 70 FR 5351, Feb. 2, 2005, unless otherwise noted.

Subpart A—General Requirements

§ 1709.1 Purpose.

The purpose of the Rural Utilities Service (RUS) Assistance to High Energy Cost Rural Communities Program is to help local communities meet their energy needs through direct loans and grants for energy facilities in qualifying extremely high energy cost communities, grants and loans to the Denali Commission for extremely high energy cost communities in Alaska, and grants to States to support revolving funds to finance more cost effective means of acquiring fuel in qualifying communities. This subpart sets forth definitions and requirements which are common to all grant and loan programs in this part administered by the RUS Electric Program under section 19 of the Rural Electrification Act of 1936, as amended (RE Act) (7 U.S.C. 918a).

§ 1709.2 [Reserved]

§ 1709.3 Definitions.

Administrator means the Administrator of the Rural Utilities Service (RUS), United States Department of Agriculture (USDA).

Agency means the Rural Utilities Service (RUS), an agency of the United States Department of Agriculture (USDA), or a successor agency.

Census block means the smallest geographic entity for which the U.S. Census Bureau collects and tabulates decennial census information and which are defined by boundaries shown on census maps.

Census designated place (CDP) means a statistical entity recognized by the

U.S. Census Bureau comprising a dense concentration of population that is not within an incorporated place but is locally identified by a name and which has boundaries defined on census maps.

Electric program means the office within RUS, and its successor organization, that administers rural electrification programs authorized by the Rural Electrification Act of 1936 (RE Act) (7 U.S.C. 901 *et seq.*) and such other programs so identified in USDA regulations.

Extremely high energy costs means community average residential energy costs that are at least 275 percent of one or more home energy cost benchmarks identified by RUS and based on the latest available information on national average residential energy expenditures as reported by the Energy Information Administration (EIA) of the United States Department of Energy.

Financial assistance means a grant, loan, or grant-loan combination issued under this part.

Funding opportunity announcement (FOA) means a publicly available document by which a Federal agency makes know its intentions to award discretionary grants or cooperative agreements, usually as a result of competition for funds. FOA announcements may be known as program announcements, notices of funding availability, solicitations, or other names depending on the agency and type of program. FOA announcements can be found at *www.Grants.gov* in the Search Grants tab and on the funding agency's or program's website.

Home energy means any energy source or fuel used by a household for purposes other than transportation, including electricity, natural gas, fuel oil, kerosene, liquified petroleum gas (propane), other petroleum products, wood and other biomass fuels, coal, wind and solar energy. Fuels used for subsistence activities in remote rural areas are also included.

High energy cost benchmarks means the criteria established by RUS for eligibility as an extremely high energy cost community. Extremely high energy cost benchmarks are calculated as 275 percent of the relevant national average household energy benchmarks.

Indian Tribe means a Federally recognized tribe as defined under section 4 of the Indian Self-Determination and Education Assistance Act (25 U.S.C. 450b) to include "* * * any Indian tribe, band, nation, or other organized group or community, including any Alaska Native village or regional or village corporation as defined in or established pursuant to the Alaska Native Claims Settlement Act (43 U.S.C. 1601 *et seq.*), that is recognized as eligible for the special programs and services provided by the United States to Indians because of their status as Indians."

Person means any natural person, firm, corporation, association, or other legal entity, and includes Indian tribes and tribal entities.

State means any of the several States of the United States, and, where provided by law, any Territory of the United States or other area authorized to receive the services and programs of the Rural Utilities Service or the Rural Electrification Act of 1936, as amended.

Target area means the geographic area to be served by the grant.

Target community means the unit or units of local government in which the target area is located.

[70 FR 5351, Feb. 2, 2005, as amended at 83 FR 45032, Sept. 5, 2018]

§ 1709.4 Allocation of available funds among programs.

The Administrator, in his sole discretion, shall allocate available funds among the programs administered under this part and determine the grant application periods under each program. In making fund allocations for each fiscal year, the Administrator may consider the amount of available funds, the nature and amount of unfunded grant applications and prior awards, Agency resources, Agency priorities, and any other pertinent information.

§ 1709.5 Determination of energy cost benchmarks.

(a) The Administrator shall establish, using the most recent data available, and periodically revise, the home energy cost benchmarks and the high energy cost benchmarks used to determine community eligibility for high energy cost grant and loan programs and the Denali Commission high energy cost grants and loans. In setting these energy cost benchmarks, the Administrator shall review the latest available information on home energy costs published by the EIA. High energy cost benchmarks will be set at 275 percent of the applicable national average home energy cost benchmark as determined by the Administrator from the published EIA data. Eligibility benchmarks shall be published in each grant announcement.

(b) For use in determining eligibility for High Energy Cost Grants, the Administrator may establish benchmarks for national average annual household expenditures and for national average household per unit energy expenditures for major home energy sources or fuels, including, but not limited to, electricity, natural gas, fuel oil, kerosene, liquified petroleum gas (propane), other petroleum products, wood and other biomass fuels, coal, wind and solar energy.

§ 1709.6 Appeals.

An applicant may appeal a decision by the Assistant Administrator, Electric Program rejecting an application for failure to meet eligibility requirements. Applicants may not appeal rating panel scores or rankings. An appeal must be made, in writing to the Administrator, within 10 days after the applicant is notified of the determination to reject the application. Appeals must state the basis for the appeal and shall be submitted to the Administrator, Rural Utilities Service, U.S. Department of Agriculture, 1400 Independence Ave., SW., STOP 1500, Washington, DC 20250–1500. Thereafter, the Administrator will review the appeal to determine whether to sustain, reverse, or modify the original determination. The Administrator's determination shall be final. A written copy of the Administrator's decision will be furnished promptly to the applicant.

§ 1709.7 Applicant eligibility.

An outstanding judgment obtained against an applicant by the United States in a Federal Court (other than in the United States Tax Court), which has been recorded, shall cause the applicant to be ineligible to receive a grant or loan under this part until the judgment is paid in full or otherwise satisfied. RUS financial assistance under this part may not be used to satisfy the judgment.

§ 1709.8 Electronic submission.

Applicants may submit applications and reports electronically if so provided in the applicable grant announcement and grant agreements or if other regulations provide for electronic submission. Any electronic submissions must be in the form prescribed in the applicable grant announcement, grant agreement, or regulation.

§ 1709.9 Grant awards and advance of funds.

The grantee must execute a grant agreement that is acceptable to the Agency. The grantee must sign and return the grant agreement to the Agency, within the time specified, before any grant funds will be advanced.

§ 1709.10 Ineligible grant purposes.

Grant funds under this part may not be used to:

(a) Pay costs of preparing the application package for funding under programs in this part, or for any finders fees or incentives for persons or entities assisting in the preparation or submission of an application.

(b) Fund political activities;

(c) Pay any judgment or debt owed to the United States; or

(d) Pay construction costs of the project incurred prior to the date of grant award except as provided herein. Construction work should not be started and obligations for such work or materials should not be incurred before the grant is approved.

(1) Applicants may request Agency approval for reimbursement of pre-award construction obligations if there are compelling reasons for proceeding with construction before grant approval. Such requests may be approved if the Agency determines that:

(i) Compelling reasons, as determined by the Agency, exist for incurring obligations before grant approval;

(ii) The obligations will be incurred for authorized grant purposes;

(iii) All environmental requirements applicable to the Agency and the applicant have been met;

(iv) The applicant has the legal authority to incur the obligations at the time proposed, and payment of the debts will remove any basis for any mechanic's, material, or other liens that may attach to the grant financed property: and

(v) The expenditure is incurred no more than 18 months before the date of the Administrator's approval of the grant award.

(2) The Agency may authorize payment of approved pre-award project construction obligations at the time of award approval. The applicant's request and the Agency's authorization for paying such obligations shall be in writing.

§1709.11 Award conditions.

In addition to all other grant requirements, all approved applicants will be required to do the following:

(a) Enter into a grant agreement with the Agency in form and substance acceptable to the Agency;

(b) Request advances or reimbursements, as applicable, as provided in the grant agreement; and

(c) Maintain a financial management system that is acceptable to the Agency.

§1709.12 Reporting requirements.

To support Agency monitoring of project performance and use of grant funds, Grantees shall file periodic reports, required under 2 CFR part 200, as adopted by USDA through 2 CFR part 400, as provided in this part, and the grant agreement as follows:

(a) A financial status report listing project expenditures by budget category in such form and at such times as provided in the grant agreement.

(b) Project performance reports in such form and at such intervals as provided in the grant agreement. The project performance report shall compare accomplishments to the objectives stated in the proposal and grant agreement. The project performance report should identify all completed tasks with supporting documentation. If the project schedule as approved in the grant agreement is not being met, the report should discuss the problems or delays that may affect completion of the project. Objectives for the next reporting period should be listed. Compliance with any special condition on the use of award funds should be discussed. Reports are due as provided in the grant agreement.

(c) A final project performance report with supporting documentation in such form and at the time specified in the grant agreement.

(d) Such other reports as the Agency determines are necessary to assure effective grant monitoring as part of the grant agreement or the grant announcement as a condition of the grant award or advances of funds.

[70 FR 5351, Feb. 2, 2005, as amended at 79 FR 76002, Dec. 19, 2014]

§1709.13 Grant administration.

The authority to approve administrative actions is vested in the Administrator except as otherwise provided in the RUS delegations of authority. Administration of RUS grants is governed by the provisions of this subpart and subpart B of this part, the terms of the grant agreement and, as applicable, the provisions of 2 CFR part 200, as adopted by USDA through 2 CFR part 400.

[70 FR 5351, Feb. 2, 2005, as amended at 79 FR 76002, Dec. 19, 2014]

§1709.14 Inspections.

The grantee will permit periodic inspection of the grant project operations by a representative of the Agency.

§1709.15 Grant closeout.

Grant closeout is when all required work is completed, administrative actions relating to the completion of work and expenditure of funds have been accomplished, the final project report has been submitted and found acceptable by RUS and RUS accepts final expenditure information. No monitoring action by RUS of the grantee is required after grant closeout. However, grantees remain responsible in accordance with the terms of the grant agreement for compliance with conditions on property acquired or derived through grant funds.

§ 1709.16 Performance reviews.

Each grant agreement shall include performance criteria and RUS will regularly evaluate the progress and performance of grantee according to such criteria. If the grantee does not comply with or does not meet the performance criteria set out in the grant agreement, the Administrator may require amendment of the grant agreement, or may suspend or terminate the grant pursuant to 7 CFR 2015, subpart N. If the grantee does not comply with or does not meet the performance criteria set out in the grant agreement, the Administrator may require amendment of the grant agreement, or may suspend or terminate the grant pursuant to 2 CFR part 200, as adopted by USDA through 2 CFR part 400.

[70 FR 5351, Feb. 2, 2005, as amended at 79 FR 76002, Dec. 19, 2014]

§ 1709.17 Environmental review.

(a) Grants made under this subpart must comply with the environmental review requirements in accordance with 7 CFR part 1970.

(b) Applicants must address environmental aspects of their projects in the grant application in sufficient detail to allow the Agency to categorize the project for purposes of compliance with environmental review requirements. The grant announcement will establish the form and content of the environmental information required for the application.

(c) Projects that are selected for grant awards by the Administrator will be reviewed by the Agency in accordance with 7 CFR part 1970 prior to final award approval. The Agency may require the selected applicant to submit additional information, as may be required, concerning the proposed project in order to complete the required reviews and to develop any project-specific conditions for the final grant agreement.

[70 FR 5351, Feb. 2, 2005, as amended at 81 FR 11025, Mar. 2, 2016]

§ 1709.18 Civil rights.

This program will be administered in accordance with applicable Federal Civil Rights Law. All grants made under this subpart are subject to the requirements of title VI of the Civil Rights Act of 1964, which prohibits discrimination on the basis of race, color or national origin. In addition, all grants made under this subpart are subject to the requirements of section 504 of the Rehabilitation Act of 1973, as amended, which prohibits discrimination on the basis of disability; the requirements of the Age Discrimination Act of 1975, which prohibits discrimination on the basis of age; and title III of the Americans with Disabilities Act, which prohibits discrimination on the basis of disability by private entities in places of public accommodations. Grantees are required to comply with certain regulations on nondiscrimination in program services and benefits and on equal employment opportunity including 7 CFR parts 15 and 15b; and 45 CFR part 90, as applicable.

§ 1709.19 Other USDA regulations.

The grant programs under this part are subject to the provisions of other departmental regulations, including but not limited to the following departmental regulations, or their successors, as applicable:

(a) Uniform Administrative Requirements, Cost Principles, and Audit Requirements for Federal Awards, 2 CFR part 200, as adopted by USDA through 2 CFR part 400;

(b) Drug-Free Workplace Act of 1998 (41 U.S.C. 8101 *et. seq.*), 2 CFR part 421;

(c) E.O.s 12549 and 12689, Debarment and Suspension, 2 CFR part 180, which is adopted by USDA through 2 CFR part 417;

(d) Byrd Anti-Lobbying Amendment (31 U.S.C. 1352), 2 CFR part 418; and

(e) Subpart F of 2 CFR 200, as adopted by USDA through 2 CFR 400.

[70 FR 5351, Feb. 2, 2005, as amended at 79 FR 76002, Dec. 19, 2014]

§ 1709.20 Member delegate clause.

Each grant agreement under this part shall provide that no member of Congress shall be admitted to any share or part of a grant program or any benefit that may arise there from, but this provision shall not be construed to bar as a contractor under a grant a publicly held corporation whose ownership might include a member of Congress.

§ 1709.21 Audit requirements.

The grantee shall provide the Agency with an audit for each year, beginning with the year in which a portion of the financial assistance is expended, in accordance with the following:

(a) If the grantee is a for-profit entity, an RUS Electric or Telecommunication borrower or any other entity not covered by paragraph (b) of this section, the recipient shall provide an independent audit report in accordance with 7 CFR part 1773, "Policy on Audits of RUS Borrowers" and the grant agreement.

(b) If the grantee is a State or local government, or a non-profit corporation (other than an RUS Electric or Telecommunication Borrower), the recipient shall provide an audit in accordance with subpart F of 2 CFR part 200, as adopted by USDA through 2 CFR part 400.

[70 FR 5351, Feb. 2, 2005, as amended at 79 FR 76002, Dec. 19, 2014]

§ 1709.22 Project changes.

The Grantee shall obtain prior written approval from the Agency for any change to the scope or objectives of the approved grant project.

§§ 1709.23–1709.99 [Reserved]

§ 1709.100 OMB control number.

The information collection requirements in this part are approved by the Office of Management and Budget and assigned OMB control number 0572–0136.

Subpart B—RUS High Energy Cost Grant Program

§ 1709.101 Purpose.

This subpart establishes policies and procedures for the Rural Utilities Service (RUS) High Energy Cost Grant Program under section 19(a)(1) of the Rural Electrification Act of 1936, as amended (7 U.S.C. 918a(a)(1)). The purpose of this grant program is to assure access to adequate and reliable energy services for persons in extremely high energy cost communities by providing financial assistance to acquire, construct, extend, upgrade, and otherwise improve energy generation, transmission, or distribution facilities serving the community.

§ 1709.102 Policy.

(a) All high energy cost grants will be awarded competitively subject to the limited exceptions in 2 CFR 415.1(d).

(b) RUS may give priority consideration to projects that benefit smaller rural communities, communities experiencing economic hardship, projects that extend service to households that lack reliable centralized or commercial energy services, and projects that correct imminent hazards to public safety, welfare, the environment or critical community energy facilities. RUS may also give priority to projects that are coordinated with State rural development initiatives or that serve a Federally-identified Empowerment Zone or Enterprise Community (EZ/EC) or a USDA-identified "Champion Community." Priority consideration will be provided through the award of additional points under the project selection criteria as specified in the grant announcement.

[70 FR 5351, Feb. 2, 2005, as amended at 79 FR 76002, Dec. 19, 2014]

§§ 1709.103–1709.105 [Reserved]

§ 1709.106 Eligible applicants.

(a) Eligible applicants for grants to fund projects serving eligible extremely high energy cost communities include Persons, States, political subdivisions of States, and other entities organized under the laws of States.

(b) Eligible applicants may be for-profit or non-profit business entities including but not limited to corporations, associations, partnerships, limited liability partnerships (LLPs), cooperatives, trusts, and sole proprietorships.

(c) Eligible government applicants include State and local governments, and agencies and instrumentalities of States and local governments.

(d) Indian tribes, other tribal entities, and Alaska Native Corporations are eligible applicants.

(e) Individuals are also eligible applicants under this program, however the proposed grant project must provide community benefits and not be for the sole benefit of the individual applicant or an individual household.

(f) As a condition of eligibility, the applicant must demonstrate the capacity:

(1) to enter into a binding grant agreement with the Federal Government at the time of the award approval; and

(2) to carry out the proposed grant project according to its terms.

§ 1709.107 Eligible communities.

(a) An eligible community under this program is one in which the average home energy costs exceed 275 percent of the national average under one or more high energy cost benchmarks established by RUS based on the latest available residential energy information from the Energy Information Administration (EIA) of the United States Department of Energy. RUS will update the national and high energy cost community benchmarks periodically to incorporate any changes in national home energy costs reported by EIA. RUS will publish the high energy cost community benchmark criteria in the grant announcement. Community eligibility will be determined by RUS at the time of application based on the criteria published in the applicable grant announcement.

(b) The Application must include information demonstrating that each community in the grant's proposed target area exceeds one or more of the RUS high energy cost community benchmarks to be eligible for assistance under this program. The smallest area that may be designated as a target area is a Census block according to the most recent decennial Census of the United States (decennial Census).

(c) The target community may include an extremely high cost to serve portion of a larger service area that does not otherwise meet the criteria, provided that the applicant can establish that the costs to serve the smaller target area exceed the benchmark.

(d) In determining the community energy costs, applicants may include additional revenue sources that lower the rates or out of pocket consumer energy costs such as rate averaging, and other Federal, State, or private cost contributions or subsidies.

(e) The applicant may propose a project that will serve high energy cost communities across a State or region, but where individual project beneficiaries will be selected at a later time. In such cases, to establish eligibility, the applicant must provide sufficient information in the application to determine that the proposed target area includes eligible high energy cost communities and proposed selection criteria to assure that grant funds are used to serve eligible communities.

[70 FR 5351, Feb. 2, 2005, as amended at 80 FR 9860, Feb. 24, 2015]

§ 1709.108 Supporting data for determining community eligibility.

The application shall include the following:

(a) *Documentation of energy costs.* Documents or references to published or other sources for information or data on home energy expenditures or equivalent measures used to support eligibility, or where such information is unavailable or does not adequately reflect the actual cost of average home energy use in a local community, reasonable estimates of commercial energy costs.

(b) *Served areas.* A comparison of the historical residential energy cost or expenditure information for the local commercial energy provider(s) serving the target community or target area with the benchmark criteria published by the Agency.

(c) *Engineering estimates.* Estimates based on engineering standards may be used in lieu of historical residential energy costs or expenditure information under the following circumstances:

(1) Where historical community energy cost data are unavailable (unserved areas), incomplete or otherwise inadequate;

(2) Where the target area is not connected to central station electric service to a degree comparable with other residential customers in the State or region.

(3) Where historic energy costs do not reflect the costs of providing a necessary upgrade or replacement of energy infrastructure that would have the effect of raising costs above one or more of the Agency benchmarks.

(d) *Independent Agency review.* Information to support high energy cost eligibility is subject to independent review by the Agency. The Agency may reject applications that are not based on credible data sources or sound engineering estimates.

§ 1709.109 Eligible projects.

Eligible projects are those that acquire, construct, extend, repair, upgrade or otherwise improve energy generation, transmission or distribution facilities serving communities with extremely high energy costs. All energy generation, transmission and distribution facilities and equipment used to provide or improve electricity, natural gas, home heating fuels, and other energy services to eligible communities are eligible. Projects providing or improving service to communities with extremely high energy costs through on-grid and off-grid renewable energy technologies, energy efficiency, and energy conservation projects and services are eligible. A grant project is eligible if it improves, or maintains energy services, or reduces the costs of providing energy services to eligible communities. Examples of eligible activities include, but are not limited to, the acquisition, construction, replacement, repair, or improvement of:

(a) Electric generation, transmission, and distribution facilities, equipment, and services serving the eligible community;

(b) Natural gas distribution or storage facilities and associated equipment and activities serving the eligible community;

(c) Petroleum product storage and handling facilities serving residential or community use.

(d) Renewable energy facilities used for on-grid or off-grid electric power generation, water or space heating, or process heating and power for the eligible community;

(e) Backup up or emergency power generation or energy storage equipment, including distributed generation, to serve the eligible community; and

(f) Implementation of cost-effective energy efficiency, energy conservation measures that are part of the implementation of a coordinated demand management or energy conservation program for the eligible community, such as, for example, weatherization of residences and community facilities, or acquisition and installation of energy-efficient or energy saving appliances and devices .

§ 1709.110 Use of grant funds.

(a) *Project development costs.* Grants may be used to fund the costs and activities associated with the development of an eligible energy project. RUS will in no case approve the use of grant funds to be used solely or primarily for project development costs. Eligible project development costs must be reasonable and directly related to the project and may include the following:

(1) Costs of conducting, or hiring a qualified consultant to conduct, a feasibility analysis of the proposed project to help establish the financial and technical sustainability of the project, provided that such costs do not exceed more than 10 percent of total project costs;

(2) Design and engineering costs, including costs of environmental and cultural surveys and consulting services necessary to the project and associated environmental review, siting and permit approvals; and

(3) Fees for legal and other professional services directly related to the project.

(b) *Construction costs.* Grant funds may be used for the reasonable costs of construction activities, including initial construction, installation, expansion, extension, repair, upgrades, and related activities, including the rental or lease of necessary equipment, to provide or improve energy generation, transmission, or distribution facilities or services;

(c) *Acquisitions and purchase.* Grant funds may be used for the acquisition of property, equipment, and materials, including the purchase of equipment, and materials, the acquisition or leasing of real or personal property, equipment, and vehicles associated with and necessary for project development, construction, and operation. Grant funds may be used for the acquisition of new or existing facilities or systems where such action is a cost-effective means to extend or maintain service to an eligible community or reduces the costs of such service for the primary benefit of community residents.

(d) *Grantee cost contributions.* Grant funds may be applied as matching funds or cost contributions under Federal or other programs where the terms of those programs so allow use of other Federal funds.

§ 1709.111 Limitations on use of grant funds.

(a) *Planning and administrative costs.* Not more than 4 percent of each grant award may be used for the planning and administrative expenses of the applicant that are unrelated to the grant project.

(b) *Unproven technology.* Only projects that utilize technology with a proven operating history, and for which there is an established industry for the design, installation, and service (including spare parts) of the equipment, are eligible for funding. Energy projects utilizing experimental, developmental, or prototype technologies or technology demonstrations are not eligible for grant funds. The determination by RUS that a project relies on unproven technology shall be final.

§ 1709.112 Ineligible grant purposes.

(a) Grant funds may not be used for the costs of preparing the grant application, finders fees, fuel purchases, routine maintenance or other operating costs, or purchase of equipment, structures or real property not directly associated with providing energy services in the target community, or, except as provided in § 1709.11(d), project construction costs incurred prior to the date of the grant award.

(b) In general, grant funds may not be used to support projects that primarily benefit areas outside of eligible target communities. However, grant funds may be used to finance an eligible target community's proportionate share of a larger energy project.

(c) Grant funds may not be used to refinance or repay the applicant's outstanding loans or loan guarantees under the Rural Electrification Act of 1936, as amended.

§ 1709.113 Limitations on grant awards.

(a) The Administrator may establish minimum or maximum amount of funds that may be awarded in a single grant application within in any grant cycle in order to distribute available grant funds as broadly as possible. If the Administrator elects to impose a minimum or maximum grant amount, the limitations will be published in the grant announcement.

(b) The Administrator may restrict eligible applicants to a single award of grant funds or to a monetary cap on grant awards within a grant cycle in order to assure that the available grant funds are distributed as broadly as possible. If the Administrator elects to impose a limit or cap on grant awards, the terms will be established in the grant announcement.

§ 1709.114 Application process.

The RUS will request applications for high energy cost grants on a competitive basis by posting a FOA on *www.Grant.gov.* The FOA will establish the amount of funds available, the application package contents and additional requirements, the availability of application materials, high energy cost community eligibility benchmarks, selection criteria and weights, priority considerations, deadlines and procedures for submitting applications. This information will also be made available in the RUS High Energy Cost Grant program application guide and the RUS High Energy Cost Grant program website.

[83 FR 45032, Sept. 5, 2018]

§ 1709.115 Availability of application materials.

Application materials, including copies of the grant announcement and all required forms and certifications will be available by request from the Agency and by such other means as the Agency may determine. In addition, the Agency may make available an application guide and other materials that may be of assistance to prospective applicants.

§ 1709.116 Application package.

The requirements for the application package will be established in the grant announcement. A complete application package will consist of the standard application for federal assistance (SF–424 series), as applicable, a narrative project proposal prepared in accordance with the grant announcement, an RUS environmental profile, and such other supporting documentation, forms, and certifications as required in the grant announcement and this part.

§ 1709.117 Application requirements.

(a) *Required forms.* The forms required for application and where to obtain them will be specified in the announcement. All required forms must be completed, signed and submitted by a person authorized to submit the proposal on behalf of the applicant. For applications and forms that are submitted electronically, the application must be authenticated as provided in the grant announcement. In the case of grant applications submitted electronically, the applicant may be required to provide signed originals of required forms prior to and as a condition of the grant award.

(b) *Narrative proposal.* Each application must include a narrative proposal describing the proposed project and addressing eligibility and selection criteria. The grant announcement will specify the contents, order, and format for the narrative proposal. The proposal must include all the required elements identified in this subsection. The grant announcement may establish additional required elements that must be addressed in the narrative project proposal.

(1) *Executive summary.* A summary of the proposal should briefly describe the project including target community, goals, tasks to be completed and other relevant information that provides a general overview of the project. The applicant must clearly state the amount of grant funds requested and identify any priority ratings for which the applicant believes it is qualified.

(2) *Applicant eligibility.* The narrative and supporting documentation must describe the applicant and establish its eligibility.

(3) *Community eligibility.* This section must describe the target area and communities to be served by the project and demonstrate eligibility. The applicant must clearly identify the:

(i) Location and population of the areas to be served by the project;

(ii) Population of the local government division to which they belong;

(iii) Identity of local energy providers; and

(iv) Sources of the high energy cost data and estimates used.

(4) *Project eligibility.* The narrative must describe the proposed project in sufficient detail to establish that it is an eligible project.

(5) *Project description.* The project description must:

(i) Describe the project design, materials, and equipment in sufficient detail to support a finding of technical feasibility;

(ii) Identify the major tasks to be performed and a proposed timeline for completion of each task; and

(iii) Identify the location of the project target area and the eligible extremely high energy cost communities to be served.

(6) *Project management.* The applicant must describe how and by whom the project will be managed during construction and operation. The description should address the applicant's organizational structure, key project personnel and the degree to which full time employees, affiliated entities or contractors will be utilized. The applicant must describe the identities, legal relationship, qualifications and experience of those persons that will perform project management functions. If the applicant proposes to use the equipment or design, construction and other services from non-affiliated entities, the applicant must describe how it plans to contract for such equipment or services.

(7) *Budget.* The budget narrative must present a detailed breakdown of all estimated costs and allocate these costs among the listed tasks in the work plan. All project costs, not just grant funds, must be accounted for in the budget. A pro forma operating budget for the first year of operations must also be included. The detailed

budget description must be accompanied by SF–424A, "Budget Information—Non-Construction Programs," or SF–424C "Budget Information—Construction Programs," as applicable.

(8) *Project goals and objectives.* The applicant must identify unambiguous measures for expected cost reduction, efficiencies or other improvements and the degree to which the incremental benefit will be enjoyed by residents of the eligible community. The description should specifically address how the project will provide or improve energy generation, transmission or distribution services in the target area. The project objectives and proposed evaluation measures will be the basis for project performance measures in the grant agreement.

(9) *Performance measures.* The application must include specific criteria for measuring project performance. These proposed criteria will be used in establishing performance measures incorporated in the grant agreement in the event the proposal receives funding under this subpart. These suggested criteria are not binding on the Agency. Appropriate measures of project performance include expected reductions in home energy costs, avoided cost increases, enhanced reliability, new households served, or economic and social benefits from improvements in energy services.

(10) *Proposal evaluation and selection criteria.* The application must address individually and in narrative form each of the proposal evaluation and selection criteria referenced in the grant announcement.

(11) *Rural development initiatives.* The proposal should describe whether and how the proposed project will support any State rural development initiatives. If the project is in support of a rural development initiative, the application should include confirming documentation from the appropriate rural development agency. The application must identify the extent to which the project is dependent upon or tied to other rural development initiatives, funding and approvals.

(12) *Environmental review requirements.* Grants made under this subpart must comply with the environmental review requirements in accordance with 7 CFR part 1970.

(13) *Regulatory and other required project approvals.* The applicant must identify all regulatory or other approvals required by other Federal, State, local, tribal or private entities (including conditions precedent to financing) that are necessary to carry out the proposed project and an estimated schedule for obtaining the necessary permits and approvals.

[70 FR 5351, Feb. 2, 2005, as amended at 81 FR 11026, Mar. 2, 2016]

§ 1709.118 Submission of applications.

Unless otherwise provided in the grant announcement, a complete original application package and two copies must be submitted by the application deadline to RUS at the address specified in the applicable announcement. Instructions for submittal of applications electronically will be established in the grant announcement.

§ 1709.119 Review of applications.

(a) RUS will review each application package received to determine whether the applicant is eligible and whether the application is timely, complete, and responsive to the requirements set forth in the grant announcement.

(b) RUS may, at its discretion, contact the applicant to clarify or supplement information in the application needed to determine eligibility, identifying information, and grant requests to allow for informed review. Failure of the applicant to provide such information in response to a written request by the Agency within the time frame established by the Agency may result in rejection of the application.

(c) After consideration of the information submitted, the Assistant Administrator, Electric Program will determine whether an applicant or project is eligible and whether an application is timely, complete, and responsive to the grant announcement and shall notify the applicant in writing. The Assistant Administrator's decision on eligibility may be appealed to the Administrator.

§ 1709.120 Evaluation of applications.

(a) The Agency will establish one or more rating panels to review and rate the grant applications. The panels may include persons not employed by the Agency.

(b) All timely and complete applications that meet the eligibility requirements will be referred to the rating panel. The rating panel will evaluate and rate all referred applications according to the evaluation criteria and weights established in the grant announcement. Panel members may make recommendations for conditions on grant awards to promote successful performance of the grant or to assure compliance with other Federal requirements.

(c) After the rating panel has evaluated and scored all proposals, in accordance with the point allocation specified in the grant announcement, the panel will prepare a list of all applications in rank order, together with funding level recommendations and recommendations for conditions, if any.

(d) The list of ranked projects and rating panel recommendations will be forwarded to the Administrator for review and selection.

§ 1709.121 Administrator's review and selection of grant awards.

(a) The final decision to make an award is at the discretion of the Administrator. The Administrator shall make any selections of finalists for grant awards after consideration of the applications, the rankings, comments, and recommendations of the rating panel, and other pertinent information.

(b) Based on consideration of the application materials, ranking panel ratings, comments, and recommendations, and other pertinent information, the Administrator may elect to award less than the full amount of grant requested by an applicant. Applicants will be notified of an offer of a reduced or partial award. If an applicant does not accept the Administrator's offer of a reduced or partial award, the Administrator may reject the application and offer an award to the next highest ranking project.

(c) The projects selected by the Administrator will be funded in rank order to the extent of available funds.

(d) In the event an insufficient number of eligible applications are received in response to a FOA and selected for funding to exhaust the funds available, the Administrator reserves the discretion to reopen the application period and to accept additional applications for consideration under the terms of the FOA. Another FOA regarding the reopening of an application period will be announced on *www.Grants.gov.*

[70 FR 5351, Feb. 2, 2005, as amended at 83 FR 45033, Sept. 5, 2018]

§ 1709.122 Consideration of eligible grant applications under later grant announcements.

At the discretion of the Administrator, the grant announcement may provide that all eligible but unfunded proposals submitted under preceding competitive grant announcements may also be considered for funding. This option is provided to reduce the burden on applicants and the Agency. The grant announcement shall indicate how applicants may request reconsideration of previously submitted, but unfunded, applications and how they may supplement their applications.

§ 1709.123 Evaluation criteria and weights.

(a) *Establishing evaluation criteria and weights.* The grant announcement will establish the evaluation criteria and weights to be used in ranking the grant proposals submitted. Unless supplemented in the grant announcement, the criteria listed in this section will be used to evaluate proposals submitted under this program. Additional criteria may be included in the grant announcement. In establishing evaluation criteria and weights, the total points that may be awarded for project design and technical merit criteria shall not be less than 65 percent of the total available points, and the total points awarded for priority criteria shall not be more than 35 percent of the total available points. The distribution of points to be awarded per criterion will be identified in the grant announcement.

(b) *Project design and technical merit.* In reviewing the grant proposal's project design and technical merit, reviewers will consider the soundness of the applicant's approach, the project's technical and financial feasibility, the adequacy of financial and other resources, the capabilities and experience of the applicant and its project management team, the project goals, and identified community needs and benefits. Points will be awarded under the following project elements:

(1) *Comprehensiveness and feasibility*. Reviewers will assess the technical and economic feasibility of the project and how well its goals and objectives address the challenges of the eligible communities. The panel will review the proposed design, construction, equipment and materials for the proposed energy facilities to determine technical feasibility. Reviewers may propose additional conditions on the grant award to assure that the project is technically sound. Budgets will be reviewed for completeness and the strength of non-Federal funding commitments. Points may not be awarded unless sufficient detail is provided to determine whether or not funds are being used for qualified purposes. Reviewers will consider the adequacy of the applicant's budget and resources to carry out the project as proposed. Reviewers will also evaluate how the applicant proposes to manage available resources such as grant funds, income generated from the facilities and any other financing sources to maintain and operate a financially viable project once the grant period has ended. Reviewers must make a finding of operational sustainability for any points to be awarded. Projects for which future grant funding is likely to be required in order to assure ongoing operations will not receive any points.

(2) *Demonstrated experience*. Reviewers will consider whether the applicant or its project team have demonstrated experience in successfully administering and carrying out projects that are comparable to that proposed in the application. The reviewers may assign a higher point score to proposals that develop the internal capacity to provide or improve energy services in the eligible communities over other proposals that rely extensively on temporary outside contractors.

(3) *Community needs*. Reviewers will consider the applicant's assessment of community energy needs to be addressed by the proposed project as well as the severity of physical and economic challenges affecting the target communities. In determining whether one proposal should receive more points than another under this criterion, reviewers will consider the relative burdens placed on the communities and individual households by extremely high energy costs, the hardships created by limited access to reliable and affordable energy services and the availability of other resources to support or supplement the proposed grant funding.

(4) *Project evaluation and performance measures*. Reviewers will consider the applicant's suggested project evaluation and performance criteria. Reviewers may award higher points to criteria that are quantifiable, directly relevant to project goals, and reflect serious consideration than to more subjective performance criteria that do not incorporate variables that reflect a reduction in energy cost or improvement in service.

(5) *Coordination with rural development initiatives*. Proposals that include documentation confirming coordination with State rural development initiatives may be credited points for this criterion.

(c) *Priority considerations*. Subject to the limitation in paragraph (a) of this section, evaluation points may also be awarded for projects that advance identified priority interests identified in the grant announcement to assist the Agency in selecting among competing projects when the amount of funding requests exceed available funds. The grant announcement may incorporate all or some of the priority criteria listed below, and as discussed in paragraph (a) of this section, the grant announcement may supplement these criteria. The announcement will also specify the points that will be awarded to qualifying applications under these priority criteria.

(1) *Community economic hardship*. Economic hardship points may be awarded where the median household income for the target community is significantly below the State average or where the target community suffers from economic conditions that severely constrain its ability to provide or improve

energy facilities serving the community. Applicants must describe in detail and document conditions creating severe community economic hardship in the proposal.

(2) *Rurality.* Priority consideration may be given to proposals that serve smaller rural communities. Applications will be scored based on the population of the largest incorporated cities, towns or villages or census designated places included within the grant's proposed target area as determined using the population figures from the most recent decennial Census. If the applicable population figure cannot be based on the most recent decennial Census, RD will determine the applicable population figure based on available population data.

(3) *Unserved energy needs.* Points may be awarded to projects that extend or improve electric or other energy services to eligible communities or areas of eligible communities that do not have reliable centralized or commercial service.

(4) *Imminent hazard.* Additional points may be awarded for projects that correct a condition posing an imminent hazard to public safety, public welfare, the environment, or to a critical community or residential energy facility in immediate danger of failure because of a deteriorated condition, capacity limitation, or damage from a natural disaster or accident.

(5) *Cost sharing.* Projects that evidence significant commitments of funds, contributed property, equipment, or other in kind support for the project may be awarded additional points for this criterion where the aggregate value of these contributions exceed ten percent of total eligible project costs.

[70 FR 5351, Feb. 2, 2005, as amended at 80 FR 9860, Feb. 24, 2015]

§ 1709.124 Grant award procedures.

(a) *Notification of applicants.* The Agency will notify all applicants in writing whether they have been selected for a grant award. Applicants that have been selected as finalists for a competitive grant award will be notified in writing of their selection and advised that the Agency may request additional information in order to complete environmental review requirements in accordance with 7 CFR part 1970, and to meet other pre-award conditions.

(b) *Letter of conditions.* The Agency will notify each applicant selected as a finalist in writing setting out the amount of grant funds and the terms and conditions under which the grant will be made and requesting that the applicant indicate in writing its intent to accept these conditions.

(c) *Applicant's intent to meet conditions.* Upon reviewing the conditions and requirements in the letter of conditions, the selected applicant must notify the agency in writing within the time period indicated, of its acceptance of the conditions, or if the proposed certain conditions cannot be met, the applicant must so advise the Agency and may propose alternate conditions. The Agency must concur with any changes proposed to the letter of conditions by the applicant before the application will be further processed.

(d) *Grant agreement.* The Agency and the grantee must sign a grant agreement acceptable to the Agency prior to the advance of funds.

[70 FR 5351, Feb. 2, 2005, as amended at 81 FR 11026, Mar. 2, 2016]

§§ 1709.125–1709.200 [Reserved]

Subpart C—Bulk Fuel Revolving Fund Grant Program

§ 1709.201 Purpose.

This subpart establishes policies and procedures for the Rural Utilities Service (RUS) State Bulk Fuel Revolving Fund Grants. The purpose of this grant program is to assist State entities in establishing and supporting a revolving fund to provide a more cost-effective means of purchasing fuel for communities where the fuel cannot be shipped by means of surface transportation.

§ 1709.202 [Reserved]

§ 1709.203 Definitions.

As used in this subpart, the following definitions apply:

Eligible area means any area that is primarily dependent on delivery of fuel by water or air for a significant part of the year and where fuel cannot be shipped routinely by means of surface transportation either because of absolute physical constraints or because surface transportation is not practical or is prohibitively expensive.

Fuel means oil, diesel fuel, gasoline and other petroleum products, coal, and any other material that can be burned to make energy.

State entity means a department, agency, or instrumentality of any State.

Surface transportation means transportation by road, rail or pipeline.

§§ 1709.204–1709.206 [Reserved]

§ 1709.207 Eligible applicants.

Eligible applicants are restricted to State entities in existence as of November 9, 2000. Eligible State entities may partner with other entities, including other government agencies, in carrying out the programs funded by this program. Each applicant must demonstrate that it has the authority to enter into a binding agreement with the Federal Government to carry out the grant activities.

§ 1709.208 Use of grant funds.

Grant funds must be used to establish and support a revolving loan fund that facilitates cost effective fuel purchases for persons, communities, and businesses in eligible areas. Where a recipient State entity's existing program is authorized to fund multiple purposes, grant funds may only be used to the extent the recipient fund finances eligible activities.

§ 1709.209 Limitations on use of grant funds.

Not more than 4 percent of the grant award may be used for the planning and administrative expenses of the grantee.

§ 1709.210 Application process.

(a) *Applications.* The Agency will solicit applications on a competitive basis by publication of a grant announcement establishing the amount of funds available, the maximum grant award, the required application materials and where to obtain them, the evaluation and selection criteria and weights, and application deadlines. Unless otherwise specified in the announcement, applicants must file an original application package and two copies. Where provided in the grant announcement, applicants may submit electronic applications.

(b) *Required forms.* The grant application will use the Standard Application for Federal Assistance (SF–424 series or its successor) and other forms as provided in the grant announcement. The required forms must be completed, signed and submitted by a person authorized to submit the proposal on behalf of the applicants. Where provided in the grant announcement, applicants may file electronic versions of the forms in compliance with the instructions in the grant announcement.

(c) *Narrative proposal and required elements.* Each grant application must include a narrative proposal describing the project and addressing the following elements. The form, contents, and order of the narrative proposal will be specified in the grant announcement. Additional elements may be published in the applicable grant announcement.

(1) *Executive summary.* This summary of the proposal must identify the State entity applying for the grant and the key agency contact information (telephone and fax numbers, mailing address and e-mail address). The applicant must clearly state the amount requested in this section. It should briefly describe the program, including the estimated number of potential beneficiaries in eligible areas, their estimated fuel needs, the projects and activities to be financed through the revolving fund and how the projects and activities will improve the cost effectiveness of fuel procured.

(2) *Applicant eligibility.* The application must establish that the applicant is a State entity that was in existence as of November 9, 2000, and has the legal authority to enter into a financial assistance relationship with the Federal Government to carry out the grant activities.

(3) *Assessment of needs and potential beneficiaries.* The application must provide estimates of the number, location and population of potentially eligible areas in the State and their estimated fuel needs and costs. The section must

also describe the criteria used to identify eligible areas, including the characteristics that make fuel deliveries by surface transport impossible or impracticable. The description of beneficiary communities should provide a detailed breakdown of the density profile of the area to be served by eligible projects. Indicate to what extent persons in eligible areas live outside of communities of 2,500 persons or more, communities of 5,000 or more or outside of communities of 20,000 or more. All population estimates should be based on the most recent decennial Census of the United States. If the applicable population estimate cannot be based on the most recent decennial Census, RD will determine the applicable population figure based on available population data. All representations should be supported with exhibits such as maps, summary tables and references to official information sources.

(4) *Project description*. The application must:

(i) Describe the legal structure and staffing of the revolving fund proposal for fuel purchase support.

(ii) Identify the objectives of the project, the proposed criteria for establishing project funding eligibility and how the project is to be staffed, managed and financed.

(iii) Describe how the potential beneficiaries will be informed of the availability of revolving fund benefits to them.

(iv) Explain how the proposed revolving fund program will help provide a more cost-effective means of meeting fuel supply needs in eligible areas, encourage the adoption of financially sustainable energy practices, the adequate planning and investment in bulk fuel facility operations and maintenance and cost-effective investments in energy efficiency.

(v) If the revolving fund program is not yet operational, a proposed implementation schedule and milestones should be provided.

(5) *Demonstrated experience*. The application shall describe past accomplishments and experiences that are relevant to determine whether the applicant is capable of administering the grant project.

(6) *Budget*. The application must include a pro forma operating budget for the proposed fund and a description of all funding sources. The level of detail must be sufficient for reviewers to determine that grant funds will be used only for eligible purposes and to determine the extent to which the program is entirely dependent on grant funding or whether it has financial support from the State or other sources.

(7) *Performance measures and project evaluation*. The application must provide unambiguous and quantifiable measures that will be used to evaluate the success and cost-effectiveness of the revolving fund in assuring adequate fuel supplies for eligible communities and for assessing the fuel supply projects financed. The grant announcement may establish additional required elements that must be addressed in the narrative proposal of the application package.

[70 FR 5351, Feb. 2, 2005, as amended at 80 FR 9860, Feb. 24, 2015]

§ 1709.211 Submission of applications.

Completed applications must be submitted to RUS at the address specified in the grant announcement on or before the deadline specified in the grant announcement. Instructions for submittal of applications electronically will be established in the grant announcement. Late applications will be rejected.

§ 1709.212 Application review.

The Agency will review all applications to determine whether the applicant is eligible and whether the application is timely, complete and sufficiently responsive to the requirements set forth in the grant announcement to allow for an informed review. Failure to address any of the required evaluation criteria or to submit all required forms will disqualify the proposal. The Agency reserves the right to contact the applicant to clarify information contained in the proposal to resolve issues related to eligibility and the grant request. Applications that are timely, complete, and responsive will be forwarded for further evaluation. Applications that are late, incomplete, or non-responsive will be rejected.

§ 1709.213 Evaluation of applications.

(a) The Agency will establish one or more rating panels to review and rate the grant applications. The panels may include persons not employed by the Agency.

(b) The rating panel will evaluate and rate all complete applications that meet the eligibility requirements according to the evaluation and selection criteria and weights established in the grant announcement. Panel members may make recommendations for conditions on grant awards to promote successful performance of the grant or to assure compliance with other Federal requirements.

(c) After all proposals have been evaluated and scored, the proposals, the rankings, recommendations, and comments of the rating panel will be forwarded to the Administrator.

§ 1709.214 Administrator's review and selection of grant awards.

(a) The final decision to make a grant award is at the discretion of the Administrator. The Administrator shall consider the applications, the ranking, comments, and recommendations of the rating panel, and any other pertinent information before making a decision about which, if any, applications to approve, the amount of funds awarded, and the order of approval. The Administrator reserves the right not to make any awards from the applications submitted. When the Administrator decides not to make any awards, the Administrator shall document in writing the reason for the decision.

(b) Decisions on grant awards will be made by the Administrator after consideration of the applications, the rankings and recommendations of the rating panel. The Administrator may elect to award less than the full amount of grant requested by an applicant.

(c) The applications selected by the Administrator will be funded in rank order to the extent of available funds.

§ 1709.215 Consideration of unfunded applications under later grant announcements.

The grant announcement may provide that all eligible but unfunded proposals submitted under preceding announcements may also be considered for funding. The announcement shall describe whether and how prior applicants may request reconsideration and supplement their application material.

§ 1709.216 Evaluation criteria and weights.

Unless supplemented in the grant announcement, the criteria listed in this section will be used to evaluate proposals submitted under this program. The total points available and the distribution of points to be awarded per criterion will be identified in the grant announcement.

(a) *Program Design.* Reviewers will consider the financial viability of the applicant's revolving fund program design, the proposed criteria for establishing eligible projects and borrowers, and how the program will improve the cost effectiveness of bulk fuel purchases in eligible areas. Programs demonstrating a strong design and the ability to improve cost effectiveness will receive more points than applications that are less detailed.

(b) *Assessment of needs.* Reviewers will award more points to programs that serve or give priority to assisting more costly areas than those that serve populations that suffer from less severe physical and economic challenges.

(c) *Program evaluation and performance measures.* Reviewers may award more points to performance measures that are relevant to the project objective and quantifiable than to performance measures that are more subjective and do not incorporate variables that reflect a reduction in fuel cost or improvement in service.

(d) *Demonstrated experience.* Applicants may be awarded points for relevant experience in administering revolving fund or other comparable programs.

(e) *Rurality.* Reviewers may award more points to proposals that give priority in access to funds to communities with low population density or that are located in remote eligible areas than to proposals that serve eligible, but less remote and higher population density communities.

(f) *Cost sharing.* Although cost-sharing is not required under this program, projects that evidence significant funding or contributed property, equipment or other in kind support for the project may be awarded points for this criterion where the aggregate value of these contributions exceed 25 percent of the annual funding operations.

(g) *Additional priority considerations.* The grant announcement may provide for additional points to be awarded to projects that advance identified Agency priority interests under this program.

§ 1709.217 Grant award.

(a) *Notification of applicants.* The Agency will notify all applicants in writing whether or not they have been selected for a grant award.

(b) *Letter of conditions.* The Agency will notify a selected applicant in writing, setting out the amount of grant approved and the conditions under which the grant will be made.

(c) *Applicant's intent to meet conditions.* Upon reviewing the conditions and requirements in the letter of conditions, the selected applicant must complete, sign and return the Agency's "Letter of Intent to Meet Conditions," or, if certain conditions cannot be met, the applicant may propose alternate conditions to the Agency. The Agency must concur with any changes proposed to the letter of conditions by the applicant before the application will be further processed.

(d) *Grant agreement.* The Agency and the grantee must execute a grant agreement acceptable to the Agency prior to the advance of funds.

§§ 1709.218–1709.300 [Reserved]

Subparts D–F [Reserved]

Subpart G—Recovery of Financial Assistance Used for Unauthorized Purposes

§ 1709.601 Policy.

This subpart prescribes the policies of the Rural Utilities Service (RUS) when it is subsequently determined that the recipient of an Assistance to High Energy Cost Rural Communities program loan or grant was not eligible for all or part of the financial assistance received or that the assistance received was used for unauthorized purposes. It is the policy of the Agency that when assistance under this part has been received by an ineligible recipient or used for unauthorized purposes the Agency shall initiate appropriate actions to recover from the recipient the sum that is determined to be ineligible or used for unauthorized purposes, regardless of amount, unless any applicable statute of limitation has expired. The Agency shall make full use of available authority and procedures, including but not limited to those available under 2 CFR part 200, as adopted by USDA through 2 CFR part 400.

[70 FR 5351, Feb. 2, 2005, as amended at 79 FR 76002, Dec. 19, 2014]

§§ 1709.602–1709.999 [Reserved]

PART 1710—GENERAL AND PRE-LOAN POLICIES AND PROCEDURES COMMON TO ELECTRIC LOANS AND GUARANTEES

Subpart A—General

Subpart B—Types of Loans and Loan Guarantees

Subpart C—Loan Purposes and Basic Policies

Subpart D—Basic Requirements for Loan Approval

Subpart E—Load Forecasts

Subpart F—Construction Work Plans and Related Studies

Subpart G—Long-Range Financial Forecasts

Subpart H—Energy Efficiency and Conservation Loan Program

Subpart I—Application Requirements and Procedures for Insured and Guaranteed Loans

AUTHORITY: 7 U.S.C. 901 *et seq.*, 1921 *et seq.*, 6941 *et seq.*

SOURCE: 57 FR 1053, Jan. 9, 1992, unless otherwise noted.

Subpart A—General

§ 1710.1 General statement.

(a) This part establishes general and pre-loan policies and requirements that apply to both insured and guaranteed loans to finance the construction and improvement of electric facilities in rural areas, including generation, transmission, and distribution facilities.

(b) Additional pre-loan policies, procedures, and requirements that apply specifically to guaranteed and/or insured loans are set forth elsewhere:

(1) For guaranteed loans in 7 CFR part 1712 and RUS Bulletins 20–22, 60–10, 86–3, 105–5, and 111–3, or the successors to these bulletins; and

(2) For insured loans in 7 CFR part 1714 and in RUS Bulletins 60–10, 86–3, 105–5, and 111–3, or the successors to these bulletins.

(c) This part supersedes those portions of the following RUS Bulletins and supplements that are in conflict.

20–5 Extensions of Payments of Principal and Interest
20–20 Deferment of Principal Repayments for Investment in Supplemental Lending Institutions
20–22 Guarantee of Loans for Bulk Power Supply Facilities
20–23 Section 12 Extensions for Energy Resources Conservation Loans
60–10 Construction Work Plans, Electric Distribution Systems
86–3 Headquarters Facilities for Electric Borrowers
105–5 Financial Forecast-Electric Distribution Systems
111–3 Power Supply Surveys
120–1 Development, Approval, and Use of Power Requirements Studies

(d) When parts 1710, 1712, and 1714 are published in final form, the bulletins cited in paragraph (b) of this section will be rescinded, in whole or in part, or revised.

[57 FR 1053, Jan. 9, 1992, as amended at 58 FR 66262, Dec. 20, 1993]

§ 1710.2 Definitions and rules of construction.

(a) *Definitions.* For the purpose of this part, the following terms shall have the following meanings:

Administrator means the Administrator of RUS or his or her designee.

Approved load forecast means a load forecast that RUS has determined is current for RUS purposes and has been approved by RUS pursuant to 7 CFR part 1710, subpart E.

Approved load forecast work plan means a load forecast work plan that RUS has determined is current for RUS' purposes and has been approved pursuant to 7 CFR part 1710, subpart E.

APRR means Average Adjusted Plant Revenue Ratio calculated as a simple average of the adjusted plant revenue ratios for 1978, 1979 and 1980 as follows:

$$\text{APRR} = \frac{A + B}{C - D}$$

where:

A = Distribution (plant), which equals Part E, Line 14(e) of RUS Form 7;
B = General Plant, which equals Part E, Line 24(e) of RUS Form 7;
C = Operating Revenue and Patronage Capital, which equals Part A, Line 1 of RUS Form 7; and
D = Cost of Power, which equals the sum of Part A, Lines 2, 3, and 4 of RUS Form 7.

Area Coverage means the provision of adequate electric service to the widest practical number of rural users in the borrower's service area during the life of the loan.

Borrower means any organization that has an outstanding loan made or guaranteed by RUS for rural electrification, or that is seeking such financing.

Bulk Transmission Facilities means the transmission facilities connecting power supply facilities to the subtransmission facilities, including both the high and low voltage sides of the transformer used to connect to the subtransmission facilities, as well as related supervisory control and data acquisition systems.

Call provision has the same meaning as "prepayment option".

Consolidation means the combination of 2 or more borrower or nonborrower organizations, pursuant to state law, into a new successor organization that takes over the assets and assumes the liabilities of those organizations.

Consumer means a retail customer of electricity, as reported on RUS Form 7, Part R, Lines 1–7.

Demand side management (DSM) means the deliberate planning and/or implementation of activities to influence Consumer use of electricity provided by a distribution borrower to produce beneficial modifications to the system load profile. Beneficial modifications to the system load profile ordinarily improve load factor or otherwise help in utilizing electric system resources to best advantage consistent with acceptable standards of service and lowest system cost. Load profile modifications are characterized as peak clipping, valley filling, load shifting, strategic conservation, strategic load growth, and flexible load profile. (See, for example, publications of the Electric Power Research Institute (EPRI), 3412 Hillview Avenue, Palo Alto, CA 94304, especially "Demand-Side Management Glossary" EPRI TR-101158, Project 1940–25, Final Report, October 1992.) DSM includes energy conservation programs.

Distribution Borrower means a borrower that sells or intends to sell electric power and energy at retail in rural areas.

Distribution Facilities means all electrical lines and related facilities beginning at the consumer's meter base, and continuing back to and including the distribution substation.

Distributed generation is the generation of electricity by a sufficiently small electric generating system as to allow interconnection of the electric generating system near the point of service at distribution voltages including points on the customer side of the meter. A distributed generating system may be operated in parallel or independent of the electric power system. A distributed generating system may be fueled by any source, including but not limited to renewable energy sources. A distributed generation project may include one or more distributed generation systems.

DSC means Debt Service Coverage of the borrower calculated as:

$$DSC = \frac{A+B+C}{D}$$

Where:

All amounts are for the same calendar year and are based on the RUS system of accounts and RUS Forms 7 and 12. References to line numbers in the RUS Forms 7 and 12 refer to the June 1994 version of RUS Form 7 and the December 1993 version of RUS Form 12, and will apply to corresponding information in future versions of the forms;

A = Depreciation and Amortization Expense of the borrower, which equals Part A, Line 12 of RUS Form 7 (distribution borrowers) or Section A, Line 20 of RUS Form 12a (power supply borrowers);

B = Interest expense on total long-term debt of the borrower, which equals Part A, Line 15 of RUS Form 7 or Section A, Line 22 of RUS Form 12a, except that interest expense shall be increased by ⅓ of the amount, if any, by which restricted rentals of the borrower (Part M, Line 3 of RUS Form 7 or Section K, Line 4 of RUS Form 12h) exceed 2 percent of the borrower's equity (RUS Form 7, Part C, Line 36 [Total Margins & Equities] less Line 26 [Regulatory Assets] or RUS Form 12a, Section B, Line 38 [Total Margins & Equities] less Line 28 [Regulatory Assets]);

C = Patronage Capital or Margins of the borrower, which equals Part A, Line 28 of RUS Form 7 or Section A, Line 35 of RUS Form 12a; and

D = Debt Service Billed (RUS + other), which equals the sum of all payments of principal and interest required to be made on account of total long-term debt of the borrower during the calendar year, plus ⅓ of the amount, if any, by which restricted rentals of the borrower (Part M, Line 3 of RUS Form 7 or Section K, Line 4 of RUS Form 12h) exceed 2 percent of the borrower's equity (RUS Form 7, Part C, Line 36 [Total Margins & Equities] less Line 26 [Regulatory Assets] or RUS Form 12a, Section B, Line 38 [Total Margins & Equities] less Line 28 [Regulatory Assets]);

DSM activities means activities of the type referred to in § 1710.354(f).

DSM plan means a plan that describes the implementation at the distribution level of the DSM activities identified in the integrated resource plan as having positive net benefits. See § 1710.357.

Electric system means all of the borrower's interests in all electric production, transmission, distribution, conservation, load management, general plant and other related facilities, equipment or property and in any mine, well, pipeline, plant, structure or other facility for the development, production, manufacture, storage, fabrication or processing of fossil, nuclear, or other fuel or in any facility or rights with respect to the supply of water, in each case for use, in whole or in major part, in any of the borrower's generating plants, including any interest or participation of the borrower in any such facilities or any rights to the output or capacity thereof, together with all lands, easements, rights-of-way, other works, property, structures, contract rights and other tangible and intangible assets of the borrower in each case used or useful in such electric system.

Eligible Energy Efficiency and Conservation Programs (Eligible EE Program) means an energy efficiency and conservation program that meets the requirements of subpart H of this part.

Equity means total margins and equities, which equals Part C, Line 33 of RUS Form 7 (distribution borrowers) or Section B, Line 34 of RUS Form 12a (power supply borrowers).

Final maturity means the final date on which all outstanding principal and accrued interest on an electric loan is due and payable.

Five percent hardship rate means an interest rate of 5 percent applicable to a hardship rate loan.

Fund advance period means the period of time during which the Government may advance loan funds to the borrower. See 7 CFR 1714.56.

Generation Facilities means the generating plant and related facilities, including the building containing the plant, all fuel handling facilities, and the stepup substation used to convert the generator voltage to transmission voltage, as well as related energy management (dispatching) systems.

Hardship rate loan means a loan made at the 5 percent hardship rate pursuant to 7 CFR 1714.8.

Insured Loan means a loan made pursuant to Section 305 of the RE Act, and may include a direct loan made under Section 4 of the RE Act.

Integrated Resources Plan (IRP) means a plan resulting from the planning and selection process for new energy resources that evaluates the benefits and costs of the full range of alternatives, including new generating capacity, power purchases, DSM programs, system operating efficiency, and renewable energy systems.

Interest rate cap means a maximum interest rate of 7 percent applicable to certain municipal rate loans as set forth in § 1710.7.

Interest rate term means a period of time selected by the borrower for the purpose of determining the interest rate on an advance of funds. See 7 CFR 1714.6.

Load forecast means the thorough study of a borrower's electric loads and the factors that affect those loads in order to determine, as accurately as practicable, the borrower's future requirements for energy and capacity.

Load forecast work plan means the plan that contains the resources, methods, schedules, and milestones to be used in the preparation and maintenance of a load forecast.

Loan means any loan made or guaranteed by RUS.

Loan Contract means the agreement, as amended, supplemented, or restated from time to time, between a borrower and RUS providing for loans made or guaranteed pursuant to the RE Act.

Loan Feasibility means that the borrower has the capability of repaying the loan in full as scheduled, in accordance with the terms of the mortgage, note, and loan contract.

Loan Guarantee means a loan guarantee made by RUS pursuant to the RE Act.

Loan period means the period of time during which the facilities will be constructed not to exceed the time identified in the Loan note, as approved.

Merger means the combining, pursuant to state law, of borrower or nonborrower organizations into an existing survivor organization that takes over the assets and assumes the liabilities of the merged organizations.

Mortgage means any and all instruments creating a lien on or security interest in the borrower's assets in connection with loans or guarantees under the RE Act.

Municipal rate loan means a loan made at a municipal interest rate pursuant to 7 CFR 1714.5.

ODSC means Operating Debt Service Coverage of the electric system calculated as:

$$ODSC = \frac{A+B+C}{D}$$

Where:

All amounts are for the same calendar year and are based on the RUS system of accounts and RUS Form 7. References to line numbers in the RUS Form 7 refer to the June 1994 version of the form, and will apply to corresponding information in future versions of the form;

A = Depreciation and Amortization Expense of the electric system, which usually equals Part A, Line 12 of RUS Form 7;

B = Interest expense on total long-term debt of the electric system, which usually equals Part A, Line 15 of RUS Form 7, except that such interest expense shall be increased by ⅓ of the amount, if any, by which restricted rentals of the electric system (usually Part M, Line 3 of RUS Form 7) exceed 2 percent of the borrower's equity (RUS Form 7, Part C, Line 36 [Total Margins & Equities] less Line 26 [Regulatory Assets]);

C = Patronage Capital & Operating Margins of the electric system, which usually equals Part A, Line 20 of RUS Form 7, plus cash received from the retirement of patronage capital by suppliers of electric power and by lenders for credit extended for the Electric System; and

D = Debt Service Billed (RUS + other), which equals the sum of all payments of principal and interest required to be made on account of total long-term debt of the electric system during the calendar year, plus ⅓ of the amount, if any, by which restricted rentals of the Electric System (usually Part M, Line 3 of RUS Form 7) exceed 2 percent of the borrower's equity (RUS Form 7, Part C, Line 36 [Total Margins & Equities] less Line 26 [Regulatory Assets]).

Off-grid renewable energy system is a renewable energy system not interconnected to an area electric power system (EPS). An off-grid renewable energy system in areas without access to an area EPS may include energy consuming devices and electric wiring to provide for more effective or more efficient use of the electricity produced by the system.

On-grid renewable energy system is a renewable energy system interconnected to an area electric power system (EPS) through a normally open or normally closed device. It can be interconnected to the EPS on either side of a customer's meter.

Ordinary Replacement means replacing one or more units of plant, called "retirement units", with similar units when made necessary by normal wear and tear, damage beyond repair, or obsolescence of the facilities.

OTIER means Operating Times Interest Earned Ratio of the electric system calculated as:

$$OTIER = \frac{A+B}{A}$$

Where:

All amounts are for the same calendar year and are based on the RUS system of accounts and RUS Form 7. References to line numbers in the RUS Form 7 refer to the June 1994 version of the form, and will apply to corresponding information in future versions of the form;

A = Interest expense on total long-term debt of the electric system, which usually equals Part A, Line 15 of RUS Form 7, except that such interest expense shall be increased by ⅓ of the amount, if any, by which restricted rentals of the electric system (usually Part M, Line 3 of RUS Form 7) exceed 2 percent of the borrower's equity (RUS Form 7, Part C, Line 36 [Total Margins & Equities] less Line 26 [Regulatory Assets]); and

B = Patronage Capital & Operating Margins of the electric system, which usually equals Part A, Line 20 of RUS Form 7, plus cash received from the retirement of patronage capital by suppliers of electric power and by lenders for credit extended for the Electric System.

Power requirements study (PRS) has the same meaning as load forecast.

Power Supply Borrower means a borrower that sells or intends to sell electric power at wholesale to distribution or power supply borrowers pursuant to RUS wholesale power contracts.

Prepayment option means a provision included in the loan documents to allow the borrower to prepay all or a portion of an advance on a municipal rate loan on a date other than a rollover maturity date. See 7 CFR 1714.9.

PRR means Plant Revenue Ratio calculated as:

$$PRR = \frac{A}{B-C}$$

where:

A = Total Utility Plant, which equals Part C, Line 3 of RUS Form 7;

B = Operating Revenue and Patronage Capital, which equals Part A, Line 1 of RUS Form 7; and

C = Cost of Power, which equals the sum of Part A, Lines 2, 3, and 4 of RUS Form 7.

PRS work plan has the same meaning as load forecast work plan.

RE Act means the Rural Electrification Act of 1936, as amended (7 U.S.C. 901 *et seq.).*

RE Act beneficiary means a person, business, or other entity that is located in a rural area.

REA means the Rural Electrification Administration formerly an agency of the United States Department of Agriculture and predecessor agency to RUS with respect to administering certain electric and telephone loan programs.

Renewable energy system is an energy conversion system fueled from any of the following energy sources: Solar, wind, hydropower, biomass, or geothermal. Any of these energy sources may be converted to heat or electricity, provided heat is a by-product of electricity generation. Non-renewable energy sources may be used by a renewable energy system for incidental and necessary means such as, but not limited to, system start up, flame stabilization, continuity of system processes, or reduction of the moisture content of renewable fuels. Energy from bio-mass may be converted from any organic matter available on a renewable basis, including dedicated energy crops and trees, agricultural food and feed crops, agricultural crop wastes and residues, wood wastes and residues, aquatic plants, animal wastes, municipal wastes, and other waste materials.

Retirement Unit means a substantial unit of property, which when retired, with or without being replaced, is accounted for by removing its book cost from the plant account.

Rollover maturity date means the last day of an interest rate term.

Rural area means—

(i) Any area of the United States, its territories and insular possessions (including any area within the Federated States of Micronesia, the Marshall Islands, and the Republic of Palau) other than a city, town, or unincorporated area that has a population of greater than 20,000 inhabitants; and

(ii) Any area within a service area of a borrower for which a borrower has an outstanding loan as of June 18, 2008, made under titles I through V of the Rural Electrification Act of 1936 (7 U.S.C. 901–950bb). For initial loans to a borrower made after June 18, 2008, the "rural" character of an area is determined at the time of the initial loan to furnish or improve service in the area.

RUS means the Rural Utilities Service, an agency of the United States Department of Agriculture established pursuant to Section 232 of the Federal Crop Insurance Reform and Department of Agriculture Reorganization Act of 1994 (Pub. L. 103–354, 108 Stat. 3178), successor to REA with respect to administering certain electric and telephone programs. See 7 CFR 1700.1.

Subtransmission Facilities means the transmission facilities that connect the high voltage side of the distribution substation to the low voltage side of the bulk transmission or generating facilities, as well as related supervisory control and data acquisition facilities.

System Improvement means the change or addition to electric plant facilities to improve the quality of electric service or to increase the quantity of electric power available to RE Act beneficiaries.

TIER means Times Interest Earned Ratio of the borrower calculated as:

$$\text{TIER} = \frac{A + B}{A}$$

Where:

All amounts are for the same calendar year and are based on the RUS system of accounts and RUS Forms 7 and 12. References to line numbers in the RUS Forms 7 and 12 refer to the June 1994 version of RUS Form 7 and the December 1993 version of RUS Form 12, and will apply to corresponding information in future versions of the forms;

A = Interest expense on total long-term debt of the borrower, which equals Part A, Line 15 of RUS Form 7 or Section A, Line 22 of RUS Form 12a, except that interest expense shall be increased by ⅓ of the amount, if any, by which restricted rentals of the borrower (Part M, Line 3 of RUS Form 7 or Section K, Line 4 of RUS Form 12h) exceed 2 percent of the borrower's equity (RUS Form 7, Part C, Line 36 [Total Margins & Equities] less Line 26 [Regulatory Assets] or RUS Form 12a,

Section B, Line 38 [Total Margins & Equities] less Line 28 [Regulatory Assets]); and

B = Patronage Capital or Margins of the borrower, which equals Part A, Line 28 of RUS Form 7 or Section A, Line 35 of RUS Form 12a.

Total Assets means Part C, Line 26 of RUS Form 7 (distribution borrowers) or Section B, Line 27 of RUS Form 12a (power supply borrowers).

Total Utility Plant means Part C, Line 3 of RUS Form 7 (distribution borrowers) or Section B, Line 27 of RUS Form 12a (power supply borrowers).

Transmission Facilities means all electrical lines and related facilities, including certain substations, used to connect the distribution facilities to generation facilities. They include bulk transmission and subtransmission facilities.

Urban area is defined as any area not considered a rural area per the definition contained in this subpart.

Urbanized area means an urbanized area as defined by the Bureau of the Census in notices published periodically in the FEDERAL REGISTER. Generally an urbanized area is characterized as an area that comprises a place and the adjacent densely settled territory that together have a minimum population of 50,000 people.

(b) *Rules of Construction.* Unless the context otherwise indicates, "includes" and "including" are not limiting, and "or" is not exclusive. The terms defined in paragraph (a) of this part include the plural as well as the singular, and the singular as well as the plural.

[57 FR 1053, Jan. 9, 1992; 57 FR 4513, Feb. 5, 1992, as amended at 58 FR 66263, Dec. 20, 1993; 59 FR 495, Jan. 4, 1994; 59 FR 66440, Dec. 27, 1994; 60 FR 3730, Jan. 19, 1995; 60 FR 67400, Dec. 29, 1995; 65 FR 14786, Mar. 20, 2000; 68 FR 37953, June 26, 2003; 74 FR 56543, Nov. 2, 2009; 78 FR 73365, Dec. 5, 2013; 84 FR 32610, July 9, 2019]

§ 1710.3 Form and bulletin revisions.

References in this part to RUS or REA forms or line numbers in RUS or REA forms are based on RUS or REA Form 7 and Form 12 dated December 1992, unless otherwise indicated. These references will apply to corresponding information in future versions of the forms. The terms "RUS form", "RUS standard form", "RUS specification", and "RUS bulletin" have the same meanings as the terms "REA form", "REA standard form", "REA specification", and "REA bulletin", respectively, unless otherwise indicated.

[59 FR 66440, Dec. 27, 1994]

§ 1710.4 Exception authority.

Consistent with the RE Act and other applicable laws, the Administrator may waive or reduce any requirement imposed by this part or other RUS regulations on an electric borrower, or a lender whose loan is guaranteed by RUS, if the Administrator determines that imposition of the requirement would adversely affect the Government's financial interest.

§ 1710.5 Availability of forms.

Information about the availability of RUS forms and publications cited in this part is available from Administrative Services Division, Rural Utilities Service, United States Department of Agriculture, Washington, DC 20250–1500. These RUS forms and publications may be reproduced.

§ 1710.6 Applicability of certain provisions to completed loan applications.

(a) Certain new or revised policies and requirements set forth in this part, which are listed in this paragraph, shall not apply to a pending loan application that has been determined by RUS to be complete as of January 9, 1992, the date of publication of such policies and requirements in the FEDERAL REGISTER. This exception does not apply to loan applications received after said date, nor to incomplete applications pending as of said date. This exception applies only to the following provisions:

(1) Paragraph 1710.115(b)—with respect to limiting loan maturities to the expected useful life of the facilities financed;

(2) Section 1710.116—with respect to the requirement to develop and follow an equity development plan;

(3) Paragraph 1710.151(f)—with respect to the borrower providing satisfactory evidence that a state regulatory authority will allow the facilities to be included in the rate base or

otherwise allow sufficient revenues to repay the loan;

(4) Paragraphs 1710.250(b), 1710.251(a), and 1710.252(a)—with respect to the requirement that improvements, replacements, and retirements of generation plant be included in a Construction Work Plan; and

(5) Paragraph 1710.300(d)(5)—with respect to the requirement that a borrower's financial forecast include a sensitivity analysis of a reasonable range of assumptions for each of the major variables in the forecast.

(b) Certain provisions of this part apply only to loans made on or after February 10, 1992. These provisions are identified in the individual sections of this part.

[57 FR 1053, Jan. 9, 1992; 57 FR 4513, Feb. 5, 1992, as amended at 58 FR 66263, Dec. 20, 1993]

§§ 1710.7-1710.49 [Reserved]

Subpart B—Types of Loans and Loan Guarantees

§ 1710.50 Insured loans.

RUS makes insured loans under section 305 of the RE Act.

(a) *Municipal rate loans.* The standard interest rate on an insured loan made on or after November 1, 1993, is the municipal rate, which is the rate determined by the Administrator to be equal to the current market yield on outstanding municipal obligations with remaining periods to maturity, up to 35 years, similar to the interest rate term selected by the borrower. In certain cases, an interest rate cap of 7 percent may apply. The interest rate term and rollover maturity date for a municipal rate loan will be determined pursuant to 7 CFR part 1714, and the borrower may elect to include in the loan documents a prepayment option (call provision).

(b) *Hardship rate loans.* RUS makes hardship rate loans at the 5 percent hardship rate to qualified borrowers meeting the criteria set forth in 7 CFR 1714.8

[58 FR 66263, Dec. 20, 1993]

§ 1710.51 Direct loans.

RUS makes direct loans under section 4 of the RE Act.

(a) *General.* Except as otherwise modified by this section, RUS will make loans under the direct Treasury rate loan program in the same manner that it makes loans under the municipal rate program. The general and preloan policies and procedures for municipal rate electric loans made by RUS may be found in this part and 7 CFR part 1714. Treasury rate electric loans are also governed by such municipal rate policies and procedures, except as follows:

(1) *Interest rates.* The standard interest rate on direct Treasury rate loans will be established daily by the United States Treasury. The borrower will select interest rate terms for each advance of funds. The minimum interest rate term shall be one year. Interest rate terms will be limited to terms published by the Treasury (i.e. 1, 2, 3, 5, 7, 10, 20, and 30). Interest rate terms to final maturity date, if other than published by Treasury, will be determined by RUS. Interest rates for terms greater than 30 years will be at the 30-year rate. There will be no interest rate cap on Treasury rate loans.

(2) *Prepayment.* A Treasury rate direct electric loan may be repaid at par on its rollover maturity date if there is one. Such a loan, or portion thereof, may also be prepaid after it has been advanced for not less than two years, at any time prior to its rollover or final maturity date at its "net present value" (NPV) as determined by RUS.

(3) *Supplemental financing.* Supplemental financing will not be required in connection with Treasury rate direct electric loans.

(4) *Transitional assistance.* A Treasury rate direct loan is not available to provide transitional assistance to borrowers.

(b) *Loan documents.* Successful applicants will be required to execute and deliver to RUS a promissory note evidencing the borrower's obligation to repay the loan. The note must be in form and substance satisfactory to RUS. RUS will require a form of note substantially in the form that it currently accepts for direct municipal rate electric loans, with such revisions as may be necessary or appropriate to reflect the different interest setting provisions and the terms of paragraphs

(a) (1) and (2) of this section. All notes will be secured in accordance with the terms of 7 CFR part 1718.

[66 FR 66294, Dec. 26, 2001]

§ 1710.52 Loan guarantees.

RUS provides financing through 100 percent loan guarantees made under sections 306 and 306A of the RE Act. RUS also provides 90 percent loan guarantees under section 311 of the RE Act to enable borrowers to secure financing from certain private lenders. The loan guarantees are made for a term of up to 35 years, and the interest rate is established at a rate agreed to by the borrower and the lender, with RUS concurrence. The guarantee applies to the repayment of both principal and interest.

[58 FR 66264, Dec. 20, 1993]

§§ 1710.53–1710.99 [Reserved]

Subpart C—Loan Purposes and Basic Policies

§ 1710.100 General.

RUS makes loans and loan guarantees to finance the construction of electric distribution, transmission and generation facilities, including system improvements and replacements required to furnish and improve electric service in rural areas, and for demand side management, efficiency and energy conservation programs, and on grid and off grid renewable energy systems. In some circumstances, RUS may finance selected operating expenses of its borrowers. Loans made or guaranteed by the Administrator of RUS will be made in conformance with the Rural Electrification Act of 1936, as amended (7 U.S.C. 901 *et seq.*), and 7 CFR chapter XVII. RUS provides certain technical assistance to borrowers when necessary to aid the development of rural electric service and to protect loan security.

[58 FR 66264, Dec. 20, 1993, as amended at 78 FR 73365, Dec. 5, 2013]

§ 1710.101 Types of eligible borrowers.

(a) RUS makes loans to corporations, states, territories, and subdivisions and agencies thereof; municipalities; people's utility districts; and cooperative, nonprofit, limited-dividend, or mutual associations that provide or propose to provide:

(1) The retail electric service needs of rural areas, or

(2) The power supply needs of distribution borrowers under the terms of power supply arrangement satisfactory to RUS, or

(3) Eligible purposes under the Rural Energy Savings Program, including energy efficiency, renewable energy, energy storage or energy conservation measures and related services, improvements, investments, financing or relending.

(b) In making loans, RUS gives preference to states, territories, and subdivisions and agencies thereof; municipalities; people's utility districts; and cooperative, nonprofit, or limited-dividend associations. RUS does not make direct loans to individual consumers.

(c) For the purpose of determining eligibility of a distribution borrower not in default on the repayment of a loan made or guaranteed under the RE Act for a loan, loan guarantee, or lien accommodation, a default by a borrower from which a distribution borrower purchases wholesale power shall not:

(1) Be considered a default by the distribution borrower;

(2) Reduce the eligibility of the distribution borrower for assistance under the RE Act; or

(3) Be the cause, directly or indirectly, of imposing any requirement or restriction on the borrower as a condition of the assistance, except such requirements or restrictions as are necessary to implement a debt restructuring agreed on by the power supply borrower and RUS.

(d) For the purpose of determining the eligibility of a distribution borrower, RUS will consider whether the distribution borrower is current on its obligations to its wholesale power supplier under the RUS wholesale power contract.

(e) Nothing in paragraph (c) of this section relieves any distribution borrower that is a member of a power supply borrower in default on its obligations to RUS or operating under a debt restructuring agreement, of requirements set forth in RUS regulations, including, without limitation,

§ 1710.112(b)(6), or of any terms and conditions that the Administrator may otherwise impose on any borrower as a condition of obtaining a loan or loan guarantee (including, in appropriate cases, member guarantees).

(f) Except as provided in paragraph (g) of this section, former borrowers that have paid off all outstanding loans may reapply for a loan to serve RE Act beneficiary loads accruing from the time the former borrower's complete loan application is received by RUS. The determination of whether an area is rural will be based on the Census designation of the area at the time of the reapplication for a loan, if the area is not served by electric facilities financed by RUS. If the area is served by electric facilities financed by RUS, it will continue to be considered rural.

(g) Former borrowers that have prepaid all, or portions of outstanding insured and direct loans in accordance with RUS regulations must comply with the provisions of 7 CFR part 1786 before being considered eligible to borrow additional funds from RUS.

[58 FR 66264, Dec. 20, 1993, as amended at 78 FR 73365, Dec. 5, 2013; 85 FR 18418, Apr. 2, 2020]

§ 1710.102 Borrower eligibility for different types of loans.

(a) *Insured loans under section 305.* Insured loans are normally reserved for the financing of distribution and subtransmission facilities of both distribution and power supply borrowers, including, under certain circumstances, the implementation of demand side management, energy efficiency and energy conservation programs, and on grid and off grid renewable energy systems. In accordance with § 1710.110, the Administrator may require the borrower to obtain no more than 30 percent of the total debt financing required for a proposed project by means of a supplemental loan from another lender without an RUS guarantee.

(b) *Direct loans under section 4.* Direct loans are normally reserved for the financing of distribution and subtransmission facilities of both distribution and power supply borrowers, including, under certain circumstances, the implementation of demand side management, energy efficiency and energy conservation programs, and on grid and off grid renewable energy systems.

(c) *One hundred percent loan guarantees under section 306.* Both distribution and power supply borrowers are eligible for 100 percent loan guarantees under section 306 of the RE Act for any or all of the purposes set forth in § 1710.106, including, under certain circumstances, the implementation of demand side management, energy conservation programs, and on grid and off grid renewable energy systems. (See 7 CFR part 1712). These guarantees are normally used to finance bulk transmission and generation facilities, but they may also be used to finance distribution and subtransmission facilities. If a borrower applies for a section 306 loan guarantee to finance all or a portion of distribution and subtransmission facilities, such request will not affect the borrower's eligibility for an insured loan to finance any remaining portion of said facilities or for any future insured loan to finance other distribution or subtransmission facilities. A section 306 loan guarantee, however, may not be used to guarantee a supplemental loan required by § 1710.110.

(d) *One hundred percent loan guarantees under section 306A.* Under section 306A of the RE Act, both distribution and power supply borrowers are eligible under certain conditions to use an existing section 306 guarantee to refinance advances made on or before July 2, 1986 from a loan made by the Federal Financing Bank. (See 7 CFR part 1786.)

(e) *Ninety percent guarantees of private-sector loans under section 311.* Under section 311 of the RE Act, both distribution and power supply borrowers in the state of Alaska are eligible under certain conditions to obtain from RUS a 90 percent guarantee of a private-sector loan to refinance their Federal Financing Bank loans. (See 7 CFR part 1786.)

[57 FR 2832, Jan. 24, 1992, as amended at 58 FR 66264, Dec. 20, 1993; 66 FR 66294, Dec. 26, 2001; 78 FR 73365, Dec. 5, 2013]

§ 1710.103 Area coverage.

(a) Borrowers shall make a diligent effort to extend electric service to all unserved persons within their service area who:

(1) Desire electric service; and

(2) Meet all reasonable requirements established by the borrower as a condition of service.

(b) If economically feasible and reasonable considering the cost of providing such service and/or the effects on all consumers' rates, such service shall be provided, to the maximum extent practicable, at the rates and minimum charges established in the borrower's rate schedules, without the payment by such persons, other than seasonal or temporary consumers, of a contribution in aid of construction. A seasonal consumer is one that demands electric service only during certain seasons of the year. A temporary consumer is a seasonal or year-round consumer that demands electric service over a period of less than five years.

(c) Borrowers may assess contributions in aid of construction provided such assessments are consistent with the policy set forth in this section.

[57 FR 1053, Jan. 9, 1992, as amended at 60 FR 67404, Dec. 29, 1995]

§ 1710.104 Service to non-RE Act beneficiaries.

(a) To the greatest extent practical, loans are limited to providing and improving electric facilities to serve consumers that are RE Act beneficiaries. When it is determined by the Administrator to be necessary in order to furnish or improve electric service in rural areas, loans may, under certain circumstances, be made to finance electric facilities to serve consumers that are not RE Act beneficiaries.

(b) Loan funds may be approved for facilities to serve non-RE Act beneficiaries only if:

(1) The primary purpose of the loan is to furnish or improve service for RE Act beneficiaries; and

(2) The use of loan funds to serve non-RE Act beneficiaries is necessary and incidental to the primary purpose of the loan.

[57 FR 1053, Jan. 9, 1992; 57 FR 4513, Feb. 5, 1992, as amended at 58 FR 66264, Dec. 20, 1993]

§ 1710.105 State regulatory approvals.

(a) In States where a borrower is required to obtain approval of a project or its financing from a state regulatory authority, RUS may require that such approvals be obtained, if feasible for the borrower to do so, before the following types of loans are approved by RUS:

(1) Loans requiring an Environmental Impact Statement;

(2) Loans to finance generation and transmission facilities, when the loan request for such facilities is $25 million or more; and

(3) Loans for the purpose of assisting borrowers to implement demand side management and energy conservation programs and on and off grid renewable energy systems.

(b) At minimum, in the case of all loans in states where state regulatory approval is required of the project or its financing, such state approvals will be required before loan funds are advanced.

(c) In cases where state regulatory authority approval has been obtained, but the borrower has failed to proceed with the project in a timely manner according to the schedule contained in the borrower's project design manual, or if there are cost overruns or other developments that threaten loan feasibility or security, RUS may require the borrower to obtain a reaffirmation of the project and its financing from the state authority before any additional loan funds are advanced.

[57 FR 1053, Jan. 9, 1992; 57 FR 4513, Feb. 5, 1992, as amended at 58 FR 66265, Dec. 20, 1993]

§ 1710.106 Uses of loan funds.

(a) Funds from loans made or guaranteed by RUS may be used to finance:

(1) *Distribution facilities.* (i) The construction of new distribution facilities or systems, the cost of system improvements and removals less salvage value, the cost of ordinary replacements and removals less salvage value, needed to meet load growth requirements, improve the quality of service, or replace existing facilities.

(ii) The purchase, rehabilitation and integration of existing distribution facilities and associated service territory when the acquisition is an incidental and necessary means of providing or improving service to persons in rural areas who are not receiving adequate central station service, and the borrower is unable to finance the acquisition from other sources. See § 1710.107.

(2) *Transmission and generation facilities.* (i) The construction of new transmission and generation facilities or systems, the cost of system improvements and removals, less salvage value, the cost of ordinary replacements and removals less salvage value, needed to meet load growth, improve the quality of service, or replace existing facilities.

(ii) The purchase of an ownership interest in new or existing transmission or generation facilities to serve RE Act beneficiaries.

(3) *Warehouse and garage facilities.* The purchase, remodeling, or construction of warehouse and garage facilities required for the operation of a borrower's system. See paragraph (b) of this section.

(4) *Interest.* The payment of interest on indebtedness incurred by a borrower to finance the construction of generation and transmission facilities during the period preceding the date such facilities are placed into service, if requested by the borrower and found necessary by RUS.

(5) Certain costs incurred in demand side management, energy conservation programs and on and off grid renewable energy systems.

(6) Eligible Energy Efficiency and Conservation Programs pursuant to Subpart H of this part.

(b) In cases of financial hardship, as determined by the Administrator, loans may also be made to finance the following items:

(1) The headquarters office and other headquarters facilities in addition to those cited in paragraph (a)(4) of this section;

(2) General plant equipment, including furniture, office, transportation, data processing and other work equipment; and

(3) Working capital required for the initial operation of a new system.

(c) RUS will not make loans to finance the following:

(1) Electric facilities, equipment, appliances, or wiring located inside the premises of the Consumer, except for assets financed pursuant to an Eligible EE Program, and qualifying items included in a loan for Demand side management or energy resource conservation programs, or renewable energy systems.

(2) Facilities to serve consumers who are not RE Act beneficiaries unless those facilities are necessary and incidental to providing or improving electric service in rural areas (See § 1710.104);

(3) Any facilities or other purposes that a state regulatory authority having jurisdiction will not approve for inclusion in the borrower's rate base, or will not otherwise allow rates sufficient to repay with interest the debt incurred for the facilities or other purposes; and

(d) A distribution borrower may request a loan period of up to 4 years. Except in the case of loans for new generating and associated transmission facilities, a power supply borrower may request a loan period of not more than 4 years for transmission and substation facilities and improvements or replacements of generation facilities. The loan period for new generating facilities and DSM activities will be determined on a case-by-case basis. The Administrator may approve a loan period shorter than the period requested by the borrower, if in the Administrator's sole discretion, a loan made for the longer period would fail to meet RUS requirements for loan feasibility and loan security set forth in §§ 1710.112 and 1710.113, respectively.

(e)(1) If, in the sole discretion of the Administrator, the amount authorized for lending for municipal rate loans, hardship rate loans, and loan guarantees in a fiscal year is substantially less than the total amount eligible for RUS financing, RUS may limit the size of all loans of that type approved during the fiscal year. Depending on the amount of the shortfall between the amount authorized for lending and the loan application inventory on hand for each type of loan, RUS may either reduce the amount on an equal proportion basis for all applicants for that type of loan based on the amount of funds for which the applicant is eligible, or may shorten the loan period for which funding will be approved to less than the maximum of 4 years. All applications for the same type of loan approved during a fiscal year will be treated in the same manner, except that RUS will not limit funding to any borrower requesting an RUS loan or loan guarantee of $1 million or less.

(2) If RUS limits the amount of loan funds approved for borrowers, the Administrator shall notify all electric borrowers early in the fiscal year of the manner in which funding will be limited. The portion of the loan application that is not funded during that fiscal year may, at the borrower's option, be treated as a second loan application received by RUS at a later date. This date will be determined by RUS in the same manner for all affected loans and will be based on the availability of loan funds. The second loan application shall be considered complete except that the borrower must submit a certification from a duly authorized corporate official stating that funds are still needed for loan purposes specified in the original application and must notify RUS of any changes in its circumstances that materially affects the information contained in the original loan application or the primary support documents. See 7 CFR 1710.401(f).

(f)(1) For borrowers having one or more loans approved on or after October 1, 1991, advances of funds will be made only for the primary budget purposes included in the loan as shown on RUS Form 740c as amended and approved by RUS, or on a construction work plan or a construction work plan amendment approved by RUS. Each advance will be charged to the oldest outstanding note(s) having unadvanced funds for the primary budget purpose for which the request for advances was made, regardless of whether such notes are associated with loans approved before or after October 1, 1991, unless any conditions on advances under any of these notes have not been met by the borrower.

(2) For borrowers whose most recent loan was approved before October 1, 1991, advances will be made on the oldest outstanding note having unadvanced funds, unless any conditions on advances under such note have not been met by the borrower.

(g) A borrower is permitted to use up to 10 percent of the amount provided under this part to construct, improve, or acquire broadband infrastructure related to the project financed, subject to the requirements of 7 CFR part 1980, subpart M.

[57 FR 1053, Jan. 9, 1992, as amended at 58 FR 66265, Dec. 20, 1993; 60 FR 3730, Jan. 19, 1995; 62 FR 7922, Feb. 21, 1997; 64 FR 33178, June 22, 1999; 78 FR 73365, Dec. 5, 2013; 84 FR 32610, July 9, 2019; 85 FR 57081, Sept. 15, 2020]

§ 1710.107 Amount lent for acquisitions.

The maximum amount that will be lent for an acquisition is limited to the value of the property, as determined by RUS. If the acquisition price exceeds this amount, the borrower shall provide the remainder without RUS financial assistance.

§ 1710.108 Mergers and consolidations.

(a) RUS encourages its borrowers to consider merging or consolidating with another electric borrower when such action will contribute to greater operating efficiency and financial soundness.

(b) After a merger or consolidation, RUS will give priority consideration per § 1710.119 to the processing of loans for the surviving system to finance the integration and rehabilitation of electric facilities, if necessary, and the improvement or extension of electric service in rural areas. Such priority consideration will also be given in the case of a borrower that has merged or consolidated with an electric system that has not previously received RUS financial assistance, if such system was serving primarily rural residents at the time of the merger or consolidation and such rural residents will continue to be served by the merged or consolidated system. RUS does not make loans for costs incurred in effectuating mergers or consolidations, such as legal expenses or feasibility study costs.

§ 1710.109 Reimbursement of general funds and interim financing.

(a) Borrowers may request that a loan include funds to reimburse general funds and/or replace interim financing used to finance equipment and facilities that were included in an RUS-approved construction work plan, energy efficiency and conservation program work plan, work plan amendment or other RUS-approved plan, and for which loan funds have not been provided by RUS. Such reimbursement and/or replacement of interim financing may include the direct costs of procurement and construction, as well as the related cost of engineering, architectural, environmental and other studies and plans needed to support the project, when such cost is capitalized as part of the cost of the facilities.

(b) If procurement and/or construction of the equipment and facilities was completed prior to the current loan period, reimbursement, including replacement of interim financing, will be limited, except in cases of extreme financial hardship as determined by the Administrator, to the cost of procurement and construction completed during the period immediately preceding the current loan period, as specified in paragraph (c) of this section. As defined in § 1710.2, the loan period begins on the date shown on page 1 of RUS Form 740c, Cost Estimates and Loan Budget for Electric Borrowers.

(c) The period immediately preceding the current loan period for which reimbursement and replacement of interim financing is authorized under paragraph (b) of this section is 48 months. Policies for reimbursement of general funds and interim financing following certain mergers, consolidations, and transfers of systems substantially in their entirety are set forth in 7 CFR 1717.154.

(d) If the reimbursement of general funds and/or replacement of interim financing is for approved expenditures for equipment and facilities whose procurement and/or construction is completed during the current loan period, the time limits of paragraph (c) of this section do not apply.

[57 FR 1053, Jan. 9, 1992, as amended at 58 FR 66265, Dec. 20, 1993; 61 FR 66870, Dec. 19, 1996; 78 FR 73366, Dec. 5, 2013; 86 FR 36196, July 9, 2021]

§ 1710.110 Supplemental financing.

(a) Except in the case of financial hardship as determined by the Administrator, and following certain mergers, consolidations, and transfers of systems substantially in their entirety as set forth in 7 CFR 1717.154, applicants for a municipal rate loan will be required to obtain a portion of their loan funds from a supplemental source without an RUS guarantee, in the amounts set forth in paragraph (c) of this section. RUS will normally grant a lien accommodation to the supplemental lender. RUS does not require supplemental financing in conjunction with an RUS guaranteed loan. However, if a borrower elects to obtain supplemental financing in conjunction with a guaranteed loan, the granting of RUS's loan guarantee may be conditioned on the borrower's obtaining supplemental financing.

(b) The terms and conditions of supplemental financing and any security offered to the supplemental lender are subject to RUS approval. Generally, supplemental loans must have the same final maturity and be amortized in the same manner as RUS loans made concurrently. Borrowers may elect to repay the loans either in substantially equal periodic installments covering interest and principal, or in periodic installments that include interest and level amortization of principal.

(c) *Supplemental financing required for municipal rate loans*—(1) *Distribution borrowers.* (i) Distribution borrowers that had, as of December 31, 1980, an average consumer density of 2 or fewer consumers per mile or an average adjusted plant revenue ratio (APRR), as defined in § 1710.2, of over 9.0 shall obtain supplemental financing equal to 10 percent of their loan request.

(ii) All other distribution borrowers must obtain supplemental financing according to their plant revenue ratio (PRR), as defined in § 1710.2, based on the most recent year-end data available on the date of loan approval, as follows:

PRR	Supplemental loan percentage
9.00 and above	10
8.01–8.99	20
8.00 and below	30

(iii) If a distribution borrower enters into a merger, consolidation, or transfer of system substantially in its entirety, and the provisions of 7 CFR 1717.154(b) do not apply, required supplemental financing will be determined as follows for loans approved by RUS after December 19, 1996. If one of the merging parties met the criteria in paragraph (c)(1)(i) of this section prior to the effective date of the merger consolidation or transfer, the borrower will be required to obtain supplemental financing equal to 10 percent of any loan funds requested for facilities to serve consumers located in the territory formerly served by the "paragraph (c)(1)(i)" borrower. The required amount of supplemental financing for the rest of the loan will be determined according to the provisions of paragraph (c)(1)(ii) of this section.

(2) *Power supply borrowers.* The supplemental loan proportion required of a power supply borrower is based on the simple arithmetic mean of the supplemental loan proportions required of the borrower's distribution members.

(3) *Subsequent loans.* (i) If more than 5 percent of an insured loan made prior to November 1, 1993, or of a municipal rate loan is terminated or rescinded, the amount of supplemental financing required in the borrower's next loan after the rescission for which supplemental financing is required, pursuant to paragraph (a) of this section, will be adjusted to average the actual supplemental financing portion on the terminated or rescinded loan with the supplemental financing portion that would have been required on the new loan according to paragraphs (c)(1) and (2) of this section, in accordance with the formulas set forth in paragraphs (c)(3)(ii) and (iii) of this section.

(ii) If a borrower's supplemental financing requirement as set forth in paragraphs (a), (c)(1), and (c)(2) of this section has not changed between the most recent loan and the loan being considered, then the amount of supplemental financing required for the new loan will be computed as follows:

Supplemental financing amount, new loan $= [(A + B) \times C] - D$

where:

A = The total funds ($) actually advanced from the first loan, including both RUS loan funds and funds from the supplemental loan, plus any unadvanced funds still available to the borrower after the rescission.

B = The total amount ($) for facilities of the new loan request, including both RUS loan funds and funds from supplemental loans.

C = The proportion (%) of supplemental financing required on the loans according to paragraphs (a), (c)(1) and (c)(2) of this section.

D = The amount ($) of supplemental funds actually advanced on the first loan, plus any unadvanced supplemental funds still available to the borrower after the rescission.

(iii) If a borrower's supplemental financing requirement as set forth in paragraphs (a), (c)(1), and (c)(2) of this section has changed between the most recent loan and the loan being considered, then the amount of supplemental financing required for the new loan will be the weighted average of the portions otherwise applicable on the two loans and will be computed as follows:

Supplemental financing amount, new loan $= (A \times C_1) + (B \times C_2) - D$

where:

A = The total funds ($) actually advanced from the first loan, including both RUS loan funds and funds from the supplemental loan, plus any unadvanced funds still available to the borrower after the rescission.

B = The total amount ($) for facilities of the new loan request, including both RUS funds and funds from supplemental loans.

C_1 = The proportion (%) of supplemental financing required on the old loan according to paragraphs (a), (c)(1) and (c)(2) of this section.

C_2 = The proportion (%) of supplemental financing required on the new loan according to paragraphs (a), (c)(1) and (c)(2) of this section.

D = The amount ($) of supplemental funds actually advanced on the first loan, plus any unadvanced supplemental funds still available to the borrower after the rescission.

(d) Supplemental financing will not be required in connection with hardship rate loans. Borrowers that qualify for hardship rate loans but elect to take municipal rate loans instead, will be required to obtain supplemental financing pursuant to this section, unless at the time of loan approval, there are no funds remaining available for hardship loans, in which case supplemental financing will not be required.

[57 FR 1053, Jan. 9, 1992, as amended at 58 FR 66265, Dec. 20, 1993; 60 FR 3730, Jan. 19, 1995; 61 FR 66870, Dec. 19, 1996]

§ 1710.111 Refinancing.

(a) RUS makes loans or loan guarantees to refinance the outstanding indebtedness of borrowers in the following cases:

(1) Loans or loan guarantees to refinance long-term debt owed by borrowers to the Tennessee Valley Authority for credit extended under the terms of the Tennessee Valley Authority Act of 1933, as amended.

(2) Loan guarantees made in accordance with the provisions of section 306A of the RE Act to prepay a loan (or any loan advance thereunder) made by the Federal Financing Bank.

(b) In certain circumstances, RUS may make a loan to replace interim financing obtained for the construction of facilities (See § 1710.109).

§ 1710.112 Loan feasibility.

(a) RUS will make a loan only if there is reasonable assurance that the loan, together with all outstanding loans and other obligations of the borrower, will be repaid in full as scheduled, in accordance with the mortgage, notes, and loan contracts. The borrower must provide evidence satisfactory to the Administrator that the loan will be repaid in full as scheduled, and that all other obligations of the borrower will be met.

(b) Based on evidence submitted by the borrower and other information, RUS will use the following criteria to evaluate loan feasibility:

(1) Projections of power requirements, rates, revenues, expenses, margins, and other factors for the present system and proposed additions are based on reasonable assumptions and adequate supporting data and analysis, including analysis of a range of assumptions for the significant variables, when required by § 1710.300(d)(5).

(2) Projected revenues from the rates proposed by the borrower are adequate to meet the required TIER and DSC ratios based on the borrower's total costs, including the projected maximum debt service cost of the new loan.

(3) The economics of the borrower's operations and service area are such that consumers can reasonably be expected to pay the proposed rates required to cover all expenses and meet RUS TIER and DSC requirements, and the borrower can reasonably compete with other utilities and other energy sources to prevent substantial load loss while providing satisfactory service to its consumers.

(4) Risks of possible loss of substantial loads from large consumers or from load concentrations in particular industries will not substantially impair loan feasibility.

(5) Risks of loss of portions of the borrower's service territory from annexation or other causes will not substantially impair loan feasibility. If there appears to be a substantial risk, RUS may require additional information from the borrower, such as a summary and analysis of the risk by the borrower; state, county or local planning reports having information on projected growth or expansion plans of local communities; annexation plans of the municipalities in question; and any other relevant information.

(6) In states where rates or investment decisions are subject to approval by state regulatory authorities, there is reasonable expectation that such approvals will be forthcoming to enable repayment of the loan in full according to its terms.

(7) The experience and performance of the system's management is acceptable.

(8) In the case of joint ventures, the borrower has sufficient management control or other contractual safeguards with respect to the construction and operation of the jointly owned facility to ensure that the borrower's interests are protected and the credit risk is minimized.

(9) The borrower has implemented adequate financial and management controls and there are and have been no significant financial or other irregularities.

(10) The borrower's projected capitalization, measured by its equity as a percentage of total assets, is adequate to enable the borrower to meet its financial needs and to provide service consistent with the RE Act. Among the factors to be considered in reviewing the borrower's projected capitalization are the economic strength of the borrower's service territory, the inherent cost of providing service to the territory, the disparity in rates between the

borrower and neighboring utilities, the intensity of competition faced by the borrower from neighboring utilities and other power sources, and the relative amount of new capital investment required to serve existing or new loads.

[57 FR 1053, Jan. 9, 1992; 57 FR 4513, Feb. 5, 1992, as amended at 60 FR 3731, Jan. 19, 1995; 63 FR 51793, Sept. 29, 1998; 84 FR 32610, July 9, 2019]

§ 1710.113 Loan security.

(a) RUS makes loans only if, in the judgment of the Administrator, the security therefor is reasonably adequate and the loan will be repaid according to its terms within the time agreed.

(b) RUS generally requires that borrowers provide it with a first lien on all of the borrower's real and personal property, including intangible personal property and any property acquired after the date of the loan. This lien shall be in the form of a mortgage by the borrower to the Government or a deed of trust between the borrower and a trustee satisfactory to the Administrator, together with such security documents as RUS may deem necessary in a particular case.

(c)(1) When a borrower is unable by reason of preexisting encumbrances, or otherwise, to furnish a first mortgage lien on its entire system the Administrator may accept other forms of security, such as a pledge of revenues, if he or she determines such security is reasonably adequate and the form and nature thereof is otherwise acceptable.

(2) The Administrator, at his or her discretion, may approve the use of an indenture patterned after those indentures commonly used by utilities engaged in private market financing, in lieu of a mortgage as the security instrument for loans to power supply borrowers. The use of an indenture will be by mutual agreement of the borrower and the Administrator. The terms of each indenture and related loan agreement will be negotiated on a case by case basis to best meet the needs of the individual borrower and the Government. The provisions of the indenture and loan contract shall control, notwithstanding any provisions of 7 CFR Chapter XVII which may be in conflict therewith.

(d) In the case of loans that include the financing of electric facilities that are operated as an integral component of a non-RUS financed system (such as generation and transmission facilities co-owned with other electric utilities), the borrower shall, in addition to the mortgage lien on all of the borrower's electric facilities, furnish adequate assurance, in the form of contractual or other security arrangements, that the system will be operated on an efficient and continuous basis. Satisfactory evidence must also be provided that the non-RUS financed system is financially sound and under capable management. Examples of such evidence include financial reports, annual reports, Security and Exchange Commission 10K reports if the system is required to file them, credit reports from Standard and Poor's, Moodys or other recognized sources, reports to state regulatory authorities and the Federal Energy Regulatory Commission, and evidence of a successful track record in related construction projects.

(e) Additional controls on the borrower's financial, investment and managerial activities appear in the loan contract and mortgage required by RUS.

[57 FR 1053, Jan. 9, 1992, as amended at 62 FR 7665, Feb. 20, 1997]

§ 1710.114 TIER, DSC, OTIER and ODSC requirements.

(a) *General.* Requirements for coverage ratios are set forth in the borrower's mortgage, loan contract, or other contractual agreements with RUS. The requirements set forth in this section apply to borrowers that receive a loan approved by RUS on or after February 10, 1992. Nothing in this section, however, shall reduce the coverage ratio requirements of a borrower that has contractually agreed with RUS to a higher requirement.

(b) *Coverage ratios.* (1) Distribution borrowers. The minimum coverage ratios required of distribution borrowers whether applied on an annual or average basis, are a TIER of 1.25, DSC of 1.25, OTIER of 1.1, and ODSC of 1.1. OTIER and ODSC shall apply to distribution borrowers that receive a loan approved on or after January 29, 1996.

(2) The minimum coverage ratios required of power supply borrowers, whether applied on an annual or average basis, are a TIER of 1.05 and DSC of 1.00.

(3) When new loan contracts are executed, the Administrator may, case by case, increase the coverage ratios of distribution and power supply borrowers above the levels cited in paragraphs (b)(1) and (b)(2), respectively, of this section if the Administrator determines that the higher ratios are required to ensure reasonable security for and/or the repayment of loans made or guaranteed by RUS. Also, the Administrator may, case by case, reduce said coverage ratios if the Administrator determines that the lower ratios are required to ensure reasonable security for and/or the repayment of loans made or guaranteed by RUS. Policies for coverage ratios following certain mergers, consolidations, and transfers of systems substantially in their entirety are in 7 CFR 1717.155.

(4) If a distribution borrower has in service or under construction a substantial amount of generation and associated transmission plant financed at a cost of capital substantially higher than the cost of funds under section 305 of the RE Act, then the Administrator may establish, in his or her sole discretion, blended levels for TIER, DSC, OTIER, and ODSC based on the respective shares of total utility plant represented by said generation and associated transmission plant and by distribution and other transmission plant.

(c) *Requirements for loan feasibility.* To be eligible for a loan, borrowers must demonstrate to RUS that they will, on a pro forma basis, earn the coverage ratios required by paragraph (b) of this section in each of the years included in the borrower's long-range financial forecast prepared in support of its loan application, as set forth in subpart G of this part.

(d) *Requirements for maintenance of coverage ratios*—(1) *Prospective requirement.* Borrowers must design and implement rates for utility service to provide sufficient revenue (along with other revenue available to the borrower in the case of TIER and DSC) to pay all fixed and variable expenses, to provide and maintain reasonable working capital and to maintain on an annual basis the coverage ratios required by paragraph (b) of this section. Rates must be designed and implemented to produce at least enough revenue to meet the requirements of this paragraph under the assumption that average weather conditions in the borrower's service territory will prevail in the future, including average system damage and outages due to weather and the related costs. Failure to design and implement rates pursuant to the requirements of this paragraph shall be an event of default upon notice provided in accordance with the terms of the borrower's mortgage or loan contract.

(2) *Retrospective requirement.* The average coverage ratios achieved by a borrower in the 2 best years out of the 3 most recent calendar years must meet the levels required by paragraph (b) of this section. If a borrower fails to achieve these average levels, it must promptly notify RUS in writing. Within 30 days of such notification or of the borrower being notified in writing by RUS, whichever is earlier, the borrower, in consultation with RUS, must provide a written plan satisfactory to RUS setting forth the actions that will be taken to achieve the required coverage ratios on a timely basis. Failure to develop and implement a plan satisfactory to RUS shall be an event of default upon notice provided in accordance with the terms of the borrower's mortgage or loan contract.

(3) *Fixed and variable expenses,* as used in this section, include but are not limited to: all taxes, depreciation, maintenance expenses, and the cost of electric power and energy and other operating expenses of the electric system, including all obligations under the wholesale power contract, all lease payments when due, and all principal and interest payments on outstanding indebtedness when due.

(e) *Requirements for advance of funds.* (1) If a borrower applying for a loan has failed to achieve the coverage ratios required by paragraph (b) of this section during the latest 12 month period immediately preceding approval of the loan, or if any of the borrower's average coverage ratios for the 2 best years out of the most recent 3 calendar years were below the levels required in paragraph (b) of this section, RUS may withhold the advance of loan funds until the borrower has adopted an annual financial plan and operating budget satisfactory to RUS and taken such other action as RUS may require to demonstrate that the required coverage ratios will be maintained in the future and that the loan will be repaid with interest within the time agreed. Such other action may include, for example, increasing system operating efficiency and reducing costs or adopting a rate design that will achieve the required coverage ratios, and either placing such rates into effect or taking action to obtain regulatory authority approval of such rates. If failure to achieve the coverage ratios is due to unusual events beyond the control of the borrower, such as unusual weather, system outage due to a storm or regulatory delay in approving rate increases, then the Administrator may waive the requirement that the borrower take the remedial actions set forth in this paragraph, provided that such waiver will not threaten loan feasibility.

(2) With respect to any outstanding loan approved by RUS on or after February 10, 1992, if, based on actual or projected financial performance of the borrower, RUS determines that the borrower may not achieve its required coverage ratios in the current or future years, RUS may withhold the advance of loan funds until the borrower has taken remedial action satisfactory to RUS.

[60 FR 67404, Dec. 29, 1995, as amended at 61 FR 66871, Dec. 19, 1996; 65 FR 51748, Aug. 25, 2000]

§ 1710.115 Final maturity.

(a) RUS is authorized to make loans and loan guarantees with a final maturity of up to 35 years. The borrower may elect a repayment period for a loan not longer than the expected useful life of the facilities, not to exceed 35 years. Most of the electric facilities financed by RUS have a long useful life, often approximating 35 years. Some facilities, such as load management equipment and Supervisory Control and Data Acquisition equipment, have a much shorter useful life due, in part, to obsolescence. Operating loans to finance working capital required for the initial operation of a new system are a separate class of loans and usually have a final maturity of less than 10 years.

(b) Loans made or guaranteed by RUS for facilities owned by the borrower generally must be repaid with interest within a period, up to 35 years, that approximates the expected useful life of the facilities financed. The expected useful life shall be based on the weighted average of the useful lives that the borrower proposes for the facilities financed by the loan, provided that the proposed useful lives are deemed appropriate by RUS. RUS Form 740c, Cost Estimates and Loan Budget for Electric Borrowers, submitted as part of the loan application must include, as a note, either a statement certifying that at least 90 percent of the loan funds are for facilities that have a useful life of 33 years or longer, or a schedule showing the costs and useful life of those facilities with a useful life of less than 33 years. If the useful life determination proposed by the borrower is not deemed appropriate by RUS, RUS will base expected useful life on an independent evaluation, the manufacturer's estimated useful-life or RUS experience with like-property, as applicable. Final maturities for loans for the implementation of programs for demand side management and energy resource conservation and on and off grid renewable energy sources not owned by the borrower will be determined by RUS. Due to the uncertainty of predictions over an extended period of time, RUS may add up to 2 years to the composite average useful life of the facilities in order to determine final maturity.

(c) The term for loans made to finance Eligible EE Programs will be determined in accordance with § 1710.408 of this part.

(d) The Administrator may approve a repayment period longer than the expected useful life of the facilities financed, up to 35 years, if a longer final maturity is required to ensure repayment of the loan and loan security is adequate.

(e) The final maturity of a loan established pursuant to the provisions of this section shall not be extended as a result of extending loan payments under section 12(a) of the RE Act.

[58 FR 66265, Dec. 20, 1993, as amended at 60 FR 3731, Jan. 19, 1995; 68 FR 54236, May 7, 2003; 78 FR 73366, Dec. 5, 2013]

§ 1710.116 [Reserved]

§ 1710.117 Environmental review requirements.

Borrowers are required to comply with the environmental review requirements in accordance with 7 CFR part 1970, and other applicable environmental laws, regulations and Executive orders.

[81 FR 11026, Mar. 2, 2016]

§ 1710.118 [Reserved]

§ 1710.119 Loan processing priorities.

(a) Generally loans are processed in chronological order based on the date the complete application is received in the Regional office.

(b) The Administrator may give priority to processing loans that are required to meet the following needs:

(1) To restore electric service following a major storm or other catastrophe;

(2) To bring existing electric facilities into compliance with any environmental requirements imposed by Federal or state law that were not in effect at the time the facilities were originally constructed;

(3) To finance the capital needs of borrowers that are the result of a merger, consolidation, or a transfer of a system substantially in its entirety, provided that the merger, consolidation, or transfer has either been approved by RUS or does not need RUS approval pursuant to the borrower's loan documents (See 7 CFR 1717.154); or

(4) To correct serious safety problems, other than those resulting from borrower mismanagement or negligence.

(c) The Administrator may also change the normal order of processing loan applications when it is necessary to ensure that all loan authority for the fiscal year is utilized.

[57 FR 1053, Jan. 9, 1992, as amended at 61 FR 66871, Dec. 19, 1996]

§ 1710.120 Construction standards and contracting.

Borrowers shall follow all RUS requirements regarding construction work plans, energy efficiency and conservation program work plans, construction standards, approved materials, construction and related contracts, inspection procedures, and bidding procedures.

[57 FR 1053, Jan. 9, 1992, as amended at 78 FR 73366, Dec. 5, 2013]

§ 1710.121 Insurance requirements.

Borrowers are required to comply with certain requirements with respect to insurance and fidelity coverage as set forth in 7 CFR part 1788.

§ 1710.122 Equal opportunity and nondiscrimination.

Borrowers are required to comply with certain regulations on nondiscrimination in program services and benefits and on equal employment opportunity as set forth in RUS Bulletins 20–15 and 20–19 or their successors; 7 CFR parts 15 and 15b; and 45 CFR part 90.

§ 1710.123 Debarment and suspension.

Borrowers are required to comply with certain requirements on debarment and suspension as set forth in 2 CFR part 180, as adopted by USDA through 2 CFR part 417.

[79 FR 76002, Dec. 19, 2014]

§ 1710.124 Uniform Relocation Act.

Borrowers are required to comply with applicable provisions of 49 CFR part 24, which sets forth the requirements of the Uniform Relocation Assistance and Real Property Acquisition Policy Act of 1970 (Pub. L. 91–646; 84 Stat. 1894), as amended by the Uniform Relocation Act Amendments of 1987 (Pub. L. 100–17; 101 Stat. 246–256) and the Intermodal Surface Transportation Efficiency Act of 1991.

§ 1710.125 Restrictions on lobbying.

Borrowers are required to comply with certain requirements with respect to restrictions on lobbying activities. See 2 CFR part 418.

[79 FR 76002, Dec. 19, 2014]

§ 1710.126 Federal debt delinquency.

(a) Prior to approval of a loan or advance of funds, a borrower must report to RUS whether or not it is delinquent on any Federal debt, such as Federal income tax obligations or a loan or loan guarantee from another Federal agency. If delinquent, the reasons for the delinquency must be explained, and RUS will take such explanation into consideration in deciding whether to approve the loan or advance of funds.

(b) Applicants for a loan or loan guarantee must also certify that they have been informed of the collection options the Federal government may use to collect delinquent debt.

§ 1710.127 Drug free workplace.

Borrowers are required to comply with the Drug Free Workplace Act of 1988 (41 U.S.C. 8101 *et. seq.*) and the Act's implementing regulations (2 CFR part 421) when a borrower receives a Federal grant or enters into a procurement contract awarded pursuant to the provisions of the Federal Acquisition Regulation (title 48 CFR) to sell to a Federal agency property or services having a value of $25,000 or more.

[79 FR 76002, Dec. 19, 2014]

§§ 1710.128–1710.149 [Reserved]

Subpart D—Basic Requirements for Loan Approval

§ 1710.150 General.

The RE Act and prudent lending practice require that the Administrator make certain findings before approving an electric loan or loan guarantee. The borrower shall provide the evidence determined by the Administrator to be necessary to make these findings.

§ 1710.151 Required findings for all loans.

(a) *Area coverage.* Adequate electric service will be made available to the widest practical number of rural users in the borrower's service area during the life of the loan. See § 1710.103.

(b) *Feasibility.* The loan is feasible and it will be repaid on time according to the terms of the mortgage, note, and loan contract. At any time after the original determination of feasibility, the Administrator may require the borrower to demonstrate that the loan remains feasible if there have been, or are anticipated to be, material changes in the borrower's costs, loads, rates, rate disparity, revenues, or other relevant factors from the time that feasibility was originally determined. See § 1710.112 and subpart G of this part.

(c) *Security.* RUS will have a first lien on the borrower's total system or other adequate security, and adequate financial and managerial controls will be included in loan documents. See § 1710.113.

(d) *Interim financing.* For loans that include funds to replace interim financing, there is satisfactory evidence that the interim financing was used for purposes approved by RUS and that the loan meets all applicable requirements of this part.

(e) *Facilities for nonrural areas.* Whenever a borrower proposes to use loan funds for the improvement, expansion, construction, or acquisition of electric facilities for non-RE Act beneficiaries, there is satisfactory evidence that such funds are necessary and incidental to furnishing or improving electric service for RE Act beneficiaries. See § 1710.104.

(f) *Facilities to be included in rate base.* In states having jurisdiction, the borrower has provided satisfactory evidence based on the information available, such as an opinion of counsel, that the state regulatory authority will not exclude from the borrower's rate base any of the facilities included in the loan request, or otherwise prevent the borrower from charging rates sufficient to repay with interest the debt incurred for the facilities. Such evidence may be based on, but not necessarily limited to, the provisions of applicable state laws; the rules and policies of the state authority; precedents in other similar cases; statements made by the state authority; any assurances given to the borrower by the state authority; and other relevant information and experience.

§ 1710.152 Primary support documents.

The following primary support documents and studies must be prepared by the borrower for approval by RUS in order to support a loan application:

(a) *Load forecast.* The load forecast provides the borrower and RUS with an understanding of the borrower's future system loads, the factors influencing those loads, and estimates of future loads. The load forecast provides a basis for projecting annual electricity (kWh) sales and revenues, and for engineering estimates of plant additions required to provide reliable service to meet the forecasted loads. Subpart E of this part contains the information to be included in a load forecast and when an approved load forecast is required.

(b) *Construction work plan (CWP).* The CWP shall specify and document the capital investments required to serve a borrower's planned new loads, improve service reliability and quality, and service the changing needs of existing loads. The requirements for a CWP are set forth in subpart F of this part.

(c) *Long-range financial forecasts.* RUS encourages borrowers to maintain on a current basis a long-range financial forecast, which should be used by a borrower's board of directors and manager to guide the system toward its financial goals. The forecast submitted in support of a loan application shall show the projected results of future actions planned by the board of directors. The requirements for a long-range financial forecast are set forth in subpart G of this part.

(d) *Environmental review requirements.* A borrower must comply with the environmental review requirements in accordance with 7 CFR part 1970.

(e) *EE Program work plan (EEWP).* In the case of a loan application to finance an Eligible Energy Efficient Program, an EE Program work plan shall be prepared in lieu of a traditional CWP required pursuant to paragraph (b) of this section. The requirements for an EEWP are set forth in § 1710.255 and in subpart H of this part.

[57 FR 1053, Jan. 9, 1992, as amended at 65 FR 14786, Mar. 20, 2000; 78 FR 73366, Dec. 5, 2013; 81 FR 11026, Mar. 2, 2016]

§ 1710.153 Additional requirements and procedures.

Additional requirements and procedures for obtaining RUS financial assistance are set forth in 7 CFR part 1712 for loan guarantees, and in 7 CFR part 1714 for insured loans.

§ 1710.154 Board of Director Resolutions.

Specific actions that require a Board of Director Resolution from a borrower:

(a) Board approval of loan documents;

(b) Major change in the terms of a loan, *i.e.* maturity;

(c) Initial access to RD Apply (or successor RUS online application systems);

(d) Requests for approval by a Board, acting as the regulatory authority, for any departure from the RUS Uniform System of Accounts with the exception of those deferrals specifically identified in § 1767.13(d); and

(e) eAuthentication requirements.

[84 FR 32610, July 9, 2019]

§§ 1710.155–1710.199 [Reserved]

Subpart E—Load Forecasts

SOURCE: 65 FR 14786, Mar. 20, 2000, unless otherwise noted.

§ 1710.200 Purpose.

This subpart contains RUS policies for the preparation, review, approval and use of load forecasts and load forecast work plans. A load forecast is a thorough study of a borrower's electric loads and the factors that affect those loads in order to estimate, as accurately as practicable, the borrower's future requirements for energy and capacity. The load forecast of a power supply borrower includes and integrates the load forecasts of its member systems. An approved load forecast, if required by this subpart, is one of the primary documents that a borrower is required to submit to support a loan application.

§ 1710.201 General.

(a) The policies, procedures and requirements in this subpart are intended to implement provisions of the loan documents between RUS and the electric borrowers and are also necessary to support approval by RUS of requests for financial assistance.

(b) Notwithstanding any other provisions of this subpart, RUS may require any power supply or distribution borrower to prepare a new or updated load forecast for RUS approval or to maintain an approved load forecast on an ongoing basis, if such documentation is necessary for RUS to determine loan feasibility, or to ensure compliance under the loan documents.

§ 1710.202 Requirement to prepare a load forecast—power supply borrowers.

(a) A power supply borrower with a total utility plant of $500 million or more must provide a load forecast in support of any request for RUS financial assistance. The borrower must also maintain a load forecast work plan on file. The borrower's load forecast must be prepared pursuant to the load forecast work plan.

(b) A power supply borrower that is a member of another power supply borrower that has a total utility plant of $500 million or more must provide an approved load forecast in support of any request for RUS financial assistance. The member power supply borrower may comply with this requirement by participation in and inclusion of its load forecasting information in the load forecast of its power supply borrower. The load forecasts must be prepared pursuant to the load forecast work plan.

(c) A power supply borrower that has total utility plant of less than $500 million and that is not a member of another power supply borrower with a total utility plant of $500 million or more must provide a load forecast that meets the requirements of this subpart in support of an application for any RUS loan or loan guarantee which exceeds $50 million.

[84 FR 32610, July 9, 2019]

§ 1710.203 Requirement to prepare a load forecast—distribution borrowers.

(a) A distribution borrower that is a member of a power supply borrower, with a total utility plant of $500 million or more must provide a load forecast in support of any request for RUS financial assistance. The distribution borrower may comply with this requirement by participation in and inclusion of its load forecasting information in the approved load forecast of its power supply borrower. The distribution borrower's load forecast must be prepared pursuant to the load forecast work plan of its power supply borrower.

(b) A distribution borrower that is a member of a power supply borrower which is itself a member of another power supply borrower that has a total utility plant of $500 million or more must provide a load forecast in support of any request for RUS financial assistance. The distribution borrower may comply with this requirement by participation in and inclusion of its load forecasting information in the load forecast of its power supply borrower. The distribution borrower's load forecast must be prepared pursuant to the load forecast work plan of the power supply borrower with total utility plant in excess of $500 million.

(c) A distribution borrower that is a member of a power supply borrower with a total utility plant of less than $500 million must provide a load forecast that meets the requirements of this subpart in support of an application for any RUS loan or loan guarantee that exceeds $3 million or 5 percent of total utility plant, whichever is greater. The distribution borrower may comply with this requirement by participation in and inclusion of its load forecasting information in the load forecast of its power supply borrower.

(d) A distribution borrower with a total utility plant of less than $500 million and that is unaffiliated with a power supply borrower must provide a load forecast that meets the requirements of this subpart in support of an application for any RUS loan or loan guarantee which exceeds $3 million or 5 percent of total utility plant, whichever is greater.

(e) A distribution borrower with a total utility plant of $500 million or more must provide a load forecast in support of any request for RUS financing assistance. The borrower must also maintain a load forecast work plan. The distribution borrower may comply with this requirement by participation in and inclusion of its load forecasting information in the load forecast of its power supply borrower.

[84 FR 32610, July 9, 2019]

§ 1710.204 [Reserved]

§ 1710.205 Minimum requirements for all load forecasts.

(a) *Contents of load forecast.* All load forecasts submitted by borrowers for approval must include:

(1) A narrative describing the borrower's system, service territory, and consumers;

(2) A narrative description of the borrower's load forecast including future load projections, forecast assumptions, and the methods and procedures used to develop the forecast;

(3) Projections of usage by consumer class, number of consumers by class, annual system peak demand, and season of peak demand for the number of years agreed upon by RUS and the borrower;

(4) A summary of the year-by-year results of the load forecast in a format that allows efficient transfer of the information to other borrower planning or loan support documents;

(5) The load impacts of a borrower's demand side management and energy efficiency and conservation program activities, if applicable;

(6) Graphic representations of the variables specifically identified by management as influencing a borrower's loads; and

(7) A database that tracks all relevant variables that might influence a borrower's loads.

(b) *Formats.* RUS does not require a specific format for the narrative, documentation, data, and other information in the load forecast, provided that all required information is included and available. All data must be in a tabular form that can be transferred electronically to RUS computer software applications. RUS will evaluate borrower load forecasts for readability, understanding, filing, and electronic access. If a borrower's load forecast is submitted in a format that is not readily usable by RUS or is incomplete, RUS will require the borrower to submit the load forecast in a format acceptable to RUS.

(c) *Document retention.* The borrower must retain its latest load forecasts and supporting documentation. Any load forecast work plan must be retained as part of the load forecast.

(d) *Consultation with RUS.* The borrower must designate and make appropriate staff and consultants available for consultation with RUS to facilitate RUS review of the load forecast when requested by RUS.

(e) *Correlation and consistency with other RUS loan support documents.* If a borrower relies on an approved load forecast or an update of an approved load forecast as loan support, the borrower must demonstrate that the approved load forecast and the other primary support documentation for the loan were reconciled. For example, both the load forecast and the financial forecast require input assumptions for wholesale power costs, distribution costs, other systems costs, average revenue per kWh, and inflation. Also, a borrower's engineering planning documents, such as the construction work plan, incorporate consumer and usage per consumer projections from the load forecast to develop system design criteria. The assumptions and data common to all the documents must be consistent.

(f) *Coordination.* A load forecast of a power supply borrower must consider the load forecasts of all its member systems.

[84 FR 32610, July 9, 2019]

§ 1710.206 Requirements for load forecasts prepared pursuant to a load forecast work plan.

(a) *Contents of load forecasts prepared under a load forecast work plan.* In addition to the minimum requirements for load forecasts under § 1710.205, load forecasts developed and submitted by borrowers required to have a load forecast work plan shall include the following:

(1) Scope of the load forecast. The narrative shall address the overall approach, time periods, and expected internal and external uses of the forecast. Examples of internal uses include providing information for developing or monitoring demand side management programs, supply resource planning, load flow studies, wholesale

power marketing, retail marketing, cost of service studies, rate policy and development, financial planning, and evaluating the potential effects on electric revenues caused by competition from alternative energy sources or other electric suppliers. Examples of external uses include meeting state and Federal regulatory requirements, obtaining financial ratings, and participation in reliability council, power pool, regional transmission group, power supplier or member system forecasting and planning activities.

(2) Resources used to develop the load forecast. The discussion shall identify and discuss the borrower personnel, consultants, data processing, methods and other resources used in the preparation of the load forecast. The borrower shall identify the borrower's member and, as applicable, member personnel that will serve as project leaders or liaisons with the authority to make decisions and commit resources within the scope of the current and future work plans.

(3) A comprehensive description of the database used in the study. The narrative shall describe the procedures used to collect, develop, verify, validate, update, and maintain the data. A data dictionary thoroughly defining the database shall be included. The borrower shall make all or parts of the database available or otherwise accessible to RUS in electronic format, if requested.

(4) A narrative for each new load forecast or update of a load forecast discussing the methods and procedures used in the analysis and modeling of the borrower's electric system loads as provided for in the load forecast work plan.

(5) A narrative discussing the borrower's past, existing, and forecast of future electric system loads. The narrative must identify and explain substantive assumptions and other pertinent information used to support the estimates presented in the load forecast.

(6) A narrative discussing load forecast uncertainty or alternative futures that may determine the borrower's actual loads. Examples of economic scenarios, weather conditions, and other uncertainties that borrowers may decide to address in their analysis include:

(i) Most-probable assumptions, with normal weather;

(ii) Pessimistic assumptions, with normal weather;

(iii) Optimistic assumptions, with normal weather;

(iv) Most-probable assumptions, with severe weather;

(v) Most-probable assumptions, with mild weather;

(vi) Impacts of wholesale or retail competition; or

(vii) new environmental requirements.

(7) A summary of the forecast's results on an annual basis. Include alternative futures, as applicable. This summary shall be designed to accommodate the transfer of load forecast information to a borrower's other planning or loan support documents. Computer-generated forms or electronic submissions of data are acceptable. Graphs, tables, spreadsheets or other exhibits shall be included throughout the forecast as appropriate.

(8) A narrative discussing the coordination activities conducted between a power supply borrower and its members, as applicable, and between the borrower and RUS.

(b) *Compliance with a load forecast work plan.* A borrower required to maintain a load forecast work plan must also be able to demonstrate that both it and its RUS borrower members are in compliance with its load forecast work plan.

[65 FR 14786, Mar. 20, 2000, as amended at 84 FR 32611, July 9, 2019]

§ 1710.207 RUS criteria for load forecasts by distribution borrowers.

Load forecasts submitted by distribution borrowers that are unaffiliated with a power supply borrower, or by distribution borrowers that are members of a power supply borrower that has a total utility plant less than $500 million and that is not itself a member of another power supply borrower with a total utility plant of $500 million or more must satisfy the following minimum criteria:

(a) The borrower considered all known relevant factors that influence the consumption of electricity and the known number of consumers served at the time the study was developed;

(b) The borrower considered and identified all loads on its system of RE Act beneficiaries and non-RE Act beneficiaries;

(c) The borrower developed an adequate supporting data base and considered a range of relevant assumptions; and

(d) The borrower provided RUS with adequate documentation and assistance to allow for a thorough and independent review.

[65 FR 14786, Mar. 20, 2000, as amended at 84 FR 32611, July 9, 2019]

§ 1710.208 RUS criteria for load forecasts by power supply borrowers and by distribution borrowers.

All load forecasts submitted by power supply borrowers and by distribution borrowers must satisfy the following criteria:

(a) The borrower objectively analyzed all known relevant factors that influence the consumption of electricity and the known number of customers served at the time the study was developed;

(b) The borrower considered and identified all loads on its system of RE Act beneficiaries and non-RE Act beneficiaries;

(c) The borrower developed an adequate supporting database and analyzed a reasonable range of relevant assumptions and alternative futures;

(d) The borrower adopted methods and procedures in general use by the electric utility industry to develop its load forecast;

(e) The borrower used valid and verifiable analytical techniques and models;

(f) The borrower provided RUS with adequate documentation and assistance to allow for a thorough and independent review; and

[65 FR 14786, Mar. 20, 2000, as amended at 84 FR 32611, July 9, 2019]

§ 1710.209 Requirements for load forecast work plans.

(a) In addition to the load forecast required under §§ 1710.202 and 1710.203, any power supply borrower with a total utility plant of $500 million or more and any distribution borrower with a total utility plant of $500 million or more must maintain a load forecast work plan. RUS borrowers that are members of a power supply borrower with a total utility plant of $500 million or more must cooperate in the preparation of and submittal of the load forecast work plan of their power supply borrower.

(b) A load forecast work plan establishes the process for the preparation and maintenance of a comprehensive database for the development of the borrower's load forecast, and load forecast updates. The load forecast work plan is intended to develop and maintain a process that will result in load forecasts that will meet the borrowers' own needs and the requirements of this subpart. A work plan represents a commitment by a power supply borrower and its members, or by a large unaffiliated distribution borrower, that all parties concerned will prepare their load forecasts in a timely manner pursuant to the load forecast work plan and they will modify the load forecast work plan as needed to address changing circumstances or enhance the usefulness of the load forecast work plan.

(c) A load forecast work plan for a power supply borrower and its members must cover all member systems, including those that are not borrowers. However, only members that are borrowers, including the power supply borrower, are required to follow the load forecast work plan in preparing their respective load forecasts. Each borrower is individually responsible for forecasting all its RE Act beneficiary and non-RE Act beneficiary loads.

(d) A load forecast work plan must outline the coordination and preparation requirements for both the power supply borrower and its members.

(e) A load forecast work plan must describe the borrower's process and methods to be used in producing the load forecast.

(f) Load forecast work plans for borrowers with residential demand of 50 percent or more of total kWh must provide for a residential consumer survey at least every 5 years to obtain data on appliance and equipment saturation and electricity demand. Any such borrower that is experiencing or anticipates changes in usage patterns shall consider surveys on a more frequent schedule. Power supply borrowers shall coordinate such surveys with their members. Residential consumer surveys may be based on the aggregation of member-based samples or on a system-wide sample, provided that the latter provides for relevant regional breakdowns as appropriate.

(g) Load forecast work plans must provide for RUS review of the load forecasts as the load forecast is being developed.

(h) A power supply borrower's work plan must have the concurrence of the majority of the members that are borrowers.

[84 FR 32611, July 9, 2019]

§ 1710.210 Waiver of requirements or approval criteria.

For good cause shown by the borrower, the Administrator may waive any of the requirements applicable to borrowers in this subpart if the Administrator determines that waiving the requirement will not significantly affect accomplishment of RUS' objectives and if the requirement imposes a substantial burden on the borrower. The borrower's general manager must request the waiver in writing.

§§ 1710.211–1710.249 [Reserved]

Subpart F—Construction Work Plans and Related Studies

§ 1710.250 General.

(a) An ongoing, integrated planning system is needed by borrowers to determine their short-term and long-term needs for plant additions, improvements, replacements, and retirements. The primary components of the system consist of long-range engineering plans, construction work plans (CWPs), CWP amendments, and special engineering and cost studies. Long range engineering plans identify plant investments required over a period of 10–20 years or more. CWPs specify and document plant requirements for the short-term, usually 4 years, and special engineering and cost studies are used to support CWPs and to identify and document requirements for specific items or purposes, such as load management equipment, System Control and Data Acquisition equipment, sectionalizing investments, and additions of generation capacity and associated transmission plant.

(b) A long range engineering plan specifies and supports the major system additions, improvements, replacements, and retirements needed for an orderly transition from the existing system to the system required 10 or more years in the future. The planned future system should be based on the most technically and economically sound means of serving the borrower's long-range loads in a reliable and environmentally acceptable manner, and it should ensure that planned facilities will not become obsolete prematurely.

(c) A CWP shall include investment cost estimates and supporting engineering and cost studies to demonstrate the need for each proposed facility or activity and the reasonableness of the investment projections and the engineering assumptions used in sizing the facilities. The CWP must be consistent with the borrower's long range engineering plan and both documents must be consistent with the borrower's RUS-approved power requirements study.

(d) Applications for a loan or loan guarantee from RUS (new loans or budget reclassifications) must be supported by a current CWP approved by RUS. RUS approval of these plans relates only to the facilities, equipment, and other purposes to be financed by RUS, and means that the plans provide an adequate basis from a planning and engineering standpoint to support RUS financing. RUS approval of the plans does not mean that RUS approves of the facilities, equipment, or other purposes for which the borrower is not seeking RUS financing. If RUS disagrees with a borrower's estimate of the cost of one or more facilities for which RUS financing is sought, RUS may adjust the estimate after consulting with the borrower and explaining the reasons for the adjustment.

(e) Except as provided in paragraph (f) of this section, to be eligible for RUS financing, the facilities, including equipment and other items, included in a CWP must be approved by RUS and receive Environmental Clearance before the start of construction. This requirement also applies to any amendments to a CWP required to add facilities to a CWP or to make significant physical changes in the facilities already included in a CWP. Provision for funding of "minor projects" under an RUS loan guarantee is permitted on the same basis as that discussed for insured loan funds in 7 CFR part 1721, Post-Loan Policies and Procedures for Insured Electric Loans.

(f) In the case of damage caused by storms and other natural catastrophes, a borrower may proceed with emergency repair work before a CWP or CWP amendment is prepared by the borrower and approved by RUS, without losing eligibility for RUS financing of the repairs. The borrower must notify the RUS regional office in writing after the natural catastrophe, of its preliminary estimates of damages and repair costs. Not later than 120 days after the natural catastrophe, the borrower must submit to RUS for approval, a CWP or CWP amendment detailing the repairs.

(g) A CWP may be amended or augmented when the borrower can demonstrate the need for the changes.

(h) A borrower's CWP or special engineering studies must be supported by the appropriate level of environmental review documentation, in accordance with 7 CFR part 1970.

(i) All engineering activities required by this subpart must be performed by qualified engineers, who may be staff employees of the borrower or outside consultants. All engineering services must be reviewed by a licensed professional engineer.

(j) Upon written request from a borrower, RUS may waive in writing certain requirements with respect to long-range engineering plans and CWPs if RUS determines that such requirements impose a substantial burden on the borrower and that waiving the requirements will not significantly affect the accomplishment of the objectives of this subpart. For example, if a borrower's load is forecast to remain constant or decline during the planning period, RUS may waive those portions of the plans that relate to load growth.

[84 FR 32611, July 9, 2019]

§ 1710.251 Construction work plans—distribution borrowers.

(a) All distribution borrowers must maintain a current CWP covering all new construction, improvements, replacements, and retirements of distribution and transmission plant, and improvements replacements, and retirements of any generation plant. Construction of new generation capacity need not be included in a CWP but must be specified and supported by specific engineering and cost studies. (See § 1710.253.)

(b) A distribution borrower's CWP shall typically cover a construction period of 4 years and includes all facilities to be constructed which are eligible for RUS financing, whether or not RUS financial assistance will be sought or be available for certain facilities. Any RUS financing provided for the facilities will be limited to a 4 year loan period. The construction period covered by a CWP in support of a loan application shall not be shorter than the loan period requested for financing of the facilities.

(c) The facilities, equipment and other items included in a distribution borrower's CWP may include:

(1) Line extensions required to connect consumers, improve service reliability or improve voltage conditions;

(2) Distribution tie lines to improve reliability of service and voltage regulation;

(3) Line conversions and changes required to improve existing services or provide additional capacity for new consumers;

(4) New substation facilities or additions to existing substations;

(5) Transmission and substation facilities required to support the distribution system;

(6) Distribution equipment required to serve new consumers or to provide adequate and dependable service to existing consumers, including replacement of existing plant facilities;

(7) Outdoor lights;

(8) Communications equipment and meters;

(9) Headquarters facilities;

(10) Improvements, replacements, and retirements of generation facilities;

(11) Load management equipment, automatic sectionalizing facilities, and

centralized System Control and Data Acquisition equipment. Load management equipment eligible for financing, including the related costs of installation, is limited to capital equipment designed to influence the time and manner of consumer use of electricity, which includes peak clipping and load shifting. To be eligible for financing, such equipment must be owned by the borrower, although it may be located inside or outside a consumer's premises;

(12) The cost of engineering, architectural, environmental, and other studies and plans needed to support the construction of facilities, when such cost is capitalized as part of the cost of the facilities; and

(13) Other items that are specifically determined by RUS as being eligible for financing prior to inclusion in the CWP.

[57 FR 1053, Jan. 9, 1992; 57 FR 4513, Feb. 5, 1992, as amended at 60 FR 3731, Jan. 19, 1995; 60 FR 67405, Dec. 29, 1995; 84 FR 32612, July 9, 2019; 86 FR 36196, July 9, 2021]

§ 1710.252 Construction work plans—power supply borrowers.

(a) All power supply borrowers must maintain a current CWP covering all new construction, improvements, replacements, and retirements of distribution and transmission plant, and improvements, replacements, and retirements of generation plant. Applications for RUS financial assistance for such facilities must be supported by a current, RUS-approved CWP. Construction of new generation capacity need not be included in a CWP but must be specified and supported by specific engineering and cost studies.

(b) Typically a power supply borrower's CWP shall cover a period of 4 years. While comprehensive CWP's are desired, if there are extenuating circumstances RUS may accept a single-purpose transmission or generation CWP in support of a loan application or budget reclassification. The construction period covered by a CWP in support of a loan application shall not be shorter than the loan period requested for financing of the facilities.

(c) Facilities, equipment, and other items included in a power supply borrower's CWP may include:

(1) Distribution and related facilities as set forth in § 1710.251(c);

(2) Transmission facilities required to deliver the power needed to serve the existing and planned new loads of the borrower and its members, and to improve service reliability, including tie lines for improved reliability of service, line conversions, improvements and replacements, new substations and substation improvements and replacements, and Systems Control and Data Acquisition equipment, including communications, dispatching and sectionalizing equipment, and load management equipment;

(3) The borrower's proportionate share of transmission facilities required to tie together the operating systems of supporting power pools and to connect with adjacent power suppliers;

(4) Improvements and replacements of generation facilities; and

(5) The cost of engineering, architectural, environmental and other studies and plans needed to support the construction of facilities, when such cost is capitalized as part of the cost of the facilities.

(d) A CWP for transmission facilities shall normally include studies of load flows, voltage regulation, and stability characteristics to demonstrate system performance and needs.

[57 FR 1053, Jan. 9, 1992, as amended at 60 FR 3731, Jan. 19, 1995; 60 FR 67405, Dec. 29, 1995; 84 FR 32612, July 9, 2019; 86 FR 36196, July 9, 2021]

§ 1710.253 Engineering and cost studies—addition of generation capacity.

(a) The construction or purchase of additional generation capacity and associated transmission facilities by a power supply or distribution borrower, including the replacement of existing capacity, shall be supported by comprehensive project-specific engineering and cost studies as specified by RUS. The studies shall cover a period from the beginning of the project to at least 10 years after the start of commercial operation of the facilities.

(b) The studies must include comprehensive economic present-value analyses of the costs and revenues of the available self-generation, load management, energy conservation, and purchased-power options, including assessments of service reliability and financing requirements and risks. An analysis of purchased power options, including an analysis of available alternate sources of power shall be included. The analysis should include the terms and conditions of any requests for proposals and responses to such requests.

(c) Generally, studies of self-generation, load management, and energy conservation options shall include, as appropriate, analyses of:

(1) Capital and operating costs;

(2) Financing requirements and risks;

(3) System reliability;

(4) Alternative unit sizes;

(5) Alternative types of generation;

(6) Fuel alternatives;

(7) System stability;

(8) Load flows; and

(9) System dispatching.

(d) At the request of a borrower, RUS, in its sole discretion, may waive specific requirements of this section if such requirements imposed a substantial burden on the borrower and if such waiver will not significantly affect the accomplishment of the objectives of this subpart.

[57 FR 1053, Jan. 9, 1992, as amended at 84 FR 32612, July 9, 2019]

§ 1710.254 [Reserved]

§ 1710.255 Energy efficiency work plans—energy efficiency borrowers.

(a) All energy efficiency borrowers must maintain a current EEWP covering in aggregate all new construction, improvements, replacements, and retirements of energy efficiency related equipment and activities;

(b) An energy efficiency borrower's EEWP shall cover a period of between 2 and 4 years, and include all facilities to be constructed or improved which are eligible for RUS financing, whether or not RUS financial assistance will be sought or be available for certain facilities. The construction period covered by an EEWP in support of a loan application shall not be shorter than the loan period requested for financing of the facilities;

(c) The borrower's EEWP may only include facilities, equipment and other activities that have been approved by RUS as a part of an Eligible Energy Efficiency and Conservation Program pursuant to subpart H of this part;

(d) The borrower's EEWP must be consistent with the documentation provided as part of the current RUS approved EE Program as outlined in § 1710.410(c); and

(e) The borrower's EEWP must include an estimated schedule for the implementation of included projects.

[78 FR 73366, Dec. 5, 2013, as amended at 84 FR 32612, July 9, 2019]

§§ 1710.256–1710.299 [Reserved]

Subpart G—Long-Range Financial Forecasts

§ 1710.300 General.

(a) RUS encourages borrowers to maintain a current long-range financial forecast. The forecast should be used by the board of directors and the manager to guide the system towards its financial goals.

(b) A borrower must prepare, for RUS review and approval, a long-range financial forecast in support of its loan application. The forecast must demonstrate that the borrower's system is economically viable and that the proposed loan is financially feasible. Loan feasibility will be assessed based on the criteria set forth in § 1710.112.

(c) The financial forecast and related projections submitted in support of a loan application shall include:

(1) The projected results of future actions planned by the borrower's board of directors;

(2) The financial goals established for margins, TIER, DSC, equity, and levels of general funds to be invested in plant;

(3) A pro forma balance sheet, statement of operations, and general funds summary projected for each year during the forecast period;

(4) A full explanation of the assumptions, supporting data, and analysis used in the forecast, including the methodology used to project loads, rates, revenue, power costs, operating expenses, plant additions, and other factors having a material effect on the balance sheet and on financial ratios such as equity, TIER, and DSC;

(5) Current and projected cash flows;

(6) Projections of future borrowings and the associated interest and principal expenses required to meet the projected investment requirements of the system;

(7) Current and projected kW and kWh energy sales;

(8) Current and projected unit prices of significant variables such as retail and wholesale power prices, average labor costs, and interest;

(9) Current and projected system operating costs, including, but not limited to, wholesale power costs, depreciation expenses, labor costs, and debt service costs;

(10) Current and projected revenues from sales of electric power and energy;

(11) Current and projected non-operating income and expense;

(12) A discussion of the historical experience of the borrower, and in the case of a power supply borrower its member systems as appropriate, with respect to the borrower's market competitiveness as it relates to the rates charged for electricity, competition from other fuels, and other factors. Additional data and analysis may be required by RUS on a case by case basis to assess the probable future competitiveness of those borrowers that have a history of serious competitive problems; and

(13) An analysis of the effects of major factors, such as projected increases in rates charged for electricity, on the ability of the borrower, and in the case of a power supply borrower its member systems, to compete with neighboring utilities and other energy sources.

(d) The following plans, studies and assumptions shall be used in developing the financial forecast:

(1) The RUS-approved CWP;

(2) RUS-approved power requirements data;

(3) RUS-approved EE Program work plan;

(4) The current rate schedules or new rates;

(5) Future plant additions and operating expenses projected at anticipated future cost levels rather than in constant dollars, with the annual rate of inflation for major items specified; and

(6) A sensitivity analysis may be required by RUS on a case-by-case basis taking into account such factors as the number and type of large power loads, projections of future borrowings and the associated interest, projected loads, projected revenues, and the probable future competitiveness of the borrower. When RUS determines that a sensitivity analysis is necessary for distribution borrowers, the variables to be tested will be determined by the General Field Representative in consultation with the borrower and the regional office. The regional office will consult with the Power Supply Division in the case of generation projects for distribution borrowers. For power supply borrowers, the variables to be tested will be determined by the borrower and the Power Supply Division.

(e) The financial forecast shall use the accrual method, as approved by RUS, for analyzing costs and revenues, and, as applicable, compare the economic results of the various alternatives on a present value basis.

[57 FR 1053, Jan. 9, 1992, as amended at 63 FR 53277, Oct. 5, 1998; 78 FR 73366, Dec. 5, 2013; 84 FR 32612, July 9, 2019]

§ 1710.301 Financial forecasts—distribution borrowers.

(a) Financial forecasts prepared by distribution borrowers shall cover at least a ten-year period, unless a shorter period is authorized by other RUS regulations.

(b) In addition to the requirements set forth in § 1710.300 of this part, financial forecasts prepared by distribution borrowers in support of a loan application shall:

(1) Include expenditures for any maintenance determined to be needed in the current system's operation and maintenance review and evaluation in order to comply with mortgage covenants and prudent utility practice;

(2) Fully explain the basis for the power cost projections used. Generally, the power supplier's most recent forecasted rates shall be used; and

(3) Use RUS Form 325 or computer-generated equivalent reports.

§ 1710.302 Financial forecasts—power supply borrowers.

(a) The requirements of this section apply only to financial forecasts submitted by power supply borrowers in support of a loan from RUS. The financial forecast prepared by power supply borrowers shall demonstrate the effects that the addition of generation, transmission and any distribution facilities will have on the power supply borrower's sales, costs, and revenues, and on the cost of power to the member distribution systems.

(b) The financial forecast shall cover a period of 10 years. RUS may request projections for a longer period of time if RUS deems necessary.

(c) Financial forecasts prepared in support of loan applications to finance additional generation capacity shall include a power cost study as set forth in § 1710.303.

(d) In addition to the requirements set forth in § 1710.300, financial forecasts prepared by power supply borrowers shall:

(1) Identify all plans for generation and transmission capital additions and system operating expenses on a year-by-year basis, beginning with the present and running for 10 years, unless a longer period of time has been requested by RUS.

(2) Integrate projections of operation and maintenance expenses associated with existing plant with those of new proposed facilities to determine total costs of system operation as well as the costs of new generation and generation-related facilities;

(3) Provide an in-depth analysis of the regional markets for power if loan feasibility depends to any degree on a borrower's ability to sell surplus power while its system loads grow to meet the planned capacity of a proposed plant;

(4) If not previously submitted, furnish RUS with all material information on operating agreements, ownership agreements, fuel contracts and any other special agreements that affect annual cost projections, as may be required by RUS on a case by case basis; and

(5) Include sensitivity analysis if required by RUS pursuant to § 1710.300(d)(6).

(e) The projections shall be coordinated in advance with RUS so that agreement can be reached on major aspects of the economic studies. These include, but are not limited to, projections of future kW and kWh requirements, RE Act beneficiary loads, electricity prices, revenues from system and off-system power sales, the cost of prospective plant additions, interest and depreciation rates, fuel costs, cost escalation factors, the discount rate, and other factors.

(f) The projections, analysis, and supporting information must be included in a report that will provide RUS with the information needed to:

(1) Understand and compare various power supply plans;

(2) Determine that the facilities to be financed will perform satisfactorily; and

(3) Determine that the overall system is economically viable and the loan is financially feasible and secure.

[57 FR 1053, Jan. 9, 1992, as amended at 63 FR 53278, Oct. 5, 1998; 78 FR 73366, Dec. 5, 2013]

§ 1710.303 Power cost studies—power supply borrowers.

(a) All applications for financing of additional generation capacity and the associated bulk transmission facilities shall be supported by a power cost study to demonstrate that the proposed generation and associated transmission facilities are the most economical and effective means of meeting the borrower's power requirements. This study usually is a separate study but it may be integrated with the financial forecast required by § 1710.302.

(b) A power cost study shall include the following basic elements:

(1) A study of all reasonably available self-generation, purchased-power, load management, and energy conservation alternatives as set forth in §§ 1710.253 and 1710.254;

(2) A present-value analysis of the costs of the alternatives and their effects on total power costs, covering a period of at least 10 years beyond the projected in-service date of the facilities;

(3) A description of proposed new power-purchase contracts or revisions to existing contracts, and an analysis of the effects on power costs;

(4) Use of sensitivity analyses to determine the vulnerability of the alternatives to a reasonable range of assumptions about fuel costs, failure to achieve projected load growth, changes in operating and financing costs, and other major factors, if the financial forecast is used in support of a loan or loan guarantee that exceeds the smaller of $25 million or 10 percent of the borrower's total utility plant. Individual sensitivity analyses need not be duplicated if they have been included in other materials submitted to RUS; and

(5) Assessment of the financial risks of the various alternatives, especially as between capital-intensive and noncapital-intensive alternatives, under the range of assumptions set forth in paragraph (b)(4) of this section.

(c) Power cost studies must use current, RUS-approved power requirements data, and all major assumptions are subject to RUS approval. Alternative assumptions about projected power requirements may be used, however, in conjunction with the sensitivity analyses required by paragraph (b)(4) of this section.

(Approved by the Office of Management and Budget under control number 0572–0032)

§§ 1710.304–1710.349 [Reserved]

Subpart H—Energy Efficiency and Conservation Loan Program

SOURCE: 78 FR 73366, Dec. 5, 2013, unless otherwise noted.

§ 1710.400 Purpose.

(a) This subpart establishes policies and requirements that apply to loans and loan guarantees to finance Energy Efficiency and Conservation programs (EE Programs) undertaken by an eligible utility system to finance Demand side management, energy efficiency and conservation, or on-grid and off-grid renewable energy system programs that will result in the better management of their system load growth, a more beneficial load profile, or greater optimization of the use of alternative energy resources in their service territory. These programs may be considered an essential utility service.

(b)(1) The goals of an eligible Energy Efficiency project eligible for funding under this program and Subpart H include:

(i) Increasing energy efficiency at the end user level;

(ii) Modifying electric load such that there is a reduction in overall system demand;

(iii) Effecting a more efficient use of existing electric distribution, transmission and generation facilities;

(iv) Attracting new businesses and creating jobs in rural communities by investing in energy efficiency; and

(v) Encouraging the use of renewable energy fuels for either Demand side management or the reduction of conventional fossil fuel use within the service territory.

(2) Although not a goal, RUS recognizes that there will be a reduction of green house gases with energy efficiency improvements.

§ 1710.401 RUS policy.

EE Programs under this subpart may be financed at the distribution level or by an electric generation and transmission provider. RUS encourages borrowers to coordinate with the relevant member systems regarding their intention to implement a program financed under this subpart. RUS also encourages borrowers to leverage funds available under this subpart with State, local, or other funding sources that may be available to implement such programs.

§ 1710.402 Scope.

This subpart adapts and modifies, but does not supplant, the requirements for all borrowers set forth elsewhere where the purpose of the loan is to finance an approved EE program. In the event there is overlap or conflict between this subpart and the provisions of this part 1710 or other parts of the Code of Federal Regulations, the provisions of this subpart will apply for loans made or guaranteed pursuant to this subpart.

§ 1710.403 General.

EE Programs financed under this subpart may be directed at all forms of energy consumed within a utility's service territory, not just electricity, where the electric utility is in a position to facilitate the optimization of the energy consumption profile within its service territory and do so in a way that enhances the financial or physical performance of the rural electric system and enables the repayment of the energy efficiency loan.

§ 1710.404 Definitions.

For the purpose of this subpart, the following terms shall have the following meanings. In the event there is overlap or conflict between the definitions contained in § 1710.2, the definitions set forth below will apply for loans made or guaranteed pursuant to this subpart.

British thermal unit (Btu) means the quantity of heat required to raise one pound of water one degree Fahrenheit.

Certified energy auditor for commercial and industrial energy efficiency improvements. (1) An energy auditor shall meet one of the following criteria:

(i) An individual possessing a current commercial or industrial energy auditor certification from a national, industry-recognized organization;

(ii) A Licensed Professional Engineer in the State in which the audit is conducted with at least 1 year experience and who has completed at least two similar type Energy Audits;

(iii) An individual with a four-year engineering or architectural degree with at least 3 years experience and who has completed at least five similar type Energy Audits; or

(iv) Beginning in calendar year 2015, an energy auditor certification recognized by the Department of Energy through its Better Buildings Workforce Guidelines project.

(2) For residential energy efficiency improvements, an energy auditor shall meet one of the following criteria: The workforce qualification requirements of the Home Performance with Energy Star Program, as outlined in Section 3 of the Home Performance with Energy Star Sponsor Guide; or an individual possessing a current residential energy auditor or building analyst certification from a national, industry-recognized organization.

Cost effective means the aggregate cost of an EE Program is less than the financial benefit of the program over time. The cost of a program for this purpose shall include the costs of incentives, measurement and verification activity and administrative costs, and the benefits shall include, without limitation, the value of energy saved, the value of corresponding avoided generation, transmission or distribution and reserve investments as may be displaced or deferred by program activities, and the value of corresponding avoided greenhouse gas emissions and other pollutants.

Demand means the electrical load averaged over a specified interval of time. Demand is expressed in kilowatts, kilovolt amperes, kilovars, amperes, or other suitable units. The interval of time is generally 15 minutes, 30 minutes, or 60 minutes.

Demand savings means the quantifiable reduction in the load requirement for electric power, usually expressed in kilowatts (kW) or megawatts (MW) such that it reduces the cost to serve the load.

Eligible borrower means a utility system that has direct or indirect responsibility for providing retail electric service to persons in a rural area. This definition includes existing borrowers and utilities who meet current RUS borrower requirements.

Energy audit means an inspection and analysis of energy flows in a building, process, or system with the goal of identifying opportunities to enhance energy efficiency. The activity should result in an objective standard-based technical report containing recommendations for improving the energy efficiency. The report should also include an analysis of the estimated benefits and costs of pursuing each recommendation and the simple payback period.

Energy efficiency and conservation measures means equipment, materials and practices that when installed and used at a Consumer's premises result in a verifiable reduction in energy consumption, measured in Btus, or demand as measured in Btu-hours, or both, at the point of purchase relative to a base level of output. The ultimate goal is the reduction of utility or consumer energy needs.

Energy efficiency and conservation program (EE Program) means a program of activities undertaken or financed by a utility within its service territory to reduce the amount or rate of energy used by Consumers relative to a base level of output.

HVAC means heating, ventilation, and air conditioning.

Load means the Power delivered to power utilization equipment performing its normal function.

Load factor means the ratio of the average load over a designated period of time to the peak load occurring in the same period.

Peak demand (or maximum demand) means the highest demand measured over a selected period of time, e.g., one month.

Peak demand reduction means a decrease in electrical demand on an electric utility system during the system's peak period, calculated as the reduction in maximum average demand achieved over a specified interval of time.

Power means the rate of generating, transferring, or using energy. The basic unit is the watt, where one Watt is approximately 3.41213 Btu/hr.

Re-lamping means the initial conversion of bulbs or light fixtures to more efficient lighting technology but not the replacement of like kind bulbs or fixtures after the initial conversion.

SI means the International System of Units: the modern metric system.

Smart Grid Investments means capital expenditures for devices or systems that are capable of providing real time, two way (utility and Consumer) information and control protocols for individual Consumer owned or operated appliances and equipment, usually through a Consumer interface or smart meter.

Ultimate recipient means a Consumer that receives a loan from a borrower under this subpart.

Utility Energy Services Contract (UESC) means a contract whereby a utility provides a Consumer with comprehensive energy efficiency improvement services or demand reduction services.

Utility system means an entity in the business of providing retail electric service to Consumers (distribution entity) or an entity in the business of providing wholesale electric supply to distribution entities (generation entity) or an entity in the business of providing transmission service to distribution or generation entities (transmission entity), where, in each case, the entities provide the applicable service using self-owned or controlled assets under a published tariff that the entity and any associated regulatory agency may adjust.

Watt means the SI unit of power equal to a rate of energy transfer (or the rate at which work is done), of one joule per second.

§ 1710.405 Eligible energy efficiency and conservation programs.

(a) *General.* Eligible EE Programs shall:

(1) Be developed and implemented by an Eligible borrower and applied within its service territory;

(2) Consist of eligible activities and investments as provided in § 1710.406

(3) Provide for the use of State and local funds where available to supplement RUS loan funds;

(4) Incorporate the applicant's policy applicable to the interconnection of distributed resources;

(5) Incorporate a business plan that meets the requirements of § 1710.407;

(6) Incorporate a quality assurance plan that meets the requirements of § 1710.408;

(7) Demonstrate that the program can be expected to be Cost effective;

(8) Demonstrate that the program will have a net positive or neutral cumulative impact on the borrower's financial condition over the time period contemplated in the analytical support documents demonstrating that the net present value of program costs incurred by the borrower are positive, pursuant to § 1710.411;

(9) Demonstrate energy savings or peak demand reduction for the service territory overall; and

(10) Be approved in writing by RUS prior to the investment of funds for which reimbursement will be requested.

(b) *Financial Structures.* Eligible EE Programs may provide for direct recoupment of expenditures for eligible activities and investment from Ultimate Recipients as follows:

(1) Loans made to Ultimate Recipients located in a rural area where —

(i) The Ultimate Recipients may be wholesale or retail;

(ii) The loans may be secured or unsecured;

(iii) The loan receivables are owned by the Eligible Borrower;

(iv) The loans are made or serviced directly by the Eligible Borrower or by a financial institution pursuant to a contractual relationship between the Eligible Borrower and the financial institution;

(v) Due diligence is performed to confirm the repayment ability of the Ultimate Recipient;

(vi) Loans are funded only upon completion of the project financed or to reimburse startup costs that have been incurred;

(vii) The rate charged the Ultimate Recipient is less than or equal to the direct Treasury rate established daily by the United States Treasury pursuant to §1710.51(a)(1) or §1710.52, as applicable, plus the borrower's interest rate from RUS and 1.5 percent . Exceptions will be made on a case-by-case basis to ensure repayment of the government's loan and must be clearly articulated in the business plan RUS will not accept an exception request if the loan is feasible at 1.5 percent; and

(viii) Loans are not used to refinance a preexisting loan.

(2) A tariff that is specific to an identified rural Consumer, premise or class of ratepayer; or

(3) On bill repayment and other financial recoupment mechanisms as may be approved by RUS.

(c) *Period of performance*—(1) *Performance standards.* (i) Eligible EE Programs activities that are listed under §1710.406(b) should be designed to achieve the applicable operating performance standards within one year of the date of installation of the facilities.

(ii) All activities other than those included in paragraph (c)(1)(i) of this section should be designed to achieve the applicable operating performance targets within the time period contemplated by the analytic support documents for the overall EE Program as approved by RUS.

(2) *Cost effectiveness.* Eligible EE Programs must demonstrate that Cost effectiveness as measured for the program overall will be achieved within ten years of initial funding, except in cases where the useful life of the technology on an aggregate basis can be demonstrated to be longer than the ten year period. RUS will evaluate the useful life assumption on a case-by-case basis.

§ 1710.406 Eligible activities and investments.

(a) *General.* Eligible program activities and investments:

(1) Shall be designed to improve energy efficiency and/or reduce peak demand on the customer side of the meter;

(2) Shall be Cost effective in the aggregate after giving effect to all activities and investments contemplated in the approved EE Program; and

(3) May apply to all Consumer classes.

(b) *Eligible activities and investments.* Eligible program activities and investments may include, but are not limited to, the following:

(1) Energy efficiency and conservation measures where assets financed at an Ultimate Recipient premises can be characterized as an integral part of the real property that would typically transfer with the title under applicable state law. Where applicable, it is anticipated that the loan obligation would also be expected to transfer with ownership of the metered account serving that property.

(2) Renewable Energy Systems, including —

(i) On or Off Grid Renewable energy systems;

(ii) Fuel cells;

(3) Demand side management (DSM) investments including Smart Grid Investments;

(4) Energy audits;

(5) Utility Energy Services Contracts;

(6) Consumer education and outreach programs;

(7) Power factor correction equipment on the Ultimate Recipient side of the meter;

(8) Re-lamping to more energy efficient lighting; and

(9) Fuel Switching as in:

(i) The replacement of existing fuel consuming equipment using a particular fuel with more efficient fuel consuming equipment that uses another fuel but which does not increase direct greenhouse gas emissions; or

(ii) The installation of non-electric fuel consuming equipment to facilitate management of electric system peak loads. Fuel switching to fossil or biomass fueled electric generating equipment is expressly excluded.

(10) Other activities and investments as approved by RUS as part of the EE Program such as, but not limited to, pre-retrofit improvements.

(c) *Intermediary lending.* EE Program loan funds may be used for direct relending to Ultimate Recipients where the requirements of § 1710.405(b) are met.

(d) *Performance standards.* Borrowers are required to use Energy Star qualified equipment where applicable or meet or exceed efficiency requirements designated by the Federal Energy Management Program.

§ 1710.407 Business plan.

An Eligible EE Program must have a business plan for implementing the program. The business plan is expected to have a global perspective on the borrower's energy efficiency plan. Therefore, energy efficiency upgrades should be identified in aggregate. The business plan must have the following elements:

(a) *Executive summary.* The executive summary shall capture the overall objectives to be met by the Eligible EE Program and the timeframe in which they are expected to be achieved.

(b) *Organizational background.* The background section shall include descriptions of the management team responsible for implementing the Eligible EE Program.

(c) *Marketing plan.* The marketing section should identify the target Consumers, promotional activities to be pursued and target penetration rates by Consumer category and investment activity.

(d) *Operations plan.* The operations plan shall include but is not limited to:

(1) A list of the activities and investments to be implemented under the EE Program and the Btu savings goal targeted for each category;

(2) An estimate of the dollar amount of investment by the utility for each category of activities and investments listed under paragraph (d)(1) of this section;

(3) A staffing plan that identifies whether and how outsourced contractors or subcontractors will be used to deliver the program;

(4) A description of the process for documenting and perfecting collateral arrangements for Ultimate Recipient loans, if applicable; and

(5) The overall Btu savings to be accomplished over the life of the EE Program.

(e) *Financial plan.* The financial plan shall include but is not limited to:

(1) A schedule showing sources and uses of funds for the program;

(2) An itemized budget for each activity and investment category listed in the operations plan;

(3) An aggregate Cost effectiveness forecast;

(4) Where applicable, provision for Ultimate Recipient loan loss reserves. These loan loss reserves will not be funded by RUS. Loan loss reserves are not required when a utility will not be relending RUS funds.

(5) Identify expected Ultimate Recipient loan delinquency and default rates and report annually on deviations from the expected rates.

(f) *Risk analysis.* The business plan shall include an evaluation of the financial and operational risk associated with the program, including an estimate of prospective Consumer loan losses consistent with the loan loss reserve to be established pursuant to paragraph (e)(4) of this section.

(g) The borrowers are strongly encouraged to follow a bulletin or such other publication as RUS deems appropriate that contains and describes best practices for energy efficiency business plans. RUS will make this bulletin or publication publicly available and revise it from time-to-time as RUS deems it necessary.

§ 1710.408 Quality assurance plan.

An eligible EE program must have a quality assurance plan as part of the program. The quality assurance plan is expected to have a global perspective on the borrower's energy efficiency plan. Therefore, energy efficiency upgrades should be identified in aggregate. Every effort is made to fund only EE programs that are administered in accordance with quality assurance plans meeting standards designed to achieve the purposes of this subpart. However, RUS and its employees assume no legal liability for the accuracy, completeness or usefulness of any information, product, service, or process funded directly or indirectly with financial assistance provided under this subpart. Nothing in the loan documents between RUS and the energy efficiency borrower shall confer upon any other person any right, benefit or remedy of any nature whatsoever. Neither RUS nor its employees makes any warranty, express or implied, including the warranties of merchantability and fitness for a particular purpose, with respect to any information, product, service, or process available from an energy efficiency borrower. The approval by RUS and its employees of an energy efficiency borrower's quality assurance plan is solely for the benefit of RUS. Approval of the quality assurance plan does not constitute an RUS endorsement. The quality assurance plan must have the following elements:

(a) Quality assurance assessments shall include the use of qualified energy managers or professional engineers to evaluate program activities and investments;

(b) Where applicable, program evaluation activities should use the protocols for determining energy savings as developed by the U.S. Department of Energy in the Uniform Methods Project.

(c) Energy audits shall be performed for energy efficiency investments involving the building envelope at an Ultimate Recipient premises;

(d) Energy audits must be performed by certified energy auditors; and

(e) Follow up audits shall be performed within one year after installation on a sample of investments made to confirm whether efficiency improvement expectations are being met.

(f) In cases involving energy efficiency upgrades to a single system (such as a ground source heat pump) the new system must be designed and installed by certified and insured professionals acceptable to the utility.

(g) Industry or manufacturer standard performance tests, as applicable, shall be required on any system upgraded as a result of an EE Program. This testing shall indicate the installed system is meeting its designed performance parameters.

(h) In some programs the utility may elect to recommend independent contractors who can perform energy efficiency related work for their customers. In these cases utilities shall monitor the work done by the contractors and confirm that the contractors are performing quality work. Utilities should remove substandard contractors from their recommended lists if the subcontractors fail to perform at a satisfactory level. RUS does not endorse or recommend any particular independent contractors.

(i) Contractors not hired by the utility may not act as agents of the utility in performing work financed under this subpart.

(j) The borrowers are strongly encouraged to follow a bulletin or other publication that RUS deems appropriate and contains and describes best practices for energy efficiency quality assurance plans. RUS will make this bulletin or publication publicly available and revise it from time-to-time as RUS deems it necessary.

§ 1710.409 Loan provisions.

(a) *Loan term.* The maximum term for loans under this subpart shall be 15 years unless the loans relate to ground source loop investments or technology on an aggregate basis that has a useful life greater than 15 years. Ground source loop investments as the term is used in this paragraph do not include ancillary equipment related to ground source heat pump systems.

(b) *Loan feasibility.* Loan feasibility must be demonstrated for all loans made under this subpart. Loans made under this subpart shall be secured.

(c) *Reimbursement for completed projects.* (1) A borrower may request an initial advance not to exceed five percent of the total loan amount for working capital purposes to implement an eligible EE Program;

(2) Except for the initial advance provided for in paragraph (c)(1) of this section, all advances under this subpart shall be used for reimbursement of expenditures relating to a completed activity or investment; and

(3) Advances shall be in accordance with RUS procedures.

(d) *Loan amounts.* (1) Cumulative loan amounts outstanding under this subpart will be determined by the Assistant Administrator of the Electric Program and based an applicant's business plan; and

(2) Financing for administrative costs may not exceed 5 percent of the total loan amount.

(3) The Rural Utilities Service reserves the right to place a cap on both the total amount of funds an eligible entity can apply for, as well as a cap on the total amount of funds the Energy Efficiency and Conservation Program can utilize in the appropriations.

§ 1710.410 Application documents.

The required application documentation listed in this section is not all inclusive but is specific to Eligible borrowers requesting a loan under this subpart and in most cases is supplemental to the general requirements for loan applications provided for in this part 1710:

(a) A letter from the Borrower's General Manager requesting a loan under this subpart.

(b) A copy of the statement establishing the EE Program that reflects an undertaking that funds collected in excess of then current amortization requirements for the related RUS loan will be redeployed for EE Program purposes or used to prepay the RUS loan.

(c) Current RUS-approved EE Program documentation that includes:

(1) A Business Plan that meets the requirements of § 1710.407;

(2) A Quality Assurance Plan that meets the requirements of § 1710.408;

(3) Analytical support documentation that meets the requirements of § 1710.411;

(4) A copy of RUS' written approval of the EE Program.

(d) An EE program work plan that meets the requirements of § 1710.255;

(e) A statement of whether an initial working capital advance pursuant to § 1710.409(c)(1) is included in the loan budget together with a schedule of how these funds will be used.

(f) A proposed draft Schedule C pursuant to 7 CFR part 1718 that lists assets to be financed under this subpart as excepted property under the RUS mortgage, as applicable.

[78 FR 73366, Dec. 5, 2013, as amended at 84 FR 32613, July 9, 2019]

§ 1710.411 Analytical support documentation.

Applications for loans under this subpart may only be made for eligible activities and investments included in an RUS-approved EE Program. In addition to a business plan and operations plan, a request for EE program approval must include analytical support documentation that demonstrates the program meets the requirements of § 1710.303 and assures RUS of the operational and financial integrity of the EE Program. This documentation must include, but is not necessarily limited to, the following:

(a) A comparison of the utility's projected annual growth in demand after incorporating the EE Program together with an updated baseline forecast on file with RUS, where each includes an estimate of energy consuming devices used by customers in the service territory and a specific time horizon as determined by the utility for meeting the performance objectives established by them for the EE Program;

(b) Demonstration that the required periods of performance under § 1710.405(c) can reasonably be expected to be met;

(c) A report of discussions and coordination conducted with the power supplier, where applicable, issues identified as a result, and the outcome of this effort.

(d) An estimate of the amount of direct investment in utility-owned generation that will be deferred as a result of the EE Program;

(e) A description of efforts to identify state and local sources of funding and, if available, how they are to be integrated in the financing of the EE Program; and

(f) Copies of sample documentation used by the utility in administering its EE Program.

(g) Such other documents and reports as the Administrator may require.

§ 1710.412 Borrower accounting methods, management reporting, and audits.

Nothing in this subpart changes a Borrower's obligation to comply with RUS's accounting, monitoring and reporting requirements. In addition thereto, the Administrator may also require additional management reports that provide the agency with a means of evaluating the extent to which the goals and objectives identified in the EE Plan are being accomplished.

§ 1710.413 Compliance with other laws and regulations.

Nothing in this subpart changes a Borrower's obligation to comply with all laws and regulations to which it is subject.

§§ 1710.414–1710.499 [Reserved]

Subpart I—Application Requirements and Procedures for Loans

SOURCE: 60 FR 3731, Jan. 19, 1995, unless otherwise noted.

§ 1710.500 Initial contact.

(a) Loan applicants that do not have outstanding loans from RUS should contact the Rural Utilities Service via Email at *RUSElectric@wdc.usda.gov*, call RUS at (202) 720–9545 or write to the Rural Utilities Service Administrator, United States Department of Agriculture, 1400 Independence Ave. SW, STOP 1560, Room 5165, Washington, DC 20250–1560. Loan Applicants may also visit RUS' website to locate a local General Field Representative at *https://www.rd.usda.gov/contact-us/electric-gfr*. A field or headquarters staff representative may be assigned by RUS to visit the applicant and discuss its financial needs and eligibility. Borrowers that have outstanding loans should contact their assigned RUS general field representative (GFR) or, in the case of a power supply borrower, Deputy Assistant Administrator, Office of Loan Origination and Approval. Borrowers may consult with RUS field representatives and headquarters staff, as necessary.

(b) Before submitting an application for an insured loan the borrower shall ascertain from RUS the amount of supplemental financing required, as set forth in § 1710.110. If the borrower is applying for either a municipal rate loan subject to the interest rate cap or a hardship rate loan, the application must provide a preliminary breakdown of residential consumers either by county or by census tract. Final data must be included with the application. See § 1710.401(a)(8).

[60 FR 3731, Jan. 19, 1995. Redesignated at 78 FR 73366, Dec. 5, 2013, as amended at 84 FR 32613, July 9, 2019]

§ 1710.501 Loan application documents.

(a) *All borrowers.* Borrowers may be eligible to submit their loan application via RUS' electronic application intake system instead of submitting a paper submission. Please consult your GFR in accordance with § 1710.500. All applications for electric loans shall include the documents listed in this paragraph (a).

(1) *Loan application letter.* A letter signed by the borrower's manager indicating the actual corporate name, the borrowers RUS Designation, the borrowers RUS Loan Designation, and taxpayer identification number of the borrower and addressing the following items:

(i) The amount of loan and loan type. The sources and amounts of any supplemental or other financing. For an insured loan, a statement of whether the application is for a municipal rate loan, with or without the interest rate cap, or a hardship loan. If the application is for a municipal rate loan, the board resolution must indicate whether the borrower intends to elect the prepayment option. See 7 CFR 1714.4(c);

(ii) The Maturity Date/Term of the Loan in number of years (useful life to determine maximum);

(iii) A short description of the purpose of the loan, *i.e.*, generation, distribution, transmission, energy efficiency, etc.;

(iv) Method of Amortization;

(v) The Borrower's DUNS Number;

(vi) The Borrower's Organization Number from its State Corporation Commission or similar entity;

(vii) The Borrower's Exact Legal Name (please state the legal name and identify the legal document used to state the name or attach such document;

(viii) List of current counties where real property is located;

(ix) Attach current property schedule;

(x) Identify any new counties with property since last loan;

(xi) Authorized/registered place of business;

(xii) Debt Limit;

(xiii) Identify any State regulatory approvals needed;

(xiv) List any subsidiaries;

(xv) Identify any material financial or other material change since last loan, including a list of any pending litigation and where there is insurance to cover such;

(xvi) Breakdown of loan funds by State;

(xvii) Construction Work Plan (CWP), if not previously submitted through RD Apply or other method;

(xviii) Environmental Report (ER), if not previously submitted through RD Apply or other method;

(xix) Statement authorizing RUS to release appropriate information and data relating to the loan application to the FFB and any existing supplemental lenders.

(2) *Special resolutions*. Included any special resolutions required by Federal or State Authorities and any others as identified and required by the RUS General Field Representative (for example, use of contractors, corrective action plans, etc.)

(3) *RUS Form 740c, Cost Estimates and Loan Budget for Electric Borrowers*. This form together with its attachments lists the construction, equipment, facilities, and other cost estimates from the construction work plan or engineering and cost studies. The projects and related costs, included on this form, shall be used to justify the loan amount and are not meant to be an exclusive list of those projects that could receive funds under this loan. In addition, to be included on this form, the project must have received written documentation of RUS concluding its environmental review. The advance of loan funds for projects shall be governed by 7 CFR part 1721. The date on page one (1) of the RUS Form 740c is the beginning date of the loan period. RUS Form 740c also includes the following information, exhibits, and attachments:

(i) *Description of funds and materials*. This description details the availability of materials and equipment, any unadvanced funds from prior loans, and any general funds the borrower designates, to determine the amount of such materials and funds to be applied against the capital requirements estimated for the loan period.

(ii) *Useful life of facilities financed by the loan*. Form 740c must include, as a note, either a statement certifying that at least 90 percent of the loan funds are for facilities that have a useful life of 33 years or longer, or a schedule showing the costs and useful life of those facilities with a useful life of less than 33 years. This statement or schedule will be used to determine the final maturity of the loan. See § 1710.115.

(iii) *Reimbursement schedule*. This schedule lists the date, amount, and identification number of each inventory of work orders and special equipment summary that form the basis for the borrower's request for reimbursement of general funds on the RUS Form 740c. See § 1710.109. If the borrower is not requesting reimbursement, this schedule need not be submitted.

(iv) *Location of consumers*. If the application is for a municipal rate loan subject to the interest rate cap, or for a loan at the hardship rate, and the average number of consumers per mile of the total electric system exceeds 17, Form 740c must include, as a note, a breakdown of funds included in the proposed loan to furnish or improve service to consumers located in an urban area. See 7 CFR 1714.7(c) and 1714.8(d). This breakdown must indicate the method used by the borrower for allocating loan funds between urban and non-urban consumers.

(4) *RUS Form 740g, Application for Headquarters Facilities*. This form lists the individual cost estimates from the

construction work plan or other engineering study that support the need for RUS financing for any warehouse and service type facilities included, and funding requested for such facilities shown on RUS Form 740c. If no loan funds are requested for headquarters facilities, Form 740g need not be submitted.

(5) *Financial and statistical report.* RUS will use the Borrower's year end filed Financial and Operating Report Electric Distribution (formerly known as the RUS Form 7) or the Financial and Operating Report Electric Power Supply (formerly known as the RUS Form 12) unless the borrower has failed to meet its applicable financial ratios, as required by its security instrument and loan contract. The reports are required to be filed electronically in the agency's Data Collection System. If the borrower's financial requirements have not been met, RUS will require a current Financial and Operating Report to be submitted with the loan application, which shall contain the most recent data available and shall not be more than 60 days old when received by RUS. In addition, for those borrowers not meeting their financial ratios, the following information shall also be provided as part of the loan application:

(i) Any other information required to be submitted by RUS;

(ii) A Plan to meet their Financial Ratios;

(iii) The Date of the Borrower's last rate change and the amount/percentage of that rate change;

(iv) A list of any Subsidiaries along with a brief summary identifying the purpose of each subsidiary and identify the percentage interest in each if less than 100%;

(v) If the issues with the Borrower not meeting its financial ratios involves the subsidiary or equity investment losses a business plan and exit strategy shall be provided;

(vi) An updated Financial and Operating Report within 60 days of actual loan approval which will be requested by RUS and can be submitted later.

(6) *Load Forecast Study.* A current Load Forecast Study will be included in the loan application which is not more than 2 years old when the loan application is submitted unless the borrower is a member of a Power Supplier which only completes a Load Forecast once every 3 years. In that case the Load Forecast shall not be more than 3 years old when the loan application is submitted.

(7) *Long Range Financial Forecast and assumptions.* Along with the loan application, the borrower shall submit to RUS a Long-Range Financial Forecast (LRFF) that meets the requirements of subpart G of this part in a form acceptable to RUS. The forecast shall include any sensitivity analysis and/or analysis of alternative scenarios only if requested by the RUS General Field Representative.

(8) *Rate disparity and consumer income data.* If the borrower is applying under the rate disparity and consumer income tests for either a municipal rate loan subject to the interest rate cap or a hardship rate loan, the application must provide a breakdown of residential consumers either by county or by census tract. In addition, if the borrower serves in 2 or more states, the application must include a breakdown of all ultimate consumers by state. This breakdown may be a copy of Form EIA 861 submitted by the Borrower to the Department of Energy or in a similar form. See 7 CFR 1714.7(b) and 1714.8(a). To expedite the processing of loan applications, RUS strongly encourages distribution borrowers to provide this information to the GFR prior to submitting the application.

(9) *Standard Form 100—Equal Employment Opportunity Employer Report EEO—1.* This form, required by the Department of Labor, sets forth employment data for borrowers with 100 or more employees. A copy of this form, as submitted to the Department of Labor, is to be included in the application for an insured loan if the borrower has more than 100 employees. See § 1710.122.

(10) *Form AD–1047, Certification Regarding Debarment, Suspension, and Other Responsibility Matters—Primary Covered Transactions.* This statement certifies that the borrower will comply with certain regulations on debarment and suspension required by Executive Order 12549, Debarment and Suspension (3 CFR, 1986 Comp., p. 189). See 2 CFR 417, and § 1710.123 of this part.

(11) *Uniform Relocation Act assurance statement.* This assurance, which need not be resubmitted if previously submitted, provides that the borrower shall comply with 49 CFR part 24, which implements the Uniform Relocation Assistance and Real Property Acquisition Policy Act of 1970, as amended by the Uniform Relocation Act Amendments of 1987 and 1991. See § 1710.124.

(12) *Lobbying.* The following information on lobbying is required pursuant to 2 CFR 418, and § 1710.125. Borrowers applying for both insured and guaranteed financing should consult RUS before submitting this information.

(13) *Federal debt delinquency requirements.* See § 1710.126. The following documents are required:

(i) *Report on Federal debt delinquency.* This report indicates whether or not a borrower is delinquent on any Federal debt.

(ii) *Certification regarding Federal Government collection options.* This statement certifies that a borrower has been informed of the collection options the Federal Government may use to collect delinquent debt. The Federal Government is authorized by law to take any or all of the following actions in the event that a borrower's loan payments become delinquent or the borrower defaults on its loans:

(A) Report the borrower's delinquent account to a credit bureau;

(B) Assess additional interest and penalty charges for the period of time that payment is not made;

(C) Assess charges to cover additional administrative costs incurred by the Government to service the borrower's account;

(D) Offset amounts owed directly or indirectly to the borrower under other Federal programs;

(E) Refer the borrower's debt to the Internal Revenue Service for offset against any amount owed to the borrower as an income tax refund;

(F) Refer the borrower's account to a private collection agency to collect the amount due; and

(G) Refer the borrower's account to the Department of Justice for collection.

(14) *Assurance regarding Felony Conviction (AD Form 3030). This form must be included with each application to document the status regarding a felony criminal violation and status of any unpaid federal tax liability;*

(15) *RD Form 400–4, Assurance Agreement. This form provides assurance to USDA that recipients of federal financial assistance are in compliance with Title VI of the Civil Rights Act of 1964, 7 CFR part 15 and other agency regulations;*

(16) *Seismic safety certifications.* This certification shall be included, if required under 7 CFR part 1792.

(17) *Other forms.* Other forms as required by law or as requested.

(b) *New or returning borrowers.* In addition to the items in paragraph (a) of this section, applications for loans submitted by new or returning borrowers shall include the items listed in this paragraph (b).

(1) A copy of the Borrower's Current Bylaws;

(2) Identify the Borrower's Type of Organizational Structure and a copy of their Articles of Incorporation;

(3) Provide evidence of where Borrower is registered to do business;

(4) Copies of the Borrower's Audited GAAP financials for the past 1–3 years, if available or other financial information, as requested on a case by case basis;

(5) A list of any secured outstanding debt including the amount and name of lender;

(6) Evidence of Collateral and/or its ability to pledge such collateral;

(7) An Attorney Opinion for the Borrower including the counties served, a property schedule, the state of incorporation, any pending litigation, the corporate debt limit, the Borrower's legal name and type of legal organization, and the borrower's legal authority to pledge its collateral or other assets.

(8) Copies of the Borrower's Power Supply Contracts and arrangements (including wholesale rate contracts);

(9) Competitive position information including its rates and rate disparity between neighboring utilities;

(10) Construction Work Plan and/or Engineering Power Cost Study, if not previously submitted;

(11) An Environmental Report related to the facilities for which financing is being requested, if not previously submitted.

(c) *Power Supply Borrowers.* In addition to the loan application, consisting of the documents required by paragraph (a) or (b) of this section, Power Supply Borrowers must also provide RUS with the following:

(1) Information on its Power Supply arrangements and/or wholesale power contracts including the maturity dates. Please note copies of the contracts may be requested on a case by cases basis;

(2) A Profile of the Power Supply Borrowers' fuel supply arrangements;

(3) The Borrowers Load Resource Table;

(4) Information on its Transmission and Interconnection arrangements. Please note that copies of the contracts related to such arrangements may be requested on a case by case basis;

(5) The Power Supply Borrowers' New/Returning membership chart profile and relationships as applicable.

(d) *Submission of documents.* (1) Generally, all information required by paragraphs (a), (b), and (c) of this section is submitted to RUS in a single application. Borrowers may be eligible to submit their loan application via RUS' electronic application intake system instead of submitting hard copies of the loan applications. Please contact your respective General Field Representative or RUS Headquarters to determine if you are eligible to utilize the electronic system.

(2) To facilitate loan review, RUS urges borrowers to ensure that their applications contain all of the information required by this section before submitting the application to RUS. Borrowers may consult with RUS field representatives and headquarters staff as necessary for assistance in preparing loan applications.

(3) RUS may, in its discretion, return an application to the borrower if the application is not materially complete to the satisfaction of RUS within 10 months of receipt of any of the items listed in paragraph (a) or (b) of this section. RUS will generally advise the borrower in writing at least 2 months prior to returning the application as to the elements of the application that are not complete.

(4) If an application is returned, an application for the same loan purposes will be accepted by RUS if satisfactory evidence is provided that all of the information required by this section will be submitted to RUS within a reasonable time. An application for loan purposes included in an application previously returned to the borrower will be treated as an entirely new application.

(e) *Complete applications.* An application is complete when all information required by RUS to approve a loan is materially complete in form and substance satisfactory to RUS.

(f) *Change in borrower circumstances.* A borrower shall, after submitting a loan application, promptly notify RUS of any changes in its circumstances that materially affect the information contained in the loan application or in the primary support documents.

(g) *Interest rate category.* For pending loans, RUS will promptly notify the borrower if its eligibility for an interest rate category changes pursuant to new information from the Department of Energy or the Bureau of the Census. See 7 CFR part 1714.

(Approved by the Office of Management and Budget under control numbers 0572–0017, 0572–0032 and 0572–1013)

[60 FR 3731, Jan. 19, 1995. Redesignated at 78 FR 73366, Dec. 5, 2013, as amended at 79 FR 76002, Dec. 19, 2014; 81 FR 11026, Mar. 2, 2016; 84 FR 32613, July 9, 2019; 86 FR 36196, July 9, 2021]

§§ 1710.502–1710.503 [Reserved]

§ 1710.504 Additional requirements.

Additional requirements for insured electric loans are set forth in 7 CFR part 1714.

[60 FR 3731, Jan. 19, 1995. Redesignated at 78 FR 73366, Dec. 5, 2013]

§ 1710.505 Supplemental financing documents.

(a) The borrower is responsible for ensuring that the loan documents required for supplemental financing pursuant to § 1710.110 are executed in a timely fashion. These documents are subject to RUS approval.

(b) *Security.* Any security offered by the borrower to a supplemental lender is subject to RUS approval.

[60 FR 3731, Jan. 19, 1995. Redesignated at 78 FR 73366, Dec. 5, 2013]

§ 1710.506 Loan approval.

(a) A loan is approved when the Administrator signs the administrative findings.

(b) If the loan is not approved, RUS will notify the borrower of the reason.

[60 FR 3731, Jan. 19, 1995. Redesignated at 78 FR 73366, Dec. 5, 2013]

§ 1710.507 Loan documents.

Following approval of a loan, RUS will forward the loan documents to the borrower for execution, delivery, recording, and filing, as directed by RUS.

[60 FR 3731, Jan. 19, 1995. Redesignated at 78 FR 73366, Dec. 5, 2013]

PART 1714—PRE-LOAN POLICIES AND PROCEDURES FOR INSURED ELECTRIC LOANS

Subpart A—General

Sec.

Subpart B—Terms of Insured Loans

AUTHORITY: 7 U.S.C. 901 *et seq.*; 1921 *et seq.*; and 6941 *et seq.*

SOURCE: 58 FR 66260, Dec. 20, 1993, unless otherwise noted.

Subpart A—General

§ 1714.1 [Reserved]

§ 1714.2 Definitions.

The definitions set forth in 7 CFR 1710.2 are applicable to this part, unless otherwise stated. References to specific RUS forms and other RUS documents, and to specific sections of such forms and documents, shall include the corresponding forms, documents, sections and lines in any subsequent revisions of these forms and documents.

§ 1714.3 Applicability of provisions.

(a) *Insured electric loans approved on or after November 1, 1993.* On November 1, 1993, the Rural Electrification Loan Restructuring Act, Pub. L. 103–129, 107 Stat. 1356, (RELRA) amended the Rural Electrification Act of 1936, 7 U.S.C. 901 *et seq.*, (RE Act) to establish a new interest rate structure for insured electric loans. Insured electric loans approved on or after this date, are either municipal rate loans or hardship rate loans. Borrowers meeting the criteria set forth in § 1714.8 are eligible for 5 percent hardship rate loans. The interest rate on loans to other borrowers is the municipal interest rate, and borrowers meeting the criteria set forth in § 1714.7 are eligible for the interest rate cap on their municipal rate loans. Interest rates for the initial interest rate term and rollover terms (§ 1714.6) will be determined pursuant to § 1714.4. Provisions for prepayment are set forth in § 1714.9. The provisions of this subpart apply to loans approved on or after November 1, 1993, unless otherwise stated.

(b) *Insured electric loans approved prior to November 1, 1993.* These loans have a single interest rate applicable to the entire loan. The rate is generally 5 percent, but, in some cases, may be as low as 2 percent. These loans have a single interest rate term and may be prepaid at face value at any time. Provisions for discounted prepayment of these loans are set forth in 7 CFR part 1786.

§ 1714.4 Interest rates.

(a) *Municipal rate loans.* Each advance of funds on a municipal rate loan shall bear interest at a single rate for each interest rate term. All interest rates applicable to municipal rate loans will be increased by one eighth of one percent (0.125 percent), if the borrower elects to include in the loan agreement a prepayment option (call provision), allowing the borrower to prepay all or a portion of an advance on a date other than a rollover maturity date. However, no interest rate for any advances of a loan to a borrower who qualifies for the interest rate cap may exceed 7 percent.

(b) *Hardship rate loans.* All advances of funds on hardship rate loans shall bear interest at a rate of 5 percent.

(c) *Application procedure.* The borrower must indicate whether the application is for a municipal rate loan, with or without the interest rate cap, or a hardship rate loan. If the application is for a municipal rate loan, the borrower must also indicate whether they intend to elect the prepayment option.

[58 FR 66260, Dec. 20, 1993, as amended at 67 FR 16969, Apr. 9, 2002; 84 FR 32615, July 9, 2019]

§ 1714.5 Determination of interest rates on municipal rate loans.

(a) RUS will post on the RUS website, Electric Program HomePage, a schedule of interest rates for municipal rate loans at the beginning of each calendar quarter. The schedule will show the year of maturity and the applicable interest rates in effect for all funds advanced on municipal rate loans during the calendar quarter and all interest rate terms beginning in the quarter. All interest rates will be adjusted to the nearest one eighth of one percent (0.125 percent).

(b) The rate for interest rate terms of 20 years or longer will be the average of the 20 year rates published in the Bond Buyer in the 4 weeks specified in paragraph (d) of this section for the "11-Bond GO Index" of Aa rated general obligation municipal bonds, or the successor to this index.

(c) The rate for terms of less than 20 years will be the average of the rates published in the Bond Buyer in the 4 weeks specified in paragraph (d) of this section in the table of "Municipal Market Data—General Obligation Yields" for Aa rated bonds, or the successor to this table, for obligations maturing in the same year as the interest rate term selected by the borrower.

(d) The interest rates on municipal rate loans shall not exceed the interest rate determined under section 307(a)(3)(A) of the Consolidated Farm and Rural Development Act (7 U.S.C. 1927(a)(3)(A)) for Water and Waste Disposal loans. The method used to determine this rate is set forth in the regulations of the Rural Housing Service at 7 CFR 1942.17(f)(1) and (4). Pursuant to the RUS rule, the interest rates are set using as guidance the average of the Bond Buyer Index for the four weeks prior to the first Friday of the last month before the beginning of the quarter.

[58 FR 66260, Dec. 20, 1993, as amended at 67 FR 16969, Apr. 9, 2002; 80 FR 9861, Feb. 24, 2015]

§ 1714.6 Interest rate term.

(a) *Municipal rate loans.* Selection of interest rate terms shall be made by the borrower for each advance of funds. The minimum interest rate term shall be one year. RUS will send the borrower written confirmation of each rollover maturity date and the applicable interest rate.

(1) The initial interest rate term will begin on the date of the advance. All rollover interest rate terms will begin on the first day of a month, and except for the last interest rate term to final maturity, shall end on the last day of a month. All terms except for the initial interest rate term on an advance, and the last term to final maturity shall be in yearly increments.

(2) The following limits apply to the number of advances of funds that may be made to the borrower on any municipal rate loan:

(i) If the loan period is 2 years or less, not more than 6 advances;

(ii) If the loan period is more than 2 years, not more than 8 advances.

(3) For the initial interest rate term of an advance, a letter from an authorized official of the borrower indicating the selection of the term shall accompany the request for the advance.

(4) At the end of any interest rate term, the borrower shall pay all accrued interest and principal balance then due, and either prepay the remaining principal of the advance at

face value, or roll over the remaining principal for a new term, provided that no interest rate term may end later than the date of the final maturity.

(i) If the borrower elects to prepay all or part of the remaining principal of the advance at face value, it must notify the Director of the appropriate Regional Division or the Power Supply Division in writing not later than 20 days before the rollover maturity date.

(ii) If the borrower wishes to elect a new interest rate term that is different from the term previously selected, it must notify RUS in writing of the new term not later than 20 days before the end of the current term. The election of the new term shall be addressed to the Director, Financial Operations Division, Rural Utilities Service, Washington, DC 20250–1500.

(iii) If the borrower fails to notify RUS within the timeframes set out in this paragraph of its intention to prepay or elect a different interest rate term, RUS will automatically roll over the remaining principal for the shorter of, and at the interest rate applicable to:

(A) A period equal in length to the term that is expiring; or

(B) The remaining period to final maturity.

(b) *Hardship rate loans.* Loans made at the 5 percent hardship rate are made for a single term that cannot exceed the final maturity as set forth in 7 CFR 1710.115. The hardship interest rate applies to the entire amount of the loan.

[58 FR 66260, Dec. 20, 1993, as amended at 60 FR 3734, Jan. 19, 1995]

§ 1714.7 Interest rate cap.

Except as provided in paragraph (c) of this section, the municipal interest rate may not exceed 7 percent on a loan advance to a borrower primarily engaged in providing retail electric service if the borrower meets, at the time of loan approval, either the consumer density test set forth in paragraph (a) of this section, or both the rate disparity test for the interest rate cap and the consumer income test set forth in paragraph (b) of this section.

(a) *Low consumer density test.* The borrower meets this test if the average number of consumers per mile of line of its total electric system, based on the most recent data available at the time of loan approval is less than 5.50.

(b)(1) *Rate disparity test for the interest rate cap.* The borrower meets this test if its average revenue per kWh sold is more than the average revenue per kWh sold by all electric utilities in the state in which the borrower provides service. To determine whether a borrower meets this test, RUS will compare the borrower's average total revenue with statewide data in the table of Average Revenue per Kilowatthour for Electric Utilities by Sector, Census Division and State, in the Electric Power Annual issued by the Energy Information Administration of the Department of Energy (DOE), or the successor to this table. The test will be based on the most recent calendar year for which full year DOE data are available at the time of loan approval and borrower data for the same year.

(2) *Consumer income test.* The borrower meets this test if either the average per capita income of the residents receiving electric service from the borrower is less than the average per capita income of residents of the state in which the borrower provides service or the median household income of the households receiving electric service from the borrower is less than the median household income of the households in the state.

(i) To qualify under the consumer income test, the borrower must include in its loan application information about the location of its residential consumers. The borrower must provide to RUS, based on the most recent data available at the time of loan application, either the number of consumers in each county it serves or the number of consumers in each census tract it serves. Using 5-year income data from the American Community Survey (ACS) or, if needed, other Census Bureau data, RUS will compare, on a weighted average basis, the average per capita and median household income of the counties or census tracts served by the borrower with state figures.

(ii) If there is reason to believe that the ACS or other Census Bureau data does not accurately represent the economic conditions of the borrower's consumers, the reasons will be documented and the borrower may furnish, or RD may obtain, additional information regarding such economic conditions. Information must consist of reliable data from local, regional, State, or Federal sources or from a survey conducted by a reliable impartial source. The Administrator has the sole discretion to determine whether such data submitted by the borrower is sufficient to determine whether the borrower qualifies under the consumer income test.

(3) *Borrowers serving 2 or more states.* If a borrower serves consumers in 2 or more states, the rate disparity test and the consumer income test will be determined on a weighted average based on the percentage of the borrower's total consumers that are served in each state.

(c) *High density test.* If the average number of consumers per mile of the borrower's total electric system exceeds 17, the interest rate cap will not apply to funds used for the purpose of furnishing or improving electric service to consumers located in an area that is an urban area at the time of loan approval, notwithstanding that the area must have been deemed a rural area for the purpose of qualifying for a loan under this part. (See the definition of "rural area" in 7 CFR 1710.2.) If the average number of consumers per mile of line of the borrower's total electric system exceeds 17, the borrower must include, as a note on RUS Form 740c, Cost Estimates and Loan Budget for Electric Borrowers, submitted as part of the loan application for a loan subject to the interest rate cap, a breakdown of funds included in the proposed loan to furnish or improve service to consumers located in such urban areas. For such borrowers only funds for those facilities serving consumers located outside an urban area are eligible for the interest rate cap.

[58 FR 66260, Dec. 20, 1993, as amended at 80 FR 9861, Feb. 24, 2015]

§ 1714.8 Hardship rate loans.

Except as provided in paragraph (d) of this section, the Administrator shall make an insured electric loan for eligible purposes at the 5 percent hardship rate to a borrower primarily engaged in providing retail electric service if the borrower meets, at the time of loan approval, both the rate disparity test for hardship and the consumer income test described in paragraph (a) of this section; or the extremely high rates test set forth in paragraph (b) of this section. A loan at the 5 percent hardship rate may also be made to any borrower pursuant to paragraph (c) of this section who, in the sole discretion of the Administrator, has experienced a severe hardship. The Administrator may not require a loan from a supplemental source in connection with a hardship rate loan.

(a)(1) *Rate disparity test for hardship.* The borrower meets this test if its average revenue per kWh sold is not less than 120 percent of the average revenue per kWh sold by all electric utilities in the state in which the borrower provides service, and its average residential revenue per kWh is not less than 120 percent of the average residential revenue per kWh sold by all electric utilities in the state in which the borrower provides service. To determine whether a borrower meets this test, RUS will compare the borrower's average total revenue and average residential revenue with statewide data in the table of Average Revenue per Kilowatthour for Electric Utilities by Sector, Census Division and State, in the Electric Power Annual issued by the Energy Information Administration of the Department of Energy (DOE), or the successor to this table. The test will be based on the most recent calendar year for which full year DOE data are available at the time of loan approval and borrower data for the same year.

(2) *Consumer income test.* The borrower meets this test if either the average per capita income of the residents receiving electric service from the borrower is less than the average per capita income of the residents of the state in which the borrower provides service or the median household income of the residents receiving electric service from the borrower is less than the median household income of the households in the state. RUS will determine whether the borrower qualifies under this test according to the procedure set forth in § 1714.7(b)(2).

(3) *Borrowers serving 2 or more states.* If a borrower serves consumers in 2 or more states, the rate disparity test and

the consumer income tests will be determined on a weighted average based on the percentage of the borrower's total consumers that are served in each state.

(b) *Extremely high rates test.* Except as provided in this paragraph, the Administrator shall make an insured electric loan at the 5 percent hardship rate to any borrower whose residential revenue exceeds 15.0 cents per kWh sold. Residential revenue shall be calculated for the most recent full calendar year for which data are available and shall include sales to both seasonal and non-seasonal consumers. If, at the time of loan approval, the area to be served is an urbanized area (notwithstanding that the area must be deemed a rural area to qualify for a loan under this part (See the definition of "rural area" in 7 CFR 1710.2)), then the borrower must satisfy the provisions of paragraphs (a) and (d) of this section to qualify to the 5 percent hardship interest rate. If at the time of loan approval, such area is outside an urbanized area, the loan shall not be subject to the conditions and limitations set forth in paragraphs (a) and (d) of this section.

(c) *Administrator's discretion.* The Administrator may make a hardship rate loan if, in the sole discretion of the Administrator, the borrower has experienced a severe hardship. The Administrator shall consider, among other matters, whether factors beyond the control or substantial influence of the borrower have had severe adverse effect on the borrower's ability to provide service consistent with the purposes of the RE Act, and which prudent management could not reasonably anticipate and either prevent or insure against. Among the factors that may be considered are system damage due to unusual weather or other natural disasters or Acts of God, loss of substantial loads, extreme rate disparity compared to a contiguous utility, and other factors that cause severe financial hardship. The Administrator will also consider whether a hardship rate loan will provide significant relief to the borrower in dealing with the severe hardship.

(d) *High density test.* Except as provided in paragraph (b) of this section, if the average number of consumers per mile of the borrower's total electric system exceeds 17, the 5 percent hardship rate will not apply to funds used for the purpose of furnishing or improving electric service to consumers located in an area that is an urban area at the time of loan approval, notwithstanding that the area must have been deemed a rural area for the purpose of qualifying for a loan under this part. (See the definition of "rural area" in 7 CFR 1710.2.) If the average number of consumers per mile of line of the borrower's total electric system exceeds 17, the borrower must include, as a note on RUS Form 740c, Cost Estimates and Loan Budget for Electric Borrowers, submitted as part of the loan application for a loan at the 5 percent hardship rate, a breakdown of funds included in the proposed loan to furnish or improve service to consumers located in urban areas. For such borrowers only funds for those facilities serving consumers located outside an urban area are eligible for the 5 percent hardship rate.

(Approved by the Office of Management and Budget under control number 0572–1013)

§ 1714.9 Prepayment of insured loans.

This section sets out provisions for prepayment of insured electric loans at face value. Provisions for discounted prepayment of RUS loans are set out in 7 CFR part 1786.

(a) *Municipal rate loans.* Loan documents for municipal rate loans shall provide for the following:

(1) *Prepayment on a rollover maturity date.* All, or a portion of, the outstanding balance on any advance from a municipal rate loan may be prepaid on any rollover maturity date pursuant to § 1714.6(a)(4).

(2) *Prepayment on a date other than a rollover maturity date.* A borrower may elect at the time of loan approval to include a prepayment option (call provision) that will allow the borrower to prepay all, or a portion of, the outstanding balance on any advance on a date other than a rollover maturity date. Interest rates on advances from loans with a prepayment provision will be increased as set forth in § 1714.4(a).

(b) *Hardship rate loans.* Loan documents for hardship loans shall provide that the loan may be prepaid at face value at any time without penalty.

§§ 1714.10–1714.49 [Reserved]

Subpart B—Terms of Insured Loans

SOURCE: 60 FR 3734, Jan. 19, 1995, unless otherwise noted.

§§ 1714.50–1714.54 [Reserved]

§ 1714.55 Advance of funds from insured loans.

The borrower shall request advances of funds as needed. Advances are subject to RUS approval and must be requested in writing on RUS Form 595 or an RUS approved equivalent. Funds will not be advanced until the Administrator has received satisfactory evidence that the borrower has met all applicable conditions precedent to the advance of funds, including evidence that the supplemental financing required under 7 CFR part 1710 and any concurrent loan guaranteed by RUS are available to the borrower under terms and conditions satisfactory to RUS.

§ 1714.56 Fund advance period.

(a) The fund advance period begins on the date of the loan note and will last no longer than five years after September 30 of the fifth year after the fiscal year of obligation. The fiscal year of obligation is identified in loan documentation associated with each loan. The Administrator may extend the fund advance period on any loan if the borrower meets the requirements of paragraph (b) of this section. However, under no circumstances shall the RUS ever make or approve an advance, regardless of the last day for an advance on the loan note or any extension by the Administrator, later than September 30 of the fifth year after the fiscal year of obligation if such date would result in the RUS obligating or permitting advance of funds contrary to the Antideficiency Act, 31 U.S.C. 1341.

(b) The Administrator may agree to an extension of the fund advance period for loans if the borrower demonstrates, to the satisfaction of the Administrator, that the loan funds continue to be needed for approved loan purposes (*i.e.*, facilities included in a RUS approved construction work plan). Policies for extension of the fund advance period following certain mergers, consolidations, and transfers of systems substantially in their entirety are set forth in 7 CFR 1717.156.

(c) RUS will rescind the balance of any loan funds not advanced to a borrower as of the final date approved for advancing funds.

[84 FR 32615, July 9, 2019, as amended at 86 FR 36196, July 9, 2021]

§ 1714.57 Sequence of advances.

(a) Except as set forth in paragraph (b) of this section, concurrent loan funds will be advanced in the following order:

(1) 50 percent of the RUS insured loan funds;

(2) 100 percent of the supplemental loan funds;

(3) The remaining amount of the RUS insured loan funds.

(b) At the borrower's request and with RUS approval, all or part of the supplemental loan funds may be advanced before funds in paragraph (a)(1) of this section.

§ 1714.58 Amortization of principal.

(a) For insured loans approved on or after February 21, 1995:

(1) Amortization of funds advanced during the first 2 years after the date of the note shall begin no later than 2 years from the date of the note. Except as set forth in paragraph (a)(2) of this section, amortization of funds advanced 2 years or more after the date of the note shall begin with the scheduled loan payment billed in the month following the month of the advance.

(2) For advances made 2 years or more after the date of the note, the Administrator may authorize deferral of amortization of principal for a period of up to 2 years from the date of the advance if the Administrator determines that failure to authorize such deferral would adversely affect either the Government's financial interest or the achievement of the purposes of the RE Act.

(b) For insured loans approved before February 21, 1995, amortization of principal shall begin 2 years after the date of the note for advances made during the first and second years of the loan, and 4 years after the date of the note for advances made during the third and fourth years.

§ 1714.59 Rescission of loans.

(a) A borrower may request rescission of a loan with respect to any funds unadvanced by submitting a certified copy of a resolution by the borrower's board of directors.

(b) RUS may rescind loans pursuant to § 1714.56.

(c) Borrowers who prepay RUS loans at a discounted present value pursuant to 7 CFR part 1786, subpart F, are required to rescind the unadvanced balance of all outstanding electric notes pursuant to 7 CFR 1786.158(j).

PART 1717—POST-LOAN POLICIES AND PROCEDURES COMMON TO INSURED AND GUARANTEED ELECTRIC LOANS

Subparts A–C [Reserved]

Subpart D—Mergers and Consolidations of Electric Borrowers

Subparts E–L [Reserved]

Subpart M—Operational Controls

Subpart N—Investments, Loans, and Guarantees by Electric Borrowers

Subpart O [Reserved]

Subpart P [Reserved]

Subpart Q [Reserved]

Subpart R—Lien Accommodations and Subordinations for 100 Percent Private Financing

Subpart S—Lien Accommodations for Supplemental Financing Required by 7 CFR 1710.110

Subpart T [Reserved]

Subpart U [Reserved]

Subpart V [Reserved]

Subpart W [Reserved]

Subpart X [Reserved]

Subpart Y—Settlement of Debt

AUTHORITY: 7 U.S.C. 901 *et seq.*, 1921 *et seq.*, 6941 *et seq.*

SOURCE: 55 FR 38646, Sept. 19, 1990, unless otherwise noted.

Subparts A–C [Reserved]

Subpart D—Mergers and Consolidations of Electric Borrowers

SOURCE: 61 FR 66871, Dec. 19, 1996, unless otherwise noted.

§ 1717.150 General.

(a) This subpart establishes RUS policies and procedures for mergers of electric borrowers. These policies and procedures are intended to provide borrowers with the flexibility to negotiate and enter into mergers that offer advantages to the borrowers and to rural communities, and adequately protect the integrity and credit quality of RUS loans and loan guarantees.

(b) Consistent with prudent lending practices, the maintenance of adequate security for RUS loans and loan guarantees, and the objectives of the Rural Electrification Act of 1936, as amended, (7 U.S.C. 901 *et seq.*) (RE Act), RUS encourages electric borrowers to consider mergers when such action is likely to contribute, in the long-term, to greater operating efficiency and financial soundness. Borrowers are specifically encouraged to explore mergers that are likely to enhance the ability of the successor to provide reliable electric service at reasonable cost to RE Act beneficiaries.

(c) Pursuant to the loan documents and RUS regulations, certain mergers are subject to RUS approval. See § 1717.615.

(d) Since RUS must take action in order to advance funds and otherwise conduct business with a successor, RUS encourages borrowers to consult RUS early in the process regardless of whether RUS approval of the merger is required. RUS will provide technical assistance and guidance to borrowers to help expedite the processing of their requests and to help resolve potential problems early in the process.

§ 1717.151 Definitions.

The definitions set forth in 7 CFR 1710.2 are applicable to this subpart unless otherwise stated. In addition, for the purpose of this subpart, the following terms shall have the following meanings:

Active borrower means an electric borrower that has, on the effective date, an outstanding insured or guaranteed loan from RUS for rural electrification, and whose eligibility for future RUS financing is not restricted pursuant to 7 CFR part 1786.

Active distribution borrower means an electric distribution borrower that has, on the effective date, an outstanding insured or guaranteed loan from RUS for rural electrification, and whose eligibility for future RUS financing is not restricted pursuant to 7 CFR part 1786.

Consolidation. See *Merger*.

Coverage ratios means collectively TIER, OTIER, DSC and ODSC, as these terms are defined in 7 CFR 1710.2.

Effective date means the date a merger is effective pursuant to applicable state law.

Former distribution borrower means any organization that

(1) Sells or intends to sell electric power and energy at retail;

(2) At one time had an outstanding loan made or guaranteed by RUS, or its predecessor the Rural Electrification Administration (REA) for rural electrification; and

(3) Either repaid such loans at face value or prepaid pursuant to 7 CFR part 1786.

Loan documents means the mortgage (or other security instrument acceptable to RUS), the loan contract, and the promissory note(s) entered into between the borrower and RUS.

Merger means:

(1) A consolidation where two or more companies are extinguished and a new successor is created, acquiring the assets, liabilities, franchises and powers of those passing out of existence;

(2) A merger where one company is absorbed by another, the former ceasing to exist as a separate business entity, and the latter retaining its own identity and acquiring the assets, liabilities, franchises and powers of the former; or

(3) A transfer of mortgaged property by one company to another where the transferee acquires substantially as an entirety the assets, liabilities, franchises, and powers of the transferor.

New loan means a loan to a successor approved by RUS on or after the effective date.

Preexisting loan means a loan to a borrower approved by RUS prior to, and outstanding on the effective date.

Successor means the entity that continues as the surviving business entity as of the effective date, and acquires all the assets, liabilities, franchises, and powers of the entity or entities ceasing to exist as of the effective date.

Transitional assistance means financial relief provided to borrowers by RUS during a limited period of time following a merger.

§ 1717.152 Required documentation for all mergers.

In order for RUS to advance funds, send bills, and otherwise conduct business with a successor, the documents listed in this section must be submitted to RUS regardless of the need for RUS approval of the merger. Borrowers are responsible for ensuring that these documents are received by RUS in timely fashion. In cases of mergers that require RUS approval, or cases where borrowers must submit requests for transitional assistance, the documents listed in this section may be combined with the documents required by § 1717.157 and/or § 1717.160 where appropriate.

(a) Prior to the effective date, borrowers must submit:

(1) A transmittal letter on corporate letterhead signed by the manager of each active borrower that is a party to the proposed merger indicating the borrower's intention to merge and tentative timeframes, including the proposed effective date;

(2) An original certified board resolution from each party to the proposed merger affirming the board's support of the merger;

(3) All documents necessary to evidence the merger pursuant to applicable law. Examples include plan of merger, articles of merger, amended articles of incorporation, bylaws, and notices and filings required by law.

These documents may be copies of documents filed elsewhere, unless otherwise specified by RUS; and

(4) A letter addressed to the Administrator from the counsel of at least one of the active borrowers briefly describing the merger and indicating the relevant statutes under which the merger will be consummated.

(b) On or after the effective date, borrowers must submit:

(1) An opinion of counsel from the successor addressing, among other things, any pending litigation, proper authorization and consummation of the merger, proper filing and perfection of RUS' security interest, and all approvals required by law. RUS will provide the form of the opinion of counsel to the successor;

(2) A letter signed by the manager of the successor advising RUS of the effective date of the merger; the corporate name, address, and phone number; the names of the officers of the successor; and the taxpayer identification number; and

(3) Evidence of proper filing and perfection of RUS' security interest, as instructed by RUS, and an executed loan contract.

§1717.153 Transitional assistance.

RUS recognizes that short-term financial stresses can follow even the most beneficial mergers. To help stabilize electric rates, enhance the credit quality of outstanding loans made or guaranteed by the Government, and otherwise ease the transition period before the long-term efficiencies and economies of a merger can be realized, RUS may approve one or more types of transitional assistance to a successor under the conditions set forth in this part.

§1717.154 Transitional assistance in connection with new loans.

Requests for transitional assistance in connection with new loans may be submitted to RUS no later than the loan application.

(a) *Loan processing priority*. (1) RUS loans are generally processed in chronological order based on the date the complete application is received in the regional or division office. At the borrower's request, RUS may offer loan processing priority for the first loan to a successor, provided that the loan is approved by RUS not later than 5 years after the effective date of the merger. In considering the request, the Administrator will take into account, among other factors, the amount of the loan application, whether there is a significant backlog in pending loan applications, the impact that loan priority would have on the backlog, the savings and efficiencies to be realized from the merger and the relative importance of loan priority to facilitating the merger. The Administrator may, in his or her sole discretion, grant or decline to grant priority, or grant priority for a limited amount of the loan application while deferring for later consideration the remainder of the application.

(2) For any subsequent loans approved during those 5 years, RUS may offer loan processing priority. In reviewing requests for loan processing priority on subsequent loans, RUS will consider the loan authority for the fiscal year, the borrower's projected cash flows, its electric rates and rate disparity, and the likely mitigation effects of priority loan processing. *See* 7 CFR 1710.108 and 1710.119.

(3) Loan processing priority is available following any merger where at least one of the merging parties is an active borrower.

(b) *Supplemental financing*. (1) RUS generally requires that an applicant for a municipal rate loan obtain a portion of its debt financing from a supplemental source without an RUS guarantee. See 7 CFR 1710.110. RUS will, at the borrower's request, waive the requirement to obtain supplemental financing for the first RUS loan approved after the effective date if that first loan is a municipal rate loan whose loan period does not exceed 2 years, and the loan is approved by RUS not later than 5 years after the effective date. For any subsequent loans approved during these 5 years, or if the borrower requests a loan period longer than 2 years, RUS may, subject to the availability of loan funds, waive or reduce the amount of supplemental financing required. In reviewing requests to reduce or waive supplemental financing on subsequent loans or on loans with a loan period longer than 2 years, RUS will consider the differences in interest rates between RUS and supplemental loans and the impacts of this difference on the borrower's projected cash flows and its electric rates and rate disparity. If significant differences would result, the waiver will be granted.

(2) Waiver of supplemental financing may be available if:

(i) All parties to the merger are active distribution borrowers, *or*

(ii) At least one of the merging parties is an active distribution borrower, all merging parties are either active distribution borrowers or former distribution borrowers, *and* the merger is effective after December 19, 1996.

(c) *Reimbursement of general funds and interim financing.* (1) Borrowers may request RUS loan funds to reimburse general funds and/or interim financing used to finance equipment and facilities included in a RUS approved construction work plan or amendment if the construction was completed immediately preceding the current loan period. This reimbursement period is generally limited to 48 months. See 7 CFR 1710.109. RUS may, in connection with the first RUS loan approved after the effective date, approve a reimbursement period of up to 48 months prior to the current loan period if the loan is approved not later than 5 years after the effective date. In reviewing requests for this longer reimbursement period, RUS will consider the stresses that the transaction and other costs of entering into the merger places on the borrower's rates and cash flows, and the mitigating effects of more generous reimbursement.

(2) A longer reimbursement period may be available if:

(i) All parties to the merger are active distribution borrowers, *or*

(ii) At least one of the merging parties is an active distribution borrower, all merging parties are either active distribution borrowers of former distribution borrowers, *and* the merger is effective after December 19, 1996.

[61 FR 66871, Dec. 19, 1996, as amended at 67 FR 58322, Sept. 16, 2002; 86 FR 36197, July 9, 2021]

§ 1717.155 Transitional assistance affecting new and preexisting loans.

Requests for transitional assistance affecting new and preexisting loans must be received by RUS no later than 2 years after the effective date.

(a) *Section 12 deferments.* (1) Section 12 of the RE Act (7 U.S.C. 912) allows RUS to extend the time of payment of interest or principal of RUS loans. Section 12 deferments do not extend the final maturity of the loan; lower payments during the deferment period result in higher payments later. Therefore, RUS may approve a Section 12 deferment of loan payments of up to 5 years only if such deferments will help to avoid substantial increases in retail electric rates during the transition period, without placing borrowers in financial stress after the deferment period.

(2) Section 12 deferment may be available following any merger where at least one of the merging parties is an active borrower.

(b) *Coverage ratios.* Required levels for coverage ratios are set forth in 7 CFR 1710.114 and in the loan documents. RUS may approve a plan, on a case by case basis, that provides for a phase-in period for these coverage ratios of up to 5 years from the effective date. Under such a plan the successor would be permitted to project and achieve lower levels for one or more of these coverage ratios during the phase-in period.

(1) A phase-in plan for coverage ratios must provide a pro forma level for each ratio during each year of the phase-in period and be supported by a financial forecast covering a period of not less than 10 years from the effective date of the merger. The plan must demonstrate that a minimum TIER level of 1.00 will be achieved in each year, that trends will be generally favorable, that the borrower will achieve the levels required in its loan documents and RUS regulations by the end of the phase-in period, and that these levels will be maintained in subsequent years.

(2) In reviewing phase-in plans for coverage ratios, RUS will review rates, rate disparity, and likely mitigating effects of the proposed phase-in plan.

(3) The borrower is responsible for obtaining approvals of supplemental lenders.

(4) Upon RUS approval of a phase-in plan, the levels in that plan will be substituted for the levels required in the borrower's preexisting loan documents and will be incorporated in any new loan or security documents.

(5) A phase in plan for coverage ratios may be available if:

(i) All parties to the merger are active distribution borrowers, *or*

(ii) At least one of the merging parties is an active distribution borrower, all merging parties are either active distribution borrowers or former distribution borrowers, and the merger is effective after December 19, 1996.

§ 1717.156 Transitional assistance affecting preexisting loans.

The fund advance period for an insured loan, which is the period during which RUS may advance loan funds to a borrower, terminates automatically after a specific period of time. See 7 CFR 1714.56. If, on the effective date the original fund advance period or the fund advance period as extended pursuant to 7 CFR 1714.56(c), on any preexisting RUS loan to any of the active borrowers involved in a merger has not terminated, such fund advance period shall be automatically lengthened by 2 years. On the borrower's request RUS will prepare documents necessary for the advance of loan funds. RUS will prepare documents for the borrower's execution that will reflect this extension and will provide the legal authority for RUS to advance funds to the successor.

§ 1717.157 Requests for transitional assistance.

(a) If the merger requires RUS approval, the borrower should, where possible, indicate that it desires transitional assistance at the time it requests approval of the merger. The formal request for transitional assistance must be received by RUS as specified in §§ 1717.155 and 171.156. Documents listed in this section may be combined with the documents required by §§ 1717.152 and/or 1717.160 where appropriate. If the request for transitional assistance is submitted at the same time as a loan application, documents listed in this section may be combined with the loan application documents where appropriate. See 7 CFR part 1710, subpart I. A request for transitional assistance must include:

(1) Transmittal letter(s) formally listing the types of transitional assistance requested. If the request is submitted before the effective date, a transmittal letter must be signed by the manager of each party to the transaction. If the request is submitted on or after the effective date, a transmittal letter must be signed by the manager of the successor. Transmittal letter(s) must be signed originals on corporate letterhead stationery;

(2) Board resolution(s). If the request is submitted before the effective date, a separate board resolution must be submitted from each entity involved in the merger. If the request is submitted on or after the effective date, a board resolution from the successor must be submitted. Each board resolution must be a certified original;

(3) A merger plan, financial forecasts, and any available studies such as net present value analyses showing the anticipated costs and benefits of the merger and likely timeframes for the merger. The merger plan must clearly identify those benefits that cannot be achieved without a merger, and those benefits that can be achieved through other means;

(4) If the transitional assistance requires RUS approval, the type and extent of the mitigation that the transitional assistance is expected to provide; and

(5) Other information that may be relevant.

(b) Borrowers are responsible for ensuring that requests for transitional assistance are complete and sound in form and substance when they are submitted to RUS. After submitting a request, borrowers shall promptly notify RUS of any changes or events that materially affect the request or any information in the request.

(c) In considering whether to approve requests for transitional assistance, RUS will evaluate the costs and benefits of the merger; the type and extent of the likely transitional stress; whether the transitional assistance requested is likely to materially mitigate such stress; and the likely impacts on electric rates and on the security of RUS loans. Review factors applicable to each type of transitional assistance are set forth in §§ 1717.154–1717.156.

§ 1717.158 Mergers with borrowers who prepaid RUS loans.

In some cases, an active distribution borrower may merge with a borrower that has prepaid RUS debt at a discount pursuant to 7 CFR part 1786, and whose eligibility for future RUS financing is thereby restricted. During the period when the restrictions on future financing are in effect, the successor will be eligible for RUS loans to finance facilities to serve consumers located in the territory that was served by the active distribution borrower immediately prior to the effective date, provided that other requirements for loan eligibility are met.

§ 1717.159 Applications for RUS approvals of mergers.

If a proposed merger requires RUS approval according to RUS regulations and/or the loan documents executed by any of the active borrowers involved, the application must be submitted to RUS not later than 90 days prior to the effective date of the proposed borrower action. A distribution borrower should consult with its assigned RUS general field representative, and a power supply borrower with the Director, Power Supply Division for general information prior to submitting the request.

§ 1717.160 Application contents.

An application for RUS approval of a merger must include the documents listed in this section. Documents listed in this section may be combined with the documents required by §§ 1717.152 and/or 1717.157 where appropriate.

(a) *Transmittal letters* signed by the managers of all borrowers and non-borrowers who are parties to the proposed merger. These letters must include the actual corporate name, address, and taxpayer identification number of all parties to the proposed merger. The transmittal letters must be signed originals on corporate letterhead stationery.

(b) *Resolutions from the boards of directors* of all borrowers and non-borrowers who are parties to the proposed merger. This document is the formal request by each entity for RUS approval of the proposed merger. The board resolution must include a description of the proposed merger, including timeframes, and authorization for RUS to release appropriate information to supplemental or other lenders, and for these lenders to release appropriate information to RUS. Each board resolution must be a certified original.

(c) *Evidence* that the proposed merger will result in a viable entity, and that the security of outstanding RUS loans will not be adversely affected by the action. This evidence shall include financial forecasts, and any available studies such as net present value analyses covering a period of not less than 10 years from the effective date of the merger, as well as information about any threatened actions by other parties that could adversely affect the financial condition of any of the parties to the proposed merger, or of the successor. Such threatened actions may include annexations or other actions affecting service territory, loads, rates or other such matters.

(d) *Regulatory information* about pending federal or state proceedings pertaining to any of the parties that could have material effects on the successor.

(e) *Rate information.* Distribution and power supply borrowers shall submit schedules of proposed rates after the merger, including the effects of the proposed action on rates and the status of any pending rate cases before a state regulatory authority. The rates of power supply borrowers are subject to RUS approval. If rates are not projected to change after the merger, a statement to that effect will suffice.

(f) *Area coverage and line extension policies.* If any distribution systems are parties to the proposed merger, a statement of proposed area coverage and line extension policies for the successor.

§ 1717.161 Application process.

(a) Borrowers are responsible for ensuring that their applications for RUS approval of a merger are complete and sound in form and substance when they are submitted to RUS. After submitting an application, borrowers shall promptly notify RUS of any changes or events that materially affect the application or any information in the application.

(b) In reviewing borrower requests for approval of mergers, RUS will consider the likely effects of the action on the ability of the successor to provide reliable electric service at reasonable cost to RE Act beneficiaries and on the security of outstanding RUS loans. Among the factors RUS will consider are whether the proposed merger is likely to:

(1) Contribute to greater operating efficiency and financial soundness;

(2) Mitigate high electric rates and or rate disparity;

(3) Help borrowers to diversify their loads or otherwise hedge risks;

(4) Have beneficial effects on rural economic development in the community served by the borrower, such as diversifying the economic base or alleviating unemployment; and

(5) Provide other benefits consistent with the purposes of the RE Act.

(c) RUS will not approve a merger if, in the sole judgment of the Administrator, such action is likely to have an adverse effect on the credit quality of outstanding loans made or guaranteed by the Government. RUS will thoroughly review each request for approval of such action, including review of the feasibility and security of outstanding Government loans according to the standards in 7 CFR 1710.112 and 1710.113, respectively, and in other RUS regulations.

(d) RUS will keep the borrowers apprised of the progress of their applications.

Subparts E–L [Reserved]

Subpart M—Operational Controls

SOURCE: 60 FR 67405, Dec. 29, 1995, unless otherwise noted.

§ 1717.600 General.

(a) *General.* The loan contract and mortgage between the Rural Utilities Service (RUS) and electric borrowers imposes certain restrictions and controls on the borrowers and gives RUS (and other co-mortgagees in the case of the mortgage) the right to approve or disapprove certain actions contemplated by the borrowers. Certain of these controls and approval rights are referred to informally as "operational controls" because they pertain to decisions or actions with respect to the operation of the borrowers' electric systems. The approval authority granted to RUS by the loan contract or mortgage regarding each decision or action subject to controls is often stated in broad, unlimited terms. This subpart lists the main operational controls affecting borrowers and establishes for each area of control the circumstances under which RUS approval of a decision or action by a borrower is either required or not required. In some cases, only the general principles or general circumstances pertaining to RUS approval or control are presented in this subpart, while the details regarding the circumstances and requirements of RUS approval or control are set forth in other RUS regulations. Since this subpart addresses only the main operational controls, failure to address a control or approval right in this subpart in no way invalidates such controls or rights established by the loan contract, mortgage, other agreements between a borrower and RUS, and RUS regulations.

(b) *Case by case amendments.* Upon written notice to a borrower, RUS may amend or annul the approvals and exceptions to controls set forth in this subpart or other RUS regulations if the borrower is in violation of any provision of its loan documents or any other agreement with RUS, or if RUS determines that loan security and/or repayment is threatened. Such amendment or annulment will apply to decisions and actions of the borrower after said written notice has been provided by RUS.

(c) *Generic notices.* By written notice to all borrowers or a group of borrowers, RUS may grant or waive approval of decisions and actions by the borrowers that are controlled under the loan documents and RUS regulations. RUS may also by written notice withdraw or cut back its grant or waiver of approval of said decisions and actions made by previous written notice, but may not by such notice extend its authority to approve decisions and actions by borrowers beyond the authority granted by the loan documents and RUS regulations.

§ 1717.601 Applicability.

(a) The approvals and exceptions to controls conveyed by this subpart apply only to controls and approval rights normally included in RUS loan documents dated prior to January 29, 1996. They do not apply to special controls and approval requirements included in loan documents or other agreements executed between a borrower and RUS that relate to individual problems or circumstances specific to an individual borrower.

(b) The approvals and exceptions to controls granted by RUS in this subpart shall not in any way affect the rights of other co-mortgagees under the mortgage or their loan contracts.

§ 1717.602 Definitions.

Terms used in this subpart that are not defined in this section have the meanings set forth in 7 CFR part 1710. In addition, for the purposes of this subpart:

Default means an event of default as defined in the borrower's loan documents or other agreement with RUS, and furthermore includes any event that has occurred and is continuing which, with notice or lapse of time and notice, would become an event of default.

Equity means the borrower's total margins and equities computed pursuant to RUS accounting requirements but excluding any regulatory created assets.

Financed or funded by RUS means financed or funded wholly or in part by a loan made or guaranteed by RUS, including concurrent supplemental loans required by 7 CFR 1710.110, loans to reimburse funds already expended by the borrower, and loans to replace interim financing.

Interchange agreement means a contractual arrangement that can include a variety of services utilities provide each other to increase reliability and efficiency, and to avoid duplicating expenses. Some examples are: transmission service (the use of transmission lines to move power and energy from one area to another); emergency service (an agreement by one utility to furnish another with power and energy to protect it in times of emergency, such as power plant outages); reserve sharing (contributions to a common pool of generating plant reserves so that each individual utility's reserves can be reduced); and economic exchanges (swapping power and energy from different plants to avoid running the most expensive units).

Interconnection agreement means a contract governing the terms for establishing or using one or more electrical connections between two or more electric systems permitting a flow of power and energy among the systems.

Loan documents means the mortgage (or other security instrument acceptable to RUS), the loan contract, and the promissory note entered into between the borrower and RUS.

Net utility plant means the amount constituting the total utility plant of the borrower, less depreciation, computed in accordance with RUS accounting requirements.

Pooling agreement means a contract among two or more interconnected electric systems to operate on a coordinated basis to achieve economies and/or enhance reliability in supplying their respective loads.

Power supply contract means any contract entered into by a borrower for the sale or purchase, at wholesale, of electric energy.

Regulatory created assets means the sum of any amounts properly recordable as unrecovered plant and regulatory study costs or as other regulatory assets, computed pursuant to RUS accounting requirements.

RUS accounting requirements means the system of accounts prescribed for electric borrowers by RUS regulations as such RUS accounting requirements exist at the date of applicability thereof.

RUS regulations mean regulations of general applicability published by RUS from time to time as they exist at the date of applicability thereof, and shall also include any regulations of other federal entities which RUS is required by law to implement.

Total assets means an amount constituting the total assets of the borrower as computed pursuant to RUS accounting requirements, but excluding any regulatory created assets.

Wheeling agreement means a contract providing for the use of the electric transmission facilities of one electric utility to transmit power and energy of another electric utility or other entity to a third party. Such transmission may be accomplished directly or by displacement.

§ 1717.603 RUS approval of extensions and additions.

(a) *Distribution borrowers.* Prior written approval by RUS is required for a distribution borrower to extend or add to its electric system if the extension or addition will be financed by RUS. For extensions and additions that will not be financed by RUS, approval is hereby given to distribution borrowers to make such extensions and additions to their electric systems, including the use of (or commitment to use) general funds of the borrower, except for the following:

(1) Construction, procurement, or leasing of generating facilities if the combined capacity of the facilities to be built, procured, or leased, including any future facilities included in the planned project, will exceed the lesser of 5 megawatts or 30 percent of the borrower's equity;

(2) Acquisition or leasing of existing electric facilities or systems in service whose purchase price, or capitalized value in the case of a lease, exceeds 10 percent of the borrower's net utility plant; and

(3) Construction, procurement, or leasing of electric facilities to serve a customer whose annual kWh purchases or maximum annual kW demand in the foreseeable future is projected to exceed 25 percent of the borrower's total kWh sales or maximum kW demand in the year immediately preceding the acquisition or start of construction.

(b) *Power supply borrowers.* Prior written approval by RUS is required for a power supply borrower to extend or add to its electric system if the extension or addition will be financed by RUS. Requirements for RUS approval of extensions and additions that will not be financed by RUS are set forth in other RUS regulations.

(c) *Additional details.* Additional details relating to RUS approval of extensions and additions of a borrower's electric system financed by RUS are set forth in other RUS regulations, e.g., in 7 CFR parts 1710 and 1726.

§ 1717.604 Long-range engineering plans and construction work plans.

(a) All borrowers are required to maintain up-to-date long-range engineering plans and construction work plans (CWPs) in form and substance as set forth in 7 CFR part 1710, subpart F.

(b) Applications for financing from RUS must be supported by a CWP approved by RUS.

(c) RUS approval is not required for CWPs if the borrower does not intend to seek RUS financing for any of the facilities, equipment, or other purposes included in those plans. However, if requested by RUS, a borrower must provide an informational copy of such plans to RUS.

[60 FR 67405, Dec. 29, 1995, as amended at 86 FR 36197, July 9, 2021]

§ 1717.605 Design standards, plans and specifications, construction standards, and RUS accepted materials.

All borrowers, regardless of the source of funding, are required to comply with applicable RUS requirements with respect to system design, construction standards, and the use of RUS accepted materials. Borrowers must comply with applicable RUS requirements with respect to plans and specifications only if the construction or procurement will be financed by RUS. These requirements are set forth in other RUS regulations, especially in 7 CFR parts 1724 and 1728.

§ 1717.606 Standard forms of construction contracts, and engineering and architectural services contracts.

All borrowers are encouraged to use the standard forms of contracts promulgated by RUS for construction, materials, equipment, engineering services, and architectural services, regardless of the source of funding for such construction and services. Borrowers are required to use these standard forms of contracts only if the construction, procurement or services are financed by RUS, and only to the extent required by RUS regulations. RUS requirements with respect to such standard forms of contract are set forth in 7 CFR part 1724 for architectural and engineering services, and in 7 CFR part 1726 for construction, materials, and equipment.

§ 1717.607 Contract bidding requirements.

Borrowers must follow RUS requirements regarding bidding for contracts for construction, materials, and equipment only if financing of the construction or procurement will be provided by RUS. These requirements are set forth in 7 CFR part 1726.

§ 1717.608 RUS approval of contracts.

(a) *Construction contracts and architectural and engineering contracts.* RUS approval of contracts for construction and procurement and for architectural and engineering services is required only when such construction, procurement or services are financed by RUS. Detailed requirements regarding RUS approval of such contracts are set forth in 7 CFR part 1724 for architectural and engineering services, and in 7 CFR part 1726 for construction and procurement.

(b) *Large retail power contracts.* RUS is required to be notified of contracts to sell electric power to retail customers if the contract is for longer than 5 years and the kWh sales or kW demand for any year covered by the contract exceeds 25 percent of the borrower's total kWh sales or maximum kW demand for the year immediately preceding execution of the contract. The requirement in this paragraph (b) applies regardless of the source of funding of any plant extensions, additions or improvements that may be involved in connection with the contract.

(c) *Power supply arrangements.* (1) Power supply contracts (including but not limited to economy energy sales and emergency power and energy sales), interconnection agreements, interchange agreements, wheeling agreements, pooling agreements, and any other similar power supply arrangements subject to approval by RUS are deemed approved if they have a term of 5 years or less. Amendments to said power supply arrangements are also deemed approved provided that the amendment does not extend the term of the arrangement for more than 5 years beyond the date of the amendment.

(2) Any amendment to a schedule or exhibit contained in any power supply arrangement subject to RUS approval, which merely has the effect of either altering a list of interconnection or delivery points or changing the value of a variable term (but not the formula itself) contained in a formulary rate or charge is deemed approved.

(3) The provisions of this paragraph (c) apply regardless of whether the borrower is a seller or purchaser of the services furnished by the contracts or arrangements, and regardless of whether or not a Federal power marketing agency is a party to any of them.

(d) *System management and maintenance contracts.* RUS approval of contracts for the management and operation of a borrower's electric system or for the maintenance of the electric system is required only if such contracts cover all or substantially all of the electric system.

(e) *Other contracts.* [Reserved]

[60 FR 67405, Dec. 29, 1995, as amended at 86 FR 36197, July 9, 2021]

§ 1717.609 RUS approval of general manager.

(a) If a borrower's mortgage or loan contract grants RUS the unconditioned right to approve the employment and/or the employment contract of the general manager of the borrower's system, such approval is hereby granted provided that the borrower is in compliance with all provisions of its loan documents and any other agreements with RUS.

(b) If a borrower is in default with respect to any provision of its loan documents or any other agreement with RUS:

(1) Such borrower, if directed in writing by RUS, shall replace its general manager within 30 days after the date of such written notice; and

(2) Such borrower shall not hire a general manager without prior written approval by RUS.

§ 1717.610 RUS approval of compensation of the board of directors.

If a borrower's mortgage or loan contract requires the borrower to obtain approval from RUS for compensation provided to members of the borrower's board of directors, such requirement is hereby waived.

§ 1717.611 RUS approval of expenditures for legal, accounting, engineering, and supervisory services.

(a) If a borrower's mortgage or loan contract requires the borrower to obtain approval from RUS before incurring expenses for legal, accounting, supervisory (other than for the management and operation of the borrower's electric system, see § 1717.608(d)), or other similar services, such approval is hereby granted. However, while expenditures for accounting do not require RUS approval, the selection of a certified public accountant by the borrower to prepare audited reports required by RUS remains subject to RUS approval.

(b) If a borrower's mortgage or loan contract requires the borrower to obtain approval from RUS before incurring expenses for engineering services, such approval is hereby granted if such services will not be financed by RUS. Approval requirements with respect to engineering services financed by RUS are set forth in other RUS regulations.

§ 1717.612 RUS approval of borrower's bank or other depository.

If a borrower's mortgage or loan contract gives RUS the authority to approve the bank or other depositories used by the borrower, such approval is hereby granted. However, without the prior written approval of RUS, a borrower shall not deposit funds from loans made or guaranteed by RUS in any bank or other depository that is not insured by the Federal Deposit Insurance Corporation or other Federal agency acceptable to RUS, or in any account not so insured.

§ 1717.613 RUS approval of data processing and system control equipment.

If a borrower's mortgage or loan contract requires the borrower to obtain approval from RUS before purchasing data processing equipment or system control equipment, such approval is hereby granted if the equipment will not be financed by RUS.

§ 1717.614 Notification of rate changes.

If a distribution borrower is required by its loan documents to notify RUS in writing of proposed changes in electric rates more than 30 days prior to the effective date of such rates, the required notification period shall be 30 days. Moreover, such notification shall be required only upon the request of RUS.

§ 1717.615 Consolidations and mergers.

A distribution or power supply borrower may without the prior approval of RUS, consolidate or merge with any other corporation or convey or transfer the mortgaged property substantially as an entirety if the following conditions are met:

(a) Such consolidation, merger, conveyance or transfer shall be on such terms as shall fully preserve the lien and security of the RUS mortgage and the rights and powers of the mortgagees;

(b) The entity formed by such consolidation or with which the borrower is merged or the corporation which acquires by conveyance or transfer the mortgaged property substantially as an entirety shall execute and deliver to the mortgagees a mortgage supplemental in recordable form and containing an assumption by such successor entity of the due and punctual payment of the principal of and interest on all of the outstanding notes and the performance and observance of every covenant and condition of the mortgage;

(c) Immediately after giving effect to such transaction, no default under the mortgage shall have occurred and be continuing;

(d) The borrower shall have delivered to the mortgagees a certificate of its general manager or other officer, in form and substance satisfactory to each of the mortgagees, which shall state that such consolidation, merger, conveyance or transfer and such supplemental mortgage comply with this section and that all conditions precedent herein provided for relating to such transaction have been complied with;

(e) The borrower shall have delivered to the mortgagees an opinion of counsel in form and substance satisfactory to each of the mortgagees; and

(f) The entity formed by such consolidation or with which the borrower is merged or the corporation which acquires by conveyance or transfer the mortgaged property substantially as an entirety shall be an entity having:

(1) Equity equal to at least 27% of its total assets on a pro forma basis after giving effect to such transaction;

(2) A pro forma TIER of not less than 1.25 and a pro forma DSC of not less than 1.25 for each of the two preceding calendar years;

(3) Net utility plant equal to or greater than 1.0 times its total long-term debt on a pro forma basis.

[60 FR 67405, Dec. 29, 1995, as amended at 65 FR 51748, Aug. 25, 2000; 67 FR 70153, Nov. 21, 2002]

§ 1717.616 Sale, lease, or transfer of capital assets.

A borrower may, without the prior approval of RUS, sell, lease, or transfer any capital asset if the following conditions are met:

(a) The borrower is not in default;

(b) In the most recent year for which data is available, the borrower has met its coverage ratios as set in 7 CFR part 1710.114(b) or other financial requirements as established by their Mortgages, Loan Contracts, and/or other Security Agreements;

(c) The sale, lease, or transfer of assets will not reduce the borrower's existing or future requirements for energy or capacity being furnished to the borrower under any wholesale power contract which has been pledged as security to the government;

(d) Fair market value is obtained for the assets;

(e) The aggregate value of assets sold, leased, or transferred in any 12-month period is less than 10 percent of the borrower's net utility plant prior to the transaction;

(f) The proceeds of such sale, lease, or transfer, less ordinary and reasonable expenses incident to such transaction, are immediately:

(1) Applied as a prepayment of all notes secured under the mortgage equally and ratably;

(2) In the case of dispositions of equipment, materials or scrap, applied to the purchase of other property useful in the borrower's utility business; or

(3) Applied to the acquisition of construction of utility plant.

[60 FR 67405, Dec. 29, 1995, as amended at 65 FR 51748, Aug. 25, 2000; 86 FR 36197, July 9, 2021]

§ 1717.617 Limitations on distributions.

If a distribution or power supply borrower is required by its loan documents to obtain prior approval from RUS before declaring or paying any dividends, paying or determining to pay any patronage refunds, or retiring any patronage capital, or making any other cash distributions, such approval is hereby given if the following conditions are met:

(a) After giving effect to the distribution, the borrower's equity will be greater than or equal to 30 percent of its total assets;

(b) The borrower is current on all payments due on all notes secured under the mortgage;

(c) The borrower is not otherwise in default under its loan documents; and

(d) After giving effect to the distribution, the borrower's current and accrued assets will be not less than its current and accrued liabilities.

§ 1717.618 Wholesale power contracts.

(a) Pursuant to the terms of the RUS documents each power supply borrower shall establish and adjust rates for the sale of electric power and energy in such a manner as to assure that the borrower will be able to make required payments on secured loans.

(b) Pursuant to the terms of the RUS wholesale power contract, the Board of Directors or Board of Trustees of the power supply borrower shall review rates not less frequently than once each calendar year and revise its rates as therein set forth. The RUS wholesale power contract further provides that the borrower shall notify the Administrator not less than 30 nor more than 45 days prior to the effective date of any adjustment and shall set forth the basis upon which the rate is to be adjusted and established. The RUS wholesale power contract provides that no final revision in rates shall be effective unless approved in writing by the Administrator.

NOTE 1 TO PARAGRAPH (b): The Wholesale Power Contract, with minor modifications which are approved by RUS on a case by case basis, provides that the rate charged for electric power and energy, shall produce revenues which shall be sufficient, but only sufficient, with the revenues of the Seller from all other sources, to meet the cost of the operation and maintenance (including without limitation, replacements, insurance, taxes and administrative and general overhead expenses) of the generating plant transmission system and related facilities of the Seller, the cost of any power and energy purchased for resale hereunder by the Seller, the cost of transmission service, make payments on account of principal and interest on all indebtedness of the Seller, and to provide for the establishment and maintenance of reasonable reserves.

(c) Pursuant to the terms of the RUS mortgage, each power supply borrower must design its rates as therein set forth and must give 90 days prior notice to RUS of any proposed change in its general rate structure.

(Approved by the Office of Management and Budget under control number 0572–0089)

[84 FR 32615, July 9, 2019]

Subpart N—Investments, Loans, and Guarantees by Electric Borrowers

AUTHORITY: 7 U.S.C. 901–950b; Pub. L. 103–354, 108 Stat. 3178 (7 U.S.C. 6941 *et seq.*); Title I, Subtitle D, Pub. L. 100–203, 101 Stat. 1330.

SOURCE: 60 FR 48877, Sept. 21, 1995, unless otherwise noted.

§ 1717.650 Purpose.

This subpart sets forth general regulations for implementing and interpreting provisions of the RUS mortgage and loan contract regarding investments, loans, and guarantees made by electric borrowers, as well as the provisions of the Rural Electrification Act of 1936, as amended, including section 312 (7 U.S.C. 901 *et seq.*) (RE Act), permitting, in certain circumstances, that electric borrowers under the RE Act may, without restriction or prior approval of the Administrator of the Rural Utilities Service (RUS), invest their own funds and make loans or guarantees.

§ 1717.651 General.

(a) *Policy.* RUS electric borrowers are encouraged to utilize their own funds to participate in the economic development of rural areas, provided that such activity does not in any way put government funds at risk or impair a borrower's ability to repay its indebtedness to RUS and other lenders. In considering whether to make loans, investments, or guarantees, borrowers are expected to act in accordance with prudent business practices and in conformity with the laws of the jurisdictions in which they serve. RUS assumes that borrowers will use the latitude afforded them by section 312 of the RE Act primarily to make needed investments in rural community infrastructure projects (such as water and waste systems, garbage collection services, etc.) and in job creation activities (such as providing technical, financial, and managerial assistance) and other activities to promote business development and economic diversification in rural communities. Nonetheless, RUS believes that borrowers should continue to give primary consideration to safety and liquidity in the management of their funds.

(b) *Applicability of this subpart.* This subpart applies to all distribution and power supply borrowers regardless of when their loan contract or mortgage was executed.

§ 1717.652 Definitions.

As used in this subpart:

Borrower means any organization that has an outstanding loan made or guaranteed by RUS for rural electrification.

Cash-construction fund-trustee account means the account described in the Uniform System of Accounts as one to which funds are deposited for financing the construction or purchase of electric facilities.

Distribution borrower means a Distribution Borrower as defined in 7 CFR 1710.2.

Electric system means all of the borrower's interests in all electric production, transmission, distribution, conservation, load management, general plant and other related facilities, equipment or property and in any mine, well, pipeline, plant, structure or other facility for the development, production, manufacture, storage, fabrication or processing of fossil, nuclear, or other fuel or in any facility or rights with respect to the supply of water, in each case for use, in whole or in major part, in any of the borrower's generating plants, including any interest or participation of the borrower in any such facilities or any rights to the output or capacity thereof, together with all lands, easements, rights-of-way, other works, property, structures, contract rights and other tangible and intangible assets of the borrower in each case used or useful in such electric system.

Equity means the Margins and Equities of the borrower as defined in the Uniform System of Accounts, less regulatory created assets.

Guarantee means to undertake collaterally to answer for the payment of another's debt or the performance of another's duty, liability, or obligation, including, without limitation, the obligations of subsidiaries. Some examples of such guarantees include guarantees of payment or collection on a note or other debt instrument (assuring returns on investments); issuing performance bonds or completion bonds; or cosigning leases or other obligations of third parties.

Invest means to commit money in order to earn a financial return on assets, including, without limitation, all investments properly recorded on the borrower's books and records in investment accounts as those accounts are used in the Uniform System of Accounts for RUS Borrowers. Borrowers may submit any proposed transaction to RUS for an interpretation of whether the action is an investment for the purposes of this definition.

Make loans means to lend out money for temporary use on condition of repayment, usually with interest.

Mortgaged property means any asset of the borrower which is pledged in the RUS mortgage.

Natural gas distribution system means any system of community infrastructure that distributes natural gas and whose services are available by design to all or a substantial portion of the members of the community.

Operating DSC means Operating Debt Service Coverage (ODSC) of the borrower's electric system calculated as:

$$ODSC = \frac{A + B + C}{D}$$

where:

All amounts are for the same year and are based on the RUS system of accounts;

A = Depreciation and Amortization Expense of the electric system;

B = Interest on Long-term Debt of the electric system, except that Interest on Long-term Debt shall be increased by ⅓ of the amount, if any, by which the rentals of Restricted Property of the electric system exceed 2 percent of Total Margins and Equities;

C = Patronage Capital & Operating Margins of the electric system (distribution borrowers) or Operating Margins of the electric system (power supply borrowers); and

D = Debt Service Billed (RUS + other) which equals all interest and principal billed or billable during the calendar year for long-term debt of the electric system plus ⅓ of the amount, if any, by which the rentals of Restricted Property of the electric system exceed 2 percent of Total Margins and Equities. Unless otherwise indicated, all terms used in defining ODSC and OTIER are as defined in RUS Bulletin 1717B–2 Instructions for the Preparation of the Financial and Statistical Report for Electric Distribution Borrowers, and RUS Bulletin 1717B–3 Instructions for the Preparation of the Operating Report for Power Supply Borrowers and for Distribution Borrowers with Generating Facilities, or the successors to these bulletins.

Operating TIER means Operating Times Interest Earned Ratio (OTIER)

of the borrower's electric system calculated as:

$$OTIER = \frac{A+B}{A}$$

where:

All amounts are for the same year and are based on the RUS system of accounts;

A = Interest on Long-term Debt of the electric system, except that Interest on Long-term Debt shall be increased by ⅓ of the amount, if any, by which the rentals of Restricted Property of the electric system exceed 2 percent of Total Margins and Equities; and

B = Patronage Capital & Operating Margins of the electric system (distribution borrowers) or Operating Margins of the electric system (power supply borrowers).

Own funds means money belonging to the borrower other than funds on deposit in the cash-construction fund-trustee account.

Power supply borrower means a Power Supply Borrower as defined in 7 CFR 1710.2.

Regulatory created assets means the sum of the amounts properly recordable in Account 182.2 Unrecovered Plant and Regulatory Study Costs, and Account 182.3 Other Regulatory Assets of the Uniform System of Accounts.

RUS means the Rural Utilities Service, an agency of the U.S. Department of Agriculture established pursuant to Section 232 of the Federal Crop Insurance Reform and Department of Agriculture Reorganization Act of 1994 (Pub. L. 103–354, 108 Stat. 3178, 7 U.S.C. 6941 *et seq.*) and, for purposes of this subpart, includes its predecessor, the Rural Electrification Administration.

RUS loan contract means the loan contract between the borrower and RUS.

RUS mortgage means any and all instruments creating a lien on or security interest in the borrower's assets in connection with loans or guarantees under the RE Act.

Solid waste disposal system means any system of community infrastructure that provides collection and/or disposal of solid waste and whose services are available by design to all or a substantial portion of the members of the community.

Subsidiary means a company which is controlled by the borrower through ownership of voting stock, and is further defined in 7 CFR 1767.10.

Supplemental lender means a lender that has provided a supplemental source of financing that is secured by the RUS mortgage.

Telecommunication and other electronic communication system means any community infrastructure that provides telecommunication or other electronic communication services and whose services are available by design to all or a substantial portion of the members of the community.

Total assets means the total assets of the borrower as calculated according to the Uniform System of Accounts, less regulatory created assets.

Total utility plant means the sum of the borrower's Electric Plant Accounts and Construction Work in Progress—Electric Accounts, as such terms are used in the Uniform System of Accounts.

Uniform System of Accounts means the system of accounts prescribed for RUS borrowers in 7 CFR part 1767.

Water and waste disposal system means any system of community infrastructure that supplies water and/or collects and treats waste water and whose services are available by design to all or a substantial portion of the members of the community.

§ 1717.653 Borrowers in default.

Any borrower not in compliance with all provisions of its mortgage, loan contract, or any other agreements with RUS must, unless the borrower's mortgage, loan contract, or other agreement with RUS specifically provides otherwise with respect to such a borrower:

(a) Obtain prior written approval from the Administrator to invest its own funds or to make loans or guarantees regardless of the aggregate amount of such investments, loans, or guarantees; and

(b) If requested by the Administrator, restructure or reduce the amount of its investments, loans, and guarantees to a level determined by the Administrator, in his or her sole discretion, to be in the financial interest of the government with respect to loan security and/or repayment. If the borrower does not so restructure or reduce its portfolio within a reasonable period of time determined by the Administrator, which shall not exceed 12 months from the date the borrower was notified of the required action, then, upon written notice from RUS, the borrower shall be in default of its RUS loan contract and mortgage.

§ 1717.654 Transactions below the 15 percent level.

(a) A borrower in compliance with all provisions of its RUS mortgage, RUS loan contract, and any other agreements with RUS may, without prior written approval of the Administrator, invest its own funds or make loans or guarantees not in excess of 15 percent of its total utility plant without regard to any provision contained in any RUS mortgage or RUS loan contract to the effect that the borrower must obtain prior approval from RUS, provided, however, that the borrower may not, without the prior written approval of the Administrator, make such investments, loans, and guarantees to extend, add to, or modify its electric system. Moreover, funds necessary to make timely payments of principal and interest on loans secured by the RUS mortgage remain subject to RUS controls on borrower investments, loans and guarantees.

(b) RUS will not consider requests from borrowers to exclude investments, loans, or guarantees made below the 15 percent level. (Categorical exclusions are set forth in § 1717.655.)

§ 1717.655 Exclusion of certain investments, loans, and guarantees.

(a) In calculating the amount of investments, loans and guarantees permitted under this subpart, there is excluded from the computation any investment, loan or guarantee of the type which by the terms of the borrower's RUS mortgage or RUS loan contract the borrower may make in unlimited amounts without RUS approval.

(b) Furthermore, the borrower may make unlimited investments, without prior approval of the Administrator, in:

(1) Securities or deposits issued, guaranteed or fully insured as to payment by the United States Government or any agency thereof;

(2) Capital term certificates, bank stock, or other similar securities of the supplemental lender which have been purchased as a condition of membership in the supplemental lender, or as a condition of receiving financial assistance from such lender, as well as any other investment made in, or loans made to, the National Rural Utilities Cooperative Finance Corporation, the Saint Paul Bank for Cooperatives, and CoBank, ACB;

(3) Patronage capital allocated from an electric power supply cooperative of which the borrower is a member; and

(4) Patronage capital allocated from an electric distribution cooperative to a power supply borrower.

(c) Without prior approval of the Administrator, the borrower may also:

(1) Invest or lend funds derived directly from:

(i) Grants which the borrower in not obligated to repay, regardless of the source or purpose of the grant; and

(ii) Loans received from or guaranteed by any Federal, State or local government program designed to promote rural economic development, provided that the borrower uses the loan proceeds for such purpose;

(2) Make loans guaranteed by an agency of USDA, up to the amount of principal whose repayment, with interest, is fully guaranteed; and

(3)(i) Make unlimited investments in and unlimited loans to finance the following community infrastructure that serves primarily consumers located in rural areas as defined in 7 CFR 1710.2, and guarantee debt issued for the construction or acquisition of such infrastructure, up to an aggregate amount of such guarantees not to exceed 20 percent of the borrower's equity:

(A) Water and waste disposal systems;

(B) Solid waste disposal systems;

(C) Telecommunication and other electronic communication systems; and

(D) Natural gas distribution systems.

(ii) In each of the four cases in paragraph (c)(3)(i) of this section, if the system is a component of a larger organization other than the borrower itself (e.g., if it is a component of a subsidiary of the borrower or a corporation independent of the borrower), to be eligible for the exemption the borrower must certify annually that a majority of the gross revenues of the larger organization during the most recent fiscal year came from customers of said system who were located in a rural area.

(d) Also excluded from the calculation of investments, loans and guarantees made by the borrower are:

(1) Amounts properly recordable in Account 142 Customer Accounts Receivable, and Account 143 Other Accounts Receivable;

(2) Any investment, loan, or guarantee that the borrower is required to make by an agency of USDA, for example, as a condition of obtaining financial assistance for itself or any other person or organization;

(3) Investments included in an irrevocable trust for the purpose of funding post-retirement benefits of the borrower's employees;

(4) Reserves required by a reserve bond agreement or other agreement legally binding on the borrower, that are dedicated to making required payments on debt secured under the RUS mortgage, not to exceed the amount of reserves specifically required by such agreements; and

(5) Investments included in an irrevocable trust approved by RUS and dedicated to the payment of decommissioning costs of nuclear facilities of the borrower.

(e) Grandfathered exclusions. All amounts of individual investments, loans, and guarantees excluded by RUS as of February 16, 1995 shall remain excluded. Such exclusions must have been based on the RUS mortgage, RUS loan contract, regulations, bulletins, memoranda, or other written notice from RUS. Profits, interest, and other returns earned (regardless of whether or not they are reinvested) on such investments, loans and guarantees after February 16, 1995 shall be excluded only if they are eligible for exclusion under paragraphs (a) through (d) of this section. Any new commitments of money to such investments, loans and guarantees shall likewise be excluded only if they are eligible under paragraphs (a) through (d) of this section.

(f) Any investment, loan or guarantee made by a borrower that is not excluded under this section or under § 1717.657(d) shall be included in the aggregate amount of investments, loans and guarantees made by the borrower, regardless of whether RUS has specifically approved the investment, loan or guarantee under § 1717.657(c), or has approved a related transaction (e.g., a lien accommodation).

§ 1717.656 Exemption of certain borrowers from controls.

(a) Any distribution or power supply borrower that meets all of the following criteria is exempted from the provisions of the RUS mortgage and loan contract that require RUS approval of investments, loans, and guarantees, except investments, loans, and guarantees made to extend, add to, or modify the borrower's electric system:

(1) The borrower is in compliance with all provisions of its RUS mortgage, RUS loan contract, and any other agreements with RUS;

(2) The average revenue per kWh for residential service received by the borrower during the two most recent calendar years does not exceed 130 percent of the average revenue per kWh for residential service during the same period for all residential consumers located in the state or states served by the borrower. This criterion applies only to distribution borrowers and does not apply to power supply borrowers. If a borrower serves customers in more than one state, the state average revenue per kWh will be based on a weighted average using the kWh sales by the borrower in each state as the weight. The calculation will be based on the two most recent calendar years for which both borrower and state-wide data are available. If a borrower fails to qualify for an exemption based solely on its failure to meet this criterion on rate disparity, at the borrower's request the Administrator may, at his or her sole discretion, exempt the borrower if he or she finds that the borrower's strengths with respect to the other criteria are sufficient to offset any weakness due to rate disparity;

(3) In the most recent calendar year for which data are available, the borrower achieved an operating TIER of at least 1.0 and an operating DSC of at least 1.0, in each case based on the average of the two highest ratios achieved in the three most recent calendar years;

(4) The borrower's ratio of net utility plant to long-term debt is at least 1.1, based on year-end data for the most recent calendar year for which data are available; and

(5) The borrower's equity is equal to at least 27 percent of its total assets, based on year-end data for the most recent calendar year for which data are available.

(b) While borrowers meeting the criteria in paragraph (a) of this section are exempt from RUS approval of investments, loans and guarantees, they are nevertheless subject to the recordkeeping, reporting, and other requirements of § 1717.658.

(c) Any borrower exempt under paragraph (a) of this section that ceases to meet the criteria for exemption shall, upon written notice from RUS, no longer be exempt and shall be subject to the provisions of this subpart applicable to non-exempt borrowers. A borrower may regain its exemption if it subsequently meets the criteria in paragraph (a) of this section, and is so notified in writing by RUS.

(d)(1) A borrower that loses its exemption and is not in compliance with all provisions of its mortgage, loan contract, or any other agreement with RUS may be required to restructure or reduce its portfolio of investments, loans and guarantees as provided in § 1717.653(b). If the borrower's portfolio exceeds the 15 percent level, the borrower will be required to restructure or reduce its portfolio to the 15 percent level or below. For example, if the borrower's mortgage or loan contract has an approval threshold, the borrower may be required to reduce its portfolio to that level, which in many cases is 3 percent of total utility plant.

(2) A borrower that loses its exemption but is in compliance with all provisions of its mortgage, loan contract, and any other agreements with RUS will be required, if its investments, loans and guarantees exceed the 15 percent level, to restructure or reduce its portfolio to the 15 percent level, unless the Administrator, in his or her sole discretion, determines that such action would not be in the financial interest of the government with respect to loan security and/or repayment. (Such borrower is eligible to ask RUS to exclude a portion of its investments under the conditions set forth in § 1717.657(d).)

(3) If a borrower required to reduce or restructure its portfolio does not fully comply within a reasonable period of time determined by the Administrator, which shall not exceed 12 months from the date the borrower was notified of its loss of exemption, then, upon written notice from RUS, the borrower shall be in default of its RUS loan contract and/or RUS mortgage.

(e) By no later than July 1 of each year, RUS will provide written notice to any borrowers whose exemption status has changed as a result of more recent data being available for the qualification criteria set forth in paragraph (a) of this section, or as a result of other reasons, such as corrections in the available data. An explanation of the reasons for any changes in exemption status will also be provided to the borrowers affected.

§ 1717.657 Investments above the 15 percent level by certain borrowers not exempt under § 1717.656(a).

(a) *General.* (1) This section applies only to borrowers that are in compliance with all provisions of their mortgage, loan contract, and any other agreements with RUS and that do not qualify for an exemption from RUS investment controls under § 1717.656(a).

(2) Nothing in this section shall in any way affect the Administrator's authority to exercise approval rights over investments, loans, and guarantees made by a borrower that is not in compliance with all provisions of its mortgage, loan contract and any other agreements with RUS.

(b) *Distribution borrowers.* Distribution borrowers not exempt from RUS investment controls under § 1717.656(a) may not make investments, loans and guarantees in an aggregate amount in excess of 15 percent of total utility plant. Above the 15 percent level, such borrowers will be restricted to excluded investments, loans and guarantees as defined in § 1717.655. (However, they are eligible to ask RUS to exclude a portion of their investments under the conditions set forth in paragraph (d) of this section.)

(c) *Power supply borrowers.* (1) Power supply borrowers not exempt from RUS investment controls under § 1717.656(a) may request approval to exceed the 15 percent level if all of the following criteria are met:

(i) Satisfactory evidence has been provided that the borrower is in compliance with all provisions of its RUS mortgage, RUS loan contract, and any other agreements with RUS;

(ii) The borrower is not in financial workout and has not had its government debt restructured;

(iii) The borrower has equity equal to at least 5 percent of its total assets; and

(iv) After approval of the investment, loan or guarantee, the aggregate of the borrower's investments, loans and guarantees will not exceed 20 percent of the borrower's total utility plant.

(2) Borrower requests for approval to exceed the 15 percent level will be considered on a case by case basis. The requests must be made in writing.

(3) In considering borrower requests, the Administrator will take the following factors into consideration:

(i) The repayment of all loans secured under the RUS mortgage will continue to be assured, and loan security must continue to be reasonably adequate, even if the entire investment or loan is lost or the borrower is required to perform for the entire amount of the guarantee. These risks will be considered along with all other risks facing the borrower, whether or not related to the investment, loan or guarantee;

(ii) In the case of investments, the investment must be made in an entity separate from the borrower, such as a subsidiary, whereby the borrower is protected from any liabilities incurred by the separate entity, unless the borrower demonstrates to the satisfaction of the Administrator that making the investment directly rather than through a separate entity will present no substantial risk to the borrower in addition to the possibility of losing all or part of the original investment;

(iii) The borrower must be economically and financially sound as indicated by its costs of operation, competitiveness, operating TIER and operating DSC, physical condition of the plant, ratio of equity to total assets, ratio of net utility plant to long-term debt, and other factors; and

(iv) Other factors affecting the security and repayment of government debt, as determined by the Administrator on a case by case basis.

(4) If the Administrator approves an investment, loan or guarantee, such investment, loan or guarantee will continue to be included when calculating the borrower's ratio of aggregate investments, loans and guarantees to total utility plant.

(d) *Distribution and power supply borrowers.* If the aggregate of the investments, loans and guarantees of a distribution or power supply borrower exceeds 15 percent of the borrower's total utility plant as a result of the cumulative profits or margins, net of losses, earned on said transactions over the past 10 calendar years (i.e., the sum of all profits earned during the 10 years on all transactions—including interest earned on cash accounts, loans, and similar transactions—less the sum of all losses experienced on all transactions during the 10 years) then:

(1) The borrower will not be in default of the RUS loan contract or RUS mortgage with respect to required approval of investments, loans and guarantees, provided that the borrower had not made additional net investments, loans or guarantees without approval after reaching the 15 percent level; and

(2) At the request of the borrower, the Administrator in his or her sole discretion may decide to exclude up to the amount of net profits or margins earned on the borrower's investments, loans and guarantees during the past 10 calendar years, if the Administrator determines that such exclusion will not increase loan security risks. The borrower must provide documentation satisfactory to the Administrator as to the current status of its investments, loans and guarantees and the net profits earned during the past 10 years. Any

exclusion approved by the Administrator may or may not reduce the level of investments, loans and guarantees to or below the 15 percent level. If such exclusion does not reduce the level to or below the 15 percent level, RUS will notify the borrower in writing that it must reduce or restructure its investments, loans and guarantees to a level of not more than 15 percent of total utility plant. If the borrower does not come within the 15 percent level within a reasonable period of time determined by the Administrator, which shall not exceed 12 months from the date the borrower was notified of the required action, then, upon written notice from RUS, the borrower shall be in default of its RUS loan contract and mortgage.

§ 1717.658 Records, reports and audits.

(a) Every borrower shall maintain accurate records concerning all investments, loans and guarantees made by it. Such records shall be kept in a manner that will enable RUS to readily determine:

(1) The nature and source of all income, expenses and losses generated from the borrower's loans, guarantees and investments;

(2) The location, identity and lien priority of any loan collateral resulting from activities permitted by this subpart; and

(3) The effects, if any, which such activities may have on the feasibility of loans made, guaranteed or lien accommodated by RUS.

(b) In determining the aggregate amount of investments, loans and guarantees made by a borrower, the borrower shall use the recorded value of each investment, loan or guarantee as reflected on its books and records for the next preceding end-of-month, except for the end-of-year report which shall be based on December 31 information. Every borrower shall also report annually to RUS, in the manner and on the form specified by the Administrator, the current status of each investment, outstanding loan and outstanding guarantee which it has made pursuant to this subpart.

(c) The records of borrowers shall be subject to the auditing procedures prescribed in part 1773 of this chapter. RUS reserves the right to review the financial records of any subsidiaries of the borrower to determine if the borrower is in compliance with this subpart, and to ascertain if the debts, guarantees (as defined in this subpart), or other obligations of the subsidiaries could adversely affect the ability of the borrower to repay its debts to the Government.

(d) RUS will monitor borrower compliance with this subpart based primarily on the annual financial and statistical report submitted by the borrower to RUS and the annual auditor's report on the borrower's operations. However, RUS may inspect the borrower's records at any time during the year to determine borrower compliance. If a borrower's most recent annual financial and statistical report shows the aggregate of the borrower's investments, loans and guarantees to be below the 15 percent level, that in no way relieves the borrower of its obligation to comply with its RUS mortgage, RUS loan contract, and this subpart with respect to Administrator approval of any additional investment, loan or guarantee that would cause the aggregate to exceed the 15 percent level.

§ 1717.659 Effect of this subpart on RUS loan contract and mortgage.

(a) Nothing in this subpart shall affect any provision, covenant, or requirement in the RUS mortgage, RUS loan contract, or any other agreement between a borrower and RUS with respect to any matter other than the prior approval by RUS of investments, loans, and guarantees by the borrower, such matters including, without limitation, extensions, additions, and modifications of the borrower's electric system. Also, nothing in this subpart shall affect any rights which supplemental lenders have under the RUS mortgage, or under their loan contracts or other agreements with their borrowers, to limit investments, loans and guarantees by their borrowers to levels below 15 percent of total utility plant.

(b) RUS will require that any electric loan made or guaranteed by RUS after October 23, 1995 shall be subject to a provision in the loan contract or mortgage restricting investments, loans and guarantees by the borrower substantially as follows: The borrower shall not make any loan or advance to, or make any investment in, or purchase or make any commitment to purchase any stock, bonds, notes or other securities of, or guaranty, assume or otherwise become obligated or liable with respect to the obligations of, any other person, firm or corporation, except as permitted by the RE Act and RUS regulations.

(c) RUS reserves the right to change the provisions of the RUS mortgage and loan contract relating to RUS approval of investments, loans and guarantees made by the borrower, on a case-by-case basis, in connection with providing additional financial assistance to a borrower after October 23, 1995.

Subpart O [Reserved]

§§ 1717.700–1717.749 [Reserved]

Subpart P [Reserved]

§§ 1717.750–1717.799 [Reserved]

Subpart Q [Reserved]

§§ 1717.800–1717.849 [Reserved]

Subpart R—Lien Accommodations and Subordinations for 100 Percent Private Financing

SOURCE: 58 FR 53843, Oct. 19, 1993, unless otherwise noted.

§ 1717.850 General.

(a) *Scope and applicability.* (1) This subpart R establishes policies and procedures for the accommodation, subordination or release of the Government's lien on borrower assets, including approvals of supporting documents and related loan security documents, in connection with 100 percent private sector financing of facilities and other purposes. Policies and procedures regarding lien accommodations for concurrent supplemental financing required in connection with an RUS insured loan are set forth in subpart S of this part.

(2) This subpart and subpart S of this part apply only to debt to be secured under the mortgage, the issuance of which is subject to the approval of the Rural Utilities Service (RUS) by the terms of the borrower's mortgage with respect to the issuance of additional debt or the refinancing or refunding of debt. If RUS approval is not required under such terms of the mortgage itself, a lien accommodation is not required. If the loan contract or other agreement between the borrower and RUS requires RUS approval with respect to the issuance of debt or making additions to or extensions of the borrower's system, such required approvals do not by themselves result in the need for a lien accommodation.

(b) *Overall policy.* (1) Consistent with prudent lending practices, the maintenance of adequate security for RUS's loans, and the objectives of the Rural Electrification Act (RE Act), it is the policy of RUS to provide effective and timely assistance to borrowers in obtaining financing from other lenders by sharing RUS's lien on a borrower's assets in order to finance electric facilities, equipment and systems, and certain other types of community infrastructure. In certain circumstances, RUS may facilitate the financing of such assets by subordinating its lien on specific assets financed by other lenders.

(2) It is also the policy of RUS to provide effective and timely assistance to borrowers in promoting rural development by subordinating RUS's lien for financially sound rural development investments under the conditions set forth in § 1717.858.

(c) *Decision factors.* In determining whether to accommodate, subordinate, or release its lien on property pledged by the borrower under the RUS mortgage, RUS will consider the effects of such action on the achievement of the purposes of the RE Act, the repayment and security of RUS loans secured by the mortgage, and other factors set forth in this subpart. The following factors will be considered in assessing the effects on the repayment and security of RUS loans:

(1) The value of the added assets compared with the amount of new debt to be secured;

(2) The value of the assets already pledged under the mortgage, and any

effects of the proposed transaction on the value of those assets;

(3) The ratio of the total outstanding debt secured under the mortgage to the value of all assets pledged as security under the mortgage;

(4) The borrower's ability to repay debt owed to the Government, as indicated by the following factors:

(i) Revenues, costs (including interest, lease payments and other debt service costs), margins, Times Interest Earned Ratio (TIER), Debt Service Coverage (DSC), and other case-specific economic and financial factors;

(ii) The variability and uncertainty of future revenues, costs, margins, TIER, DSC, and other case-specific economic and financial factors;

(iii) Future capital needs and the ability of the borrower to meet those needs at reasonable cost;

(iv) The ability of the borrower's management to manage and control its system effectively and plan for future needs; and

(5) Other factors that may be relevant in individual cases, as determined by RUS.

(d) *Environmental review requirements.* The environmental review requirements of 7 CFR part 1970 apply to applications for subordinations.

(e) *Co-mortgagees.* Other mortgagees under existing mortgages shared with RUS may have the right to approve requests for lien accommodations, subordinations and releases. In those cases, borrowers would have to obtain the approval of such mortgagees in order for the lien of the mortgage to be accommodated, subordinated or released. Any reference in this subpart to waiving by RUS of any of its rights under the mortgage shall apply only to the rights of RUS and shall not apply to the rights of any other co-mortgagee.

(f) *Safety and performance standards.* (1) To be eligible for a lien accommodation or subordination from RUS, a borrower must comply with RUS standards regarding facility and system planning and design, construction, procurement, and the use of materials accepted by RUS, as required by the borrower's mortgage, loan contract, or other agreement with RUS, and as further specified in RUS regulations.

(2) RUS "Buy American" requirements shall not apply.

(g) *Advance of funds.* (1) The advance of funds from 100 percent private loans lien accommodated or subordinated by RUS will not be subject to RUS approval. It is the private lender's responsibility to adopt reasonable measures to ensure that such loan funds are used for the purposes for which the loan was made and the lien accommodation or subordination granted. RUS encourages lenders to adopt the following measures:

(i) Remit loan advances to a separate subaccount of the Cash-Construction Fund-Trustee Account;

(ii) Obtain a certification from a registered professional engineer, for each year during which funds from the separate subaccount are utilized by the borrower, that all materials and equipment purchased and facilities constructed during the year from said funds comply with RUS safety and performance standards, as required by paragraph (f) of this section, and are included in an CWP or CWP amendment.

(iii) Obtain an auditor's certification from a Certified Public Accountant, for each year during which funds are advanced to or remitted from the separate subaccount, certifying:

(A) The amount of loan funds advanced to and remitted from the separate subaccount during the period of review;

(B) That based on the auditor's review of construction work orders and other records, all moneys disbursed from the separate subaccount during the period of review were used for purposes contemplated in the loan agreement and the lien accommodation; and

(iv) Immediately notify RUS in writing if the lender is unable to obtain the certifications cited in paragraphs (g)(1)(ii) and (g)(1)(iii) of this section.

(2) The measures listed in paragraph (g)(1) of this section will normally be sufficient to meet the lender's responsibility provided that additional measures are not reasonably required based on the particular circumstances of an individual case. Should a lender fail to carry out its responsibility in the manner described in this paragraph (g) or in another manner acceptable to RUS,

RUS may disqualify such lender from participation in advance approval under §§1717.854 and 1717.857 and condition the lender's receipt of a lien accommodation or subordination upon the lender providing satisfactory evidence that it will fulfill its responsibility under this paragraph (g).

(h) *Contracting and procurement procedures.* (1) Facilities financed with debt obtained entirely from non-RUS sources, without an RUS loan guarantee, are not subject to RUS post-loan requirements regarding contracting, procurement and bidding procedures; contract close-out procedures pertaining to project completion, final payment of contractor, and related matters; and standard forms of construction and procurement contracts listed in 7 CFR 1726.300.

(2) To the extent that provisions in a borrower's loan contract or mortgage in favor of RUS may be inconsistent with paragraphs (g)(1) and (h)(1) of this section, paragraphs (g)(1) and (h)(1) of this section are intended to constitute an approval or waiver under the terms of such instruments, and in any regulations implementing such instruments, with respect to facilities financed with debt obtained entirely from non-RUS sources without an RUS guarantee.

(i) *Access of handicapped to buildings and seismic safety.* A borrower must meet the following requirements to be eligible for a lien accommodation or subordination for 100 percent private financing of the construction of buildings:

(1) The borrower must provide RUS with a certification by the project architect that the buildings will be designed and constructed in compliance with Section 504 of the Rehabilitation Act of 1973 as amended (29 U.S.C. 794), as applicable under that Act, and that the facilities will be readily accessible to and usable by persons with handicaps in accordance with the Uniform Federal Accessibility Standards (UFAS), (Appendix A to 41 CFR part 101.19, subpart 101–19.6). The certification must be included in the borrower's application for a lien accommodation or subordination. In addition to these requirements, building construction may also be subject to requirements of The Americans with Disabilities Act (42 U.S.C. 12101 *et seq.*); and

(2) The borrower must comply with RUS's seismic safety requirements set forth in 7 CFR part 1792, subpart C.

(j) *Breach of warranty.* Any breach of any warranty or agreement or any material inaccuracy in any representation, warranty, certificate, document, or opinion submitted pursuant to this subpart, including, without limitation, any agreement or representation regarding the use of funds from loans lien accommodated or subordinated pursuant to this subpart, shall constitute a default by the borrower under the terms of its loan agreement with RUS.

(k) *Guaranteed loans.* The provisions of this subpart do not apply to lien accommodations or subordinations sought for loans guaranteed by RUS. Such lien accommodations and subordinations are governed by RUS regulations on guaranteed loans.

(l) *Release of lien.* To avoid repetition, release of lien is not mentioned in every instance where it may be an acceptable alternative to subordination of RUS's lien. Generally, lien subordination is favored over release of lien, and any decision to release RUS's lien is at the sole discretion of RUS.

(m) *Waiver authority.* Consistent with the RE Act and other applicable laws, any requirement, condition, or restriction imposed by this subpart, or subpart S of this part, on a borrower, private lender, or application for a lien accommodation or subordination may be waived or reduced by the Administrator, if the Administrator determines that said action is in the Government's financial interest with respect to ensuring repayment and reasonably adequate security for loans made or guaranteed by RUS.

(n) *Liability.* It is the intent of this subpart that any failure on the part of RUS to comply with any provisions hereof, including without limitation, those provisions setting forth specified timeframes for action by RUS on applications for lien accommodations or lien subordinations, shall not give rise to liability of any kind on the part of the Government or any employees of the Government including, without limitation, liability for damages, fees, expenses or costs incurred by or on behalf of a borrower, private lender or any other party.

[58 FR 53843, Oct. 19, 1993, as amended at 60 FR 67408, Dec. 29, 1995; 81 FR 11026, Mar. 2, 2016; 84 FR 32616, July 9, 2019]

§ 1717.851 Definitions.

Terms used in this subpart have the meanings set forth in 7 CFR 1710.2. References to specific RUS forms and other RUS documents, and to specific sections or lines of such forms and documents, shall include the corresponding forms, documents, sections and lines in any subsequent revisions of these forms and documents. In addition to the terms defined in 7 CFR 1710.2, the following terms have the following meanings for the purposes of this subpart:

Borrower's financial and statistical report means RUS Form 7, Parts A through D, for distribution borrowers, and RUS Form 12a for power supply borrowers.

Calendar day means any day of the year, except a Federal holiday that falls on a work day.

Capital investment. For the purposes of § 1717.860, capital investment means an original investment in an asset that is intended for long-term continued use or possession and, for accounting purposes, is normally depreciated or depleted as it is used. For example, such assets may include land, facilities, equipment, buildings, mineral deposits, patents, trademarks, and franchises. Original investments do not include refinancings or refundings.

Current refunding means any refunding of debt where the proceeds of the new debt are applied to refund the old debt within 90 days of the issuance of the new debt.

Default under the RUS mortgage, loan contract, restructuring agreement, or any other agreement between the borrower and RUS means any event of default or any event which, with the giving of notice or lapse of time or both, would become an event of default.

Equity, less deferred expenses, means Line 33 of Part C of RUS Form 7 less assets properly recordable in Account 182.2, Unrecovered Plant and Regulatory Study Costs, and Account 182.3, Other Regulatory Assets.

Front-end costs means the reasonable cost of engineering, architectural, environmental and other studies and plans needed to support the construction of facilities and other investments eligible for a lien accommodation or subordination under this subpart.

Lien accommodation means the sharing of the Government's (RUS's) lien on property, usually all property, covered by the lien of the RUS mortgage.

Lien subordination means allowing another lender to take a first mortgage lien on certain property covered by the lien of the RUS mortgage, and the Government (RUS) taking a second lien on such property.

Natural gas distribution system means any system of community infrastructure whose primary function is the distribution of natural gas and whose services are available by design to all or a substantial portion of the members of the community.

Net utility plant means Part C, Line 5 of RUS Form 7 (distribution borrowers) or Section B, Line 5 of RUS Form 12a (power supply borrowers).

Power cost study means the study defined in 7 CFR 1710.303.

Solid waste disposal system means any system of community infrastructure whose primary function is the collection and/or disposal of solid waste and whose services are available by design to all or a substantial portion of the members of the community.

Telecommunication and other electronic communication system means any system of community infrastructure whose primary function is the provision of telecommunication or other electronic communication services and whose services are available by design to all or a substantial portion of the members of the community.

Total assets, less deferred expenses means Line 26 of Part C of RUS Form 7 less assets properly recordable in Account 182.2, Unrecovered Plant and Regulatory Study Costs, and Account 182.3, Other Regulatory Assets.

Total outstanding long-term debt means Part C, Line 38 of RUS Form 7.

Transaction costs means the reasonable cost of legal advice, accounting fees, filing fees, recording fees, call premiums and prepayment penalties, financing costs (including, for example, underwriting commissions, letter of credit fees and bond insurance), and printing associated with borrower financing.

Water and waste disposal system means any system of community infrastructure whose primary function is the supplying of water and/or the collection and treatment of waste water and whose services are available by design to all or a substantial portion of the members of the community.

Weighted average life of the loan means the average life of the loan based on the proportion of original loan principal paid during each year of the loan. It shall be determined by calculating the sum of all loan principal payments, expressed as a fraction of the original loan principal amount, times the number of years and fractions of years elapsed at the time of each payment since issuance of the loan. For example, given a $5 million loan, with a maturity of 5 years and equal principal payments of $1 million due on the anniversary date of the loan, the weighted average life would be: (.2)(1 year) + (.2)(2 years) + (.2)(3 years) + (.2)(4 years) + (.2)(5 years) = .2 years + .4 years + .6 years + .8 years + 1.0 years = 3.0 years. If instead the loan had a balloon payment of $5 million at the end of 5 years, the weighted average life would be: ($5 million/$5 million)(5 years) = 5 years.

[58 FR 53843, Oct. 19, 1993, as amended at 59 FR 3986, Jan. 28, 1994; 60 FR 67409, Dec. 29, 1995]

§ 1717.852 Financing purposes.

(a) *Purposes eligible.* The following financing purposes, except as excluded in paragraph (b) of this section, are eligible for a lien accommodation from RUS, or in certain circumstances a subordination of RUS's lien on specific assets, provided that all applicable provisions of this subpart are met:

(1) The acquisition, construction, improvement, modification, and replacement (less salvage value) of systems, equipment, and facilities, including real property, used to supply electric and/or steam power to:

(i) RE Act beneficiaries; and/or

(ii) End-user customers of the borrower who are not beneficiaries of the RE Act. Such systems, equipment, and facilities include those listed in 7 CFR 1710.251(c) and 1710.252(c), as well as others that are determined by RUS to be an integral component of the borrower's system of supplying electric and/or steam power to consumers, such as, for example, coal mines, coal handling facilities, railroads and other transportation systems that supply fuel for generation, programs of demand side management and energy conservation, and on-grid and off-grid renewable energy systems;

(2) The purchase, rehabilitation and integration of existing distribution facilities, equipment and systems, and associated service territory;

(3) The following types of community infrastructure substantially located within the electric service territory of the borrower: water and waste disposal systems, solid waste disposal systems, telecommunication and other electronic communications systems, and natural gas distribution systems;

(4) Front-end costs, when and as the borrower has obtained a binding commitment from the non-RUS lender for the financing required to complete the procurement or construction of the facilities;

(5) Transaction costs included as part of the cost of financing assets or refinancing existing debt, provided, however, that the amount of transaction costs eligible for lien accommodation or subordination normally shall not exceed 5 percent of the principal amount of financing or refinancing provided, net of all transaction costs;

(6) The refinancing of existing debt secured under the mortgage;

(7) Interest during construction of generation and transmission facilities if approved by RUS, case by case, depending on the financial condition of the borrower, the terms of the financing, the nature of the construction, the treatment of these costs by regulatory authorities having jurisdiction, and such other factors deemed appropriate by RUS; and

(8) Lien subordinations for certain rural development investments, as provided in § 1717.858.

(b) *Purposes ineligible*. The following financing purposes are not eligible for a lien accommodation or subordination from RUS:

(1) Working capital, including operating funds, unless in the judgment of RUS the working capital is required to ensure the repayment of RUS loans and/or other loans secured under the mortgage;

(2) Facilities, equipment, appliances, or wiring located inside the premises of the consumer, except:

(i) Certain load-management equipment (see 7 CFR 1710.251(c));

(ii) Renewable energy systems and RUS-approved programs of Demand side management, energy efficiency and energy conservation; and

(iii) As determined by RUS on a case by case basis, facilities included as part of certain cogeneration projects to furnish electric and/or steam power to end-user customers of the borrower;

(3) Investments in a lender required of the borrower as a condition for obtaining financing; and

(4) Debt incurred by a distribution or power supply borrower to finance facilities, equipment or other assets that are not part of the borrower's electric system or one of the four community infrastructure systems cited in paragraph (a)(3) of this section, except for certain rural development investments eligible for a lien subordination under § 1717.858.

(c) *Lien subordination for electric utility investments*. RUS will consider subordinating its lien on specific electric utility assets financed by the lender, when the assets can be split off without materially reducing the stability, safety, reliability, operational efficiency, or liquidation value of the rest of the system.

[58 FR 53843, Oct. 19, 1993, as amended at 59 FR 3986, Jan. 28, 1994; 60 FR 67409, Dec. 29, 1995; 78 FR 73370, Dec. 5, 2013]

§ 1717.853 Loan terms and conditions.

(a) *Terms and conditions*. A loan, bond or other financing instrument, for which a lien accommodation or subordination is requested from RUS, must comply with the following terms and conditions:

(1) The maturity of the loan or bond used to finance facilities or other capital assets must not exceed the weighted average of the expected remaining useful lives of the assets being financed;

(2) The loan or bond must have a maturity of not less than 5 years, except for loans or bonds used to refinance debt that has a remaining maturity of less than 5 years;

(3) The principal of the loan or bond must be amortized at a rate that will yield a weighted average life not greater than the weighted average life that would result from level payments of principal and interest; and

(4) The loan, or any portion of the loan, may bear either a variable (set annually or more frequently) or a fixed interest rate.

(b) *RUS approval*. Loan terms and conditions and the loan agreement between the borrower and the lender are subject to RUS approval. However, RUS will usually waive its right of approval for distribution borrowers that meet the conditions for advance approval of a lien accommodation or subordination set forth in § 1717.854. RUS may also waive its right of approval in other cases. RUS's decision to waive its right of approval will depend on the adequacy of security for RUS's loans, the current and projected financial strength of the borrower and its ability to meet its financial obligations, RUS's familiarity with the lender and its lending practices, whether the transaction is ordinary or unusual, and the uncertainty and credit risks involved in the transaction.

§ 1717.854 Advance approval—100 percent private financing of distribution, subtransmission and headquarters facilities, and certain other community infrastructure.

(a) *Policy*. Requests for a lien accommodation or subordination from distribution borrowers for 100 percent private financing of distribution, subtransmission and headquarters facilities, and for community infrastructure listed in § 1717.852(a)(3), qualify for advance approval by RUS if they meet the conditions of this section and all other applicable provisions of this subpart. Advance approval means RUS will approve these requests once RUS is satisfied that the conditions of this section and all other applicable provisions of this subpart have been met.

(b) *Eligible purposes.* Lien accommodations or subordinations for the financing of distribution, subtransmission, and headquarters facilities and community infrastructure listed in § 1717.852(a)(3) are eligible for advance approval, except those that involve the purchase of existing facilities and associated service territory.

(c) *Qualification criteria.* To qualify for advance approval, the following requirements, as well as all other applicable requirements of this subpart, must be met:

(1) The borrower has achieved a TIER of at least 1.25 and a DSC of at least 1.25 for each of 2 calendar years immediately preceding, or any 2 consecutive 12 month periods ending within 180 days immediately preceding, the issuance of the debt;

(2) The ratio of the borrower's equity, less deferred expenses, to total assets, less deferred expenses, is not less than 20 percent, after adding the principal amount of the proposed loan to the total assets of the borrower;

(3) The borrower's net utility plant as a ratio to its total outstanding long-term debt is not less than 1.0, after adding the principal amount of the proposed loan to the existing outstanding long-term debt of the borrower;

(4) There are no actions or proceedings against the borrower, pending or overtly threatened in writing before any court, governmental agency, or arbitrator that would materially adversely affect the borrower's operations and/or financial condition;

(5) The borrower is current on all debt payments and all other financial obligations, and is not in default under the RUS mortgage, the RUS loan contract, the borrower's wholesale power contract, any debt restructuring agreement, or any other agreement with RUS;

(6) The borrower has:

(i) Submitted the annual auditor's report, report on compliance, report on internal controls, and management letter in accordance with 7 CFR part 1773;

(ii) Received an unqualified opinion in the most recent auditor's report;

(iii) Resolved all material findings and recommendations made in the most recent Loan Fund and Accounting Review;

(iv) Resolved all material findings and recommendations made in the most recent financial statement audit, including those material findings and recommendations made in the report on internal control, report on compliance, and management letter;

(v) Resolved all outstanding material accounting issues with RUS; and

(vi) Resolved any significant irregularities to RUS's satisfaction; and

(7) If the borrower has a power supply contract with a power supply borrower, the power supply borrower is current on all debt payments and all other financial obligations, and is not in default under the RUS mortgage, the loan contract, any debt restructuring agreement, or any other agreement with RUS.

(d) *Right of normal review reserved.* RUS reserves the right to review any request for lien accommodation or subordination under its normal review process rather than under advance approval procedures if RUS, in its sole discretion, determines there is reasonable doubt as to whether the requirements of paragraphs (b) and (c) of this section have been or will be met, or whether the borrower will be able to meet all of its present and future financial obligations.

[58 FR 53843, Oct. 19, 1993, as amended at 60 FR 67410, Dec. 29, 1995; 65 FR 51748, Aug. 25, 2000; 86 FR 36197, July 9, 2021]

§ 1717.855 Application contents: Advance approval—100 percent private financing of distribution, subtransmission and headquarters facilities, and certain other community infrastructure.

Applications for a lien accommodation or subordination that meet the requirements of § 1717.854 must include the following information and documents:

(a) A certification by an authorized official of the borrower that the borrower and, as applicable, the loan are in compliance with all conditions set forth in § 1717.854(c) and all applicable provisions of §§ 1717.852 and 1717.853;

(b) A statement requesting the lien accommodation or subordination and including the amount and maturity of the proposed loan, a general description of the facilities or other purposes to be financed, the name and address of the lender, and an attached term sheet summarizing the terms and conditions of the proposed loan;

(c) The borrower's financial and statistical report, the data in which shall not be more than 60 days old when the complete application is received by RUS;

(d) Draft copy of any new mortgage or mortgage amendment (supplement) required by RUS or the lender, unless RUS has notified the borrower that it wishes to prepare these documents itself;

(e) A copy of the loan agreement, loan note, bond or other financing instrument, unless RUS has notified the borrower that these documents need not be submitted;

(f) Environmental documentation, in accordance with 7 CFR part 1970;

(g) RUS Form 740c, Cost Estimates and Loan Budget for Electric Borrowers, and RUS Form 740g, Application for Headquarters Buildings;

(h) A CWP or CWP amendment covering the proposed project, in accordance with 7 CFR part 1710, subpart F, and subject to RUS approval.

(i) The certification by the project architect for any buildings to be constructed, as required by § 1717.850(i);

(j) A certification by an authorized official of the borrower that flood hazard insurance will be obtained for the full value of any buildings, or other facilities susceptible to damage if flooded, that will be located in a flood hazard area;

(k) Form AD–1047, Certification Regarding Debarment, Suspension, and Other Responsibility Matters—Primary Covered Transactions, as required by 2 CFR part 180, as adopted by USDA through 2 CFR part 417;

(l) A report by the borrower stating whether or not it is delinquent on any Federal debt, and if delinquent, the amount and age of the delinquency and the reasons therefor; and a certification, if not previously provided, that the borrower has been informed of the Government's collection options;

(m) The written acknowledgement from a registered engineer or architect regarding compliance with seismic provisions of applicable model codes for any buildings to be constructed, as required by 7 CFR 1792.104; and

(n) Other information that RUS may require to determine whether all of the applicable provisions of this subpart have been met.

[58 FR 53843, Oct. 19, 1993, as amended at 60 FR 67410, Dec. 29, 1995; 79 FR 76003, Dec. 19, 2014; 81 FR 11026, Mar. 2, 2016; 84 FR 32616, July 9, 2019]

§ 1717.856 Application contents: Normal review—100 percent private financing.

Applications for a lien accommodation or subordination for 100 percent private financing for eligible purposes that do not meet the requirements of § 1717.854 must include the following information and documents:

(a) A certification by an authorized official of the borrower that:

(1) The borrower and, as applicable, the loan are in compliance with all applicable provisions of §§ 1717.852 and 1717.853; and

(2) There are no actions or proceedings against the borrower, pending or overtly threatened in writing before any court, governmental agency, or arbitrator that would materially adversely affect the borrower's operations and/or financial condition. If this certification cannot be made, the application must include:

(i) An opinion of borrower's counsel regarding any actions or proceedings against the borrower, pending or overtly threatened in writing before any court, governmental agency, or arbitrator that would materially adversely affect the borrower's operations and/or financial condition. The opinion shall address the merits of the claims asserted in the actions or proceedings, and include, if appropriate, an estimate of the amount or range of any potential loss; and

(ii) A certification by an authorized official of the borrower as to the amount of any insurance coverage applicable to any loss that may result from the actions and proceedings addressed in the opinion of borrower's counsel;

(b) The information and documents set forth in § 1717.855 (b) through (n);

(c) A long-range financial forecast providing financial projections for at least 10 years, which demonstrates that the borrower's system is economically viable and that the proposed loan is financially feasible. The financial forecast must comply with the requirements of 7 CFR part 1710 subpart G. RUS may, in its sole discretion, waive the requirement of this paragraph that a long range financial forecast be provided, if:

(1) The borrower is current on all of its financial obligations and is in compliance with all requirements of its mortgage and loan agreement with RUS;

(2) In RUS's judgment, granting a lien accommodation or subordination for the proposed loan will not adversely affect the repayment and security of outstanding debt of the borrower owed to or guaranteed by RUS;

(3) The borrower has achieved the TIER and DSC and any other coverage ratios required by its mortgage or loan contract in each of the two most recent calendar years; and

(4) The amount of the proposed loan does not exceed the lesser of $10 million or 10 percent of the borrower's current net utility plant;

(d) [Reserved]

(e) As applicable to the type of facilities being financed, a CWP, related engineering and cost studies, a power cost study. These documents must meet the requirements of 7 CFR part 1710, subpart F and, as applicable, subpart G;

(f) Unless the requirement has been waived in writing by RUS, a current load forecast, which must meet the requirements of 7 CFR part 1710, subpart E, to the same extent as if the loan were being made by RUS; and

(g) A discussion of the borrower's compliance with RUS requirements on accounting, financial reporting, record keeping, and irregularities (see §1717.854(c)(5)). RUS will review the case and determine the effect of any noncompliance on the feasibility and security of RUS's loans, and whether the requested lien accommodation or subordination can be approved.

[58 FR 53843, Oct. 19, 1993, as amended at 60 FR 3735, Jan. 19, 1995; 60 FR 67410, Dec. 29, 1995; 84 FR 32616, July 9, 2019; 84 FR 37059, July 31, 2019]

§ 1717.857 Refinancing of existing secured debt—distribution and power supply borrowers.

(a) *Advance approval.* All applications for a lien accommodation or subordination for the refinancing of existing secured debt that meet the qualification criteria of this paragraph, except applications from borrowers in default under their mortgage or loan contract with RUS, are eligible for advance approval. Such lien accommodations and subordinations are deemed to be in the Government's interest, and RUS will approve them once RUS is satisfied that the requirements of this paragraph and paragraph (c) of this section have been met. The qualification criteria are as follows:

(1) The refinancing is a current refunding and does not involve interest rate swaps, forward delivery contracts, or similar features;

(2) The principal amount of the refinancing loan does not exceed the sum of the outstanding principal amount of the debt being refinanced plus the amount of transactions costs included in the refinancing loan that are eligible for lien accommodation or subordination under §1717.852(a)(4);

(3) The weighted average life of the refinancing loan is not greater than the weighted average remaining life of the loan being refinanced; and

(4) The present value of the cost of the refinancing loan, including all transaction costs and any required investments in the lender, is less than the present value of the cost of the loan being refinanced, as determined by a method acceptable to RUS. The discount rate used in the present value analysis shall be equal to either:

(i) The current rate on Treasury securities having a maturity equal to the weighted average life of the refunding loan, plus one-eighth percent, or

(ii) A rate approved by RUS based on documentation provided by the borrower as to its marginal long-term borrowing cost.

(b) *Other applications.* Applications for a lien accommodation or subordination for refinancing that do not meet the requirements of paragraph (a) of this section will be reviewed by RUS under normal review procedures for these applications. In the case of either advance approval or normal review, a lien subordination would be authorized only if the lien of the mortgage was subordinated with respect to the assets securing the loan being refinanced.

(c) *Application contents—advance approval of refinancing.* Applications for a lien accommodation or subordination for refinancing of existing secured debt that meet the qualification criteria for advance approval set forth in paragraph (a) of this section, must include the following information and documents:

(1) A certification by an authorized official of the borrower that the application meets the requirements of paragraph (a) of this section and all applicable provisions of §§ 1717.852 and 1717.853;

(2) Documentation and analysis demonstrating that the application meets the qualification criteria set forth in paragraph (a) of this section;

(3) A statement from the borrower requesting the lien accommodation or subordination and including the amount and maturity of the proposed loan, a general description of the debt to be refinanced, the name and address of the lender, and an attached term sheet summarizing the terms and conditions of the proposed loan;

(4) The borrower's financial and statistical report, the data in which shall not be more than 60 days old when the complete application is received by RUS;

(5) Draft copy of any new mortgage or mortgage amendment (supplement) required by RUS or the lender, unless RUS has notified the borrower that it wishes to prepare these documents itself;

(6) A copy of the loan agreement, loan note, bond or other financing instrument, unless RUS has notified the borrower that these documents need not be submitted;

(7) Form AD–1047, Certification Regarding Debarment, Suspension, and Other Responsibility Matters—Primary Covered Transactions, as required by 2 CFR part 417:

(8) A report by the borrower stating whether or not it is delinquent on any Federal debt, and if delinquent, the amount and age of the delinquency and the reasons therefor; and a certification, if not previously provided, that the borrower has been informed of the Government's collection options; and

(9) Other information, documents and opinions that RUS may require to determine whether all of the applicable provisions of this subpart have been met.

(d) *Application contents—normal review of refinancing.* Applications for a lien accommodation or subordination for refinancing of existing secured debt that do not meet the requirements for advance approval set forth in paragraph (a) of this section, must include the following information and documents:

(1) The information and documents set forth in paragraphs (c)(3) through (9) of this section;

(2) A complete description of the refinancing loan and the outstanding debt to be refinanced;

(3) An analysis comparing the refinancing loan with the loan being refinanced as to the weighted average life and the net present value of the costs of the two loans; and

(4) If the present value of the cost of the refinancing loan is greater than the present value of the cost of the debt being refinanced, financial forecasts for at least 5 years comparing the borrower's debt service and other costs, revenues, margins, cash flows, TIER, and DSC, with and without the proposed refinancing.

(e) *Application process and timeframes.* The application process and timeframes for RUS review and action for refinancings are set forth in § 1717.859(d).

(f) *Prepayments of concurrent RUS insured loans.* If the loan being refinanced was made concurrently as supplemental financing required by RUS in connection with an RUS insured loan, the refinancing will not be considered a prepayment under the RUS mortgage, and no proportional prepayment of the concurrent RUS insured loan will be required, provided that the principal amount of the refinancing loan is not less than the amount of loan principal being refinanced, and the weighted average life of the refinancing loan is materially equal to the weighted average remaining life of the loan being refinanced. The refinancing loan shall be considered a concurrent loan.

[58 FR 53843, Oct. 19, 1993, as amended at 60 FR 67410, Dec. 29, 1995; 79 FR 76003, Dec. 19, 2014; 84 FR 32616, July 9, 2019]

§ 1717.858 Lien subordination for rural development investments.

(a) *Policy.* RUS encourages borrowers to consider investing in financially sound projects that are likely to have a positive effect on economic development and employment in rural areas. In addition to the guidance set forth in § 1717.651, RUS recommends that such investments be made through a subsidiary of the borrower in order to clearly separate the financial risks and the revenues and costs of the rural development enterprise from those of the borrower's electric utility business. This should reduce credit risks to the borrower's primary business, and minimize the possibility of undisclosed cross subsidization of the rural development enterprise by electric rate payers.

(b) *Lien subordination.* RUS will consider subordinating or releasing its lien on the stock held by a borrower in a subsidiary whose primary business directly contributes to or supports economic development and employment in rural areas, as defined in section 13 of the RE Act, when requested by a lender to the subsidiary, other than the borrower. To be eligible for said lien subordination or release:

(1) The borrower must be current on all of its financial obligations and be in compliance with all provisions of its mortgage and loan agreement with RUS; and

(2) In the judgment of RUS, the borrower must be able to repay all of its outstanding debt, and the security for all outstanding loans made to the borrower by RUS, including loans guaranteed by RUS, must be adequate, after taking into account the proposed subordination or release of lien.

(c) *Application contents.* Applications for a lien subordination or release of lien for rural development investments must include the following information and documents:

(1) A statement from the borrower requesting the lien subordination or release of lien.

(2) A certification by an authorized official of the borrower that the borrower is current on all of its financial obligations and is in compliance with all provisions of its mortgage and loan agreement with RUS;

(3) A description of the facilities or other purposes to be financed and the projected effects on economic development and employment in rural areas;

(4) The borrower's financial and statistical report, the data in which shall not be more than 60 days old when the complete application is received by RUS;

(5) If requested by RUS, a long-range financial forecast providing financial projections for at least 10 years, in form and substance satisfactory to RUS, which demonstrates that the borrower's system is economically viable and that the borrower will be able to repay all of its outstanding debt and meet all other financial obligations;

(6) A discussion of the borrower's compliance with RUS requirements on accounting, financial reporting, record keeping, and irregularities (see § 1717.854(c)(5)). RUS will review the case and determine the effect of any noncompliance on the feasibility and security of RUS's loans, and whether the requested lien subordination or release of lien can be approved;

(7) If any buildings are to be constructed with the proceeds of the loan to be made to the subsidiary:

(i) A certification by the project architect that the buildings will be designed and constructed in compliance with Section 504 of the Rehabilitation Act of 1973 as amended (29 U.S.C. 794), as applicable under that Act, and that the facilities will be readily accessible to and usable by persons with handicaps in accordance with the Uniform Federal Accessibility Standards; and

(ii) A written acknowledgement from a registered engineer or architect regarding compliance with seismic provisions of applicable model codes, as required by 7 CFR 1792.104;

(8) A certification by an authorized official of the borrower that flood hazard insurance will be obtained for the full value of any buildings, or other facilities susceptible to damage if flooded, that will be located in a flood hazard area;

(9) Form AD–1047, Certification Regarding Debarment, Suspension, and Other Responsibility Matters—Primary Covered Transactions, as required by 2 CFR part 180, as adopted by USDA through 2 CFR part 417;

(10) A report by the borrower stating whether or not it is delinquent on any Federal debt, and if delinquent, the amount and age of the delinquency and the reasons therefor; and a certification, if not previously provided, that the borrower has been informed of the Government's collection options; and

(11) Other information that RUS may require to determine whether all of the applicable provisions of this subpart have been met.

[58 FR 53843, Oct. 19, 1993, as amended at 79 FR 76003, Dec. 19, 2014; 84 FR 32616, July 9, 2019]

§ 1717.859 Application process and timeframes.

(a) *General.* (1) Borrowers are responsible for ensuring that their applications for a lien accommodation or subordination are complete and sound as to substance and form before they are submitted to RUS. RUS will not accept any application that, on its face, is incomplete or inadequate as to the substantive information required by this subpart. RUS will notify borrowers in writing when their applications are complete and in form and substance satisfactory to RUS. A copy of all notifications of borrowers cited in this section will also be sent to the private lender.

(2) It is recommended that borrowers consult with RUS staff before submitting their applications to determine whether they will likely qualify for advance approval or normal review, and to obtain answers to any questions about the information and documents required for the application.

(3) A borrower shall, after submitting an application, promptly notify RUS of any changes that materially affect the information contained in its application.

(4) After submitting an application and having been notified by RUS of additional information and documents and other changes needed to complete the application, if the required information and documents are not supplied to RUS within 30 calendar days of the borrower's receipt of the notice, RUS may return the application to the borrower. The borrower may resubmit the application when the required additional information and documents are available.

(5) *Timeframes.* The timeframes for review of applications set forth in this section are based on the following conditions:

(i) The types of lien accommodations or subordinations requested are of the "standard" types that RUS has approved previously, i.e., the so-called Type I, II and III lien accommodations. Future revisions of the RUS mortgage may result in other "standard" types of lien accommodations and lien subordinations acceptable to RUS. Requests for lien accommodations or subordinations that are substantially different than the "standard" types previously approved by RUS may require additional time for review and action;

(ii) The requested lien accommodation or subordination does not require the preparation of an environmental assessment or an Environmental Impact Statement. Preparation of these documents often will require additional time beyond the timeframes cited in this section; and

(iii) The timeframes set forth in this section, except for paragraph (b)(4) of this section, which deals only with approval of a new mortgage or mortgage amendment, include RUS review and/or approval of a loan contract, if required as part of the application, and required supporting documents, such as a CWP.

(b) *Advance approval—100 percent private financing of distribution, subtransmission, and headquarters facilities.* (1) Applications that qualify under § 1717.854 for advance approval of a lien accommodation or subordination for 100 percent private financing of distribution, subtransmission, and headquarters facilities are submitted to the general field representative (GFR). The

GFR will work with the borrower to ensure that all components of the application are assembled. Once the application is satisfactory to the GFR, it will be sent promptly to the Washington office for further review and action. If a new mortgage or mortgage amendment is required, a draft of these documents must be included in the application, unless the borrower has been notified that RUS wishes to prepare the documents itself.

(2) If no additional or amended information is needed for RUS to complete its review of the application once it is received in the Washington Office, RUS will, within 45 calendar days of receiving the application in the Washington Office, either:

(i) Approve the lien accommodation or subordination if the borrower has demonstrated satisfactorily to RUS that all requirements of this subpart applicable to advance approval have been met, and send written notice to the borrower. RUS's approval, in this case and all other cases, will be conditioned upon execution and delivery by the borrower of a satisfactory security instrument, if required, and such additional information, documents, and opinions of counsel as RUS may require;

(ii) If all requirements have not been met, so notify the borrower in writing. The application will be returned to the borrower unless the borrower requests that it be reconsidered under the requirements and procedures for normal review set forth in paragraph (c) of this section and in § 1717.856; or

(iii) Send written notice to the borrower explaining why a decision cannot be made at that time and giving the estimated date when a decision is expected.

(3) If additional or amended information is needed after the application is received in the Washington Office, RUS will so notify the borrower in writing within 15 calendar days of receiving the application in the Washington Office. If RUS subsequently becomes aware of other deficiencies in the application, additional written notice will be sent to the borrower. Within 30 calendar days of receiving all of the information required by RUS to complete its review, RUS will act on the application as described in paragraphs (b)(2)(i) through (b)(2)(iii) of this section.

(4) If a new mortgage or mortgage amendment is required, within 30 days of receiving such documents satisfactory to RUS, including required execution counterparts, RUS will execute the documents and send them to the borrower, along with instructions pertaining to recording of the mortgage, an opinion of borrower's counsel, and other matters. RUS will promptly notify the borrower upon receiving satisfactory evidence that the borrower has complied with said instructions.

(c) *Normal review—100 percent private financing of distribution, transmission, and/or generation facilities*—(1) *Distribution borrowers.* (i) Applications from distribution borrowers for a lien accommodation or subordination for 100 percent private financing of distribution, transmission, and/or generation facilities (including other eligible electric utility purposes) that do not meet the criteria for advance approval, are also submitted to the GFR. Procedures at this stage are the same as in paragraph (b)(1) of this section.

(ii) If no additional or amended information is needed for RUS to complete its review of the application once it is received in the Washington office, RUS will, within 90 calendar days of receiving the application in the Washington office, send written notice to the borrower either approving the request, disapproving the request, or explaining why a decision cannot be made at that time and giving the estimated date when a decision is expected.

(iii) If additional or amended information is needed after the application is received in the Washington Office, RUS will so notify the borrower in writing within 15 calendar days of receiving the application in the Washington Office. If RUS subsequently becomes aware of other deficiencies in the application, additional written notice will be sent to the borrower. Within 90 calendar days of receiving all of the information required by RUS to complete its review, RUS will act on the application as described in paragraph (c)(1)(ii) of this section.

(iv) If a new mortgage or mortgage amendment is required, the procedures and timeframes of paragraph (b)(4) of this section will apply.

(2) *Power supply borrowers.* (i) Applications from power supply borrowers for a lien accommodation or subordination for 100 percent private financing of distribution, transmission, and/or generation facilities, and other eligible electric utility purposes, are submitted to the RUS Power Supply Division, or its successor, in Washington, DC.

(ii) Within 30 calendar days of receiving the borrower's application containing the information and documents required by § 1717.856, RUS will send written notice to the borrower of any deficiencies in its application as to completeness and acceptable form and substance. Additional written notices may be sent to the borrower if RUS subsequently becomes aware of other deficiencies in the borrower's application.

(iii) Within 90 calendar days of receiving all of the information required by RUS to complete its review, RUS will act on the application as described in paragraph (c)(1)(ii) of this section.

(iv) If a new mortgage or mortgage amendment is required, these documents will be reviewed and executed pursuant to the procedures and timeframes of paragraph (b)(4) of this section.

(d) *Refinancing of existing debt.* All requests for a lien accommodation or subordination for refinancing are sent directly to the Washington office.

(1) *Advance approval.* (i) Within 15 calendar days of receiving the borrower's application containing the information and documents required by § 1717.857(c), RUS will send written notice to the borrower of any deficiencies in its application as to completeness and acceptable form and substance. Additional written notices may be sent to the borrower if RUS subsequently becomes aware of other deficiencies in the borrower's application.

(ii) Within 15 calendar days of receiving all of the required information and documents, in form and substance satisfactory to RUS, RUS will either:

(A) Approve the lien accommodation or subordination if the borrower has demonstrated satisfactorily to RUS that all requirements of § 1717.857(a) and (c) have been met, and send written notice to the borrower;

(B) If all requirements have not been met, so notify the borrower in writing. The application will be returned to the borrower unless the borrower requests that it be reconsidered under the requirements and procedures for normal review set forth in paragraph (d)(2) of this section and in § 1717.857; or

(C) Send written notice to the borrower explaining why a decision cannot be made at that time and giving the estimated date when a decision is expected.

(iii) If a new mortgage or mortgage amendment is required, these documents will be reviewed and executed pursuant to the procedures and timeframes of paragraph (b)(4) of this section.

(2) *Normal review.* (i) Within 20 calendar days of receiving the borrower's application containing the information and documents required by § 1717.857(d), RUS will send written notice to the borrower of any deficiencies in its application as to completeness and acceptable form and substance. Additional written notices may be sent to the borrower if RUS subsequently becomes aware of other deficiencies in the borrower's application.

(ii) Within 30 calendar days of receiving all of the required information and documents, in form and substance satisfactory to RUS, RUS will notify the borrower in writing either approving the request, disapproving the request, or explaining why a decision cannot be made at that time and giving the estimated date when a decision is expected. If the proposed refinancing involves complicated transactions such as interest rate swaps or forward delivery contracts, additional time may be required for RUS review and final action.

(iii) If a new mortgage or mortgage amendment is required, these documents will be reviewed and executed pursuant to the procedures and timeframes of paragraph (b)(4) of this section.

(e) *Rural development investments.* (1) Applications for a lien subordination for rural development investments are submitted by distribution borrowers to the GFR and by power supply borrowers to the RUS Power Supply Division, or its successor, in Washington, DC.

(2) The GFR will work with the borrower to ensure that all components of the application are assembled. Once the application is satisfactory to the GFR, it will be sent promptly to the Washington Office for further review and action. After the application is received in the Washington Office, if additional or amended information is needed for RUS to complete its review, RUS will so notify the borrower in writing within 15 calendar days of receiving the application.

(3) Applications from power supply borrowers containing the information and documents required by § 1717.858(c) will be reviewed in the Washington office and the borrower given written notice within 30 calendar days of receiving the application of any deficiencies as to completeness and acceptable form and substance. Additional written notices may be sent to the borrower if RUS subsequently becomes aware of other deficiencies in the borrower's application.

(4) Within 60 calendar days of receiving in the Washington office all of the required information and documents, in form and substance satisfactory to RUS, RUS will give written notice to the borrower either approving the request, disapproving the request, or explaining why a decision cannot be made at that time and giving the estimated date when a decision is expected.

(5) If a new mortgage or mortgage amendment is required, these documents will be reviewed and executed pursuant to the procedures and timeframes of paragraph (b)(4) of this section.

§ 1717.860 Lien accommodations and subordinations under section 306E of the RE Act.

(a) *General.* Under section 306E of the RE Act, when requested by a private lender providing financing for capital investments by a borrower whose net worth exceeds 110 percent of the outstanding principal balance of all loans made or guaranteed to the borrower by RUS, the Administrator will, without delay, offer to share the government's lien on the borrower's system or subordinate the government's lien on the property financed by the private lender, provided that the security, including the assurance of repayment, for loans made or guaranteed by RUS will remain reasonably adequate. To qualify for a lien accommodation or subordination under this section, the investment must be an original capital investment, i.e., not a refinancing or refunding. (See § 1717.851 for the definition of capital investment.)

(b) *Determination of net worth to RUS debt ratio.* (1) In the case of applications for a lien accommodation, a borrower's net worth will be based on the borrower's most recent financial and statistical report, the data in which shall not be more than 60 days old at the time the application is received by RUS, and the outstanding debt owed to or guaranteed by RUS will be based on latest RUS records available. The financial and statistical reports (Form 7 for distribution borrowers and Form 12a for power supply borrowers) are subject to RUS review and revision, and they must comply with RUS's system of accounts and accounting principles set forth in 7 CFR part 1767. Since sinking fund depreciation is not approved under part 1767, net worth for borrowers using sinking fund depreciation will be calculated as if the borrower had been using straight line depreciation.

(2) Net worth shall be calculated by taking total margins and equities (Line 33 of Part C of RUS Form 7 for distribution borrowers, or Line 34 of Section B of RUS Form 12a for power supply borrowers) and subtracting assets properly recordable in account 182.2, Unrecovered Plant and Regulatory Study Costs, and account 182.3, Other Regulatory Assets, as defined in 7 CFR part 1767.

(c) *Application requirements and process.* (1) If a borrower's net worth to RUS debt ratio exceeds 110 percent, as determined by RUS, and the borrower is in compliance with all requirements of its mortgage, loan agreement with RUS, and any other agreement with RUS that have not been exempted in writing by RUS, if requested RUS will expeditiously approve a lien accommodation or subordination for 100 percent private financing of capital investments, provided that the security, including the assurance of repayment, for loans made or guaranteed by RUS will remain reasonably adequate. RUS's approval will be conditioned upon execution and delivery by the borrower of a security instrument satisfactory to RUS, if required, and such additional information, documents, and opinions of counsel as RUS may require.

(2) The application must include the following:

(i) A statement from the borrower requesting the lien accommodation and including the amount and maturity of the proposed loan, a general description of the facilities or other purposes to be financed, the name and address of the lender, and an attached term sheet summarizing the terms and conditions of the proposed loan;

(ii) A certification by an authorized official of the borrower that the borrower is in compliance with all requirements of its mortgage, loan agreement with RUS, and any other agreement with RUS that have not been exempted in writing by RUS;

(iii) The borrower's financial and statistical report, the data in which shall not be more than 60 days old when the complete application is received by RUS;

(iv) Draft copy of any new mortgage or mortgage amendment (supplement) required by RUS or the lender, unless RUS has notified the borrower that it wishes to prepare these documents itself;

(v) A copy of the loan agreement, loan note, bond or other financing instrument, unless RUS has notified the borrower that these documents need not be submitted. These documents will not be subject to RUS approval, but may be reviewed to determine whether they contain any provisions that would result in the security, including assurance of repayment, for loans made or guaranteed by RUS no longer being reasonably adequate;

(vi) The following certifications and reports required by law:

(A) The certification by the project architect for any buildings to be constructed, as required by 7 CFR 1717.850(i);

(B) A certification by an authorized official of the borrower that flood hazard insurance will be obtained for the full value of any buildings, or other facilities susceptible to damage if flooded, that will be located in a flood hazard area;

(C) Form AD–1047, Certification Regarding Debarment, Suspension, and Other Responsibility Matters—Primary Covered Transactions, as required by 2 CFR part 180, adopted by USDA through 2 CFR part 417;

(D) A report by the borrower stating whether or not it is delinquent on any Federal debt, and if delinquent, the amount and age of the delinquency and the reasons therefor; and a certification, if not previously provided, that the borrower has been informed of the Government's collection options; and

(E) The written acknowledgement from a registered engineer or architect regarding compliance with seismic provisions of applicable model codes for any buildings to be constructed, as required by 7 CFR 1792.104. All other elements of an application listed in §1717.855, §1717.856, and §1717.858(c) not listed in this paragraph (c) are exempted.

(3) Applications from distribution borrowers are submitted to the general field representative (GFR), while applications from power supply borrowers are submitted to the RUS Power Supply Division, or its successor, in Washington, DC. When an application is satisfactory to the GFR, it will be sent promptly to the Washington office. If Washington office staff determine that an application is incomplete, the borrower will be promptly notified in writing about the deficiencies. When the application is complete, and if the security, including assurance of repayment, of loans made or guaranteed by RUS will remain reasonably adequate after granting the lien accommodation or subordination, the borrower and the lender will be promptly notified in writing that the lien accommodation or subornation has been approved, subject to the conditions cited in paragraph (c)(1) of this section.

(d) *Rural development and other non-electric utility investments.* Although

RUS recommends the use of separate subsidiaries as set forth in § 1717.858, if requested by a borrower that meets the 110 percent equity test and all other applicable requirements of this section, RUS will provide a lien subordination on the specific assets financed in the case of loans made directly to the borrower for rural development and other non-electric utility purposes, provided that the outstanding balance of all such loans lien subordinated under this paragraph (d), after taking into consideration the effect of the new loan, does not exceed 15 percent of the borrower's net worth and the security, including assurance of repayment, of loans made or guaranteed by RUS will remain reasonably adequate after granting the lien subordination. Investments lien subordinated under this paragraph shall be included among those investments subject to the 15 percent of total utility plant limitation set forth in 7 CFR 1717.654(b)(1), and granting of the lien subordination will not constitute approval of the investment under 7 CFR part 1717, subpart N.

(e) *Requirements and controls not exempted.* All requirements and limitations imposed with respect to lien accommodations and subordinations by this subpart R that are not specifically exempted by this section are not exempted and shall continue to apply according to their terms.

[59 FR 3986, Jan. 28, 1994, as amended at 60 FR 3735, Jan. 19, 1995; 60 FR 67410, Dec. 29, 1995; 79 FR 76003, Dec. 19, 2014; 84 FR 32616, July 9, 2019]

§§ 1717.861–1717.899 [Reserved]

Subpart S—Lien Accommodations for Supplemental Financing Required by 7 CFR 1710.110

SOURCE: 58 FR 53851, Oct. 19, 1993, unless otherwise noted.

§ 1717.900 Qualification requirements.

Applications for a lien accommodation for supplemental financing required by 7 CFR 1710.110 must meet the same requirements as an RUS insured loan. The justification and documentation materials submitted as part of the borrower's application for an insured loan also serve as the justification and documentation of the request for a lien accommodation for the required supplemental loan. Unless early approval under § 1717.901 is requested by a borrower, these applications will be processed during the same time as RUS's review of the borrower's application for the concurrent insured loan.

§ 1717.901 Early approval.

(a) *Conditions.* If requested by a borrower in writing, RUS will review the application for a lien accommodation for required supplemental financing early in the process, before funding is available for the concurrent RUS insured loan, and approve the lien accommodation if the following conditions are met:

(1) The required supplemental loan meets the requirements for an insured loan, as set forth in 7 CFR part 1710, subparts A through G, and other RUS regulations pertaining to required supplemental loans;

(2) The borrower has demonstrated the ability to obtain the funds that would be needed to complete other portions of the project, if the portion to be constructed with private loan funds could not be used productively without completion of such other portions, in the event concurrent RUS insured loan funds are not forthcoming. Such evidence may include financial records demonstrating the availability of general funds, and/or a written commitment from the private lender to provide a loan for the remaining amount of financing required, with such commitment being conditioned upon the availability of a lien accommodation from RUS; and

(3) An authorized official of the borrower has requested early approval of the lien accommodation and explained the reasons therefor, and has certified that the funds are needed and will be drawn down before funds from the concurrent insured loan are expected to be available, assuming that the insured loan is approved.

(b) *Timeframe for RUS action.* (1) RUS will either approve or disapprove the lien accommodation within 90 days of receiving the borrower's request for early approval and the complete application for the concurrent RUS loan and required supplemental financing, in form and substance satisfactory to RUS, or notify the borrower in writing of the estimated date when a decision is expected. If an environmental assessment or an Environmental Impact Statement is required, additional time beyond the 90 days may be required to prepare these documents. RUS's approval of the lien accommodation will be conditioned upon execution and delivery by the borrower of a satisfactory security instrument, if required, and such additional information, documents, and opinions of counsel as RUS may require.

(2) If a mortgage or mortgage amendment is required, RUS will consult with the other mortgagees as to who will prepare the documents. Within 30 days of obtaining the documents satisfactory to RUS, including required execution counterparts, RUS will execute the documents and send them to the borrower, along with instructions pertaining to recording of the mortgage, an opinion of borrower's counsel, and other matters. RUS will promptly notify the borrower upon receiving satisfactory evidence that the borrower has complied with said instructions.

(c) *Approval of concurrent insured loan.* Early approval of a lien accommodation for a required supplemental loan does not ensure that the concurrent RUS insured loan will be approved. The request for the concurrent insured loan will be reviewed when funds are available to make the loan. The borrower may be requested to update certain supporting information in the loan application if substantial time has elapsed since the lien accommodation or subordination was approved.

§ 1717.902 Other RUS requirements.

Supplemental loans required by 7 CFR 1710.110 are subject to the same post-loan requirements as insured RUS loans regarding accepted materials, construction standards, contracting and procurement procedures, standard forms of contracts, RUS approval of the advance of loan funds, and other matters.

§ 1717.903 Liability.

It is the intent of this subpart that any failure on the part of RUS to comply with any provisions of this subpart, including without limitation, those provisions setting forth specified timeframes for action by RUS on applications for lien accommodations or lien subordinations, shall not give rise to liability of any kind on the part of the Government or any employees of the Government including, without limitation, liability for damages, fees, expenses or costs incurred by or on behalf of a borrower, private lender or any other party.

§ 1717.904 Exemptions pursuant to section 306E of the RE Act.

(a) *General policy.* If a borrower's net worth to RUS debt ratio exceeds 110 percent, as determined by RUS, and the borrower is in compliance with all requirements of its mortgage, loan agreement with RUS, and any other agreement with RUS that have not been exempted in writing by RUS, RUS will expeditiously approve a lien accommodation for a concurrent supplemental loan if requested in writing by the borrower, provided that the security, including assurance of repayment, of loans made or guaranteed by RUS will remain reasonably adequate. RUS's approval will be conditioned upon execution and delivery by the borrower of a security instrument satisfactory to RUS, if required, and such additional information, documents, and opinions of counsel as RUS may require.

(b) *Determination of net worth to RUS debt ratio.* A borrower's ratio of net worth to RUS debt will be determined as set forth in § 1717.860(b).

(c) *Procedures.* If a borrower meets the requirements of this section, upon receipt of a complete application RUS will promptly notify the borrower and lender in writing that the lien accommodation has been approved subject to the conditions set forth in paragraph (a) of this section.

[59 FR 3987, Jan. 28, 1994, as amended at 67 FR 70153, Nov. 21, 2002]

§§ 1717.905–1717.949 [Reserved]

Subpart T [Reserved]

§§ 1717.950–1717.999 [Reserved]

Subpart U [Reserved]

§§ 1717.1000–1717.1049 [Reserved]

Subpart V [Reserved]

§§ 1717.1050–1717.1099 [Reserved]

Subpart W [Reserved]

§§ 1717.1100–1717.1149 [Reserved]

Subpart X [Reserved]

§§ 1717.1150–1717.1199 [Reserved]

Subpart Y—Settlement of Debt

SOURCE: 62 FR 50491, Sept. 26, 1997, unless otherwise noted.

§ 1717.1200 Purpose and scope.

(a) Section 331(b) of the Consolidated Farm and Rural Development Act (Con Act), as amended on April 4, 1996 by Public Law 104–127, 110 Stat. 888 (7 U.S.C. 1981), grants authority to the Secretary of Agriculture to compromise, adjust, reduce, or charge-off debts or claims arising from loans made or guaranteed under the Rural Electrification Act of 1936, as amended (RE Act). Section 331(b) of the Con Act also authorizes the Secretary of Agriculture to adjust, modify, subordinate, or release the terms of security instruments, leases, contracts, and agreements entered into or administered by the Rural Utilities Service (RUS). The Secretary, in 7 CFR 2.47, has delegated authority under section 331(b) of the Con Act to the Administrator of the RUS, with respect to loans made or guaranteed by RUS.

(b) This subpart sets forth the policy and standards of the Administrator of RUS with respect to the settlement of debts and claims arising from loans made or guaranteed to rural electric borrowers under the RE Act. Nothing in this subpart limits the Administrator's authority under section 12 of the RE Act.

§ 1717.1201 Definitions.

Terms used in this subpart that are not defined in this section have the meanings set forth in 7 CFR part 1710. In addition, for the purposes of this subpart:

Application for debt settlement means a written application containing all of the information required by § 1717.1204(b)(2), in form and substance satisfactory to RUS.

Attorney General means the Attorney General of the United States of America.

Claim means any claim of the government arising from loans made or guaranteed under the RE Act to a rural electric borrower.

Con Act means the Consolidated Farm and Rural Development Act (7 U.S.C. 1921 *et seq.*).

Debt means outstanding debt of a rural electric borrower (including, but not necessarily limited to, principal, accrued interest, penalties, and the government's costs of debt collection) arising from loans made or guaranteed under the RE Act.

Enforced collection procedures means any procedures available to the Administrator for the collection of debt that are authorized by law, in equity, or under the borrower's loan documents or other agreements with RUS.

Loan documents means the mortgage (or other security instrument acceptable to RUS), the loan contract, and the promissory note entered into between the borrower and RUS.

RE Act means the Rural Electrification Act of 1936, as amended (7 U.S.C. 901–950b).

Restructure means to settle a debt or claim.

Settle means to reamortize, adjust, compromise, reduce, or charge-off a debt or claim.

§ 1717.1202 General policy.

(a) It is the policy of the Administrator that, wherever possible, all debt owed to the government, including but not limited to principal and interest, shall be collected in full in accordance with the terms of the borrower's loan documents.

(b) Nothing in this subpart by itself modifies, reduces, waives, or eliminates any obligation of a borrower under its loan documents. Any such modifications regarding the debt owed by a borrower may be granted under the authority of the Administrator only by means of the explicit written approval of the Administrator in each case.

(c) The Administrator's authority to settle debts and claims will apply to cases where a borrower is unable to pay its debts and claims in accordance with their terms, as further defined in § 1717.1204(b)(1), and where settlement will maximize, on a present value basis, the recovery of debts and claims owed to the government.

(d) In structuring settlements and determining the capability of the borrower to repay debt and the amount of debt recovery that is possible, the Administrator will consider, among other factors, the RE Act, the National Energy Policy Act of 1992 (Pub. L. 102–486, 106 Stat. 2776), the policies and regulations of the Federal Energy Regulatory Commission, state legislative and regulatory actions, and other market and nonmarket forces as to their effects on competition in the electric utility industry and on rural electric systems in particular. Other factors the Administrator will consider are set forth in more detail in § 1717.1204.

§ 1717.1203 Relationship between RUS and Department of Justice.

(a) The Attorney General will be notified by the Administrator whenever the Administrator intends to use his or her authority under section 331(b)of the Con Act to settle a debt or claim.

(b) If an outstanding claim has been referred in writing to the Attorney General, the Administrator will not use his or her own authority to settle the claim without the approval of the Attorney General.

(c) If an application for additional debt relief is received from a borrower whose debt has been settled in the past under the authority of the Attorney General, the Administrator will promptly notify the Attorney General before proceeding to consider the application.

§ 1717.1204 Policies and conditions applicable to settlements.

(a) *General.* Settlement of debts and claims shall be subject to the policies, requirements, and conditions set forth in this section and in § 1717.1202.

(b) *Need for debt settlement.* (1) The Administrator will not settle any debt or claim unless the Administrator has determined that the borrower is unable to meet its financial obligations under its loan documents according to the terms of those documents, or that the borrower will not be able to meet said obligations sometime within the period of 24 months following the month the borrower submits its application for debt settlement to RUS, and, in either case, such default is likely to continue indefinitely. The determination of a borrower's ability to meet its financial obligations will be based on analyses and documentation by RUS of the borrower's historical, current, and projected costs, revenues, cash flows, assets, opportunities to reduce costs and/or increase revenues, and other factors that may be relevant on a case by case basis.

(2) In its application to RUS for debt settlement, the borrower must provide, in form and substance satisfactory to RUS, an in-depth analysis supporting the borrower's contention that it is unable or will not be able to meet its financial obligations as described in paragraph (b)(1) of this section. The analysis must include:

(i) An explanation and analysis of the causes of the borrower's inability to meet its financial obligations;

(ii) A thorough review and analysis of the opportunities available or potentially available to the borrower to reduce administrative overhead and other costs, improve efficiency and effectiveness, and expand markets and revenues, including but not limited to opportunities for sharing services, merging, and/or consolidating, raising rates when appropriate, and renegotiating supplier and service contracts. In the case of a power supply borrower, the study shall include such opportunities among the members of the borrower, unless the Administrator waives this requirement;

(iii) Documentation of the actions taken, in progress, or planned by the borrower (and its member systems, if applicable) to take advantage of the opportunities cited in paragraph (b)(2)(ii) of this section; and

(iv) Other analyses and documentation prescribed by RUS on a case by case basis.

(3) RUS may require that an independent consultant provide an analysis of the efficiency and effectiveness of the borrower's organization and operations, and those of its member systems in the case of a power supply borrower. The following conditions will apply:

(i) RUS will select the independent consultant taking into account, among other matters, the consultant's experience and expertise in matters relating to electric utility operations, finance, and restructuring;

(ii) The contract with the consultant shall be to provide services to RUS on such terms and conditions as RUS deems appropriate. The consultant's scope of work may include, but shall not be limited to, an analysis of the following:

(A) How to maximize the value of the government's collateral, such as through mergers, consolidations, or sales of all or part of the collateral;

(B) The viability of the borrower's system, taking into account such matters as system size, service territory and markets, asset base, physical condition of the plant, operating efficiency, competitive pressures, industry trends, and opportunities to expand markets and improve efficiency and effectiveness;

(C) The feasibility and the potential benefits and risks to the borrower and the government of corporate restructuring, including aggregation and disaggregation;

(D) In the case of a power supply borrower, the retail rate mark-up by member systems and the potential benefits to be achieved by member restructuring through mergers, consolidations, shared services, and other alliances;

(E) The quality of the borrower's management, management advisors, consultants, and staff;

(F) Opportunities for reducing overhead and other costs, for expanding markets and revenues, and for improving the borrower's existing and prospective contractual arrangements for the purchase and sale of power, procurement of supplies and services, and the operation of plant and facilities;

(G) Opportunities to achieve efficiency gains and increased revenues based on comparisons with benchmark electric utilities; and

(H) The accuracy and completeness of the borrower's analysis provided under paragraph (b)(2) of this section;

(iii) RUS and, as appropriate, other creditors, will determine the extent to which the borrower and third parties (including the members of a power supply borrower) will be required to participate in funding the costs of the independent consultant;

(iv) The borrower will be required to make available to the consultant all corporate documents, files, and records, and to provide the consultant with access to key employees. The borrower will also normally be required to provide the consultant with office space convenient to the borrower's operations and records; and

(v) All analyses, studies, opinions, memoranda, and other documents and information produced by the independent consultant shall be provided to RUS on a confidential basis for consideration in evaluating the borrower's application for debt settlement. Such documents and information may be made available to the borrower and other appropriate parties if authorized in writing by RUS.

(4) The borrower may be required to employ a temporary or permanent manager acceptable to the Administrator, to manage the borrower's operations to ensure that all actions are taken to avoid or minimize the need for debt settlement. The employment could be on a temporary basis to manage the system during the time the debt settlement is being considered, and possibly for some time after any debt settlement, or it could be on a permanent basis.

(5) The borrower must submit, at a time determined by RUS, a resolution of its board of directors requesting debt settlement and stating that the borrower is either currently unable to meet its financial obligations to the government or will not be able to meet said obligations sometime within the next 24 months, and that, in either case, the default is likely to continue indefinitely.

(c) *Debt settlement measures.* (1) If the Administrator determines that debt settlement is appropriate, the debt settlement measures the Administrator will consider under this subpart with respect to direct, insured, or guaranteed loans include, but are not limited to, the following:

(i) Reamortization of debt;

(ii) Extension of debt maturity, provided that the maturity of the borrower's outstanding debt after settlement shall not extend more than 10 years beyond the latest maturity date prior to settlement;

(iii) Reduction of the interest rate charged on the borrower's debt, provided that the interest rate on any portion of the restructured debt shall not be reduced to less than 5 percent, unless the Administrator determines that reducing the rate below 5 percent would maximize debt recovery by the government;

(iv) Forgiveness of interest accrued, penalties, and costs incurred by the government to collect the debt; and

(v) With the concurrence of the Under Secretary for Rural Development, forgiveness of loan principal.

(2) In the event that RUS has, under section 306 of the RE Act, guaranteed loans made by the Federal Financing Bank or other third parties, the Administrator may restructure the borrower's obligations by: acquiring and restructuring the guaranteed loan; restructuring the loan guarantee obligation; restructuring the borrower's reimbursement obligations; or by such means as the Administrator deems appropriate, subject to such consents and approvals, if any, that may be required by the third party lender.

(d) *Borrower's obligations to other creditors.* The Administrator will not grant relief on debt owed to the government unless similar relief, on a pro rata basis, is granted with respect to other secured obligations of the borrower, or the other secured creditors provide other benefits or value to the debt restructuring. Unsecured creditors will also be expected to contribute to the restructuring. If it is not possible to obtain the expected contributions from other creditors, the Administrator may proceed to settle a borrower's debt if that will maximize recovery by the government and will not result in material benefits accruing to other creditors at the expense of the government.

(e) *Competitive bids for system assets.* If requested by RUS, the borrower or the independent consultant provided for in paragraph (b)(3) of this section shall solicit competitive bids from potential buyers of the borrower's system or parts thereof. The bidding process must be conducted in consultation with RUS and use standards and procedures acceptable to RUS. The Administrator may use the competitive bids received as a basis for requiring the sale of all or part of the borrower's system as a condition of settlement of the borrower's debt. The Administrator may also consider the bids in evaluating alternative settlement measures.

(f) *Valuation of system.* (1) The Administrator will consider the value of the borrower's system, including, in the case of a power supply borrower, the wholesale power contracts between the borrower and its member systems. The valuation of the wholesale power contracts shall take into account, among other matters, the rights of the government and/or third parties, to assume the rights and obligations of the borrower under such contracts, to charge reasonable rates for service provided under the contracts, and to otherwise enforce the contracts in accordance with their terms. In no case will the Administrator settle a debt or claim for less than the value (after considering the government's collection costs) of the borrower's system and other collateral securing the debt or claim.

(2) RUS may use such methods, analyses, and assessments as the Administrator deems appropriate to determine the value of the borrower's system.

(g) *Rates.* The Administrator will consider the rates charged for electric service by the borrower and, in the case of a power supply borrower, by its members, taking into account, among other factors, the practices of the Federal Energy Regulatory Commission (FERC), as adapted to the cooperative structure of borrowers, and, where applicable, FERC treatment of any investments by co-owners in projects jointly owned by the borrower.

(h) *Collection action.* The Administrator will consider whether a settlement is favorable to the government in comparison with the amount that can be recovered by enforced collection procedures.

(i) *Regulatory approvals.* Before the Administrator will approve a settlement, the borrower must provide satisfactory evidence that it has obtained all approvals required of regulatory bodies that the Administrator determines are needed to implement rates or other provisions of the settlement, or that are needed in any other way for the borrower to fulfill its obligations under the settlement.

(j) *Conditions regarding management and operations.* As a condition of debt settlement, the borrower, and in the case of a power supply borrower, its members, will be required to implement those changes in structure, management, operations, and performance deemed necessary by the Administrator. Those changes may include, but are not limited to, the following:

(1) The borrower may be required to undertake a corporate restructuring and/or sell a portion of its plant, facilities, or other assets

(2) The borrower may be required to replace senior management and/or hire outside experts acceptable to the Administrator. Such changes may include a commitment by the borrower's board of directors to restructure and/or obtain new membership to improve board oversight and leadership;

(3) The borrower may be required to agree to:

(i) Controls by RUS on the general funds of the borrower, as well as on any investments, loans or guarantees by the borrower, notwithstanding any limitations on RUS' control rights in the borrower's loan documents or RUS regulations; and

(ii) Requirements deemed necessary by RUS to perfect and protect its lien on cash deposits, securities, equipment, vehicles, and other items of real or non-real property; and

(4) In the case of a power supply borrower, the borrower may be required to obtain credit support from its member systems, as well as pledges and action plans by the members to change their operations, management, and organizational structure (e.g., shared services, mergers, or consolidations) in order to reduce operating costs, improve efficiency, and/or expand markets and revenues.

(k) *Conveyance of assets.* As a condition of a settlement, a borrower may be required to convey some or all its assets to the government.

(l) *Additional conditions.* The borrower will be required to warrant and agree that no bonuses or similar extraordinary compensation has been or will be provided, for reasons related to the settlement of government debt, to any officer or employee of the borrower or to other persons or entities identified by RUS. The Administrator may impose such other terms and conditions of debt settlement as the Administrator determines to be in the government's interests.

(m) *Certification of accuracy.* Before the Administrator will approve a debt settlement, the manager or other appropriate official of the borrower must certify that all information provided to the government by the borrower or by any agent of the borrower, in connection with the debt settlement, is true, correct, and complete in all material respects.

§ 1717.1205 Waiver of existing conditions on borrowers.

Pursuant to section 331(b) of the Con Act, the Administrator, at his or her sole discretion, may waive or otherwise reduce conditions and requirements imposed on a borrower by its loan documents if the Administrator determines that such action will contribute to enhancement of the government's recovery of debt. Such waivers or reductions in conditions and requirements under this section shall not include the exercise of any of the debt settlement measures set forth in § 1717.1204(c), which are subject to all of the requirements of said § 1717.1204.

§ 1717.1206 Loans subsequent to settlement.

In considering any future loan requests from a borrower whose debt has been settled in whole or in part (including the surviving entity of merged or consolidated borrowers, where at least one of said borrowers had its debts settled), it will be presumed that credit support for the full amount of the requested loan will be required. Such support may be in a number of forms, provided that they are acceptable to the Administrator on a case by case basis. They may include, but need not be limited to, equity infusions and guarantees of debt repayment, either from the applicant's members (in the case of a power supply borrower), or from a third party.

§ 1717.1207 RUS obligations under loan guarantees.

Nothing in this subpart affects the obligations of RUS under loan guarantee commitments it has made to the Federal Financing Bank or other lenders.

§ 1717.1208 Government's rights under loan documents.

Nothing in this subpart limits, modifies, or otherwise affects the rights of the government under loan documents executed with borrowers, or under law or equity.

PART 1718—LOAN SECURITY DOCUMENTS FOR ELECTRIC BORROWERS

Subpart A—General

Subpart B—Mortgage for Distribution Borrowers

Subpart C—Loan Contracts With Distribution Borrowers

AUTHORITY: 7 U.S.C. 901 *et seq.*, 1921 *et seq.*, 6941 *et seq.*

SOURCE: 60 FR 36888, July 18, 1995, unless otherwise noted.

Subpart A—General

§§ 1718.1–1718.49 [Reserved]

Subpart B—Mortgage for Distribution Borrowers

§ 1718.50 Definitions.

Unless otherwise indicated, terms used in this subpart are defined as set forth in 7 CFR 1710.2.

§ 1718.51 Policy.

(a) Adequate loan security must be provided for loans made or guaranteed by RUS. The loans are required to be secured by a first mortgage lien on most of the borrower's assets substantially in the form set forth in appendix A of this subpart. At the discretion of RUS, this model form of mortgage may be adapted to satisfy different legal requirements among the states and individual differences in lending circumstances, provided that such adaptations are consistent with the policies set forth in this subpart.

(b) Some borrowers, such as certain public power districts, may not be able to provide security in the form of a first mortgage lien on their assets. In these cases RUS will consider accepting other forms of security, such as resolutions and pledges of revenues.

(c) RUS may require supplemental and amending mortgages to protect its security, or in connection with additional loans.

(d) RUS may also require such other security instruments (such as loan contracts, security agreements, financing statements, guarantees, and pledges) as it deems appropriate.

(e) All distribution borrowers that receive a loan or loan guarantee from RUS on or after August 17, 1995 will be required to enter into a mortgage with RUS that meets the requirements of this subpart. The concurrence of any other lenders secured under the borrower's existing mortgage may be required before the borrower can enter into a new mortgage.

§ 1718.52 Existing mortgages.

Nothing contained in this subpart amends, invalidates, terminates or rescinds any existing mortgage entered into between the borrower and RUS and any other mortgagees.

§ 1718.53 Rights of other mortgagees.

Nothing contained in this subpart is intended to alter or affect any other mortgagee's rights under an existing mortgage.

§ 1718.54 Availability of model mortgage.

Single copies of the model mortgage (RUS Informational Publication 1718 B) are available from the Rural Utilities Service, United States Department of Agriculture, Washington, DC 20250–1500. This document may be reproduced.

[60 FR 36888, July 18, 1995, as amended at 86 FR 36197, July 9, 2021]

Subpart C—Loan Contracts With Distribution Borrowers

SOURCE: 60 FR 67410, Dec. 29, 1995, unless otherwise noted.

§ 1718.100 General.

(a) *Purpose*. The purpose of this subpart is to set forth the policies, requirements, and procedures governing loan contracts entered into between the Rural Utilities Service (RUS) and distribution borrowers or, in some cases, other electric borrowers.

(b) *Flexibility for individual circumstances*. The intent of this subpart is to provide the flexibility to address the different needs and different credit risks of individual borrowers, and other special circumstances of individual lending situations. The model loan contract contained in appendix A of this subpart provides an example of what a loan contract with an "average" or "typical" distribution borrower may look like under "average" or "typical" circumstances. Depending on the credit risks and other circumstances of individual loans, RUS may execute loan contracts with provisions that are substantially different than those set forth in the model. RUS may develop alternative model loan contract provisions. If it does, such provisions will be made available to the public.

(c) *Resolution of any differences in contractual provisions*. If any provision of the loan contract appears to be in conflict with provisions of the mortgage, the loan contract shall have precedence with respect to the contractual relationship between the borrower and RUS with respect to such provision. If either document is silent on a matter addressed in the other document, the other document shall have precedence with respect to the contractual relationship between the borrower and RUS with respect to such matter.

(d) *Certain loan contract provisions subject to subsequent rulemaking*. If a loan contract provision imposes an obligation or limitation on the borrower whose interpretation or specification is subject to RUS regulations or the discretion of the Administrator or RUS, such interpretation or specification shall be subject to subsequent rulemaking. Such interpretation or specification of the borrower's obligations or limitations may not exceed the authority granted to the Administrator or RUS in the loan contract provision.

§ 1718.101 Applicability.

(a) *Distribution borrowers*. The provisions of this subpart apply to all distribution borrowers that obtain a loan or loan guarantee from RUS approved on or after January 29, 1996. Distribution borrowers that obtain a lien accommodation or any other form of financial assistance from RUS after January 29, 1996, may be required to execute a new loan contract and new mortgage. Moreover, any distribution borrower may submit a request to RUS that a new loan contract and new mortgage be executed. Within the constraints of time and staff resources, RUS will attempt to honor such requests. Borrowers must first obtain the concurrence of any other mortgagees on their existing mortgage before a new mortgage can be executed.

(b) *Other borrowers*. Borrowers other than distribution borrowers may also submit requests for execution of a new loan contract pursuant to this subpart and a new mortgage pursuant to subpart B of this part. RUS may approve such requests if it determines that such approval is in the government's financial interest. If other mortgagees are on the borrower's existing mortgage, their concurrence would be required before a new mortgage could be executed.

§ 1718.102 Definitions.

For the purposes of this subpart:

Borrower means any organization that has an outstanding loan made or guaranteed by the Rural Utilities Service (RUS) or its predecessor, the Rural Electrification Administration, for rural electrification, or that is seeking such financing.

Distribution borrower means a borrower that sells or intends to sell electric power and energy at retail in rural areas, the latter being defined in 7 CFR 1710.2.

Loan documents means the mortgage (or other security instrument acceptable to RUS), the loan contract, and the promissory note entered into between the borrower and RUS.

§ 1718.103 Loan contract provisions.

Loan contracts executed pursuant to this subpart shall contain such provisions as RUS determines are appropriate to further the purposes of the RE Act and to ensure that the security for the loan will be reasonably adequate and that the loan will be repaid according to the terms of the promissory note. Such loan contracts will contain provisions addressing, but not necessarily limited to, the following matters:

(a) Description of the purpose of the loan;

(b) Specification of the interest to be charged on the loan, including the method for determining the interest rate if it is not fixed for the entire term of the loan;

(c) Specification of the method for repaying the loan principal, including the final maturity of the loan;

(d) The conditions under which the loan may be prepaid before its maturity date, including but not limited to requirements regarding the prepayment of loans made concurrently by RUS and another secured lender;

(e) The method for making scheduled payments on the loan;

(f) Accounting principles and system of accounts, and RUS authority to approve the accountant used by the borrower;

(g) The method and time period for advancing loan funds and the conditions precedent to the advance of funds;

(h) Representations and warranties by the borrower as a condition of obtaining the loan, including but not limited to: the legal authority of the borrower to enter into the loan contract and operate its system; that the loan documents will be a legal, valid and binding obligation of the borrower enforceable according to their terms; compliance of the borrower in all material respects with all federal, state, and local laws, regulations, codes, and orders; existence of any pending or threatened legal actions that could have a material adverse effect on the borrower's ability to perform its obligations under the loan documents; the accuracy and completeness of all information provided by the borrower in the loan application and with respect to the loan contract, and the existence of any material adverse change since the information was provided; and the existence of any material defaults under other agreements of the borrower;

(i) Representations, warranties, and covenants with respect to environmental matters;

(j) Reports and notices required to be submitted to RUS, including but not limited to: annual financial statements; notice of defaults; notice of litigation; notice of orders or other directives received by the borrower from regulatory authorities; notice of any matter that has resulted in or may result in a material adverse change in the condition or operations of the borrower; and such other information regarding the condition or operations of the borrower as RUS may reasonably require;

(k) Annual written certification that the borrower is in compliance with its loan contract, note, mortgage, and any other agreement with RUS, or if there has been a default in the fulfillment of any obligation under said agreements, specifying each such default and the nature and status thereof;

(l) Requirement that the borrower design and implement rates for utility services to meet certain minimum coverage of interest expense and/or debt service obligations;

(m) Requirement that the borrower maintain and preserve its mortgaged property in compliance with prudent utility practice and all applicable laws, which may include certain specific actions and certifications set forth in the borrower's loan contract or mortgage;

(n) Requirement that the borrower plan, design and construct its electric system according to standards and other requirements established by RUS, and if directed by the Administrator, that the borrower follow RUS planning, design and construction standards and requirements for other utility systems constructed by the borrower;

(o) Limitations on extensions and additions to the borrower's electric system without approval by RUS;

(p) Limitations on contracts and contract amendments that the borrower may enter into without approval by RUS;

(q) Limitations of the transfer of mortgaged property by the borrower;

(r) Limitations on dividends, patronage refunds, and cash distributions paid by the borrower;

(s) Limitations on investments, loans, and guarantees made by the borrower;

(t) Authority of RUS to approve a new general manager and to require that an existing general manager be replaced if the borrower is in default under its mortgage, loan contract, or any other agreements with RUS;

(u) Description of events of default under the loan contract and the remedies available to RUS;

(v) Applicability of state and federal laws;

(w) Severability of the individual provisions of the loan documents;

(x) Matters relating to the assignment of the loan contract;

(y) Requirements relating to federal laws and regulations, including but not limited to the following matters: area coverage for electric service; civil rights and equal employment opportunity; access to buildings and other matters relating to the handicapped; design and construction standards relating to earthquakes; the National Environmental Policy Act of 1969 and other environmental laws and regulations; flood hazard insurance; debarment and suspension from federal assistance programs; and delinquency on federal debt; and

(z) Special requirements applicable to individual loans, and such other provisions as RUS may require to ensure loan repayment and reasonably adequate loan security.

§ 1718.104 Availability of model loan contract.

Single copies of the model loan contract (RUS Informational Publication 1718 C) are available from the Rural Utilities Service, United States Department of Agriculture, Washington, DC 20250–1533. This document may be reproduced.

PART 1719—RURAL ENERGY SAVINGS PROGRAM

Subpart A—General Provisions

Sec.

Subpart B—Application, Submission and Administration of RESP Loans

AUTHORITY: 7 U.S.C. 8107a (Section 6407).

SOURCE: 85 FR 18418, Apr. 2, 2020, unless otherwise noted.

Subpart A—General Provisions

§ 1719.1 Purpose.

This part establishes policies and procedures for the implementation of the Rural Energy Savings Program (RESP) under Section 6407 of the Farm Security and Rural Investment Act of 2002, as amended, by the Rural Utilities Service (RUS). It is the purpose of this part to help rural families and small businesses achieve cost savings by providing loans through eligible entities to qualified consumers to implement durable cost-effective energy efficiency measures.

§ 1719.2 Definitions.

The following definitions apply to subparts A and B of this part and must have the following meanings for purposes of the Rural Energy Savings Program:

Administrator means the Administrator of the Rural Utilities Service, an agency under the Rural Development mission area of the United States Department of Agriculture.

Applicant means an Eligible entity interested in applying for a RESP loan that is planning to submit a Letter of Intent.

Commercial technology means equipment, devices, applications, or systems that have a proven, reliable performance and replicable operating history specific to the proposed application. The equipment, device, application or system is based on established patented design or has been certified by an industry-recognized organization and subject to installation, operating, and maintenance procedures generally accepted by industry practices and standards. Service and replacement parts for the equipment, device, application or system must be readily available in the marketplace with established warranty applicable to parts, labor and performance.

Completed loan application means an application containing all information required by RUS to approve a loan and that is materially complete in form and substance satisfactory to RUS within the specified time.

Conditional commitment letter means the notification issued by the Administrator to a RESP Applicant advising it of the total loan amount approved for it as a RESP borrower, the acceptable security arrangement, and such controls and conditions on the RESP borrower's financial, investment, operational and managerial activities deemed necessary by the Administrator to adequately secure the Government's interest. This notification will also describe the accounting standards and audit requirements applicable to the transaction.

Conflict of interest means a situation or situations, event or series of events, that taken together or separately undermine an individual's judgement, ability, or commitment to providing an accurate, unbiased, fair and reliable assessment, or determination about the cost effectiveness of the Energy efficiency measures, due to self-interest or if such judgement, ability, commitment or determination cannot be justified by the prevailing and sound application of the generally accepted standards and principles of the industry.

Deemed savings means the per-unit energy savings values that can be claimed from installing specific measures under specific operating situations. Savings are based on stipulated values stemming from historical and verified data, derived from research of historical savings values from typical projects.

Deemed savings calculations means standardized algorithms to calculate energy savings applicable to well-defined energy efficiency measures that have documented and consistent savings values.

Eligible entity means an entity described in § 1719.4.

Energy audit means an analysis of the current energy usage or costs of a Qualified consumer with the goal of identifying opportunities to enhance energy efficiency. The activity should result in an objective standard-based technical report containing recommendations on the Energy efficiency measures to reduce energy costs or consumption of the Qualified consumer and an analysis of the estimated benefits and costs of pursuing each recommendation in a payback period not to exceed the loan term to the Qualified consumer. The analysis must meet professional and industry standards and be commensurate to the complexity of the project.

Energy efficiency measures (EE measures) means for or at property served by an Eligible entity, structural improvements and investments in cost-effective, commercial technologies to increase energy efficiency (including cost-effective on- or off-grid renewable energy or energy storage systems).

Energy efficiency program (EE Program) means a program set up by an Eligible entity to provide financing to Qualified consumers so that they can implement durable cost-effective Energy efficiency measures.

Financial feasibility means an Eligible entity's capacity to generate enough revenues to cover its expenses, sufficient cash flow to service its debts and obligations as they come due, and meet the financial ratios set forth in the applicable loan documents.

Government means the Federal Government.

GAAP means the generally accepted accounting principles in the United States of America as issued by the Financial Accounting Standards Board (FASB) in the Accounting Standards Codification (ASC).

Implementation Work Plan or EE Program Implementation Work Plan (IWP) means an Implementation work plan that meets the requirements listed in §1719.5(b)(3)(i)(F).

Invitation to proceed means the written notification issued by RUS to the Eligible entity acknowledging that the Letter of Intent was received and reviewed, describing the next steps in the application process, and inviting the Eligible entity to submit a complete loan application.

Key performance indicators mean the set of measures that help an entity to determine if it is reaching its performance and operational goals. These indicators can be both financial and non-financial.

Letter of Intent means a signed letter issued by an Applicant notifying RUS of its intent to apply for a RESP loan and addressing all the elements identified in §1719.5(b)(2).

Loan to a Qualified consumer means a transaction by which an RUS borrower makes RESP funds available to a Qualified consumer for the purpose of implementing Energy efficiency measures at a property or for the property of a Qualified consumer to increase energy efficiency on the condition that the RUS borrower will be able to collect the funds made available to the Qualified consumer.

Manufactured home means a structure that is transportable, built on a permanent chassis and designed to be used as a dwelling that meets the U.S Department of Housing and Urban Development definition set forth in 24 CFR 3280.2 or a successor rule.

Measurement and Verification (M&V) means the process of quantifying the energy and cost savings resulting from the improvements in an energy-consuming system or systems.

Multi-tier Agreement means an agreement entered into by the RESP applicant that complies with the Rural Development's Environmental Policies and Procedures, pursuant to 7 CFR part 1970 or its successor regulation.

Qualified consumer means a consumer served by an Eligible entity that has the ability to repay a loan made by a RESP borrower under the RESP program, as determined by the Eligible entity.

RESP applicant means an Eligible entity that has received a written Invitation to proceed from RUS to apply for a RESP loan.

RESP borrower means an Eligible entity with an approved RESP loan as evidenced by duly executed RESP loan documents.

Rural, for purposes of 7 U.S.C. 8107a(a), means any area that has a population of 50,000 or less inhabitants or any other area designated eligible by statute.

Small business means an entity that is in accordance with the Small Business Administration's (SBA) small business size standards found in 13 CFR part 121.

Special advance means an advance, not to exceed 4 percent of the total approved loan amount, that a RESP borrower may request to defray the start-up costs of establishing a new EE Program.

Start-up costs mean amounts paid or incurred for:

(1) Creating or implementing an active EE program; or

(2) Investing in the integration of an active EE Program. Start-up costs may include, but are not limited to, amounts paid or incurred in the analysis or survey of potential markets,

products such as software and hardware, labor supply, consultants, salaries and other working capital directly related to the creation or enhancement of an EE Program consistent with RESP.

Technical Resource Manual (TRM) means a resource document that includes information used in program planning and reporting of EE Programs. A TRM may include savings values for measures, engineering algorithms to calculate savings, impact factors to be applied to calculated savings, foundational documentation, specified assumptions, and such other pertinent information to support the calculation of measure and program savings and the application of such values and algorithms in appropriate applications.

§ 1719.3 Policy and Federal Register Notices.

(a) Eligible entities (see § 1719.2 and § 1719.4) are permitted to participate in the Rural Energy Saving Program on the condition that loan funds will be used to make loans to Qualified consumers for the purpose of implementing EE measures.

(b) The Agency will issue annual FEDERAL REGISTER notices each year specifying the amount of funds available under this Part. Notices may also include program priorities and loan application periods. The Administrator in setting funding priorities and application periods may consider the amount of available funds, the nature and amount of unfunded loan applications, prior commitments, Agency resources, Agency priorities and policy goals, and any other pertinent information.

(c) In making loans under this Part, the Administrator may consider a proposed EE Program's effect on existing RUS borrowers and the integrity of the RUS portfolio and deny or limit approval of a specific RESP loan application on that basis if it is determined that such requested loan would have a negative effect on existing RUS or RESP borrowers or the RUS loan portfolio.

(d) The Administrator may, on a case-by-case basis, grant an exception to any requirement or provision of this subpart provided that such an exception is in the best financial interests of the Federal government. Exercise of this authority cannot be in conflict with applicable law.

(e) With regard to the rules of grammatical construction, unless the context otherwise indicates, "includes" and "including" are not limiting, and "or" is not exclusive.

Subpart B—Application, Submission and Administration of RESP Loans

§ 1719.4 Eligibility.

Under this subpart, Eligible entities for the RESP include:

(a) Any public power district, public utility district, or similar entity, or any electric cooperative described in section 501(c)(12) or 1381(a)(2) of the Internal Revenue Code of 1986, that borrowed and repaid, prepaid, or is paying an electric loan made or guaranteed by the Rural Utilities Service (or any predecessor agency);

(b) Any entity primarily owned or controlled by one (1) or more entities described in paragraph (a) of this section; or

(c) Any other entity that is an eligible borrower of the Rural Utilities Service, as determined under 7 CFR 1710.101.

§ 1719.5 Application process and required information.

(a) *General.* The following are general provisions for the application process:

(1) The RUS, from time to time and subject to appropriations, will notify the public specifying funding priorities, funding availability, and deadlines.

(2) Complete applications for loans to Eligible entities will be processed pursuant to the provisions in this Part and on a first-come-first served basis until the funding appropriated to the program is fully obligated.

(3) The submittal of a Letter of Intent is required to participate in the program. The letters of intent will be queued as they are received. If it advances program and policy goals, RUS may consider loan applications from Eligible entities that have submitted Letters of Intent under prior funding

announcements but that were not invited to proceed with a loan application.

(4) Upon review of the Letter of Intent, RUS may issue an Invitation to proceed with a loan application. RUS reserves the right to notify the Applicant in the queue that the amount of financing RUS will consider for a loan is below the level sought in the Letter of Intent. In making this consideration, RUS will consider overall RUS program objectives or budgetary constraints. An Invitation to proceed with the loan application issued by RUS is not to be deemed as an offer by RUS.

(5) A RESP applicant will have up to ninety (90) days to complete the documentation for a complete loan application. The ninety (90) day timeframe will begin on the date the RESP applicant receives RUS' Invitation to proceed. If the deadline to submit the completed loan application falls on Saturday, Sunday, or a Federal holiday, the application is due the next business day.

(6) The Administrator may grant an extension of time to complete the documentation required for an application if, in the Administrator's sole judgment, the interest of the program would be advanced by the extension.

(7) RUS may limit the number of applications it will consider in the same funding cycle from the same Applicant or combine applications from a single entity.

(b) *Application process.* The application process consists of the following two steps:

(1) An Applicant seeking financing must submit a Letter of Intent to be considered under this Part.

(2) The Letter of Intent must include the following information:

(i) Legal name and status of the entity seeking financing under this Part and its address and principal place of business.

(ii) The Applicant's tax identification number, SAM Managed Identifier (SAMMI), Dun and Bradstreet (DUNS) number, and such similar information as it may be subsequently amended or required for federal funding.

(iii) A statement indicating if the Applicant is a current or a former RUS borrower.

(iv) A description of the service territory.

(v) Value of the net assets, including any information as to whether the Applicant has been placed in receivership, liquidation, or under a workout agreement or whether the Applicant has declared bankruptcy or has had a decree or order issued for relief in any bankruptcy, insolvency or other similar action over the last 10 years. The Applicant must submit a copy of its balance sheet and income statements for the last 3 years. If applicable, the Applicant must provide the balance sheet and income statements for the last 3 years of the entity or entities providing equity or security for the RESP loan together with an explanation of the legal relationship among the entities.

(vi) Identification of a point of contact and provide contact information.

(vii) Description of the program or projects expected to be financed with the RESP loans funds. This description must not exceed five (5) pages (size 8.5 x 11). RUS reserves the right not to consider Letters of Intent where the project description exceeds five (5) pages. The description should include the following:

(A) Description of the service to be provided to Qualified consumers.

(B) Identity of the staff or contractors that will be implementing the EE Program and their credentials.

(C) A summarized version of the expected IWP addressing the following elements:

(*1*) The marketing strategy.

(*2*) The relending process.

(*3*) A brief description of the processes, procedures, and capabilities to quantify and verify the reduction in energy consumption or decrease in the energy costs of the Qualified consumers.

(*4*) A list of eligible EE measures expected to be implemented. An Applicant with an existing EE Program in place by April 8, 2014, may describe the EE measures, its IWP, and its M&V plan for the existing program in its Letter of Intent to expedite the application process.

(viii) The Applicant must provide evidence of its key performance indicators for the 5 complete years prior to the submission of the loan application if the total loan amount exceeds $5 million.

(3) Instructions on how to submit the loan application package will be included in the RUS Invitation to proceed to the RESP applicant. RUS will timely schedule an initial conference call with the RESP applicant to discuss the elements of the loan application.

(i) Content of the application package includes the following:

(A) A signed cover letter from the RESP applicant's General Manager or highest-ranking officer requesting RESP loan funds to make loans to Qualified consumers for the purpose of implementing EE measures.

(B) A signed copy of the board resolution or applicable authorizing document approving and establishing the EE Program and authorizing the Eligible entity to take a RESP loan.

(C) The RESP applicant must provide the Applicant's articles of incorporation or other applicable organizational documents currently in effect, as filed with the appropriate state office, setting forth the RESP applicant's corporate purpose; and the RESP Applicant must also provide the bylaws or other applicable governing documents currently in effect, as adopted by the RESP applicant's applicable governing body. RESP applicants that are active RUS borrowers may comply with this requirement by notifying RUS in writing that there are no material changes to the documents already on file with RUS.

(D) A copy of the duly executed Multi-Tier Action Environmental Compliance Agreement (Multi-Tier Agreement) consistent with Rural Development's Environmental Policies and Procedures, 7 CFR part 1970 or its successor regulation. A copy of the Multi-tier Agreement will be provided to the RESP applicant with the Invitation to proceed and the requirements of § 1970.55 will be discussed with the RESP applicant in the initial conference call. Activities and investments listed in the IWP must match the activities and investments identified in the Multi-tier Agreement executed between RUS and the RESP applicant. Additional RUS environmental review will be required if the RESP applicant pursues additional or different activities other than the ones listed in the Multi-tier Agreement. If funded, a RESP borrower would be responsible for performing and documenting environmental reviews consistent with § 1970.55.

(E) A financial forecast approved by the applicable governing body of the RESP applicant in support of its loan application. The financial forecast must cover a period of at least 10 years and must demonstrate that the RESP applicant's operation is economically viable and that the proposed loan is financially feasible. RUS may request additional information or projections for a longer period, if RUS deems such supplemental data necessary based on the financial structure of the RESP Applicant or necessary to make a determination regarding loan feasibility. A RESP applicant must, after submitting a loan application, promptly notify RUS of any changes in its circumstances that materially affect the information contained in the loan application. The financial forecast and related projections submitted in support of a loan application must include:

(*1*) Current and projected cash flows.

(*2*) A pro forma balance sheet, statement of operations, and general funds summary projected for each year during the forecast period. The requested RESP loan must be included in the financial forecast. Revenue from the interest charged to the Qualified consumer must also be included together with an explanation of the expected use of such proceeds.

(*3*) The financial goals established for margins, debt service coverage, equity, and levels of general funds to be invested in the EE Program. The financial forecast must use the accrual method of accounting for analyzing costs and revenues and, as applicable, compare the economic results of the various alternatives on a present value basis.

(*4*) A full explanation of the assumptions, supporting data, and analysis used in the forecast, including the methodology used to project revenues, operating expenses, and any other factors having a material effect on the balance sheet and the financial ratios such as equity and debt service coverage. RUS may require additional data and analysis on a case-by-case basis to assess the probable future competitiveness of the RESP applicant.

(*5*) Current and projected nonoperating income and expense.

(*6*) An itemized budget and schedule for the activities to be implemented with the RESP funds and a discussion on the expected delinquency and default rates and how the loan loss reserve will be set up. The RESP applicant is expected to forecast the amount of loans to be made to Qualified consumers over a 10-year timeframe. If the RESP applicant determines to charge interest, the RESP applicant must describe how it is going to use the funds generated from the interest to be received from the loans to the Qualified consumers.

(*7*) A sensitivity analysis may be required by RUS on a case-by-case basis.

(F) The RESP applicant must produce, to the satisfaction of the Administrator, an Implementation Work Plan or EE Program Implementation Work Plan (IWP), duly approved by the applicable governing body of the Eligible entity. The IWP will cross reference the Financial Forecast and must address the following core elements:

(*1*) The RESP applicant will identify the Qualified consumers by customer classes that will benefit from the proceeds of a loan made under this Part and explain the promotional activities that will be executed to carry out the energy efficiency relending program. The RESP applicant should also include the target penetration rates by market segment and expected investments in marketing the relending program. In doing so, it is expected that racial and ethnic demographics for the service area would be provided.

(*2*) The RESP applicant will describe the activities and investments (list of EE measures) to be implemented in the EE Program and the expected energy savings.

(*i*) The RESP applicant must include a schedule for implementation with an itemized list of anticipated costs for each task.

(*ii*) The RESP applicant must specify whether a Special advance will be requested and, if so, must detail the expected use of such loan proceeds.

(*iii*) In describing the EE Program, the RESP applicant must describe the intake process, including but not limited to, the underwriting criteria, if applicable, and the quantifiable elements considered in recommending energy retrofits or investments to reduce the Qualified consumer's energy cost or consumption. It is also expected that a description of the process for documenting and perfecting collateral arrangements with Qualified Consumers, when applicable, be also included in the narrative.

(*iv*) The RESP applicant will also identify the staff that will be carrying out the EE Program and will describe the tasks that will be performed by such individuals together with their expertise and credentials. Should the RESP applicant decide to outsource implementation of the EE Program, the credentials and expertise of the third party implementing the outsourced tasks must be described. Consideration must be given to the third party's ability and expertise in implementing an EE Program at the scale pursued with the RESP funding. The statement of qualifications must show the party's experience carrying out the financial and technical components of an EE Program at the desired scale. A RESP applicant with an existing EE Program as of April 8, 2014, may submit the IWP plan previously established to fulfill this requirement.

(*3*) The RESP applicant must include an evaluation of the financial and operational risk associated with the EE Program. When applicable, the RESP applicant should include an estimate of the prospective consumer loan losses consistent with the loan loss reserve.

(*4*) A Measurement and Verification (M&V) plan that meets the requirements of § 1719.10. In the alternative, a RESP applicant may provide an M&V plan approved by a state or local regulatory entity.

(G) The RESP applicant must provide a statement of compliance with the federal statutes as provided in § 1719.11.

§ 1719.6 Agency review.

(a) *General.* Loans made under this program will be made only when the

Administrator finds and certifies that in his or her judgment there is reasonably adequate security and the loan will be repaid within the time agreed.

(b) *Eligibility for other loans*. RUS will not include any debt incurred by a borrower under this program in the calculation of the debt-equity ratios of the borrower for purposes of eligibility for loans under the Rural Electrification Act of 1936 (7 U.S.C. 901 *et seq.*).

(c) *Letter of intent*. RUS will consider complete Letters of intent in the order they are received. In reviewing Letters of intent, RUS will be assessing:

(1) *Applicant eligibility*. Applicant's eligibility to participate in the program.

(2) *Project eligibility*. Eligibility of the proposed EE Program or project.

(3) *Financial status*. The financial status of the RESP applicant to determine the Applicant's likelihood to complete a loan application and successfully repay a RESP loan.

(d) *Loan application*. Prudent lending practices require that the Administrator make certain findings prior to approving a RESP loan. RESP applicants must provide the evidence, in form and substance satisfactory to the Administrator, to be able to make such findings. In making loans under this Section, the Administrator will consider, including, but not limited to, the following factors:

(1) *Loan feasibility*. The RESP applicant's ability to repay the loan in full as scheduled and all other obligations of the borrower will be met.

(2) *RESP applicant's character*. The RESP applicant's past performance and determination to satisfy its obligations; evidenced by such factors as credit history, previous experience addressing adversity, and manner of conducting business.

(3) *RESP applicant's equity*. The financial resources retained by the RESP applicant to provide a cushion against unexpected losses.

(4) *Overall condition of RESP applicant and project*. Verification that the proposed EE Program meets all the requirements of the Rural Energy Savings Program and an assessment of those factors that may affect the RESP applicant's ability to repay the RESP loan or implement the EE Program as proposed.

(5) *Loan security*. The RESP applicant's assets pledged to secure the loan. Collateral will be assessed for each applicant taking into consideration asset value, lien position, credit risk and borrower's profile. Collateral pledged should be adequate to protect the Government's interest. RUS reserves the right to require an asset appraisal.

(6) *EE program implementation and measurement and verification*. RESP applicant's IWP must be based on reasonable assumptions and adequate supporting data and the M&V plan reasonably complies with § 1719.10. However, the Administrator, in his or her sole discretion, may deem this requirement satisfied upon finding that the IWP and M&V plan from an existing EE Program as of April 8, 2014 is consistent with the purpose of the Rural Energy Savings Program. A RESP applicant with an existing EE Program as of April 8, 2014, may submit the M&V plan previously established to fulfill this requirement.

§ 1719.7 Conditional commitment letter and loan closing.

(a) *Conditional commitment letter*. A successful RESP loan applicant will receive a Conditional commitment letter from the Administrator notifying the RESP applicant of the total loan amount approved by RUS; any additional controls on the its financial, investment, operational and managerial activities; acceptable security arrangements; and such other conditions deemed necessary by the Administrator to adequately secure the Government's interest, ensure repayment, and abide by the RESP requirements as outlined in this Part. This written notification is a conditional RESP loan offer.

(1) The requirements for coverage ratios will be set forth in the Conditional commitment letter.

(2) Receipt of a Conditional commitment letter from the Administrator does not authorize the RESP applicant to commence performance under the approved loan.

(b) *Intent to meet conditions.* The RESP applicant must acknowledge receipt of the Conditional commitment letter and notify RUS in writing within 60 days or otherwise specified in the Conditional commitment letter that it has reviewed and understood the conditions set forth in the Conditional commitment letter and that it is the intent of the RESP applicant to meet all the conditions. The RESP applicant must promptly notify RUS should circumstances or its intent of meeting the conditions change. The Administrator may consider requests to amend the conditions and amend the conditions in a subsequent Conditional commitment letter, when it advances program and policy goals and is in the best interest of the Government.

(c) *Loan closing.* The loan will be closed in accordance with RUS instructions.

(1) Upon receipt of the acceptance of the loan offer from the RESP applicant, RUS, working with its legal counsel, will draft the loan documents which will include the loan conditions and other applicable legal requirements.

(2) The loan documents will be forwarded to the RESP applicant by RUS for execution by the RESP applicant's signatories and returned to RUS prior to a mutually acceptable closing date. RUS reserves the right to unilaterally set a closing date to advance program and policy goals.

(3) The loan closing date will be used to determine the RESP loan maturity date which under no circumstances will exceed 20 years.

(4) An opinion of counsel is required at closing and must be in form and substance acceptable to the Administrator. A form opinion of counsel will be included in the closing instructions.

(d) *Post-closing activities.* All RUS requirements and conditions for lending set forth in the loan agreement must be met before the loan will be advanced. RUS will notify the RUS borrower when it is authorized to commence activities to be funded by the RESP loan.

§ 1719.8 Loan provisions.

(a) *Financial ratios.* The Administrator will set financial coverage ratios based on the risk profile of the RESP applicant and specific loan terms. Those financial ratios will be included in the RESP borrower's loan documents with RUS.

(1) Unless otherwise notified, existing RUS borrowers will be subject to their current debt service coverage ratios as provided in their previously executed loan contracts with RUS.

(2) The minimum coverage ratio required for RESP borrowers, whether applied on annual or average basis is 1.05 Debt Service Coverage (DSC) unless specifically waived by the Administrator.

(3) DSC for RESP borrowers that are not existing RUS borrowers under the Rural Electrification Act will be defined as (Net Income or Total Margins) + (Interest Charges on Long Term Debt) + (Principal payments from RESP relending activities) + (Depreciation and Amortization Expenses)/Total Debt Service Billed.

(4) In reviewing and approving a RESP loan, the Administrator may increase the coverage ratio required to be met by an individual RESP borrower if the Administrator determines that higher ratios are required to ensure the repayment of the loan made by RUS, or reduce the coverage ratios if the Administrator determines that the lower ratios are in the best interest of the Government. The coverage ratios will be set forth in the loan documents.

(b) *Collateral.* RUS generally requires that borrowers provide it with a first priority lien on all of the borrower's real and personal property, including intangible personal property and any property acquired after the date of the loan. Collateral that is used to secure a loan must ordinarily be free from liens or security interests other than those permitted by RUS or existing security documents.

(1) For existing RUS borrowers, the Administrator may, in his or her sole discretion, rely on existing security arrangements with RUS.

(2) When a RESP borrower is unable, by reason of preexisting encumbrances, or otherwise, to furnish a first priority

lien on its entire system, the Administrator may accept other forms of security, including but not limited to a parent guarantee, state guarantee, an irrevocable letter of credit, surety bond, pledge of revenues, or other security if the Administrator determines such credit support is reasonably adequate to protect the government's interests and otherwise acceptable in form and substance.

(3) RUS may in certain circumstances agree to share its priority lien position with another lender provided the RESP loan is adequately secured and the security arrangements are acceptable to RUS. In such circumstances, RUS will consider entering into joint security arrangements with other lenders on a *pari passu* basis.

(c) *Equity contributions.* To be eligible for a RESP loan, a newly created Eligible entity or an entity primarily owned or controlled by one (1) or more entities as described in § 1719.4 must meet a minimum equity contribution in the proposed EE Program requirement at the time of the loan closing. The eligible entity will be required to continue to maintain the minimum equity contribution for the life of the loan or other time period as determined by the Administrator and as set forth in the loan documents. The minimum acceptable equity contribution for each RESP borrower will be determined by the Administrator as set forth below and will be included in the Conditional commitment letter and the loan documents as a condition and covenant to the RESP loan.

(1) The required equity contribution and related terms will be determined by the Administrator for the individual RESP applicant based upon the its risk profile and available collateral for the RESP loan.

(2) RUS reserves the right to require additional equity contributions from existing RUS or RESP borrowers when it is in the best interest of the Government.

(3) If the RESP applicant under this section is unable to achieve a minimal acceptable contribution, as set forth in the Conditional commitment letter, the Administrator may consider the following to meet such shortfall to the minimum acceptable equity contribution:

(i) The infusion of additional capital into the EE Program by an Investor to meet the shortfall to the minimum acceptable equity contribution. RUS may require that the additional capital be deposited into a RESP applicant's special account subject to a deposit account control agreement with RUS prior to loan closing.

(ii) An unconditional, irrevocable letter of credit, in form and substance satisfactory to the Administrator, in the amount necessary to meet the shortfall to the minimum acceptable equity contribution. RUS must be an unconditional payee under the letter of credit and the letter of credit must be in place prior to loan closing and remain in place until the loan is repaid unless specified otherwise in the loan documents.

(iii) General obligation bonds or special revenue bonds issued by tribal, state or local governments in the amount necessary to meet the shortfall to the minimum acceptable equity contribution. If the minimum acceptable equity position is satisfied in full or part with general obligation bonds or special revenue bonds, any lien securing the bonds must be subordinate to the lien of the Government securing the RESP loan.

(iv) Any other requirements or mechanisms approved by the Administrator to meet the shortfall to the minimum acceptable equity contribution.

(d) *Loan advances.* RUS will disburse loan funds to the RESP borrower in accordance with the terms and conditions of the executed loan documents.

(1) Excluding the Special Advance, all loan funds will be disbursed either as an advance in anticipation of loans to be made by the RESP borrower to the Qualified consumers; or as a reimbursement for eligible program costs, including loans already made to Qualified consumers. No disbursements will be made until the RESP borrower has complied with the loan conditions set forth in the loan documents. Any disbursement of loan funds to a RESP borrower within a 12-month consecutive period must not exceed 50 percent of the approved loan amount.

(i) The RESP borrower must provide to the Qualified consumers all RESP loan funds that the RESP borrower receives within one year of receiving them from RUS. If the RESP borrower does not re-lend the RESP loan funds within one year, the unused RESP loan funds, and any interest earned on those RESP loan funds, must be returned to the Government and will be applied to the RESP borrower's debt.

(ii) The RESP borrower will not be eligible to receive additional RESP loan funds from RUS until providing evidence, in form and substance satisfactory to the Administrator, that RESP loan funds from a previous advance have been fully relent to Qualified consumers or returned to the Government.

(iii) RUS will disburse the RESP loan funds as an advance in anticipation of loans to be made by the RESP borrower to the Qualified consumers only if the RESP borrower has established written procedures that will minimize the time elapsing between the transfer of RESP loan funds from RUS to the RESP borrower and its corresponding disbursement to the Qualified consumer.

(iv) A RESP borrower's request for an advance in anticipation of loans to Qualified consumers should be limited to the minimum amounts needed and timed to be in accordance with the actual immediate cash needs to carry out the EE Program.

(2) The RESP borrower may elect to request a Special advance to defray the appropriate start-up costs of establishing a new EE Program or modify an existing EE Program.

(i) The Special advance must not exceed 4 percent of the total approved loan amount.

(ii) Repayment of the Special advance must be required during the 10-year period beginning on the date on which the Special advance is made.

(iii) The RESP borrower may elect to defer the repayment of the Special advance to the end of the 10-year period.

(iv) All Special advances must be made during the first 10-years of the term of the loan.

(v) All amounts advanced on the loan by RUS to the RESP borrower, including the Special advance, must be paid prior to the final maturity which must not exceed 20 years.

(vi) The Special advance maximum amount must be requested by the Borrower and approved by RUS prior to loan closing.

(e) *Loans to Qualified Consumers*. RUS borrowers loans to Qualified Consumers will be subject to the following terms and for the purposes listed below.

(1) RESP borrower's loans to its Qualified consumers must be for the purpose of implementing EE measures.

(2) Loans to Qualified consumers may bear interest not to exceed 5 percent.

(3) Each loan made by the RESP borrower to a Qualified consumer may not exceed a term of 10 years.

(4) The EE measures financed with a RESP loan proceeds must be for the purpose of decreasing energy (not just electricity) usage or costs of the Qualified consumer by an amount that ensures, to the maximum extent practicable, that a loan term of not more than 10 years will not pose an undue financial burden on the Qualified consumer.

(5) RESP loan proceeds must not be used to fund purchases of, or modifications to, personal property unless the personal property is or becomes attached to real property (including a manufactured home) as a fixture.

(6) Loans made to Qualified consumers must be repaid through charges added to the recurring service bill for the property for, or, at which the EE measures have been or will be implemented. This requirement does not prohibit the voluntary prepayment of the loan by the owner of the property; or the use of any additional repayment mechanisms that are demonstrated to have appropriate risk mitigation measures, as determined by the RESP borrower, or required if the Qualified consumer is no longer a customer of the RESP Borrower.

(7) Loans made by a RESP borrower to a Qualified consumer using RESP loan funds must require an Energy audit by the RESP borrower to determine the impact of the proposed EE measures on the energy costs and consumption of the Qualified consumer. For purposes of this section, an energy audit performed by a contractor or

agent of the RESP borrower would be deemed as performed by the RESP borrower.

(8) The RESP borrower must comply with all applicable federal, state, and local laws and regulations in making loans to Qualified consumers. Approval by RUS and its employees of a loan under this section does not constitute a Government endorsement. The Government and its employees assume no legal liability for the accuracy, completeness or usefulness of any information, product, service, or process funded directly or indirectly with financial assistance provided under RESP. Nothing in the loan documents between RUS and the RESP borrower will confer upon any other person any right, benefit or remedy of any nature whatsoever. Neither the Government nor its employees make any warranty, express or implied, including the warranties of merchantability and fitness for a particular purpose, with respect to any information, product, service, or process available from a RESP borrower or its agents.

(f) *Loan term and repayment. RUS loans to an eligible borrow will be subject to the following terms and repayment conditions set forth in this section.*

(1) The RESP loans under this section will bear no interest (0 percent) and have a maturity not exceeding 20 years.

(2) The amortization schedule must be based on a loan term that does not exceed 20 years from the date on which the loan is closed.

(3) Except for the Special advance, the repayment of each advance must be amortized for a period not to exceed 10 years.

(4) The Administrator may include additional conditions on the repayment schedule if, in his or her sole discretion, it is in the best interest of the Government.

(5) The RESP borrower is responsible for fully repaying the RESP loan to RUS according to the loan documents regardless of repayment by its Qualified consumers.

(6) The RESP borrower may use the revenues from the interest charged to the Qualified consumer to establish a loan loss reserve, and to offset personnel and EE Program costs.

(7) Loans under this Section will not bear interest (0 percent), however, indebtedness not paid when due will be subject to interest, penalties, administrative costs and late fees as provided in the loan documents.

§ 1719.9 Eligible activities and energy efficiency measures.

(a) A RESP Borrower may provide financing to Qualified consumers to implement or invest in one or more set of EE measures such as those listed in this section.

(b) A RESP borrower may be able to provide financing to Qualified consumers for EE measures not listed in this section, if it can justify, to the satisfaction of the Administrator, that the proposed EE measure is consistent with the RESP statute, is cost effective, and the technology is commercially available. The Administrator must make the determination prior to the borrower implementing the EE measure.

(c) A RESP applicant with an existing EE Program as of April 8, 2014, may submit the list of the EE measures used in its program to RUS for validation and approval. The Administrator will make a finding as to whether such EE measures are consistent with the purpose of RESP.

(d) A RESP borrower, subject to the Administrator's written approval, may modify the list of EE measures if those measures are consistent with the statutory purpose of RESP.

(e) RESP loan proceeds must finance EE measures for the purpose of decreasing energy usage or costs of the Qualified consumer by an amount that ensures, to the maximum extent practicable, that the loan term will not pose an undue financial burden on the Qualified consumer.

(f) Eligible EE measures and investments include, but are not limited, to:

(1) Lighting:

(i) Lighting fixture upgrades to improve efficiency.

(ii) Lighting control technologies.

(iii) Daylighting systems.

(iv) Energy-efficient lighting technologies.

(2) Space conditioning, including Heating, Ventilation, and Air Conditioning (HVAC):

(i) Central Air Systems—Energy Star ® qualified equipment.
(ii) Room air conditioners.
(iii) Boilers.
(iv) Heat pumps.
(v) Ducts and duct sealing.
(vi) Furnaces—Energy Star® qualified equipment.
(vii) Thermostats.
(viii) Economizers.
(ix) Air handlers.
(x) Automated controls.

(3) Building Envelope Improvements:
(i) Improved insulation—adding insulation beyond existing levels, or above existing building codes.
(ii) Moisture barrier improvements and air sealing.
(iii) Caulking and weather stripping of doors and windows.
(iv) Windows upgrades—Energy Star® qualified windows.
(v) Door upgrades—including mandoors, overhead doors with integrated insulation and energy efficient windows.

(4) Motor Systems:
(i) Pumps, coupling and low-friction pipes.
(ii) Capacitors.
(iii) Variable frequency drives.
(iv) Induction motors repairs or replacements for energy efficiency.
(v) High efficiency motors—motors with a rated efficiency beyond the Energy Policy Act standards.
(vi) Permanent magnet motors.
(vii) Reluctance motors.

(5) Waste Heat Recovery:
(i) Recuperators.
(ii) Regenerators.
(iii) Waste heat boilers.
(iv) Combined heat and power (CHP) and Waste heat to power (WHP).

(6) Compressed Air Systems.

(7) Water heaters.

(8) Fuel switching.

(9) Irrigation or water system and waste disposal system efficiency improvements.

(10) On or off-grid renewable energy systems if consistent with the statutory purpose of this section.

(11) Energy storage devices if permanently installed to reduce energy cost or usage of the Qualified consumer.

(12) Energy efficient appliance upgrades if attached to real property as fixtures.

(13) Energy audits.

(14) Necessary and incidental activities and investments directly related to the implementation of an Energy efficiency measure.

§1719.10 Measurement and verification and quality control.

(a) *General.* A RESP applicant must provide a Measurement and Verification (M&V) plan, satisfactory to the Administrator, to ensure the effectiveness of the energy efficiency loans made to its Qualified Consumers and that there is no conflict of interest in carrying out the EE Program.

(1) RUS acknowledges the broad nature of energy efficiency projects and diverse scope of EE Programs that can be carried out under RESP. A RESP applicant, and its designees, must exercise professional judgment in developing their M&V plans. The nature, scope, and complexity of the EE measures and activities will dictate the level of effort needed for quantifying and verifying the savings. The effort expended should be commensurate with the project capital investment and the risk of miscalculating the savings.

(2) A RESP applicant with an existing EE Program as of April 8, 2014, may submit for consideration the M&V plan previously established to fulfill this requirement.

(3) RUS may reject a loan application or refuse to disburse loan proceeds to an RESP borrower that fails to demonstrate that the Energy audits or M&V plan have been adequately implemented and performed by qualified individuals.

(4) The M&V plan should be based on generally accepted principles and use the best practices of the industry, reliable data, reasonable assumptions and verifiable analytical methodologies.

(5) The M&V plan must describe the organized activities that the RESP applicant will implement to facilitate the adoption of the Energy efficiency measures that will result in energy use or cost savings to the Qualified consumer.

(6) Energy savings should be determined by comparing measured energy unit values (consumption or demand) before and after the implementation of

the EE measures, making appropriate adjustments for changes in conditions.

(7) The computation of the savings formula is as follow:

Savings = (Baseline Energy—Post-Installation of EE Measures Energy*) ± Adjustments

Note: * = performance period

(b) *M&V Techniques for measuring, calculating and reporting savings.* The RESP borrower may address the M&V requirements by applying any of the following techniques recognized in the International Performance Measurement and Verification Protocol.

(1) The Retrofit Isolation with Key Parameter Measurement Option (RIKPM) alternative is based on a combination of measured and estimated factors. Measurements will be taken at the component or system level for both the baseline and the retrofit equipment and should include the key performance parameters that define the energy use of the energy conservation measure. Savings will be determined by calculating the baseline and reporting period energy use predicated on the measured and estimated values. Estimated values will have to be supported by historical or manufacturer's data.

(2) The Retrofit Isolation with All Parameter Measurement Option (RIAPM) option will be based on short-term, periodic or continuous measurements of baseline and post-retrofit energy use (or proxies of energy use) taken at the component or system level. Savings will be based on the analysis of the baseline and reporting-period energy use or proxies of energy use.

(3) The Whole Facility Measurement Option (WFMO) will be based on continuous measurement of the energy use (such as utility billing data) at the whole facility or sub-facility level during the baseline and post-retrofit periods. Savings will be established from the analysis of the baseline and reporting-period energy data.

(4) The Calibrated Simulation Option (CSO) is an alternative where computer simulations can be used to model energy performance of a whole facility (or sub-facility). Models must be calibrated with actual hourly or monthly billing data from the facility. In this option, savings will be determined by comparing a simulation of the baseline (after having calibrated the model) with either a simulation of the performance period or actual utility data.

(c) *Use of deemed savings.* A RESP applicant may elect to meet the M&V plan requirements by applying deemed savings values and calculations. If choosing this option, the RESP applicant's M&V plan must:

(1) Describe the process to stipulate with the Qualified consumer the values and assumptions for determining the energy savings.

(2) Identify the TRMs upon which the deemed savings values and assumptions are based. In the alternative, identify such other technical M&V studies reasonably applicable to the conditions of the RESP applicant's service area or such other detailed M&V studies performed by similar entities to determine deemed savings for identical or similar energy programs or energy efficiency measures.

(3) Describe the mechanism to ensure that deemed savings values and related calculations will be maintained and kept up to date.

(4) The approval by RUS of a M&V plan under this section is solely for the benefit of RUS. Approval of a plan pursuant to this section does not constitute an RUS endorsement of the M&V plan or an EE Program. RUS and its employees assume no legal liability for the accuracy, completeness or usefulness of any information, product, service, or process funded directly or indirectly with financial assistance provided under RESP.

(d) *Quality control.* The RESP borrower must produce a detailed explanation, in form and substance satisfactory to the Administrator, describing the methods and processes to verify that the installation of the EE measures for the EE program, for which those measures have been implemented were properly executed.

(1) The RESP borrower and the Qualified consumer must agree on the EE measures to be implemented based on a quantifiable and verifiable assessment of the impacts that such measures will have in reducing the Qualified consumer's energy cost or consumption.

(2) A RESP borrower may elect to engage a third-party contractor to carry out the assessments required in this Section and install the EE measures as long as there is no Conflict of interest.

(3) RESP borrower employees and third-party contractors engaged to carry out activities in the EE Program must be qualified and have adequate expertise to perform energy audits, retrofit installations, and do the quality control assessments according to the applicable industry best-practices. Individual's credentials and expertise should be accredited through one of the following options:

(i) Possessing a current Home Energy Professional Certification or a similar certification from a nationally, industry-recognized organization that is consistent with the Job Task Analyses Guidelines issued by the US Department of Energy's National Renewable Energy Laboratory or its successor.

(ii) Possessing a current certification issued by an organization recognized by the U.S. Department of Energy in accordance with the Better Buildings Workforce Guidelines or its successor.

(iii) Producing evidence, in form and substance satisfactory to the Administrator, that the individual possesses proficiency in the knowledge, skills and abilities needed to perform the tasks and critical work functions relevant to the duties assigned in the EE Program.

(4) A RESP borrower that elects to carry out the EE Program with a contractor, must validate and document the following:

(i) The contractor has adequate capacity and resources to engage with customers, conduct whole-property assessments, performance testing, diagnostic reasoning, and fulfill all data collection and reporting requirements. This includes, but is not limited to, having access to satisfactory diagnostic equipment, tools, qualified staff, data systems and software, and administrative support.

(ii) The contractor is current and in good standing with all applicable registration and licensing requirements for their specific jurisdiction and trade.

(iii) The contractor employs individuals (either its own employees or subcontractors) that are qualified to install or physically oversee the installation of home improvements in compliance with local building codes and industry-accepted protocols.

(5) A RESP borrower is responsible for actions or omissions departing from the required standards under this Section by third party partners or contractors employed in connection with an EE Program funded under this Section.

(6) The RESP loan documents are solely for the benefit of RUS and the RESP Borrower and nothing in the loan documents between RUS and the RESP borrower will confer upon any third party any right, benefit or remedy of any nature whatsoever. Neither RUS nor its employees makes any warranty, express or implied, including the warranties of merchantability and fitness for a particular purpose, with respect to any information, product, service, or process available from a RESP borrower or its agents.

§ 1719.11 Compliance with USDA departmental regulations, policies and other federal laws.

(a) *Equal opportunity and nondiscrimination.* RUS will ensure that equal opportunity and nondiscriminatory requirements are met in accordance with the Equal Credit Opportunity Act and 7 CFR part 15. In accordance with federal civil rights law and U.S. Department of Agriculture (USDA) civil rights regulations and policies, the USDA, its agencies, offices, and employees, and institutions participating in or administering USDA programs are prohibited from discriminating based on race, color, national origin, religion, sex, gender identity (including gender expression), sexual orientation, disability, age, marital status, family/parental status, income derived from a public assistance program, political beliefs, or reprisal or retaliation for prior civil rights activity, in any program or activity conducted or funded by USDA (not all bases apply to all programs).

(b) *Civil rights compliance.* Recipients of federal assistance hereunder must comply with the Americans with Disabilities Act of 1990, Title VI of the Civil Rights Act of 1964, and Section

504 of the Rehabilitation Act of 1973. In general, recipients should have available the Agency racial and ethnic data showing the extent to which members of minority groups are beneficiaries of federally assisted programs. The Agency will conduct compliance reviews in accordance with 7 CFR part 15. Awardees will be required to complete Form RD 400–4, "Assurance Agreement," for each federal award received.

(c) *Discrimination complaints*. Persons believing, they have been subjected to discrimination prohibited by this section may file a complaint personally, or by an authorized representative with USDA, Director, Office of Adjudication, 1400 Independence Avenue SW, Washington, DC 20250. A complaint must be filed no later than 180 days from the date of the alleged discrimination, unless the time for filing is extended by the designated officials of USDA or the Agency.

(d) *Appeal Rights*. Applicants and RESP applicants have appeal or review rights for RUS decisions made under this part.

(1) Programmatic decisions based on clear and objective statutory or regulatory requirements are not appealable; however, such decisions are reviewable for appealability by the National Appeals Division (NAD).

(2) An Applicant and a RESP applicant can appeal any RUS decision that directly and adversely impacts it. Appeals will be conducted by USDA NAD and will be handled in accordance with 7 CFR part 11.

(e) *Federal Debt and Settlement of Debt*. It is the policy of the Administrator that, whenever possible, all debt owed to the Government shall be collected in full in accordance with the terms of the borrower's loan documents. Debt owed to RUS constitutes federal debt and is subject to collection under the Debt Collection Improvement Act. RUS can use all remedies available to it to collect the debt from the borrower, including offset in accordance with part 3 of this title. In addition, it is the intent of the Administrator, notwithstanding §1717.1200(b) of this chapter, that debt settlements under this Part will be governed by the provisions set forth in 7 CFR part 1717, subpart Y or its successor Agency policies or regulations.

§ 1719.12 Reporting.

(a) *General*. RESP borrowers must file periodic performance and financial reports as provided in the loan documents.

(b) *Frequency of reporting*. Performance and financial reports will be filed semiannually for the first 10 years of the RESP loan and annually thereafter through the term of the loan. However, RUS may require additional, or more frequent, reporting when necessary to preserve the quality and integrity of the program portfolio or advance policy goals.

(c) *Reporting elements*. RUS will identify the reporting requirements, in form and substance, in the loan documents based on the RESP borrower and EE Program profile. The RESP borrower's reports to RUS will include, but will not be limited to, the following information:

(1) Number and amount of loans to qualified consumers.

(2) Types of investments in EE measures and eligible activities.

(3) EE Program portfolio performance.

(4) Evidence of compliance with Multi-Tier Action Environmental Compliance Agreement.

(5) Status and amount of Loan Loss Reserve (when applicable).

§ 1719.13 Auditing and accounting requirements.

(a) *Accounting requirements*. RESP borrowers must follow RUS accounting requirements as set forth in the loan documents.

(1) Existing RUS borrowers must continue recording and reporting transactions pursuant to the RUS Uniform Systems of Accounts—Electric, 7 CFR part 1767. Such borrowers will continue to follow the accounting and reporting requirements set forth in the previously executed loan documents for RUS outstanding loans.

(2) New and RESP only borrowers must adopt and follow a GAAP based system of accounts acceptable to RUS, as well as compliance with the requirements of 2 CFR part 200 (for RESP Awardees, the term "grant recipient"

in 2 CFR part 200 will also mean "loan recipient.")

(3) All RESP borrowers must promptly notify RUS should a state regulatory authority with jurisdiction over it require it to apply accounting methods or principles different from the ones specified in the loan documents.

(4) RUS will consider borrowers' reasonable proposals to streamline reporting and accounting requirements only when such proposals afford RUS adequate mechanisms to ensure the full and timely repayment of the loan, as determined by RUS.

(5) The Administrator may modify the accounting requirements for RESP borrowers if, in his or her judgement, it is necessary to satisfy the statutory purpose of the program, streamline procedures, or advance policy goals.

(6) Nothing in this policy shall be construed as a limitation or waiver of any other federal statute or requirement or the Administrator's authority and discretion to implement the RESP in such a way that the Government's interest is adequately preserved.

(b) *Auditing requirements.* RESP borrowers will be required to prepare and furnish to RUS, at least once during each 12-month period, a full and complete report of its financial condition, operations, and cash flows, on a comparative basis, along with a report on internal control over financial reporting and on compliance in other matters, both reports in form and substance satisfactory to RUS, audited and certified by an independent certified public accountant, satisfactory to RUS according to the requirements set forth in 7 CFR 1773.5.

(1) Audits must follow governmental auditing standards issued by the Comptroller General of the United States (GAGAS) and the provisions of 2 CFR part 200, subpart F—Audit Requirements if applicable.

(2) RESP borrowers with outstanding RUS loans will be subject to the auditing requirements set forth in their existing RUS loan documents. RUS Policy on Audits of RUS Borrowers as provided in 7 CFR part 1773 will govern audits under this paragraph.

(3) RESP borrowers must comply with all reasonable RUS requests to support ongoing monitoring efforts. The RESP borrowers must afford RUS, through their representatives, a reasonable opportunity, at all times during business hours and upon prior notice, to have access to and the right to inspect any or all books, records, accounts, invoices, contracts, leases, payrolls, timesheets, cancelled checks, statements, and other documents, electronic or paper of every kind belonging to or in possession of the RESP borrowers or in any way pertaining to its property or business, including its parents, affiliates, and subsidiaries, if any, and to make copies or extracts therefrom.

(4) The Administrator may modify the audit requirements for RESP borrowers if, in his or her judgement, it is necessary to satisfy the statutory purpose of the program or advance policy goals.

(5) Nothing in this policy shall be construed as a limitation or waiver of any other federal statute or requirement or the Administrator's authority and discretion to implement the RESP in such a way that the Government's interest is adequately preserved.

PART 1720—GUARANTEES FOR BONDS AND NOTES ISSUED FOR ELECTRIFICATION OR TELEPHONE PURPOSES

AUTHORITY: 7 U.S.C. 901 *et seq.;* 7 U.S.C. 940C.

SOURCE: 69 FR 63049, Oct. 29, 2004, unless otherwise noted.

§ 1720.1 Purpose.

This part prescribes regulations implementing a guarantee program for bonds and notes issued for electrification or telephone purposes authorized by section 313A of the Rural Electrification Act of 1936 (7 U.S.C. 940c–1).

[75 FR 42573, July 22, 2010]

§ 1720.2 Background.

The Rural Electrification Act of 1936 (the "RE Act") (7 U.S.C. 901 *et seq.*) authorizes the Secretary to guarantee and make loans to persons, corporations, States, territories, municipalities, and cooperative, non-profit, or limited-dividend associations for the purpose of furnishing or improving electric and telephone service in rural areas. Responsibility for administering electrification and telecommunications loan and guarantee programs along with other functions the Secretary deemed appropriate have been assigned to RUS under the Department of Agriculture Reorganization Act of 1994 (7 U.S.C. 6941 *et seq.*). The Administrator of RUS has been delegated responsibility for administering the programs and activities of RUS, *see* 7 CFR 1700.25. Section 6101 of the Farm Security and Rural Investment Act of 2002 (Pub. L. 107–171) (FSRIA) amended the RE Act to include a new program under section 313A entitled Guarantees for Bonds and Notes Issued for Electrification or Telephone Purposes. This measure directed the Secretary of Agriculture to promulgate regulations that carry out the Program. The Secretary published the regulations for the program in the FEDERAL REGISTER as a final rule on October 29, 2004, adding part 1720 to title 7 of the Code of Federal Regulations. Section 6106(a)(1)(A) of the Food, Conservation, and Energy Act of 2008 (Pub. L. 110–246) amended section 313A of the RE Act by replacing the level of "concurrent loans" as a factor limiting the amount of bonds and notes that could be guaranteed and inserted "for eligible electrification or telephone purposes" as the limitation on the amount of bonds and notes that can be guaranteed under section 313A up to an annual program limit of $1,000,000,000, subject to availability of funds. Section 6106(a)(1)(B) further amended section 313A of the RE Act by removing the prohibition against the recipient using an amount obtained from the reduction in funding costs as a result of a new guarantee under section 313A to reduce the interest rate charged on a new or concurrent loan.

[75 FR 42573, July 22, 2010]

§ 1720.3 Definitions.

For the purpose of this part:

Administrator means the Administrator of RUS.

Applicant means a bank or other lending institution organized as a private, not-for-profit cooperative association, or otherwise on a non-profit basis, that is applying for RUS to guarantee a bond or note under this part.

Bond Documents means the trust indenture, bond resolution, guarantee, guarantee agreement and all other instruments and documentation pertaining to the issuance of the guaranteed bonds.

Borrower means any organization that has an outstanding loan made or guaranteed by RUS for rural electrification or rural telephone under the RE Act, or that is eligible for such financing.

Concurrent Loan means a loan that a guaranteed lender extends to a borrower for up to 30 percent of the cost of an eligible electrification or telephone purpose under the RE Act, concurrently with an insured loan made by the Secretary pursuant to section 307 of the RE Act.

Eligible loan means a loan that a guaranteed lender extends to a borrower for up to 100 percent of the cost of eligible electrification or telephone purposes consistent with the RE Act.

Federal Financing Bank (FFB) means a government corporation and instrumentality of the United States of America under the general supervision of the Secretary of the Treasury.

Guarantee means the written agreement between the Secretary and a guaranteed bondholder, pursuant to which the Secretary guarantees full repayment of the principal, interest, and call premium, if any, on the guaranteed lender's guaranteed bond.

Guarantee Agreement means the written agreement between the Secretary and the guaranteed lender which sets forth the terms and conditions of the guarantee.

Guaranteed Bond means any bond, note, debenture, or other debt obligation issued by a guaranteed lender on a fixed or variable rate basis, and approved by the Secretary for a guarantee under this part.

Guaranteed Bondholder means any investor in a guaranteed bond.

Guaranteed Lender means an applicant that has been approved for a guarantee under this part.

Loan means any credit instrument that the guaranteed lender extends to a borrower for any electrification or telephone purpose eligible under the RE Act, including loans as set forth in section 4 of the RE Act for electricity transmission lines and distribution systems (excluding generating facilities) and as set forth in section 201 of the RE Act for telephone lines, facilities and systems.

Loan documents means the loan agreement and all other instruments and documentation between the guaranteed lender and the borrower evidencing the making, disbursing, securing, collecting, or otherwise administering of a loan.

Program means the guarantee program for bonds and notes issued for electrification or telephone purposes authorized by section 313A of the RE Act as amended.

Rating Agency means a bond rating agency identified by the Securities and Exchange Commission as a nationally recognized statistical rating organization.

RE Act means the Rural Electrification Act of 1936 (7 U.S.C. 901 *et seq.*) as amended.

RUS means the Rural Utilities Service, a Rural Development agency of the U.S. Department of Agriculture.

Secretary means the Secretary of Agriculture acting through the Administrator of RUS.

Subsidy Amount means the amount of budget authority sufficient to cover the estimated long-term cost to the Federal government of a guarantee, calculated on a net present value basis, excluding administrative costs and any incidental effects on government receipts or outlays, in accordance with the provisions of the Federal Credit Reform Act of 1990 (2 U.S.C. 661 et. seq.)

[69 FR 63049, Oct. 29, 2004, as amended at 75 FR 42574, July 22, 2010]

§ 1720.4 General standards.

(a) In accordance with section 313A of the RE Act, a guarantee will be issued by the Secretary only if the Secretary determines, in accordance with the requirements set forth in this part, that:

(1) The proceeds of the guaranteed bonds will be used by the guaranteed lender to make loans to borrowers for electrification or telephone purposes eligible for assistance under this chapter, or to refinance bonds or notes previously issued by the guaranteed lender for such purposes;

(2) At the time the guarantee is executed, the total principal amount of guaranteed bonds outstanding would not exceed the principal amount of outstanding eligible loans previously made by the guaranteed lender;

(3) The proceeds of the guaranteed bonds will not be used directly or indirectly to fund projects for the generation of electricity; and

(4) The guaranteed lender will not use any amounts obtained from the reduction in funding costs provided by a loan guarantee issued prior to June 18, 2008, to reduce the interest rates borrowers are paying on new or outstanding loans, other than new concurrent loans as provided in part 1710 of this chapter.

(b) During the term of the guarantee, the guaranteed lender shall:

(1) Limit cash patronage refunds, for guaranteed lenders having a credit rating below "A−" on its senior secured debt without regard to the guarantee. For such guaranteed lenders, cash patronage refunds are limited to five percent of the total patronage refund eligible. The limit on patronage refunds must be maintained until the credit rating is restored to "A−" or above. For those guaranteed lenders subject to patronage limitations, equity securities issued as part of the patronage refund shall not be redeemable in cash during the term of any part of the guarantee, and the guaranteed lender shall not issue any dividends on any class of equity securities during the term of the guarantee.

(2) Maintain sufficient collateral equal to the principal amount outstanding, for guaranteed lenders having a credit rating below "A−" on its senior secured debt without regard to the guarantee, or in the case of a lender that does not have senior secured debt, a corporate (counterparty) credit rating below "A−" without regard to the guarantee. Collateral shall be in the form of specific and identifiable unpledged securities equal to the value of the guaranteed amount. In the case of a guaranteed lender's default, the U.S. government claim shall not be subordinated to the claims of other creditors, and the indenture must provide that in the event of default, the government has first rights on the asset. Upon application and throughout the term of the guarantee, guaranteed lenders not subject to collateral pledging requirements shall identify, with the concurrence of the Secretary, specific assets to be held as collateral should the credit rating of its senior secured debt, or its corporate credit rating, as applicable, without regard to the guarantee fall below "A−." The Secretary has discretion to require collateral at any time should circumstances warrant.

(c) The final maturity of the guaranteed bonds shall not exceed 20 years.

(d) The guaranteed bonds shall be issued to the Federal Financing Bank on terms and conditions consistent with comparable government-guaranteed bonds and satisfactory to the Secretary.

(e) The Secretary shall guarantee payment son guaranteed bonds in such forms and on such terms and conditions and subject to such covenants, representations, warranties and requirements (including requirements for audits) as determined appropriate for satisfying the requirements of this part. The Secretary shall require the guaranteed lender to enter into a guarantee agreement to evidence its acceptance of the foregoing. Any guarantee issued under this part shall be made in a separate and distinct offering.

[69 FR 63049, Oct. 29, 2004, as amended at 75 FR 42574, July 22, 2010]

§ 1720.5 Eligibility criteria.

(a) To be eligible to participate in the program, a guaranteed lender must be:

(1) A bank or other lending institution organized as a private, not-for-profit cooperative association, or otherwise organized on a non-profit basis; and

(2) Able to demonstrate to the Secretary that it possesses the appropriate expertise, experience, and qualifications to make loans for electrification or telephone purposes.

(b) To be eligible to receive a guarantee, a guaranteed lender's bond must meet the following criteria:

(1) The guaranteed lender must furnish the Secretary with a certified list of the principal balances of eligible loans then outstanding and certify that such aggregate balance is at least equal to the sum of the proposed principal amount of guaranteed bonds to be issued, and any previously issued guaranteed bonds outstanding; and

(2) The guaranteed bonds to be issued by the guaranteed lender must receive an underlying investment grade rating from a Rating Agency, without regard to the guarantee;

(c) A lending institution's status as an eligible applicant does not assure that the Secretary will issue the guarantee sought in the amount or under the terms requested, or otherwise preclude the Secretary from declining to issue a guarantee.

[69 FR 63049, Oct. 29, 2004, as amended at 75 FR 42574, July 22, 2010]

§ 1720.6 Application process.

(a) Applications shall contain the following:

(1) Background and contact information on the applicant;

(2) A term sheet summarizing the proposed terms and conditions of, and the security pledged to assure the applicant's performance under, the guarantee agreement;

(3) A statement by the applicant as to how it proposes to use the proceeds of the guaranteed bonds, and the financial benefit it anticipates deriving from participating in the program;

(4) A pro-forma cash flow projection or business plan for the next five years,

demonstrating that there is reasonable assurance that the applicant will be able to repay the guaranteed bonds in accordance with their terms;

(5) Consolidated financial statements of the guaranteed lender for the previous three years that have been audited by an independent certified public accountant, including any associated notes, as well as any interim financial statements and associated notes for the current fiscal year;

(6) Evidence of having been assigned an investment grade rating on the debt obligations for which it is seeking the guarantee, without regard to the guarantee;

(7) Evidence of a credit rating, from a Rating Agency, on its senior secured debt or its corporate credit rating, as applicable, without regard to the government guarantee and satisfactory to the Secretary; and

(8) Such other application documents and submissions deemed necessary by the Secretary for the evaluation of applicants.

(b) The application process occurs as follows:

(1) The applicant submits an application to the Secretary;

(2) The application is screened by RUS pursuant to 7 CFR 1720.7(a) of this part, to ascertain its threshold eligibility for the program;

(3) RUS evaluates the application pursuant to the selection criteria set forth in 7 CFR 1720.7(b) of this part;

(4) If RUS provisionally approves the application, the applicant and RUS negotiate terms and conditions of the bond documents, and

(5) The applicant offers its guaranteed bonds, and the Secretary upon approval of the pricing, redemption provisions and other terms of the offering, executes the guarantee.

(c) If requested by the applicant at the time it files its application, the General Counsel of the Department of Agriculture shall provide the Secretary with an opinion regarding the validity and authority of a guarantee issued to the lender under section 313A of the RE Act.

[69 FR 63049, Oct. 29, 2004, as amended at 75 FR 42574, July 22, 2010]

§ 1720.7 Application evaluation.

(a) *Eligibility screening*. Each application will be reviewed by the Secretary to determine whether it is eligible under 7 CFR 1720.5, the information required under 7 CFR 1720.6 is complete and the proposed guaranteed bond complies with applicable statutes and regulations. The Secretary can at any time reject an application that fails to meet these requirements.

(b) *Evaluation*. Pursuant to paragraph (a) of this section, applications will be subject to a substantive review, on a competitive basis, by the Secretary based upon the following evaluation factors, listed in order of importance:

(1) The extent to which the proposed provisions indicate the applicant will be able to repay the guaranteed bonds;

(2) The adequacy of the proposed provisions to protect the Federal government, based upon items including, but not limited to the nature of the pledged security, the priority of the lien position, if any, pledged by the applicant, and the provision for an orderly retirement of principal such as an amortizing bond structure or an internal sinking fund;

(3) The applicant's demonstrated performance of financially sound business practices as evidenced by reports of regulators, auditors and credit rating agencies;

(4) The extent to which the applicant is subject to supervision, examination, and safety and soundness regulation by an independent federal agency;

(5) The extent of concentration of financial risk that RUS may have resulting from previous guarantees made under section 313A of the RE Act; and

(6) The extent to which providing the guarantee to the applicant will help reduce the cost and/or increase the supply of credit to rural America, or generate other economic benefits, including the amount of fee income available to be deposited into the Rural Economic Development Subaccount, maintained under section 313(b)(2)(A) of the RE Act (7 U.S.C. 940c(b)(2)(A)), after payment of the subsidy amount.

(c) *Independent Assessment*. Before a guarantee decision is made by the Secretary, the Secretary shall request that the Federal Financing Bank review the adequacy of the determination by the Rating Agency, required under § 1720.5(b)(2) as to whether the bond or note to be issued would be below investment grade without the guarantee.

(d) *Decisions by the Secretary.* The Secretary shall approve or deny applications in a timely manner as such applications are received; provided, however, that in order to facilitate competitive evaluation of applications, the Secretary may from time to time defer a decision until more than one application is pending. The Secretary may limit the number of guarantees made to a maximum of five per year, to ensure a sufficient examination is conducted of applicant requests. RUS shall notify the applicant in writing of the Secretary's approval or denial of an application. Approvals for guarantees shall be conditioned upon compliance with 7 CFR 1720.4 and 1720.6 of this part. The Secretary reserves the discretion to approve an application for an amount less than that requested.

[69 FR 63049, Oct. 29, 2004, as amended at 75 FR 42574, July 22, 2010]

§ 1720.8 Issuance of the guarantee.

(a) The following requirements must be met by the applicant prior to the endorsement of a guarantee by the Secretary.

(1) A guarantee agreement suitable in form and substance to the Secretary must be delivered.

(2) Bond documents must be executed by the applicant setting forth the legal provisions relating to the guaranteed bonds, including but not limited to payment dates, interest rates, redemption features, pledged security, additional borrowing terms including an explicit agreement to make payments even if loans made using the proceeds of such bond or note is not repaid to the lender, other financial covenants, and events of default and remedies;

(3) Prior to the issuance of the guarantee, the applicant must certify to the Secretary that the proceeds from the guaranteed bonds will be applied to fund new eligible loans under the RE Act, to refinance concurrent loans, or to refinance existing debt instruments of the guaranteed lender used to fund eligible loans;

(4) The applicant provides a certified list of eligible loans and their outstanding balances as of the date the guarantee is to be issued;

(5) Counsel to the applicant must furnish an opinion satisfactory to the Secretary as to the applicant being legally authorized to issue the guaranteed bonds and enter into the bond documents;

(6) No material adverse change occurs between the date of the application and date of execution of the guarantee;

(7) The applicant shall provide evidence of an investment grade rating from a Rating Agency for the proposed guaranteed bond without regard to the guarantee;

(8) The applicant shall provide evidence of a credit rating on its senior secured debt or its corporate credit rating, as applicable, without regard to the guarantee and satisfactory to the Secretary; and

(9) Certification by the Chairman of the Board and the Chief Executive Officer of the applicant (or other senior management acceptable to the Secretary), acknowledging the applicant's commitment to submit to the Secretary, an annual credit assessment of the applicant by a Rating Agency, an annual review and certification of the security of the government guarantee that is audited by an independent certified public accounting firm or federal banking regulator, annual consolidated financial statements audited by an independent certified public accountant each year during which the guarantee bonds are outstanding, and other such information requested by the Secretary.

(b) The Secretary shall not issue a guarantee if the applicant is unwilling or unable to satisfy all requirements.

[69 FR 63049, Oct. 29, 2004, as amended at 75 FR 42574, July 22, 2010]

§ 1720.9 Guarantee Agreement.

(a) The guaranteed lender will be required to sign a guarantee agreement with the Secretary setting forth the terms and conditions upon which the Secretary guarantees the payment of the guaranteed bonds.

(b) The guaranteed bonds shall refer to the guarantee agreement as controlling the terms of the guarantee.

(c) The guarantee agreement shall address the following matters:

(1) Definitions and principles of construction;

(2) The form of guarantee;

(3) Coverage of the guarantee;

(4) Timely demand for payment on the guarantee;

(5) Any prohibited amendments of bond documents or limitations on transfer of the guarantee;

(6) Limitation on acceleration of guaranteed bonds;

(7) Calculation and manner of paying the guarantee fee;

(8) Consequences of revocation of payment on the guaranteed bonds;

(9) Representations and warranties of the guaranteed lender;

(10) Representations and warranties for the benefit of the holder of the guaranteed bonds;

(11) Claim procedures;

(12) What constitutes a failure by the guaranteed lender to pay;

(13) Demand on RUS;

(14) Assignment to RUS;

(15) Conditions of guarantee which may include requiring the guaranteed lender to adopt measures to ensure adequate capital levels are retained to absorb losses relative to risk in the guaranteed lender's portfolio and requirements on the guaranteed lender to hold additional capital against the risk of default;

(16) Payment by RUS;

(17) RUS payment does not discharge guaranteed lender;

(18) Undertakings for the benefit of the holders of guaranteed bonds, including: notices, registration, prohibited amendments, prohibited transfers, indemnification, multiple bond issues;

(19) Governing law;

(20) Notices;

(21) Benefit of agreement;

(22) Entirety of agreement;

(23) Amendments and waivers;

(24) Counterparts;

(25) Severability, and

(26) Such other matters as the Secretary believes to be necessary or appropriate.

§ 1720.10 Fees.

(a) *Guarantee fee.* An annual fee equal to 30 basis points (0.3 percent) of the amount of the unpaid principal of the guarantee bond will be deposited into the Rural Economic Development Subaccount maintained under section 313(b)(2)(A) of the RE Act.

(b) Subject to paragraph (c) of this section, up to one-third of the 30 basis point guarantee fee may be used to fund the subsidy amount of providing guarantees, to the extent not otherwise funded through appropriation actions by Congress.

(c) Notwithstanding subsections (c) and (e)(2) of section 313A of the RE Act, the Secretary shall, with the consent of the lender and if otherwise authorized by law, adjust the schedule for payment of the annual fee, not to exceed an average of 30 basis points per year for the term of the loan, to ensure that sufficient funds are available to pay the subsidy costs for note guarantees.

§ 1720.11 Servicing.

The Secretary, or other agent of the Secretary on his or her behalf, shall have the right to service the guaranteed bond, and periodically inspect the books and accounts of the guaranteed lender to ascertain compliance with the provisions of the RE Act and the bond documents.

§ 1720.12 Reporting requirements.

(a) As long as any guaranteed bonds remain outstanding, the guaranteed lender shall provide the Secretary with the following items each year within 90 days of the guaranteed lender's fiscal year end:

(1) Consolidated financial statements and accompanying footnotes, audited by independent certified public accountants;

(2) A review and certification of the security of the government guarantee, audited by reputable, independent certified public accountants or a federal banking regulator, who in the judgment of the Secretary, has the requisite skills, knowledge, reputation, and experience to properly conduct such a review;

(3) Pro forma projection of the guaranteed lender's balance sheet, income

statement, and statement of cash flows over the ensuing five years;

(4) Credit assessment issued by a Rating Agency;

(5) Credit rating, by a Rating Agency, on its senior secured debt or its corporate credit rating, as applicable, without regard to the guarantee and satisfactory to the Secretary; and

(6) Other such information requested by the Secretary.

(b) The bond documents shall specify such bond monitoring and financial reporting requirements as deemed appropriate by the Secretary.

[69 FR 63049, Oct. 29, 2004, as amended at 75 FR 42575, July 22, 2010]

§ 1720.13 Limitations on guarantees.

In a given year the maximum amount of guaranteed bonds that the Secretary may approve will be subject to budget authority, together with receipts authority from projected fee collections from guaranteed lenders, the principal amount of outstanding eligible loans made by the guaranteed lender, and Congressionally-mandated ceilings on the total amount of credit. The Secretary may also impose other limitations as appropriate to administer this guarantee program.

[75 FR 42575, July 22, 2010]

§ 1720.14 Nature of guarantee; acceleration of guaranteed bonds.

(a) Any guarantee executed by the Secretary under this part shall be an obligation supported by the full faith and credit of the United States and incontestable except for fraud or misrepresentation of which the guaranteed bondholder had actual knowledge at the time it purchased the guaranteed bonds.

(b) Amounts due under the guarantee shall be paid within 30 days of demand by a bondholder, certifying the amount of payment then due and payable.

(c) The guarantee shall be assignable and transferable to any purchaser of guaranteed bonds as provided in the bond documents.

(d) The following actions shall constitute events of default under the terms of the guarantee agreements:

(1) The guaranteed lender failed to make a payment of principal or interest when due on the guaranteed bonds;

(2) The guaranteed bonds were issued in violation of the terms and conditions of the bond documents;

(3) The guarantee fee required by 7 CFR 1720.10 of this part, has not been paid;

(4) The guaranteed lender made a misrepresentation to the Secretary in any material respect in connection with the application, the guaranteed bonds, or the reporting requirements listed in 7 CFR 1720.12; or

(5) The guaranteed lender failed to comply with any material covenant or provision contained in the bond documents.

(e) In the event the guaranteed lender fails to cure such defaults within the notice terms and the timeframe set forth in the bond documents, the Secretary may demand that the guaranteed lender redeem the guaranteed bonds. Such redemption amount will be in an amount equal to the outstanding principal balance, accrued interest to the date of redemption, and prepayment premium, if any. To the extent the Secretary makes any payments under the guarantee, the Secretary shall be deemed the guaranteed bondholder.

(f) To the extent the Secretary makes any payments under the guarantee, the interest rate the government will charge to the guaranteed lender for the period of default shall accrue at an annual rate of the greater of 1.5 times the 91-day Treasury-Bill rate or 200 basis points (2.00%) above the rate on the guaranteed bonds.

(g) Upon guaranteed lender's event of default, under the bond documents, the Secretary shall be entitled to take such other action as is provided for by law or under the bond documents.

§ 1720.15 Equal opportunity requirements.

Executive Order 12898, "Environmental Justice." To comply with Executive Order 12898, RUS will conduct a Civil Rights Analysis for each guarantee prior to approval. Rural Development Form 2006–28, "Civil Rights Impact Analysis", will be used to document compliance in regards to environmental justice. The Civil Rights Impact Analysis will be conducted prior to application approval or a conditional commitment of guarantee.

§ 1720.16 Environmental review requirements.

Guarantees made under this subpart are subject to the environmental review requirements in accordance with 7 CFR part 1970.

[81 FR 11026, Mar. 2, 2016]

PART 1721—POST-LOAN POLICIES AND PROCEDURES FOR INSURED ELECTRIC LOANS

Subpart A—Advance of Funds

Subpart B—Extensions of Payments of Principal and Interest

AUTHORITY: 7 U.S.C. 901 *et seq.*; 1921 *et seq.*; and 6941 *et seq.*

SOURCE: 50 FR 5368, Feb. 8, 1985, unless otherwise noted. Redesignated at 64 FR 72489, Dec. 28, 1999.

Subpart A—Advance of Funds

§ 1721.1 Advances.

(a) *Purpose and amount.* With the exception of minor projects which are addressed in paragraph (b) of this section and generation projects which need to be included on a RUS Form 740c or an amendment to a RUS Form 740c, loan funds will be advanced for projects which are included in a RUS approved construction work plan (CWP), Energy Efficiency and Conservation Program work plan (EEWP), or approved amendment to either, have received written documentation of RUS concluding its environmental reviews and have complied with all Contracting and Bidding Procedures included in 7 CFR part 1726. Loan fund advances can be requested in an amount representing actual costs incurred.

(b) *Minor project.* Minor project means a project costing $100,000 or less. Such a project qualifies for advance of loan funds even though it may not have been included in an RUS-approved borrower's CWP, amendment to such CWP, or approved loan. Total advances requested shall not exceed the total loan amount. All projects for which loan fund advances are requested must be constructed to achieve purposes permitted by terms of the loan contract between the borrower and RUS.

(c) *Certification.* Pursuant to the applicable provisions of the RUS loan contract, borrowers must certify with each request for funds to be approved for advance that such funds are for projects in compliance with this section and shall also provide for those that cost in excess of $100,000, a contract or work order number as applicable and a CWP cross-reference project coded identification number. For a minor project not included in a RUS approved borrower's CWP or CWP amendment, the Borrower shall describe the project and do one of the following to satisfy RUS' environmental review requirements in accordance with 7 CFR part 1970:

(1) If applicable, state that the project is a categorical exclusion of a type described in § 1970.53 of this title; or

(2) If applicable, state that the project is a categorical exclusion of a type that normally requires the preparation of an environmental report (see § 1970.54 of this title) and then submit the environmental report with the request for funds to be approved for advance.

(d) *Noncompliance.* Where insured loan funds are found to have been advanced in noncompliance with this section, borrowers will be required to deposit the appropriate amount of the over-advance in the construction fund-trustee account and pay any accrued and unpaid interest to RUS. The Administrator will require borrowers, in order to remedy such noncompliance, to pay an additional amount equal to the interest on the funds over-advanced for the period such funds were outstanding, calculated at a rate equal to the difference between the RUS loan interest rate and the most recent rate at which RUS sold Certificates of Beneficial Ownership (CBO's). While RUS will generally permit the amount of over-advance deposited in the construction fund-trustee account to be subsequently used by the borrower for RUS approved projects, nothing in this section shall be construed to preclude RUS from exercising any rights or remedies which RUS may have pursuant to the loan contract.

[64 FR 72489, Dec. 28, 1999, as amended at 78 FR 73370, Dec. 5, 2013; 81 FR 11026, Mar. 2, 2016; 86 FR 36197, July 9, 2021]

Subpart B—Extensions of Payments of Principal and Interest

SOURCE: 67 FR 485, Jan. 4, 2002, unless otherwise noted.

§ 1721.100 Purpose.

This subpart contains RUS procedures and conditions under which Borrowers of loans made by RUS may request RUS approval for extensions for the payment of principal and interest.

§ 1721.101 General.

(a) The procedures in this subpart are intended to provide Borrowers with the flexibility to request an extension of principal and interest as authorized under section 12(a) of the RE Act and section 236 of the Disaster Relief Act of 1970 (Public Law 91–606).

(b) The total amount of interest that has been deferred, including interest on deferred principal, will be added to the principal balance, and the total amount of principal and interest that has been deferred will be reamortized over the remaining life of the applicable note beginning in the first year the deferral period ends.

(c) Payment of principal and interest will not be extended more than 5 years after such payment is due as originally scheduled. However, in cases where the extension is being granted because, at the sole discretion of the Administrator, a severe hardship has been experienced, the Administrator may grant a longer extension provided that the maturity date of any such loan does not extend to a date beyond forty (40) years from the date of the note.

[67 FR 485, Jan. 4, 2002, as amended at 68 FR 37953, June 26, 2003]

§ 1721.102 Definitions.

The definitions contained in 7 CFR 1710.2 are applicable to this subpart unless otherwise stated.

§ 1721.103 Policy.

(a) In reviewing requests for extension of payment of principal and interest, consideration shall be given to the effect of such extensions on the security of the Government's loans, and on the ability of the Borrower to achieve program objectives. It is the policy of RUS to extend the time for payment of principal and interest on the basis of findings that such extension does not impair the security and feasibility of the Government's loans and:

(1) Is essential to the effectiveness of the Borrower's operations in achieving RUS program objectives which include providing reliable, affordable electricity to RE Act beneficiaries;

(2) Is necessary to help a Borrower place its operations on a more stable financial basis and thereby provide assurance of repayment of loans within the time when payments of such loans are due under the terms of the note or notes as extended; or

(3) Is otherwise in the best interest of the Government.

(b) Extensions will be given in the minimum amount to achieve the purpose of the extension.

(c) The maximum interest rate a RUS Borrower can charge on deferments for programs relating to consumer loans, *e.g.*, energy resource conservation (ERC) program, contribution-in-aid of construction (CIAC), etc., will not be more than 300 basis points above the average interest rate on the note(s) being deferred. For example, if the RUS Borrower's average interest rate on the note(s) being deferred is 5 percent, the RUS Borrower can charge a maximum interest rate of 8 percent.

[67 FR 485, Jan. 4, 2002, as amended at 68 FR 37953, June 26, 2003]

§ 1721.104 Eligible purposes.

(a) *Deferments for financial hardship.* (1) In cases of financial hardship, a Borrower may request that RUS defer principal or interest or both. RUS will consider whether the deferral will help a Borrower place its operations on a more stable financial basis and thereby provide assurance of repayment of loans within the time when payment of such loans are due under the terms of the note or notes as extended.

(2) RUS will determine whether a Borrower qualifies for the deferment on a case-by-case basis, considering such factors as the following:

(i) Substantial unreimbursed or uninsured expenses relating to storm damage;

(ii) Loss of large power load (as defined in § 1710.7(c)(6)(ii) of this chapter, Large retail power contracts); or

(iii) Substantial loss of consumers or load due to hostile annexations and condemnations, without adequate compensation.

(b) *Deferments for energy resource conservation (ERC) loans.* (1) A Borrower may request that RUS defer principal payments to make funds available to the Borrower's consumers to conserve energy. Amounts deferred under this program can be used to cover the cost of labor and materials for the following energy conservation measures:

(i) Caulking;

(ii) Weather-stripping;

(iii) Heat pump systems (including water source heat pumps);

(iv) Heat pumps, water heaters, and central heating or central air conditioning system replacements or modifications, which reduce energy consumption;

(v) Ceiling insulation;

(vi) Wall insulation;

(vii) Floor insulation;

(viii) Duct insulation;

(ix) Pipe insulation;

(x) Water heater insulation;

(xi) Storm windows;

(xii) Thermal windows;

(xiii) Storm or thermal doors;

(xiv) Electric system coordinated customer-owned devices that reduce the maximum kilowatt demand on the electric system;

(xv) Clock thermostats; or

(xvi) Attic ventilation fans.

(2) ERC loans will be amortized over not more than 84 months, without penalty for prepayment of principal.

(c) *Deferments for renewable energy projects.* (1) A Borrower may request that RUS defer principal payments to enable the Borrower to finance renewable energy projects. Amounts deferred under this program can be used to cover costs to install all or part of a renewable energy system including, without limitation:

(i) Energy conversion technology;

(ii) Electric power system interfaces;

(iii) Delivery equipment;

(iv) Control equipment; and

(v) Energy consuming devices.

(2) A Borrower may request that RUS defer principal payments for the purpose of enabling the Borrower to provide its consumers with loans to install all or part of customer-owned renewable energy systems up to 5kW.

(3) A renewable energy system is defined in § 1710.2 of this chapter.

(4) For the purpose of this subpart, a renewable energy project consists of one or more renewable energy systems.

(d) *Deferments for distributed generation projects.* (1) A Borrower may request that RUS defer principal payments to enable the Borrower to finance distributed generation projects. Amounts deferred under this program can be used to cover costs to install all or part of a distributed generation system that:

(i) The Borrower will own and operate, or

(ii) The consumer owns, provided the system owned by the consumer does not exceed 5KW.

(2) A distributed generation project may include one or more individual systems.

(e) *Deferments for contributions-in-aid of construction.* (1) A Borrower may request RUS to defer principal payments to enable the Borrower to make funds available to new full time residential consumers to assist them in paying their share of the construction costs (contribution-in-aid of construction) needed to connect them to the Borrower's system.

(2) Amounts available for this purpose will be limited to the amount of the construction costs that are in excess of the average cost per residential

consumer incurred by the Borrower to connect new full time residential consumers during the last calendar year for which data are available. The average cost per residential consumer is the total cost incurred by the Borrower and will not be reduced by the amounts received as a contribution-in-aid of construction.

[67 FR 485, Jan. 4, 2002, as amended at 68 FR 37954, June 26, 2003]

§ 1721.105 Application documents.

(a) *Deferments for financial hardship.* A Borrower requesting a section 12 deferment because of financial hardship must submit the following:

(1) A summary of the financial position of the Borrower, based on the latest information available (usually less than 60 days old).

(2) A copy of the board resolution requesting an extension due to financial hardship.

(3) A 10-year financial forecast of revenues and expenses on a cash basis, by year, for the period of the extension and 5 years beyond to establish that the remaining payments can be made as rescheduled.

(4) A listing of notes or portions of notes to be extended, the effective date for the beginning of the extension, and the length of the extension.

(5) A narrative description of the nature and cause of the hardship and the strategy that will be instituted to mitigate or eliminate the effects of the hardship.

(b) *Deferments for energy resource conservation loans.* A Borrower requesting principle deferments for an ERC loan program must submit a letter from the Borrower's General Manager requesting an extension of principle payments for the purpose of offering an ERC loan program to its members and describing the details of the program.

(c) *Deferments for renewable energy projects.* A Borrower requesting principle deferments for its renewable energy project must submit a letter from the Borrower's General Manager requesting an extension of principle payments for the purpose of offering an ERC loan program to its members and describing the details of the program.

(d) *Deferments for distributed generation projects.* A Borrower requesting principle deferments for distributed generation projects must submit a letter from the Borrower's General Manager requesting an extension of principle payments for the purpose of offering an ERC loan program to its members and describing the details of the program and approval is also subject to any applicable terms and conditions of the Borrower's loan contract, mortgage, or indenture.

(e) *Deferments for contribution-in-aid of construction.* A Borrower requesting principle deferments for contribution-in-aid of construction must submit the following:

(1) A letter from the Borrower's General Manager requesting an extension of principle payments for the purpose of offering an ERC loan program to its members and describing the details of the program.

(2) A summary of the calculations used to determine the average cost per residential customer. (See § 1721.104(e)(2)).

[67 FR 485, Jan. 4, 2002, as amended at 68 FR 37954, June 26, 2003; 84 FR 32616, July 9, 2019]

§ 1721.106 Repayment of deferred payments.

(a) *Deferments relating to financial hardship.* The total amount of interest that has been deferred, including interest on deferred principal, will be added to the principal balance, and the total amount of principal and interest that has been deferred will be reamortized over the remaining life of the applicable note beginning in the first year the deferral period ends. For example: the amount of interest deferred in years 2003, 2004, 2005, 2006, and 2007, will be added to the principal balance and reamortized over the life of the applicable note for repayment starting in year 2008.

(b) *Deferments relating to the ERC loan program, renewable energy project(s), distributed generation project(s), and the contribution(s)-in-aid of construction.* An extension agreement is for a term of two (2) years. The installment will be recalculated each time the Borrower defers the payment of principal and recognition of the deferred amount will begin with the next payment. For example: the amount deferred in the October payment will be reamortized over a 84 month period starting with the next payment (November if paying on a monthly basis). When a Borrower defers principal under any of these programs the scheduled payment on the account will increase by an amount sufficient to pay off the deferred amount, with interest, by the date specified in the agreement (usually 84 months (28 quarters)).

[67 FR 485, Jan. 4, 2002, as amended at 68 FR 37954, June 26, 2003]

§ 1721.107 Agreement.

After approval of the Borrower's request for a deferment of principal and interest, an extension agreement, containing the terms of the extension, together with associated materials, will be prepared and forwarded to the Borrower by RUS. The extension agreement will then be executed and returned to RUS by the Borrower.

§ 1721.108 Commencement of the deferment.

The deferment of principal and interest will not begin until the extension agreement and other supporting materials, in form and substance satisfactory to RUS, have been executed by the Borrower and returned to RUS. Examples of other supporting materials are items such as approving legal opinions from the Borrower's attorney and approvals from the relevant regulatory body for extending the maturity of existing debt and for the additional debt service payment incurred.

§ 1721.109 OMB control number.

The information collection requirements in this part are approved by the Office of Management and Budget and assigned OMB control number 0572–0123.

PART 1724—ELECTRIC ENGINEERING, ARCHITECTURAL SERVICES AND DESIGN POLICIES AND PROCEDURES

Subpart A—General

Subpart B—Architectural Services

Subpart C—Engineering Services

Subpart D—Electric System Planning

Subpart E—Electric System Design

Subpart F—RUS Contract Forms

AUTHORITY: 7 U.S.C. 901 *et seq.*, 1921 *et seq.*, 6941 *et seq.*

SOURCE: 63 FR 35314, June 29, 1998, unless otherwise noted.

Subpart A—General

§ 1724.1 Introduction.

(a) The policies, procedures and requirements in this part implement certain provisions of the standard form of loan documents between the Rural Utilities Service (RUS) and its electric borrowers.

(b) All borrowers, regardless of the source of financing, shall comply with RUS' requirements with respect to design, construction standards, and the use of RUS accepted material on their electric systems.

(c) Borrowers are required to use RUS contract forms only if the facilities are financed by RUS. Borrowers have three options:

(1) Submit the actual contract used for review and approval;

(2) Submit a certification that the required contract was used for the electric project or;

(3) Submit a certification that the contract was not used but the essential and identical provisions specifically listed in the certification were used in the contract for constructing the electric facilities.

[63 FR 35314, June 29, 1998, as amended at 84 FR 32617, July 9, 2019]

§ 1724.2 Waivers.

The Administrator may waive, for good cause on a case-by-case basis, requirements and procedures of this part.

§ 1724.3 Definitions.

Terms used in this part have the meanings set forth in § 1710.2 of this chapter. References to specific RUS forms and other RUS documents, and to specific sections or lines of such forms and documents, shall include the corresponding forms, documents, sections and lines in any subsequent revisions of these forms and documents. In addition to the terms defined in § 1710.2 of this chapter, the following terms have the following meanings for the purposes of this part:

Architect means a registered or licensed person employed by the borrower to provide architectural services for a project and duly authorized assistants and representatives.

Engineer means a registered or licensed person, who may be a staff employee or an outside consultant, to provide engineering services and duly authorized assistants and representatives.

Force account construction means construction performed by the borrower's employees.

GPO means Government Printing Office.

NESC means the National Electrical Safety Code.

RE Act means the Rural Electrification Act of 1936 as amended (7 U.S.C. 901 *et seq.*).

Repowering means replacement of the steam generator or the prime mover or both at a generating plant.

RUS means Rural Utilities Service.

RUS approval means written approval by the Administrator or a representative with delegated authority. RUS approval must be in writing, except in emergency situations where RUS approval may be given orally followed by a confirming letter.

RUS financed means financed or funded wholly or in part by a loan made or guaranteed by RUS, including concurrent supplemental loans required by § 1710.110 of this chapter, loans to reimburse funds already expended by the borrower, and loans to replace interim financing.

[63 FR 35314, June 29, 1998, as amended at 63 FR 58284, Oct. 30, 1998]

§ 1724.4 Qualifications.

The borrower shall ensure that:

(a) All selected architects and engineers meet the applicable registration and licensing requirements of the States in which the facilities will be located;

(b) All selected architects and engineers are familiar with RUS standards and requirements; and

(c) All selected architects and engineers have had satisfactory experience with comparable work.

§ 1724.5 Submission of documents to RUS.

(a) *Where to send documents.* Documents required to be submitted to RUS under this part are to be sent to the Office of Loan Origination & Approval.

(b) *Contracts requiring RUS approval.* The borrower shall submit to RUS three copies of each contract that is subject to RUS approval under subparts B and C of this part. At least one copy of each contract must be an original signed in ink (*i.e.*, no facsimile signature).

(c) *Contract amendments requiring RUS approval.* The borrower shall submit to RUS three copies of each contract amendment (at least one copy of which must be an original signed in ink) which is subject to RUS approval.

[84 FR 32617, July 9, 2019]

§ 1724.6 Insurance requirements.

(a) Borrowers shall ensure that all architects and engineers working under contract with the borrower have insurance coverage as required by part 1788 of this chapter.

(b) Borrowers shall also ensure that all architects and engineers working under contract with the borrower have insurance coverage for Errors and Omissions (Professional Liability Insurance) in an amount at least as large as the amount of the architectural or engineering services contract but not less than $500,000.

§ 1724.7 Debarment and suspension.

Borrowers shall comply with the requirements on debarment and suspension in connection with procurement activities set forth in 2 CFR part 180, as adopted by USDA through 2 CFR part 417, particularly with respect to lower tier transactions, *e.g.*, procurement contracts for goods or services.

[79 FR 76003, Dec. 19, 2014]

§ 1724.8 Restrictions on lobbying.

Borrowers shall comply with the restrictions and requirements in connection with procurement activities as set forth in 2 CFR part 418.

[79 FR 76003, Dec. 19, 2014]

§ 1724.9 Environmental review requirements.

Borrowers must comply with the environmental review requirements in accordance with 7 CFR part 1970.

[81 FR 11027, Mar. 2, 2016]

§ 1724.10 Standard forms of contracts for borrowers.

The standard loan agreement between RUS and its borrowers provides that, in accordance with applicable RUS regulations in this chapter, the borrower shall use standard forms of contracts promulgated by RUS for construction, procurement, engineering services, and architectural services financed by a loan made or guaranteed by RUS. This part implements these provisions of the RUS loan agreement. Subparts A through E of this part prescribe when and how borrowers are required to use RUS standard forms of contracts for engineering and architectural services. Subpart F of this part prescribes the procedures that RUS follows in promulgating standard contract forms and identifies those contract forms that borrowers are required to use for engineering and architectural services.

[63 FR 58284, Oct. 30, 1998]

§§ 1724.11–1724.19 [Reserved]

Subpart B—Architectural Services

§ 1724.20 Borrowers' requirements—architectural services.

The provisions of this section apply to all borrower electric system facilities regardless of the source of financing.

(a) Each borrower shall select a qualified architect to perform the architectural services required for the design and construction management of headquarters facilities. The selection of the architect is not subject to RUS approval unless specifically required by RUS on a case by case basis. Architect's qualification information need not be submitted to RUS unless specifically requested by RUS on a case by case basis.

(b) The architect retained by the borrower shall not be an employee of the building supplier or contractor, except in cases where the building is prefabricated and pre-engineered.

(c) The architect's duties are those specified under the Architectural Services Contract and under subpart E of this part, and, as applicable, those duties assigned to the "engineer" for competitive procurement procedures in part 1726 of this chapter.

(d) If the facilities are RUS financed, the borrower shall submit or require the architect to submit one copy of each construction progress report to RUS upon request.

(e) Additional information concerning RUS requirements for electric borrowers' headquarters facilities are set forth in subpart E of this part. See also RUS Bulletin 1724E–400, Guide to Presentation of Building Plans and Specifications, for additional guidance. This bulletin is available from Program Development and Regulatory Analysis, Rural Utilities Service, U.S. Department of Agriculture, Stop 1522, 1400 Independence Ave., SW., Washington, DC 20250–1522.

§ 1724.21 Architectural services contracts.

The provisions of this section apply only to RUS financed electric system facilities.

(a) RUS Form 220, Architectural Services Contract, may be used by electric borrowers when obtaining architectural services.

(b) The borrower shall ensure that the architect furnishes or obtains all architectural services related to the design and construction management of the facilities.

(c) Reasonable modifications or additions to the terms and conditions in the RUS contract form may be made to define the exact services needed for a specific undertaking. Such modifications or additions shall not relieve the architect or the borrower of the basic responsibilities required by the RUS contract form, and shall not alter any terms and conditions required by law. All substantive changes must be approved by RUS prior to execution of the contract.

(d) Architectural services contracts are not subject to RUS approval and need not be submitted to RUS unless specifically requested by RUS on a case by case basis.

(e) *Closeout.* Upon completion of all services and obligations required under each architectural services contract, including, but not limited to, submission of final documents, the borrower must closeout that contract. The borrower shall obtain from the architect a final statement of cost, which must be supported by detailed information as appropriate. For example, out-of-pocket expense and per diem types of compensation should be listed separately with labor, transportation, etc., itemized for each service involving these types of compensation. RUS Form 284, Final Statement of Cost for Architectural Service, may be used. All computations of the compensation must be made in accordance with the terms of the architectural services contract. Closeout documents need not be submitted to RUS unless specifically requested by RUS on a case by case basis.

[63 FR 35314, June 29, 1998, as amended at 84 FR 32617, July 9, 2019]

§§ 1724.22–1724.29 [Reserved]

Subpart C—Engineering Services

§ 1724.30 Borrowers' requirements—engineering services.

The provisions of this section apply to all borrower electric system facilities regardless of the source of financing.

(a) Each borrower shall select one or more qualified persons to perform the engineering services involved in the planning (including the development of an EE Program eligible for financing pursuant to subpart H of part 1710 of this chapter, design, and construction management of the system.

(b) Each borrower shall retain or employ one or more qualified engineers to inspect and certify all new construction in accordance with § 1724.32. The engineer must not be the borrower's manager.

(c) The selection of the engineer is not subject to RUS approval unless specifically required by RUS on a case by case basis. Engineer's qualification information need not be submitted to RUS unless specifically requested by RUS on a case by case basis.

(d) The engineer's duties are specified under the Engineering Services Contract and under part 1726 of this chapter. The borrower shall ensure that the engineer executes all certificates and other instruments pertaining to the engineering details required by RUS.

(e) Additional requirements related to appropriate seismic safety measures are contained in part 1792, subpart C, of this chapter, Seismic Safety of Federally Assisted New Building Construction.

(f) If the facilities are RUS financed, the borrower shall submit or require the engineer to submit one copy of each construction progress report to RUS upon RUS' request.

[63 FR 35314, June 29, 1998, as amended at 78 FR 73371, Dec. 5, 2013]

§ 1724.31 Engineering services contracts.

The provisions of this section apply only to RUS financed electric system facilities.

(a) RUS contract forms for engineering services shall be used. Reasonable modifications or additions to the terms and conditions in the RUS contract form may be made to define the exact services needed for a specific undertaking. Any such modifications or additions shall not relieve the engineer or the borrower of the basic responsibilities required by the RUS contract form, and shall not alter any terms and conditions required by law. All substantive changes to the RUS contract form shall be approved by RUS prior to execution of the contract.

(b) RUS Form 236, Engineering Service Contract—Electric System Design and Construction, may be used for all distribution, transmission, substation, and communications and control facilities. These contracts are not subject to RUS approval and need not be submitted to RUS unless specifically requested by RUS on a case by case basis.

(c) RUS Form 211, Engineering Service Contract for the Design and Construction of a Generating Plant, shall be used for all new generating units and repowering of existing units. These contracts require RUS approval.

(d) Any amendments to RUS approved engineering services contracts require RUS approval.

(e) *Closeout.* Upon completion of all services and obligations required under each engineering services contract, including, but not limited to, submission of final documents, the borrower must closeout the contract. The borrower shall obtain from the engineer a completed final statement of engineering fees, which must be supported by detailed information as appropriate. RUS Form 234, Final Statement of Engineering Fee, may be used. All computations of the compensation shall be made in accordance with the terms of the engineering services contract. Closeout documents need not be submitted to RUS unless specifically requested by RUS on a case by case basis.

[63 FR 35314, June 29, 1998, as amended at 84 FR 32617, July 9, 2019]

§ 1724.32 Inspection and certification of work order construction.

The provisions of this section apply to all borrower electric system facilities regardless of the source of financing.

(a) The borrower shall ensure that all field inspection and related services are performed within 6 months of the completion of construction, and are performed by a licensed engineer, except that a subordinate of the licensed engineer may make the inspection, provided the following conditions are met:

(1) The inspection by the subordinate is satisfactory to the borrower;

(2) This practice is acceptable under applicable requirements of the States in which the facilities are located;

(3) The subordinate is experienced in making such inspections;

(4) The name of the person making the inspection is included in the certification; and

(5) The licensed engineer signs such certification which appears on the inventory of work orders.

(b) The inspection shall include a representative and sufficient amount of construction listed on each RUS Form 219, Inventory of Work Orders (or comparable form), being inspected to assure the engineer that the construction is acceptable. Each work order that was field inspected shall be indicated on RUS Form 219 (or comparable form.) The inspection services shall include, but not be limited to, the following:

(1) Determination that construction conforms to RUS specifications and standards and to the requirements of the National Electrical Safety Code (NESC), State codes, and local codes;

(2) Determination that the staking sheets or as-built drawings represent the construction completed and inspected;

(3) Preparation of a list of construction clean-up notes and staking sheet discrepancies to be furnished to the owner to permit correction of construction, staking sheets, other records, and work order inventories;

(4) Reinspection of construction corrected as a result of the engineer's report;

(5) Noting, initialing, and dating the staking or structure sheets or as-built drawings and noting the corresponding work order entry for line construction; and

(6) Noting, initialing, and dating the as-built drawings or sketches for generating plants, substations, and other major facilities.

(c) *Certification.* (1) The following certification must appear on all inventories of work orders:

I hereby certify that sufficient inspection has been made of the construction reported by this inventory to give me reasonable assurance that the construction complies with applicable specifications and standards and meets appropriate code requirements as to strength and safety. This certification is in accordance with acceptable engineering practice.

(2) A certification must also include the name of the inspector, name of the firm, signature of the licensed engineer, the engineer's State license number, and the date of signature.

§§ 1724.33–1724.39 [Reserved]

Subpart D—Electric System Planning

§ 1724.40 General.

Borrowers shall have ongoing, integrated planning to determine their short-term and long-term needs for plant additions, improvements, replacements, and retirements for their electric systems. The primary components of the planning system consist of long-range engineering plans and construction work plans. Long-range engineering plans identify plant investments required over a long-range period, 10 years or more. Construction work plans specify and document plant requirements for a shorter term, 2 to 4 years. Long-range engineering plans and construction work plans shall be in accordance with part 1710, subpart F, of this chapter. See also RUS Bulletins 1724D–101A, Electric System Long-Range Planning Guide, and 1724D–101B, System Planning Guide, Construction Work Plans, for additional guidance. These bulletins are available from Program Development and Regulatory Analysis, Rural Utilities Service, U.S. Department of Agriculture, Stop 1522, 1400 Independence Ave., SW., Washington, DC 20250–1522.

§§ 1724.41–1724.49 [Reserved]

Subpart E—Electric System Design

§ 1724.50 Compliance with National Electrical Safety Code (NESC).

The provisions of this section apply to all borrower electric system facilities regardless of the source of financing.

(a) A borrower shall ensure that its electric system, including all electric distribution, transmission, and generating facilities, is designed, constructed, operated, and maintained in accordance with all applicable provisions of the most current and accepted criteria of the National Electrical Safety Code (NESC) and all applicable and current electrical and safety requirements of any State or local governmental entity. Copies of the NESC may be obtained from the Institute of Electrical and Electronic Engineers, Inc., 445 Hoes Lane, Piscataway, NJ 08855. This requirement applies to the borrower's electric system regardless of the source of financing.

(b) Any electrical standard requirements established by RUS are in addition to, and not in substitution for or a modification of, the most current and accepted criteria of the NESC and any applicable electrical or safety requirements of any State or local governmental entity.

(c) Overhead distribution circuits shall be constructed with not less than the Grade C strength requirements as described in Section 26, Strength Requirements, of the NESC when subjected to the loads specified in NESC Section 25, Loadings for Grades B and

C. Overhead transmission circuits shall be constructed with not less than the Grade B strength requirements as described in NESC Section 26.

§ 1724.51 Design requirements.

The provisions of this section apply to all borrower electric system facilities regardless of the source of financing.

(a) *Distribution.* All distribution facilities must conform to the applicable RUS construction standards and utilize RUS accepted materials.

(b) *Transmission lines.* (1) All transmission line design data must be approved by RUS or a licensed professional engineer may certify that the design data, plans and profiles drawings for the electric system facilities meets all applicable RUS electric design requirements, specifications, local, state and national requirements and that RUS listed materials were used.

(2) Design data consists of all significant design features, including, but not limited to, transmission line design data summary, general description of terrain, right-of-way calculations, discussion concerning conductor and structure selection, conductor sag and tension information, design clearances, span limitations due to clearances, galloping or conductor separation, design loads, structure strength limitations, insulator selection and design, guying requirements, and vibration considerations. For lines composed of steel or concrete poles, or steel towers, in which load information will be used to purchase the structures, the design data shall also include loading trees, structure configuration and selection, and a discussion concerning foundation selection.

(3) Line design data for uprating transmission lines to higher voltage levels or capacity must be approved by RUS.

(4) Transmission line design data which has received RUS approval in connection with a previous transmission line construction project for a particular borrower is considered approved by RUS for that borrower, provided that:

(i) The conditions on the project fall within the design data previously approved; and

(ii) No significant NESC revisions have occurred.

(c) *Substations.* (1) All substation design data must be approved by RUS or a licensed professional engineer may certify that the design data, plans and profiles drawings for the electric system facilities meets all applicable RUS electric design requirements, specifications, local, state and national requirements and that RUS listed materials were used.

(2) Design data consists of all significant design features, including, but not limited to, a discussion of site considerations, oil spill prevention measures, design considerations covering voltage, capacity, shielding, clearances, number of low and high voltage phases, major equipment, foundation design parameters, design loads for line support structures and the control house, seismic considerations, corrosion, grounding, protective relaying, and AC and DC auxiliary systems. Reference to applicable safety codes and construction standards are also to be included.

(3) Substation design data which has received RUS approval in connection with a previous substation construction project for a particular borrower is considered approved by RUS for that borrower, provided that:

(i) The conditions on the project fall within the design data previously approved; and

(ii) No significant NESC revisions have occurred.

(d) *Generating facilities.* (1) This section covers all portions of a generating plant including plant buildings, the generator step-up transformer, and the transmission switchyard at a generating plant. Warehouses and equipment service buildings not associated with generation plants are covered under paragraph (e) of this section. Generation plant buildings must meet the requirements of paragraph (e)(1) of this section.

(2) For all new generation units and for all repowering projects, the design outline shall be approved by RUS, unless RUS determines that a design outline is not needed for a particular project.

(3) The design outline will include all significant design criteria. During the early stages of the project, RUS will, in consultation with the borrower and its consulting engineer, identify the specific items which are to be included in the design outline.

(e) *Headquarters*—(1) *Applicable laws.* The design and construction of headquarters facilities shall comply with all applicable Federal, State, and local laws and regulations, including, but not limited to:

(i) Section 504 of the Rehabilitation Act of 1973, (29 U.S.C. 794), which states that no qualified individual with a handicap shall, solely by reason of their handicap, be excluded from participation in, be denied the benefits of, or be subject to discrimination under any program or activity receiving Federal financial assistance. The Uniform Federal Accessibility Standards (41 CFR part 101–19, subpart 101–19.6, appendix A) are the applicable standards for all new or altered borrower buildings, regardless of the source of financing.

(ii) The Architectural Barriers Act of 1968 (42 U.S.C. 4151), which requires that buildings financed with Federal funds are designed and constructed to be accessible to the physically handicapped.

(iii) The Earthquake Hazards Reduction Act of 1977 (42 U.S.C. 7701 *et seq.*), and Executive Order 12699, Seismic Safety of Federal and Federally Assisted or Regulated New Building Construction (3 CFR 1990 Comp., p. 269). Appropriate seismic safety provisions are required for new buildings for which RUS provides financial assistance. (See part 1792, subpart C, of this chapter.)

(2) The borrower shall provide evidence, satisfactory in form and substance to the Administrator, that each building will be designed and built in compliance with all Federal, State, and local requirements.

(f) *Communications and control.* (1) This section covers microwave and powerline carrier communications systems, load control, and supervisory control and data acquisition (SCADA) systems.

(2) The performance considerations for a new or replacement master system must be approved by RUS. A master system includes the main controller and related equipment at the main control point. Performance considerations include all major system features and their justification, including, but not limited to, the objectives of the system, the types of parameters to be controlled or monitored, the communication media, alternatives considered, and provisions for future needs.

[63 FR 35314, June 29, 1998, as amended at 84 FR 32617, July 9, 2019]

§ 1724.52 Permitted deviations from RUS construction standards.

The provisions of this section apply to all borrower electric system facilities regardless of the source of financing.

(a) *Structures for raptor protection.* (1) RUS standard distribution line structures may not have the extra measure of protection needed in areas frequented by eagles and other large raptors to protect such birds from electric shock due to physical contact with energized wires. Where raptor protection in the design of overhead line structures is required by RUS; a Federal, State or local authority with permit or license authority over the proposed construction; or where the borrower voluntarily elects to comply with the recommendations of the U.S. Fish and Wildlife Service or State wildlife agency, borrowers are permitted to deviate from RUS construction standards, provided:

(i) Structures are designed and constructed in accordance with "Suggested Practices for Raptor Protection on Powerlines: The State of the Art in 1996" (Suggested Practices for Raptor Protection); and,

(ii) Structures are in accordance with the NESC and applicable State and local regulations.

(2) Any deviation from the RUS construction standards for the purpose of raptor protection, which is not in accordance with the Suggested Practices for Raptor Protection, must be approved by RUS prior to construction. "Suggested Practices for Raptor Protection on Powerlines: The State of the Art in 1996," published by the Edison Electric Institute/Raptor Research Foundation, is hereby incorporated by

reference. This incorporation by reference is approved by the Director of the Office of the Federal Register in accordance with 5 U.S.C. 552(a) and 1 CFR part 51. Copies of this publication may be obtained from the Raptor Research Foundation, Inc., c/o Jim Fitzpatrick, Treasurer, Carpenter Nature Center, 12805 St. Croix Trail South, Hastings, Minnesota 55033. It is also available for inspection during normal business hours at RUS, Electric Staff Division, 1400 Independence Avenue, SW., Washington, DC, Room 1246-S, and at the National Archives and Records Administration (NARA). For information on the availability of this material at NARA, call 202-741-6030, or go to: *http://www.archives.gov/federal_register/code_of_federal_regulations/ibr_locations.html.*

(b) *Transformer neutral connections.* Where it is necessary to separate the primary and secondary neutrals to provide the required electric service to a consumer, the RUS standard transformer secondary neutral connections may be modified in accordance with Rule 97D2 of the NESC.

(c) *Lowering of neutral conductor on overhead distribution lines.* (1) It is permissible to lower the neutral attachment on standard construction pole-top assemblies an additional distance not exceeding two feet (0.6 m) for the purpose of economically meeting the clearance requirements of the NESC.

(2) It is permissible to lower the transformer and associated neutral attachment up to two feet (0.6 m) to provide adequate clearance between the cutouts and single-phase, conventional distribution transformers.

(3) It is permissible to lower the neutral attachment on standard construction pole-top assemblies an additional distance of up to six feet (2 m) for the purpose of performing construction and future line maintenance on these assemblies from bucket trucks designed for such work.

[63 FR 35314, June 29, 1998, as amended at 69 FR 18803, Apr. 9, 2004]

§ 1724.53 Preparation of plans and specifications.

The provisions of this section apply to all borrower electric system facilities regardless of the source of financing.

(a) *General.* (1) The borrower (acting through the engineer, if applicable) shall prepare plans and specifications that adequately represent the construction to be performed.

(2) Plans and specifications for distribution, transmission, or generating facilities must be based on a construction work plan (as amended, if applicable), engineering study or construction program which has been approved by RUS if financing for the facilities will at any time be requested from RUS.

(b) *Composition of plans and specifications package.* (1) Whether built by force account or contract, each set of plans and specifications must include:

(i) *Distribution lines.* Specifications and drawings, staking sheets, key map and appropriate detail maps;

(ii) *Transmission lines.* Specifications and drawings, transmission line design data manual, vicinity maps of the project, a one-line diagram, and plan and profile sheets;

(iii) *Substations.* Specifications and drawings, including a one-line diagram, plot and foundation plan, grounding plan, and plans and elevations of structure and equipment, as well as all other necessary construction drawings, in sufficient detail to show phase spacing and ground clearances of live parts;

(iv) *Headquarters.* Specifications and drawings, including:

(A) A plot plan showing the location of the proposed building plus paving and site development;

(B) A one line drawing (floor plan and elevation view), to scale, of the proposed building with overall dimensions shown; and

(C) An outline specification including materials to be used (type of frame, exterior finish, foundation, insulation, etc.); and

(v) *Other facilities (e.g., generation and communications and control facilities).* Specifications and drawings, as necessary and in sufficient detail to accurately define the scope and quality of work required.

(2) For contract work, the appropriate standard RUS construction contract form shall be used as required by part 1726 of this chapter.

§ 1724.54 Requirements for RUS approval of plans and specifications.

The provisions of this section apply only to RUS financed electric system facilities.

(a) For any contract subject to RUS approval in accordance with part 1726 of this chapter, the borrower shall obtain RUS approval of the plans and specifications, as part of the proposed bid package, prior to requesting bids. RUS may require approval of other plans and specifications on a case by case basis.

(b) *Distribution lines.* RUS approval of the plans and specifications for distribution line construction is not required if standard RUS drawings, specifications, RUS accepted material, and standard RUS contract forms (as required by part 1726 of this chapter) are used. Drawings, plans and specifications for nonstandard distribution construction must be submitted to RUS and receive approval prior to requesting bids on contracts or commencement of force account construction.

(c) *Transmission lines.* (1) Plans and specifications for transmission construction projects which are not based on RUS approved line design data or do not use RUS standard structures must receive RUS design approval or RUS certification approval prior to requesting bids on contracts or commencement of force account construction.

(2) Unless RUS approval is required by paragraph (a) of this section, plans and specifications for transmission construction which use previously approved design data and standard structures do not require RUS approval. Plans and specifications for related work, such as right-of-way clearing, equipment, and materials, do not require RUS approval unless required by paragraph (a) of this section.

(d) *Substations.* (1)(i) Plans and specifications for all new substations must receive RUS design approval or RUS certification approval prior to requesting bids on contracts or commencement of force account construction, unless:

(A) The substation design has been previously approved by RUS; and

(B) No significant NESC revisions have occurred.

(ii) The borrower shall notify RUS in writing that a previously approved design will be used, including identification of the previously approved design.

(2) Unless RUS approval is required by paragraph (a) of this section, plans and specifications for substation modifications and for substations using previously approved designs do not require RUS approval.

(e) *Generation facilities.* (1) This paragraph (e) covers all portions of a generating plant including plant buildings, the generator step-up transformer, and the transmission switchyard at a generating plant. Warehouses and equipment service buildings not associated with generation plants are covered under paragraph (f) of this section.

(2) The borrower shall obtain RUS approval, prior to issuing invitations to bid, of the terms and conditions for all generating plant equipment or construction contracts which will cost $5,000,000 or more. Unless RUS approval is required by paragraph (a) of this section, plans and specifications for generating plant equipment and construction do not require RUS approval.

(f) *Headquarters buildings.* (1) This paragraph (f) covers office buildings, warehouses, and equipment service buildings. Generating plant buildings are covered under paragraph (e) of this section.

(2) Unless RUS approval is required by paragraph (a) of this section, plans and specifications for headquarters buildings do not require RUS approval. The borrower shall submit two copies of RUS Form 740g, Application for Headquarters Facilities. This form is available from Program Development and Regulatory Analysis, Rural Utilities Service, United States Department of Agriculture, Stop 1522, 1400 Independence Ave., SW., Washington, DC 20250–1522. The application must show floor area and estimated cost breakdown between office building space and space for equipment warehousing and service facilities, and include a one line drawing (floor plan and elevation view), to scale, of the proposed building with overall dimensions shown. The information concerning the planned building may be included in the borrower's construction work plan in lieu of submitting it with the application.

(See 7 CFR part 1710, subpart F.) Prior to issuing the plans and specifications for bid, the borrower shall also submit to RUS a statement, signed by the architect or engineer, that the building design meets the Uniform Federal Accessibility Standards (See § 1724.51(e)(1)(i)).

(g) *Communications and control facilities.* (1) This paragraph (g) covers microwave and powerline carrier communications systems, load control, and supervisory control and data acquisition (SCADA) systems.

(2) The borrower shall obtain RUS approval, prior to issuing invitations to bid, of the terms and conditions for communications and control facilities contracts which will cost $1,500,000 or more. Unless RUS approval is required by paragraph (a) of this section, plans and specifications for communications and control facilities do not require RUS approval.

(h) Terms and conditions include the RUS standard form of contract, general and special conditions, and any other non-technical provisions of the contract. Terms and conditions which have received RUS approval in connection with a previous contract for a particular borrower are considered approved by RUS for that borrower.

[63 FR 35314, June 29, 1998, as amended at 65 FR 63196, Oct. 23, 2000; 77 FR 3071, Jan. 23, 2012; 84 FR 32617, July 9, 2019]

§ 1724.55 Dam safety.

(a) The provisions of this section apply only to RUS financed electric system facilities.

(1)(i) Any borrower that owns or operates a RUS financed dam must utilize the "Federal Guidelines for Dam Safety," (Guidelines), as applicable. A dam, as more fully defined in the Guidelines, is generally any artificial barrier which either:

(A) Is 25 feet (8 m) or more in height; or

(B) Has an impounding capacity at maximum water storage elevation of 55 acre-feet (68,000 m^3) or more.

(ii) The "Federal Guidelines for Dam Safety," FEMA 93, June, 1979, published by the Federal Emergency Management Agency (FEMA), is hereby incorporated by reference. This incorporation by reference is approved by the Director of the Office of the Federal Register in accordance with 5 U.S.C. 552(a) and 1 CFR part 51. Copies of the "Federal Guidelines for Dam Safety" may be obtained from the Federal Emergency Management Agency, Mitigation Directorate, PO Box 2012, Jessup, MD 20794. It is also available for inspection during normal business hours at RUS, Electric Staff Division, 1400 Independence Avenue, SW., Washington, DC, Room 1246–S, and at the National Archives and Records Administration (NARA). For information on the availability of this material at NARA, call 202–741–6030, or go to: *http://www.archives.gov/federal_register/code_of_federal_regulations/ibr_locations.html.*

(2) The borrower shall evaluate the hazard potential of its dams in accordance with Appendix E of the U.S. Army Corps of Engineers Engineering and Design Dam Safety Assurance Program, ER 1110–2–1155, July 31, 1995. A summary of the hazard potential criteria is included for information as Appendix A to this subpart. The U.S. Army Corps of Engineers Engineering and Design Dam Safety Assurance Program, ER 1110–2–1155, July 31, 1995, published by the United States Army Corps of Engineers, is hereby incorporated by reference. This incorporation by reference is approved by the Director of the Office of the Federal Register in accordance with 5 U.S.C. 552(a) and 1 CFR part 51. Copies of the U. S. Army Corps of Engineers Engineering and Design Dam Safety Assurance Program may be obtained from the U. S. Army Corps of Engineers, Publications Depot, 2803 52nd Ave., Hyattsville, MD 20781. It is also available for inspection during normal business hours at RUS, Electric Staff Division, 1400 Independence Avenue, SW., Washington, DC, Room 1246–S, and at the National Archives and Records Administration (NARA). For information on the availability of this material at NARA, call 202–741–6030, or go to: *http://www.archives.gov/federal_register/code_of_federal_regulations/ibr_locations.html.*

(3) For high hazard potential dams, the borrower must obtain an independent review of the design and critical features of construction. The reviewer must have demonstrated experience in the design and construction of dams of a similar size and nature. The reviewer must be a qualified engineer not involved in the original design of the dam or a Federal or State agency responsible for dam safety. The reviewer must be approved by RUS.

(4) The independent review of design must include, but not necessarily be limited to, plans, specifications, design calculations, subsurface investigation reports, hydrology reports, and redesigns which result from encountering unanticipated or unusual conditions during construction.

(5) The independent review of construction shall include:

(i) *Foundation preparation and treatment.* When the foundation has been excavated and exposed, and before critical structures such as earth embankments or concrete structures are placed thereon, the borrower shall require the reviewer to conduct an independent examination of the foundation to ensure that suitable foundation material has been reached and that the measures proposed for treatment of the foundation are adequate. This examination must extend to the preparation and treatment of the foundation for the abutments.

(ii) *Fill placement.* During initial placement of compacted fill materials, the borrower shall require the reviewer to conduct an independent examination to ensure that the materials being used in the various zones are suitable and that the placement and compaction procedures being used by the contractor will result in a properly constructed embankment.

(6) If the reviewer disagrees with any aspect of the design or construction which could affect the safety of the dam, then the borrower must meet with the design engineer and the reviewer to resolve the disagreements.

(7) *Emergency action plan.* For high hazard potential dams, the borrower must develop an emergency action plan incorporating preplanned emergency measures to be taken prior to and following a potential dam failure. The plan should be coordinated with local government and other authorities involved with the public safety.

(b)(1) For more information and guidance, the following publications regarding dam safety are available from FEMA:

(i)"Emergency Action Planning Guidelines for Dams,"FEMA 64.

(ii)"Federal Guidelines for Earthquake Analysis and Design of Dams,"FEMA 65.

(iii)"Federal Guidelines for Selecting and Accommodating Inflow Design Floods for Dams,"FEMA 94.

(iv)"Dam Safety: An Owner's Guidance Manual,"FEMA 145, August, 1987.

(2) These publications may be obtained from the Federal Emergency Management Agency, Mitigation Directorate, PO Box 2012, Jessup, MD 20794.

[63 FR 35314, June 29, 1998, as amended at 69 FR 18803, Apr. 9, 2004; 84 FR 32617, July 9, 2019]

§§ 1724.56–1724.69 [Reserved]

APPENDIX A TO SUBPART E OF PART 1724—HAZARD POTENTIAL CLASSIFICATION FOR CIVIL WORKS PROJECTS

The source for this appendix is U.S. Army Corps of Engineers Engineering and Design Dam Safety Assurance Program, ER 1110–2–1155, Appendix E. Appendix E is available from the address listed in § 1724.55(a)(2).

Category [1]	Low	Significant	High
Direct Loss of Life [2]	None expected (due to rural location with no permanent structures for human habitation).	Uncertain (rural location with few residences and only transient or industrial development).	Certain (one or more extensive residential, commercial or industrial development).
Lifeline Losses [3]	No disruption of services—repairs are cosmetic or rapidly repairable damage.	Disruption of essential facilities and access.	Disruption of critical facilities and access.

Category[1]	Low	Significant	High
Property Losses[4]	Private agricultural lands, equipment and isolated buildings.	Major public and private facilities.	Extensive public and private facilities.
Environmental Losses[5]	Minimal incremental damage.	Major mitigation required	Extensive mitigation cost or impossible to mitigate.

NOTES:
[1] Categories are based upon project performance and do not apply to individual structures within a project.
[2] Loss of life potential based upon inundation mapping of area downstream of the project. Analysis of loss of life potential should take into account the extent of development and associated population at risk, time of flood wave travel and warning time.
[3] Indirect threats to life caused by the interruption of lifeline services due to project failure, or operation, i.e., direct loss of (or access to) critical medical facilities or loss of water or power supply, communications, power supply, etc.
[4] Direct economic impact of value of property damages to project facilities and down stream property and indirect economic impact due to loss of project services, i.e., impact on navigation industry of the loss of a dam and navigation pool, or impact upon a community of the loss of water or power supply.
[5] Environmental impact downstream caused by the incremental flood wave produced by the project failure, beyond which would normally be expected for the magnitude flood event under a without project conditions.

Subpart F—RUS Contract Forms

§ 1724.70 Standard forms of contracts for borrowers.

(a) *General.* The standard loan agreement between RUS and its borrowers provides that, in accordance with applicable RUS regulations in this chapter, the borrower shall use standard forms of contract promulgated by RUS for construction, procurement, engineering services, and architectural services financed by a loan made or guaranteed by RUS. (See section 5.16 of appendix A to subpart C of part 1718 of this chapter.) This subpart prescribes RUS procedures in promulgating electric program standard contract forms and identifies those forms that borrowers are required to use.

(b) *Contract forms.* RUS promulgates standard contract forms, identified in the List of Required Contract Forms, § 1724.74(c), that borrowers are required to use in accordance with the provisions of this part. In addition, RUS promulgates standard contract forms identified in the List of Guidance Contract Forms contained in § 1724.74(c) that the borrowers may but are not required to use in the planning, design, and construction of their electric systems. Borrowers are not required to use these guidance contract forms in the absence of an agreement to do so.

[63 FR 58284, Oct. 30, 1998]

§ 1724.71 Borrower contractual obligations.

(a) *Loan agreement.* As a condition of a loan or loan guarantee under the RE Act, borrowers are normally required to enter into RUS loan agreements pursuant to which the borrower agrees to use RUS standard forms of contracts for construction, procurement, engineering services and architectural services financed in whole or in part by the RUS loan. Normally, this obligation is contained in section 5.16 of the loan contract. To comply with the provisions of the loan agreements as implemented by this part, borrowers must use those forms of contract (hereinafter sometimes called"listed contract forms") identified in the List of Required Standard Contract Forms contained in § 1724.74(c).

(b) *Compliance.* If a borrower is required by this part or by its loan agreement with RUS to use a listed standard form of contract, the borrower shall use the listed contract form in the format available from RUS, either paper or electronic format. Exact electronic reproduction is acceptable. The approved RUS standard forms of contract shall not be retyped, changed, modified, or altered in any manner not specifically authorized in this part or approved by RUS in writing on a case-by-case basis. Any modifications approved by RUS on a case-by-case basis must be clearly shown so as to indicate the modification difference from the standard form of contract.

(c) *Amendment.* Where a borrower has entered into a contract in the form required by this part, no change may be made in the terms of the contract, by amendment, waiver or otherwise, without the prior written approval of RUS.

(d) *Waiver.* RUS may waive for good cause, on a case by case basis, the requirements imposed on a borrower pursuant to this part. Borrowers seeking a waiver by RUS must provide RUS with a written request explaining the need for the waiver.

(e) *Violations.* A failure on the part of the borrower to use listed contracts as prescribed in this part is a violation of the terms of its loan agreement with RUS and RUS may exercise any and all remedies available under the terms of the agreement or otherwise.

[63 FR 58285, Oct. 30, 1998, as amended at 69 FR 7108, Feb. 13, 2004]

§ 1724.72 Notice and publication of listed contract forms.

(a) *Notice.* Upon initially entering into a loan agreement with RUS, borrowers will be provided with all listed contract forms. Thereafter, new or revised listed contract forms promulgated by RUS, including RUS approved exceptions and alternatives, will be sent by regular or electronic mail to the address of the borrower as identified in its loan agreement with RUS.

(b) *Availability.* Listed contract forms are published by RUS. Interested parties may obtain the forms from: Rural Utilities Service, Program Development and Regulatory Analysis, U.S. Department of Agriculture, Stop 1522, 1400 Independence Avenue, SW., Stop 1522, Washington, DC 20250–1522, telephone number (202) 720–8674. The list of contract forms can be found in § 1724.74(c), List of Required Contract Forms.

[63 FR 58285, Oct. 30, 1998]

§ 1724.73 Promulgation of new or revised contract forms.

RUS may, from time to time, undertake to promulgate new contract forms or revise or eliminate existing contract forms. In so doing, RUS shall publish notice of rulemaking in the FEDERAL REGISTER announcing, as appropriate, a revision in, or a proposal to amend § 1724.74, List of Electric Program Standard Contract Forms. The amendment may change the existing identification of a listed contract form; for example, changing the issuance date of a listed contract form or by identifying a new required contract form. The notice of rulemaking will describe the new standard contract form or the substantive change in the listed contract form, as the case may be, and the issues involved. The standard contract form or relevant portions thereof may be appended to the supplementary information section of the notice of rulemaking. As appropriate, the notice of rulemaking shall provide an opportunity for interested persons to provide comments. A copy of each such FEDERAL REGISTER document shall be sent by regular or electronic mail to all borrowers.

[63 FR 58285, Oct. 30, 1998]

§ 1724.74 List of electric program standard contract forms.

(a) *General.* The following is a list of RUS electric program standard contract forms for architectural and engineering services. Paragraph (c) of this section contains the list of required contract forms, *i.e.*, those forms of contracts that borrowers are required to use by the terms of their RUS loan agreements as implemented by the provisions of this part. Paragraph (d) of this section contains the list of guidance contract forms, *i.e.*, those forms of contracts provided as guidance to borrowers in the planning, design, and construction of their systems. All of these forms are available from RUS. See § 1724.72(b) for availability of these forms.

(b) *Issuance date.* Where required by this part to use a standard form of contract in connection with RUS financing, the borrower shall use that form identified by issuance date in the List of Required Contract Forms in paragraph (c) of this section, as most recently published as of the date the borrower executes the contract.

(c) *List of required contract forms.* (1) RUS Form 211, Rev. 4–04, Engineering Service Contract for the Design and Construction of a Generating Plant. This form is used for engineering services for generating plant construction.

(2) RUS Form 220, Rev. 6–98, Architectural Services Contract. This form is used for architectural services for building construction.

(3) RUS Form 236, Rev. 6–98, Engineering Service Contract—Electric

System Design and Construction. This form is used for engineering services for distribution, transmission, substation, and communications and control facilities.

(d) *List of guidance contract forms.* (1) RUS Form 179, Rev. 9–66, Architects and Engineers Qualifications. This form is used to document architects and engineers qualifications.

(2) RUS Form 215, Rev. 5–67, Engineering Service Contract—System Planning. This form is used for engineering services for system planning.

(3) RUS Form 234, Rev. 3–57, Final Statement of Engineering Fee. This form is used for the closeout of engineering services contracts.

(4) RUS Form 241, Rev. 3–56, Amendment of Engineering Service Contract. This form is used for amending engineering service contracts.

(5) RUS Form 244, Rev. 12–55, Engineering Service Contract—Special Services. This form is used for miscellaneous engineering services.

(6) RUS Form 258, Rev. 4–58, Amendment of Engineering Service Contract—Additional Project. This form is used for amending engineering service contracts to add an additional project.

(7) RUS Form 284, Rev. 4–72, Final Statement of Cost for Architectural Service. This form is used for the closeout of architectural services contracts.

(8) RUS Form 297, Rev. 12–55, Engineering Service Contract—Retainer for Consultation Service. This form is used for engineering services for consultation service on a retainer basis.

(9) RUS Form 459, Rev. 9–58, Engineering Service Contract—Power Study. This form is used for engineering services for power studies.

[63 FR 58285, Oct. 30, 1998, as amended at 65 FR 63196, Oct. 23, 2000; 69 FR 52595, Aug. 27, 2004]

§§ 1724.75–1724.99 [Reserved]

PART 1726—ELECTRIC SYSTEM CONSTRUCTION POLICIES AND PROCEDURES

Subpart A—General

Sec.

Subpart B—Distribution Facilities

Subpart C—Substation and Transmission Facilities

Subpart D—Generation Facilities

Subpart E—Buildings

Subpart F—General Plant

Subpart G—Procurement Procedures

AUTHORITY: 7 U.S.C. 901 *et seq.*, 1921 *et seq.*, 6941 *et seq.*

SOURCE: 60 FR 10155, Feb. 23, 1995, unless otherwise noted.

Subpart A—General

§§ 1726.1–1726.9 [Reserved]

§ 1726.10 Introduction.

The policies, procedures and requirements included in this part are intended to implement provisions of the standard form of loan documents between the Rural Utilities Service (RUS) and its electric borrowers. Unless prior written approval is received from RUS, borrowers are required to comply with RUS policies and procedures as a condition to RUS providing loans, loan guarantees, or reimbursement of general funds for the construction and improvement of electric facilities. Requirements relating to RUS approval of plans and specifications, duties and responsibilities of the engineer and architect, and engineering and architectural services contracts, are contained in other RUS regulations. The terms "RUS form", "RUS standard form", "RUS specification", "and RUS bulletin" have the same meanings as the terms "REA form", "REA standard form", "REA specification", "and REA bulletin", respectively, unless otherwise noted.

§ 1726.11 Purpose.

Each borrower is responsible for the planning, design, construction, operation and maintenance of its electric system. RUS, as a secured lender, has a legitimate interest in accomplishing RUS's programmatic objectives, and in assuring that the costs of construction, materials, and equipment are reasonable and economical and that the property securing the loans is constructed adequately to serve the purposes for which it is intended.

§ 1726.12 Applicability.

The requirements of this part apply to the procurement of materials and equipment for use by electric borrowers in their electric systems and to the construction of their electric systems if such materials, equipment, and construction are financed, in whole or in part, with loans made or guaranteed by RUS, including reimbursable projects. In order for general fund expenditures for procurement or construction to be eligible for reimbursement from loan funds, the borrower must comply with the procedures required by this part. In the case of jointly owned projects, RUS will determine on a case by case basis the applicability of the requirements of this part.

§ 1726.13 Waivers.

The Administrator may waive, for good cause on a case by case basis, certain requirements and procedures of this part. RUS reserves the right, as a condition of providing loans, loan guarantees, or other assistance, to require any borrower to make any specification, contract, or contract amendment subject to the approval of the Administrator.

§1726.14 Definitions.

Terms used in this part have the meanings set forth in 7 CFR 1710.2. References to specific RUS forms and other RUS documents, and to specific sections or lines of such forms and documents, shall include the corresponding forms, documents, sections and lines in any subsequent revisions of these forms and documents. In addition to the terms defined in 7 CFR 1710.2, the following terms have the following meanings for the purposes of this part:

Approval of proposed construction means RUS approval of a construction work plan or other appropriate engineering study and RUS approval, for purposes of system financing, of the completion of all appropriate environmental review requirements in accordance with 7 CFR part 1970.

Architect means a registered or licensed person employed by the borrower to provide architectural services for a project and duly authorized assistants and representatives.

Bona fide bid means a bid which is submitted by a contractor on the borrower's list of qualified bidders for the specific contract, prior to bid opening.

"Buy American" certificate means a certification that the contractor has complied with the "Buy American" requirement (see §1726.15).

Competitive procurement means procurement of goods or services based on lowest evaluated bid for similar products or services when three or more bids are received.

Construction unit means a specifically defined portion of a construction project containing materials, labor, or both, for purposes of bidding and payment.

Contracting committee means the committee consisting of three to five members representing the borrower's management and board of directors and the engineer. The contracting committee represents the borrower during contract clarifying discussions or negotiations under informal competitive bidding or multiparty negotiation, respectively.

Encumbrance means the process of approval for advance of loans funds by RUS.

Engineer means a registered or licensed person, who may be a staff employee or an outside consultant, to provide engineering services and duly authorized assistants and representatives.

Equipment means a major component of an electric system, e.g., a substation transformer, heat exchanger or a transmission structure.

Force account construction means construction performed by the borrower's employees.

Formal competitive bidding means the competitive procurement procedure wherein bidders submit sealed proposals for furnishing the goods or services stipulated in the specification. Bids are publicly opened and read at a predetermined time and place. If a contract is awarded, it must be to the lowest evaluated responsive bidder (see §1726.201).

Goods or services means materials, equipment, or construction, or any combination thereof.

Informal competitive bidding means the competitive procurement procedure which provides for private opening of bids and allows clarifying discussions between the contracting committee and the bidders. During the clarifying discussions any exceptions to the bid documents must be eliminated, or the bid rejected, so that the contract is awarded to the lowest evaluated responsive bidder (see §1726.202).

Material means miscellaneous hardware which is combined with equipment to form an electric system, e.g., poles, insulators, or conductors.

Minor error or irregularity means a defect or variation in a bid that is a matter of form and not of substance. Errors or irregularities are "minor" if they can be corrected or waived without being prejudicial to other bidders and when they do not affect the price, quantity, quality, or timeliness of construction. A minor error or irregularity is not an exception for purposes of determining whether a bid is responsive.

Minor modification or improvement means a project the cost of which is $150,000 or less, exclusive of the cost of owner furnished materials.

Multiparty lump sum quotations means the procurement of goods or services on a lump sum basis, based on the lowest evaluated offering, when three or more offers are received. (See § 1726.205).

Multiparty negotiation means the procurement procedure where three or more bids are received and provides for negotiations between the contracting committee and each bidder to determine the bid which is in the borrower's best interest (see § 1726.203).

Multiparty unit price quotations means the procurement of goods or services on a unit price basis, based on the lowest evaluated offering, when three or more offers are received (See § 1726.204).

Net utility plant (NUP) means Part C, Line 5 of RUS Form 7 for distribution borrowers or Section B, Line 5 of RUS Form 12a for power supply borrowers for the immediately preceding calendar year.

Procurement method means a procedure, including, but not limited to, those in subpart G of this part, that a borrower uses to obtain goods and services.

Owner furnished materials means materials or equipment or both supplied by the borrower for installation by the contractor.

Responsive bid means a bid with no exceptions or non-minor errors or irregularities on any technical requirement or in the contract terms and conditions.

RUS approval means written approval by the Administrator or a representative with delegated authority. RUS approval must be in writing, except in emergency situations where RUS approval may be given over the telephone followed by a confirming letter.

Unit prices means individual prices for specific construction units defined in accordance with RUS approved units specified in RUS standard contract forms.

[60 FR 10155, Feb. 23, 1995, as amended at 77 FR 3071, Jan. 23, 2012; 81 FR 11027, Mar. 2, 2016]

§ 1726.15 "Buy American".

The borrower must ensure that all materials and equipment financed with loans made or guaranteed by RUS complies with the "Buy American" provisions of the Rural Electrification Act of 1938 (7 U.S.C. 903 note), as amended by the North American Free Trade Agreement Implementation Act (107 Stat 2129). When a "Buy American" certificate is required by this part, this must be on RUS Form 213.

§ 1726.16 Debarment and suspension.

Borrowers are required to comply with certain requirements on debarment and suspension in connection with procurement activities set forth in 2 CFR part 180, as adopted by USDA through 2 CFR part 417, particularly with respect to lower tier transactions, *e.g.*, procurement contracts for goods or services.

[79 FR 76003, Dec. 19, 2014]

§ 1726.17 Restrictions on lobbying.

Borrowers are required to comply with certain restrictions and requirements in connection with procurement activities as set forth in 2 CFR part 418.

[79 FR 76003, Dec. 19, 2014]

§ 1726.18 Pre-loan contracting.

Borrowers must consult with RUS prior to entering into any contract for material, equipment, or construction if a construction work plan, general funds, loan or loan guarantee for the proposed work has not been approved. While the RUS staff will work with the borrower in such circumstances, nothing contained in this part is to be construed as authorizing borrowers to enter into any contract before the availability of funds has been ascertained by the borrower and all environmental review requirements in accordance with 7 CFR part 1970, have been met.

[81 FR 11027, Mar. 2, 2016]

§ 1726.19 Use of competitive procurement.

RUS borrowers' procurement is not subject to the provisions of the Federal

Acquisition Regulation (48 CFR chapter 1); however, since borrowers receive the benefit of Federal financial assistance borrowers must use competitive procurement to the greatest extent practical. The borrower must use competitive procurement for obtaining all goods or services when a RUS loan or loan guarantee is involved except:

(a) As specifically provided for in subparts B through F of this part; or

(b) A waiver is granted.

§ 1726.20 Standards and specifications.

All materials, equipment, and construction must meet the minimum requirements of all applicable RUS standards and specifications. (See part 1728 of this chapter, Electric Standards and Specifications for Materials and Construction, which is applicable regardless of the source of funding.)

[69 FR 7109, Feb. 13, 2004]

§ 1726.21 New materials.

The borrower shall purchase only new materials and equipment unless otherwise approved by RUS, on a case by case basis, prior to the purchase.

§ 1726.22 Methods of construction.

The borrower is generally responsible for determining whether construction will be by contract or force account. If construction is by contract, the borrower must determine whether materials will be supplied by the contractor or will be furnished by the borrower. RUS reserves the right to require contract construction in lieu of force account construction on a case by case basis.

§ 1726.23 Qualification of bidders.

(a) *Qualified bidder list (QBL).* The borrower shall (acting through its engineer, if applicable) review the qualifications of prospective bidders for contract construction and for material and equipment procurement, and select firms qualified for inclusion on the borrower's list of qualified bidders for each contract. (See also § 1726.16 and § 1726.17.) A bid may not be solicited from a prospective bidder or opened by the borrower unless that bidder has been determined to be a qualified bidder for the contract. When preparing the QBL, in addition to the actual experience of the borrower, if any, in dealing with a prospective bidder, the borrower may solicit information from that bidder or from other parties with firsthand experience regarding the firm's capabilities and experience. It is also important to consider the firm's performance record, safety record, and similar factors in determining whether to include that firm on the QBL, since the borrower may not evaluate these factors when evaluating a bid from a qualified and invited bidder.

(b) *Conflict of interest.* If there is a relationship between the borrower or engineer and a prospective bidder which might cause the borrower or engineer to have or appear to have a conflict of interest, that prospective bidder shall not be included on the QBL unless the engineer discloses the nature of the relationship to the borrower. In the case of the borrower, if its employees or directors have a relationship with a prospective bidder, the prospective bidder shall not be included on the qualified bidders list unless the nature of the relationship is disclosed to the board of directors, and the board of directors specifically approves the inclusion of that bidder in light of the potential for a conflict of interest.

§ 1726.24 Standard forms of contracts for borrowers.

(a) *General.* The standard loan agreement between RUS and the borrowers provides that, in accordance with applicable RUS regulations in this chapter, the borrower shall use standard forms of contracts promulgated by RUS for construction, procurement, engineering services, and architectural services financed by a loan made or guaranteed by RUS. This part implements these provisions of the RUS loan agreement. Subparts A through H and J of this part prescribe when and how borrowers are required to use RUS standard forms of contracts in procurement and construction. Subpart I of this part prescribes the procedures that RUS follows in promulgating standard contract forms and identifies those contract forms that borrowers are required to use for procurement and construction.

(b) *Amendments to contracts*—(1) *Contract forms.* The borrower must use RUS Form 238, Construction or Equipment Contract Amendment, for any change or addition in any contract for construction or equipment.

(2) *Special considerations.* Each time an amendment to a construction contract is executed, the borrower must ensure that contractor's bond is adequate, that all necessary licenses and permits have been obtained, and that any environmental requirements associated with the proposed construction have been met.

(3) *Amendment approval requirements.* (i) If a RUS approved form of contract is required by this part, an amendment must not alter the terms and conditions of the RUS approved form of contract without prior RUS approval.

(ii) The borrower must make a contract amendment subject to RUS approval if the underlying contract was made subject to RUS approval and the total amended contract price exceeds 120 percent of the original contract price (excluding any escalation provision contained in the contract).

(iii) Contract amendments, except as provided in paragraph (b)(3)(ii) of this section, are not subject to RUS approval and need not be submitted to RUS unless specifically requested by RUS on a case by case basis.

[60 FR 10155, Feb. 23, 1995, as amended at 63 FR 58286, Oct. 30, 1998; 69 FR 7109, Feb. 13, 2004]

§ 1726.25 Subcontracts.

Subcontracts are not subject to RUS approval and need not be submitted to RUS unless specifically requested by RUS on a case by case basis.

[69 FR 7109, Feb. 13, 2004]

§ 1726.26 Interest on overdue accounts.

Certain RUS contract forms contain a provision concerning payment of interest on overdue accounts. Prior to issuing the invitation to bidders, the borrower must insert an interest rate equal to the lowest "Prime Rate" listed in the "Money Rates" section of the Wall Street Journal on the date such invitation to bid is issued. If no prime rate is published on that date, the last such rate published prior to that date must be used. The rate must not, however, exceed the maximum rate allowed by any applicable state law.

[63 FR 58286, Oct. 30, 1998]

§ 1726.27 Contractor's bonds.

(a) RUS Form 168b, Contractor's Bond, shall be used when a contractor's bond is required by RUS Forms 200, 257, 786, 790, or 830 unless the contractor's surety has accepted a Small Business Administration guarantee and the contract is for $1 million or less.

(b) RUS Form 168c, Contractor's Bond, shall be used when a contractor's bond is required by RUS Forms 200, 257, 786, 790, or 830 and the contractor's surety has accepted a Small Business Administration guarantee and the contract is for $1 million or less.

(c) Surety companies providing contractor's bonds shall be listed as acceptable sureties in the U.S. Department of the Treasury Circular No. 570, Companies Holding Certificates of Authority as Acceptable Sureties on Federal Bonds and as Acceptable Reinsuring Companies. Copies of the circular and interim changes may be obtained directly from the Government Printing Office (202) 512–1800. Interim changes are published in the FEDERAL REGISTER as they occur. The list is also available through the Internet at *http://www.fms.treas.gov/c570/index.html* and on the Department of the Treasury's computerized public bulletin board at (202) 874–6887.

[63 FR 58286, Oct. 30, 1998, as amended at 69 FR 7109, Feb. 13, 2004]

§§ 1726.28–1726.34 [Reserved]

§ 1726.35 Submission of documents to RUS.

(a) *Where to send documents.* Documents required to be submitted to RUS under this part are to be sent electronically to RUS, unless otherwise directed.

(b) *Borrower certification.* When a borrower certification is required by this part, it must be made by the borrower's manager unless the board of directors specifically authorizes another person to make the required certification. In such case, a certified copy of the specific authorizing resolution must accompany the document or be on file with RUS.

(c) *Contracts requiring RUS approval.* The borrower shall submit to RUS, one copy of each contract that is subject to RUS approval under subparts B through F of this part. Any contract submitted by the borrower contract must be accompanied by:

(1) A bid tabulation and evaluation and, if applicable, a written recommendation of the architect or engineer.

(2) For awards made under the informal competitive bidding procedure or the multiparty negotiation procedure, a written recommendation of the contracting committee (See §§1726.202 and 1726.203).

(3) One copy of an executed contractor's bond on RUS approved bond forms as required in the contract form and one copy of the bid bond or copy of the certified check.

(4) A certification by the borrower or chairperson of the contracting committee, as applicable, that the appropriate bidding procedures were followed as required by this part.

(5) Evidence of clear title to the site for substations and headquarters construction contracts, if not previously submitted.

(6) Documentation that all reasonable measures were taken to assure competition if fewer than three bids were received.

(d) *Contract amendments requiring RUS approval.* The borrower must submit to RUS, one copy of each contract amendment which is subject to RUS approval under §1726.24(b). Each contract amendment submittal to RUS must be accompanied by a bond extension, where necessary.

(e) *Encumbrance of loan or loan guarantee funds.* (1) For contracts subject to RUS approval, the submittals required under paragraph (c) of this section will initiate RUS action to encumber loan or loan guarantee funds for such contracts.

(2) For contracts not subject to RUS approval (except for generation projects), loan or loan guarantee funds will normally be encumbered using RUS Form 219, Inventory of Work Orders, after closeout of the contracts. In cases where the borrower can show good cause for a need for immediate cash, the borrower may request encumbrance of loan or loan guarantee funds based on submittal of a copy of the executed contract, provided it meets all applicable RUS requirements.

(3) For generation project contracts not subject to RUS approval, the borrower must submit to RUS the following documentation:

(i) A brief description of the scope of the contract, including contract identification (name, number, etc.);

(ii) Contract date;

(iii) Contractor's name;

(iv) Contract amount;

(v) Bidding procedure used;

(vi) Borrower certification that:

(A) The bidding procedures and contract award for each contract were in conformance with the requirements of Part 1726, Electric System Construction Policies and Procedures;

(B) If a RUS approved form of contract is required by this part, the terms and conditions of the RUS approved form of contract have not been altered;

(C) If RUS has approved plans and specifications for the contract, the contract was awarded on the basis of those plans and specifications; and

(D) No restriction has been placed on the borrower's right to assign the contract to RUS or its successors.

(4) *Contract amendments.* (i) For amendments subject to RUS approval, the submittals required under paragraph (c) of this section will initiate RUS action to encumber loan or loan guarantee funds for contract amendments requiring RUS approval.

(ii) For amendments not subject to RUS approval (except generation projects), loan or loan guarantee funds will normally be encumbered using RUS Form 219, Inventory of Work Orders, after closeout of the contracts. In cases where the borrower can justify a need for immediate cash, the borrower may request encumbrance of loan or loan guarantee funds based on submittal of a copy of the executed amendment, providing it meets all applicable RUS requirements.

(iii) For each generation project contract amendment not subject to RUS approval, the borrower must submit to

RUS the following information and documentation:

(A) The contract name and number;

(B) The amendment number;

(C) The amendment date;

(D) The dollar amount of the increase or the decrease of the amendment;

(E) Borrower certification that:

(*1*) The amendment was approved in accordance with the policy of the board of directors;

(*2*) If a RUS approved form of contract is required by this part, the terms and conditions of the RUS approved form of contract has not been altered; and

(*3*) No restriction has been placed on the borrower's right to assign the contract to RUS or its successors.

[60 FR 10155, Feb. 23, 1995, as amended at 84 FR 32617, July 9, 2019; 86 FR 36197, July 9, 2021]

§ 1726.36 Documents subject to RUS approval.

Unless otherwise indicated, the borrower shall make all contracts and amendments that are subject to RUS approval effective only upon RUS approval.

§ 1726.37 OMB control number.

The collection of information requirements in this part have been approved by the Office of Management and Budget and assigned OMB control number 0572–0107.

§§ 1726.38–1726.49 [Reserved]

Subpart B—Distribution Facilities

§ 1726.50 Distribution line materials and equipment.

(a) *Contract forms.* (1) The borrower shall use RUS Form 198, Equipment Contract, for purchases of equipment where the total cost of the contract is more than $1,000,000.

(2) The borrower may, in its discretion, use RUS Form 198, Equipment Contract, or a written purchase order equal to $1,000,000 or less for purchases of equipment, and for all materials.

(b) *Standards and specifications.* Distribution line materials and equipment must meet the minimum requirements of RUS standards as determined in accordance with the provisions of part 1728 of this chapter, Electric Standards and Specifications for Materials and Construction. The borrower must obtain RUS approval prior to purchasing any unlisted distribution line material or equipment of the types listed in accordance with the provisions of part 1728 of this chapter.

(c) *Procurement procedures.* It is the responsibility of each borrower to determine the procurement method that best meets its needs for the purchase of material and equipment to be used in distribution line construction.

(d) *Contract approval.* Contracts for purchases of distribution line materials and equipment are not subject to RUS approval and need not be submitted to RUS unless specifically requested by RUS on a case by case basis.

[60 FR 10155, Feb. 23, 1995, as amended at 69 FR 7109, Feb. 13, 2004; 77 FR 3072, Jan. 23, 2012]

§ 1726.51 Distribution line construction.

(a) *Contract forms.* The borrower must use RUS Form 790, or 830, as outlined in this paragraph (a), for distribution line construction, except for minor modifications or improvements.

(1) The borrower may use RUS Form 790, Electric System Construction Contract—Non-Site Specific Construction, under the following circumstances:

(i) For contracts for which the borrower supplies all materials and equipment; or

(ii) For non-site specific construction contracts accounted for under the work order procedure; or

(iii) If neither paragraph (a)(1)(i) or (a)(1)(ii) of this section are applicable, the borrower may use RUS Form 790 for contracts, up to a cumulative total of $500,000 or one percent of net utility plant (NUP), whichever is greater, per calendar year of distribution line construction, exclusive of the cost of owner furnished materials and equipment.

(2) The borrower must use RUS Form 830, Electric System Construction Contract—Project Construction, for all other distribution line construction.

(b) *Procurement procedures.* (1) It is the responsibility of each borrower to determine the procurement method that best meets its needs to award contracts in amounts of up to a cumulative total of $750,000 or three percent of NUP (not to exceed $6,000,000), whichever is greater, per calendar year of distribution line construction (including minor modifications or improvements), exclusive of the cost of owner furnished materials and equipment. Borrowers may award Cost-Plus/Hourly contracts as part of these borrower responsibility limits up to a cumulative total of $250,000 or one percent of NUP (not to exceed $2,000,000), whichever is greater, per calendar year of distribution line construction (including minor modifications or improvements), exclusive of the cost of owner furnished materials and equipment.

(2) The borrower shall use formal competitive bidding for all other distribution line contract construction unless the RUS specifically approves an alternative method. The dollar amounts of contracts bid using the formal competitive bidding procedure do not apply to the cumulative total stipulated in paragraph (b)(1) of this section.

(3) An amendment which increases the scope of the contract by adding a project is not considered competitively bid, therefore, the dollar amount of that amendment does apply to the cumulative total stipulated in paragraph (b)(1) of this section.

(c) *Contract approval.* Contracts for distribution line construction are not subject to RUS approval and need not be submitted to RUS unless specifically requested by RUS on a case by case basis.

[60 FR 10155, Feb. 23, 1995, as amended at 69 FR 7109, Feb. 13, 2004; 77 FR 3072, Jan. 23, 2012; 86 FR 36198, July 9, 2021]

§§ 1726.52–1726.74 [Reserved]

Subpart C—Substation and Transmission Facilities

§ 1726.75 General.

As used in this part, "substations" includes substations, switching stations, metering points, and similar facilities.

§ 1726.76 Substation and transmission line materials and equipment.

(a) *Contract forms.* (1) The borrower shall use RUS Form 198, Equipment Contract, for purchases of equipment where the total cost of the contract is $1,000,000 or more.

(2) The borrower may, in its discretion, use RUS Form 198, Equipment Contract, or a written purchase order for purchases of equipment of less than $1,000,000 and for all materials.

(b) *Standards and specifications.* Substation and transmission line materials and equipment must meet the minimum requirements of RUS standards as determined in accordance with the provisions of part 1728 of this chapter, Electric Standards and Specifications for Materials and Construction. The borrower must obtain RUS approval prior to purchasing of any unlisted substation or transmission line material or equipment of the types listed in accordance with the provisions of part 1728 of this chapter.

(c) *Procurement procedures.* It is the responsibility of each borrower to determine the procurement method that best meets its needs for purchase of material and equipment to be used in substation and transmission line construction.

(d) *Contract approval.* Contracts for purchases of substation and transmission line materials and equipment are not subject to RUS approval and need not be submitted to RUS unless specifically requested by RUS on a case by case basis.

[60 FR 10155, Feb. 23, 1995, as amended at 69 FR 7109, Feb. 13, 2004; 77 FR 3072, Jan. 23, 2012]

§ 1726.77 Substation and transmission line construction.

(a) *Contract forms.* The borrower must use RUS Form 830, Electric System Construction Contract—Project Construction, for construction of substations, except for minor modifications or improvements.

(b) *Procurement procedures.* (1) It is the responsibility of each borrower to determine the procurement method that best meets its needs to award contracts in amounts of up to a cumulative total of $750,000 or three percent of NUP (not to exceed $6,000,000), whichever is greater, per calendar year of substation and transmission line construction (including minor modifications or improvements), exclusive of the cost of owner furnished materials and equipment. Borrowers may award Cost-Plus/Hourly contracts as part of these borrower responsibility limits up to a cumulative total of $250,000 or one percent of NUP (not to exceed $2,000,000), whichever is greater, per calendar year of substation and transmission line construction (including minor modifications or improvements), exclusive of the cost of owner furnished materials and equipment.

(2) The borrower shall use formal competitive bidding for all other contract construction unless RUS specifically approves an alternative method. The dollar amount of contracts bid using the formal competitive bidding procedure do not apply to the cumulative total stipulated in paragraph (b)(1) of this section.

(3) An amendment which increases the scope of the contract by adding a project is not considered competitively bid, therefore, the dollar amount of that amendment does apply to the cumulative total stipulated in paragraph (b)(1) of this section.

(c) *Contract approval.* Individual contracts in the amount of $750,000 or more or three percent of NUP (not to exceed $6,000,000), whichever is greater, exclusive of the cost of owner furnished materials and equipment, are subject to RUS approval.

[60 FR 10155, Feb. 23, 1995, as amended at 69 FR 7109, Feb. 13, 2004; 77 FR 3072, Jan. 23, 2012; 86 FR 36198, July 9, 2021]

§§ 1726.78–1726.124 [Reserved]

Subpart D—Generation Facilities

§ 1726.125 Generating plant facilities.

This section covers the construction of all portions of a generating plant, including plant buildings and the generator step-up transformer. Generally, the transmission switchyard will be covered under this section during initial construction of the plant. Subpart C of this part covers subsequent modifications to transmission switchyards. Warehouses and equipment service type buildings are covered under subpart E of this part.

(a) *Contract forms.* (1) The borrower shall use RUS Form 198, Equipment Contract, for the purchase of generating plant equipment in the amount of $5,000,000 or more and for any generating plant equipment contract requiring RUS approval.

(2) The borrower shall use RUS Form 200, Construction Contract—Generating, for generating project construction contracts in the amount of $5,000,000 or more and for any generating project construction contract requiring RUS approval.

(3) The borrower may, in its discretion, use other contract forms or written purchase order forms for those contracts in amounts of less than $5,000,000 and that do not require RUS approval.

(b) *Procurement procedures.* (1) It is the responsibility of each borrower to determine the procurement method that best meets its needs to award contracts in amounts of less than $5,000,000 each.

(2) If the amount of the contract is $5,000,000 or more or if the contract requires RUS approval, the borrower must use formal or informal competitive bidding to award the contract.

(3) Where formal or informal competitive bidding is not applicable, or does not result in a responsive bid, multiparty negotiation may be used only after RUS approval is obtained.

(c) *Contract approval.* During the early stages of generating plant design or project design, RUS will, in consultation with the borrower and its consulting engineer, identify the specific contracts which require RUS approval based on information supplied in the plant design manual. The following are typical contracts for each type of generating project which will require RUS approval. Although engineering services are not covered by this part, they are listed in this paragraph (d) to emphasize that RUS approval is required for all major generating station engineering service contracts in accordance with applicable RUS rules. For types of projects not shown, such as nuclear and alternate energy projects, RUS will identify the specific contracts which will require RUS approval on a case by case basis.

(1) *Fossil generating stations.* Engineering services, steam generator, turbine generator, flue gas desulfurization system, particulate removal system, electric wiring and control systems, mechanical equipment installation (including turbine installation and plant piping), power plant building (foundation and superstructure), site preparation, coal unloading and handling facilities, main step-up substation, cooling towers, and dams or reservoirs.

(2) *Diesel and combustion turbine plants.* Engineering services, prime mover and generator, building (foundation and superstructure), and electrical control systems.

(3) *Hydro installations.* Engineering services, turbine/generator, civil works and powerhouse construction, electrical control system, and mechanical installation.

[60 FR 10155, Feb. 23, 1995, as amended at 69 FR 7109, Feb. 13, 2004; 77 FR 3072, Jan. 23, 2012]

§§ 1726.126–1726.149 [Reserved]

Subpart E—Buildings

§ 1726.150 Headquarters buildings.

This section includes headquarters buildings such as warehouses and equipment service type buildings. Generating plant buildings are covered under subpart D of this part.

(a) *Contract forms.* The borrower must use RUS Form 257, Contract to Construct Buildings, for all contracts for construction of new headquarters facilities, and additions to, or modifications of existing headquarters facilities (except for minor modifications or improvements).

(b) *Procurement procedures.* A borrower may use Multiparty Lump Sum Quotations to award contracts in amounts of up to a cumulative total of $1,500,000 or three percent of NUP (not to exceed $10,000,000), whichever is greater, per calendar year of headquarters construction (including minor modifications or improvements). The borrower shall use formal competitive bidding for all other headquarters contract construction unless RUS specifically approves an alternative method.

(c) *Contract approval.* Contracts for headquarters construction are not subject to RUS approval and need not be submitted to RUS unless specifically requested by RUS on a case by case basis.

[60 FR 10155, Feb. 23, 1995, as amended at 77 FR 3072, Jan. 23, 2012; 86 FR 36198, July 9, 2021]

§§ 1726.151–1726.174 [Reserved]

Subpart F—General Plant

§ 1726.175 General plant materials.

This section covers items such as office furniture and equipment; transportation equipment and accessories, including mobile radio systems, stores and shop equipment, laboratory equipment, tools and test equipment.

(a) *Contract forms.* The borrower may, in its discretion, use RUS Form 198, Equipment Contract, or a written purchase order.

(b) *Procurement procedures.* It is the responsibility of each borrower to determine the procurement method that best meets its needs for purchase of general plant material and equipment.

(c) *Contract approval.* Contracts for the purchase of general plant items are not subject to RUS approval and need not be submitted to RUS unless specifically requested by RUS on a case by case basis.

[60 FR 10155, Feb. 23, 1995, as amended at 69 FR 7109, Feb. 13, 2004]

§ 1726.176 Communications and control facilities.

This section covers the purchase of microwave, fiber, power line carrier, and other communications technologies or systems, including load control and supervisory control and data acquisition (SCADA) systems, automated meter reading/automated metering infrastructure (AMR/AMI), or other smart grid technologies. Mobile radio systems are covered as general plant materials in § 1726.175.

(a) *Power line carrier systems.* Power line carrier equipment will frequently be purchased as part of a substation and will be included in the complete substation plans and specifications. When purchased in this manner, the requirements of subpart C of this part, Substation and Transmission Facilities, will apply. If obtained under a contract for only a power line carrier system, the requirements of paragraph (b) of this section apply.

(b) *Load control systems, communications systems, and SCADA systems*—(1) *Contract forms.* The borrower must use RUS Form 786, Electric System Communication and Control Equipment Contract. This form may be modified to be a "purchase only" contract form.

(2) *Procurement procedures.* (i) It is the responsibility of each borrower to determine the procurement method that best meets its needs to award contracts not requiring RUS approval in amounts of up to a cumulative total of $750,000 or one percent of NUP (not to exceed $5,000,000), whichever is greater, per calendar year of communications and control facilities construction (including minor modifications or improvements), exclusive of the cost of owner furnished materials and equipment.

(ii) The borrower must use multiparty negotiation for all other communications and control facilities contract construction, including all contracts requiring RUS approval. The amount of contracts bid using the multiparty negotiation procedure do not apply to the cumulative total stipulated in paragraph (b)(2)(i) of this section.

(iii) An amendment which increases the scope by adding a project is not considered competitively bid, therefore, the amount of that amendment does apply to the cumulative total stipulated in paragraph (b)(2)(i) of this section.

(3) *Contract approval.* Individual contracts in amounts of $750,000 or more or one percent of NUP (not to exceed $5,000,000 for all borrowers), whichever is greater, exclusive of the cost of owner furnished materials and equipment, are subject to RUS approval.

[60 FR 10155, Feb. 23, 1995, as amended at 77 FR 3072, Jan. 23, 2012; 86 FR 36198, July 9, 2021]

§§ 1726.177–1726.199 [Reserved]

Subpart G—Procurement Procedures

§ 1726.200 General requirements.

The borrower must use the procedures described in this subpart where such procedures are required under subparts B through F of this part. The borrower must ensure that arrangements prior to announcement of the award of the contract are such that all bidders are treated fairly and no bidder is given an unfair advantage over other bidders.

§ 1726.201 Formal competitive bidding.

Formal competitive bidding is used for distribution, transmission, and headquarters facilities, and may be used for generation facilities. The borrower must use the following procedure for formal competitive bidding:

(a) *Selection of qualified bidders.* The borrower (acting through its engineer, if applicable) will compile a list of qualified bidders for each proposed contract. The borrower will send invitations to bid only to persons or organizations on its QBL for the specific project (see § 1726.23).

(b) *Invitations to bid.* The borrower (acting through its engineer, if applicable) is responsible for sending out invitations to prospective bidders, informing them of scheduled bid openings and taking any other action necessary to procure full, free and competitive bidding. The borrower should send out a sufficient number of invitations in order to assure adequate competition and so that at least three bids will be received. Subject to the foregoing criteria, the determination of how many and which bidders will be permitted to bid will be the responsibility of the borrower.

(c) *Evaluation basis.* Any factors, other than lowest dollar amount of the bid, which are to be considered in evaluating the proposals of qualified bidders (e.g., power consumption, losses, etc.) must be stated in the "Notice and Instructions to Bidders." The borrower will not evaluate a bidder's performance record, safety record, and similar factors when evaluating a bid from a qualified and invited bidder. Such factors are to be considered when determining whether to include a particular bidder on the qualified bidders list.

(d) *Handling of bids received.* The borrower or the engineer, as applicable, will indicate, in writing, the date and time of receipt by the borrower or the engineer on the outside envelope of each bid and all letters and other transmittals amending or modifying the bids. Any bid received at the designated location after the time specified must be returned to the bidder unopened.

(e) *Bid openings.* Bid openings are generally conducted by the engineer in the presence of bidders and a representative of the borrower and the borrower's attorney. Each bona fide bid must be opened publicly and reviewed for any irregularities, errors, or exceptions. It must be verified that any addendum or supplement to the specification has been acknowledged by the bidder. The adequacy of bid bonds or certified checks must be verified at this time.

(f) *Conditions affecting acceptability of bids.* The borrower must take the following specified action if any of the following exist:

(1) *Fewer than three bona fide bids received.* If fewer than three bona fide bids are received for the contract project, the borrower must determine that all reasonable measures have been taken to assure competition prior to awarding the contract. This determination must be documented and such documentation submitted to RUS where required by subpart A of this part. The borrower may, however, elect to reject all bids, make changes in the specification or the qualified bidders list or both and invite new bids.

(2) *Significant error or ambiguity in the specification.* If a significant error or ambiguity in the specification is found which could result in the bidders having varying interpretations of the requirements of the bid, the borrower must either issue an addendum to each prospective bidder correcting the error or ambiguity before bids are received, or reject all bids and correct the specification. If a significant error or ambiguity in the specification is discovered after the bids are opened, the borrower must reject all bids, correct the specification and invite new bids.

(3) *Minor errors or omissions in the specification.* If minor errors or omissions in the specification are found, the borrower must issue an addendum to each prospective bidder correcting the error or omission prior to opening any bids. After bid opening, the error or omission must be corrected in the executed contract.

(4) *Minor errors or irregularities in bid.* The borrower may waive minor errors or irregularities in any bid, if the borrower determines that such minor errors or irregularities were made through inadvertence. Any such minor errors or irregularities so waived must be corrected on the bid in which they occur prior to the acceptance thereof by the borrower.

(5) *Non-minor error or irregularity in bid.* If a bid contains a non-minor error or irregularity, the bid must be rejected and the bid price must not be disclosed.

(6) *Unbalanced bid.* If a bid contains disproportionate prices between labor and materials or between various construction units, the borrower may reject the bid.

(7) *No acceptable price quoted.* If none of the bidders quote an acceptable price, the borrower may reject all bids.

(g) *Evaluating bids.* The borrower (acting through the engineer, if applicable) must conduct the evaluation of bids on the basis of the criteria set out in the "Notice and Instructions to Bidders." The contract, if awarded, must be awarded to the bidder with the lowest evaluated responsive bid.

(h) *Announcement of bids.* If possible, the borrower will announce bids at the bid opening. However, where extensive evaluation is required, the borrower may elect to adjourn and make formal written announcement to all bidders at a later time. Any discrepancy in a rejected bid must be indicated in the bid announcement.

(i) *Award of contract.* Upon completion of the bid evaluations and based upon the findings and recommendations of the borrower's management and engineer, the borrower's board of directors will either:

(1) Resolve to award the contract to the lowest evaluated responsive bidder; or

(2) Reject all bids.

(j) *Certification by the borrower and its engineer.* The borrower shall certify and the engineer shall certify as follows: "The procedures for formal competitive bidding, as described in 7 CFR 1726.201, were followed in awarding this contract." The certification executed by and on behalf of the borrower and its engineer shall be submitted to RUS in writing where required by subpart A of this part.

§ 1726.202 Informal competitive bidding.

Informal competitive bidding may be used for equipment purchases and generation construction. The borrower must use the following procedure for informal competitive bidding:

(a) *Selection of qualified bidders.* The borrower (acting through its engineer, if applicable) will compile a list of qualified bidders for each proposed contract. The borrower will send invitations to bid only to persons or organizations on its qualified bidder list for the specific project (see § 1726.23).

(b) *Invitations to bid.* The borrower (acting through its engineer, if applicable) is responsible for sending out invitations to prospective bidders, informing them of scheduled bid openings and any other action necessary to procure full, free and competitive bidding. In any event, however, sufficient invitations need to be sent out to assure competition and that at least three bids will be received. Subject to the criteria in the preceding sentence, the determination of how many and which bidders will be permitted to bid will be the responsibility of the borrower.

(c) *Notice and instructions to bidders.* The borrower must indicate in the "Notice and Instructions to Bidders" section of the bid documents that bids will be opened privately. The borrower may elect to conduct clarifying discussions with the bidders. If such clarifying discussions are held, at least the three apparent low evaluated bidders must be given an equal opportunity to resolve any questions related to the substance of the bidder's proposal and to arrive at a final price for a responsive bid.

(d) *Evaluation basis.* Any factors, other than lowest dollar amount of the bid, which are to be considered in evaluating the proposals of qualified bidders (e.g., power consumption, losses, etc.) must be stated in the "Notice and Instructions to Bidders." The borrower will not evaluate a bidder's performance record, safety record, and similar factors when evaluating a bid from a qualified and invited bidder. Such factors are to be considered when determining whether to include a particular bidder on the qualified bidders list.

(e) *Handling of bids received.* The borrower or the engineer, as applicable, will indicate, in writing, the date and time of receipt by the borrower or the engineer on the outside envelope of each bid and all letters and other transmittals amending or modifying the bids. Any bid received at the designated location after the time specified must be returned to the bidder unopened.

(f) *Bid opening.* The contracting committee will conduct the bid opening in private. The contracting committee will open each bona fide bid which has been received prior to the deadline, and review it for any irregularities, errors, or exceptions. It must be verified that any addendum to the specification has been acknowledged by each bidder. The adequacy of bid bonds or certified checks must also be verified.

(g) *Conditions affecting acceptability of bids.* The borrower must take the following specified action if any of the following exist:

(1) *Fewer than three bona fide bids received.* If fewer than three bona fide bids are received for the contract project, the borrower must determine that all reasonable measures have been taken to assure competition prior to awarding the contract. This determination must be documented and such documentation submitted to RUS where required by subpart A of this part. The borrower may, however, elect to reject all bids, make changes in the specification or the qualified bidders list or both and invite new bids.

(2) *Significant error or ambiguity in the specification.* If a significant error or ambiguity in the specification is found which could result in the bidders having varying interpretations of the requirements of the bid, the borrower must either issue an addendum to each prospective bidder correcting the error or ambiguity before bids are received, or reject all bids and correct the specification. If a significant error or ambiguity in the specification is discovered after the bids are opened, the borrower must reject all bids, correct the specification and invite new bids.

(h) *Clarification of proposals*. The contracting committee may elect not to hold any clarifying discussions and recommend awarding the contract to the low responsive bidder. Otherwise, the contracting committee must give at least each of the three apparent lowest evaluated bidders an equal opportunity to participate in discussions for the purpose of resolving questions regarding the specification and contract terms and to arrive at a final price. Neither prices of other bids nor relative ranking of any bidder are to be revealed under any circumstances. Such discussions may be held by telephone or similar means provided at least each of the three apparent lowest evaluated bidders have an equal opportunity to participate. Upon completion of the clarifying discussions, the contracting committee will determine the lowest evaluated responsive bid. If no bids are responsive after the contracting committee has completed clarifying discussions, no contract award can be made under the informal bidding procedure.

(i) *Award of the contract*. Upon completion of the bid evaluations, the contracting committee will promptly report all findings and recommendations to the borrower's board of directors. The board will either:

(1) Resolve to award the contract to the lowest evaluated responsive bidder; or

(2) Reject all bids.

(j) *Certifications by the contracting committee*. The chairperson of the contracting committee shall certify as follows: "The procedures for informal competitive bidding as described in 7 CFR 1726.202 were followed in awarding this contract." The certification executed by the chairperson of the contracting committee shall be submitted to RUS in writing where required by subpart A of this part.

§ 1726.203 Multiparty negotiation.

Multiparty negotiation may only be used where permitted under subpart F of this part or where prior RUS approval has been obtained. The borrower must use the following procedure for multiparty negotiation:

(a) *Selection of qualified bidders*. The borrower (acting through its engineer, if applicable) will compile a list of qualified bidders for each proposed contract. The borrower will send invitations to bid only to persons or organizations on its qualified bidder list for the specific project (see § 1726.23).

(b) *Invitations to bid*. The borrower (acting through its engineer, if applicable) is responsible for sending out invitations to prospective bidders, informing them of scheduled bid openings and any other action necessary to procure full, free and competitive bidding. In any event, however, sufficient invitations need to be sent out to assure competition and so that at least three bids will be received. Subject to the criteria in the preceding sentence, the determination of how many and which bidders will be permitted to bid will be the responsibility of the borrower.

(c) *Notice and instructions to bidders*. The borrower must indicate in the "Notice and Instructions to Bidders" section of the bid documents that bids will be opened privately. The borrower may elect to conduct negotiations with the bidders. If such negotiations are held, at least the three apparent low evaluated bidders must be given an equal opportunity to resolve any questions related to the substance of the bidder's proposal and to arrive at a final price.

(d) *Evaluation basis*. Any factors, other than lowest dollar amount of the bid, which are to be considered in evaluating the proposals of qualified bidders (e.g., power consumption, losses, etc.) must be stated in the "Notice and Instructions to Bidders." The borrower will not evaluate a bidder's performance record, safety record, and similar factors when evaluating a bid from a qualified and invited bidder. Such factors are to be considered when determining whether to include a particular bidder on the qualified bidders list.

(e) *Handling of bids received*. The borrower or the engineer, as applicable, will indicate, in writing, the date and time of receipt by the borrower or the engineer on the outside envelope of each bid and all letters and other transmittals amending or modifying the bids. Any bid received at the designated location after the time specified must be returned to the bidder unopened.

(f) *Bid opening*. The contracting committee will conduct the bid opening in private. The contracting committee will open each bona fide bid which has been received prior to the deadline, and review it for any irregularities, errors, or exceptions. It must be verified that any addendum to the specification has been acknowledged by each bidder. The adequacy of bid bonds or certified checks must also be verified.

(g) *Conditions affecting acceptability of bids*. The borrower must take the following specified action if any of the following exist:

(1) *Fewer than three bona fide bids received*. If fewer than three bona fide bids are received for the contract project, the borrower must determine that all reasonable measures have been taken to assure competition prior to awarding the contract. This determination must be documented and such documentation submitted to RUS where required by subpart A of this part. The borrower may, however, elect to reject all bids, make changes in the specification or the qualified bidders list or both and invite new bids.

(2) *Significant error or ambiguity in the specification*. If a significant error or ambiguity in the specification is found which could result in the bidders having varying interpretations of the requirements of the bid, the borrower must either issue an addendum to each prospective bidder correcting the error or ambiguity before bids are received, or reject all bids and correct the specification. If a significant error or ambiguity in the specification is discovered after the bids are opened, the borrower must reject all bids, correct the specification and invite new bids.

(h) *Negotiations*. The contracting committee may elect not to hold any negotiations and recommend award of the contract. Otherwise, the contracting committee must give at least each of the three apparent lowest evaluated bidders an equal opportunity to participate in negotiations for the purpose of resolving questions regarding the specification and contract terms and to arrive at a final price. Neither prices of other bids nor relative ranking of any bidder are to be revealed under any circumstances. Such discussions may be held by telephone or similar means provided at least each of the three apparent lowest evaluated bidders have an equal opportunity to participate. Upon completion of the negotiations, the contracting committee will determine the bid that is in the borrower's best interest.

(i) *Award of the contract*. Upon completion of the bid evaluations, the contracting committee will promptly report all findings and recommendations to the borrower's board of directors. The board will either:

(1) Resolve to award the contract to the selected bidder; or

(2) Reject all bids.

(j) *Certifications by the contracting committee*. The chairperson of the contracting committee shall certify as follows: "The procedures for multiparty negotiation as described in 7 CFR 1726.203 were followed in awarding this contract." The certification executed by the chairperson of the contracting committee shall be submitted to RUS in writing where required by subpart A of this part.

§ 1726.204 Multiparty unit price quotations.

The borrower or its engineer must contact a sufficient number of suppliers or contractors to assure competition and so that at least three bids will be received. On the basis of written unit price quotations, the borrower will select the supplier or contractor based on the lowest evaluated cost.

§ 1726.205 Multiparty lump sum quotations.

The borrower or its engineer must contact a sufficient number of suppliers or contractors to assure competition and so that at least three bids will be received. On the basis of written lump sum quotations, the borrower will select the supplier or contractor based on the lowest evaluated cost.

§§ 1726.206–1726.249 [Reserved]

Subpart H—Modifications to RUS Standard Contract Forms

§ 1726.250 General.

RUS provides standard forms of contract for the procurement of materials, equipment, and construction and for contract amendments and various related forms for use by RUS borrowers. See §1726.304 for a listing of these forms and how to obtain them. The standard forms of contract shall be used by the borrowers in accordance with the provisions of this part. RUS will give prior approval to certain modifications to these forms without changing the applicable requirements for RUS approval. Such approved modifications are set forth in this subpart. These are the only modifications given prior RUS approval.

[69 FR 7109, Feb. 13, 2004]

§ 1726.251 Prior approved contract modification related to price escalation on transmission equipment, generation equipment, and generation construction contracts.

(a) *General.* Where the borrower encounters reluctance among manufacturers, suppliers, and contractors to bid a firm price on transmission equipment or generation equipment, materials or construction, modifications may be made in the RUS standard form of contracts. These modifications, if applicable, may include, as an alternative to the standard form, provisions for adjusting a base price either upward or downward as determined by changes in specified indexes between the time of the bid and the time the work is performed or materials are procured by the contractor for such work. A large number of labor and materials indexes are published monthly by the Bureau of Labor Statistics (BLS). The borrower (acting through its engineer, if applicable) will select the indexes for the particular item to be used in the price adjustment clause. Suppliers' corporate indexes may not be used. Labor and materials indexes are reported in the BLS's monthly publications entitled "Employment and Earnings" and "Producer Prices and Price Indexes." These publications may be ordered through the Superintendent of Documents, U.S. Government Printing Office, Washington, DC 20402, or any of the BLS regional offices.

(b) *Material and equipment contracts.* The approved provisions needed to reflect the modifications to provide for price escalation in the material or equipment contract forms for generation facilities are as follows:

(1) Insert new paragraphs in the Notice and Instructions to Bidders as follows:

"Proposals are invited on the basis of firm prices (or prices with a stated maximum percentage escalation) or on the basis of nonfirm prices to be adjusted as provided for below or on both bases. The owner may award the contract on either basis.

Nonfirm prices. The prices are subject to adjustment upward or downward based on change in the Bureau of Labor Statistics labor and material indexes.

A proportion of ___ percent [the borrower will enter the appropriate percentage amount] of the contract price shall be deemed to represent labor cost and shall be adjusted based on changes in the Bureau of Labor Statistics, Average Hourly Earnings Rate___ [the borrower will enter the appropriate BLS index] from the month in which the bids are opened to the month in which the labor is incorporated in the equipment or materials. The adjustment for labor costs shall be obtained by applying the percentage of increase or decrease in such index, calculated to the nearest one-tenth of one percent, to the percentage of the contract prices deemed to represent labor costs. A portion of ___ percent [the borrower will enter the appropriate percentage amount] of the contract price shall be deemed to represent material costs and shall be adjusted based on changes in the Bureau of Labor Statistics, material index ___ [the borrower will enter the appropriate BLS index] for the period and in a manner similar to the labor cost adjustment."

(2) Insert the following in the contract documents under the "Proposal" section:

"Firm Price $_______
Nonfirm Price $_______"

(3) For equipment that uses a large quantity of insulating oil, the borrower may insert the following in the contract documents under the "Proposal" section:

"The price for insulating oil shall be adjusted upward or downward based on the change in the Bureau of Labor Statistics Refined Petroleum Rate (057) from the month in which the bids are opened to the month in which the oil is purchased by the equipment supplier. Contracts shall be evaluated based on an estimated cost of ___ cents per gallon [the borrower will enter the appropriate cost] for oil. Such adjustment, if any, shall not change the contract amount for purpose of applying any other adjustments to the contract prices."

(c) *Construction contracts.* The approved provisions needed to reflect the modifications to provide for price escalation in the construction contract forms for generation facilities are as follows:

(1) Insert new paragraphs in the "Notice and Instructions" to Bidders as follows:

"Proposals are invited on the basis of firm prices (or prices with a stated maximum percentage escalation) or on the basis of nonfirm prices to be adjusted as provided for below or on both bases. The owner may award the contract on either basis.

Nonfirm Prices—The prices are subject to adjustment upward or downward based on changes in the Bureau of Labor Statistics labor and material indexes.

A proportion of ___ percent [the borrower will enter the appropriate percentage amount] of the contract price shall be deemed to represent shop labor costs and shall be adjusted based on changes in the Bureau of Labor Statistics, Average Hourly Earnings Rate ___ [the borrower will enter the appropriate BLS index] from the month in which bids are opened to the month in which the work is accomplished. The adjustment for shop labor costs shall be obtained by applying the percentage increase or decrease in such index, to the percentage of each partial payment deemed to represent shop labor costs. A portion of ___ percent [the borrower will enter the appropriate percentage amount] of the contract prices shall be deemed to represent material costs and shall be adjusted based on changes in the Bureau of Labor Statistics, Producer Price Index, ___ [the borrower will enter the appropriate BLS index] for the period and in a manner similar to the shop labor costs adjustment. A portion of ___ percent [the borrower will enter the appropriate percentage amount] of the contract price shall be deemed to represent field labor costs and shall be adjusted based on changes in the Bureau of Labor Statistics, Average Hourly Earnings Rate ___ [the borrower will enter the appropriate BLS index], for the period and in a manner similar to the shop labor costs adjustment."

(2) Insert the following in the contract documents under the "Proposal" section:

"Firm Price $_______
Nonfirm Price $_______"

§ 1726.252 Prior approved contract modification related to liability for special and consequential damages.

This section applies only to transmission equipment purchases and generation contracts. Where the borrower anticipates difficulty in obtaining responsive bids on RUS standard contract forms due to a lack of limitation with respect to special and consequential damages, and where the borrower believes that such a modification will encourage competition through the receipt of an alternative bid which limits the bidder's liability for special and consequential damages, the borrower may make the following approved phrase modifications in the RUS standard contract form on which the borrower solicits bids:

(a) Insert new paragraphs in the "Notice and Instructions to Bidders" as follows:

"Proposals are invited on the basis of alternative Liability Clause Numbers 1 and 2. The Owner will determine on which Liability Clause basis the award will be made. Any other liability clauses in the proposal or any other modifications will be considered not responsive and unacceptable. These Liability Clauses are defined as follows:

Liability Clause Number 1. This will include unmodified all of the standard terms and conditions of the form of contract furnished by the Owner and attached hereto.

Liability Clause Number 2. This will include the following paragraph, in addition to all of the standard terms and conditions, otherwise unmodified, of the form of contract furnished by the Owner and attached hereto:

"Except for the Bidder's willful delay or refusal to perform the contract in accordance with its terms, the Bidder's liability to the Owner for special or consequential damages on account of breach of this contract shall not exceed in total an amount equal to ___ percent [the borrower will insert an appropriate percentage between 0 and 100 percent, inclusive] of the contract price."

(b) Insert the following in the contract documents under the "Proposal" section:

"Price $ (Based on Liability Clause 1)_______
Price $ (Based on Liability Clause 2)_______"

(c) Insert the following in the acceptance section of the standard contract form:

"This contract is based on Liability Clause Number_______."

[60 FR 10155, Feb. 23, 1995, as amended at 69 FR 7109, Feb. 13, 2004]

§ 1726.253 Prior approved contract modification related to alternative bid provision for payment to contractor for bulk purchase of materials.

When construction is to be performed over an extended period of time, but large quantities of material are to be purchased by the contractor at the beginning of the project (e.g., cable for URD installations), the borrower may allow alternative bids providing for payment to the contractor of 90 percent of the cost of such materials within 30 days of delivery of those materials at the job site. The borrower will retain the right to award the contract with or without the alternative payment provision, however, the contract still must be awarded on the basis of the lowest evaluated responsive bid for the alternative accepted.

§ 1726.254 [Reserved]

§ 1726.255 Prior approved contract modifications related to indemnification.

(a) As an alternative to the indemnification provision required in RUS standard construction contract forms in those jurisdictions requiring specific language concerning the requirement that the indemnitor indemnify the indemnitee for the indemnitee's own negligence, the borrower may add the words "otherwise this provision shall apply to any alleged negligence or condition caused by the Owner" so that the first paragraph reads as follows:

"i. To the maximum extent permitted by law, Bidder shall defend, indemnify, and hold harmless Owner and Owner's directors, officers, and employees from all claims, causes of action, losses, liabilities, and expenses (including reasonable attorney's fees) for personal loss, injury, or death to persons (including but not limited to Bidder's employees) and loss, damage to or destruction of Owner's property or the property of any other person or entity (including but not limited to Bidder's property) in any manner arising out of or connected with the Contract, or the materials or equipment supplied or services performed by Bidder, its subcontractors and suppliers of any tier. But nothing herein shall be construed as making Bidder liable for any injury, death, loss, damage, or destruction caused by the sole negligence of Owner, otherwise this provision shall apply to any negligence or condition caused by the Owner."

(b) As an alternative to the indemnification provision required in RUS standard construction contract forms in those jurisdictions that have a legal prohibition against one party indemnifying another for the other's negligence, the borrower may replace the words "defend, indemnify, and hold harmless" with the words " shall pay on behalf of" so that the first paragraph reads as follows:

"i. To the maximum extent permitted by law, Bidder shall pay on behalf of Owner and Owner's directors, officers, and employees from all claims, causes of action, losses, liabilities, and expenses (including reasonable attorney's fees) for personal loss, injury, or death to persons (including but not limited to Bidder's employees) and loss, damage to or destruction of Owner's property or the property of any other person or entity (including but not limited to Bidder's property) in any manner arising out of or connected with the Contract, or the materials or equipment supplied or services performed by Bidder, its subcontractors and suppliers of any tier. But nothing herein shall be construed as making Bidder liable for any injury, death, loss, damage, or destruction caused by the sole negligence of Owner, otherwise this provision shall apply to any negligence or condition caused by the Owner."

(c) If the alternative indemnification provision in paragraph (a) or (b) of this section is chosen by the borrower, the language of paragraph (a) or (b) of this section would be inserted in lieu of paragraph (i) of the section indicated in the RUS standard construction contract forms as follows:

(1) RUS Form 198, Equipment Contract, article IV, section 1(d).

(2) RUS Form 200, Construction Contract—Generating, article IV, section 1(d).

(3) RUS Form 257, Contract to Construct Buildings, article IV, section 1(d).

(4) RUS Form 786, Electric System Communications and Control Equipment Contract, article IV, section 1(d).

(5) RUS Form 790, Electric System Construction Contract—Non-Site Specific Construction, article IV, section 1(g).

(6) RUS Form 830, Electric System Construction Contract—Project Construction, article IV, section 1(g).

[60 FR 10155, Feb. 23, 1995, as amended at 69 FR 7110, Feb. 13, 2004]

§§ 1726.256–1726.299 [Reserved]

Subpart I—RUS Standard Forms

§ 1726.300 Standard forms of contracts for borrowers.

(a) *General.* The standard loan agreement between RUS and its borrowers provides that, in accordance with applicable RUS regulations in this chapter, the borrower shall use standard forms of contract promulgated by RUS for construction, procurement, engineering services, and architectural services financed by a loan made or guaranteed by RUS. (See section 5.16 of appendix A to subpart C of part 1718 of this chapter.) This subpart prescribes RUS procedures in promulgating standard contract forms and identifies those forms that borrowers are required to use.

(b) *Contract forms.* RUS promulgates standard contract forms, identified in the List of Required Contract Forms, § 1726.304(c), that borrowers are required to use in accordance with the provisions of this part. In addition, RUS promulgates standard contract forms contained in § 1726.304(d) that the borrowers may but are not required to use in the construction of their electric systems. Borrowers are not required to use these guidance contract forms in the absence of an agreement to do so.

[63 FR 58286, Oct. 30, 1998]

§ 1726.301 Borrower contractual obligations.

(a) *Loan agreement.* As a condition of a loan or loan guarantee under the Rural Electrification Act, borrowers are normally required to enter into RUS loan agreements pursuant to which the borrower agrees to use RUS standard forms of contracts for construction, procurement, engineering services and architectural services financed in whole or in part by the RUS loan. Normally, this obligation is contained in section 5.16 of the loan contract. To comply with the provisions of the loan agreements as implemented by this part, borrowers must use those forms of contract (hereinafter sometimes called "listed contract forms") identified in the List of Required Contract Forms, § 1724.304(c).

(b) *Compliance.* If a borrower is required by this part or by its loan agreement with RUS to use a listed standard form of contract, the borrower shall use the listed contract form in the format available from RUS, either paper or electronic format. Exact electronic reproduction is acceptable. The approved RUS standard forms of contract shall not be retyped, changed, modified, or altered in any manner not specifically authorized in this part or approved by RUS in writing on a case-by-case basis. Any modifications approved by RUS on a case-by-case basis must be clearly shown so as to indicate the modification difference from the standard form of contract.

(c) *Amendment.* Where a borrower has entered into a contract in the form required by this part, no change may be made in the terms of the contract, by amendment, waiver or otherwise, without the prior written approval of RUS.

(d) *Waiver.* RUS may waive for good cause, on a case by case basis, the requirements imposed on a borrower pursuant to this part. Borrowers seeking a waiver by RUS must provide RUS with a written request explaining the need for the waiver. Waiver requests should be made prior to issuing the bid package to bidders.

(e) *Violations.* A failure on the part of the borrower to use listed contracts as prescribed in this part is a violation of the terms of its loan agreement with RUS and RUS may exercise any and all remedies available under the terms of the agreement or otherwise.

[63 FR 58286, Oct. 30, 1998, as amended at 69 FR 7110, Feb. 13, 2004]

§ 1726.302 Notice and publication of listed contract forms.

(a) *Notice.* Upon initially entering into a loan agreement with RUS, borrowers will be provided with all listed contract forms. Thereafter, new or revised listed contract forms promulgated by RUS, including RUS approved exceptions and alternatives, will be sent by regular or electronic mail to the address of the borrower as identified in its loan agreement with RUS.

(b) *Availability.* Listed standard forms of contract are available from: Rural Utilities Service, Program Development and Regulatory Analysis, U.S. Department of Agriculture, Stop 1522, 1400 Independence Avenue, SW., Washington DC 20250–1522, telephone number (202) 720–8674. The listed standard forms of contract are also available on the RUS Web site at: *http://www.usda.gov/rus/electric/forms/index.htm.* The listed standard forms of contract can be found in §1724.304(c), List of Required Contract Forms.

[63 FR 58287, Oct. 30, 1998, as amended at 69 FR 7110, Feb. 13, 2004]

§ 1726.303 Promulgation of new or revised contract forms.

RUS may, from time to time, undertake to promulgate new contract forms or revise or eliminate existing contract forms. In so doing, RUS shall publish notice of rulemaking in the FEDERAL REGISTER announcing, as appropriate, a revision in, or a proposal to amend §1726.304, List of Electric Program Standard Contract Forms. The amendment may change the existing identification of a listed contract form; for example, changing the issuance date of a listed contract form or by identifying a new required contract form. The notice of rulemaking will describe the new standard contract form or the substantive change in the listed contract form, as the case may be, and the issues involved. The standard contract form or relevant portions thereof may be appended to the supplementary information section of the notice of rulemaking. As appropriate, the document shall provide an opportunity for interested persons to provide comments. A copy of each such FEDERAL REGISTER document will be sent by regular or electronic mail to all borrowers.

[63 FR 58287, Oct. 30, 1998]

§ 1726.304 List of electric program standard contract forms.

(a) *General.* This section contains a list of RUS electric program standard contract forms. Paragraph (c) of this section contains the list of required contract forms, *i.e.*, those forms of contracts that borrowers are required to use by the terms of their RUS loan agreements as implemented by the provisions of this part. Paragraph (d) of this section sets forth the list of guidance contract forms, i.e., those forms of contracts provided as guidance to borrowers in the construction of their systems. See §1726.302(b) for availability of these forms.

(b) *Issuance date.* Where required by this part to use a standard form of contract in connection with RUS financing, the borrower shall use that form identified by issuance date in the List of Required Contract Forms in paragraph (c) of this section, as most recently published as of the date the borrower issues the bid package to bidders.

(c) List of required contract forms.

(1) RUS Form 168b, Rev. 2–04, Contractor's Bond. This form is used to obtain a surety bond and is used with RUS Forms 200, 257, 786, 790, and 830.

(2) RUS Form 168c, Rev. 2–04, Contractor's Bond (less than $1 million). This form is used in lieu of RUS Form 168b to obtain a surety bond when contractor's surety has accepted a Small Business Administration guarantee.

(3) RUS Form 187, Rev. 2–04, Certificate of Completion, Contract Construction. This form is used for the closeout of RUS Forms 200, 257, 786, and 830.

(4) RUS Form 198, Rev. 4–04, Equipment Contract. This form is used for equipment purchases.

(5) RUS Form 200, Rev. 2–04, Construction Contract—Generating. This form is used for generating plant construction or for the furnishing and installation of major items of equipment.

(6) RUS Form 213, Rev. 2–04, Certificate ("Buy American"). This form is used to document compliance with the "Buy American" requirement.

(7) RUS Form 224, Rev. 2–04, Waiver and Release of Lien. This form is used for the closeout of RUS Forms 198, 200, 257, 786, 790, and 830.

(8) RUS Form 231, Rev. 2–04, Certificate of Contractor. This form is used for the closeout of RUS Forms 198, 200, 257, 786, 790, and 830.

(9) RUS Form 238, Rev. 2–04, Construction or Equipment Contract Amendment. This form is used for amendments.

(10) RUS Form 254, Rev. 2–04, Construction Inventory. This form is used for the closeout of RUS Form 830. Minor electronic modifications are acceptable for RUS Form 254.

(11) RUS Form 257, Rev. 2–04, Contract to Construct Buildings. This form is used to construct headquarters buildings and other structure construction.

(12) RUS Form 307, Rev. 2–04, Bid Bond. This form is used to obtain a bid bond.

(13) RUS Form 786, Rev. 2–04, Electric System Communications and Control Equipment Contract (including installation). This form is used for delivery and installation of equipment for system communications.

(14) RUS Form 790, Rev. 2–04, Electric System Construction Contract—Non-Site Specific Construction. This form is used for limited distribution construction accounted for under work order procedure.

(15) RUS Form 792b, Rev. 2–04, Certificate of Construction and Indemnity Agreement. This form is used for the closeout of RUS Form 790.

(16) RUS Form 830, Rev. 2–04, Electric System Construction Contract—Project Construction. This form is used for distribution and transmission line project construction.

(d) List of guidance contract forms. RUS does not currently publish any guidance forms for electric borrowers.

[63 FR 58287, Oct. 30, 1998, as amended at 69 FR 7110, Feb. 13, 2004; 69 FR 52595, Aug. 27, 2004]

§§ 1726.305–1726.399 [Reserved]

Subpart J—Contract Closeout

§ 1726.400 Final contract amendment.

As needed, a final contract amendment will be prepared and processed in accordance with § 1726.24(b) prior to or in conjunction with the closeout of the contract.

§ 1726.401 Material contract closeout.

(a) *Delivery inspection.* The borrower (acting through its engineer, if applicable) will verify that all materials are delivered in proper quantities, in good condition, and in compliance with applicable specifications.

(b) *Closeout documents.* The borrower (acting through its engineer, if applicable) will obtain from the supplier a "Buy American" certificate, RUS Form 213, any manufacturer's guarantee(s) and, if applicable, a copy of RUS Form 224, Waiver and Release of Lien. Closeout documents for materials contracts need not be submitted to RUS unless specifically requested by RUS on a case by case basis.

(c) *Final payment.* Upon completion of the actions required under paragraphs (a) and (b) of this section, the borrower shall make final payment to the supplier in accordance with the provisions of the material contract or written purchase order.

[60 FR 10155, Feb. 23, 1995, as amended at 69 FR 7110, Feb. 13, 2004]

§ 1726.402 Equipment contract closeout.

This section is applicable to contracts executed on RUS Form 198.

(a) *Final inspection and testing of equipment.* The borrower (acting through its engineer, if applicable) will perform the final inspection and testing of equipment as appropriate for the specific equipment. The borrower (acting through its engineer, if applicable) will schedule such inspection and testing at a time mutually agreeable to the borrower, engineer, and the supplier or manufacturer. Within thirty (30) days after completion of the inspection and testing, the borrower (acting through its engineer, if applicable) will prepare a report of the inspection and testing, obtain a copy of the report from the engineer, and submit a copy to the supplier or manufacturer. This report must include a detailed description of the methods of conducting the test(s), observed data, comparison of guaranteed and actual performance, and recommendations concerning acceptance. The borrower will obtain from the engineer a written certification stating that the equipment has been installed, placed in satisfactory operation and tested, and meets the contract requirements. Where more than one-hundred and eighty (180) days have elapsed since the delivery of the equipment and the equipment has not been installed or tested, the contract may be closed out upon certification by the engineer that the equipment has been inspected and appears to be in accordance with the contract requirements.

(b) *Closeout documents.* (1) The borrower (acting through its engineer, if applicable) will obtain the following executed documents:

(i) Certification by the project engineer in accordance with paragraph (a) of this section.

(ii) All guarantees or warranties.

(iii) A "Buy American" certificate, RUS Form 213, from the supplier or manufacturer.

(2) Closeout documents for materials contracts need not be submitted to RUS unless specifically requested by RUS.

(c) *Final payment.* Upon completion of the actions required under paragraphs (a) and (b) of this section, the borrower will make final payment to the supplier or manufacturer in accordance with the provisions of the equipment contract.

§ 1726.403 Project construction contract closeout.

This section is applicable to contracts executed on RUS Forms 200, 257, 786, and 830.

(a) *Final test of equipment supplied under a construction contract.* If equipment is supplied under a construction contract, the borrower (acting through its architect or engineer, if applicable) will perform the final inspection and testing of equipment as appropriate for the specific equipment. The borrower (acting through its architect or engineer, if applicable) will schedule such inspection and testing at a time mutually agreeable to the borrower, architect or engineer, and the contractor. Within thirty (30) days after completion of the inspection and testing, the borrower (acting through its architect or engineer, if applicable) will prepare a report of the inspection and testing, obtain a copy of the report from its architect or engineer, and submit a copy to the contractor. This report must include a detailed description of the methods of conducting the test(s), observed data, comparison of guaranteed and actual performance, and recommendations concerning acceptance. The borrower will obtain from its architect or engineer a written certification stating that the equipment has been installed, placed in satisfactory operation and tested, and meets the contract requirements. Where more than one-hundred and eighty (180) days have elapsed since the delivery of the equipment and the equipment has not been installed or tested, the contract may be closed out upon certification by its architect or engineer that the equipment has been inspected and appears to be in accordance with the contract requirements.

(b) *Final inspection of construction.* The borrower will require the contractor to notify the architect or engineer when construction is complete. The borrower (acting through the architect or engineer, if applicable) will schedule such final inspection at a time mutually agreeable to the borrower, architect or engineer, contractor, and the respective RUS General Field Representative (GFR), if the GFR has notified the borrower or its architect or engineer of a desire to observe the final inspection. The borrower (acting through its architect or engineer, if applicable) will perform a final inspection of the construction and notify the contractor of any required changes or corrections.

(c) *Closeout documents.* (1) Upon satisfactory completion of construction (including all changes and corrections by the contractor), the borrower (acting through its architect or engineer, if applicable) will obtain executed copies of the following documents:

(i) RUS Form 187, Certificate of Completion, Contract Construction.

(ii) RUS Form 213, "Buy American" certificate.

(iii) RUS Form 224, Waiver and Release of Lien, from each manufacturer, supplier, and contractor which has furnished material or services or both in connection with the construction.

(iv) RUS Form 231, Certificate of Contractor.

(v) RUS Form 254, Construction Inventory, including all supporting documents, such as RUS Forms 254a-c, construction change orders, and amendments for contracts executed on RUS Form 830.

(vi) Certification by the project architect or engineer in accordance with §1726.403(a), if applicable.

(vii) Final design documents, as outlined in part 1724 of this chapter.

(2) *Distribution of closeout documents.* (i) The borrower will retain one copy of each of the documents identified in paragraph (c)(1) of this section in accordance with applicable RUS requirements regarding retention of records.

(ii) For contracts subject to RUS approval, the borrower will submit either a certification or the following closeout documents for RUS approval:

(A) RUS Form 187, Certificate of Completion, Contract Construction.

(B) RUS Form 231, Certificate of Contractor.

(C) RUS Form 254, Construction Inventory, including all supporting documents, such as RUS Forms 254a-c and construction change orders, for contracts executed on RUS Form 830.

(iii) For contracts not subject to RUS approval, the closeout is not subject to RUS approval. The borrower will send one copy of RUS Form 187 to RUS for information prior to or in conjunction with the applicable RUS Form 219, Inventory of Work Orders. The remaining closeout documents need not be sent to RUS unless specifically requested by RUS.

(d) *Final payment.* (1) The borrower will make final payment to the contractor upon completion of approval of all closeout documents by the parties to the contract, in accordance with the terms of the construction contract.

(2)(i) Upon receipt of final payment by the contractor, the borrower will obtain from the contractor a certification of receipt of final payment in the following form:

"The undersigned acknowledges receipt of the final contract payment of $____ as satisfaction in full of all claims of the undersigned under the construction contract between the undersigned and ____ (borrower), dated as amended, and as complete performance by the latter of all obligations to be performed by it pursuant thereto. The total amount received under this contract is shown above."

(ii) The certification in paragraph (d)(2)(i) of this section is to be executed for the contractor by: The sole owner, a partner, or an officer of the corporation.

[60 FR 10155, Feb. 23, 1995, as amended at 69 FR 7110, Feb. 13, 2004; 84 FR 32618, July 9, 2019; 86 FR 36198, July 9, 2021]

§ 1726.404 Non-site specific construction contract closeout.

This section is applicable to contracts executed on RUS Form 790.

(a) *Final test of equipment supplied under a construction contract.* If equipment is supplied under a construction contract, the borrower (acting through its engineer, if applicable) will perform the final inspection and testing of equipment as appropriate for the specific equipment. The borrower (acting through its engineer, if applicable) will schedule such inspection and testing at a time mutually agreeable to the borrower, its engineer, and the contractor. Within thirty (30) days after completion of the inspection and testing, the borrower (acting through its engineer, if applicable) will prepare a report of the inspection and testing, obtain a copy of the report from its engineer, and submit a copy to the contractor. This report must include a detailed description of the methods of conducting the test(s), observed data, comparison of guaranteed and actual performance, and recommendations concerning acceptance. The borrower will obtain from the engineer a written certification stating that the equipment has been installed, placed in satisfactory operation and tested, and meets the contract requirements. Where more than one-hundred and eighty (180) days have elapsed since the delivery of the equipment and the equipment has not been installed or tested, the contract may be closed out upon certification by the engineer that the equipment has been inspected and appears to be in accordance with the contract requirements.

(b) *Final inspection of construction.* The borrower will require the contractor to notify its engineer when construction of a section of the project is complete. The borrower (acting through its engineer, if applicable) will schedule such final inspection at a time mutually agreeable to the borrower, its engineer, contractor, and the respective GFR, if the GFR has notified the borrower or its engineer of a desire to observe the final inspection. The borrower (acting through its engineer, if applicable) will perform a final inspection of the construction of that section of the project and notify the contractor of any required changes or corrections.

(c) *Closeout documents.* (1) Upon satisfactory completion of construction of a section of the project (including all changes and corrections by the contractor), the borrower (acting through its engineer, if applicable) will obtain executed copies of the following documents:

(i) RUS Form 792b, Certificate of Contractor and Indemnity Agreement

(ii) RUS Form 213, "Buy American" certificate.

(iii) Certification by the project engineer in accordance with paragraph (a) of this section, if applicable.

(iv) Final design documents, as outlined in part 1724 of this chapter.

(2) *Distribution of closeout documents.* (i) The borrower will retain one copy of each of the documents identified in paragraph (c)(1) of this section in accordance with applicable RUS requirements regarding retention of records.

(ii) For contracts not subject to RUS approval, the closeout is not subject to RUS approval and the closeout documents need not be sent to RUS unless specifically requested by RUS.

[60 FR 10155, Feb. 23, 1995, as amended at 69 FR 7111, Feb. 13, 2004]

§ 1726.405 Inventory of work orders (RUS Form 219).

Upon completion of the contract closeout, the borrower shall complete RUS Form 219, Inventory of Work Orders, in accordance with part 1717, Post-Loan Policies and Procedures Common to Insured and Guaranteed Electric Loans, of this chapter.

PART 1728—ELECTRIC STANDARDS AND SPECIFICATIONS FOR MATERIALS AND CONSTRUCTION

AUTHORITY: 7 U.S.C. 901 *et seq.*, 1921 *et seq.*, 6941 *et seq.*

SOURCE: 48 FR 31853, July 12, 1983, unless otherwise noted. Redesignated at 55 FR 39395, Sept. 27, 1990.

§ 1728.10 General purpose and scope.

(a) The requirements of this part are based on contractual provisions between RUS and the organizations which receive financial assistance from RUS.

(b) RUS will establish certain specifications and standards for materials, equipment, and construction units that will be acceptable for RUS financial assistance for the electric program. Materials and equipment purchased by the electric borrowers or accepted as contractor-furnished material must conform to RUS standards and specifications where they have been established and, if included in RUS Bulletin 43–5, "List of Materials Acceptable for Use on Systems of RUS Electrification Borrowers" (List of Materials), must be selected from that list or must have received technical acceptance from RUS. RUS, through its Technical Standards

Committees, will evaluate certain materials, equipment and construction units, and will determine acceptance.

[50 FR 47710, Nov. 20, 1985. Redesignated at 55 FR 39395, Sept. 27, 1990]

§ 1728.20 Establishment of standards and specifications.

(a) *National and other standards.* RUS will utilize standards of national standardizing groups, such as the American National Standards Institute (ANSI), American Wood Preservers' Association (AWPA), the various national engineering societies and the National Electrical Safety Code (NESC), to the greatest extent practical. When there are no national standards or when RUS determines that the existing national standards are not adequate for rural electric systems, RUS will prepare standards for material and equipment to be used on systems of electric borrowers. RUS standards and specifications will be codified or listed in § 1728.97, Incorporation by Reference of Electric Standards and Specifications. RUS will also prepare specifications for materials and equipment when it determines that such specifications will result in reduced costs, improved materials and equipment, or in the more effective use of engineering services.

(b) *Deviations from Standards.* No member of the RUS staff will be permitted to authorize deviations from the standard specifications, or to establish or change the technical standards, or to authorize the use of items that have not received acceptance by the Technical Standards Committees, except as provided for under § 1728.70, or by authorization and/or delegation of authority by the Administrator of RUS.

(c) *Category of Items.* Items appearing in the List of Materials are listed by categories of generic items which are used in RUS construction standards incorporated by reference in § 1728.97. RUS will establish and define these categories and will establish all criteria for acceptability within these categories.

[50 FR 47710, Nov. 20, 1985. Redesignated at 55 FR 39395, Sept. 27, 1990, as amended at 55 FR 53487, Dec. 31, 1990]

§ 1728.30 Inclusion of an item for listing or technical acceptance.

(a) *Scope.* RUS, through its Technical Standards Committees "A" and "B" will determine the acceptability of certain standards, standard specifications, standard drawings, and items of materials and equipment to be used in transmission, distribution and general plant (excluding office equipment, tools, and work equipment, and consumer-owned electric wiring facilities).

(b) *Addresses of Committees.* The address of Technical Standards Committee "A" is: Chairman, Technical Standards Committee "A" (Electric), Rural Utilities Service, U.S. Department of Agriculture, Washington, DC 20250–1500. The address of Technical Standards Committee "B" is: Chairman, Technical Standards Committee "B" (Electric), Rural Utilities Service, U.S. Department of Agriculture, Washington, DC 20250–1500.

(c) *Review by Technical Standards Committee "A".* All proposals for listing a product in the List of Materials must be addressed to Technical Standards Committee "A." This committee will consider all proposals made by sponsors of specifications, drawings, materials, or equipment in categories for which RUS has established criteria for acceptability. A sponsor may be a manufacturer, supplier, contractor or any other person or organization which has made an application for listing or has requested an action by the committee. Committee "A" will consider all relevant information presented in determining whether an item should be accepted by Technical Standards Committee "A." Formal rules of evidence and procedure shall not apply to proceedings before this committee.

(d) *Action by Technical Standards Committee "A".* (1) Committee "A" may take one of the following actions:

(i) Accept an item for listing without conditions (domestic items only),

(ii) Reject an item (domestic or nondomestic),[1]

(iii) Accept an item for listing with conditions (domestic items only),

[1] Nondomestic items are items which do not qualify as domestic products pursuant to RUS "Buy American" requirement.

(iv) Table an item for a time period sufficient to allow the sponsor to be notified and furnish additional information (domestic or nondomestic),

(v) Grant technical acceptance with or without conditions for a period of one year from the date of notification by RUS (nondomestic items only).

(2) All committee decisions regarding the actions listed above must be unanimous. If the vote is not unanimous, the item shall be referred to Technical Standards Committee "B." Written notice of Technical Standards Committee "A's" decision, stating the basis for the decision, will be provided to the sponsor.

(3) Items accepted without conditions by the Technical Standards Committees will be considered to be accepted on a general basis. No restrictions as to quantity or application will be placed on items which have received general acceptance. Items accepted subject to certain conditions, such as limited use to gain service experience, or limited use appropriate to certain areas and conditions, will be considered to be accepted on a conditional basis. The conditions will be cited as a part of the listing provided for in §1728.60, or as part of the technical acceptance for nondomestic items.

(e) *Appeal to Technical Standards Committee "B"*. A sponsor may request a review of an adverse decision by Technical Standards Committee "A" within ten (10) days of notification of such decision by submitting a letter requesting such review to Technical Standards Committee "B" (Electric).

(f) *Action by Technical Standards Committee "B"*. Committee "B" may take any of the actions listed for Committee "A" in §1728.30(d). However, for a Committee "B" action to be effective it must be by majority vote. Failure to obtain a majority on one of the proposed actions shall mean that the product will not be listed or accepted. Committee "B's" determination shall be based on the record developed before Committee "A" and such additional information as Committee "B" may request. Formal rules of procedure and evidence shall not apply to proceedings before Committee "B." Written notice of Committee "B's" decision, stating the basis of the decision, will be provided to the sponsor.

(g) *Appeal to the Administrator*. In the event of an adverse decision by Committee "B," the sponsor may, within ten (10) days of notification of such decision, request a review of this decision by submitting a letter to the Administrator requesting such a review.

(h) *Change in Design*. RUS acceptance of an item will be conditioned on the understanding that no design changes (material or dimensions) affecting the quality, strength, or electrical characteristics of the item shall be made without prior concurrence of Technical Standards Committee "A."

[50 FR 47711, Nov. 20, 1985. Redesignated at 55 FR 39395, Sept. 27, 1990]

§ 1728.40 Procedure for submission of a proposal.

(a) *Written Request*. Consideration of an item of material or equipment will be obtained by the sponsor through the submission of a written request in an original and five copies addressed to the Chairman, Technical Standards Committee "A" (Electric). The letter must include the catalog number or other identifying number or code as well as a description of the item. In the event that an item being submitted is also intended for consideration by Technical Standards Committee "A" (Telephone), a separate request must be made to the telephone committee. (See part 1755 of this chapter).

(b) *Technical and Performance Data*. Six copies of the specification of manufacture, drawings and test data must be submitted to the committee. Six copies of the performance history shall also be submitted unless RUS determines that such performance history is not reasonably available.

(c) *Sample*. One sample of the item must be submitted to the Chairman, Technical Standards Committee "A," unless RUS waives the requirements of the sample. In case of large, bulky or extremely heavy samples, the sponsor should contact the Chairman, Technical Standards Committee "A" (Electric), at the above address, before any sample is shipped.

(d) *Action on Proposal.* RUS will inform a sponsor of the action taken on the sponsor's proposal.

[50 FR 47711, Nov. 20, 1985. Redesignated at 55 FR 39395, Sept. 27, 1990]

§ 1728.50 Removal of an item from listing or technical acceptance.

(a) *Removal Actions.* An item of material or equipment may be removed from the listing or technical acceptance in accordance with the following procedures upon determination that the item is unsatisfactory or has been misrepresented to the owner or RUS.

(b) *Notification by the Committee.* The sponsor of an item of material or equipment will be notified in writing of a proposal to remove such item from the listing or technical acceptance.

(c) *Supplemental Information.* Within ten (10) days of receipt of such notification, the sponsor may submit to Committee "A" a letter expressing the sponsor's intent to submit written supplemental technical information relevant to Committee "A's" determination. The sponsor must submit such information within twenty (20) days from the submission of its letter to Committee "A." Committee "A" will have the discretion of making a decision following the expiration of the time periods provided in this paragraph.

(d) *Review by the Technical Standards Committee "A".* Committee "A" will consider all relevant information presented in determining whether an item should be removed from the listing or technical acceptance. Formal rules of evidence and procedure shall not apply to proceedings before Technical Standards Committee "A."

(e) *Action by the Technical Standards Committee "A".* Committee "A" may take one of the following actions:

(1) Order the immediate removal of the item from the listing, or technical acceptance,

(2) Condition the item's continued listing, or technical acceptance,

(3) Recommend a basis of settlement which will adequately protect the interest of the Government, or

(4) Delay the effectiveness of its decision for a time period sufficient to allow the sponsor to appeal to Technical Standards Committee "B."

All committee "A" decisions regarding the actions listed above must be by unanimous vote. If the vote is not unanimous, the item will be referred to Technical Standards Committee "B."

Written notice of Technical Standards Committee "A's" decision, stating the basis for the decision, will be provided to the sponsor.

(f) *Additional Opportunity to Present Information.* At the request of the sponsor, RUS may afford additional opportunity for consideration of relevant information. Such additional opportunity may include, without limitation, a meeting between RUS and the sponsor in such a forum that RUS may determine. In making this decision, RUS will consider, among other things, the best interests of RUS, its borrowers, and the sponsor, and the best manner to develop sufficient information relating to the proposed action.

(g) *Appeal to the Technical Standards Committee "B".* Within ten (10) days of notification of Committee "A's" decision, a sponsor may appeal in writing to Technical Standards Committee "B" to review Committee "A's" decision, specifying the reasons for such a request. Committee "B's" determination, in response to such request, shall be based on the record developed before Committee "A" and such additional information as Committee "B" may request. Formal rules of procedure and evidence shall not apply to proceedings before Committee "B."

(h) *Action by Technical Standards Committee "B".* Committee "B," by majority vote, may take one of the following actions:

(1) Order the immediate removal of the item from listing, or technical acceptance,

(2) Condition the item's continued listing, or technical acceptance,

(3) Recommend a basis of settlement which adequately protects the interests of the Government, or

(4) Delay the effectiveness of its decision for a time period sufficient to allow the sponsor to appeal to the Administrator of RUS.

Failure to obtain a majority vote on any of the above actions shall mean that the product will continue to be listed or accepted.

Written notice of Committee "B's" decision stating the basis of the decision will be provided to the sponsor.

(i) *Appeal to the Administrator.* Within ten (10) days of the receipt of Committee "B's" decision, a sponsor may appeal to the Administrator to review Committee "B's" decision. If an appeal is made, the sponsor shall submit a written request to the Administrator, Rural Utilities Service, Room 4053, South Building, U.S. Department of Agriculture, Washington, DC 20250–1500 specifying the reasons to request reconsideration. The Administrator will have the option to decline the request, in which case the decision of Committee "B" shall stand. If a review is granted, the determination by the Administrator or the Administrator's designee shall be based on the record developed before Committee "A" and Committee "B" and such additional information as the Administrator may request. Formal rules of procedure and evidence shall not apply to the actions of the Administrator.

(j) *Action by the Administrator.* The Administrator may take one of the following actions:

(1) Order the immediate removal of the item from the listing, or technical acceptance,

(2) Condition its continued listing, or technical acceptance, or

(3) Recommend a basis of settlement which adequately protects the interests of the Government.

Written notice of the Administrator's determination, stating the basis for the decision, will be provided to the sponsor.

The Administrator's actions are final.

[50 FR 47711, Nov. 20, 1985. Redesignated at 55 FR 39395, Sept. 27, 1990]

§ 1728.60 List of materials and equipment.

(a) *General.* Those items of material or equipment accepted by Technical Standards Committee "A" or "B," with the exception of technically accepted nondomestic items, will be listed in the List of Materials. Items which do not qualify as domestic products may be accepted on a technical basis only (technical acceptance) for a period of one year as provided in § 1728.30(c)(1) and will not be included in the List of Materials.

(b) *Publishing and Revisions.* RUS will reissue the List of Materials every year, dated July, and issue supplements, if needed, dated October, January, and April of every year. An RUS office copy, which is the official current copy, of the List of Materials, will be updated every time changes are made by the Technical Standards Committees.

(c) *Dual Listings.* RUS, through its Technical Standards Committees, will accept for listing only one item of a particular type of material or equipment for each manufacturer. If a manufacturer submits an item to perform the identical function of a listed item, RUS, through its Technical Standards Committees, may accept that item and remove the one previously listed. RUS will list only new items of material and equipment in the List of Materials. Used items will not be considered for listing.

[50 FR 47712, Nov. 20, 1985. Redesignated at 55 FR 39395, Sept. 27, 1990]

§ 1728.70 Procurement of materials.

(a) *By Owner.* When purchasing the type of materials included in the List of Materials, RUS borrowers shall purchase only materials listed in the List of Materials, or materials which have a current technical acceptance by RUS and meet the "Buy American" requirement.

(b) *By Contractor.* When performing work for an RUS borrower, contractors shall supply only items from the general acceptance pages of the List of Materials, or obtain the borrower's concurrence prior to purchase and use of a technically nondomestic item or any item listed on a conditional basis.

(c) *Procurement of Unlisted Items.* (1) The borrower shall request prior approval from RUS for use of an item that does not fall in categories established by RUS in the List of Materials for which acceptability has been established by the Technical Standards Committees.

(2) RUS will also determine, on a case-by-case basis, whether to allow use of an unlisted item in emergency situations and for experimental use or to meet a specific need. For purposes of

this part 1728, an emergency shall mean a situation wherein the supply of listed material and equipment from the industry is not readily available, or the standard designs are not applicable to the borrower's specific problem under consideration.

(3) RUS will make arrangements for test or experimental use of newly developed items requiring limited trial use. RUS, working with the borrower and the manufacturer, will establish test locations for the items to facilitate installation and observation.

[50 FR 47712, Nov. 20, 1985. Redesignated at 55 FR 39395, Sept. 27, 1990]

§ 1728.97 Incorporation by reference of electric standards and specifications.

Certain material is incorporated by reference into this part with the approval of the Director of the Federal Register under 5 U.S.C. 552(a) and 1 CFR part 51. All approved material is available for inspection at the Rural Utilities Service, U.S. Department of Agriculture, Room 5170–S, Washington, DC 20250–1522, call (202) 720–8674 and is available from the sources listed in this section. It is also available for inspection at the National Archives and Records Administration (NARA). For information on the availability of this material at NARA, email *fr.inspection@nara.gov* or go to *www.archives.gov/federal-register/cfr/ibr-locations.html.*

(a) Rural Utilities Service, U.S. Department of Agriculture, Room 5170–S–S, U.S. Department of Agriculture, Washington, DC 20250. For information on the availability of this material, call (202) 720–8674 or go to: *https://www.rd.usda.gov/publications/regulations-guidelines/bulletins.*

(1) Bulletin 50–4 (D–801), Specification and Drawings for 34.5/19.9 kV Distribution Line Construction (11–86), incorporation approved for § 1728.98.

(2) Bulletin 50–15 (DT–3), RUS Specifications for Pole Top Pins with 1⅜′ Diameter Lead Thread (1–51), incorporation approved for § 1728.98.

(3) Bulletin 50–16 (DT–4), RUS Specifications for Angle Suspension Brackets (3–52), incorporation approved for § 1728.98.

(4) Bulletin 50–19 (DT–7), RUS Specifications for Clevis Bolts (8–53), incorporation approved for § 1728.98.

(5) Bulletin 50–23 (DT–18), RUS Specifications for 60″ Wood Crossarm Braces (2–71), incorporation approved for § 1728.98.

(6) Bulletin 50–31 (D–3), RUS Specifications for Pole Top Pins with 1'' Diameter Lead Threads (2–79), incorporation approved for § 1728.98.

(7) Bulletin 50–32 (D–4), RUS Specifications for Steel Crossarm Mounted Pins with 1″ Diameter Lead Threads (10–50), incorporation approved for § 1728.98.

(8) Bulletin 50–33 (D–5), RUS Specifications for Single and Double Upset Spool Bolts (2–51), incorporation approved for § 1728.98.

(9) Bulletin 50–34 (D–6), RUS Specifications for Secondary Swinging Clevises (12–70), incorporation approved for § 1728.98.

(10) Bulletin 50–35 (D–7), RUS Specifications for Service Swinging Clevises (9–52), incorporation approved for § 1728.98.

(11) Bulletin 50–36 (D–8), RUS Specifications for Service Deadend Clevises (9–52), incorporation approved for § 1728.98.

(12) Bulletin 50–40 (D–14), RUS Specifications for Pole Top Brackets for Channel Type Pins (9–51), incorporation approved for § 1728.98.

(13) Bulletin 50–41 (D–15), RUS Specifications for Service Wireholders (11–51), incorporation approved for § 1728.98.

(14) Bulletin 50–55 (T–2), RUS Specifications for Overhead Ground Wire Support Brackets (5–53), incorporation approved for § 1728.98.

(15) Bulletin 50–56 (T–3), RUS Specifications for Steel Plate Anchors for Transmission Lines (12–53), incorporation approved for § 1728.98.

(16) Bulletin 50–60 (T–9), RUS Specification—Single Pole Steel Structures, Complete with Arms (12–71), incorporation approved for § 1728.98.

(17) Bulletin 50–72 (U–4), RUS Specification for Electrical Equipment Enclosures (5–35 kV) (10–79), incorporation approved for § 1728.98.

(18) Bulletin 50–73 (U–5), RUS Specifications for Pad-Mounted Transformers (Single and Three-Phase) (1–77), incorporation approved for § 1728.98.

(19) Bulletin 50–74 (U–6), RUS Specification for Secondary Pedestals (600 Volts and Below) (10–79), incorporation approved for § 1728.98.

(20) Bulletin 50–91 (S–3), RUS Specifications for Step-Down Distribution Substation Transformers (34.4–138 kV) (1–78), incorporation approved for § 1728.98.

(21) Bulletin 1728F–700, RUS Specification for Wood Poles, Stubs and Anchor Logs, September 9, 2021, incorporation approved for §§ 1728.98 and 1728.202.

(22) Bulletin 1728F–803, Specifications and Drawings for 24.9/14.4 kV Line Construction (10–98), incorporation approved for § 1728.98.

(23) Bulletin 1728F–804 (D–804), Specification and Drawings for 12.47/7.2 kV Line Construction, October 2005, incorporation approved for § 1728.98.

(24) Bulletin 1728F–806 (D–806) Specifications and Drawings for Underground Electric Distribution, October 11, 2018, incorporation approved for § 1728.98.

(25) Bulletin 1728F–810, Electric Transmission Specifications and Drawings, 34.5 kV to 69 kV (3–98), incorporation approved for §§ 1728.98 and 1728.201.

(26) Bulletin 1728F–811, Electric Transmission Specifications and Drawings, 115 kV to 230 kV (3–98), incorporation approved for §§ 1728.98 and 1728.201.

(b) American Institute of Timber Construction (AITC), 7012 S Revere Park Way, Englewood, Colorado 80112, telephone (303) 792–9559, web address: *https://www.aitc-glulam.org/index.asp*.

(1) AITC 200–2009, Manufacturing Quality Control Systems Manual For Structural Glued Laminated Timber, copyright 2009, incorporation by reference approved for §§ 1728.201 and 1728.202.

(2) [Reserved]

(c) American National Standards Institute (ANSI), 25 West 43rd Street, New York, New York 10036, telephone (212) 642–4900, Web address: *http://www.ansi.org*.

(1) ANSI O5.2–2020, Structural Glued Laminated Timber for Utility Structures, approved January 10, 2020, incorporation by reference approved for §§ 1728.201 and 1728.202.

(2) ANSI O5.3–2015, Solid Sawn Wood Crossarms & Braces: Specifications & Dimensions, approved January 9, 2015, incorporation by reference approved for § 1728.201.

(d) ASTM International, 100 Barr Harbor Drive, West Conshohocken, PA 19428–2959, Telephone: (610) 832–9585, website: *www.astm.org*.

(1) ASTM B 3–01 (Reapproved 2007)—Standard Specification for Soft or Annealed Copper Wire, (ASTM B 3–01) approved March 15, 2007, incorporated by reference approved for § 1728.204.

(2) ASTM B 8–04—Standard Specification for Concentric-Lay-Stranded Copper Conductors, Hard, Medium-Hard, or Soft (ASTM B 8–04), approved April 1, 2004, incorporated by reference approved for § 1728.204.

(3) ASTM B 230/B 230M–07—Standard Specification for Aluminum 1350–H19 Wire for Electrical Purposes (ASTM B 230/B 230M–07), approved March 15, 2007, incorporated by reference approved for § 1728.204.

(4) ASTM B 231/B 231M–04—Standard Specification for Concentric-Lay-Stranded Aluminum 1350 Conductors (ASTM B 231/B 231M–04), approved April 1, 2004, incorporated by reference approved for § 1728.204.

(5) ASTM B 400–08—Standard Specification for Compact Round Concentric-Lay-Stranded Aluminum 1350 Conductors (ASTM B 400–08), approved September 1, 2008, incorporated by reference approved for § 1728.204.

(6) ASTM B 496–04—Standard Specification for Compact Round Concentric-Lay-Stranded Copper Conductors (ASTM B 496–04), approved April 1, 2004, incorporated by reference approved for § 1728.204.

(7) ASTM B 609/B 609M–99—Standard Specification for Aluminum 1350 Round Wire, Annealed and Intermediate Tempers, for Electrical Purposes (ASTM B 609/B 609M–99), approved April 1, 2004, incorporated by reference approved for § 1728.204.

(8) ASTM B 786–08—Standard Specification for 19 Wire Combination Unilay-Stranded Aluminum 1350 Conductors for Subsequent Insulation (ASTM B 786–08), approved September 1, 2008, incorporated by reference approved for § 1728.204.

(9) ASTM B 787/B 787M–04—Standard Specification for 19 Wire Combination Unilay-Stranded Copper Conductors for

Subsequent Insulation (ASTM B 787/B 787M–04), approved September 1, 2004, incorporated by reference approved for § 1728.204.

(10) ASTM B 835–04—Standard Specification for Compact Round Stranded Copper Conductors Using Single Input Wire Construction (ASTM B 835–04), approved September 1, 2004, incorporated by reference approved for § 1728.204.

(11) ASTM B902–04a—Standard Specification for Compressed Round Stranded Copper Conductors, Hard, Medium-Hard, or Soft Using Single Input Wire Construction (ASTM B902–04a), approved September 1, 2004, incorporated by reference approved for § 1728.204.

(12) ASTM D 1248–05—Standard Specification for Polyethylene Plastics Extrusion Materials for Wire and Cable (ASTM D 1248–05), approved March 1, 2005, incorporated by reference approved for § 1728.204.

(13) ASTM D 2275–01 (Reapproved 2008)—Standard Test Method for Voltage Endurance of Solid Electrical Insulating Materials Subjected to Partial Discharges (Corona) on the Surface (ASTM D 2275–01), approved May 1, 2008, incorporated by reference approved for § 1728.204.

(14) ASTM E 96/E 96M–05—Standard Test Methods for Water Vapor Transmission of Materials (ASTM E 96/E 96M–05), approved May 1, 2005, incorporated by reference approved for § 1728.204.

(e) American Wood Protection Association (AWPA), P.O. Box 361784, Birmingham, AL 35236–1784, telephone 205–733–4077, *www.awpa.com.*

(1) AWPA A6–20, Standard for the Determination of Retention of Oil-Type Preservatives from Small Samples, Revised 2020, incorporation by reference approved for § 1728.202.

(2) AWPA A9–20, Standard Method for Analysis of Treated Wood and Treating Solutions By X-Ray Spectroscopy, Revised 2020, incorporation by reference approved for § 1728.202.

(3) AWPA A15–19, Referee Methods, Revised 2019, incorporation by reference approved for § 1728.202.

(4) AWPA A30–18, Standard Method for the Determination of 4,5 Dichloro-2-n-octyl-4-isothiazolin-3 one (DCOI) in Wood and Solutions by High Performance Liquid Chromatography (HPLC), Revised 2018, incorporation by reference approved for § 1728.202.

(5) AWPA A69–18, Standard Method to Determine the Penetration of Copper Containing Preservatives, Reaffirmed 2018, incorporation by reference approved for § 1728.202.

(6) AWPA A70–18, Standard Method to Determine the Penetration of Pentachlorophenol Using a Silver-Copper Complex Known as Penta-Check, Reaffirmed in 2018, incorporation by reference approved for § 1728.202.

(7) AWPA A71–18, Standard Method to Determine the Penetration of Solvent Used with Oil-Soluble Preservatives, Reaffirmed 2018, incorporation by reference approved for § 1728.202.

(8) AWPA A83–18, Standard Method for Determination of Chloride for Calculating Pentachlorophenol in Solution or Wood, Reaffirmed 2018, incorporation by reference approved for § 1728.202.

(9) AWPA M2–19, Standard for the Inspection of Preservative Treated Products for Industrial Use, Revised 2019, incorporation by reference approved for § 1728.202.

(10) AWPA T1–20, Use Category System: Processing and Treatment Standard, Revised 2020, incorporation by reference § 1728.201.

(11) AWPA U1–20, Use Category System: User Specification for Treated Wood, Revised 2020, incorporation by reference approved for §§ 1728.201 and 1728.202.

(f) Insulated Cable Engineers Association (ICEA). The following material may be purchased from: IHS Global Engineering Documents, 15 Inverness Way East, Englewood, CO 80112, Phone: (303) 397–7956; (800) 854–7179, Fax: (303) 397–2740, email: *global@ihs.com,* website: *http://global.ihs.com.*

(1) ANSI/ICEA S–94–649–2004—Standard for Concentric Neutral Cables Rated 5 Through 46 KV (ANSI/ICEA S–94–649–2004), approved September 20, 2005, incorporation by reference approved for § 1728.204.

(2) ANSI/ICEA T–31–610–2007—Test Method for Conducting Longitudinal Water Penetration Resistance Tests on Blocked Conductors (ANSI/ICEA T–31–610–2007), approved October 31, 2007, incorporated by reference approved for § 1728.204.

(3) ICEA T–32–645–93—Guide for Establishing Compatibility of Sealed Conductor Filler Compounds with Conducting Stress Control Materials (ICEA T–32–645–93), approved February 1993, incorporated by reference approved for §1728.204.

(g) Southern Pine Inspection Bureau Standards, 4709 Scenic Highway, Pensacola, Florida 32504–9094, telephone (850) 434–2611. The web address for the Southern Pine Inspection Bureau is *http://www.spib.org/*.

(1) Standard Grading Rules for Southern Pine Lumber, 2014 Edition, effective January 25, 2014, incorporation by reference approved for §1728.201.

(2) [Reserved]

(h) West Coast Lumber Inspection Bureau, P.O. Box 23145, Portland, Oregon 97281, telephone (503) 639–0651, fax (503) 684–8928. The web address for is *http://www.wclib.org/*.

(1) Standard No. 17, Grading Rules for West Coast Lumber, Revised September 1, 2018, incorporation by reference approved for §1728.201.

(2) [Reserved]

[76 FR 36963, June 24, 2011, as amended at 77 FR 19528, Apr. 2, 2012; 83 FR 55467, Nov. 6, 2018; 84 FR 28190, June 18, 2019; 86 FR 57020, Oct. 14, 2021]

§1728.98 Electric standards and specifications.

(a) To comply with this part, you must follow the requirements contained in the following REA/RUS bulletins. These bulletins are incorporated by reference in §1728.97 of this part.

(1) Bulletin 50–4 (D–801), Specification and Drawings for 34.5/19.9 kV Distribution Line Construction (11–86).

(2) Bulletin 50–15 (DT–3), RUS Specifications for Pole Top Pins with 1⅜′ Diameter Lead Thread (1–51).

(3) Bulletin 50–16 (DT–4), RUS Specifications for Angle Suspension Brackets (3–52).

(4) Bulletin 50–19 (DT–7), RUS Specifications for Clevis Bolts (8–53).

(5) Bulletin 50–23 (DT–18), RUS Specifications for 60″ Wood Crossarm Braces (2–71).

(6) Bulletin 50–31 (D–3), RUS Specifications for Pole Top Pins with 1″ Diameter Lead Threads (2–79).

(7) Bulletin 50–32 (D–4), RUS Specifications for Steel Crossarm Mounted Pins with 1″ Diameter Lead Threads (10–50).

(8) Bulletin 50–33 (D–5), RUS Specifications for Single and Double Upset Spool Bolts (2–51).

(9) Bulletin 50–34 (D–6), RUS Specifications for Secondary Swinging Clevises (12–70).

(10) Bulletin 50–35 (D–7), RUS Specifications for Service Swinging Clevises (9–52).

(11) Bulletin 50–36 (D–8), RUS Specifications for Service Deadend Clevises (9–52).

(12) Bulletin 50–40 (D–14), RUS Specifications for Pole Top Brackets for Channel Type Pins (9–51).

(13) Bulletin 50–41 (D–15), RUS Specifications for Service Wireholders (11–51).

(14) Bulletin 50–55 (T–2), RUS Specifications for Overhead Ground Wire Support Brackets (5–53).

(15) Bulletin 50–56 (T–3), RUS Specifications for Steel Plate Anchors for Transmission Lines (12–53).

(16) Bulletin 50–60 (T–9), RUS Specification—Single Pole Steel Structures, Complete with Arms (12–71).

(17) Bulletin 50–72 (U–4), RUS Specification for Electrical Equipment Enclosures (5–35 kV) (10–79).

(18) Bulletin 50–73 (U–5), RUS Specifications for Pad-Mounted Transformers (Single and Three-Phase) (1–77).

(19) Bulletin 50–74 (U–6), RUS Specification for Secondary Pedestals (600 Volts and Below) (10–79).

(20) Bulletin 50–91 (S–3), RUS Specifications for Step-Down Distribution Substation Transformers (34.4–138 kV) (1–78).

(21) Bulletin 1728F–700, RUS Specification for Wood Poles, Stubs and Anchor Logs, September 9, 2021.

(22) Bulletin 1728F–803, Specifications and Drawings for 24.9/14.4 kV Line Construction (10–98).

(23) Bulletin 1728F–804 (D–804), Specification and Drawings for 12.47/7.2 kV Line Construction, October 2005.

(24) Bulletin 1728F–806 (D–806) Specifications and Drawings for Underground Electric Distribution), October 11, 2018.

(25) Bulletin 1728F–810, Electric Transmission Specifications and Drawings, 34.5 kV to 69 kV (3–98).

(26) Bulletin 1728F–811, Electric Transmission Specifications and Drawings, 115 kV to 230 kV (3–98).

(b) The terms "RUS form", "RUS standard form", "RUS specification", and "RUS bulletin" have the same meanings as the terms "REA form", "REA standard form", "REA specification", and "REA bulletin", respectively unless otherwise indicated.

[76 FR 36964, June 24, 2011, as amended at 83 FR 55467, Nov. 6, 2018; 84 FR 28191, June 18, 2019; 86 FR 57020, Oct. 14, 2021]

§ 1728.201 Bulletin 1728H–701, Specification for Wood Crossarms (Solid and Laminated), Transmission Timbers and Pole Keys.

(a) *Scope.* (1) The specification in this section describes the minimum acceptable quality of wood transmission and distribution crossarms (hereinafter called arms) purchased by or for RUS borrowers. Where there is conflict between the specification in this section and any other specification referred to in this section, the specification in this section shall govern.

(2) The requirements of the specification in this section implement contractual provisions between RUS and borrowers receiving financial assistance from RUS. The contractual agreement between RUS and a RUS borrower requires the borrower to construct its system in accordance with RUS accepted plans and specifications. Each RUS electric and telecommunications borrower shall purchase only arms produced in accordance with the specification in this section. Each RUS electric and telecommunications borrower shall require a written confirmation from their selected contractor that all material utilized shall be produced in accordance with the specifications in this section.

(b) *General stipulations.* (1) Conformance of arms to RUS specifications is the responsibility of the producer. A member of the producer's staff shall be designated as quality control supervisor and charged with the responsibility for the exercise of proper quality control procedures throughout the production process. The primary responsibility of third-party inspection agencies is to verify that producers involved in the manufacture of RUS treated wood products have functional in-house quality control systems in place that result in the shipment of materials meeting applicable RUS specification requirements to borrowers.

(2) Treated wood products intended for RUS borrowers shall not be inspected when in the opinion of the inspector, unsafe conditions are present.

(3) Various requirements relating to quality control and inspection that are contained in § 1728.202 and ANSI O5.2 and ANSI O5.3 (both incorporated by reference in § 1728.97) shall be followed exactly and shall not be interpreted or subject to judgment by the producer's quality control personnel or by the third party inspector.

(4) The requirements of AWPA M3 (incorporated by reference in § 1728.97) pertaining to record keeping, pre-treatment storage, analytical laboratories, plant gauges and other plant facilities, shall be followed.

(5) The producer shall maintain its own properly staffed and equipped analytical laboratory or contract with an independent testing laboratory at or near the treating plant to provide the required analytical service. On a case-by-case basis, with written permission from RUS, a producer with more than one treatment facility may be allowed to use a central laboratory.

(6) Arms can be purchased under either of two purchase plans; a RUS approved Quality Assurance Plan or an Independent Inspection Plan. The method of inspection described in this section shall be used no matter which plan timber products are purchased under.

(7) All third-party inspectors involved in the inspection of RUS products shall maintain their impartiality when providing their inspection service. This requires that these individuals and their employers, as well as producers and suppliers involved in providing RUS borrowers with treated wood products, maintain a professional separation during the performance of their respective functions to eliminate any possible conflict of interest.

(8) With the exception of financial agreements for inspection services, inspection agencies shall neither accept nor provide gratuities or free services to suppliers.

(9) Inspection agencies shall not offer product warranties on inspected material.

(10) Arms shall be warranted to conform to this specification. Arms shall meet or exceed their minimum allowable dimensions for at least one year from time of delivery to the borrower. If any arm is determined to be defective or does not conform to this specification within 1 year from the date of delivery to the borrower, it shall be replaced as promptly as possible by the supplier. In the event of failure to do so, the purchaser may make such replacement and the cost of the arm, at destination, shall be recovered from the supplier.

(11) Arm producers shall have and maintain liability insurance in the amount of $1 million. Evidence of compliance to this requirement shall be forwarded to the RUS annually. The evidence shall be in the form of a certificate of insurance or a bond signed by a representative of the insurance company or Surety Bonding company and include a provision that no change in, or cancellation of, will be made without the prior written notice to the Chairman, Technical Standards Committee "A" (Electric), 1400 Independence Ave. SW, Stop 1569, Washington, DC 20250–1569.

(c) *Definitions.* The following definitions apply to this section:

Agency refers to Rural Utilities Service (RUS), United States Department of Agriculture.

Certificate of compliance is a written certification by an authorized employee of the producer that the material shipped meets the requirements of this specification and any supplemental requirements specified in a purchase order from a borrower or the borrower's contractor.

Crossarm refers to the structural wood member used to support electrical conductors and equipment. The word arm is used interchangeably with crossarm.

Independent inspection refers to examination of material by a trained inspector employed by a commercial inspection agency.

Inspection means an examination of material in sufficient detail to ensure conformity to all requirements of the specification under which it was purchased.

Lot is a certain number of pieces of a given item submitted for inspection at one time.

Producer is the party who manufactures arms. In some cases the producer may also be the treating plant.

Purchaser refers to either the RUS borrower or contractors acting as the borrower's agent, except where a part of the specification in this section specifically refers to only the borrower or the contractor.

Quality control supervisor refers to an employee of the producer designated to be responsible for quality control procedures carried out by said producer.

Reserve treated stock consists of treated material held in storage by a producer for purchase and immediate shipment to a borrower.

Supplier may refer to the producer, the treater, or to a third-party broker or distributorship involved in supplying RUS products to the borrowers.

Treating plant is the facility that applies the preservative treatment to the arms.

(d) *Material requirements*—(1) *Material and grade.* All arms furnished under the specification in this section shall be free of brashy wood, decay, and shall meet additional requirements as shown on specific drawings in this section. Arms shall be made of one of the following:

(i) Douglas-fir which conforms to the applicable provisions of paragraphs 170 and 170a, or the applicable transmission arm provisions of paragraphs 169 and 169a of the West Coast Lumber Standard No. 17 (incorporated by reference in § 1728.97). Only coastal origin Douglas-fir shall be used for Douglas-fir arms manufactured under the specification in this section;

(ii) Southern Yellow Pine which conforms to the provisions of Dense Industrial Crossarm 65, as described in Southern Pine Inspection Bureau's Standard Grading Rules for Southern Pine Lumber (incorporated by reference at § 1728.97); or

(iii) Laminated wood arms shall conform to ANSI O5.2 and have at least the same load carrying capacity as the solid sawn arms being replaced. The load carrying capacity of the laminated arms shall be determined by one of the procedures outlined in ANSI O5.2. The testing and inspection of laminated arms shall be in accordance with AITC 200 (incorporated by reference at § 1728.97).

(2) *Alternative arms.* Borrowers may use alternative arms that are listed in Informational Publication 202–1, *List of Materials Acceptable for Use on Systems of USDA Rural Utilities Service Borrowers.* For information on the availability of such material, contact the Chairman, Technical Standards Committee "A" (Electric), 1400 Independence Ave. SW, Stop 1569, Washington, DC 20250–1569, or go to: *https://www.rd.usda.gov/files/UEP_LoM.pdf.*

(3) *Knots.* Well-spaced sound, firm, and tight knots are permitted.

(i) Slightly decayed knots are permitted, except on the top face, provided the decay extends no more than ¾ of an inch into the knot and provided the cavities will drain water when the arm is installed. For knots to be considered well-spaced, the sum of the sizes of all knots in any 6 inches of length of a piece shall not exceed twice the size of the largest knot permitted. More than one knot of maximum permissible size shall not be in the same 6 inches of length. Slightly decayed, firm, or sound "pin knots" (⅜ of an inch or less) are not considered in size, spacing, or zone considerations.

(ii) Knots are subject to limits on size and location as detailed in Tables 1 and 2 to this paragraph (d)(3)(ii).

TABLE 1 TO PARAGRAPH (d)(3)(ii)—KNOT LIMITS FOR DISTRIBUTION ARMS (SEE FIGURE 1 TO THIS SECTION)

[All dimensions in inches]

Class of knot and location	Maximum knot diameter	
	Close grain	Dense grain
Round Knots:		
Single Knot: Maximum Diameter Center Section[1]		
Upper Half	¾	1
Lower Half	1	1¼
Elsewhere	1¼	1½
Sum of Diameters in 6-Inch Length: Maximum Center Section:		
Upper Half	1½	2
Lower Half	2	2½
Elsewhere	2½	3

[1] No knot shall be closer than its diameter to the pole mounting hole.

TABLE 2 TO PARAGRAPH (d)(3)(ii)—KNOT LIMITS FOR TRANSMISSION ARMS (SEE FIGURE 2 TO THIS SECTION)

[All dimensions in inches]

Pole mounting hole zone[1]	Maximum diameter for single knot
Upper Half (inner zone)	¾.
Upper Half (outer zone)	1 for close grain. 1¼ dense grain.

Other locations transmission arm size[2]	Narrow face	Wide face (two sides)	
		Edge	Along centerline
4⅝ × 5⅝ or less	1	1¼	1¼
5⅝ × 7⅜	1¼	1⅜	1⅞
3⅝ × 9⅜	¾	1¾	2¼

[1] No knot shall be closer than its diameter to the pole mounting hole.
[2] For cross sections not shown, refer to grading rules.

(iii) Knot clusters shall be prohibited unless the entire cluster, measured on the worst face, is equal to or less than the round knot allowed at the specific location.

(iv) Spike knots shall be prohibited in deadend arms. Any spike knot across the top face shall be limited to the equivalent displacement of a knot ⅜ of an inch deep on one face and the maximum round knot for its particular location on the worst face, with a maximum width of 1 inch measured at the midpoint of the spiked section. Elsewhere across the bottom or side faces, spike knots shall not exceed ½ the equivalent displacement of a round knot permitted at that location, provided that the depth of the knot on the worst face shall not exceed the maximum round knot allowed at that location.

(v) Loose knots shall be prohibited in deadend arms. Loose knots and knot holes shall be permitted only if they allow water to drain when the arm is installed in its normal position. In the center section, upper half, loose knots shall not be greater than ½ the dimensions of round knots. Elsewhere, loose knots shall not be greater than the round knot dimension.

(vi) All knots except those "spike" knots intersecting a corner shall be measured on the least diameter of the knot.

(vii) A knot shall be considered to occupy a specific zone or section if the center of the knot (*i.e.*, pith of knot) is within the zone or on the zone's boundary.

(viii) If a round or oval knot appears on two faces and is in two zones, each face shall be judged independently. When this does not occur, average the least dimension showing on both faces. Knots which occur on only one face of a free of heart center (FOHC) arm shall be permitted to be 25 percent larger than the stated size.

(ix) Two or more knots opposite each other on any face shall be limited by a sum not to exceed the size of a maximum single knot permitted for the location. On all four faces, all knots shall be well spaced.

(x) No knot over ⅝ inch in diameter may intersect pin holes in the center section. One-inch diameter knots may intersect insulator pin holes elsewhere.

(e) *Miscellaneous characteristics, features and requirements.* (1) The top face of distribution arms shall not have more than four medium pitch and bark pockets in 8-foot arms, and not more than five pitch and bark pockets in 10-foot arms. Elsewhere a maximum of six medium pitch and bark pockets in 8-foot arms and eight in 10-foot arms shall be permitted. Equivalent smaller pockets shall be permissible. An occasional large pocket is permissible.

(2) Shakes shall be prohibited.

(3) Prior to treatment on properly seasoned arms, single face checks shall not exceed an average penetration of ¼ the depth from any face and shall be limited to 10 inches long on the top face, and ⅓ the arm length on the other faces. Checks shall not be repeated in the same line of grain in adjacent pin holes. The sum of the average depths of checks occurring in the same plane on opposite faces shall be limited to ¼ the face depth.

(4) Compression wood shall be prohibited on any face. Compression wood is permitted if wholly enclosed in the arm, more than six annual rings from the surface, and not over ⅜ of an inch in width.

(5) Insect holes 3⁄32 of an inch and larger shall be prohibited. Insect pin holes (*i.e.*, holes not over 1⁄16 of an inch diameter) shall be allowed if scattered and not exceeding 10 percent of the arm girth.

(6) Wane shall be allowed on one edge, limited to approximately 1 inch measured across the corner. Outside of the top center section, an aggregate length not to exceed 2 feet may have wane up to 1½ inches on an occasional piece on one or both edges. Bark shall be removed.

(7) Prior to and after preservative treatment, crook, bow, or twist shall not exceed ½ of an inch in 8-foot arms and ⅝ of an inch in 10-foot arms.

(f) *Manufacturing*—(1) *Quality of work.* All arms shall be of the highest quality production. Arms shall be dressed on all four sides, although "hit and miss skips" may occur on two adjacent faces on occasional pieces.

(2) *Dimensions and tolerances.* All dimensions and tolerances shall conform

to those shown on the drawings in this section or drawings supplied with the purchase order. Arms supplied shall meet or exceed minimum dimensions shown on the drawings in this section. Cross-sectional dimensions shall be measured and judged at about 1/4 the arm length, except when the defects of "skip dressing" or "machine bite or offset" are involved.

(3) *Shape.* The shape of the arms at any cross section, except for permissible wane, shall be as shown on the respective drawings in this section or supplied with the order. The two top edges may be either chamfered or rounded 3/8 of an inch radius. The two bottom edges shall be slightly eased 1/8 of an inch radius for the entire length.

(4) *Lamination techniques.* Lamination techniques shall comply with ANSI O5.2.

(5) *Pin and bolt holes.* Pin and bolt holes shall be smoothly bored without undue splintering where drill bits break through the surface. The center of any hole shall be within 1/8 of an inch of the center-line locations on the face in which it appears. Holes shall be perpendicular to the starting and finishing faces.

(6) *Incising.* The lengthwise surfaces of Douglas-fir arms shall be incised a minimum of 1/4 of an inch deep. The incision shall be reasonably clean cut with a spacing pattern that ensures uniform penetration of preservative.

(g) *Conditioning prior to treatment.* AWPA T1 (incorporated by reference at §1728.97) shall be followed.

(1) All solid sawn arms shall be made of lumber which has been kiln-dried. Douglas-fir arms shall have an average moisture content of 19 percent or less, with a maximum not to exceed 22 percent in a single arm. Southern Yellow Pine arms shall have an average moisture content of 22 percent or less, with a maximum not to exceed 30 percent in a single arm.

(2) Moisture content levels shall be measured at about 1/4 the length and at a depth of about 1/5 the arm's thickness. Additionally, the moisture content gradient between the shell (*i.e.*, 1/4 of an inch deep) and the core (*i.e.*, about 1 inch deep) shall not exceed 5 percentage points.

(3) A minimum of at least 20 solid sawn arms per treating charge shall be measured and the individual results recorded by the producer to verify moisture content.

(4) The moisture content of lumber used in laminating shall, at the time of gluing, be within the range of 8 to 12 percent, inclusive.

(h) *Preservatives.* (1) Creosote, waterborne preservatives, pentachlorophenol, DCOI, and copper naphthenate shall conform to the requirements of AWPA U1 (incorporated by reference at §1728.97). Oxide formulations of waterborne preservatives shall be supplied. If CCA is the selected preservative, CCA–C shall be the type required.

(2) Douglas-fir arms shall not be treated with CCA.

(i) *Preservative treatment.* (1) All timber products manufactured under the specification in this section shall be pressure treated. AWPA T1 shall be followed.

(2) These materials may be further conditioned by steaming, or by heating in hot oil (Douglas-fir), within the following time and temperature limits:

	Max. time (hours)	Temperature
(i) Steam	3	220 °F
(ii) Heating in Preservative	3	210 °F

(3) A final steam or hot oil bath may be used only to meet cleanliness requirements. Total duration of the final steam bath shall not exceed 2 hours and the temperature shall not exceed 240 °F.

(j) *Results of treatment*—(1) *Penetration and retention.* The quality control supervisor shall test or supervise the testing of each treated charge for penetration and retention.

(2) *Method of sampling.* When testing penetration and retention, a borer core shall be taken from a minimum of 20 arms in each treating charge. The borings shall be taken from any face except the top face at a point as close to the end as possible, being at least 3 inches from the end of the arm and no closer than 3 inches from the edge of the holes. The bored holes shall be plugged with treated plugs. Borings

from laminated arms shall not be taken from the same laminate unless there is an end joint separation.

(3) *Preservative penetration.* All of the sapwood present in Douglas-fir and southern yellow pine arms shall be completely penetrated with preservative. Preservative penetration in the heartwood of Douglas-fir arms shall be not less than 3 inches longitudinally from the edge of holes and ends, and at least 3⁄16 inch from the surface of any face.

(4) *Preservative retention.* Preservative retention in the outer 0.6 inch for Douglas-fir arms and in the outer one inch of southern yellow pine arms shall be not less than the following:

Preservative	Retention (pcf)
(i) Creosote	8.0
(ii) Pentachlorophenol	[1] 0.4/0.36
(iii) ACA, ACZA, or CCA–C	0.4
(iv) Copper Naphthenate	0.04
(v) DCOI	0.13

[1] If the copper pyridine method is used when timbers may have been in contact with salt water, a penta retention of 0.36 pcf is required for all species native to the Pacific Coast region.

(5) *Arms surfaces.* The surfaces of all arms shall be free from oil exudation (bleeding) and pentachlorophenol crystallization (blooming), and other surface deposits.

(6) *Retreatment of arms.* Arms may be retreated no more than twice. Initial treatment steaming time plus re-treatment steaming time, combined, shall not exceed total steaming time allowed.

(k) *Marking/branding.* (1) Before treatment, arms shall be legibly branded (hot brand) or die-stamped to a depth of approximately 1⁄16 of an inch, with the top of the brand oriented to the top of the arm. The brand shall be placed on either of the wide surfaces of the arm, approximately one foot from the midpoint of the piece.

(2) The letters and figures shall be not less than ½ of an inch in height.

(3) The brand or die-stamp shall include:

(i) The manufacturer's identification symbol;

(ii) Month and year of manufacture;

(iii) Species (DF for Douglas-fir and SP for southern yellow pine);

(iv) Preservative (C for creosote, DA for DCOI, PA for penta, SK for CCA, SZ for ACZA, N for Copper Naphthenate); and

(v) Required retention. An example of required retention is: M–6–16 Manufacturer—Month—Year and DF–PA–.4 Douglas-fir–penta treated—.40 pcf retention.

(4) Brands and quality assurance/inspection marks shall be removed from arms that do not meet these specifications.

(l) *Storage.* (1) Producers may manufacture/treat RUS arms for reserve treated stock under either of the allowable purchase plans. (See paragraph (b)(6) of this section).

(2) Arms treated with creosote or oil-borne preservatives, and which have been held in storage for more than 1 year before purchase and shipment to the borrower shall be re-assayed before shipment. Any such arms found to be nonconforming for retention shall be retreated and reassayed per the requirements of this section of the specification.

(m) *Drawings.* (1) The drawings of Figure 3 to this section, Crossarm Drilling Guide, have a type number and show in detail the hole size, shape, and pattern desired for arms ordered under the specification in this section.

(2) Purchase orders shall indicate the type arm required.

(3) Arms shall be furnished in accordance with the details of the drawings in this section or in accordance with drawings attached to the purchase order.

(4) Appropriate drawings for transmission arms are to be specified and included with purchase orders. Technical drawings for transmission arms are published in Bulletin 1728F–811 (incorporated by reference at § 1728.97) and Bulletin 1728F–810 (incorporated by reference at § 1728.97).

(n) *Destination inspection.* The RUS borrower shall have the prerogative to inspect materials at destination. All provisions of the specification in this section shall apply to material inspected at destination. If a disagreement arises over conformance of materials received at destination, it shall be the responsibility of the supplier to resolve the matter with the purchaser.

(o) *Purchase of related specifications and standards.* (1) All ANSI and AWPA standards may be purchased from: American Wood Protection' Association (AWPA), P.O. Box 361784, Birmingham, AL 35236–1784, Telephone (205)733–4077, Web address: *http://www.awpa.com.*

(2) Standard Grading Rules for Southern Pine Lumber and Special Products Rules for Structural, Industrial, and Railroad Freight Car Lumber may be purchased from: Southern Pine Inspection Bureau, 4709 Scenic Highway, Pensacola, Florida 32504–9094, Telephone (850) 434–2611, Web address: *http://www.spib.org.*

(3) Standard Grading Rules for West Coast Lumber may be purchased from: West Coast Lumber Inspection Bureau, P.O. Box 23145, Portland, Oregon 97281, Telephone (503) 639–0651, Web address: *http://www.wclib.org.*

(4) AITC 200 may be purchased from: American Institute of Timber Construction, 7012 S Revere Park Way, Englewood, Colorado 80112, Telephone (303) 792–9559, Web address: *http://aitc-glulam.org.*

(p) *Information to be completed by the borrower.* When using the specification in this section, the borrower or borrower's representative should enter into a written agreement with a material supplier by way of a contract or purchase order. This agreement should state that all arms shall be manufactured in strict accordance with the specifications in this section.

Figures 1 and 2 to § 1728.201
Distribution and Transmission Arms

DISTRIBUTION ARMS
Figure 1

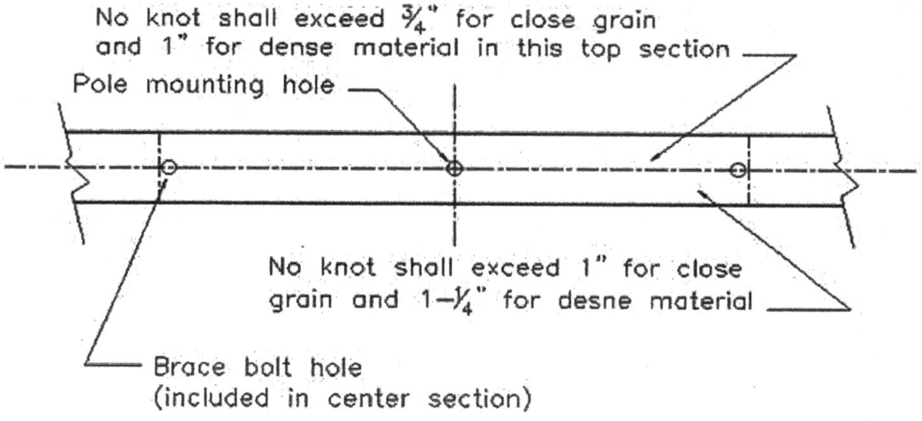

TRANSMISSION ARMS
POLE MOUNTING HOLE ZONE
Figure 2

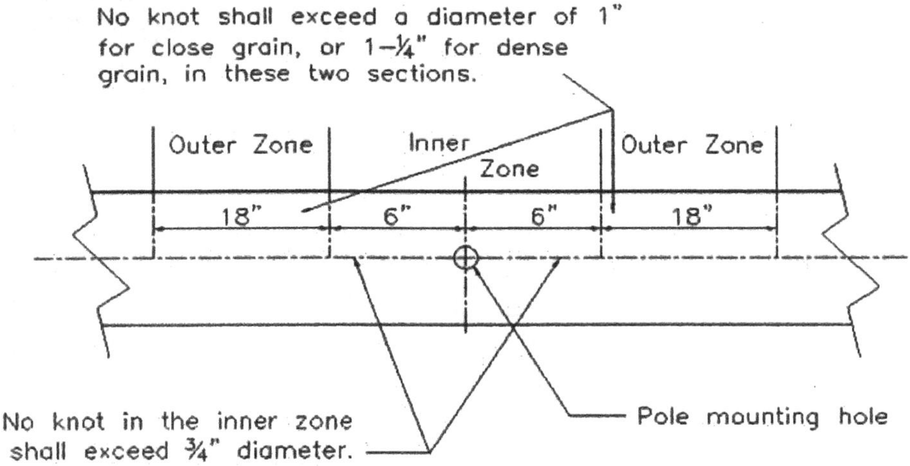

Figure 3 to §1728.201 – Crossarm Drilling Guide

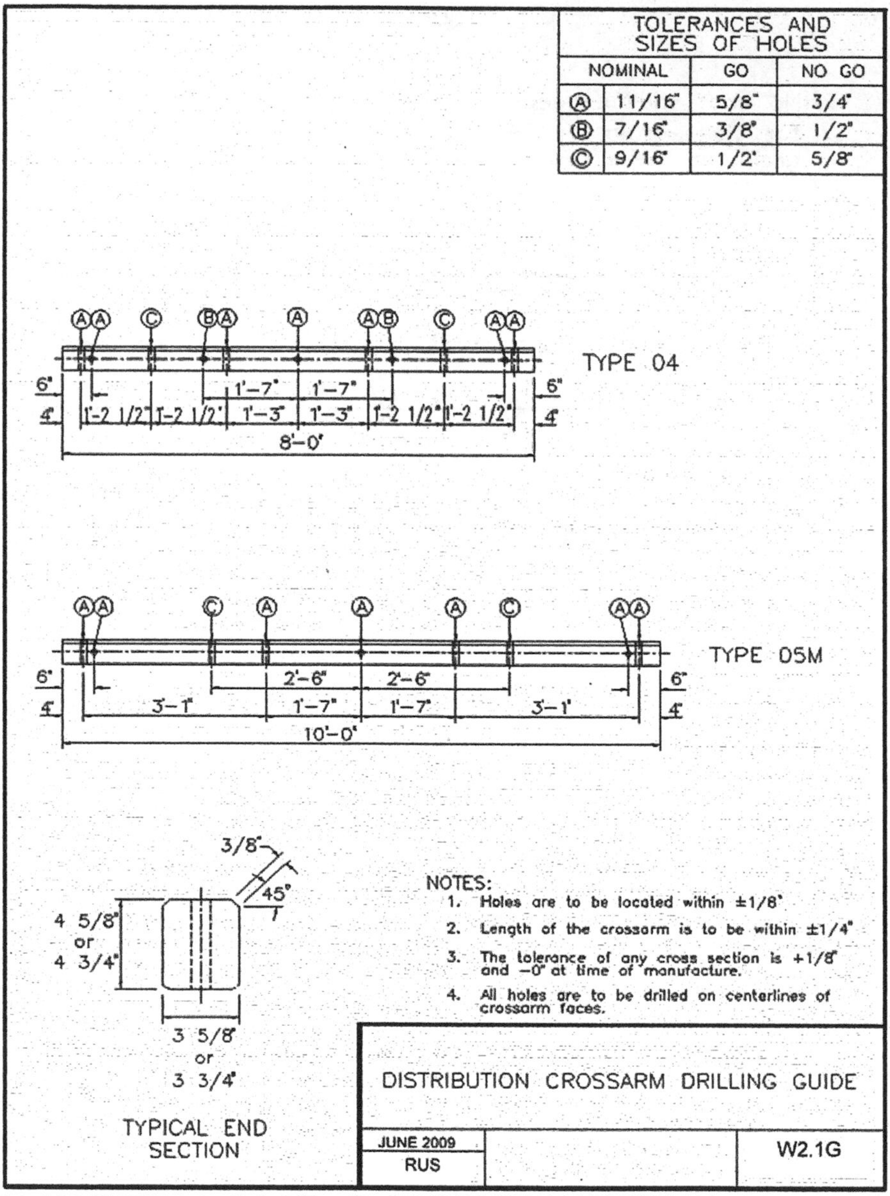

NOMINAL		GO	NO GO
Ⓐ	11/16"	5/8"	3/4"
Ⓑ	7/16"	3/8"	1/2"
Ⓒ	9/16"	1/2"	5/8"

[84 FR 28191, June 18, 2019, as amended at 86 FR 57021, Oct. 14, 2021]

§ 1728.202 Bulletin 1728H–702, Specification for Quality Control and Inspection of Timber Products.

(a) *Scope.* (1) The specification in this section describes the responsibilities and procedures pertaining to the quality control by producers and pertaining to inspection of timber products produced in accordance with the following RUS specifications in § 1728.201, and poles, covered in Bulletin 1728F–700 (incorporated by reference in § 1728.97) and in § 1755.97 of this chapter.

(2) Where there is conflict between the specification in this section and any other specification referred in this section, the specification in this section shall govern.

(3) The specification in this section also describes and designates responsibilities of RUS borrowers in regard to their purchases under the specifications referenced in paragraph (a)(1) of this section.

(b) *General stipulations.* (1) Conformance of poles and crossarms to RUS specifications is the responsibility of the producer. A member of the producer's staff shall be designated as quality control supervisor and charged with the responsibility for the exercise of proper quality control procedures throughout the production process. The primary responsibility of third party inspection agencies is to verify that producers involved in the manufacture of RUS treated wood products have functional in-house quality control systems in place that result in the shipment of materials meeting applicable RUS specification requirements to borrowers.

(2) The requirements of AWPA M3 (incorporated by reference at § 1728.97), pertaining to recordkeeping, pretreatment storage, analytical laboratories, plant gauges, and other plant facilities, shall be followed.

(3) Treated wood products intended for RUS borrowers shall not be inspected when in the opinion of the inspector, unsafe conditions are present.

(4) Poles and crossarms can be purchased under either of two purchase plans; a RUS approved Quality Assurance Plan or an Independent Inspection Plan. The method of inspection described in this section shall be used no matter which plan timber products are purchased under.

(5) Under the Independent Inspection Plan, the borrower should designate in the purchase order which inspection agency it has selected. Unless the borrower contracts for inspection as a separate transaction, the treating company shall obtain the services of the borrower's designated inspection agency. For reserve treated stock held in inventory by the producer, the producer shall obtain the services of the appropriate inspection program.

(6) All third-party inspectors involved in the inspection of RUS products shall maintain their impartiality when providing their inspection service. This requires that these individuals and their employers, as well as producers and suppliers involved in providing RUS borrowers with treated wood products, maintain the greatest degree of professional separation during the performance of their respective functions to eliminate any possible conflict of interest.

(7) With the exception of financial agreements for inspection services, inspection agencies shall not accept nor provide gratuities or free services to suppliers.

(8) Inspection agencies shall not offer product warranties on inspected material.

(9) Inspection agencies shall have and maintain liability insurance in the amount of $500,000 and a surety bond or miscellaneous Errors and Omission insurance for consequential damages for not less than $250,000. Evidence of compliance to the requirement in this paragraph (b)(9) shall be forwarded to the RUS annually. The evidence shall be in the form of a certificate of insurance or a Bond signed by a representative of the insurance or Surety Bonding company and include a provision that no change in, or cancellation of, will be made without the prior written notice to Chairman, Technical Standards Committee "A" (Electric).

(10) Inspection agencies shall maintain their own properly equipped laboratory that, at a minimum, is able to run the referee methods listed in table 1 to this paragraph (b)(10) for retention

analysis for all preservatives being inspected. This laboratory shall be independent from any treating plant laboratory. Inspection Agencies may use one central laboratory. All XRF units maintained by third party inspection agencies as part of their RUS required laboratories shall be calibrated at least quarterly by said agency utilizing the referee method for each preservative treatment being analyzed or via comparison with a set of graduated treated wood standards. Each agency shall keep an up-to-date written record of these quarterly calibration results. AWPA A83 or AWPA A9 (incorporated by reference at §1728.97) shall be followed for Pentachlorophenol testing, AWPA–A30 or AWPA A9 (incorporated by reference at §1728.97) shall be followed for DCOI testing, AWPA A6 (incorporated by reference at §1728.97) shall be followed for Creosote testing, and AWPA A9 (incorporated by reference at §1728.97) shall be followed for XRF, as illustrated in the following table:

TABLE 1 TO PARAGRAPH (b)(10)

Preservative	Analytical method	Referee method
Pentachlorophenol	XRF, Lime Ignition, Copper Pyridine	Lime Ignition, Copper Pyridine.
Creosote	Toluene Extraction	Toluene Extraction.
Waterborne	XRF	XRF.
Copper Naphthenate	XRF	ICP, GC.
DCOI	XRF, HPLC	HPLC.

Note 1 to table 1 to paragraph (b)(10): XFR means X-ray fluorescence; HPLC means High Performance Liquid Chromatography; ICP means Inductively coupled plasma; and GC means Gas Chromatography.

(11) If used for analysis, plant XRF units shall be accurate and generate reproducible results per AWPA A9. At least once monthly, their accuracy and precision shall be checked by the third-party inspector. This verification shall consist of the inspector taking a retention sample previously analyzed by the plant quality control supervisor on-site and rerunning it in the inspection agency's own laboratory using said agency's XRF unit or the referee method for a specific preservative. If the inspection agency's analytical result is within +5% of the plant's retention result on that sample, the plant XRF unit needs no further calibration.

(12) Individual inspectors in the employ of inspection agencies shall be properly trained and experienced. See §1728.203, for details of an inspector's minimum qualifications. Upon request, inspection agencies must provide RUS with detailed written documentation verifying that each of their employees inspecting RUS materials has the minimum experience and training described in §1728.203. Failure of an individual inspector to follow proper procedures or failure of an inspection agency to properly train and supervise their inspectors or follow applicable RUS specifications constitutes grounds for RUS debarment of said inspector and said inspection agency from future inspection of RUS financed material.

(c) *Quality control and inspection procedures.* It is the responsibility of the plant quality control supervisor to perform the following procedures to ensure that a particular lot of material conforms to the requirements of the applicable Agency specification prior to treatment. After the plant quality control supervisor has performed these procedures, a particular lot of material shall be released to the inspector for verification of conformance.

(1) For poles, inspection prior to treatment shall include:

(i) Ample space and assistance shall be provided by the treating plant for handling and turning poles. Regardless of the purchase plan poles are being purchased under, all poles in a lot shall be inspected by the plant quality control supervisor prior to offering the lot for verification by a third party.

(ii) When limited by the purchaser in a written purchase order, moisture content shall be measured with a calibrated electronic moisture meter. Calibration of the moisture meter shall include not only the zero settings for the X and Y readings, but also two resistance standards for 12 and 22 percent moisture content. Material failing to conform for moisture content may be retested upon request after recalibration of the moisture meter.

(iii) Dimensions, length, and circumference shall be measured by a standard steel tape to determine that they meet specification requirements and that they agree with the details for class and length found in the face brand/tag and butt of each pole. If it is obvious by visual comparison with a measured pole that the brand information regarding class and length is correct, individual poles need not be measured. Pole circumference dimensions measured prior to treatment shall govern acceptance. Reduction in dimension due to treatment and shipping shall be no more than 2 percent below the minimum for the pole class.

(iv) Poles in a lot shall be of the same seasoning condition and all shall be inspected for decay. If the plant quality control supervisor suspects that decay is present in a pole, a slice from both ends shall be cut for closer examination. If 3 percent or more of the poles in the lot inspected by the plant quality control supervisor show evidence of decay, the entire lot shall be unconditionally withdrawn without further sorting.

(v) Under the Independent Inspection Plan, all poles shall be examined by the third-party Inspector for verification of conformance. Under a RUS approved Quality Assurance Plan, the number of poles inspected for verification of conformance may vary according to the terms of the approved plan.

(vi) Whenever it becomes evident during third party inspection of any lot of poles offered by the producer that non-conforming pieces exceed 3 percent for any one defect or 5 percent for all defects, the inspector shall withhold further inspection and reject the balance of the lot. After the producer has acted to eliminate all defective pieces, the rejected balance may be inspected as a new lot. Sorting, however, shall not be permitted when a lot has been rejected for decay.

(vii) Re-examination for mechanical damage or deterioration and for original acceptance shall be conducted on timber products not treated within 10 days after the original third-party inspection.

(2) For crossarms, inspection prior to treatment shall include:

(i) Regardless of the purchase plan arms are being purchased under, all arms in a lot shall be inspected by the plant quality control supervisor prior to offering the lot for verification by a third party. *After* the plant quality control supervisor has performed the procedures in paragraphs (c)(2)(ii) through (vii) of this section, a particular lot of arms shall be released to the inspector for verification of conformance.

(ii) Moisture content of the arms in the lot shall be checked with a calibrated moisture meter.

(iii) Surface inspection of both ends and the side surfaces of each arms. Particular attention shall be paid to visible defects such as compression wood, red heart, honeycomb and other forms of decay, shakes, splits, through checks, low density, wane, undersize, and pitch pockets.

(iv) Inspection of bolt and insulator pin holes for proper location, dimension and excessive splintering.

(v) Inspection of brands for proper location, required content and legibility.

(vi) Under the Independent Inspection, both ends of all crossarms and a random representative sample of the lengthwise side faces of all crossarms shall be inspected. The sample size shall equal 20 percent of the lot size or 200 arms, whichever is smaller. Under a RUS approved Quality Assurance Plan, the number of crossarms inspected for verification of conformance may vary according to the terms of the approved plan.

(vii) Whenever it becomes evident during third party inspection of any lot of arms offered by the producer that non-conforming pieces exceed 2 percent of the sample size, the entire lot shall be rejected. After the producer has acted to eliminate all defective pieces, the rejected balance may be inspected as a new lot.

(d) *Preservatives.* Creosote, waterborne preservatives, pentachlorophenol, DCOI, and copper naphthenate shall conform to current AWPA U1 (incorporated by reference in § 1728.97).

(e) *Results of treatment—Poles.* (1) Following treatment, poles shall be sampled for preservative retention and penetration utilizing a calibrated increment borer 0.2 inches +0.02 inches in diameter in accordance with procedures listed in AWPA M2 (incorporated by reference in § 1728.97).

(2) Inspectors may take their own retention samples and analyze them concurrently with those taken by the quality control supervisor, but each shall work independently. The results of the plant's analysis shall be presented before verification and acceptance of the charge by the third-party inspector.

(3) Unless otherwise specified, borings shall be taken from the section of the pole extending from 1 foot below the face brand/tag to 1 foot above the face brand/tag. For pressure treated Western Red Cedar, Alaska Yellow Cedar and all butt treated poles, borings shall be taken from the section of the pole approximately 1 foot below groundline.

(4) For all species, core samples shall be taken from 20 poles in charges of 20 or more poles. If a charge consists of less than 20 poles, each pole shall be bored and then individual poles shall be bored a second time to obtain a minimum of 20 core samples. Any additional borings required to obtain the required 20 core samples shall be taken in a manner that represents the lot of material with respect to variations in size, seasoning condition, or other features that may affect the results of treatment.

(5) Retention and penetration samples shall consist of borings representative of pole volumes for each class and length in the charge, as illustrated in the following table:

TABLE 2 TO PARAGRAPH (e)(5)

Number of poles	Class/length	Vol. in ft^3	% of total vol.	Number of borings [2]
20	5/40	550	22	4
30	4/40	840	34	7
20	4/45	510	20	4
20	3/45	600	24	5
Total	....	2,500	....	

Note 2 to table 2 to paragraph (e)(5): Retention and penetration requirements for each different species and preservative are listed in Table 8 of Appendix A, RUS Bulletin 1728F–700, Specification for Wood Poles, Stubs and Anchor Logs (incorporated by reference at § 1728.97).

(6) Preservative retention analyses shall be performed per the standard AWPA U1, (incorporated by reference at § 1728.97).

(7) Penetration compliance of both poles and crossarms shall be determined in accordance with the standard AWPA A15 (incorporated by reference at § 1728.97). Chrome Azurol S shall be used to determine the penetration of copper containing preservatives AWPA A69 (incorporated by reference at § 1728.97), Penta-Check shall be used to determine the penetration of penta AWPA A70 (incorporated by reference at § 1728.97), and Red-O dye for penetration of DCOI AWPA A71 (incorporated by reference at § 1728.97), respectively.

(8) All bored holes created by penetration and retention sampling shall be promptly filled with tight fitting treated plugs.

(9) Penetration sampling of poles shall be carried out as follows:

(i) Group A poles (Those poles with a circumference of 37.5 inches or less at 6 feet from butt.):

(A) Bore 20 poles or 20 percent of the poles in the charge, whichever is greater. Accept all poles in the charge for penetration if every boring in the sample conforms. If any sample fails penetration, bore all poles in the charge.

(B) If more than 15% of the poles in the charge are found to be nonconforming, the entire charge shall be retreated. If 15% or less are found to be nonconforming, remove and retreat only those that are nonconforming.

(ii) Group B poles (Those poles with a circumference greater than 37.5 inches at 6 feet from the butt.):

(A) For Group B poles 45 feet and under, bore each pole in the charge. If more than 15% of these poles are found to be nonconforming, the entire charge shall be retreated. If 15% or less are found to be nonconforming, remove and retreat only those that are nonconforming.

(B) For Group B poles 50 feet and over, bore each pole twice at 90 degrees apart and accept only those poles conforming to penetration in both borings.

(iii) Nonconforming poles may be treated only twice. The letter "R" shall be added to the original charge number in the butts of all poles that are retreated. Poles failing to meet treatment requirements after two retreatments shall be permanently rejected and all brand and butt information removed.

(f) *Results of treatment—Crossarms.* Retention and penetration samples shall be taken from not less than 20 crossarms in each charge. The sampling method and retention and penetration requirements for both Douglas-fir and Southern Yellow Pine crossarms are listed in §1728.201.

(g) *Product acceptance.* (1) Third party inspectors shall verify their acceptance of untreated poles that have been offered by the producer as conforming by marking each accepted piece in the tip with a clear, legible hammer stamp. Following treatment, inspectors shall verify their acceptance of treated poles that have been offered by the producer as conforming by marking each accepted piece in the butt with a clear, legible hammer stamp. Inspection marks are not to be placed in the butt surfaces of any poles prior to treatment and proper retention analysis and penetration testing being completed. The inspector shall personally mark each piece for acceptance and shall not delegate this responsibility to any other individual.

(2) Third-party inspectors shall verify their acceptance of untreated crossarms that have been offered by the producer as conforming by marking each accepted piece in one end with a clear, legible hammer stamp. Following treatment, inspectors shall verify their acceptance of treated crossarms that have been offered by the producer as conforming by marking each accepted piece in the opposing end with a clear, legible hammer stamp. The inspector shall personally mark each piece for acceptance and shall not delegate this responsibility to any other individual.

(3) Each inspector or inspection agency shall retain for a period of at least one year a copy or transcript of each pre-treatment inspection report and a copy of analytical worksheets covering retention and penetration test results for each treated charge of material inspected. On request, a copy or transcript of these reports shall be furnished to the Chairman, Technical Standards Committee "A", Rural Utilities Service, Washington, DC 20250-1569.

(i) On each inspection report the third-party inspector and the plant quality control supervisor shall certify in writing that the material listed on the report has been properly inspected both before and after treatment and that the preservative used met the requirements of this section. Inspection reports shall also include the following information:

(A) Conditioning details of the material prior to treatment.

(B) Total number of pieces offered by the producer.

(C) Number of pieces rejected by the inspector, cause for rejection.

(D) Copy of preservative analysis (usually supplied by the preservative supplier).

(E) Treating sheet containing details of treatment for each charge.

(F) Separate worksheets for retention analyses done by the plant quality control supervisor and by the inspector.

(G) Penetration result on each individual core boring taken from poles in the charge.

(ii) [Reserved]

(h) *Laminated materials.* (1) All lumber used to fabricate laminated materials shall be inspected and its grade verified by a qualified lumber grader, then marked appropriately.

(2) Laminated materials shall comply with manufacturing requirements specified in ANSI O5.2 (incorporated by reference at §1728.97). Melamine urea adhesives shall not be used. Plant quality control procedures and any third-party inspection shall be conducted in accordance with AITC 200 (incorporated

by reference at § 1728.97), and § 1728.201 (Bulletin 1728H–701).

(3) Following treatment, laminated material shall be checked for proper preservative retention and penetration, and for any evidence of delamination. All conforming laminated materials shall be clearly marked with either an American Institute of Timber Construction (AITC) or American Plywood Association (APA) quality stamp.

(i) *Safety provisions.* Poles intended for agency borrowers shall not be inspected when, in the opinion of the inspector, unsafe conditions are present.

[84 FR 28198, June 18, 2019, as amended at 86 FR 57021, Oct. 14, 2021]

§ 1728.203 Inspector's qualifications.

Inspection agencies must assure borrowers that employees assigned to the inspection of timber products and preservative treatments are competent and experienced. In general, any of the following examples are considered as minimum qualifying experience before an individual may be permitted to inspect timber products for borrowers:

(a) Three years of direct experience inspecting untreated and treated utility products; or

(b) Three years of direct experience conducting in-plant quality control work at a treating plant producing treated utility products; or

(c) Under the direct supervision of an experienced, qualified inspector, the individual shall have performed the following:

(1) For poles, inspected at least 10,000 individual untreated poles, and checked preservative penetration on at least 3,000 individual poles;

(2) For crossarms, inspected at least 5,000 individual untreated arms and checked penetration on at least 500 individual arms;

(3) Conducted at least 100 retention assays, including at least 25 analyses for each different preservative treatment being inspected.

(d) In both paragraphs (a) and (b) of this section, the experience should be not less than that required in paragraph (c) of this section.

(e) Individuals involved in the inspection of more than one commodity must have the minimum experience required in paragraph (c) of this section for each respective product.

[84 FR 28200, June 18, 2019]

§ 1728.204 Electric standards and specifications for materials and construction.

(a) *General specifications.* This section details requirements for 15 and 25 kV single phase, V-phase, and three-phase power cables for use on 12.5/7.2 kV (15 kV rated) and 24.9/14.4 kV (25 kV rated) underground distribution systems with solidly multi-grounded neutral. Cable complying with this specification shall consist of solid or strand-filled conductors which are insulated with tree-retardant cross-linked polyethylene (TR–XLPE) or ethylene propylene rubber (EPR), with concentrically wound copper neutral conductors covered by a nonconducting or semiconducting jacket. 35 kV rated cables may be used in 24.9/14.4 kV application where additional insulation is desired.

(1) The cable may be used in single-phase, two (V)-phase, or three-phase circuits.

(2) Acceptable conductor sizes are: No. 2 AWG (33.6 mm^2) through 1000 kcmil (507 mm^2) for 15 kV cable, No. 1 AWG (42.4 mm^2) through 1000 kcmil (507 mm^2) for 25 kV, and 1/0 (53.5 mm^2) through 1000 kcmil (507 mm^2) for 35 kV cable.

(3) Except where provisions therein conflict with the requirements of this specification, the cable shall meet all applicable provisions of ANSI/ICEA S–94–649–2004 (incorporated by reference in § 1728.97). Where provisions of the ANSI/ICEA specification conflict with this section, § 1728.204 shall apply.

(b) *Definitions.* As used in this section:

Agency refers to the Rural Utilities Service (RUS), an agency of the United States Department of Agriculture's (USDA), hereinafter referred to as the Agency.

EPR Insulating Compound is a mixture of ethylene propylene base resin and selected ingredients.

TR–XLPE Insulating Compound is a tree retardant crosslinked polyethylene (TR–XLPE) insulation compound containing an additive, a polymer modification filler, which helps to retard the growth of electrical trees in the compound.

(c) *Phase conductors.* (1) Central phase conductors shall be copper or aluminum as specified by the borrower within the limit of § 1728.204(a)(2).

(2) Central copper phase conductors shall be annealed copper in accordance with ASTM B 3–01 (incorporated by reference in § 1728.97). Concentric-lay-stranded phase conductors shall conform to ASTM B 8–04 (incorporated by reference in § 1728.97) for Class B stranding. Compact round concentric-lay-stranded phase conductors shall conform to ASTM B 496–04 (incorporated by reference in § 1728.97). Combination unilay stranded phase conductors shall conform to ASTM B 787/B 787M–04 (incorporated by reference in § 1728.97). Compact round atranded copper conductors using single input wire construction shall conform to ASTM B835–04 (incorporated by reference in § 1728.97). Compressed round stranded copper conductors, hard, medium-hard, or soft using single input wire construction shall conform to ASTM B902–04a (incorporated by reference in § 1728.97). If not specified, stranded phase conductors shall be Class B stranded.

(3) Central aluminum phase conductors shall be one of the following:

(i) Solid: Aluminum 1350 H12 or H22, H14 or H24, H16 or H26, in accordance with ASTM B 609/B 609M–99 (incorporated by reference in § 1728.97).

(ii) Stranded: Aluminum 1350 H14 or H24, H142 or H242, H16, or H26, in accordance with ASTM B 609/B 609M–99 (incorporated by reference in § 1728.97) or Aluminum 1350–H19 in accordance with ASTM B 230/B 230M–07 (incorporated by reference in § 1728.97). Concentric-lay-stranded (includes compacted and compressed) phase conductors shall conform to ASTM B 231/B 231M–04 (incorporated by reference in § 1728.97) for Class B stranding. Compact round concentric-lay-stranded phase conductors shall conform to ASTM B 400–08 (incorporated by reference in § 1728.97). Combination unilay stranded aluminum phase conductors shall conform to ASTM B 786–08 (incorporated by reference in § 1728.97). If not specified, stranded phase conductors shall be class B stranded.

(4) The interstices between the strands of stranded conductors shall be filled with a material designed to fill the interstices and to prevent the longitudinal migration of water that might enter the conductor. This material shall be compatible with the conductor and conductor shield materials. The surfaces of the strands that form the outer surface of the stranded conductor shall be free of the strand fill material. Compatibility of the strand fill material with the conductor shield shall be tested and shall be in compliance with ICEA T–32–645–93 (incorporated by reference in § 1728.97). Water penetration shall be tested and shall be in compliance with ANSI/ICEA T–31–610–2007 (incorporated by reference in § 1728.97).

(5) The center strand of stranded conductors shall be indented with the manufacturer's name and year of manufacture at regular intervals with no more than 12 inches (0.3 m) between repetitions.

(d) *Conductor shield (stress control layer).* A non-conducting (for discharge resistant EPR) or semi-conducting shield (stress control layer) meeting the applicable requirements of ANSI/ICEA S–94–649–2004 (incorporated by reference in § 1728.97) shall be extruded around the central conductor. The minimum thickness at any point shall be in accordance with ANSI/ICEA S–94–649–2004. The void and protrusion limits on the conductor shield shall be in compliance with ANSI/ICEA S–94–649–2004. The shield shall have a nominal operating temperature equal to, or higher than, that of the insulation.

(e) *Insulation.* (1) The insulation shall conform to the requirements of ANSI/ICEA S–94–649–2004 (incorporated by reference in § 1728.97) and may either be tree retardant cross-linked polyethylene (TR–XLPE) or ethylene propylene rubber (EPR), as specified by the borrower. The void and protrusion limits on the insulation shall be in compliance with ANSI/ICEA S–94–649–2004.

(2) The thickness of insulation shall be as follows:

ABLE RATED VOLTAGE

Cable rated voltage	Nominal thickness	Minimum thickness	Maximum thickness
15 kV	220 mils (5.59 mm)	210 mils (5.33 mm)	250 mils (6.35 mm).
25 kV	260 mils (6.60 mm)	245 mils (6.22 mm)	290 mils (7.37 mm).
35 kV	345 mils (8.76 mm)	330 mils (8.38 mm)	375 mils (9.53 mm).

(f) *Insulation shield.* (1) A semi-conducting thermosetting polymeric layer meeting the requirements of ANSI/ICEA S–94–649–2004 (incorporated by reference in §1728.97) shall be extruded tightly over the insulation to serve as an electrostatic shield and protective covering. The shield compound shall be compatible with, but not necessarily the same material composition as, that of the insulation (e.g., cross-linked polyethylene shield may be used with EPR insulation). The void and protrusion limits on the semi-conducting shields shall be in compliance with the ANSI/ICEA S–94–649–2004.

(2) The thickness of the extruded insulation shield shall be in accordance with ANSI/ICEA S–94–649–2004 (incorporated by reference in §1728.97).

(3) The shield shall be applied such that all conducting material can be easily removed without the need for externally applied heat. Stripping tension values shall be 3 through 18 pounds (1.36 through 8.16 kg) for TR–XLPE and EPR discharge free cables. Discharge resistant cables shall have strip tension of 0 through 18 pounds (0 through 8.16 kg).

(4) The insulation shield shall meet all applicable tests of ANSI/ICEA S–94–649–2004 (incorporated by reference in §1728.97).

(g) *Concentric neutral conductor.* (1) Concentric neutral conductor shall consist of annealed round, uncoated copper wires in accordance with ASTM B 3–01 (incorporated by reference in §1728.97) and shall be spirally wound over the shielding with uniform and equal spacing between wires. The concentric neutral wires shall remain in continuous intimate contact with the extruded insulation shield. Full neutral is required for single phase and ⅓ neutral for three phase applications unless otherwise specified. The minimum wire size for the concentric neutral is 16 AWG (1.32 mm^2).

(2) When a strap neutral is specified by the borrower, the neutral shall consist of uncoated copper straps applied concentrically over the insulation shield with uniform and equal spacing between straps and shall remain in intimate contact with the underlying extruded insulation shield. The straps shall not have sharp edges. The thickness of the flat straps shall be not less than 20 mils (0.5 mm).

(h) *Overall outer jacket.* (1) An electrically nonconducting (insulating) or semi-conducting outer jacket shall be applied directly over the concentric neutral conductors.

(2) The jacket material shall fill the interstice area between conductors, leaving no voids. The jacket shall be free stripping. The jacket shall have three red stripes longitudinally extruded into the jacket surface 120° apart.

(3) Nonconducting jackets shall consist of low density, linear low density, medium density, or high density HMW black polyethylene (LDPE, LLDPE, MDPE, HDPE) compound meeting the requirements of ANSI/ICEA S–94–649–2004 (incorporated by reference in §1728.97) and ASTM D 1248–05 (incorporated by reference in §1728.97) for Type I, Class C, Category 4 or 5, Grade J3 before application to the cable. Polyvinyl chloride (PVC) and chlorinated polyethylene (CPE) jackets are not acceptable.

(4) Semi-conducting jackets shall have a maximum radial resistivity of 100 ohm-meter and a maximum moisture vapor transmission rate of 1.5 g/m^2/24 hours at 38 °C (100 °F) and 90 percent relative humidity in accordance with ASTM E 96/E96M–05 (incorporated by reference in §1728.97).

(5) The minimum thickness of the jacket over metallic neutral wires or straps shall comply with the thickness specified in ANSI/ICEA S–94–649–2004 (incorporated by reference in §1728.97).

(i) *Tests.* (1) As part of a request for Agency consideration for acceptance and listing, the manufacturer shall submit certified test data results to the Agency that detail full compliance with ANSI/ICEA S–94–649–2004 (incorporated by reference in § 1728.97) for each cable design.

(i) Test results shall confirm compliance with each of the material tests, production sampling tests, tests on completed cable, and qualification tests included in ANSI/ICEA S–94–649–2004 (incorporated by reference in § 1728.97).

(ii) The testing procedure and frequency of each test shall be in accordance with ANSI/ICEA S–94–649–2004 (incorporated by reference in § 1728.97).

(iii) Certified test data results shall be submitted to the Agency for any test, which is designated by ANSI/ICEA S–94–649–2004 (incorporated by reference in § 1728.97) as being "for Engineering Information Only," or any similar designation.

(2) *Partial discharge tests.* Manufacturers shall demonstrate that their cable is not adversely affected by excessive partial discharge. This demonstration shall be made by completing the procedures described in paragraphs (i)(2)(i) and (i)(2)(ii) of this section.

(i) Each shipping length of completed cable shall be tested and have certified test data results available indicating compliance with the partial discharge test requirements in ANSI/ICEA S–94–649–2004 (incorporated by reference in § 1728.97).

(ii) Manufacturers shall test production samples and have available certified test data results indicating compliance with ASTM D 2275–01 (incorporated by reference in § 1728.97) for discharge resistance as specified in the ANSI/ICEA S–94–649–2004 (incorporated by reference in § 1728.97). Samples of insulated cable shall be prepared by either removing the overlying extruded insulation shield material, or using insulated cable before the extruded insulation shield material is applied. The sample shall be mounted as described in ASTM D 2275–01 and shall be subjected to a voltage stress of 250 volts per mil of nominal insulation thickness. The sample shall support this voltage stress, and not show evidence of degradation on the surface of the insulation for a minimum of 100 hours. The test shall be performed at least once on each 50,000 feet (15,240 m) of cable produced, or major fractions thereof, or at least once per insulation extruder run.

(3) *Jacket tests.* Tests described in paragraph (i)(3)(i) of this section shall be performed on cable jackets from the same production sample as in paragraphs (i)(2)(i) and (i)(2)(ii) of this section.

(i) A Spark Test shall be performed on nonconducting jacketed cable in accordance with ANSI/ICEA S–94–649–2004 (incorporated by reference in § 1728.97) on 100 percent of the completed cable prior to its being wound on shipping reels. The test voltage shall be 4.5 kV AC for cable diameters <1.5 inches and 7.0 kV for cable diameters >1.5 inches, and shall be applied between an electrode at the outer surface of the nonconducting (insulating) jacket and the concentric neutral for not less than 0.15 second.

(ii) [Reserved]

(4) Frequency of sample tests shall be in accordance with ANSI/ICEA S–94–649–2004 (incorporated by reference in § 1728.97).

(5) If requested by the borrower, a certified copy of the results of all tests performed in accordance with this section shall be furnished by the manufacturer on all orders.

(j) *Miscellaneous.* (1) All cable provided under this specification shall have suitable markings on the outer surface of the jacket at sequential intervals not exceeding 2 feet (0.61 m). The label shall indicate the name of the manufacturer, conductor size, type and thickness of insulation, center conductor material, voltage rating, year of manufacture, and jacket type. There shall be no more than 6 inches (0.15 m) of unmarked spacing between texts label sequence. The jacket shall be marked with the symbol required by Rule 350G of the National Electrical Safety Code and the borrower shall specify any markings required by local safety codes. This is in addition to extruded red stripes required in this section.

(2) Watertight seals shall be applied to all cable ends to prevent the entrance of moisture during transit or storage. Each end of the cable shall be firmly and properly secured to the reel.

(3) Cable shall be placed on shipping reels suitable for protecting it from damage during shipment and handling. Reels shall be covered with a suitable covering to help provide physical protection to the cable.

(4) A durable label shall be securely attached to each reel of cable. The label shall indicate the purchaser's name and address, purchase order number, cable description, reel number, feet of cable on the reel, tare and gross weight of the reel, and beginning and ending sequential footage numbers.

[77 FR 19529, Apr. 2, 2012]

PART 1730—ELECTRIC SYSTEM OPERATIONS AND MAINTENANCE

Subpart A—General

Subpart B—Operations and Maintenance Requirements

Subpart C—Interconnection of Distributed Resources

AUTHORITY: 7 U.S.C. 901 *et seq.*, 1921 *et seq.*, 6941 *et seq.*

SOURCE: 63 FR 3450, Jan. 23, 1998, unless otherwise noted.

Subpart A—General

§ 1730.1 Introduction.

(a) This part contains the policies and procedures of the Rural Utilities Service (RUS) related to electric borrowers' operation and maintenance practices and RUS' review and evaluation of such practices.

(b) The policies and procedures included in this part apply to all electric borrowers (both distribution borrowers and power supply borrowers) and are intended to clarify and implement certain provisions of the security instrument and loan contract between RUS and electric borrowers regarding operations and maintenance. This part is not intended to waive or supersede any provisions of the security instrument and loan contract between RUS and electric borrowers.

(c) The Administrator may waive, for good cause, on a case by case basis, certain requirements and procedures of this part.

§ 1730.2 RUS policy.

It is RUS policy to require that all property of a borrower be operated and maintained properly in accordance with the requirements of each borrower's loan documents. It is also RUS policy to provide financial assistance only to borrowers whose operations and maintenance practices and records are satisfactory or to those who are taking corrective actions expected to make their operations and maintenance practices and records satisfactory to RUS.

§ 1730.3 RUS addresses.

(a) Persons wishing to obtain forms referred to in this part should contact: Program Support and Regulatory Analysis, Rural Utilities Service, U.S. Department of Agriculture, Stop 1522, 1400 Independence Ave., SW., Washington, DC 20250–1522, telephone (202) 720–8674. Borrowers or others may reproduce any of these forms in any number required.

(b) Documents required to be submitted to RUS under this part are to be sent to the office of the borrower's assigned RUS General Field Representative (GFR) or such other office as designated by RUS.

§ 1730.4 Definitions.

Terms used in this part have the meanings set forth in 7 CFR Part 1710.2. References to specific RUS forms and other RUS documents, and to specific sections or lines of such forms and documents, shall include the corresponding forms, documents, sections and lines in any subsequent revisions of these forms and documents. In addition to the terms defined in 7 CFR Part 1710.2, the term *Prudent Utility Practice* has the meaning set forth in Article 1, Section 1.01 of Appendix A to Subpart B of 7 CFR Part 1718—Model Form of Mortgage for Electric Distribution Borrowers, for the purposes of this Part.

§§ 1730.5–1730.19 [Reserved]

Subpart B—Operations and Maintenance Requirements

§ 1730.20 General.

Each electric program distribution, transmission and generation borrower (as defined in § 1710.2) shall operate and maintain its system in compliance with prudent utility practice, in compliance with its loan documents, and in compliance with all applicable laws, regulations and orders, shall maintain its systems in good repair, working order and condition, and shall make all needed repairs, renewals, replacements, alterations, additions, betterments and improvements, in accordance with applicable provisions of the borrower's security instrument. Each borrower is responsible for on-going operations and maintenance programs, individually or regionally performing a system security Vulnerability and Risk Assessment (VRA), establishing and maintaining an Emergency Restoration Plan (ERP), maintaining records of the physical, cyber and electrical condition and security of its electric system and for the quality of services provided to its customers. The borrower is also responsible for all necessary inspections and tests of the component parts of its system, and for maintaining records of such inspections and tests. Each borrower shall budget sufficient resources to operate and maintain its system and annually exercise its ERP in accordance with the requirements of this part. An actual manmade or natural event on the borrowers system in which a borrower utilizes a significant portion of its ERP shall count as an annual exercise for that calendar year, provided that after conclusion of the event, the borrower verifies accuracy of the emergency points-of-contact (POC) and the associated contact numbers as listed in their ERP. For portions of the borrower's system that are not operated by the borrower, if any, the borrower is responsible for ensuring that the operator is operating and maintaining the system properly in accordance with the operating agreement.

[69 FR 60540, Oct. 12, 2004]

§ 1730.21 Inspections and tests.

(a) Each borrower shall conduct all necessary inspections and tests of the component parts of its electric system, annually exercise its ERP, and maintain records of such inspections and tests. For the purpose of this part, "Exercise" means a borrower's Tabletop execution of, or actual implementation of, the ERP to verify the operability of the ERP. Such Exercise may be performed singly by an individual borrower, or as an active participant in a multi-party (to include utilities, government agencies and other participants or combination thereof) Tabletop execution or actual full implementation of the ERP. For the purpose of this part, "Tabletop" means a hypothetical emergency response scenario in which participants will identify the policy, communication, resources, data, coordination, and organizational elements associated with an emergency response.

(b) The frequency of inspection and testing will be determined by the borrower in conformance with applicable laws, regulations, national standards, and Prudent Utility Practice. The frequency of inspection and testing will be determined giving due consideration to the type of facilities or equipment, manufacturer's recommendations, age, operating environment and hazards to which the facilities are exposed, consequences of failure, and results of previous inspections and tests. The records of such inspections and tests will be retained in accordance with applicable regulatory requirements and Prudent Utility Practice. The retention period should be of a sufficient time period to identify long-term trends. Records must be retained at least until the applicable inspections or tests are repeated.

(c) Inspections of facilities must include a determination of whether the facility complies with the National Electrical Safety Code, National Electrical Code (as applicable), and applicable State or local regulations and whether additional security measures are considered necessary to reduce the vulnerability of those facilities which, if damaged or destroyed, would severely impact the reliability and security of the electric power grid, cause significant risk to the safety and health of the public and/or impact the ability to provide service to consumers over an extended period of time. The electric power grid, also known as the transmission grid, consists of a network of electrical lines and related facilities, including certain substations, used to connect distribution facilities to generation facilities, and includes bulk transmission and subtransmission facilities as defined in §1710.2 of this title. Any serious or life-threatening deficiencies shall be promptly repaired, disconnected, or isolated in accordance with applicable codes or regulations. Any other deficiencies found as a result of such inspections and tests are to be recorded and those records are to be maintained until such deficiencies are corrected or for the retention period required by paragraph (b) of this section, whichever is longer.

[63 FR 3450, Jan. 23, 1998, as amended at 69 FR 60540, Oct. 12, 2004]

§ 1730.22 Borrower analysis.

(a) Each borrower shall periodically analyze and document its security, operations and maintenance policies, practices, and procedures to determine if they are appropriate and if they are being followed. The records of inspections and tests are also to be reviewed and analyzed to identify any trends which could indicate deterioration in the physical or cyber condition or the operational effectiveness of the system or suggest a need for changes in security, operations or maintenance policies, practices and procedures. For portions of the borrower's system that are not operated by the borrower, if any, the borrower's written analysis would also include a review of the operator's performance under the operating agreement.

(b) When a borrower's security, operations and maintenance policies, practices, and procedures are to be reviewed and evaluated by RUS, the borrower shall:

(1) Conduct the analysis required by paragraph (a) of this section not more than 90 days prior to the scheduled RUS review;

(2) Complete RUS Form 300, Review Rating Summary, and other related forms, prior to RUS' review and evaluation; and

(3) Make available to RUS the borrower's completed RUS Form 300 (including a written explanation of the basis for each rating) and records related to the operations and maintenance of the borrower's system.

(c) For those facilities not included on the RUS Form 300 (e.g., generating plants), the borrower shall prepare and complete an appropriate supplemental form for such facilities.

[63 FR 3450, Jan. 23, 1998, as amended at 69 FR 60541, Oct. 12, 2004]

§ 1730.23 Review rating summary, RUS Form 300.

The RUS Form 300 is available from RUS and shall be used when required by this part.

[86 FR 36198, July 9, 2021]

§ 1730.24 RUS review and evaluation.

RUS will initiate and conduct a periodic review and evaluation of the operations and maintenance practices of each borrower for the purpose of assessing loan security and determining borrower compliance with RUS policy as outlined in this part. This review will normally be done at least once every three years. The borrower will make available to RUS the borrower's policies, procedures, and records related to the operations and maintenance of its complete system. Reports made by other inspectors (e.g., other Federal agencies, State inspectors, etc.) will also be made available, as applicable. RUS will not duplicate these other reviews but will use their reports to supplement its own review. RUS may inspect facilities, as well as records, and may also observe construction and maintenance work in the field. Key borrower personnel responsible for the facilities being inspected are to accompany RUS during such inspections, unless otherwise determined by RUS. RUS personnel may prepare an independent summary of the operations and maintenance practices of the borrower. The borrower's management will discuss this review and evaluation with its Board of Directors.

§ 1730.25 Corrective action.

(a) For any items on the RUS Form 300 rated unsatisfactory (i.e., 0 or 1) by the borrower or by RUS, the borrower shall prepare a corrective action plan (CAP) outlining the steps (both short term and long term) the borrower will take to improve existing conditions and to maintain an acceptable rating. The CAP must include a time schedule and cost estimate for corrective actions, and must be approved by the borrower's Board of Directors. The CAP must be submitted to RUS for approval within 90 days after the completion of RUS' evaluation noted in § 1730.24.

(b) The borrower must periodically report to RUS in writing progress under the CAP. This report must be submitted to RUS every six months until all unsatisfactory items are corrected unless RUS prescribes a different reporting schedule.

§ 1730.26 Certification.

(a) *Engineer's certification.* Where provided for in the borrower's loan documents, RUS may require the borrower to provide an "Engineer's Certification" as to the condition of the borrower's system (including, but not limited to, all mortgaged property.) Such certification shall be in form and substance satisfactory to RUS and shall be prepared by a professional engineer satisfactory to RUS. If RUS determines that the Engineer's Certification discloses a need for improvements to the condition of its system or any other operations of the borrower, the borrower shall, upon notification by RUS, promptly undertake to accomplish such improvements.

(b) *Emergency Restoration Plan certification.* The borrower's Manager or Chief Executive Officer shall provide written certification to RUS stating that a VRA has been satisfactorily completed that meets the criteria of § 1730.27 (a), (b), (c), or (d), as applicable and § 1730.27(e)(1) through (e)(8), and that the borrower has an ERP that meets the criteria of § 1730.28 (a), (b), (c), or (d), as applicable, and § 1730.28 (e), (f), and (g). The written certification shall be in letter form. Applicants for new RUS electric loans, loan guarantees or grants shall include the written certification in the application package submitted to RUS. If the self-certification of an ERP and VRA are not received as set forth in this section, approval of the loan, loan guarantees or grants will not be considered until the certifications are received by RUS.

[63 FR 3450, Jan. 23, 1998, as amended at 69 FR 60541, Oct. 12, 2004]

§ 1730.27 Vulnerability and Risk Assessment (VRA).

(a) Each borrower with an approved RUS electric program loan as of October 12, 2004 shall perform an initial VRA of its electric system no later than July 12, 2005. Additional or periodic VRA's may be necessary if significant changes occur in the borrower's system, and records of such additional assessments shall be maintained by the borrower.

(b) Each applicant that has submitted an application for an RUS electric program loan or grant prior to October 12, 2004, but whose application has not been approved by RUS by such date, shall perform an initial VRA of its electric system in accordance with § 1730.27(a).

(c) Each applicant that submits an application for an RUS electric program loan or grant between October 12, 2004 and July 12, 2005 shall perform an initial VRA of its electric system in accordance with § 1730.27(a).

(d) Each applicant that submits an application for an RUS electric program loan or grant on or after July 12, 2005 shall include with its application package a letter certification that such applicant has performed an initial VRA of its electric system. Additional or periodic VRA's may be necessary if significant changes occur in the borrower's system, and records of such additional assessments shall be maintained by the borrower.

(e) The VRA shall include identifying:

(1) Critical assets or facilities considered necessary for the reliability and security of the electric power grid as described in § 1730.21(c);

(2) Facilities that if damaged or destroyed would cause significant risk to the safety and health of the public;

(3) Critical assets or infrastructure owned or served by the borrower's electric system that are determined, identified and communicated as elements of national security by the consumer, State or Federal government;

(4) External system impacts (interdependency) with loss of identified system components;

(5) Threats to facilities and assets identified in paragraphs (e)(1), (e)(2), (e)(3), and (e)(4) of this section;

(6) Criticality and risk level of the borrower's system;

(7) Critical asset components and elements unique to the RUS borrower's system; and

(8) Other threats, if any, identified by an individual borrower.

[69 FR 60541, Oct. 12, 2004]

§ 1730.28 Emergency Restoration Plan (ERP).

(a) Each borrower with an approved RUS electric program loan as of October 12, 2004 shall have a written ERP no later than January 12, 2006. The ERP should be developed by the borrower individually or in conjunction with other electric utilities (not all having to be RUS borrowers) through the borrower's unique knowledge of its system, prudent utility practices (which includes development of an ERP) and the borrower's completed VRA. If a joint electric utility ERP is developed, each RUS borrower shall prepare an addendum to meet the requirements of paragraphs (e), (f), and (g) of this section as it relates to its system.

(b) Each applicant that has submitted an application for an RUS electric program loan or grant prior to October 12, 2004, but whose application has not been approved by RUS by such date, shall have a written ERP in accordance with § 1730.28(a).

(c) Each applicant that submits an application for an RUS electric program loan or grant between October 12, 2004 and January 12, 2006, shall have a written ERP in accordance with § 1730.28(a).

(d) Each applicant that submits an application for an RUS electric program loan or grant on or after January 12, 2006 shall include with its application package a letter certification that such applicant has a written ERP.

(e) The ERP shall include:

(1) A list of key contact emergency telephone numbers (emergency agencies, borrower management and other key personnel, contractors and equipment suppliers, other utilities, and others that might need to be reached in an emergency);

(2) A list of key utility management and other personnel and identification of a chain of command and delegation of authority and responsibility during an emergency;

(3) Procedures for recovery from loss of power to the headquarters, key offices, and/or operation center facilities;

(4) A Business Continuity Section describing a plan to maintain or re-establish business operations following an event which disrupts business systems (computer, financial, and other business systems);

(5) A section describing a plan to comply with the eligibility requirements to qualify for the FEMA Public Assistance Grant Program; and

(6) Other items, if any, identified by the borrower as essential for inclusion in the ERP.

(f) The ERP must be approved and signed by the borrower's Manager or Chief Executive Officer, and approved by the borrower's Board of Directors.

(g) Copies of the most recent approved ERP must be made readily available to key personnel at all times.

(h) The ERP shall be Exercised at least annually to ensure operability and employee familiarity. Completion of the first exercise of the ERP must occur on or before January 12, 2007.

(i) If modifications are made to an existing ERP:

(1) The modified ERP must be prepared in compliance with the provisions of paragraphs (e), (f), and (g) of this section; and

(2) Additional Exercises may be necessary to maintain employee operability and familiarity.

(j) Each borrower shall maintain records of such Exercises.

[69 FR 60541, Oct. 12, 2004, as amended at 76 FR 47056, Aug. 4, 2011]

§ 1730.29 Grants and Grantees.

For the purposes of this part, the terms "borrower" shall include recipients of RUS electric program grants, and "applicant" shall include applicants for such grants. References to "security documents" shall, with respect to recipients of RUS electric program grants, include grant agreements and other grant-related documents.

[69 FR 60541, Oct. 12, 2004]

§§ 1730.30–1730.59 [Reserved]

Subpart C—Interconnection of Distributed Resources

SOURCE: 74 FR 32409, July 8, 2009, unless otherwise noted.

§ 1730.60 General.

Each electric program distribution borrower (as defined in § 1710.2) is responsible for establishing and maintaining a written standard policy relating to the Interconnection of Distributed Resources (IDR) having an installed capacity of not more than 10 megavolt amperes (MVA) at the point of common coupling.

§ 1730.61 RUS policy.

The Distributed Resource facility must not cause significant degradation of the safety, power quality, or reliability on the borrower's electric power system or other electric power systems interconnected to the borrower's electric power system. The Agency encourages borrowers to consider model policy templates developed by knowledgeable and expert institutions, such as, but not limited to the National Association of Regulatory Utility Commissioners, the Federal Energy Regulatory Commission and the National Rural Electric Cooperative Association. The Agency encourages all related electric borrowers to cooperate in the development of a common Distributed Resource policy.

§ 1730.62 Definitions.

"Distributed resources" as used in this subpart means sources of electric power that are not directly connected to a bulk power transmission system, having an installed capacity of not more than 10 MVA, connected to the borrower's electric power system through a point of common coupling. Distributed resources include both generators and energy storage technologies.

"Responsible party" as used in this subpart means the owner, operator or any other person or entity that is accountable to the borrower under the borrower's interconnection policy for Distributed Resources.

§ 1730.63 IDR policy criteria.

(a) *General.* (1) The borrower's IDR policy and procedures shall be readily available to the public and include, but not limited to, a standard application, application process, application fees, and agreement.

(2) All costs to be recovered from the applicant regarding the application process or the actual interconnection and the process to determine the costs are to be clearly explained to the applicant and authorized by the applicant prior to the borrower incurring these costs. The borrower may require separate non-refundable deposits sufficient to insure serious intent by the applicant prior to proceeding either with the application or actual interconnection process.

(3) IDR policies must be approved by the borrower's Board of Directors.

(4) The borrower may establish a new rate classification for customers with Distributed Resources.

(5) IDR policies must provide for reconsideration and updates every five years or more frequently as circumstances warrant.

(b) *Technical requirements.* (1) IDR policies must be consistent with prudent electric utility practice.

(2) IDR policies must incorporate the Institute of Electrical and Electronic Engineers (IEEE): IEEE 1547™—Standard for Interconnecting Distributed Resources with Electric Power Systems, approved June 12, 2003, and IEEE 1547.1™—Standard Conformance Test Procedures for Equipment Interconnecting Distributed Resources with Electric Power Systems, approved June 9, 2005. Copies of the IEEE Standards 1547™ and 1547.1™ may be obtained from the IEEE Operations Center, 445 Hoes Lane, Piscataway, NJ 08854–4141, telephone 1–800–678–4333 or online at *http://www.standards.ieee.org*. Copies of the material are available for inspection during normal business hours at RUS, Room 1265, U.S. Department of Agriculture, Washington, DC 20250. Telephone (202) 720–3720, e-mail *Donald.Junta@wdc.usda.gov*, or at the National Archives and Records Administration (NARA). For information on the availability of this material at NARA, call 202–741–6030, or go to: *http://www.archives.gov/federal_register/code_of_federal_regulations/ibr_locations.html*.

(3) IDR policies must provide for appropriate electric power system disconnect facilities, as determined by the borrower, which shall include a lockable disconnect and a visible open, that are readily accessible to and operable by authorized personnel at all times.

(4) IDR policies must provide for borrower access to the Distributed Resources facility during normal business hours and all emergency situations.

(c) *Responsible Party obligations.* IDR policies must provide for appropriate Responsible Parties to assume the following risks and responsibilities:

(1) A Responsible Party must agree to maintain appropriate liability insurance as outlined in the borrower's interconnection policy.

(2) A Responsible Party must be responsible for the Distributed Resources compliance with all national, State, local government requirements and electric utility standards for the safety of the public and personnel responsible for utility electric power system operations, maintenance and repair.

(3) A Responsible Party must be responsible for the safe and effective operation and maintenance of the facility.

(4) Only Responsible Parties may apply for interconnection and the Responsible Party must demonstrate that the facility will be capably developed, constructed and operated, maintained, and repaired.

§ 1730.64 Power purchase agreements.

Nothing in this subpart requires the borrower to enter into purchase power arrangements with the owner of the Distributed Resources.

§ 1730.65 Effective dates.

(a) All electric program borrowers with an approved electric program loan as of July 8, 2009 shall have an IDR policy board approved and in effect no later than July 8, 2011.

(b) All other electric program borrowers that have pending applications or submit an application to the Agency for financial assistance on or after July 8, 2009 shall provide a letter of certification executed by the General Manager that the borrower meets the requirements of this subpart before such loan may be approved.

§ 1730.66 Administrative waiver.

The Administrator may waive in all or part, for good cause, the requirements and procedures of this subpart.

§§ 1730.67–1730.99 [Reserved]

§ 1730.100 OMB Control Number.

The Information collection requirements in this part are approved by the Office of Management and Budget and assigned OMB control number 0572–0141.

PART 1734—DISTANCE LEARNING AND TELEMEDICINE LOAN AND GRANT PROGRAMS

Subpart A—Distance Learning and Telemedicine Loan and Grant Programs—General

Sec.

Subpart B—Distance Learning and Telemedicine Grant Program

Subpart C—Distance Learning and Telemedicine Combination Loan and Grant Program

Subpart D—Distance Learning and Telemedicine Loan Program

AUTHORITY: 7 U.S.C. 901 *et seq.* and 950aaa *et seq.*

SOURCE: 82 FR 55925, Nov. 27, 2017, unless otherwise noted.

Subpart A—Distance Learning and Telemedicine Loan and Grant Program—General

§ 1734.1 Purpose.

The purpose of the Distance Learning and Telemedicine (DLT) Loan and Grant Program is to encourage and improve telemedicine services and distance learning services in rural areas through the use of telecommunications, computer networks, and related advanced technologies by students, teachers, medical professionals, and rural residents. This subpart describes the general policies for administering the DLT program. Subpart B of this part contains the policies and procedures related to grants; subpart C contains the policies and procedures related to a combination loan and grant; and subpart D contains the policies and procedures related to loans.

§ 1734.2 Policy.

(a) The transmission of information is vital to the economic development, education, and health of rural Americans. To further this objective, the Rural Utilities Service (RUS) will provide financial assistance to distance learning and telemedicine projects that will improve the access for people residing in rural areas to educational, learning, training, and health care services.

(b) In providing financial assistance, RUS will give priority to rural areas that it believes have the greatest need for distance learning and telemedicine services. RUS believes that generally the need is greatest in areas that are economically challenged, costly to serve, and experiencing outward migration. This program is consistent with the provisions of the Telecommunications Act of 1996 that designate telecommunications service discounts for schools, libraries, and rural health care centers. RUS will take into consideration the community's involvement in the proposed project and the applicant's ability to leverage grant funds.

(c) In administering this subpart, RUS will not favor or mandate the use of one particular technology over another.

(d) Rural institutions are encouraged to cooperate with each other, with applicants, and with end-users to promote the program being implemented under this subpart.

(e) RUS staff will make diligent efforts to inform potential applicants in rural areas of the programs being implemented under this subpart.

(f) The Administrator may provide loans under this subpart to an entity that has received a telecommunications or electric loan under the Rural Electrification Act of 1936. A borrower receiving a loan shall:

(1) Make the funds provided available to entities that qualify as distance learning or telemedicine projects satisfying the requirements of this subpart, under any terms it so chooses as long as the terms are no more stringent than the terms under which it received the financial assistance.

(2) Use the loan to acquire, install, improve, or extend a distance learning or telemedicine system referred to in this subpart.

(g) The Administrator will allocate funds that are appropriated each fiscal year for subparts B, C, and D, of this part respectively. Not more than 30 days before the end of the fiscal year, the Administrator may transfer any funds not committed to grants in the combination loan and grant program to the grant program.

(h) Financial assistance may be provided for end user sites. Financial assistance may also be provided for hubs located in rural or non-rural areas if they are necessary to provide distance learning or telemedicine services to rural residents at end user sites.

(i) The Administrator will publish, at the end of each fiscal year, a notice on the RUS Telecommunications Program Web site of all applications receiving financial assistance under this subpart. Subject to the provisions of the Freedom of Information Act (5 U.S.C. 552), applications will be available for public inspection at the U.S. Department of Agriculture, 1400 Independence Avenue SW., Washington, DC 20250.

§ 1734.3 Definitions.

As used in this part:

1996 Act means the Federal Agriculture Improvement Act of 1996.

Act means the Rural Electrification Act of 1936 (7 U.S.C. 901 *et seq.*).

Administrator means the Administrator of the Rural Utilities Service, or designee or successor.

Applicant means an eligible organization that applies for financial assistance under this subpart.

Approved purposes means project purposes for which grant, loan, or combination loan and grant financial assistance may be expended.

Broadband facilities means facilities that transmit, receive, or carry voice, video, or data between the terminal equipment at each end of the circuit or path. Such facilities include microwave antennae, relay stations and towers, other telecommunications antennae, fiber-optic cables and repeaters, coaxial cables, communication satellite ground station complexes, copper cable electronic equipment associated with telecommunications transmissions, and similar items.

Combination loan and grant means a grant in combination with a loan made under the DLT program.

Completed application means an application that includes all those items specified in §§ 1734.125, 1734.134, and in form and substance satisfactory to the Administrator.

Consortium means a combination or group of entities undertaking the purposes for which the distance learning and telemedicine financial assistance is provided. At least one of the entities in a consortium must meet the requirements of § 1734.4.

Construct means to acquire, construct, extend, improve, or install a facility or system.

Distance learning means a telecommunications link to an end user through the use of eligible equipment to provide educational programs, instruction, or information originating in one area, whether rural or not, to students and teachers who are located in rural areas.

DLT borrower means an entity that has an outstanding loan under the provisions of the DLT program.

DLT program means the Distance Learning and Telemedicine Loan and Grant Program administered by RUS.

Economic useful life as applied to equipment and facilities financed under the DLT program is calculated based on Internal Revenue Service depreciation rules or recognized telecommunications industry guidelines.

Eligible equipment means computer hardware and software, audio and video equipment, computer networking components, telecommunications terminal equipment, terminal equipment, inside wiring, interactive video equipment.

Eligible facilities means land, buildings, or building construction needed to carry out an eligible distance learning or telemedicine project for loan financial assistance only.

End user is one or more of the following:

(1) Rural elementary, secondary schools, and other educational institutions, such as institutions of higher education, vocational and adult training and educational centers, libraries and teacher training centers, and students, teachers and instructors using such rural educational facilities, that participate in distance learning telecommunications program through a project funded under this subpart;

(2) Rural hospitals, primary care centers or facilities, such as medical centers, nursing homes, and clinics, and physicians and staff using such rural medical facilities, that participants in a rural distance learning telecommunications program through a project funded under this part; and

(3) Other rural community facilities, institutions, or entities that receive distance learning or telemedicine services.

End user site means a facility that is part of a network or telecommunications system that is utilized by end users. An end user site can also be the residence of someone living in a rural area that is receiving telemedicine or distant learning services.

Financial assistance means a grant, combination loan and grant, or loan.

GFR means RUS telecommunications program General Field Representative.

Grant documents means the grant agreement, including any amendments and supplements thereto, between RUS and the grantee.

Grantee means a recipient of a grant from RUS to carry out the purposes of the DLT program.

Guarantee means a guarantee for a loan provided by a RUS borrower or other qualified third party.

Hub means a facility that is part of a network or telecommunications system that provides educational or medical services to end user sites.

Instructional programming means course material for teaching over the Distance Learning or Telemedicine network, including computer software.

Interactive equipment means equipment used to produce and prepare for transmission of audio and visual signals from at least two distant locations so that individuals at such locations can orally and visually communicate with each other. Such equipment includes, but is not limited to, monitors, other display devices, cameras or other recording devices, audio pickup devices, and other related equipment.

Loan means a loan made under the DLT program bearing interest at a rate equal to the then current cost-of-money to the government.

Loan documents mean the loan agreement, note, and security instrument, including any amendments and supplements thereto, between RUS and the DLT borrower.

Local exchange carrier (LEC) is a regulatory term in telecommunications for the local telephone company. In the United States, wireline telephone companies are divided into two large categories: Long distance (interexchange carrier, or IXCs) and local (local exchange carrier, or LECs). This structure is a result of 1984 divestiture of then regulated monopoly carrier American Telephone & Telegraph. Local telephone companies at the time of the divestiture are also known as Incumbent Local Exchange Carriers (ILEC).

Matching contribution means the applicant's contribution for approved purposes.

Project means approved purposes for which financial assistance has been provided.

Project service area means the area in which at least 90 percent of the persons to be served by the project are likely to reside.

Recipient means a grantee, borrower, or both of a DLT program grant, loan or combination loan and grant.

Rural community facility means a facility such as a school, library, learning center, training facility, hospital, or medical facility that provides benefits primarily to residents of rural areas.

RUS means the Rural Utilities Service, an agency of the United States Department of Agriculture, successor to the Rural Electrification Administration.

Secretary means the Secretary of Agriculture.

Technical assistance means:

(1) Assistance in learning to manage, operate, or use equipment or systems; and

(2) Studies, analyses, designs, reports, manuals, guides, literature, or other forms of creating, acquiring, or disseminating information.

Telecommunications carrier means any provider of telecommunications services.

Telecommunications or electric borrower means an entity that has outstanding RUS electric or telecommunications loan or loan guarantee under the provisions of the Act.

Telecommunications systems plan means the plan submitted by an applicant in accordance with §1734.25 for grants, §1734.34 for a combination loan and grant, or §1734.44 for loans.

Telemedicine means a telecommunications link to an end user through the use of eligible equipment which electronically links medical professionals at separate sites in order to exchange health care information in audio, video, graphic, or other format for the purpose of providing improved health care services primarily to residents of rural areas.

§ 1734.4 Applicant eligibility and allocation of funds.

To be eligible to receive a grant, loan and grant combination, or loan under this subpart:

(a) The applicant must be legally organized as an incorporated organization, an Indian tribe or tribal organization, as defined in 25 U.S.C. 450b(b) and (c), a state or local unit of government, a consortium, as defined in §1734.3, or other legal entity, including a private corporation organized on a for-profit or not-for profit basis. Each applicant must provide written evidence of its legal capacity to contract with RUS to obtain the grant, loan and grant combination, or the loan, and comply with all applicable requirements. If a consortium lacks the legal capacity to contract, each individual entity must contract with RUS in its own behalf.

(b) The applicant proposes to utilize the financing to:

(1) Operate a rural End-User Site for the purpose of providing Distance Learning or Telemedicine services; or

(2) Deliver distance learning or telemedicine services to entities that operate a rural community facility or to residents of rural areas at rates calculated to ensure that the benefit of the financial assistance is passed through to such entities or to residents of rural areas.

§ 1734.5 Processing of selected applications.

(a) During the period between the submission of an application and the execution of documents, the applicant must inform RUS if the project is no longer viable or the applicant no longer is requesting financial assistance for the project. When the applicant so informs RUS, the selection will be rescinded or the application withdrawn and written notice to that effect sent to the applicant.

(b) If an application has been selected and the scope of the project changes substantially, the applicant may be required to reapply in the next program window if the agency and the selected applicant cannot agree on the new scope of the award.

(c) If state or local governments raise objections to a proposed project under the intergovernmental review process that are not resolved within 90 days from the time the public is made aware of the award, the Administrator will rescind the selection and written notice to that effect will be sent to the applicant. The Administrator, in his sole discretion, may extend the 90 day period if it appears resolution is imminent.

(d) RUS may request additional information that would not change the application or scoring, in order to complete the appropriate documents covering financial assistance.

(e) *Financial assistance documents.* (1) The documents will include a grant agreement for grants; loan documents, including third party guarantees, notes and security instruments for loans; or any other legal documents the Administrator deems appropriate, including suggested forms of certifications and legal opinions.

(2) The grant agreement and the loan documents will include, among other things, conditions on the release or advance of funds and include at a minimum, a project description, approved purposes, the maximum amount of the financial assistance, supplemental funds required for the project, and certain agreements or commitments the applicant may have proposed in its application. In addition, the loan documents may contain covenants and conditions the Administrator deems necessary or desirable to provide additional assurance that loans will be repaid and the purposes of the loan will be accomplished.

(3) The recipient of a loan will be required to execute a security instrument in form and substance satisfactory to RUS and must, before receiving any advance of loan funds, provide security that is adequate, in the opinion of RUS, to assure repayment, within the time agreed, of all loans to the borrower under the DLT program. This assurance will generally be provided by a first lien upon all facilities and equipment financed by the loan. RUS may require additional security as it deems necessary.

(4) Adequate security may also be provided by third-party guarantees, letters of credit, pledges of revenue, or other forms of security satisfactory to RUS.

(5) The security instrument and other loan documents required by RUS in connection with a loan under the DLT program shall contain such pledges, covenants, and other provisions as may, in the opinion of RUS, be required to secure repayment of the loan.

(6) If the project does not constitute a complete operating system, the DLT recipient shall provide evidence, in form and substance satisfactory to RUS, demonstrating that the recipient has sufficient contractual, financing, or other arrangements to assure that the project will provide adequate and efficient service.

(f) Prior to the execution of a grant and loan document, RUS reserves the right to require any changes in the project or legal documents covering the project to protect the integrity of the DLT program and the interests of the government.

(g) If the applicant fails to submit, within 120 calendar days from the date RUS notifies the applicant that they have been selected for an award, all of the information that RUS determines to be necessary to prepare legal documents and satisfy other requirements of this subpart, RUS may rescind the selection of the application.

§ 1734.6 Disbursement of loans and grants.

(a) For financial assistance of $100,000 or greater, prior to the disbursement of a grant and a loan, the recipient, if it is not a unit of government, will provide evidence of fidelity bond coverage as required by 2 CFR part 200, which is adopted by USDA through 2 CFR part 400.

(b) Grants and loans will be disbursed to recipients on a reimbursement basis, or with unpaid invoices for the eligible purposes contained in this subpart, by the following process:

(1) An SF 270, "Request for Advance or Reimbursement," will be completed by the recipient and submitted to RUS not more frequently than once a month;

(2) RUS will review the SF 270 for accuracy when received and will schedule payment if the form is satisfactory. Payment will ordinarily be made within 30 days; and

(c) The recipient's share in the cost of the project must be disbursed in advance of the loan and grant, or if the recipient agrees, on a pro rata distribution basis with financial assistance during the disbursement period. Recipients will not be permitted to provide their contributions at the end of the project.

(d) A combination loan and grant will be disbursed on a pro rata basis based on the respective amounts of financial assistance provided.

§ 1734.7 Reporting and oversight requirements.

(a) A project performance activity report will be required of all recipients on an annual basis until the project is complete and the funds are expended by the applicant.

(b) Recipients shall diligently monitor performance to ensure that time schedules are being met, projected work by time periods is being accomplished, and other performance objectives are being achieved. Recipients are to submit all project performance reports, including, but not limited to, the following:

(1) A comparison of actual accomplishments to the objectives established for that period;

(2) A description of any problems, delays, or adverse conditions which have occurred, or are anticipated, and which may affect the attainment of overall project objectives, prevent the meeting of time schedules or objectives, or preclude the attainment of particular project work elements during established time periods. This disclosure shall be accompanied by a statement of the action taken or planned to resolve the situation; and

(3) Objectives and timetable established for the next reporting period.

A final project performance report must be provided by the recipient. It must provide an evaluation of the success of the project in meeting the objectives of the program. The final report may serve as the last annual report.

(c) RUS will monitor recipients, as it determines necessary, to ensure that projects are completed in accordance with the approved scope of work and that the financial assistance is expended for approved purposes.

§ 1734.8 Audit requirements.

A recipient of financial assistance shall provide RUS with an audit for each year, beginning with the year in which a portion of the financial assistance is expended, in accordance with the following:

(a) If the recipient is a for-profit entity, a Telecommunications or Electric borrower, or any other entity not covered by the following paragraph, the recipient shall provide an independent audit report in accordance with 7 CFR part 1773, "Policy on Audits of RUS Borrowers."

(b) If the recipient is a state or local government, or non-profit organization, the recipient shall provide an audit in accordance with subpart F of 2 CFR part 200, as adopted by USDA through 2 CFR part 400.

(c) Grantees shall comply with 2 CFR part 200, as adopted by USDA through 2 CFR part 400, and rules on the disposition of grant assets in Part 200 shall be applied regardless of the type of legal organization of the grantee.

§ 1734.9 Grant and loan administration.

RUS will conduct reviews as necessary to determine whether the financial assistance was expended for approved purposes. The recipient is responsible for ensuring that the project complies with all applicable regulations, and that the grants and loans are expended only for approved purposes. The recipient is responsible for ensuring that disbursements and expenditures of funds are properly supported by invoices, contracts, bills of sale, canceled checks, or other appropriate forms of evidence, and that such supporting material is provided to RUS, upon request, and is otherwise made available, at the recipient's premises, for review by the RUS representatives, the recipient's certified public accountant, the Office of Inspector General, U.S. Department of Agriculture, the General Accounting Office, and any other official conducting an audit of the recipient's financial statements or records, and program performance for the grants and loans made under this subpart. The recipient shall permit RUS to inspect and copy any records and documents that pertain to the project.

§ 1734.10 Changes in project objectives or scope.

The recipient shall obtain prior written approval by RUS for any material change to the scope or objectives of the project, including any changes to the scope of work or the budget submitted to RUS. Any material change shall be contained in a revised scope of work plan to be prepared by the recipient, submitted to, and approved by RUS in writing. If RUS does not approve the change and the awardee is unable to fulfill the original purposes of the award, the awardee will work with RUS to return or rescind the financial assistance.

§ 1734.11 Grant and loan termination.

(a) The financial assistance may be terminated when RUS and the recipient agree upon the conditions of the termination, the effective date of the termination, and, in the case of a partial termination of the financial assistance, any unadvanced portion of the financial assistance to be terminated and any advanced portion of the financial assistance to be returned.

(b) The recipient may terminate the financial assistance by written notification to RUS, providing the reasons for such termination, the effective date, and, in the case of a partial termination, the portion of the financial assistance to be terminated. In the case of a partial termination, if RUS believes that the remaining portion of the financial assistance will not accomplish the approved purposes, then RUS may terminate the financial assistance in its entirety, pursuant to the provisions of paragraph (a) of this section.

§ 1734.12 Expedited telecommunications loans

RUS will expedite consideration and determination of an application submitted by an RUS telecommunications borrower for a loan under the Act or an advance of such loan funds to be used in conjunction with financial assistance under subparts B, C, or D of this part. See 7 CFR part 1737 for loans and 7 CFR part 1744 for advances under this section.

§§ 1734.13–1734.19 [Reserved]

Subpart B—Distance Learning and Telemedicine Grant Program

§ 1734.20 [Reserved]

§ 1734.21 Approved purposes for grants.

For distance learning and telemedicine projects, grants shall finance only the costs for approved purposes. Grants shall be expended only for the costs associated with the capital assets associated with the project. The following are approved grant purposes:

(a) Acquiring and installing, by lease or purchase, eligible equipment as defined in § 1734.3;

(b) Purchases of extended warranties, site licenses, and maintenance contracts, for a period not to exceed 3 years from installation date, so long as such purchases are in support of eligible equipment included in the project and made concurrently;

(c) Acquiring or developing instructional programming; but shall not include salaries, benefits, and overhead of medical, educational, or any personnel employed by the applicant. The funded development and acquisition of instructional programming must be done through an independent 3rd party, and may not be performed using the applicant's employees.

(d) Providing technical assistance and instruction for using eligible equipment, including any related software; developing instructional programming; or providing engineering and environmental studies relating to the establishment or expansion of the phase of the project that is being financed with the grant. These purposes shall not exceed 10 percent of the grant; and

(e) Purchasing and installing broadband facilities. This purpose is limited to a maximum of 20 percent of the request grant amount and must be used for providing distance learning or telemedicine services.

§ 1734.22 Matching contributions.

(a) The grant applicant's minimum matching contribution must equal 15 percent of the grant amount requested and shall be used for approved purposes for grants listed in § 1734.21. Matching contributions generally must be in the form of cash. However, in-kind contributions solely for the purposes listed in § 1734.21 may be substituted for cash.

(b) In-kind items listed in § 1734.21 must be non-depreciated or new assets with established monetary values. Use of specific manufacturers' equipment or services, or discounts thereon, are not considered eligible in-kind matching if the manufacturer, or its authorized reseller, is a vendor on the project, the grant writer for the grant application, or has undertaken any responsibility on the grant application, including payment.

(c) Costs incurred by the applicant, or others on behalf of the applicant, for facilities or equipment installed, or other services rendered prior to submission of a completed application, shall not be considered as an eligible in-kind matching contribution.

(d) Costs incurred for non-approved purposes for grant outlined in § 1734.23 shall not be used as an in-kind matching contribution.

(e) Any financial assistance from Federal sources will not be considered as matching contributions under this subpart unless there is a Federal statutory exception specifically authorizing the Federal financial assistance to be considered as a matching contribution.

§ 1734.23 Nonapproved purposes for grants.

(a) A grant made under this subpart will not be provided or used:

(1) To pay for medical or educational equipment not having telemedicine or distance learning as its essential function;

(2) To pay for Electronic Medical Records (EMR) systems;

(3) To pay salaries, wages, or employee benefits to medical or educational personnel;

(4) To pay for the salaries or administrative expenses of the applicant or the project;

(5) To purchase equipment that will be owned by the local exchange carrier or another telecommunications service provider unless that service provider is the applicant.

(6) To duplicate facilities providing distance learning or telemedicine services in place or to reimburse the applicant or others for costs incurred prior to RUS' receipt of the completed application;

(7) To pay costs of preparing the application package for financial assistance under this program;

(8) For projects whose sole objective is to provide links between teachers and students or between medical professionals who are located at the same facility or campus environment;

(9) For site development and the destruction or alteration of buildings;

(10) For the purchase of land, buildings, or building construction;

(11) For projects located in areas covered by the Coastal Barrier Resources Act (16 U.S.C. 3501 *et seq.*);

(12) For any purpose that the Administrator has not specifically approved;

(13) Except for leases provided for in § 1734.21, to pay the cost of recurring or operating expenses for the project; or

(14) For any other purposes not specifically contained in § 1734.21.

(b) Except as otherwise provided in § 1734.12, grants shall not be used to finance a project, in part, when the success of the project is dependent upon the receipt of additional financial assistance under this subpart or is dependent upon the receipt of other financial assistance that is not assured.

§ 1734.24 Maximum and minimum grant amounts.

Applications for grants under this subpart will be subject to limitations on the proposed amount of grant funds. The Administrator will establish the maximum and minimum amounts of a grant to be made available to an individual recipient for each fiscal year under this subpart by publishing notice of the maximum and minimum amounts in the RUS DLT Program Application Guide and/or the RUS DLT Program Web site and in the funding opportunity posted on *www.Grants.gov* on an annual basis.

§ 1734.25 Completed application.

The following items are required to be submitted to RUS in support of an application for grant funds:

(a) An application for Federal Assistance. A completed Standard Form 424.

(b) An executive summary of the project. The applicant must provide RUS with a general project overview that addresses the following 9 categories:

(1) A description of why the project is needed;

(2) An explanation of how the applicant will address the need cited in paragraph (b)(1) of this section, why the applicant requires financial assistance, the types of educational or medical services to be offered by the project, and the benefits to rural residents;

(3) A description of the applicant, documenting eligibility in accordance with § 1734.4;

(4) An explanation of the total project cost including a breakdown of the grant required and the source of matching contribution and other financial assistance for the remainder of the project;

(5) A statement specifying whether the project is either a distance learning or telemedicine facility as defined in § 1734.3. If the project provides both distance learning and telemedicine services, the applicant must identify the predominant use of the system;

(6) A general overview of the telecommunications system to be developed, including the types of equipment, technologies, and facilities used;

(7) A description of the participating hubs and end user sites and the number of rural residents that will be served by the project at each end user site

(8) A certification by the applicant that facilities constructed with grants do not duplicate adequate established telemedicine or distance learning services; and

(9) A listing of the location of each end user site (city, town, village, borough, or rural areas) plus the State.

(c) Scoring criteria documentation. Each grant applicant must address and provide documentation on how it meets each of the scoring criteria contained in § 1734.26, and as supplemented in the listing on *grants.gov*, the DLT Application Guide and the agency's Web site.

(d) A scope of work. The scope of work must include, at a minimum:

(1) The specific activities to be performed under the project;

(2) Who will carry out the activities;

(3) The time-frames for accomplishing the project objectives and activities; and

(4) A budget for all capital expenditures reflecting the line item costs for approved purposes for both the grant funds and other sources of funds for the project. Separately, the budget must specify any line item costs that are non-approved purposes for grants as contained in § 1734.23.

(e) Financial information and sustainability. The applicant must provide a narrative description demonstrating: Feasibility of the project, including having sufficient resources and expertise necessary to undertake and complete the project; and, how the project will be sustained following completion of the project.

(f) A statement of experience. The applicant must provide a written narrative (not exceeding three single spaced pages) describing its demonstrated capability and experience, if any, in operating an educational or health care endeavor and any project similar to the project. Experience in a similar project is desirable but not required.

(g) Funding commitment from other sources. The applicant must provide evidence, in form and substance satisfactory to RUS, that funding agreements have been obtained to ensure completion of the project. These agreements shall be sufficient to ensure:

(1) Payment of all proposed expenditures for the project;

(2) All required matching contributions in § 1734.22; and

(3) Any other funds necessary to complete the project.

(h) A telecommunications system plan. A telecommunications system plan consisting of the following:

(1) The capabilities of the telecommunications terminal equipment, including a description of the specific equipment which will be used to deliver the proposed service. The applicant must document discussions with various technical sources which could include consultants, engineers, product vendors, or internal technical experts, provide detailed cost estimates for operating and maintaining the end user equipment and provide evidence that

alternative equipment and technologies were evaluated.

(2) A listing of the proposed telecommunications terminal equipment, telecommunications transmission facilities, data terminal equipment, interactive video equipment, computer hardware and software systems, and components that process data for transmission via telecommunications, computer network components, communication satellite ground station equipment, or any other elements of the telecommunications system designed to further the purposes of this subpart, that the applicant intends to build or fund using RUS financial assistance. If funds are being requested for broadband facilities, a description of the use of these facilities and how they will be used to deliver distance learning or telemedicine services.

(3) A description of the consultations with the appropriate telecommunications carriers (including other interexchange carriers, cable television operators, enhanced service providers, providers of satellite services and telecommunications equipment manufacturers and distributors) and the anticipated role of such providers in the proposed telecommunications system.

(i) Compliance with other Federal statutes. The applicant must provide evidence of compliance with other Federal statutes and regulations including, but not limited to the following:

(1) E.O. 11246, Equal Employment Opportunity, as amended by E.O. 11375 and as supplemented by regulations contained in 41 CFR part 60;

(2) Architectural barriers;

(3) Flood hazard area precautions;

(4) Assistance and Real Property Acquisition Policies Act of 1970;

(5) Drug-Free Workplace Act of 1998 (41 U.S.C. 8101 *et seq.*), 2 CFR part 421;

(6) E.O.s 12549 and 12689, Debarment and Suspension, 2 CFR part 180, which is adopted by USDA through 2 CFR part 417;

(7) Byrd Anti-Lobbying Amendment (31 U.S.C. 1352), 2 CFR part 418.

(j) Environmental review requirements.

(1) The applicant must provide details of the project's impact on the human environment and historic properties, in accordance with 7 CFR part 1970. The application must contain a separate section entitled "Environmental Impact of the Project."

(2) The applicant must use any programmatic environmental agreements, available from RUS, in effect at the time of filing to assist in complying with the requirements of this section.

(k) Evidence of legal authority and existence. The applicant must provide evidence of its legal existence and authority to enter into a grant agreement with RUS and perform the activities proposed under the grant application.

(l) Federal debt certification. The applicant must provide a certification that it is not delinquent on any obligation owed to the government (31 U.S.C. 3720B).

(m) Consultation with USDA State Director, Rural Development. The applicant must provide evidence that it has consulted with the USDA State Director, Rural Development, concerning the availability of other sources of funding available at the State or local level.

(n) Supplemental information. The applicant should provide any additional information it considers relevant to the project and likely to be helpful in determining the extent to which the project would further the purposes of the 1996 Act.

§ 1734.26 Criteria for scoring grant applications.

The criteria by which applications will be scored will be published in the RUS DLT Program application guide and/or the RUS DLT Program Web site and in the funding opportunity posted on *www.Grants.gov* Web site on an annual basis. The criteria will be used to determine and evaluate: Rurality; economic need; service need and benefit; and special considerations as determined by the Administrator

§ 1734.27 Application selection provisions.

(a) Applications will be evaluated competitively by the Agency and will be ranked in accordance with § 1734.26. Applications will then be awarded generally in rank order until all grant funds are expended, subject to paragraphs (b), (c), and (d) of this section.

RUS will make determinations regarding the reasonableness of all numbers; dollar levels; rates; the nature and design of the project; costs; location; and other characteristics of the application and the project to determine the number of points assigned to a grant application for all selection criteria.

(b) Regardless of the number of points an application receives in accordance with § 1734.26, the Administrator may, based on a review of the applications in accordance with the requirements of this subpart:

(1) Limit the number of applications selected for projects located in any one State during a fiscal year;

(2) Limit the number of selected applications for a particular type of project;

(3) Select an application receiving fewer points than another higher scoring application if there are insufficient funds during a particular funding period to select the higher scoring application. In this case, however, the Administrator will provide the applicant of the higher scoring application the opportunity to reduce the amount of its grant request to the amount of funds available. If the applicant agrees to lower its grant request, it must certify that the purposes of the project can be met, and the Administrator must determine the project is financially feasible at the lower amount in accordance with § 1734.25(e). An applicant or multiple applicants affected under this paragraph will have the opportunity to be considered for loan financing in accordance with subparts C and D of this part.

(c) RUS will not approve a grant if RUS determines that:

(1) The applicant's proposal does not indicate financial feasibility or is not sustainable in accordance with the requirements of § 1734.25(e);

(2) The applicant's proposal indicates technical flaws, which, in the opinion of RUS, would prevent successful implementation, operation, or sustainability of the project;

(3) Other applications would provide more benefit to rural America based on a review of the financial and technical information submitted in accordance with § 1734.25(e).

(4) Any other aspect of the applicant's proposal fails to adequately address any requirement of this subpart or contains inadequacies which would, in the opinion of RUS, undermine the ability of the project to meet the general purpose of this subpart or comply with policies of the DLT Program contained in § 1734.2.

(d) RUS may reduce the amount of the applicant's grant based on insufficient program funding for the fiscal year in which the project is reviewed. RUS will discuss its findings informally with the applicant and make every effort to reach a mutually acceptable agreement with the applicant. Any discussions with the applicant and agreements made with regard to a reduced grant amount will be confirmed in writing.

§ 1734.28 Submission of applications.

(a) Applications will be accepted as announced in the RUS DLT Program application guide and/or the RUS DLT Program Web site and in the funding opportunity posted on *www.Grants.gov* on an annual basis.

(b) When submitting paper applications:

(1) Applications for grants shall be submitted to the RUS, U.S. Department of Agriculture, 1400 Independence Avenue SW., STOP 1590, Washington, DC 20250–1590. Applications should be marked "Attention: Assistant Administrator, Telecommunications Program".

(2) Applications must be submitted to RUS postmarked no later than the application filing deadline established by the Administrator if the applications are to be considered during the period for which the application was submitted. The deadline for submission of applications each fiscal year will be announced in the RUS DLT Program application guide and/or the RUS DLT Program Web site and in the funding opportunity posted on *www.Grants.gov* on an annual basis.

(3) All applicants must submit an original and a digital copy of a completed application.

§ 1734.29 Appeals.

RUS Telecommunications and Electric Borrowers may appeal the decision to reject their application. Any appeal must be made, in writing, within 10 days after the applicant is notified of the determination to deny the application. Appeals shall be submitted to the Administrator, RUS, U.S. Department of Agriculture, 1400 Independence Ave. SW., STOP 1590, Washington, DC 20250–1590. Thereafter, the Administrator will review the appeal to determine whether to sustain, reverse, or modify the original determination. Final determinations will be made after consideration of all appeals. The Administrator's determination will be final. A copy of the Administrator's decision will be furnished promptly to the applicant.

Subpart C—Distance Learning and Telemedicine Combination Loan and Grant Program

§ 1734.30 Use of combination loan and grant.

(a) A combination loan and grant may be used by eligible organizations as defined in § 1734.4 for distance learning and telemedicine projects to finance 100 percent of the cost of approved purposes contained in § 1734.31 provided that no financial assistance may exceed the maximum amount for the year in which the combination loan and grant is made published in the funding opportunity posted on *www.grants.gov* on an annual basis.

(b) Applicants must meet the minimum eligibility requirement for determining the extent to which the project serves rural areas as determined in § 1734.26(b)

§ 1734.31 Approved purposes for a combination loan and grant.

The approved purposes for a combination loan and grant are:

(a) Acquiring, by lease or purchase, eligible equipment or facilities as defined in § 1734.3;

(b) Acquiring instructional programming;

(c) Providing technical assistance and instruction for using eligible equipment, including any related software; developing instructional programming; providing engineering or environmental studies relating to the establishment or expansion of the phase of the project that is being financed with a combination loan and grant (this purpose shall not exceed 10 percent of the total requested financial assistance);

(d) Paying for medical or educational equipment and facilities that are shown to be necessary to implement the project, including vehicles utilizing distance learning and telemedicine technology to deliver educational and health care services. The applicant must demonstrate that such items are necessary to meet the purposes under this subpart and financial assistance for such equipment and facilities is not available from other sources at a cost which would not adversely affect the economic viability of the project;

(e) Providing links between teachers and students or medical professionals who are located at the same facility, provided that such facility receives or provides distance learning or telemedicine services as part of a distance learning or telemedicine network which meets the purposes of this subpart;

(f) Providing for site development and alteration of buildings in order to meet the purposes of this subpart. Financial assistance for this purpose must be necessary and incidental to the total amount of financial assistance requested;

(g) Purchasing of land, buildings, or building construction determined by RUS to be necessary and incidental to the project. The applicant must demonstrate that financial assistance funding from other sources is not available at a cost that does not adversely impact the economic viability of the project as determined by the Administrator. Financial assistance for this purpose must be necessary and incidental to the total amount of financial assistance requested; and

(h) Acquiring telecommunications or broadband facilities provided that no telecommunications carrier will install such facilities under the Act or through other financial procedures within a reasonable time period and at a cost to the applicant that does not impact the economic viability of the project, as determined by the Administrator.

§ 1734.32 Nonapproved purposes for a combination loan and grant.

(a) Without limitation, a combination loan and grant made under this subpart shall not be expended:

(1) To pay salaries, wages, or employee benefits to medical or educational personnel;

(2) To pay for the salaries or administrative expenses of the applicant or the project;

(3) To purchase equipment that will be owned by the local exchange carrier or another telecommunications service provider, unless the applicant is the local exchange carrier or other telecommunications service provider;

(4) To duplicate facilities providing distance learning or telemedicine services in place or to reimburse the applicant or others for costs incurred prior to RUS' receipt of the completed application;

(5) For projects located in areas covered by the Coastal Barrier Resources Act (16 U.S.C. 3501 *et seq.*);

(6) For any purpose that the Administrator has not specifically approved;

(7) Except for leases (see § 1734.31), to pay the cost of recurring or operating expenses for the project; or,

(8) For any other purposes not specifically outlined in § 1734.31.

(b) Except as otherwise provided in § 1734.12, funds shall not be used to finance a project, in part, when the success of the project is dependent upon the receipt of additional financial assistance under this subpart or is dependent upon the receipt of other funding that is not assured.

§ 1734.33 Maximum and minimum amounts.

Applications for a combination loan and grant under this subpart will be subject to limitations on the proposed amount of loans and grants. The Administrator will establish the maximum and minimum amount of loans and grants and the portion of grant funds as a percentage of total assistance for each project to be made available to an individual recipient for each fiscal year under this subpart, by posting a funding opportunity in the RUS DLT Program Application Guide and/or the RUS DLT Program Web site and in the funding opportunity posted on *www.Grants.gov* on an annual basis.

§ 1734.34 Completed application.

The following items are required to be submitted to RUS in support of an application for a combination loan and grant:

(a) *An application for federal assistance:* A completed Standard Form 424.

(b) *An executive summary of the project:* The applicant must provide RUS with a general project overview that addresses each of the following 9 categories:

(1) A description of why the project is needed;

(2) An explanation of how the applicant will address the need cited in paragraph (b)(1) of this section, why the applicant requires financial assistance, the types of educational or medical services to be offered by the project, and the benefits to the rural residents;

(3) A description of the applicant, documenting eligibility in accordance with § 1734.4;

(4) An explanation of the total project cost including a breakdown of the combination loan and grant required and the source of funding, if applicable, for the remainder of the project;

(5) A statement specifying whether the project provides predominantly distance learning or telemedicine services as defined in § 1734.3. If the project provides both distance learning and telemedicine services, the applicant must identify the predominant use of the system;

(6) A general overview of the telecommunications system to be developed, including the types of equipment, technologies, and facilities used;

(7) A description of the participating hubs and end user sites and the number of rural residents that will be served by the project at each end user site;

(8) A certification by the applicant that facilities constructed with a combination loan and grant do not duplicate adequately established telemedicine or distance learning services.

(9) A listing of the location of each end user site (city, town, village, borough, or rural area plus the State).

(c) A scope of work. The scope of work must include, at a minimum:

(1) The specific activities to be performed under the project;

(2) Who will carry out the activities;

(3) The time-frames for accomplishing the project objectives and activities; and

(4) A budget for capital expenditures reflecting the line item costs for both the combination loan and grant and any other sources of funds for the project.

(d) Financial information. The applicant must show its financial ability to complete the project; show project feasibility; and provide evidence that it can execute a note for a loan with a maturity period greater than one year. For educational institutions participating in a project application (including all members of a consortium), the financial data must reflect revenue and expense reports and balance sheet reports, reflecting net worth, for the most recent annual reporting period preceding the date of the application. For medical institutions participating in a project application (including all members of a consortium), the financial data must include income statement and balance sheet reports, reflecting net worth, for the most recent completed fiscal year preceding the date of the application. When the applicant is a partnership, company, corporation, or other entity, current balance sheets, reflecting net worth, are needed from each of the entities that has at least a 20 percent interest in such partnership, company, corporation or other entity. When the applicant is a consortium, a current balance sheet, reflecting net worth, is needed from each member of the consortium and from each of the entities that has at least a 20 percent interest in such member of the consortium.

(1) Applicants must include sufficient pro-forma financial data that adequately reflects the financial capability of project participants and the project as a whole to continue a sustainable project for a minimum of 10 years and repay the loan portion of the combination loan and grant. This documentation should include sources of sufficient income or revenues to pay operating expenses including telecommunications access and toll charges, system maintenance, salaries, training, and any other general operating expenses, provide for replacement of depreciable items, and show repayment of interest and principal for the loan portion of the combination loan and grant.

(2) A list of property which will be used as collateral to secure repayment of the loan. The applicant shall purchase and own collateral that secures the loan free from liens or security interests and take all actions necessary to perfect a security interest in the collateral that secures the loan. RUS considers as adequate security for a loan, a guarantee by a RUS Telecommunications or Electric borrower or by another qualified party. Additional forms of security, including letters of credit, real estate, or any other items will be considered. RUS will determine the adequacy of the security offered.

(3) As applicable, a depreciation schedule covering all assets of the project. Those assets for which a combination loan and grant are being requested should be clearly indicated.

(4) For each hub and end user site, the applicant must identify and provide reasonable evidence of each source of revenue. If the projection relies on cost sharing arrangements among hub and end user sites, the applicant must provide evidence of agreements made among project participants.

(5) For applicants eligible under §1734.4(1), an explanation of the economic analysis justifying the rate structure to ensure that the benefit, including cost saving, of the financial assistance is passed through to the other persons receiving telemedicine or distance learning services.

(e) A statement of experience. The applicant must provide a written narrative (not exceeding three single spaced pages) describing its demonstrated capability and experience, if any, in operating an educational or health care endeavor similar to the project. Experience in a similar project is desirable but not required.

(f) A telecommunications system plan. A telecommunications system plan, consisting of the following (the items in paragraphs (f)(4) and (5) of this section are required only when the applicant is requesting a combination loan and grant for telecommunications transmission facilities):

(1) The capabilities of the telecommunications terminal equipment, including a description of the specific equipment which will be used to deliver the proposed service. The applicant must document discussions with various technical sources which could include consultants, engineers, product vendors, or internal technical experts, provide detailed cost estimates for operating and maintaining the end user equipment and provide evidence that alternative equipment and technologies were evaluated.

(2) A listing of the proposed purchases or leases of telecommunications terminal equipment, telecommunications or broadband transmission facilities, data terminal equipment, interactive video equipment, computer hardware and software systems, and components that process data for transmission via telecommunications, computer network components, communication satellite ground station equipment, or any other elements of the telecommunications system designed to further the purposes of this subpart, that the applicant intends to build or fund using a combination loan and grant.

(3) A description of the consultations with the appropriate telecommunications carriers (including other interexchange carriers, cable television operators, enhanced service providers, providers of satellite services, and telecommunications equipment manufacturers and distributors) and the anticipated role of such providers in the proposed telecommunications system.

(4) Results of discussions with local exchange carriers serving the project area addressing the concerns contained in § 1734.31(h).

(5) The capabilities of the telecommunications or broadband transmission facilities, including bandwidth, networking topology, switching, multiplexing, standards, and protocols for intra-networking and open systems architecture (the ability to effectively communicate with other networks). In addition, the applicant must explain the manner in which the transmission facilities will deliver the proposed services. For example, for medical diagnostics, the applicant might indicate whether or not a guest or other diagnosticians can join the network from locations off the network. For educational services, indicate whether or not all hub and end-user sites are able to simultaneously hear in real-time and see each other or the instructional material in real-time. The applicant must include detailed cost estimates for operating and maintaining the network, and include evidence that alternative delivery methods and systems were evaluated.

(g) Compliance with other Federal statutes. The applicant must provide evidence of compliance with other federal statutes and regulations including, but not limited to the following:

(1) E.O. 11246, Equal Employment Opportunity, as amended by E.O. 11375 and as supplemented by regulations contained in 41 CFR part 60;

(2) Architectural barriers;

(3) Flood hazard area precautions;

(4) Assistance and Real Property Acquisition Policies Act of 1970;

(5) Drug-Free Workplace Act of 1998 (41 U.S.C. 8101 *et seq.*), 2 CFR part 421;

(6) E.O.s 12549 and 12689, Debarment and Suspension, 2 CFR part 180, which is adopted by USDA through 2 CFR part 417;

(7) Byrd Anti-Lobbying Amendment (31 U.S.C. 1352), 2 CFR part 418.

(h) Environmental review requirements.

(1) The applicant must provide details of the project's impact on the human environment and historic properties, in accordance with 7 CFR part 1970. The application must contain a separate section entitled "Environmental Impact of the Project."

(2) The applicant must use any programmatic environmental agreements, available from RUS, in effect at the time of filing to assist in complying with the requirements of this section.

(i) Evidence of legal authority and existence. The applicant must provide evidence of its legal existence and authority to enter into a grant and incur debt with RUS.

(j) Federal debt certification. The applicant must provide evidence that it is not delinquent on any obligation owed to the government (31 U.S.C. 3720B).

(k) Supplemental information. The applicant should provide any additional information it considers relevant to the project and likely to be helpful in determining the extent to which the project would further the purposes of this subpart.

(l) Additional information required by RUS. The applicant must provide any additional information RUS may consider relevant to the application and necessary to adequately evaluate the application. RUS may also request modifications or changes, including changes in the amount of funds requested, in any proposal described in an application submitted under this subpart.

§ 1734.35 Application selection provisions.

(a) A combination loan and grant will be approved based on availability of funds, the financial feasibility of the project in accordance with § 1734.34(d), the services to be provided which demonstrate that the project meets the general requirements of this subpart, the design of the project; costs; location; and other characteristics of the application.

(b) RUS will determine, from the information submitted with each application for a combination loan and grant, whether the application achieves sufficient priority, based on the criteria set forth in the 1996 Act, to receive a combination loan and grant from funds available for the fiscal year. If such priority is achieved, RUS will process the combination loan and grant application on a first-in, first-out basis, provided that the total amount of applications on-hand for combination loans and grants does not exceed 90 percent of the total loan and grant funding available for the fiscal year. At such time as the total amount of applications eligible for combination loans and grants, if such applications were approved, exceeds 90 percent of amount of combination loan and grant funding available, RUS will process the remaining applications using the evaluation criteria referenced in § 1734.26.

(c) RUS will not approve a combination loan and grant if RUS determines that:

(1) The applicant's proposal does not indicate financial feasibility, or will not be adequately secured in accordance with the requirements contained in § 1734.34(d);

(2) The applicant's proposal indicates technical flaws, which, in the opinion of RUS, would prevent successful implementation, or operation of the project; or

(3) Any other aspect of the applicant's proposal fails to adequately address any requirements of this subpart or contains inadequacies which would, in the opinion of RUS, undermine the ability of the project to meet the general purpose of this subpart or comply with policies of the DLT program contained in § 1734.2.

(d) RUS will provide the applicant with a statement of any determinations made with regard to paragraphs (c)(1) through (c)(3) of this section. The applicant will be provided 15 days from the date of RUS' letter to respond, provide clarification, or make any adjustments or corrections to the project. If, in the opinion of the Administrator, the applicant fails to adequately respond to any determinations or other findings made by the Administrator, the project will not be funded, and the applicant will be notified of this determination. If the applicant does not agree with this finding, an appeal may be filed in accordance with § 1734.37.

§ 1734.36 Submission of applications.

(a) RUS will accept applications for a combination loan and grant submitted by RUS Telecommunications General Field Representatives (GFRs), by Rural Development State Directors, or by applicants themselves. Applications for a combination loan and grant under this subpart may be filed at any time and will be evaluated as received.

(b) Applications submitted to the State Director, Rural Development, in the State serving the headquarters of the project will be evaluated as they are submitted. All applicants must submit an original and an electronic copy of a completed application. The applicant must also submit a copy of the application to the State government point of contact, if one has been designated for the State, at the same time it submits an application to the State Director. The State Director will:

(1) Review each application for completeness in accordance with § 1734.34, and notify the applicant, within 15 working days of receiving the application, of the results of this review, acknowledging a complete application, or citing any information that is incomplete. To be considered for a combination loan and grant, the applicant must submit any additional information requested to complete the application within 15 working days of the date of the State Director's written response. If the applicant fails to submit such information, the application will be returned to the applicant.

(2) Within 30 days of the determination of a completed application in accordance with paragraph (b)(1) of this section, review the application to determine suitability for financial assistance in accordance with § 1734.35, and other requirements of this subpart. Based on its review, the State Director will work with the applicant to resolve any questions or obtain any additional information. The applicant will be notified, in writing, of any additional information required to allow a financial assistance recommendation and will be provided a reasonable period of time to furnish the additional information.

(3) Based on the review in accordance with § 1734.35 and other requirements of this subpart, make a preliminary determination of suitability for financial assistance. A combination loan and grant recommendation will be prepared by the State Director with concurrence of the RUS telecommunications GFR that addresses the provisions of § 1734.34 and § 1734.35 and other applicable requirements of this subpart.

(4) If the application is determined suitable for further consideration by RUS, forward an original and electronic version of the application with a financial assistance recommendation, signed jointly, to the Assistant Administrator, Telecommunications Program, Rural Utilities Service, Washington, DC. The applicant will be notified by letter of this action. Upon receipt of the application from the State Director, RUS will conduct a review of the application and the financial assistance recommendation. A final determination will be made within 15 days. If the Administrator determines that a combination loan and grant can be approved, the State Director will be notified and the State Director will notify the applicant. A combination loan and grant will be processed, approved, and serviced in accordance with §§ 1734.5 through 1734.12.

(5) If the State Director determines that the application is not suitable for further consideration by RUS,. notify the applicant with the reasons for this determination.

(c) Applications submitted by RUS Telecommunications GFRs or directly by applicants will be evaluated as they are submitted. All applicants must submit an original and an electronic version a completed application. The applicant must also submit a copy of the application to the State government point of contact, if one has been designated for the State, at the same time it submits an application to RUS. RUS will:

(1) Review each application for completeness in accordance with § 1734.34, and notify the applicant, within 15 working days of receiving the application, of the results of this review, acknowledging a complete application, or citing any information that is incomplete. To be considered for a combination loan and grant assistance, the applicant must submit any additional information requested to complete the application within 15 working days of the date of the RUS written response. If the applicant fails to submit such information, the application will be returned to the applicant.

(2) Within 30 days of the determination of a completed application in accordance with paragraph (c)(1) of this section, review the application to determine suitability for financial assistance in accordance with § 1734.35, and other requirements of this subpart. Based on its review, RUS will work with the applicant to resolve any questions or obtain any additional information. The applicant will be notified, in writing, of any additional information required to allow a financial assistance recommendation and will be provided a

reasonable period of time to furnish the additional information.

(3) If the application is determined suitable for further consideration by RUS, conduct a review of the application and financial assistance recommendation. A final determination will be made within 15 days. If the Administrator determines that a combination loan and grant can be approved, the applicant will be notified. A combination loan and grant will be processed, approved, and serviced in accordance with §§ 1734.5 through 1734.12.

(4) If RUS determines that the application is not suitable for further consideration, notify the applicant with the reasons for this determination. The applicant will be able to appeal in accordance with § 1734.37.

§ 1734.37 Appeals.

RUS Electric and Telecommunications Borrowers may appeal a decision to reject their application. Any appeal must be made, in writing, within 10 days after the applicant is notified of the determination to deny the application. Appeals shall be submitted to the Administrator, RUS, U.S. Department of Agriculture, 1400 Independence Ave. SW., STOP 1590, Washington, DC 20250–1590. Thereafter, the Administrator will review the appeal to determine whether to sustain, reverse, or modify the original determination. Final determinations will be made after consideration of all appeals. The Administrator's determination will be final. A copy of the Administrator's decision will be furnished promptly to the applicant.

§§ 1734.38–1734.39 [Reserved]

Subpart D—Distance Learning and Telemedicine Loan Program

§ 1734.40 Use of loan funds.

A loan may be used by eligible organizations as defined in § 1734.4 for distance learning and telemedicine projects to finance 100 percent of the cost of approved purposes contained in § 1734.41 provided that no financial assistance may exceed the maximum amount for the year in which the loan is made. Entities seeking a loan must be able to provide security and execute a note with a maturity period greater than one year. The following entities are eligible for loans under this subpart:

(a) Organizations as defined in § 1734.4. If a RUS Telecommunications Borrower is seeking a loan, the borrower does not need to submit all of the financial security information required by § 1734.44(d). The borrower's latest financial report (Form 479) filed with RUS and any additional information relevant to the project, as determined by RUS, will suffice;

(b) Any non-profit or for-profit entity, public or private entity, urban or rural institution, or rural educational broadcaster, which proposes to provide and receive distance learning and telemedicine services to carry out the purposes of this subpart; or

(c) Any entity that contracts with an eligible organization in paragraphs (a) or (b) of this section for constructing distance learning or telemedicine facilities for the purposes contained in § 1734.41, except for those purposes in § 1734.41(h).

(d) Applicants must meet the minimum eligibility requirement for determining the extent to which the project serves rural areas as contained in § 1734.26(b)

§ 1734.41 Approved purposes for loans.

The following are approved purposes for loans:

(a) Acquiring, by lease or purchase, eligible equipment or facilities as defined in § 1734.3;

(b) Acquiring instructional programming;

(c) Providing technical assistance and instruction for using eligible equipment, including any related software; developing instructional programming; providing engineering or environmental studies relating to the establishment or expansion of the phase of the project that is being financed with the loan (financial assistance for this purpose shall not exceed 10 percent of the requested financial assistance);

(d) Paying for medical or educational equipment and facilities which are shown to be necessary to implement the project, including vehicles utilizing distance learning and telemedicine technology to deliver educational and health care services. The applicant must demonstrate that such items are necessary to meet the purposes under this subpart and financial assistance for such equipment and facilities is not available from other sources at a cost which would not adversely affect the economic viability of the project;

(e) Providing links between teachers and students or medical professionals who are located at the same facility, provided that such facility receives or provides distance learning or telemedicine services as part of a distance learning or telemedicine network which meets the purposes of this subpart;

(f) Providing for site development and alteration of buildings in order to meet the purposes of this subpart. Loans for this purpose must be necessary and incidental to the total amount of financial assistance requested;

(g) Purchasing of land, buildings, or building construction, where such costs are demonstrated necessary to construct distance learning and telemedicine facilities. The applicant must demonstrate that funding from other sources is not available at a cost which does not adversely impact the economic viability of the project as determined by the Administrator. Financial assistance for this purpose must be necessary and incidental to the total amount of financial assistance requested;

(h) Acquiring of telecommunications or broadband facilities provided that no telecommunications carrier will install such facilities under the Act or through other financial procedures within a reasonable time period and at a cost to the applicant that does not impact the economic viability of the project, as determined by the Administrator;

(i) Any project costs, except for salaries and administrative expenses, not included in paragraphs (a) through (h) of this section, incurred during the first two years of operation after the financial assistance has been approved. The applicant must show that financing such costs are necessary for the establishment or continued operation of the project and that financing is not available for such costs elsewhere, including from the applicant's financial resources. The Administrator will determine whether such costs will be financed based on information submitted by the applicant. Loans shall not be made exclusively to finance such costs, and financing for such costs will not exceed 20 percent of the loan provided to a project under this section; and

(j) All of the costs needed to provide distance learning broadcasting to rural areas. Loans may be used to cover the costs of facilities and end-user equipment dedicated to providing educational broadcasting to rural areas for distance learning purposes. If the facilities are not 100 percent dedicated to broadcasting, a portion of the financing may be used to fund such facilities based on a percentage of use factor that approximates the distance learning broadcasting portion of use.

§ 1734.42 Non-approved purposes for loans.

(a) Loans made under this subpart will not be provided to pay the costs of recurring or operating expenses incurred after two years from approval of the project except for leases (see § 1734.41).

(b) Loans made under this subpart will not be provided for any of the following costs:

(1) To purchase equipment that will be owned by the local exchange carrier or another telecommunications service provider, unless the applicant is the local exchange carrier or other telecommunications service provider;

(2) To duplicate facilities providing distance learning or telemedicine services in place or to reimburse the applicant or others for costs incurred prior to RUS' receipt of the completed application;

(3) For projects located in areas covered by the Coastal Barrier Resources Act (16 U.S.C. 3501 *et seq.*); or

(4) To pay for salaries, wages, or administrative expenses; or

(5) For any purpose that the Administrator has not specifically approved.

(c) Except as otherwise provided in § 1734.12, funds shall not be used to finance a project, in part, when the success of the project is dependent upon the receipt of additional financial assistance under this subpart D or is dependent upon the receipt of other funding that is not assured.

§ 1734.43 Maximum and minimum amounts.

Applications for loans under this subpart will be subject to limitations on the proposed amount of loans. The Administrator will establish the maximum amount of a loan available to an applicant under this subpart.

§ 1734.44 Completed application.

The following items are required to be submitted in support of an application for a loan:

(a) *An application for federal assistance:* A completed standard form 424.

(b) An executive summary of the project. The applicant must provide RUS with a general project overview that addresses each of the following 9 categories:

(1) A description of why the project is needed;

(2) An explanation of how the applicant will address the need (see paragraph (b)(1) of this section), why the applicant requires financial assistance, the types of educational or medical services to be offered by the project, and the benefits to the rural residents;

(3) A description of the applicant, documenting eligibility in accordance with § 1734.4;

(4) An explanation of the total project cost including a breakdown of the loan required and the source of funding, if applicable, for the remainder of the project;

(5) A statement specifying whether the project provides predominantly distance learning or telemedicine services as defined in § 1734.3. If the project provides both distance learning and telemedicine services, the applicant must identify the predominant use of the system;

(6) A general overview of the telecommunications system to be developed, including the types of equipment, technologies, and facilities used;

(7) A description of the participating hubs and end user sites and the number of rural residents which will be served by the project at each end user site;

(8) A certification by the applicant that facilities funded by a loan do not duplicate adequate established telemedicine or distance learning services;

(9) A listing of the location of each end user site (city, town, village, borough, or rural area plus the State).

(c) A scope of work. The scope of work must include, at a minimum:

(1) The specific activities to be performed under the project;

(2) Who will carry out the activities;

(3) The time-frames for accomplishing the project objectives and activities; and

(4) A budget for capital expenditures reflecting the line item costs for the loan and any other sources of funds for the project.

(d) Financial information. The applicant must show its financial ability to complete the project; show project feasibility; and provide evidence that it can execute a note for a loan for a maturity period greater than one year. For educational institutions participating in a project application (including all members of a consortium), the financial data must reflect revenue and expense reports and balance sheet reports, reflecting net worth, for the most recent annual reporting period preceding the date of the application. For medical institutions participating in a project application (including all members of a consortium), the financial data must include income statement and balance sheet reports, reflecting net worth, for the most recent completed fiscal year preceding the date of the application. When the applicant is a partnership, company, corporation, or other entity, current balance sheets, reflecting net worth, are needed from each of the entities that has at least a 20 percent interest in such partnership, company, corporation or other entity. When the applicant is a consortium, a current balance sheet, reflecting net worth, is needed from each member of the consortium and from each of the entities that has at least a 20 percent interest in such member of the consortium.

(1) Applicants must include sufficient pro-forma financial data which adequately reflects the financial capability of project participants and the

project as a whole to continue a sustainable project for a minimum of 10 years and repay the requested loan. This documentation should include sources of sufficient income or revenues to pay operating expenses including telecommunications access and toll charges, system maintenance, salaries, training, and any other general operating expenses, provide for replacement of depreciable items, and show repayment of interest and principal for the loan.

(2) A list of property which will be used as collateral to secure repayment of the proposed loan. The applicant shall purchase and own collateral that secures the loan free from liens or security interests and take all actions necessary to perfect a first lien in the collateral that secures the loan. RUS will consider as adequate security a loan guarantee by a telecommunications or electric borrower or by another qualified party. Additional forms of security, including letters of credit, real estate, or any other items will be considered. RUS will determine the adequacy of the security offered.

(3) As applicable, a depreciation schedule covering all assets of the project. Those assets for which a loan is being requested should be clearly indicated.

(4) For each hub and end user site, the applicant must identify and provide reasonable evidence of each source of revenue. If the projection relies on cost sharing arrangements among hub and end user sites, the applicant must provide evidence of agreements made among project participants.

(5) For applicants eligible under §1734.4(a)(1), an explanation of the economic analysis justifying the rate structure to ensure that the benefit, including cost saving, of the financial assistance is passed through to the other persons receiving telemedicine or distance learning services.

(e) A statement of experience. The applicant must provide a written narrative (not exceeding three single spaced pages) describing its demonstrated capability and experience, if any, in operating an educational or health care endeavor and any project similar to the project. Experience in a similar project is desirable but not required.

(f) A telecommunications system plan. A telecommunications system plan, consisting of the following (the items in paragraphs (f)(4) and (5) of this section are required only when the applicant is requesting a loan for telecommunications transmission facilities):

(1) The capabilities of the telecommunications terminal equipment, including a description of the specific equipment which will be used to deliver the proposed service. The applicant must document discussions with various technical sources which could include consultants, engineers, product vendors, or internal technical experts, provide detailed cost estimates for operating and maintaining the end user equipment and provide evidence that alternative equipment and technologies were evaluated.

(2) A listing of the proposed purchases or leases of telecommunications terminal equipment, telecommunications transmission facilities, data terminal equipment, interactive video equipment, computer hardware and software systems, and components that process data for transmission via telecommunications, computer network components, communication satellite ground station equipment, or any other elements of the telecommunications system designed to further the purposes of this subpart, that the applicant intends to build or fund using a loan.

(3) A description of the consultations with the appropriate telecommunications carriers (including other interexchange carriers, cable television operators, enhanced service providers, providers of satellite services, and telecommunications equipment manufacturers and distributors) and the anticipated role of such providers in the proposed telecommunications system.

(4) Results of discussions with local exchange carriers serving the project area addressing the concerns contained in §1734.41(h).

(5) The capabilities of the telecommunications transmission facilities, including bandwidth, networking topology, switching, multiplexing,

standards, and protocols for intra-networking and open systems architecture (the ability to effectively communicate with other networks). In addition, the applicant must explain the manner in which the transmission facilities will deliver the proposed services. For example, for medical diagnostics, the applicant might indicate whether or not a guest or other diagnosticians can join the network from locations off the network. For educational services, indicate whether or not all hub and end-user sites are able to simultaneously hear in real-time and see each other or the instructional material in real-time. The applicant must include detailed cost estimates for operating and maintaining the network, and include evidence that alternative delivery methods and systems were evaluated.

(g) Compliance with other Federal statutes. The applicant must provide evidence of compliance with other Federal statutes and regulations including, but not limited to the following:

(1) E.O. 11246, Equal Employment Opportunity, as amended by E.O. 11375 and as supplemented by regulations contained in 41 CFR part 60;

(2) Architectural barriers;

(3) Flood hazard area precautions;

(4) Assistance and Real Property Acquisition Policies Act of 1970;

(5) Drug-Free Workplace Act of 1998 (41 U.S.C. 8101 *et seq.*), 2 CFR part 421;

(6) E.O.s 12549 and 12689, Debarment and Suspension, 2 CFR part 180, which is adopted by USDA through 2 CFR part 417;

(7) Byrd Anti-Lobbying Amendment (31 U.S.C. 1352), 2 CFR part 418.

(h) Environmental review requirements.

(1) The applicant must provide details of the project's impact on the environment and historic properties, in accordance with 7 CFR part 1970. The application must contain a separate section entitled "Environmental Impact of the Project."

(2) The applicant must use any programmatic environmental agreements, available from RUS, in effect at the time of filing to assist in complying with the requirements of this section.

(i) Evidence of legal authority and existence. The applicant must provide evidence of its legal existence and authority to enter into debt with RUS and perform the activities proposed under the loan application.

(j) Federal debt certification. The applicants must provide a certification that it is not delinquent on any obligation owed to the government (31 U.S.C. 3720B).

(k) Supplemental information. The applicant should provide any additional information it considers relevant to the project and likely to be helpful in determining the extent to which the project would further the purposes of this subpart.

(l) Additional information required by RUS. The applicant must provide any additional information RUS determines is necessary to adequately evaluate the application. Modifications or changes, including changes in the loan amount requested, may be requested in any project described in an application submitted under this subpart.

§ 1734.45 Application selection provisions.

(a) Loans will be approved based on availability of funds, the financial feasibility of the project in accordance with § 1734.44(d), the services to be provided which demonstrate that the project meets the general requirements of this subpart, the design of the project; costs; location; and other characteristics of the application.

(b) RUS will determine, from the information submitted with each application for a loan, whether the application achieves sufficient priority, based on the criteria set forth in the 1996 Act, to receive a loan from funds available for the fiscal year. If such priority is achieved, RUS will process the loan application on a first-in, first-out basis, provided that the total amount of applications on-hand for loans does not exceed 90 percent of the total loan funding available for the fiscal year. At such time as the total amount of applications eligible for loans, if such applications were approved, exceeds 90 percent of amount of loan funding available, RUS will process the remaining applications using the evaluation criteria referenced in § 1734.26.

(c) A loan will not be approved if it is determined that:

(1) The applicant's proposal does not indicate financial feasibility, or is not adequately secured in accordance with the requirements of § 1734.44(d);

(2) The applicant's proposal indicates technical flaws, which, in the opinion of RUS, would prevent successful implementation, or operation of the project; or

(3) Any other aspect of the applicant's proposal fails to adequately address any requirements of this subpart or contains inadequacies which would, in the opinion of RUS, undermine the ability of the project to meet the general purpose of this subpart or comply with policies of the DLT program contained in § 1734.2.

(d) RUS will provide the applicant with a statement of any determinations made with regard to paragraphs (c)(1) through (c)(3) of this section. The applicant will be provided 15 days from the date of the RUS letter to respond, provide clarification, or make any adjustments or corrections to the project. If, in the opinion of the Administrator, the applicant fails to adequately respond to any determinations or other findings made by the Administrator, the loan will not be approved, and the applicant will be notified of this determination. If the applicant does not agree with this finding an appeal may be filed in accordance with § 1734.47.

§ 1734.46 Submission of applications.

(a) RUS will accept applications for loans submitted by RUS Telecommunications GFRs, by Rural Development State Directors, or by applicants themselves. Applications for loans under this subpart may be filed at any time and will be evaluated as received on a non-competitive basis.

(b) Applications submitted to the State Director, Rural Development, in the State serving the headquarters of the project will be evaluated as they are submitted. All applicants must submit an original and an electronic version of a completed application. The applicant must also submit a copy of the application to the State government point of contact, if one has been designated for the State, at the same time it submits an application to the State Director. The State Director will:

(1) Review each application for completeness in accordance with § 1734.44, and notify the applicant, within 15 working days of receiving the application, of the results of this review, acknowledging a complete application, or citing any information that is incomplete. To be considered for a loan, the applicant must submit any additional information requested to complete the application within 15 working days of the date of the State Director's written response. If the applicant fails to submit such information, the application will be returned to the applicant.

(2) Within 30 days of the determination of a completed application in accordance with paragraph (b)(1) of this section, review the application to determine suitability for financial assistance in accordance with § 1734.45, and other requirements of this subpart. Based on its review, the State Director will work with the applicant to resolve any questions or obtain any additional information. The applicant will be notified, in writing, of any additional information required to allow a financial assistance recommendation and will be provided a reasonable period of time to furnish the additional information.

(3) Based on the review in accordance with § 1734.45 and other requirements of this subpart, make a preliminary determination of suitability for financial assistance. A loan recommendation will be prepared by the State Director with concurrence of the RUS telecommunications GFR that addresses the provisions of §§ 1734.44 and 1734.45 and other applicable requirements of this subpart.

(4) If the application is determined suitable for further consideration by RUS, forward an original and an electronic version of the application with a loan recommendation, signed jointly, to the Assistant Administrator, Telecommunications Program, Rural Utilities Service, Washington, DC. The applicant will be notified by letter of this action. Upon receipt of the application from the State Director, RUS will conduct a cursory review of the application and the recommendation. A final determination will be made within 15 days. If the Administrator determines that a loan can be approved, the State Director will be notified and the State

Director will notify the applicant. Applications for loans will be processed, and approved loans serviced, in accordance with §§ 1734.5 through 1734.12.

(5) If the State Director determines that the application is not suitable for further consideration by RUS, notify the applicant with the reasons for this determination.

(c) Applications submitted by RUS Telecommunications GFRs or directly by applicants will be evaluated as they are submitted. All applicants must submit an original and an electronic version of a completed application. The applicant must also submit a copy of the application to the State government point of contact, if one has been designated for the State, at the same time it submits an application to the RUS. RUS will:

(1) Review each application for completeness in accordance with § 1734.44, and notify the applicant, within 15 working days of receiving the application, of the results of this review, acknowledging a complete application, or citing any information that is incomplete. To be considered for a loan, the applicant must submit any additional information requested to complete the application within 15 working days of the date of the RUS written response. If the applicant fails to submit such information, the application will be returned to the applicant.

(2) Within 30 days of the determination of a completed application in accordance with paragraph (c)(1) of this section, review the application to determine suitability for financial assistance in accordance with this subpart. Based on its review, RUS will work with the applicant to resolve any questions or obtain any additional information. The applicant will be notified, in writing, of any additional information required to allow a financial assistance recommendation and will be provided a reasonable period of time to furnish the additional information.

(3) If the application is determined suitable for further consideration by RUS, conduct a review of the application and financial assistance recommendation. A final determination will be made within 15 days. If the Administrator determines that a loan can be approved, the applicant will be notified. Applications will be processed, and approved loans serviced, in accordance with §§ 1734.5 through 1734.12.

(4) If RUS determines that the application is not suitable for further consideration, notify the applicant with the reasons for this determination. The applicant will be offered appeal rights in accordance with § 1734.47.

§ 1734.47 Appeals.

RUS Electric and Telecommunications Borrowers may appeal a decision to reject their application. Any appeal must be made, in writing, within 10 days after the applicant is notified of the determination to deny the application. Appeals shall be submitted to the Administrator, RUS, U.S. Department of Agriculture, 1400 Independence Ave. SW., STOP 1590, Washington, DC 20250–1590. Thereafter, the Administrator will review the appeal to determine whether to sustain, reverse, or modify the original determination. Final determinations will be made after consideration of all appeals. The Administrator's determination will be final. A copy of the Administrator's decision will be furnished promptly to the applicant.

PART 1735—GENERAL POLICIES, TYPES OF LOANS, LOAN REQUIREMENTS—TELECOMMUNICATIONS PROGRAM

Subpart A—General

Subpart B—Loan Purposes and Basic Policies

AUTHORITY: 7 U.S.C. 901 *et seq.*, 1921 *et seq.*, and 6941 *et seq.*

SOURCE: 54 FR 13352, Apr. 3, 1989, unless otherwise noted. Redesignated at 55 FR 39395, Sept. 27, 1990.

Subpart A—General

§ 1735.1 General statement.

(a) Subparts A through E of this part set forth the general policies, types of loans and loan requirements under the Telephone loan program.

(b) The standard RUS security documents (see 7 CFR 1744 subpart D or RUS Bulletins 320–4, 320–22, 321–2, 322–2, 323–1, 326–1) contain provisions regarding acquisitions, mergers, and consolidations. Subparts F through J of this part implement those provisions by setting forth the policies, procedures, and requirements for telephone borrowers planning to acquire existing telephone lines, facilities, or systems with RUS loan or other funds, or planning to merge or consolidate with another system. This part supersedes all RUS Bulletins that are in conflict with it.

(c) Subparts F through J of this part also detail RUS's requirements with respect to mergers and acquisitions involving RUS loan funds.

(d) Entities applying for a loan under this part may be eligible to receive a grant under 7 CFR 1738.101, for a portion of the project providing retail broadband service.

[55 FR 39395, Sept. 27, 1990; 55 FR 41170, Oct. 9, 1990, as amended at 86 FR 50608, Sept. 10, 2021]

§ 1735.2 Definitions.

As used in this part:

Access line means a transmission path between user terminal equipment and a switching center that is used for local exchange service. For multiparty service, the number of access lines equals the number of lines/paths terminating on the mainframe of the switching center.

Acquisition means the purchase of another telephone system, lines, or facilities whether by acquiring telephone plant in service or majority stock interest of one or more organizations.

Acquisition agreement means the agreement, including a sales agreement, between the seller and purchaser outlining the terms and conditions of the acquisition. Acquisition agreements also include any other agreements, such as options and subsidiary agreements relating to terms of the transaction.

Administrator means the Administrator of RUS.

Advance of funds means the transferring of funds by RUS to the borrower's construction fund.

Appropriated means funds appropriated based on subsidy.

Affiliate means an organization that directly, or indirectly through one or more intermediaries, controls or is controlled by, or is under common control with, the borrower.

Borrower means any organization which has an outstanding loan made or guaranteed by RUS, or which is seeking such financing.

Cash distribution means investments, guarantees, extensions of credit, advances, loans, non-affiliated company joint ventures, affiliated company investments, and dividend and capital credit distributions. Not included in this definition are qualified investments (see 7 CFR part 1744, subpart D).

Composite economic life as applied to facilities financed by loan funds means the weighted (by dollar amount of each class of facility in the loan) average economic life of all classes of facilities in the loan.

Consolidation means the combination of two or more borrower or nonborrower organizations, pursuant to state law, into a new successor organization that takes over the assets and assumes the liabilities of those organizations.

Construction fund means the RUS Construction Account required by § 2.4 of the standard loan contract into which all RUS loan funds are advanced.

Depreciation means the loss not restored by current maintenance, incurred in connection with the consumption or prospective retirement of telecommunications plant in the course of service from causes which are known to be in current operation, against which the company is not protected by insurance, and the effect of which can be forecast to a reasonable approach to accuracy.

Economic life as applied to facilities financed by loan funds, means the number of years resulting from dividing 100 percent by the depreciation rate (expressed as a percent) approved by the regulatory body with jurisdiction over the telephone service provided by the borrower for the class of facility involved or, if no approved rate exists, by the median depreciation rate expressed as a percent as published by RUS in its Statistical Report, Rural Telephone Borrowers for all RUS borrowers for that class of facility.

Exchange access means the offering of access to telephone exchange services or facilities for the purpose of the origination or termination of telephone toll services.

Feasibility study means the pro forma financial analysis performed by RUS to determine the economic feasibility of a loan. See 7 CFR part 1737.

Forecast period means the time period beginning on the date (base date) of the borrower's balance sheet used in preparing the feasibility study and ending on a date equal to the base date plus the number of years estimated in the feasibility study for completion of the project. Feasibility projections are usually for 5 years, see § 1737.70(a) of this chapter. For example, the forecast period for a loan based on a December 31, 1990 balance sheet and having a 5-year estimated project completion time is the period from December 31, 1990 to December 31, 1995.

Funded reserve means a separate asset account, approved by RUS, consisting of any or all of the following:

(1) Federal government securities purchased in the name of the borrower;

(2) Other securities issued by an institution whose senior unsecured debt obligations are rated in any of the top three categories by a nationally recognized rating organization; or

(3) Cash.

GFR means the RUS general field representative.

Guaranteed loan means a loan guaranteed by RUS under section 306 of the RE Act bearing interest at a rate agreed to by the borrower and the lender.

Hardship loan means a loan made by RUS under section 305(d)(1) of the RE Act bearing interest at a rate of 5 percent per year.

Interim financing means funding for a project which RUS has acknowledged could be included in a loan, should said loan be approved, but for which RUS funds have not yet been made available. See 7 CFR part 1737, subpart E.

Loan means any loan made or guaranteed by RUS.

Loan contract means the loan agreement between RUS and the borrower, including all amendments thereto.

Loan funds means funds provided by RUS through direct or guaranteed loans.

Local exchange carrier (LEC) means an organization that is engaged in the provision of telephone exchange service or exchange access.

Majority noteholders means the holder or holders of a majority in principal amount of the notes outstanding at a particular time.

Merger means the combining, pursuant to state law, of one or more borrower or nonborrower organizations into an existing survivor organization that takes over the assets and assumes the liabilities of the merged organizations. While the terms merger and consolidation have different meanings, for the purpose of this part, "mergers" also include consolidations as defined above. Furthermore, "mergers" also include acquisitions where the acquired systems, lines, or facilities and the acquiring system are operated as one system.

Mobile telecommunications service means radio communication voice service between mobile and land or fixed stations, or between mobile stations.

Modernization Plan (State Telecommunications Modernization Plan) means a State plan, which has been approved by RUS, for improving the telecommunications network of those telecommunications providers covered by the plan. A Modernization Plan must conform to the provisions of 7 CFR 1751, subpart B.

Mortgage means the security agreement between RUS and the borrower, including any amendments and supplements thereto.

Net worth means the sum of the balances of the following accounts of the borrower:

Account names	Number
(1) Capital stock	4510
(2) Additional paid-in capital	4520
(3) Treasury stock	4530
(4) Other capital	4540
(5) Retained earnings	4550

Note: For nonprofit organizations, owners' equity is shown in subaccounts of 4540 and 4550. All references regarding account numbers are to the Uniform System of Accounts (47 CFR part 32).

Public switched network means any common carrier switched network, whether by wire or radio, including local exchange carriers, interexchange carriers, and mobile telecommunications service providers, that use the North American Numbering Plan in connection with the provision of switched services.

RE Act means the Rural Electrification Act of 1936, as amended (7 U.S.C. 901 *et seq.*).

Retail broadband service means any technology identified by the Administrator as having the capacity to provide transmission facilities that enable the subscriber to receive a minimum level of service equal to at least a downstream transmission capacity of 25 megabits per second (Mbps) and an upstream transmission capacity of 3 Mbps. The agency may change the minimum transmission capacity by way of notice in the FEDERAL REGISTER. The minimum transmission capacity may be higher than 25 Mbps downstream and 3 Mbps upstream but cannot be lower.

Rural area means any area of the United States, its territories and insular possessions (including any area within the Federated States of Micronesia, the Republic of the Marshall Islands, and the Republic of Palau) not included within the boundaries of any incorporated or unincorporated city, village or borough having a population exceeding 5,000 inhabitants. The population figure is obtained from the most recent decennial Census of the United

States (decennial Census). If the applicable population figure cannot be obtained from the most recent decennial Census, RD will determine the applicable population figure based on available population data. For purposes of the "rural area" definition, the character of an area is determined as of a time the initial loan for the system is made.

RUS means the Rural Utilities Service, an agency of the United States Department of Agriculture, successor to the Rural Electrification Administration.

RUS cost-of-money loan means a loan made under section 305(d)(2) of the RE Act bearing an interest rate as determined under § 1735.31(c).

Specialized telecommunications service means any telephone service other than telephone exchange service, exchange access, or mobile telecommunications service.

Subscriber means the same as access line.

Survivor means (1) the successor corporation formed by the consolidation of one or more borrowers, (2) the corporation remaining after completion of a merger involving one or more borrowers, and (3) a corporation assuming all or a portion of an RUS loan in connection with an acquisition.

Telecommunications means the transmission or reception of voice, data, sounds, signals, pictures, writings, or signs of all kinds, by wire, fiber, radio, light, or other visual or electromagnetic means.

Telephone exchange service means: (1) Service provided primarily to fixed locations within a telephone exchange, or within a connected system of telephone exchanges within the same exchange area operated to furnish to subscribers intercommunicating service of the character ordinarily furnished by a single exchange, and which is covered by the exchange service charge; or

(2) Comparable service provided through a system of switches, transmission equipment, or other facilities (or combination thereof) by which a subscriber can originate and terminate a telecommunications service.

Telephone service means any communication service for the transmission or reception of voice, data, sounds, signals, pictures, writing, or signs of all kinds by wire, fiber, radio, light, or other visual or electromagnetic means and includes all telephone lines, facilities and systems to render such service. It does not mean:

(1) Message telegram service;

(2) Community antenna television system services or facilities other than those intended exclusively for educational purposes; or

(3) Radio broadcasting services or facilities within the meaning of section 3(o) of the Communications Act of 1934, as amended.

Times Interest Earned Ratio (TIER) means the ratio of a borrower's net income (after taxes) plus interest expense, all divided by interest expense. For the purpose of this calculation, all amounts will be annual figures and interest expense will include only interest on debt with a maturity greater than one year.

Total assets means the sum of the balances of the following accounts of the borrower:

Account names	Number
(1) Current assets	1100s through 1300s.
(2) Noncurrent Assets	1400s through 1500s.
(3) Total telecommunications plant.	2001 through 2007.
(4) Less: Accumulated depreciation.	3100 through 3300s.
(5) Less: Accumulated amortization.	3400 through 3600s.

Note: All references regarding account numbers are to the Uniform System of Accounts (47 CFR part 32).

[54 FR 13351, Apr. 3, 1989; 54 FR 16194, Apr. 21, 1989. Redesignated at 55 FR 39395, Sept. 27, 1990, as amended at 56 FR 26596, June 10, 1991; 58 FR 66253, Dec. 20, 1993; 62 FR 46869, Sept. 5, 1997; 65 FR 42619, July 11, 2000; 65 FR 54402, Sept. 8, 2000; 80 FR 9861, Feb. 24, 2015; 84 FR 59920, Nov. 7, 2019; 86 FR 50608, Sept. 10, 2021]

§ 1735.3 Availability of forms.

Single copies of RUS forms and publications cited in this part are available from Program Support Regulatory Analysis, Rural Utilities Service, STOP 1522, 1400 Independence Ave., SW., Washington, DC 20250–1522. These RUS forms and publications may be reproduced. The terms "RUS form", "RUS standard form", and "RUS specification" have the same meanings as the terms "REA form" "REA standard form", and "REA specification", respectively, unless otherwise indicated.

[54 FR 13351, Apr. 3, 1989. Redesignated at 55 FR 39395, Sept. 27, 1990, as amended at 59 FR 66441, Dec. 27, 1994; 62 FR 46870, Sept. 5, 1997]

§§ 1735.4–1735.9 [Reserved]

Subpart B—Loan Purposes and Basic Policies

SOURCE: 54 FR 13351, Apr. 3, 1989, unless otherwise noted. Redesignated at 55 FR 39395, Sept. 27, 1990.

§ 1735.10 General.

(a) Loans made or guaranteed by the Administrator of RUS will be made in conformance with the Rural Electrification Act of 1936 (RE Act), as amended (7 U.S.C. 901 *et seq.*), and 7 CFR chapter XVII. RUS provides borrowers with specialized and technical accounting, engineering, and other managerial assistance in the construction and operation of their facilities when necessary to aid in the development of rural telephone service and to protect loan security. The Rural Utilities Service (RUS) makes loans to:

(1) Furnish and improve telephone service in rural areas; and

(2) To finance facilities and equipment which expand, improve or provide:

(i) 911 access;

(ii) Integrated interoperable emergency communications, including multiuse networks that provide critical transportation-related information services in addition to emergency communications services;

(iii) Homeland security communications;

(iv) Transportation safety communications; or

(v) Location technologies used outside an urbanized area.

(b) RUS will not make hardship loans or RUS cost-of-money loans for any wireline local exchange service or similar fixed-station voice service that, in RUS' opinion, is inconsistent with the borrower achieving the requirements stated in the State's telecommunication modernization plan within the time frame stated in the plan (see 7 CFR part 1751, subpart B), unless RUS has determined that achieving the requirements as stated in such plan is not technically or economically feasible.

(c) A borrower receiving a loan to provide mobile telecommunications services or special telecommunications services shall be considered to be participating in the state telecommunications plan (TMP) with respect to the particular loan so long as the loan funds are not used in a manner that, in RUS' opinion, is inconsistent with the borrower achieving the goals set forth in the plan, except that a borrower must comply with any portion of a TMP made applicable to the borrower by a state commission with jurisdiction.

(d) RUS will not deny or reduce a loan or an advance of loan funds based on a borrower's level of general funds.

(e) The Administrator may use consultants funded by the borrower for financial, legal, engineering, and other technical advice in connection with the review of a borrower's loan application.

(f) For the purpose of paragraph (a)(2) of this section, rural areas means any area that is not located within a city, town, or incorporated area that has a population of greater than 20,000 inhabitants or within an urbanized area contiguous and adjacent to a city or town that has a population of greater than 50,000 inhabitants. For the purpose of the definition of rural area,

(1) The population figure is obtained from the most recent decennial Census of the United States (decennial Census). If the applicable population figure cannot be obtained from the most recent decennial Census, RD will determine the applicable population figure based on available population data; and

(2) An urbanized area means a densely populated territory as defined in the most recent decennial Census.

[58 FR 66253, Dec. 20, 1993, as amended at 59 FR 17464, Apr. 13, 1994; 65 FR 42619, July 11, 2000; 65 FR 54403, Sept. 8, 2000; 76 FR 56093, Sept. 12, 2011; 80 FR 9861, Feb. 24, 2015; 84 FR 59921, Nov. 7, 2019; 86 FR 50608, Sept. 10, 2021]

§ 1735.11 Area coverage.

Borrowers must make adequate telephone service available to the widest practical number of rural subscribers during the life of the loan. Both the nature of the service area and the cost per subscriber must be fully considered. The borrower must seek to provide service to all interested potential subscribers in the service area. Borrowers are not required to extend service in situations where the costs would be exorbitant. The loan contract shall contain appropriate provisions to effect this requirement. See 7 CFR 1737.11(a), Preapplication Determinations: Area to be Served.

§ 1735.12 Nonduplication.

(a) A loan will not be made unless the Administrator determines that no duplication of lines, facilities, or systems already providing reasonably adequate services shall result from such a loan.

(b) Existing borrowers that apply to upgrade existing facilities in their existing service area are exempt from the non-duplication requirement in paragraph (a) of this section.

(c) RUS shall consider the following criteria for any wireline local exchange service or similar fixed-station voice service provided by a local exchange carrier (LEC) in determining whether such service is reasonably adequate:

(1) The LEC is providing area coverage as described in § 1735.11.

(2) The LEC makes available custom calling features (at a minimum, call waiting, call forwarding, abbreviated dialing, and three-way calling).

(3) The LEC is able to provide E911 service to all subscribers, when requested by the government entity responsible for this service.

(4) The LEC is able to offer local service with blocked toll access to those subscribers who request it.

(5) The LEC's network is capable of providing retail broadband service as defined in § 1735.2 to any subscriber location.

(6) There is an absence of frequent service interruptions.

(7) The LEC is interconnected with the public switched network.

(8) No Federal or State regulatory commission having jurisdiction has determined that the quality, availability, or reliability of the service provided is inadequate.

(9) Services are provided at reasonably affordable rates.

(10) Any other criteria the Administrator determines to be applicable to the particular case.

(d) RUS shall consider the following criteria for any of mobile telecommunications service in determining whether such service is reasonably adequate:

(1) The extent to which area coverage is being provided as described in 7 CFR 1735.11.

(2) Clear and reliable call transmission is provided with sufficient channel availability.

(3) The mobile telecommunications service signal strength is at least −85dBm (decibels expressed in miliwatts).

(4) The mobile telecommunications service is interconnected with the public switched network.

(5) Mobile 911 service is available to all subscribers, when requested by the local government entity responsible for this service.

(6) No Federal or State regulatory commission having jurisdiction has determined that the quality, availability, or reliability of the service provided is inadequate.

(7) Mobile telecommunications service is not provided at rates which render the service unaffordable to a significant number of rural persons.

(8) Any other criteria the Administrator determines to be applicable to the particular case.

(e) RUS does not consider mobile telecommunications service a duplication of existing wireline local exchange service or similar fixed-station voice service. RUS may finance mobile telecommunications systems designed to provide eligible services in rural areas under the Rural Electrification Act even though the services provided by the system may incidentally overlap services of existing mobile telecommunications providers.

(f) RUS shall consider the following criteria for any provider of a specialized telecommunications service in determining whether such service is reasonably adequate:

(1) The provider of a specialized telecommunications service is providing area coverage as described in § 1735.11.

(2) An adequate signal strength is provided throughout the largest practical portion of the service area.

(3) There is an absence of frequent service interruptions.

(4) The quality and variety of service provided is comparable to that provided in nonrural areas.

(5) The service provided complies with industry standards.

(6) No Federal, State, or local regulatory commission having jurisdiction has determined that the quality, availability, or reliability of the service provided is inadequate.

(7) Services are provided at reasonably affordable rates.

(8) Any other criteria the Administrator determines to be applicable to the particular case.

(g) RUS shall consider the following criteria for loans made for the purposes described in § 1735.10(a)(2):

(1) In making a preliminary assessment and a credit decision, the RUS will take into consideration the extent to which the emergency communications capability or emergency communications benefits already exist in the affected area and the need expressed by the proposed user of the emergency communications technology.

(2) The RUS will not consider an application to finance an upgrade of 911 capabilities or other emergency communications capability by different providers serving the same geographic area to be automatically duplicative. For example, RUS will generally not consider an application from two competing wireless carriers to upgrade their E911 capabilities in overlapping geographic territories to be duplicative, however the carrier's competitive situation will be a relevant consideration in evaluating the ability of a service provider to repay their loan.

(3) Duplication considerations will be reviewed on the basis of the emergency communications benefit; the Agency encourages applicants to fully embrace interoperability to maximize the impact of RUS financed investments. In the case of dual or multi-use technologies, the extent to which the proposed non-emergency communications benefits are available from other providers within the proposed service area will be considered in determining loan feasibility.

[54 FR 13351, Apr. 3, 1989. Redesignated at 55 FR 39395, Sept. 27, 1990, as amended at 65 FR 42619, July 11, 2000; 65 FR 54403, Sept. 8, 2000; 76 FR 56093, Sept. 12, 2011; 86 FR 50608, Sept. 10, 2021]

§ 1735.13 Location of facilities and service for nonrural subscribers.

(a) When it is determined by the Administrator to be necessary in order to furnish or improve telephone service in rural areas, loans may be made for the improvement, expansion, construction, acquisition, and operation of telephone lines, facilities, or systems without regard to their geographical location.

(b) To the greatest extent practical, loans are limited to providing telephone facilities that serve subscribers in rural areas. In order to furnish and improve service to rural subscribers it may at times be necessary to provide loan funds to finance telephone facilities which (1) will also serve nonrural subscribers, or (2) are located in nonrural areas. Loans may be approved to finance such facilities if the Administrator determines, on a case-by-case basis, that (i) the primary purpose of the loan is to provide service to rural areas and (ii) the financing of facilities for nonrural subscribers is necessary and incidental to furnishing or improving telephone service in rural areas.

(c) Loan funds may be approved for facilities to serve nonrural subscribers only if (1) the principal purpose of the loan is to furnish and improve rural service and (2) the use of loan funds to serve nonrural subscribers is necessary and incidental to the principal purpose of the loan. The following are examples of purposes for which such loans may be made (such loans are not limited to these examples):

(1) In the case of construction of a new system, if the loan would not be economically feasible and self-liquidating unless the nonrural as well as the rural portions of the telephone service area are included in the proposed system, the loan may include funds for both portions.

(2) Where the acquisition of an existing system located in and serving a nonrural area is necessary to serve as the nucleus of an expanded system to

furnish area coverage service in rural areas, the loan may include funds to finance the acquisition.

(3) When a system is being converted to modern service for rural subscribers, the loan may include funds for the conversion of the nonrural facilities, if the rural service will be improved as a result of such nonrural improvements and it is impractical to finance and serve the nonrural and rural areas separately.

(4) A loan may include funds to serve nonrural subscribers located in community centers frequently called by the rural subscribers if the construction to serve such nonrural subscribers will be incidental to, and contribute substantially to, the provision of adequate service for the rural subscribers.

(d) RUS may also approve financing for facilities to serve nonrural areas if, at the time financing was first approved by RUS:

(1) The nonrural area had a population of 1,500 or less when first financed by RUS and that financing was approved prior to November 1, 1993; or

(2) The nonrural area had a population of 5,000 or less when first financed by RUS and that financing was approved on or after November 1, 1993.

[54 FR 13351, Apr. 3, 1989. Redesignated at 55 FR 39395, Sept. 27, 1990, as amended at 58 FR 66253, Dec. 20, 1993]

§ 1735.14 Borrower eligibility.

(a) RUS makes loans to:

(1) Entities providing, or who may hereafter provide, telephone service in rural areas;

(2) Public bodies providing telephone service in rural areas as of October 28, 1949; and

(3) Cooperative, nonprofit, limited dividend or mutual associations.

(4) For purposes of § 1735.10(a)(2):

(i) Any entity eligible to borrow from the RUS;

(ii) State or local governments;

(iii) Indian Tribes (as defined in § 4 of the Indian Self-Determination and Education Assistance Act (25 U.S.C. 450b); or

(iv) An emergency communications equipment provider that in the sole discretion of RUS offers adequate security for a loan where the State or local government that has jurisdiction over the proposed project is prohibited by law from acquiring debt.

(b) RUS does not make loans to individuals.

(c) RUS gives preference to those borrowers (including initial loan applicants) already providing telephone service in rural areas, and to cooperative, nonprofit, limited dividend, or mutual associations. To be eligible for a loan, a borrower:

(1) Must have sufficient authority to carryout the purposes of the RE Act; and

(2) Must be incorporated or a limited liability company.

[58 FR 66253, Dec. 20, 1993, as amended at 64 FR 50429, Sept. 17, 1999; 65 FR 42619, July 11, 2000; 76 FR 56094, Sept. 13, 2011]

§ 1735.15 Civil rights.

Borrowers are required to comply with certain regulations on nondiscrimination and equal employment opportunity. See RUS Bulletin 320–19 and RUS Bulletin 320–15, respectively.

§ 1735.16 Minimum loan amount.

Recognizing plant costs, the borrower's cost of system design, and RUS's administrative costs, RUS will not consider applications for loans of less than $50,000.

§ 1735.17 Facilities financed.

(a) RUS makes hardship and guaranteed loans to finance the improvement, expansion, construction, acquisition, and operation of systems or facilities (including station apparatus owned by the borrower, headquarters facilities, and vehicles not used primarily in construction) to furnish and improve telephone service in rural areas, except as noted under paragraph (c) of this section.

(b) RUS makes RUS cost-of-money to finance the improvement, expansion, construction, and acquisition of systems or facilities (excluding station apparatus owned by the borrower, headquarters facilities, and vehicles not used primarily in construction) to furnish and improve telephone service in rural areas, except as noted under paragraph (c) of this section.

(c) RUS will not make any type of loan to finance the following items:

(1) Station apparatus (including PBX and key systems) not owned by the borrower and any associated inside wiring;

(2) Certain duplicative facilities, see § 1735.12;

(3) Facilities to provide service other than 1-party; and

(4) System designs or facilities to provide service that cannot withstand or are not designed to minimize damage caused by storms and other natural catastrophes, including, but not limited to hurricanes, floods, tornadoes, mudslides, lightning, windstorms, hail, fire, and smoke, unless an alternate design or facility for modern telecommunications is more economically or technically feasible. Economic and technical feasibility will be determined using total long range economic costs and risk analysis.

(d) Generally, RUS will not make a loan to another entity to provide the same telecommunications service in an area served by an incumbent RUS telecommunications borrower providing such service. RUS may, however, consider an application for a loan to provide the same type of service being provided by an incumbent RUS borrower if the Administrator determines that the incumbent borrower is unable to meet its obligations to the government, including the obligation to provide service set forth in its loan documents and to repay its loans.

(e) If an unadvanced loan, or portion thereof, is rescinded, a new loan shall not be made for the same purposes as in the rescinded loan, except as provided in § 1735.47.

[54 FR 13351, Apr. 3, 1989. Redesignated at 55 FR 39395, Sept. 27, 1990, as amended at 58 FR 66253, Dec. 20, 1993; 62 FR 46870, Sept. 5, 1997; 65 FR 42619, July 11, 2000; 84 FR 59921, Nov. 7, 2019]

§ 1735.18 Additional equity.

If determined by the Administrator to be necessary for loan security, a borrower applying for an initial loan shall increase its net worth as a percentage of assets to the highest level recorded, not to exceed 40 percent, at the end of any calendar quarter in the period beginning 2 years prior to the receipt by RUS of the borrower's loan application form (RUS Form 490). This restoration to the higher level of net worth shall take place before RUS will determine the feasibility of the proposed loan.

§ 1735.19 Mergers and consolidations.

RUS does not make loans for the sole purpose of merging or consolidating telephone organizations. After a merger or consolidation, RUS will consider making loans to the telephone system to finance the improvement or extension of telephone service in rural areas. See RUS Bulletins 320–4, 321–2, 325–1, and 326–1.

§ 1735.20 Acquisitions.

(a) RUS finances the acquisition by a borrower of another system, lines, or facilities only when the acquisition is necessary and incidental to furnishing or improving rural telephone service. See 7 CFR 1735.13.

(b) RUS determines the amount it will lend for each acquisition. If the acquisition price exceeds this amount, the borrower shall provide the remainder.

(c) For additional policies on acquisitions, see subpart F through J of this part.

[54 FR 13351, Apr. 3, 1989. Redesignated at 55 FR 39395, Sept. 27, 1990, as amended at 58 FR 66253, Dec. 20, 1993]

§ 1735.21 Refinancing loans.

(a) Any new direct or guaranteed loan authority provided under the RE Act may be used to refinance an outstanding obligation of the applicant on another loan made under Titles II and VI of the RE Act, or on a non-RUS loan if that loan would have been for eligible telecommunications purposes under the RE Act provided that:

(1) The applicant is current with its payments on the RUS loan(s) to be refinanced; and

(2) The amortization period for that portion of the loan request that will be needed for refinancing will not exceed the remaining amortization period for the loan(s) to be refinanced. If multiple notes are being refinanced, an average remaining amortization period will be calculated based on the weighted dollar average of the notes being refinanced.

(b) The amount that can be refinanced will be included in the funding opportunity announcement that will open a funding window based on the funds authorized for any given fiscal year.

[86 FR 50608, Sept. 10, 2021]

§ 1735.22 Loan security.

(a) RUS makes loans only if, in the judgment of the Administrator, the security therefor is reasonably adequate and the loan will be repaid within the time agreed. See 7 CFR 1735.18 and 7 CFR 1735.51.

(b) RUS generally requires that borrowers provide it with a first lien on all of the borrower's property. See 7 CFR 1735.46.

(c) The RUS will consider Government-imposed fees related to emergency communications (including State or local 911 fees) which are pledged to the repayment of a loan as security.

(d) In the case of loans that include the financing of telephone facilities that do not constitute self-contained operating systems or units (such as lines switched by other systems), the borrower shall, in addition to the mortgage lien on all of the borrower's telephone facilities, furnish adequate assurance, in the form of contractual or other security arrangements, that continuous and efficient telephone service will be rendered.

(e) The borrower shall provide RUS with a satisfactory Area Coverage Survey. See 7 CFR 1737.30 and 1737.31.

(f) RUS makes loans only if the borrower's entire system, including the facilities to be constructed with the proceeds of the loan, is economically feasible, as determined by RUS.

(g) For purposes of determining compliance with TIER requirements, unless a borrower whose existing mortgage contains TIER maintenance requirements notifies RUS in writing differently, RUS will apply the requirements described in paragraph (h) of this section to the borrower regardless of the provisions of the borrower's existing mortgage.

(h) For Loans approved after December 22, 2008, the borrower shall be required to maintain a TIER, at the end of the Forecast Period, at least equal to the projected TIER determined by the feasibility study prepared in connection with the loan, which shall be at least 1.0 and not greater than 1.5.

(i) Nothing in this section shall affect any rights of supplemental lenders under the RUS mortgage, or other creditors of the borrower, to limit a borrower's TIER requirement to a level above that established in paragraph (h) of this section.

(j) A borrower will not be required to raise its TIER as a condition for receiving a loan. Additional financial, investment, and managerial controls appear in the loan contract and mortgage required by RUS.

[54 FR 13351, Apr. 3, 1989. Redesignated at 55 FR 39395, Sept. 27, 1990, as amended at 56 FR 26597, June 10, 1991; 58 FR 66254, Dec. 20, 1993; 62 FR 46870, Sept. 5, 1997; 63 FR 45678, Aug. 27, 1998; 73 FR 65726, Nov. 5, 2008; 76 FR 56094, Sept. 13, 2011; 86 FR 50608, Sept. 10, 2021]

§ 1735.23 Public notice.

(a) Applications for funding request in which the applicant will provide retail broadband service, the Agency's mapping tool will include the following information from each application, and will be displayed for the public:

(1) The identity of the applicant;

(2) A description of the project that can deliver retail broadband service;

(3) A map of the areas to be served, the proposed funded service area (PFSA), including identification of the associated census blocks;

(4) The amount and type of funding requested;

(5) The status of the application; and

(6) The estimated number and proportion of households and businesses in the proposed funded service area without fixed retail broadband service, whether terrestrial or wireless, excluding mobile and satellite service.

(b) For funding requests outside an area where the applicant receives Federal universal service support, the public notice filing referenced under paragraph (a) of this section will accept public notice responses from existing service providers with respect to retail broadband service already being provided in the PFSA for 45 calendar days on the Agency's web page. Existing service providers are requested to submit the following information through the Agency's mapping tool:

(1) The number of residential and business customers within the PFSA currently purchasing broadband at the minimum threshold, the rates of data transmission being offered, and the cost of each level of broadband service charged by the existing service provider;

(2) The number of residential and business customers within the applicant's service area receiving voice and video services and the associated rates for these other services;

(3) A map showing where the existing service provider's services coincide with the applicant's service area using the Agency's mapping tool; and

(4) Test results for the service area in question for a minimum of at least the prior three months demonstrating that the asserted level of broadband is being provided. The test results shall be for different times of the day.

(c) The Agency may contact service providers that respond under paragraph (b) of this section to validate their submission, and so responding service providers should be prepared to:

(1) Provide additional information supporting that the area in question has sufficient access to broadband service;

(2) Have a technician on site during the field validation by RUS staff;

(3) Run on site tests with RUS personnel being present, if requested; and

(4) Provide copies of any test results that have been conducted in the last six months and validate the information submitted in the public notice response months.

(d) If no broadband service provider submits information pursuant to a pending application or if the existing provider does not provide the information requested under paragraphs (b) and (c) of this section, RUS will consider the extent of broadband service using any other data available through reasonable efforts, including utilizing the National Telecommunications and Information Administration's National Broadband Availability Map and the Federal Communications Commission broadband availability map. That may include the agency conducting field validations so as to locate facilities in the application service area and determine, to the extent possible, if those facilities can provide the minimum threshold of broadband. Notwithstanding, conclusive evidence as to the existence of the level of broadband will be taken only through the public notice process. As a result, the Agency highly recommends that existing service providers in a PFSA submit public notice response to ensure that their service is considered in the determination of eligibility on an application.

(e) The Agency will notify respondents who are existing service providers whether their public notice response was accepted or not and allow for an opportunity to respond.

(f) The information submitted by an existing service provider under paragraphs (b) and (c) of this section will be treated as proprietary and confidential and not subject to disclosure, pursuant to 7 U.S.C. 950cc(b)(3).

(g) For all applications that are approved, the following information will be made available to the public:

(1) The information provided in paragraph (a) of this section;

(2) Each annual report required under § 1735.24 will be redacted to protect any proprietary information; and

(3) Such other information as the Administrator of the RUS deems sufficient to allow the public to understand the assistance provided.

[86 FR 50608, Sept. 10, 2021]

§ 1735.24 Additional reporting requirements.

(a) Entities receiving financial assistance from RUS that are used for retail broadband must submit annual reports for 3 years after project completion. The reports must include the following information:

(1) The purpose of the financing, including new equipment and capacity enhancements that support high-speed broadband access for educational institutions, health care providers, and public safety service providers (including the estimated number of end users who are currently using or forecasted to use the new or upgraded infrastructure); and

(2) The progress towards fulfilling the objectives for which the assistance was made, including:

(i) The number of service points that will receive new broadband service, existing network improvements, and facility upgrades resulting from the federal assistance;

(ii) The speed of the broadband services;

(iii) The average price of the most subscribed tier of retail broadband service in each PFSA;

(iv) The number of new subscribers generated from the project; and

(v) Complete, reliable, and precise geolocation information that indicates the location of new broadband service that is being provided or upgraded within the service territory supported by the grant, loan, or loan guarantee.

(b) A notice will be published on the Agency's website that will include each annual broadband improvement report, redacted as appropriate to protect any proprietary information in the report.

[86 FR 50609, Sept. 10, 2021]

§§ 1735.25–1735.29 [Reserved]

Subpart C—Types of Loans

§ 1735.30 Hardship loans.

(a) RUS makes hardship loans under section 305(d)(1) of the RE Act. These loans bear interest at a rate of 5 percent per year. To qualify for a hardship loan on or after November 1, 1993, a borrower must meet each of the following requirements:

(1) The average number of proposed subscribers per mile of line in the service area of the borrower is not more than 4;

(2) The borrower has a projected TIER (including the proposed loan or loans) of at least 1.0, but not greater than 3.0, as determined by the feasibility study prepared in connection with the loan, see 7 CFR part 1737, subpart H; and

(3) The Administrator has approved and the borrower is participating in a telecommunications modernization plan for the state, see 7 CFR part 1751, subpart B.

(b)(1) Hardship loan funds shall not be used to finance facilities located in any exchange of the borrower that has:

(i) More than 1,000 existing subscribers; and

(ii) An average number of proposed subscribers per mile of line greater than 17.

(2) Those facilities may, however, be financed with a RUS cost-of-money loans or a guaranteed loan if the borrower is eligible for such financing.

(c) The Administrator may waive the TIER requirement in paragraph (a)(2) of this section in any case in which the Administrator determines, and sets forth the reasons therefor in writing, that the requirement would prevent emergency restoration of the telephone system of the borrower or result in severe hardship to the borrower.

(d) In order to fairly and equitably approve hardship loans to ensure that borrowers most in need receive hardship financing first, RUS will prioritize for approval all applications qualifying for hardship loans. The criteria in this paragraph will be used by the Administrator to rank, from high to low, applications that have been determined to qualify for hardship financing. Subject to the availability of funds, applications receiving the highest number of points will be selected for loan approval each fiscal year quarter (the application with the most points will be approved first, the second highest next, etc.) The following ranking methodology and loan approval conditions apply:

(1) *Ranking criteria*. Borrowers will receive points based on each of the following criteria applicable to the proposed loan:

(i) *Forecasted Average Number of Subscribers Per Mile of Line (Density)*. The number of points assigned to a borrower will be the value 4 less the value of the borrower's forecasted density as determined by the Feasibility Study prepared in connection with the loan (i.e., if a borrower's forecasted system density is 2.75, the borrower would receive 4 less 2.75 points, or 1.25 points).

(ii) *Forecasted TIER*. The number of points assigned to a borrower will be the value 3 less the value of the borrower's forecasted TIER as determined by the Feasibility Study prepared in connection with the loan (i.e., if a borrower's forecasted TIER is 1.75, the borrower would receive 3 less 1.75 points, or 1.25 points).

(iii) *Unserved Territories.* Borrowers will receive points for loan funds included in the application to provide telephone service in areas previously unserved because it was considered cost prohibitive (for example, high costs resulting from the terrain, remoteness, or system design). In particular, borrowers will receive one tenth of a point, up to a maximum of 2 points, for each subscriber added (in connection with the loan) that currently resides in an unserved area.

(iv) *Plant Modernization.* Borrowers will receive 1 point for loan funds included in the application for at least one of the following basic plant modernizations or system improvements:

(A) Providing digital switching capabilities where those capabilities did not previously exist; and/or

(B) Upgrading to equal access; and/or

(C) Conversion of service to 1-party making an entire exchange all 1-party service.

(v) *Distance Learning and Medical Link Facilities.* Borrowers will receive 2 points for loan funds included in the application for the purpose of providing distance learning or medical link transmission facilities. If loan funds are included for both distance learning and medical link transmission facilities, borrowers will receive 3 points. (See 7 CFR part 1734 for definitions of distance learning and medical link.)

(vi) *Time Factor.* If a borrower's application has been ranked but cannot be approved due to the lack of funds available for loans in that quarter, the borrower will receive .25 points for each quarter in which its loan is pending but not approved.

(2) *Ranking and approval of loans.* Eligible loan applications (satisfying the requirements of 7 CFR 1737.21) will be ranked during the quarter in which the application is received. If an application is received in which insufficient time remains in that quarter to process and rank the application, it will be ranked in the next quarter. At the beginning of the quarter and as soon as practical, RUS will approve all eligible hardship loans ranked in the previous quarter to the extent loan funds are available, beginning with the borrowers that received the highest number of points and working downwards. Any qualified application that is not approved due to the lack of funds will be carried forward to the next quarter and ranked with all other eligible hardship loan applications in that quarter. Upon completion of the ranking and approval of loans, all borrowers will be informed in writing of the status of their loan applications.

(e) *Optimal use of funds.* RUS retains the right to limit the size of hardship loans made to individual borrowers in order to more equitably distribute the amount of hardship funds appropriated among the greatest number of qualified borrowers. Generally, no more than 10 percent of the funds appropriated in any fiscal year may be loaned to a single borrower. In addition, RUS retains the right to approve loans to borrowers that are ranked lower in the priority system, or without regard to when the application was received and ranked, if it is necessary to:

(1) Expedite restoration of service outages due to natural disasters; or

(2) Maximize the use of all available hardship funds appropriated for loans in that fiscal year.

(f) On request of any borrower who is eligible for a hardship loan for which funds are not available, the borrower shall be considered to have applied for a RUS cost-of-money loans under sections 305 of the RE Act.

(g) Hardship loans may be made simultaneously with a RUS cost-of-money loans or guaranteed loans.

[58 FR 66254, Dec. 20, 1993, as amended at 82 FR 55939, Nov. 27, 2017; 84 FR 59921, Nov. 7, 2019]

§ 1735.31 RUS cost-of-money.

(a) RUS makes cost-of-money loans, under section 305(d)(2) of the RE Act. To qualify for cost-of-money loans, a borrower must meet each of the following requirements:

(1) The average number of proposed subscribers per mile of line in the service area of the borrower is not more than 15, or the borrower has a projected TIER (including the proposed loans) of at least 1.0, but not greater than 5.0, as determined by the feasibility study prepared in connection with the loans, see 7 CFR part 1737, subpart H; and

(2) The Administrator has approved and the borrower is participating in a

telecommunications modernization plan for the state, see 7 CFR part 1751, subpart B.

(b) To determine the RUS cost-of-money, the total loan amount will be multiplied by the ratio of RUS cost-of-money funds appropriated for the fiscal year to the sum of RUS cost-of-money funds appropriated for the fiscal year in which the loan is approved. If during the fiscal year the amount of funds appropriated changes, the ratio will be adjusted accordingly and applied only to those loans approved afterwards.

(c) The RUS cost-of-money loan shall bear interest as described in paragraphs (c)(1) and (2) of this section.

(1) Each advance of funds included in RUS cost-of-money loans shall bear interest at a rate (the "Cost of Money Interest Rate") equal to the current cost of money to the Federal Government for loans of a similar maturity. The Cost of Money Rate is determined when the funds are advanced to the borrower but cannot exceed 7 percent per year.

(2) RUS shall use the Federal Treasury Statistical Release (the "Statistical Release") issued by the United States Treasury to determine the interest rate for each advance of RUS cost-of-money loan funds. Generally, the Statistical Release is issued each Monday to cover the preceding week. RUS shall determine the Cost of Money Interest Rate as follows:

(i) Each advance shall bear the interest rate stated in the applicable Statistical Release for Treasury constant maturities with a maturity similar to that of the advance.

(ii) RUS shall determine the interest rate for an advance bearing a maturity other than those stated in the applicable Statistical Release by straight-line interpolation between the next higher and next lower stated maturities.

(iii) The first Statistical Release published after the date of an advance shall apply to that advance.

(iv) If the interest rate determined under paragraph (c)(2)(i) or (c)(2)(ii) of this section is higher than 7 percent, then the advance shall bear interest at the rate of 7 percent per year.

(v) Advances with maturities greater than 30 years shall bear interest at the rate stated in the applicable Statistical Release for 30-year maturities.

(vi) RUS may use an alternative method to determine the Cost of Money Interest Rate if the Treasury ceases to issue the Statistical Release or changes its format or frequency of issue so that it is no longer appropriate for use in the manner described in paragraph (c)(2) of this section. In this eventuality, RUS shall immediately notify all borrowers with unadvanced RUS cost-of-money loan funds. RUS may, with the borrower's consent, determine the Cost of Money Interest Rate on a case-by-case basis for subsequent advances of RUS cost-of-money loan funds but may also decide, in its discretion, that it is unable to continue advancing funds until an alternative method is in effect.

(vii) Refer to § 1735.43(a) for additional information on maturities of RUS loans.

(viii) RUS shall provide borrowers with prompt written confirmation of the Cost of Money Interest Rate borne by each advance of funds included in a RUS cost-of-money loan.

(d) Generally, no more than 10 percent of lending authority from appropriations in any fiscal year for RUS cost-of-money loans may be loaned to a single borrower. RUS will publish by notice in the FEDERAL REGISTER the dollar limit that may be loaned to a single borrower in that particular fiscal year based on approved RUS lending authority.

(e) On request of any borrower who is eligible for RUS cost-of-money loans for which funds are not available, the borrower shall be considered to have applied for a loan guarantee under section 306 of the RE Act.

(f) RUS cost-of-money loans may be made simultaneously with hardship loans or guaranteed loans.

[58 FR 66255, Dec. 20, 1993, as amended at 62 FR 46870, Sept. 5, 1997; 84 FR 59921, Nov. 7, 2019]

§ 1735.32 Guaranteed loans.

(a) *General.* Loan guarantees under this section will be considered for only those borrowers specifically requesting a guarantee. Borrowers may also specify that the loan to be guaranteed shall be made by the Federal Financing Bank (FFB). RUS provides loan guarantees pursuant to section 306 of the

RE Act. Guaranteed loans may be made simultaneously with hardship loans or RUS cost-of-money loans. No fees or charges are assessed for any guarantee of a loan provided by RUS. In view of the Government's guarantee, RUS generally obtains a first lien on all assets of the borrower (see § 1735.46).

(b) *Requirements.* To qualify for a guaranteed loan, a borrower must have a projected TIER (including the proposed loan or loans) of at least 1.2 as determined by the feasibility study prepared in connection with the loan. In addition, a borrower must meet all requirements set forth in the regulations applicable to a loan made by RUS with the exception that it is not required to participate in a state telecommunications modernization plan and is not subject to a subscriber per mile eligibility requirement, as provided in § 1735.31(a).

(c) *Net worth requirements.* RUS generally requires that borrowers seeking guaranteed loans have a net worth in excess of 20 percent of assets. RUS will, however, consider loan guarantees for borrowers with a net worth less than 20 percent.

(d) *Full amount guaranteed.* Loans are guaranteed in the full amount of principal and interest. Because of the Government's full faith and credit 100 percent guarantee of these loans, only RUS obtains a mortgage on the borrower's assets.

(e) *Federal Register notice.* After RUS has reviewed an application and determined that it shall consider guaranteeing a loan for the proposed project and if the borrower has not specified that the loan be made from the FFB, RUS shall publish a notice in the FEDERAL REGISTER. The Notice will include a description of the proposed project, the estimated total cost, the estimated amount of the guaranteed loan, a statement that the Federal Financing Bank (FFB) has a standing loan commitment agreement with RUS, and the name and address of the borrower to which financing proposals may be submitted.

(f) *Qualified lenders.* RUS considers loan guarantees on a case by case basis for loans made by the FFB and any other legally organized lending agency or by a combination of lenders that the Administrator determines to be qualified to make, hold and service the loan. "Legally organized lending agency" and "lender" include commercial banks, trust companies, mortgage banking firms, insurance companies, and any other institutional investor authorized by law to loan money. The borrower is responsible for evaluating all proposals received from lenders other than FFB. The borrower furnishes RUS with a report on the evaluations and its choice of proposals. However, at the request of the borrower, the guaranteed loan shall be made by the FFB.

(g) *Interest rate.* Guaranteed loans shall bear interest at the rate agreed upon by the borrower and lender. Guaranteed FFB loans shall be at a rate of interest that is not more than the rate of interest applicable to other similar loans then being made or purchased by FFB.

(h) *Condition of guarantee.* RUS will not guarantee a loan if the income from the loan or the income from obligations issued by the holder of the loan, when the obligations are created by the loan, is excluded from gross income for the purpose of chapter I of the Internal Revenue Code of 1954.

(i) *Contract of guarantee.* If RUS is satisfied with the engineering and economic feasibility of the project and approves the borrower's choice of proposal, subject to the submission of satisfactory financing documents and to the satisfaction of other pertinent terms and conditions, RUS will prepare a contract of guarantee to be executed by the borrower, the lender, and RUS within a specified time. The lender, or its representative, shall have the right to examine the borrower's application and supporting data submitted to RUS in support of its request for financial assistance.

(j) *Loan servicing.* The contract of guarantee will require that arrangements satisfactory to RUS be made to service the loan. Required servicing by the lender will include:

(1) Determining that all prerequisites to each advance of loan funds by the lender under the terms of the contract of guarantee, all financing documents, and all related security instruments

have been fulfilled. Such determinations may be met by obtaining RUS approval of each advance.

(2) Billing and collecting loan payments from the borrower.

(3) Notifying the Administrator promptly of any default in the payment of principal and interest on the loan and submitting a report, as soon as possible thereafter, setting forth its views as to the reasons for the default, how long it expects the borrower will be in default, and what corrective actions the borrower states it is taking to achieve a current debt service position.

(4) Notifying the Administrator of any known violations or defaults by the borrower under the lending agreement, contract of guarantee, or related security instruments, or conditions of which the lender is aware which might lead to nonpayment, violation, or other default.

(k) *Payments under the contract of guarantee.* Upon receipt of the notification required in §1735.32(j)(3) of this section, RUS will pay the lender the amount in default with interest to the date of payment. When RUS has made a payment under a contract of guarantee, it will establish in its accounts the amount of the payment as due and payable from the borrower, with interest at the rate of interest specified in the lending agreement. RUS will work with the borrower and the lender in an effort to eliminate the borrower's default as soon as possible. RUS may also proceed with other remedies available under its security instruments.

(l) *Pledging of contract of guarantee.* Subject to applicable law, RUS will consider, on a case by case basis, permitting pledging of the contract of guarantee in order to facilitate the obtaining of funds by the lending agency to make the guaranteed loan.

[54 FR 13351, Apr. 3, 1989; 54 FR 16194, Apr. 21, 1989. Redesignated at 55 FR 39395, Sept. 27, 1990, as amended at 56 FR 26597, June 10, 1991; 58 FR 66255, Dec. 20, 1993; 62 FR 46870, Sept. 5, 1997; 84 FR 59921, Nov. 7, 2019]

§ 1735.33 Variable interest rate loans.

After June 10, 1991, and prior to November 1, 1993, RUS made certain variable rate loans at interest rates less than 5 percent but not less than 2 percent. For those borrowers that received variable rate loans, this section describes the method by which interest rates are adjusted. The interest rate used in determining feasibility is the rate charged to the borrower until the end of the Forecast Period for that loan. At the end of the Forecast Period, the interest rate for the loan may be annually adjusted by the Administrator upward to a rate not greater than 5 percent, or downward to a rate not less than the rate determined in the feasibility study on which the loan was based, based on the borrower's ability to pay debt service and maintain a minimum TIER of 1.0. Downward and upward adjustments will be rounded down to the nearest one-half or whole percent. To make this adjustment, projections set forth in the loan feasibility study will be revised annually by RUS (beginning within four months after the end of the Forecast Period) to reflect updated revenue and expense factors based on the borrower's current operating condition. Any such adjustment will be effective on July 1 of the year in which the adjustment was determined. If the Administrator determines that the borrower is capable of meeting the minimum TIER requirements of §1735.22(f) at a loan interest rate of 5 percent on a loan made as described in this section, then the loan interest rate shall be fixed, for the remainder of the loan repayment period, at the standard interest rate of 5 percent.

[62 FR 46870, Sept. 5, 1997]

§§ 1735.34–1735.39 [Reserved]

Subpart D—Terms of Loans

SOURCE: 54 FR 13351, Apr. 3, 1989, unless otherwise noted. Redesignated at 55 FR 39395, Sept. 27, 1990.

§ 1735.40 General.

Terms and conditions of loans are set forth in a mortgage, note, and loan contract. Provisions of the mortgage and loan contract are implemented by provisions in RUS Bulletins and Regulations. Forms of the mortgage, note, and loan contract can be obtained from RUS.

§ 1735.41 Notes.

Loans are represented by one or more notes. Interest accrues only on funds advanced. There are no loan commitment fees or charges. See RUS Bulletin 320–12 for additional information. This CFR part supersedes those portions of RUS Bulletin 320–12 "Loan Payments and Statements" with which it is in conflict.

§ 1735.42 [Reserved]

§ 1735.43 Payments on loans.

(a) Except as described in this paragraph (a), RUS loans approved after October 6, 1997 must be repaid with interest within a period that, rounded to the nearest whole year, equals the expected composite economic life of the facilities to be financed, as calculated by RUS; expected composite economic life means the depreciated life plus three years. The expected composite economic life shall be based on the depreciation rates for the facilities financed by the loan. In states where the borrower must obtain state regulatory commission approval of depreciation rates, the depreciation rates used shall be the rates currently approved by the state commission or rates for which the borrower has received state commission approval. In cases where a state regulatory commission does not approve depreciation rates, the expected composite economic life shall be based on the most recent median depreciation rates published by RUS for all borrowers (see 7 CFR 1737.70). Borrowers may request a repayment period that is longer or shorter than the expected composite economic life of the facilities financed. If the Administrator determines that a repayment period based on the expected composite economic life of the facilities financed is likely to cause the borrower to experience hardship, the Administrator may agree to approve a period longer than requested. A shorter period may be approved as long as the Administrator determines that the loan remains feasible.

(b) Borrowers that have demonstrated to the satisfaction of the Administrator an inability to maintain the funded reserve or net plant to secured debt ratio requirements, if any, contained in their mortgage, may elect to replace notes with an original maturity that exceeded the composite economic life of the facilities financed with notes bearing a shorter maturity approximating the expected composite economic life of the facilities financed, if this will result in a shorter maturity for the loan. The principal balance of the notes (hereinafter in this section called the "refunding notes") issued to refund and substitute for the original notes would be the unpaid principal balance of the original notes. The refunding notes would mature at a date no later than the remaining economic life of the facilities financed by the loan, plus three years, as determined by the original feasibility study prepared in connection with the loan. Interest on the original note must continue to be paid through the closing date. All other payment terms, including the rate of interest on the refunding notes, would remain unchanged. Disposition of funds in the funded reserve will be determined by RUS at the closing date. RUS will notify the borrower in writing of the amendment of loan payment requirements and the terms and conditions thereof.

(c) A borrower qualifying under paragraph (b) of this section shall not be required to pay a prepayment premium on such portion of the payments under its new notes as exceeds the payments required under the notes being replaced.

(d) To apply for refunding notes, borrowers must send to the Area Office the following:

(1) A certified copy of a board resolution requesting an amendment of loan payment requirements and that certain notes be replaced;

(2) If applicable, evidence of approval by the regulatory body with jurisdiction over the telecommunications service provided by the borrower to issue refunding notes; and

(3) Such other documents as may be required by the RUS.

(e) Principal and interest will be repaid in accordance with the terms of the notes. Generally, interest is payable each month as it accrues. Principal payments on each note generally are scheduled to begin 2 years after the date of the note. After this deferral period, interest and principal payments on all funds advanced during this 2-year period are scheduled in equal monthly installments. Principal payments on funds advanced 2 years or more after the date of the note will begin with the first billing after the advance. The interest and principal payments on each of these advances will be scheduled in equal monthly installments. This CFR part supersedes those portions of RUS Bulletin 320–12, "Loan Payments and Statements" with which it is in conflict.

[56 FR 26598, June 10, 1991, as amended at 62 FR 46871, Sept. 5, 1997; 84 FR 59921, Nov. 7, 2019]

§ 1735.44 Prepayment premiums.

The loan documents normally provide that RUS insured loans may be repaid in full at any time without prepayment premiums. Depending upon the lender, there may be prepayment premiums on loans guaranteed by RUS. See RUS Bulletin 320–12 for additional information. This CFR part supersedes those portions of RUS Bulletin 320–12, "Loan Payments and Statements," with which it is in conflict.

[84 FR 59922, Nov. 7, 2019]

§ 1735.45 Extension of payments.

RUS may extend the time of payment of principal or interest on a loan. Under section 12 of the Rural Electrification Act, as amended, this extension may be up to 5 years after such payment is due. Under section 236 of the Disaster Relief Act of 1970 (Pub. L. 91–606) payment may be deferred by the Secretary of Agriculture as long as necessary in disaster situations so long as the final maturity date is not later than 40 years after the date of the loan. See RUS Bulletin 320–2 for additional information.

§ 1735.46 Loan security documents.

(a) Loans are to be repaid according to their terms. RUS generally obtains a first lien on all assets of the borrower. This lien shall be in the form of a mortgage by the borrower to the Government or a deed of trust made by and between the borrower and a trustee, satisfactory to the Administrator, together with such security agreements, financing statements, or other security documents as RUS may deem necessary in a particular case. Where a borrower is unable by reason of pre-existing encumbrances, or otherwise, to furnish a first mortgage lien on its entire system the Administrator may, if he determines such security to be reasonably adequate and the form and nature thereof otherwise appropriate, accept other forms of security. See RUS Bulletins 320–4, 320–22, 321–2, 322–2, 323–1, and 326–1 for details. See 7 CFR part 1744, subpart B for information on lien accommodations and subordinations.

(b) Loan security documents of borrowers with loans approved after October 6, 1997 will provide limits on allowable cash distributions in any calendar year as follows:

(1) No more than 25 percent of the prior calendar year's net earnings or margins if the borrower's net worth is at least 1 percent of its total assets after the distribution is made;

(2) No more than 50 percent of the prior calendar year's net earnings or margins if the borrower's net worth is at least 20 percent of its total assets after the distribution is made;

(3) No more than 75 percent of the prior calendar year's net earnings or margins if the borrower's net worth is at least 30 percent of its total assets after the distribution is made; or

(4) No limit on distributions if the borrower's net worth is at least 40 percent of its total assets after the distribution is made.

(c) Borrowers that have not received a loan after October 6, 1997 may request the Administrator to apply these requirements to them. Borrowers may request in writing that RUS substitute the new requirements described in paragraphs (b)(1) through (b)(4) of this section. Upon request by the borrower, the provisions of the borrower's loan documents restricting cash distributions or investments shall not be enforced to the extent that such provisions are inconsistent with this section.

(d) Rural development investments meeting the criteria set forth in 7 CFR part 1744, subpart D, will not be counted against a borrower's allowable cash distributions in any calendar year (7 U.S.C. 926).

(e) References to a borrower's mortgage in this section include deeds of trust and any other loan document applying the same requirements to a borrower.

(f) This section does not limit the rights of any parties to the mortgage other than RUS.

[54 FR 13351, Apr. 3, 1989. Redesignated at 55 FR 39395, Sept. 27, 1990, as amended at 59 FR 29537, June 8, 1994; 62 FR 46871, Sept. 5, 1997; 84 FR 59922, Nov. 7, 2019]

§ 1735.47 Rescissions of loans.

(a) Rescission of a loan may be requested by a borrower at any time. To rescind a loan, the borrower must demonstrate to RUS that:

(1) The purposes of the loan being rescinded have been completed;

(2) Sufficient funds are available from sources other than RUS or FFB to complete the purposes of the loan being rescinded; or

(3) The purposes of the loan are no longer required to extend or improve telephone service in rural areas.

(b) Borrowers submitting loan applications containing purposes previously covered by a loan that has been rescinded shall include in the application an explanation, satisfactory to RUS, of the change of conditions since the rescission that re-establishes the need for those purposes.

(c) RUS shall not initiate the rescission of a loan unless all of the purposes for which telephone loans have been made to the borrower under the Act have been accomplished with funds provided under the Act.

[56 FR 26598, June 10, 1991, as amended at 84 FR 59922, Nov. 7, 2019]

§§ 1735.48–1735.49 [Reserved]

Subpart E—Basic Requirements For Loan Approval

SOURCE: 54 FR 13351, Apr. 3, 1989, unless otherwise noted. Redesignated at 55 FR 39395, Sept. 27, 1990.

§ 1735.50 Administrative findings.

The RE Act requires that the Administrator make certain findings to approve a telephone loan or loan guarantee. The borrower shall provide the evidence determined by the Administrator to be necessary to make these findings. Details on the information required to support these findings are included in 7 CFR part 1737.

§ 1735.51 Required findings.

(a) *Feasibility of and security for the Loan.* The borrower shall provide RUS with satisfactory evidence to enable the Administrator to determine that the security for the loan is reasonably adequate and the loan will be repaid on time. This finding is based on the following factors:

(1) Self-liquidation of the loan within the loan amortization period; this requires that there be sufficient revenues from the borrower's system, in excess of operating expenditures (including maintenance and replacement), to repay the loan with interest.

(2) Reasonable assurance of achieving the telephone market projections upon which the loan is based.

(3) Economic feasibility (based on projected revenues, expenses, net income, maximum debt service, and rate of return on investment) for the proposed system using local service rate schedules appropriate for the area served.

(4) Impact of the proposed loan and construction on the ratio of the borrower's secured debt to assets.

(5) Projected growth in the borrower's equity.

(6) Satisfactory experience and reputation of the system's principal owners and manager.

(7) A first lien on the borrower's total system or other adequate security.

(8) Fair market value of the borrower's assets as represented in its financial reports to RUS.

(9) Appropriate financial and managerial controls included in the loan documents.

(10) Other factors determined to be relevant by RUS.

(b) *Area coverage.* The borrower shall provide RUS with satisfactory evidence to enable the Administrator to determine that adequate telephone service will be made available to the widest practical number of rural users during the life of the loan.

(c) *Nonduplication or certificate requirement.* The borrower shall provide RUS with satisfactory evidence to enable the Administrator to determine that no duplication of telephone service shall result from a particular loan.

[54 FR 13351, Apr. 3, 1989. Redesignated at 55 FR 39395, Sept. 27, 1990, as amended at 56 FR 26598, June 10, 1991; 86 FR 50609, Sept. 10, 2021]

§ 1735.52 Findings required for particular loan purposes.

Whenever a borrower proposes to use loan funds for the improvement, expansion, construction, or acquisition of telephone facilities within or for nonrural areas, the borrower shall provide RUS with satisfactory evidence to enable the Administrator to determine that such funds shall be necessary and incidental to furnishing or improving telephone service in rural areas.

[54 FR 13351, Apr. 3, 1989. Redesignated at 55 FR 39395, Sept. 27, 1990, as amended at 86 FR 50609, Sept. 10, 2021]

§§ 1735.53–1735.59 [Reserved]

Subpart F—Mortgage Controls on Acquisitions and Mergers

SOURCE: 54 FR 14626, Apr. 12, 1989, unless otherwise noted. Redesignated at 55 FR 39395, Sept. 27, 1990.

§ 1735.60 Specific provisions.

(a) The standard form of RUS mortgage contains certain provisions concerning mergers and acquisitions:

(1) Article II, section 4(a) requires the borrower to obtain the written approval of the majority noteholders before taking any action to reorganize, or to consolidate with or merge into any other corporation.

(2) Article II, section 4(b), if made applicable, provides certain exceptions to the requirements of section 4(a).

(b) Similar provisions are contained in other forms of documents executed by borrowers that have not entered into the standard form of mortgage.

(c) Mortgages and loan contracts may contain other provisions concerning mergers and acquisitions.

[54 FR 14626, Apr. 12, 1989. Redesignated at 55 FR 39395, Sept. 27, 1990, as amended at 62 FR 46871, Sept. 5, 1997]

§ 1735.61 Approval criteria.

(a) If a borrower is required by the terms of its mortgage or loan contract to obtain RUS approval of a merger or acquisition, the borrower shall request RUS approval and shall provide RUS with such data as RUS may request.

(b) If loan funds are requested, the borrower shall comply with subpart G of this part. If no additional loan funds are involved, the borrower shall comply with subpart H of this part.

(c) In considering whether to approve the request, RUS will take into account, among other matters:

(1) Whether the operation, management, and the economic and loan-repayment feasibility characteristics of the proposed system are satisfactory;

(2) Whether the merger or acquisition may result in any relinquishment, impairment, or waiver of a right or power of the Government;

(3) Whether the proposed merger or acquisition is in the best interests of the Government as note holder; and

(4) Whether the proposed purchase price and terms of an acquisition are reasonable, regardless of the source of funds used to pay for the purchase. RUS will consider the purchase price unreasonable if, in RUS's opinion, it will endanger financial feasibility.

§ 1735.62 Approval of acquisitions and mergers.

(a) If a proposal is unsatisfactory to RUS, then RUS shall inform the borrower in writing of those features it considers objectionable and, as appropriate, recommend corrective action.

(b) If a proposal is satisfactory to RUS, then RUS shall inform the borrower in writing of its approval and any conditions of such approval. Among the conditions of approval are the following:

(1) RUS shall require a compensating benefit in return for any relinquishment, impairment, or waiver of its rights or powers.

(2) If the survivor is an affiliate of another company, RUS shall require any investments in, advances to, accounts receivable from, and accounts payable to the affiliated company contrary to mortgage provisions shall be eliminated in a manner satisfactory to the Administrator.

(3) RUS requires that the borrower agree not to extend credit to, perform services for, or receive services from any affiliated company unless specifically authorized in writing by the Administrator or pursuant to contracts satisfactory in form and substance to the Administrator.

(4) RUS may require the borrower to execute additional mortgages, loan agreements, and associated documentation.

§§ 1735.63–1735.69 [Reserved]

Subpart G—Acquisitions Involving Loan Funds

SOURCE: 54 FR 14626, Apr. 12, 1989, unless otherwise noted. Redesignated at 55 FR 39395, Sept. 27, 1990.

§ 1735.70 Use of loan funds.

(a) See 7 CFR part 1735 and 1737 for RUS's general loan policies and requirements.

(b) RUS will finance an acquisition by a borrower only when the acquisition is necessary and incidental to furnishing or improving rural telephone service and the service area is eligible for RUS assistance.

(c) RUS does not make loans for the sole purpose of merging or consolidating telephone organizations. After a merger or consolidation, RUS will consider making loans to the telephone system to finance the improvement or extension of telephone service in rural areas.

(d) Generally, RUS will not make a loan for the acquisition of an existing borrower unless, in addition to all other requirements, such acquisition will improve the likelihood of repayment of an outstanding RUS loan and all outstanding balances of the previous RUS loans are paid in full.

(e) In determining the amount it will lend for each acquisition, RUS shall place a valuation on all telephone facilities that are to be acquired with loan funds. RUS may consider fair market value, the original cost less depreciation of the facilities, income generating potential, any improvement in the financial strength of the borrower as a result of the acquisition, and any other factors deemed relevant by RUS to determine the reasonableness of the acquisition price and the amount of loan funds RUS will provide for an acquisition. RUS shall not consider the acquisition price reasonable or approve a loan if, in the Administrator's opinion, the acquisition price will endanger financial feasibility. If the acquisition price exceeds the amount RUS will lend, the borrower provides the remainder.

(f) When a borrower intends to request RUS loan funds for an acquisition, it shall present a proposal in writing to the Area Office as soon as possible. The borrower must either obtain RUS approval prior to making any binding commitments with the seller or make the commitments subject to RUS's approval. Failure to comply with these requirements will disqualify the borrower from obtaining an RUS loan for the acquisition unless the Administrator determines there were extenuating circumstances.

§ 1735.71 Nonrural areas.

Loan funds may be approved for the acquisition and improvement of facilities to serve nonrural subscribers only if the principal purpose of the loan is to furnish and improve rural service and only if the use of loan funds to serve nonrural subscribers is necessary and incidental to the principal purpose of the loan. For example, when the acquisition of an existing system located in and serving a nonrural area is necessary to serve as the nucleus of an expanded system to furnish area coverage service in rural areas, the loan may include funds to finance the acquisition. Approval for the use of loan funds in these circumstances shall be made only on a case by case basis by the Administrator.

§ 1735.72 Acquisition agreements.

When borrowers are seeking RUS financing, acquisition agreements between the borrower and the seller must be in form and substance satisfactory to RUS and shall be expressly conditioned on approval of the agreement by RUS and on obtaining an RUS loan. Normally, the acquisition agreement will not be approved by RUS until the loan has been approved.

§ 1735.73 Loan design.

When loan funds are requested for an acquisition, details of the proposed acquisition shall be included in the Loan Design. See 7 CFR part 1737.

§ 1735.74 Submission of data.

(a) RUS will not approve any acquisition, other than of toll facilities (see subpart J of this part), financed in whole or in part with loan funds until the borrower submits the following data to the GFR:

(1) For any nonborrowers involved, their most recent balance sheets, operating statements, detail of plant accounts, reports to the state commission, and audits, if available.

(2) Completed RUS Form 507, "Report on Telephone Acquisition," which provides system data, including the type of purchase and purchase price, a system description, and data by exchange. See § 1735.3 for information on obtaining copies of this form.

(3) A map (such as a road map) showing county lines, the boundaries of the proposed acquisition and the borrower's existing service territory, and the names of other telephone companies serving adjoining areas.

(4) A brief statement of the plans for incorporating the acquired facilities into the borrower's existing system.

(5) The number of subscribers currently receiving service in the area to be acquired and the number of new subscribers that will be served over the next 5 years as a result of the acquisition.

(6) The proposed purchase price.

(7) Two copies of any options, bills of sale, or deeds, and four copies of any acquisition agreements. All of these documents are subject to RUS approval. If the acquisition agreement is approved by RUS, two copies of it shall be returned to the borrower.

(8) An appraisal by the borrower's consulting engineer or other qualified person of the physical plant to be acquired. The appraisal shall include the following:

(i) Inspection of each central office, noting the age and condition of the switch and associated equipment, and the extent and quality of maintenance of the equipment and premises.

(ii) Inspection of the outside plant, noting the general age and condition of cable and wire, poles and related hardware, pedestals, and subscriber drops. Any joint use or ownership shall be explained.

(iii) Inspection of miscellaneous items such as commercial office facilities, vehicles, furniture, tools and work equipment, and materials and supplies in stock, noting age and condition.

(iv) Inspection of all buildings and other structures (such as radio towers), noting age and condition.

(v) Detailed description of all real estate including the present market value that local real estate dealers, bankers, insurance agents, etc., place on the property.

(vi) Any widely accepted method, approved by RUS, may be used to estimate the condition of the facilities in paragraphs (a)(8)(i) through (a)(8)(iv) of this section. The "percent condition" method is recommended, but is not required.

(9) Copies of deeds to real estate to be acquired, with an explanation of the proposed use of the land.

(10) Copies of leases to be acquired.

(11) Copies of any existing mortgages with parties other than RUS, indentures, deeds of trust, or other security documents or financing agreements relating to the property to be acquired and any contracts or other rights or obligations to be assumed as part of the acquisition.

(12) A list of all counties in which the proposed system will have facilities.

(13) If the borrower is a cooperative-type organization, a description of its plans for taking subscribers in as members, membership fees, equity payments required because of the acquisition, and extent of membership support.

(14) A certification, signed by the president of the borrower, that the borrower is participating in the State's telecommunications modernization plan (for information concerning the plan, see 7 CFR part 1751, subpart B). This certification is not required if the borrower is seeking a guaranteed loan.

(15) Any other data deemed necessary by the Administrator for an evaluation of the acquisition.

(b) For stock acquisitions, the borrower shall submit the following in addition to the items listed in (a) of this section:

(1) A list of all stockholders of the company to be acquired and the number of shares each owns.

(2) Guarantees and indemnifications to be obtained from the sellers of the stock.

(Approved by the Office of Management and Budget under control number 0572–0084)

[54 FR 14626, Apr. 12, 1989. Redesignated at 55 FR 39395, Sept. 27, 1990; 58 FR 66256, Dec. 20, 1993]

§ 1735.75 Interim financing.

(a) A borrower may submit a written request for RUS approval of interim financing if it is necessary to close an acquisition before the loan to finance the acquisition is approved. Loan funds shall not be used to reimburse acquisition costs unless RUS has granted approval of interim financing prior to the closing of the acquisition.

(b) RUS will approve interim financing of acquisitions only in cases where loan funds cannot be made available in time for the closing.

(c) RUS will not approve interim financing unless the following information is acceptable:

(1) A written request for approval of interim financing, including a brief description of the acquisition, an explanation of the urgency of proceeding with the acquisition, and the source of funds to be used.

(2) A completed RUS Form 490, "Application for Telephone Loan or Loan Guarantee." See 7 CFR part 1737.

(3) The portions of the Loan Design that cover the proposed acquisition, including cost estimates and information on any investments in nonrural areas. See 7 CFR 1737.

(4) The information required in § 1735.74 (a)(1) through (a)(8), (a)(14) and (b)(1).

(5) Any other data deemed necessary by the Administrator to approve the interim financing of the acquisition.

(d) Furthermore, RUS will not approve interim financing if, in RUS's judgment, the proposed acquisition will not qualify for RUS financing or the proposed interim financing presents unacceptable loan security risks to RUS.

(e) Because RUS approval of interim financing is not a commitment to make a loan, RUS will not approve interim financing unless the borrower is prepared to assume responsibility for financing all obligations incurred.

(f) If the borrower plans to proceed with the closing after receiving RUS approval of interim financing, it must first receive preliminary approval from RUS. See § 1735.90

(g) See 7 CFR part 1737 for regulations on interim financing for construction.

(h) See 7 CFR part 1744, subpart B for conditions under which RUS will provide shared first lien and/or a lien accommodation for non-RUS lenders.

§ 1735.76 Acquisition of affiliates.

A borrower shall not use RUS loan funds to acquire any stock or any telephone plant of an affiliate.

[54 FR 14626, Apr. 12, 1989. Redesignated at 55 FR 39395, Sept. 27, 1990, as amended at 62 FR 46871, Sept. 5, 1997]

§ 1735.77 Release of loan funds, requisitions, advances.

(a) RUS will not approve the advance of loan funds until the borrower has fulfilled all loan contract provisions to the extent deemed necessary by RUS.

(b) The first advance of loan funds pursuant to the loan contract normally shall provide funds needed for the acquisition. Unless the borrower has received approval of interim financing, it must submit the requisition in time for the advance to be made by the closing date.

(c) After the borrower has closed the acquisition, it shall furnish RUS all documents necessary to demonstrate to RUS's satisfaction that the transaction has been closed.

(d) Advances for improvements or expansion of the acquired facilities will not be approved until RUS has determined that the transaction has been closed and the borrower has obtained satisfactory title to the acquired facilities.

(e) See 7 CFR part 1737 (or RUS Bulletin 320–4) for additional requirements for releases of loan funds and 7 CFR part 1744, subpart C for additional requirements for requisitions and advances.

§§ 1735.78–1735.79 [Reserved]

Subpart H—Acquisitions or Mergers Not Involving Additional Loan Funds

SOURCE: 54 FR 14626, Apr. 12, 1989, unless otherwise noted. Redesignated at 55 FR 39395, Sept. 27, 1990

§ 1735.80 Submission of data.

When a borrower is not requesting loan funds for an acquisition or merger, the borrower shall first notify RUS and submit for review by RUS the documents and information listed in (a) through (l) of this section required by RUS.

(a) For any nonborrowers involved, their most recent balance sheets, operating statements, detail of plant accounts, reports to the state commission, and audits, if available.

(b) Completed RUS Form 507, "Report on Telephone Acquisition."

(c) A map (such as a road map) showing county lines, the boundaries of the proposed acquisition and the borrower's existing service territory, and the names of other telephone companies serving adjoining areas.

(d) A brief statement of the plans for incorporating the acquired facilities into the borrower's existing system.

(e) The number of subscribers currently receiving service in the areas involved in the acquisition or merger and the number of new subscribers that will be served over the next 5 years as a result of the acquisition or merger.

(f) Copies of deeds of real estate to be acquired, with an explanation of the proposed use of the land.

(g) Copies of security documents of any other lenders involved and any contracts or other rights of obligations to be assumed by the survivor.

(h) A list of all counties in which the proposed system will have facilities.

(i) If Article II, section 4(b) of the standard mortgage has not been made applicable, plans for operating the unified system.

(j) In the case of a merger, the proposed articles of merger that are to be used.

(k) In the case of an acquisition, the proposed purchase price, plus two copies of any options, bills of sale, or deeds, and two copies of any acquisition agreements. All of these documents are subject to RUS approval. If the acquisition agreement is approved by RUS, two copies of it shall be returned to the borrower.

(l) Any other data deemed necessary by the Administrator for an evaluation of the acquisition or merger.

(Approved by the Office of Management and Budget under control number 0572–0084)

[54 FR 14626, Apr. 12, 1989. Redesignated at 55 FR 39395, Sept. 27, 1990]

§§ 1735.81–1735.89 [Reserved]

Subpart I—Requirements for All Acquisitions and Mergers

SOURCE: 54 FR 14626, Apr. 12, 1989, unless otherwise noted. Redesignated at 55 FR 39395, Sept. 27, 1990.

§ 1735.90 Preliminary approvals.

(a) In cases where the borrower's schedule for completion of the proposed action leaves insufficient time for RUS to prepare and process the required documentation, including new mortgages and replacement notes, the borrower may request RUS to give preliminary approval to the acquisition or merger. However, the borrower may not obtain additional loan funds until the documentation is completed to RUS's satisfaction.

(b) Consideration of preliminary approvals generally will not be practicable in cases in which compensating benefits are required.

(c) RUS will not give preliminary approval when the lien of the mortgage on after-acquired property may be affected.

(d) Before RUS will grant preliminary approval, the borrower shall submit:

(1) Merger or acquisition documents required by state law;

(2) Acquisition agreements covering the transaction;

(3) Any required franchises, licenses, and permits;

(4) All required regulatory body approvals;

(5) All required corporate actions;

(6) Leases, contracts, and evidence of titles to be assigned to the purchaser; and

(7) The latest audited financial statements for any nonborrowers involved.

(e) If the information in (d) of this section is acceptable to RUS, the borrower may proceed with the closing.

(Approved by the Office of Management and Budget under control number 0572–0084)

[54 FR 14626, Apr. 12, 1989, unless otherwise noted. Redesignated at 55 FR 39395, Sept. 27, 1990.]

§ 1735.91 Location of facilities.

Telephone facilities to be acquired must be located so that they can be efficiently operated by the borrower and provide adequate security for the RUS loan.

[54 FR 14626, Apr. 12, 1989, unless otherwise noted. Redesignated at 55 FR 39395, Sept. 27, 1990.]

§ 1735.92 Accounting considerations.

(a) Proper accounting shall be applied to all acquisitions and mergers, as required by the regulatory commission having jurisdiction, or in the absence of such a commission, as required by RUS based on Generally Accepted Accounting Principles or other accounting conventions as deemed necessary by RUS.

(b) If RUS determines that the plant accounts are not properly depreciated, the borrower should adjust its depreciation rates. Depending upon the characteristics of the case, commission jurisdiction and requirements, and similar factors, one of the following actions shall be taken:

(1) In states where commission approval of depreciation rates is required, a covenant shall be included in the loan contract that requires the borrower to:

(i) Have the consulting engineer make an original cost less depreciation inventory and appraisal of retained plant as part of the final inventory, and

(ii) Request commission approval of adjustments to its records on the basis of this inventory.

(2) In states where commission approval is not required, informal discussions between RUS and the borrower may be undertaken to reach satisfactory voluntary adjustments. If this does not resolve the situation to RUS's satisfaction, a covenant similar to that in paragraph (b)(1)(i) of this section shall be included in the loan contract and the borrower shall agree to submit evidence satisfactory to the Administrator that it has adjusted its records on the basis of the inventory.

[54 FR 14626, Apr. 12, 1989, unless otherwise noted. Redesignated at 55 FR 39395, Sept. 27, 1990.]

§ 1735.93 Notes.

Substitute notes may be required in the case of an acquisition or merger, regardless of the source of funds.

[54 FR 14626, Apr. 12, 1989, unless otherwise noted. Redesignated at 55 FR 39395, Sept. 27, 1990.]

§ 1735.94 Final approval and closing procedure.

(a) Legal documents relating to the acquisition or merger, including copies of required franchises, commission orders, permits, licenses, leases, title evidence, corporate proceedings, and contracts to be assigned to the purchaser shall be forwarded to the Area Office prior to closing.

(b) The Administrator will not give final approval to any acquisition or merger until all RUS requirements relating to the transactions are satisfied.

(c) Following the Administrator's final approval of the proposal, the Area Office shall inform the borrower in writing of the necessary legal and other actions required for the advance of loan funds to finance the acquisition, including the submission, in form and substance satisfactory to the Administrator, of (1) all information and documents necessary to demonstrate that the transaction has been completed, and (2) all loan contracts, notes, mortgages, and related documents and materials required by RUS.

(d) Deeds reflecting the change in ownership, executed bills of sale, and opinions of counsel shall be forwarded to the Area Office following closing.

(e) RUS will not advance loan funds to furnish or improve service in the acquired or merged areas until the Administrator has given final approval and the transaction has been closed. RUS may, however, advance funds if it determines that loan security will not be jeopardized.

(f) At the discretion of RUS, a GFR may be present at the closing to assist the borrower and protect the interests of RUS. Under certain circumstances the closing may take place prior to RUS granting final approval for the transaction and the execution of amended loan security documents.

[54 FR 14626, Apr. 12, 1989, unless otherwise noted. Redesignated at 55 FR 39395, Sept. 27, 1990.]

§ 1735.95 Unadvanced loan funds.

(a) The unadvanced loan funds of a borrower that will not be a survivor of an acquisition or merger shall be advanced only to the survivor and only under the following circumstances.

(1) If the funds are to be used for purposes approved in prior loans, the funds shall be advanced after the effective date of the proposed action only when all loan contract prerequisites have been met and documents have been submitted in form and substance satisfactory to the Administrator.

(2) If the funds are to be used for new purposes, then in addition to the requirements in (a)(1) of this section, RUS must also approve the change in purpose.

(b) No loan or other money in the construction fund shall be used to finance facilities outside areas to be served by projects approved by RUS.

[54 FR 14626, Apr. 12, 1989, unless otherwise noted. Redesignated at 55 FR 39395, Sept. 27, 1990.]

§§ 1735.96–1735.99 [Reserved]

Subpart J—Toll Line Acquisitions

SOURCE: 54 FR 14626, Apr. 12, 1989, unless otherwise noted. Redesignated at 55 FR 39395, Sept. 27, 1990.]

§ 1735.100 Use of loan funds.

An acquisition of toll line facilities financed with loan funds must be necessary and incidental, as determined by the Administrator, to furnishing or improving telephone service in rural areas. The borrower shall submit to RUS the acquisition agreement, the original cost less depreciation of the facilities, any concurrences with the connecting companies involved, and a detailed inventory of the facilities to be purchased. The borrower must submit to RUS evidence, satisfactory to the Administrator, of the borrower's ownership of the toll line facilities before loan funds for improvement of those facilities will be advanced.

[54 FR 14626, Apr. 12, 1989. Redesignated at 55 FR 39395, Sept. 27, 1990]

§ 1735.101 With nonloan funds.

When an acquisition is limited to toll line facilities and loan funds are not involved, RUS approval of the acquisition is not required. The borrower, however, shall submit to RUS for its approval all concurrences with the connecting companies involved and any other proof of ownership of the toll facilities required by RUS.

[54 FR 14626, Apr. 12, 1989. Redesignated at 55 FR 39395, Sept. 27, 1990]

PART 1737—PRE-LOAN POLICIES AND PROCEDURES COMMON TO INSURED AND GUARANTEED TELECOMMUNICATIONS LOANS

Subpart A—General

Subpart B—Preapplication Stage

Subpart C—The Loan Application

AUTHORITY: 7 U.S.C. 901 *et seq.*, 1921 *et seq.*; Pub. L. 103–354, 108 Stat. 3178 (7 U.S.C. 6941 *et. seq.*).

SOURCE: 54 FR 13356, Apr. 3, 1989, unless otherwise noted. Redesignated at 55 FR 39396, Sept. 27, 1990.

Subpart A—General

§ 1737.1 General statement.

(a) This part prescribes policies, procedures and responsibilities relating to applications for RUS loans to finance the improvement and extension of telephone service in rural areas. Requirements for both initial and subsequent loans are discussed, with differences pointed out.

(b) This part sets forth the policies, procedures, and requirements of RUS during the period from the receipt of a completed loan application until the advance of funds. This part sets forth the factors RUS considers in determining the characteristics of a loan, such as the amount of the loan, and conditions to the advance of funds. Involved in this determination are:

A loan budget, feasibility study, characteristics letter, loan recommendation, and release of funds. This CFR part supersedes all RUS Bulletins that are in conflict with it.

(c) See 7 CFR part 1735 on general loan policies, 7 CFR part 1737 for details on submitting a loan application, and 7 CFR part 1744 on the advance of funds.

§ 1737.2 Definitions.

As used in this part:

Access line means a transmission path between user terminal equipment and a switching center that is used for local exchange service. For multiparty service, the number of access lines equals the number of lines/paths terminating on the mainframe of the switching center.

Acquisition means the purchase of another telephone system, lines, or facilities whether by acquiring telephone plant in service or majority stock interest of one or more organizations.

Administrator means the Administrator of RUS.

Area Coverage means the provision of adequate telephone service to the widest practical number of rural users during the life of the loan.

Advance of funds means the transferring of funds by RUS to the borrower's construction fund.

Borrower means any organization which has an outstanding loan made or guaranteed by RUS, on which is seeking such financing.

Characteristics letter means the letter informing the borrower of the characteristics of the proposed loan before the loan is recommended.

Feasibility study means the pro forma financial analysis performed by RUS to determine the economic feasibility of a loan.

Forecast period means the time period beginning on the date (base date) of the borrower's balance sheet used in preparing the feasibility study and ending on a date equal to the base date plus the number of years estimated in the feasibility study for the completion of the project. Feasibility projections are usually for 5 years, see §1737.70(a). For example, the forecast period for a loan based on a December 31, 1990 balance sheet and having a 5-year estimated project completion time is the period from December 31, 1990 to December 31, 1995.

Guaranteed loan means a loan guaranteed by RUS under section 306 of the RE Act bearing interest at a rate agreed to by the borrower and the lender.

Hardship loan means a loan made by RUS under section 305(d)(1) of the RE Act bearing interest at a rate of 5 percent per year.

Initial loan means the first loan made to a borrower.

Interim construction means the purchase of equipment or the conduct of construction under an RUS-approved plan of interim financing.

Interim financing means funding for a project which RUS has acknowledged will be included in a loan, should said loan be approved, but for which RUS loan funds have not yet been made available.

Loan means any loan made or guaranteed by RUS.

Project means the improvements and telephone facilities financed by a particular RUS loan.

RE Act means the Rural Electrification Act of 1936, as amended (7 U.S.C. 901 *et seq.*).

Release of funds means determination by RUS that a borrower has complied with all of the conditions prerequisite to the advances as set forth in the loan contract to the extent deemed necessary by RUS for approval of the use of loan funds and any required equity or other nonloan funds.

Reserves means loan or nonloan funds that have not been encumbered. Funds are encumbered when they have been set aside for by RUS for a particular loan purpose.

Rural area means any area of the United States, its territories and possessions (including any area within the Federated States of Micronesia, the Republic of the Marshall Islands, and the Republic of Palau) not included within the boundaries of any incorporated or unincorporated city, village or borough having a population exceeding 5,000 inhabitants. The population figure is obtained from the most recent decennial Census of the United States. If the applicable population figure cannot be obtained from the most recent decennial Census, RD will determine the applicable population figure based on available population data. For purposes of the "rural area" definition, the character of an area is determined as of a time the initial loan for the system is made.

RUS cost-of-money loan means a loan made under section 305(d)(2) of the RE Act bearing an interest rate as determined under 7 CFR 1735.31(c).

Special project means facilities involving investment in excess of $100,000 for any single subscriber.

Subscriber means the same as access line.

Subsequent Loan means any loan to a borrower which has already received a loan.

Telephone service means any communication service for the transmission or reception of voice, data, sounds, signals, pictures, writing, or signs of all kinds by wire, fiber, radio, light, or other visual or electromagnetic means and includes all telephone lines, facilities and systems to render such service. It does not mean:

(1) Message telegram service;

(2) Community antenna television system services or facilities other than those intended exclusively for educational purposes; or

(3) Radio broadcasting services or facilities within the meaning of section 3(o) of the Communications Act of 1934, as amended.

Times Interest Earned Ratio (TIER) means the ratio of a borrower's net income (after taxes) plus interest expense, all divided by interest expense. For the purpose of this calculation, all amounts will be annual figures and interest expense will include only interest on debt with a maturity greater than one year.

[54 FR 13356, Apr. 3, 1989. Redesignated at 55 FR 39396, Sept. 27, 1990, as amended at 56 FR 26598, June 10, 1991; 58 FR 66256, Dec. 20, 1993; 80 FR 9861, Feb. 24, 2015; 84 FR 59922, Nov. 7, 2019]

§ 1737.3 Availability of RUS forms.

Single copies of RUS forms and publications cited in this part are available from Administrative Services Division, Rural Utilities Service, United States Department of Agriculture, Washington, DC 20250. These RUS forms and publications may be reproduced. The terms "RUS form", "RUS standard form", and "RUS specification" have the same meanings as the terms "REA form" "REA standard form", and "REA specification", respectively, unless otherwise indicated.

[54 FR 13356, Apr. 3, 1989. Redesignated at 55 FR 39396, Sept. 27, 1990, as amended at 59 FR 66441, Dec. 27, 1994]

§§ 1737.4–1737.9 [Reserved]

Subpart B—Preapplication Stage

§ 1737.10 Initial contact.

Initial loan applicants seeking assistance should write the Rural Utilities Service, United States Department of Agriculture, Washington, DC 20250. A field representative will be assigned by RUS to visit the applicant and discuss its financial needs and eligibility. Existing borrowers initiate the contact directly with their assigned field representative. Borrowers consult with RUS field representatives and headquarters staff, as necessary.

§ 1737.11 Preapplication determinations.

Before submitting an application to RUS, the borrower should consider the following:

(a) *Area to be served.* The proposed service area should neither include subscribers already receiving adequate service from another telephone system nor leave out unserved pockets of potential subscribers who have indicated an interest in service and are located between the proposed system and neighboring systems. See 7 CFR 1735.11 on Area Coverage and 7 CFR 1735.12 on Nonduplication. In establishing service area boundaries, borrowers should consider the location of adjoining systems, natural boundaries such as rivers and mountains, and economic and cultural features such as trading and community centers.

(b) *Number of subscribers.* The borrower must estimate the number of subscribers that will request service from the proposed system.

(c) *Acquisitions.* A borrower considering an acquisition should refer to 7 CFR 1735.20 and RUS Bulletins 320–4, 321–2, 325–1, and 326–1.

(d) *Mergers and consolidations.* A borrower considering a merger or consolidation should refer to 7 CFR 1735.19.

(e) *Refinancing.* Restrictions on the use of loan funds for refinancing are contained in 7 CFR 1735.21.

(f) *Service for nonrural subscribers.* In some situations, RUS loan funds may be used to finance facilities to serve nonrural subscribers. See 7 CFR 1735.13.

(g) *Loan amount.* The initial loan request is based on the borrower's best estimate of financing needs. RUS requires detailed studies by the borrower to complete the application and the initial estimate is subject to revision.

(h) *Loans for a portion of a system.* If it is impractical to finance facilities to provide adequate service throughout the borrower's entire telephone service area, RUS will consider a loan application to finance improvements to a portion of a borrower's system.

(i) *Telecommunications modernization plan.* A borrower applying for hardship or concurrent RUS cost-of-money loans should refer to 7 CFR part 1751, subpart B.

[54 FR 14626, Apr. 12, 1989. Redesignated at 55 FR 39395, Sept. 27, 1990, as amended at 58 FR 66256, Dec. 20, 1993; 84 FR 59922, Nov. 7, 2019]

§§ 1737.12–1737.19 [Reserved]

Subpart C—The Loan Application

§ 1737.20 [Reserved]

§ 1737.21 The completed loan application.

(a) The completed loan application consists of four parts:

(1) A completed RUS Form 490.

(2) A market survey called the Area Coverage Survey (ACS).

(3) The plan and associated costs for the proposed construction, called the Loan Design (LD).

(4) Various supplementary information specified in 7 CFR 1737.22.

(b) The RUS field representative assists the borrower in assembling this information. Certain information is required from initial loan applicants but usually not from borrowers seeking subsequent loans. Borrowers are to submit all information in paragraph (a) of this section to their RUS field representatives, who will review and then forward the packages to RUS headquarters.

(c) RUS will make a determination of completeness of the application package and will notify the borrower of this determination within 10 working days of receipt of the information at RUS headquarters. If the application package is not complete, RUS will notify the borrower of what information is needed in order to complete the application package. If the information required to complete the application package is not received by RUS within 90 working days from the date the borrower was notified of the information needed, RUS may return the application package to the borrower. Returned applications are without prejudice and borrowers may resubmit the completed application.

(Approved by the Office of Management and Budget under control number 0572–0079)

[54 FR 13356, Apr. 3, 1989. Redesignated at 55 FR 39396, Sept. 27, 1990, as amended at 56 FR 26598, June 10, 1991]

§ 1737.22 Supplementary information.

RUS requires additional information in support of the loan application form. The information listed in paragraphs (a), (b), and (c) of this section must be submitted as part of the loan application as specified in 7 CFR 1737.21.

(a) The following must be submitted by all initial loan applicants. Borrowers seeking subsequent loans must submit any changes in these items since they were last submitted.

(1) Name of attorney and manager, and certified copies of board resolutions selecting them.

(2) Certified copy of articles of incorporation showing evidence of filing with the Secretary of State and in county records.

(3) Certified copies of bylaws and board minutes showing their adoption.

(4) Certified sample stock certificates.

(5) Amounts of common and preferred stock issued and outstanding.

(6) Names, addresses, business affiliations, and stockholdings of the manager, officers, directors, and other principal stockholders (those owning at least 20 percent of borrower's voting stock).

(7) Certified copies of real estate deeds showing all recording information.

(8) Service agreements, such as for management or system maintenance.

(9) Certified copies of existing leases, except those for vehicles, furniture and office equipment, and computer equipment.

(10) Certified copies of existing franchises.

(11) Information on any franchises required as a result of the proposed loan project.

(12) Federal Communications Commission (FCC) authorizations.

(13) For toll, operator office, traffic, and EAS agreements, the names of all parties to the agreement, the type of agreement, and the effective and termination dates of the agreement and annexes, and the exchanges involved.

(14) Copies of rate schedules. (A copy of the tariff must be available for review by the RUS field representative.)

(15) Executed copy of RUS Form 291, "Certification of Nonsegregated Facilities".

(16) A sketch or map showing the existing and proposed service areas.

(17) Executed assurance that the borrower will comply with the Uniform

Relocation Assistance and Real Property Acquisitions Policies Act of 1970, as amended (see 49 CFR 24.4).

(18) A certification (which is included on RUS Form 490, "Application for Telephone Loan or Guarantee") that the borrower has been informed of the collection options listed below that the Federal government may use to collect delinquent debt. RUS and other government agencies are authorized to take any or all of the following actions in the event that a borrower's loan payments become delinquent or the borrower defaults (OMB Circular A–129 defines "delinquency" for direct or guaranteed loans as debt more than 31 days past due on a scheduled payment):

(i) Report the borrower's delinquent account to a credit bureau.

(ii) Assess additional interest and penalty charges for the period of time that payment is not made.

(iii) Assess charges to cover additional administrative costs incurred by the Government to service the borrower's account.

(iv) Offset amounts owed to the borrower under other Federal programs.

(v) Refer the borrower's debt to the Internal Revenue Service for offset against any amount owed to the borrower as an income tax refund.

(vi) Refer the borrower's account to a private collection agency to collect the amount due.

(vii) Refer the borrower's account to the Department of Justice for litigation in the courts.

(19)(i) A certification, signed by the president of the borrower, that the borrower is participating in the State's telecommunications modernization plan (for additional information concerning the plan, see 7 CFR part 1751, subpart B). This certification is not required if the borrower is seeking a guaranteed loan.

(ii) All of the actions in paragraph (a)(18) of this section can and will be used to recover any debts owed when it is determined to be in the interest of the Government to do so. The notification and the required form of certification in paragraph (a)(19) of this section are included on RUS Form 490, Application for Telephone Loan or Guarantee.

(b) The following must be submitted by borrowers seeking subsequent loans:

(1) Certified financial statements for the last 3 years.

(2) Toll settlement statements and related data.

(3) Present exchange rates and any pending changes.

(4) Environmental review documentation in accordance with 7 CFR part 1970.

(5) A "Certification Regarding Lobbying" for loans, or a "Statement for Loan Guarantees and Loan Insurance" for loan guarantees, and when required, an executed Standard Form LLL, "Disclosure of Lobbying Activities," (see section 319, Public Law 101–121 (31 U.S.C. 1352)).

(6) Executed copy of Form AD–1047, "Certification Regarding Debarment, Suspension, and Other Responsibility Matters—Primary Covered Transactions."

(7) Borrower's determination of loan maturity, including information noted in § 1735.43(a) of this chapter as required.

(8) Approved depreciation rates for items under regulatory authority jurisdiction.

(9) A statement that the borrower is or is not delinquent on any Federal debt, such as income tax obligations or a loan or loan guarantee from another Federal agency. If delinquent, the reasons for the delinquency must be explained and RUS will take such explanation into consideration in deciding whether to approve the loan. RUS Form 490, "Application for Telephone Loan or Guarantee," contains a section for providing the required statement and any appropriate explanation.

(10) Any other supporting data required by the Administrator.

(c) For borrowers requesting funds for construction or refinancing, in addition to the information included in paragraphs (a) and (b) of this section, the following must be submitted:

(1) Copies of all bonds, notes, mortgages, and contracts covering outstanding indebtedness proposed to be refinanced.

(2) For each note or bond, the name of the creditor, original amount of debt

and amount as of last year-end, purpose of debt, dates incurred and due, interest rates, and repayment terms.

(3) Justification for refinancing and evidence that the underlying loan to be refinanced would have been eligible for RUS financing under the RE Act.

(d) Loan requests whose sole purpose is to refinance loans under Titles II and VI of the RE Act must submit the following:

(1) Certified financial statements for the last 3 years.

(2) Five-year financial projections consisting of Income Statement, Balance Sheet, and Cash Flow Statement.

(3) A "Certification Regarding Lobbying" for loans, or a "Statement for Loan Guarantees and Loan Insurance" for loan guarantees, and when required, an executed Standard Form LLL, "Disclosure of Lobbying Activities," (see section 319, Pub. L. 101–121 (31 U.S.C. 1352)).

(4) Executed copy of Form AD–1047, "Certification Regarding Debarment, Suspension, and Other Responsibility Matters—Primary Covered Transactions."

(5) Borrower's determination of loan maturity.

(6) A statement that the borrower is or is not delinquent on any Federal debt, such as income tax obligations or a loan or loan guarantee from another Federal agency. If delinquent, the reasons for the delinquency must be explained and RUS will take such explanation into consideration in deciding whether to approve the loan. RUS Form 490, "Application for Telephone Loan or Guarantee," contains a section for providing the required statement and any appropriate explanation.

(7) Any other supporting data required by the Administrator.

(e) Borrowers requesting loan funds for acquisitions should refer to RUS bulletins 320–4, 321–2, 325–1, and 326–1 for requirements.

(f) For all applications that request funding for retail broadband as defined in 7 CFR 1735.2, the application must include all information required for the public notice as stated in 7 CFR 1735.23(a).

(Approved by the Office of Management and Budget under control number 0572–0079)

[54 FR 13356, Apr. 3, 1989. Redesignated at 55 FR 39396, Sept. 27, 1990, as amended at 56 FR 26599, June 10, 1991, 58 FR 66256, Dec. 20, 1993; 79 FR 76003, Dec. 19, 2014; 81 FR 11027, Mar. 2, 2016; 86 FR 50609, Sept. 10, 2021]

§§ 1737.23–1737.29 [Reserved]

Subpart D—Preloan Studies—Area Coverage Survey and Loan Design

§ 1737.30 General.

In support of a loan application, the borrower shall prepare and submit to RUS: (a) A market forecast to determine service requirements (the Area Coverage Survey) and (b) engineering studies to determine the system design that provides service most efficiently (the Loan Design). The RUS field representative confers with the borrower and its engineer to schedule the completion and submission of these studies.

(Approved by the Office of Management and Budget under control number 0572–0079)

§ 1737.31 Area Coverage Survey (ACS).

(a) The Area Coverage Survey (ACS) is a market forecast of service requirements of subscribers in a proposed service area.

(b) The objective of the ACS is to determine the location, number and telephone service requirements of subscribers in a service area. RUS will use the ACS to appraise the proposed plan for area coverage and to determine the largest practical number of rural subscribers which can be served on an economically feasible basis. Preparation of the ACS requires:

(1) A field survey of the service area to locate and identify on maps all business and residential establishments, whether currently served or not. The location and identification of future establishments are also recorded on the maps.

(2) A forecast of the number of telephone subscribers, in the entire service area, by exchange, grade and class of

service, projected for the end of the 5-year study period.

(c) The results of the survey and forecast shall be:

(1) Shown on maps (maps for those service areas previously financed by RUS do not have to be included in the ACS provided that the borrower's records contain sufficient information as to subscriber development to enable cost estimates for the proposed facilities to be prepared);

(2) Tabulated on RUS Form 569 "Area Coverage Survey Report," or its equivalent; and

(3) Supported by a narrative (see § 1737.32(f)(1)(ii)) containing information on the bases for the service requirement forecasts in each exchange.

(d) Guidelines on preparing an ACS are provided in RUS Telecommunications Engineering and Construction Manual section 205.

(e) The RUS field representative reviews and approves the borrower's ACS. The borrower should make sure this is done before proceeding with the Loan Design in order to prevent unnecessary expense should the ACS not be approved. The borrower's engineer must use the RUS-approved ACS in preparing the Loan Design.

(Approved by the Office of Management and Budget under control number 0572–0079)

§ 1737.32 Loan Design (LD).

(a) A loan application requires supporting data collectively called a "Loan Design." The LD contains a forecast of service requirements and a narrative with supporting exhibits. Most of the items included in the LD are similar for all loan applications. However, as noted below, there are certain additional requirements for initial loans and for any exchange areas not previously financed by RUS, and other additional requirements for subsequent loans for areas previously financed by RUS. The LD must conform to the borrower's state telecommunications modernization plan unless the borrower is seeking a guaranteed loan (for additional information concerning the plan, see 7 CFR part 1751, subpart B).

(b) Because of the importance and complexity of the engineering studies necessary for the LD, it should be prepared by a competent experienced telecommunications engineer. While the LD is subject to RUS approval, the borrower's selection of an engineer to perform preloan work is not. Note: The borrower's selection of an engineer to perform postloan work *is* subject to RUS approval. This should be considered when selecting a preloan engineer, if the same individual or company is to perform both services. See 7 CFR 1753.17.

(c) An LD for initial loans or for any exchange areas not previously financed by RUS requires an Outside Plant Design that provides:

(1) The most economical and practical design for a telephone system that meets immediate service demands; and

(2) The basis for orderly expansion of the system to serve the widest practical number of rural establishments.

(d) The LD for a subsequent loan (which only includes areas previously financed by RUS) does not require a detailed Outside Plant Design. The detailed Outside Plant Design for these subsequent loans may be completed for RUS review and approval after loan approval, but before staking is started and plans and specifications are prepared. By scheduling preparation of the outside plant design closer to preparation for construction, the need for redesign resulting from changing conditions and its attendant costs are reduced.

(e) Guidelines on preparing an LD are provided in RUS Telecommunications Engineering and Construction Manual section 205.

(f) The LD shall include a narrative, several exhibits, and a certification, as explained below:

(1) *Narrative.* This section discusses the following topics, as appropriate.

(i) *General.* The purposes and amount of the proposed construction and both immediate and long range plans must be covered. The source and amount of any nonloan funds to be used for this construction must be discussed.

(ii) *Subscriber data.* The basis for the subscriber forecast, including any unusual factors expected to influence growth, must be discussed. Reasons for growth projections which vary from historic trends must be explained.

(iii) *Proposed construction.* All proposed construction must be described fully. Reference to the BER must be made here.

(iv) *Service area.* For subsequent loans only, proposed construction which is not within the boundaries of prior loan projects must be discussed. New areas to be served (even if from existing exchanges) must be shown on maps submitted with the proposal.

(v) *Toll and EAS.* Proposed new toll or extended area service (EAS) facilities, including any changes from the existing trunking arrangements, must be described fully. Minutes of meetings and correspondence with connecting companies, and connecting company concurrences, if any, must be included.

(vi) *Radio telephone service.* Proposed radio telephone service must be discussed. Results of studies demonstrating demand and/or need most be included as an exhibit.

(vii) *Special projects.* Facilities involving investment in excess of $100,000 for any single subscriber must be discussed fully. Contractual arrangements with the subscriber, including a termination agreement providing for (A) the full recovery by the borrower of its capital costs of the facilities no later than the maturity date of the note representing the loan, (B) the immediate repayment of all remaining capital costs, if terminated, and (C) repayment to RUS of the outstanding amount of the special note shall be submitted. Usually a separate short-term note is prepared for loans to finance Special Projects.

(viii) *Investment in nonrural areas.* (A) For initial loans, or loans for areas not previously financed by RUS, the borrower must fully discuss proposed improvements or expansions in an exchange serving a community over 5,000 population. The name of the community, the number of existing and projected new subscribers by grades of service within the community, detailed cost estimates of the facilities involved, and information sufficient to establish the necessity for the use of loan funds must be provided.

(B) For subsequent loans, the borrower must fully discuss as specified in paragraph (f)(1)(viii)(A) of this section proposed improvements or expansions in an exchange serving a community over 5,000 population which had a population of more than 5,000 at the time the facilities to serve the community were first financed by RUS. The population determination is based on the corporate limits or boundaries of unincorporated areas in existence at the time the facilities to serve the community were first financed by RUS.

(C) For subsequent loans, the borrower shall state whether the population of a community, which is currently more than 5,000, was considered rural at the time RUS first financed the facilities to serve the community. Detailed cost estimates are not required if the population was considered rural at the time RUS first financed facilities to serve the community, see 7 CFR 1735.13(d).

(ix) *Prior loan project.* For subsequent loans only, the reason for and amount of additional loan funds needed to complete construction in progress which was part of a prior loan project in central office areas not included in the current LD must be discussed fully.

(x) *Route miles.* Route miles of outside plant in central office areas not shown on RUS Form 495 must be provided.

(xi) *Future plans.* Where the loan application is to finance part of a systemwide upgrading plan, plans for those remaining exchanges not included in the current loan proposal must be discussed.

(2) *Exhibits.* (i) An RUS Form 569, "Area Coverage Survey Report," or its equivalent shall be included for the total system and for each exchange in which system improvements or additions are proposed.

(ii) An RUS Form 495, "Construction Cost Estimates," or its equivalent shall be prepared for each exchange in which system improvements or additions are proposed. An explanation of the method used in developing these cost estimates must be included.

(iii) RUS Form 494, "Loan Design Summary," or its equivalent shall be prepared for each loan. This must show all expected 5-year construction costs, loan and nonloan.

(iv) A schematic trunking diagram shall be included showing the number and type, length, ownership and make-up of existing and proposed toll and

EAS trunks, plus transmission and traffic data for each trunk group.

(v) Detailed outside plant design maps must be submitted for all central office areas of initial loan applicants and for areas not previously served by existing borrowers or financed by RUS. These design maps must be in sufficient detail to substantiate the construction cost estimates.

(vi) For subsequent loans only, if a change in system boundaries is proposed, a map must be furnished showing present and proposed boundaries, and existing establishments and subscribers in the new areas.

(vii) Any other special exhibits needed to support particular items in the loan proposal must be included.

(3) *Certification.* The following certification shall be signed by a principal of the engineering firm and the borrower:

We, the undersigned, certify that the data in this Loan Design are correct to the best of our knowledge and belief and reasonably reflect the cost to serve the subscribers as proposed on the Forms 569, "Area Coverage Survey Report," which are integral parts hereof, and that this Loan Design adheres to RUS engineering and construction standards and practices.

(g) The RUS field representative shall review and make a recommendation on each LD.

(1) After completion of the LD, the borrower arranges a meeting with its engineer and RUS's field representative to review:

(i) Design and cost estimates.

(ii) Reserves available from prior loans, if any, or internally generated funds which may be applied against the requirements of the current application.

(2) One copy of RUS Form 567, "Checklist for Review of Loan Design," completed and signed by the borrower's engineer must be attached to the LD submitted to the RUS field representative.

(3) The RUS field representative recommends acceptance of the LD as the basis for RUS financing.

(4) Three copies of the final LD with the RUS field representative's recommendation are then sent to the relevant Area Office in RUS. A fourth copy is retained by the RUS field representative.

(5) A transmittal letter from the borrower must accompany the LDs, requesting that the application previously submitted be amended so as to be consistent with the approved LD.

(6) Final approval of the LD is given by the relevant Area Office in RUS. To be approved, the LD must be cost effective, include appropriate technology, and provide area coverage.

(7) Upon receipt of the LD and any other required information, RUS makes a preliminary analysis of the loan proposal. Before final consideration of the loan, RUS reviews the results of its preliminary analysis with the borrower.

(Approved by the Office of Management and Budget under control number 0572–0079)

[54 FR 13356, Apr. 3, 1989; 54 FR 16194, Apr. 21, 1989. Redesignated at 55 FR 39396, Sept. 27, 1990, as amended at 58 FR 66256, Dec. 20, 1993]

§§ 1737.33–1737.39 [Reserved]

Subpart E—Interim Financing of Construction of Telephone Facilities

§ 1737.40 General.

(a) Under special circumstances a borrower may request that RUS approve interim financing for interim construction. This subpart describes the circumstances in which RUS will consider approving interim financing of construction, the information to be submitted to RUS to support the borrower's request, RUS's requirements relating to interim construction, and related matters.

(b) For a borrower to preserve the option of obtaining loan funds for reimbursement of interim financing, it must obtain prior RUS approval of its interim financing plan and follow the procedures in 7 CFR 1737.41 and 7 CFR 1737.42.

(c) RUS will approve interim financing only for projects which must be performed immediately.

(d) RUS approval of interim financing is not a commitment that RUS will make loan funds available.

(e) Equal employment opportunity requirements apply to interim construction. See RUS Bulletin 320–15.

§ 1737.41 Procedure for obtaining approval.

(a) The borrower shall submit to the RUS Area Office a written request for approval of interim financing. This request shall include:

(1) A description of the construction proposed under interim financing.

(2) An explanation of the urgency of proceeding with the proposed construction.

(3) An estimate of the cost.

(4) The source of funds to be used for interim financing.

(b) RUS will not approve interim financing until it has reviewed and found acceptable:

(1) All of the information required under § 1737.21; or

(2) The following documents:

(i) The loan application (RUS Form 490) clearly marked "in support of interim financing request."

(ii) The Loan Design (LD), or the portion thereof that covers the proposed construction if the completed LD is not available. See 7 CFR 1737.32.

(iii) Evidence that the borrower has complied with the environmental review requirements in accordance with 7 CFR part 1970.

(iv) A statement that the borrower is or is not delinquent on any Federal debt, such as income tax obligations or a loan guarantee from another Federal agency. If delinquent, the reasons for the delinquency must be explained and RUS will take such explanation into consideration in deciding whether to approve the interim financing, see 7 CFR 1737.22(b)(9).

(v) A "Certification Regarding Lobbying" for loans, or a "Statement for Loan Guarantees and Loan Insurance" for loan guarantees, and when required, an executed Standard Form LLL, "Disclosure of Lobbying Activities," (see section 319, Pub. L. 101–121 (31 U.S.C. 1352)).

(vi) Executed copy of Form AD–1047, "Certification Regarding Debarment, Suspension, and Other Responsibility Matters—Primary Covered Transactions."

(vii) Any other supporting data required by the Administrator.

(c) RUS will not approve a borrower's request for approval of interim financing if, in RUS's judgment:

(1) The proposed interim financing does not comply with the requirements of this subpart.

(2) The proposed interim construction will not qualify for RUS financing.

(3) The proposed interim financing presents unacceptable loan security risks to RUS, or otherwise is not in the best interests of RUS.

(Approved by the Office of Management and Budget under control number 0572–0079)

[54 FR 13356, Apr. 3, 1989. Redesignated at 55 FR 39396, Sept. 27, 1990, as amended at 56 FR 26599, June 10, 1991; 59 FR 54381, Oct. 31, 1994; 79 FR 76003, Dec. 19, 2014; 81 FR 11027, Mar. 2, 2016]

§ 1737.42 Procedure for construction.

(a) If RUS approves the interim financing, interim construction shall be conducted in accordance with 7 CFR part 1753, 7 CFR part 1788, RUS Bulletin 320–15, and RUS Bulletins 381–1, 381–2, 381–4, 381–7, 381–8, 381–9, 381–10, 381–11, 381–13, 382–1, 382–2, 382–3, 383–1, 383–4, 384–1, 384–2, 384–3, 385–1, 385–2, 385–3, 385–4, 385–5, 385–6, 387–1, 387–2, 387–3, 387–4, and 387–5) except for the following:

(1) All sellers and contractors invited to bid must be informed that funds from sources other than RUS will be used to pay for construction.

(2) Contracts involving the interim construction must contain a provision, in form and substance satisfactory to RUS, stating that RUS is not committed to lend or advance funds to finance the project.

(3) Contracts will not be approved by RUS until the borrower demonstrates to RUS's satisfaction that funds from sources other than RUS will be available when needed to pay invoices submitted in accordance with contract payment terms.

(4) The borrower shall not begin interim construction until all necessary licenses, permits, and other governmental approvals have been obtained.

(b) After RUS loan funds are released, the borrower can obtain reimbursement for interim financing by submitting a Financial Requirement Statement. See 7 CFR part 1744, subpart C (or RUS Bulletin 327–1).

(1) The first advance of loan funds to a borrower that has received interim financing approval generally will be limited to funds to repay any interim financing indebtedness and such additional amounts as RUS deems necessary. RUS will make no further advances of loan funds until the borrower has submitted evidence, in form and substance satisfactory to the Administrator, that (i) any indebtedness created by the interim financing and any liens associated therewith have been fully discharged of record and (ii) the borrower has satisfied all other conditions on the advance of additional loan funds.

(2) If the source of funds for interim financing is the borrower's internally generated funds, the borrower may request reimbursement of those funds along with advances for other purposes on the first Financial Requirement Statement.

[54 FR 13356, Apr. 3, 1989; 54 FR 16194, Apr. 21, 1989. Redesignated at 55 FR 39396, Sept. 27, 1990]

§§ 1737.43–1737.49 [Reserved]

Subpart F—Review of Application Procedures

§ 1737.50 Review of completed loan application.

(a) The completed loan application consists of:

(1) A completed RUS Form 490, "Application for Telephone Loan or Loan Guarantee;"

(2) A completed certification Form AD–1047, "Certification Regarding Debarment, Suspension, and Other Responsibility Matters—Primary Covered Transactions;"

(3) A market survey called the Area Coverage Survey (ACS);

(4) The plan and associated costs for the proposed construction, called the Loan Design (LD);

(5) Evidence that the borrower is participating in a telecommunications modernization plan in the state where the proposed construction will occur, unless the borrower is seeking a guaranteed loan; and

(6) Various supplementary information. See 7 CFR part 1737 for additional information.

(b) RUS shall review the completed loan application, particularly noting subscriber data, grades of service, extended area service (EAS), connecting company commitments, commercial facilities, system and exchange boundaries, and proposed acquisitions. RUS shall review the LD to determine that the system design is acceptable to RUS, that the design is technically correct, that the cost estimates are reasonable, and that the design provides for area coverage service. RUS shall also review the population and incorporation status of all communities served or to be served by the borrower to determine if any nonrural areas are served and if municipal franchises are required. Any RUS lending for nonrural areas must be in accordance with 7 CFR part 1735. RUS shall also check the "List of Parties Excluded from Federal Procurement of Nonprocurement Programs", compiled, maintained and distributed by General Services Administration, to determine whether the borrower is debarred, suspended, ineligible, or voluntarily excluded (see 2 CFR 180.430).

(c) RUS will notify the borrower if RUS recommends major changes in subscriber projections, design, cost estimates, or other significant matters. RUS will not continue loan processing until RUS and the borrower agree on all major changes.

[54 FR 13356, Apr. 3, 1989; 54 FR 16194, Apr. 21, 1989. Redesignated at 55 FR 39396, Sept. 27, 1990, as amended at 58 FR 66256, Dec. 20, 1993; 79 FR 76003, Dec. 19, 2014]

§ 1737.51 Approval of loan design.

RUS shall notify the borrower when the preloan data concerning the system design and costs and subscriber projections have been approved. If found acceptable, RUS will approve the LD with any required changes. A copy of the approved LD, with any significant changes, as determined by RUS, will be returned to the borrower.

§§ 1737.52–1737.59 [Reserved]

Subpart G—Project Cost Estimation Procedures

§ 1737.60 Telephone loan budget.

(a) RUS shall prepare a "Telephone Loan Budget" (RUS Form 493) showing all costs for the proposed project and the amount of loan and nonloan funds to be used. The budget shall show, as applicable, amounts for central offices, outside plant and station equipment, right-of-way procurement, land, buildings, removal costs, special projects, engineering, vehicles and work equipment, office equipment, operating funds, refinancing with loan funds, debt retirement with nonloan funds, acquisitions, and contingencies. The amount of funds included in any loan shall be limited for certain items:

(1) Operating funds for working capital or current operating deficiencies shall be included only in cases of financial hardship as determined by the Administrator.

(2) Contingencies shall not exceed 3 percent of the total amount of loan funds to be used for construction, engineering, operating equipment and operating funds.

(b) RUS shall prepare the cost estimates based on the data in RUS Form 494, "Loan Design Summary," and RUS Form 495, "Construction Cost Estimates," or their equivalents, and other parts of the LD submitted by the borrower, and on other pertinent information. See subpart D of this part. The amounts included in the proposed budget shall be the estimated costs, less the value of materials and supplies on hand or acquired that can be used in the proposed construction. The cost estimates in the LD may be adjusted by RUS in consultation with the borrower. See § 1737.50(c).

(c) Generally, the new loan shall be reduced by any required equity funds and funds available in reserves no longer needed for prior loan purposes to determine the proposed loan requirement.

(d) When amounts are available in reserves no longer needed for prior loan purposes, RUS may, at its option, deny further advances of these funds if they will be used to finance projects in the proposed loan.

(e) The budget shall also show, if applicable, the reserves for each budget item as of the date of the latest RUS Form 481, "Financial Requirement Statement," submitted by the borrower. To ensure that sufficient funds are included in the budget to finance all proposed construction, RUS includes in the budget any funds deposited by the borrower for approved interim financing.

[54 FR 13356, Apr. 3, 1989, as amended at 84 FR 59922, Nov. 7, 2019]

§ 1737.61 Cost allocation for rural and nonrural areas.

(a) Pursuant to the requirements in 7 CFR part 1735, if loan funds are proposed for facilities to serve subscribers in nonrural areas, RUS shall allocate costs between rural and nonrural areas. This allocation will be used to determine whether the use of loan funds in nonrural areas is necessary and incidental to furnishing and improving telephone service in rural areas. Cost estimates shall be provided by the borrower in the LD. See subpart D of this part. RUS will use the following method to review the cost breakdowns and to determine their appropriateness:

(1) The costs of facilities associated directly with particular subscribers shall be allocated to those subscribers.

(2) The costs of facilities that serve both rural and nonrural subscribers shall be allocated based on the relative number of rural and nonrural subscribers receiving service from those facilities.

(3) When a borrower's exchange that includes a nonrural community will have an extended area of service (EAS) with other exchanges of the borrower, the breakdown of subscribers and funds in the allocation for rural and nonrural areas included in the proposed loan shall show the number of rural and nonrural subscribers and the costs to serve each group, as determined per paragraphs (a)(1) and (a)(2) of this section, in the subject exchange and in all exchanges connected by EAS.

(b) If RUS determines that costs cannot be adequately allocated using the procedures in paragraphs (a)(1) through (a)(3) of this section, RUS shall, on a case by case basis, allocate costs between the rural and nonrural subscribers using whatever methodology it deems reasonable. All allocations in paragraphs (a) and (b) of this section shall be documented.

§§ 1737.62–1737.69 [Reserved]

Subpart H—Feasibility Determination Procedures

§ 1737.70 Description of feasibility study.

(a) In connection with each loan RUS shall prepare a feasibility study that includes sections on consolidated loan estimates, operating statistics, projected telecommunications, plant, projected retirement computations, and projected revenue and expense estimates (including detailed estimates of depreciation and amortization expense, scheduled debt service payments, toll and access charge revenues, and local service revenues). Normally, projections will be for a 5-year period and used to determine the ability of the borrower to repay its loans in accordance with the terms thereof. RUS will not require borrowers to raise local service rates. Local service revenue projections will be based on the borrower's existing local service rates or regulatory body approved rates not yet in effect but to be implemented within the Forecast period. In the latter case, if a borrower is not required to obtain regulatory body approval for the implementation of such rates, RUS will require a resolution of the board of directors indicating when those rates will be in effect.

(b) RUS makes loans only to rural telephone systems that are financially feasible. RUS shall consider the factors discussed in paragraphs (c) through (j) of this section in determining feasibility.

(c) The revenue and expense estimates for the feasibility study generally will be based on the borrower's operating experience provided that:

(1) Adjustments are made for any nonrecurring revenues and expenses that are not representative of the borrower's past operations and would thus make the borrower's experience data inappropriate for the forecast; and

(2) Adjustments are made for any special or new characteristics or other considerations deemed necessary by the Administrator.

(d) [Reserved]

(e) Depreciation expense will be determined using depreciation rates appropriate to the normal operation of the borrower, based on:

(1) The borrowers regulatory body approved depreciation rates; and

(2) Where such rates as described in paragraph (e)(1) of this section do not exist for items which the borrower is seeking financing, the most recent median depreciation rates published by RUS for all borrowers. RUS will publish such depreciation rates annually in RUS's "Statistical Report, Rural Telephone Borrowers."

(f) Projected scheduled debt service payments will generally be based on all of the borrower's outstanding and proposed loans from RUS and all other lenders as of the end of the feasibility Forecast period (i.e. for a 5-year Forecast period, the amount of debt outstanding in year 5).

(g) The financial and statistical data are derived from RUS Form 479, "Financial and Statistical Data for Telephone Borrowers," or for initial loans, the data may be obtained from the borrower's financial statements and other reports, and from other information supplied with the completed loan applications (see 7 CFR 1737.21 and 1737.22).

(h) When, in RUS's opinion, the borrower's operating experience is not adequate or the borrower's current operations are not representative, the estimates in the feasibility study normally will be developed from state and regional standards based on the experience of RUS borrowers. These standards are included in the Borrower's Statistical Profile (BSP), which is revised annually by RUS. If the borrower's operating experience is not the basis for one or more per-subscriber estimates used in the feasibility study, the estimates generally may not vary from the standard by more than 20 percent to reflect the particular characteristics of the loan applicant. Any variation from the standard shall be documented.

(i) In cases where these per-subscriber standards do not represent a reasonable forecast of a particular borrower's operations (for example, when a variation greater than 20 percent is necessary), estimates based upon a special analysis of the borrower's projected operations shall be used. The special analysis will accompany the feasibility study.

(j) When it is reasonably expected that a subscriber, classified as a special project, may discontinue service, a second feasibility study will be prepared, for comparison purposes, omitting revenues and expenses from this subscriber.

(k) RUS may obtain and review commercially available credit reports on applicants for a loan or loan guarantee to verify income, assets, and credit history, and to determine whether there are any outstanding delinquent Federal or other debts. Such reports will also be reviewed for parties that are or propose to be joint owners of a project with a borrower.

(l) If it is determined that loan feasibility cannot be proven as described in this section, the loan application will be returned to the borrower with an explanation. A borrower whose application has been returned will have 90 working days, from the date the application was returned, to revise and resubmit its application. If a revised application is not received by RUS within the 90-day period described above, the application will be canceled and a new application will need to be submitted if the borrower wishes further consideration.

[54 FR 13356, Apr. 3, 1989. Redesignated at 55 FR 39396, Sept. 27, 1990, as amended at 56 FR 26599, June 10, 1991; 58 FR 66256, Dec. 20, 1993; 62 FR 46872, Sept. 5, 1997]

§ 1737.71 Interest rate to be considered for the purpose of assessing feasibility for loans.

(a) For purposes of determining the creditworthiness of a borrower for RUS cost-of-money, the Administrator shall assume that the loan, if made, would bear interest at the Treasury rate on the date of determination as described in paragraph (b) of this section. If the Treasury rate exceeds 7 percent, the interest rate used to determine eligibility for the RUS cost-of-money loan will be 7 percent.

(b) The 30-year Treasury rate will be used in all feasibility studies for loans with a final maturity of at least 30 years. A straight-line interpolation between other Treasury rates will be used to determine the rate used in feasibility studies for loans with final maturities of less than 30 years.

(c) The Treasury rate will be obtained each Tuesday, or as soon as possible thereafter, from the Federal Reserve. The rate for the current week, from the column labeled "This week" in the Federal Reserve statistical release, will be used from that Wednesday through the following Tuesday.

(d) As used in this section, the "date of determination" means the date of the feasibility study used in support of the loan recommendation.

[58 FR 66257, Dec. 20, 1993, as amended at 84 FR 59922, Nov. 7, 2019]

§§ 1737.72–1737.79 [Reserved]

Subpart I—Characteristics Letter

§ 1737.80 Description of characteristics letter.

(a) After all of the studies and exhibits for the proposed loan have been prepared, but before the loan is recommended, RUS shall inform the borrower, in writing, of the characteristics of the proposed loan. The purpose of the characteristics letter is to inform the borrower and obtain its concurrence, before further consideration by RUS of the loan approval and the preparation of legal documents relating to the loan, in such matters as the amount of the proposed loan, its purposes, rate of interest, loan security requirements, and other prerequisites to the advance of loan funds. The letter, whether or not concurred in by the borrower, does not commit RUS to approve the loan on these or any other terms.

(b) The Forecast of Revenues and Expenses and a copy of RUS Form 493, "Telephone Loan Budget," shall be enclosed with the characteristics letter. This copy of the budget shall be subject to change by RUS with the borrower's agreement.

[54 FR 13356, Apr. 3, 1989. Redesignated at 55 FR 39396, Sept. 27, 1990, as amended at 56 FR 26600, June 10, 1991]

§§ 1737.81–1737.89 [Reserved]

Subpart J—Final Loan Approval Procedures

§ 1737.90 Loan approval requirements.

(a) In addition to requirements set forth in 7 CFR part 1735, 7 CFR part 1737 and other applicable parts of 7 CFR chapter XVII, the following are certain additional requirements that must be met before RUS will approve a loan:

(1) If the borrower had 100 or more employees as of the prior December 31, it must submit the current annual Employer Information Report EEO–1, Standard Form 100, as required by the Department of Labor; see 29 CFR 1602.7 through 1602.14.

(2) The borrower must be in compliance with regulations on nondiscrimination. See 7 CFR part 1790 (or RUS Bulletin 320–19).

(3) For subsequent loans, RUS must determine whether the borrower's accounting records are adequate. If the records are not adequate, as determined by RUS based on Generally Accepted Accounting Principles or other accounting conventions as deemed necessary by RUS, a provision will be included in the loan contract requiring the borrower to improve its records to an adequate level.

(4) The borrower must not have any receivables, loans, guarantees, investments, or other obligations that are contrary to the mortgage provisions or any RUS regulations including, but not limited to, 7 CFR part 1758 (or RUS Bulletins 320–4, 320–22, 321–2, 322–2, 323–1, or 326–1). If the borrower has any of these items, the loan contract shall contain a provision requiring that they be eliminated prior to the release of funds. See 7 CFR part 1744 for conditions under which RUS will provide a shared first lien and/or a lien accommodation for non-RUS lenders.

(5) RUS must make a determination on flood insurance requirements. In accordance with the National Flood Insurance Act of 1968, as amended by the Flood Disaster Protection Act of 1973, as amended (the "Flood Insurance Act"), RUS shall not approve financial assistance for the acquisition, construction, repair or improvement of any building or any machinery, equipment, fixtures or furnishings contained or to be contained in any such building located in an area which has been identified by the Director of the Federal Emergency Management Agency (the "Director of FEMA") pursuant to the Flood Insurance Act as an area having special flood hazards unless:

(i) Flood insurance has been made available, pursuant to the Flood Insurance Act, in the area in which the acquisition, construction, repair or improvement is proposed to occur; and

(ii) The borrower has obtained flood insurance coverage with respect to such building, machinery, equipment, fixtures or furnishings as may be required pursuant to the Flood Insurance Act.

Accordingly, a finding shall be made on whether loan funds will be used to finance buildings, machinery, fixtures or furnishings located in an identified special flood hazard area. If loan funds are to be used in such a special flood hazard area, a provision will be included in the loan contract restricting the release of funds until all the requirements of the Flood Insurance Act have been satisfied.

(6) All environmental review requirements must be met in accordance with 7 CFR part 1970. The Agency may obligate, but not disperse, funds under the program pursuant to 7 U.S.C. 950cc–1, before the completion of the otherwise required environmental, historical, or other types of reviews if the Secretary of Agriculture determines that subsequent site-specific review shall be adequate and easily accomplished for the location of towers, poles, or other broadband facilities in the service area of the awardee without compromising the project or the required reviews.

(b) [Reserved]

[54 FR 13356, Apr. 3, 1989. Redesignated at 55 FR 39396, Sept. 27, 1990, as amended at 56 FR 26600, June 10, 1991; 81 FR 11027, Mar. 2, 2016; 86 FR 50610, Sept. 10, 2021]

§ 1737.91 Approval.

(a) A loan is approved when the Administrator, or whoever is delegated authority, signs the administrative findings and the letter to the borrower announcing the loan.

(b) If the loan is not approved, RUS shall notify the borrower, in writing, of the reasons.

§ 1737.92 Loan documents.

Following approval of the loan, RUS shall forward the necessary loan documents to the borrower for execution, delivery, recording, and filing, as directed by RUS. See 7 CFR part 1758 for details (or RUS Bulletins 320–4, 320–22, 321–2, 322–2, 323–1, or 326–1).

§§ 1737.93–1737.99 [Reserved]

Subpart K—Release of Funds Procedure

§ 1737.100 Prerequisites to the release and advance of funds.

(a) Standard prerequisites to the advance of funds, generally applied to all loans, are set forth in Article II of the form of loan contract attached as appendix A to 7 CFR part 1758. Additional prerequisites may be added on a case by case basis to the loan contract.

(b) Before any loan funds can be advanced, RUS must approve a release of funds.

(c) RUS approves the release of funds only after it determines that all prerequisites to the advance of loan funds have been met or funds should be advanced even though certain loan contract prerequisites remain unsatisfied.

(d) Following release approval, loan funds and related nonloan funds may be advanced in accordance with 7 CFR part 1744.

(e) The borrower may be required to discharge indebtedness and/or to close acquisitions before advances are made for construction purposes. In such cases, the borrower shall submit evidence that these actions have been completed. If the evidence is satisfactory to RUS, RUS shall allow the remaining loan funds to be advanced in accordance with 7 CFR part 1744.

(Approved by the Office of Management and Budget under control number 0572–0085)

§ 1737.101 Amounts spent for preloan activities.

If the borrower desires to credit amounts spent for preloan activities against any equity or general funds required by the loan contract, it shall submit an itemized statement of such expenditures to the Area Office. These expenditures will be accounted for on RUS Form 503, "Release of Telephone Loan Funds," if RUS determines that the amounts spent are reasonable based on normal industry practice and that the procedures set forth in 7 CFR part 1737, subpart D, have been complied with. Statements of preloan expenditures will be verified as to accuracy by loan fund audits.

(Approved by the Office of Management and Budget under control number 0572–0085)

§§ 1737.102–1737.109 [Reserved]

PART 1738—RURAL BROADBAND LOANS, LOAN/GRANT COMBINATIONS, AND LOAN GUARANTEES

Subpart A—General

Subpart B—Eligibility Requirements

Subpart C—Award Requirements

Subpart D—Loan and Loan/Grant Combination Award Terms

Subpart E—Loan and Loan/Grant Combination Application Review and Underwriting

Subpart F—Closing, Servicing, and Reporting for Loan and Loan/Grant Combination Awards

Subpart G—Loan Guarantee

AUTHORITY: 7 U.S.C. 901 *et seq.*

SOURCE: 85 FR 14398, Mar. 12, 2020, unless otherwise noted.

Subpart A—General

§ 1738.1 Overview.

(a) The Rural Broadband Program furnishes loans, loan/grant combinations, and loan guarantees for the costs of construction, improvement, or acquisition of facilities and equipment needed to provide service at the broadband lending speed in eligible rural areas. This part sets forth the general policies, eligibility requirements, types and terms of loans, loan/grant combinations and loan guarantees, and program requirements under 7 U.S.C. 901 *et seq.*

(b) Additional information and application materials regarding the Rural Broadband Program can be found on the Rural Development website.

§ 1738.2 Definitions.

(a) The following definitions apply to this part:

Acquisition means the purchase of assets by an eligible entity as defined in § 1738.51 to acquire facilities, equipment, operations, licenses, or majority stock interest of one or more organizations. Stock acquisitions must be arm's-length transactions.

Administrator means the Administrator of the Rural Utilities Service (RUS).

Advance means the transfer of loan or grant funds from the Agency to the Awardee.

Affiliate or *affiliated company* of any specified person or entity means any other person or entity directly or indirectly controlling of, controlled by, under direct or indirect common control with, or related to, such specified entity, or which exists for the sole purpose of providing any service to one company or exclusively to companies which otherwise meet the definition of affiliate. For the purpose of this definition, "control" means the possession directly or indirectly, of the power to direct or cause the direction of the management and policies of a company, whether such power is exercised through one or more intermediary companies, or alone, or in conjunction with or pursuant to an agreement with, one or more other companies, and whether such power is established through a majority or minority ownership voting of securities, common directors, officers, or stockholders, voting trust, or holding trusts (other than money exchanged) for property or services.

Agency means the Rural Utilities Service (RUS).

Applicant means an entity requesting approval of assistance under this part.

Assistance means a request for a loan, loan/grant combination, or loan guarantee.

Associated loan means any loan that is granted in association with a grant. Every grant will have an associated loan.

Award means a loan, loan/grant combination, or loan guarantee made under this part.

Award documents means, as applicable, all associated loan agreements, loan/grant combination agreements, or loan guarantee documents.

Award term means the term of the loan as defined in the Award documents. The Award term shall be equal to the composite economic life of the facilities being financed with RUS loan or grant funding plus 3 years.

Awardee means an entity that has applied for and been awarded assistance under this part.

Borrower means an entity that has applied for and been awarded loan funding under this part.

Broadband grant means a Community Connect, Broadband Initiatives Program, ReConnect Program, or Rural Broadband Program grant approved by the Agency.

Broadband lending speed means the minimum bandwidth requirements, as published by the Agency in its latest notice in the FEDERAL REGISTER that Applicants must propose to deliver to every customer in the proposed funded service area in order for the Agency to approve a broadband Award. Broadband lending speeds will vary depending on the technology proposed and the term of the average composite economic life of the facilities. Initially, the broadband lending speed for terrestrial service, whether fixed or wireless, as well as mobile broadband serving ranches and farmland is 25 megabits per second (Mbps) downstream and 3 Mbps upstream, until further amended by notice. If a new broadband lending speed is published in the FEDERAL REGISTER while an application is pending, the pending application will be processed based on the broadband lending speed that was in effect when the application was submitted.

Broadband loan means any loan approved under Title VI of the Rural Electrification Act of 1936, as amended (RE Act).

Broadband service means any technology identified by the Administrator as having the capacity to provide transmission facilities that enable the subscriber to receive a minimum level of service equal to at least a downstream transmission capacity of 25 Mbps and an upstream transmission capacity of 3 Mbps. The Agency will publish the minimum transmission capacity with respect to terrestrial service that will qualify as broadband service in a notice in the FEDERAL REGISTER. If a new minimum transmission capacity is published in the FEDERAL REGISTER while an application is pending, broadband service for the purpose of reviewing the application will be defined by the minimum transmission capacity that was required at the time the application was received by the Agency.

Build-out means the construction, improvement, or acquisition of facilities and equipment, except for customer premises equipment (CPE).

Competitive analysis means a study that identifies service providers and products in the service area that will compete with the Applicant's operations.

Composite economic life means the weighted (by dollar amount of each class of facility in the requested assistance) average economic life as determined by the Agency of all classes of facilities financed by the award.

Current Ratio (CR) means the current assets divided by the current liabilities.

Customer premises equipment (CPE) means any network-related equipment used by a customer to connect to a service provider's network.

Debt Service Coverage Ratio (DSCR) *means* the ratio of the sum of the Awardee's total net income or margins, depreciation and amortization expense, and interest expense, minus an allowance for funds used during construction and amortized grant revenue, all divided by the sum of interest on funded debt, other interest, and principal payment on debt and capital leases.

Density means the total population to be served by the project divided by the total number of square miles to be served by the project. If multiple service areas are proposed, the density calculation will be made on the combined areas as if they were a single area, and not the average densities.

Development costs mean the pre-application costs associated with construction, design of the system, and other professional labor, as approved by the Agency. Further guidance on what constitutes approved development costs will be outlined in the Agency's application guide.

Economic life means the estimated useful service life of an asset financed by the loan or grant, as determined by the Agency.

Feasibility study means the pro forma financial analysis performed by the Agency, based on the financial projections prepared by the Applicant, to determine the financial feasibility of a loan or loan/grant combination request.

Financial feasibility means the Applicant's ability to generate sufficient revenues to cover its expenses, sufficient cash flow to service its debts and obligations as they become due and meet the Net worth and minimum Times Interest Earned Ratio (TIER), CR, or DSCR requirements of §1738.206(b)(2)(i) by the end of the forecast period. Financial feasibility of an application is based on a projection that spans the forecast period and the entire operation of the Applicant, not just the proposed project.

Fiscal year refers to the Applicant or awardee's fiscal year, unless otherwise indicated.

Forecast period means the time period used in the feasibility study to determine if an application is financially feasible.

GAAP means generally accepted accounting principles in the United States of America.

Grant documents means the grant contract and security agreement between the Agency and the Awardee securing the grant.

Grantee means an entity that has an outstanding broadband grant made by the Agency, with outstanding obligations under the Award documents.

Incumbent service provider means a service provider that provides terrestrial broadband service to at least 5 percent of the households in the proposed funded service area at the time of application submission. Resellers are not considered incumbent service providers. If an Applicant proposes an acquisition, the Applicant will be considered a service provider for that area. The Agency will not consider mobile or satellite providers when determining the incumbent service providers in the area.

Indefeasible right to use (IRU) means the long-term agreement of the rights to capacity, or a portion thereof specified in terms of a certain amount of bandwidth or number of fibers.

Interim financing means funds used for eligible Award purposes after an Award offer has been extended to the Applicant by the Agency. Such funds may be eligible for reimbursement from Award funds if an Award is made.

Loan guarantee means Federal assistance in the form of a guarantee of a loan, or a portion thereof, made by another lender.

Loan funds means funds provided pursuant to a broadband loan made or guaranteed under this part by the Agency.

Market survey means the collection of information on the supply, demand, usage, and rates for proposed services to be offered by an Applicant in support of the Applicant's financial projections.

Net worth means the difference between an entity's total assets and total liabilities.

Project means all work to be performed to bring broadband service to all premises in the proposed funded service area under the Application that is approved for assistance. This includes the construction, purchase and installation of equipment, and professional services including engineering and accountant/consultant fees. A project may be funded with Federal assistance or other funds.

Project completion means that all Award funds for construction of the broadband system, excluding those funds for subscriber connections and CPE, have been advanced to the Awardee by RUS.

Proposed funded service area means the geographic service territory within which the Applicant is proposing to offer service at the broadband lending speed.

RE Act means the Rural Electrification Act of 1936, as amended (7 U.S.C. 901 *et seq.*).

Reseller means a company that purchases network services from service providers in bulk and resells them to commercial businesses and residential households. Resellers are not considered incumbent service providers.

Rural area(s) means any area which is not located within:

(i) A city, town, or incorporated area that has a population of greater than 20,000 inhabitants; or

(ii) An urbanized area contiguous and adjacent to a city or town that has a population of greater than 50,000 inhabitants. For purposes of this definition, an urbanized area means a densely populated territory as defined in the latest decennial census of the U.S. Census Bureau; and

(iii) Which excludes certain populations pursuant to 7 U.S.C. 1991(a)(13), or as otherwise provided by law.

RUS Borrower or *RUS Grantee* means any recipient of a loan or grant administered by the RUS Telecommunications Program that has a loan outstanding, or a grant which still has unadvanced funds available.

Security documents means any mortgage, deed of trust, security agreement, financing statement, or other document which grants or perfects to the Agency a security interest in collateral given as security for the assistance under this part.

Service area or *Service territory* means the geographic area within which a service provider offers broadband service.

Service provider means an entity providing broadband service.

System of accounts means the Agency's system of accounts for maintaining financial records as described in 7 CFR part 1770, subpart B.

TIER means times interest earned ratio. TIER is the ratio of an Applicant's net income (after taxes) plus interest expense, all divided by interest expense and with all financial terms customarily-required by GAAP or by the Uniform System of Accounts (USOA).

Total project cost means all eligible costs associated with the project that are laid out in the application budget schedule, including RUS loan and grant funding and non-RUS funds, as approved by the Agency.

(b) Accounting terms not otherwise defined in this part shall have the commonly-accepted meaning under GAAP and shall be recorded using the Agency's system of accounts.

§ 1738.3 Funding parameters.

(a) The amount of funds available for assistance, as well as the maximum and minimum Award amounts, will be published in the FEDERAL REGISTER. Applicants may apply for loans, loan/grant combinations, and loan guarantees.

(b) An Applicant that provides telecommunications or broadband service to at least 20 percent of the households in the United States is limited to an Award amount that is no more than 15 percent of the funds available to the Rural Broadband Program for the Federal fiscal year.

§§ 1738.4–1738.50 [Reserved]

Subpart B—Eligibility Requirements

§ 1738.51 Eligible entities.

(a) To be eligible for funding, an Applicant may be either a nonprofit or for-profit organization, and must take one of the following forms:

(1) Corporation;

(2) Limited liability company (LLC);

(3) Cooperative or mutual organization;

(4) Indian tribe or tribal organization as defined in 25 U.S.C. 5304; or

(5) State or local government, including any agency, subdivision, or instrumentality thereof.

(b) For loan guarantees, the underlying loan must be issued to an entity that meets the requirements in this part.

§ 1738.52 Eligible projects.

To be eligible for assistance under this part, the Applicant must:

(a) Agree to complete the build-out of the broadband system described in the application within 5 years from the day the Applicant is notified that funds are available. Under the terms of the Award documents, this 5-year period will commence from the date that the legal documents are cleared, and funds are made available to the Awardee. The application must demonstrate that all proposed construction can be completed within this 5-year period with the exception of CPE;

(b) Demonstrate an ability to provide service at the broadband lending speed to all premises in the proposed funded service area; and

(c) Provide additional equity, if necessary, to ensure financial feasibility (see § 1738.204) as determined by the Administrator.

(d) For loan guarantees, the underlying loan must be issued on a project that meets all eligibility requirements required in this part.

§ 1738.53 Eligible service area.

(a) A service area may be eligible for assistance as follows:

(1) For loan and loan/grant combinations, the proposed funded service area is completely contained within a rural area. For loan guarantee applications, the proposed funded service area must be contained within an area with a population of 50,000 or less, as defined in 7 U.S.C. 1991(a)(13);

(2) For loan/grant combinations, at least 90 percent of the households in the proposed service area must not have access to broadband service. For loans and loan guarantees, at least 50 percent of the households in the proposed service area must not have access to broadband service;

(3) No part of the proposed funded service area has three or more incumbent service providers; and

(4) No part of the proposed funded service area overlaps with the service area of current RUS borrowers or grantees with outstanding obligations. Notwithstanding, after October 1, 2020, the service areas of grantees that are providing service that is less than 10 Mbps downstream or less than 1 Mbps upstream will be considered unserved unless, at the time of the proposed application, the grantee has begun to construct broadband facilities that will meet the minimum acceptable level of service established in § 1738.55.

(b) Non-contiguous areas in the same application will be considered separate service areas and must be treated separately for the purpose of determining service area eligibility. If one or more non-contiguous areas within an application are is determined to be ineligible, the Agency may consider the remaining areas in the application for eligibility.

(c) When determining the eligibility of a proposed funded service area, the Agency will use the information submitted through the public notice response (see § 1738.106) as well as all available information collected through various means by the Agency, including but not limited to consultation with other Federal and State agencies and RUS' own site-specific assessment of the level of service in an area.

(d) Mobile and satellite services will not be considered in making the determination that households in the proposed service area do not have access to broadband service.

§ 1738.54 Eligible service area exceptions for broadband facility upgrades.

(a) Applicants upgrading existing broadband facilities in their existing service area are exempt from the requirement concerning the limit of incumbent service providers in § 1738.53(a)(3). Additionally, applicants for loans or loan guarantee funding that have received a broadband loan under Section 601 of the RE Act are exempt from the requirement concerning the number of households in § 1738.53(a)(2) without access to broadband service.

(b) Applicants submitting one application to upgrade existing broadband facilities and to expand service beyond their existing service area must segregate the upgrade and expansion into two service areas, even if the upgrade and expansion areas are contiguous. The expansion service area will not be subject to any exemptions.

(c) Applicants will be asked to remove areas determined to be ineligible from their proposed funded service area. The application will then be evaluated based on what remains if the resultant service territory is *de minimis* in change. Otherwise, the Applicant will be requested to provide additional information to the Agency relating to the ineligible areas, such as updated pro forma financials. If the Applicant fails to respond, the application may be returned.

§ 1738.55 Broadband lending speed requirements.

(a) Projects must meet the broadband build-out standards in paragraphs (a)(1) through (5) of this section in order to be considered for assistance.

(1) Projects with an Award term of less than 5 years must provide service at the broadband lending speed;

(2) Projects with an Award term of 5 to 10 years must provide service at four times the broadband lending speed;

(3) Projects with an Award term of 11 to 15 years must provide service at six times the broadband lending speed;

(4) Projects with an Award term of 16 to 20 years must provide service at eight times the broadband lending speed; and

(5) Projects with an Award term over 20 years must provide service at ten times the broadband lending speed.

(b) If an Applicant demonstrates that it would be cost prohibitive to meet the broadband lending speed in paragraph (a) of this section in the proposed funded service area due to the unique characteristics of the service territory, the Administrator may agree to utilize substitute service standards. In such cases, Applicants must document in their application why the unique characteristics of such an area make it cost prohibitive to provide service at the broadband lending speed. Note that the proof of burden on Applicants will be extremely high.

§ 1738.56 Eligible assistance purposes.

Assistance under this part may be used to pay for any of the following expenses:

(a) To fund the construction, improvement, or acquisition of all facilities required to provide service at the broadband lending speed to rural areas, including facilities required for providing other services over the same facilities.

(b) To fund the cost of leasing facilities required to provide service at the broadband lending speed if such lease qualifies as a capital/finance lease under GAAP. Notwithstanding, assistance can only be used to fund the cost of the capital/finance lease for no more than the first three years of the lease period. If an IRU qualifies as a capital/finance lease, the entire cost of the lease will be amortized over the life of the lease and only the first 3 years of the amortized cost can be funded.

(c) To fund an acquisition, provided that:

(1) The acquisition is necessary for furnishing or improving service at the broadband lending speed;

(2) The acquired service area, if any, meets the eligibility requirements set forth in § 1738.53;

(3) The acquisition cost does not exceed 50 percent of the broadband assistance; and

(4) For the acquisition of another entity, the purchase provides the Applicant with a controlling majority interest in the entity acquired.

(d) To refinance an outstanding obligation of the Applicant on another telecommunications loan made under the RE Act or on a non-RUS loan if that loan would have been for an eligible purpose under the Rural Broadband Program provided that:

(1) No more than 50 percent of the broadband assistance amount is used to refinance a non-RUS loan;

(2) The Applicant is current with its payments on the RUS telecommunications loan(s) to be refinanced; and

(3) The amortization period for that portion of the broadband loan that will be needed for refinancing will not exceed the remaining amortization period for the loan(s) to be refinanced. If multiple notes are being refinanced, an average remaining amortization period will be calculated based on the weighted dollar average of the notes being refinanced.

(e) To fund development costs in an amount not to exceed 5 percent of the total Award amount excluding amounts requested to refinance outstanding telecommunications loans. Development costs may be reimbursed only if they are incurred prior to the date on which notification of a complete application is issued (see § 1738.203) and a loan contract is entered into with RUS. Entities that meet the requirements in § 1738.101(d) may request this funding be provided as a grant. Otherwise, the funding will be provided in the form of a loan.

§ 1738.57 Ineligible assistance purposes.

Assistance under this part must not be used for any of the following purposes:

(a) To fund operating expenses of the Applicant except for eligible development costs under § 1738.56(e).

(b) To fund any costs associated with the project incurred prior to the date on which notification of a complete application is issued (see § 1738.203), except for eligible development costs under § 1738.56(e).

(c) To fund the acquisition of the stock of an affiliate.

(d) To fund the purchase or acquisition of any facilities or equipment of an affiliate.

(e) To fund the purchase of CPE and the installation of associated inside wiring, unless the CPE will be owned by the Applicant throughout its economic life.

(f) To fund the purchase or lease of any vehicle unless it is used primarily in construction or system improvements.

(g) To fund the cost of systems or facilities that have not been designed and constructed in accordance with the Award contract and other applicable requirements.

(h) To fund broadband facilities leased under the terms of an operating lease, a short-term lease, or more than 3 years of a capital/finance lease.

(i) To fund merger or consolidation of entities.

(j) To fund non-capitalized labor in accordance with 2 CFR part 200 except for eligible development costs under § 1738.56(e).

(k) To provide grant funding, a subsidized loan or payment assistance to cover the costs to refinance an outstanding loan.

§§ 1738.58–1738.100 [Reserved]

Subpart C—Award Requirements

§ 1738.101 Grant assistance.

(a) To be eligible for grant funding, the Applicant must:

(1) Submit an application for an associated loan component under Title I, Title II, or Title VI of the RE Act; and

(2) Not be the recipient of any other broadband grant from RUS with unadvanced grant funds.

(b) The amount of grant funding on any project shall not exceed:

(1) 75 percent of the total project cost when the proposed funded service area has a density of fewer than 7 people per square mile;

(2) 50 percent of the total project cost when the proposed funded service area has a density of 7 or more and fewer than 12 people per square mile; and

(3) 25 percent of the total project cost with respect to an area with a density of 12 or more and 20 or fewer people per square mile.

(c) Subsequent density determinations, as well as density requirements for projects on tribal lands will be set by notice in the FEDERAL REGISTER.

(d) The Agency may provide additional grant funding of up to 75 percent of the development costs of projects requesting funding under Title VI that serve rural areas that:

(1) Lack access to broadband service with speeds of at least 10 Mbps downstream and 1 Mbps upstream; and

(2) Meet any one of the priorities set forth in § 1738.105(a)(3)(i).

§ 1738.102 Payment assistance for loans.

(a) Grant funding may also be used to provide assistance to Title VI Awardees in the form of subsidized loans at such rates as the Agency will issue from time to time by notice in the FEDERAL REGISTER, or in the form of a payment assistance loan, which shall require no interest and principal payments or require nominal periodic payments as determined by the Agency and published in the FEDERAL REGISTER.

(b) Subsidized loans shall only be available to projects which will serve rural areas lacking access to service with speeds of at least 10 Mbps downstream and 1 Mbps upstream and meets any one of the priorities set forth in § 1738.105(a)(3)(i).

(c) The Agency may determine, at its sole discretion, to provide a payment assistance loan which shall require no interest and principal payments or such nominal payments as the Secretary determines to be appropriate. Such loans will only be provided to projects which will serve rural areas lacking access to service of speeds of 10 Mbps downstream and 1 Mbps upstream and meets any two of the priorities set forth in § 1738.105(a)(3)(i). When considering the authority to provide a payment assistance loan, the Agency will consider how such assistance will:

(1) Improve the Applicant's compliance with the commitments of the Agency's standard Award agreement, in addition to any additional requirements imposed by the Agency specific to the project;

(2) Promote the completion of the broadband project;

(3) Protect taxpayer resources; and

(4) Support the integrity of the Agency's broadband programs.

(d) The Agency and recipients of payment assistance loans must agree to specific milestones and objectives for the project which must be met, in addition to the other requirements of this part. Such terms may be amended by mutual agreement for good cause. Failure to meet the agreed upon terms, upon the Agency's determination that such failure was a direct result of the Awardee's own actions, may result in the Agency's request to the return of all, or any portion, of the grant funds used for the payment assistance loan.

(e) Additionally, Applicants with an associated loan under Title I and Title II of the RE Act and which are seeking any grant assistance under this part, are not eligible for a subsidized loan or payment assistance loans.

§ 1738.103 Substantially Underserved Trust Areas (SUTA).

Applicants seeking assistance may request consideration under the SUTA provisions in 7 U.S.C. 936f.

(a) If the Administrator determines that a community within "trust land" (as defined in 38 U.S.C. 3765) has a high need for the benefits of the Rural Broadband Program, he/she may designate the community as a "substantially underserved trust area" (as defined in section 306F of the RE Act).

(b) To receive consideration under SUTA, the Applicant must submit to the Agency a completed application that includes all of the information requested in 7 CFR part 1700, subpart D. In addition, the Applicant must notify the Agency in writing that it seeks consideration under SUTA and identify the discretionary authorities of 7 CFR part 1700, subpart D, it seeks to have applied to its application. Note, however, that the two years of historical audited financial statements and Net worth requirement for loan and loan/grant combination Applicants in § 1738.206(b)(2)(i) cannot be waived.

§ 1738.104 Technical assistance.

Projects which will serve communities that meet, at least, three of the priorities as identified in § 1738.105(a)(3)(i) may request technical assistance and training from the Agency to:

(a) Prepare reports and surveys necessary to request grants, loans, and loan guarantees for broadband deployment;

(b) Improve management, including financial management, relating to the proposed broadband deployment;

(c) Prepare applications for grants, loans, and loan guarantees; and

(d) Assist with other areas of need as identified by the Agency through a notice in the FEDERAL REGISTER.

§ 1738.105 Priorities for approving assistance.

(a) The Agency will compare and evaluate all applications for assistance and shall give priority to applications in the manner set out in paragraphs (a)(1) through (4) of this section, which shall be scored as outlined in a notice published in the FEDERAL REGISTER. (Note that for applications containing multiple proposed funded service areas, the percentage will be calculated combining all proposed funded service areas.)

(1) Applicant's providing broadband service to rural areas that do not have access to service of at least 10 Mbps upstream and 1 Mbps downstream.

(2) Projects that provide the maximum level of broadband service to the greatest proportion of rural households.

(3) Projects that:

(i) Serve rural areas:

(A) With a population of less than 10,000 permanent residents;

(B) Are experiencing outmigration and have adopted a strategic community investment plan under section 379H(d) of the Consolidated Farm and Rural Development Act (7 U.S.C. 2008v) that includes considerations for improving and expanding broadband service;

(C) With a high percentage of low-income families or persons (as defined in section 501(b) of the Housing Act of 1949 (42 U.S.C. 1471(b)));

(D) That are isolated from other significant population centers; or

(E) That provide rapid and expanded deployment of fixed and mobile broadband on cropland and ranchland within a service territory for use in various applications of precision agriculture; and

(ii) Were developed with the participation of, and will receive a substantial portion of the funding for the project from two or more stakeholders, including:

(A) State, local, and tribal governments;

(B) Nonprofit institutions; and

(C) Community anchor institutions, such public libraries, schools, institutions of higher education, health care facilities, private entities, philanthropic organizations and cooperatives.

(4) New construction projects requesting no refinancing.

(b) The Agency may assign special consideration priority points that will be issued in a notice in the FEDERAL REGISTER with respect to any funding opportunity.

(c) With respect to two or more applications that have the same priority, as outlined in paragraphs (a) and (b) of this section, the Agency shall give priority to the application that requests the least amount of grant funding as calculated based on the total amount of grant funds requested.

§ 1738.106 Public notice.

(a) The Agency will publish a public notice of each application requesting assistance under this part. The application must provide a summary of the information required for such public notice including all of the following information:

(1) The identity of the Applicant;

(2) A map of each proposed funded service area showing the rural area boundaries and the areas without broadband service using the Agency's mapping tool;

(3) The amount and type of support requested;

(4) The estimated number of households in each proposed funded service area without broadband service, excluding mobile and satellite service; and

(5) A description of all the types of services that the Applicant proposes to offer in each proposed funded service area.

(b) The public notice will remain available for 45 calendar days on the Agency's website, and will request existing service providers to submit to the Agency, within the same period, the following information:

(1) The number of residential and business customers within the Applicant's proposed funded service area that are currently offered, and that are purchasing, broadband service by the existing service provider, and the cost of each level of broadband service charged by the existing service provider;

(2) The number of residential and business customers within the Applicant's proposed funded service area that receive non-broadband services from the existing service provider, and the associated rates for these other services; and

(3) A map showing where the existing service provider's services coincide with the Applicant's proposed funded service area using the Agency's mapping tool.

(c) For purposes of 5 U.S.C. 552, information received from existing service providers under paragraph (b) of this section shall be exempt from disclosure.

(d) If an application is approved, an additional notice will be published on the Agency's website that will include the following information:

(1) The name of the entity receiving the financial assistance;

(2) The amount and type of assistance being received;

(3) The purpose of the assistance; and

(4) Each annual report submitted under § 1738.107, redacted as appropriate to protect any proprietary information in the report.

§ 1738.107 Additional reporting requirements for Awardees.

(a) Entities receiving assistance from the USDA to provide retail broadband service must submit annual reports for 3 years after project completion. The reports must include the following information:

(1) The purpose of the financing, including new equipment and capacity enhancements that support high-speed broadband access for educational institutions, health care providers, and public safety service providers (including the estimated number of end users who are currently using or forecasted to use the new or upgraded infrastructure); and

(2) The progress towards fulfilling the objectives for which the assistance was granted, including:

(i) The number of service points that will receive new broadband service, existing network service improvements, and facility upgrades resulting from the Federal assistance;

(ii) The speed of broadband services;

(iii) The average price of the most subscribed tier of broadband service in each proposed service area; and

(iv) The number of new subscribers generated from the project.

(b) Awardees must provide complete, reliable, and precise geolocation information that indicates the location of new broadband service that is being provided or upgraded within the service territory supported by the assistance no later than 30 days after the earlier of the date of:

(1) Completion of the project milestone established in the applicable assistance contract; or

(2) Project completion.

(c) Any other reporting requirements established by the Administrator by notice in the FEDERAL REGISTER before an application is submitted.

§ 1738.108 Environmental reviews.

(a) Federal agencies are required to analyze the potential environmental impacts, as required by the National Environmental Policy Act (NEPA) and the National Historic Preservation Act (NHPA) for Applicant projects or proposals seeking funding. Please refer to 7 CFR part 1970 for all of Rural Development's environmental policies. All Applicants are required to provide environmental review documents, provide a description of program activities, and to submit all other required environmental documentation as requested in the application system or by the Agency after the application is submitted. It is the Applicant's responsibility to obtain all necessary Federal, tribal, State, and local governmental permits and approvals necessary for the proposed work to be conducted. Applicants are expected to design their projects so that they minimize the potential for adverse impacts to the environment. Applicants also will be required to cooperate with the granting agencies in identifying feasible measures to reduce or avoid any identified adverse environmental impacts of their proposed projects. The failure to do so may be grounds for not making an Award.

(b) The Agency may obligate, but not disperse, funds under Title VI of the Rural Electrification Act of 1936, before the completion of the otherwise required environmental historical, or other types of reviews if the Secretary determines that subsequent site-specific review shall be adequate and easily accomplished for the location of towers, poles, or other broadband facilities in the service area of the awardee without compromising the project or the required reviews.

§ 1738.109 Civil rights procedures and requirements.

(a) *Equal opportunity and non-discrimination.* The agency will ensure that equal opportunity and non-discriminatory requirements are met in accordance with the Equal Credit Opportunity Act and 7 CFR part 15. In accordance with Federal civil rights law and USDA civil rights regulations and policies, the USDA, its agencies, offices, and employees, and institutions participating in or administering USDA programs are prohibited from discriminating based on race, color, national origin, religion, sex, gender identity (including gender expression), sexual orientation, disability, age, marital status, family/parental status, income derived from a public assistance program, political beliefs, or reprisal or retaliation for prior civil rights activity, in any program or activity conducted or funded by USDA (not all bases apply to all programs).

(b) *Civil rights compliance.* Recipients of Federal assistance under this part must comply with the Americans with Disabilities Act of 1990, Title VI of the Civil Rights Act of 1964, and Section 504 of the Rehabilitation Act of 1973. In general, recipients should have available for the Agency racial and ethnic data showing the extent to which members of minority groups are beneficiaries of federally assisted programs. The Agency will conduct compliance reviews in accordance with 7 CFR part 15. Awardees will be required to complete Form RD 400–4, "Assurance Agreement," for each Federal Award received.

(c) *Discrimination complaints.* Persons believing they have been subjected to discrimination prohibited by this section may file a complaint personally or by an authorized representative with USDA, Director, Office of Adjudication, 1400 Independence Avenue SW, Washington, DC 20250. A complaint must be filed no later than 180 days from the date of the alleged discrimination, unless the time for filing is extended by the designated officials of USDA or the Agency.

§§ 1738.110–1738.150 [Reserved]

Subpart D—Loan and Loan/Grant Combination Award Terms

§ 1738.151 General.

Direct loans shall be in the form of a cost-of-money loan except as detailed in § 1738.152.

§ 1738.152 Interest rates.

(a) Direct cost-of-money loans shall bear interest at a rate equal to the cost of borrowing to the Department of Treasury for obligations of comparable maturity unless the project qualifies for a reduced interest rate as detailed in § 1738.102. The applicable interest rate will be set at the time of each advance.

(b) The interest rate for Applicants receiving payment assistance or Substantially Underserved Trust Areas (SUTA) consideration will be set at the time of the Award.

§ 1738.153 Terms and conditions.

Terms and conditions of the loan and loan/grant combinations are set forth in a mortgage, note, and loan contract. Samples of the mortgage, note, and loan contract can be found on the Agency's website.

(a) Unless requested to be shorter by the Applicant, loans must be repaid with interest within a period that, rounded to the nearest whole year, is equal to the expected composite economic life of the assets to be financed, as determined by the Agency based upon acceptable depreciation rates. Expected composite economic life means the weighted average economic life of all classes of facilities necessary to complete construction of the broadband facilities plus 3 years.

(b) Principal payments for each advance are amortized over the remaining term of the loan and are due monthly. Principal payments will be deferred until 3 years after the date of the first advance of loan funds. Interest begins accruing when the first advance of loan funding is made and interest payments are due monthly, with no deferral period.

(c) Awardees are required to carry fidelity bond coverage. Generally, this amount will be 15 percent of the loan or loan/grant combination Award amount, not to exceed $5 million. The Agency may reduce the percentage required if it determines that the amount is not commensurate with the risk involved.

§ 1738.154 Security.

(a) The broadband loan or loan/grant combination must be secured by the assets purchased with the loan or loan/grant combination funds, as well as all other assets of the Applicant and any other cosigner of the Award documents except as allowed under section 601(h)(2) of the RE Act. With respect to loan/grant combinations, all grant assets must also be covered by a security interest in favor of the Government for the average composite economic life of all project assets financed with assistance, regardless of whether the loan is paid off before the maturity date. Additionally, the sale of all such grant assets shall be governed by 2 CFR part

200, regardless of the entity type of the Awardee.

(b) The Agency must be given an exclusive first lien, in form and substance satisfactory to the Agency, on all of the Applicant's property and revenues and such additional security as the Agency may require. The Agency may share its first lien position with another lender on a *pari passu*, prorated basis if security arrangements are acceptable to the Agency.

(c) Unless otherwise designated by the Agency, all property purchased with loan and loan/grant combination funds must be owned by the Applicant.

(d) In the case of loan and loan/grant combinations that include financing of facilities that do not constitute self-contained operating systems, the Applicant shall furnish assurance, satisfactory to the Agency, that continuous and efficient service that meets the broadband build-out requirements as noted in § 1738.55 will be rendered.

(e) The Agency will require adequate financial, investment, operational, reporting, and managerial controls in the Award documents.

§ 1738.155 Advance of funds.

RUS loan and grant advances are made at the request of the Awardee according to the procedures stipulated in the Award documents. For loan and loan/grant combination Awards, all non-RUS funds must be expended first, followed by loan funds and then grant funds, except for RUS approved development costs. Grant funds for eligible development costs, if any, will be used only on the first advance request.

§ 1738.156 Buy American requirement.

Awardees shall use in connection with the expenditure of loan and grant funds only such unmanufactured articles, materials, and supplies, as have been mined or produced in the United States or in any eligible country, and only such manufactured articles, materials, and supplies as have been manufactured in the United States or in any eligible country, substantially all from articles, materials, or supplies mined, produced, or manufactured, as the case may be, in the United States or in any eligible country. For purposes of this section, an "eligible country" is any country that applies with respect to the United States an agreement ensuring reciprocal access for United States products and services and United States suppliers to the markets of that country, as determined by the United States Trade Representative. The Buy American regulations may be found at, and any requests for waiver must be submitted pursuant to, 7 CFR part 1787.

§§ 1738.157–1739.200 [Reserved]

Subpart E—Loan and Loan/Grant Combination Application Review and Underwriting

§ 1738.201 Application submission.

(a) Loan and loan/grant combination applications must be submitted through the Agency's online application system.

(b) The Agency may publish additional application submission requirements in the FEDERAL REGISTER.

§ 1738.202 Elements of a complete application.

(a) *Online application system*. Loan and loan/grant combination applications must be submitted through RUS' online application system and include all information as required by that system and detailed in the Rural Broadband Program Application Guide (the Application Guide), available on the Agency's website, so that applications can be uniformly evaluated and compared.

(b) *DUNS registration*. All Applicants must register for a Dun and Bradstreet Universal Numbering System (DUNS) number as part of the application. The Applicant can obtain the DUNS number free of charge by calling Dun and Bradstreet. Go to *http://fedgov.dnb.com/webform* for more information on assignment of a DUNS number or confirmation.

(c) *SAM registration*. Prior to submitting an application, all Applicants requesting loan/grant combination funds must register in the System for Award Management (SAM) at *https://www.sam.gov/SAM/* and supply a Commercial and Government Entity (CAGE) code number as part of the application. SAM registration must be active with current data at all times, from the application review throughout the active Federal Award funding period. To maintain active SAM registration, the Applicant must review and update the information in the SAM database annually from the date of initial registration or from the date of the last update. The Applicant must ensure that the information in the database is current, accurate, and complete.

(d) *Contents of the application.* A complete application will include the following information as requested in the RUS online application system and Application Guide:

(1) General information on the Applicant and the project including:

(i) A description of the project that will be made public consistent with the requirements in this part; and

(ii) The estimated dollar amount of the funding request.

(2) An executive summary of the proposed project. The summary shall include, but not be limited to, a detailed description of the existing operations, discussion of key management, description of the workforce and a detailed description of the proposed project.

(3) A description of the proposed funded service area including the number of premises passed.

(4) Subscriber projections including the number of subscribers for broadband, video and voice services and any other service that may be offered. A description of the proposed service offerings and the associated pricing plan that the Applicant proposes to offer, and an explanation showing that the proposed service offerings are affordable.

(5) A map, utilizing the RUS mapping tool, of the proposed funded service areas identifying the areas lacking access to broadband service and the areas lacking access to service of speeds of at least 10 Mbps downstream and 1 Mbps upstream and any non-funded service areas of the Applicant

(6) A competitive analysis of the entire proposed service territory(ies) as required by § 1738.205.

(7) A network design which includes a description of the proposed technology used to deliver service at the required broadband lending speed (see § 1738.55) to all premises in the proposed funded service area, a network diagram, a build-out timeline and milestones for implementation of the project, and a capital investment schedule showing that the system can be built within 5 years from the date funds are made available to the Awardee. All of which must be certified by a professional engineer who is certified in at least one of the states where the project is to be constructed. The certification from the professional engineer must clearly state that the proposed network can deliver service at the required broadband lending speed (see § 1738.55) to all premises in the proposed funded service area.

(8) All environmental information as required by § 1738.108.

(9) Resumes of key management personnel, a description of the organization's readiness to manage a broadband services network, and an organizational chart showing all parent organizations and/or holding companies (including parents of parents, etc.) and all subsidiaries and affiliates.

(10) A legal opinion that addresses the Applicant's ability to enter into loan or loan/grant combination as requested in the application for financial assistance, to pledge security as required by the Agency, to describe all pending litigation matters, and such other requirements as are detailed in the Application Guide.

(11) A summary and itemized budgets of the infrastructure costs of the proposed project, including if applicable, the ratio of loans to grants, and any other sources of outside funding.

(12) A detailed description of working capital requirements and the sources of those funds.

(13) Complete copies of audited financial statements for the two years preceding the application submission as detailed in § 1738.206.

(14) The historical and projected financial information required in § 1738.206.

(15) Documentation proving that all required licenses and regulatory approvals for the proposed operation have been obtained, or the status of obtaining such licenses or approvals.

(16) If service is being proposed on tribal land, a certification from the proper tribal official that they are in

support of the project and will allow construction to take place on tribal land. The certification must:

(i) Include a description of the land proposed for use as part of the proposed project;

(ii) Identify whether the land is owned, held in Trust, land held in fee simple by the Tribe, or land under a long-term lease by the Tribe;

(iii) If owned, identify the land owner; and

(iv) Provide a commitment in writing from the land owner authorizing the Applicant's use of that land for the proposed project.

(17) Scoring sheet, analyzing the scoring criteria set forth in this part and most recent funding opportunity announcement.

(18) Additional items that may be required by the Administrator through a notice in the FEDERAL REGISTER.

(e) *Material representations.* The application, including certifications and all forms submitted as part of the application, will be treated as material representations upon which RUS will rely in awarding loans and loan/grant combinations.

§ 1738.203 Notification of completeness.

If all proposed funded service areas in a loan or loan/grant combination application are eligible, the Agency will review the application for completeness. The completeness review will include an assessment of whether all required documents and information have been submitted and whether the information provided is of adequate quality to allow further analysis.

(a) If the application contains all documents and information required by this part and is sufficient, in form and substance acceptable to the Agency, the Agency will notify the Applicant, in writing, that the application is complete. A notification of completeness is not a commitment that assistance will be approved. By submitting an application, the Applicant acknowledges that no obligation to enter into an agreement exists until the actual Award documents have been executed.

(b) If the application is considered to be incomplete or inadequate, the Agency will notify the Applicant, in writing, with detailed information regarding the reasons the applications was found to be incomplete or inadequate.

§ 1738.204 Evaluation for feasibility.

After a loan or loan/grant combination Applicant is notified that the application is complete, the Agency will evaluate the application's financial and technical feasibility. Only applications that, as determined by the Agency, are technically and financially feasible will be considered for funding.

(a) The Agency will determine financial feasibility by evaluating the impact of the facilities financed with the proceeds of the loan and the associated debt, the Applicant's equity, competitive analysis, financial information—including the Applicant's ability to meet the Agency's Net worth and TIER, DSCR, or CR requirements in § 1738.206(b)(2)(i)—and other relevant information in the application.

(b) The Agency will determine technical feasibility by evaluating the Applicant's network design and other relevant information in the application.

§ 1738.205 Competitive analysis.

The Applicant must submit a competitive market analysis for each service area regardless of projected penetration rates. Each analysis must identify all existing service providers and all resellers in each service area regardless of the provider's market share, for each type of service the Applicant proposes to provide. The analysis must compare the rates, services, and the quality of that service being offered by competitors against those that will be offered by the Applicant. The analysis must also discuss strategies the Applicant will use to compete, as well as the impacts of the competitors on the projected penetration rates for the project.

§ 1738.206 Financial information.

(a) The Applicant must submit financial information acceptable to the Agency that demonstrates that the Applicant has the financial capacity to fulfill the loan or loan/grant combination requirements in this part and to successfully complete the proposed project.

(1) Applicants must provide complete copies of audited financial statements (opinion letter, balance sheet, income statement, statement of changes in financial position, and notes to the financial statement) for the two years preceding the application submission. For governmental entities financial statements must be accompanied with certifications identifying unrestricted cash that will be available on a yearly basis to the Applicant. Subsidiary operations formed from existing utility providers may provide audited financial statements for the two previous years from the parent company, as long as the parent will be a cosigner of the loan or loan/grant documents, pledging its assets in accordance with §1738.154(a), or will guarantee the debt.

(2) If the Applicant relies on services provided by a parent or affiliated operation, it must also provide complete copies of audited financial statements for those entities for the fiscal year preceding the application submission. If audited statements are not available, unaudited statements and tax returns for the previous year must be submitted.

(3) Applicants must provide detailed information for all outstanding obligations. Copies of existing notes, loan agreements, security agreement, or other legal documents covering loans, grants, leases, or other loan guarantees must be included in the application.

(4) Applicants must provide a detailed description of working capital requirements and the source of these funds, if internally generated funds are insufficient.

(b) Applicants must submit the following documents that demonstrate the proposed project's financial viability and ability to repay the requested loan.

(1) Customer projections for the 5-year forecast period that substantiate the projected revenues for each service that is to be provided. The projections must be provided on at least an annual basis and must be developed separately for each service area and must be clearly supported by evidence such as market surveys or current company take rates.

(2) Pro forma financial forecast, including a balance sheet, income statement, and statement of cash flows. For non-regulated entities, the pro forma should be prepared in conformity with U.S. GAAP and the Agency's guidance on grant accounting found at *https://www.rd.usda.gov/files/AccountingGuidance10.pdf*. Regulated telecommunications providers may follow the USOA and RUS accounting standards for their pro forma, including accounting for grant-funded assets as a contribution, in accordance with 47 CFR 32.2, if the project assets will be treated as regulated plant. The pro forma should validate the sustainability of the project by including subscriber estimates related to all proposed service offerings; annual financial projections with balance sheets, income statements, and cash flow statements; supporting assumptions for a 5-year forecast period and a depreciation schedule for existing facilities and those funded with Federal assistance, and other funds. This pro forma should indicate the committed sources of capital funding and include a bridge year prior to the start of the forecast period. This bridge year is the year in which the application is submitted and serves as a buffer between the historical financial information and the forecast period. Including the bridge year, the pro forma statements span a 6-year period.

(i) The financial projections submitted by Applicants must demonstrate that their entire operation will be able to meet two of the following three ratio requirements: A minimum TIER, CR, or DSCR equal to 1.25 by the end of the 5-year forecast period. Additionally, the projections must demonstrate the Applicant's ability to maintain a Net worth of at least 20 percent throughout the forecast period. Demonstrating that the operation can achieve a projected Net worth of 20 percent and TIER, CR, or DSCR of 1.25 does not ensure that the Agency will approve the loan or loan/grant combination.

(ii) If the financial analysis suggests that the operation will not be able to achieve the Net worth requirement or two of the required TIER, CR, or DSCR in paragraph (b)(2)(i) of this section, the Agency will not approve the loan or loan/grant combination Award without additional capital, additional cash, additional security, and/or a change in the Award terms.

(c) Based on the financial evaluation, the Award documents will specify the Net worth and TIER, CR, or DSCR requirements in paragraph (b)(2)(i) of this section that must be met throughout the amortization period.

§ 1738.207 Network design.

(a) Applications for loan or loan/grant combinations must include a network design that demonstrates the project's technical feasibility. The network design must fully support the delivery of service to meet the broadband build-out requirements specified in §1738.55, together with any other services to be provided. In measuring speed, the Agency will take into account industry and regulatory standards. The design must demonstrate that the project will be complete within the 5-year forecast period and must include the following items:

(1) A detailed description of the proposed technology that will be used to provide service at the broadband lending speed. This description must clearly demonstrate that all premises in the proposed funded service area will be able to receive service at the broadband lending speed;

(2) A detailed description of the existing network. This description should provide a synopsis of the current network infrastructure;

(3) A detailed description of the proposed network. This description should provide a synopsis of the proposed network infrastructure;

(4) A description of the approach and methodology for monitoring ongoing service delivery and service quality for the services being deployed;

(5) Estimated project costs detailing all facilities that are required to complete the project. These estimated costs must be broken down to indicate costs associated with each proposed service area and must specify how Agency and non-Agency funds will be used to complete the project;

(6) A construction build-out schedule of the proposed facilities by service area on an annual basis. The build-out schedule must include:

(i) A description of the workforce that will be required to complete the proposed construction;

(ii) A timeline demonstrating project completion within the forecast period; and

(iii) Detailed information showing that all premises within the proposed funded service area will be offered service at the broadband lending speed when the system is complete;

(7) A depreciation schedule for all facilities financed with loan and loan/grant combination funds;

(8) An environmental report prepared in accordance with 7 CFR part 1970; and

(9) Any other system requirements required by the Administrator through a notice published in the FEDERAL REGISTER.

(b) The network design must be prepared by a registered Professional Engineer with telecommunications experience who is certified in at least one of the states where a project is to be constructed or by qualified personnel on the Applicant's staff. If the network design is prepared by the Applicant's staff, the application must clearly demonstrate the staff's qualifications, experience, and ability to complete the network design. To be considered qualified, staff must have at least 3 years of experience in designing the type of broadband system proposed in the application.

§ 1738.208 Award determinations.

(a) If the loan or loan/grant combination application meets all statutory and regulatory requirements and the feasibility study demonstrates that the Net worth and TIER, CR, or DSCR requirements in §1738.206(b)(2)(i) can be satisfied and the business plan is sustainable, the application will be submitted to the Agency's credit committees for consideration according to the priorities in §1738.105. Such submission of an application to the Agency's credit committees does not guarantee that a loan or loan/grant combination will be approved. In making a loan and/or loan/grant combination Award determination, the Administrator shall consider the recommendations of the credit committees.

(b) The Applicant will be notified of the Agency's decision in writing. If the

Agency does not approve the loan or loan/grant combination, a rejection letter will be sent to the Applicant, and the application will be returned with an explanation of the reasons for the rejection.

§§ 1738.209–1738.250 [Reserved]

Subpart F—Closing, Servicing, and Reporting for Loan and Loan/ Grant Combination Awards

§ 1738.251 Offer and closing.

The Agency will notify the Applicant of the loan or loan/grant combination offer in writing, and the date by which the Applicant must accept the offer. If the Applicant accepts the terms of the offer, a loan or loan/grant combination contract, note, security agreement, and any other necessary documents will be executed by the Agency and sent to the Applicant. The Applicant must execute the Award documents and satisfy all conditions precedent to closing within the timeframe specified by the Agency. If the conditions are not met within this timeframe, the loan or loan/grant combination offer may be terminated, unless the Applicant requests and the Agency approves, an extension. The Agency may approve such a request if the Applicant has diligently sought to meet the conditions required for closing and has been unable to do so for reasons outside its control.

§ 1738.252 Construction.

(a) Construction paid for with loan or loan/grant combination funds must comply with 7 CFR parts 1787, 1788, and 1970, the RUS Broadband Construction Procedures located at *https://reconnect.usda.gov,* and any other guidance from the Agency.

(b) Once the Agency has extended a loan or loan/grant combination offer, the Applicant, at its own risk, may start construction that is included in the application on an interim financing basis. For this construction to be eligible for reimbursement with loan or loan/grant combination funds, all construction procedures contained in this part must be followed. Note, however, that the Agency's extension of a loan or loan/grant combination offer is not a guarantee that a loan or loan/grant combination will be made, unless and until a contract has been entered into between the Applicant and RUS.

(c) All Awardees must complete build-out within 5 years from the date that funds have been made available. Build-out is considered complete when the network design has been fully implemented, the service operations and management systems infrastructure is operational, and the awardee is ready to support the activation and commissioning of individual customers to the new system.

§ 1738.253 Servicing of loan and loan/ grant combinations.

(a) Borrowers must make payments on the broadband loan as required in the note.

(b) Awardees must comply with all terms, conditions, affirmative covenants, and negative covenants contained in the Award documents.

(c) In the event of default of any required payment or other term or condition:

(1) The Agency may exercise the default remedies provided in the Award documents and any remedy permitted by law but is not required to do so.

(2) If the Agency chooses not to exercise its default remedies, it does not waive its right to do so in the future.

§ 1738.254 Accounting, reporting, and monitoring requirements.

(a) Loan and loan/grant combination Awardees must adopt a system of accounts for maintaining financial records acceptable to the Agency, as described in 7 CFR part 1770, subpart B.

(b) Loan and loan/grant combination Awardees must submit annual audited financial statements along with a report on compliance and on internal control over financial reporting and management letter in accordance with the requirements of 7 CFR part 1773. The Certified Public Accountant (CPA) conducting the annual audit is selected by the awardee and must be approved by RUS as set forth in 7 CFR 1773.4.

(c) Loan and loan/grant combination Awardees must submit to RUS 30 calendar days after the end of each calendar year quarter, balance sheets, income statements, statements of cash flow, rate package summaries, and the number of customers subscribing to broadband service from the Awardee utilizing RUS' online reporting system. These reports must be submitted throughout the loan amortization period.

(d) Loan and loan/grant combination Awardees must submit annually updated service area maps through the RUS mapping tool showing the areas where construction has been completed and premises are receiving service until the entire proposed funded service area can receive service at the broadband lending speed. At the end of the project, Awardees must submit a service area map indicating that all construction has been completed as proposed in the application. If parts of the proposed funded service area have not been constructed, RUS may require a portion of the Award to be rescinded and/or paid back.

(e) Loan and loan/grant combination Awardees must comply with all reasonable Agency requests to support ongoing monitoring efforts. The Awardee shall afford RUS, through its representatives, reasonable opportunity, at all times during business hours and upon prior notice, to have access to and the right to inspect the broadband system, and any other property encumbered by the mortgage or security agreement, and any or all books, records, accounts, invoices, contracts, leases, payrolls, timesheets, cancelled checks, statements, and other documents, electronic or paper of every kind belonging to or in the possession of the Awardee or in any way pertaining to its property or business, including its subsidiaries, if any, and to make copies or extracts therefore.

(f) Awardee records shall be retained and preserved in accordance with the provisions of 7 CFR part 1770, subpart A.

§ 1738.255 Default and deobligation.

If a default under the loan or loan/grant combination documents occurs and such default has not been cured within the timeframes established in the Award documents, the Applicant acknowledges that the Agency may, depending on the seriousness of the default, take any of the following actions:

(a) To the greatest extent possible recover the maximum amount of grant and loan funds;

(b) De-obligate all funds that have not been advanced or demonstrate an insufficient level of performance or fraudulent spending; and

(c) Reallocate recovered funds to the extent possible.

§§ 1738.256–1738.300 [Reserved]

Subpart G—Loan Guarantee

§ 1738.301 General.

(a) To be eligible for a loan guarantee, the Applicant must submit an application that meets the requirements in this part along with the requirements as stated in 7 CFR part 4279, subparts A and B, as well as any additional requirements published in the FEDERAL REGISTER.

(b) The Agency may approve Rural Broadband Program loan guarantees in excess of $10 million but less than $25 million when the project meets one of the priorities in § 1738.105(a)(3)(i).

(c) The lender will service the loan in accordance with 7 CFR part 4287, subpart B.

(d) Any reference to priorities in 7 CFR part 4279 or 4287 shall have the meaning as stated in § 1738.105 and any reference to Administrator or Agency shall have the meaning as defined in § 1738.2.

§ 1738.302 Fees.

The Agency shall charge and collect from the lender fees in such amounts as to bring down the costs of subsidies for guaranteed loans, except that such fees shall not act as a bar to participation in the programs nor be inconsistent with current practices in the marketplace.

§§ 1738.303–1738.349 [Reserved]

§ 1738.350 OMB control number.

The information collection requirements in this part are approved by the Office of Management and Budget (OMB) and assigned OMB control number 0572–0154.

PART 1739—BROADBAND GRANT PROGRAM

Subpart A—Community Connect Grant Program

AUTHORITY: Title III, Pub. L. 108–199, 118 Stat. 3.

SOURCE: 78 FR 25791, June 3, 2013, unless otherwise noted.

Subpart A—Community Connect Grant Program

§ 1739.1 Purpose.

(a) The provision of broadband service is vital to the economic development, education, health, and safety of rural Americans. The purpose of the Community Connect Grant Program is to provide financial assistance in the form of grants to eligible applicants that will provide, on a "community-oriented connectivity" basis, broadband service that fosters economic growth and delivers enhanced educational, health care, and public safety benefits. The Agency will give priority to rural areas that have the greatest need for broadband services, based on the criteria contained herein and in the RUS Community Connect Program application guide and/or the Community Connect Program website and in the funding opportunity announcement (FOA) posted on *www.Grants.gov*.

(b) Grant authority will be used for the deployment of service to all premises in eligible rural areas at the Broadband Grant Speed on a "community-oriented connectivity" basis. In addition to providing service to all premises the "community-oriented connectivity" concept will stimulate practical, everyday uses and applications of broadband by cultivating the deployment of new broadband services that improve economic development and provide enhanced educational and health care opportunities in rural areas. Such an approach will also give rural communities the opportunity to benefit from the advanced technologies that are necessary to achieve these goals.

[78 FR 25791, June 3, 2013, as amended at 83 FR 45033, Sept. 5, 2018]

§ 1739.2 Funding availability and application dates and submission.

(a) The Agency will post a FOA on *www.Grants.gov* that will set forth the total amount of funding available; the maximum and minimum funding for each grant; funding priority; the application submission dates; and the appropriate addresses and agency contact information. The FOA will also outline and explain the procedures for submission of applications, including electronic submissions. The Agency may publish more than one FOA should additional funding become available. This information will also be made available in the RUS Community Connect Grant program application guide and on the RUS Community Connect Grant program website.

(b) Notwithstanding paragraph (a) of this section, the Agency may, in response to a surplus of qualified eligible applications which could not be funded from the previous fiscal year, decline to post a FOA for the following fiscal year and fund said applications without further public notice.

[83 FR 45033, Sept. 5, 2018]

§ 1739.3 Definitions.

As used in this subpart:

Agency or *RUS* shall mean the Rural Utilities Service, which administers the United States Department of Agriculture (USDA) Rural Development Utilities Programs.

Broadband Grant Speed means the minimum bandwidth described in the funding opportunity that an applicant must propose to deliver to every customer in the proposed funded service area in order for the Agency to approve a broadband grant. The Broadband Grant Speed may be different for fixed and mobile broadband services and from the minimum rate of data transmission required to determine the availability of broadband service when qualifying a service area.

Broadband service means any terrestrial technology having the capacity to provide transmission facilities that enable subscribers of the service to originate and receive high-quality voice, data, graphics, and video at the minimum rate of data transmission described in the funding opportunity. Satellite and mobile services are not considered broadband service. The broadband service speed may be different from the broadband grant speed for the Community Connect program.

Community Center means a building within the Proposed Funded Service Area that provides access to the public, or a section of a public building with at least two (2) Computer Access Points and wireless access, that is used for the purposes of providing free access to and/or instruction in the use of broadband Internet service, and is of the appropriate size to accommodate this purpose. The community center must be open and accessible to area residents before, during, and after normal working hours and on Saturdays or Sunday.

Computer Access Point means a new computer terminal with access to service at the Broadband Grant Speed.

Critical Community Facilities means an essential community facility as defined pursuant to section 306(a) of the Consolidated Farm and Rural Development Act (7 U.S.C. 1926(a)).

Eligible applicant shall have the meaning as set forth in § 1739.10.

Eligible grant purposes shall have the meaning as set forth in § 1739.12.

Funding opportunity announcement (FOA) means a publicly available document by which a Federal agency makes know its intentions to award discretionary grants or cooperative agreements, usually as a result of competition for funds. FOA announcements may be known as program announcements, notices of funding availability, solicitations, or other names depending on the agency and type of program. FOA announcements can be found at *www.Grants.gov* in the Search Grants tab and on the funding agency's or program's website.

Matching contribution means the applicant's qualified contribution to the Project, as outlined in § 1739.14 of this part.

Project means the delivery of service at the Broadband Grant Speed financed by the grant and Matching Contribution for the Proposed Funded Service Area.

Proposed Funded Service Area (PFSA) means the contiguous geographic area within an eligible Rural Area or eligible Rural Areas, in which the applicant proposes to provide service at the Broadband Grant Speed.

Rural area means any area, as confirmed by the most recent decennial Census of the United States (decennial Census), which is not located within:

(1) A city, town, or incorporated area that has a population of greater than 20,000 inhabitants; or

(2) An urbanized area contiguous and adjacent to a city or town that has a population of greater than 50,000 inhabitants. For purposes of the definition of rural area, an urbanized area means a densely populated territory as defined in the most recent decennial Census.

[78 FR 25791, June 3, 2013, as amended at 80 FR 9862, Feb. 24, 2015; 83 FR 45033, Sept. 5, 2018; 85 FR 14408, Mar. 12, 2020]

§§ 1739.4–1739.7 [Reserved]

§ 1739.8 Buy American requirement.

Awardees shall use in connection with the expenditure of grant funds only such unmanufactured articles, materials, and supplies, as have been mined or produced in the United States or in any eligible country, and only such manufactured articles, materials, and supplies as have been manufactured in the United States or in any eligible country, substantially all from articles, materials, or supplies mined, produced, or manufactured, as the case may be, in the United States or in any eligible country. For purposes of this section, an "eligible country" is any country that applies with respect to the United States an agreement ensuring reciprocal access for United States products and services and United States suppliers to the markets of that country, as determined by the United States Trade Representative. The Buy American regulations may be found at, and any requests for waiver must be submitted pursuant to, 7 CFR part 1787.

[85 FR 14408, Mar. 12, 2020]

§ 1735.9 USDA Rural Development State Director notification.

Applicants shall complete a notification form which will be a public document that the RUS provides to USDA Rural Development State Directors and others in the state(s) of the PFSA. The notification shall include a brief project description and the location of the PFSA.

§ 1739.10 Eligible applicant.

To be eligible for a Community Connect competitive grant, the applicant must:

(a) Be legally organized as an incorporated organization, an Indian tribe or tribal organization, as defined in 25 U.S.C. 450b(e), a state or local unit of government, or other legal entity, including cooperatives or private corporations or limited liability companies organized on a for-profit or not-for-profit basis.

(b) Have the legal capacity and authority to own and operate the broadband facilities as proposed in its application, to enter into contracts and to otherwise comply with applicable federal statutes and regulations.

(c) As required by the Office of Management and Budget (OMB), all applicants for grants must supply a Dun and Bradstreet Data Universal Numbering System (DUNS) number when applying. The Standard Form 424 (SF–424) contains a field for you to use when supplying your DUNS number. Obtaining a DUNS number costs nothing and requires a short telephone call to Dun and Bradstreet. Please see *http://www.grants.gov/applicants/request_duns_number.jsp* for more information on how to obtain a DUNS number or how to verify your organization's number.

(d) Register in the System for Award Management (SAM) (formerly Central Contractor Registry (CCR)).

(1) In accordance with 2 CFR part 25, applicants, whether applying electronically or by paper, must be registered in the SAM prior to submitting an application. Applicants may register for the SAM at *https://www.sam.gov/*.

(2) The SAM registration must remain active, with current information, at all times during which an entity has an application under consideration by an agency or has an active Federal Award. To remain registered in the SAM database after the initial registration, the applicant is required to review and update, on an annual basis from the date of initial registration or subsequent updates, its information in the SAM database to ensure it is current, accurate and complete.

§ 1739.11 Eligible Community Connect Competitive Grant Project.

To be eligible for a Community Connect competitive grant, the Project must:

(a) Serve a PFSA in which Broadband Service does not currently exist;

(b) Offer service at the Broadband Grant Speed to all residential and business customers within the PFSA ;

(c) Offer free service at the Broadband Grant Speed to all Critical Community Facilities located within the PFSA for at least 2 years starting from the time service becomes available to each Critical Community Facility;

(d) Provide a Community Center with at least two (2) Computer Access Points and wireless access at the Broadband Grant Speed, free of all charges to all users for at least 2 years; and

(e) Not overlap with the service areas of current RUS borrowers and grantees.

§ 1739.12 Eligible grant purposes.

Grant funds may be used to finance the following:

(a) The construction, acquisition, or leasing of facilities, including spectrum, land or buildings, used to deploy service at the Broadband Grant Speed to all residential and business customers located within the PFSA and all participating Critical Community Facilities, including funding for up to ten Computer Access Points to be used in the Community Center. Buildings constructed with grants funds must reside on property owned by the awardee. Leasing costs will only be covered through the advance of funds period included in the award documents;

(b) The improvement, expansion, construction, or acquisition of a Community Center and provision of Computer Access Points. Grant funds for the Community Center will be limited to ten percent of the requested grant amount. If a community center is constructed with grant funds, the center must reside on property owned by the awardee;

(c) The cost of providing the necessary bandwidth for service free of charge to the Critical Community Facilities for 2 years.

§ 1739.13 Ineligible grant purposes.

Operating expenses not specifically permitted in § 1739.12.

§ 1739.14 Matching contributions.

(a) At the time of closing of the award, the awardee must contribute or demonstrate available cash reserves in an account(s) of the awardee equal to at least 15% of the grant. Matching contributions must be used solely for the Project and shall not include any financial assistance from federal sources unless there is a federal statutory exception specifically authorizing the federal financial assistance to be considered as such. An applicant must provide evidence of its ability to comply with this requirement in its application.

(b) At the end of every calendar quarter, the award must submit a schedule to RUS that identifies how the match contribution was used to support the project until the total contribution is expended.

§ 1739.15 Completed application.

Applications should be prepared in conformance with the provisions of this part and all applicable regulations, including 2 CFR part 200, as adopted by USDA through 2 CFR part 400. Applicants must also conform to the requirements of the FOA posted on *www.Grants.gov*, the RUS Community Connect Grant program application guide, and the Community Connect Grant program website. Applicants should refer to the FOA and the application guide for submission directions. The application guide contains instructions and forms, as well as other important information needed to prepare an application and is updated on an annual basis. Paper copies of the application guide can be requested by contacting the Loan Origination and Approval Division at 202–720–0800. Completed applications must include the following documentation, studies, reports and information, in form and substance satisfactory to the Agency:

(a) *An Application for Federal Assistance*. A completed Standard Form 424;

(b) *An executive summary of the Project*. A general project overview that addresses the following categories:

(1) A description of why the Project is needed;

(2) A description of the applicant;

(3) An explanation of the total Project costs;

(4) A general overview of the broadband telecommunications system to be developed, including the types of equipment, technologies, and facilities to be used;

(5) Documentation describing the procedures used to determine the unavailability of existing Broadband Service; and

(6) A list of the Critical Community Facilities that will take service from the Applicant at the Broadband Grant Speed, and evidence that any remaining Critical Community Facility located in the PFSA has rejected the offer;

(c) *Scoring Criteria Documentation*. A narrative, with documentation where necessary, addressing the elements listed in the scoring criteria of § 1739.17;

(d) *System design*. A system design of the Project that is economical and practical, including a detailed description of the facilities to be funded, technical specifications, data rates, and costs. In addition, a network diagram detailing the proposed system must be provided. The system design must also comply with the environmental review requirements in accordance with 7 CFR part 1970 and as supplemented by 7 CFR 1738.108;

(e) *Service Area Demographics*. The following information about the PFSA:

(1) A map, submitted electronically through RUS' web-based Mapping Tool, which identifies the Rural Area boundaries of the PFSA; and

(2) The total population, number of households, and number of businesses located within the PFSA;

(f) *Scope of work*. A description of the scope of work, which at a minimum must include:

(1) The specific activities and services to be performed under the Project;

(2) Who will carry out the activities and services;

(3) A construction build-out schedule and project milestones, showing the time-frames for accomplishing the Project objectives and activities on a quarterly basis; and

(4) A budget for all capital and administrative expenditures reflecting the line item costs for Eligible Grant Purposes and other sources of funds necessary to complete the Project;

(g) *Community-oriented connectivity plan*. A community-oriented connectivity plan consisting of the following:

(1) A listing of all participating Critical Community Facilities to be connected. The applicant must also provide documentation that it has consulted with the appropriate agent of every Critical Community Facility in the PFSA, and must provide statements from each one as to its willingness to participate, or not to participate, in the proposed Project;

(2) A description of the services the applicant will make available to local residents and businesses; and

(3) A list of any other telecommunications provider (including interexchange carriers, cable television operators, enhanced service providers, wireless service providers and providers of satellite services) that is participating in the delivery of services and a description of the consultations and the anticipated role of such provider in the Project;

(h) *Financial information and sustainability*. A narrative description demonstrating the sustainability of the Project: from the commencement of construction to completion, and beyond the grant period; the sufficiency of resources; how and when the matching requirement is met; and the expertise necessary to undertake and complete the Project. The following financial information is required:

(1) If the applicant is an existing company, it must provide complete copies of audited financial statements, if available, for the two fiscal years preceding the application submission. If audited statements are unavailable, the applicant must submit unaudited financial statements for those fiscal years. Applications from start-up entities must, at minimum, provide an opening balance sheet dated within 30 days of the application submission date; and

(2) Annual financial projections in the form of balance sheets, income statements, and cash flow statements for a forecast period of five years, which prove the sustainability of the Project for that period and beyond. These projections must be inclusive of the applicant's existing operations and the Project, and must be supported by a detailed narrative that fully explains the methodology and assumptions used to develop the projections, including details on the number of subscribers projected to take the applicant's services. Applicants submitting multiple applications for funding must demonstrate that each Project is feasible and sustainable on its own, funds are available to cover each of the matching requirements and that all Projects for which funding is being requested are financially feasible as a whole;

(i) *Statement of experience*. A statement of experience which includes information on the owners' and principal employees' relevant work experience that would ensure the success of the Project. The applicant must also provide a written narrative demonstrating its capability and experience, if any, in operating a broadband telecommunications system;

(j) *Legal authority*. Evidence of the applicant's legal authority and existence, and its ability to enter into a grant agreement with the RUS, and to perform the activities proposed under the grant application;

(k) *Additional funding.* Evidence that funding agreements have been attained, if the Project requires funding commitment(s) from sources other than the grant. An applicant submitting multiple applications for funding must demonstrate its financial wherewithal to support all applications, if accepted, and that it can simultaneously complete and operate all of the Projects under consideration. Additionally, commitments for outside funding must be explicit that they will be available if all applications are not funded;

(l) *Public notice.* The Agency will publish a public notice of each application requesting assistance under this part. The application must provide a summary of the information required for such public notice. The information required can be found in 7 CFR 1738.106.

(m) *Federal compliance.* Evidence of compliance with other federal statutes and regulations including, but not limited to the following:

(1) 7 CFR part 15, subpart A—Nondiscrimination in Federally Assisted Programs of the Department of Agriculture—Effectuation of Title VI of the Civil Rights Act of 1964;

(2) 2 CFR part 200, as adopted by USDA through 2 CFR part 400.

(3) 2 CFR part 417—Nonprocurement Debarment and Suspension;

(4) 2 CFR part 418—New Restrictions on Lobbying;

(5) 2 CFR part 421—Requirements for Drug-Free Workplace (Financial Assistance);

(6) Certification regarding Architectural Barriers;

(7) Certification regarding Flood Hazard Precautions;

(8) Environmental review documentation prepared in accordance with 7 CFR part 1970 and as supplemented by 7 CFR 1738.108.

(9) A certification that grant funds will not be used to duplicate lines, facilities, or systems providing Broadband Service;

(10) Federal Obligation Certification on Delinquent Debt; and

(11) Assurance Regarding Felony Conviction or Tax Delinquent Status for Corporate Applicants.

[78 FR 25791, June 3, 2013, as amended at 79 FR 76004, Dec. 19, 2014; 81 FR 11027, Mar. 2, 2016; 83 FR 45033, Sept. 5, 2018; 85 FR 14409, Mar. 12, 2020]

§ 1739.16 Review of grant applications.

(a) All applications for grants must be delivered to the Agency at the address and by the date specified in the FOA, the Community Connect Grant program application guide and the Community Connect Grant program website (*see* § 1739.2) to be eligible for funding. The Agency will review each application for conformance with the provisions of this part, and may contact the applicant for clarification of information in the application.

(b) Incomplete applications as of the deadline for submission will not be considered. If an application is determined to be incomplete, the applicant will be notified in writing and the application will be returned with no further action.

(c) If the Agency determines that the Project is technically or financially infeasible or unsustainable, the Agency will notify the applicant, in writing, and the application will be returned with no further action.

(d) Applications conforming with this part will be evaluated competitively by the Agency and will be ranked in accordance with § 1739.17. Applications will then be awarded generally in rank order until all grant funds are expended, subject to paragraphs (e) and (f) of this section.

(e) In addition to scoring, the Agency may take geographic distribution into consideration when making final award determinations.

(f) An award may be made out of rank order if a higher ranked application would require an award that exceeded available funding or would consume a disproportionate amount of funds available relative to its ranking.

(g) The Agency reserves the right to offer an applicant a lower amount than proposed in the application.

[78 FR 25791, June 3, 2013, as amended at 83 FR 45033, Sept. 5, 2018]

§ 1739.17 Scoring of applications.

The ranking of the "community-oriented connectivity" benefits of the Project will be based on documentation in support of the need for services, benefits derived from the proposed services, characteristics of the PFSA, local community involvement in planning and implementation of the Project, and the level of experience of the management team. In ranking applications the Agency will consider the following criteria based on a scale of 100 possible points:

(a) An analysis of the challenges of the following criteria, laid out on a community-wide basis, and how the Project proposes to address these issues (up to 50 points):

(1) The economic characteristics;

(2) Educational challenges;

(3) Health care needs; and

(4) Public safety issues;

(b) The extent of the Project's planning, development, and support by local residents, institutions, and Critical Community Facilities. Documentation must include evidence of community-wide involvement, as exemplified by community meetings, public forums, and surveys. In addition, applicants should provide evidence of local residents' participation in the Project planning and development (up to 40 points).

(c) The level of experience and past success of operating broadband systems for the management team. (up to 10 points)

(d) In making a final selection among and between applications with comparable rankings and geographic distribution, the Administrator may take into consideration the characteristics of the PFSA. Only information provided in the application will be considered. Applicants should therefore specifically address each of the following criteria to differentiate their applications:

(1) Persistent poverty counties that will be served within the PFSA;

(2) Out-migration Communities that will be served within the PFSA;

(3) The rurality of the PFSA;

(4) The speed of service provided by the project;

(5) Substantially underserved trust areas that will be served within the PFSA;

(6) Community members with disabilities that will be served within the PFSA; and

(7) Any other additional factors that may be outlined in the FOA, the Community Connect Grant program application guide, and the Community Connect Grant program website.

[78 FR 25791, June 3, 2013, as amended at 83 FR 45033, Sept. 5, 2018]

§ 1739.18 Grant documents.

The terms and conditions of grants shall be set forth in grant documents prepared by the Agency. The documents shall require the applicant to own all equipment and facilities financed by the grant. Among other matters, the Agency may prescribe conditions to the advance of funds that address concerns regarding the Project feasibility and sustainability. The Agency may also prescribe terms and conditions applicable to the construction and operation of the Project and the delivery of service at the Broadband Grant Speed to eligible Rural Areas, as well as other terms and conditions applicable to the individual Project. Dividend distributions will not be allowed until all grant funds and matching contributions have been expended.

§ 1739.19 Reporting and oversight requirements.

(a) A project performance activity report will be required of all recipients on an annual basis until the Project is complete and the funds are expended by the applicant. The reporting period will start with the calendar year the award is made and continue for every calendar year through the term of the award. The report must be submitted by January 31 of the following year of the reporting period. Recipients are to submit an original and one copy of all project performance reports, including, but not limited to, the following:

(1) A comparison of actual accomplishments to the objectives established for that period;

(2) A description of any problems, delays, or adverse conditions which have occurred, or are anticipated, and

which may affect the attainment of overall Project objectives, prevent the meeting of time schedules or objectives, or preclude the attainment of particular Project work elements during established time periods. This disclosure shall be accompanied by a statement of the action taken or planned to resolve the situation; and

(3) Objectives and timetable established for the next reporting period.

(b) A final project performance report must be provided by the recipient. It must provide an evaluation of the success of the Project in meeting the objectives of the program. The final report may serve as the last annual report.

(c) The Agency will monitor recipients, as it determines necessary, to assure that Projects are completed in accordance with the approved scope of work and that the grant is expended for Eligible Grant Purposes.

(d) Recipients shall diligently monitor performance to ensure that time schedules are being met, projected work within designated time periods is being accomplished, and other performance objectives are being achieved.

(e) The applicant must have the necessary processes and systems in place to comply with the reporting requirements for first-tier sub-awards and executive compensation under the Federal Funding Accountability and Transparency Act of 2006 in the event the applicant receives funding unless such applicant is exempt from such reporting requirements pursuant to 2 CFR 170.110(b). The reporting requirements under the Transparency Act pursuant to 2 CFR part 170 are as follows:

(1) First Tier Sub-Awards of $25,000 or more in non-Recovery Act funds (unless they are exempt under 2 CFR part 170) must be reported by the Recipient to *http://www.fsrs.gov* no later than the end of the month following the month the obligation was made.

(2) The Total Compensation of the Recipient's Executives (5 most highly compensated executives) must be reported by the Recipient (if the Recipient meets the criteria under 2 CFR part 170) to *http://www.sam.gov* by the end of the month following the month in which the award was made.

(3) The Total Compensation of the Subrecipient's Executives (5 most highly compensated executives) must be reported by the Subrecipient (if the Subrecipient meets the criteria under 2 CFR part 170) to the Recipient by the end of the month following the month in which the subaward was made.

(f) Entities that receive assistance from the Agency under this part to provide retail broadband service must submit annual reports for 3 years after project completion. The information required can be found in 7 CFR 1738.107(a) and (c).

[78 FR 25791, June 3, 2013, as amended at 85 FR 14409, Mar. 12, 2020]

§ 1739.20 Audit requirements.

A grant recipient shall provide the Agency with an audit for each year in which a portion of the financial assistance is expended, in accordance with the following:

(a) If the recipient is a for-profit entity, an existing Telecommunications or Electric Borrower with the Agency, or any other entity not covered by the following paragraph, the recipient shall provide an independent audit report in accordance with 7 CFR part 1773, "Policy on Audits of the Agency's Borrowers." Please note that the first audit submitted to the Agency and all subsequent audits must be comparative audits as described in 7 CFR part 1773.

(b) If the recipient is a Tribal, State or local government, or non-profit organization, the recipient shall provide an audit in accordance with subpart F of 2 CFR part 200, as adopted by USDA through 2 CFR part 400.

[78 FR 25791, June 3, 2013, as amended at 79 FR 76004, Dec. 19, 2014]

§ 1739.21 OMB control number.

The information collection requirements in this part are approved by the Office of Management and Budget (OMB) and assigned OMB control number 0572–0127.

Subpart B [Reserved]

PART 1740—RURAL ECONNECTIVITY PROGRAM

AUTHORITY: 7 U.S.C. 1981(b)(4), 7 U.S.C. 901 *et seq.*, 7 U.S.C. 950aaa *et seq.*, and 7 U.S.C. 950cc.

SOURCE: 86 FR 11609, Feb. 26, 2021, unless otherwise noted.

Subpart A—General

§ 1740.1 Overview.

(a) The Rural eConnectivity Program, hereinafter referred to as Program, provides funding in the form of loans, grants, and loan/grant combinations for the costs of construction, improvement, or acquisition of facilities and equipment needed to facilitate broadband deployment in rural areas. One of the essential goals of the Program is to expand broadband service to rural areas that do not have sufficient access to broadband. This part sets forth the general policies, eligibility requirements, types and terms of loans, grants, and loan/grant combinations and program requirements.

(b) Additional information and application materials regarding the Program can be found on the Rural Development website.

§ 1740.2 Definitions.

(a) The following definitions apply to this part:

Administrator means the Administrator of the Rural Utilities Service, or the Administrator's designee.

Agency means the Rural Utilities Service (RUS).

Applicant means an entity requesting funding under this part.

Application means the Applicant's request for federal funding, which may be approved in whole or in part by RUS.

Award documents mean, as applicable, all associated grant agreements, loan agreements, or loan/grant agreements.

Award means a grant, loan, or loan/grant combination made under this part.

Awardee means a grantee, borrower, or borrower/grantee that has applied and been awarded federal assistance under this part.

Broadband loan means, for purposes of this regulation, a loan that has been approved or is currently under review by RUS after the beginning of Fiscal Year 2000 in the Telecommunications Infrastructure Program, Farm Bill Broadband Program, Broadband Initiatives Program or this Program.

Broadband loans that were rescinded or defaulted on, or the terms and conditions of which were not met, are not included in this definition, so long as the entity under consideration for an award under this part has not previously defaulted on, or failed to meet the terms and conditions of, an RUS loan or had an RUS loan rescinded.

Broadband service means any fixed terrestrial technology, including fixed wireless, having the capacity to transmit data to enable a subscriber to the service to originate and receive high quality voice, data, graphics and video.

CALEA means the Communications Assistance for Law Enforcement Act, 47 U.S.C. 1001 *et seq.*

Composite economic life means the weighted (by dollar amount of each class of facility) average economic life of all classes of facilities necessary to complete construction of the broadband facilities in the proposed funded service area.

Current ratio means the current assets divided by the current liabilities.

Debt Service Coverage Ratio (DSCR) means the ratio of the sum of the Awardee's total net income or margins, depreciation and amortization expense, and interest expense, minus an allowance for funds used during construction and amortized grant revenue, all divided by the sum of interest on funded debt, other interest and principal payment on debt and capital leases.

Economic life means the estimated useful service life of an asset as determined by RUS.

Eligible service area means any contiguous proposed funded service area where 90 percent of the households to be served do not have sufficient access to broadband service. For eligibility purposes, if an applicant is applying for multiple proposed funded service areas, each service area will be evaluated on a stand-alone basis.

Equity means total assets minus total liabilities as reflected on the Applicant's balance sheet.

Fixed wireless service means a wireless system between two fixed locations (*e.g.*, fixed transmitting tower to fixed customer premise equipment).

Forecast period means the five-year period of projections in an application, which shall be used by RUS to determine financial and technical feasibility of the application.

GAAP means accounting principles generally accepted in the United States of America.

Grant means any federal assistance in the form of a grant made under this part.

Grant agreement means the grant contract and security agreement between RUS and the Awardee securing the Grant awarded under this part, including any amendments thereto, available for review on the Agency's web page.

Indefeasible Right to Use (IRU) means the long-term agreement of the rights to capacity, or a portion thereof specified in the terms of a certain amount of bandwidth or number of fibers.

Loan means any federal assistance in the form of a loan made under this part.

Loan agreement means the loan contract and security agreement between RUS and the Awardee securing the Loan, including all amendments thereto, available for review on the Agency's web page.

Loan/grant means any federal assistance in the form of a loan/grant combination made under this part.

Loan/grant agreement means the loan/grant contract and security agreement between RUS and the Awardee securing the loan/grant, including all amendments thereto, available for review on the Agency's web page.

Non-funded service area (NFSA) means any area in which the applicant offers broadband service or intends to offer broadband service during the forecast period but is not a part of its proposed funded service area.

Pre-application expenses means any reasonable expenses, as determined by RUS, incurred after the release of a FEDERAL REGISTER notice opening an application window to prepare an Application or to respond to RUS inquiries about the Application.

Premises means households, farms, and businesses.

Project means all of the work to be performed to bring broadband service to all premises in the proposed funded service area under the Application, including construction, the purchase and installation of equipment, and professional services including engineering and accountant/consultant fees, whether funded by federal assistance, matching, or other funds.

Proposed funded service area (PFSA) means the area (whether all or part of an existing or new service area) where the applicant is requesting funds to provide broadband service. Multiple service areas will be treated as separate standalone service areas for the purpose of determining how much of the PFSA does not have sufficient access to broadband. Each service area must meet the minimum requirements for the appropriate funding category to be an eligible area.

RE Act means the "Rural Electrification Act of 1936," as amended (7 U.S.C. 901 *et seq.*).

Rural area means any area that is not located within: (1) A city, town, or incorporated area that has a population of greater than 20,000 inhabitants; or (2) an urbanized area contiguous and adjacent to a city or town that has a population of greater than 50,000 inhabitants as defined in the Agency mapping tool.

RUS Accounting Requirements shall mean compliance with GAAP, acceptable to RUS, the system of accounting prescribed by RUS Bulletin 1770B–1 and the Uniform Administrative Requirements, Cost Principles, and Audit Requirements for Federal Awards, found at 2 CFR part 200. For all Awardees the term "grant recipient" in 2 CFR 200 shall also be read to encompass "loan recipient" and "loan/grant recipient", such that 2 CFR 200 shall be applicable to all Awardees under this part.

Sufficient access to broadband means a rural area in which households have *broadband service* at the minimum acceptable level of broadband, as set forth in the latest FEDERAL REGISTER notice announcing funding for the program. This definition will be used to determine the eligibility of a proposed service area and cannot be lower than 10 megabit per second (Mbps) downstream and 1 Mbps upstream. Mobile/Cellular and satellite services, which include systems that use satellite backbone facilities to connect to the internet, will not be considered in making the determination of sufficient access to broadband.

TIER means times interest earned ratio. TIER is the ratio of an Applicant's net income (after taxes) plus interest expense, all divided by interest expense and with all financial terms defined by GAAP.

(b) Unless otherwise provided in the award documents, all financial terms not defined herein shall have the meaning as defined by GAAP.

§ 1740.3 Funding parameters.

(a) For the purposes of this part:

(1) Ninety (90) percent of the PFSA must not have sufficient access to broadband service;

(2) Applicants must propose to build a network that is capable of providing broadband service to every premises located in the PFSA at the time the application is submitted at a speed defined in the latest FEDERAL REGISTER notice announcing funding for the Program; and

(3) The Agency reserves the right to make funding offers or seek consultations to resolve partially overlapping applications. RUS may contact the applicant for additional information during the review process. If additional information is requested, the applicant will have up to 30 calendar days to submit the information. If such information is not timely submitted, RUS may reject the application.

(b) The amount and types of funds available for assistance, as well as the maximum and minimum award amounts will be published in the FEDERAL REGISTER. Applicants may apply for grants, loans and loan/grant combinations.

§ 1740.4 Certifications.

The Applicant must certify to the following within the online application system:

(a) That it is authorized to submit the application on behalf of the eligible entity(ies) listed in the Application;

(b) That the Applicant has examined the Application;

(c) That all information in the Application, including certifications and forms submitted are, at the time furnished, true and correct in all material respects;

(d) That the entity requesting funding will comply with the terms, conditions, purposes, and federal requirements of the program;

(e) That a false, fictitious, or fraudulent statement or claim on the Application is grounds for denial or termination of an award, and/or possible punishment by a fine or imprisonment as provided in 18 U.S.C. 1001 and civil violations of the False Claims Act (31 U.S.C. 3729 *et seq.*);

(f) That the Applicant will comply with all applicable federal, tribal, state, and local laws, rules, regulations, ordinances, codes, orders, and programmatic rules and requirements relating to the project, and acknowledges that failure to do so may result in rejection or de-obligation of the award, as well as civil liability or criminal prosecution, if applicable, by the appropriate law enforcement authorities.

§§ 1740.5–1740.8 [Reserved]

Subpart B—Eligibility Requirements

§ 1740.9 Eligible and ineligible entities.

(a) To be eligible for funding, an Applicant may be either a nonprofit or for-profit organization, and must take one of the following forms:

(1) Corporation;

(2) Limited Liability Company and Limited Liability Partnership;

(3) Cooperative or mutual organization;

(4) States or local governments, including any agency, subdivision, instrumentality, or political subdivision thereof;

(5) A territory or possession of the United States; or

(6) An Indian tribe, as defined in section 4 of the Indian Self-Determination and Education Assistance Act (25 U.S.C. 450b).

(b) Individuals and legal general partnerships that are formed with individuals are not eligible entities.

(c) Co-Applicants are not eligible entities. If two entities would like to partner with each other in delivering broadband to areas without sufficient access, then one entity must take the lead on submitting an application. Inter-company agreements can be used to account for revenues and expenses on the applicant's financial projections. However, based on the existing financial and security arrangements, the Agency may require that both, or other entities, be parties to the award documents, or guarantee the award.

§ 1740.10 Eligible projects.

To be eligible for funding assistance under the part, the Applicant must:

(a) Submit a complete application and provide all supporting documentation including unqualified, comparative, audited financial statements for the previous year from the date the application is submitted as detailed in § 1740.63.

(b) Demonstrate that the project can be completely built out within five years from the date funds are first made available.

(c) Demonstrate that the project is technically feasible as detailed in § 1740.64.

(d) Demonstrate that all project costs can be fully funded or accounted for as detailed in § 1740.63.

(e) Submit documentation which enables RUS to determine that the project is financially feasible and sustainable as detailed in § 1740.61.

(f) Demonstrate that the following service requirements will be met:

(1) Facilities funded with grant funds will provide broadband service proposed in the application for the composite economic life of the facilities, as approved by RUS, or as provided in the Award Documents.

(2) Facilities funded with loan funds must provide broadband service through the amortization period of the loan.

§ 1740.11 Eligible and ineligible service areas.

(a) *Eligible service areas.* (1) Applicants must propose to provide broadband service directly to all premises in the PFSA.

(2) If any part of the applicant's PFSA is ineligible, RUS, in its sole discretion, may request that an applicant modify its application, if RUS believes

the modification is feasible. Otherwise, RUS will reject the application.

(b) *Ineligible service areas*—(1) *Overlapping service areas.* RUS will not fund more than one project that serves any one given geographic area. Invariably, however, applicants will propose service areas that overlap, varying from small *de minimis* areas of the territory, but which may be significant with respect to households involved, to larger areas of the service territory, but which may contain few households or businesses, if any. As a result, devising a procedure that will cover every overlap circumstance is not practicable. Nevertheless, it is the agency's intent to make as many eligible applications viable for consideration as possible. That may mean the agency may:

(i) Determine the overlap to be so insignificant that no agency action is necessary;

(ii) Request one or more applications to be revised to eliminate the overlapping territory;

(iii) Choose one application over another given the amount of assistance requested, the number of awards already chosen in the area or State, or the need for the project in the specific area due to other factors; or

(iv) Simply choose the project that scores higher or in the judgement of the agency is more financially feasible.

(2) *Prior funded service areas to include:* (i) RUS Broadband loans. Service areas of borrowers that have RUS Broadband loans, as defined in this part, are ineligible for all other applicants, and can be found on the Agency web page for the program. However, RUS Broadband Borrowers that have built out their service areas consistent with their application and award documents, but were not required to provide, and are currently not providing, sufficient access to broadband pursuant to this regulation are eligible to apply for funding for these service areas; provided that they have not defaulted on, and have materially complied with, in the sole discretion of RUS, their prior Broadband loan award requirements. Current RUS Broadband Borrowers that have received funding to provide sufficient access to broadband but have not yet built out their system are ineligible to apply for funding for these service areas.

(ii) RUS Community Connect Grants. Service areas that received grants under the RUS Community Connect Grant Program are eligible if they do not have sufficient access to broadband, except for those grants still under construction. Service areas still under construction can be found on the Agency's web page.

(iii) RUS BIP Grants. Service areas that received a 100 percent grant under the RUS Broadband Initiatives Program are eligible if they do not have sufficient access to broadband.

(c) *Service areas with other funding.* (1) Applicants are encouraged to work with the Governor's office for the states, and tribal governments for the tribal areas where they are proposing to provide broadband service and submit information detailing where state funding has been provided.

(2) Service areas that have received federal grant funds, or funds from the Federal Communications Commission, to provide broadband service will be restricted from funding, if such funding is principally to construct facilities throughout the service area that provide broadband service at the threshold level of service. If additional service areas are restricted from funding, these areas will be identified in the funding opportunity announcement that opens an application window.

§ 1740.12 Eligible and ineligible cost purposes.

Award and any matching funds must be used to pay only eligible costs incurred post award, except for approved pre-application expenses. Eligible costs must be consistent with the cost principles identified in 2 CFR 200, Subpart E, Cost Principles. In addition, costs must be reasonable, allocable, and necessary to the project. Any application that proposes to use any portion of the award or matching funds for any ineligible costs may be rejected.

(a) *Eligible award costs.* Award funds under this part may be used to pay for the following costs:

(1) To fund the construction or improvement of facilities, including buildings and land, required to provide

fixed terrestrial broadband service, including fixed wireless service, and any other facilities required for providing other services over the same facilities, such as equipment required to comply with CALEA;

(2) To fund reasonable preapplication expenses in an amount not to exceed five percent of the award. Preapplication expenses must be included in the first request for advance of award funds and will be funded with either grant or loan funds. If the funding category applied for has a grant component, then grant funds will be used for this purpose. If preapplication expenses are not included in the first request for advance of award funds, they will become an ineligible purpose; and

(3) To fund the acquisition of an existing system that does not currently provide sufficient access to broadband for upgrading that system to meet the requirements of this regulation. The cost of the acquisition is limited to 40 percent of the award amount requested. Acquisitions can be considered for 100 percent loans.

(b) *Ineligible award costs.* Award funds under this part may not be used for any of the following purposes:

(1) To fund operating expenses of the Awardee;

(2) To fund costs incurred prior to the date on which the application was submitted other than eligible preapplication expenses;

(3) To fund an acquisition of an affiliate, or the purchase or acquisition of any facilities or equipment of an affiliate. Note that if affiliated transactions are contemplated in the application, approval of the application does not constitute approval to enter into affiliated transactions, nor acceptance of the affiliated arrangements that conflict with the obligations under the award documents;

(4) To fund the acquisition of a system previously funded by RUS without prior written approval of RUS before an application is submitted;

(5) To fund the purchase or lease of any vehicle other than those used primarily in construction or system improvements;

(6) To fund broadband facilities leased under the terms of an operating lease or an indefeasible right of use (IRU) agreement;

(7) To fund the merger or consolidation of entities;

(8) To fund costs incurred in acquiring spectrum as part of a Federal Communication Commission (FCC) auction or in a secondary market acquisition. Spectrum that is part of a system acquisition may be considered;

(9) To fund facilities that provide mobile services;

(10) To fund facilities that provide satellite service including satellite backbone services;

(11) To fund the acquisition of a system that is providing sufficient access to broadband; or

(12) To refinance outstanding debt.

§§ 1740.13–1740.24 [Reserved]

Subpart C—Award Requirements

§ 1740.25 Substantially Underserved Trust Areas (SUTA).

Applicants seeking assistance may request consideration under the SUTA provisions in 7 U.S.C. 936f.

(a) If the Administrator determines that a community within "trust land" (as defined in 38 U.S.C. 3765) has a high need for the benefits of the Program, the Administrator may designate the community as a "substantially underserved trust area" (as defined in section 306F of the RE Act).

(b) To receive consideration under SUTA, the applicant must submit to the Agency a completed application that includes all information requested in 7 CFR part 1700, subpart D. In addition, the application must identify the discretionary authorities within subpart D that it seeks to have applied to its application. Note, however, the following:

(1) Given the prohibition on funding operating expenses in the Program, requests for waiver of the equity requirements cannot be considered; and

(2) Due to the statutory requirements that established the Program, waiver of the nonduplication requirements cannot be considered.

§ 1740.26 Public notice.

(a) To ensure transparency for the Program, the Agency's mapping tool will include the following information from each application, and be displayed for the public:

(1) The identity of the applicant;

(2) The areas to be served, including identification of the associated census blocks;

(3) The type of funding requested;

(4) The status of the application; and

(5) The number of households without sufficient access to broadband.

(b) The Agency will publish a public notice of each application requesting assistance under this part in accordance with the requirements of 7 U.S.C. 950cc. All applicants must provide the following information, which will be posted publicly on RUS' fully searchable website, in addition to the status of the application:

(1) A description of the proposed broadband project;

(2) A map of the PFSA;

(3) The amount and type of support requested by the applicant;

(4) The estimated number and proportion of service points in the proposed service territory without fixed broadband service, whether terrestrial or wireless; and

(5) Any other information required of the applicant in a funding notice.

(c) The public notice referenced under paragraph (b) of this section will be published after application submission and will remain available for 45 calendar days on the Agency's web page. During this period, existing service providers are requested to submit the following information through the Agency's mapping tool:

(1) The number of residential and business customers within the applicant's service area currently purchasing sufficient access to broadband, the rates of data transmission being offered, and the cost of each level of broadband service charged by the existing service provider;

(2) The number of residential and business customers within the applicant's service area receiving voice and video services and the associated rates for these other services;

(3) A map showing where the existing service provider's services coincide with the applicant's service area using the Agency's Mapping Tool; and

(4) Test results for the service area in question for a minimum of at least the prior three months demonstrating that sufficient access to broadband is being provided. The test results shall be for different times of the day.

(d) The Agency may contact service providers that respond under paragraph (b) of this section to validate their submission, and so responding service providers should be prepared to:

(1) Provide additional information supporting that the area in question has sufficient access to broadband service;

(2) Have a technician on site during the field validation by RUS staff;

(3) Run on site tests with RUS personnel being present, if requested; and

(4) Provide copies of any test results that have been conducted in the last six months and validate the information submitted in the public notice response months.

(e) If no broadband service provider submits information pursuant to a pending application or if the existing provider does not provide the information requested under paragraphs (b) and (c) of this section, RUS will consider the number of providers and extent of broadband service using any other data available through reasonable efforts, including utilizing the National Telecommunications and Information Administration National Broadband Availability Map and FCC broadband availability map. That may include the agency conducting field validations so as to locate facilities in the PFSA and determine, to the extent possible, if those facilities can provide sufficient access to broadband. Notwithstanding, conclusive evidence as to the existence of sufficient access to broadband will be taken only through the public notice process. As a result, the Agency highly recommends that existing service providers in a proposed funded service territory submit responses to the public notice to ensure that their service is considered in the determination of eligibility on an application.

(f) The Agency will notify respondents who are existing service providers whether their challenge was successful or not and allow for an opportunity to respond.

(g) The information submitted by an existing service provider under paragraph (c) of this section will be treated as proprietary and confidential and not subject to disclosure, pursuant to 7 U.S.C. 950cc(b)(3).

(h) For all applications that are approved, the following information will be made available to the public:

(1) The information provided in paragraph (a) of this section;

(2) Each annual report required under §1740.80(g) will be redacted to protect any proprietary information; and

(3) Such other information as the Administrator of the RUS deems sufficient to allow the public to understand the assistance provided.

§ 1740.27 Environmental and related reviews.

(a) Federal Agencies are required to analyze the potential environmental impacts, as required by the National Environmental Policy Act (NEPA), for Applicant projects or proposals seeking funding. Please refer to 7 CFR part 1970 for all of Rural Development's environmental policies. All Applicants must follow the requirements in 7 CFR part 1970 and are required to complete an Environmental Questionnaire, to provide a description of program activities, and to submit all other required environmental documentation as requested in the application system or by the Agency after the application is submitted. It is the Applicant's responsibility to obtain all necessary federal, tribal, state, and local governmental permits and approvals necessary for the proposed work to be conducted.

(b) Applications will be reviewed to ensure that they contain sufficient information to allow Agency staff to conduct a NEPA analysis so that appropriate NEPA documentation can be submitted to the appropriate federal and state agencies, along with the recommendation that the proposal is in compliance with applicable environmental and historic preservation laws.

(c) Applicants proposing activities that cannot be covered by existing environmental compliance procedures will be informed whether NEPA requirements and other environmental requirements can otherwise be expeditiously met so that a project can proceed within the timeframes anticipated under the Program.

(d) If additional information is required after an application is accepted for funding, funds can be withheld by the agency under a special award condition requiring the Awardee to submit additional environmental compliance information sufficient for the Agency to assess any impacts that a project may have on the environment.

§ 1740.28 Civil rights procedures and requirements.

(a) *Equal opportunity and non-discrimination.* The agency will ensure that equal opportunity and non-discriminatory requirements are met in accordance with the Equal Credit Opportunity Act and 7 CFR part 15. In accordance with federal civil rights law and USDA civil rights regulations and policies, the USDA, its agencies, offices, and employees, and institutions participating in or administering USDA programs are prohibited from discriminating based on race, color, national origin, religion, sex, gender identity (including gender expression), sexual orientation, disability, age, marital status, family/parental status, income derived from a public assistance program, political beliefs, or reprisal or retaliation for prior civil rights activity, in any program or activity conducted or funded by USDA (not all bases apply to all programs).

(b) *Civil rights compliance.* Recipients of federal assistance under this part must comply with the Americans with Disabilities Act of 1990, Title VI of the Civil Rights Act of 1964, and Section 504 of the Rehabilitation Act of 1973. In general, recipients should have available for the Agency, racial and ethnic data showing the extent to which members of minority groups are beneficiaries of federally assisted programs. The Agency will conduct compliance reviews in accordance with 7 CFR part 15. Awardees will be required to complete RD 400–4, "Assurance Agreement," for each Federal Award received.

(c) *Discrimination complaints.* Persons believing they have been subjected to discrimination prohibited by this section may file a complaint personally or by an authorized representative with

USDA, Director, Office of Adjudication, 1400 Independence Avenue SW, Washington, DC 20250. A complaint must be filed no later than 180 days from the date of the alleged discrimination, unless the time for filing is extended by the designated officials of USDA or the Agency.

§§ 1740.29–1740.41 [Reserved]

Subpart D—Award Terms

§ 1740.42 Interest rates.

Interest rates for the different funding options that will become available will be included in the FEDERAL REGISTER as part of the funding announcement opening a funding window.

(a) Direct cost-of-money loans shall bear interest at a rate equal to the cost of borrowing to the Department of Treasury for obligations of comparable maturity.

(b) The agency may offer 100 percent loans at a reduced interest rate, and in such cases, the applicable interest rate will be stated in the FEDERAL REGISTER or applicable funding opportunity notice.

§ 1740.43 Terms and conditions.

Terms and conditions of loans, grants, or loan/grant combinations are set forth in the non-negotiable standard loan, grant, or loan/grant agreements and the corresponding note, and/or mortgage, if applicable, which may be found on the Agency's web page.

(a) Unless the Applicant requests a shorter repayment period, loans must be repaid with interest within a period that, rounded to the nearest whole year, is equal to the expected Composite Economic Life of the project assets, as determined by RUS based upon acceptable depreciation rates, plus three years. Acceptable depreciation rates can be found in the Program Construction Procedures found on the Agency's web page.

(b) Interest begins accruing on the date of each loan advance. Any deferral period for loans will be set in the FEDERAL REGISTER notice opening a funding window.

(c) All proposed construction (including construction with matching and other funds) and all advance of funds must be completed no later than five years from the time funds are made available.

(d) No funds will be disbursed under this program until all other sources of funding have been obtained and any other pre-award conditions have been met. Failure to obtain one or more sources of funding committed to in the Application or to fulfill any other pre-award condition within 90 days of award announcement may result in withdrawal of the award. The RUS may modify this requirement in the FEDERAL REGISTER or applicable funding opportunity notice.

§ 1740.44 Security.

(a) *Loans and loan/grant combinations.* The loan portion of the award must be adequately secured, as determined by RUS.

(1) For Corporations and limited liability entities, the loan and loan/grant combinations must be secured by all assets of the Awardee.

(i) RUS must be given an exclusive first lien, in form and substance satisfactory to RUS, on all assets of the Awardee, including all revenues.

(ii) RUS may share its first lien position with one or more lenders on a *pari passu* basis, except with respect to grant funds, if security arrangements are acceptable to RUS.

(iii) Applicants must submit a certification that their prior lender or lienholder on any Awardee assets has already agreed to sign the RUS' standard intercreditor agreement or co-mortgage found on the Agency's web page.

(iv) RUS will not share a lien position on assets with any related party or affiliate of the Awardee.

(2) For Tribal entities and municipalities, RUS will develop appropriate security arrangements.

(3) Unless otherwise approved by RUS in writing, all property and facilities purchased with award funds must be owned by the Awardee.

(b) *Grant security.* The grant portion of the award must also be adequately secured, as determined by RUS.

(1) The government must be provided an exclusive first lien on all grant funded assets during the service obligation of the grant, and thereafter any

sale or disposition of grant assets must comply with the Uniform Administrative Requirements, Cost Principles, and Audit Requirements for Federal Awards, codified in 2 CFR part 200. Note that this part will apply to ALL grant funds of an Awardee, regardless of the entity status or type of organization.

(2) All Awardees must repay the grant if the project is sold or transferred without receiving written approval from RUS during the service obligation of the grant.

(c) *Substitution of Collateral and Irrevocable Letter of Credit*—(1) *Loans and combination loan and grant.* The Agency's standard loan/grant documents require that applicants pledge all assets and revenues of their operations as collateral. Applicants may propose other forms of collateral as long as the amount of the collateral is equal to the full amount of the loan. The collateral must be pledged to the Agency. Acceptable forms of substitute collateral are limited to following: Certificates of Deposit, with the Agency named as the beneficiary on the certificate, or Bonds with a AAA rating from an accredited rating agency. All other conditions of the standard loan documents will apply. A copy of the Substitution Documents can be found on the Agency's web page.

(2) *Grants.* For grant-only applications, applicants may request that standard grant security arrangements be replaced with an Irrevocable Letter of Credit (ILOC), to ensure that the project is completed. The ILOC must be for the full amount of funding requested and must remain in place until project completion. If an ILOC is offered as security, applicants will not be required to provide financial projections, meet any financial ratios requirements as part of the application process, or submit the maps for their NFSAs. Although the ILOC will replace security for the grant security arrangements, all other requirements of the standard grant agreement will remain the same. A copy of the ILOC award documents can be found on the Agency's web page.

§ 1740.45 Advance of funds.

RUS loan and grant advances are made at the request of the Awardee according to the procedures stipulated in the Award Documents. All non-RUS funds, to include matching funds and cash provided in lieu of RUS loan funds, must be expended first, followed by loan funds and then grant funds, except for RUS-approved pre-application expenses. RUS may modify this requirement in the FEDERAL REGISTER or applicable funding opportunity notice. Grant funds, if any, will be used for eligible preapplication expenses only on the first advance request. Applications that do not account for such advance procedures in the pro forma five-year forecast may be rejected.

§ 1740.46 Buy American requirement.

Awardees shall use in connection with the expenditure of loan and grant funds only such unmanufactured articles, materials, and supplies, as have been mined or produced in the United States or in any eligible country, and only such manufactured articles, materials, and supplies as have been manufactured in the United States or in any eligible country, substantially all from articles, materials, or supplies mined, produced, or manufactured, as the case may be, in the United States or in any eligible country. For purposes of this section, an "eligible country" is any country that applies with respect to the United States an agreement ensuring reciprocal access for United States products and services and United States suppliers to the markets of that country, as determined by the United States Trade Representative. The Buy American regulations may be found at, and any requests for waiver must be submitted pursuant to, 7 CFR part 1787.

§§ 1740.47–1740.58 [Reserved]

Subpart E—Application Submission and Evaluation

§ 1740.59 Application submission.

(a) Applications must be submitted through the Agency's online application system.

(b) The Agency may publish additional application submission requirements in a notice in the FEDERAL REGISTER.

(c) Unless otherwise identified in the notice, applicants can only submit one application under any funding window.

§ 1740.60 Elements of a complete application.

(a) *Online application system.* All applications under this regulation must be submitted through the RUS Online Application System located on the Agency's web page. Additional information can be found in the Application Guide found on the Agency's web page.

(b) *Dun and Bradstreet Universal Numbering System (DUNS) Number.* All applicants must register for a DUNS number, or other Government non-proprietary identifier as part of the application process. The applicant can obtain the DUNS number free of charge by calling Dun and Bradstreet. Go to *https://fedgov.dnb.com/webform* for more information on assignment of a DUNS number or confirmation. DUNS numbers of parent or affiliated operations cannot be substituted for the applicant. If a DUNS number is not provided, the application cannot be considered for an award.

(c) *System for Award Management (SAM).* Prior to submitting an application, the applicant must also register in SAM at *https://www.sam.gov/SAM/* and supply a Commercial and Government Entity (CAGE) Code number as part of the application. SAM registration must be active with current data at all times, from the application review throughout the active Federal award funding period. To maintain active SAM registration, the applicant must review and update the information in the SAM database annually from the date of initial registration or from the date of the last update. The applicant must ensure that the information in the database is current, accurate, and complete. If the CAGE Code of the applicant is not included in the application, the application will not be considered for an award.

(d) *Contents of the application.* A complete application will include the following information as requested in the RUS Online Application System and application guide:

(1) General information on the applicant and the project including:

(i) A description of the project, that will be made public, consistent with the requirements herein; and

(ii) The estimated dollar amount of the funding request.

(2) An executive summary that includes, but is not be limited to, a detailed description of existing operations, discussion about key management, description of the workforce, description of interactions between any parent, affiliated or subsidiary operation, a detailed description of the proposed project, and the source of the matching and other funds;

(3) A description of the PFSA including the number of premises passed;

(4) Subscriber projections including the number of subscribers for broadband, video and voice services and any other service that may be offered. A description of the proposed service offerings and the associated pricing plan that the applicant proposes to offer;

(5) A map, utilizing the RUS mapping tool located on the Agency's web page, of the PFSAs identifying the areas without sufficient access to broadband and any NFSA of the applicant. If an applicant has multiple NFSAs, they can elect to submit each NFSA individually or they can submit them as a single file through the mapping tool;

(6) A description of the advertised prices of service offerings by competitors in the same area;

(7) A network design and all supporting information as detailed in § 1740.64.

(8) Resumes of key management personnel, a description of the organization's readiness to manage a broadband services network, and an organizational chart showing all parent organizations and/or holding companies (including parents of parents, etc.), and all subsidiaries and affiliates;

(9) A legal opinion that:

(i) Addresses the applicant's ability to enter into the award documents;

(ii) Describes all material pending litigation matters;

(iii) Addresses the applicant's ability to pledge security as required by the award documents; and

(iv) Addresses the applicant's ability to provide broadband service under state or tribal law.

(10) Summary and itemized budgets of the infrastructure costs of the proposed project, including if applicable, the ratio of loans to grants, and any other sources of outside funding. The summary must also detail the amount of matching and other funds and the source of these funds. If the matching and other funds are coming from a third party, a commitment letter and support that the funds are available must also be submitted. Matching and other funds must be deposited into the RUS Pledged Deposit Account at the closing of the award;

(11) A detailed description of working capital requirements and the sources of those funds;

(12) Unqualified, comparative audited financial statements for the previous calendar year from the date the application is submitted as detailed in §1740.63;

(13) The historical and projected financial information required in §1740.63;

(14) All information and attachments required in the RUS Online application system;

(15) A scoring sheet, analyzing any scoring criteria set forth in the funding announcement opening the application window;

(16) A list of all the applicant's outstanding and contingent obligations as required in §1740.63;

(17) All environmental information as required by §1740.27;

(18) Certification from the applicant that agreements with, or obligations to, investors do not breach the obligations to the government under the standard Award Documents located on the Agency's web page, especially distribution requirements, and that any such agreements will be amended so that such obligations are made contingent to compliance with the Award Documents. Such certification should also specifically identify which, if any, provisions would need to be amended;

(19) If service is being proposed on tribal land, a certification from the proper tribal official that they are in support of the project and will allow construction to take place on tribal land. The certification must:

(i) Include a description of the land proposed for use as part of the proposed project;

(ii) Identify whether the land is owned, held in Trust, land held in fee simple by the Tribe, or land under a long-term lease by the Tribe;

(iii) If owned, identify the landowner; and

(iv) Provide a commitment in writing from the landowner authorizing the applicant's use of that land for the proposed project; and

(20) Additional items that may be required by the Administrator through a notice in the FEDERAL REGISTER.

(e) *Material representations.* The application, including certifications, and all forms submitted as part of the application will be treated as material representations upon which RUS will rely in awarding grants and loans.

§1740.61 Evaluation for technical and financial feasibility.

(a) A project is financially feasible when the applicant demonstrates to the satisfaction of RUS that it will be able to generate sufficient revenues to cover expenses; will have sufficient cash flow to service all debts and obligations as they come due; will have a positive ending cash balance as reflected on the cash flow statement for each year of the forecast period; and, by the end of the forecast period, will meet at least two of the following requirements: A minimum TIER requirement of 1.2, a minimum DSCR requirement of 1.2, and a minimum current ratio of 1.2. In addition, applicants must demonstrate positive cash flow from operations at the end of the forecast period.

(b) For any funding option that includes grant funds, evaluation criteria for scoring the application will be included in the FEDERAL REGISTER notice that opens an application window. Grant applications submitted for a certain category will be ranked and awarded based only on those applications included in that category.

(c) The Agency will determine technical feasibility by evaluating the Applicant's network design and other relevant information in the application.

§ 1740.62 Evaluation of Awardee operations.

(a) RUS may send a team to the awardee's facilities to complete a Management Analysis Profile (MAP) of the entire operation. MAPs are used by RUS as a means of evaluating an Awardee's strengths and weaknesses and ensuring that awardees are prepared to fulfil the terms of the award. Once an applicant accepts an award offer, RUS may schedule a site visit as soon as possible.

(b) RUS reserves the right not to advance funds until the MAP has been completed. If the MAP identifies issues that can affect the operation and completion of the project, those issues must be addressed to the satisfaction of RUS before funds can be advanced. Funding may be rescinded if following a MAP, the agency determines that the awardee will be unable to meet the requirements of the award.

§ 1740.63 Financial information.

(a) The Applicant must submit financial information acceptable to the Agency that demonstrates that the Applicant has the financial capacity to fulfill the grant, loan, and loan/grant combination requirements in this part and to successfully complete the proposed project.

(1) Applicants must submit unqualified, comparative, audited financial statements for the previous year from the date the application is submitted. If an application is submitted and the most recent year-end audit has not been completed, the applicant can submit the previous unqualified audit that has been completed. If qualified audits containing a disclaimer or adverse opinion are submitted, the application will not be considered.

(i) An applicant can use the consolidated audit of a parent as long as the parent fully guarantees the loan, or in the case of a grant, guarantees that construction will be completed as approved in the application or will repay the grant to RUS.

(ii) If the applicant has more than one parent, then each parent's audits must be submitted, and each parent must fully guarantee the award.

(iii) For governmental entities, financial statements must be accompanied with certifications as to unrestricted cash that may be available on a yearly basis to the applicant.

(2) Applicants must provide detailed information for all outstanding and contingent obligations. Copies of existing notes, loan and security agreements, guarantees, any existing management or service agreements, and any other agreements with parents, subsidiaries and affiliates, including but not limited to debt instruments that use the applicant's assets, revenues or stock as collateral must be included in the application.

(3) Applicants must provide evidence of all funding, other than the RUS award, necessary to support the project, such as bank account statements, firm letters of commitment from equity participants, or outside loans, which must evidence the timely availability of funds. If outside loans are used to cover any matching requirement, they may only be secured by assets other than those used for collateral under this regulation. Equity partners that are not specifically identified by name will not be considered in the financial analysis of the application. If the application states that other funds are required for the broadband project in addition to the Program funding requested, evidence must be included in the application identifying the source of funds and when the funds will be available. If the additional funding is not clearly identified, the application may not be considered for an award. If the applicant is providing non-telecommunication services and is proposing expansion to those services and states that additional funds are required to support sustainability of the overall operation of the applicant, then evidence must be submitted supporting the availability of these funds or the application may not be considered for funding.

(4) Historical financial statements for the last four years consisting of a balance sheet, income statement, and cash flow statement must be provided.

If an entity has not been operating for four years, historical statements for the period of time the entity has been operating are acceptable.

(5) Pro Forma financial analysis prepared in conformity with GAAP and the Agency's guidance on grant accounting can be found at *https://www.rd.usda.gov/files/AccountingGuidance10.pdf.* The Pro Forma should validate the sustainability of the project by including subscriber estimates related to all proposed service offerings; annual financial projections with balance sheets, income statements, and cash flow statements; supporting assumptions for a five-year forecast period and a depreciation schedule for existing facilities, those facilities funded with federal assistance, matching funds, and other funds. This pro forma should indicate the committed sources of capital funding and include a bridge year prior to the start of the forecast period. This bridge year shall be used as a buffer between the historical financial information and the forecast period and is the year in which the application is submitted.

(i) The financial projections must demonstrate that by the end of the forecast period, the project will meet at least two of the requirements described in § 1740.61(a).

(ii) The financial projections must also demonstrate positive cash flow from operations at the end of the forecast period.

(iii) Based on the financial evaluation, additional conditions may be added to the Award documents to ensure financial feasibility and security on the award.

(b) Publicly traded companies that have a bond rating from Moody's, Standard and Poor's, or Fitch of Investment Grade at the time an application is submitted do not have to complete the pro forma financial projections. In addition, applicants with this classification that elect not to submit financial projections do not need to submit NFSAs.

§ 1740.64 Network design.

(a) Only projects that RUS determines to be technically feasible will be eligible for an award.

(b) The network design must include a description of the proposed technology used to deliver the broadband service, demonstrating that all premises in the PFSA can be offered broadband service; a network diagram, identifying cable routes, wireless access points, and any other equipment required to operate the network; a buildout timeline and milestones for implementation of the project; and a capital investment schedule showing that the system can be built within five years. All of these items must be certified by a professional engineer who is certified in at least one of the states where there is or will be project construction. The certification from the professional engineer must clearly state that the proposed network can deliver the broadband service to all premises in the PFSA at the minimum required service level. In addition, a list of all required licenses and regulatory approvals needed for the proposed project and how much the applicant will rely on contractors or vendors to deploy the network facilities must be submitted. Note that in preparing budget costs for equipment and materials, RUS' Buy American requirements apply, as referenced in § 1740.46.

§§ 1740.65–1740.76 [Reserved]

Subpart F—Closing, Servicing, and Reporting

§ 1740.77 Offer and closing.

Successful applicants will receive an offer letter and award documents from RUS following award notification. Applicants may view sample award documents on the Agency's web page.

§ 1740.78 Construction.

(a) All project assets must comply with 7 CFR part 1788 and 7 CFR part 1970, the Program Construction Procedures located on the Agency's web page, any successor regulations found on the agency's website, and any other guidance from the Agency.

(b) The build-out of the project must be completed within five years from the date funds are made available. Build-out is considered complete when the network design has been fully implemented, the service operations and management systems infrastructure is operational, and the awardee is ready to support the activation and commissioning of individual customers to the new system.

§ 1740.79 Servicing of grants, loans and loan/grant combinations.

(a) Awardees must make payments on the loan as required in the note and Award Documents.

(b) Awardees must comply with all terms, conditions, affirmative covenants, and negative covenants contained in the Award Documents.

(c) The sale or lease of any portion of the Awardee's facilities must be approved in writing by RUS prior to initiating the sale or lease.

§ 1740.80 Accounting, monitoring, and reporting requirements.

(a) Awardees must adopt a system of accounts for maintaining financial records acceptable to the Agency, as described in 7 CFR part 1770, subpart B.

(b) Awardees must submit annual comparable audited financial statements along with a report on compliance and on internal control over financial reporting, and management letter in accordance with the requirements of 7 CFR part 1773 using the RUS' on-line reporting system. The Certified Public Accountant (CPA) conducting the annual audit is selected by the borrower and must be satisfactory to RUS as set forth in 7 CFR 1773, subpart B, "RUS Audit Requirements."

(c) Thirty (30) calendar days after the end of each calendar year quarter, Awardees must submit to RUS, balance sheets, income statements, statements of cash flow, rate package summaries, and the number of customers taking broadband service on a per community basis utilizing RUS' on-line reporting system. These reports must be submitted throughout the loan amortization period or for the economic life of the facilities funded with a grant.

(d) Awardees will be required to submit annually updated service area maps through the RUS mapping tool showing the areas where construction has been completed and premises are receiving service until the entire PFSA can receive the broadband service. At the end of the project, Awardees must submit a service area map indicating that all construction has been completed as proposed in the application. If parts of the PFSA have not been constructed, RUS may require a portion of the award to be rescinded or paid back.

(e) Awardees must comply with all reasonable Agency requests to support ongoing monitoring efforts. The Awardee shall afford RUS, through its representatives, reasonable opportunity, at all times during business hours and upon prior notice, to have access to and the right to inspect: The Broadband System, any other property encumbered by the Award Documents, any and all books, records, accounts, invoices, contracts, leases, payrolls, timesheets, cancelled checks, statements, and other documents (electronic or paper, of every kind) belonging to or in the possession of the Awardee or in any way pertaining to its property or business, including its subsidiaries, if any, and to make copies or extracts thereof.

(f) Awardee records shall be retained and preserved in accordance with the provisions of 7 CFR part 1770, subpart A.

(g) Awardees receiving assistance under this part will be required to submit annual reports for three (3) years after the completion of construction. The reports must include the following information:

(1) Existing network service improvements and facility upgrades, as well as new equipment and capacity enhancements that support high-speed broadband access for educational institutions, health care providers, and public safety service providers;

(2) The estimated number of end users who are currently using or forecasted to use the new or upgraded infrastructure;

(3) The progress towards fulfilling the objectives for which the assistance was granted;

(4) The number and geospatial location of residences and businesses that will receive new broadband service;

(5) The speed and price of the Awardee's broadband service offerings; and

(6) The average price of broadband service in the Project's service area.

§ 1740.81 Default and de-obligation.

RUS reserves the right to deobligate awards to Awardees under this part that demonstrate an insufficient level of performance, wasteful or fraudulent spending, or noncompliance with environmental and historic preservation requirements.

§§ 1740.82–1740.93 [Reserved]

Subpart G—Other Information and Federal Requirements

§ 1740.94 Confidentiality of Applicant information.

Applicants are encouraged to identify and label any confidential and proprietary information contained in their applications. The Agency will protect confidential and proprietary information from public disclosure to the fullest extent authorized by applicable law, including the Freedom of Information Act, as amended (5 U.S.C. 552), the Trade Secrets Act, as amended (18 U.S.C. 1905), the Economic Espionage Act of 1996 (18 U.S.C. 1831 *et seq.*), and CALEA (47 U.S.C. 1001 *et seq.*). Applicants should be aware, however, that this program requires substantial transparency. For example, RUS is required to make publicly available on the internet a list of each entity that has applied for a loan or grant, a description of each application, the status of each application, the name of each entity receiving funds, and the purpose for which the entity is receiving the funds.

§ 1740.95 Compliance with applicable laws.

Any recipient of funds under this regulation shall be required to comply with all applicable federal, tribal and state laws, including but not limited to:

(a) The Architectural Barriers Act of 1968, as amended (42 U.S.C. 4151 *et seq.*);

(b) The Uniform Federal Accessibility Standards (UFAS) (Appendix A to 41 CFR subpart 101–19.6); and

(c) All applicable federal, tribal and state communications laws and regulations, including, for example, the Communications Act of 1934, as amended, (47 U.S.C. 151 *et seq.*) the Telecommunications Act of 1996, as amended (Pub. L. 104–104, 110 Stat. 56 (1996), and CALEA. For further information, *see http://www.fcc.gov*.

§§ 1740.96–1740.99 [Reserved]

§ 1740.100 OMB control number.

The information collection requirements in this part are approved by the Office of Management and Budget (OMB) and assigned OMB control number 0572–0152.

PART 1741—PRE-LOAN POLICIES AND PROCEDURES FOR INSURED TELEPHONE LOANS [RESERVED]

PART 1744—POST-LOAN POLICIES AND PROCEDURES COMMON TO GUARANTEED AND INSURED TELEPHONE LOANS

Subpart A [Reserved]

Subpart B—Lien Accommodations and Subordination Policy

Subpart C—Advance and Disbursement of Funds

Subpart D [Reserved]

Subpart E—Borrower Investments

AUTHORITY: 7 U.S.C. 901 *et seq.*, 1921 *et seq.*, and 6941 *et seq.*

SOURCE: 55 FR 39396, Sept. 27, 1990, unless otherwise noted.

Subpart A [Reserved]

Subpart B—Lien Accommodations and Subordination Policy

SOURCE: 51 FR 32430, Sept. 12, 1986, unless otherwise noted. Redesignated at 55 FR 39396, Sept. 27, 1990.

§ 1744.20 General.

(a) Recent changes in the telecommunications industry, including deregulation and technological developments, have caused Rural Utilities Service (RUS) borrowers and other organizations providing telecommunications services to consider undertaking projects that provide new telecommunications services and other telecommunications services not ordinarily financed by RUS. Although some of these services may not be eligible for financing under the Rural Electrification Act of 1936 (RE Act), these services may nevertheless advance RE Act objectives where the borrower obtains financing from private lenders. The borrower's financial strength and the assurance of repayment of outstanding Government debt may be improved as a result of providing such telecommunications services.

(b) To facilitate the financing of new services and other services not ordinarily financed by RUS, RUS is willing to consider accommodating the Government's lien on telecommunications borrowers' systems or accommodating or subordinating the Government's lien on after-acquired property of telecommunications borrowers. To expedite this process, requests for lien accommodations meeting the requirements of § 1744.30 will receive automatic approval from RUS.

(c) This subpart establishes RUS policy with respect to all requests for lien accommodations and subordinations for loans from private lenders. For borrowers that do not qualify for automatic lien accommodations in accordance with § 1744.30, RUS will consider lien accommodations for RE Act purposes under § 1744.40 and non-Act purposes under § 1744.50.

[66 FR 41758, Aug. 9, 2001]

§ 1744.21 Definitions.

The following definitions apply to this subpart:

Administrator means the Administrator of RUS.

Advance means transferring funds from RUS or a lender guaranteed by RUS to the borrower's construction fund.

After-acquired property means property which is to be acquired by the borrower and which would be subject to the lien of the Government mortgage when acquired.

Amortization expense means the sum of the balances of the following accounts of the borrower:

Account names	Number
(1) Amortization expense	6560.2
(2) Amortization expense—tangible	6563
(3) Amortization expense—intangible	6564
(4) Amortization expense—other	6565

NOTE: All references to account numbers are to the Uniform System of Accounts (7 CFR part 1770, subpart B).

Asset means a future economic benefit obtained or controlled by the borrower as a result of past transactions or events.

Automatic lien accommodation means the approval, by RUS, of a request to share the Government's lien on a pari passu or pro-rata basis with a private lender in accordance with the provisions of §1744.30.

Borrower means any organization that has an outstanding telecommunications loan made or guaranteed by RUS, or that is seeking such financing. See 7 CFR part 1735.

Construction Fund means the RUS Construction Fund Account into which all advances of loan funds are deposited pursuant to the provisions of the loan documents.

Debt Service Coverage (DSC) ratio means the ratio of the sum of the borrower's net income, depreciation and amortization expense, and interest expense, all divided by the sum of all payments of principal and interest required to be paid by the borrower during the year on all its debt from any source with a maturity greater than 1 year and capital lease obligations.

Default means any event or occurrence which, unless corrected, will, with the passage of time and the giving of proper notices, give rise to remedies under one or more of the loan documents.

Depreciation expense means the sum of the balances of the following accounts of the borrower:

Account names	Number
(1) Depreciation expense	6560.1
(2) Depreciation expense—telecommunications plant in service	6561
(3) Depreciation expense—property held for future telecommunications use	6562

NOTE: All references to account numbers are to the Uniform System of Accounts (7 CFR part 1770, subpart B).

Disbursement means a transfer of money by the borrower out of the construction fund in accordance with the provisions of the fund.

Equity percentage means the total equity or net worth of the borrower expressed as a percentage of the borrower's total assets.

FFB means the Federal Financing Bank.

Financial Requirement Statement (FRS) means RUS Form 481 (OMB—No. 0572—0023). (This RUS Form is available from RUS, Program Development and Regulatory Analysis, Washington, DC 20250–1522).

Government mortgage means any instrument to which the Government, acting through the Administrator, is a party and which creates a lien or security interest in the borrower's property in connection with a loan made or guaranteed by RUS whether the Government is the sole mortgagee or is a co-mortgagee with a private lender.

Hardship loan means a loan made by RUS under section 305(d)(1) of the RE Act.

Interim construction means the purchase of equipment or the conduct of construction under an RUS-approved plan of interim financing. See 7 CFR part 1737.

Interest expense means the sum of the balances of the following accounts of the borrower:

Account names	Number
(1) Interest and related items	7500
(2) Interest on funded debt	7510
(3) Interest expense—capital leases	7520
(4) Amortization of debt issuance expense	7530
(5) Less Allowance for funds used during construction	7340/ 7300.4
(6) Other interest deductions	7540

NOTE: All references to account numbers are to the Uniform System of Accounts (7 CFR part 1770, subpart B).

Interim financing means funding for a project which RUS has acknowledged may be included in a loan, should said loan be approved, but for which RUS loan funds have not yet been made available.

Lien accommodation means sharing the Government's lien on a pari passu or pro-rata basis with a private lender.

Loan means any loan made or guaranteed by RUS.

Loan documents means the loan contract, note and mortgage between the borrower and RUS and any associated document pertinent to a loan.

Loan funds means the proceeds of a loan made or guaranteed by RUS.

Material and supplies means any of the items properly recordable in the following account of the borrower:

Account names	Number
(1) Material and Supplies	1220.1

NOTE: All references to account numbers are to the Uniform System of Accounts (7 CFR part 1770, subpart B).

Net income/Net margins means the sum of the balances of the following accounts of the borrower:

Account names	Number
(1) Local Network Services Revenues	5000 through 5069
(2) Network Access Services Revenues	5080 through 5084
(3) Long Distance Network Services Revenues.	5100 through 5169
(4) Miscellaneous Revenues	5200 through 5270
(5) Nonregulated Revenues	5280
(6) Less Uncollectible Revenues	5200 through 5302
(7) Less Plant Specific Operations Expense.	6110 through 6441
(8) Less Plant Nonspecific Operations Expense.	6510 through 6565
(9) Less Customer Operations Expense	6610 through 6623
(10) Less Corporate Operations Expense.	6710 through 6790
(11) Other Operating Income and Expense.	7100 through 7160
(12) Less Operating Taxes	7200 through 7250/ 7200.5
(13) Nonoperating Income and Expense	7300 through 7370
(14) Less Nonoperating Taxes	7400 through 7450/ 7400.5
(15) Less Interest and Related Items	7500 through 7540
(16) Extraordinary Items	7600 through 7640/ 7600.4
(17) Jurisdictional Differences and Nonregulated Income Items.	7910 through 7990

NOTE: All references to account numbers are to the Uniform System of Accounts (7 CFR part 1770, subpart B).

Net plant means the sum of the balances of the following accounts of the borrower:

Account names	Number
(1) Property, Plant and Equipment	2001 through 2007
(2) Less Depreciation and Amortization	3100 through 3600

NOTE: All references to account numbers are to the Uniform System of Accounts (7 CFR part 1770, subpart B).

Notes means evidence of indebtedness secured by or to be secured by the Government mortgage.

Pari Passu means equably; ratably; without preference or precedence.

Plant means any of the items properly recordable in the following accounts of the borrower:

Account names	Number
(1) Property, Plant and Equipment	2001 through 2007

NOTE: All references to account numbers are to the Uniform System of Accounts (7 CFR part 1770, subpart B).

Private lender means any lender other than the RUS or the lender of a loan guaranteed by RUS.

Private lender notes means the notes evidencing a private loan.

Private loan means any loan made by a private lender.

RE Act (Act) means the Rural Electrification Act of 1936 (7 U.S.C. 901 *et. seq.*).

RUS means the Rural Utilities Service, and includes its predecessor, the Rural Electrification Administration.

RUS cost-of-money loan means a loan made under section 305(d)(2) of the RE Act.

Subordination means allowing a private lender to have a lien on specific property which will have priority over the Government's lien on such property.

Tangible plant means any of the items properly recordable in the following accounts of the borrower:

Account names	Number
(1) Telecommunications Plant in Service—General Support Assets.	2110 through 2124
(2) Telecommunications Plant in Service—Central Office Assets.	2210 through 2232
(3) Telecommunications Plant in Service—Information Origination/Termination Assets.	2310 through 2362
(4) Telecommunications Plant in Service—Cable and Wire Facilities Assets.	2410 through 2441
(5) Amortizable Tangible Assets	2680 through 2682
(6) Nonoperating Plant	2006

NOTE: All references to account numbers are to the Uniform System of Accounts (7 CFR part 1770, subpart B).

Telecommunication services means any service for the transmission, emission, or reception of signals, sounds, information, images, or intelligence of any nature by optical waveguide, wire, radio, or other electromagnetic systems and shall include all facilities used in providing such service as well as the development, manufacture, sale, and distribution of such facilities.

Times interest earned ratio (TIER) means the ratio of the borrower's net income or net margins plus interest expense, divided by said interest expense.

Total assets means the sum of the balances of the following accounts of the borrower:

Account names	Number
(1) Current Assets	1100s through 1300s
(2) Noncurrent Assets	1400s through 1500s
(3) Total telecommunications plant	2001 through 2007
(4) Less accumulated depreciation	3100 through 3300s
(5) Less accumulated amortization	3400 through 3600s

NOTE: All references to account numbers are to the Uniform System of Accounts (7 CFR part 1770, subpart B).

Total equity or net worth means the excess of a borrower's total assets over its total liabilities.

Total liabilities means the sum of the balances of the following accounts of the borrower:

Account names	Number
(1) Current Liabilities	4010 through 4130.2
(2) Long-Term Debt	4210 through 4270.3
(3) Other Liabilities and Deferred Credits.	4310 through 4370

NOTE: All references to account numbers are to the Uniform System of Accounts (7 CFR part 1770, subpart B).

Total long-term debt means the sum of the balances of the following accounts of the borrower:

Account names	Number
(1) Long-Term Debt	4210 through 4270.3

NOTE: All references to account numbers are to the Uniform System of Accounts (7 CFR part 1770, subpart B).

Weighted-average life of the loans or notes means the average life of the loans or notes based on the proportion of original loan principal paid during each year of the loans or notes. It shall be determined by calculating the sum of all loan or note principal payments expressed as a fraction of the original loan or note principal amount, times the number of years and fractions of years elapsed at the time of each payment since issuance of the loan or note. For example, given a $5 million loan, with a maturity of 5 years and equal principal payments of $1 million due on the anniversary date of the loan, the weighted-average life would be: (.2)(1 year) + (.2)(2 years) + (.2)(3 years) + (.2)(4 years) + (.2)(5 years) = .2 years + .4 years + .6 years + .8 years + 1.0 years = 3.0 years. If instead the loan had a balloon payment of $5 million at the end of 5 years, the weighted-average life would be: ($5 million/$5 million)(5 years) = 5 years.

Weighted-average remaining life of the loans or notes means the remaining average life of the loans or notes based on the proportion of remaining loan or note principal expressed in years remaining to maturity of the loans or notes. It shall be determined by calculating the sum of the remaining principal payments of each loan or note expressed as a fraction of the total remaining loan or note amounts times the number of years and fraction of years remaining until maturity of the loan or note.

Weighted-average remaining useful life of the assets means the estimated original average life of the assets to be acquired with the proceeds of the private lender notes expressed in years based on depreciation rates less the number of years those assets have been in service (or have been depreciated). It shall be determined by calculating the sum of each asset's remaining value expressed as a fraction of the total remaining value of the assets, times the estimated number of years and fraction of years remaining until the assets are fully depreciated.

Wholly-owned subsidiary means a corporation owned 100 percent by the borrower.

[66 FR 41758, Aug. 9, 2001, as amended at 84 FR 59922, Nov. 7, 2019]

§§ 1744.22–1744.29 [Reserved]

§ 1744.30 Automatic lien accommodations.

(a) *Purposes and requirements for approval.* Automatic lien accommodations are available only for refinancing and refunding of notes secured by the borrower's existing Government mortgage; financing assets, to be owned by the borrower, to provide telecommunications services; or financing assets, to be owned by a wholly-owned subsidiary of the borrower, to provide telecommunications services in accordance with the procedures set forth in this section.

(b) *Private lender responsibility.* The private lender is responsible for ensuring that its notes, for which an automatic lien accommodation has been approved as set forth in this section, are secured under the mortgage. The private lender is responsible for ensuring that the supplemental mortgage is a valid and binding instrument enforceable in accordance with its terms, and recorded and filed in accordance with applicable law. If the private lender determines that additional documents are required or that RUS must take additional actions to secure the notes under the mortgage, the private lender shall follow the procedures set forth in § 1744.40 or § 1744.50, as appropriate.

(c) *Refinancing and refunding.* The Administrator will automatically approve a borrower's execution of private lender notes and the securing of such notes on a pari passu or pro-rata basis with all other notes secured under the Government mortgage, when such private lender notes are issued for the purpose of refinancing or refunding any notes secured under the Government mortgage, provided that all of the following conditions are met:

(1) No default has occurred and is continuing under the Government mortgage;

(2) The borrower has delivered to the Administrator, at least 10 business days before the private lender notes are to be executed, a certification and agreement executed by the President of the borrower's Board of Directors, such certification and agreement to be substantially in the form set forth in Appendix A of this subpart, providing that:

(i) No default has occurred and is continuing under the Government mortgage;

(ii) The principal amount of such refinancing or refunding notes will not be greater than 112 percent of the then outstanding principal balance of the notes being refinanced or refunded;

(iii) The weighted-average life of the private loan evidenced by the private lender notes will not exceed the weighted-average remaining life of the notes being refinanced or refunded;

(iv) The private lender notes will provide for substantially level debt service or level principal amortization over a period not less than the original remaining years to maturity;

(v) Except as provided in the Government mortgage, the borrower has not agreed to any restrictions or limitations on future loans from RUS; and

(vi) If the private lender determines that a supplemental mortgage is necessary, the borrower will comply with those procedures contained in paragraph (h) of this section for the preparation, execution, and delivery of a supplemental mortgage and take such additional action as may be required to secure the notes under the Government mortgage.

(d) *Financing assets to be owned directly by a borrower.* The Administrator will automatically approve a borrower's execution of private lender notes and the securing of such notes on a pari passu or pro-rata basis with all other notes secured under the Government mortgage, when such private lender notes are issued for the purpose of financing the purchase or construction of plant and material and supplies to provide telecommunication services and when such assets are to be owned and the telecommunications services are to be offered by the borrower, provided that all of the following conditions are met:

(1) The borrower has achieved a TIER of not less than 1.5 and a DSC of not less than 1.25 for each of the borrower's

two fiscal years immediately preceding the issuance of the private lender notes;

(2) The ratio of the borrower's net plant to its total long-term debt at the end of any calendar month ending not more than 90 days prior to execution of the private lender notes is not less than 1.2, on a pro-forma basis, after taking into account the effect of the private lender notes and additional plant on the total long-term debt of the borrower;

(3) The borrower's equity percentage, as of the most recent fiscal year-end, was not less than 25 percent;

(4) No default has occurred and is continuing under the Government mortgage;

(5) The borrower has delivered to the Administrator, at least 10 business days before the private lender notes are to be executed, a certification by an independent certified public accountant that the borrower has met each of the requirements in paragraphs (d)(1) and (d)(3) of this section, such certification to be substantially in the form in appendix B of this subpart; and

(6) The borrower has delivered to the Administrator, at least 10 business days before the private lender notes are to be executed, a certification and agreement executed by the President of the borrower's Board of Directors, such certification and agreement to be substantially in the form in appendix C of this subpart: provided, that:

(i) The borrower has met each of the requirements in paragraphs (d)(2) and (d)(4) of this section;

(ii) The proceeds of the private lender notes are to be used for the construction or purchase of the plant and materials and supplies to provide telecommunications services in accordance with this section and such construction or purchase is expected to be completed not later than 4 years after execution of such notes;

(iii) The weighted-average life of the private loan evidenced by the private lender notes does not exceed the weighted-average remaining useful life of the assets being financed;

(iv) The private lender notes will provide for substantially level debt service or level principal amortization over a period not less than the original remaining years to maturity;

(v) All of the assets financed by the private loans will be purchased or otherwise procured in bona fide arm's length transactions;

(vi) The financing agreement with the private lender will provide that the private lender shall cease the advance of funds upon receipt of written notification from RUS that the borrower is in default under the RUS loan documents;

(vii) Except as provided in the Government mortgage, the borrower has not agreed to any restrictions or limitations on future loans from RUS; and

(viii) If the private lender determines that a supplemental mortgage is necessary, the borrower will comply with those procedures set forth in paragraph (h) of this section for the preparation, execution, and delivery of a supplemental mortgage and take such additional action as may be required to secure the notes under the Government mortgage.

(e) *Financing assets to be owned by a wholly-owned subsidiary of the borrower.* The Administrator will automatically approve a borrower's execution of private lender notes and the securing of such notes on a pari passu or pro-rata basis with all other notes secured under the Government mortgage, when such private lender notes are issued for the purpose of financing the purchase or construction of tangible plant and material and supplies to provide telecommunication services and when such services are to be offered and the associated tangible assets are to be owned by a wholly-owned subsidiary of the borrower, provided that all of the following conditions are met:

(1) The borrower has achieved a TIER of not less than 2.5 and a DSC of not less than 1.5 for each of the borrower's two fiscal years immediately preceding the issuance of the private lender notes;

(2) The ratio of the borrower's net plant to its total long-term debt at the end of any calendar month ending not more than 90 days prior to execution of the private lender notes is not less than 1.6, on a pro-forma basis, after taking into account the effect of the private lender notes and additional

plant on the total long-term debt of the borrower;

(3) The borrower's equity percentage, as of the most recent fiscal year-end, was not less than 45 percent;

(4) No default has occurred and is continuing under the Government mortgage;

(5) The borrower has delivered to the Administrator, at least 10 business days before the private lender notes are to be executed, a certification by an independent certified public accountant that the borrower has met each of the requirements in paragraphs (e)(1) and (e)(3) of this section, such certification to be substantially in the form in appendix D of this subpart; and

(6) The borrower has delivered to the Administrator, at least 10 business days before the private lender notes are to be executed, a certification and agreement executed by the President of the borrower's Board of Directors, such certification and agreement to be substantially in the form in appendix E of this subpart; providing that:

(i) The borrower has met each of the requirements in paragraphs (e)(2) and (e)(4) of this section;

(ii) The proceeds of the private lender notes are to be used for the construction or purchase of the tangible plant and materials and supplies to provide telecommunications services in accordance with this section and such construction or purchase is expected to be completed not later than 4 years after execution of such notes;

(iii) The weighted-average life of the private loan evidenced by the private lender notes does not exceed the weighted-average remaining useful life of the assets being financed;

(iv) The private lender notes will provide for substantially level debt service or level principal amortization over a period not less than the original remaining years to maturity;

(v) All of the assets financed by the private loans will be purchased or otherwise procured in bona fide arm's length transactions;

(vi) The proceeds of the private lender notes will be lent to a wholly-owned subsidiary of the borrower pursuant to terms and conditions agreed upon by the borrower and subsidiary;

(vii) The borrower will, whenever requested by RUS, provide RUS with a copy of the financing or guarantee agreement between the borrower and the subsidiary or any similar or related material including security instruments, loan contracts, or notes issued by the subsidiary to the borrower;

(viii) The borrower will promptly report to the Administrator any default by the subsidiary or other actions that impair or may impair the subsidiary's ability to repay its loans;

(ix) The financing agreement with the private lender will provide that the private lender shall cease the advance of funds upon receipt of written notification from RUS that the borrower is in default under the RUS loan documents;

(x) Except as provided in the Government mortgage, the borrower has not agreed to any restrictions or limitations on future loans from RUS; and

(xi) If the private lender determines that a supplemental mortgage is necessary, the borrower will comply with those procedures contained in paragraph (h) of this section for the preparation, execution, and delivery of a supplemental mortgage and take such additional action as may be required to secure the notes under the Government mortgage.

(f) *Borrower notification.* The borrower shall notify RUS of its intention to obtain an automatic lien accommodation under §1744.30 by providing the following:

(1) The board resolution cited in §1744.55(b)(1) and the opinion of counsel cited in §1744.55(b)(2);

(2) The applicable certification or certifications required by paragraph (c)(2); paragraphs (d)(5) and (d)(6); or paragraphs (e)(5) and (e)(6), respectively, of this section, in substantially the form contained in the applicable appendices to this subpart.

(g) *RUS acknowledgment.* Within 5 business days of receipt of the completed certifications and any other information required under this section, RUS will review the information and provide written acknowledgment to the borrower and the private lender of its qualification for an automatic lien accommodation. Upon receipt of the acknowledgment, the borrower may execute the private lender notes.

(h) *Supplemental mortgage.* If the private lender determines that a supplemental mortgage is required to secure the private lender notes on a pari passu or pro-rata basis with all other notes secured under the Government mortgage, the private lender may prepare the supplemental mortgage using the form attached as appendix F to this subpart or the borrower may request RUS to prepare such supplemental mortgage in accordance with the following procedures:

(1) The private lender preparing the supplemental mortgage shall execute and forward the completed document to RUS. Upon ascertaining the correctness of the form and the information concerning RUS, RUS will execute and forward the supplemental mortgage to the borrower.

(2) When requested by the borrower, RUS will expeditiously prepare the supplemental mortgage, using the form in appendix F to this subpart, upon submission by the private lender of:

(i) The name of the private lender;

(ii) The Property Schedule for inclusion as supplemental mortgage Schedule B, containing legally sufficient description of all real property owned by the borrower; and

(iii) The amount of the private lender note.

(3) The government is not responsible for ensuring that the supplemental mortgage has been executed by all parties and is a valid and binding instrument enforceable in accordance with its terms, and recorded and filed in accordance with applicable law. If the private lender determines that additional security instruments or other documents are required or that RUS must take additional actions to secure the private lender notes under the mortgage, the private lender shall follow the procedures established in §§1744.40 or 1744.50, as appropriate. Except for the actions of the government expressly established in §1744.40, the government undertakes no obligation to effectuate an automatic lien accommodation. When processing of the supplemental mortgage has been completed to the satisfaction of the private lender, the borrower shall provide RUS with the following:

(i) A fully executed counterpart of the supplemental mortgage, including all signatures, seals, and acknowledgements; and

(ii) Copies of all opinions rendered by borrower's counsel to the private lender.

(i) *Other approvals.* (1) The borrower is responsible for meeting all requirements necessary to issue private lender notes and to accommodate the lien of the Government mortgage to secure the private lender notes including, but not limited to, those of the private lender, of any other mortgagees secured under the existing RUS mortgage, and of any governmental entities with jurisdiction over the issuance of notes or the execution and delivery of the supplemental mortgage.

(2) To the extent that the borrower's existing mortgage requires RUS approval before the borrower can make an investment in an affiliated company, approval is hereby given for all investments made in affiliated companies with the proceeds of private lender notes qualifying for an automatic lien accommodation under paragraph (e) of this section. Any reference to an approval by RUS under the mortgage shall apply only to the rights of RUS and not to any other party.

[66 FR 41760, Aug. 9, 2001]

§ 1744.40 Act purposes.

(a) Borrowers are encouraged to submit requests for accommodation of the Government's lien on the borrower's system in order to facilitate obtaining financing from private lenders for purposes provided in the RE Act.

(b) The Administrator will consider requests for the subordination of the Government's lien on after-acquired property which will enable borrowers to obtain financing from private lenders for purposes provided in the Act: Provided, however, that property integral to the operation of projects financed with loans made or guaranteed by RUS shall be financed with funds obtained through lien accommodations instead of lien subordinations, unless the Administrator determines that it is in the Government's interest to do otherwise.

[51 FR 32430, Sept. 12, 1986. Redesignated at 55 FR 39396, Sept. 27, 1990, and further redesignated at 66 FR 41760, Aug. 9, 2001]

§§ 1744.41–1744.49 [Reserved]

§ 1744.50 Non-Act purposes.

(a) The Administrator will consider requests for the accommodation of the Government's lien on the borrower's system or the subordination of the Government's lien on after-acquired property which will enable the borrowers to obtain financing from private lenders for the purpose of providing new telecommunication services which may not be eligible for financing under the Act if the Administrator is satisfied that:

(1) The borrower will have the ability to repay its existing and proposed indebtedness;

(2) The security for outstanding Government loans and guarantees is reasonably adequate and will not be adversely affected by the accommodation or subordination; and

(3) Approval of the request is in the interests of the Government with respect to the financial soundness of the borrower and other matters, such as assuring that the borrower's system is constructed cost-effectively using sound engineering practices.

(b) In determining that the security for outstanding Government loans and guarantees is reasonably adequate and will not be adversely affected by the accommodation or subordination the Administrator will consider, among other matters, when applicable, the following:

(1) Market forecasts for the project;

(2) Projected revenues, expenses and net income of the borrower's existing system and the project;

(3) Maximum debt service on indebtedness of both the borrower's system and the project;

(4) Projected rate of return on the borrower's investment in the project;

(5) Fair market value of property acquired by the borrower as part of the project;

(6) Impact of the project on the ratio of the borrower's secured debt to assets;

(7) Projected growth in borrower's system and project equity; and

(8) Amount of funds available for plant additions, replacements and other similar costs of the system and the project.

(c) In determining whether the accommodation or subordination is in the interests of the Government, the Administrator may consider, among other matters, whether the project will improve the borrower's financial strength and the assurance of repayment of Government debt.

[51 FR 32430, Sept. 12, 1986. Redesignated at 55 FR 39396, Sept. 27, 1990, as amended at 59 FR 43716, Aug. 25, 1994. Redesignated at 66 FR 41760, Aug. 9, 2001, as amended at 66 FR 41763, Aug. 9, 2001]

§§ 1744.51–1744.54 [Reserved]

§ 1744.55 Application procedures.

(a) Requests for information regarding applications for lien accommodations or subordination under this part should be addressed to the Assistant Administrator, Telecommunications Program, Rural Utilities Service, Washington, DC 20250–1590.

(b) An application for a lien accommodation or subordination shall include the following supporting information:

(1) A board Resolution from the applicant requesting the lien accommodation or subordination and stating the general purpose for which the funds from the private lender will be used, the proposed amount of the loan, and the proposed terms and conditions of the loan;

(2) An opinion from counsel representing the applicant that the applicant has the authority under its articles of incorporation, bylaws, and under applicable state law to undertake the project;

(3) Engineering and pertinent studies related to the projects or purposes to be financed, when applicable;

(4) Feasibility studies with pro forma financial statements showing the ability to repay the loan and provide an appropriate margin or net income;

(5) Any other information or documentation deemed pertinent by the borrower or the Administrator in support of the application.

(c) When the Administrator makes a determination that an application for an accommodation or subordination will not be approved the Administrator shall set forth the reasons therefor in writing and furnish such determination and reasons to the borrower within 30 days of the determination.

[51 FR 32430, Sept. 12, 1986. Redesignated at 55 FR 39396, Sept. 27, 1990, and further redesignated at 66 FR 41760, Aug. 9, 2001, as amended at 66 FR 41763, Aug. 9, 2001]

§§ 1744.56–1744.59 [Reserved]

APPENDIX A TO SUBPART B OF PART 1744—STATEMENT, CERTIFICATION, AND AGREEMENT OF BORROWER'S PRESIDENT OF BOARD OF DIRECTORS REGARDING REFINANCING AND REFUNDING NOTES PURSUANT TO 7 CFR 1744.30(C)

I ______(Name of President)______, am President of ______(Name of Borrower)______ (the "borrower"). The borrower proposes to issue notes (the "private lender notes"), to be dated on or about ______________ and delivered to ______(Name of Private Lender)______ (the "private lender"). I am duly authorized to make and enter into the following statements, certifications, and agreements for the purpose of inducing the United States of America (the "government"), to give automatic approval to the issuance of the private lender notes pursuant to 7 CFR 1744.30(c).

(a) The private lender:

___ is a mortgagee under the existing mortgage securing the government's loan to the borrower (the "government mortgage"); or

___ is not a mortgagee under the government mortgage and the borrower has executed the attached form of supplemental mortgage as provided in 7 CFR 1744.30(h).

(b) I hereby certify that all other requirements of 7 CFR 1744.30(c) are met; said requirements being as follows:

(1) No default has occurred and is continuing under the government mortgage;

(2) The principal amount of such refinancing or refunding notes, which is ________ dollars, will not be greater than 112 percent of the then outstanding principal balance of the notes being refinanced or refunded; such outstanding principal balance being ________ dollars;

(3) The weighted-average life of the private loan evidenced by the private lender notes, which is _______ years, will not exceed the weighted-average remaining life of the notes being refinanced or refunded, which is _______ years;

(4) Except as provided in the government mortgage, the borrower has not agreed to any restrictions or limitations on future loans from the Rural Utilities Service (RUS); and

(5) This certificate is being delivered to RUS at least 10 business days before the private lender notes are to be executed.

(c) The borrower agrees that the private lender notes will provide for substantially level debt service or level principal amortization.

(d) All terms not defined herein shall have the meaning set forth in 7 CFR 1744, subpart B.

______________________________ Signed

______________________________ Date

______________________________ Name

Name and Address of Borrower:

[66 FR 41763, Aug. 9, 2001]

APPENDIX B TO SUBPART B OF PART 1744—CERTIFICATION OF INDEPENDENT CERTIFIED PUBLIC ACCOUNTANT REGARDING NOTES TO BE ISSUED PURSUANT TO 7 CFR 1744.30(C)

I/We, __(Name of Independent Certified Public Accountant)__, hereby certify the following with respect to the note or notes (the "private lender notes") to be issued by __(Name of Borrower)__ ("the borrower") on or about __(Date private lender notes are to be Signed)__, evidencing a total loan principal of ____________ dollars:

(a) The borrower has achieved a TIER of not less than 1.5 and a DSC of not less than 1.25 for each of the borrower's 2 fiscal years immediately preceding the issuance of the private lender notes. The TIER and DSC ratios achieved are as follows:

Year	TIER	DSC
____	____	____
____	____	____

(b) The borrower's equity percentage, as of the most recent fiscal year-end, was not less than 25 percent:

Year	Total Equity
____	____

________________________ Signed

________________________ Date

Name and address of CPA Firm:

All terms not defined herein shall have the meaning set forth in 7 CFR 1744, Subpart B.

[66 FR 41763, Aug. 9, 2001]

APPENDIX C TO SUBPART B OF PART 1744—STATEMENT, CERTIFICATION, AND AGREEMENT OF BORROWER'S PRESIDENT OF BOARD OF DIRECTORS REGARDING NOTES TO BE ISSUED PURSUANT TO 7 CFR 1744.30(d)

I ________(Name of President)_______, am President of ______(Name of Borrower) (the "borrower"). The borrower proposes to issue notes (the "private lender notes"), to be dated on or about ________________ and delivered to______(Name of Private Lender)______ (the "private lender"). I am duly authorized to make and enter into the following statements, certifications, and agreements for the purpose of inducing the United States of America (the "government"), to give automatic approval to the issuance of the private lender notes pursuant to 7 CFR 1744.30(d).

(a) The private lender:
___ is a mortgagee under the existing mortgage securing the government's loan to the borrower (the "government mortgage"); or
___ is not a mortgagee under the government mortgage and the borrower has executed the attached form of supplemental mortgage as provided in 7 CFR 1744.30(h).

(b) I have reviewed the certificate of the independent certified public accountant also being delivered to the government in connection with the private lender notes to be issued pursuant to 7 CFR 1744.30(d) and concur with the conclusions expressed therein.

(c) I hereby certify that all other requirements of 7 CFR 1744.30(d) are met as follows:

(1) The ratio of the borrower's net plant to its total long-term debt at the end of any calendar month ending not more than 90 days prior to execution of the private lender notes is ____, which is not less than 1.2, on a pro-forma basis, after taking into account the effect of the private lender notes on the total long-term debt of the borrower;

(2) No default has occurred and is continuing under the government mortgage;

(3) The weighted-average life of the private loan evidenced by the private lender notes, which is _____ years, does not exceed the weighted-average remaining useful lives of the assets being financed, which is _____ years;

(4) Except as provided in the Government mortgage, the borrower has not agreed to any restrictions or limitations on future loans from the Rural Utilities Service (RUS); and

(5) This certificate is being delivered to RUS at least 10 business days before the private lender notes are to be executed.

(d) The borrower agrees that:

(1) The proceeds of the private lender notes are to be used for the construction or purchase of the plant and materials and supplies to provide telecommunications services in accordance with 7 CFR 1744.30 and such construction or purchase is expected to be completed not later than 4 years after execution of such notes;

(2) The private lender notes will provide for substantially level debt service or level principal amortization;

(3) All of the assets financed by the private lender notes will be purchased or otherwise procured in bona fide arm's length transactions; and

(4) The financing agreement with the private lender will provide that the private lender shall cease the advance of funds upon receipt of written notification from RUS that the borrower is in default under the RUS loan documents.

(e) All terms not defined herein shall have the meaning set forth in 7 CFR 1744, Subpart B.

________________________ ________________
Signed Date

____
Name

Name and Address of Borrower:

[66 FR 41763, Aug. 9, 2001]

APPENDIX D TO SUBPART B OF PART 1744—CERTIFICATION OF INDEPENDENT CERTIFIED PUBLIC ACCOUNTANT REGARDING NOTES TO BE ISSUED PURSUANT TO 7 CFR 1744.30

I/We, __(Name of Independent Certified Public Accountant)__, hereby certify the following with respect to the note or notes (the "private lender notes") to be issued by __(Name of Borrower)__ ("the borrower") on or about __(Date private lender notes are to be Signed)__, evidencing a total loan principal of ______________ dollars:

(a) The borrower has achieved a TIER of not less than 2.5 and a DSC of not less than 1.5 for each of the borrower's 2 fiscal years immediately preceding the issuance of the private lender notes. The TIER and DSC ratios achieved are as follows:

Year	TIER	DSC
____	____	____
____	____	____

(b) The borrower's equity percentage, as of the most recent fiscal year-end, was not less than 45 percent.

Year	Total Equity
____	____

______________________________ Signed

______________________________ Date

Name and address of CPA Firm:

All terms not defined herein shall have the meaning set forth in 7 CFR 1744, Subpart B.

[66 FR 41763, Aug. 9, 2001]

APPENDIX E TO SUBPART B OF PART 1744—STATEMENT, CERTIFICATION, AND AGREEMENT OF BORROWER'S PRESIDENT OF BOARD OF DIRECTORS REGARDING NOTES TO BE ISSUED PURSUANT TO 7 CFR 1744.30(e)

I ________(Name of President)________, am President of ____________(Name of Borrower)____________ (the "borrower"). The borrower proposes to issue notes (the "private lender notes"), to be dated on or about ________________ and delivered to ______(Name of Private Lender)______ (the "private lender"). I am duly authorized to make and enter into the following statements, certifications, and agreements for the purpose of inducing the United States of America (the "government"), to give automatic approval to the issuance of the private lender notes pursuant to 7 CFR 1744.30(e).

(a) The private lender:

___ is a mortgagee under the existing mortgage securing the government's loan to the borrower (the "government mortgage"); or

___ is not a mortgagee under the government mortgage and the borrower has executed the attached form of supplemental mortgage as provided in 7 CFR 1744.30(h).

(b) I have reviewed the certificate of the independent certified public accountant also being delivered to the government in connection with private lender notes to be issued pursuant to said § 1744.30(e) and concur with the conclusions expressed therein.

(c) I hereby certify that all other requirements of 7 CFR 1744.30(e) are met; said requirements being as follows:

(1) The ratio of the borrower's net plant to its total long-term debt at the end of any calendar month ending not more than 90 days prior to execution of the private lender notes is ____, which is not less than 1.6, on a pro-forma basis, after taking into account the effect of the private lender notes on the total long-term debt of the borrower;

(2) No default has occurred and is continuing under the government mortgage;

(3) The weighted-average life of the private loan evidenced by the private lender notes, which is _____ years, does not exceed the weighted-average remaining useful lives of the assets being financed, which is _____ years;

(4) Except as provided in the government mortgage, the borrower has not agreed to any restrictions or limitations on future loans from the Rural Utilities Service "RUS"; and

(5) This certificate is being delivered to RUS at least 10 business days before the private lender note or notes are to be executed.

(d) The borrower agrees that:

(1) The proceeds of the private lender notes are to be used for the construction or purchase of the tangible plant and materials and supplies to provide telecommunications services in accordance with 7 CFR 1744.30 and such construction or purchase is expected to be completed not later than 4 years after execution of such notes;

(2) The private lender notes will provide for substantially level debt service or level principal amortization;

(3) All of the assets financed by the private lender notes will be purchased or otherwise procured in bona fide arm's length transactions;

(4) The proceeds of the private lender notes will be lent to, __(Name of Subsidiary)__, a wholly-owned subsidiary of the borrower pursuant to terms and conditions agreed upon by the borrower and subsidiary;

(5) The borrower will, whenever requested by RUS, provide RUS with a copy of the financing or guarantee agreement between the borrower and the subsidiary or any similar or related material including security instruments, loan contracts, or notes issued by the subsidiary to the borrower;

(6) The borrower will promptly report to RUS any default by the subsidiary or other actions that impair or may impair the subsidiary's ability to repay its private loans; and

(7) The financing agreement with the private lender will provide that the private lender shall cease the advance of funds upon receipt of written notification from RUS that the borrower is in default under the RUS loan documents.

(e) All terms not defined herein shall have the meaning set forth in 7 CFR 1744, Subpart B.

______________________________ ______________________

Signed Date

____

Name Name and Address of Borrower:

[66 FR 41763, Aug. 9, 2001]

APPENDIX F TO SUBPART B OF PART 1744—FORM OF SUPPLEMENTAL MORTGAGE

Supplemental Mortgage and Security Agreement, dated as of_____________, (hereinafter sometimes called this "Supplemental Mortgage") is made by and among ____ (hereinafter called the "Mortgagor"), a corporation existing under the laws of the State of _________________________________, and the UNITED STATES OF AMERICA acting by and through the Administrator of the Rural Utilities Service (hereinafter called the "Government"[1]), ____ (Supplemental Lender[2]) (hereinafter called ______________________), a _____________________ existing under the laws of ___________________, and is intend to confer rights and benefits on both the Government and____ and ____ in accordance with this Supplemental Mortgage and the Original Mortgage (hereinafter defined) (the Government and the Supplemental Lenders being hereinafter sometimes collectively referred to as the "Mortgagees").

Recitals

Whereas, the Mortgagor, the Government and __________________________________ are parties to that certain Restated Mortgage (the "Original Mortgage" as identified in Schedule "A" of this Supplemental Mortgage) originally entered into between the Mortgagor, the Government acting by and through the Administrator of the Rural Utilities Service (hereinafter called "RUS"), and ____; and

Whereas, the Original Mortgage as the same may have been previously supplemented, amended or restated is hereinafter referred to as the "Existing Mortgage"; and

Whereas, the Mortgagor deems it necessary to borrow money for its corporate purposes and to issue its promissory notes and other debt obligations therefor, and to mortgage and pledge its property hereinafter described or mentioned to secure the payment of the same, and to enter into this Supplemental Mortgage pursuant to which all secured debt of the Mortgagor hereunder shall be secured on parity, and to add ________________________ as a Mortgagee and secured party hereunder and under the Existing Mortgage (the Supplemental Mortgage and the Existing Mortgage, hereinafter sometimes collectively referred to the "Mortgage"); and

Whereas, all of the Mortgagor's Outstanding Notes listed in Schedule "A" hereto is secured pari passu by the Existing Mortgage for the benefit of all of the Mortgagees under the Existing Mortgage; and

Whereas, by their execution and delivery of this Supplemental Mortgage the parties hereto do hereby secure the Additional Notes listed in Schedule "A" ((hereinafter called the Supplemental Lender Notes[3])) pari passu with the Outstanding Notes under the Existing Mortgage {and do hereby add ________________ as a Mortgagee and a secured party under the Existing Mortgage}; and

1 If the Rural Telephone Bank is a party to the original Mortgage, then "Rural Telephone Bank (herein after called the "Bank")" should be added here and the words "and the Bank" should be added after each reference to the Government.

2 If the Existing Mortgage already defines a Supplemental Lender, then the supplemental lender in the present transaction is to be called the "Second Supplemental Lender" and the supplemental mortgage should refer to both the supplemental lender and the second supplemental lender.

3 If the Second Supplemental Lender is being added to the mortgage, the reference here should be to the "Second Supplemental Lender's Notes."

Whereas, all acts necessary to make this Supplemental Mortgage a valid and binding legal instrument for the security of such notes and related obligations under the terms of the Mortgage, have been in all respects duly authorized:

Now, Therefore, This Supplemental Mortgage Witnesseth: That to secure the payment of the principal of (and premium, if any) and interest on all Notes issued hereunder according to their tenor and effect, and the performance of all provisions therein and herein contained, and in consideration of the covenants herein contained and the purchase or guarantee of Notes by the guarantors or holders thereof, the Mortgagor has mortgaged, pledged and granted a continuing security interest in, and by these presents does hereby grant, bargain, sell, alienate, remise, release, convey, assign, transfer, hypothecate, pledge, set over and confirm, pledge and grant to the Mortgagees, for the purposes hereinafter expressed, a continuing security interest in all property, rights, privileges and franchises of the Mortgagor of every kind and description, real, personal or mixed, tangible and intangible, of the kind or nature specifically mentioned herein or any other kind or nature, in accordance with the Existing Mortgage owned or hereafter acquired by the Mortgagor (by purchase, consolidation, merger, donation, construction, erection or in any other way) wherever located, including (without limitation) all and singular the following:

A. all of those fee and leasehold interests in real property set forth in Schedule "B" hereto, subject in each case to those matters set forth in such Schedule; and

B. all of those fee and leasehold interests in real property set forth in __________ the Existing Mortgage or in any restatement, amendment or supplement thereto,__________; and

C. all of the kinds, types or items of property, now owned or hereafter acquired, described as Mortgaged Property in the Existing Mortgage or in any restatement, amendment to supplement thereto as Mortgaged Property.

It is Further Agreed and Covenanted That the Original Mortgage, as previously restated, amended or supplemented, and this Supplement shall constitute one agreement and the parties hereto shall be bound by all of the terms thereof and, without limiting the foregoing:

1. All terms not defined herein shall have the meaning given in the Existing Mortgage.
2. The Supplemental Lender Notes are "notes" and "Additional Notes" under the terms of the Existing Mortgage and the Supplemental Mortgage is a supplemental mortgage under the terms of the Existing Mortgage.
3. The holders of the Supplemental Lenders Notes shall be considered as a class, so that in those instances where the Existing Mortgage providers that the holders of majority of the notes issued to other Mortgagees, voting as a class, may approve certain actions or make certain demands, so shall the holders of the Supplemental Lender Notes be considered to be a class with rights and authority equal to those of the holders of notes issued to such other Mortgagees.
4. The Maximum Debt Limit for the Existing Mortgage shall be as set forth in Schedule "A" hereto.
5. The [Second] Supplemental Lender shall immediately cease transfer of funds covered by the Supplemental Lender Notes if it receives notice that RUS has determined that the borrower's financial condition has deteriorated to a level that impairs the security or feasibility of the government's loans to the borrower.

In Witness Whereof, ____ as Mortgagor[4]

4 Spaces are to be provided for the execution by all other parties, together with the printed name and office of the executing individual and the name of the organization represented. Each execution must be acknowledged.

Supplemental Mortgage Schedule A

Maximum Debt Limit and Other Information

1. The Maximum Debt Limit is $______________________________ .

2. The Original Mortgage as referred to in the first WHEREAS clause above is more particularly described as follows:__________________________________ .

3. The Outstanding Notes referred to in the fourth WHEREAS clause above are more particularly described as follows:

4. The Additional Notes described in the fifth WHEREAS clause above are more particularly described as follows:

Supplemental Mortgage Schedule B

Property Schedule

The fee and leasehold interests in real property referred to in clause A of the granting clause are more particularly described as follows:

[66 FR 41763, Aug. 9, 2001]

Subpart C—Advance and Disbursement of Funds

SOURCE: 54 FR 12186, Mar. 24, 1989, unless otherwise noted. Redesignated at 55 FR 39396, Sept. 27, 1990.

§ 1744.60 General.

(a) The standard loan documents (as defined in 7 CFR part 1758) contain provisions regarding advances and disbursements of loan funds by telephone borrowers. This part implements certain of the provisions by setting forth requirements and procedures to be followed by borrowers in obtaining advances and making disbursements of loan and nonloan funds.

(b) This part supersedes any sections of RUS Bulletins with which it is in conflict.

§ 1744.61 [Reserved]

§ 1744.62 Introduction.

RUS is under no obligation to make or approve advances of loan funds unless the borrower is in compliance with all terms and conditions of the loan documents. The borrower shall use funds in its construction fund only to make disbursements approved by RUS.

§ 1744.63 The telephone loan budget.

When the loan is made, RUS provides the borrower a Telephone Loan Budget, RUS Form 493. This budget divides the loan into budget accounts such as "Engineering." When a contract or other document is approved by RUS, funds are encumbered from the appropriate budget account. See 7 CFR part 1753.

§ 1744.64 Budget adjustment.

(a) If more funds are required than are available in a budget account, the borrower may request RUS's approval of a budget adjustment to use funds from another account. The request shall include an explanation of the change, the budget account to be used, and a description of how the adjustment will affect loan purposes. RUS will not approve a budget adjustment that affects other loan purposes unless the borrower satisfies RUS that the additional funds are available from another source, requests a deficiency loan, or scales back the project.

(b) RUS may make a budget adjustment without a formal request by the borrower when a budget account is insufficient to encumber funds for a contract that otherwise would be approved by RUS. See 7 CFR part 1753.

§ 1744.65 The construction fund.

(a) The construction fund is used by the borrower primarily to hold advances until disbursed.

(b) All advances shall be deposited in the construction fund.

(c) RUS may require that other funds be deposited in the construction fund. These may include equity or general fund contributions to construction, service termination payments, proceeds from the sale of property, amounts recovered from insurance for losses during the construction period, and interest received on loan funds in savings or interest bearing checking accounts, and similar receipts. Deposit slips for any deposit to the construction fund shall show the source and amount of funds deposited and be executed by an authorized representative of the bank.

(d) Funds shall be disbursed only up to the amount approved for advance on the FRS as described in § 1744.66. No funds may be withdrawn from the fund except for loan purposes approved by RUS.

(e) The disbursement of nonloan funds requires the same RUS approvals as loan funds.

(f) Disbursements must be evidenced by canceled checks. The invoices and supporting documentation needed for construction contracts are specified in the contracts and in 7 CFR part 1753. Disbursements to reimburse the borrower's general funds shall be documented by a reimbursement schedule, *to be retained in the borrower's files,* that lists the construction fund check number, date, and an explanation of amounts reimbursed by budget account.

§ 1744.66 The financial requirement statement (FRS).

(a) To request advances, the borrower must submit to RUS an FRS, a description of the advances desired, and other information related to the transactions when required by RUS.

(b) The FRS is used by RUS and the borrower to record and control transactions in the construction fund. Approved contracts and other items are shown on the FRS under "Approved Purposes." Except as noted below, the amount approved for advance is 100 percent of the amount encumbered for that item. Funds are approved for advance as follows:

(1) *Construction*—(i) *Construction contracts and force account proposals.* Ninety percent of the encumbered amount (95 percent for outside plant), with the final 10 percent (5 percent) approved when RUS approves the closeout documents. When a contract contains supplement "A" (See 7 CFR part 1753), 90 percent (95 percent) of the contract is approved less materials supplied by the borrower. For the Supplement "A" materials, which are a separate entry on the FRS, 100 percent of the material cost is approved.

(ii) *Work orders.* The portion of the work order summary (See 7 CFR part 1753) determined by RUS to be for approved loan purposes.

(iii) *Work order fund.* Based on a borrower's request as described in 7 CFR part 1753.

(iv) *Real estate.* Upon request by the borrower after submission of evidence of a valid title.

(v) *Right of way procurement.* Based on the borrower's itemized costs.

(vi) *Joint use charges.* Based on copies of invoices from the other utility.

(2) *Engineering*—(i) *Preloan engineering.* Based on a final itemized invoice from the engineer.

(ii) *Postloan engineering contracts.* The amount shown on the engineering estimate, RUS Form 506, less the amount estimated for construction contract closeouts. The balance is approved when the engineering contract is closed.

(iii) *Force account engineering.* Ninety percent of the total amount of the RUS approved force account engineering proposal. The balance is approved when the force account engineering proposal is closed.

(3) *Office equipment, vehicles and work equipment.* Based on copies of invoices for the equipment.

(4) *General*—(i) *Organization and loan expenditures.* Based on an itemized list of requirements prepared by the borrower.

(ii) *Construction overhead.* Based on an itemized list of expenditures. If funds are required for employee salaries, the itemization shall include the employee's position, the period covered, total compensation for the period, and the portion of compensation attributable to the itemized construction.

(iii) *Legal fees.* Based on itemized invoices from the attorney.

(iv) *Bank stock.* Based on the requirements for purchase of class B Rural Telephone Bank stock established in the loan. Funds for class B stock will be advanced in an amount equal to 5 percent of the amount, exclusive of the amount for class B stock, of each loan advance, at the time of such advance.

(5) *Operating expenses*—(i) *Working capital—new system.* Based on the borrower's itemized estimate.

(ii) *Current operating deficiencies.* Based on a current and projected balance sheet submitted by the borrower.

(6) *Debt retirement and refinancing.* Upon release of the loan, based on the amount in the approved budget.

(7) *Acquisitions.* Based on final itemized costs, but cannot exceed the amount in the approved loan budget.

(c) Funds other than loan funds deposited in the construction fund, which shall include proceeds from the sale of property on which RUS has a lien, (lines 10 and 11 on the FRS) are reported as a credit under total disbursements. Disbursements of these funds are subject to the same RUS approvals as loan funds.

(d) The borrower shall request advances as needed to meet its obligations promptly. Generally, RUS does not approve an advance requested more than 60 days before the obligation is payable.

(e) Funds should be disbursed for the item for which they were advanced. If the borrower needs to pay an invoice for which funds have not been advanced, and disbursement of advanced funds for another item has been delayed, the latter funds may be disbursed to pay the invoice up to the amount approved for advance for that item on the FRS. The borrower shall make erasable entries on the next FRS showing the changes under "Total Advances to Date" and shall explain the changes in writing before RUS will process the next FRS.

(f) Advances will be rounded down to the nearest thousands of dollars except for final amounts.

(g) The certification on each of the three copies of the FRS sent to RUS shall be signed by a corporate officer of manager authorized by resolution of the board of directors to sign such statements. At the time of such authorization a certified copy of the resolution and one copy of RUS Form 675, Certificate of Authority, shall be submitted to RUS.

(h) The documentation required for the FRS transactions are the deposit slips, the canceled construction fund checks and the supporting invoices or reimbursement schedules. These shall be kept in the borrower's files for periodic audits by RUS.

[54 FR 12186, Mar. 24, 1989. Redesignated at 55 FR 39396, Sept. 27, 1990, as amended at 56 FR 26600, June 10, 1991]

§ 1744.67 Temporary excess construction funds.

(a) When unanticipated events delay the borrower's disbursement of advanced funds, the funds may be used as follows:

(1) With RUS loan funds for loans approved prior to November 1, 1993, or hardship loan funds, the borrower may invest the funds in 5 percent Treasury Certificates of Indebtedness—RUS Series.

(2) With RUS cost-of-money or FFB loan funds, the following apply:

(i) The borrower may invest the funds in short term securities issued by the United States Treasury.

(ii) If permitted by state law, the borrower may deposit the funds in savings accounts, including certificates of deposit, of federally insured savings institutions.

(3) Funds advanced by a guaranteed lender other than the FFB may, if so permitted by such lender, be invested under the terms and conditions described above for FFB advances.

(4) Any security or investment made under this authorization shall identify the borrower by its corporate name followed by the words "Trustee, Rural Utilities Service."

(5) All temporary investments and all income derived from them shall be considered part of the construction fund and be subject to the same controls as cash in that account.

(6) Securities and other investments shall have maturity dates or liquidating provisions that ensure the availability of funds as required for the completion of projects and the payment of obligations.

(7) Any instrument evidencing a security or other investment herein authorized to be purchased or made, may not be sold, discounted, or pledged as collateral for a loan or as security for the performance of an obligation or for any other purpose.

(8) The Administrator may, at his sole discretion, require a borrower to pledge any security or other evidence of investment authorized hereby by forwarding to him all pertinent instruments and related documentation as he may reasonably require.

(9) Borrowers shall be responsible for the safekeeping of securities and other investments.

(b) All interest and income received from investments of temporary excess funds, as described in this section, shall be deposited in the Construction Fund.

(c) The borrower shall account for investment proceeds on the next FRS submitted to RUS. RUS will make the necessary adjustments on budgetary records.

(d) The Administrator reserves the right to suspend any borrower's authorization to invest temporary excess funds contained herein if the borrower does not comply with the requirements.

(e) For RUS loans approved prior to October 1, 1991, the borrower may return advanced funds to RUS as a refund of an advance. Interest stops accruing on the refunded advance upon receipt by RUS. A refunded advance may be readvanced. A refund of an advance shall be sent to the Rural Utilities Service, United States Department of Agriculture, Collections and Custodial Section, Washington, DC, 20250. The borrower should clearly indicate that this is a refund of an advance, and not a loan payment or prepayment.

[54 FR 12186, Mar. 24, 1989. Redesignated at 55 FR 39395, Sept. 27, 1990, as amended at 58 FR 66257, Dec. 20, 1993; 84 FR 59922, Nov. 7, 2019]

§ 1744.68 Order and method of advances of telephone loan funds.

(a) Borrowers may specify the sequence of advances of funds under any combination of approved telephone loans from RUS or FFB, except that for all loans approved on or after November 1, 1993, the borrower may use loan funds:

(1) Only for purposes for which that type of loan (*i.e.* Hardship, RUS cost-of-money, or FFB) may be made; and

(2) Only in exchanges that qualify for the type of loan from which the funds are drawn.

(b) The first or subsequent advances of loan funds may be conditioned on the satisfaction of certain requirements stated in the borrower's loan contract.

(c) Normally, only one payment is made by the Automatic Clearing House (ACH) for an advance of funds.

(d) Borrowers of RUS funds may request advances by wire service only for amounts greater than $500,000 or for advances to borrowers outside the Continental United States. FFB advances in any amount over $100,000 can be sent by wire service.

(e) The following information shall be included with the FRS:

(1) Name and address of borrower's bank.

(2) If borrower's bank is not a member of the Federal Reserve System, the name and address of its correspondent bank that is a member of the Federal Reserve System.

(3) American Bankers Association (ABA) nine digit identifier of the receiving banks (routing number and check digit).

(4) Borrower's bank account title and number.

(5) Any other necessary identifying information.

[54 FR 12186, Mar. 24, 1989. Redesignated at 55 FR 39395, Sept. 27, 1990, as amended at 58 FR 66257, Dec. 20, 1993, as amended at 84 FR 59922, Nov. 7, 2019]

§ 1744.69 [Reserved]

Subpart D [Reserved]

Subpart E—Borrower Investments

SOURCE: 58 FR 52642, Oct. 12, 1993, unless otherwise noted.

§ 1744.200 General statement.

(a) RUS telephone borrowers are encouraged to utilize their own funds to participate in the economic development of rural areas, provided that such activity does not impair a borrower's ability to provide modern telecommunications services at reasonable rates or to repay its indebtedness to RUS and other lenders. When considering loans, investments, or guarantees, borrowers are expected to act in accordance with prudent business practices and in conformity with the laws of the jurisdictions in which they serve.

(b) [Reserved]

[58 FR 52642, Oct. 12, 1993, as amended at 84 FR 59923, Nov. 7, 2019]

§ 1744.201 Definitions.

As used in this subpart:

Administrator means the Administrator of the Rural Utilities Service (RUS).

Advance means any funds provided of which repayment is expected.

Affiliated company means any organization that directly, or indirectly through one or more intermediaries, controls or is controlled by, or is under common control with, the borrower.

Borrower means any organization which has an outstanding loan made by RUS or guaranteed by RUS, or which is seeking such financing.

Extension of credit means to make loans or advances.

Guarantee means to undertake collaterally to answer for the payment of another's debt or the performance of another's duty, liability, or obligation, including, without limitation, the obligations of affiliated companies. Some examples of such guarantees would include:

(1) Guarantees of payment or collection on a note or other debt instrument;

(2) Issuing performance bonds or completion bonds; or

(3) Cosigning leases or other obligations of third parties.

Maximum investment ratio means that the aggregate of all qualified investments by the borrower including the proposed qualified investment shall not be more than one-third of the net worth of the borrower.

Minimum total assets ratio means the borrower's net worth is at least twenty percent of its total assets including the proposed qualified investment.

Net plant means the sum of the balances of the following accounts of the borrower:

Account Names	Number
(1) Telecommunications plant in service	2001
(2) Property held for future telecommunications use	2002
(3) Telecommunications plant under construction-short term	2003
(4) Telecommunications plant under construction-long term	2004
(5) Telecommunications plant adjustment	2005
(6) Nonoperating plant	2006
(7) Goodwill	2007
(8) Less accumulated depreciation	3100 through 3300s
(9) Less accumulated amortization	3400 through 3600s

NOTE: All references to account numbers are to the Uniform System of Accounts (47 CFR part 32).

Net worth means the sum of the balances of the following accounts of the borrower:

Account Names	Number
(1) Capital stock	4510
(2) Additional paid-in capital	4520
(3) Treasury stock	4530
(4) Other capital	4540
(5) Retained earnings	4550

NOTE: For nonprofit organizations, owners' equity is shown in subaccounts of 4540 and 4550. All references regarding account numbers are to the Uniform System of Accounts (47 CFR part 32).

Qualified investment is defined in §1744.202(b).

RE Act means the Rural Electrification Act of 1936, as amended (7 U.S.C. 901 *et seq.*).

REA means the Rural Electrification Administration formerly an agency of the United States Department of Agriculture and predecessor agency to RUS with respect to administering certain electric and telephone loan programs.

Rural development investment is defined in §1744.202(d).

RUS means the Rural Utilities Service, an agency of the United States Department of Agriculture established pursuant to Section 232 of the Federal Crop Insurance Reform and Department of Agriculture Reorganization Act of 1994 (Pub. L. 103–354, 108 Stat. 3178), successor to REA with respect to administering certain electric and telephone programs. See 7 CFR 1700.1.

RUS mortgage means the instrument creating a lien on or security interest in the borrower's assets in connection with a loan made or guaranteed under the RE Act.

Total assets means the sum of the balances of the following accounts of the borrower:

Account Names	Number
(1) Current assets	1100s through 1300s
(2) Noncurrent assets	1400s through 1500s
(3) Total telecommunications plant	2001 through 2007
(4) Less accumulated depreciation	3100 through 3300s
(5) Less accumulated amortization	3400 through 3600s

NOTE: All references regarding account numbers are to the Uniform System of Accounts (47 CFR part 32).

Uniform System of Accounts means the Federal Communications Commission Uniform System of Accounts for Telecommunications Companies (47 CFR part 32) as supplemented by 7 CFR Part 1770, Accounting Requirements for RUS Telephone Borrowers.

[58 FR 52642, Oct. 12, 1993, as amended at 59 FR 66440, Dec. 27, 1994; 84 FR 59923, Nov. 7, 2019]

§ 1744.202 Borrowers may make qualified investments without prior approval of the Administrator.

(a) A borrower that equals or exceeds the minimum total assets ratio may make a qualified investment, defined in paragraph (b) of this section without prior written approval of the Administrator.

(b) A qualified investment is a rural development investment, defined in paragraph (d) of this section meeting the following criteria:

(1) Unless the borrower's commitment is a guarantee, extension of credit, or advance, the borrower receives any financial return accruing to such investment, or the borrower's proportionate share of such return;

(2) Unless the borrower's commitment is a guarantee, extension of credit, or advance, the borrower retains title to any asset acquired with such investment, or the borrower's proportionate share of such title; and

(3) The funds committed are the borrower's own funds. As used in this subpart, the term own funds shall not include proceeds of loans made, guaranteed or lien accommodated by RUS; funds necessary to make timely payments of principal and interest on loans made, guaranteed or lien accommodated by RUS; and funds on deposit in the cash construction fund-trustee account, as defined in the borrower's loan contract with RUS.

(c) A rural development investment will not be considered to be a qualified investment to the extent that the amount of such investments exceeds the borrower's maximum investment ratio.

(d) A rural development investment is an investment, extension of credit, advance, or guarantee by a borrower for a period longer than one year and for one or more of the following purposes:

(1) Improve the economic well-being of rural residents and alleviate the

problems of low income, elderly, minority, and otherwise disadvantaged rural residents;

(2) Improve the business and employment opportunities, occupational training and employment services, health care services, educational opportunities, energy utilization and availability, housing, transportation, community services, community facilities, water supplies, sewage and solid waste management systems, credit availability, and accessibility to and delivery of private and public financial resources in the maintenance and creation of jobs in rural areas;

(3) Improve state and local government management capabilities, institutions, and programs related to rural development and expand educational and training opportunities for state and local officials, particularly in small rural communities;

(4) Strengthen the family farm system; or

(5) Maintain and protect the environment and natural resources of rural areas.

(e) As used in paragraph (d) of this section, the term rural development investment shall include investments by a borrower in its own name, in affiliated companies, and in entities not affiliated with the borrower.

§ 1744.203 Establishing amount of rural development investment.

For purposes of determining whether a rural development investment is within the limits of the borrower's maximum investment ratio or the minimum total assets ratio, the amount of the qualified investment shall be the total amount of funds committed to the rural development project as of the date of determination. The total amount of funds committed to the rural development project includes:

(a) The principal amount of loans and advances made by the borrower;

(b) Guarantees made by the borrower; and

(c) A reasonable estimate of the amount the borrower is committed to provide to the rural development project in future years.

§ 1744.204 Rural development investments that do not meet the ratio requirements.

(a) Each borrower is authorized to make investments other than qualified investments only in accordance with the provisions of the borrower's mortgage with RUS. Without RUS's approval, the portion of any investment of funds or commitment to invest funds for any rural development investment that will exceed the borrower's maximum investment ratio or cause the borrower to fall below the minimum total assets ratio, must comply with the provisions of the RUS mortgage.

(b) RUS will consider, on a case-by-case basis, requests for approval of rural development investments not constituting qualified investments. RUS may condition such approval, if granted, on such requirements and restrictions as RUS may determine to be in the best interests of the Government, including, without limitation, the borrower's agreement to limit dividends or distributions of capital by an amount specified by RUS. Requests for such approvals must be submitted in writing to the relevant RUS regional office and shall include:

(1) A description of the rural development project and the type of investment to be made, such as a loan, guarantee, stock purchase or equity investment;

(2) A reasonable estimate of the amount the borrower is committed to provide to the rural development project including investments that may be required in the future; and

(3) A pro forma balance sheet and cash flow statement for the period covering the borrower's future commitments to the rural development project.

(c) In determining whether to approve a rural development investment that may cause the borrower to exceed the maximum investment ratio or to fall below the minimum total assets ratio in the future, RUS will consider annual increases to the borrower's net worth and total assets as might be reasonably anticipated from the borrower's normal operations.

§ 1744.205 Determinations and application of limitations described in § 1744.202.

(a) RUS will not include qualified investments, including qualified investments in affiliated companies, in calculating the amount of dividend or capital distributions a borrower may make under its RUS mortgage.

(b) A borrower's investment in its net plant shall not be considered a rural development investment for purposes of calculating the maximum investment ratio or the minimum total assets ratio.

(c) The borrower's net worth and total assets shall be determined using the balances of the respective accounts of the borrower as of December 31 of the last complete calendar year preceding the date on which the borrower's maximum investment ratio and minimum total assets ratio are calculated.

(d) All determinations required to be made under 7 U.S.C. 926 or this subpart will be made in accordance with the Uniform System of Accounts (USoA)(47 CFR part 32). References to specific USoA accounts shall include revised or replacement accounts.

§ 1744.206 Effect of subsequent failure to maintain ratios.

If an expenditure constitutes a qualified investment under the terms of this subpart, it does not cease to be a qualified investment merely because subsequently the borrower fails to maintain the maximum investment ratio or the minimum total assets ratio.

§ 1744.207 Investment not to jeopardize loan security.

A borrower shall not make a qualified investment or a rural development investment which jeopardizes:

(a) The security of loans made or guaranteed by RUS; or

(b) The borrower's ability to repay such loans under the terms and conditions as agreed.

§ 1744.208 Rural development investments before November 28, 1990.

All investments made by a borrower shall be subject to the provisions of this subpart, regardless of when the investment was made or whether it has been approved by RUS. Any restrictions required by RUS as a condition to approving a rural development investment before November 28, 1990, shall continue to be in effect to the extent that such investment exceeds the maximum investment ratio or causes the borrower to fall below the minimum total assets ratio.

§ 1744.209 Records.

(a) The records of borrowers, including records relating to qualified investments, shall be subject to the auditing procedures prescribed in part 1773 of this chapter. RUS reserves the right to review the records of the borrower relating to qualified investments to determine if the borrower is in compliance with this subpart.

(b) Borrowers shall report to RUS on the end-of-year operating report, RUS Form 479, the current status and principal amount of each qualified investment it has made or is committed to make pursuant to § 1744.202.

(Approved by the Office of Management and Budget under control number 0572–0098)

§ 1744.210 Effect of this subpart on RUS loan contract and mortgage.

(a) Except as expressly provided in this subpart, the borrower shall comply with all provisions of its loan contract with RUS, its notes issued to RUS, and the RUS mortgage, including all provisions thereof relating to investments not covered by this subpart.

(b) Nothing in this subpart shall affect any rights of supplemental lenders under the RUS mortgage, or other creditors of the borrower, to limit a borrower's investments, loans and guarantees to levels below those permitted in § 1744.202.

(c) As used in paragraph (b) of this section, supplemental lender means a creditor of the borrower, other than RUS, whose loan to the borrower is secured by the RUS mortgage.

PART 1748—POST-LOAN POLICIES AND PROCEDURES FOR INSURED TELEPHONE LOANS [RESERVED]

PART 1751—TELECOMMUNICATIONS SYSTEM PLANNING AND DESIGN CRITERIA, AND PROCEDURES

Subpart A [Reserved]

Subpart B—State Telecommunications Modernization Plan

AUTHORITY: 7 U.S.C. 901 *et seq.*, 1921 *et seq.*; Pub. L. 103–354, 108 Stat. 3178 (7 U.S.C. 6941 *et seq.*).

SOURCE: 60 FR 8174, Feb. 13, 1995, unless otherwise noted.

Subpart A [Reserved]

§§ 1751.1–1751.99 [Reserved]

Subpart B—State Telecommunications Modernization Plan

§ 1751.100 Definitions.

As used in this subpart:

Bit rate. The rate of transmission of telecommunications signals or intelligence in binary (two state) form in bits per unit time, e.g., Mb/s (megabits per second), kb/s (kilobits per second), etc.

Borrower. Any organization that has received a RUS loan designation number and which has an outstanding telephone loan made by RUS or guaranteed by RUS, or which has a completed loan application with RUS.

Emerging technologies. New or not fully developed methods of telecommunications.

Modernization Plan (State Telecommunications Modernization Plan). A State plan, which has been approved by RUS, for improving the telecommunications network of those Telecommunications Providers covered by the plan. A Modernization Plan must conform to the provisions of this subpart.

New facilities. Facilities which are wholly or partially constructed or reconstructed after a short- or medium-term requirements start date, as appropriate. This does not include connections or capacity extensions within the wired capacity of existing plant such as adding line cards to existing equipment.

Plan Developer. The entity creating the Modernization Plan for the State, which may be the State PUC, the State legislature, or a numeric majority of the RUS Borrowers within the State. When this part refers to the PUC as the Plan Developer, this includes the State legislature.

PUC (Public Utilities Commission). The public utilities commission, public service commission or other State body with such jurisdiction over rates, service areas or other aspects of the services and operation of providers of telecommunications services as vested in the commission or other body authority, to the extent provided by the State, to guide development of telecommunications services in the State. When this part refers to the PUC as the Plan Developer, this includes the State legislature.

RE Act. The Rural Electrification Act of 1936, as amended (7 U.S.C. 901 *et seq.*).

REA. The Rural Electrification Administration, formerly an agency of the United States Department of Agriculture and predecessor agency to RUS with respect to administering certain electric and telephone loan programs.

RELRA. The Rural Electrification Loan Restructuring Act of 1993 (107 Stat. 1356).

RUS. The Rural Utilities Service, an agency of the United States Department of Agriculture established pursuant to Section 232 of the Federal Crop Insurance Reform and Department of Agriculture Reorganization Act of 1994 (Pub. L. 103–354, 108 Stat. 3178 (7 U.S.C. 6941 *et seq.*)), successor to REA with respect to administering certain electric and telephone programs. See 7 CFR 1700.1.

RUS hardship loan. A loan made by RUS under section 305(d)(1) of the RE Act bearing interest at a rate of 5 percent per year.

State. Each of the 50 states of the United States, the District of Columbia, and the territories and insular possessions of the United States. This does not include countries in the Compact of Free Association.

Telecommunications. The transmission or reception of voice, data, sounds, signals, pictures, writings, or signs of all kinds, by wire, fiber, radio, light, or other visual or electromagnetic means.

Telecommunications providers. RUS Borrowers and if the Plan Developer is a PUC, such other entities providing telecommunications services as the developer of the Modernization Plan (See § 1751.101) may determine.

Wireline Service. Telecommunications service provided over telephone lines. It is characterized by a wire or wirelike connection carrying electricity or light between the subscriber and the rest of the telecommunications network. Wireline Service implies a physical connection. Although radio may form part of the circuit, it is not the major method of transmission as in radiotelephone.

[60 FR 8174, Feb. 13, 1995, as amended at 84 FR 59923, Nov. 7, 2019]

§ 1751.101 General.

(a) It is the policy of RUS that every State have a Modernization Plan which provides for the improvement of the State's telecommunications network.

(b) A proposed Modernization Plan must be submitted to RUS for approval. RUS will approve the proposed Modernization Plan if it conforms to the provisions of this subpart. Once obtained, RUS's approval of a Modernization Plan cannot be rescinded.

(c) The Modernization Plan shall not interfere with RUS's authority to issue such other telecommunications standards, specifications, requirements, and procurement rules as may be promulgated from time to time by RUS including, without limitation, those set forth in 7 CFR part 1755.

(d) The Modernization Plan must, at a minimum, apply to RUS Borrowers' wireline service areas. If a Modernization Plan is developed by the PUC, RUS encourages, but does not require, that the Modernization Plan's requirements apply to the rural service areas of all providers of telecommunications services in the State. A PUC's decision not to include non-RUS Borrowers will not prejudice RUS approval of that PUC's Modernization Plan. The PUC may also, at its option, extend coverage of the Modernization Plan to all service areas of all providers of telecommunications services in the State. In addition, while the requirements and goals contained in § 1751.106 apply only to wireline services, the PUC, at its discretion, may extend coverage of Modernization Plans to wireless or other communications services in the State as it deems appropriate. Borrower-developed Modernization Plans apply only to Borrowers.

§ 1751.102 Modernization Plan Developer; eligibility.

(a) Each PUC is eligible until February 13, 1996 to develop a proposed Modernization Plan and deliver it to RUS. RUS will review and consider for approval all PUC-developed Modernization Plans received by RUS within this one year period. The review and approval, if any, may occur after the one year period ends even though the PUC is no longer eligible to submit a proposed Modernization Plan.

(b) The PUC must notify all Telecommunications Providers in the State and other interested parties of its intent to develop a proposed Modernization Plan. The PUC is encouraged to consider all Telecommunications Providers' and interested parties' views and incorporate these views into the Modernization Plan. In the event that the PUC does not intend to develop a proposed Modernization Plan, RUS requests that the PUC inform RUS of this decision as soon as possible.

(c)(1) If the PUC is no longer eligible to develop a Modernization Plan or has informed RUS that it will not develop a Modernization Plan, as described in paragraphs (a) and (b) of this section, a majority of the Borrowers within the State may develop the Modernization Plan. If a majority of Borrowers develops the Modernization Plan, the following apply:

(i) All Borrowers shall be given reasonable notice of and shall be encouraged to attend and contribute to all meetings and other proceedings relating to the development of the Modernization Plan; and

(ii) Borrowers developing a Modernization Plan are encouraged to solicit the views of other providers of telecommunications services and interested parties in the State.

(2) There is no time limit placed on Borrowers to develop a Modernization Plan. Borrowers should be aware that certain types of loans may be restricted until a Modernization Plan is approved. See § 1751.103.

§ 1751.103 Loan and loan advance requirements.

(a) For information about loan eligibility requirements in relation to the Modernization Plan, see 7 CFR part 1735. In particular, beginning February 13, 1996, RUS will make RUS hardship and RUS cost-of-money loans for facilities and other RE Act purposes in a State only if:

(1) The State has an RUS approved Modernization Plan; and

(2) The Borrower to whom the loan is to be made is participating in the Modernization Plan for the State. A Borrower is considered to be participating if, in RUS's opinion, the purposes of the loan requested by the Borrower are consistent with the Borrower achieving the requirements stated in the Modernization Plan within the timeframe stated in the Modernization Plan unless RUS has determined that achieving the requirements is not technically or economically feasible.

(b) With regard to the three types of loans discussed in paragraph (a), only loans approved after the date the State has an RUS approved Modernization Plan are subject to complying with the Modernization Plan.

(c) For loans subject to complying with the Modernization Plan, advances will not be made if, in RUS's opinion, the advances are not consistent with achieving the requirements of the Modernization Plan.

[60 FR 8174, Feb. 13, 1995, as amended at 84 FR 59923, Nov. 7, 2019]

§ 1751.104 Obtaining RUS approval of a proposed Modernization Plan.

(a) To obtain RUS approval of a proposed Modernization Plan, the Plan Developer must submit the following to RUS:

(1) A certified copy of the statute or PUC order, if the PUC is the Plan Developer, or a written request for RUS approval of the proposed Modernization Plan signed by an authorized representative of the Plan Developer, if a majority of Borrowers is the Plan Developer; and

(2) Three copies of the proposed Modernization Plan.

(b) Generally, RUS will review the proposed Modernization Plan within (30) days and either:

(1) Approve the Modernization Plan if it conforms to the provisions of this subpart in which case RUS will return a copy of the Modernization Plan with notice of approval to the Plan Developer; or

(2) Not approve the proposed Modernization Plan if it does not conform to the provisions of this subpart. In this event, RUS will return the proposed Modernization Plan to the Plan Developer with specific written comments and suggestions for modifying the proposed Modernization Plan so that it will conform to the provisions of this subpart. If the Plan Developer remains eligible, RUS will invite the Plan Developer to submit a modified proposed Modernization Plan for RUS consideration. This process can continue until the Plan Developer gains approval of a proposed Modernization Plan unless the Plan Developer is a PUC whose eligibility has expired. If a PUC's eligibility has expired, RUS will return the proposed Modernization Plan unapproved. Because RUS does not have authority to extend the term of a PUC's eligibility, RUS recommends that the PUC submit a proposed Modernization Plan at least 90 days in advance of February 13, 1996 to allow time for this process.

§ 1751.105 Amending a Modernization Plan.

(a) RUS understands that changes in standards, technology, regulation, and the economy could indicate that an RUS-approved Modernization Plan should be amended.

(b) The Plan Developer of the Modernization Plan may amend the Modernization Plan if RUS finds the proposed changes continue to conform to the provisions of this subpart.

(c) The procedure for requesting approval of an amended Modernization Plan is identical to the procedure for a proposed Modernization Plan except that there are no time limits on the eligibility of the Plan Developer.

(d) The existing Modernization Plan remains in force until RUS has approved the proposed amended Modernization Plan.

(e) RUS may from time to time revise these regulations to incorporate newer technological and economic standards that RUS believes represent more desirable goals for the future course of telecommunications services. Such revisions will be made in accordance with the Administrative Procedure Act. These revisions shall not invalidate Modernization Plans approved by RUS but shall be used by RUS to determine whether to approve amendments to Modernization Plans presented for RUS approval after March 15, 1995.

§ 1751.106 Modernization Plan; requirements.

(a) The requirements for a Modernization Plan as stated in RELRA are:

(1) The plan must provide for the elimination of party line service.

(2) The plan must provide for the availability of telecommunications services for improved business, educational, and medical services.

(3) The plan must encourage and improve computer networks and information highways for subscribers in rural areas.

(4) The plan must provide for—

(i) Subscribers in rural areas to be able to receive through telephone lines—

(A) Conference calling;

(B) Video images; and

(C) Data at a rate of at least 1,000,000 bits of information per second; and

(ii) The proper routing of information to subscribers.

(5) The plan must provide for uniform deployment schedules to ensure that advanced services are deployed at the same time in rural and nonrural areas.

(6) The plan must provide for such additional requirements for service standards as may be required by the Administrator.

(b) To implement the requirements of the law described in paragraph (a) of this section, RUS has set minimum requirements as described in paragraphs (i) and (j) of this section. They are grouped into short-term and medium-term requirements. RUS has also included long-term goals which are not requirements. The Modernization Plan must meet all of the statutory requirements of RELRA and shall provide that short- and medium-term requirements be implemented as set forth in this section of the regulation except that the PUC, if it is the Plan Developer, or RUS, if a majority of Borrowers is the Plan Developer, may approve extensions of time if the required investment is not economically feasible or if the best available telecommunications technology lacks the capability to enable the Telecommunications Provider receiving the extension to comply with the Modernization Plan. Extensions shall be granted only on a case-by-case basis and generally shall not exceed a total of five years from the first such extension granted to the Telecommunications Provider.

(c) Each State's Modernization Plan shall be a strategic development proposal for modernizing the telecommunications network of the Telecommunications Providers covered by the Modernization Plan. In addition to implementing the requirements described in paragraphs (a), (i), and (j) of this section, the Modernization Plan shall include a short engineering description of the characteristics of a future telecommunications structure that would enable all Telecommunications Providers to achieve the requirements and goals of the Modernization Plan.

(d) Within the scope of § 1751.101(d), if the Plan Developer is the PUC, the Modernization Plan shall name the Telecommunications Providers in the State, in addition to Borrowers, that are covered by the Modernization Plan.

(e) The Modernization Plan must require that the design of the network provided by Telecommunications Providers allow for the expeditious deployment and integration of such emerging technologies as may from time to time become commercially feasible.

(f) The Modernization Plan must provide guidelines to Telecommunications Providers for the development of affordable tariffs for medical links and distance learning services.

(g) With regard to the uniform deployment requirement of the law restated in paragraph (a)(5) of this section, if services cannot be deployed at the same time, only the minimum feasible interval of time shall separate availability of the services in rural and nonrural areas.

(h) The Modernization Plan must make provision for reliable powering of ordinary voice telephone service operating over those portions of the telecommunications network which are not network powered. In the event of electric utility power outages, an alternative source of power must be available to ensure reliable voice service.

(i) *Short-term requirements.* (1) The "short-term requirements start date" is the date one year after the date RUS approves the Modernization Plan for the State.

(2) All New Facilities providing Wireline Service after the short-term requirements start date, even if the construction began before such date, shall be constructed so that:

(i) Every subscriber can be provided 1-party service.

(ii) The New Facilities are suitable, as built or with additional equipment, to provide transmission and reception of data at a rate no lower than 1 Mb/sec.

(3) All switching equipment installed by a Telecommunications Provider after the short-term requirements start date shall be capable of:

(i) Providing custom calling features. At a minimum, custom calling features must include call waiting, call forwarding, abbreviated dialing, and three-way calling; and

(ii) Providing E911 service for areas served by the Telecommunication Provider when requested by the government responsible for this service.

(j) *Medium-term requirements.* (1) The "medium-term requirements start date" is the date six years after the date RUS approves the Modernization Plan for the State, or such earlier date as the Modernization Plan shall provide.

(2) All New Facilities providing Wireline Service after the medium-term requirements start date, even if the construction began before such date, shall be capable, as built or with additional equipment, of transmitting video to a subscriber. The video must be capable of depicting a reasonable representation of motion. The frame rate, resolution, and other measures of audio and video quality shall be determined by the Plan Developer.

(3) No later than the medium-term requirements start date, all switching equipment of Telecommunications Providers covered by the Modernization Plan must be capable of providing E911 service when requested by the government responsible for this service.

(4) No later than five years after the medium-term requirements start date, one-party service must be provided upon demand to any subscriber of a Telecommunications Provider covered by the Modernization Plan.

(k) *Long-term goals.* RUS suggests, but does not require, that the provisions of each Modernization Plan be consistent with the accomplishment of the following:

(1) The elimination of party line service.

(2) For subscribers that desire the service, universal availability of:

(i) Digital voice and data service (56–164 kb/sec).

(ii) Service that provides transmission and reception of high bit rate (no less than 1 Mb/sec) data.

(iii) Service that provides reception of video as described in paragraph (j)(2) of this section.

PART 1752—SERVICING OF TELECOMMUNICATIONS PROGRAMS

Sec.

AUTHORITY: 7 U.S.C. 1981(b)(4), 7 U.S.C. 901 *et seq.* and 7 U.S.C. 950aaa *et seq.*

SOURCE: 85 FR 10558, Feb. 25, 2020, unless otherwise noted.

§ 1752.1 Purpose.

This part prescribes the policies and procedures for loan and grant servicing for financial assistance made under the Rural Utilities Service (RUS) Telecommunications Infrastructure Loan Program, Rural Broadband Program, Distance Learning and Telemedicine Program, Broadband Initiatives Program, and the Rural e-Connectivity Pilot Program (in this part collectively referred to as the "RUS Telecommunications Programs").

§ 1752.2 Objectives.

The purpose of loan and grant servicing functions is to assist recipients to meet the objectives of loans and grants, repay loans on schedule, comply with agreements and protect RUS' financial interests. The provisions of this part will ensure recipients comply with any revised terms in repayment on loans and ensures serving actions are handled consistently by the Agency.

§ 1752.3 Definitions.

The terms and conditions provided in this section are applicable to this part only. All financial terms not defined in this section shall have the commonly-accepted meaning under Generally Accepted Accounting Principles.

Acceleration. A written notice informing the Borrower that the total unpaid principal and interest is due and payable immediately.

Administrator. Administrator of the Rural Utilities Service.

Agency. The Rural Utilities Service, an agency of the United States Department of Agriculture's Rural Development mission area.

Assumption of debt. Agreement by one party to legally bind itself to repay the debt of the RUS borrower.

Borrower. Recipient of loan funding under a RUS Telecommunications Program.

Broadband system. The telecommunications or broadband network financed with RUS loan and/or grant funding or maintained by the Borrower and contained as part of the collateral to the loan.

Cancellation. Final discharge of debt with a release of liability.

Charge-off. Write-off of a debt and termination of servicing activity without release of liability. A charge-off is a decision by the Agency to remove debt from Agency receivables, however, future payments may be received.

Collateral. Means the assets, equipment and/or revenues pledged as security for the loan as defined in the loan documents.

Disposition of facility. Relinquishing control of a facility to another entity.

Loan Documents. All associated loan agreements, loan and security agreements, loan/grant agreements, mortgages, and promissory notes, as applicable.

Liquidation. Satisfaction of a debt through the sale of a Borrower's assets and cancellation of liabilities.

Parity lien. A lien having an equal lien position to another lender's lien on a Borrower's asset.

Rural Utilities Service (RUS). An agency of the United States Department of Agriculture's Rural Development mission area.

Settlement. Compromise, adjustment, cancellation, or charge-off of a debt owed to the Agency. The term "settlement" is used to refer to any of these actions, whether individually or collectively.

Unliquidated obligations. Obligated loan funds that have not been advanced.

Voluntary conveyance. A method by which title to security is voluntarily transferred to the Federal Government.

§ 1752.4 Availability of forms, bulletins, and procedures.

Forms, bulletins, and procedures referenced in this part are available online at *https://www.rd.usda.gov/publications/regulations-guidelines.*

§ 1752.5 Monetary default by Borrower.

A defaulting Borrower's primary responsibility is to expeditiously bring the delinquent account current. If a monetary default exceeds 60 days, RUS will attempt to discuss the situation with the Borrower and make the Borrower aware of options that may be available. In considering options, the prospects for providing a permanent cure without adversely affecting the risk to the Agency is the paramount objective. RUS will also work with entities that are not in monetary default but whose financial position is such that, without RUS action, a monetary default is imminent within the next 24 months, as evidenced by a financial forecast provided by the Borrower. RUS receives quarterly financial reports and annual audits from borrowers and actively monitors the borrower's Times Interest Earned Ratio (TIER), Current Ration, Debt Service Coverage Ratio, and Net Worth.

§ 1752.6 Request for special servicing action.

(a) Special servicing actions include, but are not limited to, one or more of the following:

(1) Consent to additional, unsecured debt;

(2) Parity lien;

(3) Reamortization or rescheduling of debt payments;

(4) Deferment of principal and/or interest;

(5) Interest rate adjustment;

(6) Transfer of collateral and assumption of debt;

(7) Sale or exchange of loan collateral;

(8) Sale of the note; and

(9) Debt settlement.

(b) In order for the Agency to consider one or more of the curative actions cited in paragraph (a) of this section, the Borrower must submit a written request to RUS.

(1) The written request must contain the following items:

(i) A detailed explanation of the request and why it is needed.

(ii) Most recent audited financial statements for the Borrower.

(iii) Borrower's Pro Forma 5-year financial forecast, which includes an Income Statement, Balance Sheet, and Statement of Cash Flows, 2 years of historical data, current year data and a 5-year forecast, with detailed supporting assumptions. Additionally, in order to request assistance under this paragraph (b)(1)(iii), the Borrower must make a showing that the account is delinquent and cannot be brought current within one year, or that the Borrower will become delinquent within 24 months, as demonstrated in the Pro Forma.

(iv) Existing and projected subscriber numbers and service tiers, along with pricing for each tier. Additionally, for companies receiving support from the Federal Communications Commission, a detailed forecast of the support revenue, certified by a cost consultant, must be included.

(v) Current organizational chart for the Borrower, related entities, and affiliated companies, as well as information relating to ownership interest in the Borrower and its related entities.

(vi) A complete list of all collateral and steps the Borrower is taking to preserve the collateral.

(2) The Agency may request the additional documents in paragraphs (b)(2)(i) through (iv) of this section after reviewing the Borrower's servicing request:

(i) An appraisal in order to determine the adequacy of loan security or repayment ability;

(ii) An itemized list of estimated liquidation expenses expected to be incurred along with justification for each expense;

(iii) A legal opinion regarding RUS' interests in the impacted collateral and supporting evidence, in the form of Uniform Commercial Code Statements and filed Mortgages, that RUS maintains a first lien position on all assets of the Borrower, or such collateral as mandated by the Loan Documents; and

(iv) Such other documents that may be relevant in individual cases, as determined by RUS.

(3) When submitting a request for a servicing action, the distressed Borrower must consent to the following during the request and for the duration of the servicing action:

(i) *On-site visit.* A Management Analysis Profile (MAP) visit of the Borrower's entire operation;

(ii) *RUS priority payment.* Borrowers must agree that no other creditors will be paid without RUS consent, if RUS is not receiving full principal and interest payments;

(iii) *Additional reporting and monitoring.* Throughout the term of the servicing action(s), RUS will require increased frequency and/or additional details to the reporting and monitoring required under the terms of the Loan Documents; and

(iv) *Additional controls and limitations.* RUS may require additional controls and limitations such as segregation of accounts, RUS review of expenditures, etc.

(c) False information provided by a Borrower, or by entities acting on behalf of the Borrower, will give rise to the immediate termination of any servicing action(s).

§ 1752.7 Civil rights and requirements.

(a) *Equal opportunity and nondiscrimination.* The Agency will ensure that equal opportunity and nondiscriminatory requirements are met in accordance with the Equal Credit Opportunity Act and 7 CFR part 15. In accordance with Federal civil rights law and U.S. Department of Agriculture (USDA) civil rights regulations and policies, the USDA, its agencies, offices, and employees, and institutions participating in or administering USDA programs are prohibited from discriminating based on race, color, national origin, religion, sex, gender identity (including gender expression), sexual orientation, disability, age, marital status, family/parental status, income derived from a public assistance program, political beliefs, or reprisal or retaliation for prior civil rights activity, in any program or activity conducted or funded by USDA (not all bases apply to all programs).

(b) *Civil rights compliance.* Recipients of Federal assistance under this part must comply with the Americans with Disabilities Act of 1990, Title VI of the Civil Rights Act of 1964, and Section 504 of the Rehabilitation Act of 1973. Prior to determining eligibility of any servicing action under this part, the Agency will determine that the Borrower is in compliance with all civil rights requirements of the latest Civil Rights Compliance Review conducted by the Agency.

(c) *Discrimination complaints.* Persons believing they have been subjected to discrimination prohibited by this section may file a complaint personally, or by an authorized representative with USDA, Director, Office of Adjudication, 1400 Independence Avenue SW, Washington, DC 20250. A complaint must be filed no later than 180 days from the date of the alleged discrimination, unless the time for filing is extended by the designated officials of USDA or the Agency.

§§ 1752.8–1752.10 [Reserved]

§ 1752.11 Consent to additional, unsecured debt.

(a) An additional, unsecured loan from another lender to the Borrower may be approved subject to the conditions set forth in this section. In order to request assistance under this section, the Borrower must make a showing that the additional debt will cure any existing or projected delinquency. Additionally, the following requirements must be met, as determined by RUS:

(1) The additional debt will not disadvantage RUS's standing or lien on any of the collateral already pledged to RUS;

(2) The additional debt will not adversely impact the continued financial viability of the Borrower or the Borrower's ability to carry out the purposes of the RUS loan;

(3) The debt is needed to resolve short-term, negative cashflow problems; and

(4) The Borrower is in good standing with the Agency or will become so with the additional debt.

(b) In the case where all assets of the Borrower are not secured by the Government's debt, the Borrower may request additional debt that is secured by collateral that is not subject to the Government's security interest.

§ 1752.12 Parity lien.

A Borrower's request for parity may be approved subject to the conditions set forth in this section. In order to request assistance under this section, the Borrower must make a showing that the amount of new debt is at least equal to the amount of the collateral being added and will cure any existing or projected delinquency. The following factors will be considered in assessing whether the request is in the Government's best interest:

(a) The value of the added assets compared with the amount of new debt to be secured;

(b) The value of the assets already pledged under the Loan Documents, and any effects of the proposed transaction on the value of those assets;

(c) The ratio of the total outstanding debt secured under the Loan Documents to the value of all assets pledged as security under the Loan Documents;

(d) The Borrower's ability to repay its debt owed to the Government;

(e) The overall financial viability of the Borrower; and

(f) That the Borrower is in good standing with the Agency or will become so with the parity lien; and

(g) Such other conditions as may be imposed by the Agency on a case-by-case basis, as determined by RUS.

§ 1752.13 Reamortization of or rescheduling of the debt payments.

A reamortization or rescheduling of debt payments may be approved subject to the conditions set forth in this section. In order to request a reamortization or rescheduling of debt payments, the Borrower must make a showing that the Borrower does not have access to other sources of capital or alternatives for resolving the delinquency, and that the reamortization or rescheduling of debt payment will cure any existing or projected delinquency. Reamortizations or rescheduling of debt will be limited to 10 years beyond the original maturity date. Additionally, the following requirements must be met, as determined by RUS:

(a) The Borrower has cooperated with RUS in exploring alternative servicing options and has acted in good faith with regard to eliminating the delinquency and complying with its loan agreements and Agency regulations;

(b) Any management deficiencies identified by RUS have been corrected or the Borrower has submitted a plan acceptable to RUS to correct any deficiencies;

(c) The Borrower has presented a budget which clearly indicates that it is able to meet the proposed payment schedule and the reamortization or rescheduling of debt payments will ensure the continued financial viability of the Borrower;

(d) The Agency will consider the useful life of the facilities along with the level of debt service payments that the Borrower can contribute when determining the appropriate term to place on any reamortized loan; and

(e) Such other conditions as may be imposed by the Agency on a case-by-case basis, as determined by RUS.

§ 1752.14 Deferment of principal and/or interest payments.

A deferment of principal and/or interest payments which will continue the original purpose of the loan may be approved subject to the conditions set forth in this section.

(a) *Principal-only deferrals.* In order to request a principal deferral, the Borrower must make a showing that at the end of the deferment period the Borrower's financial position has improved and the Borrower is able to make full principal and interest payments, curing any delinquency or projected delinquency. Deferments of principal will be limited to no more than 36 months. Additionally, the following requirements must be met, as determined by RUS:

(1) Any management deficiencies identified by RUS have been corrected or the Borrower has submitted a plan

acceptable to RUS to correct any deficiencies;

(2) The Borrower has presented a budget which clearly indicates that it is able to meet the new proposed payment schedule and after the end of the deferral period is able to resume making full principal and interest payments while maintaining a positive cashflow position;

(3) Unless authorized by prior RUS written consent, the Borrower will only use funds otherwise due and payable under the RUS Note for the benefit of the broadband system. Such expenditures include, but are not limited to, costs to complete any necessary construction of the Project, costs to connect additional subscribers, marketing and sales costs, and other such costs that are necessary to maximize the value of the broadband system; and

(4) The Borrower will comply with such other conditions as may be imposed by the Agency on a case-by-cases basis, as determined by RUS.

(b) *Principal and interest deferrals.* A principal and interest deferral shall only be approved when the Borrower has demonstrated that it is the only option for the Agency to avoid foreclosure and is in the best interest of the Government to avoid a substantial loss to the Government. Additionally, principal and interest deferrals may be approved if the Borrower and RUS have agreed to a public sale of the broadband system and such a deferral is needed to provide time to complete the sale of the broadband system. Principal and interest deferrals will be limited to no more than 24 months, unless extended by the Agency for good cause and full cooperation of the Borrower. Additionally, the following requirements must be met, as determined by RUS:

(1) The Borrower has cooperated with RUS in exploring alternative servicing options and has acted in good faith with regard to eliminating the delinquency and complying with its Loan Documents and Agency regulations;

(2) Any management deficiencies identified by RUS have been corrected or the Borrower has submitted a plan acceptable to RUS to correct any deficiencies; and

(3) Unless authorized by prior RUS written consent, the Borrower will only use funds otherwise due and payable under the RUS Note for the benefit of the broadband system. Such expenditures include, but are not limited to, costs to execute a sale of the broadband system, costs to complete any necessary construction of the Project, costs to connect additional subscribers, marketing and sales costs, and other such costs that are necessary to maximize the value of the broadband system for an eventual sale;

(4) In cases when the Borrower and RUS have agreed to a public sale of the broadband system, the Borrower agrees that within 30 days of the execution of the deferral agreement, the Borrower will develop a process and timeline for the sale of the broadband system, in form and substance satisfactory to RUS, and will continually execute on those plans in order to effectuate a public sale of the broadband system; and

(5) The Borrower will comply with such other conditions as may be imposed by the Agency on a case-by-cases basis, as determined by RUS.

§ 1752.15 Interest rate adjustments.

Interest rate reductions may be approved subject to the conditions set forth in this section. In order to request an interest rate reduction, the Borrower must make a showing that the Borrower does not have access to other sources of capital or alternatives to resolve the delinquency, and the interest rate adjustment will cure any existing or projected delinquency. Additionally, the following requirements must be met, as determined by RUS:

(a) The Borrower has cooperated with RUS in exploring alternative servicing options and has acted in good faith with regard to eliminating the delinquency and complying with its loan agreements and Agency regulations;

(b) Any management deficiencies identified by RUS have been corrected or the Borrower has submitted a plan acceptable to RUS to correct any deficiencies;

(c) The Borrower has presented a budget which clearly indicates that it is able to meet the proposed payment schedule and the interest rate reduction will improve the financial viability of the Borrower;

(d) The Borrower has agreed to not maintain cash or cash reserves beyond what is reasonable at the time of interest rate adjustment to meet debt service, operating, and reserve requirements; and

(e) The Borrower will comply with such other conditions as may be imposed by the Agency on a case-by-cases basis, as determined by RUS.

§ 1752.16 Transfer of collateral and assumption of debt.

A transfer of collateral and assumption of debt may be approved subject to the conditions set forth in this section. In order to request assistance under this section, the Borrower must make a showing that the transfer of collateral and assumption of debt will improve the likelihood that the government will be repaid and maximize the Agency's recovery on such loans. Such actions will be subject to the following requirements:

(a) The transfer will not be disadvantageous to the Government, as determined by RUS;

(b) The Agency has concurred to plans for disposition of funds in any reserve account, including project construction bank accounts;

(c) The transferee will assume all of the Borrower's responsibilities regarding the loan(s) and will accept the original loan conditions, as well as any others that may be imposed by the Agency;

(d) There must be no lien, judgement, or similar claims of other parties against the loan collateral being transferred, and once transferred, such collateral may not be subject to the lien, judgement, or similar claims of other parties of the transferee;

(e) Title to all assets must be conveyed to the transferee; and

(f) The Borrower will comply with such other conditions as may be imposed by the Agency on a case-by-case basis, as determined by RUS.

§ 1752.17 Sale or exchange of loan collateral.

A cash sale of all or a portion of a Borrower's assets or an exchange of security property for Borrowers in a distressed situation may be approved subject to the conditions set forth in this section. In order to request assistance under this section, the Borrower must make a showing that the sale or exchange of collateral is in the best interest of the Government to avoid a substantial loss to the Government. Additionally, the following requirements must be met, as determined by RUS:

(a) If a sale of all of the assets, that the consideration is for the full amount of the debt or the present fair market value as determined by an independent appraiser that has been approved by RUS, and which addresses any conditions of the appraisal as may be imposed by RUS; and

(b) If the sale is for a portion of the assets, that the remaining property is adequate security for the loan and that the transaction will not adversely affect the Agency's security position; and provided that any proceeds remaining after paying reasonable and necessary selling expenses, as approved by the Agency in advance, are to be used for the following purposes:

(1) Repayment of the RUS debt, and other non-RUS debt if secured by a parity lien with the Agency; and/or

(2) Improvement of the broadband network or other facilities of the Borrower, including customer premise equipment and other equipment needed to upgrade the broadband network, if necessary to improve the Borrower's ability to repay the loan; and

(c) Any grant assets in the sale of collateral that were financed with Agency grants must follow the disposition rules as stated in the Loan Documents or Grant/Loan Documents.

§ 1752.18 Sale of the note.

In the event of one or more incidents of default by the Borrower that cannot or will not be cured within a reasonable period of time, the Agency may sell the note. A decision to sell the note may be made when the Agency determines that the monetary default cannot be cured through the other actions as outlined in this part, or it has been determined that it is in the best interest of the Agency. The decision to sell the note should be made as soon as possible when one or more of the following exist:

(a) A loan is 90 days behind on any scheduled payment and the Agency and Borrower have not been able to cure the delinquency through actions such as those contained in this part;

(b) It is determined that delaying sale of the note will jeopardize full recovery on the loan; or

(c) The Borrower is uncooperative in resolving the delinquency or the Agency has reason to believe the Borrower is not acting in good faith, and it would improve the position of the Agency to sell the note immediately.

§ 1752.19 Debt settlement.

Debts will not be settled directly by the Agency if:

(a) Referral to the Office of Inspector General and/or to Office of General Counsel is contemplated or pending because of suspected criminal violation;

(b) Civil action to protect the interest of the Government is contemplated or pending;

(c) An investigation for suspected fiscal irregularity is contemplated or pending;

(d) The Borrower is uncooperative in resolving the delinquency or the Agency has reason to believe the Borrower is not acting in good faith; or

(e) The debt has been referred to the Department of Justice, such as in bankruptcy proceedings.

§§ 1752.20–1752.24 [Reserved]

§ 1752.25 Special terms.

If the Administrator determines the servicing actions in this part would not protect the Government's interest due to unique circumstances of the debtor, the Agency reserves the right to negotiate special terms to maximize the Government's recovery on the debt.

§ 1752.26 No rights to special servicing actions.

Nothing in this part should be assumed guaranteed as a right to the Borrower for any of the special servicing actions noted in this part.

§ 1752.27 Confidentiality of borrower information.

Borrowers are encouraged to identify and label any confidential and proprietary information contained in their applications. The Agency will protect confidential and proprietary information from public disclosure to the fullest extent authorized by applicable law, including the Freedom of Information Act, as amended (5 U.S.C. 552), the Trade Secrets Act, as amended (18 U.S.C. 1905), the Economic Espionage Act of 1996 (18 U.S.C. 1831 *et seq.*), and Communications Assistance for Law Enforcement Act (CALEA) (47 U.S.C. 1001 *et seq.*).

§ 1752.28 Interest accrual.

(a) The Agency may determine to stop accruing interest if the account has remained delinquent for a period of 18 months or more, and the Agency has determined, in its sole discretion, that it will not recover the full outstanding principal balance on the loan.

(b) Notwithstanding paragraph (a) of this section, the Administrator may waive the accrual of interest on any outstanding delinquent debt, if in the sole determination of the Administrator, such waiver facilitates and maximizes the Government's recovery of the debt, such as under a voluntary foreclosure by the Borrower.

§ 1752.29 Communications laws.

Borrowers must comply with all applicable Federal and state communications laws and regulations, including, for example, the Communications Act of 1934, as amended (47 U.S.C. 151 *et seq.*), the Telecommunications Act of 1996, as amended (Pub. L. 104–104, 110 Stat. 56 (1996), and CALEA. For further information see *http://www.fcc.gov.*

§ 1752.30 Information collection and reporting requirements.

Copies of all forms and instructions referenced in this part may be obtained from RUS. Data furnished by Borrowers will be used to determine eligibility for certain servicing actions. Furnishing the data is voluntary; however, the failure to provide data could result in a servicing action being denied or the Agency taking adverse action against the Borrower to collect funds. The collection of information is vital to RUS to ensure compliance with the provisions of this part. The information collection requirements contained in this part have been approved by the Office of Management and Budget (OMB) under OMB Control Number 0572–0153.

§ 1752.31 Authorized signatories.

Only the RUS Administrator can bind the Government to the expenditure of funds. Notwithstanding anything contained in this part, however, any settlement resulting in the reduction of $500,000 or more in payment to the Government, inclusive of attorney's fees, shall be approved by the Under Secretary for Rural Development, or higher.

PART 1753—TELECOMMUNICATIONS SYSTEM CONSTRUCTION POLICIES AND PROCEDURES

Subpart A—General

Subpart B—Engineering Services

Subpart C [Reserved]

Subpart D—Construction of Buildings

Subpart E—Purchase and Installation of Central Office Equipment

Subpart F—Outside Plant Major Construction by Contract

Subpart G—Outside Plant Major Construction by Force Account

Subpart H—Purchase and Installation of Special Equipment

Subpart I—Minor Construction

Subpart J—Construction Certification Program

AUTHORITY: 5 U.S.C. 501, 7 U.S.C. 901 *et seq.*

Subpart A—General

SOURCE: 54 FR 39267, Sept. 25, 1989, unless otherwise noted. Redesignated at 55 FR 39396, Sept. 27, 1990.

§ 1753.1 General.

(a) The standard RUS Telecommunications Loan Documents contain provisions regarding procurement of materials and equipment and construction of telecommunications facilities by telecommunications borrowers. This part implements certain of the provisions by setting forth requirements and procedures. Borrowers shall follow these requirements and procedures whenever using loan funds to purchase materials and equipment or perform construction, unless they have received the Administrator's written approval to do otherwise.

(b) The typical procedure followed in constructing a project financed by an RUS loan begins with the prospective borrower obtaining the necessary preloan engineering and developing a complete loan application, including an LD (See 7 CFR part 1737). If a loan is approved and all prerequisites to advance of funds are satisfied, the borrower may proceed with the purchase and installation of materials and equipment and the construction of telephone facilities pursuant to this part 1753. Subpart A describes

(1) RUS's general requirements with respect to steps to be taken after the loan is approved and before construction begins (See § 1753.3),

(2) RUS requirements with respect to methods of construction (See §§ 1753.5 and 1753.6),

(3) RUS requirements regarding sealed competitive bidding and negotiated bidding of construction contracts (See §§ 1753.6 and 1753.9),

(4) RUS standards for materials, equipment, and construction financed with loan funds (See § 1753.7), and

(5) RUS requirements for subcontracts and contract amendments covering construction financed with loan funds (See §§ 1753.10 and 1753.12).

(c) Each borrower is responsible for the construction of its facilities and for the procurement of materials and equipment that are best suited to its needs.

(d) If contracts, P&S, or other methods of procurement are subject to RUS approval pursuant to the provisions of the loan contract, as implemented by this part, RUS will review the documents or proposals submitted and notify the borrower in writing of approval or disapproval. RUS may withhold approval if, in RUS's judgment:

(1) The P&S or contract will not accomplish loan purposes.

(2) Provisions of the P&S or contract will add unnecessary expense to the project.

(3) The proposal, method of procurement, or P&S do not conform to RUS engineering criteria or construction standards, or if they present unacceptable loan security risks to RUS.

(4) The P&S or contract have been modified.

(e) The requirements and procedures covering procurement of architectural and engineering services are described in subpart B of this part.

(f) Single copies of RUS forms cited in this part are available from Administrative Services Division, Rural Utilities Service, United States Department of Agriculture, Washington, DC 20250–1500. These RUS forms may be reproduced.

[54 FR 39267, Sept. 25, 1989. Redesignated at 55 FR 39396, Sept. 27, 1990, as amended at 64 FR 16604, Apr. 6, 1999]

§ 1753.2 Definitions.

For the purpose of this part 1753:

Alternate— A solicitation for a bid adjustment for a specified deviation from the Plans and Specifications.

Architect— A person registered as an architect in the state where construction is performed, or a person on the borrower's staff, approved by RUS, authorized to perform architectural services.

Bid guarantee— A bid bond or certified check required of contractors bidding on construction work to ensure that the bidder, if successful, will furnish a satisfactory performance bond ensuring completion of work.

Central office building— The facility housing the central office equipment.

Central office equipment— Switching and signaling equipment that performs call origination and completion functions for subscribers.

Closeout documents— The documents required to certify satisfactory completion of all obligations under a contract or force account proposal.

Construction— Purchase and installation of telecommunications facilities in a borrower's system using loan funds.

Contract—"Contract" means collectively the Seller's Proposal, Seller's Technical Proposal including any general or feature descriptions, equipment lists, Seller's responses to the Buyer's specifications and Performance Requirements, the Buyer's Acceptance, Articles I through X herein, the Performance Requirements, and the Contractor's Bond when required by the Buyer.

Contract construction— Construction and installations performed using an RUS contract form. See 7 CFR 1755.93.

Engineer— A person registered as an engineer in the state where construction is performed, or a person on the borrower's staff, approved by RUS, authorized to perform engineering services.

FAP (force account proposal)—The borrower's detailed plans submitted to RUS for force account construction.

Force account construction—Construction performed by the borrower's employees under an RUS approved FAP, with the borrower furnishing all materials, equipment, tools, and transportation.

FRS—RUS Form 481 (OMB control number 0572–0023), Financial Requirement Statement.

GFR—RUS General Field Representative.

Installation—The act of setting up or placing in position equipment for service or use in the borrower's system.

Interim construction—The purchase of equipment or the conduct of construction under an RUS-approved plan of interim financing. See 7 CFR part 1737.

Interim financing—Funding for a project which RUS has acknowledged may be included in a loan, should said loan be approved, but for which RUS loan funds have not yet been made available.

Labor and materials—All the labor and materials required for construction.

LD (loan design)—Supporting data for a loan application. See 7 CFR part 1737.

Loan—Any loan made or guaranteed by RUS. See 7 CFR part 1735.

Loan funds—Funds provided by RUS through direct or guaranteed loans. See 7 CFR part 1744 subpart C.

Loan purposes—The high level objectives of the loan are to fund the construction. These purposes are first stated in the characteristics letter described in 7 CFR 1737.80, which is sent to the applicant to offer a loan after RUS has completed its preloan studies.

Major construction—A telecommunications plant project estimated to cost more than $250,000, including all labor and materials.

Minor construction—A telecommunications plant project estimated to cost $250,000 or less, including all labor and materials.

Minor errors or irregularities—A defect or variation in a seller's bid that is a matter of form and not of substance. Errors or irregularities are "minor" if they can be corrected or waived without being prejudicial to other bidders and when they do not affect the price, quantity, quality, or timeliness of construction. Unless otherwise noted, the borrower determines whether an error or irregularity is "minor."

Modernization plan—A State plan, which has been approved by RUS, for improving the telecommunications network of those Telecommunications Providers covered by the plan. A Modernization Plan must conform to the provisions of 7 CFR part 1751, subpart B.

Negotiation—Any form of purchasing or contracting other than sealed competitive bidding. Any contract awarded without using the sealed competitive bidding procedure is a negotiated contract.

Outside plant—The facilities that conduct electrical or optical signals between the central office and the subscriber's network interface or between central offices.

Performance bond—A surety bond on a form satisfactory to RUS guaranteeing the contractor's faithful performance of a contract.

P&S (plans and specifications)—An RUS contract form, the appropriate specifications, and such additional information and documents needed to provide a clear, accurate, and complete understanding of the installations to be made or construction to be performed.

Project—"Project" means any equipment, including but not limited to switching, routing, access, video, and/or transport equipment, which will be used in the delivery of voice, video, or data services, which are listed under Column 2, "Equipment," in Schedule 1 hereto. A Project will have a single delivery and completion schedule listed under Column 7 and Column 8. The Contract may consist of one or more Projects.

Responsive bid—A bid that complies with the requirements of the plans and specifications.

RUS—the Rural Utilities Service, an agency of the United States Department of Agriculture established pursuant to Section 232 of the Federal Crop Insurance and Reform and Department of Agriculture Reorganization Act of 1994 (Pub. L. 103–354, 108 Stat. 3178), successor to Rural Electrification Administration with respect to administering certain electric and telecommunications program. See 7 CFR 1700.1.

Sealed competitive bidding—A method of contracting that employs sealed competitive bids, public opening of bids, and award of the contract to the bidder submitting the lowest responsive bid. See §1753.8.

Single source negotiation—Negotiating with a single source (contractor or seller).

Special equipment—Equipment used primarily for the transmission and enhancement of voice, data, carrier, radio and light signals, and other equipment and facilities, including incidental cable and other transmission equipment.

Subcontract—A secondary contract undertaking some of the obligations of a primary contract. Under all RUS forms of contract, the primary contractor bears full responsibility for the performance of the subcontractor.

Unbalanced bid—A bid which contains pricing for a task or material that is significantly higher or lower than pricing for similar tasks or materials.

Work order construction—Minor construction performed by the borrower's employees, pursuant to its work order procedure, with the borrower furnishing all materials, equipment, tools, and transportation.

[54 FR 39267, Sept. 25, 1989. Redesignated at 55 FR 39396, Sept. 27, 1990, as amended at 58 FR 66259, Dec. 20, 1993; 59 FR 17464, Apr. 13, 1994; 64 FR 16604, Apr. 6, 1999; 81 FR 71582, Oct. 18, 2016]

§ 1753.3 Preconstruction review.

(a) Advance RUS approval must be obtained for any construction that does not conform to RUS standards and specifications or the approved LD, such as construction of extensions to serve subscribers in areas not included in the LD (See 7 CFR part 1737). For loans approved after RUS approval of the modernization plan in the borrower's state, the proposed construction must conform to the modernization plan, as required by 7 CFR part 1751, subpart B. To obtain approval, the borrower shall submit a written proposal containing:

(1) A description of the work, indicating any deviations from the approved LD or RUS standards and specifications.

(2) An engineering study covering the deviations if there are changes in the design.

(3) A cost estimate for labor, engineering, materials, and overheads.

(4) If applicable, a brief analysis from the borrower demonstrating that the proposed changes conform to the modernization plan.

(b) Before any construction, including interim construction, is initiated, the GFR shall meet with the borrower to review the LD to determine if any significant changes have occurred since its approval by RUS. It is important that the design and construction of the proposed facilities be based on the latest information on subscriber needs.

(c) If the borrower and GFR agree that there have been no significant changes, the borrower may proceed.

(d) If the GFR finds that the LD is no longer satisfactory, the borrower shall prepare an amendment to the LD incorporating the necessary revisions (See 7 CFR part 1737). The borrower must obtain RUS approval of the LD amendment before proceeding with engineering activities on any project to be financed with loan funds.

[54 FR 39267, Sept. 25, 1989. Redesignated at 55 FR 39396, Sept. 27, 1990, as amended at 58 FR 66259, Dec. 20, 1993; 59 FR 17464, Apr. 13, 1994; 64 FR 16604, Apr. 6, 1999]

§ 1753.4 Major and minor construction.

RUS's general requirements for construction are set forth in this subpart A. Additional requirements and procedures for different types of major construction are presented in subparts D, E, F, G, and H (OMB control number 0572–0062). The requirements and procedures for minor construction are presented in subpart I. Borrowers may, at their option, follow the procedures in subparts D, E, F, G, and H for any minor construction.

§ 1753.5 Methods of major construction.

(a) All major construction projects financed by loan funds shall be performed pursuant to a contract approved by RUS and awarded through sealed competitive bidding unless

(1) A specific exception is granted in subparts D, E, F, G, or H, or

(2) Written RUS approval is obtained.

(b) *Contract construction.* (1) RUS approval of the borrower's award of the contract is not required if the contractor is selected through sealed competitive bidding, the bid amount is $500,000 or less and the contractor is not a company or organization affiliated with the borrower. This does not relieve the borrower of the requirements of bidding or bid evaluation set contained in this part.

(2) RUS approval of the borrower's award of the contract is required for all other competitively-bid and for negotiated major construction contracts.

(3) The requirements and procedures for sealed competitive bidding are presented in § 1753.8(a). The requirements and procedures for negotiation are presented in § 1753.8(b).

(c) *Force account construction.* To obtain RUS approval of the force account method for major construction the borrower must demonstrate its ability to perform major construction based on past force account construction which fully met RUS construction standards and was as cost-effective as contract construction in the area. If the borrower has no record of past performance to support its request, but has adequate equipment and experienced personnel to perform the proposed construction, RUS may approve a small trial project. The requirements and procedures for force account construction are presented in subparts D, E, G, and H.

[54 FR 39267, Sept. 25, 1989. Redesignated at 55 FR 39396, Sept. 27, 1990, as amended at 59 FR 43716, Aug. 25, 1994; 64 FR 16604, Apr. 6, 1999]

§ 1753.6 Standards, specifications, and general requirements.

(a) Materials, equipment, and construction financed with loan funds must meet the standards and specifications established by RUS in 7 CFR 1755.97 which lists the RUS Bulletins containing the standards and specifications for telephone facilities.

(b) The borrower may use RUS loan funds to finance nonstandard construction materials or equipment only if approved by RUS in writing prior to purchase or commencement of construction.

(c) Only new materials and equipment may be financed with loan funds, unless otherwise approved by RUS. The materials and equipment must be year 2000 compliant, as defined in 7 CFR 1735.22(e).

(d) All materials and equipment financed with loan funds are subject to the "Buy American" provision (7 U.S.C. 901 *et seq.* as amended in 1938).

(e) All software, software systems, and firmware financed with loan funds must be year 2000 compliant, as defined in 7 CFR 1732.22(e).

[54 FR 39267, Sept. 25, 1989. Redesignated at 55 FR 39396, Sept. 27, 1990, as amended at 59 FR 43716, Aug. 25, 1994; 63 FR 45679, Aug. 27, 1998; 64 FR 16605, Apr. 6, 1999; 81 FR 71582, Oct. 18, 2016]

§ 1753.7 Plans and specifications (P&S).

(a) The P&S consist of an RUS contract form, the appropriate RUS specifications, and such additional information and documents needed to provide a clear, accurate, and complete understanding of what is included in the construction.

(b) 7 CFR 1755.93 provides a list of the RUS forms of telecommunications contracts for use in purchasing telephone materials and equipment and for constructing telephone facilities with loan funds. Also listed is the source where copies may be obtained.

(c) The appropriate standards and specifications listed in 7 CFR part 1755 shall be included in the P&S. When RUS has not prepared standards and specifications, the borrower shall use all appropriate project specific engineering requirements and specifications prepared by the borrower's engineer. The specifications prepared by the borrower's engineer and based on appropriate project specific engineering requirements shall be subject to review and approval by RUS for all major construction, including major projects which would be exempted from RUS approval under paragraph (e) of this section.

(d) The P&S shall be based on the LD approved by RUS. Section 1753.3 presents the requirements and procedures for obtaining RUS approval for construction that does not conform to the LD approved by RUS.

(e) RUS approval of P&S is required for construction that is estimated to cost over $500,000 or 25% of the total loan, whichever is less, and for all building construction. P&S for all other construction are exempt from RUS review and approval except that, at the time of contract approval, RUS will examine the plans and specifications for conformity with the loan purposes and to determine that they comply with other requirements of this part.

(f) RUS will approve only contracts that will provide for at least the following requirements.

(1) *Equal employment opportunity provision.* If this provision is not already in the contract, RUS Contract Form 270, Equal Opportunity Addendum, shall be attached and made a part of the contract.

(2) *Liquidated damages provision.* (i) If not covered by the contract, an appropriate liquidated damages provision, in a form prescribed by RUS, shall be included and made a part of the contract

(ii) The liquidated damages must be based upon the borrower's best estimate of the damages it would incur as a result of the contractor's default.

(3) *Insurance and bond requirements.* (i) The insurance provision shall provide coverage as required by 7 CFR 1788.

(ii) A contractor's bond shall be furnished as required by 7 CFR part 1788.

(iii) The borrower is responsible for ensuring that its contractor complies with the insurance and bond requirements.

[54 FR 39267, Sept. 25, 1989. Redesignated at 55 FR 39396, Sept. 27, 1990, as amended at 59 FR 17679, Apr. 14, 1994; 64 FR 16605, Apr. 6, 1999; 81 FR 71582, Oct. 18, 2016]

§ 1753.8 Contract construction procedures.

(a) *Sealed, competitive bidding*—(1) *Bid opening date.* The borrower is responsible for scheduling the bid opening date. If RUS review of P&S is required by § 1753.7, the borrower shall wait until approval has been received before setting the date. In setting the date, sufficient time should be allowed for the bidders to examine the project site and prepare their bids. The borrower shall notify GFR of the bid date and invite GFR to attend.

(2) *Invitations to bid.* The borrower is responsible for sending invitations to prospective bidders and taking any other action necessary to procure full, free, and competitive bidding. The borrower should obtain from its engineer a list of prospective bidders and a recommendation indicating which bidders are considered qualified. The minimum number of contractors to be invited to bid on contracts for various types of facilities is set forth in subparts D, E, F, or H.

(3) *Qualifying bidders.* If the notice and instructions to bidders require that bidders show evidence of meeting certain requirements, the borrower shall qualify bidders before issuing P&S to them. Procedures for qualifying bidders are contained in subparts D, E, and F.

(4) *Receipt of bids.* The borrower shall write on the outside envelope of any bid or bid amendment, the date and

time the bid was received. Any bid received from an unqualified bidder or after the time specified for opening shall be returned promptly to the bidder unopened.

(5) *Procedure when fewer than three bids are received.* If fewer than three valid bids are received, the borrower shall consult with RUS to determine whether the bids are to be opened or returned unopened. RUS requires that the project be rebid if fewer than three bids are received and RUS determines that one or more other bidders with an express interest in bidding is available and could meet the bid requirements, but was not invited to bid. RUS shall also require rebidding if it is found that qualified bidders were discouraged from bidding by unreasonable bid requirements (such as late notification to bidders) or if the borrower fails to follow the bid procedure.

(6) *Conduct of bid openings.* The borrower shall conduct bid openings open to the public. The borrower should be able to contact its attorney for immediate consultation.

(7) *Review of bids.* The borrower shall review all bids prior to reading any bid results to determine that:

(i) The bid guarantees are adequate.

(ii) All minor errors or irregularities made through inadvertence are corrected or waived. Failing this, the bid shall be rejected as nonresponsive.

(iii) In the event of non-minor errors or irregularities, the bid is rejected and the bid price not disclosed.

(8) *Reading of bids.* Bid prices shall not be read until the borrower has reviewed all bids to determine if there are any minor errors or irregularities that may affect the recommendation as to award. These shall be made public at the same time the bid price is announced.

(9) *Evaluating bids.* The borrower shall consider the same alternates in all bids in determining the low bid.

(10) *Rejection.* The borrower shall reject:

(i) All bids if quoted prices are not acceptable or if the specifications were ambiguous and resulted in bidders having different interpretations of the requirements.

(ii) Any bid that is not responsive, or is incomplete, or submitted by an unqualified bidder, or unbalanced between labor and materials or other respects.

(11) *Award of contract.* (i) The borrower shall obtain from the engineer the determination of the lowest responsive bid, a tabulation of all bids and the engineer's recommendation for award of the contract. Contract award is subject to RUS approval if either the cost of the project is over $500,000 or the contract is with an organization affiliated with the borrower. Contract award of all other projects is not subject to RUS approval.

(ii) If an award is made, the borrower shall award the contract to the lowest responsive bidder. The borrower may award the contract immediately upon determination of the lowest responsive bidder if the following conditions are met:

(A) The project is included in an approved loan and adequate funds were budgeted in the loan and are available.

(B) All applicable RUS procedures were followed, including those in the Notice and Instructions to Bid in the standard forms of contract.

(iii) If RUS approval of the award of contract is required under this paragraph (a)(11), the borrower shall send to RUS for consideration of approval of the award:

(A) Two copies of the low bid.

(B) The engineer's recommendation and the tabulation of all bids.

(C) Evidence of acceptance of the low bid by the borrower, such as:

(*1*) Certified copy of board resolution or

(*2*) letter or telegram to RUS signed by a properly authorized corporate official.

(iv) If RUS approval of the award of contract is not required under this paragraph (a)(11), the borrower shall keep a file available for inspection by RUS. The file shall be kept for at least two years and shall include:

(A) One copy of all received bids.

(B) The engineer's recommendation and tabulation of all bids including "Buy American" evaluations, if any, and all other evaluations required by law.

(C) Evidence of acceptance of the low bid by the borrower, such as a copy of the board resolution certified by the Secretary of the board.

(12) *Execution of contract.* (i) The borrower shall submit to RUS three original counterparts of the contract executed by the contractor and borrower.

(ii) If RUS approves the contract, it shall return one copy to the borrower and send one copy to the contractor.

(b) *Negotiated construction contracts.* (1) For the construction of certain facilities the borrower may negotiate a contract rather than solicit sealed competitive bids. Refer to the appropriate subparts E, F, or H for specific requirements and procedures.

(2) For negotiated purchases, borrowers shall use RUS contract forms, standards, and specifications.

(3) For all contract forms except RUS Form 773:

(i) After a satisfactory negotiated proposal has been obtained, the borrower shall submit it to RUS for approval, along with the engineer's recommendation, and evidence of acceptance by the borrower.

(ii) If RUS approves the negotiated proposal, the borrower shall submit three copies of the contract, executed by the contractor and borrower, to RUS for approval.

(iii) If RUS approves the contract, RUS shall return one copy of the contract to the borrower and one copy to the contractor.

(4) For RUS Form 773, the borrower is responsible for negotiating a satisfactory proposal, executing contracts, and closing the contract. See subparts F and I of this part for requirements for major and minor construction, respectively, on Form 773.

[54 FR 39267, Sept. 25, 1989. Redesignated at 55 FR 39396, Sept. 27, 1990, as amended at 59 FR 43716, Aug. 25, 1994; 64 FR 16605, Apr. 6, 1999]

§ 1753.9 Subcontracts.

(a) RUS construction contract Forms 257, 395, and 515, contain provisions for subcontracting. Reference should be made to the individual contracts for the amounts and conditions under which a contractor may subcontract work under the contract.

(b) RUS Form 282, Subcontract, shall be used for subcontracts under construction and installation contracts.

(1) Minor modifications or additions may be made to the subcontract form, as long as they do not change the intent of the primary contract. Any alterations to the subcontract shall be initialed and dated by the persons executing the subcontract.

(2) Subcontracts shall be prepared in quadruplicate and all copies executed by the contractor and subcontractor and consented to by the borrower and surety, if any.

(3) Four executed copies of the subcontract shall be forwarded to RUS for approval. Upon approval, one copy each will be sent to the borrower, contractor, and subcontractor.

(c) As stated in contract Forms 257, 395, and 515, the contractor shall bear full responsibility for the acts and omissions of the subcontractor and is not relieved of any obligations to the borrower and to the Government under the contract.

(d) As stated in the contract, construction shall not be performed by the subcontractor before approval of the subcontract by RUS.

[54 FR 39267, Sept. 25, 1989. Redesignated at 55 FR 39396, Sept. 27, 1990, as amended at 59 FR 43716, Aug. 25, 1994; 81 FR 71582, Oct. 18, 2016]

§ 1753.10 Preconstruction conference.

The borrower shall conduct a conference, attended by the borrower, contractor, and resident engineer prior to the beginning of construction to provide an opportunity to discuss and agree on responsibilities, procedures, practices, and methods before the work begins. The borrower shall provide each participant with a copy of the conference results. The GFR shall be invited to attend this conference.

§ 1753.11 Contract amendments.

(a) The borrower must obtain RUS approval before execution of any amendment to a contract if

(1) The amendment alters the terms and conditions of the contract or changes the scope of the project covered by the contract regardless of the amount of the contract before amendment,

(2) The amendment increases the amount to be paid under the contract by 20% or more, or

(3) The amendment causes an unbonded contract to require a contractor's performance bond. This would occur when a contract that is executed in an amount below that requiring a performance bond by 7 CFR part 1788, subpart C, is amended to an amount above that amount.

(b) Advance RUS approval to execute other contract amendments is not required. These amendments may be submitted to RUS at any time prior to closeout. If a borrower wishes to receive an advance of funds based on an amended contract amount (i.e., amendments that increase a contract by less than 20%), the borrower may initiate an increase in the amount approved for advance by submitting three copies of the amendment to RUS for approval.

(c) For each amendment executed, the borrower shall make certain that:

(1) The contractor's bond covers the additional work to be performed. If the amendment by itself (or together with preceding amendments) increases the original contract price by 20% or more, a bond extension will be required to bring the penal sum of the bond to the total amended contract price.

(2) If an amendment covers construction in a county or state not included in the original contract, the borrower and contractor are licensed to do business in that location.

(d) Upon execution of any amendment that causes the amended contract amount to exceed the original contract amount by 20% or more, three copies of the amendment shall be submitted to RUS for approval.

[54 FR 39267, Sept. 25, 1989. Redesignated at 55 FR 39396, Sept. 27, 1990, as amended at 64 FR 16605, Apr. 6, 1999]

§§ 1753.12–1753.14 [Reserved]

Subpart B—Engineering Services

SOURCE: 54 FR 3984, Jan. 27, 1989, unless otherwise noted. Redesignated at 55 FR 39397, Sept. 27, 1990.

§ 1753.15 General.

(a)(1) The standard RUS loan documents contain provisions regarding engineering and architectural services performed by or for RUS telecommunications borrowers. This part implements certain of the provisions by setting forth the requirements and procedures to be followed by borrowers in selecting architects and engineers and obtaining architectural and engineering services by contract or by force account.

(2) Borrowers shall obtain architectural and engineering services only from persons or firms which are not affiliated with, and have not represented, a contractor, vendor or manufacturer who may provide labor, materials, or equipment to the borrower under any current loan.

(3) Preloan architectural and engineering services may be provided by qualified personnel on the borrower's staff or by consultants. Neither the selection of a preloan architect or engineer by a borrower, nor the contractual arrangements with them, requires RUS approval.

(4) Postloan architectural and engineering services shall be obtained by borrowers from registered architects and engineers licensed in the State in which the facilities will be located, except where RUS has approved the borrower to provide these services by the force account method. When the extent of the proposed major or minor construction is such that the postloan engineering involved is within the capabilities of employees on the borrower's staff, the borrower may request RUS approval to provide such services. This method of providing engineering services is referred to as force account engineering. Refer to § 1753.17(c).

(5)(i) For major construction, services provided by architects and engineers not on the borrower's staff must be provided under Form 220, Architectural Service Contract, or Form 217, Postloan Engineering Service Contract—Telecommunications. These contracts require RUS approval.

(ii) For minor construction, borrowers may use the contracts in paragraph (a)(5)(i) of this section for postloan architectural or engineering services or any other form of contract, such as Form 245, Engineering Service Contract, Special Services—Telephone. RUS approval of contracts for postloan architectural or engineering services associated with minor construction, except for buildings covered in paragraph (a)(6) of this section, is not required.

(6) For buildings to be constructed with RUS funds, postloan architectural or engineering services shall be obtained if (1) the construction cost exceeds $50,000 (prefab buildings using manufacturer's specifications approved by RUS are exempt from this requirement) or (2) soil or seismic conditions require special design considerations.

(b) For the purpose of this subpart B:

(1) *Contract*—The services contract between the borrower and its architect or engineer.

(2) *Force Account Engineering*—Any preloan or postloan engineering services performed by the borrower's staff.

(3) *Postloan engineering services*—The design, procurement, and inspection of construction to accomplish the objectives of a loan as stated in a LD approved by RUS.

(4) *Preloan engineering services*—The planning and design work performed in preparing a LD. This consists of helping the borrower determine the objectives for a loan, including consideration of RUS's requirements relating to the modernization plan, selecting the most effective and efficient methods of meeting loan objectives, and preparing the LD which describes the objectives and presents the method selected to meet them.

(c) Single copies of RUS forms and publications cited in this part are available free from Administrative Services Division, Rural Utilities Service, United States Department of Agriculture, Washington, DC 20250–1500. These forms and publications may be reproduced.

(d)(1) All outside architects and engineers employed by RUS telephone borrowers shall have insurance coverage as required by 7 CFR part 1788.

(2) Borrowers shall ensure that their architects and engineers comply with the insurance requirements of their contracts. See 7 CFR 1788.54.

(e)(1) Borrowers shall make prompt payments to architects and engineers as required by the contract.

(2) RUS shall not make loan funds available for late payment interest charges.

[54 FR 3984, Jan. 27, 1989. Redesignated at 55 FR 39397, Sept. 27, 1990, as amended at 58 FR 66259, Dec. 20, 1993; 59 FR 17464, Apr. 13, 1994; 64 FR 16605, Apr. 6, 1999]

§ 1753.16 Architectural services.

(a) The borrower shall be responsible for selecting an architect to perform the architectural services required in the design and construction of buildings.

(b)(1) The borrower shall use Form 220 when contracting for architectural services for major construction, except that the borrower may use either Form 220 or Form 217 if the building is an unattended central office building.

(2) The borrower and the architect negotiate the fees for services under Form 220.

(3) Reasonable modifications or additions to the terms and provisions in Form 220 may be made, subject to RUS approval, to obtain the specific services needed for a building.

(4)(i) Three copies of Form 220, executed by the borrower and the architect, shall be sent to GFR to be forwarded to RUS for approval. RUS will review the contract terms and conditions. RUS will not approve the contract if, in RUS's judgment:

(A) Unacceptable modifications have been made to the contract form.

(B) The contract will not accomplish loan purposes.

(C) The architectural service fees are unreasonable.

(D) The contract presents unacceptable loan security risk to RUS.

(ii) If RUS approves the contract, RUS will send one copy to the architect and one copy to the borrower.

(5) Loan funds will not be available to pay for the preliminary architectural services if a loan is not made for the construction project, or if the construction project is abandoned.

(6) Subpart D of 7 CFR part 1753 sets forth the requirements and procedures to be followed by borrowers constructing central office, warehouse, and garage buildings with RUS loan funds.

(c)(1) RUS telephone borrowers shall obtain two copies of a completed Form

284, Final Statement of Architect's Fees, when all services and obligations required under the architectural services contract have been completed. All fees shown on the statement shall be supported by detailed information where appropriate. For example: out-of-pocket expense, cost plus, and per diem types of compensation shall be listed separately with labor, transportation, etc., itemized for each service involving these types of compensation.

(2) If Form 284 and supporting data are satisfactory, the borrower shall approve the statement, sign both copies, and send one copy to the GFR.

(3) Upon approval of Form 284 by RUS, the borrower shall promptly make final payment to the architect.

[54 FR 3984, Jan. 27, 1989. Redesignated at 55 FR 39397, Sept. 27, 1990, as amended at 59 FR 43717, Aug. 25, 1994; 64 FR 16606, Apr. 6, 1999]

§ 1753.17 Engineering services.

(a)(1) All engineering services required by a borrower to support its application for a loan shall be rendered by a qualified engineer selected by the borrower or by qualified employees on the borrower's staff. The selection of the preloan engineer, the form of preloan engineering service contract, and the contract itself, are not subject to RUS approval. Borrowers, however, should discuss their proposed method of obtaining preloan engineering services with the GFR before proceeding with any arrangements.

(2) Form 835, Preloan Engineering Service Contract, Telephone System Design, is a suggested form of preloan engineering service contract. While use of this form of contract is not required, it will be helpful in determining the tasks to be performed. Any form of contract used shall specify that preloan engineering services conform to RUS requirements for preloan studies. See subpart D of 7 CFR part 1737.

(b)(1) *Major construction.* (i) Three copies of Form 217 executed by the borrower and the engineer shall be sent to the GFR to forward to RUS for approval. The engineer's estimate of the engineering fees, on Form 506, shall be included.

(ii) RUS will review the contract terms and conditions. RUS will not approve the contract if, in RUS's judgement:

(A) Unacceptable modifications have been made to the contract form.

(B) The contract will not accomplish loan purposes.

(C) The engineering service fees are unreasonable.

(D) The contract presents unacceptable loan security risk to RUS.

(E) The consulting engineering firm is affiliated with or has represented a contractor, vendor, or manufacturer who may provide labor, materials, or equipment to the borrower under any current loan.

(2) *Minor construction.* When a borrower contracts for an engineering firm to inspect and certify construction accounted for under the work order procedure or the Contract for Miscellaneous Construction Work and Maintenance Services, Form 773 (See 7 CFR part 1753 subpart I), the borrower shall require that the certification be signed by a licensed engineer.

(c)(1) *Major construction.* When the extent and complexity of the proposed construction is such that the engineering involved is within the capabilities of employees on the borrower's staff, borrowers may request RUS approval to provide such services.

(i) The request shall include:

(A) A description of services to be performed.

(B) The name and qualifications of the employee to be in charge. RUS requires this employee to meet the State experience requirements for registered engineers. In the absence of specific State experience requirements, the employee must have at least eight years experience in the design and construction of telecommunication facilities, with at least two years of the work experience at a supervisory level. RUS does not require professional registration of this employee, but this does not relieve the borrower from compliance with applicable State registration requirements which may require a licensed individual to perform such services.

(C) The names, qualifications, and responsibilities of other principal employees who will be associated with providing the engineering services.

(D) A letter signed by an authorized representative of the borrower authorizing the engineering services to be performed by force account and certifying the information supporting the request.

(ii) RUS shall notify the borrower by letter of approval or disapproval to perform force account engineering. The letter shall set forth any conditions associated with an approval or the reasons for disapproval.

(iii) RUS's approval of force account engineering for major construction shall be only for the specific projects named in the notice of approval.

(2) *Minor construction.* (i) When the borrower proposes to perform the inspection and certification of minor construction, the following shall be submitted to the RUS:

(A) A copy of the employee's qualifications and experience record, unless previously submitted. RUS requires a minimum of four years of construction and inspection experience. The employee cannot be engaged in the actual construction.

(B) A letter signed by an authorized representative of the borrower authorizing the performance of these services by the employee, subject to RUS approval, and certifying the supporting information.

(ii) RUS shall notify the borrower by letter of approval or disapproval of the borrower's staff employee to perform the inspection and certification of construction. The approval shall be limited to the employee's area of expertise.

(d)(1) Subject to the requirements of this part and other applicable regulations, RUS will make loan funds available for the architectural and engineering services up to the amounts included in the approved loan.

(2) Advance of funds shall be requested on an FRS as set forth in 7 CFR part 1744 subpart C.

(e) The borrower shall obtain status of contract and force account proposal reports from the engineer once each month. The report shall show for each contract or FAP the approved contract or FAP amount, the date of approval, the scheduled date construction was to begin and the actual date construction began, the scheduled completion date, the estimated or actual completion date, the estimated or actual date of submission of closeout documents, and an explanation of delays or other pertinent data relative to progress of the project. One copy of this report shall be submitted to the GFR.

(f)(1) Upon completion of all services required under the engineering service contract Form 217, the borrower shall obtain from the engineer four copies of the Final Statement of Engineering Fee, Form 506.

(2) If the statement is satisfactory, the borrower shall sign all copies and send three to the GFR.

(3) After RUS approval of Form 506, one copy shall be sent to the borrower and one copy sent to the engineer.

(4) The borrower shall promptly make final payment to the engineer.

[54 FR 3984, Jan. 27, 1989. Redesignated at 55 FR 39397, Sept. 27, 1990, as amended at 59 FR 43717, Aug. 25, 1994; 64 FR 16606, Apr. 6, 1999]

§ 1753.18 Engineer and architect contract closeout certifications.

A certification of completion and inspection of construction signed by the borrower and countersigned in accordance with accepted professional engineering and architectural practice, by the engineer or architect, shall be prepared as evidence of completion of a major construction project. This certification shall make reference to the contract number and contract amount, and shall include the following:

(a) A statement that the construction is complete and was done in accordance with the RUS approved system design or layout or subsequent RUS approved changes.

(b) A statement that the construction was for loan purposes.

(c) A statement that construction used was in accordance with specifications published by RUS covering the construction which were in effect when the contract was executed, or in the absence of such specifications, that it meets other applicable specifications and standards and that it meets all applicable national and local code requirements as to strength and safety.

(d) A statement that the construction complies with the "Buy American" provision (7 U.S.C. 903 note) of the Rural Electrification Act of 1936 (7 U.S.C. 901 *et seq.*).

(e) A statement that all necessary approvals have been obtained from regulatory bodies and other entities with jurisdiction over the project.

(f) A statement that all closeout documents required by this part have been examined and found complete such that the Contractor has fulfilled all obligations under the contract except for warranty coverage.

(g) A statement that the engineer or architect is not affiliated with and does not represent the contractor, vendor, or manufacturer who is a participant in the contract.

[64 FR 16606, Apr. 6, 1999, as amended at 81 FR 71582, Oct. 18, 2016]

§§ 1753.19–1753.20 [Reserved]

Subpart C [Reserved]

Subpart D—Construction of Buildings

SOURCE: 54 FR 39267, Sept. 25, 1989, unless otherwise noted. Redesignated at 55 FR 39396, Sept. 27, 1990.

§ 1753.25 General.

(a) This subpart implements and explains the provisions of the Loan Documents setting forth the requirements and the procedures to be followed by borrowers in constructing headquarters, commercial office, central office, warehouse, and garage buildings with loan funds.

(b) Terms used in this subpart are defined in § 1753.2.

(c) All plans and specifications for buildings to be constructed with loan funds are subject to the approval of RUS. In addition, preliminary plans and specifications for headquarters and commercial office buildings to be constructed with loan funds are subject to RUS approval.

(d) RUS Form 257, Contract to Construct Buildings, shall be used for the construction of all headquarters, commercial office, central office, warehouse, and garage buildings with loan funds. Refer to § 1753.26 for further instructions.

(e) The borrower shall use the sealed competitive bid procedure for all building construction, except for:

(1) Minor construction using subpart I procedures.

(2) Major construction, where the borrower has received advanced approval to perform the construction by force account.

Refer to §§ 1753.27 and 1753.29 for further instructions.

(f) The site location, design, and construction of the facilities must comply with all applicable laws and regulations, including:

(1) Pub. L. 90–480 (42 U.S.C. 4151) (Access to Physically Handicapped), which requires certain buildings financed with Federal funds be designed and constructed to be accessible to the physically handicapped.

(2) Pub. L. 91–596 (29 U.S.C. 651) the Occupational Safety and Health Act of 1970. OSHA issues rules and regulations covering occupational safety and health standards for buildings. These regulations are codified in 29 CFR chapter XVII.

(3) 7 CFR part 1970.

(4) 7 CFR part 1792, subpart C, which requires that the building design comply with applicable seismic design criteria. Prior to the design of buildings, borrowers shall submit to RUS a written acknowledgement from the architect or engineer that the design will comply.

(g) All construction pertaining to the building structure shall be performed under one contract. Separate contracts may be used for planting shrubbery, surfacing of roads and parking areas, and other identifiable parts of the project not pertaining to the building structure. These separate contracts shall also be subject to RUS approval as described in this subpart D.

(h) The borrower is responsible for submitting evidence, satisfactory to RUS, establishing that clear title to the building site has been obtained. RUS will not approve the construction contract until it has given title clearance.

[54 FR 39267, Sept. 25, 1989. Redesignated at 55 FR 39396, Sept. 27, 1990, as amended at 59 FR 43717, Aug. 25, 1994; 64 FR 16606, Apr. 6, 1999; 81 FR 11028, Mar. 2, 2016]

§ 1753.26 Plans and specifications (P&S).

(a) For headquarters and commercial office buildings only, the borrower shall prepare preliminary P&S showing the floor plan and general architectural details of the building to be constructed using loan funds. In particular, the preliminary P&S shall address the requirements of § 1753.25(f) and the Uniform Relocation Assistance and Real Property Acquisition Policies Act of 1970 (42 U.S.C. 4601 *et seq.*). The P&S shall be submitted to the GFR and are subject to RUS approval.

(b) The borrower shall prepare P&S for construction of all buildings. Each set of P&S shall include:

(1) RUS Contract Form 257, Contract to Construct Buildings, completed to the extent explained in (c) of this section.

(2) Complete and detailed specifications covering materials and workmanship.

(3) A detailed building plan. Where the building is to house electronic apparatus, the detailed plan or specifications shall include the equipment environmental requirements and special equipment required.

(4) A site plan for each building showing the building location and giving the legal description of the site. Sufficient information must be provided for the site so that it can be identified as the same property on which title opinion was submitted to RUS. The legal description shall be typed on the site plan. The borrower shall also furnish topographical information and a description of any proposed site development work and show proposed connections for public utilities.

(c) RUS Contract Form 257 shall be completed as follows:

(1) *List of names or kinds of buildings and locations*—Site plan and specifications must be identified with the appropriate building.

(2) *Alternates*—The borrower shall keep the number of alternates to a minimum. Items for which alternates are to be taken shall be fully described on a separate sheet in the specifications and the details shown on the plans, when necessary, and identified by the alternate number. The Notice and Instructions to Bidders shall explain how bids will be evaluated with respect to alternates.

(3) *Time for construction*—A reasonable time for completion of construction, considering local conditions, shall be determined by the borrower and inserted in the space provided. Too short a construction period may discourage bidders or influence their bids. Completion of the building, where central office equipment is involved, shall be coordinated with delivery of the equipment. The time of completion shall allow adequate drying time before the central office equipment is stored or installed in the building.

(d) The plans and specifications shall show the identification and date of the model code used for seismic safety design considerations, and the seismic factor used. See 7 CFR part 1792, subpart C.

(e) Two sets of the building plans and specifications shall be prepared and submitted to the GFR.

[54 FR 39267, Sept. 25, 1989. Redesignated at 55 FR 39396, Sept. 27, 1990, as amended at 59 FR 43717, Aug. 25, 1994; 64 FR 16606, Apr. 6, 1999]

§ 1753.27 Bidding procedure.

Upon RUS approval of the P&S, the borrower shall proceed as follows:

(a) Bid documents shall consist of a copy of the approved P&S, including RUS Contract Form 257, completed in accordance with the instructions on the cover of the form and the plot plans showing site development details. For contracts in amounts of $100,000 or less, the borrower must specify in the Notice and Instructions to Bidders whether the contractor will be required to furnish a performance bond or a builder's risk policy.

(b) The borrower shall determine that title to the real estate has been approved by RUS before the invitations to bid are released.

(c) The borrower shall set the time for opening of bids, allowing ample time for bidders to prepare bids.

(d) The borrower shall solicit bids as set forth in § 1753.8(a)(2). Invitations shall be sent to at least six prospective bidders.

(e) The borrower shall conduct bid opening and award of contract in accordance with the procedure set forth in §1753.8(a).

§1753.28 Contract amendments.

(a) The general requirements for contract amendments are set forth in §1753.11.

(b) The borrower shall prepare construction contract amendments on RUS Contract Form 238, Construction or Equipment Contract Amendments. See 7 CFR 1755.93 to obtain copies of Form 238.

§1753.29 Force account procedures.

(a) The borrower must obtain RUS approval of the force account method of construction of buildings in advance in order to obtain RUS financing.

(b) The borrower shall prepare the P&S in accordance with §1753.26.

(c) Prior to any construction activity or the purchase of materials or equipment, the borrower shall submit the FAP in duplicate to RUS, accompanied by a resolution indicating approval of the board of directors of the borrower or a letter signed by an authorized corporate official. The proposal shall include:

(1) A Copy of the P&S.

(2) An itemized list of all items of materials required for construction.

(3) A construction schedule showing the estimated construction period for each major construction item.

(4) An estimate of the material and labor and other costs for any construction item not provided for in the approved loan.

(d) Force Account construction to be financed with loan funds shall not be started until RUS approval has been received by the borrower.

[54 FR 39267, Sept. 25, 1989. Redesignated at 55 FR 39396, Sept. 27, 1990, as amended at 59 FR 43717, Aug. 25, 1994]

§1753.30 Closeout procedures.

(a) This section outlines the procedure to be followed to close out RUS Contract Form 257 (Contract to Construct Buildings) and construction or rehabilitation performed by the force account method.

(b) *RUS Form 257 Contract.* (1) Whenever changes were made in the plans and specifications which did not require immediate submission to RUS of an amendment under §1753.11, a final contract amendment showing the changes shall be prepared.

(2) Upon completion of the project, the borrower shall obtain certifications from the architect or engineer that the project and all required documentation are satisfactory and complete. The requirements for this certification are contained in §1753.18.

(3) The engineer's or architect's contract closeout certification and the final amendment shall be submitted to RUS as a basis for the final advance of funds for the contract.

(4) After all required RUS approvals are obtained, final payment is made in accordance with article III of RUS Form 257 once the borrower has received the architect's or engineer's certifications regarding satisfactory completion of the project.

(c) Upon completion of force account construction, the borrower shall:

(1) Arrange with its architect or engineer and the GFR for final inspection of the project.

(2) Complete, with the assistance of its architect or engineer, the documents listed in the following table that are required for the closeout of force account construction.

DOCUMENTS REQUIRED TO CLOSEOUT CONSTRUCTION OF BUILDINGS

RUS Form No.	Description	Use with		No. of copies prepared by		Distribution	
		Contract	Force account	Contractor	Architect/engineer	Borrower	Contractor
238	Construction or Equipment Contract Amendment (if not previously submitted, send to RUS for approval).	X	....	....	(3)	(to RUS)	(to RUS)
181	Certificate of Completion (contract construction)[1]	X	....	....	2	1	1

DOCUMENTS REQUIRED TO CLOSEOUT CONSTRUCTION OF BUILDINGS—Continued

RUS Form No.	Description	Use with		No. of copies prepared by		Distribution	
		Contract	Force account	Contractor	Architect/engineer	Borrower	Contractor
231	Certificate of Contractor	X	....	1	....	1	
224	Waiver and Release of Lien From Each Supplier	X	....	1	....	1	
213	Certificate (buy American)	X	....	1	....	1	
None [2]	"As Built" Plans and Specifications	X	X	....	1	1	
None	Guarantees, Warranties, Bonds, Operating or Maintenance Instructions, etc.	X	....	1	....	1	
None	Architect/Engineer seismic safety certification	X	X	....	2	1	1

[1] Cost of materials and services furnished by borrower are not to be included in Total Cost on RUS Form 181.
[2] When only minor changes were made during construction, two copies of a statement to that effect from the Architect will be accepted instead of the "as built" Plans and Specifications.

(3) Make distribution of the completed documents as indicated in the table in this section.

(d) Final payment shall not be made until RUS has approved the closeout documents.

[54 FR 39267, Sept. 25, 1989. Redesignated at 55 FR 39396, Sept. 27, 1990, as amended at 59 FR 43717, Aug. 25, 1994; 64 FR 16606, Apr. 6, 1999]

§§ 1753.31–1753.35 [Reserved]

Subpart E—Purchase and Installation of Central Office Equipment

SOURCE: 54 FR 39267, Sept. 25, 1989, unless otherwise noted. Redesignated at 55 FR 39397, Sept. 27, 1990.

§ 1753.36 General.

(a) This subpart implements and explains the provisions of the Loan Documents setting forth the requirements and the procedures to be followed by borrowers in purchasing and installing central office equipment financed with loan funds.

(b) Terms used in this subpart are defined in § 1753.2 and Equipment Contract, RUS Contract Form 395 (RUS Contract Form 395).

(c) Borrowers shall use RUS Contract Form 395, and associated RUS Form 395a, Equipment Contract Certificate of Completion (Including Installation), when the firm supplying the equipment will install it and RUS Contract Form 395 and associated RUS Form 395b, Equipment Contract Certificate of Completion (Not Including Installation) when the supplier of the equipment will not be installing it. In either case the appropriate specifications shall be included in the contract.

(d) Alternates, if any, specified in the P&S shall be kept to a minimum.

(e) The borrower shall take sealed competitive bids for all central office equipment to be purchased under RUS Contract Form 395 using the procedure set forth in Sec. 1753.38(a), unless RUS approval to negotiate is obtained.

(f) The borrower may request permission to negotiate with a single supplier for additional central offices to standardize equipment on a system basis. RUS approval to negotiate must be obtained before release of the plans and specifications to the supplier. Except for remote switching terminals associated with an existing central office, RUS will not approve negotiation with a non-domestic manufacturer for the purpose of standardization because such a purchase does not meet the RE Act "Buy American" provisions.

(g) Materials and equipment must meet the standards and general specifications approved by RUS.

(h) Only new equipment shall be purchased unless otherwise approved by RUS.

(i) All purchases of materials and equipment are subject to the "Buy American" requirements.

(j) If the sealed competitive bid procedure is followed, negotiation after bid opening will not be permitted.

[54 FR 39267, Sept. 25, 1989. Redesignated at 55 FR 39397, Sept. 27, 1990, as amended at 64 FR 16607, Apr. 6, 1999; 81 FR 71582, Oct. 18, 2016]

§ 1753.37 Plans and specifications (P&S).

(a) *General.* (1) Prior to the preparation of P&S, the borrower shall review with the GFR the current and future requirements for central office equipment.

(2) The P&S shall specify the delivery and completion time required for each exchange.

(3) P&S for equipment to be provided under an Equipment Contract, RUS Contract Form 395 (RUS Contract Form 395) contract without installation shall require the supplier to provide specific installation information and a detailed bonding and grounding plan to be utilized by the engineer, borrower, and others responsible for the installation of the equipment.

(b) *Preparation of P&S.* The P&S shall include RUS Contract Form 395, Notice and Instructions to Bidders, specifications for the required equipment for each exchange, provision for spare parts, and all other pertinent data needed by the bidder to complete its proposal.

(c) RUS review of P&S is required for construction estimated to cost over $500,000 total or estimated to cost more than 25% of the total loan, whichever is less.

(1) If RUS review is required, the borrower shall submit one copy of the P&S to the GFR for RUS review.

(2) RUS will review the P&S and notify the borrower in writing of approval or disapproval.

[54 FR 39267, Sept. 25, 1989. Redesignated at 55 FR 39397, Sept. 27, 1990, as amended at 64 FR 16607, Apr. 6, 1999; 81 FR 71582, Oct. 18, 2016]

§ 1753.38 Procurement procedures.

(a) *Sealed competitive bidding.* Sealed competitive bidding of central office equipment shall be in two steps: presentation and evaluation of suppliers' technical proposals, and compliance with the sealed competitive bidding procedure set forth in § 1753.8(a). The procedure is as follows:

(1) *Solicitation of bids.* (i) After RUS approval of the specifications and equipment requirements (required only for projects expected to exceed $500,000 or 25% of the loan, whichever is less), the borrower shall send "Notice and Instructions to Bidders" to suppliers with central office equipment.

(ii) The "Notice" must set forth the method of evaluating bids and must require the submission of equipment lists and traffic calculations with the bids.

(iii) Equipment Contract, RUS Contract Form 395 (RUS Contract Form 395) shall be used, except that the "Notice" shall state that prior to the bid opening a technical session will be conducted with each supplier to resolve any questions related to the technical proposal submitted by the supplier. The suppliers' technical proposals should be requested for presentation 30 days in advance of the bid opening to enable sufficient time to make the technical evaluation.

(iv) The borrower shall solicit bids as set forth in § 1753.8(a)(2). The "Notice" shall be sent to at least three prospective bidders. A copy of the "Notice" and a list of such bidders shall be sent to RUS.

(v) At the request of an invited supplier, the borrower shall provide two copies of the P&S.

(2) *Technical Sessions.* (i) The borrower shall schedule individual technical sessions by the suppliers, notify each supplier of its scheduled date and time, notify the GFR of all scheduled dates and times, and request the following be available at the technical session:

(A) Lists of equipment, material and software.

(B) Proposed floor plan.

(C) Power and heat dissipation calculations.

(D) List of exceptions to plans and specifications.

(E) Protection and grounding requirements.

(F) Description of how office administration, maintenance and traffic collection are handled with step-by-step examples and printouts.

(G) Explanation of processor and/or memory expansion required to meet ultimate size. This shall include discussions of software, processor memory, and hardware additions needed for line additions and the introduction of various future services; the relative costs of installing the necessary hardware and software initially as compared with the anticipated cost if installed at the time when the future services are to be offered.

(H) Description of how special equipment such as loop tests, volunteer fire alarm circuit, line load control, etc., will function.

(I) Description of method for translating initial office administration information into machine language, and proposal as to whether it will be done by the borrower or by the supplier.

(J) Some types of equipment contain software. RUS Contract Form 395 indicates whether the equipment contains software and whether the software contract stipulations are applicable.

(K) Any other items pertinent to the technical proposal, such as information regarding changes that have been made in hardware and software of the equipment that is of like manufacture to that presently in operation in the borrower's system. This shall include requirements for additional spare parts or training which have developed as a result of significant change in system device technology.

(ii) The borrower shall review in detail all exceptions to the P&S. No exceptions will be accepted unless all bidders are notified, in writing, of the change in the specifications and permitted to incorporate the change in their proposal.

(iii) If the technical proposal is not responsive, the borrower shall notify the supplier, in writing, that its proposal will not be given further consideration and why.

(iv) Changes in the P&S resulting from the technical sessions shall be subject to RUS's review and approval.

(v) After evaluation of the technical proposals and RUS approval of the changes to P&S (required only for projects that are expected to exceed $500,000 or 25% of the loan, whichever is less), sealed bids shall be solicited from only those bidders whose technical proposals meet P&S requirements. When fewer than three bidders are adjudged qualified by the borrower to bid, RUS approval must be obtained to proceed. Generally, RUS will grant such approval only if the borrower can demonstrate to the satisfaction of RUS that a good faith effort was made to obtain at least three competitive bids. This would be demonstrated if all suppliers currently listed in I.P. 300–4 were invited to submit technical proposals.

(vi) The borrower shall invite the GFR to attend the technical sessions.

(3) *Bidding and award of contract.* (i) All bids must be completed, dated, and signed prior to submission.

(ii) The bid opening and award of contract shall be conducted in accordance with the procedure set forth in §1753.8(a).

(iii) The spare parts bid shall always be priced separately and added to the base bid when determining the low bidder.

(b) *Single source negotiated procurement.* If RUS has approved the borrower's request to procure central office equipment through single source negotiation in accordance with requirements contained in §1753.36(f), the borrower shall proceed in accordance with this subsection.

(1) After RUS approval of the P&S and equipment requirements (required only for contracts expected to exceed $500,000 or 25% of the loan, whichever is less), the borrower shall send two complete copies of the approved P&S to the supplier and request that a proposal be submitted.

(2) The borrower shall schedule a time and date for a technical session by the supplier and request that the items listed in §1753.38(a)(2)(i) be available at the technical session. In addition to these items, the supplier shall be requested to provide a description of the exact differences in hardware and software between the borrower's existing equipment and the proposed equipment so that the borrower can determine spare parts interchangeability, need for retraining, and the compatibility of administration of the old and new equipment.

(3) If the contract is expected to exceed $500,000 or 25% of the loan, whichever is less, changes in the P&S resulting from the technical session shall be subject to RUS review and approval.

(4) The submitted proposal shall be based on the agreed-upon results of the technical evaluation and must be complete, dated, and signed.

(5) The borrower shall obtain an award recommendation from its engineer.

(6) The following shall be sent to RUS for review and approval:

(i) A copy of the engineer's recommendation to the borrower, and

(ii) Evidence of acceptance of the proposal by the borrower, such as

(A) A certified copy of the board resolution, or

(B) A letter to RUS signed by an authorized corporate official.

(7) RUS approval of the proposal will be conditioned upon the borrower obtaining prices that are consistent with current competitive prices. Upon RUS approval of the proposal, three copies of the contract shall be prepared with all specifications and proposal documents, and performance bonds, to be executed by the supplier and borrower.

(8) The three complete, executed contracts shall be sent to the RUS Area Engineering Branch Chief for approval.

(9) If RUS approves the contract, one copy will be returned to the borrower and one copy will be sent to the supplier.

(10) Installation of the central office equipment and materials provided under RUS Contract Form 395 may be made in accordance with subpart I, if applicable, or by an approved Force Account Proposal (FAP).

(c) *Contract amendments*. (1) The general requirements for contract amendments are set forth in § 1753.11.

(2) Equipment contract amendments shall be prepared on RUS Contract Form 238, Construction or Equipment Contract Amendments.

(d) *Additions*. When additions to existing central office equipment are required:

(1) A proposal shall be requested from the supplier.

(2) The borrower shall prepare a plan containing an outline of the proposed use of the equipment, the proposal from the supplier and an estimate of the installation cost. If the total cost exceeds $500,000, RUS approval of the award of contract is required. The borrower shall in this case submit its plan and the supplier's proposal to GFR. If the cost does not exceed $500,000, the borrower's award of contract is not subject to RUS approval.

(3) If RUS approval was required by paragraph (d)(2) of this section, upon RUS approval the purchase may be made using RUS Contract Form 395, or when applicable, the procedures contained in subpart I of this part.

(4) If the purchase is to be made by contract, three executed copies of the contract with attachments are to be submitted to the RUS.

(5) Installation of the central office equipment and materials procured by RUS Contract Form 395 without installation may be made in accordance with subpart I, if applicable, or by an approved FAP.

(e) *Preinstallation conference*. RUS recommends, but does not require, that the borrower hold a preinstallation conference, attended by the borrower, its engineer, equipment installers, and if possible the GFR, prior to the beginning of the installation of the central office equipment.

[54 FR 39267, Sept. 25, 1989. Redesignated at 55 FR 39397, Sept. 27, 1990, as amended at 59 FR 17679, Apr. 14, 1994; 64 FR 16607, Apr. 6, 1999; 81 FR 71583, Oct. 18, 2016]

§ 1753.39 Closeout documents.

Closeout of Equipment Contract, RUS Contract Form 395 (RUS Contract Form 395) (including or not including installation) shall be conducted as follows:

(a) *Contract amendments*. Amendments that must be submitted to RUS for approval, as required by § 1753.11, shall be submitted promptly. All other amendments may be submitted to RUS with the engineer's contract closeout certification.

(b) *Acceptance tests*. The borrower will perform acceptance tests as part of the partial closeout and final closeout of RUS Contract Form 395 that will demonstrate compliance with the requirements as specified by the borrower's engineer in the Performance Requirements. Other tests demonstrating compliance will be acceptable.

(c) *Grounding system audit*. A grounding system audit shall be performed and found acceptable for equipment provided under RUS Contract Form 395, (including or not including installation), prior to placing a central office or remote switching terminal into full service operation. The audits are to be conducted in accordance with the requirements specified by the borrower's engineer in the Performance Requirements. The audits shall be performed by the contractor and borrower when using the RUS Contract Form 395 with equipment installation and by the borrower when using the RUS Contract Form 395 without equipment installation.

(d) *Partial Closeout Procedure.* Under conditions set forth in RUS Contract Form 395, a contractor may, when approved by the borrower, receive payment in full for central offices and their respective associated remote switching terminals upon completion of the installation without awaiting completion of the project. The contractor is to receive such payment, according to procedures contained in the applicable sections of RUS Contract Form 395. In addition to complying with the appropriate partial closeout procedure contained in RUS Contract Form 395, the borrower shall:

(1) Obtain from the engineer a certification of partial closeout; and

(2) Submit one copy of the summary to RUS with an FRS.

(e) *Final contract closeout procedure.* The documents required for the final closeout of the equipment contract, RUS Contract Form 395 with or without installation, are listed in the Table 1 to paragraph (e), which also indicates the number of copies and their distribution.

TABLE 1 TO PARAGRAPH (e)—DOCUMENTS REQUIRED TO CLOSE OUT EQUIPMENT CONTRACT, RUS CONTRACT FORM 395

RUS Form No.	Description	Number of copies prepared by		Distribution	
		Seller	Engineer	Buyer	Seller
213	Certificate (Buy American)	1		1.	
238	Construction or Equipment Contract Amendment (If not previously submitted, send to RUS for approval).		3 sent to RUS.	1 from RUS	1 from RUS.
395a	Certificate of Completion for Equipment Contract (Including Installation).		2	1	1.
395b	Certificate of Completion for Equipment Contract (Not Including Installation).		2	1	1.
395c	Certificate of Contractor and Indemnity Agreement (Use only for installation contracts).	1		1.	
None	Report in writing, including all measurements, any acceptance test report and other information required under Part II of the applicable specifications (Form 395d may be used).		1	1.	
None	Set of maintenance recommendations for all equipment furnished under the contract.	1		1.	

(f) Once RUS approval has been obtained for any required amendments, the borrower shall obtain certifications from the engineer that the project and all required documentation are satisfactory and complete. The requirements for the final contract certification are contained in § 1753.18.

(g) Once these certifications have been received, final payment shall be made according to the payment terms of the contract. Copies of the certifications shall be submitted with the FRS, requesting the remaining funds on the contract.

[81 FR 71583, Oct. 18, 2016]

§§ 1753.40–1753.45 [Reserved]

Subpart F—Outside Plant Major Construction by Contract

SOURCE: 54 FR 39267, Sept. 25, 1989, unless otherwise noted. Redesignated at 55 FR 39397, Sept. 27, 1990.

§ 1753.46 General.

(a) This subpart implements and explains the provisions of the loan documents setting forth the requirements and procedures to be followed by borrowers when outside plant major construction by contract is financed by loan funds. Terms used in this subpart are defined in § 1753.2 and RUS Contract Form 515.

(b) The contract method for major construction is described in § 1753.5(b).

(c) The two contract forms which may be used for major outside plant construction are Form 515 and Form 773. Limitations on the applicability of these forms shall be as follows:

(1) Form 515 shall be used for major outside plant construction projects which will be competitively bid. The contract contains plans and specifications and has no dollar limitation. See §§ 1753.47, 1753.48 and 1753.49.

(2) Contract Form 515, which is for $250,000 or less, may, at the borrower's option, be negotiated. See § 1753.48(b).

(3) RUS Form 773 may be used for minor outside plant projects which are not competitively bid because they cannot be designed and staked at the time of contract execution. Projects of this nature include routine line extensions and placement of subscriber drops. See subpart I of this part.

[54 FR 39267, Sept. 25, 1989. Redesignated at 55 FR 39397, Sept. 27, 1990, as amended at 59 FR 43717, Aug. 25, 1994; 64 FR 16608, Apr. 6, 1999]

§ 1753.47 Plans and specifications (P&S).

(a) *General.* (1) Prior to the preparation of P&S for the construction project:

(i) A review shall be made of the outside plant requirements, and the Loan Design (LD) shall be revised to reflect any needed changes (See § 1753.3).

(ii) Deviations from the approved LD (7 CFR part 1737) must be approved by RUS (See § 1753.3).

(2) The standard RUS specifications required for construction of outside plant facilities are:

(i) RUS Form 515a (Bulletin 345–150)—Specifications and Drawings for Construction of Direct Buried Plant.

(ii) RUS Form 515c (Bulletin 345–151)—Specifications and Drawings for Conduit and Manhole Construction.

(iii) RUS Form 515d (Bulletin 345–152)—Specifications and Drawings for Underground Cable Installation.

(iv) RUS Form 515f (Bulletin 345–153)—Specifications and Drawings for Construction of Pole Lines and Aerial Cables.

(v) RUS Form 515g (Bulletin 345–154)—Specifications and Drawings for Service Entrance and Station Protector Installation.

(b) *Preparation of plans and specifications.* Each set of plans and specifications shall include:

(1) RUS Contract Form 515, "Telephone System Construction Contract (Labor and Materials)."

(2) The specifications described in paragraph (a)(2) of this section as specified by the borrower in the RUS Contract Form 515.

(3) Description of special assembly units and guide drawings, if any.

(4) Key, detail, and cable layout maps.

(5) RUS Contract Form 787, "Supplement A to Construction Contract, RUS Contract Form 515," when the borrower proposes to provide any materials to the contractor. The borrower shall not order materials for a contractor without RUS approval. In such cases the borrower must attach Form 787 and a "List of Owner's Materials on Hand" and/or a "List of Materials Ordered by Owner but Not Delivered" to contract Form 515 (See § 1753.48(f) of this part). Any materials furnished under Supplement A shall be listed in RUS Bulletin 344–2 unless special RUS approval has been received by the borrower to use unlisted materials.

(c) *Submission of plans and specifications to RUS.* (1) If the project does not exceed $500,000 or 25% of the loan, whichever is less, the borrower shall furnish GFR one set of P&S. The borrower may then proceed with procurement in accordance with § 1753.48.

(2) If the project exceeds $500,000 or 25% of the loan, whichever is less, RUS approval of P&S is required. Two sets of P&S shall be furnished to GFR. RUS will return one set to the borrower upon notice of approval. The borrower may then proceed with procurement in accordance with § 1753.48.

[54 FR 39267, Sept. 25, 1989. Redesignated at 55 FR 39397, Sept. 27, 1990, 55 FR 53488, Dec. 31, 1990, as amended at 64 FR 16609, Apr. 6, 1999]

§ 1753.48 Procurement procedures.

(a) *Sealed competitive bidding*—(1) *Qualifying bidders.* (i) The borrower is responsible for selecting qualified contractors to bid on the project. See

§ 1753.8(a)(3). Questions relating to bidders' qualifications shall be resolved prior to the pre-bid conference.

(ii) RUS Form 274 or its equivalent, supplemented by RUS Form 276, shall be used for the submission of bidders' qualifications for all types of construction and for the required information on the bidder and subcontractors.

(2) *Invitations to bid*— The borrower shall solicit bids as set forth in § 1753.8(a)(2). Invitations shall be sent to at least 6 prospective bidders.

(3) *Pre-bid conference.* (i) Representatives of the borrower and its engineer shall be present at the pre-bid conference at the time and place designated in the Notice to Bidders. The borrower shall invite the GFR to attend the pre-bid conference.

(ii) The purpose of the pre-bid conference is to acquaint the bidders with the scope and special considerations of the project and to clarify any concerns the bidders may have.

(iii) No proposals shall be considered from bidders that do not attend the pre-bid conference unless the bidder has been notified by the engineer that such bidder's attendance has been waived. Attendance can be waived if, in the judgment of the engineer, the bidder would gain no additional understanding of the construction project by attending the pre-bid conference.

(iv) The borrower shall obtain from the engineer the minutes of the pre-bid conference and shall distribute them to all potential bidders.

(v) When fewer than three bidders have been qualified to submit bids, RUS written approval must be obtained to proceed with requesting bids.

(4) *Bid openings.* (i) Bid openings and award of the contract shall be conducted in accordance with §§ 1753.5(b)(1) and 1753.8(a).

(ii) If § 1753.8 requires RUS approval of award of the bid, the borrower shall submit to RUS two copies of the assembly unit sections of the apparent lowest responsive bid accepted by the borrower.

(b) *Negotiated procurement.* (1) Competitive bids are not required for outside plant construction that is estimated to cost $250,000, or less, inclusive of labor and materials.

(2) The procedures to be followed are contained in § 1753.8(b) and paragraphs (3) and (4) of this section.

(3) *Negotiation conference.* (i) The borrower shall schedule a conference to be attended by representatives of the engineer, the borrower and the contractor selected for negotiations. The borrower shall invite the GFR to attend this conference.

(ii) The purpose of the negotiation conference is to acquaint the contractor with the scope and special considerations of the project and to answer any questions.

(iii) The borrower shall obtain from the engineer notes covering the negotiation conference and shall distribute them to all attendees.

(4) Two copies of the assembly unit sections of the negotiated contractor's proposal shall be sent to the GFR for approval.

(c) *Contract amendments.* The borrower shall prepare contract amendments in accordance with § 1753.11 on RUS Contract Form 526, Construction Contract Amendment.

(d) *Subcontracts.* The RUS requirements for subcontracts and the procedures to be followed are set forth in § 1753.9.

(e) *Preconstruction conference.* The borrower shall conduct a conference, attended by the borrower, contractor, subcontractors, resident engineer, and the GFR, prior to the beginning of cable placement, to resolve any questions pertaining to the construction. Results of the conference shall be provided to each conference participant (See § 1753.10).

(f) *Owner-furnished materials.* When the borrower furnishes materials under RUS Contract Form 787, Supplement A to Construction Contract, these steps shall be followed:

(1) Materials on hand to be furnished by the borrower shall be released to the contractor at the start of construction. Materials on order but not received shall be provided to the contractor as they become available. The borrower shall obtain from the contractor a written receipt for all such materials delivered.

(2) Materials on hand, until released to the contractor, shall be covered by fire and either wind-storm or extended

coverage insurance, exclusive of materials stored in the open and not within 100 feet of any building. Poles, wherever stored, shall be covered by fire insurance. All insured values must be at least 80 percent of the cash value of the property insured.

(3) Subject to adjustment at the time of final settlement, the borrower shall obtain from the contractor monthly invoices that show credit to the borrower, at the prices quoted in Form 787, Supplement A, for all materials furnished by the borrower and installed by the contractor during the preceding month.

(4) Any materials furnished by the borrower remaining as surplus at the completion of construction shall be returned to the borrower. For such materials, the borrower shall furnish a written receipt to the contractor and credit the contractor at the prices quoted in Supplement A.

(g) *Changes or corrections in construction.* (1) When changes or corrections in construction are necessary, and the cost of such changes or corrections is properly chargeable to the borrower, the borrower shall have its engineer prepare and sign four copies of a Construction Change Order, RUS Form 216, obtain borrower's approval and forward the four copies to the contractor. Receipt of the executed Construction Change Order by the contractor will constitute authorization to proceed with the changes or corrections.

(2) When the changes or corrections have been made, the borrower shall have the contractor complete the form, itemizing the costs in accordance with the terms of the contract, and return three copies to the borrower's engineer. A copy of each change order shall be attached to each copy of the construction inventory required to close out the contract.

[54 FR 39267, Sept. 25, 1989. Redesignated at 55 FR 39397, Sept. 27, 1990, as amended at 64 FR 16609, Apr. 6, 1999]

§ 1753.49 Closeout documents.

(a) *General.* The borrower shall be responsible for preparing the closeout documents with, if necessary, the assistance of the GFR.

(b) *Documents required.* The following table lists the documents required to closeout the RUS contract Form 515.

DOCUMENTS REQUIRED TO CLOSEOUT CONSTRUCTION CONTRACT

[RUS Form 515]

RUS Form No.	Description	No. of copies prepared by		Distribution	
		Contractor	Engineer	Borrower	Contractor
724	Final Inventory—Certificate of Completion		2	1	1
724a	Final Inventory—Assembly Units		2	1	1
None	Contractor's Bond Extension (send to RUS when required).	(3)		(to RUS)	(to RUS)
281	Tabulation of Materials Furnished by Borrower	2		1	1
213	Certificate—"Buy American"	1		1	
None	Listing of Construction Change Orders		1	1	
224	Waiver and Release of Lien (from each supplier)	1		1	
231	Certificate of Contractor	1		1	
527	Final Statement of Construction		2	1	1
None	Reports on Results of Acceptance Tests		1	1	1
None	Set of Final Staking Sheets		1	1	
None	Tabulation of Staking Sheets		1	1	
None	Correction Summary (legible copy)		1	1	
None	Treated Forest Products Inspection Reports or Certificates of Compliance (prepared by inspection company or supplier).			1	
None	Final Key Map (when applicable)		1	1	
None	Final Central Office Area and Town Maps		1	1	

(c) *Closeout procedure.* (1) After construction has been completed in accordance with the plans and specifications, and acceptance tests have been made, the borrower shall arrange the time for a final inspection to be made by the borrower's engineer, the contractor, the GFR and a representative of the borrower.

(2) *Final inventory documents.* (i) The borrower shall obtain certifications from the engineer that the project and all required documentation are satisfactory and complete. Requirements for these contract closeout certifications are contained in § 1753.18.

(ii) The borrower shall prepare and distribute the final inventory documents in accordance with the tables contained in this section. The documents listed for RUS shall be retained by the borrower for inspection by RUS for at least two years from the date of the engineer's contract closeout certification.

STEP-BY-STEP PROCEDURE FOR CLOSEOUT OF CONSTRUCTION CONTRACT

[RUS Form 515]

Sequence		By	Procedure
Step No.	When		
1	Upon Completion of Construction	Borrower's Engineer.	Prepares the following: a set of Detail Maps and a set (when applicable) of Key Maps which show in red the work done under the 515 contract; a Tabulation of Staking Sheet; and a tentative Final Inventory, RUS Forms 724 and 724a.
2	After acceptance tests made	Borrower's Engineer.	Forwards letter to the borrower with copies to the GFR stating that the project is ready for final inspection. Schedules inspection date.
3	Upon receipt of letter from Borrower's Engineer.	GFR	Advises borrower whether attending the final inspection will be possible.
4	By inspection date	Borrower's Engineer.	Obtains and makes available the following documents: a set of "as constructed" detail maps and (when applicable) "as built" key maps; a list of construction change orders; the final staking sheets; the tabulation staking sheets; the treated forest products inspection reports or certificates of compliance; the tentative final inventory, RUS Forms 724 and 724a; the tentative tabulation, RUS Form 231 (if borrower furnished part of material); and, a report of results of acceptance tests.
5	During inspection	Borrower's Engineer.	Issues instructions to contractor covering corrections to be made in construction as a result of inspection.
6	During inspection	Contractor	Corrects construction on basis of instructions from the borrower's engineer. The corrections should proceed closely behind the inspection in order that the borrower's engineer can check the corrections before leaving the system.
7	During inspection	Borrower's Engineer.	Inspects and approves corrected construction. Marks inspected areas on the key map, if available, otherwise on the detail maps.
8	Upon completion of inspection	Borrower's Engineer.	Prepares or obtains all the closeout documents listed in Table 3.
9	After signing final inventory	Borrower	Prepares and submits to RUS the engineer's certifications of completion and a Financial Requirement Statement, RUS Form 481, requesting amount necessary to make final payment due under contract.
10	On receipt of final advance	Borrower	Promptly forwards check for final payment to contractor.
11	During subsequent loan fund audit review following final payment.	RUS Field Accountant.	Examines borrower's construction records for compliance with the construction contract and Subpart F, and examines RUS Form 281 (Tabulation of Materials Furnished by Borrower) if any, for appropriate costs.

(iii) When the total inventory price exceeds the maximum contract by more than 20 percent, an extension to the contractor's bond is required.

(iv) The borrower shall submit the engineer's contract closeout certification with FRS for the final advance of funds.

(3) Final payment shall be made according to the payment provisions of article III of RUS Form 515, except that certificates and other documents required to be submitted to or approved by the Administrator shall be

submitted to and approved by the Owner.

[54 FR 39267, Sept. 25, 1989. Redesignated at 55 FR 39397, Sept. 27, 1990, as amended at 59 FR 43717, Aug. 25, 1994; 64 FR 16609, Apr. 6, 1999]

§§ 1753.50–1753.55 [Reserved]

Subpart G—Outside Plant Major Construction by Force Account

SOURCE: 55 FR 3572, Feb. 2, 1990, unless otherwise noted. Redesignated at 55 FR 39397, Sept. 27, 1990.

§ 1753.56 General.

(a) This subpart implements and explains the provisions of the loan documents setting forth the requirements and the procedures to be followed by borrowers for outside plant major construction by the force account method with RUS loan funds. Terms used in this subpart are defined in § 1753.2 and RUS Contract Form 515.

(b) A borrower shall not use the force account method for construction financed with loan funds unless prior RUS approval has been obtained.

(c) Generally, RUS will not approve the force account method for major outside plant construction for the initial loan to a borrower.

(d) The Force Account Proposals (FAPs) are subject to review and approval by RUS.

(e) The FAP is approved by RUS on the basis of estimated labor and material costs. The FAP is closed based on the borrower's actual cost of performing the construction. RUS will provide loan funds only up to the amount determined by the completed assembly units priced at the unit prices in the approved FAP.

(Approved by the Office of Management and Budget under control number 0572–0062)

§ 1753.57 Procedures.

(a) *The request.* (1) The borrower shall submit to RUS a certified copy of the board resolution or a letter signed by an authorized corporate official requesting approval to use the force account method of construction. The request shall state the advantages of the force account method of construction and provide the following information:

(i) The scope of the construction to be undertaken, stating briefly the facilities and equipment to be installed and other pertinent data.

(ii) The name and qualifications of the construction supervisor who will be directly in charge of construction, the names and qualifications of the construction foremen, and the availability of qualified construction personnel. The construction supervisor must have at least 5 years outside plant construction experience with at least 2 years at the supervisory level on RUS financed projects. Construction foremen must have at least 3 years of outside plant construction experience.

(iii) The availability of equipment for construction, exclusive of equipment needed for normal operation and maintenance.

(2) [Reserved]

(b) *Force Account Proposal (FAP).* Upon receiving RUS approval to use the force account method, the borrower, prior to any construction activity or the purchase of materials or equipment, shall submit to RUS two copies of its FAP. The FAP shall consist of:

(1) The RUS Contract Form 515 and appropriate supporting attachments that normally would be provided as plans and specifications for contract construction. See § 1753.47.

(2) The cost estimate, using Form 515 as a convenient means of showing the following:

(i) The quantity and cost estimates of the various assembly units required. "Labor and other" cost will not include the cost of engineering, legal, and other professional services, interest during construction, preliminary survey and investigation charges, and right-of-way easement procurement costs.

(ii) A list identifying materials or construction for which loan funds will not be requested.

(3) The estimated completion time.

(c) *Storage of materials.* All materials ordered for the construction shall be stored separate from normal maintenance materials.

(d) *Construction*—(1) *Preconstruction conference.* The borrower shall arrange a conference, attended by the manager, construction supervisor, construction foremen, resident engineer and the GFR prior to the beginning of construction to clarify any questions pertaining to the construction. Notes of the conference shall be provided to each conference participant.

(2) *Construction schedule and progress reports.* The borrower shall obtain from the engineer a construction schedule and submit one copy to the GFR. The schedule shall include the starting date and a statement indicating that materials are either delivered or deliveries are assured to permit construction to proceed in accordance with the construction schedule. The borrower shall obtain from the engineer progress reports and submit one copy of each to the GFR. RUS Form 521 may be used for the construction schedule and the progress report.

(3) *Borrower's management responsibilities.* (i) Obtain all right-of-way easements, permits, etc., prior to construction.

(ii) Maintain records on all expenditures for materials, labor, transportation, and other costs of construction, in order that all costs may be fully accounted for upon completion of construction.

(iii) Ensure that all the required inspections and tests are made.

(4) *Engineer's responsibilities.* (i) Inspect and inventory construction as completed.

(ii) Require timely corrections and cleanup.

(iii) Perform acceptance tests as construction is completed.

(iv) Provide "as built" staking sheets of completed construction when the final inspections are made.

(v) Maintain accurate and current inventories of completed construction.

(5) *Construction supervisor's responsibilities.* (i) Correct construction errors as construction progresses.

(ii) Maintain an accurate inventory of completed construction.

(iii) Perform cleanup as construction is completed.

(iv) Perform all the inspections and acceptance tests a contractor would be required to make under the construction contract.

(v) Promptly perform cleanup required after final inspection.

§ 1753.58 Closeout documents.

(a) *General.* (1) This section outlines the procedure to be followed in the preparation of closeout documents for the FAP.

(2) The period between the completion of construction and submission of the closeout documents to RUS should not exceed 60 days.

(b) *Documents.* The documents required to close the FAP are listed in the following table. The following is a brief description of the closeout documents:

DOCUMENTS REQUIRED TO CLOSE OUT FORCE ACCOUNT OUTSIDE PLANT CONSTRUCTION

RUS Form No.	Description
817, 817a, 817b	Final Inventory Force Account Construction and Certificate of Engineer. Submit one copy to RUS, if required [1]
213	Certificate—"Buy American" (as applicable from each supplier).
None	Detail Maps.
None	Key map, if applicable.
None	Staking Sheets.
None	Tabulation of staking sheets.
None	Treated Forest Products Inspection Reports or Certificates of Compliance (prepared by inspection company or supplier).

[1] RUS Forms 817, 817a, and 817b are to be submitted to GFR only if required in paragraph (c)(5) of this section. Otherwise, the final inventory documents are to be assembled and retained by the borrower for at least two years.

(c) *Closeout procedures.* (1) The borrower shall notify the GFR when the project is ready for final inspection.

(2) The GFR shall be invited to make the final inspection accompanied by the engineer and the borrower.

(3) The borrower shall correct all deficiencies found during the final inspection.

(4) The borrower may request the assistance of an RUS field accountant to

review the borrower's record of construction expenditures and assist the borrower with any accounting problems in connection with construction expenditures.

(5) After inspection, the final inventory documents shall be assembled as indicated in the table in this section. RUS Forms 817, 817a, and 817b are to be submitted to GFR only if the amount of the closeout exceeds the original force account proposal by 20% or more. Otherwise, the final inventory documents are to be assembled and retained by the borrower for at least two years.

(6) Upon approval of the closeout documents, RUS will notify the borrower of approval and of any adjustments to be made in funds advanced in connection with the construction.

(d) The above are not intended to be a complete description of the requirements of the documents relating to RUS's closeout procedure. Refer to the documents for additional requirements.

[55 FR 3572, Feb. 2, 1990. Redesignated at 55 FR 39397, Sept. 27, 1990, as amended at 64 FR 16610, Apr. 6, 1999]

§§ 1753.59–1753.65 [Reserved]

Subpart H—Purchase and Installation of Special Equipment

SOURCE: 54 FR 39267, Sept. 25, 1989, unless otherwise noted. Redesignated at 55 FR 39397, Sept. 27, 1990.

§ 1753.66 General.

(a) This subpart implements and explains the provisions of the Loan Documents setting forth the requirements and the procedures to be followed by borrowers in purchasing and installing special equipment financed with loan funds.

(b) Terms used in this subpart are defined in § 1753.2 and Equipment Contract, RUS Contract Form 395 (RUS Contract Form 395).

(c) Borrowers must obtain RUS review and approval of the LD for their telephone systems. Applications of equipment not included in an approved LD must conform to the modernization plan as required by 7 CFR part 1751, subpart B, and must be submitted to RUS for review and approval.

(d) RUS Contract Form 395 and applicable specifications shall be used for the purchase of special equipment for major construction on a furnish-and-install basis, as well as on a furnish-only basis.

(e) The procedures provided in subpart I, if applicable, or a FAP approved by RUS may be used for the installation of special equipment purchased with a RUS Contract Form 395 contract not including installation.

(f) For special equipment purchases for minor construction, the borrower may at its option use the Methods of Minor Construction procedures contained in subpart I or the purchase procedures contained in this subpart H.

(g) Some types of special equipment contain software. RUS Contract Form 395 indicates whether the equipment contains software and whether the software contract stipulations are applicable.

[81 FR 71584, Oct. 18, 2016]

§ 1753.67 Contracts and specifications.

(a) Equipment Contract, RUS Contract Form 395 shall be used to purchase equipment on a furnish-and-install basis, as well as on a furnish-only basis.

(b) The equipment specifications must accompany the equipment contract form and each specification consists of performance specifications, installation requirements (if applicable), and application engineering requirements.

[81 FR 71584, Oct. 18, 2016]

§ 1753.68 Purchasing special equipment.

(a) *General.* (1) Equipment purchases are categorized as initial equipment purchase, equipment additions to existing systems and new system additions.

(i) An initial equipment purchase is a first time purchase by a borrower of a complete system of special equipment.

(ii) Equipment additions to existing systems are additions of components to complete operating systems to increase system capacity that require components made by the manufacturer of the existing system.

(iii) New system additions are purchases of complete systems of special equipment when the purpose can be accomplished either with equipment of the same type and manufacture as other complete operating systems in the borrower's system, or with complete systems of special equipment from other manufacturers.

(iv) Where equipment is obtained under a Equipment Contract, RUS Contract Form 395 (RUS Contract Form 395) without installation, the borrower shall require the supplier to provide a detailed proposed bonding and grounding plan and detailed installation information. The installation information is to enable acceptance testing by the borrower upon completion of the installation.

(2) For initial equipment purchases that qualify as major construction, the borrower shall obtain proposals from at least three suppliers of equipment of different manufacturers.

(3) For equipment additions to increase the capacity of existing systems, the borrower may negotiate for equipment of a specific type and manufacture. RUS approval to negotiate in this instance is not required if these additions were specifically described in the LD approved by RUS

(4) For new system additions, the borrower may request RUS approval to negotiate for additional equipment for the purpose of standardization on a system basis, provided RUS approved the procurement method used for the initial equipment purchase. RUS approval to negotiate must be obtained before release of the P&S to the seller.

(5) RUS will not approve negotiation with a seller of non-domestic equipment for the purpose of standardization, because such a purchase does not meet the "Buy American" provision.

(6) RUS recommends, but does not require, that borrowers include installation by the seller for initial installations of special equipment that qualify as major construction.

(7) Special equipment may be installed by the borrower if it has qualified personnel and test equipment available to install the equipment and make the required acceptance tests, and written approval is given by RUS.

(8) Installations, whether by the borrower or the seller, must meet the installation requirements of RUS Contract Form 395specifications. A copy of the acceptance tests results must be attached to the closeout documents or work order summary.

(9) The specifications for the various applications of equipment is prepared by the RUS borrower's engineer and based on generally accepted engineering considerations and practices found in the Telecommunications Industry.

(10) The borrower must obtain authorization from the Federal Communications Commission (FCC) to construct and operate radio transmitting equipment. Evidence of FCC authorization is required for RUS contract approval. Where required, the borrower must obtain approval of state regulatory bodies regarding tariffs and related matters.

(b) *Procurement procedures*—(1) *General.* The following are the procurement procedure steps required for the purchase of special equipment by borrowers.

(2) *Initial equipment purchase.* (i) The borrower prepares P&S and, for projects estimated to exceed $500,000 or 25% of the loan, whichever is less, sends two copies to GFR for approval.

(ii) For projects estimated to exceed $500,000 or 25% of the loan, whichever is less, RUS will either approve P&S in writing or notify the borrower of the reasons for withholding approval.

(iii) For projects estimated to cost less than $500,000 or 25% of the loan, whichever is less, the borrower may proceed with procurement upon completion of the P&S.

(iv) If the borrower has employed full competitive bidding in the selection, a contract may be executed with the successful bidder and the borrower may proceed to paragraph (b)(2)(vi) of this section.

(v) If the borrower did not follow a fully competitive bidding process as described in §1753.8, the selection, along with a summary of all proposals and an engineer's recommendation, shall be sent to RUS. RUS shall approve the proposal selection in writing or notify the borrower of any reason for withholding approval.

(vi) The borrower sends three executed contracts including specifications to RUS for approval.

(vii) After RUS approval of the contract, one copy will be returned to the borrower and one copy will be sent to the seller.

(3) *Equipment additions to existing systems.* Purchase procedures for equipment additions to existing systems are the same as for initial system purchase except that the borrower may negotiate for equipment of a specific type and manufacture instead of obtaining proposals from three or more sellers.

(4) *New system additions.* (i) The borrower prepares the P&S and, if the project is estimated to exceed $500,000 or 25% of the loan, whichever is less, sends two copies to the GFR for approval. The borrower may request RUS approval to negotiate for the purpose of standardization on a system basis prior to preparing the P&S.

(ii) RUS notifies the borrower in writing as to whether the borrower may negotiate for specific equipment. If P&S were required to be submitted to RUS under paragraph (b)(4)(i) of this section, RUS notifies the borrower in writing of P&S approval (or notifies the borrower of the reasons for withholding approval).

(iii) The remainder of the purchase procedure for new system additions is the same as for initial equipment purchase.

(c) *Contract amendments.* (1) The general requirements for contract amendments are set forth in § 1753.11.

(2) The borrower shall prepare any required amendments to the special equipment contract, arrange for the execution by all parties, and submit these amendments to RUS in accordance with § 1753.11(d). RUS Form 238, Construction or Equipment Contract Amendment, shall be used for this purpose.

(d) *Closeout procedures*—(1) *Acceptance tests for RUS Contract Form 395 with installation.* (i) Immediately upon completion of the installation and alignment of the equipment, the borrower shall arrange with the contractor's installer and the GFR for acceptance tests.

(ii) The borrower shall obtain from the contractor, in writing, the results of all inspections and tests made by the contractor as required in the specifications. The borrower will analyze the test results and determine whether the performance of the equipment meets the contract specifications.

(2) Acceptance tests for RUS Contract Form 395 without installation. (Upon completion of the installation and alignment of the equipment (under this contract the installation alignment will be by other than the seller) the borrower shall perform all the inspections and tests outlined in the specifications.

(3) *Closeout documents.* When the acceptance tests have been completed and all deficiencies have been corrected, the borrower:

(i) Assembles and distributes the documents listed in the following table that are required for the closeout of the equipment contract. The documents listed for RUS shall be retained by the borrower for inspection by RUS for at least two years from the date of the engineer's contract closeout certification.

TABLE 1 TO PARAGRAPH (d)(3)(i)—DOCUMENTS REQUIRED TO CLOSE OUT EQUIPMENT CONTRACT, RUS CONTRACT FORM 395

RUS Form No.	Description	Number of copies prepared by		Distribution	
		Seller	Engineer	Buyer	Seller
213	Certificate (Buy American)	1		1.	
238	Construction or Equipment Contract Amendment (If not previously submitted, send to RUS for approval).		3 sent to RUS.	1 sent to RUS.	1 from RUS.
395a	Certificate of Completion for Equipment Contract (Including Installation).		2	1	1.
395b	Certificate of Completion for Equipment Contract (Not Including Installation).		2	1	1.
395c	Certificate of Contractor and Indemnity Agreement (Use only for installation contracts).			1.	

TABLE 1 TO PARAGRAPH (d)(3)(i)—DOCUMENTS REQUIRED TO CLOSE OUT EQUIPMENT CONTRACT,—Continued
RUS CONTRACT FORM 395

RUS Form No.	Description	Number of copies prepared by		Distribution	
		Seller	Engineer	Buyer	Seller
None	Report in writing, including all measurements, any acceptance test report and other information required under Part II of the applicable specifications. (Form 395d may be used.).	....	1	1.	
None	Set of maintenance recommendations for all equipment furnished under the contract.	1	....	1.	

(ii) Obtains certifications from the engineer that the project and all required documentation are satisfactory and complete. Requirements for this contract closeout certification are contained in § 1753.18.

(iii) Submits copies of the engineer's certifications to RUS with the FRS requesting the remaining funds on the contract.

(iv) Makes final payment in accordance with the payment terms of the contract.

[54 FR 39267, Sept. 25, 1989. Redesignated at 55 FR 39397, Sept. 27, 1990, as amended at 59 FR 43718, Aug. 25, 1994; 64 FR 16611, Apr. 6, 1999; 81 FR 71584, Oct. 18, 2016]

§§ 1753.69–1753.75 [Reserved]

Subpart I—Minor Construction

SOURCE: 55 FR 3573, Feb. 2, 1990, unless otherwise noted. Redesignated at 55 FR 39397, Sept. 27, 1990.

§ 1753.76 General.

(a) This subpart implements and explains the provisions of the Loan Documents containing the requirements and procedures to be followed by borrowers for minor construction of telecommunications facilities using RUS loan funds. Terms used in this subpart are defined in § 1753.2.

(b) [Reserved]

(Approved by the Office of Management and Budget under control number 0572–0062)

[55 FR 3573, Feb. 2, 1990. Redesignated at 55 FR 39397, Sept. 27, 1990, as amended at 64 FR 16611, Apr. 6, 1999]

§ 1753.77 Methods of minor construction.

Minor construction may be performed by contract using RUS Contract Form 773, ''Miscellaneous Construction Work and Maintenance Services'', by RUS Contract Form 515, or by work order construction. The rules for using Form 515 for minor construction are contained in subpart F of this part.

[64 FR 16612, Apr. 6, 1999]

§ 1753.78 Construction by contract.

(a) RUS Form 773 shall be used for minor construction by contract. Compensation may be based upon unit prices, hourly rates, or another basis agreed to in advance by the borrower and the contractor. A single work project may require more than one contractor.

(b) The borrower shall prepare the contract form and attach any diagrams, sketches and tabulations necessary to specify clearly the work to be performed and who shall provide which materials. Neither the selection of the contractor nor the contract requires RUS approval.

(c) Borrowers are urged to obtain quotations from several contractors before entering into a contract to be assured of obtaining the lowest cost. The borrower must ensure that the contractor selected meets all Federal and State licensing and bonding requirements, and that the contractor maintains the insurance coverage required by the contract for the duration of the work. (See 7 part CFR 1788)

(d) Upon completion and final inspection of the construction the borrower shall obtain from the Contractor a final invoice and an executed copy of

RUS Form 743, Certificate of Contractor and Indemnity Agreement.

(e) RUS Contract Form 773 may also be used to contract for the maintenance and repair of telephone equipment and facilities. Generally, RUS will not finance maintenance and repair contracts.

[55 FR 3573, Feb. 2, 1990. Redesignated at 55 FR 39397, Sept. 27, 1990, as amended at 59 FR 43718, Aug. 25, 1994]

§ 1753.79 Construction by force account.

The borrower shall require that:

(a) Minor construction by the force account method be supervised by a competent foreman. The work shall be performed in accordance with all regulatory and safety codes.

(b) Daily time and material reports, referenced by the work project number, shall be kept to record labor and materials used as construction is performed.

(c) The construction foreman shall maintain a tabulation of all construction units installed.

§ 1753.80 Minor construction procedure.

(a) If the borrower performs minor construction financed with loan funds, the borrower's regular work order procedure shall be used to administer construction activities that may be performed entirely by a contractor under Form 773 contract, by work order, or jointly by work order and one or more contractors under Form 773 contracts.

(b) RUS financing under Form 773 contracts dated in the same calendar year is limited to the following amounts for the following discrete categories of minor construction. The date of the Form 773 contract is the date the Form 773 contract is executed.

(1) For outside plant construction, the limit is $500,000 or ten per cent (10%) of the borrower's previous calendar year's outside plant total construction, whichever is greater.

(2) For central office equipment, the limit is $500,000.

(3) For equipment and buildings, the limit is $250,000 in each category.

(c) A single minor construction project may be a discrete element of a somewhat larger overall project, such as the provision and installation of a standby power generator or heating/air conditioning equipment in connection with a building modification or expansion project or the splicing on a major cable placement project. It cannot be a portion, by dividing into smaller segments, of a discrete major construction project, such as the placement of a continuous cable facility.

(d) RUS approval must be obtained in advance for minor construction unless all of the following conditions are met:

(1) RUS has approved the engineering design.

(2) All standard RUS procedures are followed, including the application of RUS construction practices (see § 1753.6).

(3) The Standard Form 773 contract is used without modification.

(e) The borrower shall determine the scope of each proposed construction project and decide how it will be constructed. A work project number shall be assigned to which all charges for that project are referenced.

(f) The borrower shall maintain accounting and plant records sufficient to document the cost and location of all construction and to support loan fund advances and disbursements.

(g) Normally the borrower will finance minor construction with general funds and obtain reimbursement with loan funds when construction is completed and executed Form 771 has been submitted to RUS. If a borrower satisfies RUS of its inability to finance the construction temporarily with general funds, RUS may establish, on a case by case basis, a work order fund for specific construction projects. The work order fund will be closed upon receipt of an FRS and the executed Forms 771 for the specific projects for which the work order fund was established.

(h) RUS will advance funds to finance minor construction work projects only if all necessary documents, including an FRS and supporting data covering the project, are received within one year of the date construction of the project is completed.

[55 FR 3573, Feb. 2, 1990. Redesignated at 55 FR 39397, Sept. 27, 1990, as amended at 59 FR 43718, Aug. 25, 1994; 64 FR 16612, Apr. 6, 1999; 81 FR 71585, Oct. 18, 2016]

§ 1753.81 Inspection and certification.

(a) Upon completion and prior to closeout, minor construction must be inspected and certified to be in compliance with RUS construction standards, to be reasonable in cost, and to meet applicable codes. The certification is made by an experienced telephone engineer who is either licensed in the state where the inspection will be performed, or is a borrower's staff engineer, who meets the requirements of the "employee in charge" of force account engineering as described in subpart B of this part. The GFR will periodically audit the inspection of minor construction to ensure integrity of the procedure. RUS borrowers with less than 2000 subscribers may use the above procedure or have construction inspection performed by the GFR.

(b) Engineering services for minor construction may be contracted using RUS Form 245, Engineering Service Contract—Special Services. Costs for these services may be included in the costs for construction on the Form 771. (See subpart B of this part.

(c) Upon completion of construction, the borrower shall obtain the engineer's certification on RUS Form 771. An official of the borrower, designated by the board of directors, shall also execute the borrower's certification on Form 771.

[54 FR 39267, Sept. 25, 1989. Redesignated at 55 FR 39397, Sept. 27, 1990, as amended at 55 FR 53488, Dec. 31, 1990]

§ 1753.82 Minor construction closeout.

(a) For minor construction inspected by the borrower's engineer, an original and two copies of Form 771 shall be sent to the GFR. The GFR will initial and return the original and one copy.

(b) When funds are requested for minor construction, the original Form 771 signed or initialed by the GFR, shall be submitted with the FRS. Forms 771 should be submitted only with the FRS which they support. RUS does not encumber funds pursuant to Forms 771 unless an advance is made to the borrower. (See 7 CFR part 1744 subpart C).

§§ 1753.83–1753.90 [Reserved]

Subpart J—Construction Certification Program

SOURCE: 55 FR 3574, Feb. 2, 1990, unless otherwise noted. Redesignated at 55 FR 39397, Sept. 27, 1990.

§ 1753.91 General.

(a) This subpart implements and explains the provisions of the loan documents setting forth the requirements and procedures to be followed by borrowers accepting nomination for the construction certification program. Terms used in this subpart are defined in § 1753.2.

(b) [Reserved]

(Approved by the Office of Management and Budget under control number 0572–0062)

§ 1753.92 Policies and requirements.

(a) It is RUS policy that, as borrowers gain in experience and maturity, the advice and assistance rendered by RUS shall progressively diminish. Prior to approval of a loan, RUS may nominate certain borrowers to fulfill the responsibilities for administration and construction of projects financed with RUS loans. Borrowers who accept this nomination will be known as "certification borrowers," and the program in which they participate will be known as the "certification program."

(b) Generally, initial loan borrowers are not eligible for the certification program.

(c) Generally, the factors which RUS will consider in selecting borrowers for the certification program will include:

(1) The experience of the staff of the borrower.

(2) The RUS assessment of the borrower's ability to handle the certification program requirements considering the size and complexity of the proposed construction in the LD.

(3) The history of the borrower in following RUS's policies and procedures.

(4) Other factors deemed relevant by RUS.

(d) Except as specifically stated in this subpart, certification borrowers must comply with all requirements applicable to other borrowers.

(e) RUS reserves the right at any time to require submission of construction documents or to remove the borrower from the certification program.

§ 1753.93 Responsibilities.

(a) *Responsibilities transferred to certification borrowers.* (1) Approval of engineering and architectural service contracts.

(2) Approval of P&S.

(3) Approval of price quotations and bids, except where the low price bid is not accepted.

(4) Approval of award of construction contracts and amendments.

(5) Approval of FAP's if RUS has approved the force account method of construction for the construction project.

(6) Inspection and certification of construction.

(7) Approval of closeout documents.

(8) Other responsibilities as may be specifically granted in writing by RUS.

(b) *Responsibilities retained by RUS.* (1) Approval to deviate from RUS requirements, except as provided in (a) above.

(2) Approval of use of loan funds for projects other than those included in the loan construction budget. See 7 CFR part 1744 subpart C.

(3) Approval of use of loan funds in excess of amounts included in the loan budget.

(4) Approval of force account methods of engineering and construction.

(5) Approval to make significant deviations from the work plan approved by RUS.

(6) Approval of interim construction.

(7) Approval to modify or alter standard forms and contracts.

(8) Approval to open bids when fewer than the required number have been received.

(90) Approval of outside plant layouts.

(10) "Buy American" determinations.

(11) Other responsibilities not specifically transferred by this subpart or in writing by RUS.

[55 FR 3574, Feb. 2, 1990. Redesignated at 55 FR 39397, Sept. 27, 1990, as amended at 81 FR 71585, Oct. 18, 2016]

§ 1753.94 Procedures.

(a) Certification borrowers shall appoint three certification officials. These appointments shall be subject to RUS approval.

(1) The "Certifying Officer" shall be an officer or employee of the borrower who is authorized to execute binding agreements. This officer shall sign all contracts, amendments, closeout documents and the certification on RUS Form 158, Certification of Contract or Force Account Proposal Approval, and RUS Form 159, Summary of Completed Construction.

(2) The "Construction Certifier" shall be an experienced telephone engineer who is either licensed in the state where the inspection will be performed, or is a borrower's staff engineer who meets the requirements of the "employee in charge" of force account engineering as described in subpart B of this part. RUS may determine that it will accept the certification only for matters within the staff engineer's area of specialization. In such cases the position of "Construction Certifier" shall be filled by more than one engineer. This official is responsible for certifying that the construction complies with all technical and code requirements.

(3) The "Certification Coordinator" shall administer the certification program and serve as the official point of contact for RUS. The certifying officer or construction certifier may also serve as the certification coordinator.

(b) Certification borrowers, shall submit and obtain RUS approval of a work plan before construction and related engineering begin.

(1) The work plan shall provide a description of the proposed construction and methods of purchasing in such detail as to enable RUS to monitor the construction program to ensure to its satisfaction that loan purposes are accomplished in an organized construction program.

(2) The work plan shall include the following:

(i) The names and qualifications of the proposed certification officials defined in § 1753.94(a).

(ii) A listing of the proposed work projects to accomplish the loan purposes showing the estimated cost, method of performing the construction, and the proposed commencement and completion dates for each work project. The proposed work projects shall be summarized on RUS Form 157, Construction Work Plan and Cost Distribution, or a form providing essentially the same information.

(iii) The proposed source of funds for meeting cost overruns if the total estimated cost of work projects exceeds the loan budget.

(iv) A statement signed by the borrower's certification officials and the GFR that the work plan is accurate and complete.

(c) Under the certification program, the borrower shall follow all standard RUS postloan engineering and construction procedures except that the approvals shown in §1753.93(a) will be made by certification officials rather than RUS. The approvals noted in §1753.93(a)(1), (4) and (5) will be reported immediately to RUS using RUS Form 158. Approval of closeouts, §1753.93(a) (6) and (7), will be reported immediately on RUS Form 159.

(d) As the construction program progresses, the certification borrower shall request, by letter, RUS approval of any significant changes in work plan schedules and budgets and in certification officials.

§ 1753.95 Advance of loan funds.

Advance of loan funds needed to meet the certification borrower's current financial obligations are to be requested on RUS Form 481 for construction and engineering items supported by appropriate RUS Forms 158 and 159. For items other than construction or engineering, other supporting data shall be submitted. (See 7 CFR part 1744 subpart C.)

§ 1753.96 Certification addendum.

The certification borrower shall modify standard RUS forms of contract for use under the certification program by inserting an executed copy of the following certification addendum in each copy of the contract.

CERTIFICATION ADDENDUM

Permission has been obtained by the Owner to proceed with this contract under 7 CFR part 1753 subpart J, pursuant to which the references in the RUS construction document requiring approvals and other actions of the RUS Administrator will not apply unless RUS gives specific notice in writing to the affected parties that designated approval(s) or action(s) will be required. Certifications by the Contractor of amounts due and certifications of completions of work under the contract are to be construed to be rendered for the purpose of inducing the Rural Utilities Service to advance funds to the Owner to make, or reimburse the Owner for, payments under this contract.

Date

____

Owner

By ____________________

Certifying Officer

Date ____________________

____

Contractor

By ____________________

____

Title

[55 FR 3574, Feb. 2, 1990. Redesignated at 55 FR 39397, Sept. 27, 1990, as amended at 81 FR 71585, Oct. 18, 2016]

§§ 1753.97–1753.99 [Reserved]

PART 1755—TELECOMMUNICATIONS POLICIES ON SPECIFICATIONS, ACCEPTABLE MATERIALS, AND STANDARD CONTRACT FORMS

Sec.

AUTHORITY: 7 U.S.C. 901 *et seq.*, 1921 *et seq.*, 6941 *et seq.*

SOURCE: 55 FR 39397, Sept. 27, 1990, unless otherwise noted.

§§ 1755.1–1755.2 [Reserved]

§ 1755.3 Field trials.

(a) Except as covered in Bulletin 345–3, no loan funds shall be advanced for any product if any item to be included in the project is not included in the "List of Materials Acceptable for Use on Telephone Systems of RUS Borrowers," RUS Bulletin 344–2. When new items of materials or equipment are considered for acceptance by RUS or when a previously accepted item has been subjected to such major modifications that its suitability cannot be determined based on laboratory data and/or field experience, a field trial shall be required if RUS so determines. This field trial consists of limited field installations of the materials or equipment in closely monitored situations designed to determine, to RUS's satisfaction, their operational effectiveness under actual field conditions. Field trials are to be used only as a means for determining, to RUS's satisfaction, the operational effectiveness of a new or revised product under actual field conditions. Both the manufacturer and borrower are responsible for assuring that the field trial is carried out and that the required information on the product's performance is received by RUS in a timely manner. The use of materials or equipment derived from new inventions or concepts untried within the telephone industry is defined as "an experiment" and shall be handled as a special case using procedures considered appropriate by RUS to meet the individual experiment.

(b) To qualify for a field trial, the new and improved materials and equipment must appear to RUS to offer one or more of the following benefits:

(1) Improved performance.

(2) Decreased cost.

(3) Broader application.

(c) The item of material or equipment subject to field trial may be only part of the total amount of materials or equipment included in a bid or it may be the key component of the facility or system provided; therefore, RUS shall have authority to require that a satisfactory plan be provided to maintain or restore service in the event that the materials and equipment fail to meet established performance requirements. RUS shall limit the quantity of new materials and equipment installed on any field trial and shall also limit the number of field trials for a given product to what RUS considers reasonable to provide the necessary information.

(d) A borrower may participate in a field trial only if, in RUS's opinion, the borrower possesses:

(1) Adequate financial resources so that no delay in the project will result from lack of funds.

(2) The financial stability to overcome difficulties which may result from an unsuccessful field trial. The borrower must be able to restore and maintain service until the manufacturer meets its financial obligations with respect to the field trial.

(3) Qualified personnel to enable it to discharge its responsibilities.

(4) A record satisfactory to RUS for maintaining equipment and plant facilities and for providing RUS with information when requested.

(5) Willingness to participate in the field trial and awareness of the effort and responsibility this entails.

(e) The test site for the field trial shall be, in RUS's opinion, readily accessible and provide the conditions, such as temperature extremes, high probability of lightning damage, etc., for which the product is being evaluated. The material or equipment involved shall be covered by an RUS specification or a suitable standard acceptable to RUS. The supplier is required to submit test data to show conformance with the applicable specification or standard. Further testing shall be performed if required by RUS personnel.

(f) A field trial shall normally continue for a minimum of six months, or for a longer period of time determined by RUS to be required to obtain conclusive data that the item either fulfills all requirements or is unacceptable. Either the borrower or supplier may terminate a field trial at any time, in accordance with their contractual agreement. Such termination, if prior to the time required by RUS, shall constitute withdrawal of the product from consideration by RUS. RUS has authority to terminate field trials based on its determination that the equipment is not performing satisfactorily and that this lack of performance may, in RUS's opinion, cause service degradation or hazards to life or property.

(g) Field trials shall be conducted in accordance with the instructions set forth in this regulation and the agreement relating to the specific application. Both the supplier and the borrower shall agree, and obtain RUS approval before the start of the trial, on the following:

(1) The specific purpose of the field trial;

(2) Ownership of items during trial;

(3) Starting date and duration;

(4) Responsibility for costs and removal of items in the event of noncompliance with the specification or purpose intended and arrangements for service continuity or restoration;

(5) Responsibility for testing, test equipment and normal operation and maintenance during the trial period;

(6) Availability of test equipment on site during the trial period; and

(7) Responsibility for spare parts and components consumed during the trial period.

(h) Both the supplier and the borrower shall keep RUS informed of the status of a field trial. These reports shall not be limited to details of problems of failures encountered during installation and subsequent operation but shall include information on progress of the field trial. If these reports are not received in accordance with the requirements of the RUS Form 399b, RUS shall have the authority to deny or suspend loan funds related to these products until the delinquent reports are received.

(i) Before a borrower purchases materials or equipment that require a field trial, prior approval must be obtained from RUS and RUS Form 399b, RUS Telecommunications Equipment Field Trial (available from the Director, Administrative Services Division, Rural Utilities Service, Room 0175, South Building, U.S. Department of Agriculture, Washington, DC 20250) will be completed by RUS and must be signed by both the borrower and supplier as an indication that they understand their responsibilities in the field trial. Assurance must also be obtained from RUS that the "particular item" that is the subject of the field test is eligible for a field trial. To obtain this assurance, any proposal for use of an item on a field trial basis shall be forwarded to the Chief, Area Engineering Branch, for review and approval.

(j) Procedures for establishing field trials for the various categories of equipment after RUS has approved the 399b:

(1) *Electronic transmission equipment.* The procedure set forth in Bulletin 385–2 "Purchasing and Installing Special Electronic Equipment" shall be followed except that the Special Equipment Contract (Including Installation), RUS Form 397, shall be used in all purchases of electronic equipment for field trials. In addition, the borrower and supplier shall execute three copies of a "Supplemental Agreement to Equipment Contract for Field Trial," RUS Form 399, or a "Supplemental Agreement to Equipment Contract for Field Trial (Secondary—Delivery, Installation, Operation)", RUS Form 399a, as well as three copies of the RUS Form 399b, "RUS Telecommunications Equipment Field Trial", and forward them, together with three copies of the executed contract and specifications, to the Chief, Area Engineering Branch. A limited number of copies of RUS Forms 399, 399a, and 399b are available from RUS upon request from the Director, Administrative Services Division, Rural Utilities Service, Room 0175, South Building, U.S. Department of Agriculture, Washington, DC 20250. Additional copies may be reproduced by the user as needed. This category includes:

(i) Voice frequency repeaters;

(ii) Trunk carriers;

(iii) Subscriber carrier;

(iv) Point-to-point radio (Microwave);

(v) Coaxial cable system electronics;

(vi) Fiber optic cable system electronics;

(vii) Multiplex equipment;

(viii) Mobile and fixed radiotelephone; and

(ix) Other items of electronic equipment associated with transmission.

(2) *Central office equipment.* The procedure set forth in Bulletin 384–1 "Purchasing and Installing Central Office Equipment" shall be followed except that "The Central Office Equipment Contract (Including Installation)", RUS Form 525, shall be used to purchase switching equipment for field trials. In addition, the borrower and supplier shall execute three copies of a "Supplemental Agreement to Equipment Contract for Field Trial," RUS Form 399, or a "Supplemental Agreement to Equipment Contract for Field Trial (Secondary—Delivery, Installation, Operation)", RUS Form 399a, as the case may be, as well as three copies of the RUS Form 399b, "RUS Telecommunications Equipment Field Trial", and forward them, together with three copies of the executed contract and specification to the Chief, Area Engineering Branch. This category includes:

(i) Central office dial equipment;

(ii) Direct distance dialing equipment;

(iii) Automatic number identification equipment;

(iv) Line concentrators;

(v) Remote switching equipment; and

(vi) All other items of equipment associated with switching equipment, such as loop extenders.

(3) Protection equipment and materials, outside plant equipment and materials, and all other equipment and materials, which includes all items not covered in paragraph (j) (1) or (2) of this section, shall be handled as described in Bulletin 344–1 "Methods of Purchasing Materials and Equipment for Use on Systems of Telephone Borrowers" except that the borrower's purchase order form is to be used for purchasing materials and equipment in these categories. In addition, the borrower and supplier shall execute three copies of the "Supplemental Agreement to Equipment Contract for Field Trial," RUS Form 399, or a "Supplemental Agreement to Equipment Contract for Field Trial (Secondary—Delivery, Installation, Operation)", RUS Form 399a, as the case may be, as well as three copies of the RUS Form 399b, "RUS Telecommunications Field Trial", and forward them, together

with three copies of the purchase order to the Chief, Area Engineering Branch.

(k) For all items except Electronic Central Office Equipment, suppliers and manufacturers must furnish warranties or guarantees satisfactory to RUS against the failure of the material and equipment used in the field trial. Terms of this warranty must not be less than the provisions of the standard warranty included in the "Telephone System Construction Contract", RUS Form 515, or the warranty provided for similar materials and equipment included in the "List of Materials Acceptable for Use on Telephone Systems of RUS Borrowers", RUS Bulletin 344–2. In lieu of a warranty, materials and equipment are sometimes furnished to RUS borrowers on a reduced or no cost basis. Terms of such arrangements are subject to RUS approval and should be fully covered in field trial proposals forwarded by borrowers to the Chief, Area Engineering Branch for review and approval. For the purchase of electronic central office equipment, suppliers and manufacturers are to provide warranties as provided in the applicable RUS contract form: RUS Form 397 for electronic equipment and RUS Form 525 for central office equipment. Forms 399 and 399a, which apply to field trials of these devices, specify that the term of the warranty does not begin until the satisfactory conclusion of the field trial.

[49 FR 28394, July 12, 1984. Redesignated at 55 FR 39397, Sept. 27, 1990]

§§ 1755.4–1755.25 [Reserved]

§ 1755.26 RUS standard contract forms.

(a) The standard loan agreement between RUS and its borrowers provides that, in accordance with applicable RUS regulations, borrowers shall use standard contract forms promulgated by RUS for construction, procurement, engineering services, and architectural services financed by a loan or guaranteed by RUS. This part implements these provisions of the RUS loan agreement and prescribes the procedures that RUS follows in promulgating standard contract forms that borrowers are required to use. Part 1753 prescribes when and how borrowers are required to use these standard forms of contracts.

(b) *Contract forms*. RUS promulgates standard contract forms, identified in § 1755.30(c), List of Standard Contract Forms, that borrowers are required to use.

[64 FR 6500, Feb. 10, 1999]

§ 1755.27 Borrower contractual obligations.

(a) *Loan agreement*. As a condition of a loan or loan guaranteed under the RE Act, borrowers are normally required to enter into RUS loan agreements pursuant to which the borrowers agree to use RUS standard contract forms for construction, procurement, engineering services, and architectural services financed in whole or in part by the RUS loan. To comply with the provisions of the loan agreements as implemented by this part, borrowers must use those contract forms identified in the list of telecommunications standard contract forms, set forth in § 1755.30(c) of this part.

(b) *Compliance*. (1) If a borrower is required by part 1753 to use a listed contract form, the borrower shall use the listed contract form in the format available from RUS. The forms shall not be retyped, changed, modified, or altered in any manner not specifically authorized in this part or approved by RUS in writing. Any modifications approved by RUS must be clearly shown so as to indicate the difference from the listed contract form.

(2) The borrower may use electronic reproductions of a contract form if the contract documents submitted for RUS approval are exact reproductions of the RUS form and include the following certification by the borrower: I (Insert name of the person.), certify that the attached (Insert name of the contract form.), between (Insert name of the parties.), dated (Insert contract date.) is an exact reproduction of RUS Form (Insert form number), dated (Insert date of RUS form).

____
(Signature)

____
(Title)

____
(Employer's Address)

(c) *Amendment*. Where a borrower has entered into a contract in the form required by 7 CFR part 1753, no change may be made in the terms of the contract, by amendment, waiver or otherwise, without the prior written approval of RUS.

(d) *Waiver*. RUS may waive for good cause, on a case-by-case basis, the requirements imposed on a borrower pursuant to this part. Borrowers seeking an RUS waiver must provide RUS with a written request explaining the need for the waiver.

(e) *Violations*. A failure on the part of the borrower to use listed contracts as prescribed in 7 CFR part 1753 is a violation of the terms of the loan agreement with RUS and RUS may exercise any and all remedies available under the terms of the agreement or otherwise.

[64 FR 6500, Feb. 10, 1999]

§ 1755.28 Notice and publication of listed contract forms.

(a) *Notice*. Upon initially entering a loan agreement with RUS, borrowers will be provided with all listed contract forms. Thereafter, new or revised listed contract forms promulgated by RUS, including RUS approved exceptions and alternatives, will be sent by regular or electronic mail to the borrower's address as identified in its loan agreement with RUS.

(b) *Availability*. Listed contract forms are published by RUS. Interested parties may obtain the forms from the Rural Utilities Service, Program Development and Regulatory Analysis, U.S. Department of Agriculture, Stop 1522, Washington DC 20250–1522, telephone number (202) 720–8674. The list of contract forms can be found in § 1755.30(c).

[64 FR 6500, Feb. 10, 1999]

§ 1755.29 Promulgation of new or revised contract forms.

RUS may, from time to time, promulgate new contract forms or revise or eliminate existing contract forms. In so doing, RUS shall publish a notice of rulemaking in the FEDERAL REGISTER announcing, as appropriate, a revision in, or a proposal to amend § 1755.30(c), List of telecommunications standard contract forms. The amendment may change the existing identification of a listed contract form by, for example, changing the issuance date of the listed contract form or identifying a new required contract form. The notice of rulemaking will describe the new standard contract form or substantive change in the listed contract form, as the case may be, and the issues involved. The standard contract form or relevant portions thereof may be appended to the supplementary information section of the notice of rulemaking. As appropriate, the notice of rulemaking shall provide an opportunity for interested persons to provide comments. RUS shall send, by regular or electronic mail, a copy of each such FEDERAL REGISTER document to all borrowers.

[64 FR 6500, Feb. 10, 1999]

§ 1755.30 List of telecommunications standard contract forms.

(a) *General*. The following is a list of RUS telecommunications program standard contract forms for procurement, construction, engineering services, and architectural services. Borrowers are required to use these contract forms by the terms of their RUS loan agreements implemented by part 1753 and this part.

(b) *Issuance Date*. Where part 1753 requires the use of a standard contract form in connection with RUS financing, the borrower shall use the appropriate form identified in § 1755.30(c), List of Telecommunications Standard Contract Forms, published as of the date the borrower releases the plans and specifications to solicit bids or price quotes.

(c) *List of telecommunications standard contract forms*. (1) RUS Form 157, issued 10–77, Construction Work Plan and cost Distribution—Telephone.

(2) RUS Form 158, issued 10–77, Certification of Contract or Force Account Approval.

(3) RUS Form 159, issued 10–77, Summary of Completed Construction.

(4) RUS Form 168b, issued 2–04, Contractor's Bond.

(5) RUS Form 168c, issued 2–04, Contractor's Bond.

(6) RUS Form 181a, issued 3–66, Certificate of Completion (Force Account Construction).

(7) RUS Form 187, issued 2–04, Certificate of Completion, Contract Construction.

(8) RUS Form 213, issued 2–04, Certificate (Buy American).

(9) RUS Form 216, issued 7–67, Construction Change Order.

(10) RUS Form 217, issued 3–97, Postloan Engineering Services Contract—Telecommunications Systems.

(11) RUS Form 220, issued 6–98, Architectural Services Contract.

(12) RUS Form 224, issued 2–04, Waiver and Release of Lien.

(13) RUS Form 231, issued 2–04, Certificate of Contractor.

(14) RUS Form 238, issued 2–04, Construction or Equipment Contract Amendment.

(15) RUS Form 242, issued 11–58, Assignment of Engineering Service Contract.

(16) RUS Form 245, issued 11–75, Engineering Services Contract, Special Services—Telephone.

(17) RUS Form 257, issued 2–04, Contract to Construct Buildings.

(18) RUS Form 257a, issued 10–69, Contractor's Bond.

(19) RUS Form 274, issued 6–81, Bidder's Qualifications.

(20) RUS Form 276, issued 5–59, Bidder's Qualifications for Buried Plant Construction.

(21) RUS Form 281 issued 5–61, Tabulation of Materials Furnished by Borrower.

(22) RUS Form 282, issued 11–53, Subcontract (Under Construction or Equipment Contracts).

(23) RUS Form 284, issued 4–72, Final Statement of Cost for Architectural Service and Certificate of Architect.

(24) RUS Form 307, issued 2–04, Bid Bond.

(25) RUS Form 395, October 18, 2016, Equipment Contract.

(26) RUS Form 395a, October 18, 2016, Equipment Contract Certificate of Completion (Including Installation).

(27) RUS Form 395b, October 18, 2016, Equipment Contract Certificate of Completion (Not Including Installation).

(28) RUS Form 395c, October 18, 2016, Certificate of Contractor and Indemnity Agreement.

(29) RUS Form 395d, October 18, 2016, Results of Acceptance Tests.

(30) RUS Form 506, issued 3–97, Statement of Engineering Fee—Telecommunications.

(31) RUS Form 515, issued September 17, 2001, Telecommunications Systems Construction Contract (Labor and Materials).

(32) RUS Form 526, issued 8–66, Construction Contract Amendment.

(33) RUS Form 527, issued 3–71, Statement of Construction, Telephone System "Outside Plant".

(34) RUS Form 553, issued 5–67, Check List for Review of Plans and Specifications.

(35) RUS Form 724, issued 10–63, Final Inventory, Telephone Construction Contract.

(36) RUS Form 724a, issued 4–61, Final Inventory, Telephone Construction—Telephone Construction Contract (Labor and Materials), columns 1–8.

(37) RUS Form 724b, issued 3–61, Final Inventory, Telephone Construction Contract (Labor and Materials), columns 9–14.

(38) RUS Form 771, issued 10–75, Summary of Work Orders (Inspected by RUS Field Engineer).

(39) RUS Form 771a, issued 10–75, Summary of Work Orders (Inspected by Licensed Engineer or Borrower's Staff Engineer).

(40) RUS Form 773, issued 12–90, Miscellaneous Construction Work and Maintenance Services Contract.

(41) RUS Form 787, issued 8–63, Supplement A to Construction Contract.

(42) RUS Form 817, issued 6–60, Final Inventory, Telephone Force Account Construction.

(43) RUS Form 817a, issued 6–60, Final Inventory, Telephone Force Account Construction, columns 1–8.

(44) RUS Form 817b, issued 6–60, Final Inventory, Telephone Force Account Construction, Columns 9–14.

(45) RUS Form 835, issued 3–66, Preloan Engineering Service Contract, Telephone System Design.

[64 FR 6501, Feb. 10, 1999, as amended at 64 FR 53887, Oct. 5, 1999; 65 FR 51750, Aug. 25, 2000; 69 FR 7111, Feb. 13, 2004; 81 FR 71585, Oct. 18, 2016]

§§ 1755.31–1755.96 [Reserved]

§ 1755.97 Telephone standards and specifications.

(a)(1) Certain material is incorporated by reference into this part with the approval of the Director of the Federal Register under 5 U.S.C. 552(a) and 1 CFR part 51. All approved material is available for inspection at the Rural Utilities Service, U.S. Department of Agriculture, Room 5170–S, Washington, DC 20250–1522, call (202) 720–8674 and is available as listed in this section. It is also available for inspection at the National Archives and Records Administration (NARA). For information on the availability of these materials at NARA, call (202) 741–6030 or go to: *www.archives.gov/federal-register/cfr/ibr-locations.html.*

(2) To comply with the provisions of this part, you must follow the requirements set out in the RUS telecommunications bulletins incorporated by reference. These materials are incorporated as they exist on the date of the approval and notification of any change in these materials will be published in the FEDERAL REGISTER. The terms "RUS form", "RUS standard form", "RUS specification", and "RUS bulletin" have the same meaning as the terms "REA form", "REA standards form", "REA specification", and "REA bulletin", respectively, unless otherwise indicated. For information on other standards incorporated by reference into this part see § 1755.901.

(b) Rural Utilities Service, U.S. Department of Agriculture, Room 5170–S, U.S. Department of Agriculture, Washington, DC 20250, *https://www.rd.usda.gov/publications/regulations-guidelines/bulletins.*

(1) Bulletin 345–39, RUS specification for telephone station protectors, August 19, 1985.

(2) Bulletin 345–50 PE–60, RUS specification for trunk carrier systems, September 1979.

(3) Bulletin 345–54 PE–52, RUS specification for telephone cable splicing connectors, December 1971.

(4) Bulletin 345–55 PE–61, RUS specification for central office loop extenders and loop extender voice frequency repeater combinations, December 1973.

(5) Bulletin 345–65, PE–65, Specification for shield bonding connectors, March 22, 1985.

(6) Bulletin 345–66 PE–64, RUS specification for subscriber carrier systems, September 1979.

(7) Bulletin 345–69 PE–29, RUS specification for two-wire voice frequency repeater equipment, January 1978.

(8) Bulletin 345–72 PE–74, RUS specification for filled splice closures, October 1985.

(9) Bulletin 345–78 PE–78, RUS specification for carbon arrester assemblies for use in protectors, February 1980.

(10) Bulletin 345–180 Form 397a, RUS specifications for voice frequency repeaters and voice frequency repeatered trunks, January 1963.

(11) Bulletin 345–183 Form 397d, RUS design specifications for point-to-point microwave radio systems June 1970.

(12) Bulletin 345–184 Form 397e, RUS design specifications for mobile and fixed dial radio telephone equipment May 1971.

(13) Bulletin 1728F–700, RUS Specification for Wood Poles, Stubs and Anchor Logs, September 9, 2021.

(14) Bulletin 1753F–150 Form 515a, Specifications and Drawings for Construction of Direct Buried Plant, September 30, 2010.

(15) Bulletin 1753F–151 Form 515b, Specifications and Drawings for Construction of Underground Plan, September 12, 2001.

(16) Bulletin 1753F–152 Form 515c, Specifications and Drawings for Construction of Aerial Plant, September 17, 2001.

(17) Bulletin 1753F–153 Form 515d, Specifications and Drawings for Service Installation at Customer Access Locations, September 17, 2001.

[84 FR 28201, June 18, 2019, as amended at 86 FR 57022, Oct. 14, 2021]

§ 1755.98 List of telecommunications specifications included in other 7 CFR parts.

The following specifications are included throughout 7 CFR chapter XVII. These specifications are not incorporated by reference elsewhere in the chapter. The terms "RUS form," "RUS standard form," "RUS specification," and "RUS bulletin" have the same meaning as the terms "REA form,"

"REA standard form," "REA specification," and "REA bulletin," respectively, unless otherwise indicated. The list of specifications follows:

Section	Issue date	Title
(a) 1728.202	9.9.2021	RUS Specification for Quality Control and Inspection of Timber Products.
(b) [Reserved].		

[55 FR 39397, Sept. 27, 1990, as amended at 84 FR 28201, June 18, 2019; 86 FR 57022, Oct. 14, 2021]

§§ 1755.99–1755.199 [Reserved]

§ 1755.200 RUS standard for splicing copper and fiber optic cables.

(a) *Scope.* (1) This section describes approved methods for splicing plastic insulated copper and fiber optic cables. Typical applications of these methods include aerial, buried, and underground splices.

(2) American National Standard Institute/National Fire Protection Association (ANSI/NFPA) 70, 1993 National Electrical Code (NEC) referenced in this section is incorporated by reference by RUS. This incorporation by reference was approved by the Director of the Federal Register in accordance with 5 U.S.C. 552(a) and 1 CFR part 51. A copy of the ANSI/NFPA 1993 NEC standard is available for inspection during normal business hours at RUS, room 2845, U.S. Department of Agriculture, Washington, DC 20250–1500, or at the National Archives and Records Administration (NARA). For information on the availability of this material at NARA, call 202–741–6030, or go to: *http://www.archives.gov/federal_register/code_of_federal_regulations/ibr_locations.html.* Copies are available from NFPA, Batterymarch Park, Quincy, Massachusetts 02269, telephone number 1 (800) 344–3555.

(3) American National Standard Institute/Institute of Electrical and Electronics Engineers, Inc. (ANSI/IEEE), 1993 National Electrical Safety Code (NESC) referenced in this section is incorporated by reference by RUS. This incorporation by reference was approved by the Director of the Federal Register in accordance with 5 U.S.C. 552(a) and 1 CFR part 51. A copy of the ANSI/IEEE 1993 NESC standard is available for inspection during normal business hours at RUS, room 2845, U.S. Department of Agriculture, Washington, DC 20250–1500, or at the National Archives and Records Administration (NARA). For information on the availability of this material at NARA, call 202–741–6030, or go to: *http://www.archives.gov/federal_register/code_of_federal_regulations/ibr_locations.html.* Copies are available from IEEE Service Center, 455 Hoes Lane, Piscataway, New Jersey 08854, telephone number 1 (800) 678–4333.

(b) *General.* (1) Only Rural Utilities Service (RUS) accepted filled cable and splicing materials shall be used on outside plant projects financed by RUS.

(2) The installation instructions provided by the manufacturer of splicing materials shall be followed except where those instructions conflict with the procedures specified in this section.

(3) Precautions shall be taken to prevent the ingress of moisture and other contaminants during all phases of the splicing installation. When an uncompleted splice must be left unattended, it shall be sealed to prevent the ingress of moisture and other contaminants.

(4) Minor sheath damage during construction may be repaired if the repair is completed immediately and approved by the borrower's resident project representative. Minor damage is typically repaired by:

(i) Scuffing the cable sheath associated with the damaged area;

(ii) Applying several layers of DR tape over the scuffed and damaged area;

(iii) Applying several layers of plastic tape over the DR tape; and

(iv) If damage is severe enough to rupture the cable shield, a splice closure shall be installed.

(5) All splice cases installed on RUS toll trunk and feeder cables shall be filled, whether aerial, buried, or underground.

(c) *Splicing considerations for copper cables*—(1) *Preconstruction testing.* It is desirable that each reel of cable be tested for grounds, opens, shorts, crosses, and shield continuity before the cable is installed. However, manufacturer supplied test results are acceptable. All cable pairs shall be free from electrical defects.

(2) *Handling precautions.* The cable manufacturer's instructions concerning pulling tension and bending radius shall be observed. Unless the cable manufacturer's recommendation is more stringent, the minimum bending radius shall be 10 times the cable diameter for copper cables and 20 times the cable diameter for fiber optic cables.

(3) *Cable sheath removal.* (i) The length of cable sheath to be removed shall be governed by the type of splicing hardware used. Follow the splice case manufacturer's recommendations. For pedestals or large pair count splice housings, consider removing enough cable sheath to allow the conductors to extend to the top of the pedestal and then to hang downward to approximately 15 centimeters (cm) (6 inches (in.)) above the baseplate.

(ii) Caution shall be exercised to avoid damaging the conductor insulation when cutting through the cable shield and removing the shield. Sharp edges and burrs shall be removed from the cut end of the shield.

(4) *Shield bonding and grounding.* For personnel safety, the shields of the cables to be spliced shall be bonded together and grounded before splicing activities are started. (See paragraphs (g)(2), and (g)(5)(i) through (g)(5)(iii) of this section for final bonding and grounding provisions.)

(5) *Binder group identification.* (i) Color coded plastic tie wraps shall be placed loosely around each binder group of cables before splicing operations are attempted. The tie wraps shall be installed as near the cable sheath as practicable and shall conform to the same color designations as the binder ribbons. Twisted wire pigtails shall not be used to identify binder groups due to potential transmission degradation.

(ii) The standard insulation color code used to identify individual cable pairs within 25-pair binder groups shall be as shown in Table 1:

TABLE 1—CABLE PAIR IDENTIFICATION WITHIN BINDER GROUPS

Pair No.	Color	
	Tip	Ring
1	White	Blue.
2	White	Orange.
3	White	Green.
4	White	Brown.
5	White	Slate.
6	Red	Blue.
7	Red	Orange.
8	Red	Green.
9	Red	Brown.
10	Red	Slate.
11	Black	Blue.
12	Black	Orange.
13	Black	Green.
14	Black	Brown.
15	Black	Slate.
16	Yellow	Blue.
17	Yellow	Orange.
18	Yellow	Green.
19	Yellow	Brown.
20	Yellow	Slate.
21	Violet	Blue.
22	Violet	Orange.
23	Violet	Green.
24	Violet	Brown.
25	Violet	Slate.

(iii) The standard binder ribbon color code used to designate 25-pair binder groups within 600-pair super units shall be as shown in Table 2:

TABLE 2—CABLE BINDER GROUP IDENTIFICATION

Group No.	Color of bindings	Group pair count
1	White-Blue	1–25
2	White-Orange	26–50
3	White-Green	51–75
4	White-Brown	76–100
5	White-Slate	101–125
6	Red-Blue	126–150
7	Red-Orange	151–175
8	Red-Green	176–200
9	Red-Brown	201–225
10	Red-Slate	226–250
11	Black-Blue	251–275
12	Black-Orange	276–300
13	Black-Green	301–325
14	Black-Brown	326–350
15	Black-Slate	351–375
16	Yellow-Blue	376–400
17	Yellow-Orange	401–425
18	Yellow-Green	426–450
19	Yellow-Brown	451–475
20	Yellow-Slate	476–500
21	Violet-Blue	501–525
22	Violet-Orange	526–550
23	Violet-Green	551–575
24	Violet-Brown	576–600

(iv) Super-unit binder groups shall be identified in accordance with Table 3:

TABLE 3—SUPER-UNIT BINDER COLORS

Pair numbers	Binder color
1–600	White.
601–1200	Red.
1201–1800	Black.
1801–2400	Yellow.
2401–3000	Violet.
3001–3600	Blue.
3601–4200	Orange.
4201–4800	Green.
4801–5400	Brown.
5401–6000	Slate.

(v) Service pairs in screened cables shall be identified in accordance with Table 4:

TABLE 4—SCREENED CABLE SERVICE PAIR IDENTIFICATION

Service pair No.	Color	
	Tip	Ring
1	White	Red.
2	White	Black.
3	White	Yellow.
4	White	Violet.
	Red	Black.
6	Red	Yellow.
7	Red	Violet.
8	Black	Yellow.
9	Black	Violet.

(6) *Cleaning conductors.* It is not necessary to remove the filling compound from cable conductors before splicing. However, it is permissible to wipe individual conductors with clean paper towels or clean cloth rags. No cleaning chemicals, etc., shall be used. Caution shall be exercised to maintain individual cable pair and binder group identity. Binder group identity shall be maintained by using color coded plastic tie wraps. Individual pair identification shall be maintained by carefully twisting together the two conductors of each pair.

(7) *Expanded plastic insulated conductor (PIC) precautions.* Solid PIC and expanded (foam or foam skin) PIC are spliced in the same manner, using the same tools and materials and, in general, should be treated the same. However, the insulation on expanded PIC is much more fragile than solid PIC. Twisting or forming expanded PIC into extremely compact splice bundles and applying excessive amounts of tension when tightening tie wraps causes shiners and, thus shall be avoided.

(8) *Splice connectors.* (i) Only RUS accepted filled splice connectors shall be used on outside plant projects financed by RUS.

(ii) Specialized connectors are available for splicing operations such as butt splices, in line splices, bridge taps, clearing and capping, and multiple pair splicing operations. The splice connector manufacturer's recommendations shall be followed concerning connector selection and use.

(iii) Caution shall be exercised to maintain conductor and pair association both during and after splicing operations.

(iv) Splicing operations that involve pairs containing working services shall utilize splice connectors that permit splicing without the interruption of service.

(9) *Piecing out conductors.* Conductors may be pieced-out to provide additional slack or to repair damaged conductors. However, the conductors shall be pieced-out with conductors having the same gauge and type and color of insulation. The conductors used for piecing-out shall be from cables having RUS acceptance.

(10) *Splice organization.* Spliced pair bundles shall be arranged in firm lay-ups with minimum conductor tension in accordance with the manufacturer's instructions.

(11) *Binder tape.* Perforated nonhygroscopic and nonwicking binder tape should be applied to splices housed in filled splice cases. The binder tape allows the flow of filling compound while holding the splice bundles near the center of the splice case to allow adequate coverage of filling compound.

(12) *Cable tags.* Cables shall be identified by a tag indicating the cable manufacturer's name, cable size, date of placement, and generic route information. Information susceptible to changes caused by future cable throws and rearrangements should not be included. Tags on load coil stubs shall include the serial number of the coil case, the manufacturer's name, and the inductance value.

(13) *Screened cable.* Screened PIC cable is spliced in the same manner as nonscreened PIC cable. However, special considerations are necessary due to differences in the cable design. The transmit and receive bundles of the cable shall be separated and one of the

bundles shall be wrapped with shielding material in accordance with the cable manufacturer's recommendations. When acceptable to the cable manufacturer, it is permissible to use either the scrap screening tape removed from the cable during the sheath opening process provided the screening tape is edge coated or new pressure sensitive aluminum foil tape over polyethylene tape.

(14) *Service wire connections.* (i) Buried service wires may be spliced directly to cable conductors inside pedestals using the same techniques required for branch cables. Buried service wires may also be terminated on terminal blocks inside pedestals in areas where high service order activity or fixed count cable administration policies require terminal blocks. However, only RUS accepted terminal blocks equipped with grease or gel filled terminations to provide moisture and corrosion resistance shall be used.

(ii) Only filled terminal blocks having RUS acceptance shall be used on aerial service wire connections.

(15) *Copper cable testing.* Copper cable testing shall be performed in accordance with RUS Bulletin 345–63, "RUS Standard for Acceptance Tests and Measurements of Telephone Plant," PC–4, (Incorporated by reference at §1755.97).

(16) *Cable acceptance.* Installed cable shall be tested and pass the inventory and acceptance testing specified in the Telephone System Construction Contract (Labor and Materials), RUS Form 515. The tests and inspections shall be witnessed by the borrower's resident project representative. All conductors shall be free from grounds, shorts, crosses, splits, and opens.

(d) *Splice arrangements for copper cables*—(1) *Service distribution closures.* (i) Ready access closures permit cable splicing activities and the installation of filled terminal blocks for service wire connections in the same closure. Ready access designs shall allow service technicians direct access ,to the cable core as well as the terminal block.

(ii) Fixed count terminals shall restrict service technician access to the cable core. Predetermined cable pairs shall be spliced to the terminal leads or stub cable in advance of service assignments.

(2) *Aerial splices.* Aerial splice cases accommodate straight splices, branch splices, load coils, and service distribution terminals. Aerial splicing arrangements having more than 4 cables spliced in the same splice case are not recommended. Stub cabling to a second splice case to avoid a congested splice is acceptable.

(3) *Buried splices.* (i) Direct buried splice cases accommodate straight splices, branch splices, and load coils. Direct buried splices shall be filled and shall be used only when above ground splicing in pedestals is not practicable.

(ii) A treated plank or equivalent shall be placed 15 cm (6 in.) above the buried splice case to prevent damage to the splice case from future digging. Where a firm base for burying a splice cannot be obtained, a treated plank or equivalent shall be placed beneath the splice case.

(iii) Each buried splice shall be identified for future locating. One method of marking the splice point is the use of a warning sign. Another method is the burying of an electronic locating device.

(4) *BD-type pedestals.* (i) BD-type pedestals are housings primarily intended to house, organize, and protect cable terminations incorporating splice connectors, ground lugs, and load coils. Activities typically performed in pedestals are cable splicing, shield bonding and grounding, loading, and connection of subscriber service drops.

(ii) The recommended splice capacities for BD-type pedestals are shown in Table 5. However, larger size pedestals are permissible if service requirements dictate their usefulness. Table 5 is as follows:

TABLE 5—SPLICE CAPACITIES FOR BD-TYPE PEDESTALS

Pedestal type	Maximum straight splice pair capacity using single pair connectors or multiple pair splice modules	Maximum load splice pair capacity using single pair connectors or multiple pair splice modules (see note 1)
BD3, BD3A	100 Pair	50 Pair.
BD4, BD4A	200 Pair	100 Pair.
BD5, BD5A	600 Pair	300 Pair.
BD7	1200 Pair	600 Pair.
BD14, BD14A	100 Pair	50 Pair.

TABLE 5—SPLICE CAPACITIES FOR BD-TYPE PEDESTALS—Continued

Pedestal type	Maximum straight splice pair capacity using single pair connectors or multiple pair splice modules	Maximum load splice pair capacity using single pair connectors or multiple pair splice modules (see note 1)
BD15, BD15A	400 Pair	200 Pair.
BD16, BD16A	600 Pair	300 Pair.

Note 1: This table refers to load coil cases that are to be direct buried with stub cables extending into the pedestal for splicing. Requirements involving individual coil arrangements inside the pedestal should be engineered on a case-by-case basis.

(iii) Special distribution pedestals having a divider plate for mounting filled terminal blocks are available. Distribution pedestals are also equipped with service wire channels for installation of buried service wires without disturbing the cabling and gravel inside the base of the pedestal. Distribution pedestals are recommended in locations where the connection of service wires is required.

(5) *Large pair count splice housings.* Large pair count splice housings are recommended for areas not suitable for man- holes. The recommended capacities are shown in Table 6:

TABLE 6—SPLICE CAPACITIES FOR LARGE COUNT HOUSINGS

Housing type	Maximum straight splice pair capacity using single pair connectors or multiple pair splice modules	Maximum load splice pair capacity using single pair connectors or multiple pair splice modules (see note 1)
BD 6000	6,000 Pair	3,000 Pair.
BD 8000	8,000 Pair	4,000 Pair.
BD 10000	10,000 Pair	5,000 Pair.

(6) *Pedestal restricted access inserts.* Restricted access inserts may be used to protect splices susceptible to unnecessary handling where subsequent work activities are required or expected to occur after splices have been completed. Restricted access inserts also provide moisture protection in areas susceptible to temporary flooding. A typical restricted access insert is shown in Figure 1:

FIGURE 1

PEDESTAL RESTRICTED ACCESS INSERT

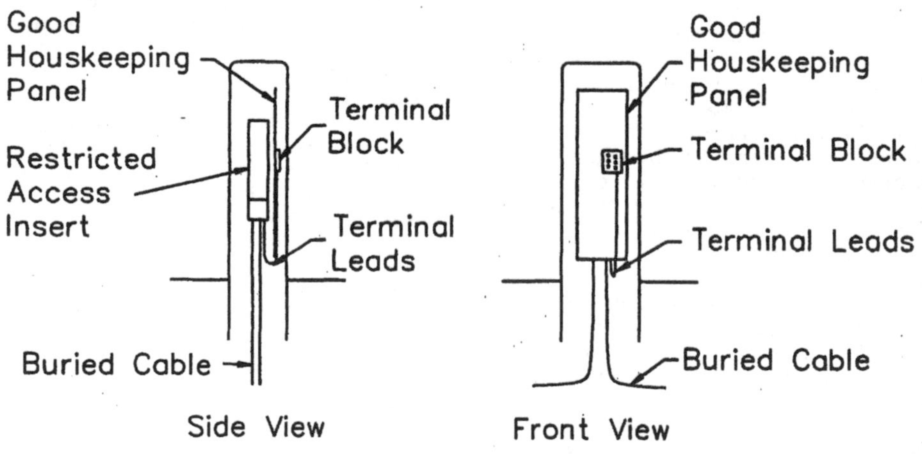

(7) *Serving Area Interface (SAI) Systems.* SAI systems provide the cross-connect point between feeder and distribution cables. Connection of feeder to distribution pairs is accomplished by placing jumpers between connecting

blocks. Only RUS accepted connecting blocks having grease or gel filled terminations to provide moisture and corrosion resistance shall be used.

(8) *Buried cable splicing arrangements.* Typical buried cable splicing arrangements are illustrated in Figures 2 through 5:

FIGURE 2

SERVICE WIRE CONNECTION TO BURIED CABLE

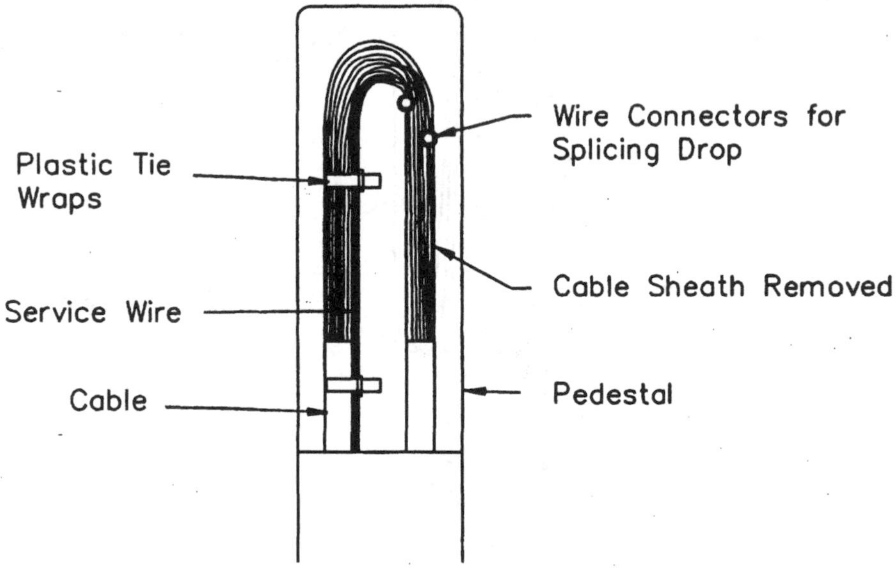

Note: See Figures 13 through 16 for cable tags, tie wraps, and bonding and grounding details.

FIGURE 3

TYPICAL SPLICE USING SINGLE PAIR CONNECTORS

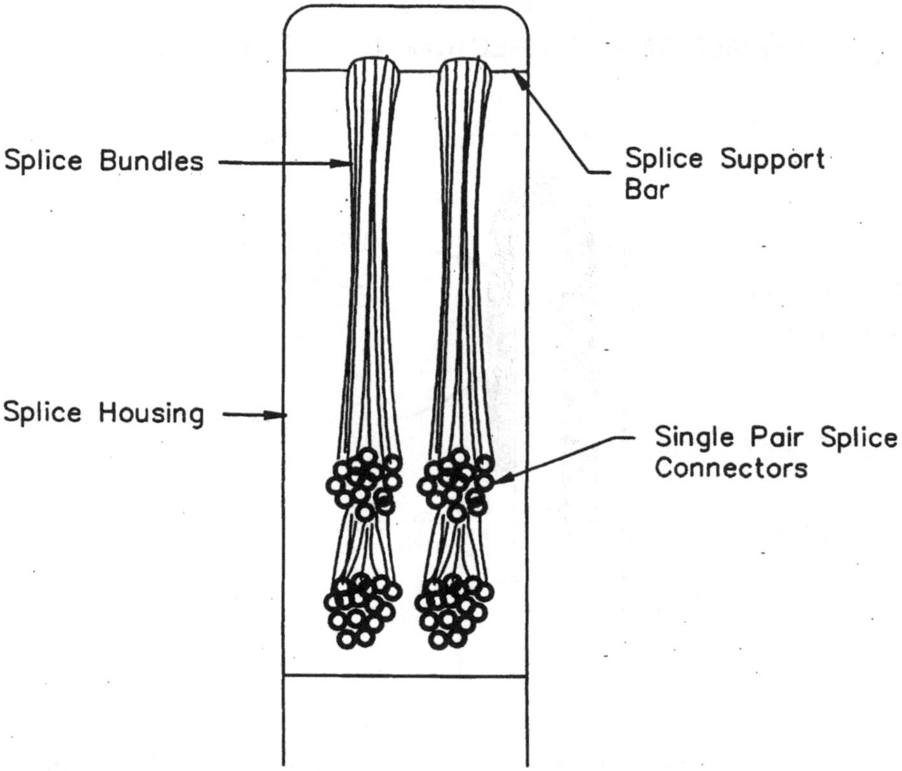

Note: Cable tags, bonding and grounding details, and plastic tie wraps have been omitted for clarity. See Figures 13 through 16 for cable tags, tie wraps, and bonding and grounding details.

FIGURE 4

LARGE SPLICE USING MULTIPLE PAIR CONNECTORS

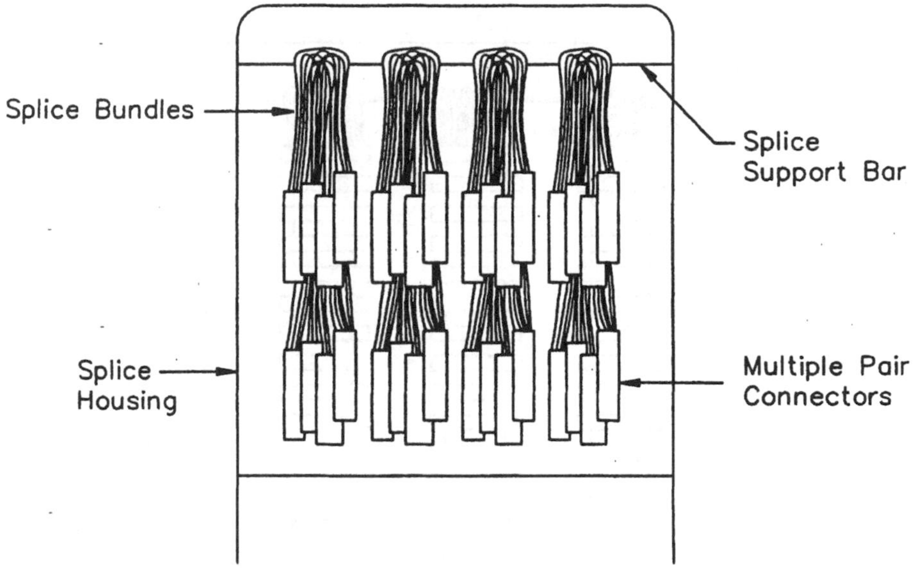

Note: Cable tags, bonding and grounding details, and plastic tie wraps have been omitted for clarity. See Figures 13 through 16 for cable tags, tie wraps, and bonding and grounding details.

FIGURE 5

LARGE SPLICE USING MULTIPLE PAIR CONNECTORS MOUNTED ON ORGANIZER RACKS

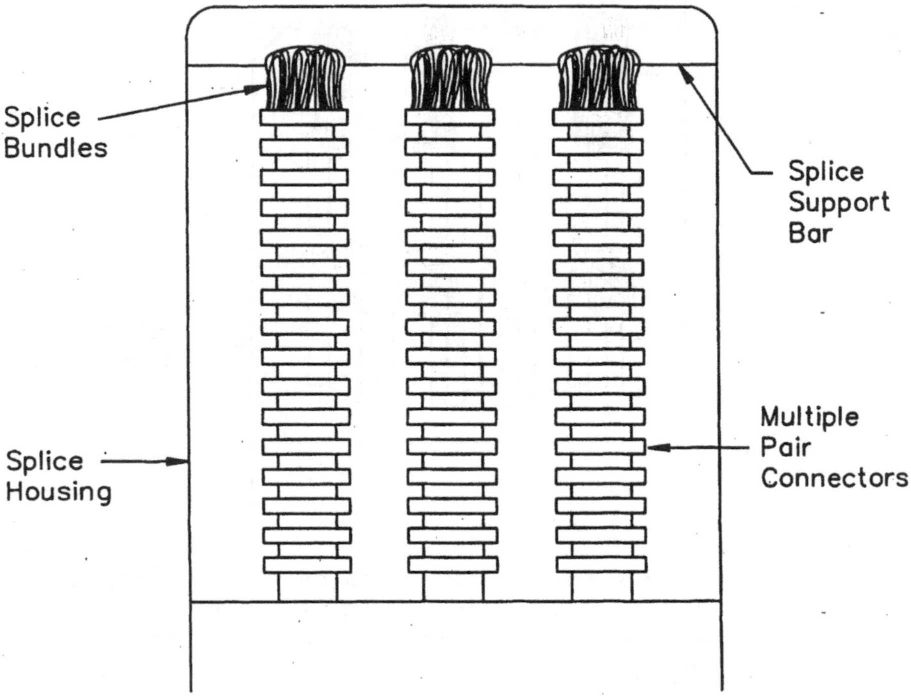

Note: Cable tags, bonding and grounding details, and plastic tie wraps have been omitted for clarity. See Figures 13 through 16 for cable tags, tie wraps, and bonding and and grounding details.

(9) *Underground splices (manholes).* Underground splice cases accommodate straight splices, branch splices, and load coils. Underground splices shall be filled.

(10) *Central office tip cable splices.* (i) Filled cable or filled splices are not recommended for use inside central offices, except in cable vault locations. Outside plant cable sheath and cable filling compound are susceptible to fire and will support combustion. Fire, smoke, and gases generated by these materials during burning are detrimental to telephone switching equipment.

(ii) Tip cables should be spliced in a cable vault. However, as a last resort, tip cables may be spliced inside a central office if flame retardant splice cases or a noncombustible central office splice housing is used to contain the splice.

(iii) Splices inside the central office shall be made as close as practical to the point where the outside plant cables enter the building. Except in cable vault locations, outside plant cables within the central office shall be wrapped with fireproof tape or enclosed in noncombustible conduit.

(e) *Splicing considerations for fiber optic cables*—(1) *Connection characteristics*. Splicing efficiency between optical fibers is a function of light loss across the fiber junctions measured in decibels (dB). A loss of 0.2 dB in a splice corresponds to a light transmission efficiency of approximately 95.5 percent.

(2) *Fiber core alignment*. Fiber splicing techniques shall be conducted in such a manner that the cores of the fibers will be aligned as perfectly as possible to allow maximum light transmission from one fiber to the next. Without proper alignment, light will leave the fiber core and travel through the fiber cladding. Light outside the fiber core is not a usable light signal. Core misalignment is illustrated in Figure 6:

FIGURE 6
CORE MISALIGNMENT

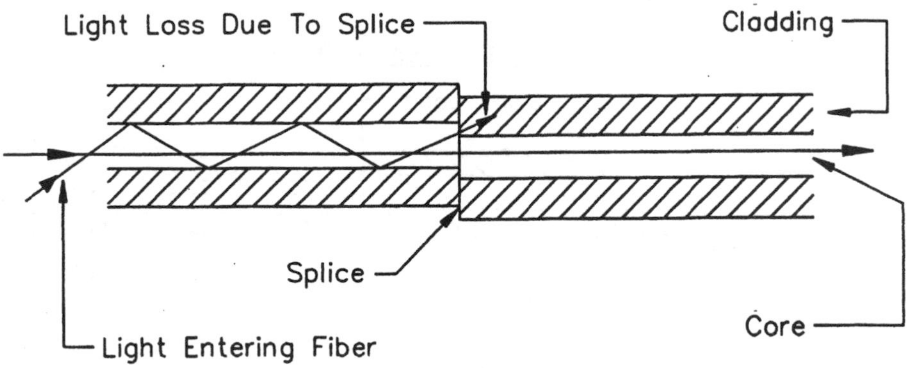

(3) *Splice loss*. (i) Splice loss can also be caused by fiber defects such as nonidentical core diameters, cores not in center of the fiber, and noncircular cores. Such defects are depicted in Figure 7:

FIGURE 7

. SPLICE LOSS CAUSED BY FIBER MANUFACTURE

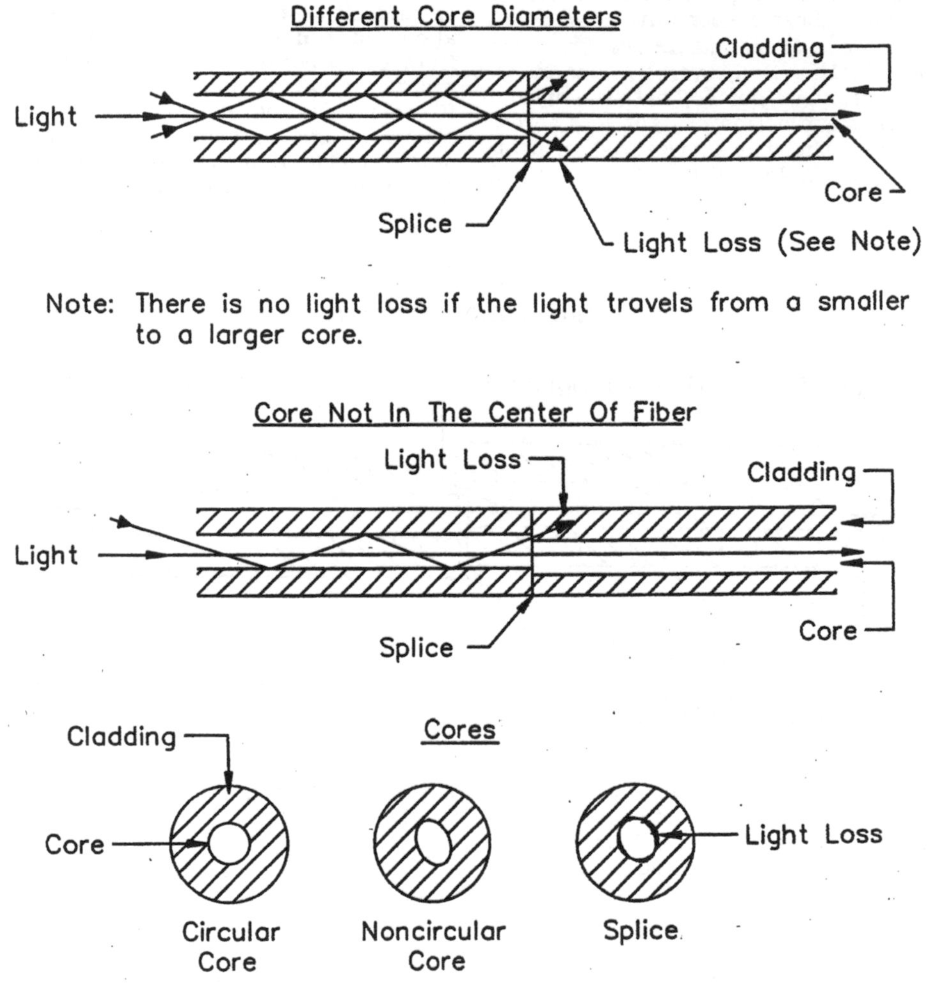

(ii) Undesirable splice losses are caused by poor splicing techniques including splicing irregularities such as improper cleaves and dirty splices. Typical cleave problems are illustrated in Figure 8:

FIGURE 8

IMPROPER CLEAVES VERSUS PROPER CLEAVE

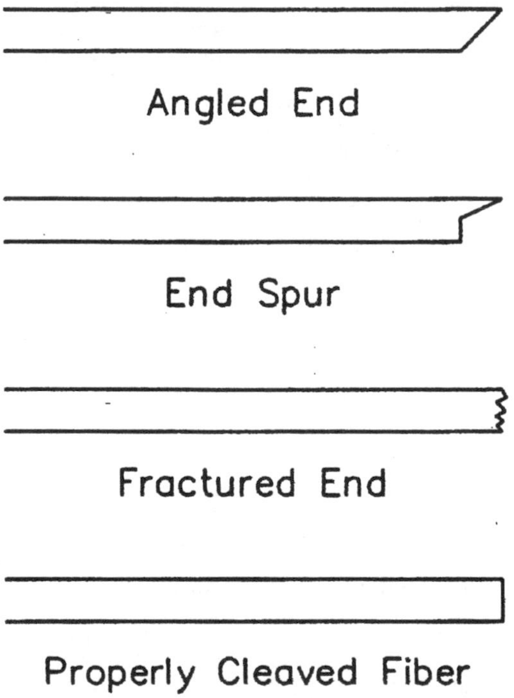

(4) *Handling precautions.* The following precautions shall be observed:

(i) Avoid damaging the cable during handling operations prior to splicing. Minor damage may change the transmission characteristics of the fibers to the extent that the cable section will have to be replaced;

(ii) The cable manufacturer's recommendations concerning pulling tension shall be observed. The maximum pulling tension for most fiber optic cable is 2669 newtons (600 pound-force);

(iii) The cable manufacturer's recommendations concerning bending radius shall be observed. Unless the cable manufacturer's recommendation is more stringent, the minimum bending radius for fiber optic cable shall be 20 times the cable diameter;

(iv) The cable manufacturer's recommendations concerning buffer tube bending radius shall be observed. Unless the cable manufacturer's recommendation is more stringent, the minimum bending radius for buffer tubes is usually between 38 millimeters (mm) (1.5 in.) and 76 mm (3.0 in.). The bending limitations on buffer tubes are intended to prevent kinking. Buffer tube kinking may cause excessive optical loss or fiber breakage; and

(v) Handle unprotected glass fibers carefully to avoid introducing flaws such as scratched or broken fibers.

(5) *Personnel safety.* The following safety precautions shall be observed:

(i) Safety glasses shall be worn when handling glass fibers;

(ii) Never view open-ended fibers with the naked eye or a magnifying device. Improper viewing of a fiber end that is transmitting light may cause irreparable eye damage; and

(iii) Dispose of bare scrap fibers by using the sticky side of a piece of tape to pick up and discard loose fiber ends. Fiber scraps easily penetrate the skin and are difficult to remove.

(6) *Equipment requirements.* (i) Fiber optic splices shall be made in areas where temperature, humidity, and cleanliness can be controlled. Both fusion and mechanical splicing techniques may require a splicing vehicle equipped with a work station that will allow environmental control.

(ii) Both fusion and mechanical splicing techniques are permitted on RUS financed projects. When using the mechanical splicing technique, only RUS accepted mechanical fiber optic splice connectors can be used.

(iii) Fusion splicing machines shall be kept in proper working condition. Regular maintenance in accordance with the machine manufacturer's recommendations shall be observed.

(iv) Mechanical splicing tools shall be in conformance with the tool manufacturer's recommendations.

(v) An optical time domain reflectometer (OTDR) shall be used for testing splices. The OTDR shall be stationed at the central office or launch point for testing individual splices as they are made and for end-to-end signature tests for the fiber optic link.

(vi) An optical power meter shall be used for end-to-end cable acceptance tests.

(vii) A prerequisite for the successful completion of a fiber optic splicing endeavor is the presence of a talk circuit between the splicing technician in the splicing vehicle and the operator of the OTDR in the central office. The splicing technician and the OTDR operator shall have access to communications with each other in order to inform each other as to:

(A) Which splices meet the loss objectives;

(B) The sequence in which buffer tubes and fibers are to be selected for subsequent splicing operations; and

(C) The timing required for the performance of OTDR testing to prevent making an OTDR test at the same time a splice is being fused.

(7) *Cable preparation.* (i) Engineering work prints shall prescribe the cable slack needed at splice points to reach the work station inside the splicing vehicle. Consideration should be given to the slack required for future maintenance activity as well as initial construction activities. The required slack may be different for each splice point, depending on the site logistics. However, the required slack is seldom less than 15 meters (50 feet). The amount of slack actually used shall be recorded for each splice point to assist future maintenance and restoration efforts.

(ii) The splice case manufacturer's recommendations concerning the amount of cable sheath to be removed shall be followed to facilitate splicing operations. The length of the sheath opening shall be identified with a wrap of plastic tape.

(iii) If the cable contains a rip cord, the cable jacket shall be ring cut approximately 15 cm (6 in.) from the end and the 15 cm (6 in.) of cable jacket shall be removed to expose the rip cord. The rip cord shall be used to slit the jacket to the tape mark.

(iv) If the cable does not contain a rip cord, the cable jacket shall be slit using a sheath splitter. No cuts shall be made into the cable core nor shall the buffer tubes be damaged.

(v) If the cable contains an armor sheath, the outer jacket shall be opened along the slit and the jacket shall be removed exposing the armor sheath. The armor shall be separated at the seam and pulled from the cable exposing the inner jacket. The armor shall be removed making allowances for a shield bond connector. The inner sheath shall be slit using a sheath splitter or rip cord. The cable core shall not be damaged nor shall there be any damage to the buffer tubes. The jacket shall be peeled back and cut at the end of the slit. The exposed buffer tubes shall not be cut, kinked, or bent.

(vi) After the cable sheath has been removed, the binder tape shall be removed from the cable. The cable shall not be crushed or deformed.

(vii) The buffer tubes shall be unstranded one at a time. The buffer tubes shall not be kinked.

(viii) If the cable is equipped with a strength member, the strength member shall be cut to the length recommended by the splice case manufacturer.

(ix) Each buffer tube shall be inspected for kinks, cuts, and flat spots. If damage is detected, an additional length of cable jacket shall be removed and all of the buffer tubes shall be cut off at the point of damage.

(x) The cable preparation sequence shall be repeated for the other cable end.

(8) *Shield bonding and grounding*. For personnel safety, the shields and metallic strength members of the cables to be spliced shall be bonded together and grounded before splicing activities are started. (See paragraphs (g)(4), and (g)(5)(i) through (g)(5)(iii) of this section for final bonding and grounding provisions).

(9) *Fiber optic color code*. The standard fiber optic color code for buffer tubes and individual fibers shall be as shown in Table 7:

TABLE 7—FIBER AND BUFFER TUBE IDENTIFICATION

Buffer tube and fiber No.	Color
1	Blue.
2	Orange.
3	Green.
4	Brown.
5	Slate.
6	White.
7	Red.
8	Black.
9	Yellow.
10	Violet.
11	Rose.
12	Aqua.
13	Blue/Black Tracer.
14	Orange/Black Tracer.
15	Green/Black Tracer.
16	Brown/Black Tracer.
17	Slate/Black Tracer.
18	White/Black Tracer.
19	Red/Black Tracer.
20	Black/Yellow Tracer.
21	Yellow/Black Tracer.
22	Violet/Black Tracer.
23	Rose/Black Tracer.
24	Aqua/Black Tracer.

(10) *Buffer tube removal*. (i) The splice case manufacturer's recommendation shall be followed concerning the total length of buffer tube to be removed. Identify the length to be removed with plastic tape.

(ii) Experiment with a scrap buffer tube to determine the cutting tool adjustment required to ring cut a buffer tube without damaging the fibers.

(iii) Buffer tubes shall be removed by carefully ring cutting and removing approximately 15 to 46 cm (6 to 18 in.) of buffer tube at a time. The process shall be repeated until the required length of buffer tube has been removed, including the tape identification marker.

(11) *Coated fiber cleaning*. (i) Each coated fiber shall be cleaned. The cable manufacturer's recommendations shall be followed concerning the solvent required to clean the coated fibers. Reagent grade isopropyl alcohol is a commonly used cleaning solvent.

(ii) A tissue or cotton ball shall be soaked in the recommended cleaning solvent and the coated fibers shall be carefully wiped one at a time using a clean tissue or cotton ball for each coated fiber. Caution shall be exercised to avoid removing the coloring agent from the fiber coating.

(12) *Fiber coating removal*. (i) Fiber coatings shall be removed. In accordance with the splicing method used, the splice case manufacturer's recommendation shall be followed concerning the length of fiber coating to be removed.

(ii) The recommended length of fiber coating shall be removed only on the two fibers to be spliced. Fiber coating removal shall be performed on a one-fiber-at-a-time basis as each splice is prepared.

(13) *Bare fiber cleaning*. After the fiber coating has been removed, the bare fibers shall be cleaned prior to splicing. Each fiber shall be wiped with a clean tissue or cotton ball soaked with the cleaning solvent recommended by the cable manufacturer. The bare fiber shall be wiped one time to minimize fiber damage. Aggressive wiping of bare fiber shall be avoided as it lowers the fiber tensile strength.

(14) *Fiber cleaving*. Cleaving tools shall be clean and have sharp cutting edges to minimize fiber scratches and improper cleave angles. Cleaving tools that are recommended by the manufacturer of the splicing system shall be used.

(15) *Cleaved fiber handling*. The cleaved and cleaned fiber shall not be allowed to touch other objects and

shall be inserted into the splicing device.

(16) *Completion of the splice.* (i) In accordance with the method of splicing selected by the borrower, the splice shall be completed by either fusing the splice or by applying the mechanical connector.

(ii) Each spliced fiber shall be routed through the organizer tray one at a time as splices are completed. The fibers shall be organized one at a time to prevent tangled spliced fibers. The splice case manufacturer's recommendation shall be followed concerning the splice tray selection.

(17) *Fiber optic testing.* Fiber optic testing shall be performed in accordance with RUS Bulletin 345–63, "RUS Standard for Acceptance Tests and Measurements of Telephone Plant," PC–4, (Incorporated by reference at § 1755.97).

(18) *Cable acceptance.* Installed cable shall be tested and pass the inventory and acceptance testing specified in the Telephone System Construction Contract (Labor and Materials), RUS Form 515. The tests and inspections shall be witnessed by the borrower's resident project representative.

(f) *Splice arrangements for fiber optic cables*—(1) *Aerial splices.* Cable slack at aerial splices shall be stored either on the messenger strand, on the pole, or inside a pedestal at the base of the pole. A typical arrangement for the storage of slack cable at aerial splices is shown in Figure 9:

FIGURE 9

AERIAL SPLICE STORED INSIDE PEDESTAL

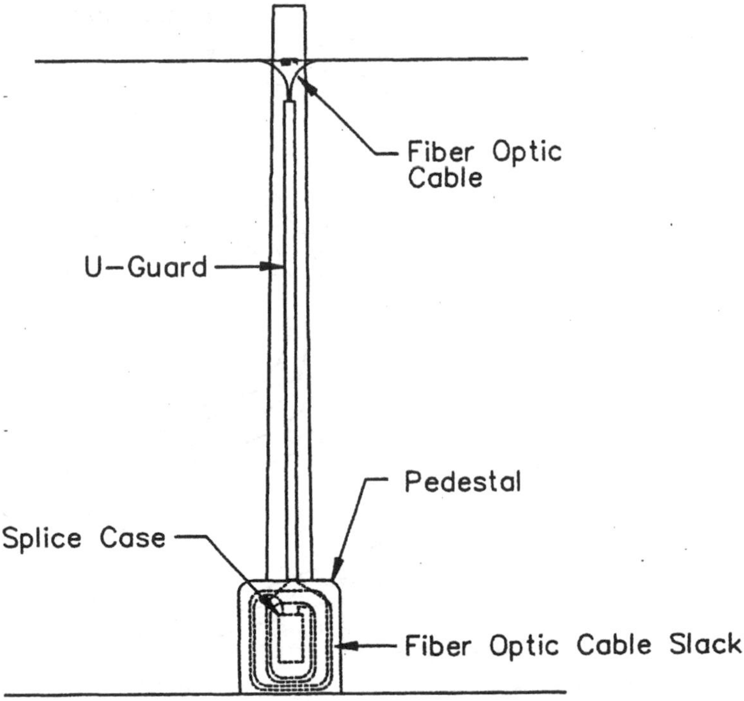

Note: See Figure 11 for details concerning storage of splice case inside pedestal.

(2) *Buried splices*. Buried splices shall be installed in handholes to accommodate the splice case and the required splicing slack. An alternative to the handhole is a pedestal specifically designed for fiber optic splice cases. Typical arrangements for buried cable splices are shown in Figures 10 and 11:

FIGURE 10

BURIED SPLICE STORED INSIDE HANDHOLE

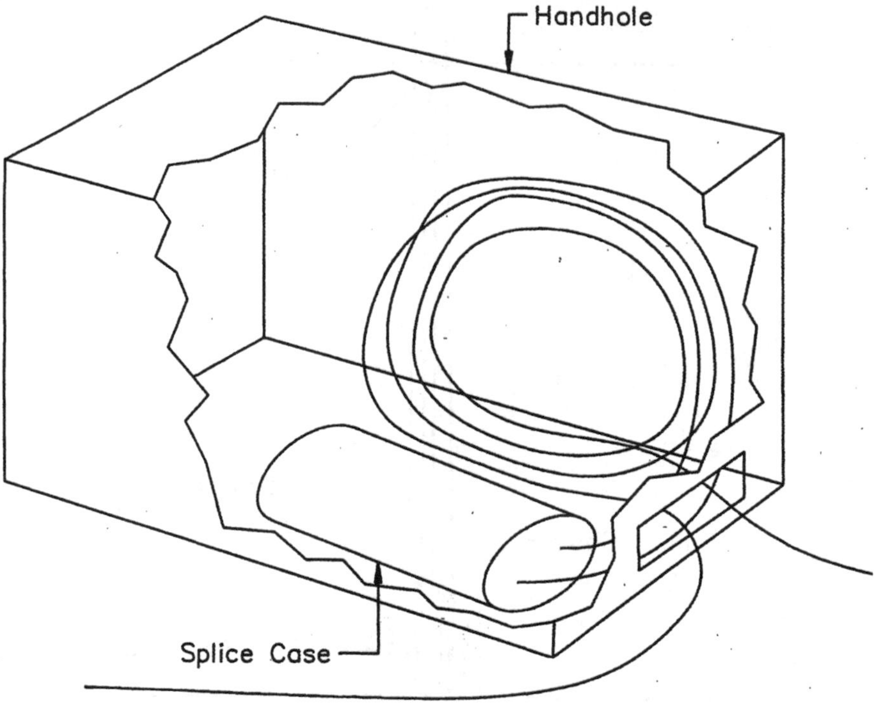

Note: Ground wires omitted for clarity. See Figure 19 for bonding and grounding details.

FIGURE 11

BURIED SPLICE STORED INSIDE PEDESTAL HOUSING

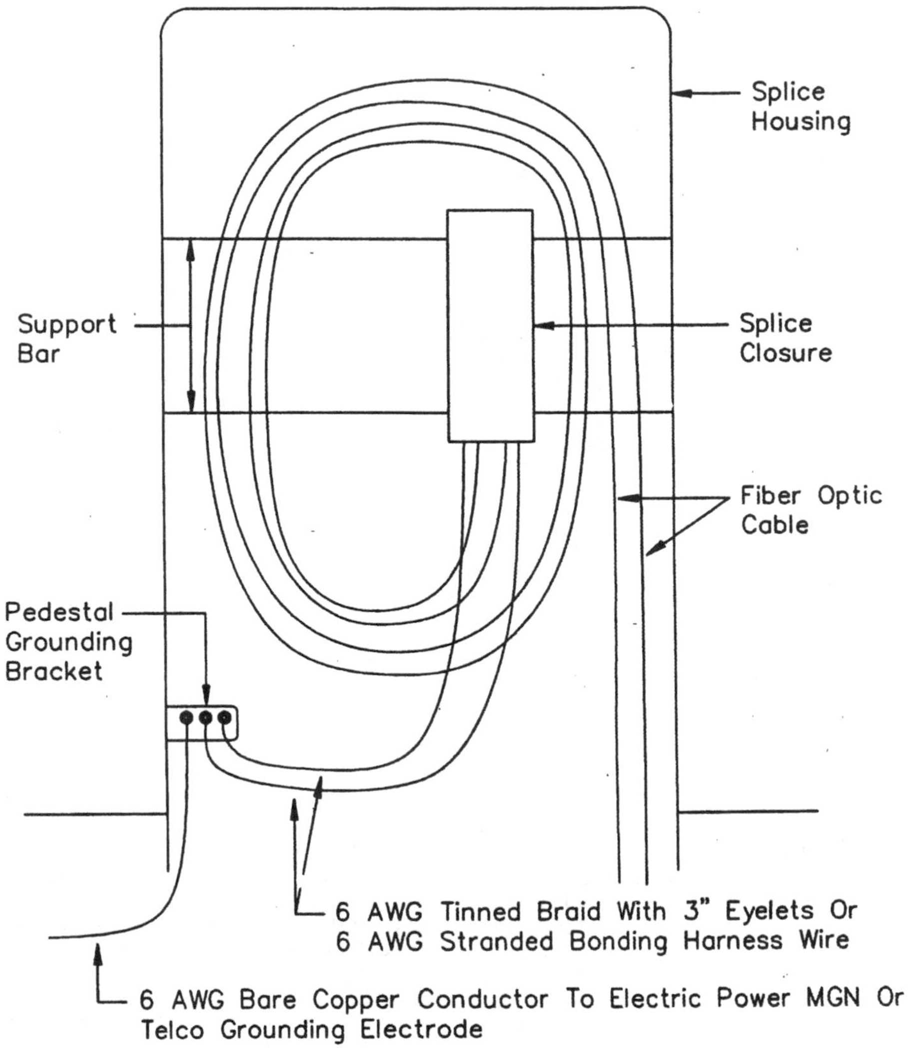

(3) *Underground manhole splices.* Underground splices shall be stored in manholes on cable hooks and racks fastened to the manhole wall. The cable slack shall be stored on cable hooks and racks as shown in Figure 12:

FIGURE 12
MANHOLE SPLICE STORAGE

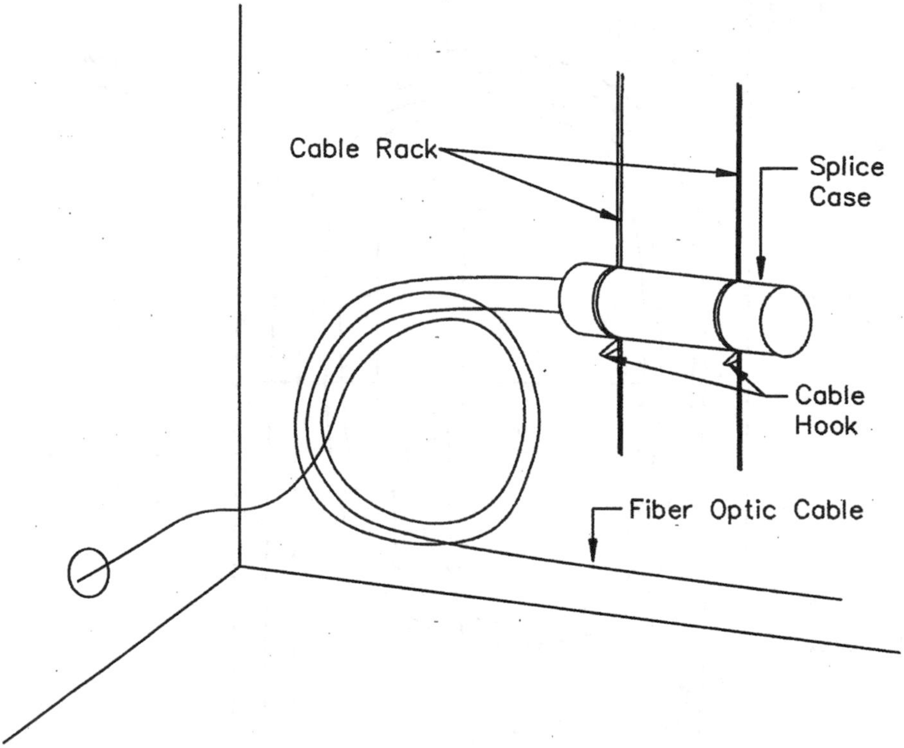

(4) *Central office cable entrance.* (i) Filled cable or filled splices are not recommended for use inside central offices except in cable vault locations. Outside plant cable sheath and cable filling compound are susceptible to fire and will support combustion. Fire, smoke, and gases generated by these materials during burning are detrimental to telephone switching equipment.

(ii) As a first choice, the outside plant fiber optic cable shall be spliced to an all-dielectric fire retardant cable in a cable vault with the all-dielectric cable extending into the central office and terminating inside a fiber patch panel.

(iii) As a second choice, the outside plant cable may be spliced inside the central office if a flame retardant fiber optic splice case or a noncombustible central office splice housing equipped with organizer trays is used to contain the splice.

(iv) In cases referenced in paragraphs (f)(4)(ii) and (f)(4)(iii) of this section, as a minimum the fire retardant all-dielectric cable used to provide the connection between the cable entrance splice and the fiber patch panel shall be listed as Communication Riser Cable (Type CMR) in accordance with Sections 800–50 and 800–51(b) of the 1993 National Electrical Code.

(v) Splices inside the central office shall be made as close as practicable to the point where the outside plant cables enter the building. Except in cable vault locations, outside plant cables within the central office shall be wrapped with fireproof tape or enclosed in noncombustible conduit.

(g) *Bonding and grounding fiber optic cable, copper cable, and copper service wire*—(1) *Bonding*. Bonding is electrically connecting two or more metallic items of telephone hardware to maintain a common electrical potential. Bonding may involve connections to another utility.

(2) *Copper cable shield bond connections*. (i) Cable shields shall be bonded at each splice location. Only RUS accepted cable shield bond connectors shall be used to provide bonding and grounding connections to metallic cable shields. The shield bond connector manufacturer's instructions shall be followed concerning installation and use.

(ii)(A) Shield bonding conductors shall be either stranded or braided tinned copper wire equivalent to a minimum No. 6 American Wire Gauge (AWG) and shall be RUS accepted. The conductor connections shall be tinned or of a compatible bimetallic design to avoid corrosion problems associated with dissimilar metals. The number of shield bond connectors required per pair size and gauge shall be as shown in Table 8:

TABLE 8—SHIELD BOND CONNECTORS PER PAIR SIZE AND GAUGE

19 AWG	Pair size and gauge			No. of shield bond connectors
	22 AWG	24 AWG	26 AWG	
0–25	0–100	0–150	0–200	1
50–100	150–300	200–400	300–600	2
150–200	400–600	600–900	900–1500	3
300–600	900–1200	1200–2100	1800–3600	4

(B) It is permissible to strap across the shield bond connectors of several cables with a single length of braided wire. However, both ends of the braid shall be terminated on the pedestal ground bracket to provide a bonding loop. Shield bond connection methods for individual cables are shown in Figures 13 through 15, and the bonding of several cables inside a pedestal using the bonding loop is shown in Figure 16:

FIGURE 13

BONDING AND GROUNDING CABLES INSIDE PEDESTALS

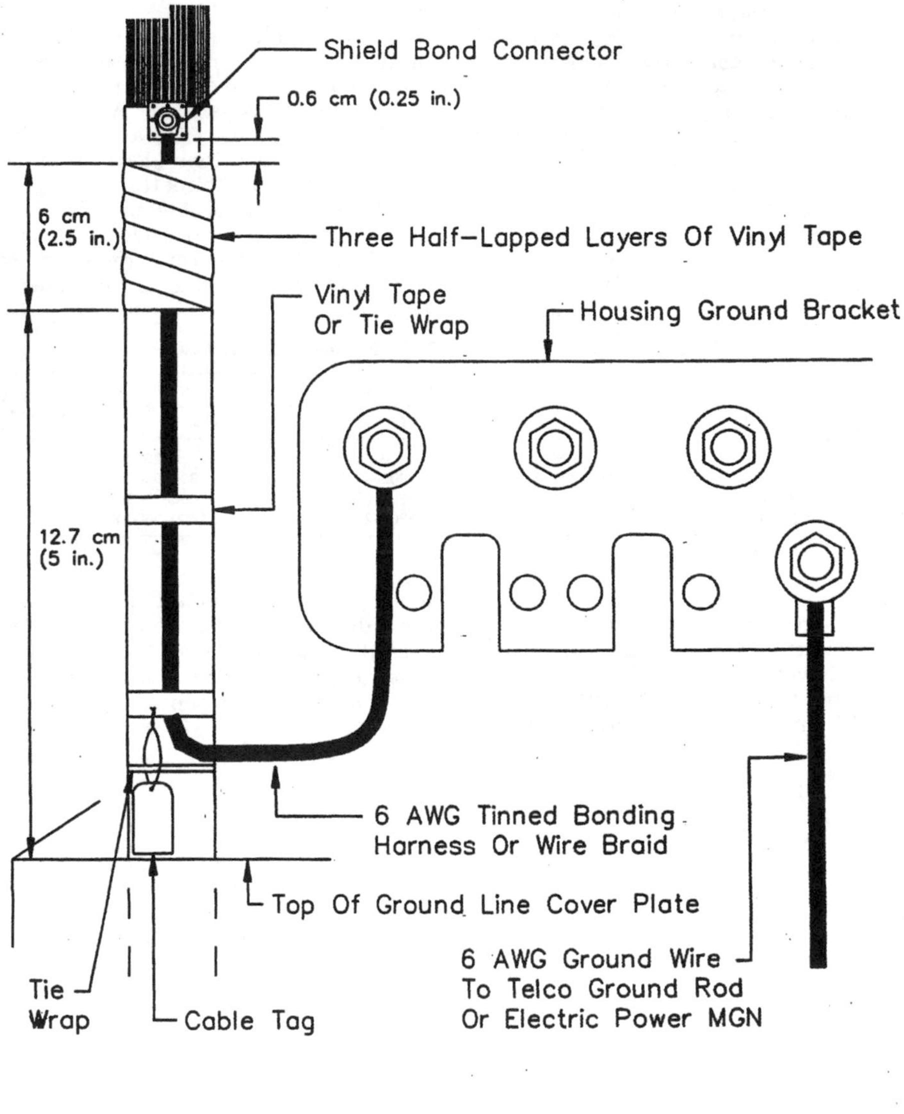

FIGURE 14

BONDING AND GROUNDING OF LARGE CABLES INSIDE PEDESTALS USING MULTIPLE SHIELD BOND CONNECTORS AND HARNESS WIRES

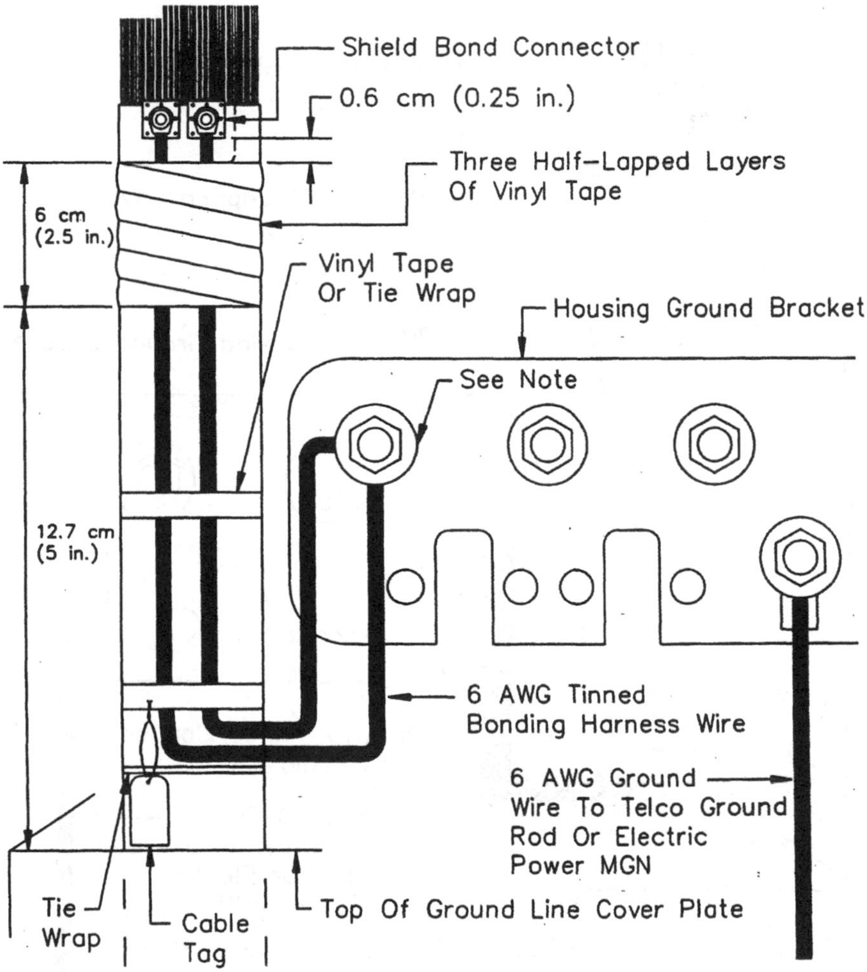

Note: The maximum number of harness wires that can be installed on each stud of the ground bracket shall be in accordance with the manufacturer's instructions.

FIGURE 15

ALTERNATIVE METHOD OF BONDING AND GROUNDING LARGE CABLES IN PEDESTALS USING MULTIPLE SHIELD BOND CONNECTORS AND 6 AWG WIRE BRAID

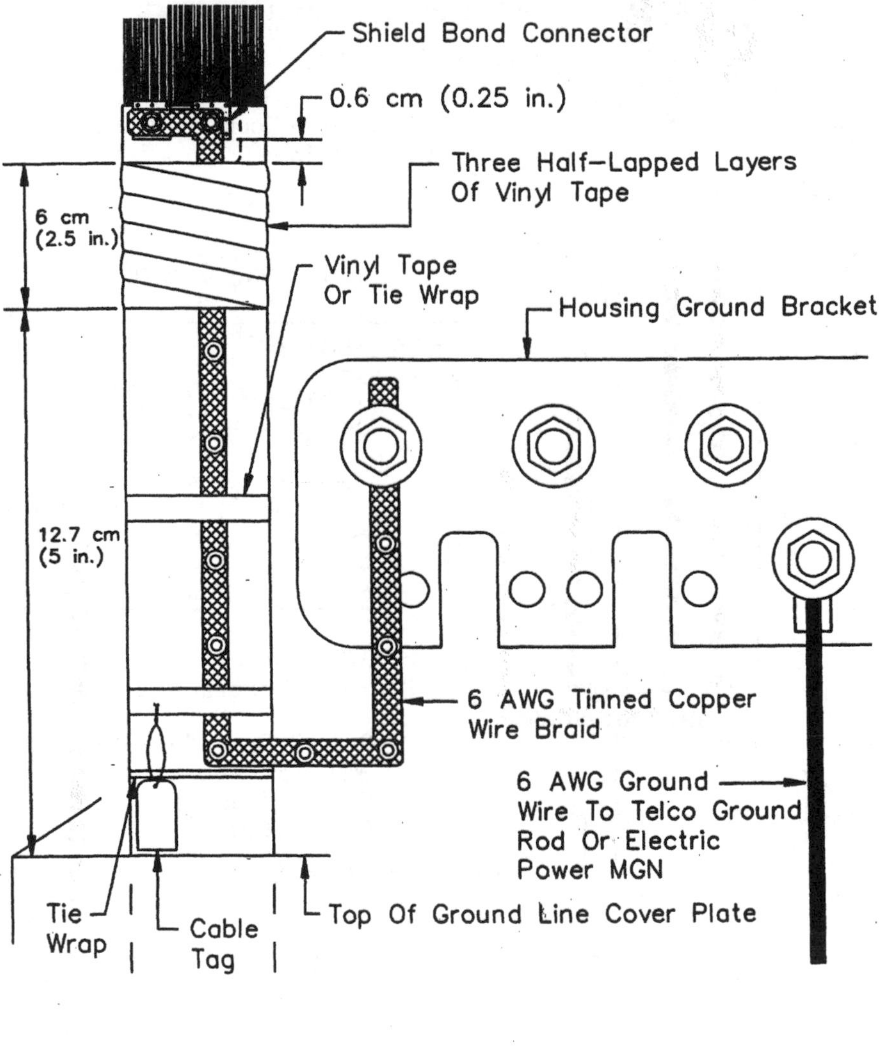

FIGURE 16

ALTERNATIVE METHOD OF BONDING AND GROUNDING SEVERAL CABLES IN PEDESTALS USING SHIELD BOND CONNECTORS AND 6 AWG WIRE BRAID LOOP

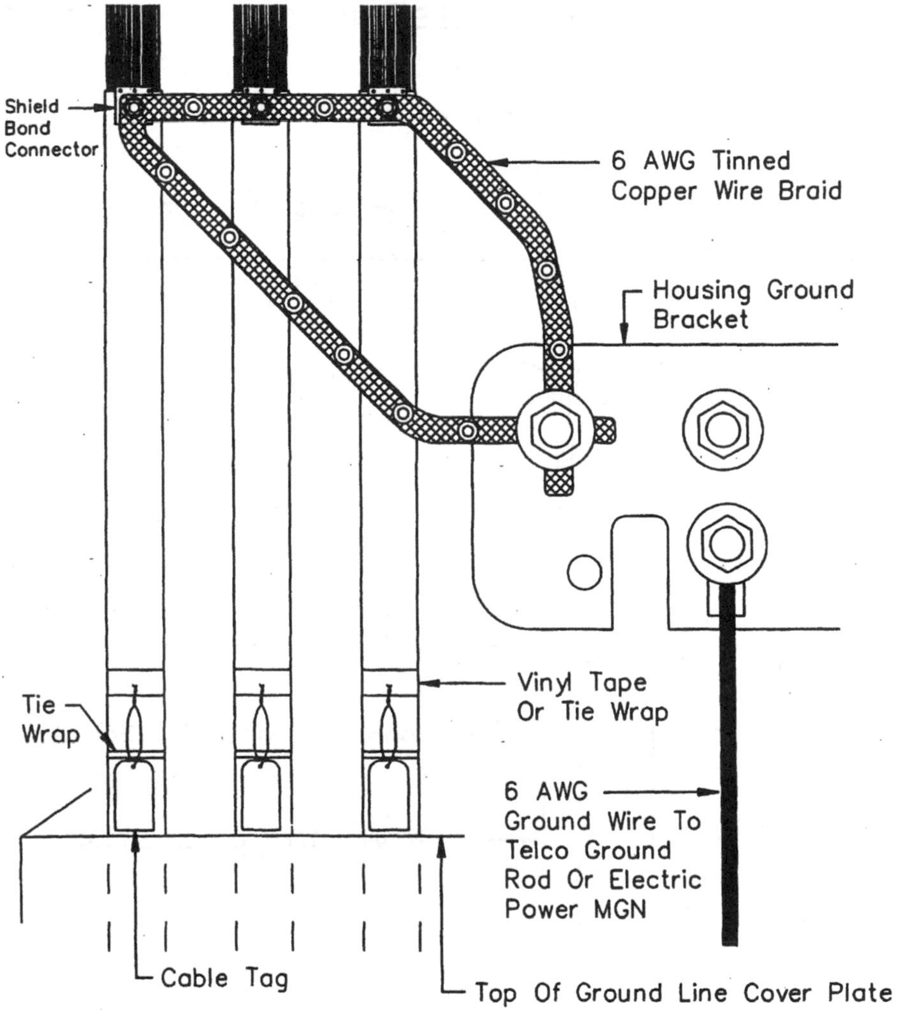

(3) *Buried service wire shield bond connections.* Buried service wire shields shall be connected to the pedestal bonding and grounding system. Typical buried service wire installations are shown in Figures 17 and 18. In addition to the methods referenced in Figures 17 and 18, the shields of buried service wires may also be connected to the pedestal bonding and grounding system using buried service wire bonding harnesses listed on Page 3.3.1, Item "gs-b," of RUS Bulletin 1755I–100. RUS Bulletin 1755I–100 may be purchased from the

Superintendent of Documents, U.S. Government Printing Office, Washington, DC 20402. When those harnesses are used they shall be installed in accordance with the manufacturer's instructions. Figures 17 and 18 are as follows:

FIGURE 17

GROUNDING SERVICE WIRE SHIELDS USING SERVICE WIRE CLAMP

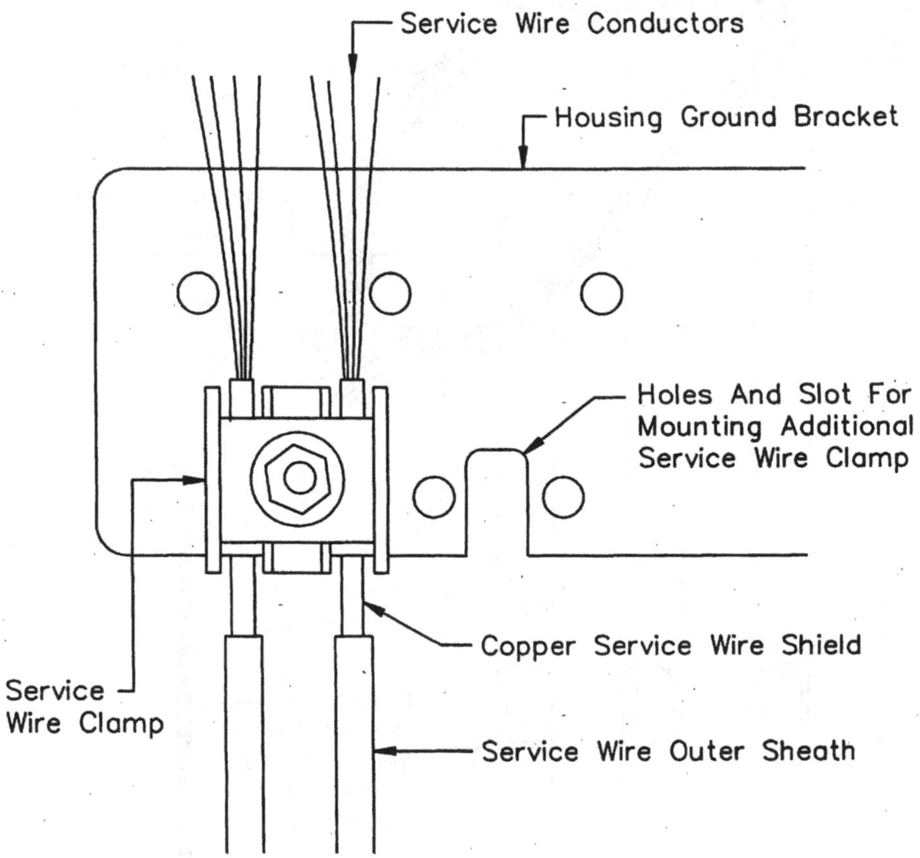

Note: Provide a loop in service drops to allow for movement of the drops without damage to the grounding connection.

FIGURE 18

ALTERNATIVE METHOD OF GROUNDING BURIED SERVICE WIRES INSIDE PEDESTALS

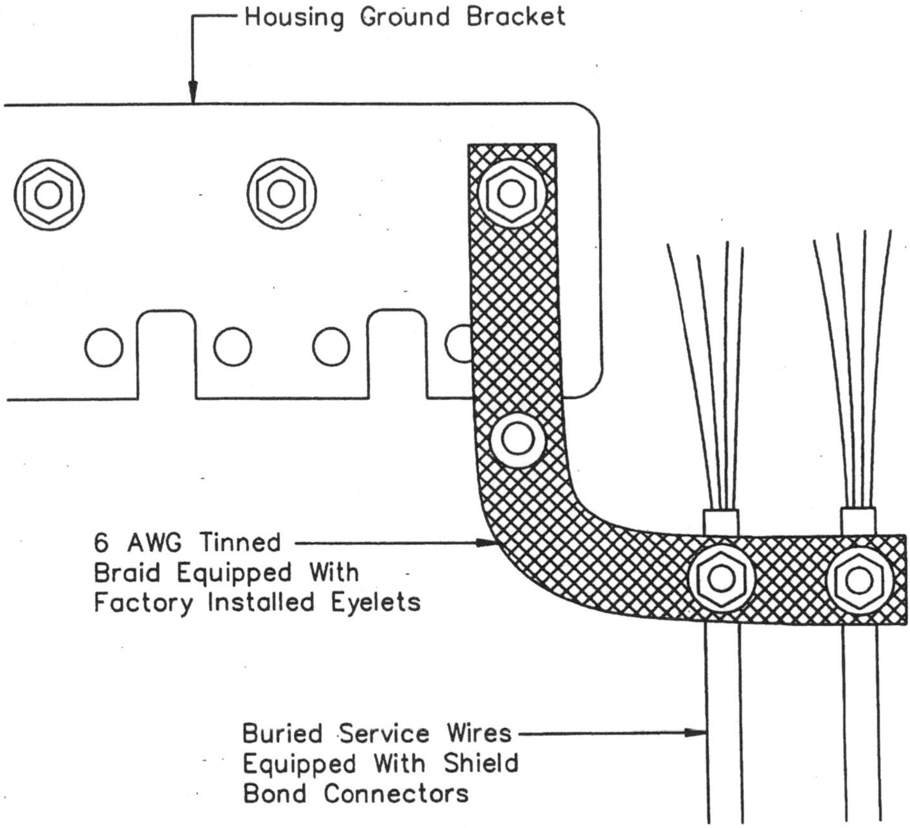

(4) *Fiber optic cable bond connections.* (i) The cable shield and metallic strength members shall be bonded at each splice location. Only RUS accepted fiber optic cable shield bond connectors shall be used to provide bonding connections to the metallic cable shields. The shield bond connector manufacturer's instructions shall be followed concerning installation and use.

(ii) Shield bonding conductors shall be either stranded or braided tinned copper wire equivalent to a minimum No. 6 American Wire Gauge (AWG) and shall be RUS accepted. The conductor connections shall be tinned or of a compatible bimetallic design to avoid corrosion problems associated with dissimilar metals.

(5) *Grounding.* (i) Grounding is electrically connecting metallic telephone hardware to a National Electrical Safety Code (NESC) acceptable grounding electrode. Acceptable grounding electrodes are defined in the Rule 99A of the NESC.

(ii) The conductor used for grounding metallic telephone hardware shall be a minimum No. 6 AWG solid, bare, copper conductor.

(iii) For copper and fiber optic cable plant, all cable shields, all metallic strength members, and all metallic hardware shall be:

(A) Grounded at each splice location to a driven grounding electrode (ground rod) of:

(*1*) At least 1.5 meters (5 feet) in length where the local frost level is normally less than 0.30 meters (1 foot) deep; or

(*2*) At least 2.44 meters (8 feet) in length where the local frost level is normally 0.30 meters (1 foot) or deeper; and

(B) Bonded to a multi-grounded power system neutral when the splice is within 1.8 meters (6 feet) of access to the grounding system of the multi-grounded neutral system. Bonding to the multi-grounded neutral of a parallel power line may help to minimize telephone interference on long exposures with copper cable plant. Consideration, thus, should be given to completing such bonds, at least four (4) times each mile, when splices are greater than 1.8 meters (6 feet) but less than 4.6 meters (15 feet) from access to the multi-grounded neutral.

(6) *Bonding and grounding splice cases.* (i) Splice cases are equipped with bonding and grounding devices to ensure that cable shields and metallic strength members maintain electrical continuity during and after cable splicing operations. The splice case manufacturer's recommendations shall be followed concerning the bonding and grounding procedures. Conductors used for bonding shall be either stranded or braided tinned copper wire equivalent to 6 AWG. Conductors used for grounding shall be a solid, bare, copper wire equivalent to minimum No. 6 AWG.

(ii) Buried splice cases installed in either handholes or pedestals shall be grounded such that the cable shield grounds are attached to a common ground connection that will allow the lifting of a ground on the cable shield in either direction to permit efficient cable locating procedures. As a first choice, buried grounding conductor(s) shall be bare. However, if two or more grounding conductors are buried in the s they shall be insulated to avoid shorts when a locating tone is applied.

(iii) A typical bonding and grounding method for fiber optic splices is shown in Figure 19:

FIGURE 19

BONDING AND GROUNDING BURIED FIBER OPTIC SPLICES

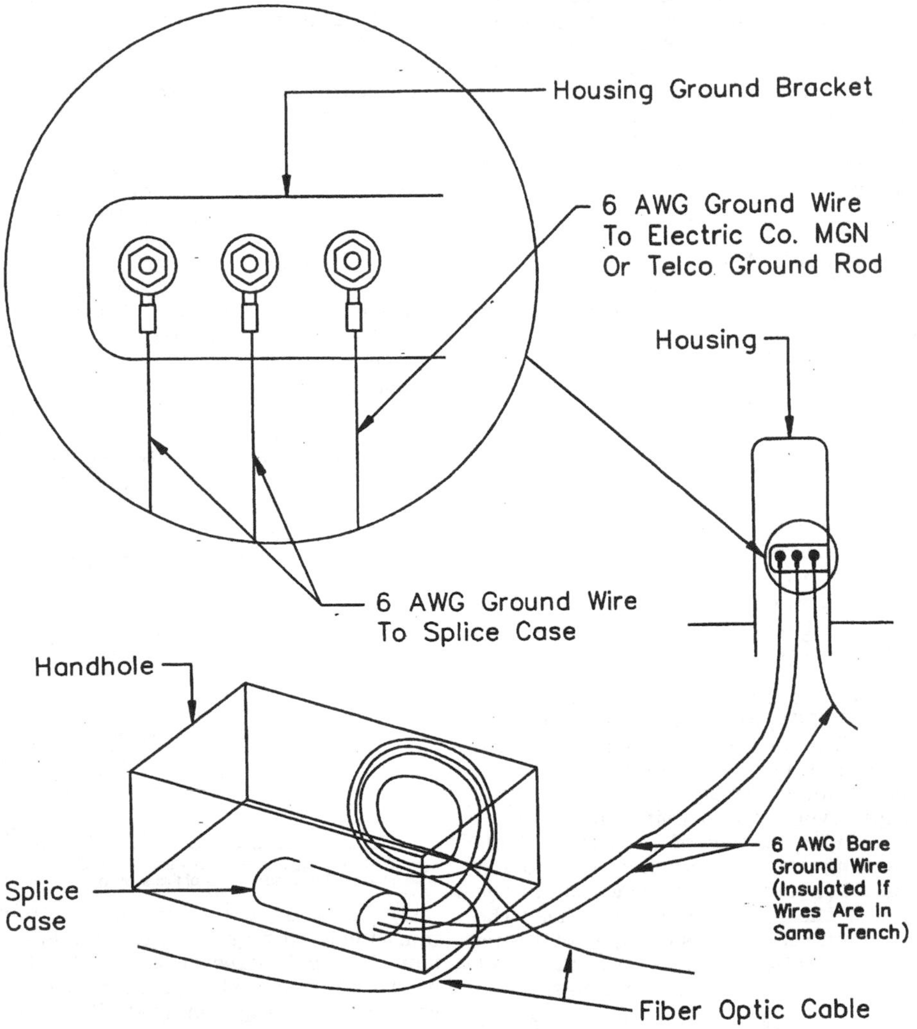

(7) *Bonding and grounding central office cable entrances.* The RUS Telecommunications Engineering and Construction Manual (TE&CM) Section 810 provides bonding and grounding guidance for central office cable entrances. Splicing operations shall not be attempted before all metallic cable shield and strength members are bonded and grounded.

[60 FR 5097, Jan. 26, 1995; 60 FR 9079, Feb. 16, 1995]

§§ 1755.201–1755.369 [Reserved]

§ 1755.370 RUS specification for seven wire galvanized steel strand.

(a) RUS incorporates by reference ASTM A475–78, Standard Specification for Zinc-Coated Steel Wire Strand, issued May 1978. All seven wire galvanized steel strand purchased after April 1, 1990, for use on telecommunications systems financed by RUS loan funds must conform to this standard. This incorporation by reference was approved by the Director of the Federal Register in accordance with 5 U.S.C. 552(a) and 1 CFR part 51 on January 19, 1990). Copies of ASTM A475–78 are available for inspection during normal business hours at the National Archives and Records Administration (NARA) and the Rural Utilities Service, Administrative Services Division, room 0175–S, U.S. Department of Agriculture, Washington, DC 20250, telephone 202–382–9551. For information on the availability of this material at NARA, call 202–741–6030, or go to: *http://www.archives.gov/federal_register/code_of_federal_regulations/ibr_locations.html*. Copies are available from the American Society for Testing and Materials, 1916 Race Street, Philadelphia, PA 19103, telephone 215–299–5400.

(b) In addition to the requirements of ASTM 475–78, all coils and reels having Class B or C coatings shall be marked with a 3-inch wide and 6-inch long deep-colored stripe, green or orange, respectively, to identify the class of galvanized coating of the strand. This marking shall be applied to the exposed convolutions of the strand in the eye of the coils and located near the midpoint on the outside layer of strand on the reels. The marking shall not cover any welded joint markings.

[55 FR 1792, Jan. 19, 1990; 55 FR 3685, Feb. 2, 1990. Redesignated at 55 FR 39397, Sept. 27, 1990, as amended at 69 FR 18803, Apr. 9, 2004]

§§ 1755.371–1755.389 [Reserved]

§ 1755.390 RUS specification for filled telephone cables.

(a) *Scope.* (1) This section covers the requirements for filled telephone cables intended for direct burial installation either by trenching or by direct plowing, for underground application by placement in a duct, or for aerial installations by attachment to a support strand.

(i) The conductors are solid copper, individually insulated with an extruded solid insulating compound.

(ii) The insulated conductors are twisted into pairs which are then stranded or oscillated to form a cylindrical core.

(iii) For high frequency applications, the cable core may be separated into compartments with screening shields.

(iv) A moisture resistant filling compound is applied to the stranded conductors completely covering the insulated conductors and filling the interstices between pairs and units.

(v) The cable structure is completed by the application of suitable core wrapping material, a flooding compound, a shield or a shield/armor, and an overall plastic jacket.

(2) The number of pairs and gauge size of conductors which are used within the RUS program are provided in the following table:

AWG	19	22	24	26
Pairs	6	6	6	
	12	12	12	
	18	18	18	
	25	25	25	25
		50	50	50
		75	75	75
		100	100	100
		150	150	150
		200	200	200
		300	300	300
		400	400	400
			600	600
				900

NOTE: Cables larger in pair sizes than those shown in this table must meet all requirements of this section.

(3) Screened cable, when specified, must meet all requirements of this section. The pair sizes of screened cables used within the RUS program are referenced in paragraph (e)(2)(i) of this section.

(4) All cables sold to RUS borrowers for projects involving RUS loan funds under this section must be accepted by RUS Technical Standards Committee "A" (Telephone). For cables manufactured to the specification of this section, all design changes to an accepted design must be submitted for acceptance. RUS will be the sole authority on what constitutes a design change.

(5) Materials, manufacturing techniques, or cable designs not specifically addressed by this section may be allowed if accepted by RUS. Justification for acceptance of modified materials, manufacturing techniques, or cable designs must be provided to substantiate product utility and long-term stability and endurance.

(6) The American National Standard Institute/Insulated Cable Engineers Association, Inc. (ANSI/ICEA) S-84-608-1988 Standard For Telecommunications Cable, Filled, Polyolefin Insulated, Copper Conductor Technical Requirements referenced throughout this section is incorporated by reference by RUS. This incorporation by reference was approved by the Director of the Federal Register in accordance with 5 U.S.C. 552(a) and 1 CFR part 51. Copies of ANSI/ICEA S-84-608-1988 are available for inspection during normal business hours at RUS, room 2845, U.S. Department of Agriculture, Washington, DC 20250, or at the National Archives and Records Administration (NARA). For information on the availability of this material at NARA, call 202-741-6030, or go to: *http://www.archives.gov/federal_register/code_of_federal_regulations/ibr_locations.html.* Copies are available from ICEA, P. O. Box 440, South Yarmouth, MA 02664, telephone number (508) 394-4424.

(7) American Society for Testing and Materials specifications (ASTM) A 505-87, Standard Specification for Steel, Sheet and Strip, Alloy, Hot-Rolled and Cold-Rolled, General Requirements For; ASTM B 193-87, Standard Test Method for Resistivity of Electrical Conductor Materials; ASTM B 224-80, Standard Classification of Coppers; ASTM B 694-86, Standard Specification for Copper, Copper Alloy, and Copper-Clad Stainless Steel Sheet and Strip for Electrical Cable Shielding; ASTM D 4565-90a, Standard Test Methods for Physical and Environmental Performance Properties of Insulations and Jackets for Telecommunications Wire and Cable; and ASTM D 4566-90, Standard Test Methods for Electrical Performance Properties of Insulations and Jackets for Telecommunications Wire and Cable referenced in this section are incorporated by reference by RUS. These incorporations by references were approved by the Director of the Federal Register in accordance with 5 U.S.C. 552(a) and 1 CFR part 51. Copies of the ASTM standards are available for inspection during normal business hours at RUS, room 2845, U.S. Department of Agriculture, Washington, DC 20250, or at the National Archives and Records Administration (NARA). For information on the availability of this material at NARA, call 202-741-6030, or go to: *http://www.archives.gov/federal_register/code_of_federal_regulations/ibr_locations.html.* Copies are available from ASTM, 1916 Race Street, Philadelphia, PA 19103-1187, telephone number (215) 299-5585.

(b) *Conductors and conductor insulation.* (1) The gauge sizes of the copper conductors covered by this specification must be 19, 22, 24, and 26 American Wire Gauge (AWG).

(2) Each conductor must comply with the requirements specified in ANSI/ICEA S-84-608-1988, paragraph 2.1.

(3) Factory joints made in conductors during the manufacturing process must comply with the requirements specified in ANSI/ICEA S-84-608-1988, paragraph 2.2.

(4) The raw materials used for conductor insulation must comply with the requirements specified in ANSI/ICEA S-84-608-1988, paragraphs 3.1 through 3.1.3.

(5) The finished conductor insulation must comply with the requirements specified in ANSI/ICEA S-84-608-1988, paragraphs 3.2.1 and 3.3.

(6) Insulated conductors must not have an overall diameter greater than 2 millimeters (mm) (0.081 inch (in.)).

(7) A permissible overall performance level of faults in conductor insulation must average not greater than one fault per 12,000 conductor meters (40,000 conductor feet) for each gauge of conductor.

(i) All insulated conductors must be continuously tested for insulation faults during the twinning operation with a method of testing acceptable to RUS. The length count and number of faults must be recorded. The information must be retained for a period of 6 months and be available for review by RUS when requested.

(ii) The voltages for determining compliance with the requirements of this section are as follows:

AWG	Direct Current Voltages (kilovolts)
19	8.0
22	6.0
24	5.0
26	4.0

(8) Repairs to the conductor insulation during manufacture are permissible. The method of repair must be accepted by RUS prior to its use. The repaired insulation must be capable of meeting the relevant electrical requirements of this section.

(9) All repaired sections of insulation must be retested in the same manner as originally tested for compliance with paragraph (b)(7) of this section.

(10) The colored insulating material removed from or tested on the conductor, from a finished cable, must meet the performance requirements specified in ANSI/ICEA S-84-608-1988, paragraphs 3.4.1, 3.4.2, 3.4.4, 3.4.5, and 3.4.6.

(c) *Identification of pairs and twisting of pairs.* (1) The insulation must be colored to identify:

(i) The tip and ring conductor of each pair; and

(ii) Each pair in the completed cable.

(2) The colors to be used in the pairs in the 25 pair group, together with the pair numbers must be in accordance with the table specified in ANSI/ICEA S-84-608-1988, paragraph 3.5.

(3) Positive identification of the tip and ring conductors of each pair by marking each conductor of a pair with the color of its mate is permissible. The method of marking must be accepted by RUS prior to its use.

(4) Other methods of providing positive identification of the tip and ring conductors of each pair may be employed if accepted by RUS prior to its use.

(5) The insulated conductors must be twisted into pairs.

(6) In order to provide sufficiently high crosstalk isolation, the pair twists must be designed to enable the cable to meet the capacitance unbalance and crosstalk loss requirements of paragraphs (k)(5), (k)(6), and (k)(8) of this section.

(7) The average length of pair twists in any pair in the finished cable, when measured on any 3 meter (10 foot) length, must not exceed the requirement specified in ANSI/ICEA S-84-608-1988, paragraph 3.5.

(d) *Forming of the cable core.* (1) Twisted pairs must be assembled in such a way as to form a substantially cylindrical group.

(2) When desired for lay-up reasons, the basic group may be divided into two or more subgroups called units.

(3) Each group, or unit in a particular group, must be enclosed in bindings of the colors indicated for its particular pair count. The pair count, indicated by the colors of insulation, must be consecutive as indicated in paragraph (d)(6) of this section through units in a group.

(4) The filling compound must be applied to the cable core in such a way as to provide as near a completely filled core as is commercially practical.

(5) Threads and tapes used as binders must comply with the requirements specified in ANSI/ICEA S-84-608-1988, paragraphs 4.2 and 4.2.1.

(6) The colors of the bindings and their significance with respect to pair count must be as follows:

Group No.	Color of Bindings	Group Pair Count
1	White-Blue	1–25
2	White-Orange	26–50
3	White-Green	51–75
4	White-Brown	76–100
5	White-Slate	101–125
6	Red-Blue	126–150
7	Red-Orange	151–175
8	Red-Green	176–200
9	Red-Brown	201–225
10	Red-Slate	226–250
11	Black-Blue	251–275
12	Black-Orange	276–300
13	Black-Green	301–325
14	Black-Brown	326–350
15	Black-Slate	351–375
16	Yellow-Blue	376–400
17	Yellow-Orange	401–425
18	Yellow-Green	426–450
19	Yellow-Brown	451–475
20	Yellow-Slate	476–500
21	Violet-Blue	501–525
22	Violet-Orange	526–550
23	Violet-Green	551–575
24	Violet-Brown	576–600

(7) The use of the white unit binder in cables of 100 pairs or less is optional.

(8) When desired for manufacturing reasons, two or more 25 pair groups

may be bound together with nonhygroscopic and nonwicking threads or tapes into a super-unit. Threads or tapes must meet the requirements specified in paragraph (d)(5) of this section. The group binders and the super-unit binders must be color coded such that the combination of the two binders must positively identify each 25 pair group from every other 25 pair group in the cable. Super-unit binders must be of the color shown in the following table:

SUPER-UNIT BINDER COLORS

Pair Numbers	Binder Color
1–600	White
601–1200	Red
1201–1800	Black
1801–2400	Yellow
2401–3000	Violet

(9) Color binders must not be missing for more than 90 meters (300 feet) from any 25 pair group or from any subgroup used as part of a super-unit. At any cable cross-section, no adjacent 25 pair groups and no more than one subgroup of any super-unit may have missing binders. In no case must the total number of missing binders exceed three. Missing super-unit binders must not be permitted for any distance.

(10) Any reel of cable which contains missing binders must be labeled indicating the colors and location of the binders involved. The labeling must be applied to the reel and also to the cable.

(e) *Screened cable.* (1) Screened cable must be constructed such that a metallic, internal screen(s) must be provided to separate and provide sufficient isolation between the compartments to meet the requirements of this section.

(2) At the option of the user or manufacturer, identified service pairs providing for voice order and fault location may be placed in screened cables.

(i) The number of service pairs provided must be one per twenty-five operating pairs plus two for a cable size up to and including 400 pairs, subject to a minimum of four service pairs. The pair counts for screened cables are as follows:

SCREENED CABLE PAIR COUNTS

Carrier Pair Count	Service Pairs	Total Pair Count
24	4	28
50	4	54
100	6	106
150	8	158
200	10	210
300	14	314
400	18	418

(ii) The service pairs must be equally divided among the compartments. The color sequence must be repeated in each compartment.

(iii) The electrical and physical characteristics of each service pair must meet all the requirements set forth in this section.

(iv) The colors used for the service pairs must be in accordance with the requirements of paragraph (b)(5) of this section. The color code used for the service pairs together with the service pair number are shown in the following table:

COLOR CODE FOR SERVICE PAIRS

Service Pair No.	Color	
	Tip	Ring
1	White	Red
2	"	Black
3	"	Yellow
4	"	Violet
5	Red	Black
6	"	Yellow
7	"	Violet
8	Black	Yellow
9	"	Violet

(3) The screen tape must comply with the requirements specified in ANSI/ICEA S-84-608-1988, paragraphs 5.1 through 5.4.

(4) The screen tape must be tested for dielectric strength by completely removing the protective coating from one end to be used for grounding purposes.

(i) Using an electrode, over a 30 centimeter (1 foot) length, apply a direct current voltage at the rate of rise of 500 volts/second until failure.

(ii) No breakdown should occur below 8 kilovolts.

(f) *Filling compound.* (1) After or during the stranding operation and prior to application of the core wrap, filling compound must be applied to the cable core. The compound must be as nearly colorless as is commercially feasible and consistent with the end product requirements and pair identification.

(2) The filling compound must comply with the requirements specified in ANSI/ICEA S-84-608-1988, paragraphs 4.4 through 4.4.4.

(3) The individual cable manufacturer must satisfy RUS that the filling compound selected for use is suitable for its intended application. The filling compound must be applied to the cable in such a manner that the cable components will not be degraded.

(g) *Core wrap.* (1) The core wrap must comply with the requirements specified in ANSI/ICEA-S-84-608-1988, paragraph 4.3.

(2) If required for manufacturing reasons, white or colored binders of nonhygroscopic and nonwicking material may be applied over the core and/or wrap. When used, binders must meet the requirements specified in paragraph (d)(5) of this section.

(3) Sufficient filling compound must be applied to the core wrap so that voids or air spaces existing between the core and the inner side of the core wrap are minimized.

(h) *Flooding compound.* (1) Sufficient flooding compound must be applied on all sheath interfaces so that voids and air spaces in these areas are minimized. When the optional armored design is used, the flooding compound must be applied between the core wrap and shield, between the shield and armor, and between the armor and the jacket so that voids and air spaces in these areas are minimized. The use of floodant over the outer metallic substrate is not required if uniform bonding, per paragraph (i)(7) of this section, is achieved between the plastic-clad metal and the jacket.

(2) The flooding compound must comply with the requirements specified in ANSI/ICEA S-84-608-1988, paragraph 4.5 and the jacket slip test requirements of appendix A, paragraph (III)(5) of this section.

(3) The individual cable manufacturer must satisfy RUS that the flooding compound selected for use is acceptable for the application.

(i) *Shield and optional armor.* (1) A single corrugated shield must be applied longitudinally over the core wrap.

(2) For unarmored cable the shield overlap must comply with the requirements specified in ANSI/ICEA S-84-608-1988, paragraph 6.3.2. Core diameter is defined as the diameter under the core wrap and binding.

(3) For cables containing the coated aluminum shield/coated steel armor (CACSP) sheath design, the coated aluminum shield must be applied in accordance with the requirements specified in ANSI/ICEA S-84-608-1988, paragraph 6.3.2, Dual Tape Shielding System.

(4) General requirements for application of the shielding material are as follows:

(i) Successive lengths of shielding tapes may be joined during the manufacturing process by means of cold weld, electric weld, soldering with a nonacid flux, or other acceptable means.

(ii) Shield splices must comply with the requirements specified in ANSI/ICEA S-84-608-1988, paragraph 6.3.3.

(iii) The corrugations and the application process of the coated aluminum and copper bearing shields must comply with the requirements specified in ANSI/ICEA S-84-608-1988, paragraph 6.3.1.

(iv) The shielding material must be applied in such a manner as to enable the cable to pass the cold bend test specified in paragraph (l)(3) of this section.

(5) The following is a list of acceptable materials for use as cable shielding. Other types of shielding materials may also be used provided they are accepted by RUS prior to their use.

Standard Cable	Gopher Resistant Cable
8-mil Coated Aluminum [1] 5-mil Copper	10-mil Copper 6-mil Copper-Clad Stainless Steel 5 mil Copper-Clad Stainless Steel 5 mil Copper-Clad Alloy Steel 7-mil Alloy 194 6-mil Alloy 194 8-mil Coated Aluminum [1] and 6-mil Coated Steel [1]

[1] Dimensions of uncoated metal

(i) The 8-mil aluminum tape must be plastic coated on both sides and must comply with the requirements of ANSI/ICEA S-84-608-1988, paragraph 6.2.2.

(ii) The 5-mil copper tape must comply with the requirements specified in ANSI/ICEA S-84-608-1988, paragraph 6.2.3.

(iii) The 10-mil copper tape must comply with the requirements specified in ANSI/ICEA S-84-608-1988, paragraph 6.2.4.

(iv) The 6-mil copper clad stainless steel tape must comply with the requirements specified in ANSI/ICEA S-84-608-1988, paragraph 6.2.5.

(v) The 5-mil copper clad stainless steel tape must be in the fully annealed condition and must conform to the requirements of American Society for Testing and Materials (ASTM) B 694-86, with a cladding ratio of 16/68/16.

(A) The electrical conductivity of the clad tape must be a minimum of 28 percent of the International Annealed Copper Standard (IACS) when measured per ASTM B 193-87.

(B) The tape must be nominally 0.13 millimeter (0.005 inch) thick with a minimum thickness of 0.11 millimeter (0.0045 inch).

(vi) The 5-mil copper clad alloy steel tape must be in the fully annealed condition and the copper component must conform to the requirements of ASTM B 224-80 and the alloy steel component must conform to the requirements of ASTM A 505-87, with a cladding ratio of 16/68/16.

(A) The electrical conductivity of the copper clad alloy steel tape must comply with the requirement specified in paragraph (i)(5)(v)(A) of this section.

(B) The thickness of the copper clad alloy steel tape must comply with the requirements specified in paragraph (i)(5)(v)(B) of this section.

(vii) The 6-mil and 7-mil 194 copper alloy tapes must comply with the requirements specified in ANSI/ICEA S-84-608-1988, paragraph 6.2.6.

(6) The corrugation extensibility of the coated aluminum shield must comply with the requirements specified in ANSI/ICEA S-84-608-1988, paragraph 6.4.

(7) When the jacket is bonded to the plastic coated aluminum shield, the bond between the jacket and shield must comply with the requirements specified in ANSI/ICEA S-84-608-1988, paragraph 7.2.6.

(8) A single plastic-coated steel corrugated armor must be applied longitudinally directly over the coated aluminum shield listed in paragraph (i)(5) of this section with an overlap complying with the requirements specified in ANSI/ICEA S-84-608-1988, paragraph 6.3.2, Outer Steel Tape.

(9) Successive lengths of steel armoring tapes may be joined during the manufacturing process by means of cold weld, electric weld, soldering with a nonacid flux, or other acceptable means. Armor splices must comply with the breaking strength and resistance requirements specified in ANSI/ICEA S-84-608-1988, paragraph 6.3.3.

(10) The corrugations and the application process of the coated steel armor must comply with the requirements specified in ANSI/ICEA S-84-608-1988, paragraph 6.3.1.

(i) The corrugations of the armor tape must coincide with the corrugations of the coated aluminum shield.

(ii) Overlapped portions of the armor tape must be in register (corrugations must coincide at overlap) and in contact at the outer edge.

(11) The armoring material must be so applied to enable the cable to pass the cold bend test as specified in paragraph (l)(3) of this section.

(12) The 6-mil steel tape must be electrolytic chrome-coated steel (ECCS) plastic coated on both sides and must comply with the requirements specified in ANSI/ICEA S-84-608-1988, paragraph 6.2.8.

(13) When the jacket is bonded to the plastic-coated steel armor, the bond between the jacket and armor must comply with the requirement specified in ANSI/ICEA-S-84-608-1988, paragraph 7.2.6.

(j) *Cable jacket.* (1) The jacket must comply with the requirements specified in ANSI/ICEA S-84-608-1988, paragraph 7.2.

(2) The raw materials used for the cable jacket must comply with the requirements specified in ANSI/ICEA S-84-608-1988, paragraph 7.2.1.

(3) Jacketing material removed from or tested on the cable must meet the performance requirements specified in ANSI/ICEA S-84-608-1988, paragraphs 7.2.3 and 7.2.4.

(4) The thickness of the jacket must comply with the requirements specified

in ANSI/ICEA S-84-608-1988, paragraph 7.2.2.

(k) *Electrical requirements*—(1) *Conductor resistance.* The direct current resistance of any conductor in a completed cable and the average resistance of all conductors in a Quality Control Lot must comply with the requirements specified in ANSI/ICEA S-84-608-1988, paragraph 8.1.

(2) *Resistance unbalance.* (i) The direct current resistance unbalance between the two conductors of any pair in a completed cable and the average resistance unbalance of all pairs in a completed cable must comply with the requirements specified in ANSI/ICEA S-84-608-1988, paragraph 8.2.

(ii) The resistance unbalance between tip and ring conductors shall be random with respect to the direction of unbalance. That is, the resistance of the tip conductors shall not be consistently higher with respect to the ring conductors and vice versa.

(3) *Mutual capacitance.* The average mutual capacitance of all pairs in a completed cable and the individual mutual capacitance of any pair in a completed cable must comply with the requirements specified in ANSI/ICEA S-84-608-1988, paragraph 8.3.

(4) *Capacitance difference.* (i) The capacitance difference for completed cables having 75 pairs or greater must comply with the requirement specified in ANSI/ICEA S-84-608-1988, paragraph 8.4.

(ii) When measuring screened cable, the inner and outer pairs must be selected from both sides of the screen.

(5) *Pair-to-pair capacitance unbalance*—(i) *Pair-to-pair.* The capacitance unbalance as measured on the completed cable must comply with the requirements specified in ANSI/ICEA S-84-608-1988, paragraph 8.5.

(ii) *Screened cable.* In cables with 25 pairs or less and within each group of multigroup cables, the pair-to-pair capacitance unbalance between any two pairs in an individual compartment must comply with the requirements specified in ANSI/ICEA S-84-608-1988, paragraph 8.5. The pair-to-pair capacitance unbalances to be considered must be:

(A) Between pairs adjacent in a layer in an individual compartment;

(B) Between pairs in centers of 4 pairs or less in an individual compartment; and

(C) Between pairs in adjacent layers in an individual compartment when the number of pairs in the inner (smaller) layer is 6 or less. The center is counted as a layer.

(iii) In cables with 25 pairs or less, the root-mean-square (rms) value must include all the pair-to-pair unbalances measured for each compartment separately.

(iv) In cables containing more than 25 pairs, the rms value must include the pair-to-pair unbalances in the separate compartments.

(6) *Pair-to-ground capacitance unbalance*—(i) *Pair-to-ground.* The capacitance unbalance as measured on the completed cable must comply with the requirements specified in ANSI/ICEA S-84-608-1988, paragraph 8.6.

(ii) When measuring pair-to-ground capacitance unbalance all pairs except the pair under test are grounded to the shield and/or shield/armor except when measuring cables containing super units in which case all other pairs in the same super unit must be grounded to the shield.

(iii) The screen tape must be left floating during the test.

(iv) Pair-to-ground capacitance unbalance may vary directly with the length of the cable.

(7) *Attenuation.* (i) For nonscreened and screened cables, the average attenuation of all pairs on any reel when measured at 150 and 772 kilohertz must comply with the requirements specified in ANSI/ICEA S-84-608-1988, paragraph 8.7, Solid Column.

(ii) For T1C type cables over 12 pairs, the maximum average attenuation of all pairs on any reel must not exceed the values listed below when measured at a frequency of 1576 kilohertz at or corrected to a temperature of 20 ±1 °C. The test must be conducted in accordance with ASTM D 4566–90.

AWG	Maximum Average Attenuation decibel/kilometer (dB/km) (decibel/mile)
19	13.4 (21.5)
22	18.3 (29.4)
24	23.1 (37.2)

(8) *Crosstalk loss.* (i) The equal level far-end power sum crosstalk loss (FEXT) as measured on the completed cable must comply with the requirements specified in ANSI/ICEA S-84-608-1988, paragraph 8.8, FEXT Table.

(ii) The near-end power sum crosstalk loss (NEXT) as measured on completed cable must comply with the requirements specified in ANSI/ICEA S-84-608-1988, paragraph 8.8, NEXT Table.

(iii) *Screened cable.* (A) For screened cables the NEXT as measured on the completed cable must comply with the requirements specified in ANSI/ICEA S-84-608-1988, paragraphs 8.9 and 8.9.1.

(B) For T1C screened cable the NEXT as measured on the completed cable must comply with the requirements specified in ANSI/ICEA S-84-608-1988, paragraphs 8.9 and 8.9.2.

(9) *Insulation resistance.* The insulation resistance of each insulated conductor in a completed cable must comply with the requirement specified in ANSI/ICEA S-84-608-1988, paragraph 8.11.

(10) *High voltage test.* (i) In each length of completed cable, the insulation between conductors must comply with the requirements specified in ANSI/ICEA S-84-608-1988, paragraph 8.12, Solid Column.

(ii) In each length of completed cable, the dielectric between the shield and/or armor and conductors in the core must comply with the requirements specified in ANSI/ICEA S-84-608-1988, paragraph 8.13, Single Jacketed, Solid Column. In screened cable the screen tape must be left floating.

(iii) *Screened cable.* (A) In each length of completed screened cable, the dielectric between the screen tape and the conductors in the core must comply with the requirement specified in ANSI/ICEA S-84-608-1988, paragraph 8.14.

(B) In this test, the cable shield and/or armor must be left floating.

(11) *Electrical variations.* (i) Pairs in each length of cable having either a ground, cross, short, or open circuit condition will not be permitted.

(ii) The maximum number of pairs in a cable which may vary as specified in paragraph (k)(11)(iii) of this section from the electrical parameters given in this section are listed below. These pairs may be excluded from the arithmetic calculation.

Nominal Pair Count	Maximum Number of Pairs With Allowable Electrical Variation
6–100	1
101–300	2
301–400	3
401–600	4
601 and above	6

(iii) *Parameter variations.* (A) *Capacitance unbalance-to-ground.* If the cable fails either the maximum individual pair or average capacitance unbalance-to-ground requirement and all individual pairs are 3937 picofarad/kilometer (1200 picofarad/1000 feet) or less, the number of pairs specified in paragraph (k)(11)(ii) of this section may be eliminated from the average and maximum individual calculations.

(B) *Resistance unbalance.* Individual pair of 7 percent for all gauges.

(C) *Conductor resistance, maximum.* The following table shows maximum conductor resistance:

AWG	ohms/kilometer	(ohms/1000 feet)
19	29.9	(9.1)
22	60.0	(18.3)
24	94.5	(28.8)
26	151.6	(46.2)

NOTE: RUS recognizes that in large pair count cable (600 pair and above) a cross, short or open circuit condition occasionally may develop in a pair which does not affect the performance of the other cable pairs. In these circumstances rejection of the entire cable may be economically unsound or repairs may be impractical. In such circumstances the manufacturer may desire to negotiate with the customer for acceptance of the cable. No more than 0.5 percent of the pairs may be involved.

(l) *Mechanical requirements*—(1) *Compound flow test.* All cables manufactured in accordance with the requirements of this section must be capable of meeting the compound flow test specified in ANSI/ICEA S-84-608-1988, paragraph 9.1 using a test temperature of 80 ±1 °C.

(2) *Water penetration.* All cables manufactured in accordance with the requirements of this section must be capable of meeting the water penetration test specified in ANSI/ICEA S-84-608-1988, paragraph 9.2.

(3) *Cable cold bend test.* All cables manufactured in accordance with the

requirements of this section must be capable of meeting the cable cold bend test specified in ANSI/ICEA S-84-608-1988, paragraph 9.3.

(4) *Cable impact test.* All cables manufactured in accordance with the requirements of this section must be capable of meeting the cable impact test specified in ANSI/ICEA S-84-608-1988, paragraph 9.4.

(5) *Jacket notch test (CACSP sheath only).* All cables utilizing the coated aluminum/coated steel sheath (CACSP) design manufactured in accordance with the requirements of this section must be capable of meeting the jacket notch test specified in ANSI/ICEA S-84-608-1988, paragraph 9.5.

(6) *Cable torsion test (CACSP sheath only).* All cables utilizing the coated aluminum/coated steel sheath (CACSP) design manufactured in accordance with the requirements of this section must be capable of meeting the cable torsion test specified in ANSI/ICEA S-84-608-1988, paragraph 9.6.

(m) *Sheath slitting cord (optional).* (1) Sheath slitting cords may be used in the cable structure at the option of the manufacturer unless specified by the end user.

(2) When a sheath slitting cord is used it must be nonhygroscopic and nonwicking, continuous throughout a length of cable and of sufficient strength to open the sheath without breaking the cord.

(n) *Identification marker and length marker.* (1) Each length of cable must be identified in accordance with ANSI/ICEA S-84-608-1988, paragraphs 10.1 through 10.1.4. The color of the ink used for the initial outer jacket marking must be either white or silver.

(2) The markings must be printed on the jacket at regular intervals of not more than 0.6 meter (2 feet).

(3) The completed cable must have sequentially numbered length markers in accordance with ANSI/ICEA S-84-608-1988, paragraph 10.1.5. The color of the ink used for the initial outer jacket marking must be either white or silver.

(o) *Preconnectorized cable (optional).* (1) At the option of the manufacturer and upon request by the purchaser, cables 100 pairs and larger may be factory terminated in 25 pair splicing modules.

(2) The splicing modules must meet the requirements of RUS Bulletin 345-54, PE-52, RUS Specification for Telephone Cable Splicing Connectors (Incorporated by reference at § 1755.97), and be accepted by RUS prior to their use.

(p) *Acceptance testing and extent of testing.* (1) The tests described in appendix A of this section are intended for acceptance of cable designs and major modifications of accepted designs. What constitutes a major modification is at the discretion of RUS. These tests are intended to show the inherent capability of the manufacturer to produce cable products having long life and stability.

(2) For initial acceptance, the manufacturer must submit:

(i) An original signature certification that the product fully complies with each section of the specification;

(ii) Qualification Test Data, per appendix A of this section;

(iii) To periodic plant inspections;

(iv) A certification that the product does or does not comply with the domestic origin manufacturing provisions of the "Buy American" requirements of the Rural Electrification Act of 1938 (7 U.S.C. 901 *et seq.*);

(v) Written user testimonials concerning field performance of the product; and

(vi) Other nonproprietary data deemed necessary by the Chief, Outside Plant Branch (Telephone).

(3) For requalification acceptance, the manufacturer must submit an original signature certification that the product fully complies with each section of the specification, excluding the Qualification Section, and a certification that the product does or does not comply with the domestic origin manufacturing provisions of the "Buy American" requirements of the Rural Electrification Act of 1938 (7 U.S.C. 901 *et seq.*), for acceptance by August 30 of each year. The required data must have been gathered within 90 days of the submission. If the initial acceptance of a product to this specification was within 180 days of August 30, then requalification for that product will not be required for that year.

(4) Initial and requalification acceptance requests should be addressed to:

Chairman, Technical Standards Committee "A" (Telephone), Telecommunications Standards Division, Rural Utilities Service, Washington, DC 20250–1500.

(5) *Tests on 100 percent of completed cable.* (i) The shield and/or armor of each length of cable must be tested for continuity in accordance with ANSI/ICEA S-84–608–1988, paragraph 8.16.

(ii) The screen tape of each length of screened cable must be tested for continuity in accordance with ANSI/ICEA S-84–608–1988, paragraph 8.16.

(iii) Dielectric strength between conductors and shield and/or armor must be tested to determine freedom from grounds in accordance with paragraph (k)(10)(ii) of this section.

(iv) Dielectric strength between conductors and screen tape must be tested to determine freedom from grounds in accordance with paragraph (k)(10)(iii) of this section.

(v) Each conductor in the completed cable must be tested for continuity in accordance with ANSI/ICEA S-84–608–1988, paragraph 8.16.

(vi) Dielectric strength between conductors must be tested to insure freedom from shorts and crosses in each length of completed cable in accordance with paragraph (k)(10)(i) of this section.

(vii) Each conductor in the completed preconnectorized cable must be tested for continuity.

(viii) Each length of completed preconnectorized cable must be tested for split pairs.

(ix) The average mutual capacitance must be measured on all cables. If the average mutual capacitance for the first 100 pairs tested from randomly selected groups is between 50 and 53 nanofarad/kilometer (nF/km) (80 and 85 nanofarad/mile), the remainder of the pairs need not be tested on the 100 percent basis (See paragraph (k)(3) of this section).

(6) *Capability tests.* Tests on a quality assurance basis must be made as frequently as is required for each manufacturer to determine and maintain compliance with:

(i) Performance requirements for conductor insulation, jacketing material, and filling and flooding compounds;

(ii) Bonding properties of coated or laminated shielding and armoring materials and performance requirements for screen tape;

(iii) Sequential marking and lettering;

(iv) Capacitance difference, capacitance unbalance, crosstalk, and attenuation;

(v) Insulation resistance, conductor resistance and resistance unbalance;

(vi) Cable cold bend and cable impact tests;

(vii) Water penetration and compound flow tests; and

(viii) Jacket notch and cable torsion tests.

(q) *Summary of records of electrical and physical tests.* (1) Each manufacturer must maintain suitable summary records for a period of at least 3 years of all electrical and physical tests required on completed cable by this section as set forth in paragraphs (p)(5) and (p)(6) of this section. The test data for a particular reel must be in a form that it may be readily available to the purchaser or to RUS upon request.

(2) Measurements and computed values must be rounded off to the number of places or figures specified for the requirement according to ANSI/ICEA S-84–608–1988, paragraph 1.3.

(r) *Manufacturing irregularities.* (1) Repairs to the shield and/or armor are not permitted in cable supplied to end users under this section.

(2) Minor defects in jackets (defects having a dimension of 3 millimeters (0.125 inch) or less in any direction) may be repaired by means of heat fusing in accordance with good commercial practices utilizing sheath grade compounds.

(s) *Preparation for shipment.* (1) The cable must be shipped on reels. The diameter of the drum must be large enough to prevent damage to the cable from reeling or unreeling. The reels must be substantial and so constructed as to prevent damage to the cable during shipment and handling.

(2) The thermal wrap must comply with the requirements of ANSI/ICEA S-84–608–1988, paragraph 10.3. When a thermal reel wrap is supplied, the wrap must be applied to the reel and must be suitably secured in place to minimize thermal exposure to the cable during

storage and shipment. The use of the thermal reel wrap as a means of reel protection will be at the option of the manufacturer unless specified by the end user.

(3) The outer end of the cable must be securely fastened to the reel head so as to prevent the cable from becoming loose in transit. The inner end of the cable must be securely fastened in such a way as to make it readily available if required for electrical testing. Spikes, staples, or other fastening devices which penetrate the cable jacket must not be used. The method of fastening the cable ends must be accepted by RUS prior to its use.

(4) Each length of cable must be wound on a separate reel unless otherwise specified or agreed to by the purchaser.

(5) The arbor hole must admit a spindle 63 millimeters (2.5 inches) in diameter without binding. Steel arbor hole liners may be used but must be accepted by RUS prior to their use.

(6) Each reel must be plainly marked to indicate the direction in which it should be rolled to prevent loosening of the cable on the reel.

(7) Each reel must be stenciled or labeled on either one or both sides with the information specified in ANSI/ICEA S-84-608-1988, paragraph 10.4 and the RUS cable designation:

Cable Designation

BFC

Cable Construction

Pair Count

Conductor Gauge

A = Coated Aluminum Shield
C = Copper Shield
Y = Gopher Resistant Shield
X = Armored, Separate Shield
H = T1 Screened Cable
H1C = T1C Screened Cable
P = Preconnectorized

Example: BFCXH100–22

Buried Filled Cable, Armored (w/separate shield), T1 Screened Cable, 100 pair, 22 AWG.

(8) When cable manufactured to the requirements of this section is shipped, both ends must be equipped with end caps acceptable to RUS.

(9) When preconnectorized cables are shipped, the splicing modules must be protected to prevent damage during shipment and handling. The protection method must be acceptable to RUS and accepted prior to its use.

(10) All cables ordered for use in underground duct applications must be equipped with a factory-installed pulling-eye on the outer end in accordance with ANSI/ICEA S-84-608-1988, paragraph 10.5.2.

(The information and recordkeeping requirements of this section have been approved by the Office of Management and Budget under the control number 0572–0059)

APPENDIX A TO § 1755.390—QUALIFICATION TEST METHODS

(I) The test procedures described in this appendix are for qualification of initial designs and major modification of accepted designs. Included in (V) of this appendix are suggested formats that may be used in submitting the test results to RUS.

(II) *Sample selection and preparation.* (1) All testing must be performed on lengths removed sequentially from the same 25 pair, 22 gauge jacketed cable. This cable must not have been exposed to temperatures in excess of 38 °C since its initial cool down after sheathing. The lengths specified are minimum lengths and if desirable from a laboratory testing standpoint longer lengths may be used.

(a) Length A shall be 10 ±0.2 meters (33 ±0.5 feet) long and must be maintained at 23 ±3 °C. One length is required.

(b) Length B shall be 12 ±0.2 meters (40 ±0.5 feet) long. Prepare the test sample by removing the jacket, shield or shield/armor and core wrap for a sufficient distance on both ends to allow the insulated conductors to be flared out. Remove sufficient conductor insulation so that appropriate electrical test connections can be made at both ends. Coil the sample with a diameter of 15 to 20 times its sheath diameter. Three lengths are required.

(c) Length C shall be one meter (3 feet) long. Four lengths are required.

(d) Length D shall be 300 millimeters (1 foot) long. Four lengths are required.

(e) Length E must be 600 millimeters (2 feet) long. Four lengths are required.

(f) Length F shall be 3 meters (10 feet) long and must be maintained at 23 ±3 °C for the duration of the test. Two lengths are required.

(2) *Data reference temperature.* Unless otherwise specified, all measurements must be made at 23 ±3 °C.

(III) *Environmental tests*—(1) *Heat aging test*—(a) *Test samples.* Place one sample each of lengths B, C, D and E in an oven or environmental chamber. The ends of Sample B must exit from the chamber or oven for electrical tests. Securely seal the oven exit holes.

(b) *Sequence of tests.* The samples are to be subjected to the following tests after conditioning:

(i) Water Immersion Test outlined in (III)(2) of this appendix;

(ii) Water Penetration Test outlined in (III)(3) of this appendix;

(iii) Insulation Compression Test outlined in (III)(4) of this appendix; and

(iv) Jacket Slip Strength Test outlined in (III)(5) of this appendix.

(c) *Initial measurements.* (i) For Sample B measure the open circuit capacitance for each odd numbered pair at 1, 150, and 772 kilohertz, and the attenuation at 150 and 772 kilohertz after conditioning the sample at the data reference temperature for 24 hours. Calculate the average and standard deviation for the data of the 13 pairs on a per kilometer or (on a per mile) basis.

(ii) The attenuation at 150 and 772 kilohertz may be calculated from open circuit admittance (Yoc) and short circuit impedance (Zsc) or may be obtained by direct measurement of attenuation.

(iii) Record on suggested formats in (V) of this appendix or on other easily readable formats.

(d) *Heat conditioning.* (i) Immediately after completing the initial measurements, condition the sample for 14 days at a temperature of 65 ±2 °C.

(ii) At the end of this period note any exudation of cable filler. Measure and calculate the parameters given in (III)(1)(c) of this appendix. Record on suggested formats in (V) of this appendix or on other easily readable formats.

(iii) Cut away and discard a one meter (3 foot) section from each end of length B.

(e) *Overall electrical deviation.* (i) Calculate the percent change in all average parameters between the final parameters after conditioning and the initial parameters in (III)(1)(c) of this appendix.

(ii) The stability of the electrical parameters after completion of this test must be within the following prescribed limits:

(A) *Capacitance.* The average mutual capacitance must be within 5 percent of its original value;

(B) The change in average mutual capacitance must be less than 5 percent over frequency 1 to 150 kilohertz; and

(C) *Attenuation.* The 150 and 772 kilohertz attenuation must not have increased by more than 5 percent over their original values.

(2) *Water immersion electrical test*—(a) *Test sample selection.* The 10 meter (33 foot) section of length B must be tested.

(b) *Test sample preparation.* Prepare the sample by removing the jacket, shield or shield/armor, and core wrap for sufficient distance to allow one end to be accessed for test connections. Cut out a series of 6 millimeter (0.25 inch) diameter holes along the test sample, at 30 centimeters (1 foot) intervals progressing successively 90 degrees around the circumference of the cable. Assure that the cable core is exposed at each hole by slitting the core wrapper. Place the prepared sample in a dry vessel which when filled will maintain a one meter (3 foot) head of water over 6 meters (20 feet) of uncoiled cable. Extend and fasten the ends of the cable so they will be above the water line and the pairs are rigidly held for the duration of the test.

(c) *Capacitance testing.* Measure the initial values of mutual capacitance of all odd pairs in each cable at a frequency of 1 kilohertz before filling the vessel with water. Be sure the cable shield or shield/armor is grounded to the test equipment. Fill the vessels until there is a one meter (3 foot) head of water on the cables.

(i) Remeasure the mutual capacitance after the cables have been submerged for 24 hours and again after 30 days.

(ii) Record each sample separately on suggested formats in (V) of this appendix or on other easily readable formats.

(d) *Overall electrical deviation.* (i) Calculate the percent change in all average parameters between the final parameters after conditioning with the initial parameters in (III)(2)(c) of this appendix.

(ii) The average mutual capacitance must be within 5 percent of its original value.

(3) *Water penetration testing.* (a) A watertight closure must be placed over the jacket of length C. The closure must not be placed over the jacket so tightly that the flow of water through pre-existing voids of air spaces is restricted. The other end of the sample must remain open.

(b) Test per Option A or Option B—(i) *Option A.* Weigh the sample and closure prior to testing. Fill the closure with water and place under a continuous pressure of 10 ±0.7 kilopascals (1.5 ±0.1 pounds per square inch gauge) for one hour. Collect the water leakage from the end of the test sample during the test and weigh to the nearest 0.1 gram. Immediately after the one hour test, seal the ends of the cable with a thin layer of grease and remove all visible water from the closure, being careful not to remove water that penetrated into the core during the test. Reweigh the sample and determine the weight of water that penetrated into the core. The weight of water that penetrated into the core must not exceed 8 grams.

(ii) *Option B.* Fill the closure with a 0.2 gram sodium fluorscein per liter water solution and apply a continuous pressure 10 ±0.7 kilopascals (1.5 ±0.1 pounds per square inch gauge) for one hour. Catch and weigh any water that leaks from the end of the cable during the one hour period. If no water leaks from the sample, carefully remove the water from the closure. Then carefully remove the jacket, shield or shield/armor and core wrap one at a time, examining with an ultraviolet light source for water penetration. After removal of the core wrap, carefully dissect the core and examine for water penetration within the core. Where water penetration is observed, measure the penetration distance. The distance of water penetration into the core must not exceed 127 millimeters (5.0 inches).

(4) *Insulation compression test*—(a) *Test Sample D.* Remove jacket, shield or shield/armor, and core wrap being careful not to damage the conductor insulation. Remove one pair from the core and carefully separate, wipe off core filler, and straighten the insulated conductors. Retwist the two insulated conductors together under sufficient tension to form 10 evenly spaced 360 degree twists in a length of 10 centimeters (4 inches).

(b) *Sample testing.* Center the mid 50 millimeters (2 inches) of the twisted pair between 2 smooth rigid parallel metal plates that are 50 millimeters × 50 millimeters (2 inches × 2 inches). Apply a 1.5 volt direct current potential between the conductors, using a light or buzzer to indicate electrical contact between the conductors. Apply a constant load of 67 newtons (15 pound-force) on the sample for one minute and monitor for evidence of contact between the conductors. Record results on suggested formats in (V) of this appendix or on other easily readable formats.

(5) *Jacket slip strength test*—(a) *Sample selection.* Test Sample E from (III)(1)(a) of this appendix.

(b) *Sample preparation.* Prepare test sample in accordance with the procedures specified in ASTM D 4565–90a.

(c) *Sample conditioning and testing.* Remove the sample from the tensile tester prior to testing and condition for one hour at 50 ±2 °C. Test immediately in accordance with the procedures specified in ASTM D 4565–90a. A minimum jacket slip strength of 67 newtons (15 pound-force) is required. Record the highest load attained.

(6) *Humidity exposure.* (a) Repeat steps (III)(1)(a) through (III)(1)(c)(iii) of this appendix for separate set of samples B, C, D, and E which have not been subjected to prior environmental conditioning.

(b) Immediately after completing the measurements, expose the test sample to 100 temperature cyclings. Relative humidity within the chamber must be maintained at 90 ±2 percent. One cycle consists of beginning at a stabilized chamber and test sample temperature of 52 ±1 °C, increasing the temperature to 57 ±1 °C, allowing the chamber and test samples to stabilize at this level, then dropping the temperature back to 52 ±1 °C.

(c) Repeat steps (III)(1)(d)(ii) through (III)(5)(c) of this appendix.

(7) *Temperature cycling.* (a) Repeat steps (III)(1)(a) through (III)(1)(c)(iii) of this appendix for separate set of samples B, C, D, and E which have not been subjected to prior environmental conditioning.

(b) Immediately after completing the measurements, subject the test sample to the 10 cycles of temperature between a minimum of −40 °C and + 60 °C. The test sample must be held at each temperature extreme for a minimum of 1½ hours during each cycle of temperature. The air within the temperature cycling chamber must be circulated throughout the duration of the cycling.

(c) Repeat steps (III)(1)(d)(ii) through (III)(5)(c) of this appendix.

(IV) *Control sample*—(1) *Test samples.* A separate set of lengths A, C, D, E, and F must have been maintained at 23 ±3 °C for at least 48 hours before the testing.

(2) Repeat steps (III)(2) through (III)(5)(c) of this appendix except use length A instead of length B.

(3) *Surge Test.* (a) One length of sample F must be used to measure the breakdown between conductors while the other length of F must be used to measure the core to shield breakdown.

(b) The samples must be capable of withstanding without damage, a single surge voltage of 20 kilovolts peak between conductors, and a 35 kilovolts peak surge voltage between conductors and the shield or shield/armor as hereinafter described. The surge voltage must be developed from a capacitor discharged through a forming resistor connected in parallel with the dielectric of the test sample. The surge generator constants must be such as to produce a surge of 1.5 × 40 microsecond wave shape.

(c) The shape of the generated wave must be determined at a reduced voltage by connecting an oscilloscope across the forming resistor with the cable sample connected in parallel with the forming resistor. The capacitor bank is charged to the test voltage and then discharged through the forming resistor and test sample. The test sample will be considered to have passed the test if there is no distinct change in the wave shape obtained with the initial reduced voltage compared to that obtained after the application of the test voltage.

(V) The following suggested formats may be used in submitting the test results to RUS:

ENVIRONMENTAL CONDITIONING______________

FREQUENCY 1 KILOHERTZ

Pair Number	Capacitance nF/km (nanofarad/mile) Initial	Final
1	______	______
3	______	______
5	______	______
7	______	______
9	______	______
11	______	______
13	______	______
15	______	______
17	______	______
19	______	______
21	______	______
23	______	______
25	______	______
Average x̄	______	______

Overall Percent Difference in Average x̄ ______

ENVIRONMENTAL CONDITIONING______________

FREQUENCY 772 KILOHERTZ

Pair Number	Capacitance nF/km (nanofarad/mile)		Attenuation dB/km (decibel/mile)	
	Initial	Final	Initial	Final
1	____	____	____	____
3	____	____	____	____
5	____	____	____	____
7	____	____	____	____
9	____	____	____	____
11	____	____	____	____
13	____	____	____	____
15	____	____	____	____
17	____	____	____	____
19	____	____	____	____
21	____	____	____	____
23	____	____	____	____
25	____	____	____	____
Average x̄	____	____	____	____

Overall Percent Difference in Average x̄ Capacitance:______ Conductance:______

ENVIRONMENTAL CONDITIONING______________

FREQUENCY 150 KILOHERTZ

Pair Number	Capacitance nF/km (nanofarad/mile)		Attenuation dB/km (decibel/mile)	
	Initial	Final	Initial	Final
1	____	____	____	____
3	____	____	____	____
5	____	____	____	____
7	____	____	____	____
9	____	____	____	____
11	____	____	____	____
13	____	____	____	____
15	____	____	____	____
17	____	____	____	____
19	____	____	____	____
21	____	____	____	____
23	____	____	____	____
25	____	____	____	____
Average x̄	____	____	____	____

Overall Percent Difference in Average x̄ Capacitance:______ Conductance:______

ENVIRONMENTAL CONDITIONING______________

WATER IMMERSION TEST (1 KILOHERTZ)

Pair Number	Capacitance nF/km (nanofarad/mile) Initial	24 Hours	Final
1	____	____	____
3	____	____	____
5	____	____	____
7	____	____	____
9	____	____	____
11	____	____	____
13	____	____	____
15	____	____	____
17	____	____	____
19	____	____	____
21	____	____	____
23	____	____	____
25	____	____	____
Average x̄	____	____	____

Overall Percent Difference in Average x̄ ______

WATER PENETRATION TEST

	Option A		Option B	
	End Leakage grams	Weight Gain grams	End Leakage grams	Penetration mm (in.)
Control.				
Heat Age.				
Humidity Exposure.				
Temperature Cycling.				

INSULATION COMPRESSION

	Failures
Control	______
Heat Age	______
Humidity Exposure	______
Temperature Cycling	______

JACKET SLIP STRENGTH @ 50 °C

	Load in newtons (pound-force)
Control	______
Heat Age	______
Humidity Exposure	______

JACKET SLIP STRENGTH @ 50 °C—Continued

	Load in newtons (pound-force)
Temperature Cycling	______________

FILLER EXUDATION (GRAMS)

Heat Age	______________
Humidity Exposure	______________
Temperature Cycle	______________

SURGE TEST (KILOVOLTS)

Conductor to Conductor	______________
Shield to Conductors	______________

[58 FR 29338, May 20, 1993; 58 FR 32749, June 11, 1993, as amended at 60 FR 1711, Jan. 5, 1995; 69 FR 18803, Apr. 9, 2004]

§§ 1755.391–1755.396 [Reserved]

§ 1755.397 RUS performance specification for line concentrators.

(a) *General.* (1) This section covers general requirements for a line concentrator (LC) system. This system shall operate in accordance with the manufacturer's specifications. Reliability shall be of prime importance in the design, manufacture and installation of the equipment. The equipment shall automatically provide for:

(i) Terminating subscriber lines at a location remote from the serving central office;

(ii) Concentrating the subscriber lines over a few transmission and supervisory paths to the serving central office; and

(iii) Terminating the lines at the central office without loss of individual identity. A subscriber connected to a line concentrator shall be capable of having essentially the same services as a subscriber connected directly to the central office equipment (COE). Intraunit calling among subscribers connected to the concentrator may be provided, but is not required.

(2) Industry standards, or portions thereof, referred to in this paragraph (a) are incorporated by reference by RUS. This incorporation by reference was approved by the Director of the Federal Register in accordance with 5 U.S.C. 552 (a) and 1 CFR part 51. Copies of these standards are available for inspection during normal business hours at RUS, room 2838, U.S. Department of Agriculture, Washington, DC 20250, or at the National Archives and Records Administration (NARA). For information on the availability of this material at NARA, call 202–741–6030, or go to: *http://www.archives.gov/federal_register/code_of_federal_regulations/ibr_locations.html.*

(3) American National Standards Institute (ANSI) standards are available from ANSI Inc., 11 West 42nd Street, 13th floor, New York, NY 10036, telephone 212–642–4900.

(i) ANSI Standard S1.4–1983, Specification for Sound Level Meters, including Amendment S1.4A–1985.

(ii) [Reserved]

(4) American Society for Testing Materials (ASTM) are available from 1916 Race Street, Philadelphia, PA 19103, telephone 215–299–5400.

(i) ASTM Specification B33–91, Standard Specifications for Tinned Soft or Annealed Copper Wire for Electrical Purposes.

(ii) [Reserved]

(5) Bell Communications Research (Bellcore) standards are available from Bellcore Customer Service, 8 Corporate Place, Piscataway, NJ 08854, telephone 1–800–521–2673.

(i) TR-TSY-000008, Issue 2, August 1987, Digital Interface between the SLC 96 Digital Loop Carrier System and a Local Digital Switch.

(ii) Bell Communications Research (Bellcore) document TR-TSY-000057, Issue 1, April 1987, including Revision 1, November 1988, Functional Criteria for Digital Loop Carrier Systems.

(iii) Bell Communications Research (Bellcore) Document TR-NWT-000303, Issue 2, December 1992, including Revision 1, December 1993, Integrated Digital Loop Carrier System Generic Requirements, Objectives, and Interface.

(6) Federal Standard H28, Screw-Thread Standards for Federal Services, March 31, 1978, including Change Notice 1, May 28, 1986; Change Notice 2, January 20, 1989; and Change Notice 3, March 12, 1990. Copies may be obtained from the General Services Administration, Specification Section, 490 East L'Enfant Plaza SW, Washington, DC 20407, telephone 202–755–0325.

(7) IEEE standards are available from IEEE Service Center, 445 Hoes Lane,

P.O. Box 1331, Piscataway, NJ 08854, telephone 1–800–521–2673.

(i) IEEE Standard 455–1985, Standard Test Procedure for Measuring Longitudinal Balance of Telephone Equipment Operating in the Voice Band.

(ii) [Reserved]

(8) RUS standards are available from Publications and Directives Management Branch, Administrative Services Division, Rural Utilities Service, room 0180, South Building, U.S. Department of Agriculture, Washington, DC 20250–1500.

(i) RUS Bulletin 345–50, PE–60 (Sept 1979), RUS Specification for Trunk Carrier Systems.

(ii) [Reserved]

(b) *Types of requirements.* (1) Unless otherwise indicated, the requirements listed in this section are considered to be fixed requirements.

(2) The concentrator system shall communicate with standard T1 digital transmission format at a minimum between the concentrator and central office terminals. Analog conversion functions at remote and central office terminals shall be capable of being eliminated to accommodate end-to-end digital transmission.

(3) The LC shall operate properly as an integral part of the telephone network when connected to physical or carrier derived circuits and central offices meeting RUS specifications and other generally accepted telecommunications practices, such as Bellcore documents TR-NWT-000303, Integrated Digital Loop Carrier System Generic Requirements, Objectives and Interface; TR-TSY-000008, Digital Interface between the SLC 96 Digital Loop Carrier System and a Local Digital Switch; and TR-TSY-000057, Functional Criteria for Digital Loop Carrier Systems.

(4) For RUS acceptance consideration of a LC, the manufacturer must certify and demonstrate that all requirements specified in this section are available and in compliance with this section.

(5) Certain requirements are included in this section for features which may not be needed for every application. Such features are identifiable by the inclusion in the requirements of some such phrase as "when specified by the owner" or "as specified by the owner." In some cases where an optional feature will not be required by an owner, either now or in the future, a system which does not provide this feature shall be considered to be in compliance with the specification for the specific installation under consideration, but not in compliance with the entire specification.

(6) The owner may properly request bids from any supplier of an RUS accepted LC whose system provides all the features which will be required for a specific installation.

(7) When required by the owner, the supplier shall state compliance to the Carrier Serving Area (CSA) requirements, as stated in Bell Communications Research (Bellcore) Standard TR-TSY-000057, Functional Criteria for Digital Loop Carrier Systems.

(c) *Reliability.* (1) The failure rate of printed circuit boards shall not exceed an average of 2.0 percent per month of all equipped cards in all system terminals during the first 3 months after cutover, and shall not exceed an average of 1.0 percent per month of all equipped cards in all system terminals during the second 3-month period. The failure rate for the equipment shall be less than 0.5 percent per month of all equipped cards in all system terminals after 6 months. A failure is considered to be the failure of a component on the PC board which requires it to be repaired or replaced.

(2) The line concentrator terminal units shall be designed such that there will be no more than 4 hours of total outages in 20 years.

(d) *System type acceptance tests.* General test results will be required on each system type. Any system provided in accordance with this section shall be capable of meeting any requirement in this section on a spot-check basis.

(e) *Features required.* The network control equipment and peripheral equipment shall be comprised of solid-state and integrated circuitry components as far as practical and in keeping with the state-of-the-art and economics of the subject system.

(f) *Subscriber lines*—(1) *General.* (i) The remote LC units shall operate satisfactorily with subscriber lines which meet all of the conditions under the bidder's specifications and all the requirements of this section. This section recognizes that the loop limit of the line concentrator is dependent upon the transmission facility between the LC central office termination and the LC remote unit. When voice frequency (physical) circuits are used, the loop limit from the COE to the subscriber shall be 1900 ohms (including the telephone set). When electronically derived circuits (carrier, lightwave, etc.) are used, the loop limits of the electronic system will control. The bidder shall identify the loop limits of the equipment to be supplied.

(ii) There should be provisions for such types of lines as ground start, loop start, regular subscriber, pay stations, etc.

(2) *Dialing.* (i) *General.* The line concentrator remote and central office terminal equipment shall satisfactorily transmit dialing information when used with subscriber dials having a speed of operation between 8 and 12 dial pulses per second and a break period of 55 to 65% of the total signaling period.

(ii) *Subscriber dial interdigital time.* The remote and central office LC equipment shall permit satisfactory telecommunications operation when used with subscriber rotary dial interdigital times of 200 milliseconds minimum, and pushbutton dialing with 50 milliseconds minimum.

(iii) *Subscriber line pushbutton dialing frequencies.* The frequency pairs assigned for pushbutton dialing when provided by the central office shall be as listed in this paragraph (f)(2)(iii), with an allowable variation of ±1.5 percent:

Low group frequencies (Hz)	High group frequencies (Hz)			
	1209	1336	1477	1633
697	1	2	3	Spare.
770	4	5	6	Spare.
852	7	8	9	Spare.
941	*	0	#	Spare.

(3) *Ringing.* (i) When LC ringing is generated at the remote end, it shall be automatic and intermittent and shall be cut off from the called line upon removal of the handset at the called station during either the ringing or silent period.

(ii) When ringing generators are provided in the LC on an ancillary basis, they shall be accepted or technically accepted by RUS.

(iii) Where ringing is generated at the remote end, the ringing system shall provide sufficient ringing on a bridged basis over the voltage and temperature limits of this specification and over subscriber loops within the limits stated by the manufacturer. The manufacturer shall state the minimum number (not less than two) of main station ringers that can be used for each ringing option available.

(g) *Traffic.* (1)(i) The minimum grade of service for traffic in the line concentrator shall be B =. 005 using the Traffic Table, based on the Erlang Lost-Calls-Cleared Formula. Required grade of service, traffic assumptions and calculations for the particular application being implemented shall be supplied by the bidder.

(ii) Service to customers served by a traffic sensitive LC should not be noticeably different than the service to customers served by the dedicated physical pairs from the central office so that uniform grade of service will be provided to all customers in any class of service. Reference §1755.522(p)(1)(i), RUS General Specification for Digital, Stored Program Controlled Central Office Equipment.

(2) *Traffic and Plant Registers.* Traffic measurements consist of three types—peg count, usage, and congestion. A peg count register scores one count per call attempt per circuit group such as trunks, digit receivers, senders, etc. Usage counters measure the traffic density in networks, trunks and other circuit groups. Congestion registers score the number of calls which fail to find an idle circuit in a trunk group or to find an idle path through the switching network when attempting to connect two given end points. These conditions constitute "network blocking."

(3) When required, traffic data will be stored in electronic storage registers or a block of memory consisting of one or more traffic counters for each item to be measured. The bidder shall indicate what registers are to be supplied, their purpose and the means for displaying the information locally (or at a remote location when available).

(h) *Transmission requirements*—(1) *General.* Unless otherwise stated, the requirements in paragraphs (h) (2) through (20) of this section are specified in terms of analog measurements made from Main Distributing Frame (MDF) terminals to MDF terminals excluding cabling loss.

(2) *Telephone transmitter battery supply.* A minimum of 20 milliamperes, dc, shall be provided for the transmitter of the telephone set at the subscriber station under all loop conditions specified by the bidder. The telephone set is assumed to have a resistance of 200 ohms.

(3) *Impedance—subscriber loops.* For the purpose of this section, the input impedance of all subscriber loops served by the equipment is arbitrarily considered to be 900 ohms in series with 2.16 microfarad capacitor at voice frequencies.

(4) *Battery noise.* Noise across the remote terminal battery at power panel distribution bus terminals shall not exceed 35 dBrnC during the specified busy hour.

(5) *Stability.* The long-term allowable variation in loss through the line concentrator system shall be ±0.5 dB from the loss specified by the bidder.

(6) *Return loss.* The specified return loss values are determined by the service and type of port at the measuring end. Two-wire ports are measured at 900 ohms in series with 2.16 microfarads, and 4-wire ports are measured at 600 ohms resistive. When other balance networks are supplied, test equipment arranged for operation with the supplied network(s) may be used. The requirement given shall meet the following cited values on each balance network available in the system:

Line-to-Line or Line-to-Trunk (2-Wire)
Echo Return Loss (ERL)—18 dB, Minimum
Singing Return Loss (SRL)—Low—15 dB, Minimum
Singing Return Loss (SRL)—High—18 dB, Minimum

(7) *Longitudinal balance.* The minimum longitudinal balance, with dc loop currents between 20 to 70 mA, shall be 60 dB at all frequencies between 60 and 2000 Hz, 55 dB at 2700 Hz and 50 dB at 3400 Hz. The method of measurement shall be as specified in the IEEE standard 455, "Standard Testing Procedure for Measuring Longitudinal Balance of Telephone Equipment Operating in the Voice Band." Source voltage level shall be 10 volts root mean square (rms) where conversation battery feed originates at the remote end.

(8) *60 hz longitudinal current immunity.* The LC 60 Hz longitudinal current immunity shall be measured in accordance with Figure 1 of this section. Under test conditions cited on Figure 1 of this section, the system noise shall be 23 dBrnC or less as follows:

Figure 1

Measuring the Effects of Low Frequency Induction

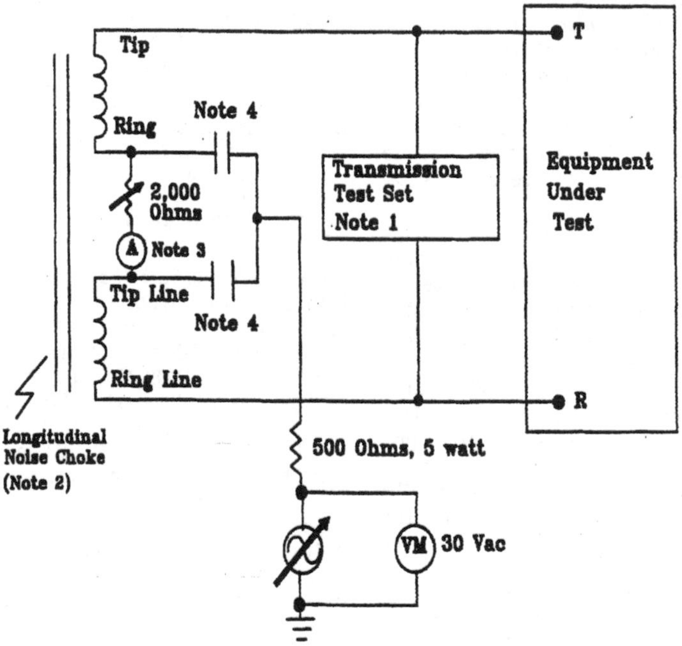

Notes:

1. **Wilcom T194C or Equivalent (900 ohm termination, C-message weighting, hold coil off)**
2. **SNC Noise Choke 35 W, or equivalent**
3. **Test at 0.020 Adc and 0.070 Adc**
4. **2 ± 0.001 microfarad, 150 Vdc**

(9) *Steady noise (idle channel at 900 ohm impedance).* Steady noise: Measure on terminated call. Noise measurements shall comply with the following:

Maximum—23 dBrnC0
Average—18 dBrnC0 or Less
3KHz Flat—Less than 35 dBrnO as an Objective

(10) *Impulse noise.* LC central office terminal equipment shall have an impulse noise limit of not more than five counts exceeding 54 dBrnC0 voice band weighted in a 5-minute period on six such measurements made during the busy hour. A WILCOM T–194C Transmission Test Set, or equivalent, should be used for the measurements. The

measurement shall be made by establishing a normal connection from the noise counter through the switching equipment in its off-hook condition to a quiet termination of 900 ohms impedance. Office battery and signaling circuit wiring shall be suitably segregated from voice and carrier circuit wiring, and frame talking battery filters provided, if and as required, in order to meet these impulse noise limits.

(11) *Crosstalk coupling.* Worst case equal level crosstalk shall be 65 dB minimum in the range 200 to 3400 Hz. This shall be measured between any two paths through the system by connecting a 0 dBm0 level tone to the disturbing pair.

(12) *Digital error rate.* The digital line concentrator shall not introduce more than one error in 10^8 bits averaged over a 5-minute period, excluding the least significant bit.

(13) *Quantizing distortion.* (i) The system shall meet the following requirements:

Input level (dBm0) 1004 or 1020 Hz	Minimum signal to distortion with C-message weighting
0 to −30	33 dB
−30 to −40	27 dB
−40 to −45	22 dB

(ii) Due to possible loss of the least significant bit on direct digital connections, a signal to distortion degradation of up to 2 dB may be allowed where adequately justified by the bidder.

(14) *Overload level.* The overload level shall be + 3 dBm0.

(15) *Gain tracking (linearity)* shall meet the following requirements:

Input signal level [1]	Maximum gain deviation
+ 3 to −37 dBm0	±0.5 dB
−37 to −50 dBm0	±1 dB

[1] 1004 Hz reference at 0 dBm0.

(16) *Frequency response (loss relative to 1004 Hz)* for line-to-line (via trunk group or intra-link) connections shall meet the following requirements:

Frequency (Hz)	Loss at 0 dBm0 input [1]
60	20 dB Min. [2]
300	−1 to + 3 dB
600 to 2400	+ 1 dB
3400	−1 to + 3 dB

[1] (−) means less loss and (+) means more loss.
[2] Transmit End.

(17) *Envelope delay distortion.* On any properly established connection, the envelope delay distortion shall not exceed the following limits:

Frequency (Hz)	Microseconds
1000 to 2600	190
800 to 2800	350
600 to 3000	500
400 to 3200	700

(18) *Absolute delay.* The absolute one-way delay through the line concentrator, excluding delays associated with the central office switching equipment, shall not exceed 1000 microseconds analog-to-analog measured at 1800 Hz.

(19) *Insertion loss.* The insertion loss in both directions of transmission at 1004 Hz shall be included in the insertion loss requirements for the connected COE switch and shall not increase the overall losses through the combined equipment beyond the values for the COE alone, when operated through a direct digital interface. Systems operated with a (VF) line circuit interface may introduce up to 3 dB insertion loss. Reference § 1755.522(q)(3).

(20) *Detailed requirements for direct digital connections.* (i) This paragraph (h)(20) covers the detailed requirements for the provision of interface units which will permit direct digital connection between the host central office and line concentrator subscriber terminals over digital facilities. The digital transmission system shall be compatible with T1 type span lines using a DS1 interface and other digital interfaces that may be specified by the owner. The RUS specification for the T1 span line equipment is PE–60. Other span line techniques may also be used. Diverse span line routing may be used when specified by the owner.

(ii) The output of a digital-to-digital port shall be Pulse Code Modulation (PCM), encoded in eight-bit words using the mu–255 encoding law and D3 encoding format, and arranged to interface with a T1 span line.

(iii) Signaling shall be by means of Multifrequency (MF) or Dual Pulsing (DP) and the system which is inherent in the A and B bits of the D3 format. In the case where A and B bits are not used for signaling or system control, these bits shall only be used for normal voice and data transmission.

(iv) When a direct digital interface between the span line and the host central office equipment is to be implemented, the following requirements shall be met:

(A) The span line shall be terminated in a central office as a minimum a DS1 (1.544Mb/s) shall be provided;

(B) The digital central office equipment shall be programmed to support the operation of the digital port with the line concentrator subscriber terminal;

(C) The line concentrator subscriber terminal used with a direct digital interface shall be interchangeable with the subscriber terminal used with a central office terminal.

(i) *Alarms.* The system shall send alarms for such conditions as blown fuses, blocked controls, power failure in the remote terminal, etc., along with its own status indication and status of dry relay contact closures or solid-state equivalent to the associated central office alarm circuits. Sufficient system alarm points shall be provided from the remote terminal to report conditions to the central office alarm system. The alarms shall be transmitted from the remote terminal to the central office terminal as long as any part of the connecting link is available for this transmission. Fuses shall be of the alarm and indicator type, and their rating designated by numerals or color code on fuse positions.

(j) *Electrical protection*—(1) *Surge protection.* (i) Adequate electrical protection of line concentrator equipment shall be included in the design of the system. The characteristics and application of protection devices must be such that they enable the line concentrator equipment to withstand, without damage or excessive protector maintenance, the dielectric stresses and currents that are produced in line-to-ground and tip-to-ring circuits through the equipment as a result of induced or conducted lightning or power system fault-related surges. All wire terminals connected to outside plant wire or cable pairs shall be protected from voltage and current surges.

(ii) Equipment must pass laboratory tests, simulating a hostile electrical environment, before being placed in the field for the purpose of obtaining field experience. For acceptance consideration RUS requires manufacturers to submit recently completed results (within 90 days of submittal) of data obtained from the prescribed testing. Manufacturers are expected to detail how data and tests were conducted. There are five basic types of laboratory tests which must be applied to exposed terminals in an effort to determine if the equipment will survive. Figure 2 of this section, Summary of Electrical Requirements and Tests, identifies the tests and their application as follows:

FIGURE 2—SUMMARY OF ELECTRICAL REQUIREMENTS AND TESTS

Test	Application criteria	Peak voltage or current	Surge waveshape	Number of applications and maximum time between	Comments
Current surge ...	Low impedance paths exposed to surges.	500A or lesser current (see fig. 4).	10 × 1000 μs	5 each polarity at 1 minute intervals.	None.
60 Hz current carrying.	High or low impedance paths exposed to surges.	10A rms or lesser current (see fig. 6).	11 Cycles of 60 Hz (0.183 Sec.).	3 each at 1 minute intervals.	None.
AC Power service surge voltage.	AC power service connection.	2500V or + 3 σ clamping V of arrester employed at 10kV/μs.	1.2 × 50 μs	5 each polarity at 1 minute intervals.	AC arrester, if used, must be removed. Communications line arresters, if used, remain in place.
Voltage surge ...	High impedance paths exposed to surges.	1000V or + 3 σ dc breakdown of arrester employed.	10 × 1000 μs	5 each polarity at 1 minute intervals.	All primary arresters, if used, must be removed.

FIGURE 2—SUMMARY OF ELECTRICAL REQUIREMENTS AND TESTS—Continued

Test	Application criteria	Peak voltage or current	Surge waveshape	Number of applications and maximum time between	Comments
Arrester response delay.	Paths protected by arresters, such as gas tubes, with breakdown dependent on V. rate of rise.	+ 3 σ breakdown of arrester employed at 100V/μs of rise.	100V/μs rise decay to ½ V. in tube's delay time.	5 each polarity at 1 minute intervals.	All primary arrestors, if used, must be removed.

(iii) Electrical protection requirements for line concentrator equipment can be summarized briefly as follows:

(A) Current surge tests simulate the stress to which a relatively low impedance path may be subjected before main frame protectors break down. Paths with a 100 Hz impedance of 50 ohms or less shall be subjected to current surges, employing a 10 × 1000 microsecond waveshape as defined in Figure 3 of this section, Surge Waveshape. For the purpose of determining this impedance, arresters which are mounted within the equipment are to be considered zero impedance. The crest current shall not exceed 500A; however, depending on the impedance of the test specimen this value of current may be lower. The crest current through the sample, multiplied by the sample's 100 Hz impedance, shall not exceed 1000 V. Where sample impedance is less than 2 ohms, peak current shall be limited to 500A as shown in Figure 4 of this section, Current Surge Tests. Figures 3 and 4 follow:

Figure 3

Explanation of Surge Waveshape

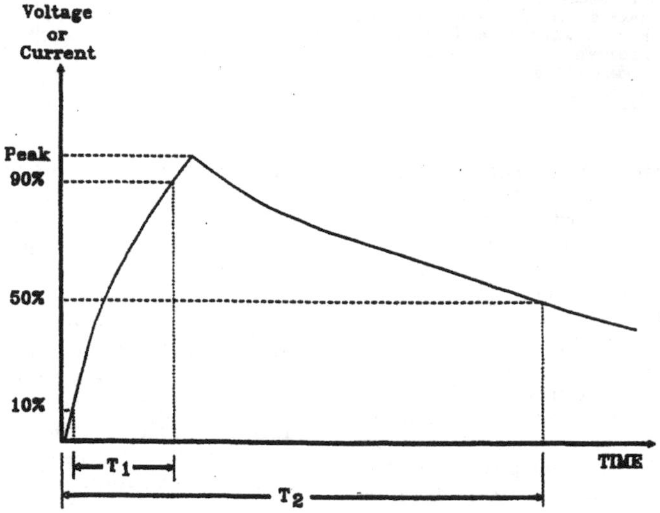

Surge Waveshape is defined as follows:

Rise Time x Time to Decay to Half Crest Value
(For example, 10 x 1000 μs)

Notes: T_1 = Time to determine the rate of rise. The rate of rise is determined as the slope between 10% and 90% of peak voltage or current.

T_2 = Time to 50% of peak voltage (decay to half value).

Figure 4

Current Surge Test

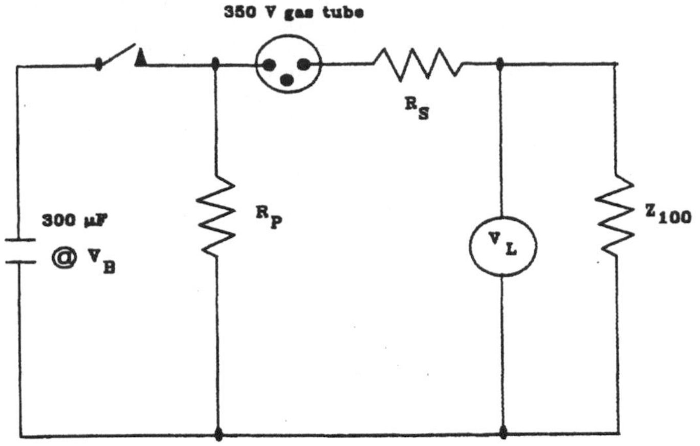

V_L = Not to exceed 1000V
V_B = Charging Voltage
Z_{100} = Test Specimen Impedance to be measured at 100 Hz.
R_P = Parallel Resistance (Waveshape)
R_S = Series Resistance (Current Limiting)

Z_{100}	R_S	R_P	V_B
0	5	∞	2500
1	4	∞	2500
2	3	∞	2500
3	2	∞	1670
4	1	∞	1250
5	0	∞	1000
7.5	0	15	1000
10	0	10	1000
15	0	7.5	1000
20	0	6.7	1000
25	0	6.25	1000
30	0	6	1000
40	0	5.7	1000
50	0	5.5	1000

(B) Sixty Hertz (60 Hz) current carrying tests shall be applied to simulate an ac power fault which is conducted to the unit over the cable pairs. The test shall be limited to 10 amperes Root Mean Square (rms) of 60 Hz ac for a period of 11 cycles (0.1835 seconds) and shall be applied longitudinally from line to ground.

(C) AC power service surge voltage tests shall be applied to the power input terminals of ac powered devices to simulate switching surges or lightning-induced transients on the ac power system. The test shall employ a 1.2 × 50 microsecond waveshape with a crest voltage of 2500 V. Communications line protectors may be left in place for these tests.

(D) Voltage surge tests which simulate the voltage stress to which a relatively high impedance path may be subjected before primary protectors break down and protect the circuit. To ensure coordination with the primary protection while reducing testing to the minimum, voltage surge tests shall be conducted at a 1000 volts with primary arresters removed for devices protected by carbon blocks, or the + 3 sigma dc breakdown voltage of other primary arresters. Surge waveshape should be 10 × 1000 microseconds.

(E) Arrester response delay tests are designed to stress the equipment in a manner similar to that caused by the delayed breakdown of gap type arresters when subjected to rapidly rising voltages. Arresters shall be removed for these tests, the peak surge voltage shall be the + 3 sigma breakdown voltage of the arrester in question on a voltage rising at 100 V per microsecond, and the time for the surge to decay to half voltage shall equal at least the delay time of the tube as explained in Figure 5 of this section, Arrester Response Delay Time as follows:

Figure 5

Explanation of Arrester Response Delay Time

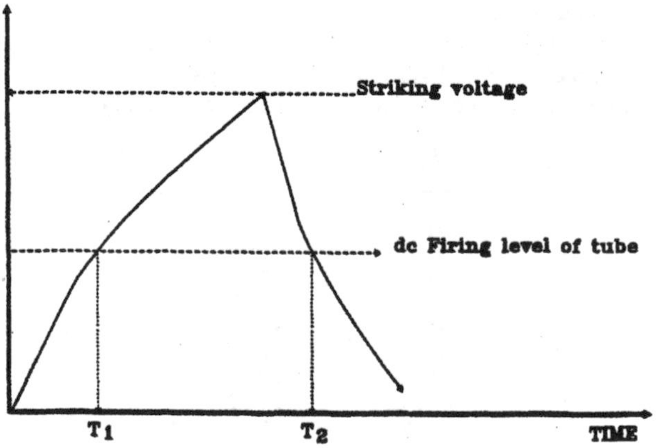

$D = T_2 - T_1$ = Delay time of tube

Note: The delay time is that period of time when the potential across an arrester exceeds its dc firing level.

(iv) Tests shall be conducted in the following sequence. As not all tests are required in every application, non-applicable tests should be omitted:

(A) Current Impulse Test;

(B) Sixty Hertz (60 Hz) Current Carrying Tests;

(C) AC Power Service Impulse Voltage Test;

(D) Voltage Impulse Test; and

(E) Arrester Response Delay Time Test.

(v) A minimum of five applications of each polarity for the surge tests and three for the 60 Hz Current Carrying Tests are the minimum required. All tests shall be conducted with not more than 1 minute between consecutive applications in each series of three or five applications to a specific configuration so that heating effects will be cumulative. See Figure 6 of this section, 60 Hz Current Surge Tests as follows:

Figure 6

60 Hz Current Surge Test

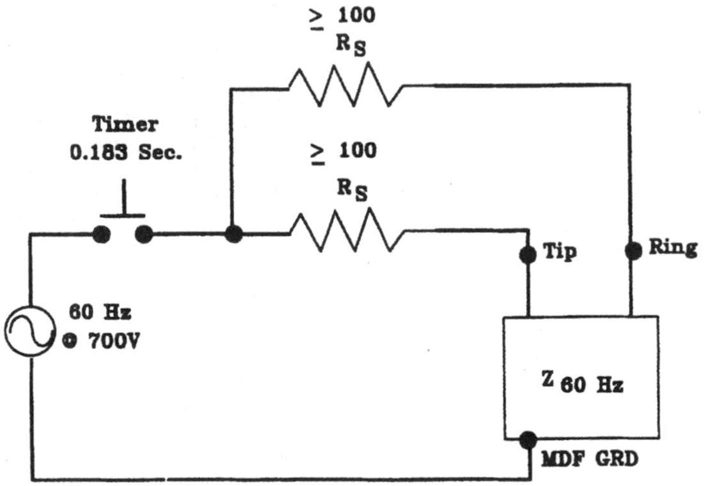

V - 700 Volts root mean square (rms) (Approximately 1000V Peak).

Z_{60} - Test specimen impedance to be measured at 60 Hz.

R_S - Series Resistance (current limiting) in each side of line. (Source impedance never less than 50 Ω longitudinal.)

Z_{60} Hz	R_S
0	140
10	120
20	100
50	100
Over 50	100

(vi) Tests shall be applied between each of the following terminal combinations for all line operating conditions:

(A) Line tip to ring;

(B) Line ring to ground;

(C) Line tip to ground; and

(D) Line tip and ring tied together to ground.

(2) *Dielectric strength*. (i) Arresters shall be removed for all dielectric strength tests.

(ii) Direct current potentials shall be applied between all line terminals and the equipment chassis and between these terminals and grounded equipment housings in all instances where the circuitry is dc open circuit from the chassis, or connected to the chassis through a capacitor. The duration of all dielectric strength tests shall be at least 1 second. The applied potential shall be at a minimum equal to the plus 3 sigma dc breakdown voltage of the arrester, provided by the line concentrator manufacturer.

(3) *Insulation resistance*. Following the dielectric tests, the insulation resistance of the installed electrical circuits between wires and ground, with the normal equipment grounds removed, shall not be less than 10 megohms at 500 volts dc at a temperature of 68 °F (20 °C) and at a relative humidity of approximately 50 percent. The measurement shall be made after the meter stabilizes, unless the requirement is met sooner. Arresters shall be removed for these tests.

(4) *Self-protection*. (i) All components shall be capable of being continuously energized at rated voltage without injury. Design precautions must be taken to prevent damage to other equipment components when a particular component fails.

(ii) Printed circuit boards or similar equipment employing electronic components should be self-protecting against external grounds applied to the connector terminals. Board components and coatings applied to finished products shall be of such material or so treated that they will not support combustion.

(iii) Every precaution shall be taken to protect electrostatically sensitive components from damage during handling. This shall include written instructions and recommendations.

(k) *Miscellaneous*—(1) *Interconnect wire*. All interconnect wire shall be of soft annealed tinned copper wire meeting the requirements of ASTM Specification B33–91 and of suitable cross-section to provide safe current carrying capacity and mechanical strength. The insulation of installed wire, connected to its equipment and frames, shall be capable of withstanding the same insulation resistance and dielectric strength requirements as given in paragraphs (j)(2) and (j)(3) of this section at a temperature of 120 °F (49 °C), and a relative humidity of 90 percent.

(2) *Wire wrapped terminals*. These terminals are preferred and where used shall be of a material suitable for wire wrapping. The connections to them shall be made with a wire wrapping tool with the following minimum number of successive non-overlapping turns of bare tinned copper wire in contact with each terminal:

(i) 6 turns of 30 gauge;

(ii) 6 turns of 26 gauge;

(iii) 6 turns of 24 gauge; or

(iv) 5 turns of 22 gauge.

(3) *Protection against corrosion*. All metal parts of equipment frames, distributing frames, cable supporting framework and other exposed metal parts shall be constructed of corrosion resistant materials or materials plated or painted to render them adequately corrosion resistant.

(4) *Screws and bolts*. Screw threads for all threaded securing devices shall be of American National Standard form in accordance with Federal Standard H28, unless exceptions are granted to the manufacturer of the switching equipment. All bolts, nuts, screws, and washers shall be of nickel-copper alloy, steel, brass or bronze.

(5) *Environmental requirements*. (i) The bidder shall specify the environmental conditions necessary for safe storage and satisfactory operation of the equipment being bid. If requested, the bidder shall assist the owner in planning how to provide the necessary environment for the equipment.

(ii) To the extent practicable, the following temperature range objectives shall be met:

(A) For equipment mounted in central office and subscriber buildings, the carrier equipment shall operate satisfactory within an ambient temperature range of 32 °F to 120 °F (0 °C to 49 °C) and at 80 percent relative humidity between 50 °F and 100 °F (10 °C and 38 °C); and

(B) Equipment mounted outdoors in normal operation (with cabinet doors closed) shall operate satisfactorily within an ambient temperature range (external to cabinet) of −40 °F to 140 °F (−40 °C to 60 °C) and at 95 percent relative humidity between 50 °F to 100 °F (10 °C to 38 °C). As an alternative to the (60 °C) requirement, a maximum ambient temperature of 120 °F (49 °C) with equipment (cabinet) exposed to direct sunlight may be substituted.

(6) *Stenciling.* Equipment units and terminal jacks shall be adequately designated and numbered. They shall be stenciled so that identification of equipment units and leads for testing or traffic analysis can be made without unnecessary reference to prints or descriptive literature.

(7) *Quantity of equipment bays.* Consistent with system arrangements and ease of maintenance, space shall be provided on the floor plan for an orderly layout of future equipment bays. Readily accessible terminals will be provided for connection to interbay and frame cables to future bays. All cables, interbay and intrabay (excluding power), if technically feasible, shall be terminated at both ends by connectors.

(8) *Radio and television interference.* Measures shall be employed by the bidders to limit the radiation of radio frequencies generated by the equipment so as not to interfere with radio, television receivers, or other sensitive equipment.

(9) *Housing.* (i) When housed in a building supplied by the owner, a complete floor plan including ceiling height, floor loading, power outlets, cable entrances, equipment entry and travel, type of construction, and other pertinent information shall be supplied.

(ii) In order to limit corrosion, all metal parts of the housing and mounting frames shall be constructed of suitable corrosion resistant materials or materials protectively coated to render them adequately resistant to corrosion under the climatic and atmospheric conditions existing in the area in which the housing is to be installed.

(10) *Distributing frame.* (i) The line concentrator terminal equipment located at the central office shall be protected by the central office main distribution frame. The bidder may supply additional protection capability as appropriate. All protection devices (new or existing) shall be arranged to operate in a coordinated manner to protect equipment, limit surge currents, and protect personnel.

(ii) The distributing frame shall provide terminals for terminating all incoming cable pairs. Arresters shall be provided for all incoming cable pairs, or for a smaller number of pairs if specified.

(iii) The current carrying capacity of each arrester and its associated mounting shall coordinate with a #22 gauge copper conductor without causing a self-sustaining fire or permanently damaging other arrester positions. Where all cable pairs entering the housing are #24 gauge or finer, the arresters and mountings need only coordinate with #24 gauge cable conductors.

(iv) Remote terminal protectors may be mounted and arranged so that outside cable pairs may be terminated on the left or bottom side of protectors (when facing the vertical side of the MDF) or on the back surface of the protectors. Means for easy identification of pairs shall be provided.

(v) Protectors shall have a "dead front" (either insulated or grounded) where live metal parts are not readily accessible.

(vi) Protectors shall be provided with an accessible terminal of each incoming conductor which is suitable for the attachment of a temporary test lead. They shall also be constructed so that auxiliary test fixtures may be applied to open and test the subscriber's circuit in either direction. Terminals shall be suitable for wire wrapped connections or connectorized.

(vii) If specified, each protector group shall be furnished with a factory assembled tip cable for splicing to the outside cable; the tip cable shall be 20 feet (6.1 m) in length, unless otherwise specified. Tip cable used shall be RUS accepted.

(viii) Protector makes and types used shall be RUS accepted.

(l) *Power equipment*—(1) *General.* When specified, batteries and charging equipment shall be supplied for the remote terminal of the line concentrator.

(2) *Operating voltage.* (i) The nominal operating voltage of the central office and remote terminal shall be 48 volts dc, provided by a battery with the positive side tied to system ground.

(ii) Where equipment is dc powered, it must operate satisfactorily over a range of 50 volts ±6 volts dc.

(iii) Where equipment is ac powered, it must operate satisfactorily over a range of 120±10 volts or 220±10 volts ac.

(3) *Batteries.* (i) Unless otherwise specified by the owner, sealed batteries shall be supplied for the remote line concentrator terminal.

(ii) The batteries shall have an ampere hour load capacity of no less than 8 busy hours. When an emergency ac supply source is available, the battery reserve may be reduced to 3 busy hours.

(iii) The batteries shall be sealed when they are mounted in the cabinet with the concentrator equipment.

(iv) When specified by the owner, battery heaters shall be supplied in a bidder-furnished housing.

(4) *Charging equipment.* (i) One charger capable of carrying the full dc power load of the remote terminal shall be supplied unless otherwise specified by the owner.

(ii) Charging shall be on a full float basis. The rectifiers shall be of the full wave, self-regulating, constant voltage, solid-state type and shall be capable of being turned on and off manually.

(iii) When charging batteries, the voltage at the battery terminals shall be adjustable and shall be set at the value recommended for the particular battery being charged, provided it is not above the maximum operating voltage of the central office switching equipment. The voltage shall not vary more than ±0.02 volt dc per cell between 10% load and 100% load. Between 3% and 10% load, the output voltage shall not vary more than ±0.04 volt dc per cell. Beyond full load current the output voltage shall drop sharply. The above output voltage shall be maintained with input line voltage variations of plus or minus 10 percent. Provision shall be made to manually change the output voltage of the rectifier to 2.25 volts per cell to provide an equalization charge on the battery.

(iv) The charger noise, when measured with a suitable noise measuring set and under the rated battery capacitance and load conditions, shall not exceed 22 dBrnC. See Figure 7 of this section, Charger Noise Test as follows:

Figure 7

Charger Noise Test

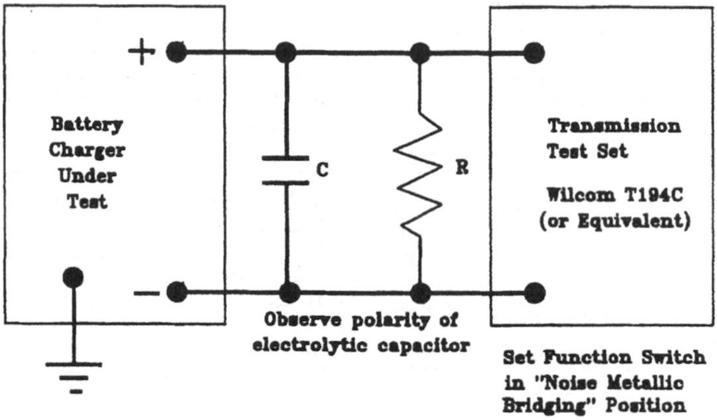

Note (1) The manufacturer may elect to eliminate the capacitor C from the measurement.

Capacitance C in μF = 30,000 μF per ampere-hour per cell. For example, 25 cells at 100 ampere-hour would be equivalent to a capacitance of:

(30,000 x 100)/25 = 120,000 μF

(2) The value of the resistive load R is determined by the nominal battery voltage in volts divided by the full load rating in amperes. For example, for a 48 volt battery and a full load current of 24 amperes, the load resistance R is 48/24 = 2 ohms of appropriate power handling capacity.

(v) The charging equipment shall be provided with a means for indicating a failure of charging current whether due to ac power failure, an internal failure in the charger, or to other circumstances which might cause the output voltage of the charger to drop below the battery voltage. Where a supplementary constant current charger is used, an alarm shall be provided to indicate a failure of the charger.

(vi) Audible noise developed by the charging equipment shall be kept to a minimum. Acoustic noise resulting from operation of the rectifier shall be expressed in terms of dB indicated on a sound level meter conforming to American National Standards Institute S1.4, and shall not exceed 65 dB (A-weighting) measured at any point 5 feet (1.5m) from any vertical surface of the rectifier.

(vii) The charging equipment shall be designed so that neither the charger nor the central office equipment is subject to damage in case the battery circuit is opened for any value of load within the normal limits.

(5) *Power panel.* (i) Battery and charger control switches, dc voltmeters, dc ammeters, fuses and circuit breakers, supervisory and timer circuits shall be provided as required. Portable or panel mounted frequency meters or voltmeters shall be provided as specified by the owner.

(ii) Power panels, cabinets and shelves, and associated wiring shall be designed initially to handle the line concentrator terminal when it reaches its ultimate capacity as specified by the owner.

(iii) The power panel shall be of the "dead front" type.

(6) *Ringing equipment.* The ringing system shall provide sufficient ringing on a bridged basis over the voltage and temperature limits of this section and over subscriber drops within the limits stated by the bidder. The ringing system shall be without operational problems such as bell tapping during dialing. The bidder shall state the minimum number (not less than two) of main station ringers that can be used for each ringing option available.

(7) *Interrupter equipment.* The interrupter may be an integral part of the system or may be part of the associated central office equipment connected to the line concentrator central office terminal.

(8) *Special systems.* Manufacturers of LC systems that operate by extending ringing current from the central office shall state their required input ringing (voltage and frequency) and the limitations on the connected subscriber loop.

(m) *Fusing requirements*—(1) *General.* (i) The equipment shall be completely wired and equipped with fuses, trouble signals, and all associated equipment for the wire capacity of the frames or cabinets provided.

(ii) Design precautions shall be taken to prevent the possibility of equipment damage arising from the insertion of an electronic package into the wrong connector or the removal of a package from any connector or improper insertion of the correct card in its connector.

(2) *Fuses.* Fuses and circuit breakers shall be of an alarm and indicator type, except where the fuse or breaker location is indicated on the alarm printout. Their rating shall be designated by numerals or color codes on the fuse or the panel.

(n) *Trouble location and test*—(1) *Equipment.* (i) Trouble indications in the system may be displayed in the form of lights on the equipment units or printed circuit boards.

(ii) When required, a jack or other connector shall be provided to connect a fault or trouble recorder (printer or display).

(2) *Maintenance system.* (i) The maintenance system shall monitor and maintain the system operation without interruption of call processing except for major failures.

(ii) The maintenance system shall be arranged to provide the ability to determine trouble to an individual card, functional group of cards, or other equipment unit.

(o) *Spare parts.* Lists of spare parts and maintenance tools as recommended by the bidder shall be provided. The cost of such tools and spare parts shall be indicated and shall not be included in the base price.

(p) *Drawings and printed material.* (1) The bidder shall supply instructional material for each line concentrator system involved at the time of delivery of the equipment. It is not the intent of this section to require system documentation necessary for the repair of individual circuit boards.

(2) Three complete sets of legible drawings shall be provided for each central office to be accessed. Each set shall include all of the following:

(i) Drawings of major equipment items such as frames, with the location of major component items of equipment shown therein;

(ii) Wiring diagrams indicating the specific method of wiring used on each item of equipment and interconnection wiring between items of equipment;

(iii) Maintenance drawings covering each equipment item that contains replaceable parts, appropriately identifying each part by name and part number; and

(iv) Job drawings including all drawings that are individual to the particular line concentrator involved such as mainframe, power equipment, etc.

(3) The following information shall also be furnished:

(i) A complete index of required drawings;

(ii) An explanation of electrical principles of operation of overall concentrator system;

(iii) A list of tests which can be made with each piece of test equipment furnished and an explanation of the method of making each test;

(iv) A sample of each form recommended for use in keeping records;

(v) The criteria for analyzing results of tests and determining appropriate corrective action;

(vi) A set of general notes on methods of isolating equipment faults to specific printed circuit cards in the equipment;

(vii) A list of typical troubles which might be encountered, together with general indications as to probable location of each trouble; and

(viii) All special line concentrator system grounding requirements.

(4) When installation is to be done by the bidder a complete set of drawings shall be provided by the owner, such as floor plans, lighting, grounding and ac power access.

(q) *Installation and acceptance*—(1) *General.* Paragraphs (q)(2)(i) through (q)(3)(xxi) of this section covers the general requirements for the installation of line concentrator equipment by the bidder, and outlines the general conditions to be met by the owner in connection with such installation work. The responsibilities apply in both the central office installation and remote terminal installations, unless otherwise noted.

(2) *Responsibilities of owner.* The owner shall:

(i) Allow the bidder and its employees free access to the premises and facilities at all hours during the progress of the installation;

(ii) Provide access to the remote site and any other site for development work needed during the installation;

(iii) Take such action as necessary to ensure that the premises are dry and free from dust and in such condition as not to be hazardous to the installation personnel or the material to be installed (not required when remote terminal is not installed in a building);

(iv) Provide heat or air conditioning when required and general illumination in rooms in which work is to be performed or materials stored;

(v) Provide suitable openings in buildings to allow material to be placed in position (not required when a remote terminal is not installed in a building);

(vi) Provide the necessary conduit and commercial and dc-ac inverter output power to the locations shown on the approved floor plan drawings;

(vii) Provide 110 volts a.c., 60 Hz commercial power equipped with a secondary arrester and a reasonable number of outlets for test, maintenance and installation equipment;

(viii) Provide suitable openings or channels and ducts for cables and conductors from floor to floor and from room to room;

(ix) Provide suitable ground leads, as designated by the bidder (not required when remote terminal is not installed in a building);

(x) Provide the necessary wiring, central office ground and commercial power service, with a secondary arrester, to the location of an exterior remote terminal installation based on the voltage and load requirements furnished voltage and load requirements furnished by the bidder;

(xi) Test at the owners expense all lines and trunks for continuity, leakage and loop resistance and ensure that all lines and trunks are suitable for operation with the central office and remote terminal equipment specified;

(xii) Make alterations and repairs to buildings necessary for proper installation of material, except to repair damage for which the bidder or its employees are responsible;

(xiii) Connect outside cable pairs on the distributing frame (those connected to protectors);

(xiv) Furnish all line, class of service assignment, and party line assignment information to permit bidder to program the data base memory within a reasonable time prior to final testing;

(xv) Release for the bidder's use, as soon as possible, such portions of the existing plant as are necessary for the proper completion of such tests as require coordination with existing facilities including facilities for T1 span lines with properly installed repeaters between the central office and the remote terminal installations;

(xvi) Make prompt inspections as it deems necessary when notified by the bidder that the equipment, or any part thereof, is ready for acceptance;

(xvii) Provide adequate fire protection apparatus at the remote terminal, including one or more fire extinguishers or fire extinguishing systems of the gaseous type, that has low toxicity and effect on equipment;

(xviii) Provide necessary access ports for cable, if underfloor cabling is selected;

(xix) Install equipment and accessory plant devices mounted external to the central office building and external to the repeater and other outside housings including filters, repeater housings, splicing of repeater cable stubs, externally mounted protective devices and other such accessory devices in accordance with written instructions provided by the bidder; and

(xx) Make all cross connections (at the MDF or Intermediate Distribution Frame IDF) between the physical trunk or carrier equipment and the central office equipment unless otherwise specified in appendix A of this section.

(3) *Responsibilities of bidder.* The bidder shall:

(i) Allow the owner and its representatives access to all parts of the building at all times;

(ii) Obtain the owner's permission before proceeding with any work necessitating cutting into or through any part of the building structure such as girders, beams, concrete or tile floors, partitions or ceilings (does not apply to the installation of lag screws, expansion bolts, and similar devices used for fastening equipment to floors, columns, walls, and ceilings);

(iii) Be responsible for and repair all damage to the building due to carelessness of the bidder's workforce, exercise reasonable care to avoid any damage to the owner's switching equipment or other property, and report to the owner any damage to the building which may exist or may occur during its occupancy of the building;

(iv) Consult with the owner before cutting into or through any part of the building structure in all cases where the fireproofing or moisture proofing may be impaired;

(v) Take necessary steps to ensure that all fire fighting apparatus is accessible at all times and all flammable materials are kept in suitable places outside the building;

(vi) Not use gasoline, benzene, alcohol, naphtha, carbon tetrachloride or turpentine for cleaning any part of the equipment;

(vii) Be responsible for delivering the CO and remote terminal equipment to the sites where they will be needed;

(viii) Install the equipment in accordance with the specifications for the line concentrator;

(ix) Have all leads brought out to terminal blocks on the MDF (or IDF if stated in appendix A of this section) and have all terminal blocks identified and permanently labeled;

(x) Use separate shielded type leads grounded at one end only unless otherwise specified by the owner or bidder or tip cables meeting RUS cable crosstalk requirements for carrier frequencies inside the central office;

(xi) Group the cables to separate carrier frequency, voice frequency, signaling, and power leads;

(xii) Make the necessary power and ground connections (location as shown in appendix A of this section) to the purchaser's power terminals and ground bus unless otherwise stated in appendix A of this section (ground wire shall be 6 AWG unless otherwise stated);

(xiii) Place the battery in service in compliance with the recommendations of the battery manufacturer;

(xiv) Make final charger adjustments using the manufacturer's recommended procedure;

(xv) Run all jumpers, except line and trunk jumpers (those connected to protectors) unless otherwise specified in appendix A of this section;

(xvi) Establish and update all data base memories with subscriber information as supplied by the owner until an agreed turnover time;

(xvii) Give the owner notice of completion of the installation at least one week prior to completion;

(xviii) Permit the owner or its representative to conduct tests and inspections after installation has been completed in order that the owner may be assured the requirements for installation are met;

(xix) Allow access, before turnover, by the owner or its representative, upon request, to the test equipment which is to be turned over as a part of the delivered equipment, to permit the checking of the circuit features which are being tested and to permit the checking of the amount of connected equipment to which the test circuits have access;

(xx) Notify the owner promptly of the completion of work of the central office terminals, remote terminals or such portions thereof as are ready for inspection; and

(xxi) Correct promptly all defects for which the bidder is responsible.

(4) *Information to be furnished by bidder.* The bidder shall accompany its bid with the following information:

(i) Two copies of the equipment list and the traffic calculations from which the quantities in the equipment list are determined;

(ii) Two copies of the traffic tables from which the quantities are determined, if other than the Erlang B traffic tables;

(iii) A block diagram of the line concentrator and associated maintenance equipment will be provided;

(iv) A prescribed method and criteria for acceptance of the completed line concentrator which will be subject to review;

(v) This special grounding requirements including the recommended configuration, suggested equipment and installation methods to be used to accomplish them;

(vi) The special handling and equipment requirements to avoid damage resulting from the discharge of static electricity (see paragraph (j)(4)(iii) of this section) or mechanical damage during transit installation and testing;

(vii) The location of technical assistance service, its availability and conditions for owner use and charges for the service by the bidder; and

(viii) The identification of the subscriber loop limits available beyond the line concentrator.

(5) *Installation requirements.* (i) All work shall be done in a neat, workmanlike manner. Equipment frames or cabinets shall be correctly located, carefully aligned, anchored, and firmly braced. Cables shall be carefully laid with sufficient radius of curvature and protected at corners and bends to ensure against damage from handling or vibration. Exterior cabinet installations for remote terminals shall be made in a permanent, eye-pleasing manner.

(ii) All multiple and associated wiring shall be continuous, free from crosses, reverses, and grounds and shall be correctly wired at all points.

(iii) An inspection shall be made by the owner or its representatives prior to performing operational and performance tests on the equipment, but after all installing operations which might disturb apparatus adjustments have been completed. The inspection shall be of such character and extent as to disclose with reasonable certainty any unsatisfactory condition of apparatus or equipment. During these inspections, or inspections for apparatus adjustments, or wire connections, or in testing of equipment, a sufficiently detailed examination shall be made throughout the portion of the equipment within which such condition is observed, or is likely to occur, to disclose the full extent of its existence, where any of the following conditions are observed:

(A) Apparatus or equipment units failing to compare in quantity and type to that specified for the installation;

(B) Apparatus or equipment units damaged or incomplete;

(C) Apparatus or equipment affected by rust, corrosion or marred finish; and

(D) Other adverse conditions resulting from failure to meet generally accepted standards of good workmanship.

(6) *Operational tests.* (i) Operational tests shall be performed on all circuits and circuit components to ensure their proper functioning in accordance with appropriate explanation of the operation of the circuit.

(ii) All equipment shall be tested to ensure proper operation with all components connected in all possible combinations and each line shall be tested for proper ring, ring trip and supervision.

(iii) All fuses shall be verified for continuity and correct rating. Alarm indication shall be demonstrated for each equipped fuse position. An already failed fuse compatible with the fuse position may be used.

(iv) Each alarm or signal circuit shall be checked for correct operation.

(v) A sufficient quantity of locally originating and incoming calls shall be made to demonstrate the function of the line concentrator including all equipped transmission paths. When intra-link calling is supplied, all intra-link transmission paths shall be demonstrated.

(7) *Acceptance tests and data required.* (i) Data shall be supplied to the owner by the bidder in writing as a part of the final documents in closing out the contract as follows:

(A) A detailed cross connect drawing of alarm to power board, central office battery to physical trunks or carrier system, wiring options used in terminals, channels, filters, repeaters, etc., marked in the owner's copy of the equipment manual or supplied separately;

(B) The measured central office supply voltages applied to the equipment terminals or repeaters at the time the jack and test point readings are made and ac supply voltages where equipment is powered from commercial ac sources;

(C) A list of all instruments, including accessories, by manufacturer and type number, used to obtain the data; and

(D) The measurements at all jack or test points recommended by the manufacturer, including carrier frequency level measurements at all carrier terminals and repeaters where utilized.

(ii) Data in the form of a checklist or other notations shall be supplied showing the results of the operational tests.

(iii) The bidder shall furnish to the owner a record of the battery cell or multicell unit voltages measured at the completion of the installation of the switching system before it is placed in commercial service. This is not required at a site where the owner furnishes dc power.

(8) *Joint inspection requirements.* (i) The bidder shall notify the owner in writing at least one week before the date the complete system will be ready for inspection and tests. A joint inspection shall be made by the bidder and the owner (or owner's engineer) to determine that the equipment installation is acceptable. The inspection shall include physical inspection, a review of acceptance test data, operational tests, and sample measurements.

(A) The owner shall review the acceptance test data and compare it to the requirements of this section.

(B) Sample measurements shall be made on all systems installed under this contract. Test methods should follow procedures described in paragraph (g)(5) of this section.

(C) A check shall be made of measured test point and jack readings for compliance with the manufacturer's specifications. This applies also to channels, terminals, carrier frequency repefault locating circuits.

(ii) In the event that the measured data or operational tests show that equipment fails to meet the requirements quirements of this section, the deficiencies are to be resolved as set forth in Article II of the 397 Special Equipment Contract. (Copies are available from RUS, room 0174, U.S. Department of Agriculture, Washington, DC 20250–1500.) The reports of the bidder and the owner shall be detailed as to deficiencies, causes, corrective action necessary, corrective action to be taken, completion time, etc.

(The information and recordkeeping requirements of this section have been approved by the Office of Management and Budget under the control number 0572–0059)

APPENDIX A TO § 1755.397—SPECIFICATION FOR LINE CONCENTRATOR DETAILED EQUIPMENT REQUIREMENTS

(INFORMATION TO BE SUPPLIED BY OWNER)

Telephone Company (Owner)

Name: ________________________

Location: ________________________

Number of LC's Required: ________

Line Concentrator Locations:

Location	No. of Lines	Central Office
....	....	....
....	....	....
....	....	....
....	....	....

1. General

1.1 Notwithstanding the bidder's equipment lists, the equipment and materials furnished by the bidder must meet the requirements of paragraphs (a) through (p) of this section, and this appendix A.

1.2 Paragraph (a) through (p) of this section cover the minimum general requirements for line concentrator equipment.

1.3 Paragraph (q) of this section covers the requirements for installation, inspection and testing when such service is included as part of the contract.

1.4 This appendix A covers the technical data for application engineering and detailed equipment requirements insofar as they can be established by the owner. This appendix A shall be filled in by the owner.

1.5 Appendix B of this section covers detailed information on the line concentrator equipment, information on system reliability and traffic capacity as proposed by the bidder. Appendix B of this section is to be filled in by the bidder and must be presented with the bid.

Office Name

(By Location) ____________

LC Designation ____________

2. Number of Subscriber Lines

	Equipped	Wired only
Single-Party. Pay Station (Type:_______). Other (Describe:_______). Total	....	....

3. Loop Resistance

3.1 Number of non-pay station lines having a loop resistance, including the telephone set as follows:

3.1.1 For physical trunks between the remote and the office units, the loop resistance is to include the resistance of the trunk.

	No. of lines
1200–1900 ohms	....
1901–3200 ohms	....
3201–4500 ohms	....

3.1.2 Number of pay station lines having a loop resistance, excluding the telephone set, greater than:

	No. of lines
1200 ohms (Prepay)	....
1000 ohms (Semi-Postpay)	....

When physical trunks are used, these resistances include that of the facility between the CO and the remote.

3.1.3 Range extension equipment, if required, is to be provided:

_______ By Bidder

_______ By Owner

(Quantity and Type) ____________

4. Traffic Data

4.1 Average combined originating and terminating hundred call seconds (CCS) per line in the busy hour:

_____ CCS/Line. (Assume originating & terminating equal.)

4.2 Percent Intra-Calling _______

4.3 Total Busy Hour Calls _______

5. TYPE or RINGING

5.1 Frequency No.	1.	2.	3.	4.
Frequency (Hz)	....	....	....	....
Max. No. of Phones/Freq.	....	....	....	....

5.2 Minimum ringing generator capacity to be supplied shall be sufficient to serve _______ lines (each frequency).

6. Central Office Equipment Interface

6.1 COE will be:

6.1.1 COE Manufacturer

____

Type ____________

Year ____________

Generic ____________

6.1.2 _______ See digital central office specification for the switchboard at ____________.

6.2 Interface will be:

6.2.1 _______ Line Circuit(s)

6.2.2 _______ Direct Digital Interface

6.2.3 _______ Other (Describe)

____

6.3 Mounting rack for line concentrator furnished by:

_______ Bidder

_______ Owner

(Specify width and height of rack available)
(Width) (Height)

6.4 Equipment to be installed in existing building:

_______ Yes (Attach detailed plan)

_______ No

7. Transmission Facilities

7.1 Transmission facilities between the central office and remote terminals shall be:

7.1.1 Type:

_______ VF Carrier Derived Circuits

_______ Digital Span Line (DS1)

_______ Other

____

(Attach a layout of the transmission facilities between the central office and the remote terminals describing transmission and signaling parameters, routing and resistance where applicable.)

7.1.2 Utilizes physical plant

_______ Cable Pairs (Existing/New)

_______ Other

____

NOTE: Unless otherwise stated, physical plant will be supplied by the owner.

7.1.3 Terminal equipment for transmission facility to be supplied by:

_______ Owner

_______ Bidder

7.1.3.1 Carrier e/w voice terminations _______ Yes _______ No

Manufacturer and type _______________

Central office voice terminations Equipped _______, Wired Only _______

7.1.3.2 Digital span line (DS1) supplied by

_______ Owner

_______ Bidder

Manufacturer and Type _______________

7.1.3.3 Number of repeaters (per span line) _______

7.1.3.4 Diverse (alternate) span line routing required

_______ Yes (Describe in Item 11) _______ No

7.1.3.5 Span line terminations only _______ Yes _______ No

7.1.3.6 Span line power required (CO and Remote Terminals) _______ Yes _______ No

7.1.3.7 Physical facility between CO and remote Loop Resistance _______ ohms, Length _______ meters

8. Power Equipment Requirements

8.1 Central Office Terminal

8.1.1 Owner-furnished −48 volt dc power _______ Yes _______ No

8.1.2 Other (Describe)

____

8.1.3 Standby power is available _______ Yes _______ No

8.2 Remote Terminal

8.2.1 Owner-furnished −48 vdc power _______ Yes _______ No

8.2.2 Bidder-furnished power supply _______ Yes _______ No

8.2.3 AC power available at site:

_______ 110 vac, 60 Hz, single-phase

_______ Other (Describe in Item 11)

8.2.4 A battery reserve of _______ busy hours shall be provided for this line concentrator terminal when it reaches _______ lines at the traffic rates specified.

8.2.5 Batteries supplied shall be:

_______ Lead Calcium

_______ Stabilized Electrolyte

_______ Sealed Lead Acid

_______ Other (Describe in item 11)

8.2.6 Standby power is available _______ Yes _______ No

9. Remote Terminal

9.1 Mounting

9.1.1 _______ Outside Housing (To be furnished by bidder)

9.1.2 _______ Concrete Slab to be furnished by owner (Bidder to supply construction details after award.)

9.1.3 _______ Manhole, environmentally controlled (Describe in Item 11)

9.1.4 _______ Pedestal Mounting

9.1.5 _______ Pole Mounting (Owner-furnished installed pole)

9.1.6 _______ Prefab Building (Owner-furnished site)

9.2 Equipment is to be installed in an existing building.

_______ Yes _______ No

(Attach detailed plan.)

9.3 Other (Describe)

____

____

10. Alternates

11. Explanatory Notes

APPENDIX B TO § 1755.397—SPECIFICATION FOR LINE CONCENTRATORS DETAILED REQUIREMENTS; BIDDER SUPPLIED INFORMATION

Telephone Company (Owner)

Name: _______________________

Location: _______________________

Line Concentrator Equipment Locations

Central Office Terminal: _______________

Remote Terminal: _______________

1. General

1.1 The equipment and materials furnished by the bidder must meet the requirements of paragraphs (a) through (p) of this section.

1.2 Paragraph (a) through (p) of this section cover the minimum general requirements for line concentrator equipment.

1.3 Paragraph (q) of this section covers requirements for installation, inspection and testing when such service is included as part of the contract.

1.4 Appendix A of this section covers the technical data for application engineering and detailed equipment requirements insofar as they can be established by the owner. Appendix A of this section is to be filled in by the owner.

1.5 This appendix B covers detailed information on the line concentrator equipment, information as to system reliability and traffic capacity as proposed by the bidder. This appendix B shall be filled in by the bidder and must be presented with the bid.

2. Performance Objectives

2.1 Reliability (See paragraph (c) of this section)

______

______

______

______

2.2 Busy Hour Load Capacity and Traffic Delay (See Paragraph (g) of this section)

______

______

______

3. Equipment Quantities Dependent on System Design

3.1 Transmission Facilities between the Central Office and Remote Terminals

Type	Quantity equipped	Quantity wired only
....	....	....
....	....	....
....	....	....

4. Power Requirements

4.1 Central Office Terminal

Voltage ______

Current Drain (Amps) Normal ______, Peak ______

Fuse Qty ______, Size ______, Type ______

Heat Dissipation (BTU/Hr.) ______

______

4.2 Remote Terminal

AC or DC ______

Voltage ______

Current Drain (Amps) Normal ______, Peak ______

Fuse Qty ______, Size ______, Type ______

Heat Dissipation (BTU/Hr.) ______

______

Power required for heating or cooling equipment in remote bidder-furnished housing

______

______

5. Temperature and Humidity Limitations

5.1 Temperature

	Central office	Remote*
Maximum °F (°C)	....	....
Minimum °F (°C)	....	....

5.2 Relative Humidity

	Central office	Remote*
Maximum	....	....
Minimum	....	....

* Show conditions outside bidder-furnished housing.

6. Explanatory Notes

[60 FR 44729, Aug. 29, 1995, as amended at 69 FR 18803, Apr. 9, 2004]

§§ 1755.398–1755.399 [Reserved]

§ 1755.400 RUS standard for acceptance tests and measurements of telecommunications plant.

Sections 1755.400 through 1755.407 cover the requirements for acceptance tests and measurements on installed copper and fiber optic telecommunications plant and equipment.

[62 FR 23960, May 2, 1997]

§ 1755.401 Scope.

(a) Acceptance tests outlined in §§ 1755.400 through 1755.407 are applicable to plant constructed by contract or force account. This testing standard provides for the following:

(1) Specific types of tests or measurements for the different types of telecommunications plant and equipment;

(2) The method of measurement and types of measuring equipment;

(3) The expected results and tolerances permitted to meet the acceptable standards and objectives;

(4) Suggested formats for recording the results of the measurements and tests; and

(5) Some probable causes of nonconformance and methods for corrective action, where possible.

(b) Alternative methods of measurements that provide suitable alternative results shall be permitted with the concurrence of the Rural Utilities Service (RUS).

(c) For the purpose of this testing standard, a "measurement" shall be defined as an evaluation where quantitative data is obtained (e.g., resistance in ohms, structural return loss in decibels (dB), etc.) and a "test" shall be defined as an evaluation where no quantitative data is obtained (e.g., a check mark indicating conformance is usually the result of the test).

(d) The sequence of tests and measurements described in this standard have been prepared as a guide. Variations from the sequence may be necessary on an individual application basis.

(e) There is some overlap in the methods of testing shown; also, the extent of each phase of testing may vary on an individual basis. The borrower shall determine the overall plan of testing, the need and extent of testing, and the responsibility for each phase of testing.

[62 FR 23960, May 2, 1997]

§ 1755.402 Ground resistance measurements.

(a) The resistance of the central office (CO) and the remote switching terminal (RST) ground shall be measured before and after it has been bonded to the master ground bar (MGB) where it is connected to the building electric service ground.

(b) The ground resistance of electronic equipment such as span line repeaters, carrier terminal equipment, concentrators, etc. shall be measured.

(c) *Method of measurement.* The connection of test equipment for the ground resistance measurement shall be as shown in Figure 1. Refer to RUS Bulletin 1751F–802, "Electrical Protection Grounding Fundamentals," for a comprehensive discussion of ground resistance measurements.

(d) *Test equipment.* The test equipment for making this measurement is shown in Figure 1 as follows:

FIGURE 1

GROUND RESISTANCE MEASUREMENT①, ②

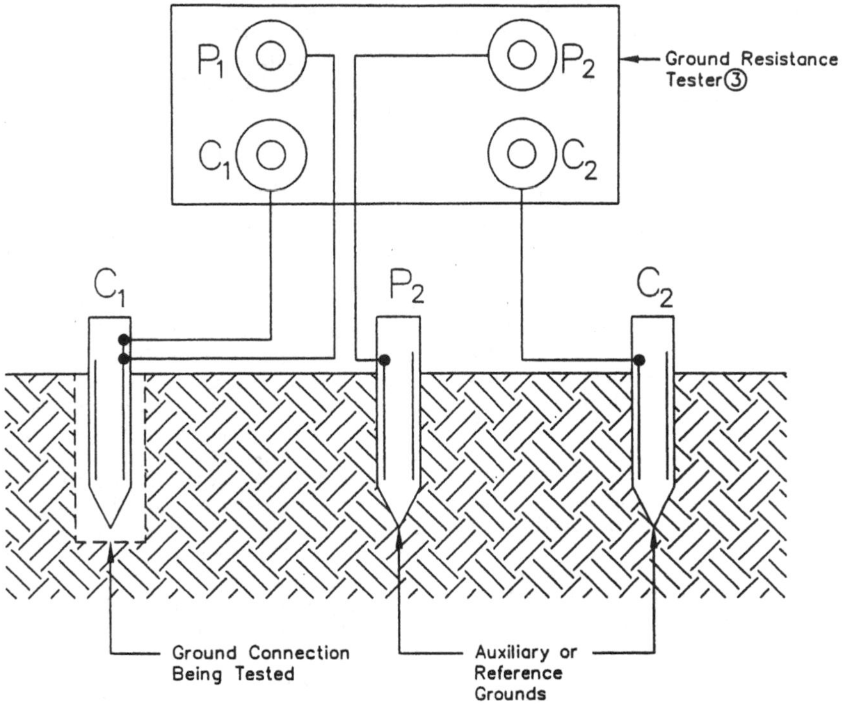

Notes:

① Measurement procedure for COs, RSTs, and electronic equipment housings approximately 10 ft by 10 ft (3 m by 3m) or smaller shall be as follows: The minimum distance between the CO ground (C_1) being tested and C_2 = 100 ft (30.5 m). Take several measurements moving P_2 from 50 ft to 75 ft (15.2 m to 23 m) away from CO ground C_2. Resistance should initially rise then level off and then start rising again. The value to record for CO ground resistance is the value where it levels off which usually should occur with P_2 at 62 % of the distance between the CO ground and C_2.

② Measurement procedure for COs, RSTs, and electronic equipment housings larger than 10 ft by 10 ft (3 m by 3 m) shall be in accordance with the test equipment manufacturer's instructions.

③ Dynatel Research–Vibroground, General Radio–Megger Bridge, Associate Research–Megohm Meter or equivalent.

(e) *Applicable results.* (1) For the CO and RST, the resistance after the bond has been made to the MGB electric service ground shall not exceed 5 ohms. Where the measured ground resistance exceeds 5 ohms, the borrower shall determine what additional grounding, if any, shall be provided.

(2) For electronic equipment, the ground resistance shall not exceed 25 ohms. Where the measured ground resistance exceeds 25 ohms, the borrower shall determine what additional grounding, if any, shall be provided.

(3) When ground resistance measurements exceed the ground resistance requirements of paragraphs (e)(1) and (e)(2) of this section, refer to RUS Bulletin 1751F–802, "Electrical Protection Grounding Fundamentals," for suggested methods of reducing the ground resistance.

(f) *Data record.* Results of the CO and RST ground resistance measurements shall be recorded. A suggested format similar to Format I, Outside Plant Acceptance Tests—Subscriber Loops, in § 1755.407 or a format specified in the applicable construction contract may be used. Results of the electronic equipment ground resistance measurements shall be recorded. A suggested format similar to Format II, Outside Plant Acceptance Tests—Trunk Circuits, in § 1755.407 or a format specified in the applicable construction contract may be used. Data showing approximate moisture content of the soil at the time of measurement, the temperature, the type of soil and a description of the test equipment used shall also be included.

(g) *Probable causes for nonconformance.* Refer to RUS Bulletin 1751F–802, "Electrical Protection Grounding Fundamentals," and Telecommunications Engineering and Construction Manual (TE&CM) Section 810, "Electrical Protection of Electronic Analog and Digital Central Office Equipment," for possible causes of nonconformance and suggested methods for corrective action.

[62 FR 23960, May 2, 1997]

§ 1755.403 Copper cable telecommunications plant measurements.

(a) *Shield or shield/armor continuity.* (1) Tests and measurements shall be made to ensure that cable shields or shield/armors are electrically continuous. There are two areas of concern. The first is shield or shield/armor bonding within a pedestal or splice and the second is shield or shield/armor continuity between pedestals or splices.

(2) Measurement techniques outlined here for verification of shield or shield/armor continuity are applicable to buried cable plant. Measurements of shield continuity between splices in aerial cable plant should be made prior to completion of splicing. Conclusive results cannot be obtained on aerial plant after all bonds have been completed to the supporting strand, multigrounded neutral, etc.

(3) *Method of measurement.* (i) The shield or shield/armor resistance measurements shall be made between pedestals or splices using either a Wheatstone bridge or a volt-ohm meter. For loaded plant, measurements shall be made on cable lengths that do not exceed one load section. For nonloaded plant, measurements shall be made on cable lengths that do not exceed 5,000 feet (ft) (1,524 meters (m)). All bonding wires shall be removed from the bonding lugs at the far end of the cable section to be measured. The step-by-step measurement procedure shall be as shown in Figure 2.

(ii) Cable shield or shield/armor continuity within pedestals or splices shall be measured with a cable shield splice continuity test set. The step-by-step measurement procedure outlined in the manufacturer's operating instructions for the specific test equipment being used shall be followed.

(4) *Test equipment.* (i) The test equipment for measuring cable shield or shield/armor resistance between pedestals or splices is shown in Figure 2 as follows:

FIGURE 2

SHIELD OR SHIELD/ARMOR RESISTANCE MEASUREMENT

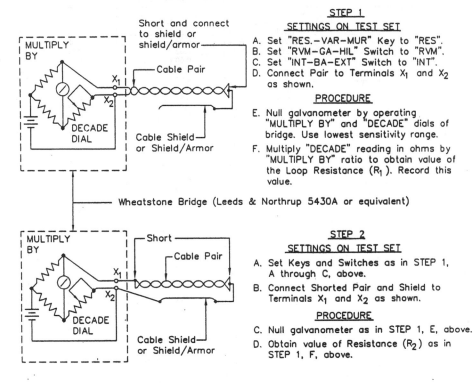

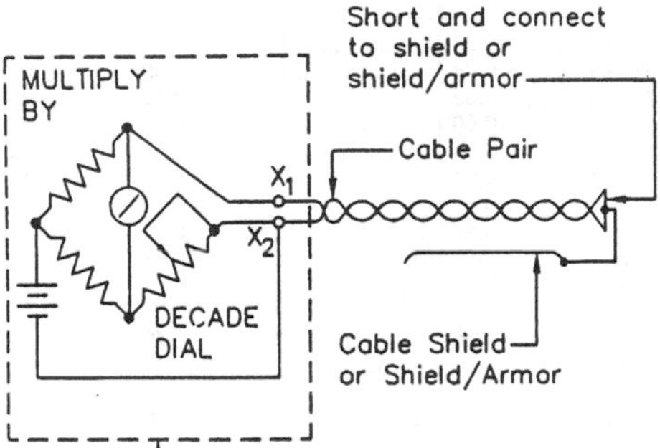

STEP 1

SETTINGS ON TEST SET

A. Set "RES.-VAR-MUR" Key to "RES".
B. Set "RVM-GA-HIL" Switch to "RVM".
C. Set "INT-BA-EXT" Switch to "INT".
D. Connect Pair to Terminals X_1 and X_2 as shown.

PROCEDURE

E. Null galvanometer by operating "MULTIPLY BY" and "DECADE" dials of bridge. Use lowest sensitivity range.
F. Multiply "DECADE" reading in ohms by "MULTIPLY BY" ratio to obtain value of the Loop Resistance (R_1). Record this value.

Wheatstone Bridge (Leeds & Northrup 5430A or equivalent)

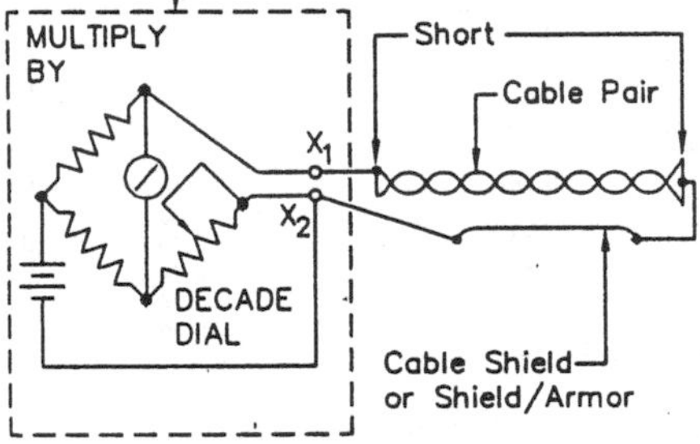

STEP 2

SETTINGS ON TEST SET

A. Set Keys and Switches as in STEP 1, A through C, above.
B. Connect Shorted Pair and Shield to Terminals X_1 and X_2 as shown.

PROCEDURE

C. Null galvanometer as in STEP 1, E, above.
D. Obtain value of Resistance (R_2) as in STEP 1, F, above.

STEP 3

COMPUTE THE SHIELD OR SHIELD/ARMOR RESISTANCE (R_S)

$$R_S = R_2 - \frac{R_1}{4}$$

(ii) A cable shield splice continuity tester shall be used to measure shield or shield/armor continuity within pedestals or splices.

(5) *Applicable results.* (i) The shield or shield/armor resistance per 1000 ft and per kilometer (km) for cable diameters and types of shielding materials are given in Table 1 (English Units) and Table 2 (Metric Units), respectively as follows:

TABLE 1—SHIELD RESISTANCE @ 68 °F (20 °C) CABLE DIAMETERS VERSUS SHIELD TYPES

[English Units]

Outside diameter inches (in.)	Nominal resistance ohm/1000 ft.					
	A	B	C	D	E	F
0.40–0.49	0.77	1.54	1.65	1.96	2.30	5.51
0.50–0.59	0.64	1.28	1.37	1.63	1.91	4.58
0.60–0.69	0.51	1.03	1.10	1.31	1.53	3.67
0.70–0.79	0.44	0.88	0.94	....	1.31	3.14

TABLE 1—SHIELD RESISTANCE @ 68 °F (20 °C) CABLE DIAMETERS VERSUS SHIELD TYPES—Continued

[English Units]

Outside diameter inches (in.)	Nominal resistance ohm/1000 ft.					
	A	B	C	D	E	F
0.80–0.89	0.38	0.77	0.82	....	1.14	2.74
0.90–0.99	0.35	0.69	0.74	....	1.03	2.47
1.00–1.09	0.31	0.62	0.66	....	0.92	2.20
1.10–1.19	0.28	0.56	0.60	....	0.84	2.00
1.20–1.29	0.26	0.51	0.55	....	0.77	1.84
1.30–1.39	0.24	0.48	0.51	....	0.71	1.70
1.40–1.49	0.22	0.44	0.47	....	0.65	1.57
1.50–1.59	0.21	0.41	0.44	....	0.61	1.47
1.60–1.69	0.19	0.38	0.41	....	0.57	1.37
1.70–1.79	0.18	0.37	0.39	....	0.54	1.30
1.80–1.89	0.17	0.35	0.37	....	0.51	1.24
1.90–1.99	0.16	0.33	0.35	....	0.49	1.17
2.00–2.09	0.15	0.31	0.33	....	0.46	1.10
2.10–2.19	0.15	0.29	0.31	....	0.43	1.03
2.20–2.29	0.14	0.28	0.30	....	0.42	1.00
2.30–2.39	0.14	0.27	0.29	....	0.40	0.97
2.40–2.49	0.13	0.25	0.27	....	0.38	0.90
2.50–2.59	0.12	0.24	0.26	....	0.36	0.87
2.60–2.69	0.12	0.23	0.25	....	0.35	0.83
2.70–2.79	0.11	0.22	0.24	....	0.33	0.80
2.80–2.89	0.11	0.22	0.24	....	0.33	0.80
2.90–2.99	0.11	0.22	0.23	....	0.32	0.77
3.00–3.09	0.10	0.21	0.22	....	0.31	0.73
3.10–3.19	0.10	0.20	0.21	....	0.29	0.70
3.20–3.29	0.10	0.20	0.21	....	0.29	0.70
3.30–3.39	0.09	0.19	0.20	....	0.28	0.67
3.40–3.49	0.09	0.18	0.19	....	0.26	0.63
3.50–3.59	0.09	0.18	0.19	....	0.26	0.63
3.60–3.69	0.08	0.17	0.18	....	0.25	0.60
3.70–3.79	0.08	0.17	0.18	....	0.25	0.60
3.80–3.89	0.08	0.16	0.17	....	0.24	0.57
3.90–3.99	0.08	0.16	0.17	....	0.24	0.57
4.00–4.99	0.07	0.15	0.16	....	0.22	0.53

Where: Column A–10 mil Copper shield.
Column B—5 mil Copper shield.
Column C—8 mil Coated Aluminum and 8 mil Coated Aluminum/6 mil Coated Steel shields.
Column D—7 mil Alloy 194 shield.
Column E—6 mil Alloy 194 and 6 mil Copper Clad Stainless Steel shields.
Column F—5 mil Copper Clad Stainless Steel and 5 mil Copper Clad Alloy Steel shields.

TABLE 2—SHIELD RESISTANCE @ 68 °F (20 °C) CABLE DIAMETERS VERSUS SHIELD TYPES

[Metric Units]

Outside diameter millimeters (mm)	Nominal Resistance ohm/km					
	A	B	C	D	E	F
10.2—12.5	2.53	5.05	5.41	6.43	7.55	18.08
12.7—15.0	2.10	4.20	4.49	5.35	6.27	15.03
15.2—17.5	1.67	3.38	3.61	4.30	5.02	12.04
17.8—20.1	1.44	2.89	3.08	....	4.30	10.30
20.3—22.6	1.25	2.53	2.69	....	3.74	8.99
22.9—25.1	1.15	2.26	2.43	....	3.38	8.10
25.4—27.7	1.02	2.03	2.16	....	3.02	7.22
27.9—30.2	0.92	1.84	1.97	....	2.76	6.56
30.5—32.8	0.85	1.67	1.80	....	2.53	6.04
33.0—35.3	0.79	1.57	1.67	....	2.33	5.58
35.6—37.8	0.72	1.44	1.54	....	2.13	5.15
38.1—40.4	0.69	1.34	1.44	....	2.00	4.82
40.6—42.9	0.62	1.25	1.34	....	1.87	4.49
43.2—45.5	0.59	1.21	1.28	....	1.77	4.26
45.7—48.0	0.56	1.15	1.21	....	1.67	4.07
48.3—50.5	0.52	1.08	1.15	....	1.61	3.84
50.8—53.1	0.49	1.02	1.08	....	1.51	3.61
53.3—55.6	0.49	0.95	1.02	....	1.41	3.38
55.9—58.2	0.46	0.92	0.98	....	1.38	3.28
58.4—60.7	0.46	0.89	0.95	....	1.31	3.18
61.0—63.2	0.43	0.82	0.89	....	1.25	2.95

TABLE 2—SHIELD RESISTANCE @ 68 °F (20 °C) CABLE DIAMETERS VERSUS SHIELD TYPES—Continued

[Metric Units]

Outside diameter millimeters (mm)	Nominal Resistance ohm/km					
	A	B	C	D	E	F
63.5—65.8	0.39	0.79	0.85	....	1.18	2.85
66.0—68.3	0.39	0.75	0.82	....	1.15	2.72
68.6—70.9	0.36	0.72	0.79	....	1.08	2.62
71.1—73.4	0.36	0.72	0.79	....	1.08	2.62
73.7—75.9	0.36	0.72	0.75	....	1.05	2.53
76.2—78.5	0.33	0.69	0.72	....	1.02	2.39
78.7—81.0	0.33	0.66	0.69	....	0.95	2.30
81.3—83.6	0.33	0.66	0.69	....	0.95	2.30
83.6—86.1	0.29	0.62	0.66	....	0.92	2.20
86.4—88.6	0.29	0.59	0.62	....	0.85	2.07
88.9—91.2	0.29	0.59	0.62	....	0.85	2.07
91.4—93.7	0.26	0.56	0.59	....	0.82	1.97
94.0—96.3	0.26	0.56	0.59	....	0.82	1.97
96.5—98.8	0.26	0.52	0.56	....	0.79	1.87
99.1—101.3	0.26	0.52	0.56	....	0.79	1.87
101.6—103.9	0.23	0.49	0.52	....	0.72	1.74

Where: Column A—10 mil Copper shield.
Column B—5 mil Copper shield.
Column C—8 mil Coated Aluminum and 8 mil Coated Aluminum/6 mil Coated Steel shields.
Column D—7 mil Alloy 194 shield.
Column E—6 mil Alloy 194 and 6 mil Copper Clad Stainless Steel shields.
Column F—5 mil Copper Clad Stainless Steel and 5 mil Copper Clad Alloy Steel shields.

(ii) All values of shield and shield/armor resistance provided in Tables 1 and 2 in (a)(5)(i) of this section are considered approximations. If the measured value corrected to 68 °F (20 °C) is within #30 percent (%) of the value shown in Table 1 or 2, the shield and shield/armor shall be assumed to be continuous.

(iii) To correct the measured shield resistance to the reference temperature of 68 °F (20 °C) use the following formulae:

$R_{68} = Rt/[1 + A(t - 68)]$ for English Units

$R_{20} = Rt/[1 + A(t - 20)]$ for Metric Units

Where:

R_{68} = Shield resistance corrected to 68 °F in ohms.

R_{20} = Shield resistance corrected to 20 °C in ohms.

R_t = Shield resistance at measurement temperature in ohms.

A = Temperature coefficient of the shield tape.

t = Measurement temperature in °F or (°C).

(iv) The temperature coefficients (A) for the shield tapes to be used in the formulae referenced in paragraph (a)(5)(iii) of this section are as follows:

(A) 5 and 10 mil copper = 0.0021 for English units and 0.0039 for Metric units;

(B) 8 mil coated aluminum and 8 mil coated aluminum/6 mil coated steel = 0.0022 for English units and 0.0040 for Metric units;

(C) 5 mil copper clad stainless steel and 5 mil copper clad alloy steel = 0.0024 for English units and 0.0044 for Metric units;

(D) 6 mil copper clad stainless steel = 0.0019 for English units and 0.0035 for Metric units; and

(E) 6 and 7 mil alloy 194 = 0.0013 for English units and 0.0024 for Metric units.

(v) When utilizing shield continuity testers to measure shield and shield/armor continuity within pedestals or splices, refer to the manufacturer's published information covering the specific test equipment to be used and for anticipated results.

(6) *Data record.* Measurement data from shield continuity tests shall be recorded together with anticipated Table 1 or 2 values (see paragraph (a)(5)(i) of this section) in an appropriate format to permit comparison. The recorded data shall include specific location, cable size, cable type, type of shield or shield/armor, if known, etc.

(7) *Probable causes for nonconformance.* Among probable causes for nonconformance are broken or damaged

shields or shield/armors, bad bonding harnesses, poorly connected bonding clamps, loose bonding lugs, etc.

(b) *Conductor continuity.* After placement of all cable and wire plant has been completed and joined together in continuous lengths, tests shall be made to ascertain that all pairs are free from grounds, shorts, crosses, and opens, except for those pairs indicated as being defective by the cable manufacturer. The tests for grounds, shorts, crosses, and opens are not separate tests, but are inherent in other acceptance tests discussed in this section. The test for grounds, shorts, and crosses is inherent when conductor insulation resistance measurements are conducted per paragraph (c) of this section, while tests for opens are inherent when tests are conducted for loop resistance, insertion loss, noise, or return loss measurements, per paragraphs (d), (e), or (f) of this section. The borrower shall make certain that all defective pairs are corrected, except those noted as defective by the cable manufacturer in accordance with the marking provisions of the applicable cable and wire specifications. All defective pairs that are not corrected shall be reported in writing with details of the corrective measures attempted.

(c) *Dc insulation resistance (IR) measurement.* (1) IR measurements shall be made on completed lengths of insulated cable and wire plant.

(2) *Method of measurement.* (i) The IR measurement shall be made between each conductor and all other conductors, sheath, shield and/or shield/armor, and/or support wire electrically connected together and to the main distributing frame (MDF) ground. The measurement shall be made from the central office with the entire length of the cable under test and, where used with all protectors and load coils connected. For COs containing solid state arresters, the solid state arresters shall be removed before making the IR measurements. Field mounted voice frequency repeaters, where used, may be left connected for the IR test but all carrier frequency equipment, including carrier repeaters and terminals, shall be disconnected. Pairs used to feed power remote from the CO shall have the power disconnected and the tip and ring conductors shall be opened before making IR tests. All conductors shall be opened at the far end of the cable being measured.

(ii) IR tests are normally made from the MDF with all CO equipment disconnected at the MDF, but this test may be made on new cables at field locations before they are spliced to existing cables. The method of measurement shall be as shown in Figure 3 as follows:

FIGURE 3

DC INSULATION RESISTANCE MEASUREMENT

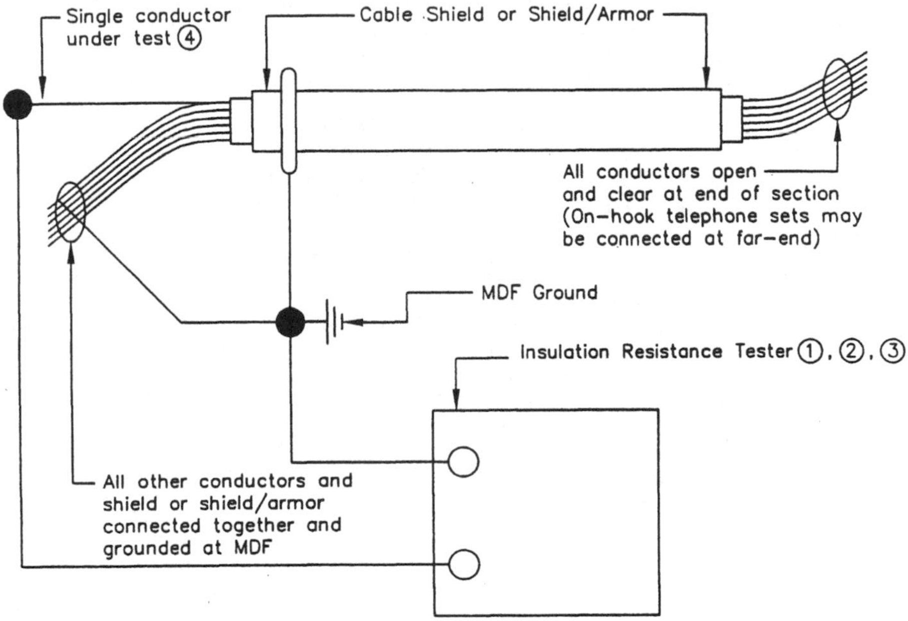

Notes:

(1.) For hand cranked or battery operated Insulation Resistance Testers, the output voltage should not exceed 500 volts dc.

(2.) For dc bridge type Megohmmeters, the voltage applied to the conductors under test should not exceed 250 volts dc when using instruments having adjustable test voltage levels.

(3.) Biddle CO.—Model BM 200, Associate Research—Model 263, General Radio—1864 Megohm Meter, or equivalent.

(4.) Repeat test for each conductor in cable.

(iii) If the IR of the conductor cannot be measured because of breakdown of lightning arresters by the test voltage, the arrester units shall be removed and the conductor IR retested. If the IR then meets the minimum requirements, the conductor will be considered satisfactory. Immediately following the IR tests, all arrester units which have been removed shall be reinstalled.

(3) *Test equipment.* (i) IR measurements shall be made with either an insulation resistance test set or a direct current (dc) bridge type megohmmeter.

(ii) The IR test set shall have an output voltage not to exceed 500 volts dc and shall be of the hand cranked or battery operated type.

(iii) The dc bridge type megohmmeter, which may be alternating current (ac) powered, shall have scales and multiplier which make it possible to accurately read IR from 1 megohm to 1 gigohm. The voltage applied to the conductors under test shall not exceed "250 volts dc" when using an instrument having adjustable test voltage levels. This will help to prevent breakdown of lightning arresters.

(4) *Applicable results.* (i) For all new insulated cable or wire facilities, the expected IR levels are normally greater than 1,000 to 2,000 megohm-mile (1,609 to 3,218 megohm-km). A value of 500 megohm-mile (805 megohm-km) at 68 °F (20 °C) shall be the minimum acceptable value of IR. IR varies inversely with the length and the temperature.

(ii) The megohm-mile (megohm-km) value for a conductor may be computed by multiplying the actual scale reading in megohms on the test set by the length in miles (km) of the conductor under test.

(iii) The objective insulation resistance may be determined by dividing 500 by the length in miles (805 by the length in km) of the cable or wire conductor being tested. The resulting value shall be the minimum acceptable meter scale reading in megohms.

(iv) Due to the differences between various insulating materials and filling compounds used in manufacturing cable or wire, it is impractical to provide simple factors to predict the magnitude of variation in insulation resistance due to temperature. The variation can, however, be substantial for wide excursions in temperature from the ambient temperature of 68 °F (20 °C).

(v) Borrowers should be certain that tip and ring IR measurements of each pair are approximately the same. Borrowers should also be certain that IR measurements are similar for cable or wire sections of similar length and cable or wire type. If some pairs measure significantly lower, borrowers should attempt to improve these pairs in accordance with cable manufacturer's recommendations.

NOTE: Only the megohm-mile (megohm-km) requirement shall be cause for rejection, not individual measurement differences.

(5) *Data record.* The measurement data shall be recorded. Suggested formats similar to Format I, Outside Plant Acceptance Tests—Subscriber Loops, or Format II, Outside Plant Acceptance Tests—Trunk Circuits, in §1755.407 or formats specified in the applicable construction contract may be used.

(6) *Probable causes for nonconformance.* (i) When an IR measurement is below 500 megohm-mile (805 megohm-km), the cable or wire temperature at the time of testing must then be taken into consideration. If this temperature is well above 68 °F (20 °C), the measurement shall be disregarded and the cable or wire shall be remeasured at a time when the temperature is approximately 68 °F (20 °C). If the result is then 500 megohm-mile (805 megohm-km) or greater, the cable or wire shall be considered satisfactory.

(ii) Should the cable or wire fail to meet the 500 megohm-mile (805 megohm-km) requirement when the temperature is known to be approximately 68 °F (20 °C) there is not yet justification for rejection of the cable or wire. Protectors, lightning arresters, etc., may be a source of low insulation resistance. These devices shall be removed from the cable or wire and the cable or wire IR measurement shall be repeated. If the result is acceptable, the cable or wire shall be considered acceptable. The removed devices which caused the low insulation resistance value shall be identified and replaced, if found defective.

(iii) When the cable or wire alone is still found to be below the 500 megohm-mile (805 megohm-km) requirement after completing the steps in paragraph (c)(6)(i) and/or paragraph (c)(6)(ii) of this section, the test shall be repeated to measure the cable or wire in sections to isolate the piece(s) of cable or wire responsible. The cable or wire section(s) that is found to be below the 500 megohm-mile (805 megohm-km) requirement shall be either repaired in accordance with the cable or wire manufacturer's recommended procedure or shall be replaced as directed by the borrower.

(d) *Dc loop resistance and dc resistance unbalance measurement.* (1) When specified by the borrower, dc loop resistance and dc resistance unbalance measurements shall be made on all cable pairs used as trunk circuits. The dc loop resistance and dc resistance unbalance measurements shall be made between CO locations. Measurements shall include all components of the cable path.

(2) Dc loop resistance and dc resistance unbalance measurements shall be made on all cable pairs used as subscriber loop circuits when:

(i) Specified by the borrower;

(ii) A large number of long loops terminate at one location (similar to trunk circuits); or

(iii) Circuit balance is less than 60 dB when computed from noise measurements as described in paragraph (e) of this section.

(3) Dc resistance unbalance is controlled to the maximum possible degree by the cable specification. Allowable random unbalance is specified between tip and ring conductors within each reel. Further random patterns should occur when the cable conductor size changes. Cable meeting the unbalance requirements of the cable specification may under some conditions result in unacceptable noise levels as discussed in paragraph (d)(6)(iii) of this section.

(4) *Method of measurement.* The method of measurement shall be as detailed in Figures 4 and 5.

(5) *Test equipment.* The test equipment is shown in Figures 4 and 5 as follows:

FIGURE 4

DC LOOP RESISTANCE MEASUREMENT

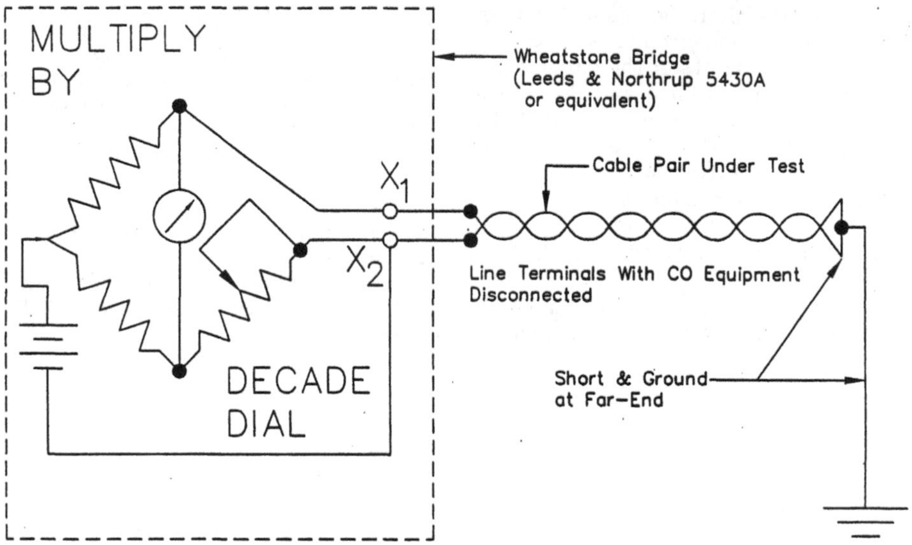

SETTINGS ON TEST SET

1. Set "RES.–VAR–MUR" Key to "RES".
2. Set "RVM–GA–HIL" Switch to "RVM".
3. Set "INT–BA–EXT" Switch to "INT".
4. Connect Pair to Terminals X_1 and X_2 as shown.

PROCEDURE

1. Null galvanometer by operating "MULTIPLY BY" and "DECADE" dials of bridge. Use lowest sensitivity range.

2. Multiply "DECADE" reading in ohms by "MULTIPLY BY" ratio to obtain value of the Loop Resistance.

FIGURE 5

DC LOOP RESISTANCE UNBALANCE MEASUREMENT

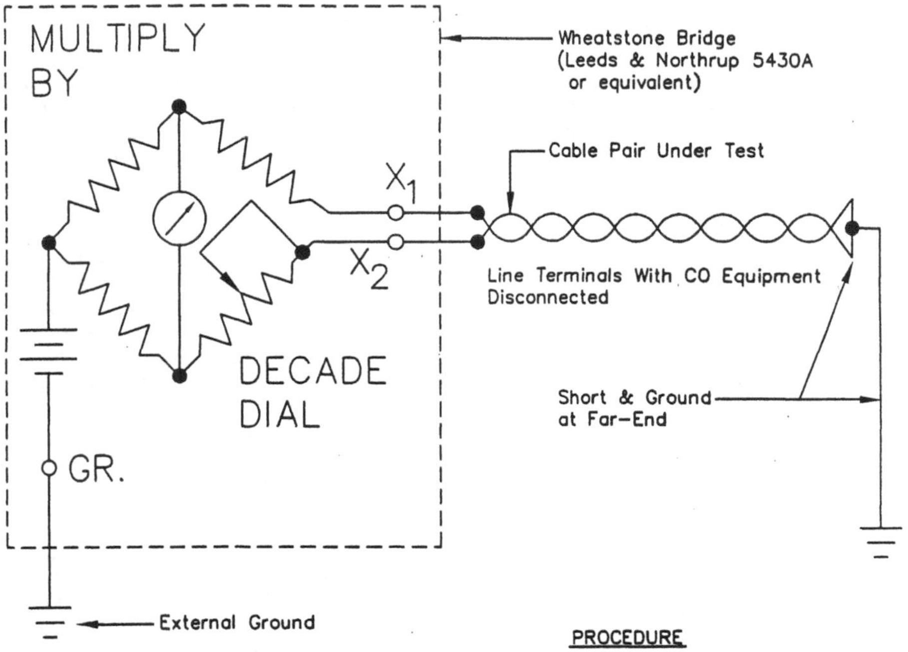

PROCEDURE

1. Null galvanometer by operating "MULTIPLY BY" and "DECADE" dials of bridge. Use lowest sensitivity range.
2. If continuously varying 1, 10, or 100 ohm switches from 1 to 999 ohms produces a deflection consistently to the left on the galvanometer, reverse the conductors of the cable pair under test to the X_1 & X_2 terminals of the bridge.
3. Vary 1, 10, or 100 ohm switches again until deflection approaches zero. Read "DECADE" dial for Resistance Unbalance in ohms.

SETTINGS ON TEST SET

1. Connect Terminals X_1 & X_2 as shown.
2. Set "RES.–VAR–MUR" Key to "VAR".
3. "MULTIPLY BY" Switch to 1/1.
4. Set "RVM–GA–HIL" Switch to "RVM".
5. Set "INT–BA–EXT" Switch to "INT".

(6) *Applicable results.* (i) The measured dc loop resistance shall be within ±5% of the calculated dc loop resistance when corrected for temperature.

(ii) The calculated dc loop resistance is computed as follows:

(A) Multiply the length of each different gauge by the applicable resistance per unit length as shown in Table 3 as follows:

TABLE 3—DC LOOP RESISTANCE @ 68 °F (20 °C)

American wire gauge (AWG)	Loop resistance	
	ohms/1000 ft	ohms/km
19	16.1	52.8
22	32.4	106.3
24	51.9	170.3
26	83.3	273.3

(B) Add the individual resistances for each gauge to give the total calculated dc loop resistance at a temperature of 68 °F (20 °C).

(C) Correct the total calculated dc loop resistance at the temperature of 68 °F (20 °C) to the measurement temperature by the following formulae:

$R_t = R_{68} \times [1 + 0.0022 \times t - 68)]$ for English Units

$R_t = R_{20} \times [1 + 0.0040 \times (t - 20)]$ for Metric Units

Where:

R_t = Loop resistance at the measurement temperature in ohms.

R_{68} = Loop resistance at a temperature of 68 °F in ohms.

R_{20} = Loop resistance at a temperature of 20 °C in ohms.

t = Measurement temperature in °F or (°C).

(D) Compare the calculated dc loop resistance at the measurement temperature to the measured dc loop resistance to determine compliance with the requirement specified in paragraph (d)(6)(i) of this section.

(iii) Resistance varies directly with temperature change. For copper conductor cables, the dc resistance changes by ±1% for every ±5 °F (2.8 °C) change in temperature from 68 °F (20 °C).

(iv) The dc resistance unbalance between the individual conductors of a pair shall not exceed that value which will result in a circuit balance of less than 60 dB when computed from noise measurements as described in paragraph (e) of this section. It is impractical to establish a precise limit for overall circuit dc resistance unbalance due to the factors controlling its contribution to circuit noise. These factors include location of the resistance unbalance in relation to a low impedance path to ground (close to the central office) and the magnitude of unbalance in short lengths of cable making up the total circuit length. The objective is to obtain the minimum unbalance throughout the entire circuit when it is ascertained through noise measurements that dc resistance unbalance may be contributing to poor cable balance.

(v) Pairs with poor noise balance may be improved by reversing tip and ring conductors of pairs at cable splices. Where dc resistance unbalances are systematic over the total trunk circuit or loop circuit length, tip and ring reversals may be made at frequent intervals. Where the unbalances are concentrated in a shorter section of cable, only one tip and ring reversal should be required. Concentrated dc resistance unbalance produces maximum circuit noise when located adjacent to the central office. Concentrated dc resistance unbalance will contribute to overall circuit noise at a point approximately two-thirds (⅔) of the distance to the subscriber. All deliberate tip and ring reversals shall be tagged and identified to prevent plant personnel from removing the reversals when resplicing these connections in the future. The number of tip and ring reversals shall be held to a minimum.

(vi) A systematic dc resistance unbalance can sometimes be accompanied by other cable parameters that are marginal. Among these are pair-to-pair capacitance unbalance, capacitance unbalance-to-ground, and 150 kilohertz (kHz) crosstalk loss. Engineering judgment has to be applied in each case. Rejection of cable for excessive dc resistance unbalance shall only apply to a single reel length, or shorter.

(7) *Data record*. The measurement data for dc loop resistance and dc resistance unbalance shall be recorded. Suggested formats similar to Format I for subscriber loops and Format II for trunk circuits in §1755.407 or formats

specified in the applicable construction contract may be used.

(8) *Probable causes for nonconformance.* Dc loop resistance and dc resistance unbalance are usually the result of the resistance of individual conductors used in the manufacture of the cable. Resistance unbalance can be worsened by defective splicing of the conductors (splicing connectors, improper crimping tool, etc.).

(e) *Subscriber loop measurement (loop checking).* (1) When specified by the borrower, insertion loss and noise measurements shall be performed on subscriber loops after connection of a line circuit to the loop by the one person method using loop checking equipment from the customer access location. For this method, the central office should be equipped with a 900 ohm plus two microfarad quiet termination and a milliwatt generator having the required test frequencies; or a portable milliwatt generator having the desired frequencies may be used, especially, where several small offices are involved.

(2) At a minimum, insertion loss and frequency response of subscriber loop plant shall be measured at 1,000, 1,700, 2,300, and 2,800 Hertz (Hz). When additional testing frequencies are desired, the additional frequencies shall be specified in the applicable construction contract.

(3) Measurements of insertion loss and noise shall be made on five percent or more of the pairs. A minimum of five pairs shall be tested on each route. Pairs shall be selected on a random basis with greater consideration in the selection given to the longer loops. Consideration shall be given to measuring a large percentage, up to 100 percent, of all loops.

(4) *Method of measurement*—(i) *Insertion loss.* The step-by-step measurement procedure shall be as shown in Figure 6. The output level of the milliwatt generator tones shall be determined prior to leaving the CO. This shall be accomplished by dialing the milliwatt generator number from a spare line at the MDF and measuring with the same equipment to be used in the tests at customer access locations. The output levels shall be recorded for reference later. Insertion loss measurements shall be made across the tip and ring terminals of the pair under test. Figure 6 is as follows:

FIGURE 6

INSERTION LOSS AND FREQUENCY RESPONSE MEASUREMENT AT SUBSCRIBER LOCATION USING LOOP CHECKING EQUIPMENT

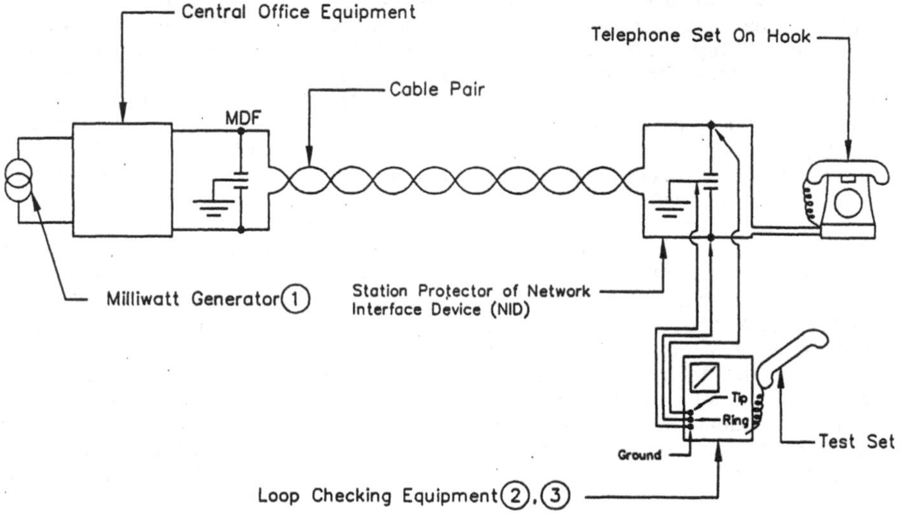

Calibration

1. Before leaving CO connect Loop Checking equipment to idle line at MDF.
   A. Dial number of Milliwatt Generator.
   B. Read and record output level of all tones in dBm for reference.

Notes:

(1.) H.P.–204B, H.P.–204C, General Radio–1335, or equivalent.

(2.) N.E.C.–125, N.E.C.–37B, Wilcom–136, Wilcom–336, Wilcom–337, or equivalent.

(3.) Do not leave test equipment connected and exposed to ringing voltage of incoming call. Ringing voltage could damage test equipment.

Measurement Procedure

1. Connect Loop Checking equipment at subscriber's NID as shown.
2. Dial number of Milliwatt Generator at central office.
3. Verify by listening on the test set that the tones are being received.
4. Switch test set to Circuit Loss mode.
5. Read loss in dBm at each frequency.
6. Record results of loss at each frequency.
7. Subtract the output levels observed at the CO for each tone by the values observed at the subscriber location. The resultant values are the Insertion Loss.
8. Disconnect leads of test equipment from NID when tests are completed.

(ii) *Noise.* The step-by-step measurement procedure shall be as shown in Figure 7. Prior to leaving the CO for testing, dial the 900 ohm plus two microfarad quiet termination from a spare pair and measure the termination to determine that it actually is quiet. Circuit noise (noise-metallic) shall be measured at the customer access location across the tip and ring terminals of the pair under test. Power influence (direct reading with loop checking equipment) shall be measured at the customer access location from tip and ring conductors-to-ground (this connection is completed via the test unit). The power influence measurement includes the entire talking connection from the quiet termination to the customer. (That is, the power influence measurement includes all the CO equipment which normally makes up the connection.) Figure 7 is as follows:

FIGURE 7

NOISE MEASUREMENT AT SUBSCRIBER LOCATION USING LOOP CHECKING EQUIPMENT

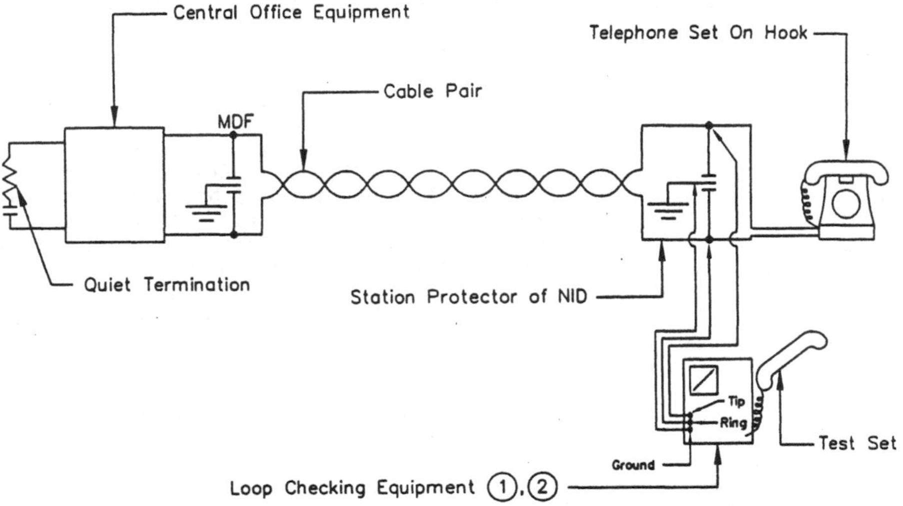

Calibration

1. Before leaving CO connect Loop Checking equipment to idle line at MDF (no outside plant attached).

   A. Dial number of Quiet Termination.

   B. Read and record Circuit Noise in dBrnc.

Note:

① N.E.C.–125, N.E.C.–37B, Wilcom–136, Wilcom–336, Wilcom–337, or equivalent.

② Do not leave test equipment connected and exposed to ringing voltage of incoming call. Ringing voltage could damage test equipment.

Measurement Procedure

1. Connect Loop Checking equipment at subscriber's NID as shown.
2. Dial number of Quiet Termination in cental office.
3. Switch test set to Circuit Noise (NM) mode.
4. Read and record Circuit Noise value in dBrnc.
5. Switch test set to Power Influence (PI) mode.
6. Read and record Power Influence value in dBrnc.
7. Compute and record apparent Balance (Balance = PI – NM).
8. Disconnect leads of test equipment from NID when tests are completed.

(5) *Test equipment.* (i) Loop checking equipment which is available from several manufacturers may be used for these measurements. The equipment

should have the capability of measuring loop current, insertion loss, circuit noise (NM) and power influence (PI). The test equipment manufacturer's operating instructions shall be followed.

(ii) There should be no measurable transmission loss when testing through loop extenders.

(6) *Applicable results*—(i) *Insertion loss.* (A) For D66 loaded cables (a specific loading scheme using a 66 millihenry inductor spaced nominally at 4,500 ft [1,371 m] intervals) measured at a point one-half section length beyond the last load point, the measured nonrepeated insertion loss shall be within ±10% at 1000, 1700, 2300, and 2800 Hz, ±15% at 3400 Hz and ±20% at 4000 Hz of the calculated insertion loss at the same frequencies and temperature.

(B) For H88 loaded cables (a specific loading scheme using an 88 millihenry inductor spaced nominally at 6,000 ft [1,829 m] intervals) measured at a point one-half section length beyond the last load point, the measured nonrepeatered insertion loss shall be within ±10% at 1000, 1700, and 2300 Hz, ±15% at 2800 Hz, and ±20% at 3400 Hz of the calculated insertion loss at the same frequencies and temperature.

(C) For nonloaded cables, the measured insertion loss shall be within ±10% at 1000, 1700, 2300, and 2800 Hz, ±15% at 3400 Hz and ±20% at 4000 Hz of the calculated insertion loss at the same frequencies and temperature.

(D) For loaded cables, the calculated loss at each desired frequency shall be computed as follows:

(*1*) Multiply the length in miles (km) of each different gauge in the loaded portion of the loop (between the office and a point one-half load section beyond the furthest load point) by the applicable decibel (dB)/mile (dB/km) value shown in Table 4 or 5. This loss represents the total loss for each gauge in the loaded portion of the loop;

(*2*) Multiply the length in miles (km) of each different gauge in the end section or nonloaded portion of the cable (beyond a point one-half load section beyond the furthest load point) by the applicable dB/mile (dB/km) value shown in Table 6. This loss represents the total loss for each gauge in the nonloaded portion of the loop; and

(*3*) The total calculated insertion loss is computed by adding the individual losses determined in paragraphs (e)(6)(i)(D)(*1*) and (e)(6)(i)(D)(*2*) of this section.

(E) For nonloaded cables, the calculated loss at each desired frequency shall be computed by multiplying the length in miles (km) of each different gauge by the applicable dB/mile (dB/km) value shown in Table 6 and then adding the individual losses for each gauge to determine the total calculated insertion loss for the nonloaded loop.

(F) The attenuation information in Tables 4, 5, and 6 are based on a cable temperature of 68 °F (20 °C). Insertion loss varies directly with temperature. To convert measured losses for loaded cables to a different temperature, use the following value for copper conductors: For each ±5 °F (±2.8 °C) change in the temperature from 68 °F (20 °C), change the insertion loss at any frequency by ±1%. To convert measured losses for nonloaded cables to a different temperature, use the following value for copper conductors: For each ±10 °F (±5.6 °C) change in the temperature from 68 °F (20 °C), change the insertion loss at any frequency by ±1%. Tables 4, 5, and 6 are as follows:

TABLE 4—FREQUENCY ATTENUATION @ 68 °F (20 °C) D66 LOADED EXCHANGE CABLES 83 NANOFARAD (NF)/MILE (52 NF/KM) (SEE NOTE)

Frequency (Hz)	Attenuation dB/mile (dB/km) AWG			
	19	22	24	26
200	0.41 (0.26)	0.67 (0.42)	0.90 (0.56)	1.21 (0.75)
400	0.43 (0.26)	0.77 (0.48)	1.09 (0.68)	1.53 (0.95)
600	0.44 (0.27)	0.80 (0.49)	1.17 (0.73)	1.70 (1.06)
800	0.44 (0.27)	0.81 (0.50)	1.21 (0.75)	1.80 (1.12)
1000	0.44 (0.27)	0.82 (0.51)	1.23 (0.76)	1.86 (1.15)
1200	0.45 (0.28)	0.83 (0.52)	1.24 (0.77)	1.91 (1.19)
1400	0.45 (0.28)	0.83 (0.52)	1.26 (0.78)	1.94 (1.20)
1600	0.45 (0.28)	0.84 (0.52)	1.26 (0.78)	1.96 (1.22)
1800	0.45 (0.28)	0.84 (0.52)	1.27 (0.78)	1.98 (1.23)

TABLE 4—FREQUENCY ATTENUATION @ 68 °F (20 °C) D66 LOADED EXCHANGE CABLES 83 NANOFARAD (NF)/MILE (52 NF/KM) (SEE NOTE)—Continued

Frequency (Hz)	Attenuation dB/mile (dB/km) AWG			
	19	22	24	26
2000	0.46 (0.29)	0.85 (0.53)	1.28 (0.79)	1.99 (1.24)
2200	0.46 (0.29)	0.85 (0.53)	1.29 (0.80)	2.01 (1.25)
2400	0.47 (0.29)	0.86 (0.53)	1.30 (0.81)	2.02 (1.26)
2600	0.47 (0.29)	0.87 (0.54)	1.31 (0.81)	2.04 (1.27)
2800	0.48 (0.30)	0.88 (0.55)	1.32 (0.82)	2.07 (1.29)
3000	0.49 (0.30)	0.89 (0.55)	1.34 (0.83)	2.10 (1.30)
3200	0.50 (0.31)	0.91 (0.57)	1.36 (0.84)	2.13 (1.32)
3400	0.52 (0.32)	0.93 (0.58)	1.40 (0.87)	2.19 (1.36)
3600	0.54 (0.34)	0.97 (0.60)	1.45 (0.90)	2.26 (1.40)
3800	0.57 (0.35)	1.02 (0.63)	1.52 (0.94)	2.36 (1.47)
4000	0.62 (0.38)	1.10 (0.68)	1.63 (1.01)	2.53 (1.57)

NOTE: Between end-section lengths of 2,250 ft (686 m) for D66 loading.

TABLE 5—FREQUENCY ATTENUATION @ 68 °F (20 °C) H88 LOADED EXCHANGE CABLES 83 NF/ MILE (52 NF/KM) (SEE NOTE)

Frequency (Hz)	Attenuation dB/mile (dB/km) AWG			
	19	22	24	26
200	0.40 (0.25)	0.66 (0.41)	0.90 (0.56)	1.20 (0.75)
400	0.42 (0.26)	0.76 (0.47)	1.08 (0.67)	1.53 (0.95)
600	0.43 (0.27)	0.79 (0.49)	1.16 (0.72)	1.70 (1.06)
800	0.43 (0.27)	0.80 (0.50)	1.20 (0.75)	1.80 (1.12)
1000	0.43 (0.27)	0.81 (0.50)	1.23 (0.76)	1.86 (1.15)
1200	0.44 (0.27)	0.82 (0.51)	1.24 (0.77)	1.91 (1.19)
1400	0.44 (0.28)	0.82 (0.51)	1.25 (0.78)	1.94 (1.20)
1600	0.44 (0.27)	0.83 (0.52)	1.26 (0.78)	1.97 (1.22)
1800	0.45 (0.28)	0.84 (0.52)	1.28 (0.79)	1.99 (1.24)
2000	0.46 (0.29)	0.85 (0.53)	1.29 (0.80)	2.02 (1.26)
2200	0.47 (0.29)	0.86 (0.53)	1.31 (0.81)	2.06 (1.28)
2400	0.48 (0.30)	0.89 (0.55)	1.34 (0.83)	2.10 (1.30)
2600	0.50 (0.31)	0.92 (0.57)	1.39 (0.86)	2.18 (1.35)
2800	0.53 (0.33)	0.97 (0.60)	1.47 (0.91)	2.29 (1.42)
3000	0.59 (0.37)	1.07 (0.66)	1.60 (0.99)	2.48 (1.54)
3200	0.71 (0.44)	1.26 (0.78)	1.87 (1.16)	2.86 (1.78)
3400	1.14 (0.71)	1.91 (1.19)	2.64 (1.64)	3.71 (2.30)
3600	4.07 (2.53)	4.31 (2.68)	4.65 (2.90)	5.30 (3.29)
3800	6.49 (4.03)	6.57 (4.08)	6.72 (4.18)	7.06 (4.39)
4000	8.22 (5.11)	8.27 (5.14)	8.36 (5.19)	8.58 (5.33)

NOTE: Between end-section lengths of 3,000 ft (914 m) for H88 loading.

TABLE 6—FREQUENCY ATTENUATION @ 68 °F (20 °C) NONLOADED EXCHANGE CABLES 83 NF/ MILE (52 NF/KM) AWG

Frequency (Hz)	Attenuation dB/mile (dB/km) AWG			
	19	22	24	26
200	0.58 (0.36)	0.82 (0.51)	1.03 (0.64)	1.30 (0.81)
400	0.81 (0.51)	1.15 (0.71)	1.45 (0.90)	1.84 (1.14)
600	0.98 (0.61)	1.41 (0.87)	1.77 (1.10)	2.26 (1.40)
800	1.13 (0.70)	1.62 (1.01)	2.04 (1.27)	2.60 (1.61)
1000	1.25 (0.78)	1.80 (1.12)	2.28 (1.42)	2.90 (1.80)
1200	1.36 (0.84)	1.97 (1.22)	2.50 (1.55)	3.17 (1.97)
1400	1.46 (0.91)	2.12 (1.32)	2.69 (1.67)	3.42 (2.12)
1600	1.55 (0.96)	2.26 (1.40)	2.87 (1.78)	3.65 (2.27)
1800	1.63 (1.01)	2.39 (1.48)	3.04 (1.89)	3.87 (2.40)
2000	1.71 (1.06)	2.51 (1.56)	3.20 (1.99)	4.08 (2.53)
2200	1.78 (1.11)	2.62 (1.63)	3.35 (2.08)	4.27 (2.65)
2400	1.85 (1.15)	2.73 (1.70)	3.49 (2.17)	4.45 (2.76)
2600	1.91 (1.19)	2.83 (1.76)	3.62 (2.25)	4.63 (2.88)
2800	1.97 (1.22)	2.93 (1.82)	3.75 (2.33)	4.80 (2.98)
3000	2.03 (1.26)	3.02 (1.88)	3.88 (2.41)	4.96 (3.08)
3200	2.08 (1.29)	3.11 (1.93)	4.00 (2.48)	5.12 (3.18)
3400	2.13 (1.32)	3.19 (1.98)	4.11 (2.55)	5.27 (3.27)
3600	2.18 (1.35)	3.28 (2.04)	4.22 (2.62)	5.41 (3.36)

TABLE 6—FREQUENCY ATTENUATION @ 68 °F (20 °C) NONLOADED EXCHANGE CABLES 83 NF/ MILE (52 NF/KM) AWG—Continued

Frequency (Hz)	Attenuation dB/mile (dB/km) AWG			
	19	22	24	26
3800	2.22 (1.38)	3.36 (2.09)	4.33 (2.69)	5.55 (3.45)
4000	2.27 (1.41)	3.43 (2.13)	4.43 (2.75)	5.69 (3.53)

(G) For loaded subscriber loops, the 1 kHz loss shall be approximately 0.45 dB per 100 ohms of measured dc loop resistance. This loss shall be the measured loss less the net gain of any voice frequency repeaters in the circuit. Testing shall also be conducted to verify that the loss increases gradually as the frequency increases. The loss on H88 loaded loops should be down only slightly at 2.8 kHz but drop rapidly above 2.8 kHz. The loss on D66 loaded loops shall be fairly constant to about 3.4 kHz and there shall be good response at 4.0 kHz. When voice frequency repeaters are in the circuit there will be some frequency weighting in the build-out network and the loss at the higher frequencies will be greater than for nonrepeatered loops.

(H) For nonloaded subscriber loops, the 1 kHz loss shall be approximately 0.9 dB per 100 ohms of measured dc loop resistance. Testing shall also be conducted to verify that the loss is approximately a straight line function with no abrupt changes. The 3 kHz loss should be approximately 70% higher than the 1 kHz loss.

(ii) *Noise.* The principal objective related to circuit noise (noise-metallic) and the acceptance of new plant is that circuit noise levels be 20 dBrnc or less (decibels above reference noise, C-message weighted (a weighting derived from listening tests, to indicate the relative annoyance or speech impairment by an interfering signal of frequency (f) as heard through a "500-type" telephone set)). For most new, properly installed, plant construction, circuit noise will usually be considerably less than 20 dBrnc unless there are unusually long sections of telephone plant in parallel with electric power facilities and/or power influence of paralleling electric facilities is abnormally high. When circuit noise is 20 dBrnc or less, the loop plant shall be considered acceptable. When measured circuit noise is greater than 20 dBrnc, loop plant shall still be considered acceptable providing circuit balance (power influence reading minus circuit noise readings) is 60 dB or greater and power influence readings are 85 dBrnc or greater. When circuit noise is greater than 20 dBrnc and circuit balance is less than 60 dB and/or power influence is less than 85 dBrnc, loop plant shall not be considered acceptable and the loop plant shall be remedied to make circuit balance equal to or greater than 60 dB.

(7) *Data record.* Measurement data shall be recorded. A suggested format similar to Format I for subscriber loops in § 1755.407 or a format specified in the applicable construction contract may be used.

(8) *Probable causes for nonconformance*—(i) *Insertion loss.* Some of the more common causes for failing to obtain the desired results may be due to reversed load coil windings, missing load coils, bridge taps between load coils, load coil spacing irregularities, excessive end sections, cables having high or low mutual capacitance, load coils having the wrong inductance, load coils inadvertently installed in nonloaded loops, moisture or water in cable, split pairs, and improperly spliced connections. The above factors can occur singularly or in combination. Experience to date indicates that the most common problems are missing load coils, reversed load coil windings or bridge taps.

(ii) *Noise.* Some of the common causes for failing to obtain the desired results may be due to high power influence from paralleling electrical power systems, poor telephone circuit balance, discontinuous cable shields, inadequate bonding and grounding of cable shields, high capacitance unbalance-to-ground of the cable pairs, high dc loop resistance unbalance, dc loop current less than 20 milliamperes, etc. The above factors can occur singularly or in combination. See TE&CM Section 451, Telephone Noise Measurement and Mitigation, for steps to be taken in reducing telecommunications line noise.

(f) *One-person open circuit measurement (subscriber loops).* (1) When specified by the borrower, open circuit measurements shall be made on all loaded and nonloaded subscriber loops upon completion of the cable work to verify that the plant is free from major impedance irregularities.

(2) For loaded loops, open circuit measurements shall be made using one of the following methods:

(i) Impedance or pulse return pattern, with cable pair trace compared to that of an artificial line of the same length and gauge. For best results, a level tracer or fault locator with dual trace capability is required;

(ii) Return loss using a level tracer, with cable pair compared to an artificial line of the same length and gauge connected in lieu of a Precision Balance Network (PBN). This method can be made with level tracers having only single trace capability; or

(iii) Open circuit structural return loss using a level tracer. This method can be made with level tracer having only single trace capability.

(3) Of the three methods suggested for loaded loops, the method specified in paragraph (f)(2)(ii) of this section is the preferred method because it can yield both qualitative and quantitative results. The methods specified in paragraphs (f)(2)(i) and (f)(2)(iii) of this section can be used as trouble shooting tools should irregularities be found during testing.

(4) For nonloaded loops, open circuit measurements shall be made using the method specified in paragraph (f)(2)(i) of this section.

(5) *Method of measurement.* Open circuit measurements shall be made at the CO on each loaded and nonloaded pair across the tip and ring terminals of the pair under test. All CO equipment shall be disconnected at the MDF for this test. For loaded loops containing voice frequency repeaters installed in the CO or field mounted, the open circuit measurement shall be made after the repeaters have been disconnected. Where field mounted repeaters are used, the open circuit measurement shall be made at the repeater location in both directions.

(i) *Impedance or pulse return pattern.* The step-by-step measurement procedure using the impedance or pulse return pattern for loaded and nonloaded loops shall be as shown in Figure 8. An artificial line of the same makeup as the cable to be tested shall be set up. The traces of the impedance or pulse return pattern from the cable pair and the artificial line shall be compared and should be essentially identical. If the impedance or pulse return traces from the cable pair are different than the artificial line trace, cable faults are possible. When the cable pair trace indicates possible defects, the defects should be identified and located. One method of identifying and locating defects involves introducing faults into the artificial line until its trace is identical with the cable trace.

(ii) *Return loss balanced to artificial line.* The step-by-step measurement procedure using the return loss balanced to artificial line for loaded loops shall be as shown in Figure 9. An artificial line of the same makeup as the cable to be tested shall be set up. The artificial line is connected to the external network terminals of the test set. The cable pair under test is compared to this standard. When defects are found, they should be identified and located by introducing faults into the artificial line. This is more difficult than with the method referenced in paragraph (f)(5)(i) of this section since this measurement is more sensitive to minor faults and only a single trace is used.

(iii) *Open circuit structural return loss using level tracer.* The step-by-step measurement procedure using the level tracer for loaded loops shall be as shown in Figure 10. The cable pair is compared to a PBN.

(6) *Test equipment.* Equipment for performing these tests is shown in Figures 8 through 10. For loaded loops, artificial loaded lines must be of the same gauge and loading scheme as the line under test. For nonloaded loops, artificial nonloaded lines must be of the same gauge as the line under test. Artificial lines should be arranged using

switches or other quick connect arrangements to speed testing and troubleshooting. Figures 8 through 10 are as follows:

FIGURE 8

ONE–PERSON OPEN CIRCUIT MEASUREMENT IMPEDANCE OR PULSE RETURN PATTERN

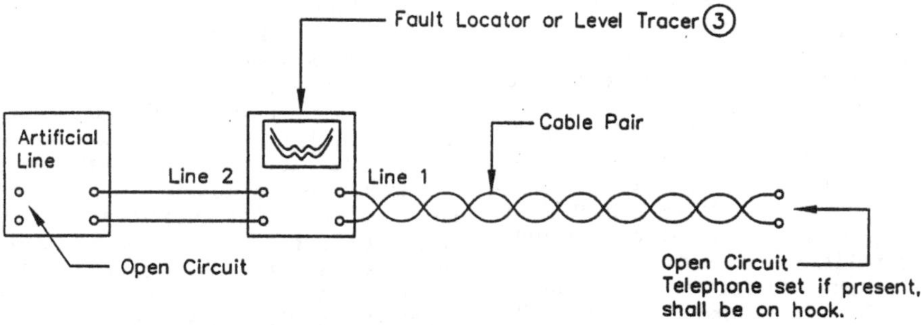

Measurement Procedure

1. Set up Artificial Line to same make–up [Length & Gauge(s)] as the cable pair.
2. Connect to test set (See Note (1)).
3. Connect cable pair to test set (See Note (1)).
4. Compare traces of Artificial Line and cable pair (2). They should be essentially identical. Differences indicate cable faults.
5. Location and type of fault may be determined by introducing faults in the Artificial Line until its trace is identical to that of the cable pair.

Notes:

(1.) Terminals to which cable pair and artificial line are attached shall be determined from the manufacturer's operating instructions. Proper settings for various switches and adjustments on the test set shall also be determined from the same source.

(2.) With test sets having trace storage capability only one set of terminals need be used. Connect Artificial Line to test set, store trace and disconnect line. Connect cable pair and compare trace to stored trace. To identify fault, store cable pair trace and connect Artificial Line. Introduce faults in the Artificial Line until traces are identical.

(3.) N.E.C.–17A, Biddle–CME110A–1, Dolcom–490, Tektronix–1503, Wilcom–T195, Wilcom–T132, or equivalent.

FIGURE 9

ONE–PERSON OPEN CIRCUIT MEASUREMENT
RETURN LOSS BALANCED TO ARTIFICIAL LINE

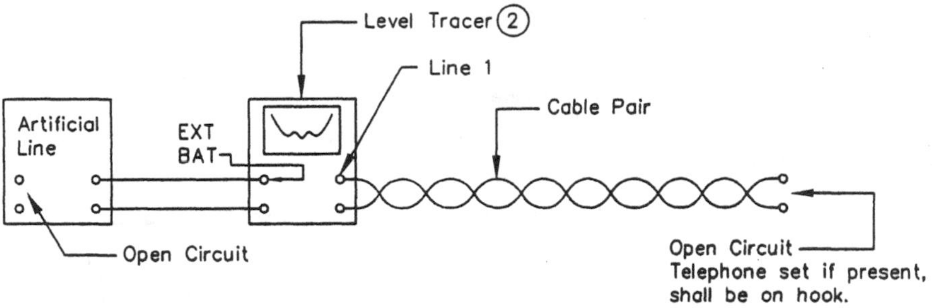

Measurement Procedure

1. Connect the test equipment and cable pair under test as shown above (See Note 1). Set up Artificial Line to same make–up [Length & Gauge(s)] as the cable pair.

2. Observe Return Loss from 200 to 3500 Hz (D66) or 200 to 3000 Hz (H88) noting maximum and minimum values. Note the value and frequency of the poorest (Lowest Numerical Value) SRL. (SRL becomes better as the readings become more negative). Record this value and frequency.

Notes:

(1.) Terminals to which cable pair and Artificial Line are attached shall be determined from the manufacturer's operating instructions. Proper settings for various switches and adjustments on the test set shall also be determined from the same source.

(2.) Wilcom–T132, Wilcom–T195, or equivalent.

FIGURE 10

ONE–PERSON OPEN CIRCUIT MEASUREMENT STRUCTURAL RETURN LOSS USING LEVEL TRACER

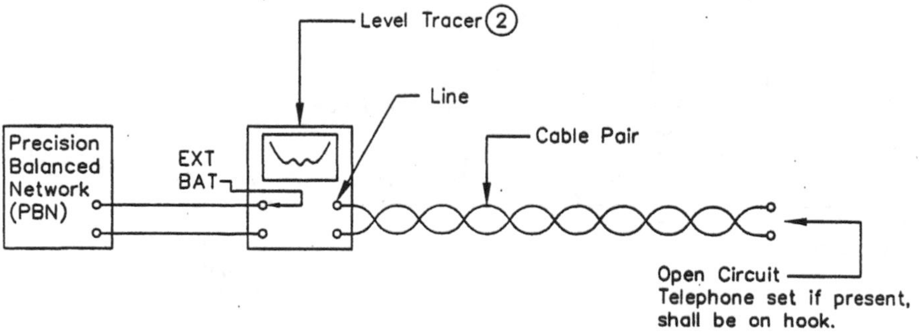

Measurement Procedure

1. Connect the test equipment and cable pair under test as shown above (See Note (1)). Set gauge of PBN for: Single Gauge – Same gauge as cable being measured; Mixed Gauge – Most predominant gauge adjacent to test set.

2. Observe Return Loss between 1000 and 3500 Hz (D66) or 1000 and 3000 Hz (H88) observing maximum and minimum values. Note the value and frequency of the poorest (Lowest Numerical Value) SRL. Single Gauge: Record this value. Mixed Gauge: Change gauge of PBN and note if SRL becomes better. (SRL becomes better as readings become more negative). If it does, record this value and frequency; if not, record value obtained with original gauge setting. (Varying gauge will be necessary, depending on actual cable layout, to obtain best SRL).

Notes:

(1) Terminals to which cable pair and Artificial Line are attached shall be determined from the manufacturer's operating instructions. Proper settings for various switches and adjustments on the test set shall also be determined from the same source.

(2) Wilcom–T132, Wilcom–T195, or equivalent.

(7) *Applicable results.* (i) For loaded and nonloaded loops, the two traces in the pulse return pattern or impedance method (paragraph (f)(5)(i) of this section) shall be essentially identical. The degree of comparison required of the two traces is to be determined by experience.

(ii) For loaded loops, results for return loss measurements using a level tracer, with artificial line, in lieu of a PBN (paragraph (f)(5)(ii) of this section) shall meet the following requirements:

(A) For D66 and H88 loaded cables the structural return loss (SRL) values shall range between 28 and 39 dB, respectively, at the critical frequency of structural return loss (CFSRL) within the pass band of the loading system being used. The minimum SRL value for uniform gauge shall be 25 dB CFSRL. These SRL values apply for loaded cables of uniform gauge for the entire length of the subscriber loop circuit. Subscriber loop circuits shall meet the loading spacing deviations and the cable mutual capacitance requirements in the applicable RUS cable specifications;

(B) For mixed gauge loaded cables the SRL values shall be 25 and 27 dB CFSRL, respectively, and the minimum SRL value shall be 22 dB CFSRL; and

(C) The two traces in the pulse return pattern should be essentially identical. The degree of comparison required of the two traces is determined by experience.

(iii) For loaded loops, the results of open circuit structural return loss measurements using a level tracer (paragraph (f)(5)(iii) of this section) shall meet the following requirements. For D66 and H88 loaded cables with uniform or mixed gauges, the worst value allowed for measured open circuit structural return loss between 1,000–3,500 Hz and 1,000–3,000 Hz, respectively, shall be approximately 0.9 dB (round trip) for each 100 ohms outside plant dc loop resistance including the resistance of the load coils. The value of 0.9 dB per 100 ohms for the round trip loss remains reasonably accurate as long as:

(A) The subscriber end section of the loaded pair under test is approximately 2,250 ft (685 m) for D66 loading or 3,000 ft (914 m) for H88 loading in length; and

(B) The one-way 1,000 Hz loss does not exceed 10 dB.

(iv) For loaded loops, the measured value of open circuit structural return loss can only be as accurate as the degree to which the dc loop resistance of the loaded pair under test is known. Most accurate results shall be obtained when the dc loop resistance is known by actual measurements as described in paragraph (d) of this section. Furthermore, where the dc loop resistance is measured at the same time as the open circuit structural return loss, no correction for temperature is needed because the loss is directly proportional to the loop resistance. Where it is not practical to measure the dc loop resistance, it shall be calculated and corrected for temperature as specified in paragraph (d)(6)(ii) of this section. When measuring existing plant, care shall be taken to verify the accuracy of the records, if they are used for the calculation of the dc loop resistance. For buried plant, the temperature correction shall be based at the normal depth of the cable in the ground. (Temperature can be measured by boring a hole to cable depth with a ground rod, placing a thermometer in the ground at the cable depth, and taking and averaging several readings during the course of the resistance measurements.) For aerial cable it shall be based on the temperature inside the cable sheath.

(v) For loaded loops, the best correlation between the measured and the expected results shall be obtained when the cable is of one gauge, one size, and the far end section is approximately 2,250 ft (685 m) for D66 loading or 3,000 ft (914 m) for H88 loading. Mixing gauges and cable sizes will result in undesirable small reflections whose frequency characteristics and magnitude cannot be accurately predicted. In subscriber loop applications, cable gauge may be somewhat uniform but the cable pair size most likely will not be uniform as cable pair sizes taper off toward the customer access location and a downward adjustment of 1 dB of the allowed value shall be acceptable. "Long" end sections (as defined in TE&CM Section 424, "Guideline for Telecommunications Subscriber Loop Plant") lower the expected value, a further downward adjustment of 3 dB in the allowed value shall be acceptable.

(vi) For loaded loops, the limiting factor when making open circuit structural return loss measurements is when the 1,000 Hz one-way loss of the loaded cable pair under test becomes 10 dB or greater; it becomes difficult to detect the presence of irregularities beyond the 10 dB point on the loop. To overcome this difficulty, loaded loops having a one-way loss at 1,000 Hz greater than 10 dB shall be opened at some convenient point (such as a pedestal or ready access enclosure) and loss measurements at the individual portions measuring less than 10 dB one-way shall be made separately. When field mounted voice frequency repeaters are used, the measurement shall be made at the repeater location in both directions.

(8) *Data record.* (i) When performing a pulse return pattern or impedance open

circuit measurement on loaded and nonloaded loops, a "check mark" indicating that the pair tests good or an "X" indicating that the pair does not test good shall be recorded in the SRL column. A suggested format similar to Format I for subscriber loops in § 1755.407 or a format specified in the applicable construction contract may be used.

(ii) When performing open circuit return loss measurements using the return loss balanced to an artificial line or return loss using a level tracer on loaded loops, the value of the poorest (lowest numerical value) SRL and its frequency in the proper column between 1,000 and 3,500 Hz for D66 loading or between 1,000 and 3,000 Hz for H88 loading shall be recorded. A suggested format similar to Format I for subscriber loops in § 1755.407 or a format specified in the applicable construction contract may be used.

(9) *Probable causes for nonconformance.* Some of the more common causes for failing to obtain the desired results may be due to reversed load coil windings, missing load coils, bridge taps between load coils, load coil spacing irregularities, excessive end sections, cables having high or low mutual capacitance, load coils inadvertently installed in nonloaded loops, moisture or water in the cable, load coils having the wrong inductance, split pairs, and improperly spliced connectors. The above can occur singularly or in combination. Experience to date indicates that the most common problems are missing load coils, reversed load coil windings or bridge taps.

(g) *Cable insertion loss measurement (carrier frequencies).* (1) When specified by the borrower, carrier frequency insertion loss measurements shall be made on cable pairs used for T1, T1C, and/or station carrier systems. Carrier frequency insertion loss shall be made on a minimum of three pairs. Select at least one pair near the outside of the core unit layup. If the three measured pairs are within 10% of the calculated loss in dB corrected for temperature, no further testing is necessary. If any of the measured pairs of a section are not within 10% of the calculated loss in dB, all pairs in that section used for carrier transmission shall be measured.

(2) *Method of measurement.* The step-by-step method of measurement shall be as shown in Figure 11.

(3) *Test equipment.* The test equipment is shown in Figure 11 as follows:

FIGURE 11

CARRIER FREQUENCY INSERTION LOSS MEASUREMENT
CABLE FACILITIES

Measurement

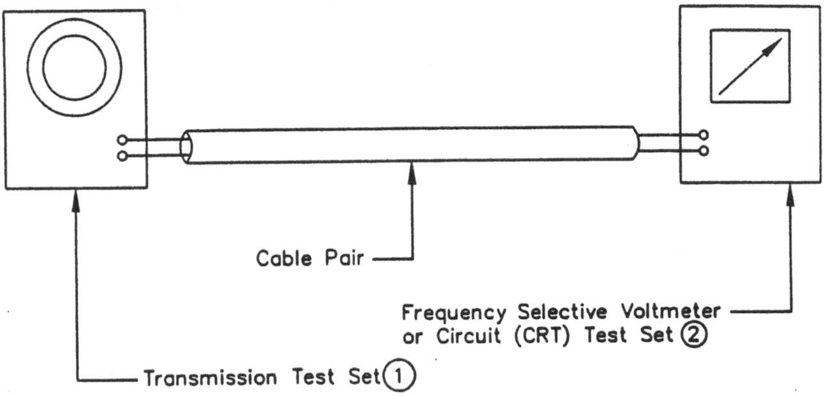

Measurement Procedure

1. Connect the transmission test set to one end of the length of cable to be measured and either the frequency selective voltmeter (FSVM) or CRT test set to the other end as shown.
2. Record the Insertion Loss in dB of the cable at each specified frequency.
3. The measured Insertion Loss of the cable should be within ± 10 percent of the calculated loss in dB when the loss is corrected for temperature.
4. Transmission test sets having an impedance between 100 and 135 ohms on the cable side are acceptable.

Notes:

① H.P.-204B, H.P.-204C, H.P.-355, Siemens-W2057, or equivalent.

② Wilcom-T136, Wilcom-T336, Wilcom-T337, Wilcom-T132B, Siemens-D2057, or equivalent.

(4) *Applicable results.* (i) The highest frequency to be measured is determined by the type of carrier system. For T1 type carrier, the highest frequency is normally 772 kHz. For T1C type carrier, the highest frequency is normally 1576 kHz. The highest frequency to be measured for station carrier is 140 kHz.

(ii) The measured insertion loss of the cable shall be within ±10% of the calculated loss in dB when the loss is corrected for temperature.

(iii) The calculated insertion loss is computed as follows:

(A) Multiply the length of each different gauge by the applicable dB per unit length as shown in Table 7 or 8 as follows:

TABLE 7—CABLE ATTENUATION @ 68 °F (20 °C) FILLED CABLES—SOLID INSULATION

Frequency (kHz)	Attenuation dB/mile (dB/km) Gauge (AWG)			
	19	22	24	26
10	2.8 (1.7)	4.8 (2.9)	6.4 (3.9)	8.5 (5.3)
20	3.2 (2.0)	5.8 (3.6)	8.2 (5.1)	11.2 (6.9)
40	3.6 (2.2)	6.5 (4.0)	9.6 (6.0)	13.9 (8.6)
60	4.0 (2.5)	6.9 (4.2)	10.3 (6.4)	15.2 (9.4)
80	4.5 (2.8)	7.3 (4.5)	10.7 (6.6)	16.0 (9.9)
100	4.9 (3.0)	7.7 (4.7)	11.1 (6.8)	16.5 (10.2)
112	5.2 (3.2)	8.0 (4.9)	11.3 (7.0)	16.8 (10.5)
120	5.4 (3.3)	8.1 (5.0)	11.5 (7.1)	17.0 (10.6)
140	5.8 (3.6)	8.6 (5.3)	11.9 (7.4)	17.4 (10.8)
160	6.2 (3.8)	9.0 (5.6)	12.3 (7.6)	17.8 (11.1)
180	6.6 (4.1)	9.5 (5.9)	12.7 (7.9)	18.2 (11.3)
200	7.0 (4.3)	10.0 (6.2)	13.2 (8.2)	18.6 (11.5)
300	8.7 (5.4)	12.2 (7.5)	15.4 (9.6)	20.6 (12.8)
400	10.0 (6.2)	14.1 (8.8)	17.7 (11.0)	22.9 (14.2)
500	11.2 (6.9)	15.9 (9.8)	19.8 (12.3)	25.2 (15.6)
600	12.2 (7.5)	17.5 (10.9)	21.8 (13.6)	27.4 (17.0)
700	13.2 (8.2)	19.0 (11.8)	23.6 (14.7)	29.6 (18.4)
772	13.8 (8.5)	19.9 (12.4)	24.8 (15.4)	31.4 (19.5)
800	14.2 (8.8)	20.1 (12.5)	27.4 (17.1)	31.7 (19.7)
900	14.8 (9.2)	21.6 (13.4)	29.0 (18.0)	33.8 (21.0)
1000	15.8 (9.8)	22.7 (14.1)	31.1 (19.3)	35.9 (22.3)
1100	16.4 (10.2)	23.8 (14.8)	32.7 (20.3)	38.0 (23.6)
1200	17.4 (10.8)	24.8 (15.4)	34.3 (21.3)	40.0 (24.9)
1300	17.9 (11.1)	25.9 (16.1)	35.4 (22.0)	41.7 (25.9)
1400	19.0 (11.8)	26.9 (16.7)	37.0 (23.0)	43.3 (26.9)
1500	19.5 (12.1)	28.0 (17.4)	38.0 (23.6)	44.3 (27.6)
1576	20.1 (12.4)	29.0 (18.0)	39.0 (24.3)	44.4 (28.2)

TABLE 8—CABLE ATTENUATION @ 68 °F (20 °C) FILLED CABLES—EXPANDED INSULATION

Frequency (kHz)	Attenuation dB/mile (dB/km) Gauge (AWG)			
	19	22	24	26
10	3.0 (1.8)	4.9 (3.0)	6.5 (4.0)	8.6 (5.3)
20	3.5 (2.1)	6.0 (4.1)	8.5 (5.2)	11.5 (7.1)
40	4.0 (2.5)	7.0 (4.3)	10.2 (6.3)	14.4 (8.9)
60	4.5 (2.8)	7.5 (4.6)	11.1 (6.8)	16.0 (9.9)
80	5.2 (3.3)	7.9 (4.9)	11.3 (6.9)	16.2 (10.1)
100	5.8 (3.6)	8.4 (5.2)	11.6 (7.2)	16.4 (10.2)
112	6.0 (3.8)	8.8 (5.4)	11.9 (7.4)	16.6 (10.3)
120	6.2 (3.9)	9.0 (5.6)	12.1 (7.5)	16.9 (10.5)
140	6.6 (4.1)	9.5 (5.9)	12.7 (7.9)	17.2 (10.7)
160	6.9 (4.3)	10.0 (6.2)	13.2 (8.2)	17.4 (10.8)
180	7.4 (4.6)	10.6 (6.6)	13.7 (8.5)	17.9 (11.1)
200	7.9 (4.9)	11.1 (6.9)	14.2 (8.8)	18.5 (11.5)
300	9.5 (5.9)	13.2 (8.2)	16.8 (10.5)	21.6 (13.4)
400	11.1 (6.9)	15.3 (9.5)	19.5 (12.1)	24.3 (15.1)
500	12.1 (7.5)	17.9 (11.1)	22.2 (13.8)	27.4 (17.1)
600	13.7 (8.5)	19.5 (12.1)	24.3 (15.1)	29.6 (18.4)
700	14.8 (9.2)	21.1 (13.1)	26.4 (16.4)	32.2 (20.0)
772	15.3 (9.5)	21.6 (13.4)	27.4 (17.1)	33.8 (21.90)
800	15.8 (9.8)	22.2 (13.8)	28.0 (17.4)	34.4 (21.3)
900	17.0 (10.5)	23.8 (14.8)	29.6 (18.4)	36.4 (22.6)
1000	17.4 (10.8)	24.8 (15.4)	31.1 (19.3)	38.5 (23.9)
1100	17.9 (11.1)	26.4 (16.4)	33.3 (20.7)	40.6 (25.3)
1200	19.0 (11.8)	27.4 (17.1	34.3 (21.3)	42.2 (26.2)
1300	19.5 (12.1)	28.5 (17.7)	35.9 (22.3)	43.8 (27.2)
1400	20.1 (12.5	29.6 (18.4)	37.0 (23.0)	45.9 (28.5)
1500	20.6 (12.8)	30.6 (19.0)	38.5 (23.9)	47.5 (29.5)
1576	21.6 (13.4)	31.1 (19.3)	39.1 (24.3)	48.6 (30.2)

(B) Add the individual losses for each gauge to give the total calculated insertion loss at a temperature of 68 °F (20 °C);

(C) Correct the total calculated insertion loss at the temperature of 68 °F (20 °C) to the measurement temperature by the following formulae:

$A_t = A_{68} \times [1 + 0.0012 \times (t - 68)]$ for English Units

$A_t = A_{20} \times [1 + 0.0022 \times (t - 20)]$ for Metric Units

Where:

A_t = Insertion loss at the measurement temperature in dB.

A_{68} = Insertion loss at a temperature of 68 °F in dB.

A_{20} = Insertion loss at a temperature of 20 °C in dB.

t = Measurement temperature in °F or (°C); and

(D) Compare the calculated insertion loss at the measurement temperature to the measured insertion loss to determine compliance with the requirement specified in paragraph (g)(4)(ii) of this section. (NOTE: Attenuation varies directly with temperature. For each ±10 °F (5.6 °C) change in temperature increase or decrease the attenuation by ±1%.)

(iv) If the measured value exceeds the ±10% allowable variation, the cause shall be determined and corrective action shall be taken to remedy the problem.

(5) *Data record.* Results of carrier frequency insertion loss measurements for station, T1, and/or T1C type carrier shall be recorded. Suggested formats similar to Format III, Outside Plant Acceptance Tests—T1 or T1C Carrier Pairs, and Format IV, Outside Plant Acceptance Tests—Station Carrier Pairs, in § 1755.407 or formats specified in the applicable construction contract may be used.

(6) *Probable causes for nonconformance.* If the measured loss is low, the cable records are likely to be in error. If the measured loss is high, there may be bridge taps, load coils or voice frequency build-out capacitors connected to the cable pairs or the cable records may be in error. Figures 12 and 13 are examples that show the effects of bridge taps and load coils in the carrier path. Figures 12 and 13 are as follows:

FIGURE 12

EFFECTS OF BRIDGE TAPS ON ATTENUATION

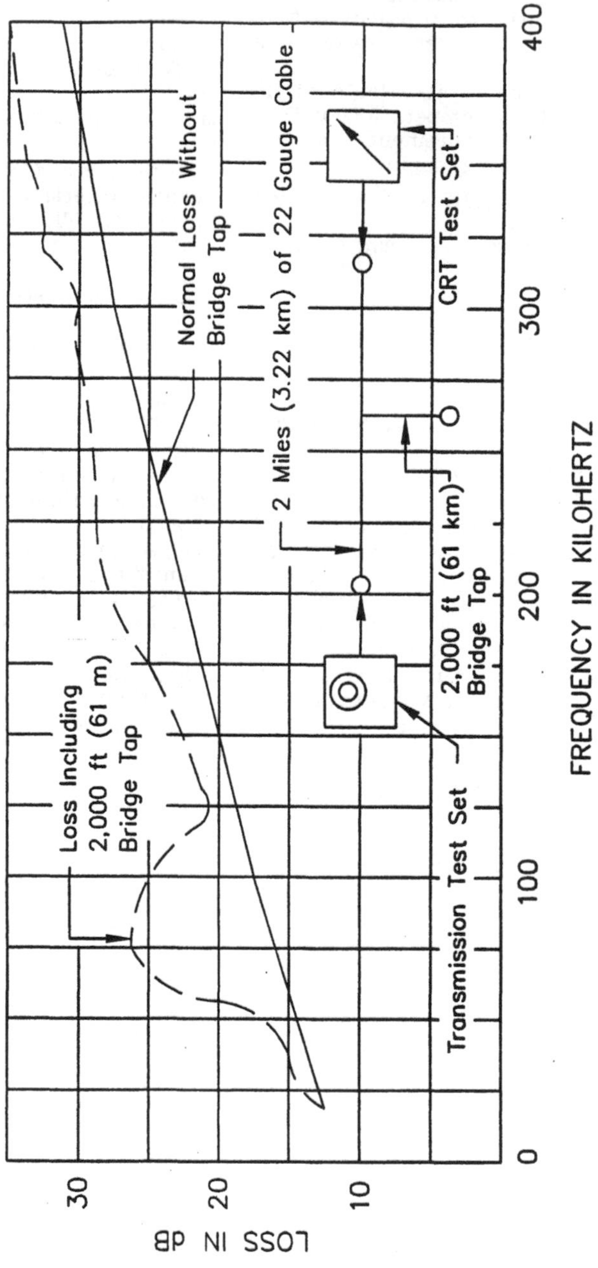

FIGURE 13

EFFECTS OF LOADING COILS ON ATTENUATION

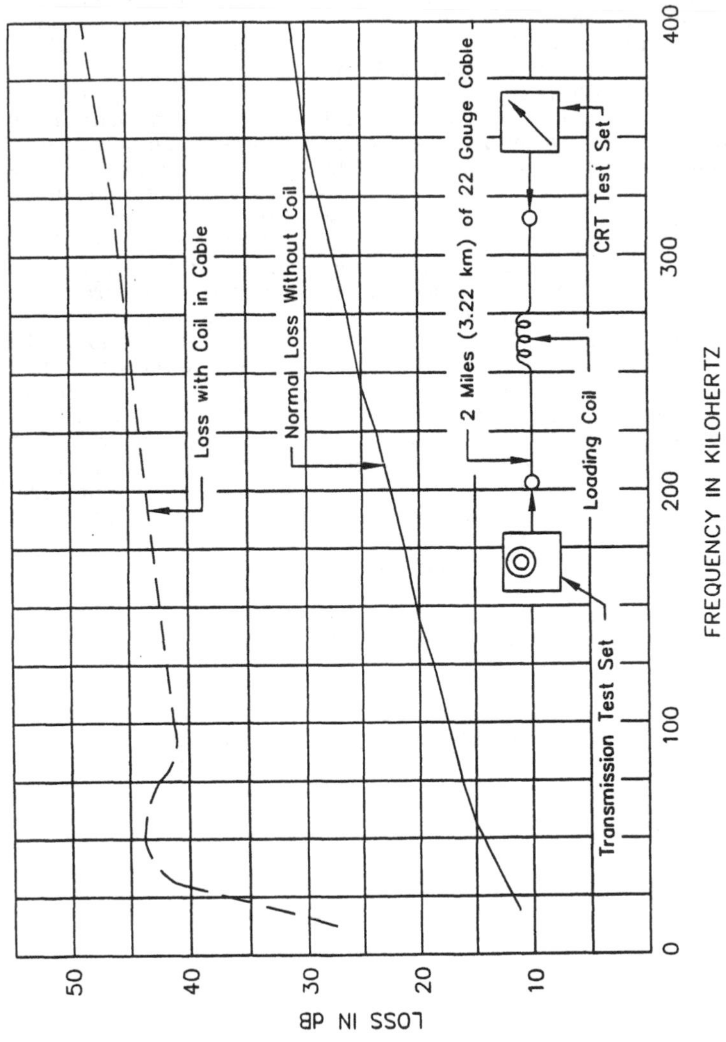

[62 FR 23962, May 2, 1997]

§ 1755.404 Fiber optic cable telecommunications plant measurements.

(a) *Armor continuity*. (1) Tests and measurements shall be made to ensure that the armor of fiber optic cables is continuous. There are two areas of concern. The first is armor bonding within a splice and the second is armor continuity between splices.

(2) Measurement techniques outlined here for verification of armor continuity are applicable to buried fiber optic cable plant. Measurements of armor continuity between splices in aerial, armored, fiber optic cable should be made prior to completion of splicing. Conclusive results cannot be obtained on aerial plant after all bonds have been completed to the supporting strand, multigrounded neutral, etc.

(3) *Method of measurement*. Armor continuity within splices shall be measured with a cable shield splice

continuity test set. The step-by-step measurement procedure outlined in the manufacturer's operating instructions for the specific test equipment being used shall be followed.

(4) *Test equipment.* A cable shield splice continuity tester shall be used to measure armor continuity within splices.

(5) *Applicable results.* When utilizing shield continuity testers to measure armor continuity within splices, refer to the manufacturer's published information covering the specific test equipment to be used and for anticipated results.

(6) *Data record.* Measurement data from armor continuity tests shall be recorded together with anticipated values in an appropriate format to permit comparison. The recorded data shall include specific location, cable size, and cable type, if known, etc.

(7) *Probable causes for nonconformance.* Among probable causes for nonconformance are broken or damaged armors, bad bonding harnesses, poorly connected bonding clamps, loose bonding lugs, etc.

(b) *Fiber optic splice loss measurement.* (1) After placement of all fiber optic cable plant has been completed and spliced together to form a continuous optical link between end termination points, splice loss measurements shall be performed on all field and central office splice points.

(2) *Method of measurement.* (i) Field splice loss measurements shall be made between the end termination points at 1310 and/or 1550 nanometers for single mode fibers and in accordance with Figure 14. Two splice loss measurements shall be made between the end termination points. The first measurement shall be from termination point A to termination point B. The second measurement shall be from termination point B to termination point A.

(ii) CO splice loss measurements shall be made at 1310 and/or 1550 nanometers for single mode fibers and in accordance with Figure 15. Two splice loss measurements shall be made between the end termination points. The first measurement shall be from termination point A to termination point B. The second measurement shall be from termination point B to termination point A.

(3) *Test equipment.* The test equipment is shown in Figures 14 and 15. The optical time domain reflectometer (OTDR) used for the testing should have dual wave length capability. Figures 14 and 15 are as follows:

FIGURE 14

FIBER OPTIC FIELD SPLICE LOSS MEASUREMENT

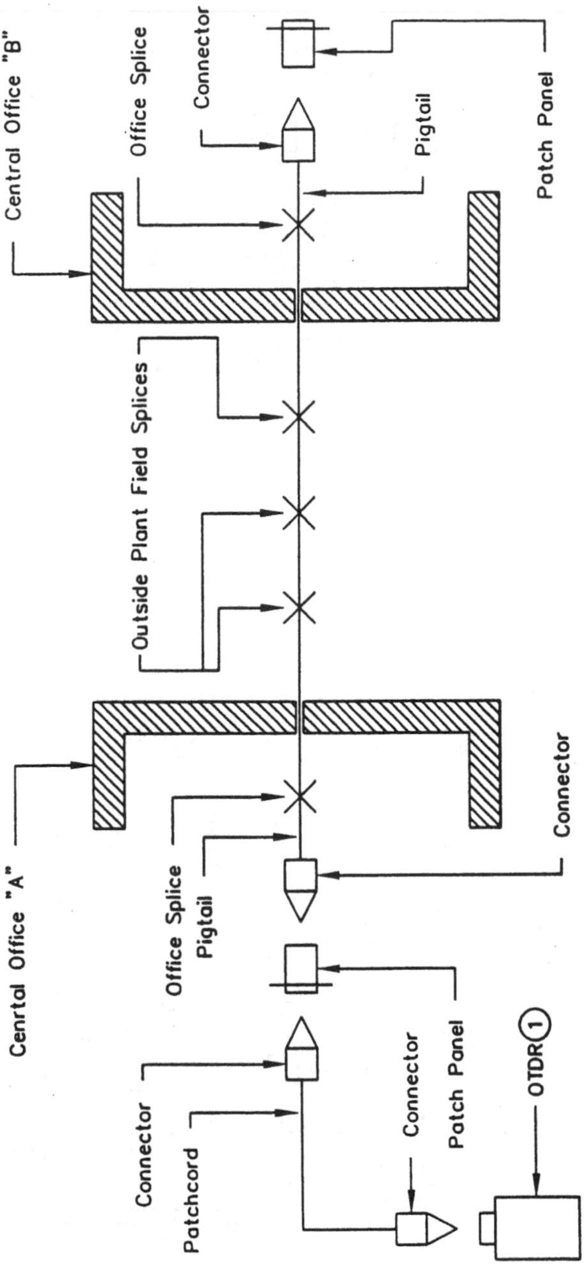

Note:

1. Tektronix–TFP2, H.P.–8146A, Opto–Electronics–DFM10, Photo Kinetics–6000, or equivalent.

FIGURE 15

FIBER OPTIC CENTRAL OFFICE SPLICE LOSS MEASUREMENT

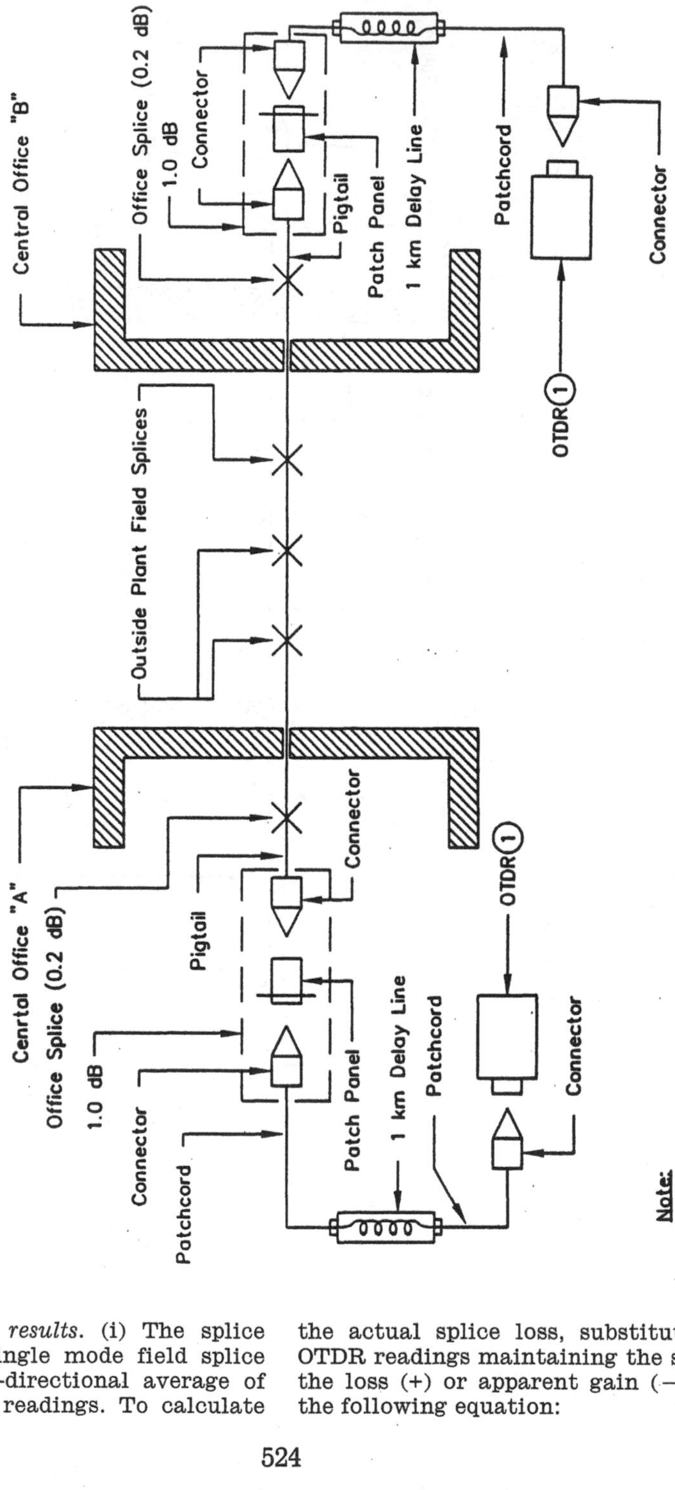

Note:

1. Tektronix–TFP2, H.P.–8146A, Opto–Electonics–DFM10, Photo Kinetics–6000, or equivalent.

(4) *Applicable results.* (i) The splice loss for each single mode field splice shall be the bi-directional average of the two OTDR readings. To calculate the actual splice loss, substitute the OTDR readings maintaining the sign of the loss (+) or apparent gain (−) into the following equation:

$$\text{Actual Splice Loss (dB)} = \frac{\begin{matrix}\text{OTDR Reading} \\ \text{From A to B}\end{matrix} + \begin{matrix}\text{OTDR Reading} \\ \text{From B to A}\end{matrix}}{2}$$

(ii) When specified in the applicable construction contract, the splice loss of each field splice at 1310 and/or 1550 nanometers shall not exceed the limit specified in the contract.

(iii) When no limit is specified in the applicable construction contract, the splice loss of each field splice shall not exceed 0.2 dB at 1310 and/or 1550 nanometers.

(iv) The splice loss for each single mode CO splice shall be the bi-directional average of the two OTDR reading. To calculate actual splice loss, substitute the OTDR reading, maintaining the sign of the loss (+) or apparent gain (−), into the equation specified in paragraph (b)(4)(i) of this section.

(v) When specified in the applicable construction contract, the splice loss of each central office splice at 1310 and/or 1550 nanometers shall not exceed the limit specified in the contract.

(vi) When no limit is specified in the applicable construction contract, the splice loss of each central office splice shall not exceed 1.2 dB at 1310 and/or 1550 nanometers.

(5) *Data record.* The measurement data shall be recorded. A suggested format similar to Format V, Outside Plant Acceptance Test—Fiber Optic Telecommunications Plant, in § 1755.407 or a format specified in the applicable construction contract may be used.

(6) *Probable causes for nonconformance.* When the results of the splice loss measurements exceed the specified limits the following factors should be checked:

(i) Proper end preparation of the fibers;

(ii) End separation between the fiber ends;

(iii) Lateral misalignment of fiber cores;

(iv) Angular misalignment of fiber cores;

(v) Fresnel reflection;

(vi) Contamination between fiber ends;

(vii) Core deformation; or

(viii) Mode-field diameter mismatch.

(c) *End-to-end attenuation measurement.* (1) After placement of all fiber optic cable plant has been completed and spliced together to form a continuous optical link between end termination points, end-to-end attenuation measurements shall be performed on each optical fiber within the cable.

(2) *Method of measurement.* For single mode fibers, the end-to-end attenuation measurements of each optical fiber at 1310 and/or 1550 nanometers in each direction between end termination points shall be performed in accordance with Figure 16.

(3) *Test equipment.* The test equipment is shown in Figure 16 as follows:

FIGURE 16

END–TO–END FIBER OPTIC ATTENUATION MEASUREMENT SHOWING MEASUREMENT IN ONE DIRECTION ONLY (2)

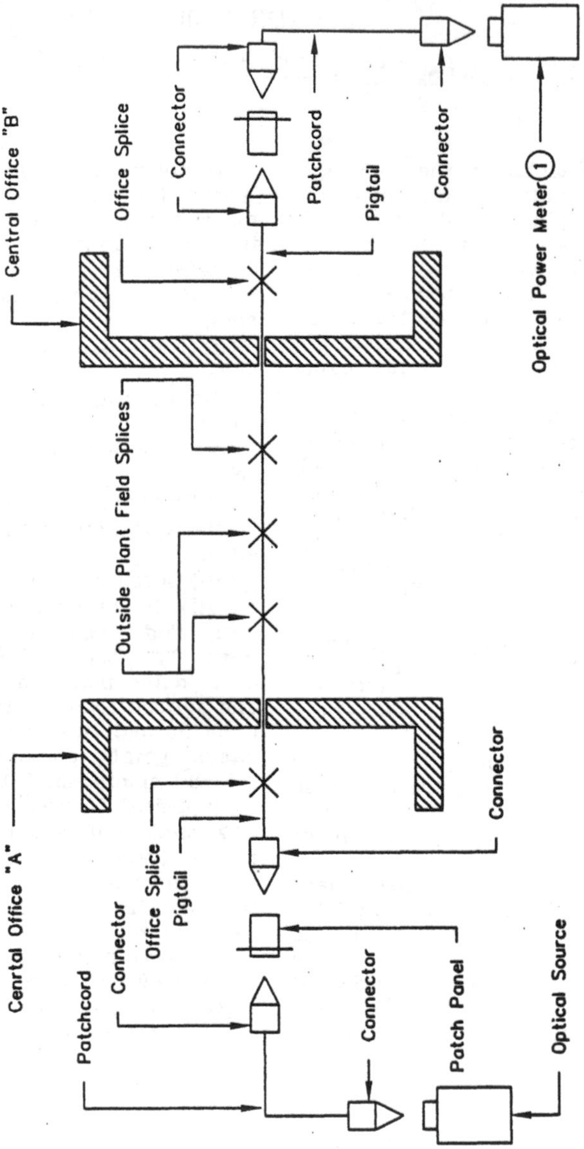

Notes:

(1) H.P.–8153A, Tektronix–OCP5002, Telecommunications Techniques Corp.–131, or equivalent.

(2) Measurement is repeated by reversing location of the optical source and optical power meter in the respective central offices.

(4) *Applicable results.* The end-to-end attenuation of each single mode optical fiber at 1310 and/or 1550 nanometers shall not exceed the limits specified in the applicable construction contract.

(5) *Data record.* The measurement data shall be recorded. A suggested format similar to Format V for fiber optic telecommunications plant in §1755.407 or on a format specified in the applicable construction contract may be used.

(6) *Probable causes for nonconformance.* Failure of each optical fiber to meet the end-to-end attenuation limit could be attributed to the following:

(i) Excessive field or central office splice loss;

(ii) Excessive cable attenuation; or

(iii) Damage to the fiber optic cable during installation.

(d) *End-to-end fiber signature measurement.* (1) After placement of all fiber optic cable plant has been completed and spliced together to form a continuous optical link between end termination points, end-to-end fiber signature testing shall be performed on each optical fiber within the cable.

(2) *Method of measurement.* For single mode fibers, the end-to-end fiber signature measurement of each optical fiber in each direction shall be performed between end termination points at 1310 and/or 1550 nanometers in accordance with Figure 17.

(3) *Test equipment.* The test equipment is shown in Figure 17 as follows:

FIGURE 17

END–TO–END FIBER OPTIC SIGNATURE MEASUREMENT
SHOWING MEASUREMENT IN ONE DIRECTION ONLY ②

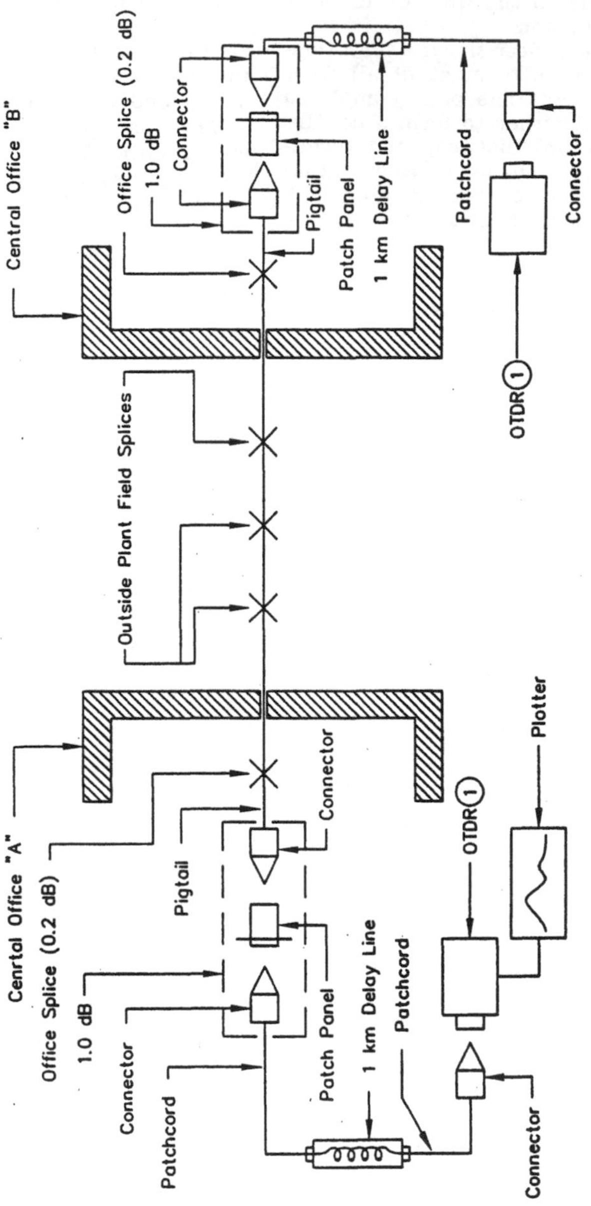

Notes:

① Tektronix–TFP2, H.P.–8146A, Opto–Electonics–DFM10, Photo Kinetics–6000, or equivalent.

② Measurement is repeated by reversing location of optical source and optical power meter in the respective central offices.

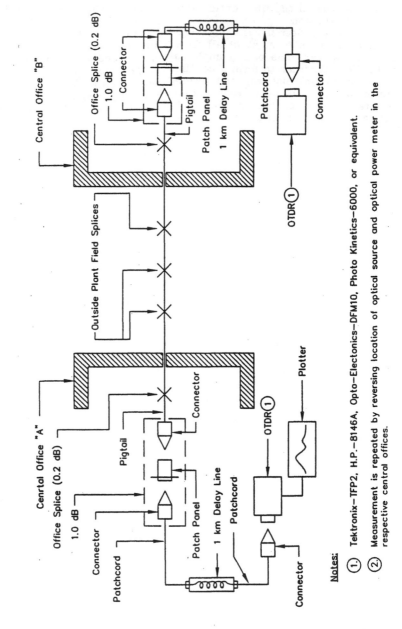

(4) *Applicable results.* The appearance of each optical fiber between end termination points.

(5) *Data record.* Plot the trace of each optical fiber and retain as a permanent record for future comparison if needed.

(6) *Probable causes for nonconformance.* None.

[62 FR 23989, May 2, 1997; 62 FR 25017, May 7, 1997]

§ 1755.405 Voiceband data transmission measurements.

(a) The data transmission measurements listed in this section shall be used to determine the acceptability of trunk and nonloaded subscriber loop circuits for data modem transmission.

(b) *Signal-to-C notched noise (S/CNN) measurement.* (1) When specified by the borrower, S/CNN measurements shall be made on trunk circuits and nonloaded subscriber loops. For trunk circuits, the measurement shall be made between CO locations. For nonloaded subscriber loops, the measurement shall be made from the CO to the station protector of the NID at the customer's access location.

(2) S/CNN is the logarithmic ratio expressed in dB of a 1,004 Hz holding tone signal compared to the C-message weighted noise level. S/CNN is one of the most important transmission parameters affecting the performance of data transmission because proper modem operation requires low noise relative to received power level. Since modulated carriers are used in data communication systems, noise measurements need to be performed with power on the connection to activate equipment having signal-level-dependent noise sources. For 4 kHz channels, a 1,004 Hz holding tone is used to activate the signal-dependent equipment on the channel or connection.

(3) *Method of measurement.* The S/CNN measurement shall be made using a 1,004 Hz holding tone at −13 dBm0 (decibels relative to one milliwatt, referred to a zero transmission level point) and performed in accordance with American National Standards Institute (ANSI) T1.506–1990, American National Standard for Telecommunications—Network Performance—Transmission Specifications for Switched Exchange Access Network including supplement ANSI T1.506a–1992, and American National Standards Institute/Institute of Electrical and Electronics Engineers (ANSI/IEEE) 743–1984, IEEE Standard Methods and Equipment for Measuring the Transmission Characteristics of Analog Voice Frequency Circuits. The ANSI T1.506–1990, American National Standard for Telecommunications—Network Performance—Transmission Specifications for Switched Exchange Access Network is incorporated by reference in accordance with 5 U.S.C. 522(a) and 1 CFR part 51. Copies of ANSI T1.506–1990 are available for inspection during normal business hours at RUS, room 2845, U.S. Department of Agriculture, STOP 1598, Washington, DC 20250–1598, or at the National Archives and Records Administration (NARA). For information on the availability of this material at NARA, call 202–741–6030, or go to: *http://www.archives.gov/federal_register/code_of_federal_regulations/ibr_locations.html.* Copies are available from ANSI, Customer Service, 11 West 42nd Street, New York, New York 10036, telephone number (212) 642–4900. The ANSI/IEEE 743–1984, IEEE Standard Methods and Equipment for Measuring the Transmission Characteristics of Analog Voice Frequency Circuits is incorporated by reference in accordance with 5 U.S.C. 522(a) and 1 CFR part 51. Copies of ANSI/IEEE 743–1984 are available for inspection during normal business hours at RUS, room 2845, U.S. Department of Agriculture, STOP 1598, Washington, DC 20250–1598, or at the National Archives and Records Administration. Copies are available from ANSI, Customer Service, 11 West 42nd Street, New York, New York 10036, telephone number (212) 642–4900.

(4) *Test equipment.* The equipment for performing the measurement shall be in accordance with ANSI/IEEE 743–1984.

(5) *Applicable results.* The S/CNN for both trunk and nonloaded subscriber loop circuits shall not be less than 31 dB.

(6) *Data record.* The measurement data shall be recorded. Suggested formats similar to Format VI, Voiceband Data Transmission Tests—Nonloaded Subscriber Loops, and Format VII, Voiceband Data Transmission Tests—Trunk Circuits, in § 1755.407 or formats specified in the applicable construction contract may be used.

(7) *Probable causes for nonconformance.* Some of the causes for failing to obtain the desired results may be due to excessive harmonic distortion, quantizing noise, phase and amplitude jitter, and loss in digital pads used for level settings.

(c) *Signal-to-intermodulation distortion (S/IMD) measurement.* (1) When specified by the borrower, S/IMD measurements shall be made on trunk circuits and nonloaded subscriber loops. For trunk circuits, the measurement shall be made between CO locations. For nonloaded subscriber loops, the measurement shall be made from the CO to the station protector of the NID at the customer's access location.

(2) S/IMD is a measure of the distortion produced by extraneous frequency cross products, known as intermodulation products, when a multi-tone tone signal is applied to a system.

(3) Intermodulation distortion (IMD) is caused by system nonlinearities acting upon the harmonic frequencies produced from an input of multiple tones. The products resulting from IMD can be more damaging than noise in terms of producing data transmission errors.

(4) IMD is measured as a signal to distortion ratio and is expressed as the logarithmic ratio in dB of the composite power of four resulting test frequencies to the total power of specific higher order distortion products that are produced. The higher order products are measured at both the 2nd order and 3rd order and are designated R2 and R3, respectively. The four frequency testing for IMD is produced with four tones of 857, 863, 1,372, and 1,388 Hz input at a composite power level of −13 dBm0.

(5) *Method of measurement.* The S/IMD measurement shall be performed in accordance with ANSI T1.506–1990 and ANSI/IEEE 743–1984.

(6) *Test equipment.* The equipment for performing the measurement shall be in accordance with ANSI/IEEE 743–1984.

(7) *Applicable results.* The 2nd order (R2) S/IMD for both trunk and nonloaded subscriber loop circuits shall not be less than 40 dB. The 3rd order (R3) S/IMD for both trunk and nonloaded subscriber loop circuits shall not be less than 40 dB.

(8) *Data record.* The measurement data shall be recorded. Suggested formats similar to Format VI for nonloaded subscriber loops and Format VII for trunk circuits in §1755.407 or formats specified in the applicable construction contract may be used.

(9) *Probable causes for nonconformance.* Some of the causes for failing to obtain the desired results may be due to channel nonlinearities, such as compression and clipping, which cause harmonic and intermodulation distortion in a voiceband signal.

(d) *Envelope delay distortion (EDD) measurement.* (1) When specified by the borrower, EDD measurements shall be made on trunk circuits and nonloaded subscriber loops. For trunk circuits, the measurement shall be made between CO locations. For nonloaded subscriber loops, the measurement shall be made from the CO to the station protector of the NID at the customer's access location.

(2) EDD is a measure of the linearity or uniformity of the phase versus frequency characteristics of a transmission facility. EDD is also known as relative envelope delay (RED).

(3) EDD is specifically defined as the delay relative to the envelope delay at the reference frequency of 1,704 Hz. EDD is typically measured at two frequencies, one low and one high in the voiceband. The low frequency measurement is made at 604 Hz. The high frequency measurement is made at 2,804 Hz.

(4) *Method of measurement.* The EDD measurement shall be performed in accordance with ANSI T1.506–1990 and ANSI/IEEE 743–1984.

(5) *Test equipment.* The equipment for performing the measurement shall be in accordance with ANSI/IEEE 743–1984.

(6) *Applicable results.* The EDD for both trunk and nonloaded subscriber loop circuits at the low frequency of 604 Hz shall not exceed 1,500 microseconds. The EDD for both trunk and nonloaded subscriber loop circuits at the high frequency of 2,804 Hz shall not exceed 1,000 microseconds.

(7) *Data record.* The measurement data shall be recorded. Suggested formats similar to Format VI for nonloaded subscriber loops and Format VII for trunk circuits in §1755.407 or formats specified in the applicable construction contract may be used.

(8) *Probable causes for nonconformance.* Some of the causes for failing to obtain the desired results may be due to nonlinearity of the phase versus frequency

characteristic of the transmission facility. This nonlinear phase versus frequency characteristic of the transmission facility causes the various frequency components to travel at different transit times which results in successively transmitted data pulses to overlap at the receive end. The overlapping of the pulses at the receive end results in distortion of the received signal. Excessive EDD on the transmission facility may be reduced using data modems with equalization or by conditioning the transmission line.

(e) *Amplitude jitter (AJ) measurement.* (1) When specified by the borrower, AJ measurements shall be made on trunk circuits and nonloaded subscriber loops. For trunk circuits, the measurement shall be made between CO locations. For nonloaded subscriber loops, the measurement shall be made from the CO to the station protector of the NID at the customer's access location.

(2) AJ is any fluctuation in the peak amplitude value of a fixed tone signal at 1,004 Hz from its nominal value. AJ is expressed in peak percent amplitude modulation.

(3) AJ is measured in two separate frequency bands, 4–300 Hz and 20–300 Hz. The 4–300 Hz band is important for modems employing echo canceling capabilities. The 20–300 Hz band is used for modems that do not employ echo cancelers.

(4) Amplitude modulation can affect the error performance of voiceband data modems. The measurement of amplitude jitter indicates the total effect on the amplitude of the holding tone of incidental amplitude modulation and other sources including quantizing and message noise, impulse noise, gain hits, phase jitter, and additive tones such as single-frequency interference.

(5) *Method of measurement.* The AJ measurement shall be performed in accordance with ANSI T1.506–1990 and ANSI/IEEE 743–1984.

(6) *Test equipment.* The equipment for performing the measurement shall be in accordance with ANSI/IEEE 743–1984.

(7) *Applicable results.* The AJ for both trunk and nonloaded subscriber loop circuits in the 4–300 Hz frequency band shall not exceed 6%. The AJ for both trunk and nonloaded subscriber loop circuits in the 20–300 Hz frequency band shall not exceed 5%.

(8) *Data record.* The measurement data shall be recorded. Suggested formats similar to Format VI for nonloaded subscriber loops and Format VII for trunk circuits in § 1755.407 or formats specified in the applicable construction contract may be used.

(9) *Probable causes for nonconformance.* Some of the causes for failing to obtain the desired results may be due to excessive S/CNN, impulse noise, and phase jitter.

(f) *Phase jitter (PJ) measurement.* (1) When specified by the borrower, PJ measurements shall be made on trunk circuits and nonloaded subscriber loops. For trunk circuits, the measurement shall be made between CO locations. For nonloaded subscriber loops, the measurement shall be made from the CO to the station protector of the NID at the customer's access location.

(2) PJ is any fluctuation in the zero crossings of a fixed tone signal (usually 1,004 Hz) from their nominal position in time within the voiceband. PJ is expressed in terms of either degrees peak-to-peak (° p-p) or in terms of a Unit Interval (UI). One UI is equal to 360° p-p.

(3) PJ measurements are typically performed in two nominal frequency bands. The frequency bands are 20–300 Hz band and either the 2–300 Hz band or the 4–300 Hz band. The 20–300 Hz band is important to all phase-detecting modems. The 4–300 Hz band or the 2–300 Hz band is important for modems employing echo canceling capabilities.

(4) Phase jitter can affect the error performance of voiceband data modems that use phase detection techniques. The measurement of phase jitter indicates the total effect on the holding tone of incidental phase modulation and other sources including quantizing and message noise, impulse noise, phase hits, additive tones such as single-frequency interference, and digital timing jitter.

(5) *Method of measurement.* The PJ measurement shall be performed in accordance with ANSI T1.506–1990 and ANSI/IEEE 743–1984.

(6) *Test equipment.* The equipment for performing the measurement shall be in accordance with ANSI/IEEE 743–1984.

(7) *Applicable results.* The PJ for both trunk and nonloaded subscriber loop circuits in the 4–300 Hz frequency band shall not exceed 6.5° p-p. The PJ for both trunk and nonloaded subscriber loop circuits in the 20–300 Hz frequency band shall not exceed 10.0° p-p.

(8) *Data record.* The measurement data shall be recorded. Suggested formats similar to Format VI for nonloaded subscriber loops and Format VII for trunk circuits in §1755.407 or formats specified in the applicable construction contract may be used.

(9) *Probable causes for nonconformance.* Some of the causes for failing to obtain the desired results may be due to excessive S/CNN, impulse noise, and amplitude jitter.

(g) *Impulse noise measurement.* (1) When specified by the borrower, impulse noise measurements shall be made on trunk circuits and nonloaded subscriber loops. For trunk circuits, the measurement shall be made between CO locations. For nonloaded subscriber loops, the measurement shall be made from the CO to the station protector of the NID at the customer's access location.

(2) Impulse noise is a measure of the presence of unusually large noise excursions of short duration that are beyond the normal background noise levels on a facility. Impulse noise is typically measured by counting the number of occurrences beyond a particular noise reference threshold in a given time interval. The noise reference level is C-message weighted.

(3) *Method of measurement.* The impulse noise measurement shall be performed using a 1,004 Hz tone at −13 dBm0 and in accordance with ANSI T1.506–1990 and ANSI/IEEE 743–1984.

(4) *Test equipment.* The equipment for performing the measurement shall be in accordance with ANSI/IEEE 743–1984.

(5) *Applicable results.* The impulse noise for both trunk and nonloaded subscriber loop circuits shall not exceed 65 dBrnC0 (decibels relative to one picowatt reference noise level, measured with C-message frequency weighting, referred to a zero transmission level point). The impulse noise requirement shall be based upon a maximum of 5 counts in a 5 minute period at equal to or greater than the indicated noise thresholds.

(6) *Data record.* The measurement data shall be recorded. Suggested formats similar to Format VI for nonloaded subscriber loops and Format VII for trunk circuits in §1755.407 or formats specified in the applicable construction contract may be used.

(7) *Probable causes for nonconformance.* Some of the causes for failing to obtain the desired results may be due to excessive transient signals originating from the various switching operations.

[62 FR 23996, May 2, 1997, as amended at 69 FR 18803, Apr. 9, 2004]

§ 1755.406 Shield or armor ground resistance measurements.

(a) Shield or armor ground resistance measurements shall be made on completed lengths of copper cable and wire plant and fiber optic cable plant.

(b) *Method of measurement.* (1) The shield or armor ground resistance measurement shall be made between the copper cable and wire shield and ground and between the fiber optic cable armor and ground, respectively. The measurement shall be made either on cable and wire lengths before splicing and before any ground connections are made to the cable or wire shields or armors. Optionally, the measurement may be made on cable and wire lengths after splicing, but all ground connections must be removed from the section under test.

(2) The method of measurement using either an insulation resistance test set or a dc bridge type megohmmeter shall be as shown in Figure 18 as follows:

FIGURE 18

SHIELD OR ARMOR GROUND RESISTANCE MEASUREMENT

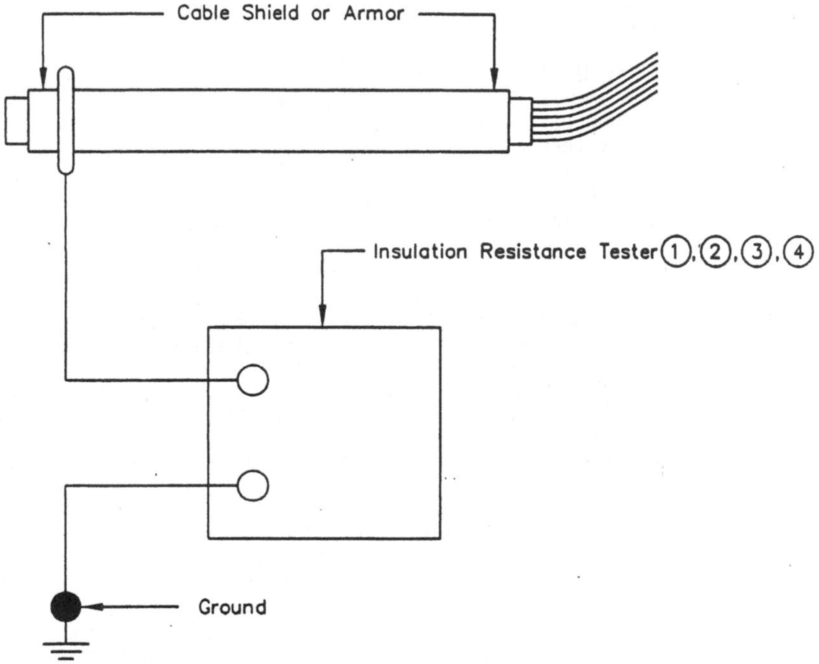

Notes:

(1) For hand cranked or battery operated Insulation Resistance Testers, the output voltage should not exceed 500 volts dc.

(2) For dc bridge type Megohmmeters, the voltage applied to the shield or armor under test should not be less than 250 volts dc nor greater than 1000 volts dc when using instruments having adjustable test voltage levels.

(3) When the distance between test points results in a measurement beyond the range of the test equipment, extended range devices recommended by the test equipment manufacturer may be used to assist in making the measurement.

(4) Biddle CO.—Model BM 200, Associate Research—Model 263, General Radio—1864 Megohm Meter, or equivalent.

(c) *Test equipment.* (1) The shield or armor ground resistance measurements may be made using an insulation resistance test set, a dc bridge type megohmmeter, or a commercially available fault locator.

(2) The insulation resistance test set should have an output voltage not to exceed 500 volts dc and may be hand cranked or battery operated.

(3) The dc bridge type megohmmeter, which may be ac powered, should have

scales and multipliers which make it possible to accurately read resistance values of 50,000 ohms to 10 megohms. The voltage that is applied to the shield or armor during the test should not be less than "250 volts dc" nor greater than "1,000 volts dc" when using an instrument having adjustable test voltage levels.

(4) Commercially available fault locators may be used in lieu of the above equipment, if the devices are capable of detecting faults having resistance values of 50,000 ohms to 10 megohms. Operation of the devices and method of locating the faults should be in accordance with manufacturer's instructions.

(d) *Applicable results.* (1) For all new copper cable and wire facilities and all new fiber optic cable facilities, the shield or armor ground resistance levels normally exceed 1 megohm-mile (1.6 megohm-km) at 68 °F (20 °C). A value of 100,000 ohm-mile (161,000 ohm-km) at 68 °F (20 °C) shall be the minimum acceptable value of the shield or armor ground resistance.

(2) Shield or armor ground resistance varies inversely with length and temperature. In addition other factors which may affect readings could be soil conditions, faulty test equipment and incorrect test procedures.

(3) For the resistance test method and dc bridge type megohmmeter, the ohm-mile (ohm-km) value for the shield or armor ground resistance shall be computed by multiplying the actual scale reading in ohms on the test set by the length in miles (km) of the cable or wire under test.

(4)(i) The objective shield or armor ground resistance may be determined by dividing 100,000 by the length in miles (161,000 by the length in km) of the cable or wire under test. The resulting value is the minimum acceptable meter scale reading in ohms. Examples for paragraphs (d)(3) and (d)(4) of this section are as follows:

Equation 1. Test Set: Scale Reading * Length = Resistance-Length

75,000 ohms * 3 miles = 225,000 ohm-mile
(75,000 ohms * 4.9 km = 367,000 ohm-km)

Equation 2. 100,000 ohm-mile ÷ Length = Minimum Acceptable Meter Scale Reading

100,000 ohm-mile ÷ 3 miles = 33,333 ohms
(161,000 ohm-km ÷ 4.9 km = 32,857 ohms)

(ii) Since the 33,333 ohms (32,857 ohms) is the minimum acceptable meter scale reading and the meter scale reading was 75,000 ohms, the cable is considered to have met the 100,000 ohm-mile (161,000 ohm-km) requirement.

(5) Due to the differences between various jacketing materials used in manufacturing cable or wire and to varying soil conditions, it is impractical to provide simple factors to predict the magnitude of variation in shield or armor to ground resistance due to temperature. The variations can, however, be substantial for wide excursions in temperature from the ambient temperature of 68 °F (20 °C).

(e) *Data record.* The data shall be corrected to the length requirement of ohm-mile (ohm-km) and a temperature of 68 °F (20 °C) and shall be recorded on a form specified in the applicable construction contract.

(f) *Probable causes for nonconformance.* (1) When results of resistance measurements are below the 100,000 ohm-mile (161,000 ohm-km) requirement at 68 °F (20 °C), the jacket temperature, soil conditions, test equipment and method shall be reviewed before the cable or wire is considered a failure. If the temperature is approximately 68 °F (20 °C) and soil conditions are acceptable, and a reading of less than 100,000 ohm-mile (161,000 ohm-km) is indicated, check the calibration of the equipment; as well as, the test method. If the equipment was found to be out of calibration, recalibrate the equipment and remeasure the cable or wire. If the temperature was 86 °F (30 °C) or higher, the cable or wire shall be remeasured at a time when the temperature is approximately 68 °F (20 °C). If the test was performed in unusually wet soil, the cable or wire shall be retested after the soil has reached normal conditions. If after completion of the above steps, the resistance value of 100,000 ohm-mile (161,000 ohm-km) or greater is obtained, the cable or wire shall be considered acceptable.

(2) When the resistance value of the cable or wire is still found to be below 100,000 ohm-mile (161,000 ohm-km) requirement after completion of the steps listed in paragraph (f)(1) of this

section, the fault shall be isolated by performing shield or armor ground resistance measurements on individual cable or wire sections.

(3) Once the fault or faults have been isolated, the cable or wire jacket shall be repaired in accordance with §1755.200, RUS Standard for Splicing Copper and Fiber Optic Cables or the entire cable or wire section may be replaced at the request of the borrower.

[62 FR 23998, May 2, 1997]

§ 1755.407 Data formats.

The following suggested formats listed in this section may be used for recording the test data:

FORMAT I

OUTSIDE PLANT ACCEPTANCE TESTS - SUBSCRIBER LOOPS

PROJECT: ______ Date of Test: ______

CO NAME OF LOCATION: ______ Time Measured: ______ Tester (Contractor): ______

CO or RST GROUND RESISTANCE: ______ Ohms (Before tying to elect. neutral) Temperature: ______ Tester (Engineer): ______

CO or RST GROUND RESISTANCE: ______ Ohms (After tying to elect. neutral) Soil Type: ______ Tester (Borrower): ______

Test Equip.: ______ Moisture Content of Soil: ______

Route No.	Pair No.	Length (Miles or km)	DC Insul. Resist. (Megohms)			DC Loop Resist. (Ohms)			Measd. DC Res. Unbal. (Ohms)	Open Ckt Meas. Type (a,b,c,d)		Insertion Loss at kHz Shown						Line Noise			
			T-GD	R-GD	Min. Permit	Comp. 68°F (20°C)	Measd. ___°F (___°C)	Corr. 68°F (20°C)		SRL (dB)	Freq. (kHz)	1.0	1.7	2.3	2.8	3.4	4.0	NMC	NG-C	NGF	Bal.

Shield or Shield/Armor Continuity Data has been attached. Yes ___ No ___

FORMAT II

OUTSIDE PLANT ACCEPTANCE TESTS – TRUNKS CIRCUITS

PROJECT: ______ Date of Test: ______

CO NAME OF LOCATION: ______ Tester (Contractor): ______

OFFICE A: ______ Tester (Engineer): ______

OFFICE B: ______ Tester (Borrower): ______

ELECTRONIC EQUIPMENT GROUND RESISTANCE: ______ Ohms

Time Measured: ______ Soil Type: ______ Test Equip: ______

Temperature: ______ Moisture Content of Soil: ______

In the space below show in a simple line diagram the facility makeup including all gauges, lengths, cable types, and repeater locations if any.

Trunk No.	Pair No. Off. A	Pair No. Off. B	Length Miles or km	DC Insul. Resist. (Megohms)			DC Loop Resist. (Ohms)			Measd DC Res. Unbal. (Ohms)
				T–GD	R–GD	Min. Permit	Comp. 68°F (20°C)	Measd. __°F (__°C)	Corr. 68°F (20°C)	

FORMAT III

OUTSIDE PLANT ACCEPTANCE TESTS – T1 or T1C CARRIER PAIRS

PROJECT: ______ Type of Proposed Carrier: ______ (Trunk – Subscriber)

LOCATION: From ______ (CO Name) to ______ (CO Name) Shield or Shield/Armor Continuity has been checked: ______

Aerial: ______ Buried: ______ Weather: ______ Temp.: ______ Date: ______ Sheet ______ of ______

CARRIER FREQUENCY INSERTION LOSS MEASUREMENTS (1)

From ______ to ______					From ______ to ______				
Freq (2) (kHz)	Send Level (dBm)	Receive Level(dBm)	Measured Loss(dB)	Estimated (3) Loss (dB)	Freq (2) (kHz)	Send Level (dBm)	Receive Level(dBm)	Measured Loss(dB)	Estimated (3) Loss (dB)
20					20				
60					60				
100					100				
140					140				
180					180				
200					200				
300					300				
400					400				
600					600				
700					700				
772					772				
800					800				
1000					1000				
1200					1200				
1300					1300				
1400					1400				
1500					1500				
1576					1576				

Notes: (1.) Refer to RUS TE&CM 925 on How to Make Measurements. (2.) Go as high in frequency as required by contract.
(3.) From either Table 7 or 8 in Paragraph (g)(4)(iii)(A) of Section 1755.403; Correct loss for temperature.

FORMAT IV

OUTSIDE PLANT ACCEPTANCE TESTS – STATION CARRIER PAIRS

PROJECT: ______________________ Type of Proposed Carrier: ______________ (Trunk – Subscriber)

LOCATION: From ______________ (CO Name) to ______________ (Sub.) Shield or Shield/Armor Continuity has been checked: ______

Aerial: ________ Buried: ________ Weather: ________ Temp.: ________ Date: ________ Sheet ____ of ____

CARRIER FREQUENCY INSERTION LOSS MEASUREMENTS (1)

From______ to______					From______ to______				
Freq. (kHz)	Send Level (dBm)	Receive Level(dBm)	Measured Loss(dB)	Estimated (2) Loss (dB)	Freq. (kHz)	Send Level (dBm)	Receive Level(dBm)	Measured Loss(dB)	Estimated (2) Loss (dB)
20					20				
60					60				
100					100				
112					112				
140					140				

From______ to______					From______ to______				
Freq. (kHz)	Send Level (dBm)	Receive Level(dBm)	Measured Loss(dB)	Estimated (2) Loss (dB)	Freq. (kHz)	Send Level (dBm)	Receive Level(dBm)	Measured Loss(dB)	Estimated (2) Loss (dB)
20					20				
60					60				
100					100				
112					112				
140					140				

Notes:

(1) Refer to RUS TE&CM 925 on How to Make Measurements.

(2) From either Table 7 or 8 in Paragraph (g)(4)(iii)(A) of Section 1755.403; correct loss for temperature.

FORMAT V

OUTSIDE PLANT ACCEPTANCE TESTS

FIBER OPTIC TELECOMMUNICATIONS PLANT

PROJECT: ______________________ Date of Test: ______________________

TERMINATION POINT A: ____________ Tester (Contractor): ______________

TERMINATION POINT B: ____________ Tester (Engineer): ________________

Time Measured: ______ Tester (Borrower): ________________

Temperature: __________ Test Equip: ________________

Soil Type: __________ Moisture Content of Soil: __________

Route No.	Fiber No.	Length Miles or km	Splice Loss (dB)		End-to-End Attenuation (dB/km)	End-to-End Fiber Signature	
			FIELD	CO		Yes	No

Armor Continuity Data has been attached. Yes ___ No ___

FORMAT VI

VOICEBAND DATA TRANSMISSION TESTS – NONLOADED SUBSCRIBER LOOPS

PROJECT: ____________ Date of Test: ____________

AERIAL: ________ Tester (Contractor): ____________

LOCATION: From ____________ (CO Name) to ____________ (Sub. Name) BURIED: ________ Tester (Engineer): ____________

TEMPERATURE: ________ DATE: ________ UNDERGROUND: ________ Tester (Borrower): ____________

Test Equip.: ____________ Sheet ____ of ____

Route No.	Pair No.	Length (Miles or km)	S/CNN 1,004 Hz Tone at −13 dBm0 (dB)	S/IMD (dB)		Impulse Noise (dBrnC0)	EDD (Microseconds)		AJ (%)		PJ (°p–p)	
				R2	R3		604 Hz	2,804 Hz	4 to 300 Hz	20 to 300 Hz	4 to 300 Hz	20 to 300 Hz

FORMAT VII

VOICEBAND DATA TRANSMISSION TESTS – TRUNK CIRCUITS

PROJECT: ______ Date of Test: ______

LOCATION: From ______ (CO Name) to ______ (CO Name) AERIAL: ______ Tester (Contractor): ______

TEMPERATURE: ______ DATE: ______ BURIED: ______ Tester (Engineer): ______

UNDERGROUND: ______ Tester (Borrower): ______

Test Equip.: ______ Sheet ____ of ____

Trunk No.	Pair No. CO-A	Pair No. CO-B	Length (Miles or km)	S/CNN 1,004 Hz Tone at −13 dBm0 (dB)	S/IMD (dB)		Impulse Noise (dBrnC0)	EDD (Microseconds)		AJ (%)		PJ (°p-p)	
					R2	R3		604 Hz	2,804 Hz	4 to 300 Hz	20 to 300 Hz	4 to 300 Hz	20 to 300 Hz

[62 FR 24000, May 2, 1997]

§§ 1755.408–1755.499 [Reserved]

§ 1755.500 RUS standard for service installations at customers access locations.

(a) Sections 1755.501 through 1755.510 cover service installations at permanent or mobile home customer access locations. Sections 1755.501 through 1755.510 do not cover service installations at customer access locations associated with boat yards or marinas.

(b) Service installations for customer access locations in boat yards or marinas shall be performed in accordance with Article 800, Communications Circuits, of the American National Standards Institute/National Fire Protection Association (ANSI/NFPA) 70–1999, *National Electrical Code®* (*NEC®*). The *National Electrical Code®* and *NEC®* are registered trademarks of the National Fire Protection Association, Inc., Quincy, MA 02269. The ANSI/NFPA 70–

1999, *NEC*® is incorporated by reference in accordance with 5 U.S.C. 552(a) and 1 CFR part 51. Copies are available from NFPA, 1 Batterymarch Park, P.O. Box 9101, Quincy, Massachusetts 02269–9101, telephone number 1 (800) 344–3555. Copies of ANSI/NFPA 70–1999, *NEC*®, are available for inspection during normal business hours at Rural Utilities Service (RUS), room 2905, U.S. Department of Agriculture, 1400 Independence Avenue, SW., STOP 1598, Washington, DC 20250–1598, or at the National Archives and Records Administration (NARA). For information on the availability of this material at NARA, call 202–741–6030, or go to: *http://www.archives.gov/federal_register/code_of_federal_regulations/ibr_locations.html.*

[66 FR 43317, Aug. 17, 2001, as amended at 69 FR 18803, Apr. 9, 2004]

§ 1755.501 Definitions applicable to §§ 1755.501 through 1755.510.

For the purpose of this section and §§ 1755.502 through 1755.510, the following terms are defined as follows:

American National Standards Institute (ANSI). A private sector standards coordinating body which serves as the United States source and information center for all American National Standards.

Ampacity. As defined in the ANSI/NFPA 70–1999, *NEC*®: The current, in amperes, that a conductor can carry continuously under the conditions of use without exceeding its temperature rating. (Reprinted with permission from NFPA 70–1999, the *National Electrical Code*®, Copyright © 1998, National Fire Protection Association, Quincy, MA 02269. This reprinted material is not the complete and official position of the National Fire Protection Association, on the referenced subject which is represented only by the standard in its entirety.) The *National Electrical Code*® and *NEC*® are registered trademarks of the National Fire Protection Association, Inc., Quincy, MA 02269. The ANSI/NFPA 70–1999, *NEC*®, is incorporated by reference in accordance with 5 U.S.C. 552(a) and 1 CFR part 51. Copies are available from NFPA, 1 Batterymarch Park, P.O. Box 9101, Quincy, Massachusetts 02269–9101, telephone number 1 (800) 344–3555. Copies of ANSI/NFPA 70–1999, *NEC*®, are available for inspection during normal business hours at RUS, room 2905, U.S. Department of Agriculture, 1400 Independence Avenue, SW., STOP 1598, Washington, DC 20250–1598, or at the National Archives and Records Administration (NARA). For information on the availability of this material at NARA, call 202–741–6030, or go to: *http://www.archives.gov/federal_register/code_of_federal_regulations/ibr_locations.html.*

AWG. American Wire Gauge.

BET. Building entrance terminal.

Bonding (Bonded). As defined in the ANSI/NFPA 70–1999, *NEC*®: The permanent joining of metallic parts to form an electrically conductive path that will ensure electrical continuity and the capacity to conduct safely any current likely to be imposed. (Reprinted with permission from NFPA 70–1999, the *National Electrical Code*®, Copyright © 1998, National Fire Protection Association, Quincy, MA 02269. This reprinted material is not the complete and official position of the National Fire Protection Association, on the referenced subject which is represented only by the standard in its entirety.)

Bonding harness wire. A reliable electrical conductor purposefully connected between metal parts which are required to be electrically connected (bonded) to one another to ensure the metal parts are at similar electrical potential.

Building entrance terminal (BET). A BET is comprised of a housing suitable for indoor and outdoor installation which contains quick-connect or binding post terminals for terminating both telecommunications service cable conductors and inside wiring cable conductors. The BET also includes primary station protectors and a means of terminating the metallic shields of service entrance cables.

Demarcation point (DP). As defined in the Federal Communications Commission (FCC) rules in 47 CFR part 68: The point of demarcation or interconnection between telecommunications company communications facilities and terminal equipment, protective apparatus, or wiring at a subscriber's premises. Carrier-installed facilities at, or constituting, the demarcation point shall consist of wire or a jack conforming to subpart F of 47 CFR part 68. "Premises" as used herein generally means a dwelling unit, other building or a legal unit of real property such as a lot on which a dwelling unit is located, as determined by the telecommunications company's reasonable and nondiscriminatory standard operating practices. The "minimum point of entry" as used herein shall be either the closest practicable point to where the wiring crosses a property line or the closest practicable point to where the wiring enters a multiunit building or buildings. The telecommunications company's reasonable and nondiscriminatory standard operating practices shall determine which shall apply. The telecommunications company is not precluded from establishing reasonable clarifications of multiunit premises for determining which shall apply. Multiunit premises include, but are not limited to, residential, commercial, shopping center, and campus situations.

(1) *Single unit installations.* For single unit installations existing as of August 13, 1990, and installations installed after that date, the demarcation point shall be a point within 12 inches (in.) (305 millimeters (mm)) of the primary protector, where there is no protector, within 12 in. (305 mm) of where the telecommunications wire enters the customer's premises.

(2) *Multiunit installations.* (i) In multiunit premises existing as of August 13, 1990, the demarcation point shall be determined in accordance with the local carrier's reasonable and nondiscriminatory standard operating practices. Provided, however, that where there are multiple demarcation points within the multiunit premises, a demarcation point for a customer shall not be further inside the customer's premises than a point 12 in. (305 mm) from where the wiring enters the customer's premises.

(ii) In multiunit premises in which wiring is installed after August 13, 1990, including additions, modifications, and rearrangements of wiring existing prior to that date, the telecommunications company may establish a reasonable and nondiscriminatory practice of placing the demarcation point at the minimum point of entry. If the telecommunications company does not elect to establish a practice of placing the demarcation point at the minimum point of entry, the multiunit premises owner shall determine the location of the demarcation point or points. The multiunit premises owner shall determine whether there shall be a single demarcation point for all customers or separate such locations for each customer. Provided, however, that where there are multiple demarcation points within the multiunit premises, a demarcation point for a customer shall not be further inside the customer's premises than a point 12 in. (305 mm) from where the wiring enters the customer's premises.

DP. Demarcation point.

Eligible country. Any country that applies with respect to the United States an agreement ensuring reciprocal access for United States products and services and United States suppliers to the markets of that country, as determined by the United States Trade Representative.

FCC. Federal Communications Commission.

Fuse link. As defined in the ANSI/NFPA 70–1999, *NEC*®: A fine gauge section of wire or cable that serves as a fuse (that is, open-circuits to interrupt the current should it become excessive) that coordinates with the telecommunications cable and wire plant, and protective devices. (Reprinted with permission from NFPA 70–1999, the *National Electrical Code*®, Copyright© 1998, National Fire Protection Association, Quincy, MA 02269. This reprinted material is not the complete and official position of the National Fire Protection Association, on the referenced subject which is represented only by the standard in its entirety.)

Grounding conductor. As defined in the ANSI/NFPA 70–1999, *NEC*®: A conductor used to connect equipment or the grounded circuit of a wiring system to a grounding electrode or electrodes. (Reprinted with permission from NFPA 70–1999, the *National Electrical Code*®, Copyright© 1998, National Fire Protection Association, Quincy, MA 02269. This reprinted material is not the complete and official position of the National Fire Protection Association, on the referenced subject which is represented only by the standard in its entirety.)

Listed. As defined in the ANSI/NFPA 70–1999, *NEC*®: Equipment, materials, or services included in a list published by an organization that is acceptable to the authority having jurisdiction and concerned with evaluation of products or services, that maintains periodic inspection of production of listed equipment or materials or periodic evaluation of services, and whose listing states that either the equipment, material, or services meets identified standards or has been tested and found suitable for a specified purpose. (Reprinted with permission from NFPA 70–1999, the *National Electrical Code*®, Copyright© 1998, National Fire Protection Association, Quincy, MA 02269. This reprinted material is not the complete and official position of the National Fire Protection Association, on the referenced subject which is represented only by the standard in its entirety.)

Manufactured home. As defined in the ANSI/NFPA 70–1999, *NEC*®: A factory-assembled structure or structures that bears a label identifying it as a manufactured home that is transportable in one or more sections, that is built on a permanent chassis and designed to be used as a dwelling with or without a permanent foundation where connected to the required utilities, and includes the plumbing, heating, air conditioning, and electric systems contained therein. Unless otherwise indicated, the term "mobile home" includes manufactured homes. Fine Print Note (FPN) No. 1: See the applicable building code for definition of the term permanent foundation. FPN No. 2: See 24 CFR part 3280, Manufactured Home Construction and Safety Standards, of the Federal Department of Housing and Urban Development for additional information on the definition. (Reprinted with permission from NFPA 70–1999, the *National Electrical Code*®, Copyright© 1998, National Fire Protection Association, Quincy, MA 02269. This reprinted material is not the complete and official position of the National Fire Protection Association, on the referenced subject which is represented only by the standard in its entirety.)

Mobile home. As defined in the ANSI/NFPA 70–1999, *NEC*®: A factory-assembled structure or structures transportable in one or more sections that is built on a permanent chassis and designed to be used as a dwelling without a permanent foundation where connected to the required utilities, and includes the plumbing, heating, air-conditioning, and electric systems contained therein. Unless otherwise indicated, the term "mobile home" includes manufactured homes. (Reprinted with permission from NFPA 70–1999, the *National Electrical Code*®, Copyright© 1998, National Fire Protection Association, Quincy, MA 02269. This reprinted material is not the complete and official position of the National Fire Protection Association, on the referenced. subject which is represented only by the standard in its entirety.)

Motor home. As defined in the ANSI/NFPA 70–1999, *NEC*®: A vehicular unit designed to provide temporary living quarters for recreational, camping, or travel use built on or permanently attached to a self-propelled motor vehicle chassis or on a chassis cab or van that is an integral part of the completed vehicle. (Reprinted with permission from NFPA 70–1999, the *National Electrical Code*®, Copyright© 1998, National Fire Protection Association, Quincy, MA 02269. This reprinted material is not the complete and official position of the National Fire Protection Association, on the referenced subject which is represented only by the standard in its entirety.)

Network interface device (NID). A NID is comprised of a housing suitable for outdoor installation which contains a compartment accessible by only telecommunications employees which includes a primary station protector and the means for terminating telecommunications service wire conductors and metallic shields, and a compartment accessible by customers which includes an RJ–11 plug and jack of the type specified in the FCC rules in 47 CFR part 68.

NID. Network interface device.

Primary station protector. An assembly which complies with RUS Bulletin 345–39, RUS Specification for Telephone

Station Protectors. Copies of RUS Bulletin 345–39 are available upon request from RUS, U.S. Department of Agriculture (USDA), 1400 Independence Avenue, SW., STOP 1522, Washington, DC 20250–1522, FAX (202) 720–4120.

Qualified Installer. A person who has extensive installation experience, complete knowledge and understanding of RUS Bulletin 1751F–805, Electrical Protection At Customer Locations; RUS Bulletin 1753F–153 (RUS Form 515d), Specifications and Drawings for Service Installations at Customer Access Locations, and applicable portions of the ANSI/NFPA 70–1999, *NEC*®, and ANSI/IEEE C2–1997, NESC. Copies of RUS Bulletins 1751F–805 and 1753F–153 are available upon request from RUS/USDA, 1400 Independence Avenue, SW., STOP 1522, Washington, DC 20250–1522, FAX (202) 720–4120.

Recreational vehicle. As defined in the ANSI/NFPA 70–1999, *NEC*®: A vehicular-type unit primarily designed as temporary living quarters for recreational, camping, or travel use, which either has its own motive power or is mounted on or drawn by another vehicle. The basic entities are: travel trailer, camping trailer, truck camper, and motor home. (Reprinted with permission from NFPA 70–1999, the *National Electrical Code*®, Copyright © 1998, National Fire Protection Association, Quincy, MA 02269. This reprinted material is not the complete and official position of the National Fire Protection Association, on the referenced subject which is represented only by the standard in its entirety.)

RUS. Rural Utilities Service.

RUS accepted (material and equipment). Equipment which RUS has reviewed and determined that:

(1) Final assembly or manufacture of the equipment is completed in the United States, its territories and possessions, or in an eligible country;

(2) The cost of components within the material or equipment manufactured in the United States, its territories and possessions, or in an eligible country is more than 50 percent of the total cost of all components used in the material or equipment; and

(3) The material or equipment is suitable for use on systems of RUS telecommunications borrowers.

RUS technically accepted (material and equipment). Equipment which RUS has reviewed and determined that the material or equipment is suitable for use on systems of RUS telecommunications borrowers but the material or equipment does not satisfy both paragraphs (1) and (2) of this definition:

(1) Final assembly or manufacture of the equipment is not completed in the United States, its territories and possessions, or in an eligible country; and

(2) The cost of components within the material or equipment manufactured in the United States, its territories and possessions, or in an eligible country is 50 percent or less than the total cost of all components used in the material or equipment.

SEA. Service entrance aerial.

SEB. Service entrance buried.

Travel trailer. As defined in the ANSI/NFPA 70–1999, *NEC*®: A vehicular unit, mounted on wheels, designed to provide temporary living quarters for recreational, camping, or travel use, of such size and weight as not to require special highway movement permits when towed by a motorized vehicle, and of gross trailer area less than 320 square feet (29.7 square meters). (Reprinted with permission from NFPA 70–1999, the *National Electrical Code*®, Copyright © 1998, National Fire Protection Association, Quincy, MA 02269. This reprinted material is not the complete and official position of the National Fire Protection Association, on the referenced subject which is represented only by the standard in its entirety.)

Truck camper. As defined in the ANSI/NFPA 70–1999, *NEC*®: A portable unit constructed to provide temporary living quarters for recreational, travel or camping use, consisting of a roof, floor, and sides, designed to be loaded onto and unloaded from the bed of a pick-up truck. (Reprinted with permission from NFPA 70–1999, the *National Electrical Code*®, Copyright © 1998, National Fire Protection Association, Quincy, MA 02269. This reprinted material is not the complete and official position of the National Fire Protection Association, on the referenced subject which is represented only by the standard in its entirety.)

[66 FR 43317, Aug. 17, 2001, as amended at 69 FR 18803, Apr. 9, 2004]

§ 1755.502 Scope.

(a) Sections 1755.503 through 1755.510 cover approved methods of making service installations at customer access locations in telecommunications systems of RUS borrowers.

(b) Requirements in §§ 1755.503 through 1755.510 cover facilities of the type described in the FCC rules in 47 CFR part 68 for one and multi-party customer owned premises wiring.

[66 FR 43317, Aug. 17, 2001]

§ 1755.503 General.

(a) For the purposes of this section and §§ 1755.504 through 1755.510, a NID shall be as defined in § 1755.501 and shall contain both a fuseless primary station protector and a modular plug and jack for each conductor pair, up to a maximum of 11 pairs, and shall be provided by the telecommunications company and used by customers.

(b) For the purposes of this section and §§ 1755.504 through 1755.510, BET shall be as defined in § 1755.501 and shall contain both primary station protectors and connector terminals for each conductor pair, of 12 or more pairs, and shall be provided by the telecommunications company and used by customers. The primary station protectors may be either fuseless or fused.

(c) The requirements provided in this section and §§ 1755.504 through 1755.510 have been designed to coordinate with the provisions of the ANSI/NFPA 70–1999, *NEC*®, and the American National Standards Institute/Institute of Electrical and Electronics Engineers, Inc. (ANSI/IEEE) C2–1997, National Electrical Safety Code (NESC). The *National Electrical Code*® and *NEC*® are registered trademarks of the National Fire Protection Association, Inc., Quincy, MA 02269. The ANSI/NFPA 70–1999, NEC®, and the ANSI/IEEE C2–1997, NESC, are incorporated by reference in accordance with 5 U.S.C. 552(a) and 1 CFR part 51. Copies of ANSI/NFPA 70–1999, *NEC*®, are available from NFPA, 1 Batterymarch Park, P.O. Box 9101, Quincy, Massachusetts 02269–9101, telephone number 1 (800) 344–3555. Copies of ANSI/IEEE C2–1997, NESC, are available from IEEE Service Center, 455 Hoes Lane, Piscataway, New Jersey 08854, telephone number 1 (800) 678–4333. Copies of the ANSI/NFPA 70–1999, *NEC*®, and the ANSI/IEEE C2–1997, NESC, are available for inspection during normal business hours at RUS, room 2905, U.S. Department of Agriculture, 1400 Independence Avenue, SW., STOP 1598, Washington, DC 20250–1598, or at the National Archives and Records Administration (NARA). For information on the availability of this material at NARA, call 202–741–6030, or go to: *http://www.archives.gov/federal_register/code_of_federal_regulations/ibr_locations.html*. Most state and local authorities require that utility construction comply with either the ANSI/NFPA 70–1999, *NEC*®, and ANSI/IEEE C2–1997, NESC, or some earlier editions of the ANSI/NFPA 70, *NEC*®, and ANSI/IEEE C2, NESC. Some authorities have their own more stringent codes which may or may not be embellishments of the ANSI/NFPA 70, *NEC*®, and ANSI/IEEE C2, NESC.

(d) RUS borrowers shall make certain that all construction financed with RUS loan funds comply with:

(1) The provisions of this section and §§ 1755.504 through 1755.510 and the ANSI/NFPA 70–1999, *NEC*®, and ANSI/IEEE C2–1997, NESC codes, or any more stringent local codes; or

(2) The provisions of this section and §§ 1755.504 through 1755.510 with borrower added adjustments to bring construction into compliance with any more stringent local codes.

(e) This section and §§ 1755.504 through 1755.510 are intended primarily for the installer who will perform the work. It assumes that decisions regarding the selection of grounding electrodes, locations, and types of equipment have been made by the RUS borrower or the engineer delegated by the RUS borrower.

(f) Only a *qualified installer* as defined in § 1755.501 shall be assigned to make installations without advance planning and without direct supervision.

(g) This section and §§ 1755.504 through 1755.509 contain information which is normally not provided on the construction drawings which are included in § 1755.510.

(h) All work shall be conducted in a careful and professional manner. Service wire and cable shall not be trampled on, run over by vehicles, pulled over or around abrasive objects or otherwise subjected to abuse.

(i) When situations not covered by this section and §§ 1755.504 through 1755.510 arise, the RUS borrower or the engineer delegated by the borrower, shall specify the installation procedure to be used. The requirements of paragraph (j) of this section shall be complied with in every installation.

(j) NIDs, BETs, and fused primary station protectors shall be installed and grounded to meet the requirements of the ANSI/NFPA 70–1999, *NEC®*, or local laws or ordinances, whichever are more stringent.

(k) Battery polarity and conductor identification shall be maintained throughout the system as indicated on construction drawings 815 and 815–1 contained in § 1755.510. Color codes and other means of conductor identification of buried and aerial service wires shall conform to the requirements of this section and §§ 1755.504 through 1755.510.

(l) All materials for which RUS makes acceptance determinations, such as service wires and cables, ground rods, ground rod clamps, etc., used in service entrance installations shall be RUS accepted or RUS technically accepted. Borrowers shall require contractors to obtain the borrower's approval before RUS technically accepted materials are to be used in service entrance installations. Borrower's shall also ensure that the cost of the RUS technically accepted materials are at least 6 percent less than the cost of equivalent RUS accepted materials, as specified in "Buy American" Requirement of the Rural Electrification Act of 1938, as amended (7 U.S.C. 903 note). Materials used in service entrance installations which are of the type which RUS does not make acceptance determinations shall be of a suitable quality for their intended application as determined by the RUS borrower or the engineer delegated by the RUS borrower.

(m) On completion of an installation, borrowers shall require the installer to make all applicable tests required by §§ 1755.400 through 1755.407, RUS standard for acceptance tests and measurements of telecommunications plant.

[66 FR 43317, Aug. 17, 2001, as amended at 69 FR 18803, Apr. 9, 2004]

§ 1755.504 Demarcation point.

(a) The demarcation point (DP) provides the physical and electrical interface between the telecommunications company's facilities and the customer's premises wiring.

(b) The Federal Communications Commission (FCC) rules in 47 CFR part 68 require telecommunications providers to establish a "DP" which marks a separation of the provider's facilities from the customer's (owned) premises wiring and equipment.

(c) RUS borrowers shall observe the FCC DP requirement by installing NIDs, BETs, or fused primary station protectors when required by section 800–30(a)(2) of ANSI/NFPA 70–1999, *NEC®*, at all new or significantly modified customer access locations which are financed with RUS loan funds. *The National Electrical Code®* and *NEC®* are registered trademarks of the National Fire Protection Association, Inc., Quincy, MA 02269. The ANSI/NFPA 70–1999, *NEC®*, is incorporated by reference in accordance with 5 U.S.C. 552(a) and 1 CFR part 51. Copies are available from NFPA, 1 Batterymarch Park, P. O. Box 9101, Quincy, Massachusetts 02269–9101, telephone number 1 (800) 344–3555. Copies of ANSI/NFPA 70–1999, *NEC®*, are available for inspection during normal business hours at RUS, room 2905, U.S. Department of Agriculture, 1400 Independence Avenue, SW., STOP 1598, Washington, DC 20250–1598, or at the National Archives and Records Administration (NARA). For information on the availability of this material at NARA, call 202–741–6030, or go to: *http://www.archives.gov/federal_register/code_of_federal_regulations/ibr_locations.html.*

(d) For all customer access locations of less than 12 pairs, RUS borrowers shall establish DPs by using either NIDs or fused primary station protectors when required by section 800–30(a)(2) of ANSI/NFPA 70–1999, *NEC®*.

For customer access locations of 12 pairs or greater, RUS borrowers shall establish DPs using either NIDs, BETs, or fused primary station protectors when required by section 800–30(a)(2) of ANSI/NFPA 70–1999, *NEC®*.

[66 FR 43317, Aug. 17, 2001, as amended at 69 FR 18803, Apr. 9, 2004]

§ 1755.505 Buried services.

(a) Buried services of two or three pairs shall consist of Service Entrance Buried (SEB) assembly units, in accordance with RUS Bulletin 1753F–153 (RUS Form 515d), Specifications and Drawings for Service Installations at Customer Access Locations. The wire used for buried services shall conform to the requirements of § 1755.860, RUS specification for filled buried wires, and shall be RUS accepted or RUS technically accepted. The conductor size for two and three pair buried service wires shall be 22 American Wire Gauge (AWG). Copies of RUS Bulletin 1753F–153 are available upon request from RUS/USDA, 1400 Independence Avenue, SW., STOP 1522, Washington, DC 20250–1522, FAX (202) 690–2268.

(b) Buried services of six or more pairs shall be RUS accepted or RUS technically accepted 22 AWG filled buried cable conforming to the requirements of § 1755.390, RUS specification for filled telephone cables.

(c) Buried service wire or cable shall be terminated in buried plant housings using either splicing connectors or filled terminal blocks in accordance with the applicable paragraphs of § 1755.200, RUS standard for splicing copper and fiber optic cables.

(d) Buried service wire or cable shall be identified at buried plant housings in accordance with construction drawing 958 contained in § 1755.510.

(e) Buried service wire or cable shall be installed up to the building in the same general manner as buried exchange cable but in addition must meet the following requirements:

(1) Light weight lawn plows or trenchers shall be used;

(2) The shortest feasible route commensurate with the requirements of § 1755.508(i), (j), and (k), and paragraph (f)(1) of this section shall be followed;

(3) Buried service wire or cable shall be plowed or trenched to a depth of 12 in. (305 mm) or greater where practicable in soil, 36 in. (914 mm) in ditches, or 3 in. (76 mm) in rock. Depths shall be measured from the top of the wire or cable to the surface of the ground or rock;

(4) In the case of a layer of soil over rock either the minimum depth in rock measured to the surface of the rock, or the minimum depth in soil measured to the surface of the soil may be used; and

(5) Where adequate advance planning has been done, burial of telecommunications services jointly with electric power services may be feasible. If a decision has been reached by management to provide joint occupancy services, the services may be installed using the recommendations in RUS Bulletin 1751F–640, "Design of Buried Plant—Physical Considerations." Copies of RUS Bulletin 1751F–640 are available upon request from RUS/USDA, 1400 Independence Avenue, SW., STOP 1522, Washington, DC 20250–1522, FAX (202) 720–4120.

(f) Buried service wire or cable shall be installed on or in buildings as follows:

(1) Each buried service wire or cable shall contact the building as close to the NID, BET, or fused primary station protector as practicable. Service wire or cable runs on buildings shall normally consist of a single vertical run held to the minimum practical length. Horizontal and diagonal runs shall not be permitted.

(2) Buried service wire or cable shall be located so as to avoid damage from lawn mowers, animals, gardening operations, etc.

(3) Buried service wire or cable shall be installed against a foundation wall or pillar to provide adequate support and mechanical protection.

(4) Where it is likely that the service wire or cable shall be subjected to mechanical damage, the wire or cable shall be enclosed in a guard in accordance with assembly unit drawing BM83 contained in § 1755.510.

(5) The first above-ground attachment for a buried service wire or cable, unless it is enclosed in a guard, shall not be more than 4 in. (100 mm) above final grade.

(6) Uninsulated attachment devices may be used to attach buried service

wire and cable to masonry and other types of noncombustible buildings and on any type of building if fuseless primary station protectors incorporated in NIDs or BETs are used and installations fully comply with section 800–30(a)(1) of ANSI/NFPA 70–1999, *NEC*®. The *National Electrical Code*® and *NEC*® are registered trademarks of the National Fire Protection Association, Inc., Quincy, MA 02269. The ANSI/NFPA 70–1999, *NEC*®, is incorporated by reference in accordance with 5 U.S.C. 552(a) and 1 CFR part 51. Copies are available from NFPA, 1 Batterymarch Park, P.O. Box 9101, Quincy, Massachusetts 02269–9101, telephone number 1(800)344–3555. Copies of ANSI/NFPA 70–1999, *NEC*®, are available for inspection during normal business hours at RUS, room 2905, U.S. Department of Agriculture, 1400 Independence Avenue, SW., STOP 1598, Washington, DC 20250–1598, or at the National Archives and Records Administration (NARA). For information on the availability of this material at NARA, call 202–741–6030, or go to: *http://www.archives.gov/federal_register/code_of_federal_regulations/ibr_locations.html.*

(7) Insulated attachments shall be used to separate service wires or cables from woodwork where section 800–30(a)(2) of ANSI/NFPA 70–1999, *NEC*® requiring the use of fused primary station protectors must be observed.

(8) Minimum separation between buried service wire or cable and other facilities shall be as listed in Table 1, as follows:

TABLE 1—MINIMUM SEPARATION FOR TELECOMMUNICATIONS WIRES AND CABLES ON OR IN BUILDINGS

Foreign facility or obstruction	Minimum clearance in. [mm] [1] [2] telecommunications company's wires or cables
Electric supply wire including neutral and grounding conductors:	
Open	4 [102]
In conduit	2 [50.8]
Radio and television antennas, Lead-in and grounding conductors	4 [102]
Lightning rods and lightning conductors	[3] 72 [1830]
All foreign grounding conductors except lightning rod ground conductors	2 [50.8]
Neon signs and associated wiring	6 [150]
Metallic objects—pipes (gas, cold water, oil, sewer) and structures	[4] 2 [50.8]
Wires or cables of another communications system	2 [50.8]

[1] If minimum separation cannot be obtained, nonshielded wire and cable facilities shall be protected with either porcelain tubes or flexible tubing as modified by Notes (3) and (4) of this table.

[2] Separation applies to crossings and parallel runs.

[3] If this separation cannot be obtained, bond the telecommunications grounding conductors or grounding electrode to the lightning rod grounding conductor or grounding electrode with at least a Number (No.) 6 AWG copper, insulated, ground wire. With this provision a minimum separation of 4 in. (100 mm) is acceptable but this provision must not be utilized if the separation cited in this table can be maintained.

[4] Increase to a minimum of 3 in. (75 mm) separation from steam or hot water pipes, heating ducts, and other heat sources.

(9) Wire and cable attachments to buildings for outside mounted NIDS, BETs, or fused primary station protectors shall be in accordance with construction drawing 962 contained in § 1755.510.

(10) Appropriate devices for attaching service wire or cable on or in buildings vary with the type of building construction and the wire or cable size. Figures 1 and 2 illustrate various types of anchoring devices and their applications. The size and type of fastening device for the wire or cable size and type of surface shall be in accordance with the manufacturer's recommendation; Figures 1 and 2 are as follows:

FIGURE 1 ANCHORING DEVICES

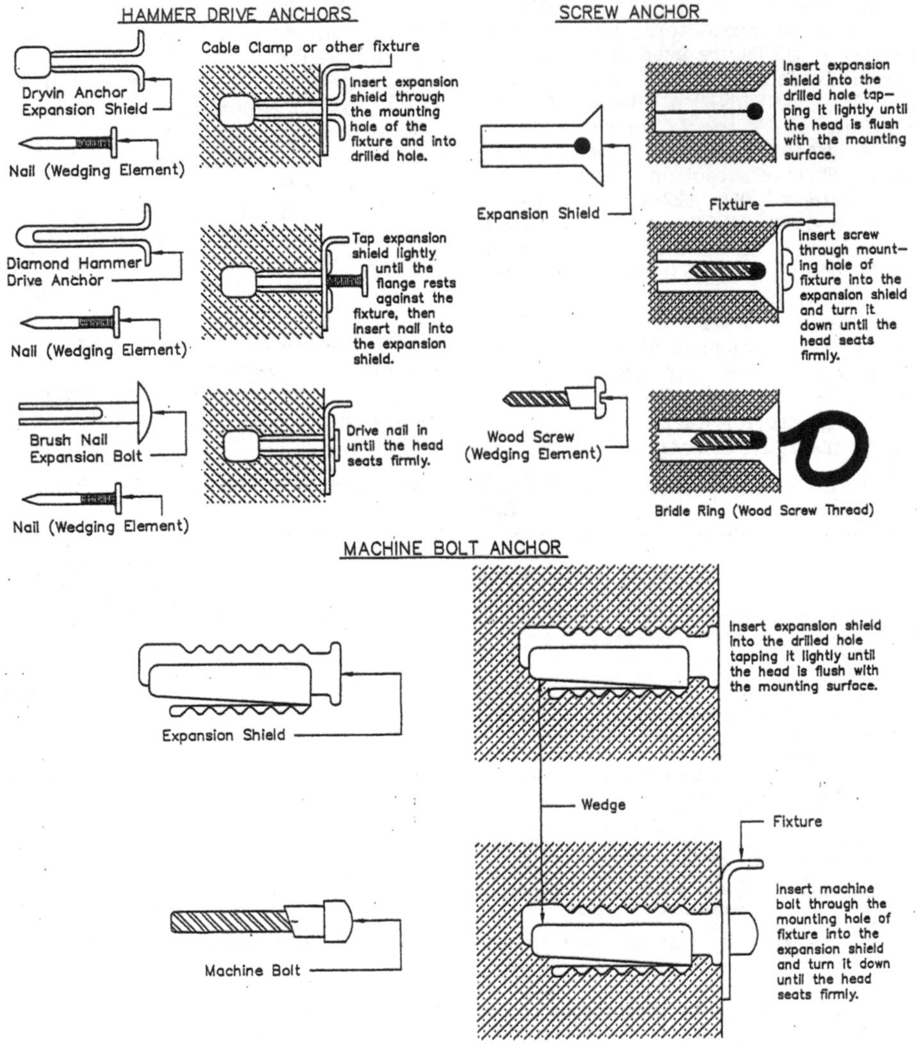

FIGURE 2 CABLE ATTACHMENT DEVICES

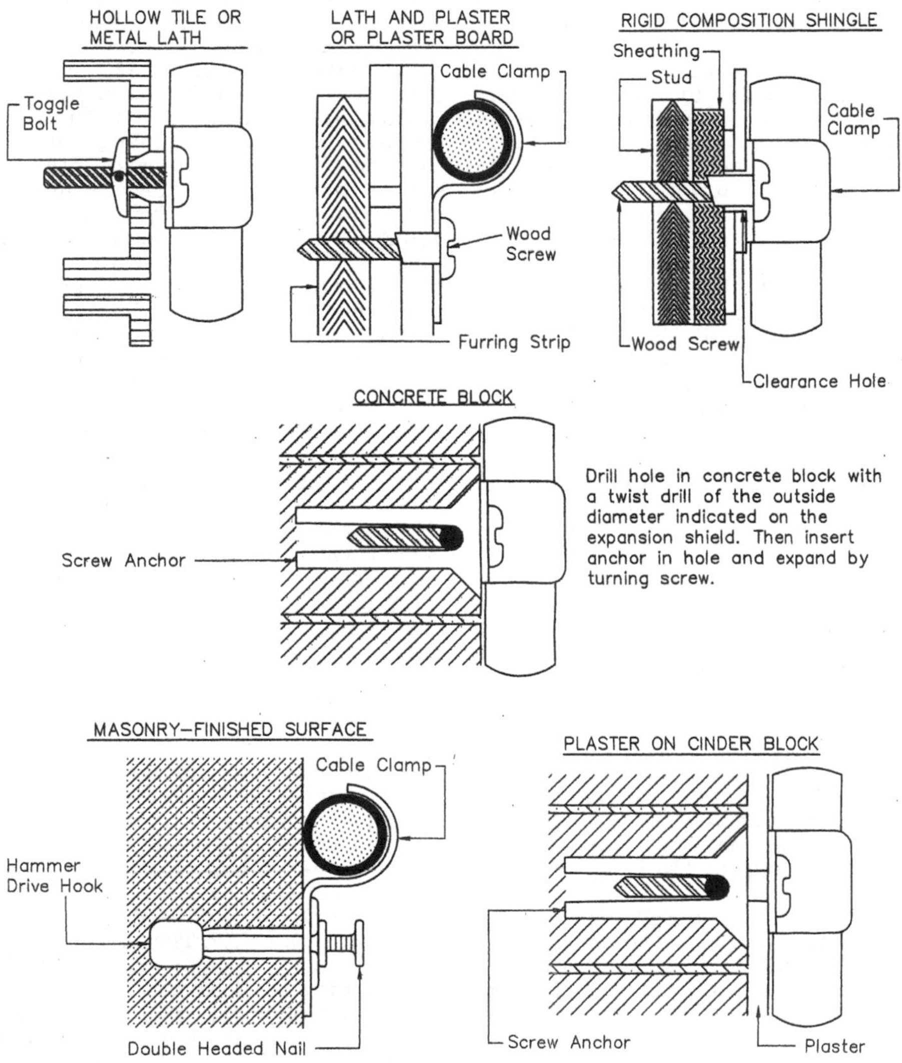

(11) Experience indicates that there are objections from many owners of buildings covered with aluminum or vinyl siding to the drilling of holes in the siding for the attachment of wires or cables, and NIDs, BETs, or fused primary station protectors. It is, therefore, important to obtain permission from the owner before drilling holes in such siding.

(12) If the NID, BET, or fused primary station protector must be mounted inside (not recommended by RUS), the service entrance into the building shall be installed in accordance with section 800–12(c) of ANSI/NFPA 70–1999, *NEC*®. After pulling-in the wire or cable, the

free space around the cable or wire shall be carefully sealed both outside and inside with a duct sealer that has RUS acceptance or RUS technical acceptance.

(13) If the customer requests an all buried installation for an alarm system or objects to above-ground facilities because of appearance and one-party service is involved, the entrance hole shall be made below grade as shown in sketch C of construction drawing 510–2 contained in § 1755.510. Care shall be exercised to prevent damage to the building foundation. The hole shall be sealed as specified in paragraph (f)(12) of this section. The installation shall comply with all the requirements of section 800–12(c) of ANSI/NFPA 70–1999, *NEC®*.

(g) When the NID, BET, or fused primary station protector is to be installed inside the building, the installation shall comply with section 800–12(c) of ANSI/NFPA 70–1999, *NEC®*, and the outside plant wire or cable shall preferably be installed in a rigid metal or intermediate metal conduit that is grounded to an electrode in accordance with section 800–40(b) of ANSI/NFPA 70–1999, *NEC®*, as shown in sketch A of Figure 3 in paragraph (h)(2) of this section. The shield of the outside plant wire or cable shall be bonded to the grounding terminal of the NID, BET, or fused primary station protector which in turn shall be connected to the closest, existing, and accessible grounding electrode, of the electrodes cited in section 800–40(b) of ANSI/NFPA 70–1999, *NEC®*.

(h) An inside NID, BET, or fused primary station protector installation may also be made without use of a rigid metal or intermediate metal conduit provided that the ingress of the outside plant wire or cable complies with section 800–12(c) of ANSI/NFPA 70–1999, *NEC®*, and provided either of the following are observed:

(1) The NID, BET, or fused primary station protector is located as close as practicable to the point where the outside plant wire or cable emerges through an exterior wall. The length of outside plant wire or cable exposed within the building shall be as short as practicable but in no case shall it be longer than 50 feet (ft) (15.2 meters (m)) in accordance with the allowable exception No. 3 of section 800–50 of ANSI/NFPA 70–1999, *NEC®*. See sketch B of Figure 3 in paragraph (h)(2) of this section. The shield of the outside plant wire or cable shall be bonded to the grounding terminal of the NID, BET, or fused primary station protector which in turn shall be connected to the closest, existing and accessible grounding electrode, of the electrodes cited in section 800–40(b) of ANSI/NFPA 70–1999, *NEC®* (Fine print Note No. 2 of ANSI/NFPA 70–1999, *NEC®*, section 800–50, warns that the full 50 ft (15.2 m) may not be authorized for outside unlisted cable (not in a metal or intermediate metal conduit) within a building if it is practicable to place the NID, BET, or fused primary station protector closer than 50 ft (15.2 m) to the cable entrance point, e.g., if there is an acceptable and accessible grounding electrode of the type cited in section 800–40(b) of ANSI/NFPA 70–1999, *NEC®*, anywhere along the proposed routing of the outside cable within the building); or

(2) Where the NID, BET, or fused primary station protector must be located within the building remote from the entrance point and the entrance point of the outside plant wire or cable cannot be designed to be closer to the NID, BET, or fused primary station protector location, the outside plant wire or cable shall be spliced, as close as practicable to the point where the outside plant wire or cable emerges through an outside wall, to an inside wiring cable that is "Listed" as being suitable for the purpose in accordance with part E of article 800 of ANSI/NFPA 70–1999, *NEC®*. The length of outside plant wire or cable exposed within the building shall be as short as practicable but in no case shall it be longer than 50 ft (15.2 m) in accordance with the allowable exception No. 3 of section 800–50 of ANSI/NFPA 70–1999, *NEC®*. See sketch C of Figure 3. The shield of the outside plant wire or cable shall be bonded to the grounding terminal of the NID, BET, or fused primary station protector which in turn shall be connected to the closest, existing, and accessible grounding electrode, of the electrodes cited in section 800–40(b) of ANSI/NFPA 70–1999, *NEC®* (Fine print Note No. 2 of the ANSI/NFPA 70–1999, *NEC®*, section 800–50, warns that the

full 50 ft (15.2 m) may not be authorized for outside unlisted cable (not in a metal or intermediate metal conduit) if it is practicable to place the NID, BET, or fused primary station protector closer than 50 ft (15.2 m) to the cable entrance point, e.g., if there is an acceptable and accessible grounding electrode of the type cited in section 800–40(b) of ANSI/NFPA 70–1999, *NEC*®, anywhere along the proposed routing of the outside cable within the building). Figure 3 is as follows:

FIGURE 3

CABLE ENTRANCES AND RUNS IN BUILDINGS

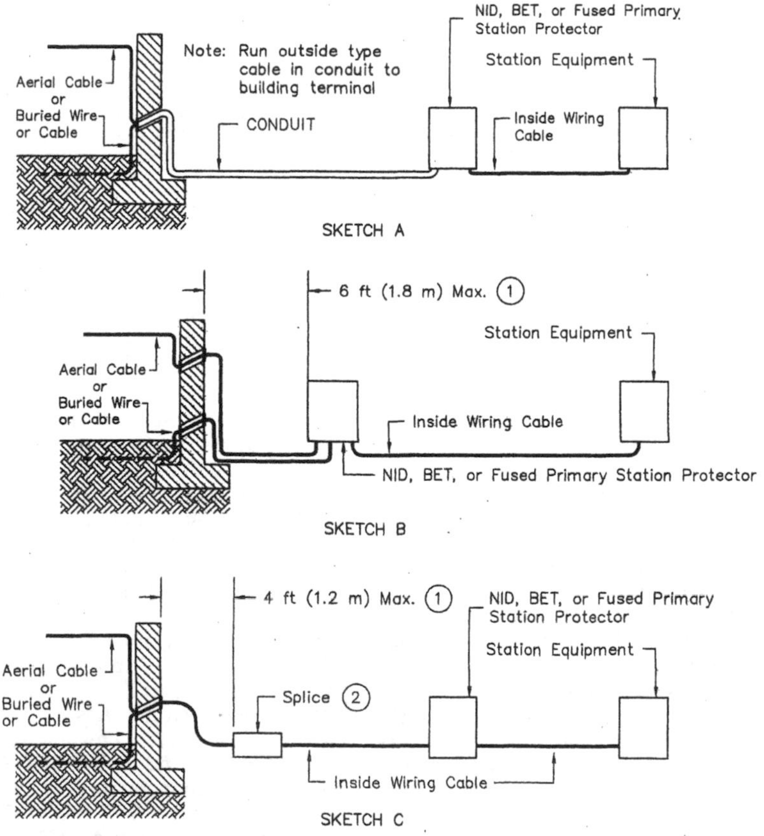

Notes:

1. Recommended maximum is shown; length cannot exceed the ANSI/NFPA 70–1999, NEC® allowable length of 50 ft (15.2 m). (See Fine Print Note No. 2 of Section 800–50 of ANSI/NFPA 70–1999, NEC®

2. Outside plant cable shield shall be connected to an acceptable grounding electrode. If splice case is metallic, the splice case shall also be connected to the same acceptable grounding electrode.

(i) The polarity of buried wire or cable "tip" and "ring" conductors shall be maintained by making the connections in accordance with Table 2, as follows:

TABLE 2—COLOR CODES FOR TIP AND RING CONNECTIONS OF INSIDE WIRING CABLE

Pair	Tip		Ring	
	Color of insulation	Color of marking	Color of insulation	Color of marking
1	White	Blue	Blue	White
2	White	Orange	Orange	White
3	White	Green	Green	White
4	White	Brown	Brown	White
5	White	Slate	Slate	White
6	Red	Blue	Blue	Red
7	Red	Orange	Orange	Red
8	Red	Green	Green	Red
9	Red	Brown	Brown	Red
10	Red	Slate	Slate	Red
11	Black	Blue	Blue	Black
12	Black	Orange	Orange	Black
13	Black	Green	Green	Black
14	Black	Brown	Brown	Black
15	Black	Slate	Slate	Black
16	Yellow	Blue	Blue	Yellow
17	Yellow	Orange	Orange	Yellow
18	Yellow	Green	Green	Yellow
19	Yellow	Brown	Brown	Yellow
20	Yellow	Slate	Slate	Yellow
21	Violet	Blue	Blue	Violet
22	Violet	Orange	Orange	Violet
23	Violet	Green	Green	Violet
24	Violet	Brown	Brown	Violet
25	Violet	Slate	Slate	Violet

[66 FR 43317, Aug. 17, 2001, as amended at 69 FR 18803, Apr. 9, 2004]

§ 1755.506 Aerial wire services.

(a) Aerial services of one through six pairs shall consist of Service Entrance Aerial (SEA) assembly units, in accordance with RUS Bulletin 1753F–153 (RUS Form 515d), Specifications and Drawings for Service Installations at Customer Access Locations. The wire used for aerial services shall conform to the requirements of §§ 1755.700 through 1755.704, RUS specification for aerial service wires, and shall be RUS accepted or RUS technically accepted. Copies of RUS Bulletin 1753F–153 are available upon request from RUS/USDA, 1400 Independence Avenue, SW., STOP 1522, Washington, DC 20250–1522, FAX (202) 720–4120.

(b) If aerial wire services are to be connected to aerial cable pairs, the NIDs or fused primary station protectors and grounds shall be installed and connected before the aerial service wires are attached to the customer's structure.

(c) Kinks or splices shall not be permitted in aerial service wire spans.

(d) Aerial service wires shall be run in accordance with the construction drawings contained in § 1755.510 and shall conform to all clearance requirements of the ANSI/NFPA 70–1999, *NEC*®, and ANSI/IEEE C2–1997, NESC, or local laws or ordinances, whichever are the most stringent. The *National Electrical Code*® and *NEC*® are registered trademarks of the National Fire Protection Association, Inc., Quincy, MA 02269. The ANSI/NFPA 70–1999, *NEC*®, and ANSI/IEEE C2–1997, NESC, are incorporated by reference in accordance with 5 U.S.C. 552(a) and 1 CFR part 51. Copies of ANSI/NFPA 70–1999, *NEC*®, are available from NFPA, 1 Batterymarch Park, P.O. Box 9101, Quincy, Massachusetts 02269–9101, telephone number 1 (800) 344–3555. Copies of ANSI/IEEE C2–1997, NESC, are available from IEEE Service Center, 455 Hoes Lane, Piscataway, New Jersey 08854, telephone number 1 (800) 678–4333. Copies of ANSI/NFPA 70–1999, *NEC*®, and ANSI/IEEE C2–1997, NESC, are available for inspection during normal business hours at RUS, room 2905, U.S. Department of Agriculture, 1400 Independence Avenue, SW., STOP 1598, Washington, DC 20250–1598, or at the National Archives and Records Administration (NARA). For information on the availability of this material at NARA, call 202–741–6030, or go to: *http://www.archives.gov/federal_register/code_of_federal_regulations/ibr_locations.html.*

(e) Aerial service wire shall be installed using the maximum practicable sag consistent with the required ground clearance and good construction practices. In no event shall the minimum sags be less than the values shown on construction drawing 505 contained in § 1755.510 for various span lengths and loading areas provided. Span lengths shall not exceed 250 ft (76 m).

(f) To reduce vibration and galloping, aerial service wire shall be twisted one complete turn for each 10 ft (3 m) of span length at the time of installation.

(g) The methods of attaching aerial service wires at poles shall be as illustrated in construction drawings 503–2 and 504 contained in § 1755.510.

(h) Horizontal and vertical climbing spaces on poles used jointly with power circuits shall be provided in conformance with the requirements of Rule 236 of ANSI/IEEE C2–1997, NESC.

(i) Not more than four aerial service wires shall be distributed from any one 7⁄16 in. (10 mm) drive hook, or more than two aerial service wires from any one 5⁄16 in. (8 mm) drive hook. Aerial service wires and drive hooks shall be arranged so that the load does not pull the drive hook out of the pole. When more than one drive hook is required, the drive hooks shall be staggered with a minimum separation of 1 in. (25.4 mm) horizontally on centers and 1.5 in. (40 mm) vertically on centers. If drive hooks are placed within 3 in. (76 mm) of the top of the pole and on the opposite side of the pole's circumference, a vertical separation of at least 3 in. (76 mm) shall be provided. A drive hook shall not be placed on the top of a pole or stub pole.

(j) When connecting aerial service wires to cable pairs at terminals, sufficient slack shall be provided so that each aerial service wire shall reach any binding post position as shown on construction drawing 312–1 contained in § 1755.510.

(k) Aerial service wire attachments on utility poles and the manner of placing bridle rings and entering cable terminals shall be as shown on construction drawing 503–2 contained in § 1755.510.

(l) Not more than two conductors shall be connected to any terminal binding post. Where it is necessary to bridge more than two aerial service wires at the same closure, the aerial service wires shall be terminated in aerial service wire terminals connected in parallel with a No. 20 AWG bridle wire which shall be terminated on the binding posts of the filled terminal block.

(m) Where aerial service wire is attached to aerial plastic cable, it shall be brought directly into a ready-access closure and shall be terminated on the binding posts of the filled terminal block as shown on construction drawing 503–2 contained in § 1755.510.

(n) The conductor of copper coated steel reinforced aerial service wires identified by tracer ridges shall be used as the ring (negative battery) conductor of the pair, and shall normally be connected to the right or lower binding post of a pair on filled terminal blocks and NIDs or fused primary station protectors.

(o) *Nonmetallic reinforced aerial service wire pair identification.* (1) The tip and ring conductors of nonmetallic reinforced aerial service wires shall be identified in accordance with Table 3, as follows:

TABLE 3—NONMETALLIC REINFORCED AERIAL SERVICE WIRE COLOR CODE

Pair number	Conductor color	
	Tip	Ring
1	White/Blue or White	Blue
2	White/Orange or White	Orange
3	White/Green or White	Green

TABLE 3—NONMETALLIC REINFORCED AERIAL SERVICE WIRE COLOR CODE—Continued

Pair number	Conductor color	
	Tip	Ring
4	White/Brown or White	Brown
5	White/Slate or White	Slate
6	Red/Blue or Red	Blue

(2) The ring (negative battery) conductor of the pair shall normally be connected to the right or lower binding post of a pair on filled terminal blocks and NIDs or fused primary station protectors.

(p) When it is necessary to avoid intervening obstacles between a pole and a building, span clamp attachments shall be used to support the aerial service wires at points between the poles that are supporting the cable on the suspension strand as indicated by construction drawings 501–1 and 501–2 contained in §1755.510.

(q) Aerial service wire strung from pole to pole shall be placed entirely below or entirely above any existing wire or cable. When adequate ground clearance can be obtained, preference shall be given to placing aerial service wire below wire and cable.

(r) When more than one aerial service wire is installed from pole to pole, the first aerial service wire shall be sagged in accordance with construction drawing 505 contained in §1755.510. Succeeding aerial service wires shall be sagged with 2 in. (50.8 mm) more sag for each aerial service wire.

(s) Aerial service wire spans from pole lines to buildings shall follow the shortest feasible route commensurate with the requirements of paragraph (t) of this section and shall be sagged in accordance with construction drawing 505 contained in §1755.510. The route shall avoid trees and other obstructions to the extent practicable. Where trees cannot be avoided, tree trimming permission shall be obtained from the owner or the owner's representative, and all limbs and foliage within 2 ft (600 mm) of the finally sagged wire shall be removed. If tree trimming permission cannot be obtained, the matter shall be referred to the borrower for resolution before proceeding with the installation.

(t) Aerial service wires shall contact buildings as closely as practicable at a point directly above the NID, or fused primary station protector. Generally, horizontal drop wire runs on buildings shall not exceed 20 ft (6 m). The warning given in §1755.505(f)(11) regarding drilling holes in aluminum and vinyl siding applies also to attaching aerial service wires.

(u) The point of the first building attachment shall be located so that the aerial service wire will be clear of roof drainage points.

(v) Where practicable, aerial service wires shall pass under electrical guys, power distribution secondaries and services, tree limbs, etc.

(w) Aerial service wire shall not pass in front of windows or immediately above doors.

(x) Aerial service wires shall be routed so as to have a minimum clearance of 2 ft (600 mm) from any part of a short wave, ham radio, etc. antenna mast and a television antenna mast in its normal vertical position and of the possible region through which it sweeps when being lowered to a horizontal position.

(y) Aerial service wires shall be installed such that all clearances and separations comply with either section 237 of ANSI/IEEE C2–1997, NESC, or ANSI/NFPA 70–1999, *NEC®*, or local laws or ordinances, whichever is the most stringent.

(z) Aerial service wire attachments to buildings shall be as follows:

(1) First attachments on buildings shall be made in accordance with construction drawings 506, 507, or 508–1 contained in §1755.510, as applicable;

(2) Intermediate attachments on buildings shall be made in accordance with construction drawings 510 or 510–1 contained in §1755.510; and

(3) Uninsulated attachments shall be permitted to be used as follows:

(i) Wherever NIDS are used as permitted by section 800–30(a)(1) of the ANSI/NFPA 70–1999, *NEC*®; and

(ii) On masonry and other types of nonflammable buildings.

(aa) Insulated attachments shall be used on wooden frame, metallic siding and other types of combustible buildings where fused primary station protectors are used, as required by section 800–30(a)(2) of ANSI/NFPA 70–1999, *NEC*®.

(bb) Aerial service wire runs on buildings shall be attached vertically and horizontally in a neat and most inconspicuous possible manner. See construction drawing 513 contained in § 1755.510. Horizontal runs on buildings are undesirable and shall be kept to a minimum. Diagonal runs shall not be made.

(cc) Aerial service wire runs on buildings shall be located so as not to be subjected to damage from passing vehicles, pedestrians, or livestock.

(dd) Minimum separation between aerial service wires and other facilities on or in buildings shall be in accordance with § 1755.505(f)(8), Table 1.

(ee) Appropriate devices for attaching aerial service wires to buildings vary with the type of building construction and with the type of customer access location equipment. Table 4 lists various types of attachments and their application with respect to construction, customer access location equipment, and proper mounting devices. Construction drawings 506 through 513 contained in § 1755.510 illustrate requirements with respect to various angles of service wire contacts and uses of various attachments. Table 4 is as follows:

Table 4
DEVICES FOR ATTACHING AERIAL SERVICE WIRES TO BUILDINGS (1), (2), (8)

TYPE OF ATTACHMENT	TYPES OF FASTENING DEVICES											
	FRAME BUILDINGS (3)								FIRE RESISTANT BUILDINGS (4)			
	FUSED STATION PROTECTOR				NID				(NID OR FUSED STATION PROTECTOR)			
	Wood Shingle–Composition (5)	Plywood–Plastic–Board Paneling	Thin Brick–Stucco–Plaster	Metal Sheath	Wood Shingle–Composition (5)	Plywood–Plastic–Board Paneling	Thin Brick–Stucco–Plaster	Metal Sheath	Concrete Block	Tile	Brick Stone Concrete	Steel
Knob, S — Under 30° Angle	2-1/2" x #18 FH Screw	3" x #18 FH Screw	3" x #18 FH Screw	2-1/2" x #18 FH Screw	2-1/2" x #18 FH Screw	3" x #18 FH Screw	3" x #18 FH Screw	2-1/2" x #18 FH Screw				
Knob, S — Over 30° Angle	5/16" Angle Screw	5/16" Angle Screw	3/8" Angle Screw	5/16" Angle Screw	5/16" Angle Screw	5/16" Angle Screw	3/8" Angle Screw	5/16" Angle Screw				
Knob, C	2-1/2" x #10 RH Screw	3" x #10 RH Screw	3-1/2" x #10 RH Screw	2-1/2" x #10 RH Screw	Note 6	Note 6	Note 6	Note 6				
Bracket, House	2" x #14 RH Screw	2" x #14 RH Screw	2-1/2" x #14 RH Screw	2" x #14 RH Screw	Note 6	Note 6	Note 6	Note 6				
Bracket, Corner	2" x #14 RH Screw	2" x #14 RH Screw	2-1/2" x #14 RH Screw	2" x #14 RH Screw	2" x #14 RH Screw	2" x #14 RH Screw	2-1/2" x #14 RH Screw	2" x #14 RH Screw	3/16" x 4" Toggle	3/16" x 4" Toggle	2" x #14 RH Screw	3/16" x 4" Toggle
Screweye, Insulated	1" Shank	1" Shank	2" Shank	1" Shank	Note 6	Note 6	Note 6	Note 6				
Ring, Bridle, Drive	Note 6	Note 6	Note 6	Note 6	Note 7	Note 7	Note 6	Note 7	Drive Anchor	Note 6	Drive Anchor	Note 6
Ring, Bridle, Screw	Note 6	Note 6	Note 6	Note 6	Note 7	Note 7	Note 6	Note 7	Expansion Anchor	Note 6	Expansion Anchor	Note 6
Hook, Drop Wire	Note 6	Note 6	Note 6	Note 6	2" x #14 RH Screw	2" x #14 RH Screw	2" x #14 RH Screw	2" x #14 RH Screw	1/4" x 4" Toggle	1/4" x 4" Toggle	2" x #18 RH Screw	1/4" x 3" Toggle
Hook, House	Note 6	Note 6	Note 6	Note 6	2" x #14 RH Screw	2" x #14 RH Screw	2" x #14 RH Screw	2" x #14 RH Screw	Expansion Anchor	Note 6	Expansion Anchor	Note 6
Ring, Bridle, Toggle									3/16" x 4" Toggle	3/16" x 4" Toggle	Note 6	3/16" x 4" Toggle
Clamp, One Hole, Offset or closed "U" Cable Strap	Note 6	Note 6	Note 6	Note 6	3/4" x #6 RH Screw	3/4" x #6 RH Screw	3/4" x #6 RH Screw	3/4" x #6 RH Screw	1" x #6 RH Screw	1/8" x 4" Toggle	1" x #6 RH Screw	1/2" x #6 SM Screw

NOTES: 1. Screw dimensions are minimum. Where appropriate, either or both dimensions shall be increased. All wood screws for exterior use shall be stainless steel. All other exterior metal devices shall be stainless steel, zinc coated steel, silicon bronze, or corrosion resistant aluminum alloy.

2. Toggle bolt dimensions are minimum. Where appropriate, either or both dimensions shall be increased.

3. All devices should be attached to studding.

4. Screw-type devices shall be secured by means of expansion-type anchors. Equivalent manual or machine-driven devices may be used. Where toggle bolts are specified equivalent devices may be used.

5. Pilot holes shall be provided for screws and bridle rings in shingles and dropsiding.

6. Attachment device not applicable.

7. Attachment device applicable but no separate fastening device required.

8. To convert English units to Metric units use 1 in. = 25.4 mm.

(ff) Fastener spacings for vertical and horizontal runs on frame or masonry buildings shall not be more than 6 ft (2 m) apart. Fasteners should be spaced close enough to prevent the aerial service wire from "slapping" against the building during windy conditions.

(gg) When it is necessary to pass behind or around obstructions such as downspouts and vertical conduits, the aerial service wire shall be supported firmly with attachment devices placed not more than 6 in. (152 mm) from the obstruction as illustrated in Figures 4 and 5 of paragraph (hh) of this section. Preferably, the aerial service wire should be routed behind obstructions to minimize the possibility of mechanical damage to the aerial service wire in the event repair work to the obstruction is required.

(hh) When passing around building projections of masonry or wood or around corners, aerial service wires shall be installed as illustrated in Figures 5 and 6. Figures 4, 5, and 6 are as follows:

FIGURE 4

AERIAL SERVICE WIRE CROSSING OBSTRUCTIONS
WOODEN BUILDING SURFACES

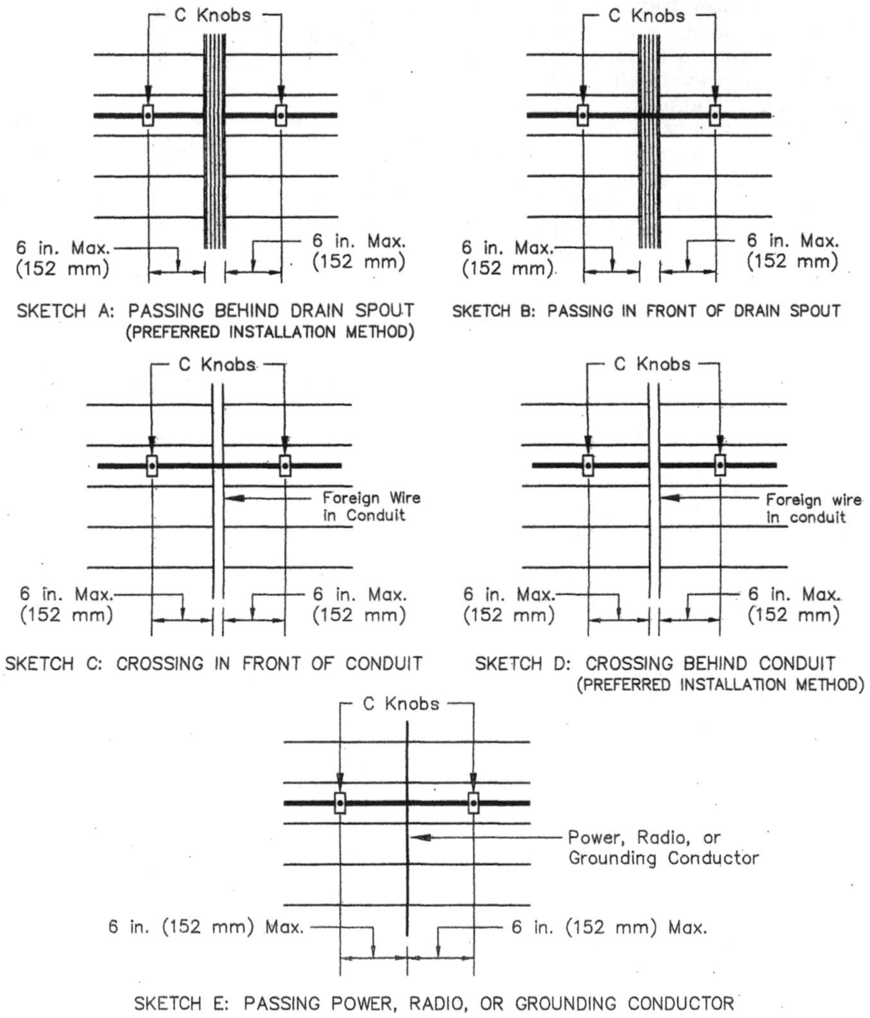

FIGURE 5

AERIAL SERVICE WIRE CROSSING OBSTRUCTIONS MASONRY BUILDING SURFACES

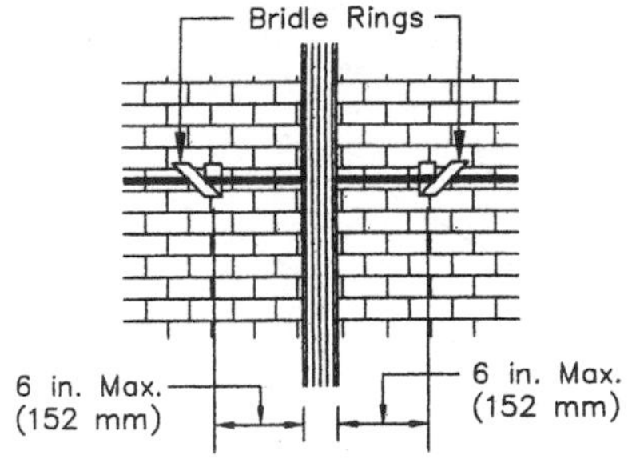

SKETCH A: PASSING BEHIND DRAIN SPOUT
(PREFERRED INSTALLATION METHOD)

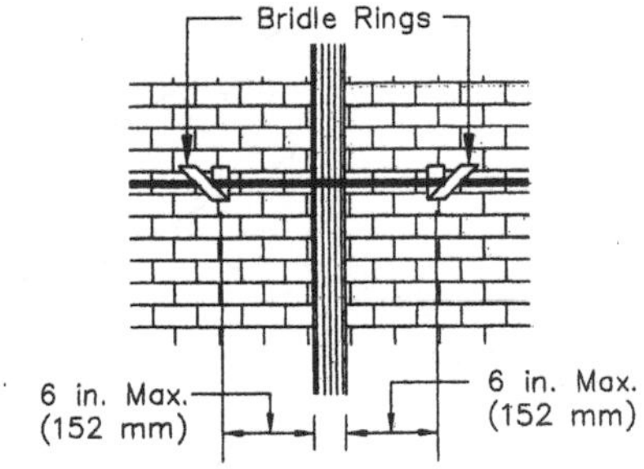

SKETCH B: PASSING IN FRONT OF DRAIN SPOUT

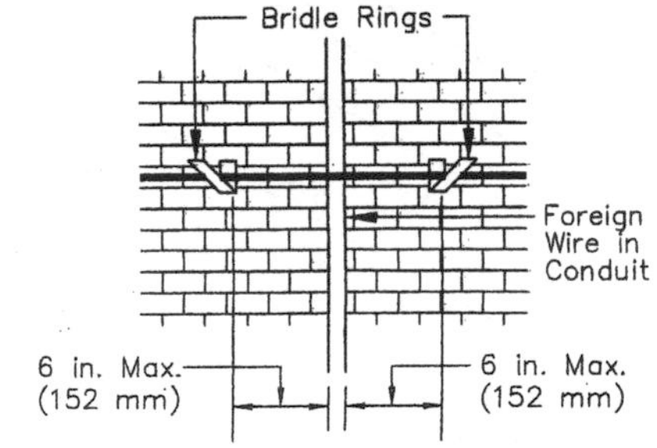

SKETCH C: CROSSING IN FRONT OF CONDUIT

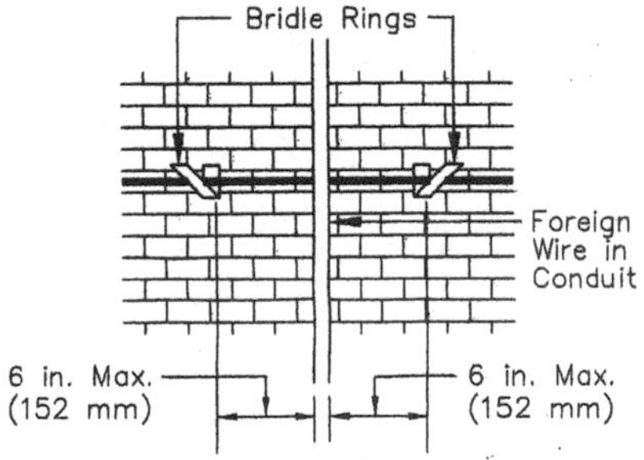

SKETCH D: CROSSING BEHIND CONDUIT
(PREFERRED INSTALLATION METHOD)

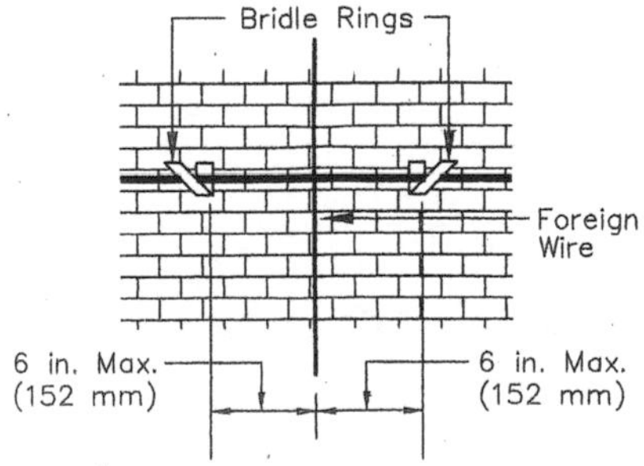

SKETCH E: PASSING BEHIND FOREIGN WIRE
(PREFERRED INSTALLATION METHOD)

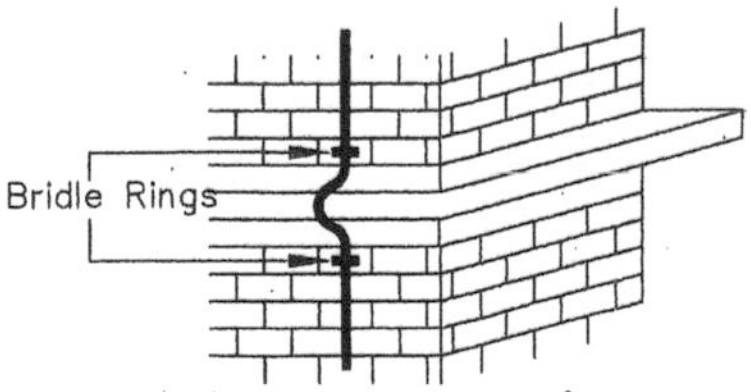

SKETCH F: MASONRY BUILDING PROJECTIONS

FIGURE 6

AERIAL SERVICE WIRE CROSSING COMBUSTIBLE BUILDING PROJECTIONS

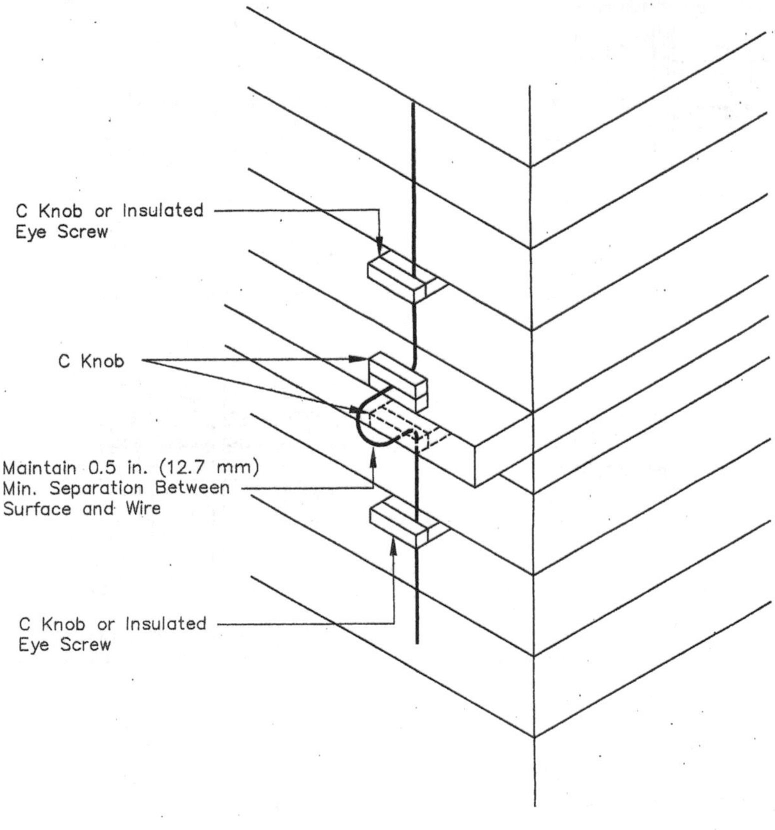

(ii) In areas where ice and snow conditions are severe, aerial service wires shall be located so that ice and snow falling from the roof will not strike the wires. However, where aerial service wires must pass under the sloping part of the roof, first attachments shall be made as close as practicable to the eaves.

(jj) If two aerial service wire spans are required to the same building, the first attachment shall be such that both aerial service wires can be attached at the same attachment device. Refer to construction drawing 508–1 contained in §1755.510. Where more than two aerial service wires are required, additional attachment devices in the same general location on the building shall be used.

(kk) When two or more aerial service wire runs are required on the same building they shall share the same type of attachment devices.

(ll) Aerial service wire entrances to buildings shall conform to sketch B of construction drawing 510–2 contained in § 1755.510, unless the entrance is made through a conduit.

(mm) When the aerial service wire approaches the entrance hole from above, a 1.5 in. (40 mm) minimum drip loop shall be formed in accordance with sketch B of construction drawing 510–2 contained in § 1755.510.

(nn) If an entrance conduit which slopes upward from outside to inside is available and suitably located, it shall be used for the aerial service wire entrance.

[66 FR 43317, Aug. 17, 2001, as amended at 69 FR 18803, Apr. 9, 2004]

§ 1755.507 Aerial cable services.

(a) Where more than six pairs are needed initially, and where an aerial service is necessary, the service shall consist of 22 AWG filled aerial cable of a pair size adequate for the ultimate anticipated service needs of the building. The cable shall comply with the requirements of § 1755.390, RUS specification for filled telephone cables, and shall be RUS accepted or RUS technically accepted.

(b) Aerial cable services shall be constructed in accordance with specific installation specifications prepared by the RUS borrower or the engineer delegated by the borrower.

(c) Unless otherwise specified in the installation specifications, aerial cable service installations shall meet the following requirements:

(1) Strand supported lashed construction shall be used.

(2) Where practicable a $\frac{5}{16}$ in. (8 mm) utility grade strand and automatic clamps shall be used in slack spans to avoid damage to the building.

(3) Construction on poles shall comply with applicable construction drawings for regular line construction. Aerial service cable shall be spliced to the main cable in accordance with § 1755.200, RUS standard for splicing copper and fiber optic cables.

(4) Where practicable, aerial cable shall pass under electrical guys, distribution secondaries, and services.

(5) The suspension strand shall be attached to the building by wall brackets as indicated in Figure 7 as follows:

FIGURE 7

SUSPENSION STRAND DEADENDING ON BUILDINGS

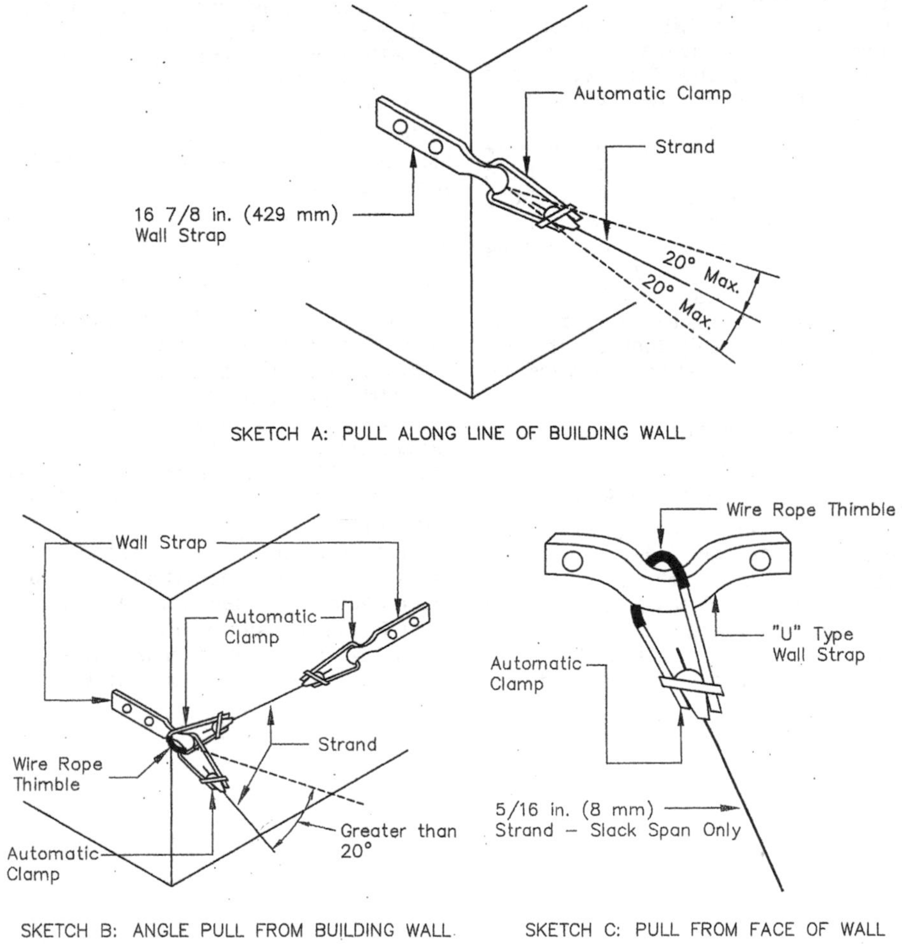

(i) If taut spans are necessary, appropriate size strand may be used if the pull is in line with one wall of the building, or within 20 degrees of being in line as illustrated in sketch A of Figure 7. If the angle of pull is greater than 20 degrees from the building, the wall bracket shall be reinforced against pullout by an arrangement equivalent to sketch B of Figure 7. Taut spans may be strung using the recommendations in RUS Bulletin 1751F–630, Design of Aerial Plant. The same tension as would be used in normal line construction so as not to exceed 60 percent of the breaking strength of the strand under maximum loading shall be used. Taut spans shall not exceed 100 ft (30.5 m) in length and the cable weight shall not exceed 1 pound/foot (lb/ft) [1.5 kilogram/meter (kg/m)] except when equivalent combinations of greater span lengths with cable weight less than 1 lb/ft (1.5 kg/m) are permissible. Copies of RUS Bulletin 1751F–630 are available upon request from RUS/USDA, 1400 Independence Avenue, SW., STOP 1522, Washington, DC 20250–1522, FAX (202) 720–4120.

(ii) When an attachment must be made to the face of a building wall away from a corner, a "U" type wall bracket shall be used as indicated in sketch C of Figure 7. Only slack span construction with 5⁄16 in. (8 mm) utility grade strand shall be permitted in this situation. The bail of the automatic clamp shall be protected by a wire rope thimble.

(6) Aerial cable shall be located on the rear or side of the building and shall be run only in a horizontal or a vertical direction. The cable route shall be selected so as to avoid building projections and obstructions to the extent practicable.

(7) Cable attachment devices shall be located on solid masonry or on studs of wood frame buildings. Cable attachment devices may be installed on sheet surface materials only when such materials are reinforced with a backing material which allows penetration and firm holding of the attachment devices through the backing material.

(8) The minimum separation on or in buildings between cable and other facilities shall be as indicated in § 1755.505(f)(8), Table 1.

(9) On horizontal runs, cable clamps shall be placed so that the attachment is below the cable. On vertical runs, cable clamps shall be placed so that the attachment is on the same side as horizontal runs. Cable clamps shall be placed on the inside of cable bends.

(10) On horizontal runs, cable clamps shall be placed not more than 16 in. (400 mm) apart for cable diameters equal to or greater than 1 in. (25.4 mm) and 24 in. (600 mm) apart for cable diameters less than 1 in. (25.4 mm).

(11) On vertical runs, cable clamps shall be approximately 24 in. (600 mm) apart for all sizes of cable.

(12) For the cable entrance, holes shall be bored slightly larger in diameter than the cable and shall slope upward from outside to inside. A duct sealer having RUS acceptance or RUS technical acceptance shall be applied to both ends of the hole after the cable is pulled in.

(13) Section 1755.505(g) and (h) shall also apply to aerial cable services.

[66 FR 43317, Aug. 17, 2001]

§ 1755.508 Customer access location protection.

(a) All customer access locations shall be protected.

(b) Customer access location protection shall consist of installing the telecommunications facilities with proper clearances and insulation from other facilities, providing primary voltage limiting protection, fuse links, NIDs, BETs, or fused primary station protectors, if required, and adequate bonding and grounding.

(c) All NIDs shall be RUS accepted or RUS technically accepted or the RUS borrower shall obtain RUS regional office approval on a case by case basis as applicable.

(d) All BETs shall be RUS accepted or RUS technically accepted.

(e) All fused primary station protectors shall be RUS accepted or RUS technically accepted.

(f) NIDs, BETs, or fused primary station protectors shall be mounted outside for all applications except for those described in paragraphs (g)(1) through (g)(3) of this section.

(g) NIDs, BETs, or fused primary station protectors may be mounted inside when:

(1) Large buildings are to be served and the customer requests an inside installation;

(2) Buried alarm circuits are requested by the subscriber; or

(3) The customer requests an all buried installation for appearance or to prevent the drilling of holes in aluminum or vinyl siding.

(h) Outside mounted NIDs, BETs, or fused primary station protectors shall be easily accessible and shall be located between 3 to 5 ft (1 to 1.5 m) above final grade.

(i) The locations of NIDs, BETs, or fused primary station protectors shall be selected with emphasis on utilizing the shortest primary station protector grounding conductor practicable and on grounding of the telecommunications primary station protector to the electric service grounding system established at the building served utilizing electrodes (c) through (g) cited in section 800–40(b)(1) of ANSI/NFPA 70–1999, *NEC*®. The *National Electrical Code*® and *NEC*® are registered trademarks of the National Fire Protection Association, Inc., Quincy, MA 02269. The ANSI/NFPA 70–1999, *NEC*®, is incorporated by reference in accordance with 5 U.S.C. 552(a) and 1 CFR part 51. Copies are available from NFPA, 1 Batterymarch Park, P. O. Box 9101, Quincy, Massachusetts 02269–9101, telephone number 1 (800) 344–3555. Copies of ANSI/NFPA 70–1999, *NEC*®, are available for inspection during normal business hours at RUS, room 2905, U.S. Department of Agriculture, 1400 Independence Avenue, SW., STOP 1598, Washington, DC 20250–1598, or at the National Archives and Records Administration (NARA). For information on the availability of this material at NARA, call 202–741–6030, or go to: *http://www.archives.gov/federal_register/code_of_federal_regulations/ibr_locations.html.*

(j) If access to the building electric service grounding system, as referenced in paragraph (i) of this section, is not possible or is not reasonable (telecommunications primary station protector grounding conductor will be longer than 10 ft (3 m)), the NID, BET, or fused primary station protector shall be located as close as practicable to electrodes (a) or (b) cited in section 800–40(b)(1) of ANSI/NFPA 70–1999, *NEC*®.

(k) In addition, the NID, BET, or fused primary station protector shall be located in, on, or immediately adjacent to the structure or building to be served as close as practicable to the point at which the telecommunications service wire attaches to the building, making sure that the telecommunications primary station protector grounding conductor is connected to the closest, existing, and accessible electrode, of the electrodes cited in paragraph (i) or (j) of this section.

(l) For the preferred customer access location installation, the ANSI/NFPA 70–1999, *NEC*®, permits the telecommunications grounding conductor to be connected to the metallic conduit, service equipment closure, or electric grounding conductor as shown in Figure 8 of paragraph (l)(2) of this section.

(1) Connections to metallic conduits shall be made by ground straps clamped over a portion of the conduit that has been cleaned by sanding down to bare metal.

(2) Connections to metallic service equipment closures shall be made by attaching a connector which is listed for the purpose by some organization acceptable to the local authority (State, county, etc.) per article 100 of ANSI/NFPA 70–1999, *NEC*®, definition for ''Listed'' (for example connectors listed for the purpose by Underwriters Laboratories (UL)). Figure 8 is as follows:

FIGURE 8

GROUNDING OF TELECOMMUNICATIONS SERVICE TO ELECTRIC SERVICE (PREFERRED METHOD)

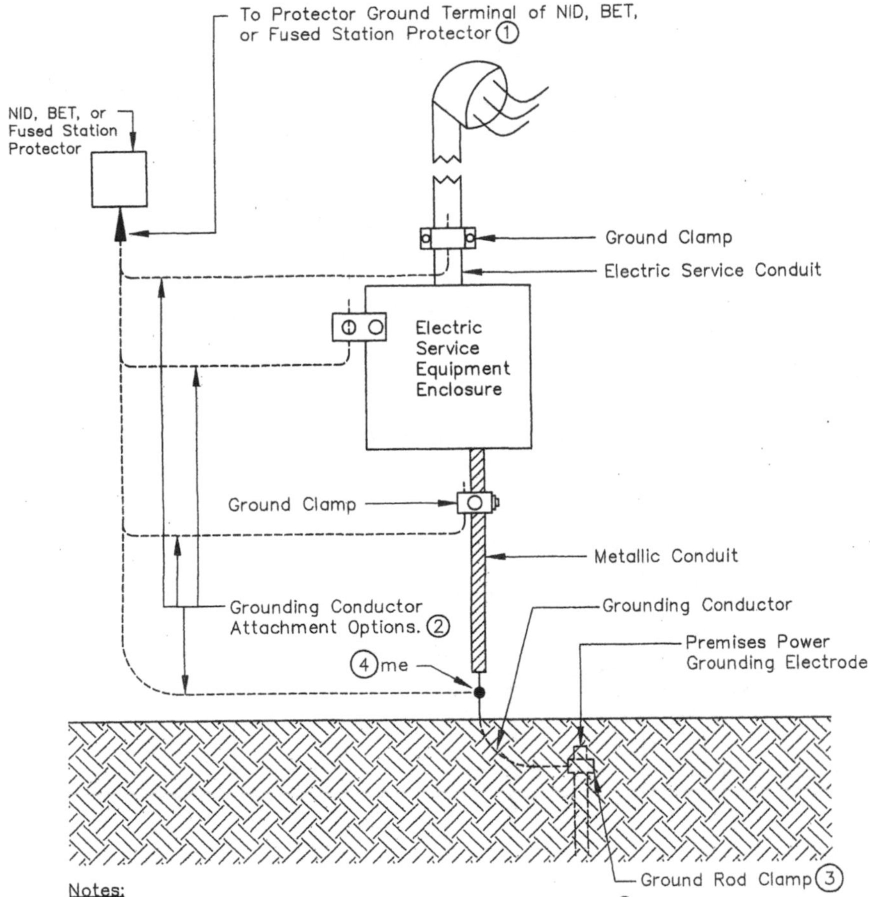

Notes:

(1.) See Section 800–40(a) of ANSI/NFPA 70–1999 NEC®. (2.) Select one of the attachment options shown above for the installation. (3.) Clamp must be accepted by Listing Agency (UL, etc.). (4.) Connector (Item "me") must be accepted by a Nationally recognized testing laboratory.

(m) Where it is not possible to accomplish the objective of paragraphs (i), (j), and (k) of this section, interior metallic pipes may be used to the maximum practicable extent to gain access to the electric service ground as shown in Figure 9. Note that the water pipe in Figure 9 is electrically continuous between electric and telecommunications bonds to the cold water pipe and it is used only as a portion of a bonding conductor and, therefore, does not have to be "acceptable" as a ground electrode but may be floating (isolated from ground by a plastic pipe section). ANSI/NFPA 70-1999, *NEC®*, requires that metal piping be used as a bonding conductor in this manner only when the connectors to the pipe are within 1.5 m (5 ft) of where the pipe enters the premises. This is not the preferred installation. The RUS preferred installation has the telecommunications primary station protector grounded directly to an accessible location near the power grounding system. See paragraph (1) of this section. Figure 9 is as follows:

FIGURE 9

ALTERNATIVE TECHNIQUE FOR BONDING TO ELECTRIC SERVICE GROUND WHERE DIRECT ATTACHMENT IS NOT POSSIBLE

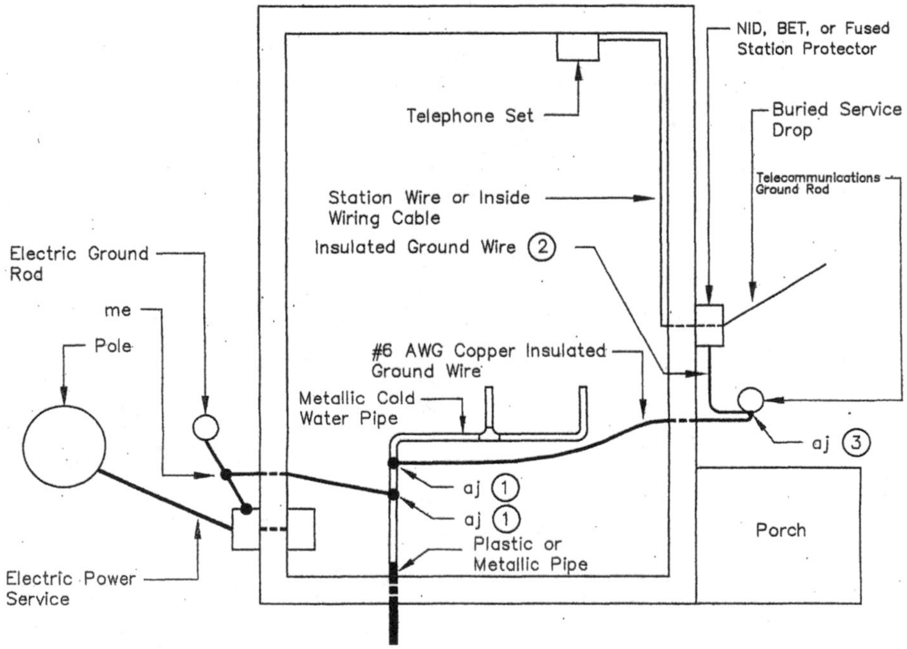

Notes:

(1.) Both electric and telephone "aj" connectors attached to the cold water pipe shall be within 5 ft (1.5 m) of where the pipe enters the premises.

(2.) Refer to Section 1755.508, Paragraph (v), Table 5 for the ground wire conductor size. Ground wire must be accepted by a Nationally recognized testing laboratory.

(3.) Connector "aj" must be accepted by a Nationally recognized testing laboratory.

(n) Where the telecommunications premises system at a customer's access location is grounded to a separate electrode (of any type) this telecommunications grounding electrode must be bonded to the electric grounding system with a No. 6 AWG or larger copper insulated grounding conductor. Bonding of separate electrodes is a requirement of the ANSI/NFPA 70–1999, *NEC®*.

(o) The NID, BET, or fused primary station protector pair size shall be selected for the number of lines anticipated within five years.

(p) When lightning damage is considered probable or customer access locations are remote from the borrower's headquarters, use of maximum duty gas tube primary station protectors incorporated in NIDs, BETs, or fused primary station protectors should be considered. (See RUS TE&CM 823, Electrical Protection by Use of Gas Tube Arresters). Copies of RUS TE&CM 823 are available upon request from RUS/USDA, 1400 Independence Avenue, SW., STOP 1522, Washington, DC 20250–1522, FAX (202) 720–4120.

(q) NIDs or BETs incorporating fuseless station protectors shall always be used in preference to fused station protectors or BETs incorporating fused protectors, when in the judgment of the RUS borrower or the engineer delegated by the RUS borrower, the requirements of ANSI/NFPA 70–1999, *NEC®*, for fuseless station protectors can be met.

(r) A fuse link consisting of a copper conductor two gauges (AWG) finer (numerically higher) conductivity than the aerial service wire shall be provided between the cable and aerial service wire where NIDs or BETs incorporating fuseless station protectors are used. Thus for a 22 AWG drop, a fuse link of No. 24 AWG or finer copper wire shall be provided. If the cable circuit is No. 24 gauge or finer, the cable conductors serve as the fuse link for the 22 AWG aerial service wire and no separate fuse link is necessary. (Note: The fuse link or the facilities serving as the fuse link must be located between the telecommunications facilities that are exposed to possible power cross and the customer drop where there is no exposure to possible power cross.)

(s) RUS's buried plant practices require buried main line plant to be protected against power contacts to aerial plant extensions and aerial inserts by No. 24 AWG fuse links at every buried-aerial junction.

(t) In aerial cable plant, fuse links are usually provided by No. 24 AWG leads on filled terminal blocks regardless of the gauge of the cable conductors. This practice is acceptable if the ampacity of the aerial service wire is sufficiently higher than the fuse link's ampacity.

(u) The grounding and bonding of each NID, BET, or fused primary station protector shall be selected by consulting paragraphs (i) through (n) of this section. The "first choice" assembly unit shall be selected whenever the prevailing conditions make its use practicable. The NID, BET, or fused primary station protector assembly unit selected shall be installed in accordance with the appropriate construction drawing specified in RUS Bulletin 1753F–153 (RUS Form 515d), Specifications and Drawings for Service Installations at Customer Access Locations (Incorporated by reference at §1755.97). Copies of RUS Bulletin 1753F–153 are available upon request from RUS/USDA, 1400 Independence Avenue, SW., STOP 1522, Washington, DC 20250–1522, FAX (202) 720–4120.

(v) The minimum size grounding conductor that can be used with a single NID; a group of NIDs; a multipair NID; fused protector; or BET shall be in accordance Table 5, as follows:

TABLE 5—GROUNDING CONDUCTOR SIZE VERSUS NUMBER OF CIRCUITS

Minimum grounding conductor size	Number of circuits	
	Fuseless (carbon or gas tube)	Fused
#12 AWG, copper, insulated	1 to 2	1 to 3.
#10 AWG, copper, insulated	3 to 5	4 to 7.
#6 AWG, copper, insulated	6 or more	8 or more.

(w) Grounding conductor runs between the NID, BET, or fused station protector and the ground electrode shall conform to the following:

(1) The shortest, most direct route practicable shall be used;

(2) Sharp bends in the grounding conductor shall be avoided during installation;

(3) No splices shall be made in the grounding conductor;

(4) Grounding conductors shall not be fished through walls, under floors, or placed in bridle rings or any metal conduit unless the grounding conductor is bonded to the conductor at both ends of the metallic conduit;

(5) Grounding conductor runs from an outside mounted NID, BET, or fused station protector to an inside ground electrode shall use the same entrance as the station wire; and

(6) Grounding conductor runs from an outside mounted NID, BET, or fused station protector to an outside ground electrode at the building shall be attached to the exterior surface of the building or buried. If buried, the grounding conductor shall be either plowed or trenched to a minimum depth of 12 in. (300 mm). When trenched, the trenches shall be as close to the side of the building as practicable, backfilled, and tamped to restore the earth to its original condition.

(x) Telecommunications grounding connectors shall be RUS accepted or RUS technically accepted. Grounding and bonding conductors shall be made of copper. Where the grounding and bonding conductors must be connected to aluminum electric service grounding conductors, bimetal grounding connectors shall be used.

(y) Grounding conductor attachments shall conform to the following:

(1) Galvanized nails or clamps, or nickel-copper alloy staples shall be used for grounding conductor attachments in accordance with Table 6 in paragraph (y)(3) of this section;

(2) Grounding conductors, station or buried service wires in parallel runs may share the same fastening device when the device is specifically designed for two wires. See Table 6 in paragraph (y)(3) of this section for station wire and grounding conductor fasteners; and

(3) Grounding conductor fasteners shall be placed 12 to 18 in. (300 to 450 mm) apart on straight runs and 2 to 4 in. (50.8 to 100 mm) apart at corners and at bends. Table 6 is as follows:

Table 6

TYPICAL FASTENING DEVICES FOR STATION WIRES AND GROUNDING CONDUCTORS (9)

TYPE AND GAUGE OF WIRE	APPROX. OVERALL DIAMETER	TYPES OF FASTENING DEVICES FOR VARIOUS TYPES OF BUILDINGS OR WALL FINISHES: Hard Woods	Soft Woods	Wallboard, Plaster on Wood, or Metal Lath, or Concrete Block(3)	Brick, Stone or Concrete(3)	Shingles and Siding(4)	Sheet Metal(5)	Wall Tile(3)
#22 AWG Station Wire	.125 in. to .155 in.	A1, D7, E1, F1, G1	A2, A3, D8, E2, F2, G2	D8, D9, E2, E3, G2, G3	D8, E2, G2	A2, A3, D7, D8, E2, F2	D7, D8, D9, G1, G2, G3, H1	D8, D9, E2, E3, G2, G3, H1
GROUNDING CONDUCTOR: #10 AWG Insulated Wire	.168 in.	A1, B1, D1	A2, A3, B1; B2, D2	B2, D2, D3	B2, D2	A2, A3, B1, B2, D1, D2	D1, D2, D3, H2	B2, D2, D3, H2
GROUNDING CONDUCTOR: #12 AWG Insulated Wire	.127 in.	A1, B1, C1, E1, F1, D7, G1	A2, B1, B2, C1, C2, D8, E2, F2, G2	B2, C2, C3, D8, D9, E2, E3, G2, G3	B1, B2, C2, D8, E2, E3, G2	A2, A3, B1, B2, C1, C2, D8, E2, F2, G2	C1, C2, C3, D7, D8, D9, E1, E2, E3, G1, G2, G3, H1	B2, B3, C3, D8, D9, E2, E3, G2, G3, H1
GROUNDING CONDUCTOR: #6 AWG Insulated Wire	.290 in.	A2, A3, B1, D4	A3, B2, D5	B2, D5, D6	B2, D5	A3, B2, D5	D4, D5, D6, H3	B2, D5, D6, H3

EXPLANATION OF FASTENER CODES

A. Staple Machine, Round Crown, Interior Use Only – (Note 6)

1. 3/16" or 1/4" Crown – 3/8" Leg
2. 3/16" or 1/4" Crown – 7/16" or 9/16" Leg
3. 3/16" or 1/4" Crown – 9/16" Leg

B. Nail, Ground Wire, Single Shank Galvanized, Interior and Exterior Use

1. 7/8" #14
2. 1–3/8" #13

C. Clamp, Ground Wire, One Hole, Galvanized, Interior and Exterior Use

1. Type B–1/2" x #6 RH Screw (1)
2. Type B–3/4" x #6 RH Screw (1)
3. Type B–1/8" x 3" Toggle Bolt (2)

D. Clamp, One Hole Offset, Galvanized or Enameled, Interior and Exterior Use – (Note 7)

	Wire Size Min.	Max.	Fasteners (1), (2)
1.	5/32"	to 7/32"	1/2" x #6 RH Screw
2.	5/32"	to 7/32"	3/4" x #6 RH Screw
3.	5/32"	to 7/32"	1/8" x 3" Toggle Bolt
4.	1/4"	to 5/16"	1/2" x #6 RH Screw
5.	1/4"	to 5/16"	1" x #6 RH Screw
6.	1/4"	to 5/16"	1/8" x 3" Toggle Bolt
7.	1/8"	to 5/32"	1/2" x #6 RH Screw
8.	1/8"	to 5/32"	3/4" x #6 RH Screw
9.	1/8"	to 5/32"	1/8" x 3" Toggle Bolt

E. Clamp, Station Wiring, One Hole, Galvanized or Enameled, Interior and Exterior Use – (Note 7)

1. Type B–1/2" x #6 RH Screw (1)
2. Type B–3/4" x #6 RH Screw (1)
3. Type B–1/8" x 3" Toggle Bolt (2)

F. Nail, Station Wiring, Galvanized or Enameled, Interior and Exterior Use – (Note 7)

1. Type B – 1/2"
2. Type B – 7/8"

G. Clamp, One Hole Double – (Note 8)

	Wire Size Min.	Max.	Fasteners
1.	Two 1/8"	to 5/32"	3/4" x #6 RH Screw(1)
2.	Two 1/8"	to 5/32"	1" x #6 RH Screw(1)
3.	Two 1/8"	to 5/32"	1/8" x 3" Toggle Bolt(2)

H. Station Wire Clip, Adhesive Backed, Interior Use Only –

Wire Size

1. 1/8" Nominal
2. 3/16" Nominal
3. 1/4" Nominal

NOTES: 1. Screw dimensions are minimum. Where appropriate, either or both dimensions shall be increased. All wood screws for exterior use shall be stainless steel. All other exterior metal devices shall be stainless steel, zinc coated steel, silicon bronze, or corrosion resistant aluminum alloy.

2. Toggle bolt dimensions are minimum. Where appropriate, either or both dimensions shall be increased.

3. Wall screw anchors may be used in wall board, plaster or tile walls. Screws and nails in masonry shall be secured by means of expansions type anchors. Equivalent manual or machine-driven devices may be used. Where toggle bolts are specified, equivalent devices may be used.

4. Lead holes shall be drilled for screws, nails, and bridle rings in shingles and dropsiding.

5. Sheet metal screws shall be used except where toggle bolts are required. Where wood sheathing under sheet metal siding is encountered, the sheet metal may be drilled or punched and a wood screw used.

6. Machine-driven staples of nickel-copper composition may be used for exterior wiring.

7. Galvanized clamps and wiring nails may be used for exterior and interior wiring. Enameled clamps shall be used for interior wiring only. Where toggle bolts or equivalent devices require holes in the structure larger than the clamp being fastened, a suitable washer of sufficient size to cover the hole must be used under the clamp.

8. Double clamp may be used where two #22 AWG station wires, two #12 AWG grounding conductors, or one #22 AWG station wire and one #12 grounding conductor parallels one another.

9. For converting English units to Metric units use 1 in. = 25.4 mm.

(z) Grounding conductors shall be separated from non-telecommunications company wires in accordance with section 800–12(b) of ANSI/NFPA 70–1999, *NEC*®.

(aa) Grounding conductors run through metal conduits shall be bonded to the conduit at each end. RUS accepted and RUS technically accepted pipe type ground clamps and grounding connectors shall be used for bonding.

(bb) Where NID, BET, or fused station protector assembly units require grounding conductor connections to pipe systems, the following apply:

(1) The connection shall be made to a cold water pipe of an operating water system;

(2) The connection point shall be preferably inside the building;

(3) Allow a minimum of 6 in. (152 mm) between the last fastener and the point where the grounding conductor first touches the water pipe;

(4) Leave 2 in. (50.8 mm) of slack in the grounding conductor to avoid breaking the conductor at the terminating point. Tape the grounding conductor to the pipe where possible to avoid movement. In no case, shall the grounding conductor be coiled or wrapped around the pipe;

(5) The pipe shall be cleaned with fine sand paper to make a good electrical connection. Care should be taken to avoid damaging the pipe while cleaning it;

(6) Attach the pipe grounding conductor connector to the cleaned area of pipe and tighten. Care shall be exercised to avoid deforming, crushing, or otherwise damaging the pipe. A simple continuity check with an ohmmeter between the connector and the pipe will indicate whether or not a good electrical contact has been made. Set the ohmmeter to "Rx1" scale to ensure that a low resistance contact is made;

(7) A warning tag shall be attached to the ground clamp with the following or equivalent statement: "Call the telecommunications company if this connector or grounding conductor is loose or must be removed;" and

(8) When the water pipe is used, the ANSI/NFPA 70–1999, *NEC*®, requires that metal piping be used as a bonding conductor in this manner only when the connections to the pipe are within 5 ft (1.5 m) of where the pipe enters the premises.

(cc) Bonding conductors shall consist of either copper or tinned copper insulated wires of appropriate sizes.

(1) Bonding conductors shall be run and attached in the same manner as grounding conductors.

(2) Attaching and terminating devices for bonding conductors shall be adequate for the size of wire involved. The No. 6 AWG copper insulated conductor or larger shall not be terminated by bending it around a threaded stud.

(dd) Where NID, BET, or fused station protector assembly units require a driven ground rod the following shall apply to the ground rod installation:

(1) Locate the ground rod at least 1 ft (300 mm) from buildings, poles, trees and other obstruction;

(2) Ground rods shall not be installed within 6 ft (2 m) of electric service ground rods (Note: This minimum separation is provided to avoid mutual impedance effects of multiple grounding electrodes that will deleteriously degrade the effective impedance-to-earth if grounding electrodes are installed any closer than 6 ft (2 m) to one another. This requirement is included for cases where the telecommunications company is not allowed, for some reason, to observe the RUS preferred grounding method of attaching the primary protector grounding conductor directly to an accessible point on the

building electric service grounding system. RUS believes that if the primary protector location can be sited within 6 ft (2 m) of the electric service ground rod then the electric service ground rod could be used as the preferred telecommunications grounding electrode and a separate telecommunications ground rod is unnecessary);

(3) A hole, 15 in. (350 mm) deep and 6 in. (150 mm) in diameter, shall be dug at the location where the ground rod is to be driven;

(4) Where "slip-on" type ground rod clamps are used instead of "clamp-around" type clamps, the ground rod clamps shall be placed onto the rod prior to driving the rod into the ground (Note there should be one clamp for the NID, BET, or fused station protector grounding conductor and one clamp for the conductor required to bond the telecommunications ground rod to the electric grounding system). However, the clamp shall not be tightened until the rod is completely driven. The end of the rod shall be placed in the bottom of the hole and the rod shall be aligned vertically adjacent to one wall of the hole prior to driving. The rod shall be driven until its tip is 12 in. (300 mm) below final grade. The grounding conductor shall then be attached, the clamp shall be tightened, and hole backfilled. Clamps employed in this manner shall be suitable for direct burial and shall be RUS accepted or RUS technically accepted; and

(5) Where rods are manually driven, a large number of blows from a light hammer (4 lbs (1.8 kg)) shall be used instead of heavy sledgehammer type blows. This should keep the rod from bending.

(ee) Terminations on fuseless primary station protectors incorporated in NIDs and on fused primary station protectors shall be as shown in Figures 10, 11, 12, and 13 of paragraph (ee)(1) of this section, Figure 14 of paragraph (ee)(4) of this section, and Figure 15 of paragraph (ee)(6) of this section. The inner jackets of buried service wires and outer jackets of cables used as service drops shall be extended into the NID or the fused primary station protector. A 10 in. (250 mm) length of each spare wire shall be left in NIDs or fused primary station protectors. The spare wires shall be coiled up neatly and stored in the NID or fused primary station protector housing.

(1) The shields of buried service wires may be connected to the ground binding post using RUS accepted or RUS technically accepted buried service shield bond connectors as shown in Figure 10 for NIDs and Figure 11 for fused primary station protectors. RUS accepted or RUS technically accepted buried service wire harness wires designed for customer access location installations may also be used for terminating buried service wire shields to the ground binding post of the NID as shown in Figure 12 and Figure 13 for fused primary station protectors. Figures 10 through 13 are as follows:

FIGURE 10

BONDING BURIED SERVICE WIRE AT STATION PROTECTOR OF NID USING SERVICE WIRE SHIELD BOND CONNECTOR

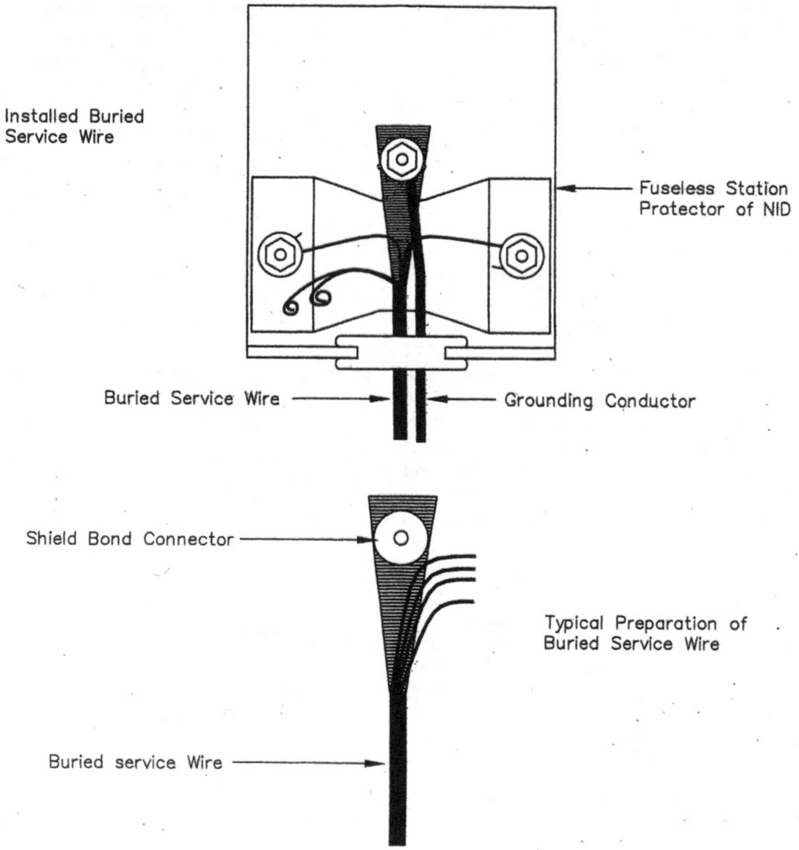

FIGURE 11

BONDING BURIED SERVICE WIRE AT FUSED STATION PROTECTOR
USING SERVICE WIRE SHIELD BOND CONNECTOR

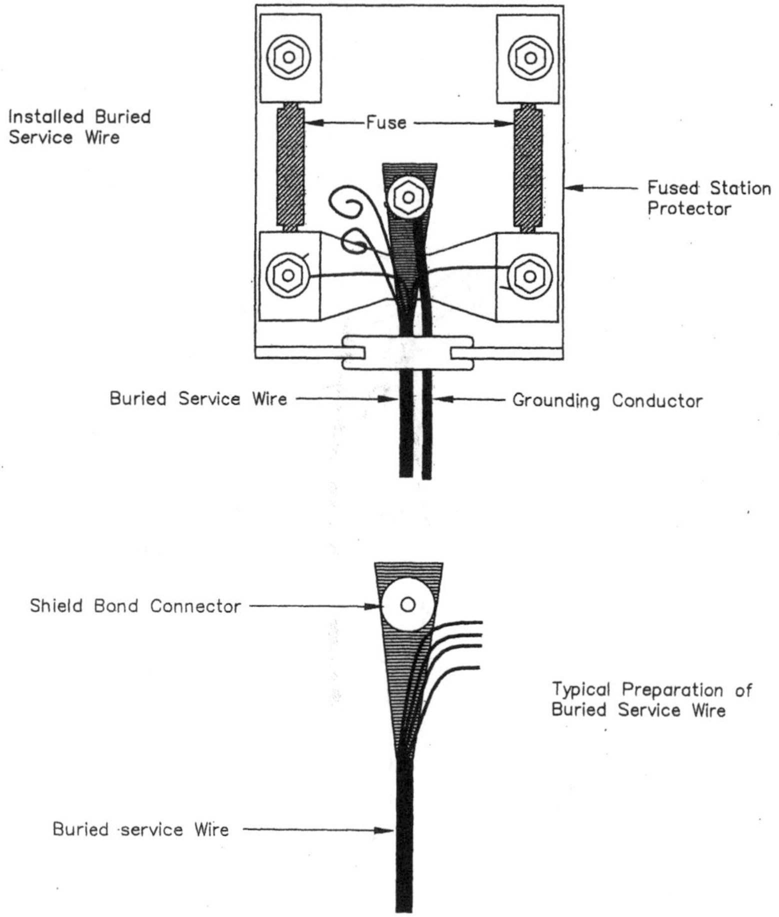

FIGURE 12

BONDING BURIED SERVICE WIRE AT STATION PROTECTOR OF NID USING SERVICE WIRE BONDING HARNESS

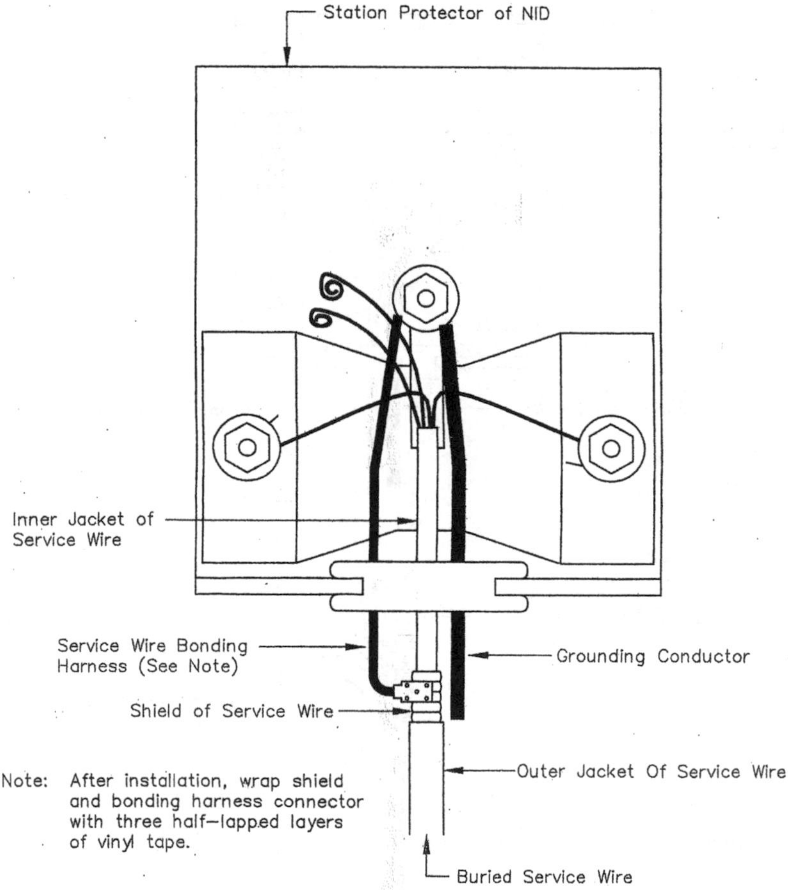

Note: After installation, wrap shield and bonding harness connector with three half–lapped layers of vinyl tape.

FIGURE 13

BONDING BURIED SERVICE WIRE AT FUSED STATION PROTECTOR USING SERVICE WIRE BONDING HARNESS

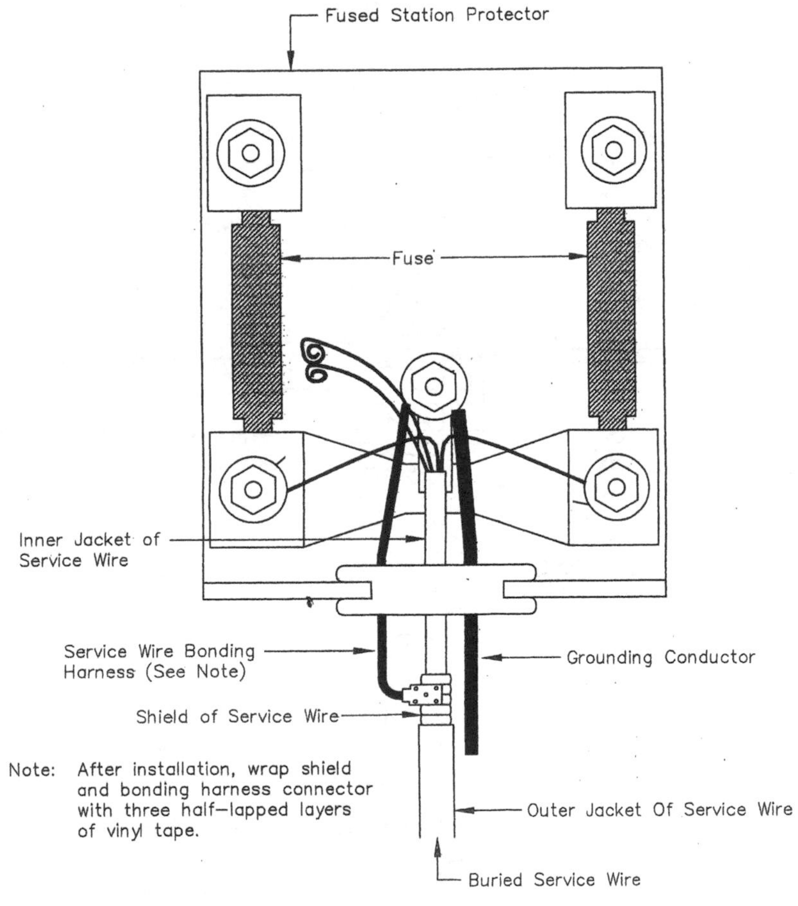

(2) On buried service drops and aerial service drops of more than 6 pairs using RUS accepted or RUS technically accepted cables, the shields shall be terminated with a RUS accepted or RUS technically accepted cable shield bonding connector and extended to the ground binding post of the NID, BET, or fused primary station protector with an RUS accepted or RUS technically accepted bonding harness wire. The installation of the shield bond connector and bonding harness wire shall be in accordance with the manufacturer's instructions.

(3) The shield and other conductors at the fuseless primary station protector incorporated in the NID shall be

terminated as shown on Figure 14 in paragraph (ee)(4) of this section. The pronged or cupped washer shall be placed above the shield. The grounding conductor shall be placed around the post on top of the pronged or cupped washer. A flat washer shall be placed above the grounding conductor.

(4) The station wire signaling ground conductor, if required, shall be placed above the first flat washer and beneath the second flat washer as indicated in Figure 14 as follows:

FIGURE 14

TERMINATION OF CONDUCTORS AND SHIELD ON STATION PROTECTOR BINDING POSTS OF NID

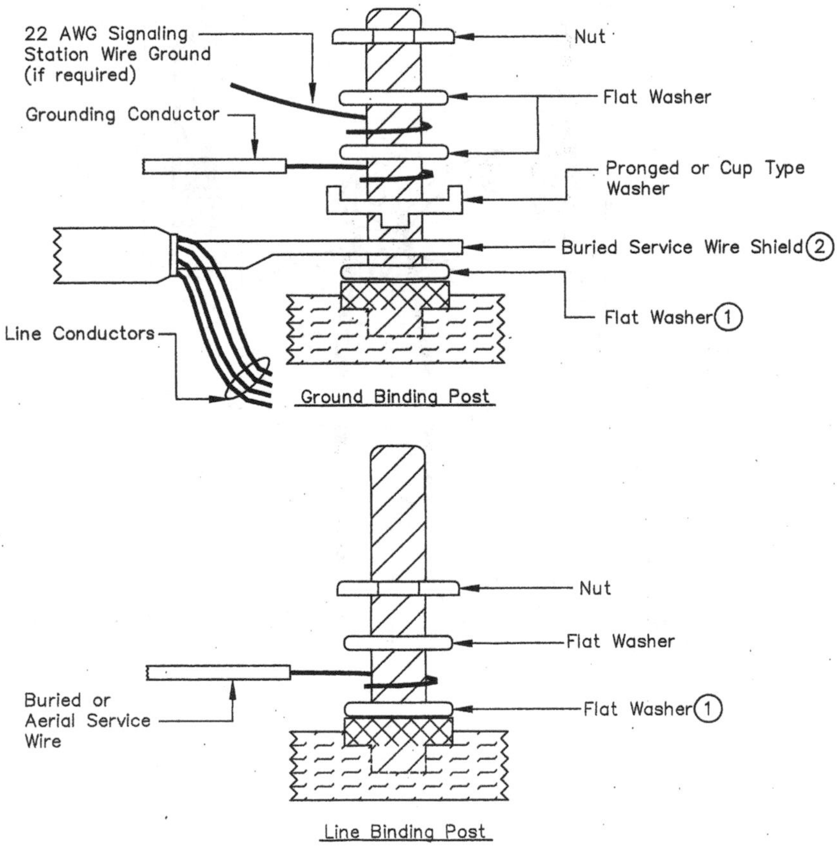

Notes:

(1.) If shoulder is inadequate to support shield or wire add a flat washer.

(2.) Terminate buried service wire shield with station protector grounding lug of NID in accordance with either Figure 10 or 12 of paragraph (ee)(1) of this section.

(5) The shield and other conductors at the fused primary station protector shall be terminated as shown on Figure 15 in paragraph (ee)(6) of this section. The pronged or cupped washer shall be placed above the shield. The grounding conductor shall be placed around the post on top of the pronged or cupped washer. A flat washer shall be placed above the grounding conductor.

(6) The station wire signaling ground conductor, if required, shall be placed above the first flat washer and beneath the second flat washer as indicated in Figure 15 as follows:

FIGURE 15

TERMINATION OF CONDUCTORS AND SHIELD ON FUSED STATION PROTECTOR BINDING POSTS

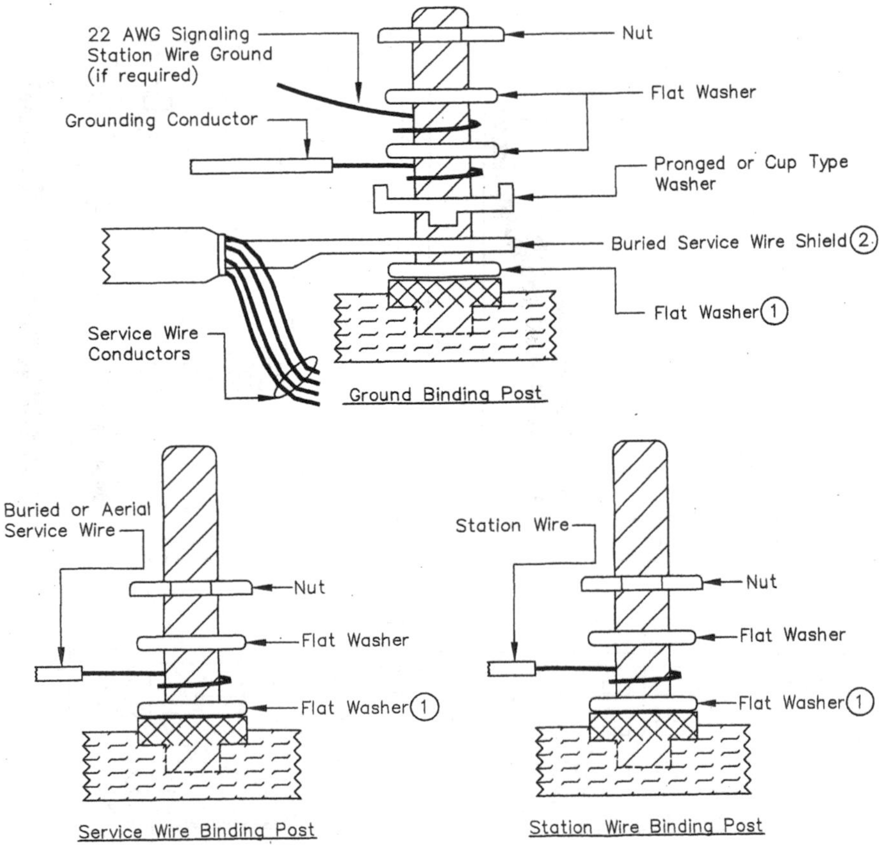

Notes:

(1) If shoulder is inadequate to support shield or wire add a flat washer.

(2) Terminate buried service wire shield on fused station protector grounding lug in accordance with either Figure 11 or 13 of paragraph (ee)(1) of this section.

(7) Indoor NIDs or BETs that are equipped with "Quick Connect" type terminals shall not have more than one wire connected per clip. No. 19 AWG copper and No. 18 AWG copper covered-steel reinforced aerial service wire conductors shall not be connected to quick connect terminals. Nonmetallic reinforced aerial service wire using No. 22 AWG copper conductors may be connected to the quick connect terminals.

(8) Tip and ring connections and other connections in multipair NIDs or BETs shall be as indicated in Figure 16 as follows:

FIGURE 16

MULTIPAIR NID OR BET TERMINAL CONNECTIONS CONTAINING FUSELESS STATION PROTECTORS

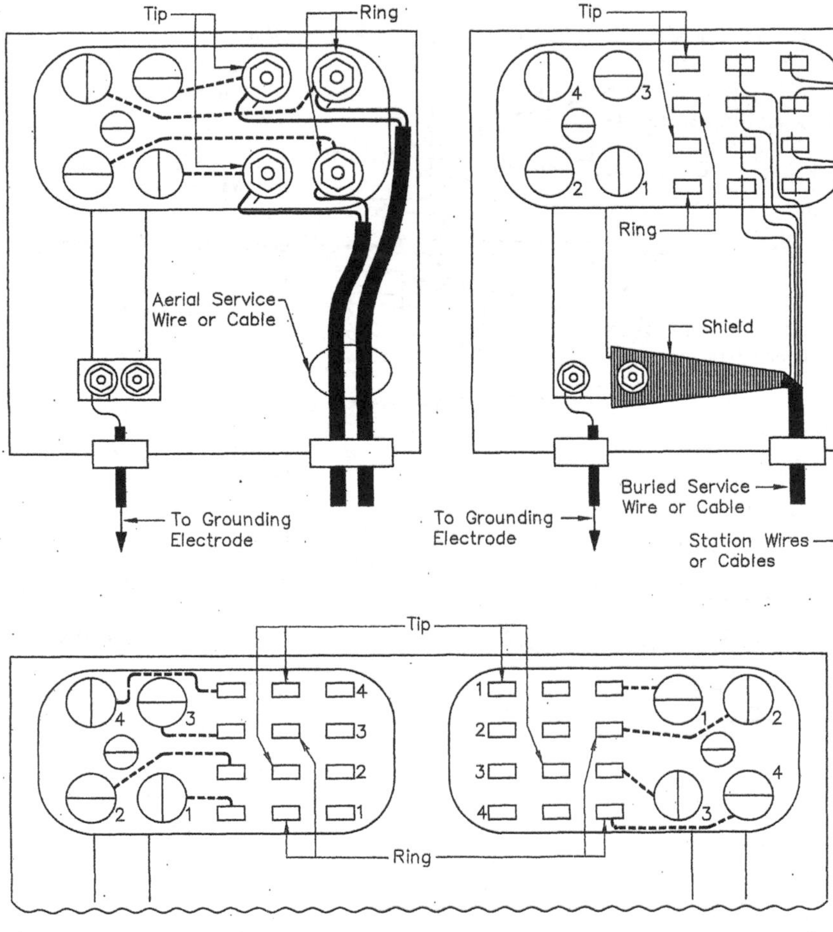

Note: #18 AWG copper–covered steel reinforced aerial service conductors shall not be connected to quick connect terminals. Nonmetallic reinforced aerial service conductors (#22 AWG copper) may be connected to quick connect terminals.

(ff) System polarity and conductor identification shall be maintained in NIDs, BETs, or fused primary station protectors in accordance with construction drawings 815 and 815–1 contained in § 1755.510.

[66 FR 43317, Aug. 17, 2001, as amended at 69 FR 18803, Apr. 9, 2004]

§ 1755.509 Mobile homes.

(a) Customer access location installations at mobile homes shall be treated the same whether the homes are mounted on permanent foundations or temporary foundations and shall be installed as specified in §§ 1755.500 through 1755.510. For the purpose of this section, mobile homes include manufactured homes, motor homes, truck campers, travel trailers, and all forms of recreational vehicles. Customer access location installations at mobile homes can be considerably different than customer access location installations at regular homes and borrowers shall be certain that the two types of installations are properly applied.

(b) The method of customer access location installation prescribed by the ANSI/NFPA 70–1999, *NEC*® for a mobile home depends on how the electric power is installed at the mobile home and it can involve considerable judgment on the part of the telecommunications installer. The *National Electrical Code*® and *NEC*® are registered trademarks of the National Fire Protection Association, Inc., Quincy, MA 02269. The ANSI/NFPA 70–1999, *NEC*®, is incorporated by reference in accordance with 5 U.S.C. 552(a) and 1 CFR part 51. Copies are available from NFPA, 1 Batterymarch Park, P. O. Box 9101, Quincy, Massachusetts 02269–9101, telephone number 1 (800) 344–3555. Copies of ANSI/NFPA 70–1999, *NEC*®, are available for inspection during normal business hours at RUS, room 2905, U.S. Department of Agriculture, 1400 Independence Avenue, SW., STOP 1598, Washington, DC 20250–1598, or at the National Archives and Records Administration (NARA). For information on the availability of this material at NARA, call 202–741–6030, or go to: *http://www.archives.gov/federal_register/code_of_federal_regulations/ibr_locations.html*. The ANSI/NFPA 70–1999, *NEC*®, requires primary station protectors to be located where specific acceptable grounding electrodes exist. The ANSI/NFPA 70–1999, *NEC*®, allows station protector installations to be at the location of the power meter or the electric disconnecting means apparatus serving the mobile home providing these electric facilities are installed in the manner specifically defined by the ANSI/NFPA 70–1999, *NEC*®. The ANSI/NFPA 70–1999, *NEC*®, requires the station protectors to be installed at the nearest of a number of other meticulously defined ANSI/NFPA 70–1999, *NEC*®, acceptable electrodes where the protector cannot be installed at the power meter or the electric disconnecting means apparatus serving the mobile home. The provisions can be confusing.

(c) NIDs shall be installed at mobile homes as follows:

(1) Where the mobile home electric service equipment (power meter, etc.,) or the electric service disconnecting means associated with the mobile home is located within 35 ft (10.7 m) of the exterior wall of the mobile homes it serves, the NID shall be installed in accordance with Figure 17 as follows:

FIGURE 17

NETWORK INTERFACE DEVICE (NID) INSTALLATION
ELECTRIC SERVICE EQUIPMENT WITHIN 35 FEET (10.7 METERS)
OF MOBILE HOME

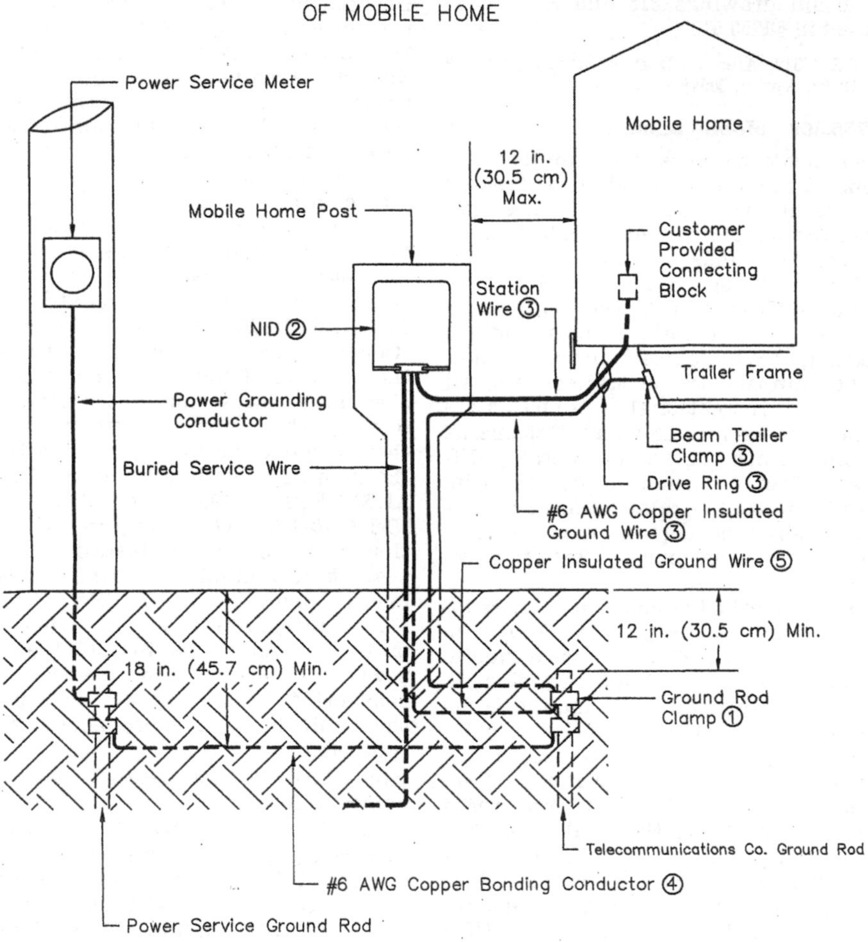

Notes: ① Clamp must be accepted by Listing Agency (UL, etc.) for two conductors, otherwise two clamps must be used. ② See Figure 19 of paragraph (e) of this section for NID terminations. ③ See Figure 20 of paragraph (e) of of this section for mobile home installation. ④ Bare if buried its entire length; insulated where human contact is possible. ⑤ See Section 1755.508, paragraph (v), Table 5 for the correct conductor size of the ground wire.

(2) Where the mobile home electric service equipment (power meter, etc.,) or the electric service disconnecting means associated with the mobile home is located more than 35 ft (10.7 m) from the exterior wall of the mobile homes it serves, the NID shall be installed in accordance with Figure 18 as follows:

FIGURE 18

NETWORK INTERFACE DEVICE (NID) INSTALLATION
ELECTRIC SERVICE EQUIPMENT MORE THAN 35 FEET (10.7 METERS)
FROM MOBILE HOME

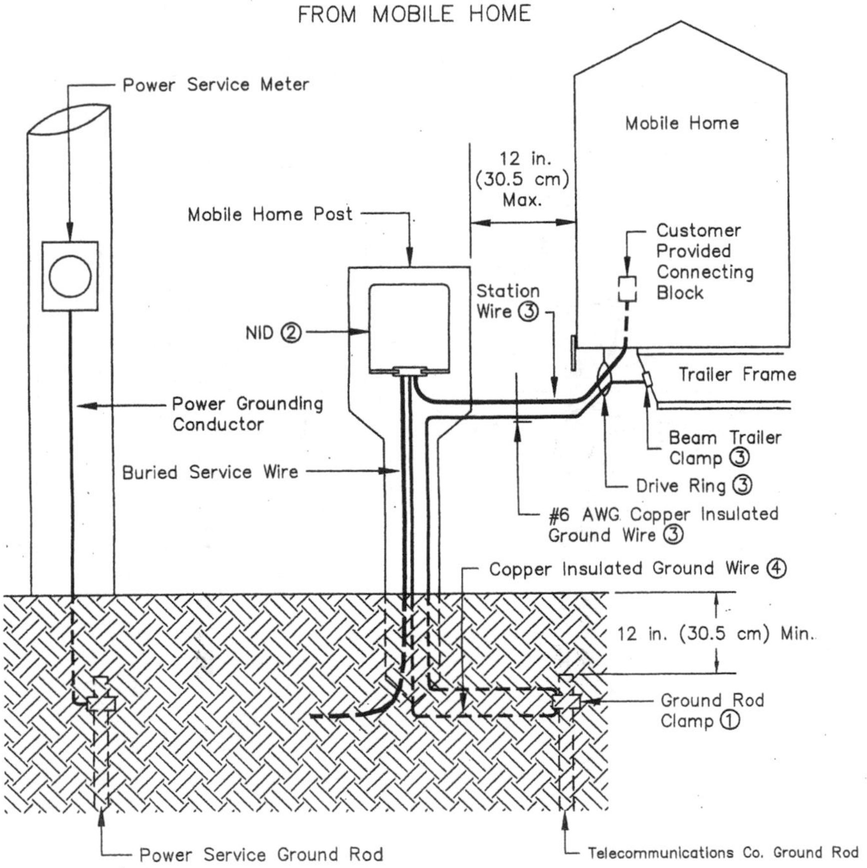

Notes: (1.) Clamp must be accepted by Listing Agency (UL, etc.) for two conductors, otherwise two clamps must be used. (2.) See Figure 19 of paragraph (e) of this section for NID terminations. (3.) See Figure 20 of paragraph (e) of this section for mobile home installation. (4.) See Section 1755.508, paragraph (v), Table for the correct conductor size of the ground wire.

(d) The service wire and station wire shall be terminated in the NID in accordance with Figure 19 in paragraph (e) of this section.

(e) Installation of the station wire and grounding conductor at the mobile home shall be in accordance with Figure 20. Figures 19 and 20 are as follows:

FIGURE 19

NID TERMINATIONS

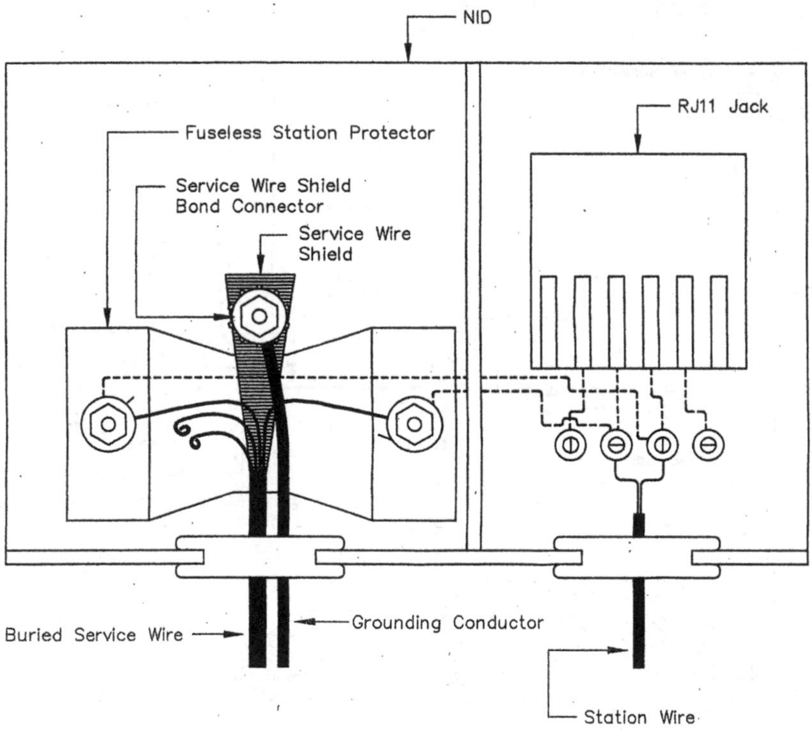

FIGURE 20
MOBILE HOME INSTALLATION

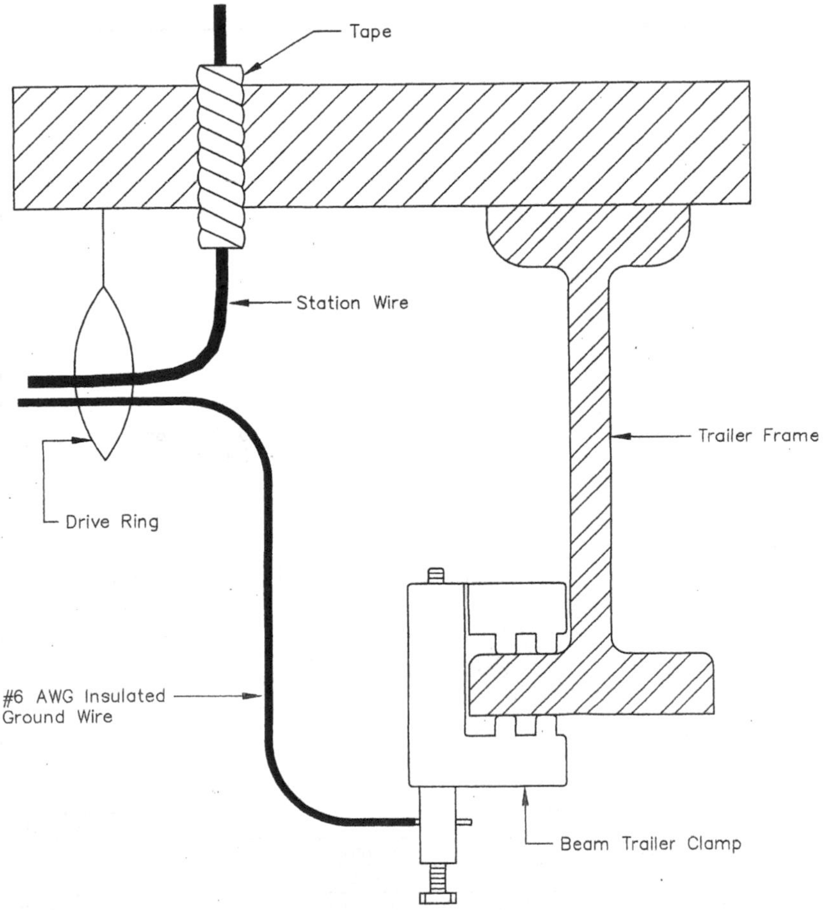

[66 FR 43317, Aug. 17, 2001, as amended at 69 FR 18803, Apr. 9, 2004]

§ 1755.510 Construction and assembly unit drawings.

(a) The construction and assembly unit drawings in this section shall be used by borrowers to assist the installer in making the customer access location installations.

(b) The asterisks appearing on the construction drawings indicate that the items are no longer listed in the RUS Informational Publication (IP) 344–2, "List of Materials Acceptable for Use on Telecommunications Systems of RUS Borrowers." RUS IP 344–2 can be obtained from the Superintendent of Documents, P. O. Box 371954, Pittsburgh, PA 15250–7954, telephone number (202) 512–1800.

(c) Drawings BM50, BM83, 312–1, 501–1, 501–2, 503–2, 504, 505, 506, 507, 508–1, 510, 510–1, 510–2, 513, 815, 815–1, 958, and 962 are as follows:

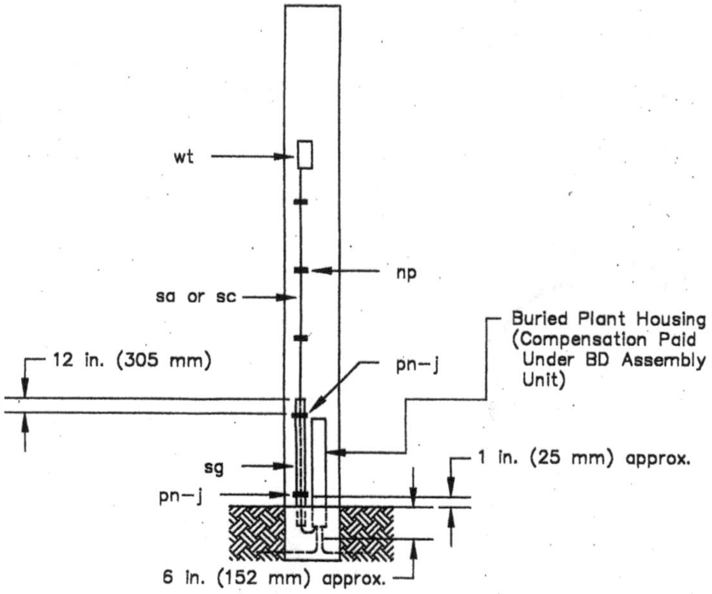

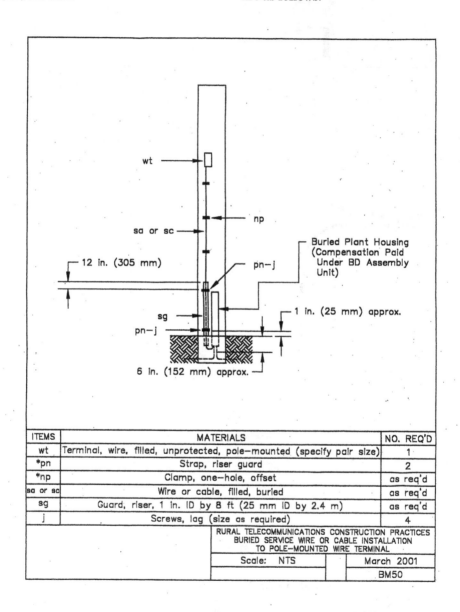

ITEMS	MATERIALS	NO. REQ'D
wt	Terminal, wire, filled, unprotected, pole–mounted (specify pair size)	1
*pn	Strap, riser guard	2
*np	Clamp, one–hole, offset	as req'd
sa or sc	Wire or cable, filled, buried	as req'd
sg	Guard, riser, 1 in. ID by 8 ft (25 mm ID by 2.4 m)	as req'd
j	Screws, lag (size as required)	4

RURAL TELECOMMUNICATIONS CONSTRUCTION PRACTICES
BURIED SERVICE WIRE OR CABLE INSTALLATION
TO POLE–MOUNTED WIRE TERMINAL

Scale: NTS | March 2001 | BM50

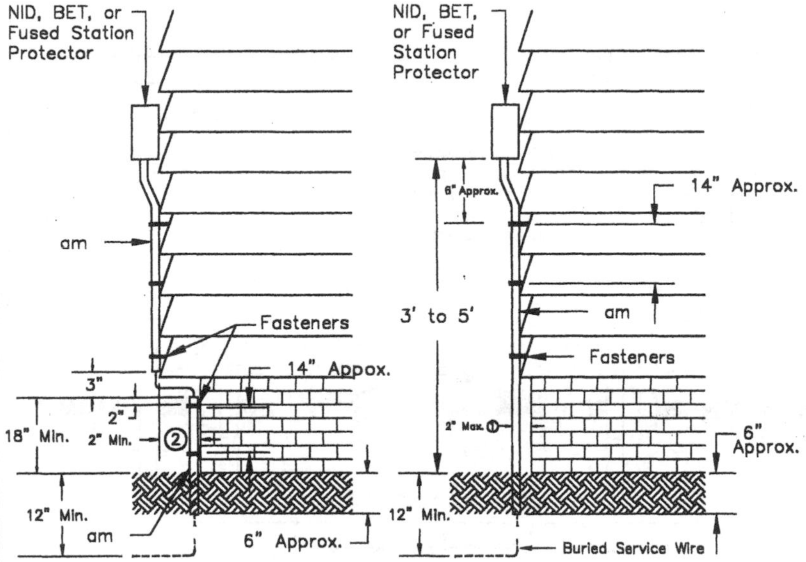

Notes:

(1.) Where an obstruction of less than 2 in. is encountered, the buried service guard (item am) shall extend from the NID, BET, or fused protector to 6 in. below the ground.

(2.) Where an obstruction of greater than 2 in. is encountered, the buried service guard (item am) shall be divided as shown (from the NID, BET, or fused protector to the obstruction, and from 3 in. below the obstruction to 6 in. below the ground). In lieu of divided service guards (item am), a continuous flexible conduit may be used from the NID, BET, or fused protector to 6 in. below the ground.

(3.) For converting English units to metric units use 1 in. = 25.4 mm and 1 ft = 0.3048 m.

ITEM	MATERIAL	NO. REQ'D
am	Guard, buried service (including fasteners)	1

RURAL TELECOMMUNICATIONS CONSTRUCTION PRACTICES BURIED SERVICE GUARD		
Scale: NTS		March 2001
		BM83

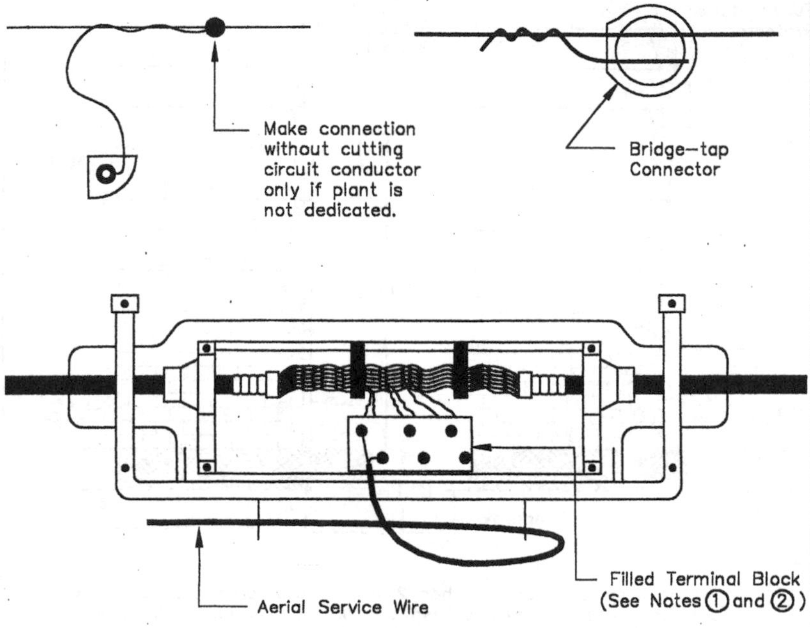

Notes:

① Where aerial service wire connections are made along aerial plastic cable, unprotected filled terminal blocks equipped with lead–out wires shall be used.

② Conductors of the aerial service wire shall be connected directly to the binding posts of the filled terminal block.

RURAL TELECOMMUNICATIONS CONSTRUCTION PRACTICES AERIAL PLASTIC CABLE DETAILS OF WIRE CONNECTIONS TO FILLED TERMINAL BLOCKS		
Scale: NTS		March 2001
		312–1

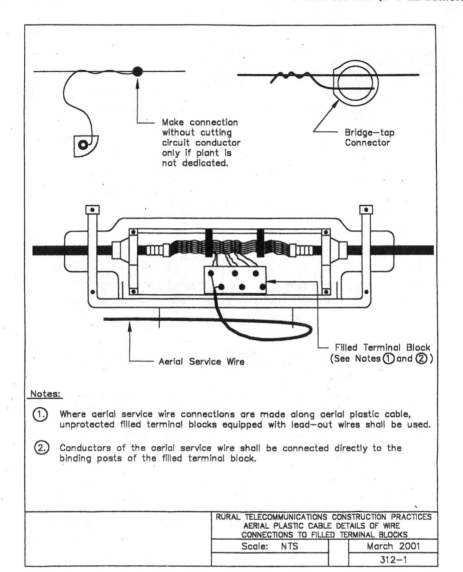

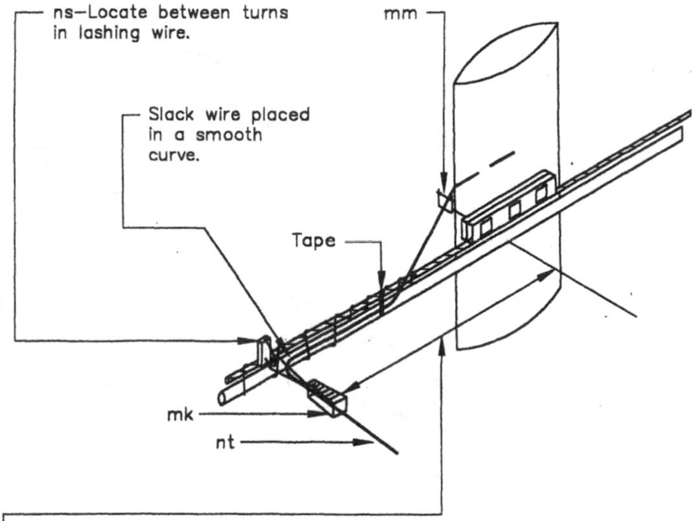

Preferably not more than 20 in. (508 mm) from cable suspension bolt.

May be increased to 3 ft (0.9 m) to provide climbing space or clearances from trees.

When greater than 3 ft (0.9 m) refer to drawing 501–2.

ITEMS	MATERIALS	NO. REQUIRED
*mm	Rings, drive	as required
*ns	Clamps, span	as required
mk	Clamps, drop wire	as required
nt	Wire, aerial service	as required

RURAL TELECOMMUNICATIONS CONSTRUCTION PRACTICES
SPAN CLAMP ATTACHMENT

Scale: NTS

March 2001

501–1

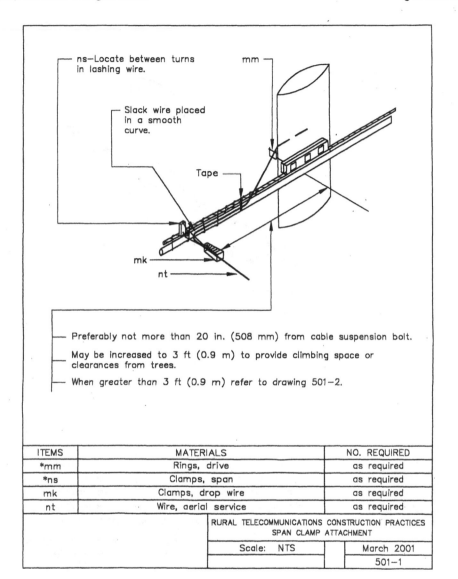

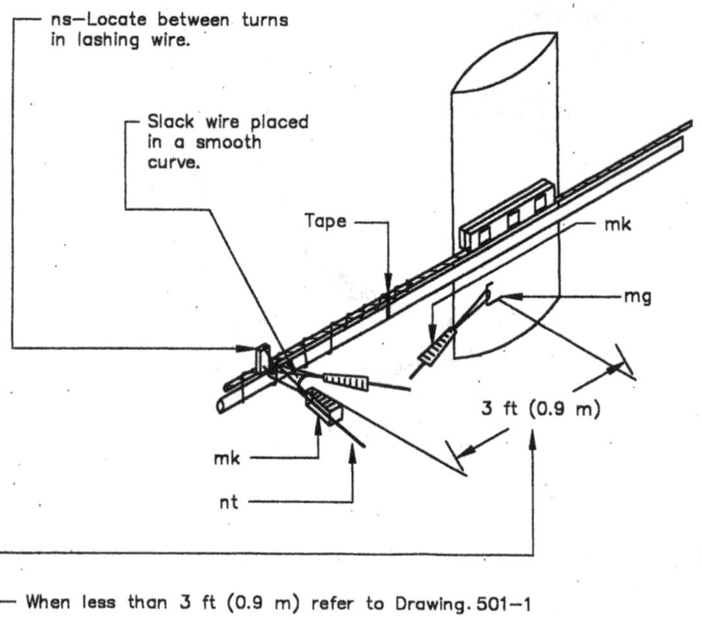

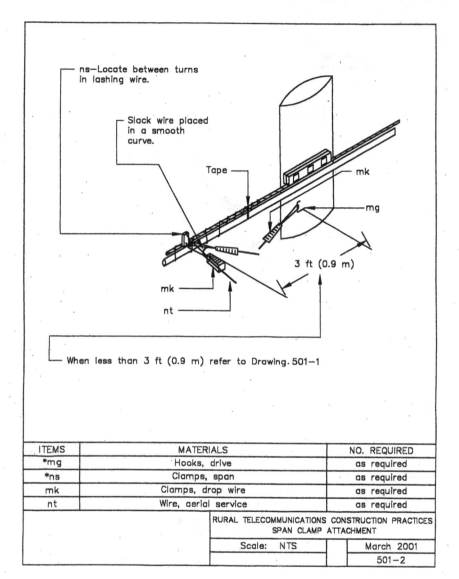

ITEMS	MATERIALS	NO. REQUIRED
*mg	Hooks, drive	as required
*ns	Clamps, span	as required
mk	Clamps, drop wire	as required
nt	Wire, aerial service	as required

RURAL TELECOMMUNICATIONS CONSTRUCTION PRACTICES
SPAN CLAMP ATTACHMENT

Scale: NTS | March 2001 | 501–2

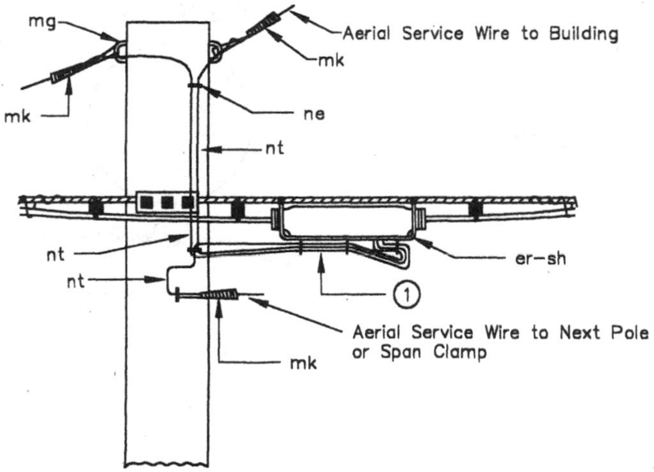

Note:

(1.) Install aerial service wiring through all rings on bottom of terminal housing. Turn wire back around last ring to assigned pair. Form wire loosely to avoid sharp bends.

ITEMS	MATERIALS	NO. REQUIRED
*mg	Hooks, drive	as required
*ne	Rings, bridle	as required
er	Enclosures, ready–access	–
sh	Blocks, filled, terminal, unprotected	–
nt	Wire, aerial service	as required
mk	Clamps, drop wire	as required

RURAL TELECOMMUNICATIONS CONSTRUCTION PRACTICES SERVICE WIRE CONNECTIONS TO AERIAL CABLE		
Scale: NTS		March 2001
		503–2

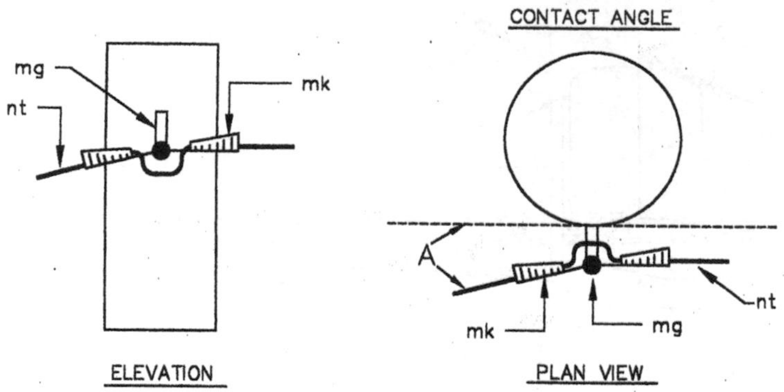

FIGURE A: Aerial service wires whose contact angle (A) exceeds five degrees and/or whose adjacent span lengths are different by 25 percent or more.

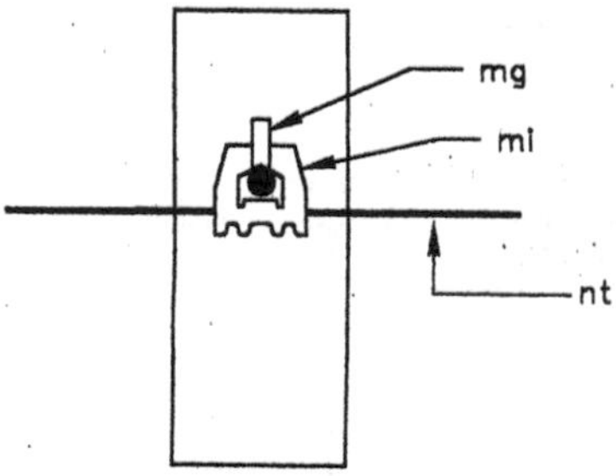

FIGURE B: Aerial service wires whose contact angle (A) is less than five degrees and/or whose adjacent span lengths are different by less than 25 percent.

ITEMS	MATERIALS	NO. REQUIRED
*mg	Hooks, drive	as required
nt	Wire, aerial service	as required
mk	Clamps, drop wire	as required
*mi	Support, drop wire	as required

RURAL TELECOMMUNICATIONS CONSTRUCTION PRACTICES
SERVICE WIRE ATTACHMENT AT INTERMEDIATE POLE

Scale: NTS		March 2001
		504

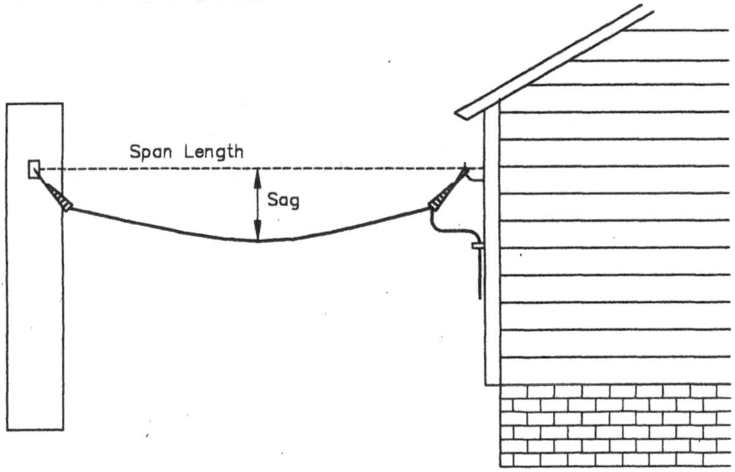

MINIMUM STRINGING SAG – COPPER COVERED STEEL REINFORCED (CCSR) and NONMETALLIC REINFORCED (NMR) AERIAL SERVICE WIRES

SPAN LENGTH ft (m)	SAG–MEDIUM AND LIGHT LOADING DISTRICTS	SAG–HEAVY LOADING DISTRICT
100 (30.5) OR LESS	20 in. (510 mm)	20 in. (510 mm)
125 (38)	34 in. (860 mm)	34 in. (860 mm)
150 (46)	4 ft (1.2 m)	4 ft (1.2 m)
175 (53)	5,5 ft (1.7 m)	7 ft (2.1 m)
200 (61)	7 ft (2.1 m)	11 ft (3.4 m)
225 (66.5)	9 ft (2.7 m)	
250 (76)	11 ft (3.4 m)	

Note: To reduce vibration and dancing, service wire shall be twisted one complete turn for each 10 ft (3 m) of span length at the time installation.

RURAL TELECOMMUNICATIONS CONSTRUCTION PRACTICES AERIAL SERVICE WIRE SAGS		
Scale: NTS		March 2001
		505

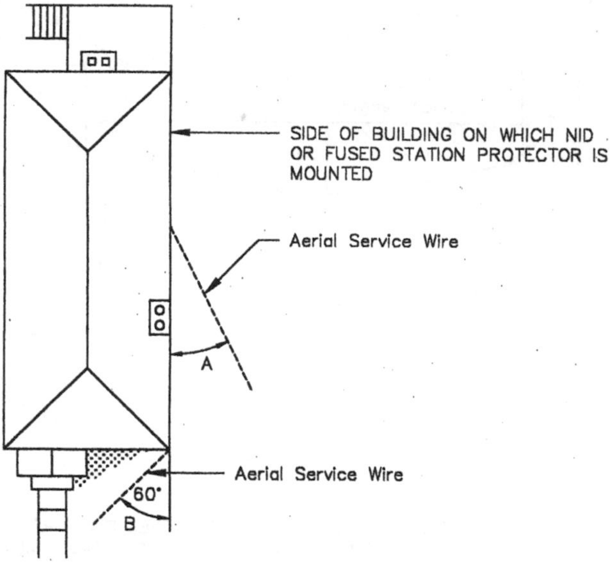

<u>Frame Buildings Where NIDs Containing Fuseless Station Protectors are Used on Fire Resistant Buildings.</u>

Use house hook or drop wire hook for any angle except angle B. When necessary to place service wire within angle B use "S" knob with corner bracket to avoid service wire attachment on front of building.

<u>Frame Buildings Where Fused Station Protectors are Used.</u>

If angle A is less than 30° use "S" knob. If angle A is greater than 30° use "S" knob with 5/16 in. (7.9 mm) angle screw. When necessary to place service wire within angle B use "S" knob with corner bracket to avoid service wire attachments on front of buildings.

RURAL TELECOMMUNICATIONS CONSTRUCTION PRACTICES
SELECTION OF SERVICE WIRE ATTACHMENT

Scale: NTS		March 2001
		506

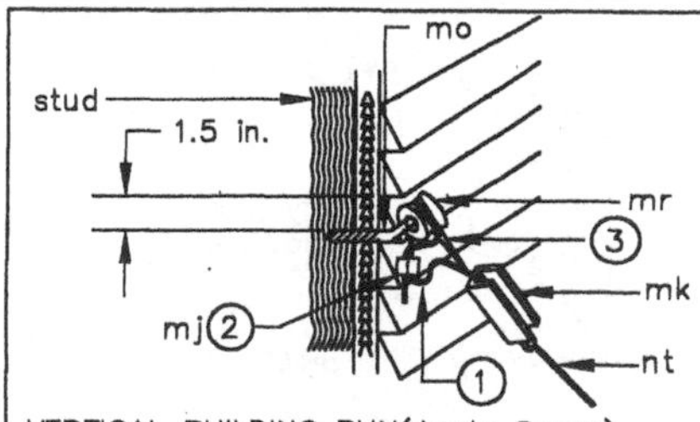

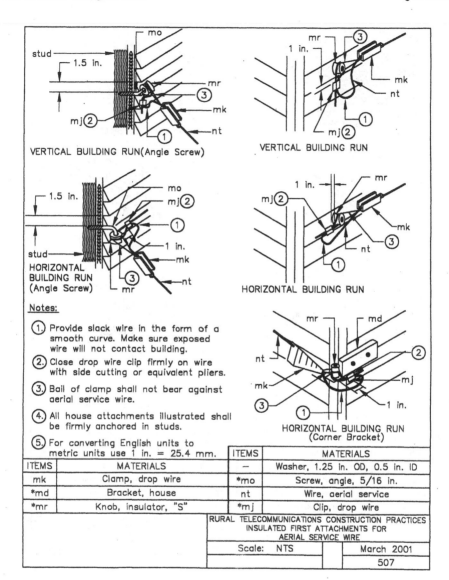

VERTICAL BUILDING RUN(Angle Screw)

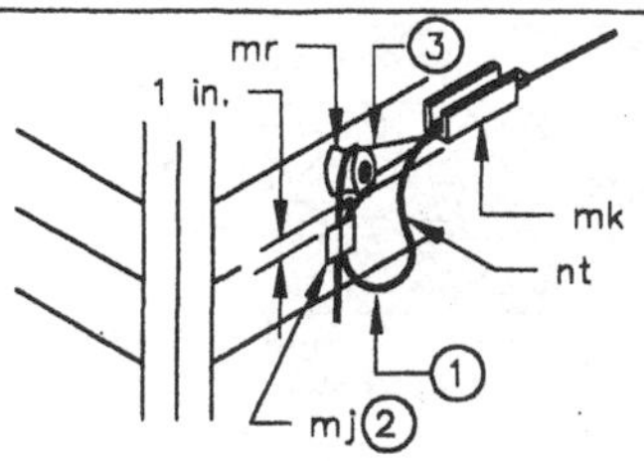

VERTICAL BUILDING RUN

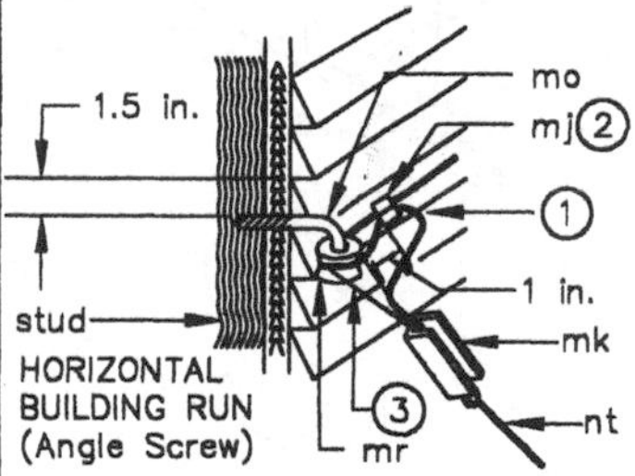

HORIZONTAL BUILDING RUN (Angle Screw)

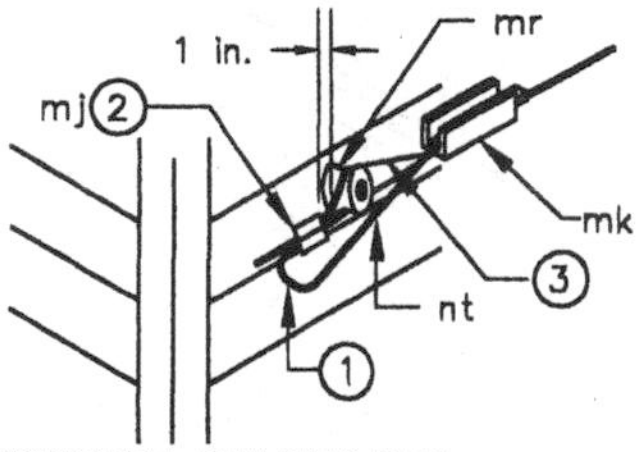

HORIZONTAL BUILDING RUN

Notes:

1. Provide slack wire in the form of a smooth curve. Make sure exposed wire will not contact building.
2. Close drop wire clip firmly on wire with side cutting or equivalent pliers.
3. Bail of clamp shall not bear against aerial service wire.
4. All house attachments illustrated shall be firmly anchored in studs.
5. For converting English units to metric units use 1 in. = 25.4 mm.

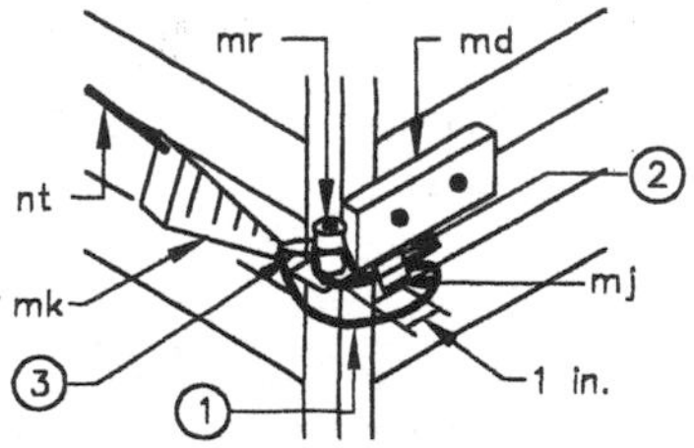

HORIZONTAL BUILDING RUN (Corner Bracket)

ITEMS	MATERIALS
mk	Clamp, drop wire
*md	Bracket, house
*mr	Knob, insulator, "S"

ITEMS	MATERIALS
—	Washer, 1.25 in. OD, 0.5 in. ID
*mo	Screw, angle, 5/16 in.
nt	Wire, aerial service
*mj	Clip, drop wire

RURAL TELECOMMUNICATIONS CONSTRUCTION PRACTICES
INSULATED FIRST ATTACHMENTS FOR
AERIAL SERVICE WIRE

Scale: NTS March 2001
507

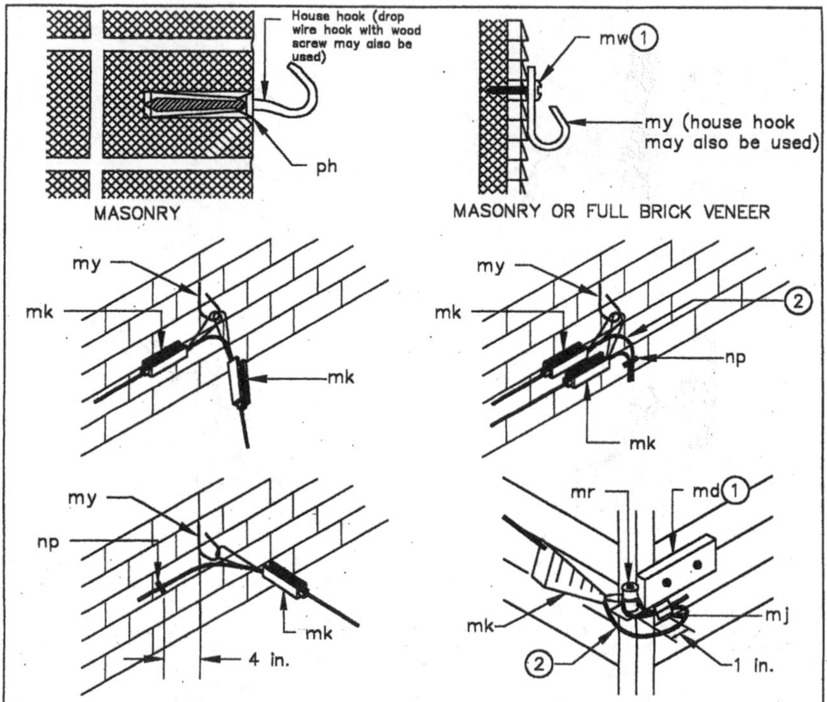

Notes:

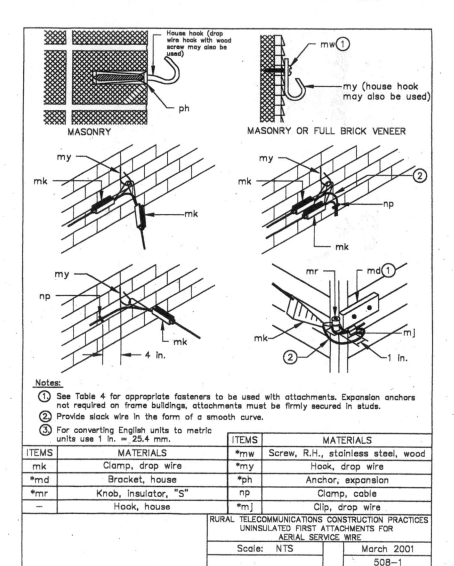

① See Table 4 for appropriate fasteners to be used with attachments. Expansion anchors not required on frame buildings, attachments must be firmly secured in studs.

② Provide slack wire in the form of a smooth curve.

③ For converting English units to metric units use 1 in. = 25.4 mm.

ITEMS	MATERIALS
mk	Clamp, drop wire
*md	Bracket, house
*mr	Knob, insulator, "S"
—	Hook, house

ITEMS	MATERIALS
*mw	Screw, R.H., stainless steel, wood
*my	Hook, drop wire
*ph	Anchor, expansion
np	Clamp, cable
*mj	Clip, drop wire

RURAL TELECOMMUNICATIONS CONSTRUCTION PRACTICES
UNINSULATED FIRST ATTACHMENTS FOR
AERIAL SERVICE WIRE

Scale: NTS		March 2001
		508–1

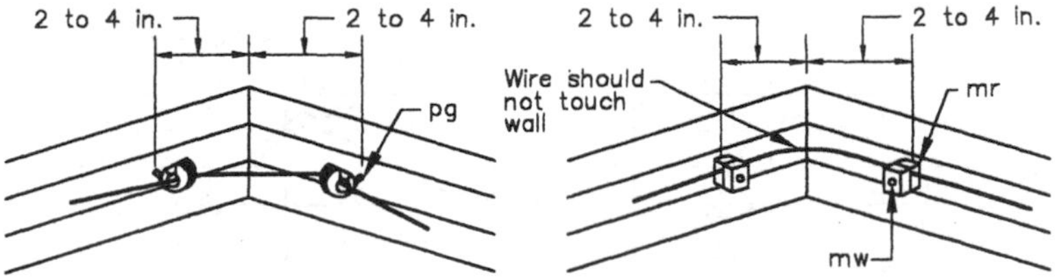

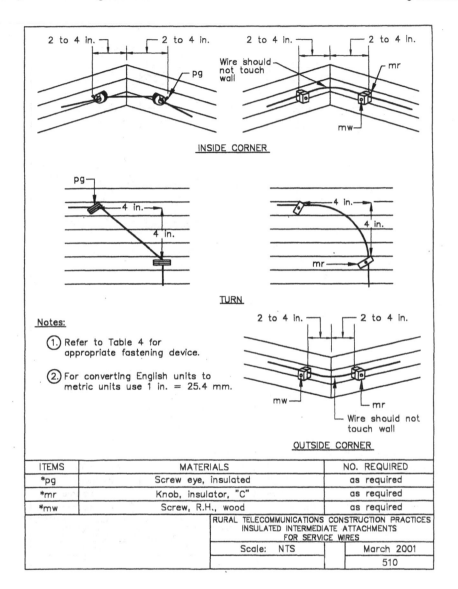

INSIDE CORNER

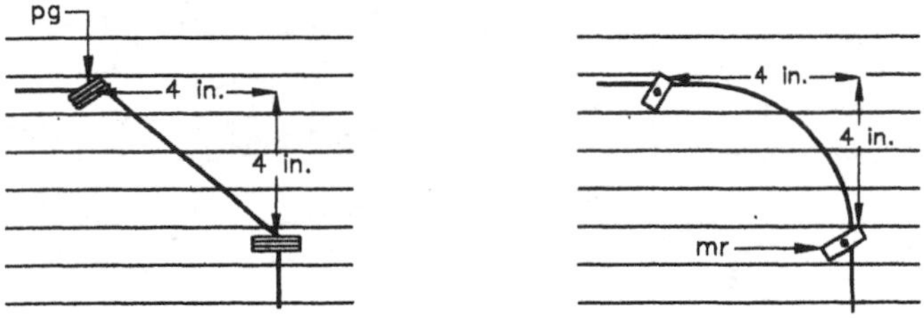

TURN

Notes:

1. Refer to Table 4 for appropriate fastening device.

2. For converting English units to metric units use 1 in. = 25.4 mm.

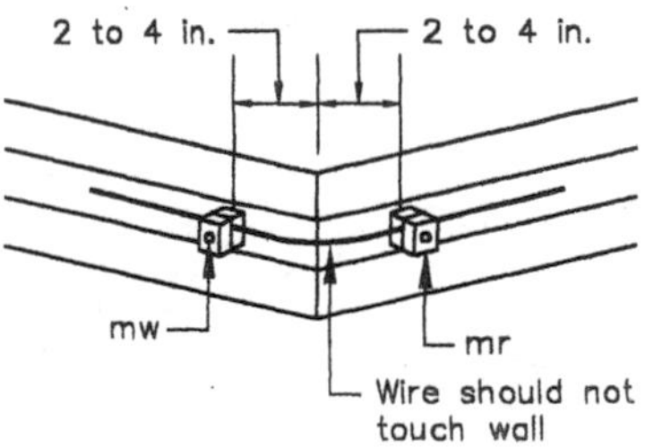

OUTSIDE CORNER

ITEMS	MATERIALS	NO. REQUIRED
*pg	Screw eye, insulated	as required
*mr	Knob, insulator, "C"	as required
*mw	Screw, R.H., wood	as required

RURAL TELECOMMUNICATIONS CONSTRUCTION PRACTICES INSULATED INTERMEDIATE ATTACHMENTS FOR SERVICE WIRES		
Scale: NTS		March 2001
		510

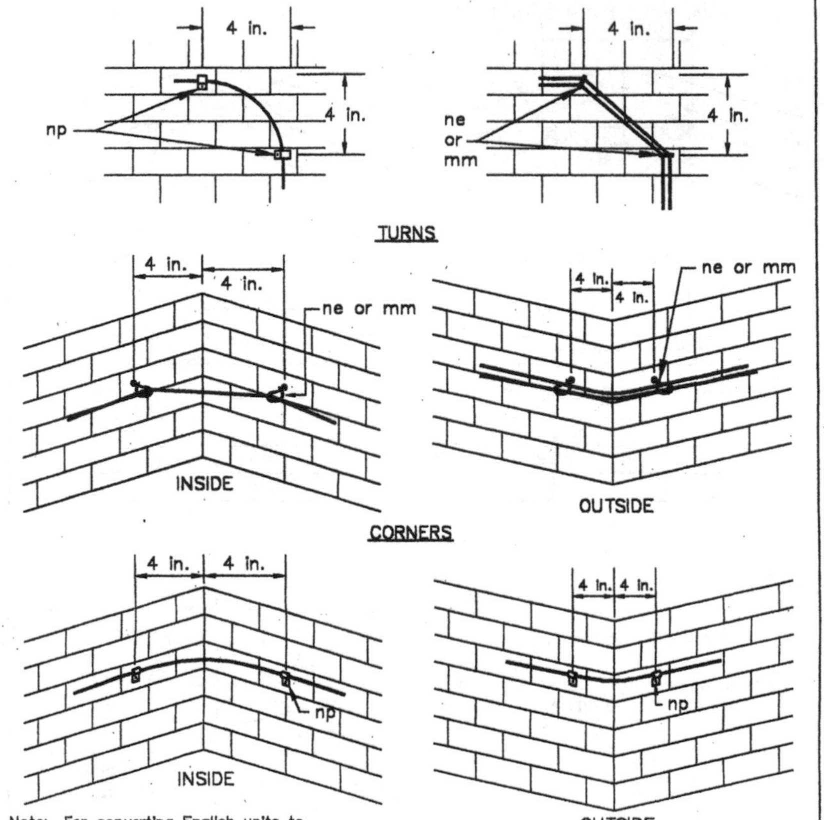

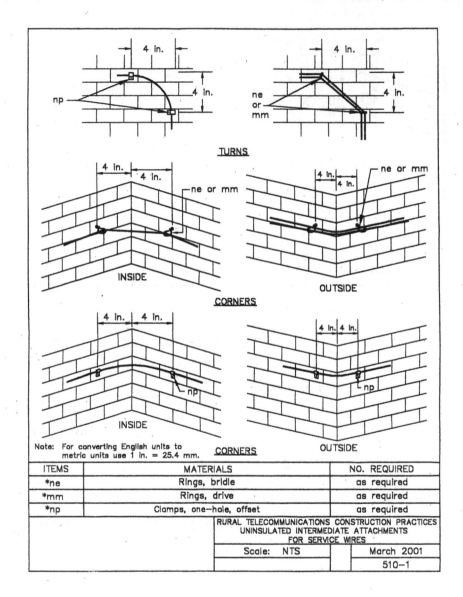

Note: For converting English units to metric units use 1 in. = 25.4 mm.

ITEMS	MATERIALS	NO. REQUIRED
*ne	Rings, bridle	as required
*mm	Rings, drive	as required
*np	Clamps, one-hole, offset	as required

	RURAL TELECOMMUNICATIONS CONSTRUCTION PRACTICES UNINSULATED INTERMEDIATE ATTACHMENTS FOR SERVICE WIRES		
	Scale: NTS		March 2001
			510–1

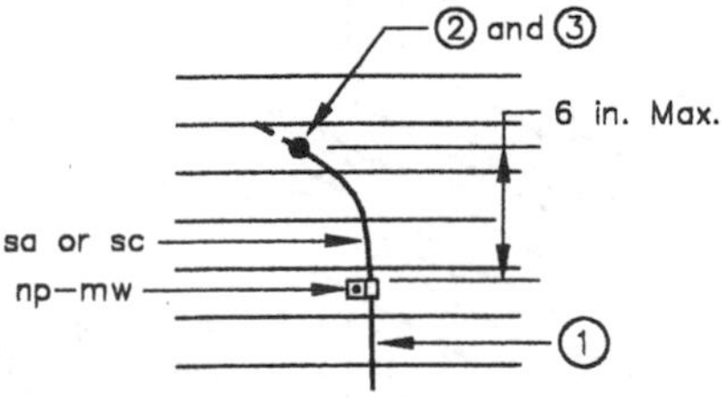

SKETCH A: Buried Service Above Grade Entrance

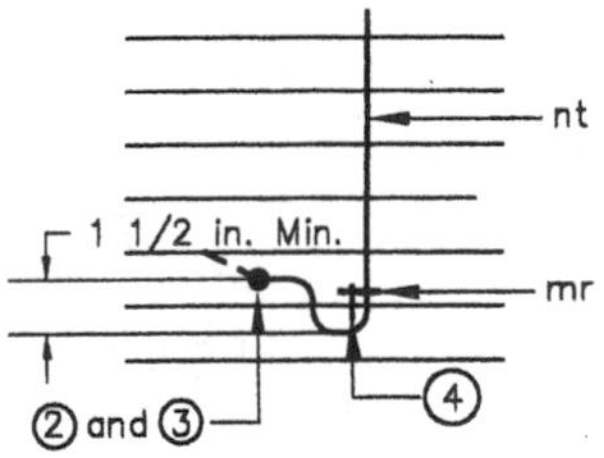

SKETCH B: Aerial Service Wire – Aerial Service Entrance

Notes:

1. The first attachment of the buried wire to the building should be located approximately 4 inches above the ground. The remaining attachments shall be spaced approximately 14 inches apart.
2. A porcelain or plastic tube shall be employed only when insulated attachments are required for support of aerial service wire on buildings.
3. Entrance hole shall be drilled to slope slightly upward. Except where a porcelain or plastic tube is required, all wires entering the hole shall be taped for a tight fit. When the aerial service wire approaches from above the entrance hole, a drip loop shall be made as shown.
4. Insert short piece of aerial service wire to cushion "C" knob.
5. Seal both ends of hole or conduit with duct seal.
6. For converting English units to metric units use 1 in. = 25.4 mm.

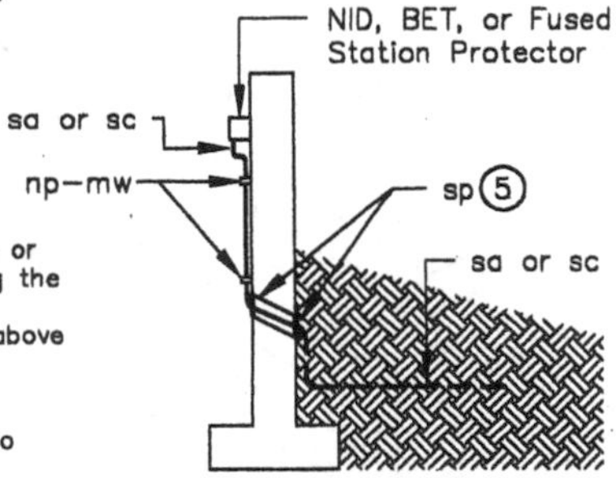

SKETCH C: Buried Service – Below Grade Entrance

ITEMS	MATERIALS	ITEMS	MATERIALS
*mr	Knob, insulator, "C"	*mw	Screw, wood
nt	Wire, aerial service	sa/sc	Wire or cable, filled, buried
–	Tube, plastic	sp	Sealer, duct
*np	Clamp, one–hole, offset		

RURAL TELECOMMUNICATIONS CONSTRUCTION PRACTICES
SERVICE ENTRANCES

Scale: NTS		March 2001
		510–2

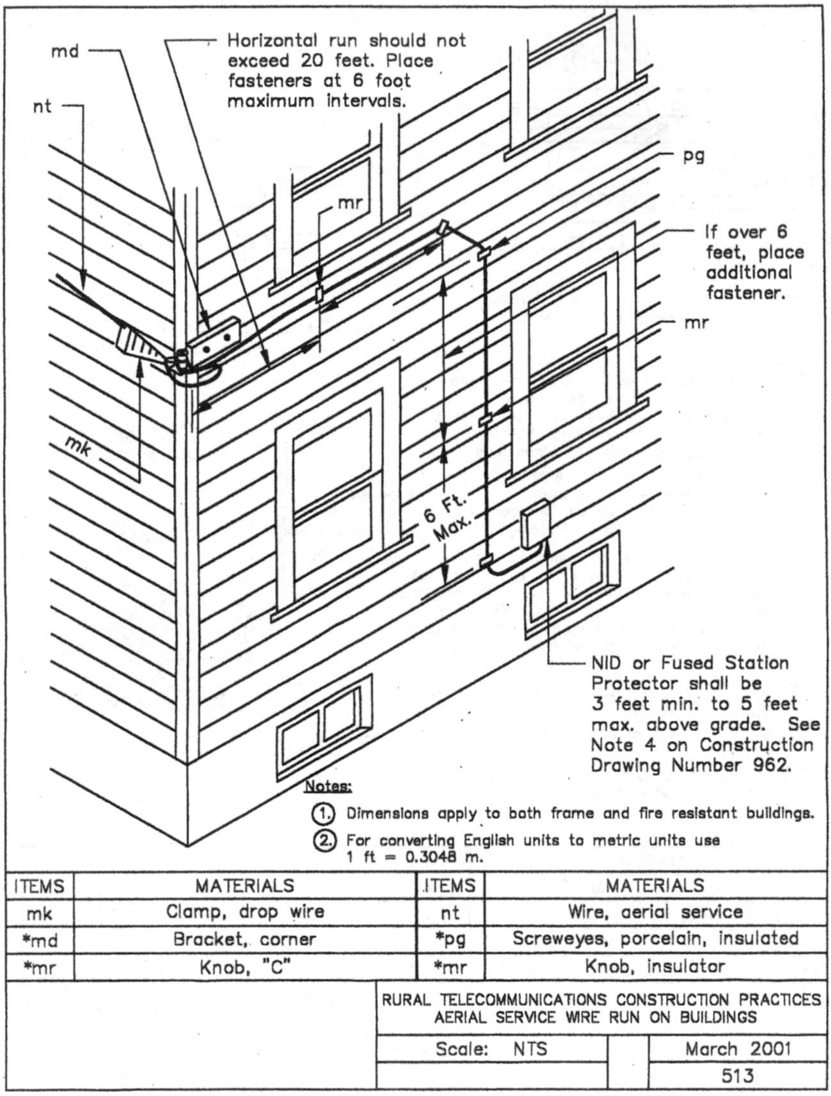
md
nt
Horizontal run should not exceed 20 feet. Place fasteners at 6 foot maximum intervals.
mr
pg
If over 6 feet, place additional fastener.
mr
mk
6 Ft. Max.
NID or Fused Station Protector shall be 3 feet min. to 5 feet max. above grade. See Note 4 on Construction Drawing Number 962.
Notes:
1. Dimensions apply to both frame and fire resistant buildings.
2. For converting English units to metric units use 1 ft = 0.3048 m.

ITEMS	MATERIALS	ITEMS	MATERIALS
mk	Clamp, drop wire	nt	Wire, aerial service
*md	Bracket, corner	*pg	Screweyes, porcelain, insulated
*mr	Knob, "C"	*mr	Knob, insulator

RURAL TELECOMMUNICATIONS CONSTRUCTION PRACTICES AERIAL SERVICE WIRE RUN ON BUILDINGS		
Scale: NTS		March 2001
		513

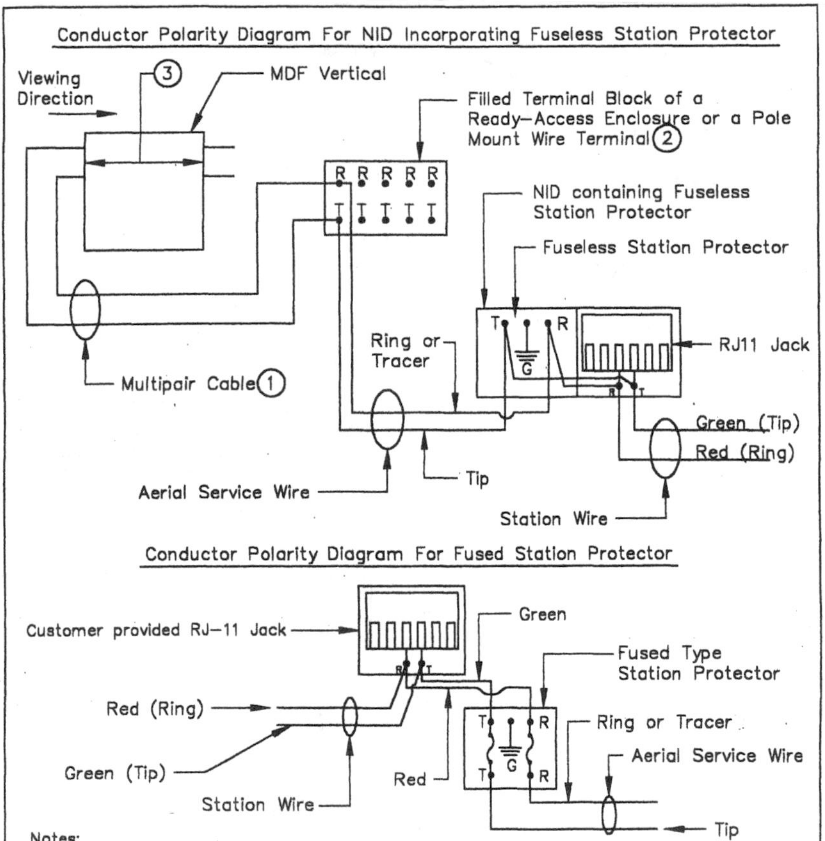

Notes:

1. Refer to appropriate cable specifications for tip and ring conductor identification.
2. When facing the cable terminal the positive (tip) is on the left and the negative (ring) is on the right side of the pair.
3. Connections to be made in accordance with the manufacturer's instructions.

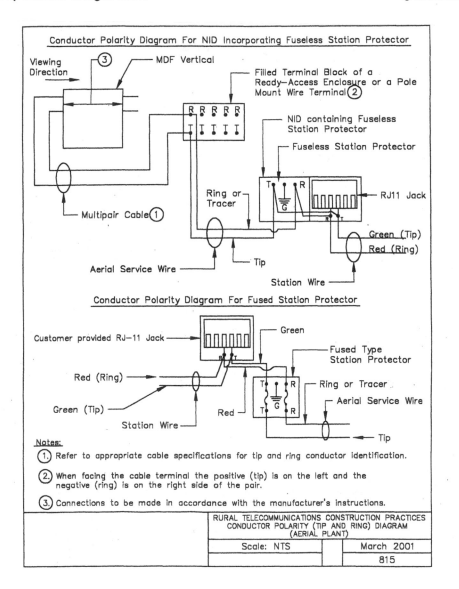

RURAL TELECOMMUNICATIONS CONSTRUCTION PRACTICES
CONDUCTOR POLARITY (TIP AND RING) DIAGRAM
(AERIAL PLANT)

Scale: NTS		March 2001
		815

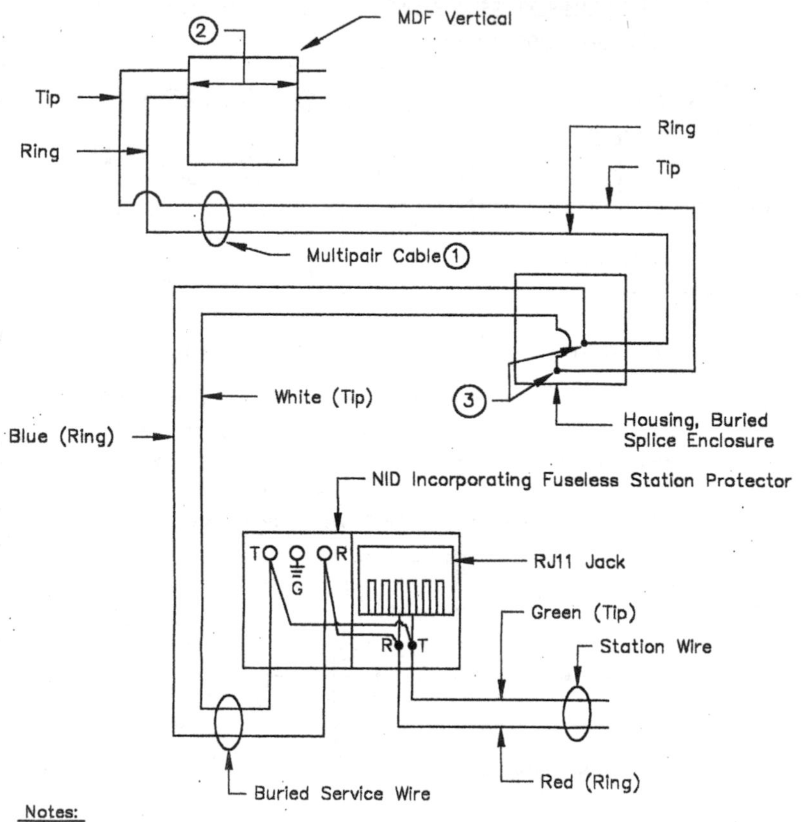

Notes:

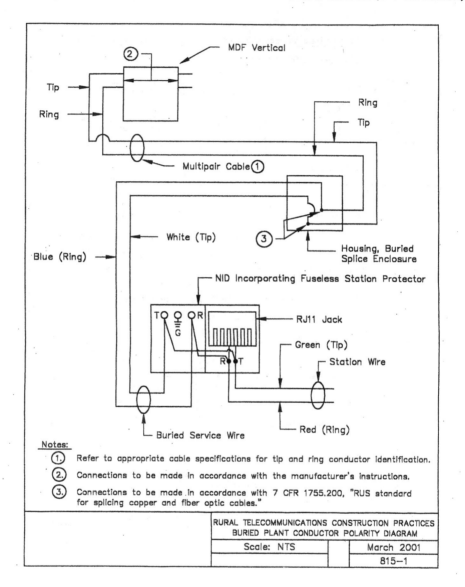

1. Refer to appropriate cable specifications for tip and ring conductor identification.
2. Connections to be made in accordance with the manufacturer's instructions.
3. Connections to be made in accordance with 7 CFR 1755.200, "RUS standard for splicing copper and fiber optic cables."

	RURAL TELECOMMUNICATIONS CONSTRUCTION PRACTICES BURIED PLANT CONDUCTOR POLARITY DIAGRAM		
	Scale: NTS		March 2001
			815–1

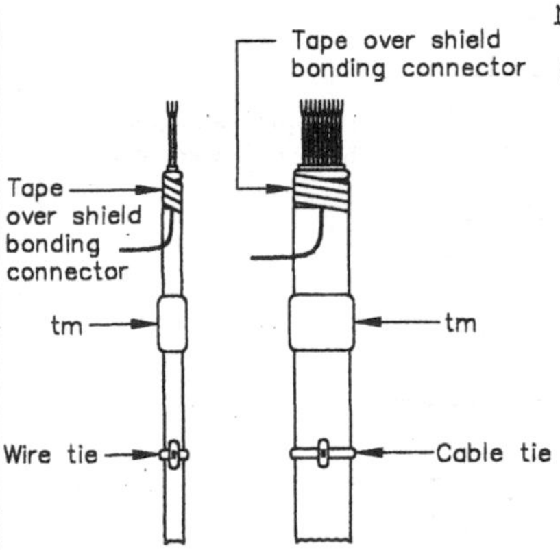

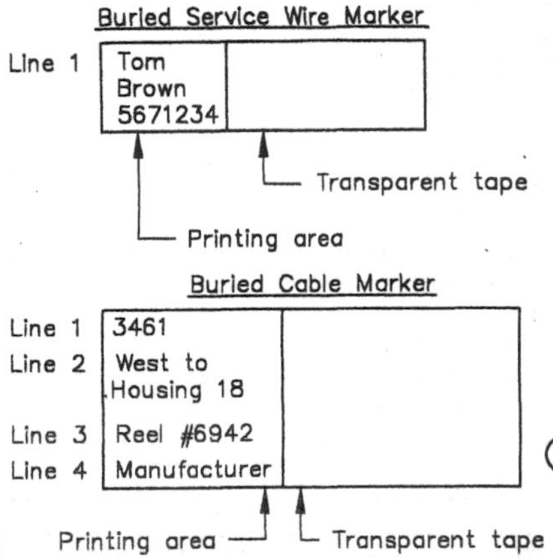

Notes:

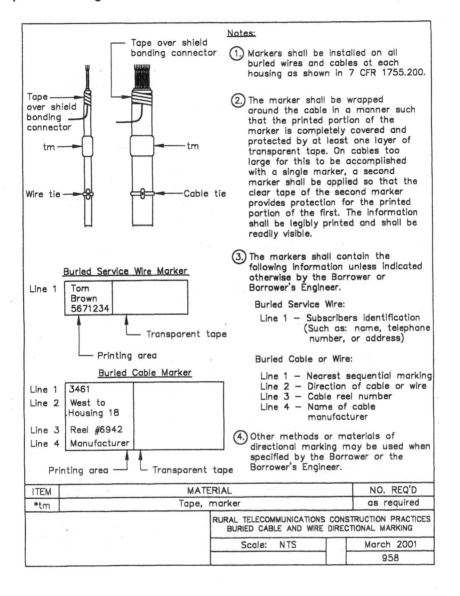

1. Markers shall be installed on all buried wires and cables at each housing as shown in 7 CFR 1755.200.

2. The marker shall be wrapped around the cable in a manner such that the printed portion of the marker is completely covered and protected by at least one layer of transparent tape. On cables too large for this to be accomplished with a single marker, a second marker shall be applied so that the clear tape of the second marker provides protection for the printed portion of the first. The information shall be legibly printed and shall be readily visible.

3. The markers shall contain the following information unless indicated otherwise by the Borrower or Borrower's Engineer.

   Buried Service Wire:

   Line 1 – Subscribers identification (Such as: name, telephone number, or address)

   Buried Cable or Wire:

   Line 1 – Nearest sequential marking
   Line 2 – Direction of cable or wire
   Line 3 – Cable reel number
   Line 4 – Name of cable manufacturer

4. Other methods or materials of directional marking may be used when specified by the Borrower or the Borrower's Engineer.

ITEM	MATERIAL	NO. REQ'D
*tm	Tape, marker	as required

RURAL TELECOMMUNICATIONS CONSTRUCTION PRACTICES
BURIED CABLE AND WIRE DIRECTIONAL MARKING

Scale: NTS		March 2001
		958

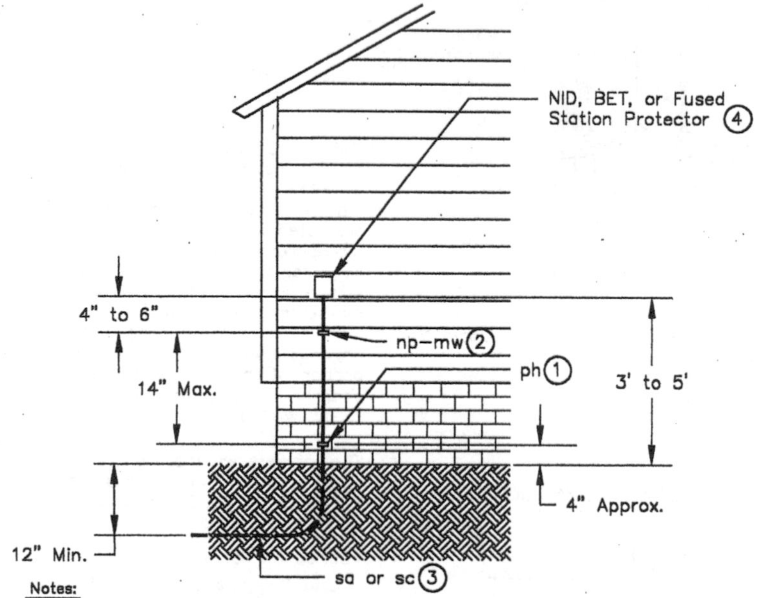

Notes:

(1) When mounting NID, BET, or fused station protector and clamps on masonry surface, use screw expansion anchors or equivalent manual or machine–driven devices.

(2) Attach filled buried service wire or cable to building with one–hole offset clamps spaced 14 in. max. apart. Where grounding conductor parallels service wire or cable, both wires may be run under the same attachment.

(3) Place filled buried service wire or cable snug against building.

(4) Details of NID, BET, or Fused station protector terminations are shown on Figures 10 through 16 of 7 CFR 1755.508, and Figure 19 of 7 CFR 1755.509.

(5) For converting English units to metric units use 1 in. = 25.4 mm and 1 ft = 0.03048 m.

ITEMS	MATERIALS	ITEMS	MATERIALS
NID	NID, protected, station, outside	*rg	Wire, station
sa	Wire, filled, buried	*mw	Screw, stainless steel, wood
*ph	Anchor, expansion, screw	sc	Cable, filled, buried
*np	Clamp, one–hole, offset		

RURAL TELECOMMUNICATIONS CONSTRUCTION PRACTICES
BURIED WIRE SERVICE INSTALLATION ON BUILDINGS

Scale: NTS		March 2001
		962

[66 FR 43327, Aug. 17, 2001]

§§ 1755.511–1755.521 [Reserved]

§ 1755.522 RUS general specification for digital, stored program controlled central office equipment.

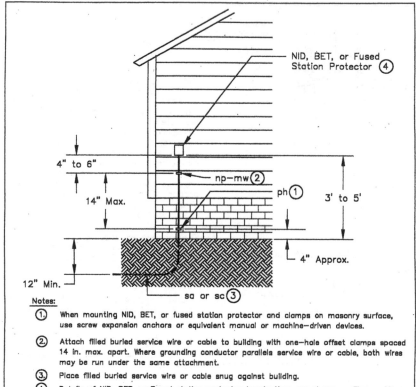

(a) *General.* (1) This section covers general requirements for a digital telephone central office switching system, which is fully electronic and controlled by stored program processors. A digital switching system transfers information which is digitally encoded from any input port to a temporarily addressed exit port. The information may enter the system in either analog or digital form and may or may not be converted to analog at the exit port depending on the facility beyond. The switching system shall operate properly as an integral part of the telephone network when connected to physical and carrier derived circuits meeting RUS specifications and other generally accepted telecommunications practices.

(2) The output of a digital-to-digital port shall be Pulse Code Modulation (PCM), encoded in eight-bit words using the mu-255 encoding law and D3 encoding format, and arranged to interface with a T1 span line.

(3) American National Standards Institute (ANSI) Standard S1.4–1983, Specification for Sound Level Meters, is incorporated by reference by RUS. This includes S1.4A-1985 that is also incorporated by reference. This incorporation by reference was approved by the Director of the Federal Register in accordance with 5 U.S.C. 552(a) and 1 CFR part 51. Copies may be obtained from ANSI Inc., 11 West 42nd Street, 13th Floor, New York, NY 10036, telephone 212–642–4900. Copies may be inspected during normal business hours at RUS, room 2838-S, U.S. Department of Agriculture, Washington, DC 20250, or at the National Archives and Records Administration (NARA). For information on the availability of this material at NARA, call 202–741–6030, or go to: *http://www.archives.gov/federal_register/code_of_federal_regulations/ibr_locations.html.*

(4) American Society for Testing Materials (ASTM) Specification B 33–91, Standard Specification for Tinned Soft or Annealed Copper Wire for Electrical Purposes, is incorporated by reference by RUS. This incorporation by reference was approved by the Director of the Federal Register in accordance with 5 U.S.C. 552(a) and 1 CFR part 51. Copies may be obtained from ASTM, 1916 Race Street, Philadelphia, PA, telephone 215–299–5400. Copies may be inspected during normal business hours at RUS, room 2838-S, U.S. Department of Agriculture, Washington, DC 20250, or at the National Archives and Records Administration (NARA). For information on the availability of this material at NARA, call 202–741–6030, or go to: *http://www.archives.gov/federal_register/code_of_federal_regulations/ibr_locations.html.*

(5) Bell Communications Research (Bellcore) document SR-TSV-002275, BOC Notes on the LEC Networks—1990, March 1991, is incorporated by reference by RUS. This incorporation by reference was approved by the Director of the Federal Register in accordance with 5 U.S.C. 552(a) and 1 CFR Part 51. Copies may be obtained from Bellcore Customer Service, 60 New England Avenue, Piscataway, NJ 08854, telephone 1–800–521–2673. Copies may be inspected during normal business hours at RUS, room 2838-S, U.S. Department of Agriculture, Washington, DC 20250, or at the National Archives and Records Administration (NARA). For information on the availability of this material at NARA, call 202–741–6030, or go to: *http://www.archives.gov/federal_register/code_of_federal_regulations/ibr_locations.html.*

(6) Bellcore TR-TSY-000508, Automatic Message Accounting, July 1987, is incorporated by reference by RUS. This incorporation by reference was approved by the Director of the Federal Register in accordance with 5 U.S.C. 552(a) and 1 CFR part 51. Copies may be obtained from Bellcore Customer Service, 60 New England Avenue, Piscataway, NJ 08854, telephone 1–800–521–2673. Copies may be inspected during normal business hours at RUS, room 2838-S, U.S. Department of Agriculture, Washington, DC 20250, or at the National Archives and Records Administration (NARA). For information on the availability of this material at NARA, call 202–741–6030, or go to: *http://www.archives.gov/federal_register/code_of_federal_regulations/ibr_locations.html.*

(7) Federal Standard H28, Screw-Thread Standards for Federal Services, March 31, 1978, is incorporated by reference by RUS. This includes: Change Notice 1, Federal Standard, Screw-Thread Standards for Federal Services, May 28, 1986; Change Notice 2, Federal Standard, Screw-Thread Standards for Federal Services, January 20, 1989; and Change Notice 3, Federal Standard, Screw-Thread Standards for Federal Services, March 12, 1990. This incorporation by reference was approved by the Director of the Federal Register in

accordance with 5 U.S.C. 552(a) and 1 CFR part 51. Copies may be obtained from the General Services Administration, Specification Section, 490 East L'Enfant Plaza SW, Washington, DC 20407, telephone 202–755–0325. Copies may be inspected during normal business hours at RUS, room 2838-S, U.S. Department of Agriculture, Washington, DC 20250, or at the National Archives and Records Administration (NARA). For information on the availability of this material at NARA, call 202–741–6030, or go to: *http://www.archives.gov/federal_register/code_of_federal_regulations/ibr_locations.html.*

(8) Institute of Electrical and Electronics Engineers (IEEE) Std 455–1985, IEEE Standard Test Procedure for Measuring Longitudinal Balance of Telephone Equipment Operating in the Voice Band, is incorporated by reference by RUS. This incorporation by reference was approved by the Director of the Federal Register in accordance with 5 U.S.C. 552(a) and 1 CFR part 51. Copies may be obtained from IEEE Service Center, 445 Hoes Lane, P. O. Box 1331, Piscataway, NJ 08854, telephone (201) 981–0060. Copies may be inspected during normal business hours at RUS, room 2838-S, U.S. Department of Agriculture, Washington, DC 20250, or at the National Archives and Records Administration (NARA). For information on the availability of this material at NARA, call 202–741–6030, or go to: *http://www.archives.gov/federal_register/code_of_federal_regulations/ibr_locations.html.*

(9) Institute of Electrical and Electronics Engineers (IEEE) Std 730–1989, IEEE Standard for Software Quality Assurance Plans, is incorporated by reference by RUS. This incorporation by reference was approved by the Director of the Federal Register in accordance with 5 U.S.C. 552(a) and 1 CFR part 51. Copies may be obtained from IEEE Service Center, 445 Hoes Lane, P. O. Box 1331, Piscataway, NJ 08854, telephone (201) 981–0060. Copies may be inspected during normal business hours at RUS, room 2838-S, U.S. Department of Agriculture, Washington, DC 20250, or at the National Archives and Records Administration (NARA). For information on the availability of this material at NARA, call 202–741–6030, or go to: *http://www.archives.gov/federal_register/code_of_federal_regulations/ibr_locations.html.*

(10) RUS Bulletin 345–50, PE-60, RUS Specification for Trunk Carrier Systems, September 1979, is incorporated by reference by RUS. This incorporation by reference was approved by the Director of the Federal Register in accordance with 5 U.S.C. 552 (a) and 1 CFR part 51. Copies may be obtained from the Rural Utilities Service, Administrative Services Division, room 0175-S, Washington, DC 20250. The bulletin may be inspected at the National Archives and Records Administration (NARA). For information on the availability of this material at NARA, call 202–741–6030, or go to: *http://www.archives.gov/federal_register/code_of_federal_regulations/ibr_locations.html.*

(11) RUS Bulletin 345–55, PE-61, Central Office Loop Extenders and Loop Extender Voice Frequency Repeater Combinations, December 1973, is incorporated by reference by RUS. This incorporation by reference was approved by the Director of the Federal Register in accordance with 5 U.S.C. 552 (a) and 1 CFR part 51. Copies may be obtained from the Rural Utilities Service, Administrative Services Division, room 0175-S, Washington, DC 20250. The bulletin may be inspected at the National Archives and Records Administration (NARA). For information on the availability of this material at NARA, call 202–741–6030, or go to: *http://www.archives.gov/federal_register/code_of_federal_regulations/ibr_locations.html.*

(12) RUS Bulletin 345–87, PE-87, RUS Specification for Terminating (TIP) Cable, December 1983, is incorporated by reference RUS. This incorporation by reference was approved by the Director of the Federal Register in accordance with 5 U.S.C. 552 (a) and 1 CFR part 51. Copies may be obtained from the Rural Utilities Service, Administrative Services Division, room 0175-S, Washington, DC 20250. The bulletin may be inspected at the National Archives and Records Administration (NARA). For information on the availability of this material at NARA, call 202–741–6030, or go to: *http://*

www.archives.gov/federal_register/code_of_federal_regulations/ibr_locations.html.

(b) *Reliability.* (1) Quality control and burn-in procedures shall be sufficient so the failure rate of printed circuit boards does not exceed an average of 1.0 percent per month of all equipped cards in the central office during the first three months after cutover, and an average of 0.5 percent per month of all equipped cards in the central office during any 6-month period thereafter. A failure is considered to be the failure of a component on the PC board which requires it to be repaired or replaced.

(2) The central office switching system shall be designed such that the expected individual line downtime does not exceed 30 minutes per year. This is the interval that the customer is out of service as a result of all failure types, excluding dispatch and travel time, i.e., hardware, software, and procedural errors.

(3) The central office switching system shall be designed such that there will be no more than 1 hour of total outages in 20 years, excluding dispatch and travel time for unattended offices.

(c) *System type acceptance tests.* (1) System type acceptance tests (general acceptance tests) are performed for the purpose of determining whether or not a type of switching system should be added or retained as an RUS accepted system. While general acceptance tests will be required on each system type, they will not be expected to cover every requirement in this section. However, any installation of a system provided in accordance with this section shall be capable of meeting any requirement in this section on a spot-check basis.

(2) A "completed call" test shall be made part of these system type acceptance tests. There shall be no more than two in 10,000 locally originating and incoming calls misdirected, unsuccessfully terminated, prematurely disconnected or otherwise failing as a result of equipment malfunction and/or equipment failures, or as a result of transients, noise or design deficiencies. This test shall be made with a load box with no less than 10 lines access and 10 subscriber numbers for completion, or equivalent, with no other traffic in the system. If there is a failure in the equipment during this test, the cause shall be repaired and the test restarted at zero calls.

(3) System type acceptance testing applies basically to factory type testing, and not to owner acceptance testing for individual installations. The overall installed and operating system shall also meet these requirements, except for unusual circumstances or where specifically excluded by this or other RUS requirements.

(d) *Types of requirements.* (1) Unless otherwise indicated, the requirements listed in this section are fixed requirements.

(2) Optional requirements are those which may not be needed for every office and are identifiable by a phrase such as, "when specified by the owner," or, "as specified by the owner."

(3) In some cases where an optional feature specified in paragraph (e) of this section will not be required by an owner, either now or in the future, a system which does not provide this feature will be considered to be in compliance with this section for the specific installation under consideration, but not in compliance with the entire section.

(4) The owner may request bids from any RUS accepted supplier whose system provides all the features which will be required for a specific installation.

(5) The Application Guide, RUS TE&CM 322, provides information about the economic and service factors involved in all optional features, as well as instructions for the completion of appendices A and B of this section.

(e) *General requirements.* (1) The equipment shall provide for terminating and automatically interconnecting subscriber lines and trunks in response to dial pulses (or push-button dialing signals, if specified) without the aid of an operator.

(2) Complete flexibility shall be provided for assigning any subscriber directory number to any central office line equipment by the use of internal programmed memory. Thus, any subscriber line and/or directory number may be moved to another terminal to distribute traffic loads, if the line

equipment hardware is compatible with the service provided.

(3) The system shall be arranged to interface with interexchange carrier trunks and networks using single digit or multi-digit access codes. The system shall be equipped to handle at least 20-digit subscriber dialed numbers. All subscriber directory numbers in the office shall be seven-digit numbers.

(4) The network and the control equipment shall be comprised of solid-state and integrated circuitry components. Peripheral equipment shall be comprised of solid-state and integrated circuitry components as far as practical and consistent with the state-of-the-art and economics of the subject system.

(5) The basic switching system shall include the provision of software programming and necessary hardware, including memory, for optional custom calling services such as call waiting, call forwarding, three-way calling, and abbreviated dialing. It shall be possible to provide these services to any individual line (single-party) subscriber. The addition of these services shall not reduce the anticipated ultimate engineered line, trunk, and traffic capacity of the switching system as specified in appendix A of this section.

(6) The requirements in this specification apply only to single party lines. Although only single frequency ringing is required, other types may be requested in appendix A of this section.

(7) Provision shall be made for local automatic message accounting (LAMA), and for traffic service position system (TSPS) trunks, or equivalent, to the operator's office when required either initially or in the future.

(8) Tandem switching features shall be provided if specified in appendix A of this section.

(9) The system shall be arranged to serve a minimum of eight All Number Calling (ANC) office codes per office, with discrimination on terminating calls by trunk group, numbering plan, or programmed memory and class mark, if specified in appendix A of this section.

(10) Busy hour load handling capacity is an important feature when an office approaches capacity. The delays which may occur in call completion during busy hour periods may prove to be excessive in some system designs. Accordingly, each bidder shall provide, in appendix C of this section, data satisfactory to RUS regarding the busy hour load handling capacity and traffic delays of the system.

(11) Provision shall be made for hotel-motel arrangements, as required by the owner, to permit the operation of message registers at the subscriber's premises to record local outdial calls by guests (see Item 10.5, appendix A of this section).

(12) Provision shall be made to identify the calling line or incoming trunk on nuisance calls (see paragraph (g)(10) of this section for details).

(13) Full access from every subscriber line to every interoffice trunk shall be provided.

(14) Facilities shall be provided to implement service orders, make traffic studies, and perform switching and transmission tests by means of remote control devices if such operations are specified in Items 11.2 and 11.3 of appendix A of this section.

(15) Provision shall be made for the addition of facilities to record all subscriber originated calls based on dialed directory number, time of day, and duration of conversation. They shall be such that the additional equipment (if any is required) may be added to an in-service system without interruption of service and a minimum of equipment, wiring and software modifications.

(16) The system shall be capable of distributed switching operation where groups of subscriber lines can be remotely located from the central office. The remotely situated units are known as "Remote Switching Terminals" (RST's) (see paragraph (w) of this section). This does not eliminate the use of pair gain devices such as direct digitally connected concentrators, regular concentrators or subscriber carrier equipment, where specifically ordered by the owner and its engineer.

(17) The switching system shall have means to synchronize its clock with switches above it in the network hierarchy, when specified by the owner in item 3, appendix A of this section (see paragraph (j) of this section).

(18) Consistent with system arrangements and ease of maintenance, space

shall be provided on the floor plan for an orderly layout of future equipment bays that will be required for anticipated traffic when the office reaches its ultimate size. Readily accessible terminals shall be provided for connection to interbay and frame cables to future bays. All cables, interbay and intrabay (excluding power), if technically feasible, shall be terminated at both ends by use of connectors.

(19) When specified in appendix A of this section, the system shall be capable of processing emergency calls to a 911 service bureau connected either by a group of one-way 911 lines or a trunk group.

(i) It shall be possible to reach the service bureau by dialing 911, 1 + 911, or a 7-digit number.

(ii) The system shall select an idle 911 line or trunk.

(iii) The system shall provide usual ringing and ringback signal until the called 911 line answers.

(iv) If the calling line goes on-hook first, the system shall hold the connection from the called 911 line and return steady low tone to the service bureau. The system shall then begin a 45-minute timeout, after which the calling line is disconnected and an alarm message is printed on a TTY. If the calling line goes off-hook before timeout, the system shall reestablish the conversation path.

(v) If the calling line does not disconnect, the service bureau attendant shall have the ability to force a disconnect of the established connection with the calling party.

(vi) When the 911 call is answered, the equipment shall be arranged so that coin lines are not charged for the call. Similarly, if some form of local call charging is used, there shall be no charge for the 911 call.

(vii) If the 911 service bureau is holding a calling line, it shall be possible for the 911 line to cause the equipment to ring back the calling line. This is done by providing a flash of on-hook signal from the 911 line lasting from 200 to 1,100 milliseconds. The signal to the calling line shall be ringing current if the line is on-hook, or receiver off-hook (ROH) tone if the line is off-hook.

(viii) Calls shall not be originated from the service bureau via the dedicated 911 lines. If an attempt is made to originate a call, it shall receive reorder tone. After 6 minutes, the system shall print an alarm message.

(ix) If 911 calls pass through intermediate switching, the forced-hold control, emergency ringback, and calling line status monitoring capabilities are lost.

(f) *Line circuit requirements*—(1) *General*. (i) The range of direct current (dc) resistances of subscriber loops, measured from the main frame in the central office and including the telephone set shall be at least 0–1900 ohms without loop extension and 1900–3600 ohms with loop extenders, or equivalent. The range when using extension equipment may be significantly reduced for straight line ringers. These limits apply under maximum adverse environmental and manufacturing variation tolerance conditions. Central office voltage shall be stabilized at a value necessary to provide at least a nominal 21 milliamperes current with a non-treated loop of at least 1900 ohms. Minimum loop insulation resistance without loop extenders shall be 25,000 ohms between conductors or from either conductor or both conductors in parallel to ground. Loop insulation resistance for loop extension devices may be 100,000 ohms minimum between conductors or from either conductor or both conductors in parallel to ground.

(ii) In addition to operating on non-loaded cable pairs and subscriber carrier, the equipment shall function properly with D-66 and H-88 loaded cable pairs, including any provisions the equipment must control for the purposes of proper transmission.

(2) *Dialing*—(i) *Subscriber dial speed*. The line equipment and central office equipment (COE) in tandem shall operate satisfactorily when used with subscriber dials having a speed of operation between eight and twelve impulses per second and a break period of 55 to 65 percent of the total impulse period.

(ii) *Subscriber dial interdigital time*. The line equipment and central office equipment shall operate satisfactorily with subscriber rotary dial interdigital times of 200 milliseconds minimum, and with pushbutton dialing interdigital times of 50 milliseconds minimum.

(iii) *Subscriber line pushbutton dialing frequencies.* (A) The frequency pairs assigned for pushbutton dialing shall be as follows, with an allowable variation of ±1.5 percent:

Low Group Frequencies (Hz)	High Group Frequencies (Hz)			
	1209	1336	1477	1633
697	1	2	3	Spare
770	4	5	6	Spare
852	7	8	9	Spare
941	*	0	#	Spare

(B) The receiver shall comply with the operating parameters of the dual-tone multifrequency (DTMF) central office receiver as described in section 6 of Bell Communications Research (Bellcore) document SR-TSV-002275, BOC Notes on the LEC Networks—1990.

(3) *Impedance.* For the purpose of this section, the input impedance of all subscriber loops served by the equipment is arbitrarily considered to be 900 ohms at voice frequencies.

(4) *Lockout.* (i) All line circuits shall be arranged for line lockout. When a permanent condition occurs prior to placing a line into lockout, a timed low level warning followed by a timed high level receiver off-hook (ROH) tone (see paragraph (i)(2)(xi) of this section) or a howler circuit (see paragraph (o)(2)(iii)(C) of this section) shall be applied to the line.

(ii) The line on lockout shall be reconnected automatically to the central office when the permanent off-hook condition is cleared.

(5) *Pay stations.* Pay stations may be prepay, or semi-postpay, as specified by the owner.

(6) *Loop extension.* (i) The number of lines which exceed 1900 ohms will be specified by the owner. When requested by the owner, the bidder shall furnish equipment to guarantee satisfactory operation of all lines.

(ii) Working limits for subscriber lines with loop extenders are covered in RUS Bulletin 345–55, PE-61, Central Office Loop Extenders and Loop Extender Voice Frequency Repeater Combinations.

(iii) Ringing from RUS accepted loop extenders, or their equivalent, shall be cut off from the called line when the handset at the called station is removed during the ringing or the silent interval.

(7) *Private branch exchange (PBX) lines.* PBX trunk hunting shall be available. It will not be necessary to segregate PBX lines to certain line groups.

(8) *Quantity.* A sufficient number of terminations shall be provided, in addition to the quantity specified by the owner for subscriber line service, to meet the requirements of the system for equipment testing, alarm checking, tone transfer, loop around test and other features.

(9) *Types.* There shall be provisions for types of lines such as ground start, loop start, regular subscriber, pay stations, etc.

(g) *Intraoffice switching requirements.* (1) The switching system shall:

(i) Provide dial tone in response to origination of a call by a subscriber, except on special lines where the application of dial tone is not applicable, such as manual and hot lines;

(ii) Remove dial tone immediately after the first digit has been dialed;

(iii) Recognize the class of service of the calling subscriber;

(iv) Register the digits dialed by the calling subscriber where the rotary dial or pushbutton dialing characteristics and the minimum interdigital times are as specified;

(v) Perform the necessary translation functions when the required number of digits have been registered, and select a channel to a proper outgoing trunk, if one is available, to the designated interexchange carrier;

(vi) Provide a transmission path from the calling subscriber line to the selected trunk, if an idle one is found;

(vii) Provide for more than one alternate route to the desired destination when specified by the owner, select an idle outgoing trunk in the first or second choice alternate route trunk group, if all trunks in the higher choice groups are busy, and provide a reorder signal (see paragraph (i)(2)(iv) of this section) to the subscriber if no trunks are available in the last choice alternate route;

(viii) Translate the proper part of the registered incoming routing data on tandem calls into an identification of an outgoing trunk group, select an idle trunk in that group, initiate the connection of the incoming trunk to the outgoing trunk, set the trunks in the proper configuration for tandem operation, and transmit information as required to permit completion to the desired destination in the distant office;

(ix) Transmit the proper stored information over the selected trunk to permit completion of outgoing calls to the desired destination by the distant office or offices, and provide multifrequency (MF) outpulsing when specified;

(x) Register all the digital information on calls incoming from a distant office, when dial or MF pulsing characteristics and interdigital times are as specified;

(xi) Translate internally a registered directory number into line equipment location, ringing code and terminating class (such as "PBX hunting") on incoming or intraoffice calls;

(xii) Test the called line for a busy condition;

(xiii) Connect the incoming trunk or locally originated call to the called line if the called line is idle;

(xiv) Permit any type of ringing voltage available in the central office to be associated with any Subscriber Directory Number (SDN), cause the proper type of ringing voltage to be connected to the called line, and remove ringing from the line upon answer whether in the ringing or silent period; and

(xv) Test and monitor the switching system continually during periods of low traffic using the maintenance and diagnostic subsystem.

(2) The switching system shall offer at least the following originating and terminating class-of-service indications on a per-line basis to subscribers, as specified by the owner:

(i) Flat rate individual line, bridged ringing;

(ii) Flat rate PBX and trunk hunting numbers, bridged ringing;

(iii) Pay station;

(iv) Message rate subscriber line;

(v) Wide Area Telephone Service (WATS);

(vi) Extended Area Service (EAS);

(vii) Data service;

(viii) Hotel-Motel capability;

(ix) Denied originating;

(x) Denied terminating;

(xi) Custom calling features;

(xii) Special interexchange carrier accesses; and

(xiii) Presubscription to designated interexchange carrier.

(3) The switching system shall provide PBX hunting.

(i) At least one trunk hunting group in each 100 SDN's equipped shall be provided. More may be provided as specified by the owner.

(ii) PBX groups shall be of a reasonable size commensurate with the ultimate size of the switching system.

(iii) Any available SDN may be used for PBX trunk hunting.

(iv) Each PBX group shall have the capability of being assigned one or more nonhunting SDN's for night service.

(v) If the called line is a PBX hunting line, the switching system shall test all assigned lines in the hunting group for a busy condition.

(vi) If the called PBX group is busy, line busy tone, as specified in paragraph (i)(2)(iii) of this section, shall be returned to the originating end of the connection.

(4) The switching system shall provide pay stations which may be prepay or semi-postpay. The system shall be arranged so that an operator and emergency service (911) may be reached from prepay or semi-postpay coin lines without the use of a coin, when the proper pay station equipment is provided.

(5) To meet dialing requirements, the switching system shall:

(i) Initiate the line lockout function after a delay, as specified in paragraph (r)(3) of this section, if dial or pushbutton dialing pulses are not received after initiation of a call, preferably routing the subscriber line to a holding circuit for tones and then automatically to lockout;

(ii) Connect 120 interruptions per minute (IPM) paths busy tone, recorded message, or other distinctive tone to the calling subscriber if an interval longer than that specified in paragraph (r)(4) of this section elapses between dialed digits;

(iii) Register the standard tone calling signals received from a subscriber station arranged for pushbutton dialing if specified by the owner, provide arrangements to function properly with 12-button pushbutton dialing sets, and return a reorder signal to the subscriber upon receipt of signal from the 11th or 12th buttons if neither of these buttons is assigned functions; and

(iv) Connect the incoming trunk to the digit register equipment within 120 milliseconds after seizure where direct dialing is received on calls from a distant office, cancel the bid for a register, and return reorder tone to the calling end if dial pulses are received before a register is attached.

(6) The switching system shall provide for appropriate circuit usage.

(i) To avoid inefficient utilization of the switching network, that portion of the common equipment that establishes the connection on intramachine calls shall not require more than 500 milliseconds, exclusive of ringing and ring trip, to complete its function under no-delay conditions.

(ii) The switching system shall provide for duplication in a load sharing or redundant configuration any circuit elements or components, the failure of which would reduce the grade of service of 100 or more lines by more than 25 percent of the traffic carrying capacity.

(iii) The switching system shall ensure that failure of access to a high choice circuit will not prevent subsequent calls from being served by lower choice circuits, wherever possible.

(iv) Where only two circuits of a type are provided, circuits shall be designed so that failure of one circuit will not permanently block any portion of the system for the duration of the failure.

(v) Where more than two circuits of a type are provided, successive usages should be on a rotational or random basis rather than the step-up selection with the possible exception of a last choice trunk.

(vi) The system shall be designed so that, in the event of a network failure, the system shall immediately or simultaneously use a redundant portion of the network to complete the call.

(7) The switching system shall provide busy verification facilities with the method of access specified by the owner.

(i) Only an operator or a switchman shall be able to override a busy line condition.

(ii) If the called line is busy, off-hook supervision shall be given the operator or switchman.

(iii) The responsibility of restricting subscribers in distant offices from having access to busy verification shall be on the distant office personnel when the toll trunks are used for both toll connecting and verification traffic.

(iv) When a verification code is used, all digits of the code must be dialed before cut-through to the called line can be accomplished.

(8) The switching system shall provide intercept facilities.

(i) All unused numbering plan area codes, home numbering plan area office codes, service codes and subscriber directory numbers (SDN's) shall be routed to intercept. All intercept administration shall be by changes in memory administrable by telephone company personnel. Maximum machine time to place a subscriber on intercept shall be 15 seconds.

(ii) Unequipped SDN's intercept shall be effective if the processor memory does not have information concerning the SDN in question.

(iii) The intercept equipment shall be arranged so that specific SDN's can be routed to a separate intercept circuit for changed numbers.

(iv) When an intercept call is answered, either by an operator or by a recorded announcement, an off-hook or charge supervision signal shall not be returned, even momentarily, to the originating end.

(v) When intercepting service is to be handled over the regular interoffice toll trunks, a distinctive identifying tone shall be transmitted when the operator answers. This tone shall be of the frequency and duration specified in paragraph (i)(2)(x) of this section.

(9) The switching system shall provide nuisance call trap facilities which, when activated, provide a permanent record of the calling and called numbers complete with date and time of day. Where the call originates over an interoffice trunk, the actual trunk number shall be recorded. There shall be provision for the called subscriber to hold the connection and for the positive trace of the call from origination to termination within the office.

(10) The switching system shall follow appropriate release procedures.

(i) The office shall be arranged so a connection to a terminating channel other than assistance operator shall be released under control of the calling party so that the channel can be reseized, unless the call is to emergency 911 service or other termination arranged for called party control.

(ii) If the called party disconnects first, the channel used in the originally established connection shall be held until the calling party disconnects or until the timing interval specified in paragraph (r)(7) of this section has elapsed. This feature shall not interfere with the normal operation of calls to intercept, fire alarm, or other special services.

(11) The switching system shall provide line load control facilities, when specified by the owner, to give preference for originating service to a limited group of subscribers during emergencies.

(i) These facilities may be activated manually by input-output (I/O) device or automatically after a manual setting of a key (or equivalent) to put line load control into effect, as determined by the bidder. The automatic procedure is preferable.

(ii) Procedures shall be established to avoid the unauthorized use of the line load control facilities.

(iii) Where automatic activation is provided, service may be provided to small groups of nonemergency subscribers on limited grade of service whenever the office load becomes low enough to permit this to be done safely.

(h) *Interoffice trunk circuit requirements*—(1) *General.* (i) The bidder shall supply, as requested by the owner, solid-state technology type trunk and signaling circuits of any of the types described in RUS TE&CM 319, Interoffice Trunking and Signaling, or, with the approval of RUS, any other more recent and desirable types not as yet covered in the manual. For dc signaling, the duplex (DX) and loop types of signaling are preferred.

(ii) Trunks shall not be directly driven from the subscriber's dial on outward calls.

(iii) In order to reduce the spares inventory and minimize incidence of improper maintenance replacement of circuit assemblies, the types of trunk circuits shall be kept to a minimum. Variation in assemblies should be mainly limited to variation in signaling modes.

(iv) Trunk circuits which connect with carrier or 4-wire transmission facilities shall be arranged for 4-wire transmission to avoid an intermediate 2-wire interface between a 4-wire switching system and trunk facilities.

(2) *Quantity.* Trunk quantities shall be as specified in appendix A of this section. Sufficient space shall be provided for an orderly layout of trunks. Trunks of a certain type going to the same destination may be grouped together on the original installation.

(3) *Requirements for interoffice connections.* (i) When operator trunks are used in common for both coin and noncoin lines, they shall be arranged to provide an indication to the operator by means of a visual signal or tone when calls are from pay stations. When a tone is used, it shall be of the type specified in paragraph (i)(2)(v) of this section and shall be connected to be heard only by the operator upon answer. It shall be possible to repeat the tone signal.

(ii) There are no requirements for trunks arranged for manual re-ring by a toll operator, either with the receiver on or off the hook, except to coin stations with the receiver on the hook.

(iii) On calls from subscribers to the assistance operator, the release of the connection shall be under control of the last party to disconnect. An exception is operator control of disconnect that is used on outgoing trunks to a TSP/TSPS system.

(iv) On calls originated by an operator, the release of the connection shall be under control of the operator.

(v) Where trunks with E and M lead signaling are used, the trunk circuits for Type I signaling shall be arranged to place ground on the M lead during the on-hook condition and battery on the M lead in the off-hook condition. For E and M Type II, only a make contact between the MA and MB lead will be required. In either type, current limiting shall be provided in the E lead of the trunk circuit itself, as required for proper operation. It shall be assumed that connection equipment in the form of trunk carrier, multiplex, or associated signaling apparatus furnishes only a contact closure to ground (Type I) or to a signal ground lead (Type II) for an off-hook condition on the E lead.

(vi) Where answer supervision is used to determine the initiation of the charging interval for a call, such answer supervision shall not be effective for charging until after the elapse of the timing interval listed in paragraph (r)(5) of this section.

(vii) When necessary, provision shall be made for reception of start and stop dial signals on toll trunk equipment.

(viii) When trunks arranged for automatic message accounting (AMA), toll ticketing, or centralized automatic message accounting (CAMA) are specified by the owner, these trunks shall provide the pertinent features described in paragraph (k) of this section applicable to such functions.

(4) *Requirements for direct digital connections.* (i) Interface units which will permit direct digital connection to other digital switches, channel banks and remote line and/or trunk circuits over digital facilities shall be provided when specified by the owner. The digital transmission system shall be compatible with T1 type span lines using a DS1 interface and other digital interfaces that may be specified by the owner. The RUS specification for the span line equipment is Bulletin 345–50, PE-60, RUS Specification for Trunk Carrier Systems.

(ii) Each interface circuit shall connect 24 voice channels to the switching system from a 1.544 megabit per second DS1 bit stream. The DS1 bit stream entering or exiting the system shall be in the D3 format and the voice signals shall be encoded in 8 bit mu-255 PCM. The format and processing of the bit stream must be compatible with characteristics of the D3 channel bank such as alarm and maintenance characteristics. Loss of receive signal (DS1) shall be detected and the equivalent of a carrier group alarm shall be executed in 2.5 ±0.5 seconds. Loss of synchronization shall be detected by slips, timing jitter, and wander in accordance with industry standards.

(iii) Signaling shall be by means of MF or dial pulse (DP) and the system which is inherent in the A and B bits of the D3 format. In the case where they are not used for signaling, the A and B bits shall be used only for normal voice and data transmission.

(i) *Tone requirements*—(1) *General.* Tones shall be provided to indicate the progress of a call through the office. Tone generators should be an integral part of the switching systems. The tones should be introduced digitally by the application of the appropriate bit stream to the line or trunk circuit via the digital switching network. The necessary precautions shall be made to ensure tone sources automatically if the primary sources fail.

(2) *Tone specifications.* (i) Dial tone shall consist of 350 Hz plus 440 Hz at a composite level of −10 dBm0 which equates to −13 dBm0 per frequency. This is the precise tone suitable for use with pushbutton dialing.

(ii) Low tone shall consist of 480 Hz plus 620 Hz at a composite level of −21 dBm0 which equates to −24 dBm0 per frequency.

(iii) Line busy tone shall be low tone interrupted at 60 IPM, with tone on 0.5 seconds and off 0.5 seconds.

(iv) Reorder, all paths busy, and no circuit tone shall be low tone interrupted at 120 IPM, with tone on 0.25 seconds and off 0.25 seconds.

(v) Identifying tone on calls from coin lines shall be uninterrupted low tone.

(vi) High tone shall consist of 480 Hz at −17 dBm0.

(vii) Audible ringback tone shall consist of 440 plus 480 Hz at a composite level of −16 dBm0 which equates to −19 dBm0 per frequency.

(viii) The call progress tones listed in this section are described in Bellcore document SR-TSV-002275, BOC Notes on the LEC Networks—1990, section 6. The 350, 440, 480, and 620 Hz tones shall be held at ±0.5 percent frequency tolerance and ±3 dB amplitude variation. The amplitude levels specified are to be measured at the main distributing frame, excluding cable loss.

(ix) Distinctive tone, when required for alarm calls, or other features, shall consist of high tone interrupted at 200 IPM with tone on 150 ms and off 150 ms.

(x) Identifying tone on intercepted calls shall consist of uninterrupted high tone impressed on the trunk circuit 300 to 600 milliseconds following the operator's answer of intercepted calls.

(xi) An ROH circuit shall have output tones which do not interfere with the pushbutton or multifrequency signaling tones. The ROH tone may be introduced digitally internal to the system near the overload level of + 3 dBm0. No power adjustment will be required. The frequency of the output shall be distinctive and urgent in order to attract the subscriber's attention to an off-hook situation. (Warning: In order to determine the signal level, a frequency selective voltmeter must be used to determine the level of each signal component and mathematical power addition used to combine these measurements into a single level value.)

(xii) During application of tones, office longitudinal balance shall be maintained within 15 dB of that specified in paragraph (q)(8) of this section.

(j) *System clock.* (1) The central office clock and network synchronization system shall have the ability to be synchronized with external clocks for network synchronization, including detection of slips, timing, jitter and wander, in a digital-to-digital environment or operate initially in an independent network (refer to Bellcore document SR-TSV-002275, BOC Notes on the LEC Networks—1990, section 11).

(2) The end office central office system clock shall be a Stratum 3 clock with:

(i) A minimum long-term accuracy of $\pm 4.6 \times 10^{-6}$ (±7 Hz @ 1.544 MHz);

(ii) A minimum stability of 3.7×10^{-7}/day upon loss of all frequency references; and

(iii) A "Pull-In Range" for the capability of synchronizing to a clock with accuracy of $\pm 4.6 \times 10^{-6}$.

(3) The access tandem central office system clock shall be a Stratum 2 clock with:

(i) A minimum long-term accuracy of $\pm 1.6 \times 10^{-8}$ (±0.025 Hz @ 1.544 MHz);

(ii) A minimum stability of 1×10^{-10}/day upon loss of all frequency references; and

(iii) A "Pull-In Range" for the capability of synchronization to a clock with accuracy of $\pm 1.6 \times 10^{-8}$.

(k) *Switched access service arrangements*—(1) *General.* The equipment shall be capable of providing Feature Group A, Feature Group B, Feature Group C, and Feature Group D switched access service arrangements, as described in Bellcore document SR-TSV-002275, BOC Notes on the LEC Networks—1990, section 6 and section 15, including arrangements for automatic number identification (ANI).

(2) *Operation.* (i) All equipment shall be arranged for Feature Group A (Line Side Connection).

(ii) All equipment shall be arranged for Feature Group B given that appendix A of this section requires the equipment of the necessary trunks (Trunk Side Connection).

(iii) The equipment shall be arranged for Feature Group C on the trunk groups specified in appendix A of this section. Even though appendix A of this section specifies Feature Group D or some other trunk group, it shall be possible through software commands available to the owner to use Feature Group C signaling protocols on a trunk group basis until such time that the trunk group in question converts to Feature Group D signaling protocols.

(iv) The equipment shall be arranged for Feature Group D on the trunk groups specified in appendix A of this section.

(v) Calls originating from coin lines toward switched access service shall be arranged either to provide signaling protocols for TSPS, or in the absence of TSPS-type service, such calls shall be blocked.

(vi) The equipment shall be arranged for forwarding routing information, calling party identification, and called party numbers in the proper feature group protocols, by trunk group as specified in appendix A of this section.

(vii) The equipment shall be arranged for AMA data collection as specified in appendix A of this section by trunk group. Unless otherwise specified by the owner, the equipment shall be arranged to collect the billing data in the Bellcore AMA format as described in Bellcore document TR-TSY-000508, Automatic Message Accounting.

(viii) If specified in Item 9.4, appendix A of this section, the equipment shall be arranged to store the billing data in a pollable system. If specified in Item 9.5, appendix A of this section, equipment shall be furnished to poll the pollable systems associated with the contract.

(l) *Fusing and protection requirements*—(1) *General.* (i) The equipment shall be completely wired and equipped with fuses, trouble signals, and arranged for printout of fault conditions, with all associated equipment for the wired capacity of the frames or cabinets provided.

(ii) Design precautions shall be taken to prevent the possibility of equipment damage arising from the insertion of an electronic package into the wrong connector, the removal of a package from any connector, or the improper insertion of the correct card in its connector.

(2) *Fuses.* Fuses and circuit breakers shall be of an alarm and indicator type, except where the fuses or breaker location is indicated on the alarm printout. Their rating shall be designated by numerals or color code on the fuse panel, where feasible.

(3) *Components.* (i) Insofar as possible, all components shall be capable of being continuously energized at rated voltage without injurious results. Insofar as possible, design precautions shall be taken to prevent damage to other equipment and components when a particular component fails.

(ii) Printed circuit boards or similar equipment employing electronic components shall be self-protecting against external grounds applied to the connector terminals, where feasible. Board components and coatings applied to finished products shall be of such material or treated so they will not support combustion.

(iii) Every precaution shall be taken to protect electrostatically sensitive components from damage during handling. This shall include written instructions and recommendations (see Item 6.1,h of appendix C of this section).

(m) *Switching network requirements*—(1) *The network.* (i) All networks shall be comprised of solid-state components.

(ii) The switching network shall employ time division digital switching and be compatible for connection to D3 type PCM channel banks without conversion to analog.

(iii) Equipment shall be available as required to connect analog lines and trunks, analog or digital service circuits, digital carriers to RST's, D3 channel banks or other digital switching units.

(2) *Network quantity.* Where the number of stages in the switching network and their control varies with the capacity of the system, sufficient equipment and wiring shall be supplied initially in order that there will be no service interruptions when additions are made up to the ultimate capacity as specified in appendix A of this section. This does not imply the necessity of supplying empty cabinets unless this is the only way the necessary wiring can be accomplished.

(n) *Stored program control (SPC) equipment requirements.* (1) The system shall provide redundancy in call processing such that the failure of a call processing unit does not degrade the call processing capabilities of the switching system nor result in the loss of established calls.

(2) Programs shall be modular, flexible and structured. In the interest of more dependable and more easily read programs, it is desirable to use a language which is more person-oriented leaving the detailed machine-oriented problems to a compiler program. Quality assurance of all software programs shall be in accordance with IEEE Std 730–1989, IEEE Standard for Software Quality Assurance Plans, or equivalent.

(3) The office administration program shall have checks within it to prevent failure due to erroneous or inconsistent input data. It shall safeguard against the possibility of upsetting machine performance with improper instructions or information. In addition, modular structure shall allow the use of a variety of human-engineered service order formats. Service changes may be performed remotely if so desired. Average machine time for service change shall be 15 seconds or less. Service changes shall not be registered in permanent memory until

verified. The access to the service change shall not have access to generic program.

(4) The switching system shall be able to offer, by request, at least the following printouts of its routine stored data for administrative purposes:

(i) A list of all assigned directory numbers, in numerical order, with their assigned class of service and line terminal numbers;

(ii) A list of all directory numbers, in numerical order, associated with a class of service;

(iii) A list of all unassigned line terminals;

(iv) Traffic data in proper form for separation studies in accordance with the revenue separations procedures current at the time of the contract;

(v) All lines on lockout;

(vi) All lines assigned to intercept;

(vii) All available (unassigned) directory numbers in the working thousands group; and

(viii) A list of equipment busied out for maintenance.

(5) The printouts in paragraph (n)(4) of this section may be delayed to times of light traffic.

(6) Maintenance diagnostics shall be performed by a fault recognition system utilizing both software and hardware, each being used where they are most effective for maintenance and reliability. In the economic interests of providing early and efficient fault detection and accurate pinpointing of faulty areas, it is desirable to have a comprehensive person-machine interface supported by extensive automatic fault detection and analysis, involving diagnostic software for fault resolution and automatic recovery mechanisms to maintain continuous service. Maintenance messages may be channeled to a remote maintenance center if so desired.

(7) Information in memory, having no requirement for changes to be introduced in the maintenance or operation of the system, may be stored in memory devices such as programmable read-only memory (PROM) or other devices that cannot be reprogrammed in the field.

(o) *Maintenance facilities*—(1) *Alarm features, including alarm sending*. (i) The equipment shall be arranged to provide audible and visual alarms indicating fuse operation or other circuit malfunctions resulting from component failure, crosses or open wiring, or any other conditions affecting service which can be detected economically.

(ii) The alarms shall be classified in accordance with their effect on the system.

(A) Catastrophic alarms demand immediate attention and require notification of the highest level of supervisory personnel. Conditions such as loss of service, loss of one or more remote line switches or line concentrators connected through Direct Digital Interface, loss of network control, and loss of computer program in all processors shall produce catastrophic alarms.

(B) Major alarms demand rapid action. Conditions such as loss of one or more groups of subscribers or trunk ports, blown fuses for common groups of channels, loss of control to groups of channels, failure of one or both redundant units, and total loss of battery charging current for more than 15 minutes shall produce major alarms.

(C) Minor alarms indicate nonemergency conditions which cause degraded service or fault conditions which causes the system to operate within less-than-optimum performance. Conditions discovered in automatic routining which have not shown in the operation of the equipment but require attention and cumulative line lockout (level adjustable) are examples of minor alarm conditions.

(iii) When the office is arranged for unattended operation, facilities shall be provided for extending the alarm indications to an attended point.

(iv) When the use of a separate outside plant facility for alarm sending is specified, the nature of the alarm may be indicated to the distant point by machine printout or other display device.

(v) When alarm sending is accomplished over a regular operator office trunk, the operator shall be apprised that the call is an alarm indication by a distinctive tone, as specified by the owner in appendix A of this section. It shall be possible for the operator to determine at any time the presence of a trouble condition by dialing a number set aside for that purpose. This number shall also be accessible from lines classmarked for this feature.

(vi) When the alarm sending circuit seizes an interoffice operator trunk, the operator must dial the alarm checking code over another trunk before the first trunk can be released except where the alarm condition has disappeared first.

(vii) The alarm sending circuit shall have access to two or more trunks if the trunks are used for subscriber traffic.

(viii) An alarm indication of higher priority shall supersede an original alarm indication and reseize an interoffice operator trunk.

(ix) In any group of offices purchased under one contract, the same codes shall be used in each office for alarm checking and test.

(x) When the alarm checking number is dialed, the alarm indications received shall be as follows:

(A) Catastrophic alarm—No tone.

(B) Major alarm—Continuous busy tone 60 IPM, unless alarm is overridden.

(C) Minor alarm—Continuous 1-ring code ringback tone, unless alarm is overridden.

(D) No trouble—Continuous 2-ring code ringback tone, unless alarm is overridden.

(xi) Audible and visual local alarms and transmitted alarms shall be provided as follows:

Classification	Delay Interval	
	Local Alarms	Alarms Transmitted
Catastrophic	0	0
Major	0	0 [1]
Minor	0	0–30 Min.

[1]Except no charge alarm delayed 15 minutes.

(xii) The central office alarm circuits shall be arranged to provide optional wiring to transmit either a minor alarm or a major alarm and a printout to accommodate various types of trunk and subscriber carrier systems, microwave, mobile radio, other transmission systems, and environmental protection systems with different priorities when a set of contacts is closed in the equipment of such systems and the alarm checking code is dialed. The alarm relay shall be furnished by the supplier of the carrier multiplex and/or mobile radio equipment. The option or options shall be specified by the owner.

(2) *Trouble location and test.* (i) *Equipment.* (A) A maintenance center shall be provided with a fault recorder (printer and/or display) for troubles. Here, system and sub-system visual trouble indications are shown for maintenance aid.

(B) The fault recorder shall provide a permanent or semi-permanent record of the circuit elements involved whenever a trouble is encountered. It shall be arranged to recognize an existing fault condition and not cause multiple printouts of the same fault, except during test routine.

(ii) *Maintenance system.* (A) The maintenance system shall monitor and maintain the system operation without interruption of call processing, except for major failures.

(B) The maintenance system shall provide both specialized maintenance hardware circuits and an extensive software package to enable maintenance to determine trouble to an individual card or functional group of cards.

(C) Maintenance programs may be both on-line and off-line. On-line maintenance programs are activated by system errors and shall be scheduled to execute call tests during low traffic periods and periodic hardware tests at specific time intervals. Programs shall provide diagnostic tools for the maintenance personnel and be initiated by them.

(D) Scheduled periodic hardware tests shall automatically detect faults and alert maintenance personnel via alarm or appropriate input/output device(s) at local and/or remote locations.

(E) Facilities shall be provided so that test calls can be set up using preselected items of switching equipment.

(F) The maintenance personnel shall be able to make tests to determine if every trunk and every item of switching equipment are functioning properly. Also, it shall be possible to make each trunk and each SPC equipment, or part thereof, busy to service calls. Where possible, equipment which is made busy to service calls shall still be accessible for test calls.

(iii) *Outside plant and subscriber stations.* (A) A subscriber loop test set or equivalent shall be provided either as a separate set or as a part of the maintenance center, as specified in item 11.2 of appendix A of this section. This circuit shall include a high resistance volt-ohm meter, wiring to tip and ring terminals to permit a portable wheatstone bridge to be used, an operator's telephone circuit, a dial circuit (and pushbutton dialing keys, if specified), outgoing trunks to dial equipment for access to lines under test without use of the main distributing frame (MDF) test shoe and the necessary test keys. No dry cell batteries shall be accepted for test potentials. Circuits shall be designed so that alternating current (ac) induction on the line will have no effect on dc measurements. All functions shall be under control of lever or pushbutton keys. As a minimum the test system shall:

(*1*) Test for bridged foreign electromotive force (EMF);

(*2*) Test for regular line battery;

(*3*) Test for booster battery voltage and polarity using the test shoe;

(*4*) Test for open circuits, short, tip ground, and ring ground;

(*5*) Test for tip or ring negative potential;

(*6*) Test for capacitance of a subscriber's line;

(*7*) Supply talking battery to the line with and without booster battery;

(*8*) Ring the subscriber through the test access circuit or through a test shoe;

(*9*) Test in and out of the central office; and

(*10*) Supply a reverse polarity key for voltage readings, except when positive or negative values are displayed directly.

(B) An acceptable arrangement for making the tests shown in paragraph (o)(2)(iii)(A) of this section is to have them under software control with results displayed at one of the system's I/O ports.

(C) A howler circuit for maintenance purposes, if ordered by the owner, shall have output tones which do not interfere with the pushbutton or multifrequency signaling tones. The harmonics of the output tones shall be attenuated at least 26 dB below the fundamental frequency for all load conditions. The frequency stability shall be 2 percent or less for all output tones when the unit is operated in the specified load and environmental range. It shall be possible to vary the output voltage (power) of the howler circuit. It shall remove tone and restore the line to service when the telephone instrument receiver is placed on-hook. The frequency of the output shall be chosen to be distinctive and urgent in order to attract the subscriber's attention to an off-hook situation.

(D) When a dial speed test facility is specified by the owner, it shall be accessed by dialing a special code and shall return to the calling station readily identifiable signals to indicate that the dial speed is slow, normal, or fast.

(E) When the office is arranged for pushbutton dialing, optional facilities shall be provided for testing the pushbutton dialing equipment at the subscriber station.

(F) When a system for testing subscriber lines in remote offices from a test position in a centrally located office is specified by the owner, it shall be capable of working with all the central offices and RST's in the remote areas. This testing equipment shall preferably be solid-state with a minimum of electromechanical devices and shall operate from central office battery. It shall be capable of working over any voice grade telephone circuit and shall not require a dedicated trunk. There shall be no interference to or from "in-band" voice channel tones. When used over a network, the verification or access shall be guarded to prevent unauthorized access by subscribers. Access to this system shall only be available to the test operator in all cases.

(3) *Transmission testing.* (i) When transmission test circuits are specified in Item 11.3 of appendix A of this section, they shall permit testing of trunks by a distant office without any assistance in the local dial office. Analog test ports shall meet appropriate trunk requirements. If Centralized Automatic Reporting on Trunks (CAROT), or equivalent, is to be used, the equipment at the end office shall comply with Bellcore document SR-

TSV-002275, BOC Notes on the LEC Networks—1990, section 8, Item 2.

(ii) Transmission test circuits are available with a variety of options. These include single frequency and multifrequency tone generators with one or more generator output terminals, quiet terminations, and loop around test arrangements for both one-way and two-way trunks.

(iii) Where multifrequency generators are used, they are usually arranged to provide a minimum of three frequencies. With some equipment, up to seven additional frequencies may be provided if needed. No industry standardization of test frequencies is as yet provided. Therefore, it is important that the selection of frequencies, the order in which they are applied and the time interval for application of each frequency be agreed upon by the connecting company and the RUS borrower and listed in appendix A of this section in those situations where connecting companies request the installation of multifrequency generators in borrowers' central offices.

(iv) The milliwatt generator shall be solid-state and generate the analog or digital equivalent of 1004 Hz. The milliwatt generator shall be assigned to a 4-wire analog test port or be digitally generated. All 2-wire and 4-wire voice frequency ports are at a nominal 0 dBm0 level. The level of the 1004 Hz tone generator shall appear at outgoing 2-wire and 4-wire ports at 0 dBm ±0.5 dB. For direct digital connections, the encoded output shall be the digital equivalent of a 0 dBm0 ±0.5 dB signal.

(v) Reference tone generators can be used individually or they can be part of a loop around test arrangement. If both single frequency and multifrequency reference tone generators are to be provided, only one can be arranged as part of a loop around test. Where a loop around arrangement is provided, the generator output can be obtained by dialing singly one of the two line terminals. By dialing the other line terminal singly, usually a 900 ohm resistor in series with a 2.16 microfarad capacitor is connected to the circuit under test to act as a "quiet termination" for noise measurements and other tests. Whenever both line terminals are held simultaneously, both the milliwatt supply and the quiet termination shall be lifted off and a "loop around" condition established. This permits the overall loss to be determined from the distant office by going out over one trunk, looping around in the end office and returning over the other trunk. The insertion loss of this test arrangement when used in a loop around configuration should not exceed 0.1 dB at the frequencies specified for the milliwatt supply. Unless otherwise specified, continuous off-hook supervision is to be provided on both line terminals to prevent collusive calling without charge. It will be permissible to accomplish the quiet termination by opening the 4-wire path internally and to accomplish the loop around by digital switching.

(vi) Provision shall be made so that the milliwatt supply can be manually patched to circuits.

(vii) Test jack access shall be provided for all interoffice trunks of the voice frequency type. The jack access shall be properly designated for line, drop, monitor, and signaling leads plus any other jacks as requested by the owner. This may be accomplished by a set of jacks located at the maintenance center which have access to each trunk on a switching basis.

(p) *Traffic*—(1) *General engineering guidelines*. (i) The Traffic Table, based on the Erlang Lost-Calls-Cleared Formula, shall be used for determining the quantity of intraoffice paths, registers, and senders where full availability conditions apply. The following table shows the traffic capacity in CCS for 1 to 200 trunks at nine grades of service.

TRAFFIC TABLE

Full Availability for Random Traffic

LOST-CALLS-CLEARED

Offered Traffic Expressed in CCS

Number of Trunks	B-.001	.002	.005	.01	.02	.05	.1	.2	.5	Number of Trunks
1	0	0	0	0	1	2	4	9	36	1

TRAFFIC TABLE—Continued
Full Availability for Random Traffic
LOST-CALLS-CLEARED
Offered Traffic Expressed in CCS

Number of Trunks	B-.001	.002	.005	.01	.02	.05	.1	.2	.5	Number of Trunks
2	2	3	4	5	8	14	22	36	98	2
3	7	9	13	17	22	32	46	69	165	3
4	16	19	25	31	39	55	74	106	234	4
5	27	32	41	49	60	80	104	144	304	5
6	41	48	58	69	82	107	135	184	374	6
7	57	65	78	90	106	135	168	224	445	7
8	74	83	98	113	131	163	202	265	516	8
9	92	103	120	136	156	193	236	307	586	9
10	111	123	143	161	183	224	270	348	656	10
11	131	145	166	186	210	255	306	391	729	11
12	152	167	190	212	238	286	341	433	801	12
13	174	190	215	238	266	318	377	476	872	13
14	196	213	240	265	295	350	413	519	944	14
15	219	237	266	292	324	383	449	562	1015	15
16	242	261	292	320	354	415	486	605	1087	16
17	266	286	318	347	384	449	523	648	1158	17
18	290	311	345	376	414	482	560	692	1230	18
19	314	337	372	404	444	515	597	735	1302	19
20	339	363	399	433	474	549	634	779	1374	20
21	364	388	427	462	505	583	671	823	1445	21
22	389	415	455	491	536	617	709	866	1517	22
23	415	441	483	521	567	651	747	910	1589	23
24	441	468	511	551	599	685	784	954	1661	24
25	467	495	540	580	630	720	822	998	1733	25
26	493	523	568	611	662	754	860	1042	1805	26
27	520	550	598	641	693	788	898	1086	1876	27
28	546	578	627	671	725	823	936	1130	1948	28
29	573	606	656	702	757	858	974	1174	2020	29
30	600	634	685	732	789	893	1012	1218	2092	30
31	628	662	715	763	822	928	1050	1263	2164	31
32	655	690	744	794	854	963	1089	1307	2236	32
33	683	719	774	825	887	998	1127	1351	2308	33
34	711	747	804	856	919	1033	1165	1395	2380	34
35	739	776	834	887	951	1068	1203	1439	2452	35
36	767	805	864	918	984	1104	1242	1484	2524	36
37	795	834	895	950	1017	1139	1281	1528	2595	37
38	823	863	925	981	1050	1174	1319	1572	2667	38
39	851	892	955	1013	1083	1210	1358	1617	2739	39
40	880	922	986	1044	1116	1246	1396	1661	2811	40
41	909	951	1016	1076	1149	1281	1435	1706	2883	41
42	937	980	1047	1108	1182	1317	1474	1750	2955	42
43	966	1010	1078	1140	1215	1352	1512	1795	3027	43
44	995	1040	1109	1171	1248	1388	1551	1839	3099	44
45	1024	1070	1140	1203	1282	1424	1590	1884	3171	45
46	1053	1099	1171	1236	1315	1459	1629	1928	3243	46
47	1083	1129	1202	1268	1349	1495	1668	1973	3315	47
48	1112	1159	1233	1300	1382	1531	1706	2017	3387	48
49	1141	1189	1264	1332	1416	1567	1745	2062	3459	49
50	1170	1220	1295	1364	1449	1603	1784	2106	3531	50
51	1200	1250	1327	1397	1483	1639	1823	2151	3603	51
52	1229	1280	1358	1429	1516	1675	1862	2195	3675	52
53	1259	1310	1390	1462	1550	1711	1901	2240	3747	53
54	1289	1341	1421	1494	1584	1747	1940	2285	3819	54
55	1319	1371	1453	1527	1618	1783	1979	2329	3891	55
56	1349	1402	1484	1559	1652	1819	2018	2374	3962	56
57	1378	1432	1516	1592	1686	1856	2057	2418	4034	57
58	1408	1463	1548	1625	1719	1892	2096	2463	4106	58
59	1439	1494	1579	1657	1753	1928	2136	2508	4178	59
60	1468	1525	1611	1690	1787	1965	2174	2552	4250	60
61	1499	1556	1643	1723	1821	2001	2214	2597	4322	61
62	1529	1587	1675	1756	1855	2037	2253	2642	4394	62

TRAFFIC TABLE—Continued
Full Availability for Random Traffic
LOST-CALLS-CLEARED
Offered Traffic Expressed in CCS

Number of Trunks	B-.001	.002	.005	.01	.02	.05	.1	.2	.5	Number of Trunks
63	1559	1617	1707	1789	1889	2073	2292	2687	4466	63
64	1590	1648	1739	1822	1923	2110	2331	2731	4538	64
65	1620	1679	1771	1855	1958	2146	2370	2776	4610	65
66	1650	1710	1803	1888	1992	2182	2409	2821	4682	66
67	1681	1742	1835	1921	2026	2219	2449	2865	4754	67
68	1711	1773	1867	1954	2060	2255	2488	2910	4826	68
69	1742	1804	1900	1987	2094	2291	2527	2955	4898	69
70	1773	1835	1932	2020	2129	2328	2566	3000	4970	70
71	1803	1867	1964	2053	2163	2364	2606	3044	5042	71
72	1834	1898	1997	2087	2197	2401	2645	3089	5114	72
73	1865	1929	2029	2120	2232	2438	2684	3134	5186	73
74	1895	1961	2061	2153	2266	2474	2723	3178	5258	74
75	1926	1992	2093	2186	2300	2511	2763	3223	5330	75
76	1957	2024	2126	2219	2335	2547	2802	3268	5402	76
77	1988	2055	2159	2253	2369	2584	2841	3313	5474	77
78	2019	2087	2191	2286	2404	2620	2881	3357	5546	78
79	2050	2118	2223	2319	2438	2657	2920	3402	5618	79
80	2081	2150	2256	2353	2473	2694	2959	3447	5690	80
81	2112	2182	2289	2386	2507	2730	2999	3492	5762	81
82	2143	2213	2321	2420	2542	2767	3038	3537	5834	82
83	2174	2245	2354	2453	2577	2803	3077	3581	5906	83
84	2206	2277	2386	2487	2611	2840	3117	3626	5977	84
85	2237	2309	2419	2521	2646	2877	3156	3671	6049	85
86	2268	2340	2452	2554	2680	2913	3196	3716	6121	86
87	2299	2372	2485	2588	2715	2950	3235	3761	6193	87
88	2331	2404	2517	2621	2750	2987	3275	3805	6265	88
89	2362	2436	2550	2655	2784	3024	3314	3850	6337	89
90	2393	2468	2583	2688	2819	3060	3353	3895	6409	90
91	2425	2500	2616	2722	2854	3097	3393	3940	6481	91
92	2456	2532	2649	2756	2889	3134	3432	3984	6553	92
93	2488	2564	2682	2790	2923	3171	3471	4029	6625	93
94	2519	2596	2715	2823	2958	3208	3511	4074	6697	94
95	2551	2628	2748	2857	2993	3244	3551	4119	6769	95
96	2582	2660	2781	2891	3028	3281	3590	4164	6841	96
97	2614	2692	2814	2925	3063	3318	3630	4209	6913	97
98	2645	2724	2847	2958	3097	3355	3669	4253	6985	98
99	2677	2757	2880	2992	3132	3392	3708	4298	7057	99
100	2709	2789	2913	3026	3167	3429	3748	4343	7129	100
105	2867	2950	3078	3196	3342	3613	3946	4567	7489	105
110	3027	3112	3244	3366	3516	3798	4143	4792	7849	110
115	3186	3275	3411	3536	3691	3983	4341	5016	8209	115
120	3347	3437	3578	3707	3867	4168	4539	5241	8569	120
125	3507	3601	3745	3878	4043	4353	4737	5465	8929	125
130	3669	3765	3912	4049	4219	4539	4935	5689	9289	130
135	3830	3929	4081	4221	4395	4724	5133	5914	9649	135
140	3992	4093	4249	4392	4571	4910	5332	6138	10009	140
145	4155	4258	4418	4564	4748	5095	5530	6363	10369	145
150	4318	4423	4586	4737	4925	5282	5728	6587	10729	150
155	4481	4589	4755	4909	5102	5467	5927	6812	11089	155
160	4644	4755	4925	5082	5279	5654	6125	7037	11449	160
165	4808	4920	5094	5255	5457	5840	6324	7261	11809	165
170	4972	5087	5264	5428	5634	6026	6523	7486	12169	170
175	5137	5253	5434	5602	5811	6213	6722	7710	12529	175
180	5301	5420	5604	5775	5989	6399	6920	7935	12889	180
185	5466	5587	5775	5949	6167	6586	7119	8160	13249	185
190	5631	5754	5945	6123	6345	6773	7318	8384	13609	190
195	5797	5922	6116	6296	6524	6960	7517	8609	13969	195
200	5962	6089	6287	6471	6702	7146	7716	8834	14329	200

(ii) The traffic capacity for all interoffice trunks shall be based on full availability, even though the distant office itself is not engineered to provide full availability access.

(iii) The Traffic Table may also be used to determine the approximate traffic capacity of high-usage intertoll trunks. The traffic offered to high-usage groups may be read at B.10, signifying that 10 percent of the traffic overflows to the alternate route. This approximates the HU12 table used by AT&T.

(iv) In reading the trunk quantity from the table, the higher quantity shall be used when the CCS load is three or more CCS over the lower quantity. For example, the number of trunks justified for 294 CCS at B.005 is 16, but for 295 CCS 17 trunks are justified.

(v) Limited availability is not permitted.

(vi) The traffic capacity in the following table should be used for small trunk groups such as pay station, special service trunks, intercept, and PBX trunks, unless otherwise specified in appendix A of this section:

Number of Circuits	Permissible CCS
1	10
2	20
3	30
4	40

(vii) The percentage of lines equipped for pushbutton dialing is to be used to determine the number of tone receivers. Local registers, if required, shall be supplied on the basis of all dial pulse.

(2) *Grade of service.* (i) Grade of service specifies the expected performance when there are adequate service facilities for an assumed volume of traffic. It is expressed as a portion of the total traffic during a busy hour that cannot be terminated immediately or within a predetermined time period due to congestion. This places responsibility on the traffic engineers to specify facilities which will be entirely satisfactory to the users and which can be equipped at a price which will be accepted as reasonable.

(ii) The number of calls encountering dial tone delay in excess of 3 seconds, measured over the busy hour of the four high-consecutive week (4HW) period, shall not be more than 1.5 percent.

(iii) The average post dialing delay objective for an intraoffice call shall not exceed 1 second. This includes all connect, operate, and translation time.

(iv) The line to line (intraoffice) network matching loss objective shall be 0.02 or less.

(v) The blocking probabilities related to trunks include both "mismatch" probability and probability of "all trunks busy." It is likely that the "mismatch" will be negligible in that many digital central offices have essentially nonblocking switching characteristics. The objectives for trunk connections are as follows:

(A) Subscriber to outgoing trunk objective 0.01 or less;

(B) Incoming trunk to subscriber objective 0.02 or less; and

(C) Local trunk tandem objective 0.01 or less.

(vi) Groups of common service circuits are to be engineered utilizing the full availability traffic tables that appear in paragraph (p)(1)(i) of this section at the following stipulated probabilities:

(A) Outgoing trunks to 2/6 MF or dial pulse senders at B.001;

(B) Incoming trunks to 2/6 MF receivers at B.001;

(C) Incoming nondelay dial trunks to receivers at B.001; and

(D) Incoming trunks with start dial at B.01.

(vii) Remote Switching Terminals (RST's) shall meet the same grade of service objectives as the host.

(3) *Holding times.* For the purpose of estimating the quantity of common control circuits, the following average holding times may be used. These holding times are conservative and represent the average effective and ineffective call. If these holding times are to be used, it must be so stated in appendix A of this section.

(i) The following average call holding times (HT) may be used.

Type of Call	HT—Seconds
Intraoffice	120
EAS	150
Special Service, Intercept, Verification	60
Toll, CLR	300
Toll, S-S	24
Toll, PPCS	270

(ii) The following average subscriber dialing holding times may be used (times used to dial digits do not include machine time).

	Digits Dialed	DP Sec.	Push-button Sec.
Operator, Non-Pay Station	1	4.7	3.4
Special Service	3	7.7	5.0
Local	7	13.7	8.2
EAS	7	13.7	8.2
DDD: 1/0 + 7	8	15.2	9.0
DDD: 1/0 + 10	11	19.7	11.4
Dialing Time Per Digit	-	1.5	0.8
Dial Tone Response	-	3.2	2.6

(iii) The following average incoming register holding times may be used (times for digit registrations do not include machine time).

	Basic		Additional Per Digit
	Holding Time (Sec.)	Digits	
MF Receiver from:			
No. 5 Crossbar—Non-LAMA	1.4	4	0.14
No. 5 Crossbar—LAMA	2.3	4	0.14
Crossbar Tandem & 4A Toll	3.1	4	0.14
No. 1 ESS	1.4	4	0.14
Key Pulsing Switchboard	5.2	4	0.60
DP Receivers—10 PPS from:			
SxS	6.0	4	1.5
Dialing Switchboard	6.6	4	1.3
4A Toll	5.6	5[1]	-
Crossbar Tandem	4.9	4	1.2

[1] No reduction for fewer digits.

(iv) The following average sender holding times may be used (does not include machine setup and release time).

	Basic		Additional Per Digit
	Holding Time (Sec.)	Digits	
MF Senders:			
No. 5 Crossbar	1.5	4	0.14
Crossbar Tandem & 4A Toll[1]	2.0	4	0.14
TSP/TSPS	2.4	7	0.14
SxS—CAMA, Called Number	3.7	7	0.14
SxS—CAMA, Calling Number	1.3	7	-
DP Senders—10 PPS:			
With Overlap Pulsing[2]	9.1	Up to 6	1.8
Without Overlap Pulsing	4.6	4	1.2

[1] Add 1.3 seconds for ANI outpulsing on special toll (0 +) calls and on DDD calls if AMA is not provided.

[2] Assumes overlap outpulsing starting on receiving of third digit; applies only to calls handled on direct trunk groups.

(4) *Traffic data requirements.* (i) Traffic measurements are composed of primarily two types—counts and usage. The following types of traffic data recording are required:

(A) *Peg count* registers shall be incremented when a successful network connection is established to a particular circuit group such as trunks, senders, digital receivers, etc.

(B) *Overflow count* registers shall be incremented when access to a particular circuit group is denied due to all resource busy condition.

(C) *Network blockage count* registers shall be incremented due to an unavailability of a path in an access or switching matrix network.

(D) *Usage* measurements of the length of time associated with a particular setup event or network connection shall be made. Usage data measurements are normally collected by scanning circuit groups resources every 10 or 100 seconds to determine busy/idle states. Measurements are accumulated and read directly in CCS (hundred call seconds).

(E) *Service delay* measurements shall provide percentage counts of the calls for a particular service that are delayed beyond a specified interval of time, e.g., calls not receiving dial tone within 3 seconds after call origination.

(ii) Traffic data shall be stored in electronic storage registers or block of memory consisting of one or more traffic counters for each item to be measured. The registers listed in paragraph (p)(4)(i) of this section shall be associated with the interoffice trunks, switching network and central control equipment in such a manner that the register readings can be used to determine the traffic load and flow to, from and within the system. Two-way trunks shall be metered to indicate inward and outward seizures. The bidder shall indicate what registers are to be supplied and their purpose.

(iii) The measured data shall be shown on a printout. It should be possible to have local or remote printout, or both. Arrangement shall be made for automatic data printout on command for 15-, 30-, or 60-minute intervals as required, and be arranged for automatic start-stop and in accordance with revenue separation procedures current at the time of contract.

(iv) All traffic records shall have dates and times and office identification.

(q) *Transmission*—(1) *General*. The transmission characteristics will be governed by the fact that the switching matrix will be based on digital operation. Unless otherwise stated, the requirements are in terms of analog measurements made from Main Distributing Frame (MDF) to MDF terminals, excluding cabling loss.

(2) *Impedance*. For the purpose of this section, the nominal input impedance of analog ports in an end office shall be 900 ohms for 2-wire ports and 600 ohms for 4-wire ports. Where the connecting facility or equipment is other than this impedance, suitable impedance matching shall be provided by the bidder when specified by the owner.

(3) *Insertion loss*. The insertion loss in both directions of transmission at 1004 Hz shall meet the following requirements when measured with a 0 dBm input signal at 900 ohms (or 600 ohms, when required) at a temperature of 77 °F ±9 °F (25 °C ±5 °C).

(i) *Trunk-to-trunk or trunk-to-line*. The loss shall be set between 0 and 0.5 dB for 2-wire to 2-wire, 2-wire to 4-wire, or 4-wire to 4-wire voice frequency connections.

(ii) *Line-to-line*. The loss shall be set between 0 and 2 dB.

(iii) *Direct digital interface*. On a direct digital interface, the loss through the office shall be adjusted to the proper level in the receive side.

(iv) *Stability*. The long-term allowable variation in loss through the office shall be ±0.5 dB from the loss specified by the bidder.

(4) *Frequency response (loss relative to 1004 hz)* shall meet the following requirements.

(i) *Trunk-to-trunk*.

Frequency (Hz)	Loss at 0 dBm0 Input[1]	
	2-Wire to 2-Wire	4-Wire to 4-Wire
60	20 dB Min.[2]	16 dB Min.[2]
200	0 to 5 dB	0 to 3 dB
300–3000	−0.5 dB to 1 dB	−0.3 to + 0.3 dB
3300	1.5 dB Max.	1.5 dB Max.
3400	0 to 3 dB	0 to 3 dB

[1](−) means less loss and (+) means more loss.
[2]Transmit End

(ii) *Line-to-line*.

Frequency (Hz)	Loss at 0 dBm0 Input[1]
60	20 dB Min.[2]
300	−1 to + 3 dB
600–2400	±1 dB
3200	−1 to + 3 dB

[1](−) means less loss and (+) means more loss.
[2]Transmit End

(iii) *Trunk-to-line*. The trunk-to-line frequency response requirements shall be a compromise between those values specified in paragraphs (q)(4)(i) and (q)(4)(ii) of this section.

(5) *Overload level*. The overload level at 900 ohm impedance shall be + 3 dBm0.

(6) *Gain tracking (linearity)* shall meet the following requirements.

Input Signal Level[1]	Maximum Gain Deviation
+ 3 to −37 dBm0	±0.5 dB
−37 to −50 dBm0	±1 dB

[1]1004 Hz reference at 0 dBm0.

(7) *Return loss*. (i) The specified return loss values are determined by the service and type of port at the measuring (near) end. Two-wire ports are measured (near end) at 900 ohms in series with 2.16 microfarads and 4-wire ports are measured at 600 ohms resistive.

(ii) Far end test terminations shall be as follows:

(A) Loaded line circuit—1650 ohms in parallel with the series combination of .005 microfarads and 100 ohms;

(B) Nonloaded line circuit—800 ohms in parallel with the series combination of .05 microfarads and 100 ohms;

(C) Special service line circuit including electronic lines and carrier lines—900 ohms in series with 2.16 microfarads;

(D) Two-wire trunk—900 ohms in series with 2.16 microfarads; and

(E) Four-wire trunk—600 ohms.

(iii) For trunk-to-trunk (2-wire or 4-wire) connections the echo return loss (ERL) shall be 27 dB, minimum and the singing return loss (SRL) shall be 20 dB, minimum low and 23 dB, minimum high.

(iv) For trunk-to-line (2-wire or 4-wire) connections the ERL shall be 24 dB, minimum and the SRL shall be 17 dB, minimum low and 20 dB, minimum high.

(v) For line-to-line or line-to-trunk (2-wire or 4-wire) connections the ERL shall be 18 dB, minimum and the SRL shall be 12 dB, minimum low and 15 dB, minimum high.

(8) *Longitudinal balance.* The minimum longitudinal balance, with dc loop currents of 20 to 70 mA, shall be 60 dB at all frequencies between 60 and 2000 Hz, 55 dB at 2700 Hz and 50 dB at 3400 Hz. The method of measurement shall be as specified in the IEEE Std 455–1985, IEEE Standard Test Procedure for Measuring Longitudinal Balance of Telephone Equipment Operating in the Voice Band. Source voltage level shall be 10 volts root-mean-square (rms).

(9) *60 hz longitudinal current immunity.* Under test conditions with 60 Hz, the system noise shall be no greater than 23 dBrnC0 as measured using the configuration in Figure 1.

FIGURE 1—MEASURING THE EFFECTS OF LOW FREQUENCY INDUCTION

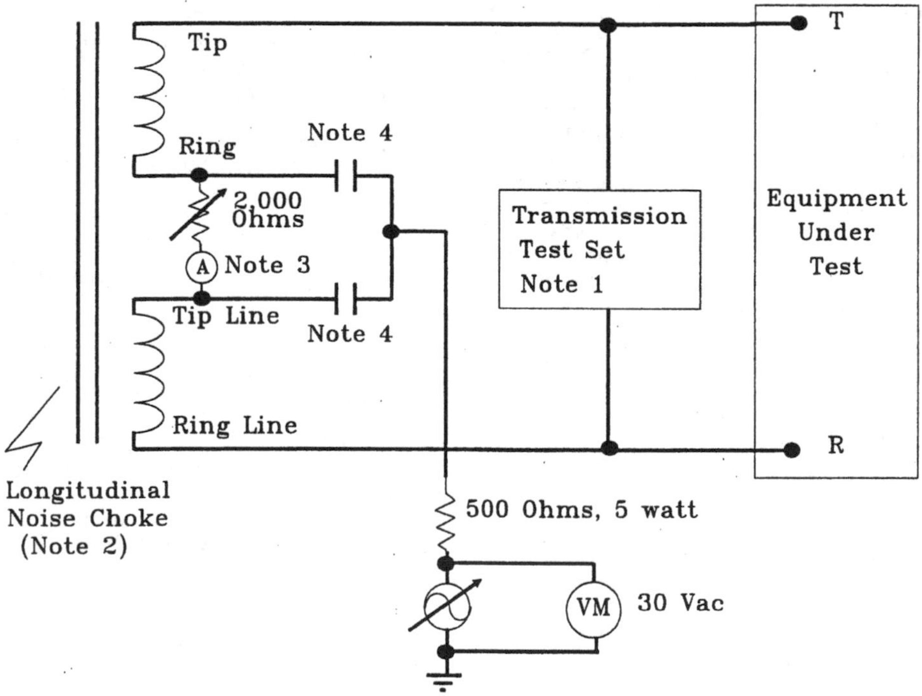

NOTES:
1. 900 ohm termination, C-message weighting, hold coil off
2. SNC Noise Choke 35 W, or equivalent
3. Test at 0.020 Adc and 0.070 Adc
4. 2 ±0.001 microfarad, 150 Vdc

(10) *Steady noise* (idle channel at 900 ohms impedance) measured on a terminated call shall be 23 dBrnC0 maximum

and average 18 dBrnC0 or less. The 3K Hz Flat noise should be less than 35 dBrnC0 as an objective.

(11) *Impulse noise*. The central office switching equipment shall be capable of meeting an impulse noise limit of not more than five counts exceeding 54 dBrnC0 voice band weighted in a 5-minute period on six such measurements made during the busy hour. A Northeast Electronics Company TTS 4002 Impulse Noise Counter, Wilcom T194C, Hewlett Packard 4945, or equivalent, should be used for the measurements. The measurement shall be made by establishing a normal connection from the noise counter through the switching equipment in its off-hook condition to a quiet termination of 900 ohms impedance. Office battery and signaling circuit wiring shall be suitably segregated from voice and carrier circuit wiring, and frame talking battery filters provided, if and as required, in order to meet these impulse noise limits.

(12) *Crosstalk coupling*. Worst case equal level crosstalk is to be 75 dB minimum in the range 200–3400 Hz. This is to be measured between any two paths through the system connecting a 0 dBm0 level tone to the disturbing pair.

(13) *Quantizing distortion*. (i) The switching system shall meet the following requirements.

Input Level (dBm0) 1004 or 1020 Hz	Minimum Signal to Distortion with C-Message Weighting
0 to −30	33 dB
−30 to −40	27 dB
−40 to −45	22 dB

(ii) Due to the possible loss of the least significant bit on direct digital connections, a signal to distortion degradation of up to 2 dB may be allowed where adequately justified by the bidder.

(14) *Absolute delay*. The absolute one-way delay through the switching system, excluding delays associated with RST switching, shall not exceed 1000 microseconds analog-to-analog measured at 1800 Hz.

(15) *Envelope delay distortion*. On any properly established connection, the envelope delay distortion shall not exceed the following limits.

Frequency Range (Hz)	Microseconds
1000 to 2600	190
800 to 2800	350
600 to 3000	500
400 to 3200	700

(16) *Digital error rate*. The digital switching system shall not introduce an error into digital connections which is worse than one error in 10^8 bits averaged over a 5-minute period.

(17) *Battery noise*. Noise across battery at power board distribution bus terminals shall not exceed 35 dBrnC during the busy hour.

(18) Radio and television interference. The central office switching equipment shall be designed and installed so that radiation of high frequency noise will be limited so as not to interfere with radio and television receivers.

(r) *Timing intervals*—(1) *Type of equipment required*. The equipment for providing the specified timing intervals shall be solid-state.

(2) *Tolerance*. Where a range of time is specified as minimum and maximum, the lower limits shall be considered as controlling and the variation between this minimum and the actual maximum shall be kept as small as practicable. In no case shall the quoted upper limit be exceeded.

(3) *Permanent signal timing*. Lockout shall occur after an interval of 20 to 30 seconds after receipt of dial tone if a "permanent" condition occurs prior to the transmission of dial pulses or pushbutton dialing signals. This interval may be reduced appreciably during periods of heavy traffic.

(4) *Partial dial timing*. Partial dial timing shall be within 15 to 37 seconds. This timing may be reduced appreciably during periods of heavy traffic.

(5) *Charge delay timing*. Charge delay timing shall be within 2 seconds.

(6) *Called party disconnect timing*. Timed disconnect of a terminating path under control of the called party shall be 10 to 32 seconds.

(7) *Timing intervals for signals involved in distance dialing*. Timing intervals shall be provided to meet the requirements for distance dialing equipment, which have been established in Bellcore document SR-TSV-002275, BOC Notes on the LEC Networks—1990.

Some of the more important times which this document specifies are for:

(i) Disconnect signal;

(ii) Wink signal;

(iii) Start dialing signal;

(iv) Pulse delay signal;

(v) Go signal;

(vi) Digit timing; and

(vii) Sender, register, and link attachment timing.

(s) *Power requirements and equipment*—(1) *Operating voltage.* The nominal operating voltage of the central office shall be 48 volts dc, provided by a battery with the positive side tied to system ground.

(2) *Batteries.* (i) When battery cells of the lead antimony type are specified, the pasted plate type shall be considered adequate.

(ii) When lead calcium cells are specified, no cell shall differ from the average voltage of the string of fully charged cells by more than ±0.03 volt when measured at a charging rate in amperes equivalent to 10 percent of the ampere hour capacity of the cells. Similarly, when cells are fully charged and floating between 2.30 and 2.33 volts per cell, the cell voltage of any cell in a given string shall not differ more than ±0.03 volt from the average. These requirements are for test purposes only and do not apply to operating conditions.

(iii) Voltage readings shall be corrected by a temperature coefficient of 0.0033 volt per degree F (0.006 per degree C), whenever temperature variations exist between cells in a given string. This correction factor shall also be applied when comparing cell voltages taken at different times and at different temperatures. The correction factor shall be added to the measured voltage when the temperature is above 77 °F (25 °C) and subtracted when the temperature is below 77 °F (25 °C).

(iv) The specific gravity readings of lead antimony cells at full charge shall be 1.210 ±.010 at 77 °F (25 °C) at maximum electrolyte height.

(v) When counter cells are supplied by the bidder, they shall be the dry counter electromotive force (CEMF) type.

(vi) When lead antimony batteries are specified, they shall be designed to last a minimum of 10 years when maintained on a full float operation between 2.15 and 2.17 volts per cell. When lead calcium batteries are specified, they shall be designed to last a minimum of 20 years when maintained on full float operation between 2.17 and 2.25 volts per cell. The battery shall be clearly designated as "antimony" or "calcium" by means of stencils, decals or other devices.

(vii) Each battery cell shall be equipped with an explosion control device.

(viii) The battery size shall be calculated in accordance with standard procedures. The battery in no case shall have a reserve capacity in ampere hours less than four times the current capacity of the largest charger.

(3) *Charging equipment.* (i) Charging shall be on a full float basis. The rectifiers shall be of the full wave, self-regulating, constant voltage, solid-state type and shall be capable of being turned on and off manually.

(ii) When charging batteries, the voltage at the battery terminals shall be adjustable and shall be set at the value recommended for the particular battery being charged, providing it is not above the maximum operating voltage of the switching system equipment. The voltage shall not vary more than plus or minus 0.02 volt per cell between 10 percent load and 100 percent load. Between 3 percent and 10 percent load, the output voltage shall not vary more than plus or minus 0.04 volt per cell. Beyond full load current, the output voltage shall drop sharply. The output voltage shall be maintained with the line voltage variations of plus or minus 10 percent. Provision shall be made to change the output voltage of the rectifier manually to 2.25 volts per cell to provide an equalization charge on the battery.

(iii) The charger noise shall not exceed 22 dBrnC when measured with a suitable noise measuring set and under the rated battery capacitance and load conditions as determined in Figure 2.

FIGURE 2—CHARGER NOISE TEST

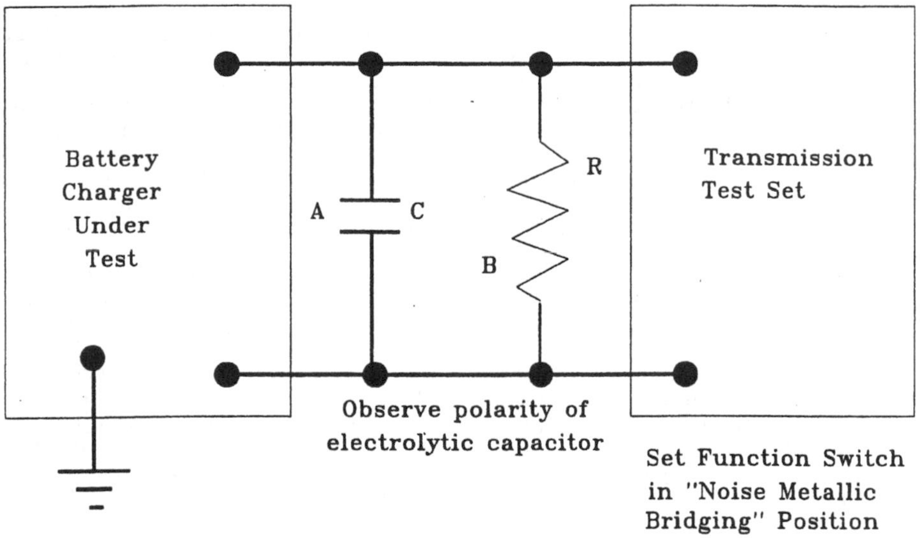

The manufacturer may elect to eliminate the capacitor C from the measurement.

A. Capacitance in μF = 30,000 μF per ampere-hour per cell. For example, 25 cells at 100 ampere-hour would be equivalent to a capacitance of:

(30,000 × 100) / 25 = 120,000 μF

B. The value of the resistive load R is determined by the nominal battery voltage in volts divided by the full load rating in amperes. For example, for a 48 volt battery and a full load current of 24 amperes, the load resistance R is 48/24 = 2 ohms of appropriate power handling capacity.

(iv) The charging equipment shall indicate a failure of charging current, whether due to ac power failure, an internal failure in the charger, or to other circumstances which might cause the output voltage of the charger to drop below the battery voltage. Where a supplementary constant current charger is used, an alarm shall be provided to indicate a failure of the charger.

(v) Audible noise developed by the charging equipment shall be kept to a minimum. Acoustic noise resulting from operation of the rectifier shall be expressed in terms of dB indicated on a sound level meter conforming to ANSI S1.4–1983, Specification for Sound Level Meters, and shall not exceed 65 dB (A-weighting) measured at any point 5 feet (152.4 cm) from any vertical surface of the rectifier.

(vi) The charging equipment shall be designed so that neither the charger nor the central office switching equipment is subject to damage in case the battery circuit is opened for any value of load within the normal limits.

(vii) The charging equipment shall have a capacity to meet the requirements of central office size and special requirements of the owner in appendix A of this section.

(viii) Minimum equipment requirement for chargers is one of the following:

(A) Two chargers either capable of carrying the full office load as specified in Item 12 of appendix A of this section; or

(B) Three chargers each capable of carrying half the office load as specified in Item 12 of appendix A of this section.

(4) *Miscellaneous voltage supplies.* (i) Any power supply required for voltages other than the primary battery voltage shall be provided by either a solid-state dc-to-dc converter or dc-to-ac inverter,

operating from the central office battery or from a separate battery and charger. These power supplies shall meet the noise limit specified for chargers in paragraph (s)(3)(iii) of this section, except the capacitor "C" shall be eliminated and the resistive load "R" shall be determined by the nominal output voltage in volts divided by the full load current rating in amperes. This requirement does not preclude the use of commercial ac power to operate input/output devices.

(ii) Power converters required for the purpose of providing various operating voltages to printed circuit boards or similar equipment employing electronic components shall be provided in duplicate with each unit capable of immediately assuming the full operating load upon failure of a unit. An exception to the duplicate power converter requirement permits nonduplicated power converter(s) to be utilized where there is full compliance with the following criteria.

(A) The failure of any single nonduplicated power converter shall not reduce the grade of service of common control and service circuits to any individual line or trunk by more than 50 percent.

(B) The failure of any single nonduplicated power converter shall not reduce the traffic carrying capacity of any interoffice trunk group by more than 50 percent.

(C) In central office switching systems of 400 or more equipped lines, any single nonduplicated power converter failure shall not cause a complete loss of service to more than 100 equipped lines.

(D) In central office switching systems of less than 400 equipped lines, any single nonduplicated power converter failure shall not cause a complete loss of service to more than 25 percent of the total equipped lines.

(5) *Ringing generators.* Ringing generators supplied on an ancillary basis shall be selected from RUS Bulletin 1755I-100, List of Materials Acceptable for Use on Telephone Systems of RUS Borrowers. Regardless of whether the ringing is generated on an ancillary basis or is generated integrally to the switching system, the ringing equipment shall meet the requirements of this section.

(i) *Ringing equipment provisioning.* (A) Redundant ringing equipment shall be provided. There shall be automatic transfer to the redundant equipment within the period of one ringing cycle, in case of failure of the equipment in use (either regular or standby). Automatic transfer shall not take place under any other conditions. Manual transfer in each direction shall be provided.

(B) An exception to the redundant ringing equipment requirement permits nonredundant ringing equipment to be utilized where there is full compliance with the following service criteria.

(*1*) In a central office switching system of 400 or more equipped lines, a single nonduplicated ringing source failure shall not cause the complete loss of ringing capability to more than 100 lines.

(*2*) In a central office switching system of less than 400 equipped lines, a single nonredundant ringing source failure shall not cause the complete loss of ringing capability to more than 25 percent of the total equipped lines.

(ii) *Output voltage.* (A) The ringing generators shall have an output voltage which approximates a sine wave and, as a minimum, shall be suitable for ringing straight-line ringers. Although not a requirement for RUS listing, decimonic, synchromonic, or harmonic ringing may also be specified in appendix A of this section.

(B) The ringing generator shall obtain its energy from the nominal 48-volt office battery.

(C) The output of each generator shall have three or more voltage taps or a single tap with associated variable control. Taps or control shall be easily accessible as installed in the field. Software control of ringing generator outputs via I/O devices may be provided in lieu of taps. The taps, or equivalent, shall be designated L, M, and H. The variable control shall have a locking device to prevent accidental readjustment. The outputs at the terminals of the generators with a voltage input of 52.1 volts and rated full resistive load shall be as follows for the ringing frequencies provided:

Frequency Range (Hz)	Output Volts rms (Tolerance 3 Volts)		
	L	M	H
16⅔ through 20	90	105	120
21 through 30	95	110	120
31 through 42	100	115	130
43 through 54	110	125	140

(D) No voltages in excess of the values in column H of the table in paragraph (s)(5)(ii)(C) of this section shall be provided at the output taps. Additional intermediate and/or lower taps may be provided without restriction.

(iii) *Voltage regulation.* (A) The output voltage for resistive, capacitive power factor of 0.8, and inductive power factor of 0.5 loads from no load to full rated output with 52.1 volts input battery shall not vary more than ±3 percent from the output voltage measured at ½ rated output, 1.0 power factor with 52.1 volts dc input applied.

(B) The output voltage for resistive, capacitive power factor of 0.8, and inductive power factor of 0.5 from no load to full rated output with input battery variations between 48–56 volts dc shall not vary more than ±10 percent from the output voltage measured at ½ rated output and 1.0 power factor with 52.1 volts dc input applied.

(C) The output voltage for resistive, capacitive power factor of 0.8, and inductive power factor of 0.5 loads from no load to full rated output and with input battery variations between 44–56 volts dc shall not vary more than + 10/ −15 percent from the output voltage measured at ½ rated output and 1.0 power factor with 52.1 volts dc input applied.

(iv) *Cross ringing.* Unwanted voltage caused by harmonic distortion or intermodulation distortion shall not exceed 15 volts rms when measured within ±5 Hz of any other assigned ringing frequency under any condition of load or input battery specified by paragraph (s)(5)(iii) of this section.

(v) *Frequency stability.* At ambient temperature of 70 ° ±5 °F (21 ° ±0.3 °C), for any combination of capacitive power factor of 0.8, inductive power factor of 0.5, and resistive loads with variations in input battery ranging from 44 to 56 volts, the output frequency shall not vary more than ±1/3 Hz or ±1 percent, whichever is less stringent. At temperatures between 15 °F (4 °C) to 130 °F (54 °C), and for any combination of resistive load and variations in input battery ranging from 44–56 volts, the output frequency shall not vary more than ±1/3 Hz or ±1 percent, whichever is less stringent.

(vi) *Self-protection on overloads.* The ringing generator equipment shall be capable of withstanding a short circuit across any pair of output terminals for a period of 5 minutes without fuse operation or damage.

(6) *Interrupter equipment.* (i) The interrupter shall be an integral part of the switching system and shall be controlled by any call processor or equivalent.

(ii) The ringing cycle provided by the interrupter equipment shall not exceed 6 seconds in length. The ringing period shall be 2 seconds.

(7) *Power panels.* (i) Battery and charger control switches, dc voltmeters, dc ammeters, fuses and circuit breakers, supervisory and timer circuits shall be provided as required. Voltmeters shall be provided as specified by the owner.

(ii) Portable or panel mounted frequency meters shall be provided as specified by the owner unless the system is equipped to measure actual ringing generator voltage and frequency outputs internally. If the system is equipped to make such measurements and print the results, the bidder is not required to provide a frequency meter.

(iii) Power panels, cabinets and shelves, and associated wiring shall be designed initially to handle the exchange when it reaches its ultimate capacity as specified by the owner.

(iv) The power panel shall be of the "dead front" type.

(t) *Main distributing frames.* (1) The main distributing frame shall provide terminals for terminating all incoming cable pairs. Arresters shall be provided for all incoming cable pairs, or for a smaller number of pairs if specified, provided an acceptable means of temporarily grounding all terminated pairs which are not equipped with arresters is furnished.

(2) The current carrying capacity of each arrester and its associated mounting shall coordinate with a #22 gauge

copper conductor without causing a self-sustaining fire or permanently damaging other arrester positions. Where all cable pairs entering the central office are #24 gauge or finer, the arresters and mountings need only coordinate with #24 gauge cable conductors. Item 13 of appendix A of this section designates the gauge of the cable conductors serving the host office. Item 7 of appendix B of this section designates the gauge of the cable conductors serving the RST(s).

(3) Central office protectors shall be mounted and arranged so that outside cable pairs may be terminated on the left side of protectors (when facing the vertical side of the MDF) or on the back surface of the protectors. Means for easy identification of pairs shall be provided.

(4) Protectors shall have a "dead front" (either insulated or grounded) whereby live metal parts are not readily accessible.

(5) Protectors shall be provided with an accessible terminal of each incoming conductor which is suitable for the attachment of a temporary test lead. They shall also be constructed so that auxiliary test fixtures may be applied to open and test the subscriber's circuit in either direction. Terminals shall be tinned or plated and shall be suitable for wire wrapped, insulation displacement or connectorized connections.

(6) If specified in appendix A of this section, each protector group shall be furnished with a factory assembled tip cable for splicing to the entrance cable; the tip cable to be 20 feet (610 cm) in length unless otherwise specified. Factory assembled tip cable shall be #22 gauge and selected from RUS Bulletin 1755I-100, List of Materials Acceptable for Use on Telephone Systems of RUS Borrowers. Tip cable requirements are provided in RUS Bulletin 345–87, PE-87, RUS Specification for Terminating (TIP) Cable. Cables having other kinds of insulation and jackets which have equivalent resistance to fire and which produce less smoke and toxic fumes may be used if specifically approved by RUS.

(7) Protectors shall be mounted on vertical supports, with centers not exceeding 9 inches (22.9 cm). The space between protector units shall be adequate for terminating conductors.

(8) Cable supporting framework shall be provided between the cable entrance and the MDF when overhead cable entrance is specified in Item 14.3.3 of appendix A of this section.

(9) The main distributing frame shall be equipped with a copper ground bus bar having the conductivity of a #6 American Wire Gauge (AWG) copper conductor or a greater conductivity, or may consist of another metal if specifically approved, provided it has adequate cross-sectional area to provide conductivity equivalent to, or better than, bare copper. A guardrail or equivalent shall also be furnished.

(10) Other features not specified in paragraph (t) of this section may be required at the option of the owner, if checked in Item 13.4 of appendix A of this section.

(11) Main frame protector makes and types shall be selected only from RUS Bulletin 1755I-100, List of Materials Acceptable for Use on Telephone Systems of RUS Borrowers. Protectors shall be capable of easy removal.

(u) *Electrical protection*—(1) *Surge protection.* (i) Adequate electrical protection of central office switching equipment shall be included in the design of the system. The characteristics and application of protection devices shall be such that they enable the central office switching equipment to withstand, without damage or excessive protector maintenance, the dielectric stresses and currents that are produced in line-to-ground and tip-to-ring circuits through the equipment as a result of induced or conducted lightning or power system fault-related surges. All wire terminals connected to outside plant wire or cable pairs shall be protected from voltage and current surges.

(ii) Central office switching equipment shall pass laboratory tests, simulating the hostile electrical environment, before being placed in the field for the purpose of obtaining field experience. There are five basic types of laboratory tests which shall be applied to exposed terminals in an effort to determine if the equipment will survive. Figure 3 summarizes these tests and the minimum acceptable levels of protection for equipment to pass them.

FIGURE 3—SUMMARY OF ELECTRICAL REQUIREMENTS AND TESTS

Test	Application Criteria	Peak Voltage or Current	Surge Waveshape	No. of Applications & Max. Time Between	Comments
Current Surge	Low Impedance Paths Exposed to Surges	500A or Lesser Current (See Fig. 5)	10 × 1000 μs	5 each Polarity at 1 minute intervals	
60 Hz Current Carrying	High or Low Impedance paths Exposed to Surges	10A rms or Lesser Current (See Fig. 6)	11 Cycles of 60 Hz (0.183 Sec.)	3 each Polarity at 1 minute intervals	
AC Power Service Surge Voltage	AC Power Service Connection	2500V or + 3 σ clamping V of arrester employed at 10kV/μs	1.2 × 50 μs	5 each Polarity at 1 minute intervals	AC arrester, if used, must be removed. Communications line arresters, if used, remain in place.
Voltage Surge	High Impedance Paths Exposed to Surges	1000V or + 3 σ dc breakdown of arrester employed	10 × 000 μs	Same	All primary arresters, if used, must be removed.
Arrester Response Delay	Paths protected by arresters, such as gas tubes, with breakdown dependent on V. rate of rise.	+ 3 σ breakdown of arrester employed at 100V/μs of rise	100V/μs rise decay to 1/2 V. in tube's delay time	Same	Same

(iii) *Two categories of surge tests.* (A) Current surge tests simulate the stress to which a relatively low impedance path may be subjected before main frame protectors break down. Paths with a 100 Hz impedance of 50 ohms or less shall be subjected to current surges, employing a 10 × 1000 microseconds waveshape as defined in Figure 4. For the purpose of determining this impedance, arresters which are mounted within the equipment are to be considered zero impedance. The crest current shall not exceed 500A; however, depending on the impedance of the test specimen this value of current may be lower. The crest current through the sample, multiplied by the sample's 100 Hz impedance, shall not exceed 1000 volts (V). Where sample impedance is less than two ohms, crest current shall be limited to 500A as shown in Figure 5.

FIGURE 4—EXPLANATION OF SURGE WAVESHAPE

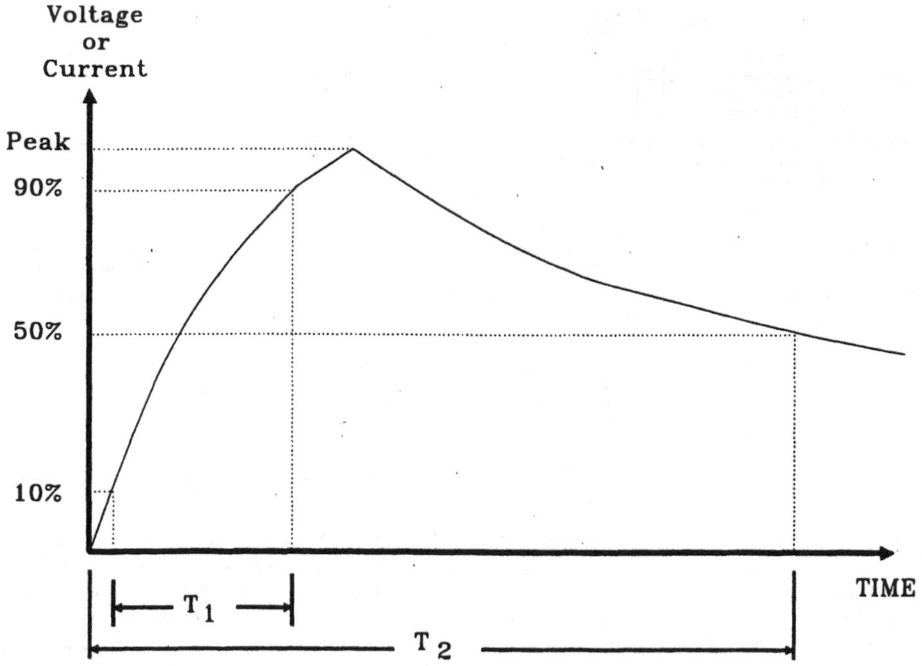

Surge Waveshape is defined as follows:

Rise Time × Time to Decay to Half Crest Value

(For example, 10 × 1000 μs)

Notes:

T_1 = Time to determine the rate of rise. The rate of rise is determined as the slope between 10% and 90% of peak voltage or current.

T_2 = Time to 50% of peak voltage (decay to half value).

FIGURE 5—EXPLANATION OF SURGE WAVESHAPE

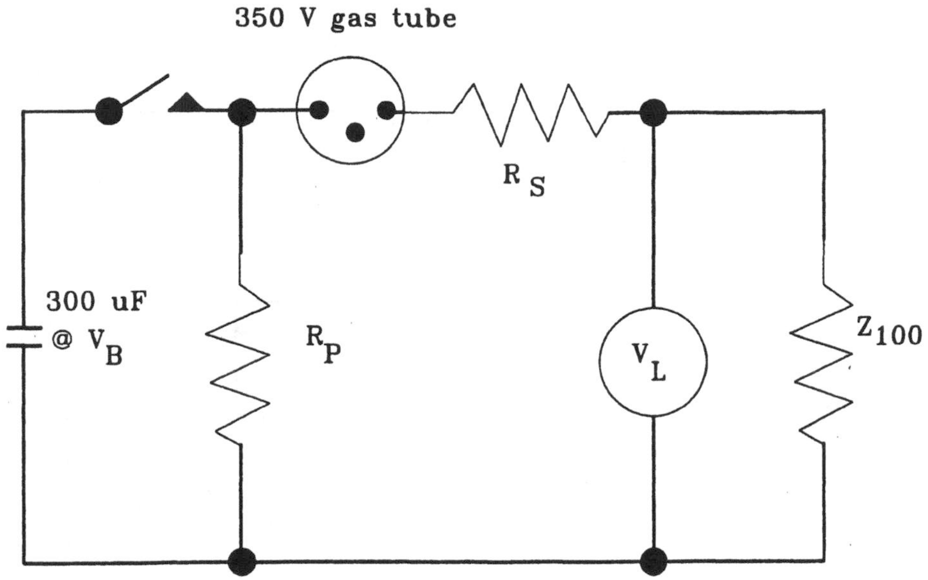

V_L = Not to exceed 1000V
V_B = Charging Voltage
Z_{100} = Test Specimen Impedance to be measured at 100 Hz.
R_P = Parallel Resistance (Waveshape)
R_S = Series Resistance (Current Limiting)

Z_{100}	R_S	R_P	V_B
0	5	∞	2500
1	4	∞	2500
2	3	∞	2500
3	2	∞	1670
4	1	∞	1250
5	0	∞	1000
7.5	0	15	1000
10	0	10	1000
15	0	7.5	1000
20	0	6.7	1000
25	0	6.25	1000
30	0	6	1000
40	0	5.7	1000
50	0	5.5	1000

(B) Sixty Hertz (60 Hz) current-carrying tests should be applied to simulate an ac power fault which is conducted to the unit over the cable pairs. The test should be limited to 10 amperes rms at 60 Hz for a period of 11 cycles (0.1835 seconds) and should be applied longitudinally from line to ground (see Figures 3 and 6 of this section).

FIGURE 6—60 HZ CURRENT SURGE TEST

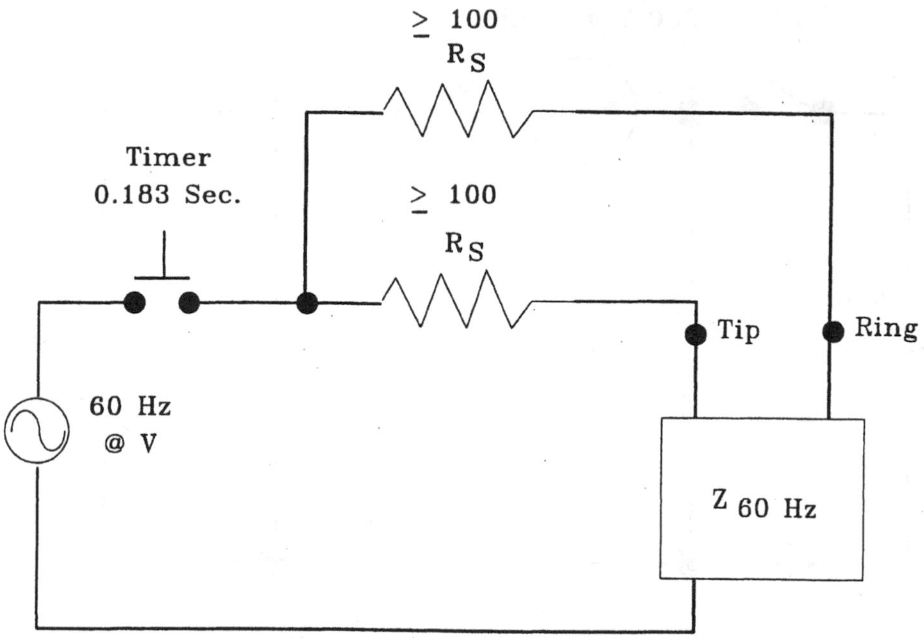

V—700 Volts RMS (Approximately 1000V Peak).

Z_{60}—Test specimen impedance to be measured at 60 Hz.

R_S—Series Resistance (current limiting) in each side of line. (Source impedance never less than 50 Ω longitudinal.)

Z_{60} Hz	R_S
0	140
10	120
20	100
50	100
Over 50	100

(C) AC power service surge voltage tests should be applied to the power input terminals of ac powered devices to simulate switching surges or lightning-induced transients on the ac power system. The test shall employ a 1.2 × 50 microseconds waveshape with a crest voltage of 2500V. Communications line protectors may be left in place for this test. Borrowers are urged to install commercially available surge protectors at the ac service entrance as part of their COE building program.

(D) Voltage surge tests simulate the voltage stress to which a relatively high impedance path may be subjected before primary protectors break down and protect the circuit. To assure coordination with the primary protection while reducing testing to the minimum, voltage surge tests should be conducted at a 1000 volts with primary arresters removed for devices protected by carbon blocks, or the + 3 sigma dc breakdown of other primary arresters. Surge waveshape should be 10 × 1000 microseconds.

(E) Arrester response delay tests are designed to stress the equipment in a manner similar to that caused by the delayed breakdown of gap type arresters when subjected to rapidly rising voltages. Arresters shall be removed for these tests, the peak surge voltage should be the + 3 sigma breakdown of the arrester in question on a voltage rising at 100V per microsecond and the time for the surge to decay to half voltage shall equal at least the delay time of the tube, as explained in Figure 7.

FIGURE 7—EXPLANATION OF ARRESTER RESPONSE DELAY TIME

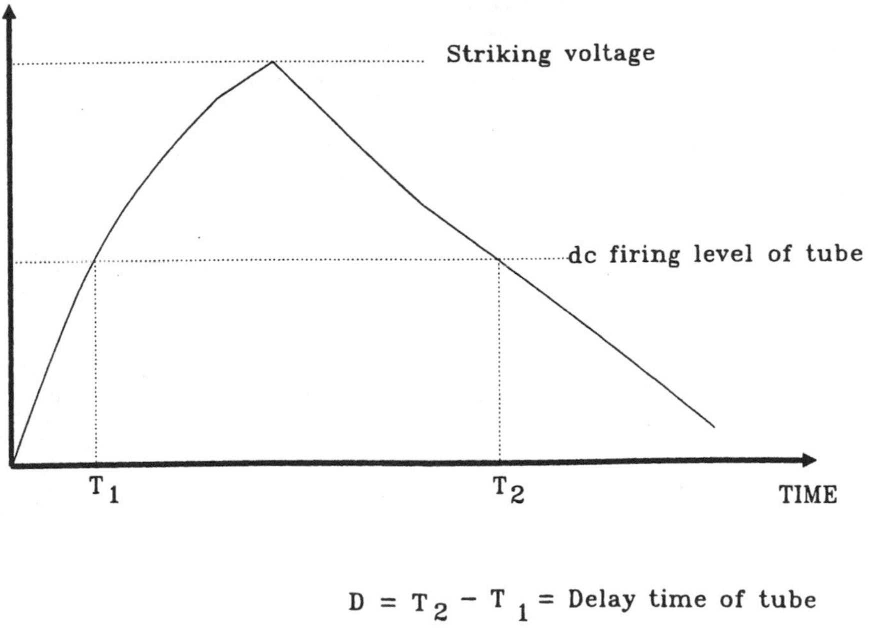

The delay time is that period of time when the potential across an arrester exceeds its dc firing level.

(iv) Five applications of each polarity for the surge tests and three for the 60 Hz Current Carrying Test are the minimum required. All tests should be conducted with not more than 1 minute between consecutive applications in each series of three or five to a specific configuration so that heating effects will be cumulative. As not all tests are required in every application, nonapplicable tests should be omitted. Tests should be conducted in the following sequence.

(A) Current Impulse Test.

(B) Sixty Hertz (60 Hz) Current Carrying Test.

(C) AC Power Service Impulse Voltage Test.

(D) Voltage Impulse Test.

(E) Arrester Response Delay Test.

(v) Tests should be applied between each of the following terminal combinations for all line operating conditions.

(A) Line tip to ring.

(B) Line ring to ground.

(C) Line tip to ground.

(D) Line tip to ring tied together to ground.

(2) *Extraordinary surge protection.* A central office or RST may be located in an area where ground conditions prevent the reasonable economic achievement of a low resistance to ground and/or there exists a greater than average probability of surge damage. Such an unusually hostile operating environment shall be recognized and taken into consideration by the bidder in the engineering and specification of the central office switching system and line protection. This subject of operating environment, ground conditions, etc., should be discussed at the time of technical presentation to assure the owner that adequate system protection will be provided by the bidder.

(3) *Dielectric strength.* Arresters shall be removed for all dielectric strength tests. The duration of all dielectric strength tests shall be at least 1 second. The applied potential shall equal or exceed the + 3 sigma dc breakdown voltage of the arrester, provided by the

COE manufacturer. Direct current potentials shall be applied between all line terminals and equipment chassis and between these terminals and grounded equipment housings in all instances where the circuitry is dc open circuit from the chassis, or connected to the chassis through a capacitor.

(4) *Insulation resistance.* Following the dielectric tests, the insulation resistance of the installed electrical circuits between wires and ground, with the normal equipment grounds removed, shall not be less than 10 megohms at 500 volts dc at approximately room temperature (68 °F (20 °C)) and at a relative humidity of approximately 50 percent. The measurement shall be made after the meter stabilizes, unless the requirement is met sooner. Arresters shall be removed for these tests.

(5) *Self-protection.* (i) All components shall be of the self-protecting type, capable of being continuously energized at rated voltage without injurious results.

(ii) The unit equipment shall not be permanently damaged by accidental short circuits of any duration across either the central office side tip and ring or the line side tip and ring. A test is to be made with the unit energized at the highest recommended voltages.

(6) *Static discharge.* Assemblies subject to damage by static discharge shall be identified and special handling instructions shall be supplied.

(v) *Miscellaneous*—(1) *Office wire.* All office wire shall be of soft annealed tinned copper wire meeting the requirements of ASTM Specification B 33–91, Standard Specification for Tinned Soft or Annealed Copper Wire for Electrical Purposes, and of suitable cross-section to provide safe current carrying capacity and mechanical strength. The insulation of installed wire, connected to its equipment and frames, shall be capable of withstanding the same insulation resistance and dielectric strength requirements as given in paragraphs (u)(3) and (u)(4) of this section at a temperature of 120 °F (49 °C) and a relative humidity of 90 percent.

(2) *Wire wrapped terminals.* These terminals are preferred and where used shall be of a material suitable for wire wrapping. The connections to them shall be made with a wire wrapping tool with the following minimum number of successive nonoverlapping turns of bare tinned copper wire in contact with each terminal.

(i) 6 Turns of 30 Gauge.

(ii) 6 Turns of 26 Gauge.

(iii) 6 Turns of 24 Gauge.

(iv) 5 Turns of 22 Gauge.

(3) *Protection against corrosion.* All metal parts of equipment frames, distributing frames, cable supporting framework, and other exposed metal parts shall be constructed of corrosion resistant materials or materials plated or painted to render them adequately corrosion resistant.

(4) *Screws and bolts.* Screw threads for all threaded securing devices shall be of American National Standard form in accordance with Federal Standard H28, Screw-Thread Standards for Federal Services, unless exceptions are granted to the manufacturer of the switching equipment. All bolts, nuts, screws, and washers shall be of nickel-copper alloy, steel, brass or bronze.

(5) *Temperature and humidity range.* The supplier shall furnish the operating temperature and humidity ranges of the equipment being provided in order that adequate heating and cooling may be supplied (see Items 5.2.1 and 5.2.2 of appendix C of this section).

(6) *Stenciling.* Equipment units and terminal jacks shall be adequately designated and numbered. They shall be stenciled so that identification of equipment units and leads for testing or traffic analysis can be made without unnecessary reference to prints or descriptive literature.

(7) *Equipment frame design.* For newly designed systems, consideration should be given to the desirability of providing frames which can be installed in rooms of normal ceiling height [up to 10 feet (305 cm)]. Where feasible, frames and equipment units shall be designed for ready portability and high salvage value.

(8) *Quantity of equipment bays.* Consistent with system arrangements and ease of maintenance, space shall be provided on the floor plan for an orderly layout of future equipment bays that will be required for anticipated traffic when the office reaches its ultimate size. Readily accessible terminals

shall be provided for connection to interbay and frame cables to future bays. All cables, interbay and intrabay (excluding power), if technically feasible, shall be terminated at both ends by use of connectors.

(w) *Remote switching terminal (RST)*—(1) *General.* The RST is a remotely located digital switching terminal which is placed at a subordinate wire center for subscriber lines and is a part of the host central office from a switching standpoint, and has hardware interchangeable with the host office, except for items that are applicable only to RST control and associated peripheral equipment. This does not preclude the use of existing in-service remote units on a new or upgraded host central office of the latest series generic or release.

(2) *Span line.* The RST is to be connected to the host central office via a means compatible with T1 type span lines using a DS-1 interface. This connection will be for control supervision and subscriber communication. The RUS equipment specification for a span line is PE-60.

(3) *Switching.* (i) The RST may have its switching functions controlled either by the host central office stored program control processors or by local subordinate processors which communicate with the host office processors.

(ii) As long as the connecting span line is intact, the subscribers served by the RST shall have all features, traffic capacity, and services including busy verification, available to all other subscribers in the system.

(iii) The RST shall have available an emergency call processing option which permits calling among all subscribers and from subscribers to emergency numbers within the RST if control link connections to the host central office are severed or otherwise disabled. The RST shall be capable of rerouting normally used emergency numbers, such as 911, to predetermined line terminations in this emergency standalone operating condition. This RST emergency call processing option shall be provided only when specified by the owner in Item 6.1 of appendix B of this section.

(4) *Subscriber line test.* (i) Means shall be available on an optional basis to the maintenance personnel to make subscriber line tests from a common location for all subscriber lines including the RST.

(ii) If tests in paragraph (w)(4)(i) of this section are not requested by the owner for a particular installation, a subscriber loop test set (see paragraph (o)(2)(iii)(A) of this section) shall be supplied at the RST with a means to access all lines.

(5) *Housing.* When housed in a building supplied by the owner, a complete floor plan including ceiling height, power outlets, cable entrances, equipment entry and travel, type of construction, and other pertinent dimensions shall be supplied with this section.

(6) *Power*—(i) *Chargers.* A single charger meeting the requirements of paragraph (s)(3) of this section (with the exception of paragraph (s)(3)(viii) of this section) is required. An additional charger capable of carrying the full load or a combination of three chargers each capable of carrying half the load shall be supplied if redundant chargers are specified in appendix B of this section.

(ii) *Ringing equipment provisioning.* (A) Ringing sources shall be supplied in duplicate.

(B) An exception to the duplicated ringing source requirement permits nonduplicated ringing source(s) to be utilized where there is full compliance with the following service criteria.

(*1*) In a remote switching terminal (RST) of 400 or more equipped lines, a single nonduplicated ringing source failure shall not cause the complete loss of ringing capability to more than 100 lines.

(*2*) In a remote switching terminal (RST) of less than 400 equipped lines, a single nonredundant ringing source failure shall not cause the complete loss of ringing capability to more than 25 percent of the total equipped lines.

(iii) *Power converter.* (A) Power converters required for the purpose of providing various operating voltages to printed circuit boards or similar equipment employing electronic components shall be provided in duplicate with each unit capable of immediately assuming the full operating load upon failure of a unit.

(B) An exception to the duplicate power converter requirement permits nonduplicated power converter(s) to be utilized where there is full compliance with the following criteria.

(*1*) The failure of any single nonduplicated power converter shall not reduce the grade of service of common control and service circuits to any individual line or trunk by more than 50 percent.

(*2*) The failure of any single nonduplicated power converter shall not reduce the traffic carrying capacity of any trunk group or service links to a host office by more than 50 percent.

(*3*) In a remote switching terminal (RST) of 400 or more equipped lines, any single nonduplicated power converter failure shall not cause a complete loss of service to more than 100 equipped lines.

(*4*) In a remote switching terminal (RST) of less than 400 equipped lines, any single nonduplicated power converter failure shall not cause a complete loss of service to more than 25 percent of the total equipped lines.

(7) *Alarm.* Sufficient system alarm points shall be provided from the RST to report conditions to the host alarm system.

(x) *Responsibilities of the bidder*—(1) *Central office layout.* (i) The successful bidder shall furnish tentative floor plan layout drawings showing the arrangement of the equipment and the dimensions of major equipment units. These drawings shall include minimum door dimensions and ceiling heights required for installation, maintenance and ventilation. If requested by the owner, the floor plan shall be such that the battery, charger, power board, main distributing frame and wire chief's test equipment are isolated from the other equipment by a partition.

(ii) The layout drawings shall also show provision for the ultimate capacity of the central office as specified by the owner.

(iii) After approval by the owner of the tentative floor plan, and within 10 calendar days after approval of the contract by the Administrator, the owner shall furnish the bidder the necessary data on the actual floor plan. Within 20 calendar days after receiving the necessary building data, the bidder shall then supply floor plan drawings showing exact locations of all equipment, both initial and ultimate, including points where connection to commercial power are required, with voltage and wattage indicated at each point. Within 20 calendar days after receiving the floor plan drawings from the bidder, the owner shall approve these drawings or take the necessary steps to have the drawings changed to meet his approval. The layout planning must be so coordinated between the owner and the bidder as not to delay the scheduled equipment installation date.

(2) *Shipment of main distributing frame (MDF).* The bidder shall ship the MDF equipment, with all necessary instructions to permit its installation by the owner, at the time requested by the owner in writing, provided such time is not earlier than 90 days prior to the date specified for the shipment of the rest of the central office equipment. If the owner or the owner's agent installs the main distributing frame, the owner shall assume the responsibility and the expense of proper installation according to information furnished by the bidder.

(3) *Drawings and printed material.* (i) The bidder shall supply instructional material for each exchange involved at the time of delivery of the equipment. It is not the intent of this section to require system documentation necessary for the repair of individual circuit boards. The bidder shall supply three complete sets of legible drawings, each set to include all of the following drawings and documentation:

(A) A floor plan showing exact dimensions and location of each equipment frame or item to a convenient scale;

(B) A block schematic drawing showing the various equipment components in the system, and their identifying circuit number (e.g., MDF, line circuits, memory, trunks, etc.);

(C) Drawings of major equipment items such as frames, with the location of major component items of equipment shown;

(D) Individual functional drawings for electrical circuits in the system;

(E) A detailed description of the operation of each circuit down to a circuit package level;

(F) Wiring diagrams indicating the specific method of wiring used on each item of equipment and interconnection wiring between items of equipment;

(G) Sufficient software documentation to maintain and service the system, including drawings showing principal aspects of the software architecture;

(H) Individual maintenance drawings covering each equipment item that contains replaceable parts, appropriately identifying each part by name and part number, or, complete ordering instructions for all replaceable parts if individual item drawings are not provided; and

(I) Job drawings including all drawings that are individual to the particular office involved, such as main frame, power panel, test board, etc.

(ii) The following information shall also be furnished:

(A) Complete index of the required drawings;

(B) Explanation of electrical principles of operation of the overall switching system;

(C) List of tests which can be performed with each piece of test equipment furnished, and explanation of the method of performing each test;

(D) Sample of each form recommended for use in keeping records of tests;

(E) Criteria for analyzing results of tests and determining appropriate corrective action;

(F) General notes on the methods of isolating equipment faults to specific printed circuit cards in the equipment;

(G) List of typical troubles which might be encountered, together with general indications as to the probable location of each trouble;

(H) Special office grounding requirements;

(I) A site specific central office ground system acceptance checklist that is consistent with industry practice; and

(J) A site specific layout of the master ground bar (MGB) showing assignment of P, A, N, and I equipment areas.

(4) *Distributing frame wire.* The bidder shall provide sufficient tinned copper conductor distributing frame wire for the initial installation. The insulation of this wire shall be such that it will not support combustion. The insulation shall have good abrasion resistance and cut-through properties, exhibit good solder heat resistance, and be suitable for wire wrap connections or insulation displacement connections.

(5) *Technical assistance service.* A technical assistance service shall be made available to assist the owner and its maintenance personnel on a 24-hour, 7 days a week basis. There is to be assistance available for both hardware and software problems. The necessary interface devices shall be supplied by the bidder.

(6) *Spare parts.* (i) The spare parts bid price shall be based upon the required quantities of spares shown in Item 6.2 of appendix C of this section, and shall be added to the base bid for comparison purposes in awarding the bid. It is the supplier's responsibility to provide all spares required by this section. If the supplier neglects to list certain spare parts in Item 6.2 of appendix C of this section, but they are necessary to comply with this section, they shall be provided by the supplier at no additional cost to the owner.

(A) "Units" are defined as user replaceable components used in the central office equipment. "Spare Parts" are direct replacements for units. Spare parts are necessary for the maintenance and diagnostic operations where the suspected faulty unit may be removed and a spare part substituted in anticipation that the trouble will be cleared.

(B) Examples of units for which spare parts should be furnished are printed circuit cards; circuit pack assemblies; fuses; and power supplies.

(C) Spare parts are not required as part of this addendum for items such as connectorized cables, nuts, bolts, and similar hardware; nor for items which can be obtained from sources other than the bidder such as battery cells, chargers, powerboards, magnetic tape transport assemblies, disk drives, ringing machines, recorded announcement machines, loop extenders and voice frequency repeaters, fire bars, teletypewriters, and video monitors.

(D) When 100 or more like units are used in the hosts and RST's to be bid, the quantity of spares to be furnished is determined by multiplying the total number of like units in the contract by .05 or .03, as applicable, and rounding off to the next lowest integer. For example, 119 Class 1 units require five spares; 120 require six.

(E) When alternates are required, the price of the spare parts for the alternates shall be included with the price of the alternate.

(F) For equipment in which the line cards consist of a number of plug-in "daughter" boards on a "mother" board, the line card is defined as the "daughter" board unit. In a similar manner for those designs which have line cards backed up by a "control card," the "control card" is not, by definition, a line card.

(G) The quantities of spare parts determined in paragraph (x)(6)(vi) of this section are a minimum quantity. The bidder may add quantities of spare parts to bring the number of spare parts up to the bidder's list of spare parts necessary for proper operation in the field.

(ii) A Class 1 unit does not have automatic transfer to a redundant or standby pool of identical units, and provides any function for 24 or more lines or trunks or for all trunks in a group. Nonredundant digital trunk interfaces are included in this category.

(iii) A Class 2 unit has automatic transfer to a redundant or standby pool of identical units, and provides any function for 24 or more lines or trunks or for all trunks in a group. Redundant digital trunk interfaces and units of a redundant stored program processor are included in this category.

(iv) A Class 3 unit does not have automatic transfer to a redundant or standby pool of identical units and provides any function for no more than 23 lines or trunks or for less than all trunks in a group. Nonredundant analog trunks are included in this category. Excluded from this category are line cards, which are in Class 4.

(v) A Class 4 unit has automatic transfer to a redundant or standby pool of identical units and provides any function for no more than 23 lines or trunks or for less than all trunks in a group. Also, any line cards are in Class 4.

(vi) The spare parts for all of the hosts and the RST's included in this contract shall be provided as follows:

Quantity of Units used in the CO's & RST's To Be Bid	Required Quantity of Spares By Class of Unit			
Class —>	1	2	3	4
1 through 9	1	1	0	0
10 through 24	2	2	1	0
25 through 49	3	2	2	0
50 through 99	4	3	2	0
100 or More	5%	3%	3%	0

(vii) As a part of the response to the bid, the supplier shall furnish a list of units used by class and a list of spare parts to be furnished with this contract. This list shall be placed in Item 6.2 of appendix C of this section for only one of the host specifications included in the entire contract.

(7) *Environmental requirements.* The bidder shall specify the environmental conditions necessary for safe storage and satisfactory operation of the equipment being bid. If requested, the bidder shall assist the owner in planning how to provide the necessary environment for the equipment.

(8) *Unit costs for cost separation purposes.* The successful bidder shall present a cost breakdown of the central office equipment on a discrete element basis 90 days after installation completion. This shall include the various frames, switching and transmission components, and software.

(9) *Single-point grounding system acceptance.* Qualified representatives of the central office system supplier and the owner are to conduct a thorough joint acceptance audit of the grounding system prior to the central office being placed into service. A grounding system acceptance checklist provided by RUS, which is consistent with standard industry practice, will be used in conducting this audit. All required grounding system corrections are to be made prior to placing the central office system into full service operation. The successful completion of this grounding system audit will constitute an acceptance on the part of both parties, the owner and the central office supplier (refer to paragraph (y)(5) of this

section, and appendix D of this section).

(y) *Installation.* The following responsibilities apply to the central office equipment installation and Remote Switching Terminal (RST) installations, unless otherwise noted.

(1) *Responsibilities of owner.* The owner shall:

(i) Allow the bidder and its employees free access to the premises and facilities at all hours during the progress of the installation;

(ii) Take such action as necessary to ensure that the premises are dry and free from dust and in such condition as not to be hazardous to the installation personnel or the material to be installed (not required for an RST installed in a self-contained environmentally controlled cabinet);

(iii) Provide heat or air conditioning when required and general illumination in rooms in which work is to be performed or materials stored (not required for an RST installed in a self-contained environmentally controlled cabinet);

(iv) Provide suitable openings in buildings to allow material to be placed in position (not required for an RST installed in a self-contained environmentally controlled cabinet);

(v) Provide the necessary conduit and commercial and dc-ac inverter output power to the locations shown on the approved floor plan drawings; provide 120 volts, 60 Hz commercial power equipped with a secondary arrester and a reasonable number of outlets for test, maintenance and installation equipment; provide suitable openings or channels and ducts for cables and conductors, from floor to floor and from room to room; provide an acceptable central office grounding system and at a ground resistance level that is reasonable for office site conditions (not required for an RST installed in a self-contained environmentally controlled cabinet);

(vi) Provide the necessary wiring, central office grade ground and commercial power service, with a secondary arrester, to the location of an exterior RST installation based on the voltage and load requirements furnished by the bidder;

(vii) Test at the owner's own expense all lines and trunks for continuity, leakage and loop resistance and ensure that all lines and trunks are suitable for operation with the central office equipment specified;

(viii) Make alterations and repairs to buildings necessary for proper installation of material, except to repair damage for which the bidder or its employees are responsible;

(ix) Connect outside cable pairs on the distributing frame and run all line and trunk jumpers (those connected to protectors);

(x) Furnish all trunk, line, and party assignment information to permit the bidder to program the data base memory within a reasonable time prior to final testing;

(xi) Release for the bidder's use such portions of the existing plant as are necessary for the proper completion of such tests as require coordination with existing facilities including facilities for T1 span lines with properly installed repeaters between the central office and the RST installations;

(xii) Make prompt inspections as it deems necessary when notified by the bidder that the equipment, or any part of the equipment, is ready for acceptance;

(xiii) Provide and install adequate fire protection apparatus, including one or more fire extinguishers or fire extinguishing systems of the gaseous type that has low toxicity and effect on equipment; and

(xiv) Provide necessary access ports for cable, if underfloor cable is selected.

(2) *Responsibilities of bidder.* The bidder shall:

(i) Allow the owner and its representatives access to all parts of the buildings at all times during the installation;

(ii) Obtain the owner's permission before cutting into or through any part of the building structure such as girders, beams, concrete or tile floors, partitions or ceilings (not applicable to the installation of lag screws, expansion bolts, and similar devices used for fastening equipment to floors, columns, walls and ceilings);

(iii) Be responsible for reporting to the owner any damage to the building which may exist or may occur during its occupancy of the building, repairing all damage to the building due to carelessness of the bidder's workforce, and exercising reasonable care to avoid any damage to the owner's property;

(iv) Consult with the owner before cutting into or through any part of the building structure where the fireproofing or moisture proofing may be impaired;

(v) Take necessary steps to ensure that all fire fighting apparatus is accessible at all times and all flammable materials are kept in suitable places outside the building;

(vi) Not use gasoline, benzene, alcohol, naphtha, carbon tetrachloride or turpentine for cleaning any part of the equipment;

(vii) Install the equipment in accordance with the specifications for the office;

(viii) Run all jumpers, except line and trunk jumpers (those connected to protectors);

(ix) Establish and update all data base memories with subscriber and trunk information as supplied by the owner until an agreed turnover time;

(x) Give the owner notice of completion of the installation at least 1 week prior to completion;

(xi) Permit the owner or its representative to conduct tests and inspections after installation has been completed in order that the owner may be assured that the requirements for installation are met;

(xii) Allow access, before turnover, by the owner or its representative, upon request, to the test equipment which is to be turned over as a part of the office equipment, to permit the checking of the circuit features which are being tested and to permit the checking of the amount of connected equipment to which the test circuits have access;

(xiii) Make final charger adjustments using the manufacturer's recommended procedure;

(xiv) Notify the owner promptly of the completion of work of the central office, or such portions as are ready for inspection;

(xv) Correct promptly all defects for which the bidder is responsible;

(xvi) Provide the owner with one set of marked prints, or strapping prints, showing which of the various options and figures are in use on each switching system as specified in paragraph (x)(3)(i) of this section;

(xvii) Place the battery in service in compliance with the recommendations of the battery manufacturer; and

(xviii) Furnish the owner with a record of the cell voltages and specific gravity readings made at the completion of the installation of the switching system and before it is placed in commercial service.

(3) *Installation requirements.* (i) All work shall be done in a neat, workmanlike manner. Equipment frames or cabinets shall be correctly located, carefully aligned, anchored and firmly braced. Cables shall be carefully laid with sufficient radius of curvature and protected at corners and bends to ensure against damage from handling or vibration. Exterior cabinet installations for RST's shall be made in a permanent, eye-pleasing manner.

(ii) All multiple and associated wiring shall be continuous, free from crosses, reverses and grounds and shall be correctly wired at all points.

(iii) An inspection shall be made by the owner or its representatives prior to performing operational and performance tests on the equipment. However, this inspection shall be made after all installing operations which might disturb apparatus adjustments have been completed. The inspection shall be of such character and extent as to disclose with reasonable certainty any unsatisfactory condition of apparatus or equipment. During these inspections, or inspections for apparatus adjustments, or soldering, or in testing of equipment, a sufficiently detailed examination shall be made throughout the portion of the equipment within which such condition is observed, or is likely to occur, to disclose the full extent of its existence, where any of the following conditions are observed:

(A) Apparatus or equipment units failing to compare in quantity and code with that specified for the installation;

(B) Apparatus or equipment units damaged or incomplete;

(C) Apparatus or equipment affected by rust, corrosion or marred finish; or

(D) Other adverse conditions resulting from failure to meet generally accepted standards of good workmanship.

(4) *Operational test requirements.* (i) Operational tests shall be performed on all circuits and circuit components to ensure their proper functioning in accordance with appropriate applicable documents supplied by the bidder.

(ii) A sufficient quantity of overall tests shall be made to ensure proper operation of all specified features.

(iii) A sufficient quantity of locally originating and incoming calls shall be made to prove the switching system can accept and process calls to completion.

(5) *Grounding system audit.* (i) A grounding system audit shall be performed to ensure that a viable single-point grounding system is in place prior to the time the switching system is placed into full service operation. It is suggested that such an audit be conducted at the time the switching system is ready for turnover to the owner.

(ii) This single-point grounding system audit is to be conducted by authorized representatives of the supplier and owner, and with the RUS general field representative participating at his discretion.

(iii) The single-point grounding system audit is to be conducted using the checklist contained in appendix D of this section.

(iv) Appendix D of this section shall be the principal single-point grounding system audit guideline document. A supplemental checklist may be prepared and provided by the switching system supplier which recognizes unique grounding requirements related to their particular switching system. The scope of this supplier checklist is to be confined to unique and specific switching system requirements only. Acceptable supplier supplemental grounding checklist must have prior approval of and be on file with the Central Office Equipment Branch of the Telecommunications Standards Division of RUS.

(v) It is the responsibility of the central office supplier to ensure that the grounding system evaluation criteria contained in the combination of the appendix D checklist of this section and their optional supplemental checklist adequately fulfill requirements for warranty coverage.

(vi) All deficiencies in the single-point grounding system are to be corrected prior to the switching system being placed into full service operation. Exceptions are permitted only by mutual agreement of the owner and supplier and with written approval of the RUS general field representative.

(vii) The acceptance statement facesheet of the audit checklist in appendix D of this section shall be signed by authorized representatives of the supplier and owner to indicate mutual approval of the single-point grounding system. Copies of all completed grounding system audit documents are to be provided to the supplier, owner and appropriate RUS telephone program regional offices.

(The information and recordkeeping requirements of this section have been approved by the Office of Management and Budget under the control number 0572–0059)

APPENDIX A TO § 1755.522—SPECIFICATION FOR DIGITAL, STORED PROGRAM CONTROLLED CENTRAL OFFICE EQUIPMENT DETAILED REQUIREMENTS (HOST)

(INFORMATION TO BE SUPPLIED BY OWNER)

Telephone Company Name ____________

____

Location

____

Central Office Name (By Location)

Town ____________

County ____________

State ____________

______ Attended

______ Unattended

______ Remotes

1. General

1.1 Notwithstanding the bidder's equipment lists, the equipment and materials furnished by the bidder must meet the requirements of paragraphs (a) through (x), Appendix A and Appendix B of § 1755.522.

1.2 Paragraphs (a) through (x) of § 1755.522 cover the minimum general requirements for digital, stored program controlled central office switching equipment.

1.3 Paragraph (y) of § 1755.522 covers requirements for installation, inspection, and testing when such service is included as part of the contract.

1.4 Appendices A and B of § 1755.522 cover the technical data for application engineering and detailed equipment requirements insofar as they can be established by the owner. These appendices are to be filled in by the owner.

1.5 Appendix C of § 1755.522 covers detailed information on the switching network equipment and the common control equipment, and information as to system reliability and heavy traffic delays as proposed by the bidder. This appendix is to be filled in by the bidder and must be presented with the bid.

1.6 Appendix D of § 1755.522 is the single-point grounding system audit checklist.

2. Numbering Scheme

2.1 This office shall be arranged to serve the following area and office code(s):

If more than one code is to be served, discrimination shall be determined by the following:

Number Translation ____

Separate Trunk Groups ____

Both (Explain in Item 16, Appendix A) ____

2.2 This office shall be arranged to provide EAS service to the following:

Connecting office	Code	Connecting office	Code

2.2.1 Seven digits shall be dialed for all local and EAS calls.

2.3 Additional dialing procedures to be provided include the following:

Feature	Required
Station Paid Toll (Including Coin):	
Home Numbering Plan Area (HNPA):	
"1" + 7 Digits	
"1" + 10 Digits	
Other (Explain in Item 16, Appendix A)	
Foreign Numbering Plan Area (FNPA):	
"1" + 10 Digits	
Other (Explain in Item 16, Appendix A)	

10XXX Dialing to Interexchange Carriers:

Name	Access code

Feature	Required
Person, Special (Including Coin):	
HNPA—"0" + 7 Digits.	
"0" + 10 Digits.	
FNPA "0" + 10 Digits.	
Other (Explain in Item 16, Appendix A).	
Directory Assistance:	
HNPA Local—411.	
"1" + 411.	
HNPA Toll "1" + 555–1212.	
FNPA Toll "1" + NPA + 555–1212.	
IDDD:	
Operator Serviced 01.	
Station-Station 011.	

Other service codes	No. to be dialed
Wire Chief.	
Repair Service.	
Business Office.	
Emergency Calls to 911 Lines.	
Emergency Calls to 911 Trunks.	
Time.	
Weather.	
100 Test Line.	
102 Test Line.	
105 Test Line.	
Other (Explain in Item 16, Appendix A).	

2.4 Assistance calls are answered: (Check appropriate items)

2.4.1 At the operator office in ________

2.4.1.1 By means of the regular interoffice toll trunks ________

2.4.1.2 By means of the regular interoffice EAS trunks ________

2.4.1.3 By means of a separate special service trunk group ________

2.4.1.4 Locally ________

Explain:

3. Office Clock

3.1 This office is to be slave clock synchronized with another office:

_______ Yes _______ No

(Explain details in Appendix A, Item 16 if "Yes".)

3.2 This office is to be a master clock office to provide synchronization timing for other offices:

_______ Yes _______ No

(Explain details in Appendix A, Item 16 if "Yes".)

4. Interoffice Trunking Diagram

4.1 A sketch showing relative location of exchanges, RST's, and number of circuits shall be included, also the office and area codes of the direct trunk points. The diagram should indicate whether toll or EAS trunk groups are "High Usage" or "Final." Alternate routes should be included. Indicate whether the trunk termination is direct digital or analog.

5. Translator Function Chart

Called point	Subscriber dials	First route			Alternate routes		
		Translator action	Send		Translator action	Send	
		Deletes	Prefixes		Deletes	Prefixes	

6. Line Circuit Requirements (Includes all lines associated with RST's.)

6.1 *Types of Lines*

	No. of lines		No. of EAS areas	Total No. of lines required
	Local service only	both local and EAS service		
6.1.1 Individual—Flat Rate	______	______	______	______
6.1.2 Individual—Message Rate	______	______	______	______
6.1.3 Pay Station	______	______	______	______
6.1.4 Telephone Company Official Lines	______	______	______	______
6.1.5 Wire Chief	______	______	______	______
6.1.6 911 Emergency Service Bureau Lines.	______	______	______	______
6.1.7 Number Hunting PBX Groups:	______	______	______	______

No. of lines in group	No. of groups	Direct in dial*	Re-stricted service at COE	Type		No. of lines		No. of EAS areas	Total No. of lines required
				Ground start	Loop start	Local service only	Both local and EAS service		
____	____	____	____	____	____	____	____	____	____
____	____	____	____	____	____	____	____	____	____
____	____	____	____	____	____	____	____	____	____
____	____	____	____	____	____	____	____	____	____
____	____	____	____	____	____	____	____	____	____

* Furnish translation information under Item 5.

6.1.8 WATS Lines (Give details in Appendix A, Item 16)

Number of Inward WATS Lines _______

Number of Outward WATS Lines _______

6.1.9 Special Lines Required _______ (Explain in Item 16, Appendix A)

6.1.10 Total Number of Lines Required

Host _____ (Incl. DDI Concentrator Lines)

RST 1 _____

RST 2 _____

RST 3 _____

Total _____

6.1.11 Total Director Numbers Required _____ (Including RST's) (see Item 7.1, Appendix A)

6.1.12 Pay Station

Type __________

New ____ Reused ____

(Describe in Item 16, Appendix A)

6.1.13 Line Concentrator

6.1.13.1 Supplied by Owner (see Item 16, Appendix A, for details)

___ Yes ___ No

6.1.13.2 Supplied by Bidder (If "Yes", attach REA Form 397g, Performance Specification for Line Concentrators)

___ Yes ___ No

6.2. *Data on Lines Required Range Extension*

6.2.1 Number of non-pay station lines having a loop resistance, including the telephone set, as follows:

	No. of lines
1901–3200 ohms	_____
3201–3600 ohms	_____

6.2.2 Number of pay station lines having loop resistance, excluding the telephone set, greater than:

	No. of lines
1200 ohms (For Prepay)	_____
1000 ohms (For Semi-Postpay Operation).	_____

6.2.3 Range extension equipment is to be provided:

6.2.3.1 Loop Extenders: Total Quantity _____

By Bidder—Quantity _____

By Owner—Quantity _____

(Explain in Item 16, Appendix A)

6.2.3.2 VF Repeaters: Total Quantity _____

By Bidder—Quantity _____

(Bidder must have information on loading and cable size.)

By Owner—Quantity _____

(Explain in Item 16, Appendix A)

6.2.3.3 Range extension may be furnished as an extended range line circuit at the option of the supplier. If this option is used, the quantities of loop extenders and VF repeaters will be different from the quantities listed above (see Item 6.1,a, Appendix C).

___ Yes ___ No

7. Traffic Data-Line Originating and Terminating Traffic

7.1 Originating Line Traffic—Estimated per Busy Hour (Includes all Lines Associated With RST's):

	(a) CCS per Main Station	(b) No. of Main Stations	(axb) Total CCS	No. of Lines Required [1]
Ind.—Res	_____	_____	_____	_____
Ind.—Bus	_____	_____	_____	_____
Special Lines	_____	_____	_____	_____
Pay Station	_____	_____	_____	_____
Telco Official	_____	_____	_____	_____
Wire Chief	_____	_____	_____	_____
No. Htg. or PBX	_____ [2]	_____ [3]	_____	_____
WATS	_____	_____	_____	_____
Data Service	_____	_____	_____	_____
911 Emerg. Service	_____	_____	_____	_____
Total		_____	_____	_____ [4]
		(c)	(d)	(e)

[1] See Appendix A, Item 6.1.

[2] This figure is the CCS per PBX trunk.

[3] This figure is the number of PBX trunks.

[4] This is the total number of line equipments required. The number to be provided will be determined by the equipment design of the system of the selected bidder. See Appendix C, Item 3.1.1.2.

7.2 Average Originating CCS per Line per Busy Hour

(d) / (e) = ___ / ___ = ___ CCS/Line

This office shall be engineered to handle an initial average originating busy hour traffic of ___ CCS per line. It is anticipated that the average originating busy hour traffic will increase to ___ CCS per line.

Originating Traffic Attributed to Host Only ___ CCS/Line

7.3 Terminating Traffic—Estimated CCS per Busy Hour

It is assumed that the total CCS for terminating traffic is the same as for originating traffic. Since digital switch networks are on a terminal per line basis, the terminating CCS per line will be the same as the originating CCS per line as shown in Item 7.2, Appendix A.

Terminating Traffic Attributed to Host Only ___ CCS/Line

7.4 Percent of Pushbutton Lines ___

7.5 Anticipated Ultimate Capacity (20 years)

7.5.1 Subscriber Lines

Host _______ (Incl. DDI Concentrator Lines)

RST 1 _______

RST 2 _______

RST 3 _______

Total _______

8. Trunk Circuit Requirements

8.1 *Interoffice Trunking*

8.1.1 Trunking Requirements

1. Connecting Office	....	....	....	....
2. Use of Trunk	....	....	....	....
3. Trk. Grp. Ntwk. Connection [1]	....	....	....	....
4. Quantity Equipped	....	....	....	....
5. Ultimate % Growth	....	....	....	....
6. CCS Capacity	....	....	....	....
7. Direction	....	....	....	....
8. No. Digits Dialed	....	....	....	....
9. No. Digits Outpulsed	....	....	....	....
10. No. Digits Inpulsed	....	....	....	....
11. Type Signaling	....	....	....	....
12. Type Pulsing	....	....	....	....
13. Carrier Type (2-Wire)	....	....	....	....
14. Carrier Type (4-Wire)	....	....	....	....
15. Physical	....	....	....	....
16. Repeat Coils [2]	....	....	....	....
17. DX Signaling Set	....	....	....	....
18. Other Type Signaling	....	....	....	....
19. Delay Dial	....	....	....	....
20. Direct Digital Interface	....	....	....	....
21. a. Feature Group B	....	....	....	....
b. Feature Group C	....	....	....	....
c. Feature Group D	....	....	....	....

[1] Designation of trunk group network connection involves the following categories:
IC—Direct Inter-LATA Connecting Trunk = (IC/POP)
TC—Tandem Connecting Trunks
IT—Intertandem Connecting Trunks
IL—Intra-LATA Connecting Trunks
TIC—Tandem Inter-LATA Connecting Trunks
Misc.—Intercept, Busy Verification, etc.

[2] Omit repeating coils for carrier derived trunks.

8.1.2 Pads for 4-Wire Carrier (7dB and 16dB)

Total Quantity _____
By Bidder Quantity _____
By Owner Quantity _____

Refer to the attached information regarding connecting company trunk circuit drawing numbers and name of manufacturer.

8.2 *Switched Traffic Data*

8.2.1 Originating Traffic

Type	CCS	H.T. secs.	BHC	No. of digits out-pulsed	Sender sig. mode	Remarks
Toll "0" – [1]	....	....	....	....	....	....
Toll "0" + 7 [1] [2]	....	....	....	....	....	....
Toll "0" + 10 [1] [2]	....	....	....	....	....	....
Toll S-S "1" + 7 [2]	....	....	....	....	....	....
Toll S-S "1" + 10 [2]	....	....	....	....	....	....
Toll Other	....	....	....	....	....	....
Special Service	....	....	....	....	....	....
Intercept	....	....	....	....	....	....
Intraoffice	....	....	....	XXXXXXX	XXXXXXX	....
EAS	....	....	....	....	....	....
EAS	....	....	....	....	....	....
EAS	....	....	....	....	....	....
Tandem	....	....	....	....	....	....
Tandem	....	....	....	....	....	....
Tandem	....	....	....	....	....	....
911 Emerg. Service	....	....	....	....	....	....
Total	....	....	....	....	....	....

[1] PPCS traffic assumed to be divided 20 percent "0" – and 80 percent "0" + if unknown.

[2] Toll calls assumed to be divided two-thirds 7 digits and one-third 10 digits.

Busy Hour Attempts = BHC Total × 1.4 = ____

8.2.2 Terminating Traffic

Type	CCS	H.T. secs.	BHC	No. of digits inpulsed	Receiver sig. mode	Remarks
Toll Compl.	....	....	....	....	....	....
Test & Ver.	....	....	....	....	....	....
Intraoffice	....	....	....	....	....	....
EAS	....	....	....	....	....	....
EAS	....	....	....	....	....	....
EAS	....	....	....	....	....	....
Tandem	....	....	....	....	....	....
Tandem	....	....	....	....	....	....
Tandem	....	....	....	....	....	....
Total	....	....	....	....	....	....

9. Checklist of Features Required

9.1 *Alternate Routing*
(Explain in Item 16, Appendix A)
9.2 *Data Service*
(Explain in Item 16, Appendix A)
9.3 This office shall be:
9.3.1 End Office Only
9.3.2 End Office and Intermediate Tandem
(Explain in Item 16, Appendix A)
9.3.3 End Office and Access Tandem
(Explain in Item 16, Appendix A)
9.4 Billing Data

	Trunk group	Send ANI feature group			Store billing data	
		B	C	D	AMA system	Pollable system
9.4.1 This office only	....	......	......	......	....	....
9.4.2 Trunks from Tributaries	....	......	......	......	....	....
9.4.3 Local Message Detail Recording:						

9.5 *Pollable Systems*

9.5.1 Polling device to be provided on this contract

____ Required
____ Not Required

(Provide details in Item 16, Appendix A)

9.5.2 Pollable system to be backed up by tape or disc standby

____ Required
____ Not Required

9.6 *AMA Format*

9.6.1 Bellcore Format

____ Required
____ Not Required

(Provide details in Item 16, Appendix A)

10. Miscellaneous Operating Features

10.1 *Busy Verification*
10.1.1 By dedicated trunk from toll operator: ____
10.1.1.1 One-Way, Inward ____
10.1.1.2 Two-Way (Busy verification inward, intercept outward) ____
10.1.2 By prefix digit over intertoll trunk ____

(Indicate digit(s) dialed) ____
10.1.3 Access by Switchman
10.1.3.1 Dedicated Trunk ____
10.1.3.2 Multiple of Operator Trunk ____

10.2 *Intercept Facilities*
10.2.1 Vacant code, disconnected number, and unassigned number intercept shall be: (Check One)
By recorded announcement:
Without cut-through to operator ____
With cut-through to operator ____
By operator ____
10.2.2 Changed number intercept shall be: (Check One)
By recorded announcement:
Without cut-through to operator ____
With cut-through to operator ____
By operator ____
By automatic intercept system (AIS) in distant office ____
10.2.3 Method of Reaching Operator, if required:
Separate trunk group ____
Regular interoffice toll trunks with idle trunk selecting over at least three trunks when three or more toll trunks are equipped ____
10.2.4 Number of separate intercept trunk circuits ____
10.3 *Line Load Control*
10.3.1 Line load control facilities are:
____ Required ____ Not Required
(Explain in Item 16, Appendix A)
10.4 *Service Observing Facilities*
10.4.1 Service observing facilities are:
____ Required ____ Not Required

(Explain in Item 16, Appendix A)
10.5 *Hotel-Motel Arrangements*
10.5.1 Hotel-motel arrangements for operation of message registers at the subscriber's premises are:
___ Required ___ Not Required
(Explain in Item 16, Appendix A)
10.5.1.1 How are message registers to be activated?

Line Reversal ___
Third Wire ___
Other ___
(Explain in Item 16, Appendix A)
10.6 *Nailed-Up Connections*
___ Required ___ Not Required
(Explain in Item 16, Appendix A)
10.7 *Vertical Services:* (RST Lines are Included)

	Initially	*Ultimate*
10.7.1 Call Waiting—No. of Lines	___	___
10.7.2 Call Forwarding—No. of Lines	___	___
___ Local ___ Remote (Explain in Item 16, Appendix A).		
10.7.3 Abbreviated Dialing No. of Lines	___	___
No. of Codes per Line ___ for ___ Lines		
No. of Codes per Line ___ for ___ Lines		
10.7.4 Three-Way Calling—No. of Lines	___	___
CCS Per Line	___	___

10.7.5 Other ___
(Explain in Item 16, Appendix A)

11. Maintenance Facility Requirements

11.1 *Alarm Signals*
11.1.1 Handled locally ___
Explain in Detail: ___

11.1.2 Transmitted to attended point
11.1.2.1 Via operator office trunks ___
11.1.2.2 Via printout or other display service ___
Explain in Detail: ___

11.1.2.3 Type of tone to operator
11.1.2.3.1 Distinctive tone (see (i)(2)(ix) of § 1755.522) ___
11.1.2.3.2 Other
Explain in Detail: ___

11.1.3 Alarm checking signals for carrier and mobile radio systems
11.1.3.1 Minor Alarm.
11.1.3.2 Major Alarm.
11.1.3.3 Terminals for both.
11.2 *Trouble Location and Test*
11.2.1 Outside plant and stations (check desired items)
11.2.1.1 Subscriber's loop test circuit:
11.2.1.1.1 As part of the maintenance center.
11.2.1.1.2 Separately.
11.2.1.2 Remote test set (Explain in Item 16, Appendix A).
11.2.1.3 Dial speed test circuit (Explain in Item 16, Appendix A).
11.2.1.4 Pushbutton dialing test circuit.
11.2.1.5 Howler (per (o)(2)(iii)(C) of § 1755.522).
11.2.1.6 Hand test sets, number required ___ (Explain in Item 16, Appendix A).

11.3 *Transmission Tests*
11.3.1 Furnish reference tone
Yes ___
No ___

Frequencies and order in which applied	Time interval for application of each frequency
___ Hz	___ Seconds
___ Hz	___ Seconds
___ Hz	___ Seconds
___ Hz	___ Seconds

11.3.2 Test Lines
11.3.2.1 Test Line 100 ___
11.3.2.2 Test Line 102 ___
11.3.2.3 Test Line 104 ___
11.3.2.4 Test Line 105 ___
(Explain in Item 16, Appendix A)
11.3.2.5 Test Line 107 ___
11.3.2.6 Remote Office Test Line ___
(Explain in Item 16, Appendix A)
11.4 *Line Testing*
11.4.1 Automatic line insulation testing
Yes ___
No ___
11.4.2 Owner supplied equipment
Yes ___
No ___
11.4.2.1 Vendor supplied interface only
Yes ___
No ___
If supplied by owner, explain in Item 16, Appendix A, including manufacturer, model, location.
11.5 *Remote Control*
11.5.1 Remote control of the system shall be provided.
Yes ___
No ___
If required, explain in Item 16, Appendix A, including number, type and location.

12. Power Equipment Requirements (Host Office Only)

12.1 *Central Office Battery*

12.1.1 A battery reserve of ___ busy hours shall be provided for this office when it reaches ___ lines at the ultimate anticipated traffic rates specified in Item 7.2, Appendix A.

12.1.1.1 The owner will furnish a standby generator, permanently installed in this office, with capacity sufficient to power air conditioning equipment required for cooling of the central office equipment and to maintain an adequate dc supply in the event of a failure of the commercial ac supply.

Yes ___
No ___

12.1.2 *Type of battery:* (Check One)
Lead Calcium ___
Lead Antimony ___

12.1.3 Voltmeter (portable 3–60–150 volt scale, 1% accuracy) shall be furnished.

Yes ___
No ___

12.1.4 Hydrometer in a hydrometer holder with glass or plastic drop cup shall be furnished.

Yes ___
No ___2112.1.5 Type of battery rack required: (Check One)

Two Tier ___
Other ___
Explain:

12.1.6 Special equipment power requirements (carrier, voice frequency repeaters, etc.). Drain in amperes ___

12.1.6.1 Supply all necessary equipment to provide the following 48-volt battery taps:

Number of circuits	Fuse (or circuit breaker) size
....	
....	
....	
....	

12.2 *Charging Equipment*

12.2.1 Charging equipment shall be provided capable of charging the office battery on a full float basis when the office reaches ___ lines at the ultimate anticipated traffic rates specified in Item 7.2, Appendix A.

12.2.2 Charger input rating shall be:

	3-Phase Connection:
Voltage ___	3–Wire ___
Phase ___	4–Wire ___
Frequency ___	Delta ___
....	Y ___

12.3 *Ringing Equipment*

12.3.1 Solid-state ringing equipment in accordance with paragraph (s)(5)(i) of § 1755.522 shall be provided for generating the frequencies specified by check marks in the following table. Ringing generator sets serving the entire office shall each be sized to carry the full office ringing load when the office size reaches ___ lines at the ultimate anticipated traffic rates specified in Item 7.2, Appendix A.

12.3.2 Ringing frequencies to be supplied:

	Frequency in Hz		Maximum No. of telephones
Single Frequency	20	......	....
Decimonic	20	......	....
	30	......	....
	40	......	....
	50	......	....
Harmonic	16⅔	......	....
	25	......	....
	33⅓	......	....
	50	......	....
Synchromonic	20	......	....
	30	......	....
	42	......	....
	54	......	....

12.3.3 Furnish frequency meter (accurate within 1.3 Hz) and voltmeter (5% accuracy) for ringing measurements (see paragraph (s)(7)(ii) of § 1755.522). Check One:

Panel Mounted ___
Portable ___
Not Required ___

12.4 *Power Board*

The power panel and associated wiring shall be of ample size to meet the load requirements when this office reaches ___ lines at the ultimate anticipated traffic rates specified in Item 7.2, Appendix A.

13. Distributing Frame Requirements (Host Office Only)

13.1 Total number of outside plant cable pairs to be terminated.

13.1.1 Gauge of outside plant cable pairs.

13.2 Number of outside plant cable pairs to be protected.

13.3 Number of additional protector pair units to be provided on MDF ....

Explain:

13.4 *Main Frame Details*

Is present MDF to be reused?
Yes ___
No ___

If "Yes," Type ___
Reused protectors are:
___ (Mfgr.)
___ (Type)

13.4.1 Number of pairs of arrester units (switching equipment) ___

13.4.2 Number of pairs of gas tube arrester units (special equipment) ___

13.4.2.1 Gas tubes to be:
___light,
___medium,
___heavy,
___max. duty units

13.4.2.2 Fail shorted/low breakdown failure mode required

Yes ____
No ____
13.4.2.3 Breakdown voltage of gas tube arresters ____
13.4.3 Number of terminated pairs to be grounded ____
13.4.4 Factory assembled tip cable
Yes ____
No ____
13.4.4.1 Tip cable length [if other than 20 feet (610 cm)]
____
13.4.4.2 Tip cable formed
Up ____
Down ____
13.4.5 Pairs per vertical ____
13.4.6 Height of vertical ____feet ____ inches

14. Building and Floor Plan Information (Host Office Only)

14.1 Equipment is to be installed in an existing building (Attach detailed plan.) ____
14.2 A new building is planned ____
14.2.1 Tentative plan (*Note to Engineer:* Show sketch without dimensions.)
14.3 *Detailed Arrangements*
14.3.1 Partition required (to isolate space containing battery, charger, power board, test panel, main distributing frame and subscriber's loop test circuit (wire chief's test desk) from that of the remaining equipment).
Yes ____
No ____
14.3.2 Vestibule required
Yes ____
No ____
14.3.3 Cable entrance
Overhead____
Underground ____
14.3.4 Additional floor space will be required for the following equipment which is being furnished by the owner or by the connecting company:

____
____
____

14.3.5 The office will be arranged for
Overhead Interbay Cabling ____
Underfloor (Computer Room Type) Interbay Cabling ____
14.3.6 Is earthquake bracing required?
Yes ____
No ____
(If "Yes," explain zone and criteria used for zone in Item 16, Appendix A.)
14.3.7 Office ground will be ____ ohms or less (Refer to Item 4.6.3 of RUS TE&CM 810.)
14.3.8 The office is considered to be in the following category for lightning damage probability based on the Figure 1 map of RUS TE&CM 823 (see paragraph (u)(2) of 1755.522).
____Very High
____Higher than Average
____Average
____Lower than Average
____Very Low
14.3.9 The following is additional information regarding operating environment conditions which should be considered in determining system protection requirements (tower in vicinity, high exposure, etc.):

15. Alternate Requests

16. Explanatory Notes (Include a detailed description of any equipment to be reused, or otherwise supplied by the owner, loop extenders, subscriber carrier, VF repeaters, etc.)

APPENDIX B TO 7 CFR 1755.522—DETAILED INFORMATION ON REMOTE SWITCHING TERMINALS (RST'S)

(Complete One Form For Each RST)

1. Number of Subscriber Lines (These lines included in totals in Item 6, Appendix A).

1.1 Single-Party: ______ Flat Rate ______ Message Rate.
1.2 Semi-Postpay Pay Station ______.
1.3 Prepay Pay Station ______.
1.4 PABX Lines ______ Loop Start______ Ground Start ______ Restricted at Office ______ Other ______ (Describe in Item 12, Appendix B)
1.5 Number of lines to be pushbutton ______
1.6 911 Emergency Lines ______
1.7 Anticipated ultimate capacity (20-Year) ______

2. Traffic

2.1 Originating traffic per line—CCS/BH: ______ Initial ______ Ultimate.
2.2 Terminating traffic per line—CCS/BH: ______ Initial ______ Ultimate
2.2.1 Terminating will be made equal to originating if it is not known to be different.

3. Subscriber Loop Resistance

3.1 Number of subscriber lines having loop resistance, including the telephone set of:

No. of Lines
1501–1900 Ohms ______
1901–3200 Ohms ______
3.2 Number of pay station lines having loop resistance, excluding the telephone set, greater than:

No. of Lines
1200 Ohms (For Prepay) ______
1000 Ohms (For Semi-Post Pay Operation) ______.

4. Range Extension

4.1 If no standby power is available at the site, loop extenders may be required on 1501 to 1900 ohms loops.

4.2 *Loop extenders:* Total Quantity ________By Bidder—Quantity _______By Owner—Quantity
(Explain in Item 12, Appendix B)
4.3 *VF repeaters:* Total Quantity ________ By Bidder—Quantity ________By Owner—Quantity ________.
(Explain in Item 12, Appendix B)

5. Power Supply

5.1 Power Board.
5.1.1 The power board and associated wiring shall be of ample size to meet the load requirements when this RST reaches _______lines at the ultimate anticipated traffic rates specified in Item 2, Appendix B.
5.2 *Charger input rating shall be:* Voltage ________Phase ________Frequency ________

3-Phase Connection:
3-Wire ________
4-Wire ________
Delta ________
Y________
5.2.1 Charger shall be capable of charging the RST battery on a full float basis when the RST reaches _______ lines at ultimate traffic rate specified in Item 2, Appendix B.
5.2.2 Charger shall be redundant _______.
5.3 Battery reserve shall be _______ busy hours when the RST reaches _______ lines at the ultimate anticipated traffic specified in Item 2, Appendix B.
5.4 Standby power is available. Yes _____ No _____.
5.5 Special equipment power requirements _______ amps.
5.6 *Ringing.*
5.6.1 Type of Ringing.

5.6.2 Frequency No.	1.	2.	3.	4.
Frequency (Hz)	....	....	....	....
Max. No. Phones/Frequency	....	....	....	....

5.6.3 Wattage to be sized for _______ lines.
5.6.4 Frequency Meter (see Item 12.3.3, Appendix A). Panel Mounted _____ Not Required ______.

6. Emergency Operation

6.1 If path to central office is opened, the RST shall be able to complete calls between subscribers in its own system: Yes _____ No _____

Further requirements should be listed under Item 12, Appendix B.

7. RST Distribution Frame Requirements

7.1 Total number of outside plant cable pairs to be terminated _____.
7.1.1 Gauge of outside plant cable pairs _______.
7.2 Number of outside plant cable pairs to be protected _______.
7.3 Number of additional protector pair units to be provided on MDF _______.
Explain:
7.4 *Main Frame Details*
7.4.1 Present MDF to be reused Yes_____ No_____.
If "Yes", Type _______.
Reused protectors are: _______ (Mfr.) _______ (Type).
7.4.2 Number of pairs of arrester units (switching equipment) _______.
7.4.3 Number of pairs of gas tube arrester units (special equipment) _______.
7.4.3.1 Gas tubes to be: _____ light, _____ medium, _____ heavy, _____ maximum duty units.
7.4.3.2 Fail shorted/low breakdown failure mode required Yes _____ No _____.
7.4.3.3 Breakdown voltage of gas tube arresters _______.
7.4.4 Number of terminated pairs to be grounded _______.
7.4.5 Factory assembled tip cable Yes _____ No _____.
7.4.5.1 Tip cable length [if other than 20 feet (610 cm)] _____.
7.4.5.2 Tip cable formed Up _____ Down _____.
7.4.6 Pairs per vertical _______.
7.4.7 Height of vertical _______ feet _______ inches.

8. Building and Floor Plan Information

8.1 RST to be mounted in building _____.
8.1.1 Earthquake bracing required Yes _____No _____(see Item 14.3.6, Appendix A).
8.1.2 Supply building floor plan.
8.2 RST to be mounted in cabinet out of doors _____.
8.2.1 Cabinet to be mounted _____on pole _____on ground.

9. Subscriber Line Test

9.1 Remote testing of subscriber lines is required Yes _____No _____.
9.2 Subscriber loop test set _____.

10. Span Lines to Host Central Office

10.1 To be supplied by Owner _____.
10.2 To be supplied by Bidder _____.
10.2.1 When the bidder is to supply the span lines, an RUS Form 397b, Trunk Carrier Systems, with the applicable parts completed must be attached with a physical layout of the span line.

11. Grounding Considerations

11.1 The RST ground will be _____ohms or less. (Refer to Item 4.6.3 of RUS TE&CM 810.)
11.2 This RST is considered to be in the following category for lightning damage probability based on the Figure 1 map of RUS TE&CM 823._______Very High _______ Higher than Average _______Average

______ Lower than Average ______ Very Low

11.3 The following is additional information regarding operating environment conditions which should be considered in determining system protection requirements (tower in vicinity, high exposure, etc.):

12. Explanatory Notes

APPENDIX C TO 7 CFR 1755.522—SPECIFICATIONS FOR DIGITAL, STORED PROGRAM CONTROLLED CENTRAL OFFICE EQUIPMENT DETAILED REQUIREMENTS—BIDDER SUPPLIED INFORMATION

Telephone Company
Name ______
Location

______

______

Central Office Name (By Location)

Town ______

County ______ State ______

Attended ______ Unattended ______

1. General

1.1 The equipment and materials furnished by the bidder must meet the requirements of paragraphs (a) through (x), Appendix A, and Appendix B of §1755.522.

1.2 Paragraphs (a) through (x) of §1755.522 cover the minimum general requirements for digital, stored program controlled central office switching equipment.

1.3 Paragraph (y) of §1755.522 covers requirements for installation, inspection, and testing when such service is included as part of the contract.

1.4 Appendices A and B of §1755.522 cover the technical data for application engineering and detailed equipment requirements insofar as they can be established by the owner. These appendices are to be filled in by the owner.

1.5 Appendix C of §1755.522 covers detailed information on the switching network equipment and the stored program controlled equipment, and information as to system reliability and heavy traffic delays as proposed by the bidder. This appendix is to be filled in by the bidder and must be presented with the bid.

1.6 Appendix D of §1755.522 is the single-point grounding system audit checklist.

2. Performance Objectives

2.1 *Reliability* (see paragraph (b) of §1755.522).

______

______

______

______

2.2 *Busy Hour Load Capacity and Traffic Delay* (see paragraph (e)(10) of §1755.522. Describe basis for traffic analysis).

______

______

______

______

3. Equipment Quantities Dependent on System Design

3.1 *Switch Frames and Circuits.*

3.1.1 Number of Lines.

3.1.1.1 The number of lines to be provided shall include the number required for the termination of subscriber lines, Item 7, Appendix A, plus the number required for routine testing plus any additional to meet the minimum switch increment of the selected system.

3.1.1.2 The number of lines provided for this office will be ______

3.1.2 Number of Ports Used for Trunks

3.1.2.1 The number of trunk ports to be provided shall be based on the trunk quantities required (Item 8, Appendix A) as modified by the minimum increment of the selected system. Provision shall be made for at least 5 percent additional inlet and outlet ports over those required initially. The additional ports shall be used for connecting additional trunks that may be required in the future.

3.1.2.2 The number of trunk ports provided for this office will be ______

3.1.3 Number of Subscriber Directory Numbers

3.1.3.1 The number of directory numbers provided shall be based on the total directory numbers required (Item 6.1.11, appendix A), as modified by the memory increment of the proposed system.

3.1.3.2 The number of subscriber directory numbers provided for this office will be ______

4. RST

4.1 Information for RST's must be supplied for each RST to be furnished.

4.2 Number of line terminals for this RST will be ______.

4.3 Number of span line terminations to the central office being supplied ______

4.4 If the emergency operation option is required, it will provide the following service when connection to the main office is severed:

______

______

4.5 The ac power drain at the remote end will be:

Initial ______ Ultimate ______
Voltage: Single-Phase ______ Three-Phase ______

4.6 Special environmental requirements for the remote end:

____

____

____

5. Power

5.1 *AC Power Drain Watts*

Initial __________ Ultimate __________

5.2 *Heat Dissipation Watts*

Provide the initial and ultimate equipment dissipation for each equipment room.

5.2.1 Operating Temperature Range

Minimum __________ Maximum __________

5.2.2 Operating Humidity Range

Minimum __________ Maximum __________

6. Additional Information to be Furnished by Bidder

6.1 The bidder shall accompany its bid with the following information:

a. Two copies of the equipment list and the calculations from which the quantities in the equipment list are determined;

b. Two copies of the traffic tables from which the quantities are determined, other than the full availability tables shown in paragraph (p)(1)(i) of §1755.522;

c. Two copies of detailed switching diagram showing the traffic on each route, the grade of service, the quantity of circuits, and main distributing frames;

d. Block diagram of stored program control and associated maintenance equipment;

e. A prescribed method and criteria for acceptance of the completed central office, which is subject to review;

f. Location of technical assistance service with 24-hour maintenance, and conditions when owner will be charged for access to the service;

g. Calculations showing the method by which ringing machine sizes were derived;

h. Precautions to be taken against static discharge;

i. Details of central office grounding requirements, recognizing local grounding conditions;

j. Details concerning traffic measurement capabilities and formats; and

k. Details concerning AMA features and formats to be provided.

6.2 As a part of the response to the bid, the bidder must also list information concerning the types and quantities of spare parts to be furnished. All units, excluding those units described in paragraph (x)(6)(i)(C) of §1755.522, must fall into one of the four classes. The information must be in the following format:

Unit No.	Unit name	Quantity of units in the CO's and RST's which are bid				Quantity of spare parts furnished with this bid			
		Class 1	Class 2	Class 3	Class 4	Class 1	Class 2	Class 3	Class 4

7. Explanatory Notes

APPENDIX D TO 7 CFR 1755.522—ACCEPTANCE CHECKLIST—SINGLE-POINT GROUNDING SYSTEM

1. Approval Statement

Telephone Company: __________

RUS Borrower Designation: __________

RUS Contract Number: __________

N/A __________

Name:

Central Office: __________

Remote: __________

Date of Inspection: __________

Names of Inspectors:

Owner Representative __________

Central Office Supplies __________

Consulting Engineer __________

Mutually Approved Exceptions:

____

____

____

____

Grounding System Approval:

Name (Owner Representative) __________

Signature __________

Title __________

Date __________

Name (Supplier Representative) __________

Signature __________

Title __________

Date __________

2. General Survey

2.1 This office is considered to be in the following category for probability of lightning damage based on the Figure 1 map in RUS TE&CM 823 (also refer to paragraph (u)(2) of §1755.522)

___Very High ___Higher than Average ___Average ___Lower than Average ___Very low

2.2 Central office ground field (COGF) to be inspected for proper bonding of conductors to ground rods, etc. COGF to earth grounding reading is _______ohms. (Refer to

RUS TE&CM 802, Appendices C and D, Measurement Techniques.) Is this resistance reading acceptable? (Refer to RUS TE&CM 810, Items 1.6, 4.6.2 and 4.6.3 for protection considerations.)

Acceptable: ___Yes ___No

Comments: ___

2.3 Ground connection to be inspected from the master ground bar (MGB) to the central office ground field (COGF) to ensure it is properly sized and installed by most direct route with no sharp bends. (Refer to RUS TE&CM 810, Item 4.3.2 and section 8.1.)

Acceptable: ___Yes ___No

Comments: ___

2.4 Building structure grounds (steel rebar in footings, ironwork, etc.) are to be properly bonded and connected to the MGB. (Refer to RUS TE&CM 810, Item 4.3.4.)

Acceptable: ___Yes ___No

Comments: ___

2.5 Metallic central office door(s) are to be painted with metallic paint with doorknobs left bare. Door(s) and frames are to be grounded to the building structural ground or the MGB.

Acceptable: ___Yes ___No

Comments: ___

2.6 Metallic fences within 6 feet (183 cm) of the exchange building, storage facilities ground field, etc. are to be properly bonded to the COGF outside of the central office building. Handhole enclosure is to be used for the COGF connection to permit inspection and disconnect for earth resistance testing. (Refer to RUS TE&CM 810, Appendix C, Item 4.6.1.)

Acceptable: ___Yes ___No

Comments: ___

2.7 Lightning rod systems are to be grounded by a separate dedicated ground field. A bond should be provided between the COGF and the lightning rod ground field. Handhole enclosure is to be used for the COGF connection to permit inspection and disconnect for earth resistance testing. (Refer to RUS TE&CM 810, Item 4.3.2.1.)

Acceptable: ___Yes ___No

Comments: ___

2.8 Radio/microwave tower ground grid is to be properly bonded to the COGF by a direct outside connection. Handhole enclosure is to be used for the COGF connection to permit inspection and disconnect for earth resistance testing. (Refer to RUS TE&CM 810, Item 4.3.2 and section 10.)

Acceptable: ___ Yes ___ No

Comments: ___

2.9 If a qualified metallic water system is present, inspect the MGB connecting conductor to ensure that it is properly sized and installed by the most direct route with no sharp bends and that it is clamped solidly on the water pipes. (Refer to RUS TE&CM 810, Item 4.3.3 for details on metallic water system grounding.)

Acceptable: ___ Yes ___ No

Comments: ___

2.10 All power and grounding conductors are to be continuous, end to end, with no splices, size discontinuity or intermediate terminations. If an exception is necessary, unusual care must be taken to assure proper bonding between the two sections. (Refer to RUS TE&CM 810, Appendix C, section 5.)

Acceptable: ___ Yes ___ No

Comments: ___

2.11 All ground conductors should be void of sharp bends along their entire lengths. (Refer to RUS TE&CM 810, Item 8.2.2.)

Acceptable: ___ Yes ___ No

Comments: ___

2.12 Ground conductors should only be placed in nonmetallic conduit. Those routed through metallic conduit require that both ends of the conduit be bonded to the ground conductor. (Refer to RUS TE&CM 810, Item 8.2.4.)

Acceptable: ___ Yes ___ No

Comments: ___

2.13 Ground conductors should not be encircled by metallic clamp. Metallic straps are to be removed and replaced with nonmetallic clamps. (Refer to RUS TE&CM 810, Item 8.2.4.)

Acceptable: ___ Yes ___ No

Comments: ___

2.14 If metallic conduit is used, it is to be insulated from all ironwork.

Acceptable: ___ Yes ___ No

Comments: ______________________

2.15 Inspect to determine if the required central office supplier electrostatic discharge plates, wrist wraps, antistatic floor mats, etc. are available and properly installed. (Refer to RUS TE&CM 810, Item 12.3.)

Acceptable: ___ Yes ___ No

Comments: ______________________

2.16 Ground conductors, except green wires, should not be routed close and parallel to other conductors so as to minimize induction on surges into equipment wiring. It is also better not to route these ground conductors through cable racks or troughs, or within the confines of any iron work. (Refer to RUS TE&CM 810, Item 8.2.3.)

Acceptable: ___ Yes ___ No

Comments: ______________________

3. Master Ground Bar (MGB)

3.1 The designated P, A, N, and I segments of the master ground bar (MGB) should be clearly identified. (Refer to RUS TE&CM 810, Figure 1 for MGB segmentation arrangement.)

Acceptable: ___ Yes ___ No

Comments: ______________________

3.2 Check for appearance and proper location of following on MGB:

(a) R—Interior radio equipment[1]

(b) C—Cable entrance ground bar[1]

(c) M—MDF ground bar[1]

(d) G—Standby power equipment frame ground[1]

(e) N—Commercial power MGN[2]

(f) B—Building structure ground[2]

(g) L—Central office ground field[2]

(h) W—Water pipe system[2]

(i) N^1—Battery Return[3]

(j) N^2—Outside IGZ: __________[3]

(k) N^3—Outside IGZ: __________[3]

(l) I^1—Ground window bar[4]

(m) I^2—Ground window bar[4]

Acceptable: ___ Yes ___ No

Comments: ______________________

[1] Surge Producer—(P)

[2] Surge Absorber—(A)

[3] Grounds to non-IGZ Equipment—(N)

[4] Grounds to IGZ Equipment (GWB's)—(I)

3.3 All connections to MGB are to be two-hole bolted down copper crimped or compression type terminal lugs. (NOTE: No solder connections are permitted.)

Acceptable: ___ Yes ___ No

Comments: ______________________

3.4 MGB is to be properly insulated from the mounting surface.

Acceptable: ___ Yes ___ No

Comments: ______________________

3.5 All connections are to be tight.

Acceptable: ___ Yes ___ No

Comments: ______________________

3.6 The MGB is to have an anticorrosion coating of the type which enhances conductivity.

Acceptable: ___ Yes ___ No

Comments: ______________________

3.7 Bar is to be clearly stenciled or legibly labeled "MGB."

Acceptable: ___ Yes ___ No

Comments: ______________________

3.8 All ground leads are to be properly sized and labeled as to point of origin. (Refer to RUS TE&CM 810, Item 8.3.1 and section 8.1.)

Acceptable: ___ Yes ___ No

Comments: ______________________

4. Ground Window Bar (GWB)

4.1 All equipment grounds that originate inside of an Isolated Ground Zone (IGZ) are to be terminated on the GWB which is preferably located physically inside the IGZ and insulated from its support. (Refer to RUS TE&CM 810, Item 5.1.)

Acceptable: ___ Yes ___ No

Comments: ______________________

4.2 Each GWB is to be connected to the MGB by the most direct route with a conductor of 2/0-gauge or coarser, or resistance of less than 0.005 ohms. Parallel conductors for redundancy if required by the supplier. (Refer to RUS TE&CM 810, Item 8.1.2.)

Acceptable: ___ Yes ___ No

Comments: ______

4.3 The metal framework grounds of only that switching equipment and associated electrical equipment located inside of the IGZ should be connected to the GWB as required by the central office equipment supplier. (Refer to RUS TE&CM 810, Item 5.5.)

Acceptable: ___ Yes ___ No

Comments: ______

4.4 GWB is to be clearly stenciled or labeled "GWB."

Acceptable: ___ Yes ___ No

Comments: ______

4.5 All connections are to be tight.

Acceptable: ___ Yes ___ No

Comments: ______

5. Isolated Ground Zone (IGZ)

5.1 IGZ areas are to be clearly marked on the floor or in some other easily recognizable manner. (Refer to RUS TE&CM 810, Item 6.1.1)

Acceptable: ___ Yes ___ No

Comments: ______

5.2 Confirm that all framework, cabinets, etc., within the IGZ are ground connected only to the GWB. (Refer to RUS TE&CM 810, Item 5.5.)

Acceptable: ___ Yes ___ No

Comments: ______

5.3 All cable racks, ground mats, switching and transmission equipment within the IGZ are to have ground leads *only* to the GWB. (Refer to RUS TE&CM 810, Item 5.5.2.)

Acceptable: ___ Yes ___ No

Comments: ______

5.4 Review ac power feed arrangement within the IGZ for acceptable receptacle type and confirm that all green wires are properly connected. (Refer to RUS TE&CM 810, Item 5.5.4.)

Acceptable: ___ Yes ___ No

Comments: ______

5.5 All ironwork, metallic conduit, and other equipment associated with the switch are to be properly insulated at the IGZ boundary as stipulated by the supplier. (Refer to RUS TE&CM 810, Item 6.2.)

Acceptable: ___ Yes ___ No

Comments: ______

5.6 With the GWB disconnected from the MGB, the resistance reading of _____ ohms between the GWB and the MGB indicates adequate isolation. (CAUTION: Test is to be conducted only with the approval and under the direction of the central office supplier.)

Acceptable: ___ Yes ___ No

Comments ______

6. Entrance and Tip Cables

6.1 When neither a cable vault nor a splicing trough exists, the outside plant cable should be brought into the central office and spliced to tip cables with a PVC outer jacket (ALVYN[R]) or equivalent as close as practical to the cable entrance. (Refer to RUS TE&CM 810, Item 7.3.4.)

Acceptable: ___ Yes ___ No

Comments: ______

6.2 All outside entrance cables and all tip cable shields are to be separated by at least a 3-inch (7.6 cm) gap between shield ends.

Acceptable: ___ Yes ___ No

Comments: ______

6.3 All entrance cable shields are to be bonded separately to #6 AWG or larger insulated wire or bonding ribbon and connected to the Cable Entrance Ground Bar (CEGB) by most direct route with minimum bends.

Acceptable: ___ Yes ___ No

Comments: ______

6.4 Outside plant cable shields are to be connected only to the CEGB, and the tip cable shields are to be connected only to the Main Distributing Frame Bar (MDFB).

Acceptable: ___ Yes ___ No

Comments: ______

7. Cable Entrance Ground Bar (CEGB)

7.1 The CEGB is to be properly insulated from the mounting surface. (Refer to TE&CM 810, Item 4.2.1.)

Acceptable: ___ Yes ___ No

Comments: ____________________

7.2 The CEGB is to be located as close as possible to the physical ends of the entrance cable shields.

Acceptable: ___ Yes ___ No

Comments: ____________________

7.3 All connections are to use two-hole bolted down copper crimped or compression type terminal lugs. (NOTE: No solder connections are permitted.)

Acceptable: ___ Yes ___ No

Comments: ____________________

7.4 All connections are to be tight.

Acceptable: ___ Yes ___ No

Comments: ____________________

7.5 Bar is to be clearly stenciled or legibly labeled "CEGB."

Acceptable: ___ Yes ___ No

Comments: ____________________

7.6 All ground leads are to be properly sized and labeled.

Acceptable: ___ Yes ___ No

Comments: ____________________

7.7 The CEGB is to have an anticorrosion coating of the type which enhances conductivity.

Acceptable: ___ Yes ___ No

Comments: ____________________

7.8 The CEGB is to be connected to the MGB by a properly sized conductor and by the most direct route. (Refer to RUS TE&CM 810, section 8.1.)

Acceptable: ___ Yes ___ No

Comments: ____________________

8. Main Distributing Frame (MDF)

8.1 RUS strongly recommends that MDF protectors be furnished without heat coils. (Refer to RUS TE&CM 810, section 7.6.)

Acceptable: ___ Yes ___ No

Comments: ____________________

8.2 Incoming cable pairs terminated on MDF protector assemblies should be protected with protector modules. These modules should contain white coded carbon blocks or orange coded gas tube arrestors that are included in the RUS List of Materials. (Refer to RUS TE&CM 810, Item 7.4)

Acceptable: ___ Yes ___ No

Comments: ____________________

7 ____________________

8.3 All incoming subscriber cable pairs are to be properly terminated at either a protector equipped terminal or connected to ground.

Acceptable: ___ Yes ___ No

Comments ____________________

8.4 MDF protector assemblies may be mounted directly on the vertical frame ironwork. Protector assemblies on each vertical are interconnected with each other and the Main Distributing Frame Bar (MDFB) with a #6 copper grounding conductor. Alternative means of connecting to the MDFB are also acceptable which do not rely on the frame ironwork for conducting surge currents to ground. (Refer to RUS TE&CM 810, section 7.)

Acceptable: ___ Yes ___ No

Comments ____________________

8.5 Protective "ground connections" should be provided between the MDFB and the frame ironwork for personnel protection regardless of the type of protector assembly used. Protective ground leads should be 14-gauge, less than 12 inches (30.5 cm) in length with paint thoroughly removed at point of connection to the ironwork. (Refer to RUS TE&CM 810, Item 7.1.3.)

Acceptable: ___ Yes ___ No

Comments ____________________

8.6 The MDFB should be insulated from the frame ironwork in all cases where it is used as a Master Ground Bar (MGB). (Refer to RUS TE&CM 810, Item 7.1.2.)

Acceptable: ___ Yes ___ No

Comments ____________________

8.7 Where the MDFB is used as the MGB in very small offices the protective "ground connections" should be connected on the N section of the bar. The MDF line protector assembly grounds should be connected to the P section of the bar. (Refer to RUS TE&CM 810, Item 7.1.4.)

Acceptable: ____Yes ____No

Comments ______________________________

8.8 The MDFB is to be connected to the MGB by the most direct path with minimum bends and proper conductor size. (Refer to RUS TE&CM 810, Item 8.1.4.)

Acceptable: ____Yes ____No

Comments ______________________________

8.9 The MDFB should be free of all other ground leads when not used as an MGB.

Acceptable: ____Yes ____No

Comments ______________________________

8.10 Alternative arrangements which insulate the line protector assemblies and MDFB from the frame ironwork may require a direct ground connection of the frame ironwork to the MGB for personnel protection. Conductor is properly sized and tightened with paint removal on main frame ironwork at point of connection.

Acceptable: ____Yes ____No

Comments ______________________________

9. Power Service Protection and Grounding

9.1 The ground conductor between the ac power system multigrounded neutral (MGN) at the main ac disconnect panel and the master ground bar (MGB) is to be properly sized and connected. (Refer to RUS TE&CM 810, Items 2.19, 4.3.1 and 8.1.3.)

Acceptable: ____ Yes ____ No

Comments: ______________________________

9.2 If there is a non-MGN ac power system, there is to be a properly sized and connected insulated conductor bond between the power service ground electrode and the MGB. (Refer to RUS TE&CM 810, Item 4.3.1.1.)

Acceptable: ____ Yes ____ No

Comments: ______________________________

9.3 AC conductors including ground conductors serving 120-volt ac electric convenience receptacles and all direct wire peripheral equipment, located in the IGZ, should be sized in accordance with normal "green wire" criteria. (Refer to RUS TE&CM 810, Items 5.5.4, 5.5.5, and 5.5.6.)

Acceptable: ____ Yes ____ No

Comments: ______________________________

9.4 Minimum protection for ac power serving the central office buildings should consist of an RUS accepted secondary arrestor at the service entrance. (Refer to RUS TE&CM 810, section 9.)

Acceptable: ____ Yes ____ No

Comments: ______________________________

9.5 A properly sized conductor for ground bonding between the standby power plant framework (not separately derived) and the MGB is to be provided to equalize framework voltages for personnel safety reasons. (Refer to RUS TE&CM 810, Item 4.2.4.)

Acceptable: ____ Yes ____ No

Comments: ______________________________

10. Miscellaneous

10.1 All non-IGZ equipment frames, relay racks, cable racks and other ironwork are to be properly connected to the MGB. (Refer to TE&CM 810, Item 4.4.)

Acceptable: ____ Yes ____ No

Comments: ______________________________

10.2 Shields on high frequency intra-office cables are to be properly isolated and connected only to an isolation ground bar in the relay rack. All shielded cables entering the IGZ should only be referenced at the IGZ termination point as given by the manufacturer. (Refer to RUS TE&CM 810, Item 7.2.1.2.)

Acceptable: ____ Yes ____ No

Comments: ______________________________

10.3 Isolation ground bars in the relay racks are to be properly connected to the MGB with appropriate sized conductor with no sharp bends.

Acceptable: ____ Yes ____ No

Comments: ______________________________

10.4 All radio equipment cabinet(s) are to be at least 10 feet (305 cm) from the IGZ.

Acceptable: ___ Yes ___ No

Comments: ______________________________

____

____

10.5 The metal spare parts cabinet is to be grounded with a #6 AWG or larger insulated wire to non-IGZ cable rack, etc. or directly to the MGB.

Acceptable: ___ Yes ___ No

Comments: ______________________________

____

____

[58 FR 30938, May 28, 1993; 58 FR 36252, July 6, 1993, as amended at 60 FR 1711, Jan. 5, 1995, 60 FR 64312, 64314, Dec. 15, 1995; 69 FR 18803, Apr. 9, 2004]

§§ 1755.523–1755.699 [Reserved]

§ 1755.700 RUS specification for aerial service wires.

§§ 1755.701 through 1755.704 cover the requirements for aerial service wires.

[61 FR 26074, May 24, 1996]

§ 1755.701 Scope.

(a) This section covers the requirements for aerial service wires intended for aerial subscriber drops.

(b) The aerial service wires can be either copper coated steel reinforced or nonmetallic reinforced designs.

(c) For the copper coated steel reinforced design, the reinforcing members are the conductors.

(1) The conductors are solid copper-covered steel wires.

(2) The wire structure is completed by insulating the conductors with an overall extruded plastic insulating compound.

(d) For the nonmetallic reinforced design, the conductors are solid copper individually insulated with an extruded solid insulating compound.

(1) The insulated conductors are either laid parallel (two conductor design only) or twisted into pairs (a star-quad configuration is permitted for two pair wires).

(2) The wire structure is completed by the application of nonmetallic reinforcing members and an overall plastic jacket.

(e) All wires sold to RUS borrowers for projects involving RUS loan funds under §§ 1755.700 through 1755.704 must be accepted by RUS Technical Standards Committee "A" (Telecommunications). For wires manufactured to the specification of §§ 1755.700 through 1755.704, all design changes to an accepted design must be submitted for acceptance. RUS will be the sole authority on what constitutes a design change.

(f) Materials, manufacturing techniques, or wire designs not specifically addressed by §§ 1755.700 through 1755.704 may be allowed if accepted by RUS. Justification for acceptance of modified materials, manufacturing techniques, or wire designs must be provided to substantiate product utility and long term stability and endurance.

[61 FR 26074, May 24, 1996]

§ 1755.702 Copper coated steel reinforced (CCSR) aerial service wire.

(a) *Conductors.* (1) Each conductor shall comply with the requirements specified in the American National Standard Institute/Insulated Cable Engineers Association, Inc. (ANSI/ICEA) S–89–648–1993, paragraphs 2.1 through 2.1.5. The ANSI/ICEA S–89–648–1993 Standard For Telecommunications Aerial Service Wire, Technical Requirements (approved by ANSI July 11, 1994) is incorporated by reference in accordance with 5 U.S.C. 552(a) and 1 CFR part 51. Copies of ANSI/ICEA S–89–648–1993 are available for inspection during normal business hours at RUS, room 2845, U.S. Department of Agriculture, Washington, DC 20250–1500, or at the National Archives and Records Administration (NARA). For information on the availability of this material at NARA, call 202–741–6030, or go to: *http://www.archives.gov/federal_register/code_of_federal_regulations/ibr_locations.html.* Copies are available from ICEA, P. O. Box 440, South Yarmouth, MA 02664, telephone number (508) 394–4424.

(2) Factory joints in conductors shall comply with the requirement specified in ANSI/ICEA S–89–648–1993, paragraph 2.1.6.

(b) *Conductor insulation.* (1) The raw materials used for the conductor insulation shall comply with the requirements specified in ANSI/ICEA S–89–648–1993, paragraph 3.1.1.

(2) The raw materials shall be accepted by RUS prior to their use.

(3) The finished conductor insulation shall be free from holes, splits, blisters, or other imperfections and shall be as smooth as is consistent with best commercial practice.

(4) The finished conductor insulation shall comply with the requirements specified in ANSI/ICEA S–89–648–1993, paragraphs 3.1.5 through 3.1.5.4.

(5) The insulation shall have a minimum spot thickness of not less than 0.9 millimeters (mm) (0.03 inches (in.)) at any point.

(c) *Wire assembly*. (1) The two conductors shall be insulated in parallel to form an integral configuration.

(2) The finished wire assembly shall be either a flat or a notched oval. Other finished wire assemblies may be used provided that they are accepted by RUS prior to their use.

(3) The overall dimensions of the finished wire assembly shall be in accordance with the following requirements:

Diameter	Dimensions	
	Minimum mm (in.)	Maximum mm (in.)
Major	5.5 (0.22)	8.0 (0.31)
Minor	3.0 (0.12)	5.0 (0.19)

(d) *Conductor marking*. The insulated conductors of a finished wire shall be marked in accordance with the requirements specified in ANSI/ICEA S–89–648–1993, paragraph 3.1.4.

(e) *Electrical requirements*—(1) *Conductor resistance*. The direct current (dc) resistance of each conductor in a completed CCSR aerial service wire shall comply with the requirement specified in ANSI/ICEA S–89–648–1993, paragraph 7.1.2.

(2) *Wet mutual capacitance*. The wet mutual capacitance of the completed CCSR aerial service wire shall comply with the requirement specified in ANSI/ICEA S–89–648–1993, paragraph 7.1.3.

(3) *Wet attenuation*. The wet attenuation of the completed CCSR aerial service wire shall comply with the requirement specified in ANSI/ICEA S–89–648–1993, paragraph 7.1.4.

(4) *Wet insulation resistance*. The wet insulation resistance of the completed CCSR aerial service wire shall comply with the requirement specified in ANSI/ICEA S–89–648–1993, paragraph 7.1.5.

(5) *Dielectric strength*. (i) The wet dielectric strength between conductors and between each conductor of the completed CCSR aerial service wire and the surrounding water shall comply with the requirement specified in ANSI/ICEA S–89–648–1993, paragraph 7.1.6.

(ii) The dry dielectric strength between conductors of the completed CCSR aerial service wire shall comply with the requirement specified in ANSI/ICEA S–89–648–1993, paragraph 7.1.7.

(6) *Fusing coordination*. The completed CCSR aerial service wire shall comply with the fusing coordination requirement specified in ANSI/ICEA S–89–648–1993, paragraph 7.1.8.

(7) *Insulation imperfections*. Each length of completed CCSR aerial service wire shall comply with the requirement specified in ANSI/ICEA S–89–648–1993, paragraph 7.1.9.

(f) *Mechanical requirements*—(1) *Impact test*. (i) All CCSR aerial service wires manufactured in accordance with this section shall comply with the unaged impact test specified in ANSI/ICEA S–89–648–1993, paragraph 8.1.2.

(ii) All CCSR aerial service wires manufactured in accordance with this section shall comply with the aged impact test specified in ANSI/ICEA S–89–648–1993, paragraph 8.1.3.

(2) *Abrasion resistance test*. All CCSR aerial service wires manufactured in accordance with this section shall comply with the abrasion resistance test specified in ANSI/ICEA S–89–648–1993, paragraph 8.1.4.

(3) *Static load test*. All CCSR aerial service wires manufactured in accordance with this section shall comply with the static load test specified in ANSI/ICEA S–89–648–1993, paragraph 8.1.5.

(4) *Plasticizer compatibility test*. All CCSR aerial service wires manufactured in accordance with this section shall comply with the plasticizer compatibility test specified in ANSI/ICEA S–89–648–1993, paragraph 8.1.8.

(g) *Environmental requirements*—(1) *Cold temperature handling test*. (i) All

CCSR aerial service wires manufactured in accordance with this section shall comply with the unaged cold temperature handling test specified in ANSI/ICEA S–89–648–1993, paragraph 8.2.1.

(ii) All CCSR aerial service wires manufactured in accordance with this section shall comply with the aged cold temperature handling test specified in ANSI/ICEA S–89–648–1993, paragraph 8.2.2.

(2) *Light absorption test.* All CCSR aerial service wires manufactured in accordance with this section shall comply with the light absorption test specified in ANSI/ICEA S–89–648–1993, paragraph 8.2.3.

(3) *Low temperature separation test.* All CCSR aerial service wires manufactured in accordance with this section shall comply with the low temperature separation test specified in ANSI/ICEA S–89–648–1993, paragraph 8.2.4.

(4) *Flammability test.* All CCSR aerial service wires manufactured in accordance with this section shall comply with the flammability test specified in ANSI/ICEA S–89–648–1993, paragraph 8.3.

(5) *Wire listing.* All CCSR aerial service wires manufactured in accordance with this section shall comply with the listing requirements specified in ANSI/ICEA S–89–648–1993, paragraph 8.4.

(h) *Identification marker.* Each length of CCSR aerial service wire shall be identified in accordance with ANSI/ICEA S–89–648–1993, paragraph 9.1.4. When surface marking is employed, the color of the initial marking shall be either white or silver.

(i) *Length marking (optional).* (1) Sequentially numbered length marking of the completed CCSR aerial service wire may be used at the option of the manufacturer unless specified by the end user.

(2) When sequentially numbered length markings are used, the length markings shall be in accordance with ANSI/ICEA S–89–648–1993, paragraph 9.1.5. The color of the initial marking shall be either white or silver.

(j) *Durability of marking.* The durability of the marking of the CCSR aerial service wire shall comply with the requirements specified in ANSI/ICEA S–89–648–1993, paragraph 9.1.6.

[61 FR 26075, May 24, 1996, as amended at 69 FR 18803, Apr. 9, 2004]

§ 1755.703 Nonmetallic reinforced (NMR) aerial service wire.

(a) *Conductors.* (1) Each conductor shall comply with the requirements specified in ANSI/ICEA S–89–648–1993, paragraphs 2.2 and 2.2.1. The ANSI/ICEA S–89–648–1993 Standard For Telecommunications Aerial Service Wire, Technical Requirements (approved by ANSI July 11, 1994) is incorporated by reference in accordance with 5 U.S.C. 552(a) and 1 CFR part 51. Copies of ANSI/ICEA S–89–648–1993 are available for inspection during normal business hours at RUS, room 2845, U.S. Department of Agriculture, Washington, DC 20250–1500, or at the National Archives and Records Administration (NARA). For information on the availability of this material at NARA, call 202–741–6030, or go to: *http://www.archives.gov/federal_register/code_of_federal_regulations/ibr_locations.html.* Copies are available from ICEA, P. O. Box 440, South Yarmouth, MA 02664, telephone number (508) 394–4424.

(2) Factory joints made in the conductors during the manufacturing process shall comply with the requirement specified in ANSI/ICEA S–89–648–1993, paragraph 2.2.2.

(b) *Conductor insulation.* (1) The raw materials used for the conductor insulation shall comply with the requirements specified in ANSI/ICEA S–89–648–1993, paragraphs 3.2 through 3.2.2.

(2) The finished conductor insulation shall comply with the requirements specified in ANSI/ICEA S–89–648–1993, paragraph 3.2.3.

(3) The dimensions of the insulated conductors shall comply with the requirements specified in ANSI/ICEA S–89–648–1993, paragraph 3.2.3.1.

(4) The colors of the insulation shall comply with the requirements specified in ANSI/ICEA S–89–648–1993, paragraph 3.2.3.2.

(5) A permissible overall performance level of faults in conductor insulation shall comply with the requirement specified in ANSI/ICEA S–89–648–1993, paragraph 3.2.4.6. The length count and

number of faults shall be recorded. The information shall be retained for a period of 6 months and be available for review by RUS when requested.

(6) Repairs to the conductor insulation during manufacture are permissible. The method of repair shall be accepted by RUS prior to its use. The repaired insulation shall comply with the requirement specified in ANSI/ICEA S-89-648-1993, paragraph 3.2.3.3.

(7) All repaired sections of insulation shall be retested in the same manner as originally tested for compliance with paragraph (b)(5) of this section.

(8) The colored insulating material removed from or tested on the conductor, from a finished wire shall comply with the requirements specified in ANSI/ICEA S-89-648-1993, paragraphs 3.2.4 through 3.2.4.5.

(c) *Identification of pairs and layup of pairs.* (1) The insulation shall be colored coded to identify:

(i) The tip and ring conductor of each pair; and

(ii) Each pair in the completed wire.

(2) The colors to be used in the pairs together with the pair numbers shall be in accordance with the table specified in ANSI/ICEA S-89-648-1993, paragraph 4.1.1.

(3) The insulated conductors shall be either layed parallel (two conductor design only) or twisted into pairs.

(4) When using parallel conductors for the two conductor design, the parallel conductors shall be designed to enable the wire to meet the electrical requirements specified in paragraph (g) of this section.

(5) When twisted pairs are used, the following requirements shall be met:

(i) The pair twists shall be designed to enable the wire to meet the electrical requirements specified in paragraph (g) of this section; and

(ii) The average length of pair twists in any pair in the finished wire, when measured on any 3 meter (10 foot) length, shall not exceed the requirement specified in ANSI/ICEA S-89-648-1993, paragraph 4.1.

(6) An alternative method of forming the two-pair wire is the use of a star-quad configuration.

(i) The assembly of the star-quad shall be such as to enable the wire to meet the electrical requirements specified in paragraph (g) of this section.

(ii) The star-quad configuration shall be assembled in accordance with ANSI/ICEA S-89-648-1993, paragraph 4.1.2.

(iii) The average length of twist for the star-quad in the finished wire, when measured on any 3 meter (10 foot) length, shall not exceed the requirement specified in ANSI/ICEA S-89-648-1993, paragraph 4.1.

(iv) The color scheme used to provide identification of the tip and ring conductors of each pair in the star-quad shall comply with the table specified in ANSI/ICEA S-89-648-1993, paragraph 4.1.2.

(d) *Strength members.* The strength members shall comply with the requirements specified in ANSI/ICEA S-89-648-1993, paragraphs 6.1 and 6.1.1.

(e) *Wire jacket.* (1) The jacket shall comply with the requirements specified in ANSI/ICEA S-89-648-1993, paragraphs 5.1 and 5.1.1.

(2) The jacket raw materials shall be accepted by RUS prior to their use.

(f) *Wire assembly.* The finished wire assembly shall be in accordance with ANSI/ICEA S-89-648-1993, paragraph 5.1.3 and Figure 5-1.

(g) *Electrical requirements*—(1) *Conductor resistance.* The dc resistance of each conductor in a completed NMR aerial service wire shall comply with the requirement specified in ANSI/ICEA S-89-648-1993, paragraph 7.2.2.

(2) *Resistance unbalance.* (i) The dc resistance unbalance between the two conductors of any pair in a completed NMR aerial service wire and the average resistance unbalance of all pairs in a Quality Control Lot shall comply with the requirements specified in ANSI/ICEA S-89-648-1993, paragraph 7.2.3.

(ii) The resistance unbalance between tip and ring conductors shall be random with respect to the direction of unbalance. That is, the resistance of the tip conductors shall not be consistently higher with respect to the ring conductors and vice versa.

(3) *Dry mutual capacitance.* The dry mutual capacitance of the completed NMR aerial service wire shall comply with the requirements specified in ANSI/ICEA S-89-648-1993, paragraph 7.2.4, Type 1.

(4) *Pair-to-pair capacitance unbalance.* The pair-to-pair capacitance unbalance as measured on the completed NMR aerial service wire shall comply with the requirements specified in ANSI/ICEA S–89–648–1993, paragraph 7.2.5.

(5) *Attenuation.* (i) The dry attenuation of the completed NMR aerial service wire shall comply with the requirement specified in ANSI/ICEA S–89–648–1993, paragraph 7.2.7.

(ii) The wet attenuation of the completed NMR aerial service wire shall comply with the requirement specified in ANSI/ICEA S–89–648–1993, paragraph 7.2.8.

(6) *Insulation resistance.* (i) The dry insulation resistance of the completed NMR aerial service wire shall comply with the requirement specified in ANSI/ICEA S–89–648–1993, paragraph 7.2.9.

(ii) The wet insulation resistance of the completed NMR aerial service wire shall comply with the requirement specified in ANSI/ICEA S–89–648–1993, paragraph 7.2.10.

(7) *Wet dielectric strength.* The wet dielectric strength between conductors and between each conductor of the completed NMR aerial service wire and the surrounding water shall comply with the requirement specified in ANSI/ICEA S–89–648–1993, paragraph 7.2.11.

(8) *Fusing coordination.* The completed NMR aerial service wire shall comply with the fusing coordination requirement specified in ANSI/ICEA S–89–648–1993, paragraph 7.2.13.

(9) *Crosstalk loss.* (i) The output-to-output far-end crosstalk loss (FEXT) for any pair of completed NMR aerial service wire shall comply with the requirement specified in ANSI/ICEA S–89–648–1993, paragraph 7.2.14.

(ii) The input-to-input near-end crosstalk loss (NEXT) for any pair of completed NMR aerial service wire shall comply with the requirement specified in ANSI/ICEA S–89–648–1993, paragraph 7.2.14.

(h) *Mechanical requirements*—(1) *Impact test.* (i) All NMR aerial service wires manufactured in accordance with this section shall comply with the unaged impact test specified in §1755.702(f)(1)(i).

(ii) All NMR aerial service wires manufactured in accordance with this section shall comply with the aged impact test specified in §1755.702(f)(1)(ii).

(2) *Abrasion resistance test.* All NMR aerial service wires manufactured in accordance with this section shall comply with the abrasion resistance test specified in §1755.702(f)(2).

(3) *Static load test.* All NMR aerial service wires manufactured in accordance with this section shall comply with the static load test specified in §1755.702(f)(3).

(4) *Elongation test.* All NMR aerial service wires manufactured in accordance with this section shall comply with the elongation test specified in ANSI/ICEA S–89–648–1993, paragraph 8.1.7.

(5) *Plasticizer compatibility test.* All NMR aerial service wires manufactured in accordance with this section shall comply with the plasticizer compatibility test specified in §1755.702(f)(4).

(i) *Environmental requirements*—(1) *Cold temperature handling test.* (i) All NMR aerial service wires manufactured in accordance with this section shall comply with the unaged cold temperature handling test specified in §1755.702(g)(1)(i).

(ii) All NMR aerial service wires manufactured in accordance with this section shall comply with the aged cold temperature handling test specified in §1755.702(g)(1)(ii).

(2) *Light absorption test.* All NMR aerial service wires manufactured in accordance with this section shall comply with the light absorption test specified in §1755.702(g)(2).

(3) *Flammability test.* All NMR aerial service wires manufactured in accordance with this section shall comply with the flammability test specified in §1755.702(g)(4).

(4) *Wire listing.* All NMR aerial service wires manufactured in accordance with this section shall comply with the listing requirements specified in §1755.702(g)(5).

(j) *Ripcord (optional).* (1) A ripcord may be used in the NMR aerial service wire structure at the option of the manufacturer unless specified by the end user.

(2) When a ripcord is used it shall comply with the requirements specified

in ANSI/ICEA S–89–648–1993, paragraphs 4.2 through 4.2.3.

(k) *Identification marker*. Each length of NMR aerial service wire shall be identified in accordance with ANSI/ ICEA S–89–648–1993, paragraphs 9.1 through 9.1.4. When surface marking is employed, the color of the initial marking shall be either white or silver.

(l) *Length marking (optional)*. (1) Sequentially numbered length marking of the completed NMR aerial service wire may be used at the option of the manufacturer unless specified by the end user.

(2) When sequentially numbered length markings are used, the length markings shall be in accordance with in accordance with § 1755.702(i)(2).

(m) *Durability of marking*. The durability of the marking of the NMR aerial service wire shall comply with the requirements specified in § 1755.702(j).

[61 FR 26076, May 24, 1996, as amended at 69 FR 18803, Apr. 9, 2004]

§ 1755.704 Requirements applicable to both CCSR and NMR aerial service wires.

(a) *Acceptance testing*. (1) The tests described in §§ 1755.700 through 1755.704 are intended for acceptance of wire designs and major modifications of accepted designs. What constitutes a major modification is at the discretion of RUS. These tests are intended to show the inherent capability of the manufacturer to produce wire products having long life and stability.

(2) For initial acceptance, the manufacturer shall:

(i) Certify that the product fully complies with each paragraph in §§ 1755.700 through 1755.704;

(ii) Agree to periodic plant inspections by RUS;

(iii) Certify whether the product complies with the domestic origin manufacturing provisions of the "Buy American" requirements of the Rural Electrification Act of 1938 (7 U.S.C. 903 note), as amended (the "REA Buy-American provision");

(iv) Submit at least three written user testimonials concerning field performance of the product; and

(v) Provide any other nonpropriety data deemed necessary by the Chief, Outside Plant Branch (Telecommunications).

(3) In order for RUS to consider a manufacturer's request that a product be requalified, the manufacturer shall certify not later than June 30 of the year in which requalification is required, that the product:

(i) Fully complies with each paragraph in §§ 1755.700 through 1755.704; and

(ii) Does or does not comply with the domestic origin manufacturing provisions of the REA Buy American provisions. The required certifications shall be dated within 90 days of the submission.

(4) Initial and requalification acceptance requests should be addresses to: Chairman, Technical Standards Committee "A" (Telecommunications), Telecommunications Standards Division, Rural Utilities Service, AG Box 1598, Washington, DC 20250–1598.

(b) *Extent of testing*—(1) *Tests on 100 percent of completed wire*. (i) Each conductor in the completed CCSR and NMR aerial service wire shall be tested for continuity in accordance with ANSI/ICEA S–89–648–1993, paragraphs 7.1.1 and 7.2.1, respectively. The ANSI/ ICEA S–89–648–1993 Standard For Telecommunications Aerial Service Wire, Technical Requirements (approved by ANSI July 11, 1994) is incorporated by reference in accordance with 5 U.S.C. 552(a) and 1 CFR part 51. Copies of ANSI/ICEA S–89–648–1993 are available for inspection during normal business hours at RUS, room 2845, U.S. Department of Agriculture, Washington, DC 20250–1500, or at the National Archives and Records Administration (NARA). For information on the availability of this material at NARA, call 202–741–6030, or go to: *http://www.archives.gov/federal_register/code_of_federal_regulations/ibr_locations.html*. Copies are available from ICEA, P. O. Box 440, South Yarmouth, MA 02664, telephone number (508) 394–4424.

(ii) Each conductor in the completed CCSR and NMR aerial service wire shall be tested for shorts in accordance with ANSI/ICEA S–89–648–1993, paragraphs 7.1.1 and 7.2.1, respectively.

(iii) Each length of completed CCSR and NMR aerial service wire shall be tested for insulation imperfections in accordance with § 1755.702(e)(7) and § 1755.703(b)(5), respectively.

(2) *Capability tests.* Tests on a quality assurance basis shall be made as frequently as is required for each manufacturer to determine and maintain compliance with:

(i) Performance of the conductors;

(ii) Performance of the conductor insulation and jacket material;

(iii) Sequential marking and lettering;

(iv) Mutual capacitance, capacitance unbalance, attenuation, and crosstalk;

(v) Conductor resistance, resistance unbalance, and insulation resistance;

(vi) Dielectric strength and fusing coordination;

(vii) Impact, abrasion, static load, elongation, and plasticizer compatibility tests; and

(viii) Cold temperature handling, light absorption, low temperature separation, and flammability tests.

(c) *Summary of records of electrical and physical tests.* (1) Each manufacturer shall maintain suitable summary records for a period of at least 3 years of all electrical and physical tests required on completed wire as set forth in paragraph (b) of this section. The test data for a particular lot of aerial service wire shall be in a form such that it may be readily available to the purchaser or to RUS upon request.

(2) Measurements and computed values shall be rounded off to the number of places or figures specified for the requirement according to ANSI/ICEA S–89–648–1993, paragraph 1.3.

(d) *Manufacturing irregularities.* (1) Repairs to the insulation of CCSR aerial service wires are not permitted in wires supplied to end users under §§ 1755.700 through 1755.704.

(2) Repairs to the jacket of NMR aerial service wires are not permitted in wires supplied to end users under §§ 1755.700 through 1755.704.

(e) *Splicing.* Splicing of completed CCSR and NMR aerial service wires shall comply with the requirement specified in ANSI/ICEA S–89–648–1993, paragraph 8.1.1.

(f) *Preparation for shipment.* (1) CCSR and NMR aerial service wire shall be shipped either in coils or on reels.

(2) When CCSR and NMR aerial service wires are shipped on reels the following provisions shall apply:

(i) The diameter of the drum shall be large enough to prevent damage to the wire from reeling or unreeling. The reels shall be substantial and so constructed as to prevent damage to the wire during shipment and handling;

(ii) A waterproof corrugated board or other suitable means of protection accepted by RUS prior to its use may be applied to the reel. If the waterproof corrugated board or other suitable material is used for protection, it shall be suitably secured in place to prevent damage to the wire during storage and handling. The use of the waterproof corrugated board or other suitable means of protection shall be at the option of the manufacturer unless specified by the end user;

(iii) The outer end of the wire shall be securely fastened to the reel head so as to prevent the wire from becoming loose in transit. The inner end of the wire shall be securely fastened in such a way as to make it readily available if required for electrical testing. Spikes, staples, or other fastening devices which penetrate the conductor insulation of the CCSR aerial service wire and the jacket of the NMR aerial service wire shall not be used. The method of fastening the wire ends shall be accepted by RUS prior to their use;

(iv) Each length of wire shall be wound on a separate reel;

(v) Each reel shall be plainly marked to indicate the direction in which it should be rolled to prevent loosening of the wire on the reel; and

(vi) Each reel shall be stenciled or labeled on either one or both sides with the following information:

(A) Customer order number;

(B) Manufacturer's name and product code;

(C) Factory reel number and year of manufacture;

(D) Gauge of conductors and pair size of wire;

(E) Length of wire; and

(F) RUS designation letter "K."

(3) When CCSR and NMR aerial service wires are shipped in coils the following provisions shall apply:

(i) The diameter of the coil shall be large enough to prevent damage to the wire from coiling or uncoiling;

(ii) The nominal length of the wire in a coil shall be 305 meters (1,000 feet). No coil shall be less than 290 meters (950 feet) long or more than 460 meters (1,500 feet) long; however, 25 percent of the total number of coils may be less than 305 meters (1,000 feet);

(iii) The coils of wire shall be wound securely with strong tape in four separate evenly spaced places;

(iv) The coils may be protected from damage by wrapping the coil with heavy paper, burlap, or other suitable material accepted by RUS prior to its use. The use of the heavy paper, burlap, or other suitable means of protection shall be at the option of the manufacturer unless specified by the end user; and

(v) Each coil shall be tagged with the following information:

(A) Customer order number;

(B) Manufacturer's name and product code;

(C) Year of manufacture;

(D) Gauge of conductors and pair size of wire;

(E) Length of wire; and

(F) RUS designation letter "K."

(4) In lieu of wrapping the coil with heavy paper, burlap, or other suitable material, the coil may be packaged in a moisture resistant carton.

(5) When the coils are shipped in moisture resistant cartons, each carton shall be marked with the information specified in paragraphs (f)(3)(v)(A) through (f)(3)(v)(F) of this section.

(6) Other methods of shipment may be used if accepted by RUS prior to their use.

(7) When NMR aerial service wire is shipped, the ends of the wire shall be sealed in accordance with ANSI/ICEA S–89–648–1993, paragraph 9.2.

[61 FR 26077, May 24, 1996, as amended at 69 FR 18803, Apr. 9, 2004]

§§ 1755.705–1755.859 [Reserved]

§ 1755.860 RUS specification for filled buried wires.

(a) *Scope.* (1) This section covers the requirements for filled buried wires intended for direct burial as a subscriber drop and/or distribution wire.

(i) The conductors are solid copper, individually insulated with an extruded solid insulating compound.

(ii) The insulated conductors are twisted into pairs (a star-quad configuration is permitted for the two pair wires) which are then stranded or oscillated to form a cylindrical core.

(iii) A moisture resistant filling compound is applied to the stranded conductors completely covering the insulated conductors and filling the interstices between the pairs.

(iv) The wire structure is completed by the application of an optional core wrapping material, an inner jacket, a flooding compound, a shield, a flooding compound, and an overall plastic jacket.

(2) The number of pairs and gauge size of conductors which are used within the RUS program are provided in the following table:

American Wire Gauge (AWG)	22	24
Pairs	2 3	2 3

(3) All wires sold to RUS borrowers for projects involving RUS loan funds under this section must be accepted by RUS Technical Standards Committee "A" (Telephone). For wires manufactured to the specification of this section, all design changes to an accepted design must be submitted for acceptance. RUS will be the sole authority on what constitutes a design change.

(4) Materials, manufacturing techniques, or wire designs not specifically addressed by this section may be allowed if accepted by RUS. Justification for acceptance of modified materials, manufacturing techniques, or wire designs must be provided to substantiate product utility and long term stability and endurance.

(5) The American National Standards Institute/Electronic Industries Association (ANSI/EIA) 359-A-84, EIA Standard Colors for Color Identification and Coding, referenced in this section is incorporated by reference by RUS. This incorporation by reference was approved by the Director of the Federal Register in accordance with 5 U.S.C. 552(a) and 1 CFR part 51. Copies of ANSI/EIA 359-A-84 are available for inspection during normal business hours at RUS, room

2845, U.S Department of Agriculture, Washington, DC 20250–1500, or at the National Archives and Records Administration (NARA). For information on the availability of this material at NARA, call 202–741–6030, or go to: *http://www.archives.gov/federal_register/code_of_federal_regulations/ibr_locations.html*. Copies are available from EIA, 2001 Pennsylvania Avenue, NW., suite 900, Washington, DC 20006, telephone number (202) 457–4966.

(6) American Society for Testing and Materials specifications (ASTM) A 505–87, Standard Specification for Steel, Sheet and Strip, Alloy, Hot-Rolled and Cold-Rolled, General Requirements for; ASTM B 3–90, Standard Specification for Soft or Annealed Copper Wire; ASTM B 193–87, Standard Test Method for Resistivity of Electrical Conductor Materials; ASTM B 224–91, Standard Classification of Coppers; ASTM B 694–86, Standard Specification for Copper, Copper Alloy, and Copper-Clad Stainless Steel Sheet and Strip for Electrical Cable Shielding; ASTM D 150–87, Standard Test Methods for A-C Loss Characteristics and Permittivity (Dielectric Constant) of Solid Electrical Insulating Materials; ASTM D 257–91, Standard Test Methods for D-C Resistance or Conductance of Insulating Materials; ASTM D 1238–90b, Standard Test Method for Flow Rates of Thermoplastics by Extrusion Plastometer; ASTM D 1248–84(1989), Standard Specification for Polyethylene Plastics Molding and Extrusion Materials; ASTM D 1535–89, Standard Test Method for Specifying Color by the Munsell System; ASTM D 3349–86, Standard Test Method for Absorption Coefficient of Carbon Black Pigmented Ethylene Plastic; ASTM D 4101–82(1988), Standard Specification for Propylene Plastic Injection and Extrusion Materials; ASTM D 4565–90a, Standard Test Methods for Physical and Environmental Performance Properties of Insulations and Jackets for Telecommunications Wire and Cable; ASTM D 4566–90, Standard Test Methods for Electrical Performance Properties of Insulations and Jackets for Telecommunications Wire and Cable; ASTM D 4568–86, Standard Test Methods for Evaluating Compatibility between Cable Filling and Flooding Compounds and Polyolefin Cable Materials; ASTM D 4872–88, Standard Test Method for Dielectric Testing of Wire and Cable Filling Compounds; ASTM E 8–91, Standard Test Methods of Tension Testing of Metallic Materials; and ASTM E 29–90, Standard Practice for Using Significant Digits in Test Data to Determine Conformance with Specifications, referenced in this section are incorporated by reference by RUS. These incorporations by references were approved by the Director of the Federal Register in accordance with 5 U.S.C. 552(a) and 1 CFR part 51. Copies of the ASTM standards are available for inspection during normal business hours at RUS, room 2845, U.S. Department Agriculture, Washington, DC 20250–1500, or at the National Archives and Records Administration (NARA). For information on the availability of this material at NARA, call 202–741–6030, or go to: *http://www.archives.gov/federal_register/code_of_federal_regulations/ibr_locations.html*. Copies are available from ASTM, 1916 Race Street, Philadelphia, Pennsylvania 19103–1187, telephone number (215) 299–5585.

(b) *Conductors and conductor insulation.* (1) Each conductor must be a solid round wire of commercially pure annealed copper. Conductors must meet the requirements of the American Society for Testing and Materials (ASTM) B 3–90 except that requirements for *Dimensions and Permissible Variations* are waived and elongation requirements are superseded by this section.

(2) The minimum conductor elongation in the final wire must comply with the following limits when tested in accordance with ASTM E 8–91.

Conductor—AWG	Minimum Elongation—Percent
22	20
24	16

(3) Joints made in conductors during the manufacturing process may be brazed, using a silver alloy solder and nonacid flux, or they may be welded using either an electrical or cold welding technique. In joints made in uninsulated conductors, the two conductor ends must be butted. Splices

made in insulated conductors need not be butted but may be joined in a manner acceptable to RUS.

(4)(i) The tensile strength of any section of a conductor containing a factory joint must not be less than 85 percent of the tensile strength of an adjacent section of the solid conductor of equal length without a joint.

(ii) *Engineering Information:* The sizes of wire used and their nominal diameters shall be as shown in the following table:

AWG	Nominal Diameter	
	Millimeters (mm)	(Inches (in.))
22	0.643	(0.0253)
24	0.511	(0.0201)

(5) Each conductor must be insulated with either a colored, solid, insulating grade, high density polyethylene or crystalline propylene/ethylene copolymer or with a solid natural primary layer and a colored, solid outer skin using one of the insulating materials listed in paragraphs (b)(5)(i) through (b)(5)(ii) of this section.

(i) The polyethylene raw material selected to meet the requirements of this section must be Type III, Class A, Category 4 or 5, Grade E9, in accordance with ASTM D 1248–84(1989).

(ii) The crystalline propylene/ethylene raw material selected to meet the requirements of this section must be Class PP 200B 40003 E11 in accordance with ASTM D 4101–82(1988).

(iii) Raw materials intended as conductor insulation furnished to these requirements must be free from dirt, metallic particles, and other foreign matter.

(iv) All insulating raw materials must be accepted by RUS prior to their use.

(6) All conductors in any single length of wire must be insulated with the same type of material.

(7) A permissible overall performance level of faults in conductor insulation must average not greater than one fault per 12,000 conductor meters (40,000 conductor feet) for each gauge of conductor.

(i) All insulated conductors must be continuously tested for insulation faults during the twinning operation with the method of test acceptable to RUS. The length count and number of faults must be recorded. The information must be retained for a period of 6 months and be available for review by RUS when requested.

(ii) The voltages for determining compliance with the requirements of this section are as follows:

AWG	Direct Current Voltages (Kilovolts)
22	6.0
24	5.0

(8) Repairs to the conductor insulation during manufacturing are permissible. The method of repair must be accepted by RUS prior to its use. The repaired insulation must be capable of meeting the relevant electrical requirements of this section.

(9) All repaired sections of insulation must be retested in the same manner as originally tested for compliance with paragraph (b)(7) of this section.

(10) Colored insulating material removed from or tested on the conductor, from a finished wire, must be capable of meeting the following performance requirements:

Property	Polyethylene	Crystalline Propylene/Ethylene Copolymer
Melt Flow Rate		
Percent increase from raw material, Maximum.		
$<$0.5 (Initial Melt Index).	50	—
0.5–2.00 (Initial Melt Index).	25	—
≤5.0 (Initial Melt Index).	—	110
Tensile Strength—Minimum		
Megapascals (MPa).	16.5	21.0
(Pounds per Square Inch (psi)).	(2,400)	(3,000)
Ultimate Elongation		
Minimum, Percent.	300	300
Cold Bend		
Failures, Maximum.	0/10	0/10

Property	Poly-ethylene	Crystalline Propylene/Ethylene Copolymer
Shrinkback		
Maximum, mm (in.).	10 (0.375)	10 (0.375)
Oxygen Induction Time		
Minimum, Minutes.	20	20

(11) *Testing procedures.* The procedures for testing the insulation samples for compliance with paragraph (b)(10) of this section must be as follows.

(i) *Melt flow rate.* The melt flow rate must be determined as described in ASTM D 1238–90b. Condition E must be used for polyethylene. Condition L must be used for crystalline propylene/ethylene copolymer. The melt flow test must be conducted prior to the filling operation.

(ii) *Tensile strength and ultimate elongation.* Samples of the insulation material, removed from the conductor, must be tested in accordance with ASTM D 4565–90a using the following conditions. The minimum length of unclamped specimen must be 50 mm (2.0 in.). The minimum speed of jaw separation must be 25 mm (1.0 in.) per minute per 25 mm (1.0 in.) of unclamped specimen. The temperature of specimens and surrounding shall be 23 ±1 °C.

NOTE: Quality assurance testing at a jaw separation speed of 500 mm/min (20 in./min) is permissible. Failures at this rate must be retested at the 50 mm/min (2 in./min) rate to determine section compliance.

(iii) *Cold bend.* Samples of the insulation material on the conductor must be tested in accordance with ASTM D 4565–90a at a temperature of −40 ±1 °C with a mandrel diameter equal to 3 times the outside diameter of the insulated conductor. There must be no cracks visible to normal or corrected-to-normal vision.

(iv) *Shrinkback.* Samples of insulation must be tested for four hours in accordance with ASTM D 4565–90a. The temperature for the type of material is listed as follows:

Material	Temperature
Polyethylene	115 ±1 °C
Crystalline propylene/ethylene Copolymer	130 ±1 °C

(v) *Oxygen induction time.* Samples of insulation, which have been conditioned in accordance with paragraph 17.3 of ASTM D 4565–90a, must be tested in accordance with the procedures of ASTM D 4565–90a using copper pans and a test temperature of 199 ±1 °C.

(12) Other methods of testing may be used if acceptable to RUS.

(c) *Identification of pairs and twisting of pairs.* (1) The insulation must be colored to identify:

(i) The tip and ring conductor of each pair; and

(ii) Each pair in the completed wire.

(2) The colors to be used to provide identification of the tip and ring conductor of each pair are shown in the following table:

Pair No.	Color	
	Tip	Ring
1	White	Blue
2	White	Orange
3	White	Green

(3) *Standards of color.* The colors of the insulated conductors supplied in accordance with this section are specified in terms of the Munsell Color System (ASTM D 1535–89) and must comply with the "Table of Wire and Cable Limit Chips" as defined in ANSI/EIA-359-A-84. (Visual color standards meeting these requirements may be obtained directly from the Munsell Color Company, Inc., 2441 North Calvert Street, Baltimore, Maryland 21218).

(4) Positive identification of the tip and ring conductors of each pair by marking each conductor of a pair with the color of its mate is permissible. The method of marking must be accepted by RUS prior to its use.

(5) Other methods of providing positive identification of the tip and ring conductors of each pair may be employed if accepted by RUS prior to its use.

(6) The insulated conductors must be twisted into pairs.

(7) In order to provide sufficiently high crosstalk isolation, the pair

twists must be designed to enable the wire to meet the capacitance unbalance and the crosstalk loss requirements of paragraphs (m)(2), (m)(3), and (m)(4) of this section.

(8) The average length of pair twists in any pair in the finished wire, when measured on any 3 meter (m) (10 foot(ft)) length, must not exceed 152 mm (6 in.).

(9) An alternative method of forming the two pair wire is the use of a star-quad configuration.

(i) The assembly of the star-quad must be such as to enable the wire to meet the capacitance unbalance and the crosstalk loss requirements of paragraphs (m)(2), (m)(3), and (m)(4) of this section.

(ii) The four individual insulated conductors must be twisted together to form a star-quad configuration with the tip and ring conductors of each pair diagonally opposite each other in the quad.

(iii) The average length of twist for the star-quad in the finished wire, when measured on any 3 m (10 ft) length, must not exceed 152 mm (6 in.).

(iv) The following color scheme must be used to provide identification of the tip and ring conductor of each pair in the star-quad:

Pair No.	Color	
	Tip	Ring
1	White with blue stripe.	Blue
2	White with orange stripe.	Orange

(v) If desired, the blue and orange conductors may contain a white stripe. The stripes in this case must be narrow enough so that the tip and ring identification is obvious.

(d) *Forming of the wire core.* (1) Twisted pairs or star-quad configuration must be assembled in such a way as to form a substantially cylindrical group.

(2) The filling compound must be applied to the wire core in such a way as to provide a completely filled core as is commercially practical.

(3) If desired for manufacturing reasons, white or colored binders of nonhygroscopic and nonwicking material may be applied over the core.

(e) *Filling compound.* (1) After or during the stranding operation and prior to application of the optional core wrap and inner jacket, a homogeneous filling compound free of agglomerates must be applied to the wire core. The compound must be as nearly colorless as is commercially feasible and consistent with the end product requirements and pair identification.

(2) The filling compound must be free from dirt, metallic particles, and other foreign matter. It must be applied in such a way as to fill the space within the wire core.

(3) The filling compound must be nontoxic and present no dermal hazards.

(4) The filling compound must exhibit the following dielectric properties at a temperature of 23 ±3 °C when measured in accordance with ASTM D 150–87 or ASTM D 4872–88.

(i) The dissipation factor must not exceed 0.0015 at a frequency of 1 megahertz (MHz).

(ii) The dielectric constant must not exceed 2.30.

(5) The volume resistivity must not be less than 10^{12} ohm-cm at a temperature of 23 ±3 °C when measured in accordance with ASTM D 257–91 or ASTM D 4872–88.

(6) The individual wire manufacturer must satisfy RUS that the filling compound selected for use is suitable for its intended application. The filling compound must be compatible with the wire components when tested in accordance with ASTM D 4568–86 at a temperature of 80 °C.

(f) *Core wrap (optional).* (1) When a core wrap is used, it must consist of a layer of nonhygroscopic and nonwicking dielectric material. The wrap must be applied with an overlap.

(2) The core wrap must provide a sufficient heat barrier to prevent visible evidence of conductor insulation deformation or adhesion between conductors, caused by adverse heat transfer during the inner jacketing operation.

(3) If required for manufacturing reasons, white or colored binders of nonhygroscopic and nonwicking material may be applied over the core wrap.

(4) Sufficient filling compound must be applied to the core wrap that voids or air spaces existing between the core

and inner side of the core wrap are minimized.

(g) *Inner jacket.* (1) An inner jacket must be applied over the core and/or core wrap.

(2) The jacket must be free from holes, splits, blisters, or other imperfections and must be as smooth and concentric as is consistent with the best commercial practice.

(3) The inner jacket material and test requirements must be as specified for the outer jacket material per paragraphs (j)(3) through (j)(5)(iv) of this section.

(4) The inner jacket thickness at any point must not be less than 0.5 mm (0.020 in.). The thickness must be determined from measurements on 50 mm (2 in.) samples taken not less than 0.3 m (1 ft) from either end of the wire. The average must be determined from 4 readings taken approximately 90 °apart on any cross section of the samples. The maximum and minimum points must be determined by exploratory measurements. The maximum thickness minus the minimum thickness at any cross section must not exceed 43 percent of the average thickness at that cross section.

(h) *Flooding compound.* (1) Sufficient flooding compound must be applied on all sheath interfaces so that voids and air spaces in these areas are minimized.

(2) The flooding compound must be compatible with the jacket when tested in accordance with ASTM D 4568–86 at a temperature of 80 °C. The floodant must exhibit adhesive properties sufficient to prevent jacket slip when tested in accordance with the requirements of appendix A, paragraph (III)(5), of this section.

(3) The individual wire manufacturer must satisfy RUS that the flooding compound selected for use is acceptable for the application.

(i) *Shield.* (1) A shield must be applied either longitudinally or helically over the inner jacket.

(i) If the shield is applied longitudinally, it must be corrugated.

(ii) If the shield is applied helically, it must be smooth.

(2) The overlap for longitudinally applied shields must be a minimum of 2 mm (0.075 in.) The overlap for helically applied shields must be a minimum of 23 percent of the tape width.

(3) General requirements for application of the shielding material are as follows:

(i) Successive lengths of shielding tapes may be joined during the manufacturing process by means of cold weld, electric weld, soldering with a nonacid flux, or other acceptable means;

(ii) Where two ends of a metal shield are to be joined together, care shall be taken to clean the metal surfaces in order to provide for a good mechanical and electrical connection;

(iii) The shields of each length of wire must be tested for continuity. A one meter (3 ft) section of shield containing a factory joint must exhibit not more than 110 percent of the resistance of a shield of equal length without a joint;

(iv) The breaking strength of any section of a shield tape containing a factory joint must not be less than 80 percent of the breaking strength of an adjacent section of the shield of equal length without a joint;

(v) The reduction in thickness of the shielding material due to the corrugating or application process must be kept to a minimum and must not exceed 10 percent at any spot; and

(vi) The shielding material must be applied in such a manner as to enable the wire to pass the bend test as specified in paragraph (n)(3) of this section.

(4) The following materials are acceptable for use as wire shielding:

Standard Wire	Gopher Resistant Wire
Copper Alloy 220 (Bronze).	Copper-Clad Stainless Steel
(0.1016 ±0.0076 mm) ...	0.1270 ±0.0127 mm
(0.0040 ±0.0003 in.)	(0.0050 ±0.0005 in.)
Copper Alloy 220 (Bronze).	Copper Alloy 664
0.1270 ±0.0127 mm	0.1397 ±0.0127 mm
(0.0050 ±0.0005 in.)	(0.0055 ±0.0005 in.)
	Copper-Clad Alloy Steel
	0.1270 ±0.0127
	(0.0050 ±0.0005 in.)

(i) The copper-clad steels and copper alloy 664 shielding tapes must be capable of meeting the following performance requirements prior to application to the wire:

Property	Requirement
Tensile Strength	
Minimum, MPa (psi)	379 (55,000)
Tensile Yield	
Minimum, MPa (psi)	241 (35,000)
Elongation	
Minimum, percent in 50 mm (2 in.).	15

(ii) *Copper alloy 220.* The shielding material, prior to application to the wire, must be in the fully annealed condition and shall conform to the requirements of ASTM B 694–86 for C22000 commercial bronze.

(iii) *Copper-clad stainless steel.* In addition to meeting the requirements of paragraph (i)(4)(i) of this section, the shielding material, prior to application to the wire, must be in the fully annealed condition and must conform to the requirements of ASTM B 694–86, with a cladding ratio of 16/68/16 and must have a minimum electrical conductivity of 28 percent IACS when measured in accordance with ASTM B 193–87.

(iv) *Copper alloy 664.* In addition to meeting the requirements of paragraph (i)(4)(i) of this section, the shielding material, prior to application to the wire, must be annealed temper and must conform to the requirements of ASTM B 694–86 and must have a minimum electrical conductivity of 28 percent IACS when measured in accordance with ASTM B 193–87.

(v) *Copper-clad alloy steel.* In addition to meeting the requirements of paragraph (i)(4)(i) of this section, the shielding material, prior to application to the wire, must be in the fully annealed condition and the copper component must conform to the requirements of ASTM B 224–91 and the alloy steel component must conform to the requirements of ASTM A 505–87, with a cladding ratio of 16/68/16, and must have a minimum electrical conductivity of 28 percent IACS when measured in accordance with ASTM B 193–87.

(j) *Outer jacket.* (1) The outer jacket must provide the wire with a tough, flexible, protective covering which can withstand exposure to sunlight, to atmospheric temperatures and stresses reasonably expected in normal installation and service.

(2) The jacket must be free from holes, splits, blisters, or other imperfections and must be as smooth and concentric as is consistent with the best commercial practice.

(3) The raw material used for the outer jacket must be one of the five types listed in paragraphs (j)(3)(i) through (j)(3)(v) of this section. The raw material must contain an antioxidant to provide long term stabilization and the materials must contain a 2.60 ±0.25 percent concentration of furnace black to provide ultraviolet shielding. Both the antioxidant and furnace black must be compounded into the material by the raw material supplier.

(i) Low density, high molecular weight polyethylene (LDHMW) must conform to the requirements of ASTM D 1248–84(1989), Type I, Class C, Category 4 or 5, Grade J3.

(ii) Low density, high molecular weight ethylene copolymer (LDHMW) must conform to the requirements of ASTM D 1248–84 (1989), Type I, Class C, Category 4 or 5, Grade J3.

(iii) Linear low density, high molecular weight polyethylene (LLDHMW) must conform to the requirements of ASTM D 1248–84(1989), Type I, Class C, Category 4 or 5, Grade J3.

(iv) High density polyethylene (HD) must conform to the requirements of ASTM D 1248–84(1989), Type III, Class C, Category 4 or 5, Grade J4.

(v) Medium density polyethylene (MD) must conform to the requirements of ASTM D 1248–84(1989), Type II, Class C, Category 4 or 5, Grade J4.

(vi) Particle size of the carbon selected for use must not average greater than 20 nanometers.

(vii) Absorption coefficient must be a minimum of 400 in accordance with the procedures of ASTM D 3349–86.

(4) The outer jacketing material removed from or tested on the wire must be capable of meeting the following performance requirements:

Property	LLDHMW, Ethylene Copolymer	LDHMW Polyethylene	HD or MD Polyethylene
Melt Flow Rate Percent increase from raw material Maximum		50	50
<0.41 (Initial Melt Index)	100	—	—
0.41–2.00 (Initial Melt Index)	50	—	—
Tensile Strength Minimum, MPa (psi)	12.0 (1,700)	12.0 (1,700)	16.5 (2,400)
Ultimate Elongation Percent, Minimum	400	400	300
Shrinkback Percent of Length, Maximum	5	5	5
Impact Failures, Maximum	2/10	2/10	2/10

(5) *Testing procedures.* The procedures for testing the jacket samples for compliance with paragraph (j)(4) of this section must be as follows:

(i) *Melt flow rate.* The melt flow rate must be as determined by ASTM D 1238–90b, Condition E. Jacketing material must be free from flooding and filling compound.

(ii) *Tensile strength and ultimate elongation.* Test in accordance with ASTM D 4565–90a, using a jaw separation speed of 500 mm/min (20 in./min) for low density material and 50 mm/min (2 in./min) for high and medium density materials.

(iii) *Shrinkback.* Test in accordance with the procedures specified in ASTM D 4565–90a using a test temperature of 100 ±1 °C for low density material and a test temperature of 115 ±1 °C for high and medium density materials.

(iv) *Impact.* The test must be performed in accordance with ASTM D 4565–90a using an impact force of 4 newton-meter (3 pound force-foot) at a temperature of −20 ±2 °C. The cylinder must strike the sample at the shield overlap. A crack or split in the jacket constitutes failure.

(6) *Jacket thickness.* The minimum jacket thickness must be 0.64 mm (0.025 in.) except that the minimum thickness over the sheath slitting cord, if present, must be 0.46 mm (0.018 in.). The minimum point must be determined by exploratory measurements. The average thickness at any cross section must be determined from four readings including the minimum point, taken approximately 90 °apart. The thickness measurement must exclude any jacket material that has formed into the corrugation. The maximum thickness at any cross section must not be greater than 155 percent of the minimum thickness.

(7) *Eccentricity.* The eccentricity of the jacket must not exceed 43 percent when calculated using the formula as follows:

$$\frac{\text{Maximum Thickness} \atop -\text{Minimum Thickness}}{\text{Average Thickness}} \times 100 \text{ Percent}$$

(k) *Sheath slitting cord (optional).* (1) Sheath slitting cords may be used in the wire structure at the option of the manufacturer.

(2) When a sheath slitting cord is used it must be nonhygroscopic and nonwicking, continuous throughout a length of wire, and of sufficient strength to open the sheath without breaking the cord.

(3) Sheath slitting cords must be capable of consistently slitting the jacket(s) and/or shield for a continuous length of 0.6 m (2 ft) when tested in accordance with the procedure specified in appendix B of this section.

(l) *Identification marker and length marker.* (1) Each length of wire must be permanently identified as to manufacturer and year of manufacture.

(2) The number of conductor pairs and their gauge size must be marked on the jacket.

(3) The marking must be printed on the jacket at regular intervals of not more than 1.5 m (5 ft).

(4) An alternative method of marking may be used if accepted by RUS prior to its use.

(5) The completed wire must have sequentially numbered length markers in FEET OR METERS at regular intervals of not more than 1.5 m (5 ft) along the outside of the jacket.

(6) The method of length marking must be such that for any single length

of wire, continuous sequential numbering must be employed.

(7) The numbers must be dimensioned and spaced to produce good legibility and must be approximately 3 mm (0.125 in.) in height. An occasional illegible marking is permissible if there is a legible marking located not more than 1.5 m (5 ft) from it.

(8) The method of marking must be by means of suitable surface markings producing a clear, distinguishable, contrasting marking acceptable to RUS. Where direct or transverse printing is employed, the characters should be indented to produce greater durability of marking. Any other method of length marking must be acceptable to RUS as producing a marker suitable for the field. Size, shape and spacing of numbers, durability, and overall legibility of the marker will be considered in acceptance of the method.

(9) The accuracy of the length marking must be such that the actual length of any wire section is never less than the length indicated by the marking and never more than one percent greater than the length indicated by the marking.

(10) The color of the initial marking must be white or silver. If the initial marking fails to meet the requirements of the preceding paragraphs, it will be permissible to either remove the defective marking and re-mark with the white or silver color or leave the defective marking on the wire and re-mark with yellow. No further re-marking is permitted. Any re-marking must be on a different portion of the wire circumference than any existing marking when possible and have a numbering sequence differing from any other existing marking by at least 5,000.

(11) Any reel of wire which contains more than one set of sequential markings must be labeled to indicate the color and sequence of marking to be used. The labeling must be applied to the reel and also to the wire.

(m) *Electrical requirements*—(1) *Mutual capacitance and conductance*. (i) The average mutual capacitance (corrected for length) of all pairs in any reel must not exceed 52 ±4 nanofarad/ kilometer (nF/km) (83 ±7 nanofarad/mile (nF/mile)) when tested in accordance with ASTM D 4566–90 at a frequency of 1.0 ±0.1 kilohertz (kHz) and a temperature of 23 ±3 °C.

(ii) The mutual conductance (corrected for length and gauge) of any pair must not exceed 2 micromhos/kilometer (micromhos/km) (3.3 micromhos/mile) when tested in accordance with ASTM D 4566–90 at a frequency of 1.0 ±0.1 kHz and a temperature of 23 ±3 °C.

(2) *Pair-to-pair capacitance unbalance*. The capacitance unbalance between any pair of the completed wire must not exceed 145 picofarad/kilometer (pF/km) (80 picofarad/1000 ft (pF/1000 ft)) when tested in accordance with ASTM D 4566–90 at a frequency of 1.0 ±0.1 kHz and a temperature of 23 ±3 °C.

(3) *Pair-to-ground capacitance unbalance*—(i) *Pair-to-ground*. The capacitance unbalance as measured on the individual pairs of the completed wire must not exceed 2625 pF/km (800 pF/1000 ft) when tested in accordance with ASTM D 4566–90 at a frequency of 1.0 ±0.1 kHz and a temperature of 23 ±3 °C.

(ii) When measuring pair-to-ground capacitance unbalance, all pairs, except the pair under test, are grounded to the shield.

(iii) Pair-to-ground capacitance unbalance may vary directly with the length of the wire.

(4) *Far-end crosstalk loss*. (i) The output-to-outputfar-end crosstalk loss (FEXT) between any pair combination of a completed wire when measured in accordance with ASTM D 4566–90 at a test frequency of 150 kHz must not be less than 58 decibel/ kilometer (dB/km) (63 decibel/1000 ft). If the loss K_o at a frequency F_o for length L_o is known, then K_x can be determined for any other frequency F_x or length L_x by:

$$\text{FEXT loss}\,(K_X) = K_O - 20\log 10\frac{F_X}{F_O} - 10\log 10\frac{L_X}{L_O}$$

(5) *Attenuation*. The attenuation of any individual pair on any reel of wire must not exceed the following limits when measured at or corrected to a temperature of 20 ±1 °C and a test frequency of 150 kHz. The test must be conducted in accordance with ASTM D 4566–90.

Conductor AWG	Individual Pair Attenuation dB/km (decibel/mile (dB/mile))	
	Maximum	Minimum
22	6.8 (11.0)	5.0 (8.1)
24	8.7 (14.0)	6.6 (10.7)

(6) *Insulation resistance.* Each insulated conductor in each length of completed wire, when measured with all other insulated conductors and the shield grounded, must have an insulation resistance of not less than 1600 megohm-kilometer (1000 megohm-mile) at 20 ±1 °C. The measurement must be made in accordance with the procedures of ASTM D 4566–90.

(7) *High voltage test.* (i) In each length of completed wire, the insulation between conductors when tested in accordance with ASTM D 4566–90 must withstand for 3 seconds a direct current (dc) potential whose value is not less than:

(A) 5.0 kilovolts for 22-gauge conductors; and

(B) 4.0 kilovolts for 24-gauge conductors.

(ii) In each length of completed wire, the dielectric strength between the shield and all conductors in the core must be tested in accordance with ASTM D 4566–90 and must withstand, for 3 seconds, a dc potential whose value is not less than 20 kilovolts.

(8) *Conductor resistance.* The dc resistance of any conductor must be measured in the completed wire in accordance with ASTM D 4566–90 and must not exceed the following values when measured at or corrected to a temperature of 20 ±1 °C.

AWG	Maximum Resistance	
	ohms/kilometer	(ohms/1000 ft)
22	57.1	(17.4)
24	90.2	(27.5)

(9) *Resistance unbalance.* (i) The difference in dc resistance between the two conductors of any pair in the completed wire must not exceed 5.0 percent when measured in accordance with the procedures of ASTM D 4566–90.

(ii) The resistance unbalance between tip and ring conductors shall be random with respect to the direction of unbalance. That is, the resistance of the tip conductors shall not be consistently higher with respect to the ring conductors and vice versa.

(n) *Mechanical requirements*—(1) *Defective wire.* Pairs in each length of wire will not be permitted to have either a ground, cross, short or open circuit condition.

(2) *Wire breaking strength.* The breaking strength of the completed wire must not be less than 890 newtons (200 pound-force) when tested in accordance with ASTM D 4565–90a using a jaw separation speed of 25 mm/min (1.0 in./min).

(3) *Wire bending test.* The completed wire must be capable of meeting the requirements of ASTM D 4565–90a after conditioning at −20 ±2 °C and at 23 ±2 °C.

(4) *Water penetration test.* (i) A one meter (3 ft) length of completed wire must be stabilized at 23 ±2 °C and tested in accordance with ASTM D 4565–90a using a one meter (3 ft) water head over the sample or placed under the equivalent continuous pressure for one hour.

(ii) After the one hour period, there must be no water leakage in the sheath interfaces, under the core wrap or between any insulated conductors in the core.

(iii) If water leakage is detected in the first sample, one 3 m (10 ft) additional adjacent sample from the same reel of wire must be tested in accordance with paragraph (n)(4)(ii) of this section. If the second sample exhibits water leakage, the entire reel of wire is to be rejected. If the second sample exhibits no leakage, the entire reel of wire is considered acceptable.

(5) *Compound flow test.* The completed wire must be capable of meeting the compound flow test specified in ASTM D 4565–90a when exposed for a period of 24 hours at a temperature of 80 ±1 °C. At the end of this test period, there must be no evidence of flowing or dripping of compound from either the core or sheath interfaces.

(o) *Acceptance testing and extent of testing.* (1) The tests described in appendix A of this section are intended for acceptance of wire designs and major modifications of accepted designs. RUS decides what constitutes a major modification. These tests are intended to show the inherent capability of the manufacturer to produce wire products having long life and stability.

(2) For initial acceptance, the manufacturer must submit:

(i) An original signature certification that the product fully complies with each requirement of this section;

(ii) Qualification Test Data, per appendix A of this section;

(iii) To periodic plant inspections;

(iv) A certification that the product does or does not comply with the domestic origin manufacturing provisions of the "Buy American" requirements of the Rural Electrification Act of 1938 (7 U.S.C. 901 *et seq.*);

(v) Written user testimonials concerning performance of the product; and

(vi) Other nonproprietary data deemed necessary by the Chief, Outside Plant Branch (Telephone).

(3) For requalification acceptance, the manufacturer must submit an original signature certification that the product fully complies with each section of the specification, excluding the Qualification Section, and a certification that the product does or does not comply with the domestic origin manufacturing provisions of the "Buy American" requirements of the Rural Electrification Act of 1938 (7 U.S.C. 901 *et seq.*) for acceptance by June 30 every three years. The required data and certification must have been gathered within 90 days of the submission.

(4) Initial and requalification acceptance requests should be addressed to: Chairman, Technical Standards, Committee "A" (Telephone), Telecommunications Standards Division, Rural Utilities Service, Washington, DC 20250–1500.

(5) Tests on 100 percent of completed wire. (i) The shield of each length of wire must be tested for continuity using the procedures of ASTM D 4566–90.

(ii) Dielectric strength between all conductors and the shield must be tested to determine freedom from grounds in accordance with paragraph (m)(7)(ii) of this section.

(iii) Each conductor in the completed wire must be tested for continuity using the procedures of ASTM D 4566–90.

(iv) Dielectric strength between conductors must be tested to ensure freedom from shorts and crosses in accordance with paragraph (m)(7)(i) of this section.

(v) The average mutual capacitance must be measured on all wires.

(6) *Capability tests.* Tests on a quality assurance basis must be made as frequently as is required for each manufacturer to determine and maintain compliance with:

(i) Performance requirements for conductor insulation and jacket material;

(ii) Performance requirements for filling and flooding compounds;

(iii) Sequential marking and lettering;

(iv) Capacitance unbalance and crosstalk;

(v) Insulation resistance;

(vi) Conductor resistance and resistance unbalance;

(vii) Wire bending and wire breaking strength tests;

(viii) Mutual conductance and attenuation; and

(ix) Water penetration and compound flow tests.

(p) *Summary of records of electrical and physical tests.* (1) Each manufacturer must maintain suitable summary of records for a period of at least 3 years for all electrical and physical tests required on completed wire by this section as set forth in paragraphs (o)(5) and (o)(6) of this section. The test data for a particular reel shall be in a form that it may be readily available to the purchaser or to RUS upon request.

(2) Measurements and computed values must be rounded off to the number of places of figures specified for the requirement according to ASTM E 29–90.

(q) *Manufacturing irregularities.* (1) Repairs to the inner jacket and shield are not permitted in wire supplied to the end user under this section.

(2) Minor defects in the outer jackets (defects having a dimension of 3 mm (0.125 in.) or less in any direction) may be repaired by means of heat fusing in accordance with good commercial practices utilizing sheath grade compound.

(r) *Preparation for shipment.* (1) The wire must be shipped on reels. The diameter of the drum must be large enough to prevent damage to the wire from reeling or unreeling. The reels must be substantial and so constructed as to prevent damage to the wire during shipment and handling.

(2) The thermal wrap must comply with the requirements of appendix C of this section. When a thermal reel wrap is supplied, the wrap must be applied to the reel and must be suitably secured in place to minimize thermal exposure to the wire during storage and shipment. The use of the thermal reel wrap as a means of reel protection will be at the option of the manufacturer unless specified by the end user.

(3) The outer end of the wire must be securely fastened to the reel head so as to prevent the wire from becoming loose in transit. The inner end of the wire must be securely fastened in such a way as to make it readily available if required for electrical testing. Spikes, staples, or other fastening devices which penetrate the wire jacket must not be used. The method of fastening the wire ends must be accepted by RUS prior to it being used.

(4) Each length of wire must be wound on a separate reel unless otherwise specified or agreed to by the purchaser.

(5) Each reel must be plainly marked to indicate the direction in which it should be rolled to prevent loosening of the wire on the reel.

(6) Each reel must be stenciled or labeled on either one or both sides with the name of the manufacturer, year of manufacture, actual shipping length, an inner and outer end sequential length marking, description of the wire, reel number and the RUS wire designation:

Wire Designation

BFW
Wire Construction
Pair Count
Conductor Gauge

N = Copper Alloy 220 (Bronze) Shield
Y = Gopher Resistant Shields

Example: BFWY 3–24

Buried Filled Wire, Gopher Resistant Shield, 3 pair, 24 AWG

(7) Both ends of the filled buried wire, manufactured to the requirements of this section, must be equipped with end caps which are acceptable to RUS.

(The information and recordkeeping requirements of this section have been approved by the Office of Management and Budget under the control number 0572–0059)

APPENDIX A TO § 1755.860—QUALIFICATION TEST METHODS

(I) The test procedures described in this appendix are for qualification of initial designs and major modifications of accepted designs. Included in (V) of this appendix are suggested formats that may be used in submitting test results to RUS.

(II) *Sample Selection and Preparation.* (1) All testing must be performed on lengths removed sequentially from the same 3 pair, 22 gauge jacketed wire. This wire must not have been exposed to temperatures in excess of 38 °C since its initial cool down after sheathing. The lengths specified are minimum lengths and if desirable from a laboratory testing standpoint longer lengths may be used.

(a) Length A shall be 10 ±0.2 meters (33 ±0.5 feet) long and must be maintained at 23 ±3 °C. One length is required.

(b) Length B shall be 12 ±0.2 meters (40 ±0.5 feet) long. Prepare the test sample by removing the inner and outer jacket, shield, and core wrap, if present, for a sufficient distance on both ends to allow the insulated conductors to be flared out. Remove sufficient conductor insulation so that appropriate electrical test connections can be made at both ends. Coil the specimen with a diameter of 15 to 20 times its sheath diameter. Three lengths are required.

(c) Length C shall be one meter (3 feet) long. Four lengths are required.

(d) Length D shall be 300 millimeters (1 foot) long. Four lengths are required.

(e) Length E shall be 600 millimeters (2 feet) long. Four lengths are required.

(f) Length F shall be 3 meters (10 feet) long and must be maintained at 23 ±3 °C for the duration of the test. Two lengths are required.

(2) *Data Reference Temperature.* Unless otherwise specified, all measurements shall be made at 23 ±3 °C.

(III) *Environmental Tests*—(1) *Heat Aging Test*—(a) *Test Samples.* Place one sample each of lengths B, C, D, and E in an oven or environmental chamber. The ends of sample B must exit from the chamber or oven for electrical tests. Securely seal the oven exit holes.

(b) *Sequence of Tests.* After conditioning the samples are to be subjected to the following tests:

(i) Water Immersion Test outlined in (III)(2) of this appendix;

(ii) Water Penetration Test outlined in (III)(3) of this appendix; .

(iii) Insulation Compression Test outlined in (III)(4) of this appendix; and

(iv) Jacket Slip Strength Test outlined in (III)(5) of this appendix.

(c) *Initial Measurements.* (i) For sample B, measure the open circuit capacitance and conductance for each pair at 1 and 150 kilohertz and the attenuation at 150 kilohertz after conditioning the sample at the data reference temperature for 24 hours. Calculate the average and standard deviation for the data of the 3 pairs on a per kilometer (per mile) basis.

(ii) The attenuation at 150 kilohertz may be calculated from open circuit admittance (Yoc) and short circuit impedance (Zsc) or may be obtained by direct measurement of attenuation.

(iii) Record on suggested formats attached in (V) of this appendix or on other easily readable formats.

(d) *Heat Conditioning.* (i) Immediately after completing the initial measurements, condition the sample for 14 days at a temperature of 65 ±2 °C.

(ii) At the end of this period note any exudation of filling compound. Measure and calculate the parameters given in (III)(1)(c) of this appendix. Record on suggested formats attached in (V) of this appendix or on other easily readable formats.

(iii) Cut away and discard a one meter (3 foot) section from each end of length B.

(e) *Overall Electrical Deviation.* (i) Calculate the percent change in all average parameters between the final parameters after conditioning with the initial parameters in (III)(1)(c) of this appendix.

(ii) The stability of the electrical parameters after completion of this test must be within the following prescribed limits:

(A) *Capacitance.* The average mutual capacitance must be within 5 percent of its original value;

(B) The change in average mutual capacitance must be less than 5 percent over the frequency range of 1 to 150 kilohertz;

(C) *Conductance.* The average mutual conductance must not exceed 2 micromhos/kilometer (3.3 micromhos/mile) at a frequency of 1 kilohertz; and

(D) *Attenuation.* The attenuation must not have increased by more than 5 percent over its original value.

(2) *Water Immersion Electrical Test*—(a) *Test Sample Selection.* The 10 meter (33 foot) section of length B must be tested.

(b) *Test Sample Preparation.* Prepare the sample by removing the inner and outer jacket, shield, and core wrap, if present, for a sufficient distance to allow one end to be accessed for test connections. Cut out a series of 2.5 millimeter by 13 millimeter (0.1 inch by 0.5 inch) rectangular slots along the test sample, at 300 millimeter (1 foot) intervals progressing successively 90 degrees around the circumference of the wire. Assure that the wire core is exposed at each slot by slitting the inner jacket and core wrap if present. Place the prepared sample in a dry vessel which when filled will maintain a one meter (3 foot) head of water over 6 meters (20 feet) of uncoiled wire. Extend and fasten the ends of the wire so they will be above the water line and the pairs are rigidly held for the duration of the test.

(c) *Capacitance and Conductance Testing.* Measure the initial values of mutual capacitance and conductance of all pairs in each wire at a frequency of 1 kilohertz before filling the vessel with water. Be sure the wire shield is grounded to the test equipment. Fill the vessel until there is a one meter (3 foot) head of water on the wires.

(i) Remeasure the mutual capacitance and conductance after the wires have been submerged for 24 hours and again after 30 days.

(ii) Record each sample separately on the suggested formats attached in (V) of this appendix or on other easily readable formats.

(d) *Overall Electrical Deviation.* (i) Calculate the percent change in all average parameters between the final parameters after conditioning with the initial parameters in (III)(2)(c) of this appendix.

(ii) The stability of the electrical parameters after of the test must be within the following prescribed limits:

(A) *Capacitance.* The average mutual capacitance must be within 5 percent of its original value; and

(B) *Conductance.* The average mutual conductance must not exceed 2 micromhos/kilometer (3.3 micromhos/mile) at a frequency of 1 kilohertz.

(3) *Water Penetration Testing.* (a) A watertight closure must be placed over the jacket of length C. The closure must not be placed over the jacket so tightly that the flow of water through preexisting voids or air spaces is restricted. The other end of the sample must remain open.

(b) Test per Option A or Option B. (i) *Option A.* Weigh the sample and closure prior to testing. Fill the closure with water and place under a continuous pressure of 10 ±0.7 kilopascals (1.5 ±0.1 pounds per square inch gauge) for one hour. Collect the water leakage from the end of the test sample during the test and weigh to the nearest 0.1 gram. Immediately after the one hour test, seal the ends of the wire with a thin layer of grease and remove all visible water from the closure, being careful not to remove water that penetrated into the core during the test. Reweigh the sample and determine the weight of water that penetrated into the core. The weight of water that penetrated into the core must not exceed 1 gram.

(ii) *Option B.* Fill the closure with a 0.2 gram sodium fluorscein per liter water solution and apply a continuous pressure of 10 ±0.7 kilopascals (1.5 ±0.1 pounds per square inch gauge) for one hour. Catch and weigh any water that leaks from the end of the wire during the one hour period. If no water leaks from the sample, carefully remove the water from the closure. Then carefully remove the outer jacket, shield, inner jacket and core wrap, if present, one at a time, examining with an ultraviolet light source for water penetration. After removal of the inner jacket and core wrap, if present, carefully dissect the core and examine for water penetration within the core. Where water penetration is observed, measure the penetration distance. The distance of water penetration into the core must not exceed 127 millimeters (5.0 inches).

(4) *Insulation Compression Test.* (a) *Test Sample D.* Remove inner and outer jacket, shield, and core wrap, if present, being careful not to damage the conductor insulation. Remove one pair from the core and carefully separate, wipe off core filler and straighten the insulated conductors. Retwist the two insulated conductors together under sufficient tension to form 10 evenly spaced 360 degree twists in a length of 100 millimeters (4 inches).

(b) *Sample Testing.* Center the mid 50 millimeters (2 inches) of the twisted pair between two smooth rigid parallel metal plates measuring 50 millimeters (2 inches) in length or diameter. Apply a 1.5 volt direct current potential between the conductors, using a light or buzzer to indicate electrical contact between the conductors. Apply a constant load of 67 newtons (15 pound-force) on the sample for one minute and monitor for evidence of contact between the conductors. Record results on suggested formats attached in (V) of this appendix or on other easily readable formats.

(5) *Jacket Slip Strength Test*—(a) *Sample Selection.* Test sample E from (III)(1)(a) of this appendix.

(b) *Sample Preparation.* Prepare test sample in accordance with the procedures specified in ASTM D 4565–90a.

(c) *Sample Conditioning and Testing.* Remove the sample from the tensile tester prior to testing and condition for one hour at 50 ±2 °C. Test immediately in accordance with the procedure specified in ASTM D 4565–90a. A minimum outer jacket slip strength of 67 newtons (15 pound-force) is required. Record the load attained.

(6) *Humidity Exposure.* (a) Repeat steps (III)(1)(a) through (III)(1)(c)(iii) of this appendix for separate set of samples B, C, D and E which have not been subjected to prior environmental conditioning.

(b) Immediately after completing the measurements, expose the test sample to 100 temperature cyclings. Relative humidity within the chamber must be maintained at 90 ±2 percent. One cycle consists of beginning at a stabilized chamber and test sample temperature of 52 ±1 °C, increasing the temperature to 57 ±1 °C, allowing the chamber and test samples to stabilize at this level, then dropping the temperature back to 52 ±1 °C.

(c) Repeat steps (III)(1)(d)(ii) through (III)(5)(c) of this appendix.

(7) *Temperature Cycling.* (a) Repeat steps (III)(1)(a) through (III)(1)(c)(iii) of this appendix for separate set of samples B, C, D and E which have not been subjected to prior environmental conditioning.

(b) Immediately after completing the measurements, subject the test sample to 10 cycles of temperature between −40 °C and +60 °C. The test sample must be held at each temperature extreme for a minimum of 1½ hours during each cycle of temperature. The air within the temperature cycling chamber must be circulated throughout the duration of the cycling.

(c) Repeat steps (III)(1)(d)(ii) through (III)(5)(c) of this appendix.

(IV) *Control Sample*—(1) *Test Samples.* A separate set of lengths for samples A, C, D, and E must have been maintained at 23 ±3 °C for at least 48 hours before the testing.

(2) Repeat steps (III)(2) through (III)(5)(c) of this appendix except use length A instead of length B.

(3) *Surge Test.* (a) One length of sample F must be used to measure the breakdown between conductors while the other length of F must be used to measure core to shield breakdown.

(b) The samples must be capable of withstanding, without damage, a single surge voltage of 20 kilovolts peak between conductors, and 35 kilovolts peak between conductors and the shield as hereinafter described. The surge voltage must be developed from a capacitor discharge through a forming resistor connected in parallel with the dielectric of the test sample. The surge generator constants must be such as to produce a surge of 1.5 × 40 microseconds wave shape.

(c) The shape of the generated wave must be determined at a reduced voltage by connecting an oscilloscope across the forming resistor with the wire sample connected in parallel with the forming resistor. The capacitor bank is charged to the test voltage and then discharged through the forming resistor and test sample. The test sample will be considered to have passed the test if there is no distinct change in the wave shape obtained with the initial reduced voltage compared to that obtained after the application of the test voltage.

(V) The following suggested formats may be used in submitting the test results to RUS:

ENVIRONMENTAL CONDITIONING ____________

FREQUENCY 1 KILOHERTZ

PAIR NUMBER	CAPACITANCE		CONDUCTANCE	
	nF/km	(nF/mile)	micromhos/km	(micromhos/mile)
	Initial	Final	Initial	Final
1	______	______	______	______
2	______	______	______	______
3	______	______	______	______
Average x̄	______	______	______	______

Overall Percent Difference in Average x̄ Capacitance:____________ Conductance: ____________

ENVIRONMENTAL CONDITIONING ____________

FREQUENCY 150 KILOHERTZ

PAIR NUMBER	CAPACITANCE		CONDUCTANCE		ATTENUATION	
	nF/km	(nF/mile)	micromhos/km	(micromhos/mile)	dB/km	(dB/mile)
	Initial	Final	Initial	Final	Initial	Final
1	______	______	______	______	______	______
2	______	______	______	______	______	______
3	______	______	______	______	______	______
Average x̄	______	______	______	______	______	______

Overall Percent Difference in Average x̄ Capacitance:____________ Conductance: ____________ Attenuation:____________

ENVIRONMENTAL CONDITIONING ____________

WATER IMMERSION TEST (1 KILOHERTZ)

PAIR NUMBER	CAPACITANCE			CONDUCTANCE		
	nF/km	(nF/mile)		micromhos/km	(micromhos/mile)	
	Initial	24 hours	Final	Initial	24 hours	Final
1	______	______	______	______	______	______
2	______	______	______	______	______	______
3	______	______	______	______	______	______
Average x̄	______	______	______	______	______	______

Overall Percent Difference in Average x̄ Capacitance:____________ Conductance: ____________

WATER PENETRATION TEST

	Option A		Option B	
	End Leakage grams	Weight Gain grams	End Leakage grams	Penetration mm (in.)
Control	______	______	______	______
Heat Age	______	______	______	______
Humidity Exposure	______	______	______	______
Temperature Cycling.	______	______	______	______

INSULATION COMPRESSION

	Failures
Control	
Heat Age	
Humidity Exposure	
Temperature Cycling	

JACKET SLIP STRENGTH @ 50 °C

	Load in newtons (pound-force)
Control	
Heat Age	
Humidity Exposure	
Temperature Cycling	

FILLER EXUDATION (GRAMS)

Heat Age	
Humidity Exposure	
Temperature Cycle	

SURGE TEST (KILOVOLTS)

Conductor to Conductor	
Shield to Conductors	

APPENDIX B TO § 1755.860—SHEATH SLITTING CORD QUALIFICATION

(I) The test procedures described in this appendix are for qualification of initial and subsequent changes in sheath slitting cords.

(II) *Sample Selection.* All testing must be performed on two 1.2 meters (4 feet) lengths of wire removed sequentially from the same 3 pair, 22 gauge jacketed wire. This wire must not have been exposed to temperatures in excess of 38 °C since its initial cool down after sheathing.

(III) *Test Procedure.* (1) Using a suitable tool, expose enough of sheath slitting cord to permit grasping with needle nose pliers.

(2) The prepared test specimens must be maintained at a temperature of 23 ±1 °C for at least 4 hours immediately prior to and during the test.

(3) Wrap the sheath slitting cord around the plier jaws to ensure a good grip.

(4) Grasp and hold the wire in a convenient position while gently and firmly pulling the sheath slitting cord longitudinally in the direction away from the wire end. The angle of pull may vary to any convenient and functional degree. A small starting notch is permissible.

(5) The sheath slitting cord is considered acceptable if the cord can slit the jacket and/or shield for a continuous length of 0.6 meter (2 feet) without breaking the cord.

APPENDIX C TO § 1755.860—THERMAL REEL WRAP QUALIFICATION

(I) The test procedures described in this appendix are for qualification of initial and subsequent changes in thermal reel wraps.

(II) *Sample Selection.* All testing must be performed on two 450 millimeter (18 inch) lengths of wire removed sequentially from the same 3 pair, 22 gauge jacketed wire. This wire must not have been exposed to temperatures in excess of 38 °C since its initial cool down after sheathing.

(III) *Test Procedure.* (1) Place the two samples on an insulating material such as wood, etc.

(2) Tape thermocouples to the jackets of each sample to measure the jacket temperature.

(3) Cover one sample with the thermal reel wrap.

(4) Expose the samples to a radiant heat source capable of heating the uncovered jacket sample to a minimum of 71 °C. A 600 watt photoflood lamp or an equivalent lamp having the light spectrum approximately that of the sun shall be used.

(5) The height of the lamp above the jacket shall be 380 millimeters (15 inches) or a height that produces the 71 °C jacket temperature on the unwrapped sample.

(6) After the samples have stabilized at the temperature, the jacket temperatures of the samples must be recorded after one hour of exposure to the heat source.

(7) Compute the temperature difference between the jackets.

(8) For the thermal reel wrap to be acceptable to RUS, the temperature differences between the jacket with the thermal reel wrap and the jacket without the reel wrap must be greater than or equal to 17 °C.

[58 FR 61004, Nov. 19, 1993, as amended at 60 FR 1711, Jan. 5, 1995; 69 FR 18803, Apr. 9, 2004]

§§ 1755.861–1755.869 [Reserved]

§ 1755.870 RUS specification for terminating cables.

(a) *Scope.* (1) This section establishes the requirements for terminating cables used to connect incoming outside plant cables to the vertical side of the main distributing frame in a telephone central office.

(i) The conductors are solid tinned copper, individually insulated with extruded solid dual insulating compounds.

(ii) The insulated conductors are twisted into pairs which are then stranded or oscillated to form a cylindrical core.

(iii) The cable structure is completed by the application of a core wrap, a shield, and a polyvinyl chloride jacket.

(2) The number of pairs and gauge size of conductors which are used within the RUS program are provided in the following table:

American Wire Gauge (AWG)	22	24
Number of Pairs	12	12
	50	50
	100	100
	200	200
	300	300
	400	400
	600	600
	800	800

NOTE: Cables larger in pair sizes from those shown in this table shall meet all the requirements of this section.

(3) All cables sold to RUS borrowers for projects involving RUS loan funds under this section must be accepted by RUS Technical Standards Committee "A" (Telephone). For cables manufactured to the specification of this section, all design changes to an accepted design must be submitted for acceptance. RUS will be the sole authority on what constitutes a design change.

(4) Materials, manufacturing techniques, or cable designs not specifically addressed by this section may be allowed if accepted by RUS. Justification for acceptance of modified materials, manufacturing techniques, or cable designs shall be provided to substantiate product utility and long term stability and endurance.

(5) The American National Standard Institute/Electronic Industries Association (ANSI/EIA) 359–A–84, EIA Standard Colors for Color Identification and Coding, referenced in this section is incorporated by reference by RUS. This incorporation by reference was approved by the Director of the Federal Register in accordance with 5 U.S.C. 552(a) and 1 CFR part 51. Copies of ANSI/EIA 359–A–84 are available for inspection during normal business hours at RUS, room 2845, U.S. Department of Agriculture, Washington, DC 20250–1500, or at the National Archives and Records Administration (NARA). For information on the availability of this material at NARA, call 202–741–6030, or go to: *http://www.archives.gov/federal_register/code_of_federal_regulations/ibr_locations.html.* Copies are available from Global Engineering Documents, 15 Inverness Way East, Englewood, CO 80112, telephone number (303) 792–2181.

(6) American Society for Testing and Materials Specifications (ASTM) B 33–91, Standard Specification for Tinned Soft or Annealed Copper Wire for Electrical Purposes; ASTM B 736–92a Standard Specification for Aluminum, Aluminum Alloy and Aluminum-Clad Steel Cable Shielding Stock; ASTM D 1248–84 (1989), Standard Specification for Polyethylene Plastics Molding and Extrusion Materials; ASTM D 1535–89, Standard Test Method for Specifying Color by the Munsell System; ASTM D 2287–81 (Reapproved 1988), Standard Specification for Nonrigid Vinyl Chloride Polymer and Copolymer Molding and Extrusion Compounds; ASTM D 2436–85, Standard Specification for Forced-Convection Laboratory Ovens for Electrical Insulation; ASTM D 2633–82 (Reapproved 1989), Standard Methods of Testing Thermoplastic Insulations and Jackets for Wire and Cable; ASTM D 4101–82 (1988), Standard Specification for Propylene Plastic Injection and Extrusion Materials; ASTM D 4565–90a, Standard Test Methods for Physical and Environmental Performance Properties of Insulations and Jackets for Telecommunications Wire and Cable; ASTM D 4566–90, Standard Test Methods for Electrical Performance Properties of Insulations and Jackets for Telecommunications Wire and Cable; and ASTM E 29–90, Standard Practice for Using Significant Digits in Test Data to Determine Conformance with Specifications, referenced in this section are incorporated by reference by RUS. These incorporations by references were approved by the Director of the Federal Register in accordance with 5 U.S.C. 552(a) and 1 CFR part 51. Copies of the ASTM standards are available for inspection during normal business hours at RUS, room 2845, U.S. Department of Agriculture, Washington, DC 20250–1500, or at the National Archives and Records Administration (NARA). For information on the availability of this material at NARA, call 202–741–6030, or go to: *http://www.archives.gov/federal_register/code_of_federal_regulations/ibr_locations.html.* Copies are available

from ASTM, 1916 Race Street, Philadelphia, Pennsylvania 19103–1187, telephone number (215) 299–5585.

(7) American National Standards Institute/National Fire Protection Association (ANSI/NFPA), NFPA 70–1993 National Electrical Code referenced in this section is incorporated by reference by RUS. This incorporation by reference was approved by the Director of the Federal Register in accordance with 5 U.S.C. 552(a) and 1 CFR part 51. A copy of the ANSI/NFPA standard is available for inspection during normal business hours at RUS, room 2845, U.S. Department of Agriculture, Washington, DC 20250–1500, or at the National Archives and Records Administration (NARA). For information on the availability of this material at NARA, call 202–741–6030, or go to: *http://www.archives.gov/federal_register/code_of_federal_regulations/ibr_locations.html.* Copies are available from NFPA, Batterymarch Park, Quincy, Massachusetts 02269, telephone number 1 (800) 344–3555.

(8) Underwriters Laboratories Inc. (UL) 1666, Standard Test for Flame Propagation Height of Electrical and Optical-Fiber Cables Installed Vertically in Shafts, dated January 22, 1991, referenced in this section is incorporated by reference by RUS. This incorporation by reference was approved by the Director of the Federal Register in accordance with 5 U.S.C. 552(a) and 1 CFR part 51. A copy of the UL standard is available for inspection during normal business hours at RUS, room 2845, U.S. Department of Agriculture, Washington, DC 20250–1500, or at the National Archives and Records Administration (NARA). For information on the availability of this material at NARA, call 202–741–6030, or go to: *http://www.archives.gov/federal_register/code_of_federal_regulations/ibr_locations.html.* Copies are available from UL Inc., 333 Pfingsten Road, Northbrook, Illinois 60062–2096, telephone number (708) 272–8800.

(b) *Conductors and conductor insulation.* (1) Each conductor shall be a solid round wire of commercially pure annealed tin coated copper. Conductors shall meet the requirements of the American Society for Testing and Materials (ASTM) B 33–91 except that requirements for *Dimensions and Permissible Variations* are waived.

(2) Joints made in conductors during the manufacturing process may be brazed, using a silver alloy solder and nonacid flux, or they may be welded using either an electrical or cold welding technique. In joints made in uninsulated conductors, the two conductor ends shall be butted. Splices made in insulated conductors need not be butted but may be joined in a manner acceptable to RUS.

(3) The tensile strength of any section of a conductor, containing a factory joint, shall not be less than 85 percent of the tensile strength of an adjacent section of the solid conductor of equal length without a joint.

(4) Engineering Information: The sizes of wire used and their nominal diameters shall be as shown in the following table:

AWG	Nominal diameter	
	Millimeters	(Inches)
22	0.643	(0.0253)
24	0.511	(0.0201)

(5) Each conductor shall be insulated with a primary layer of natural or white solid, insulating grade, high density polyethylene or crystalline propylene/ethylene copolymer and an outer skin of colored, solid, insulating grade, polyvinyl chloride (PVC) using one of the insulating materials listed in paragraphs (b)(5)(i) through (iii) of this section.

(i) The polyethylene raw material selected to meet the requirements of this section shall be Type III, Class A, Category 4 or 5, Grade E9, in accordance with ASTM D 1248–84 (1989).

(ii) The crystalline propylene/ethylene raw material selected to meet the requirements of this section shall be Class PP 200B 40003 E11 in accordance with ASTM D 4101–82 (1988).

(iii) The PVC raw material selected to meet the requirements of this section shall be either Type PVC–64751E3XO, Type PVC–76751E3XO, or Type PVC–77751E3XO in accordance with ASTM D 2287–81 (1988).

(iv) Raw materials intended as conductor insulation furnished to these requirements shall be free from dirt, metallic particles, and other foreign matter.

(v) All insulating raw materials shall be accepted by RUS prior to their use.

(6) All conductors in any single length of cable shall be insulated with the same type of material.

(7) A permissible overall performance level of faults in conductor insulation when using the test procedures in paragraph (b)(8) of this section shall average not greater than one fault per 12,000 conductor meters (40,000 conductor feet) for each gauge of conductor.

(8) The test used to determine compliance with paragraph (b)(7) of this section shall be conducted as follows:

(i) Samples tested shall be taken from finished cables selected at random from standard production cable. The samples tested shall contain a minimum of 300 conductor meters (1,000 conductor feet) for cables sizes less than 50 pairs and 1,500 conductor meters (5,000 conductor feet) for cables sizes greater than or equal to 50 pairs. No further sample need be taken from the same cable production run within 6,000 cable meters (20,000 cable feet) of the original test sample from that run.

(ii) The cable sample shall have its jacket, shield, and core wrap removed and its core shall be immersed in tap water for a minimum period of 6 hours. In lieu of removing the jacket, shield, and core wrap from the core, the entire cable may be tested. In this case, the core shall be completely filled with tap water, under pressure; then the cable assembly shall be immersed for a minimum period of 6 hours. With the cable core still fully immersed, except for end connections, the insulation resistance (IR) of all conductors to water shall be measured using a direct current (dc) voltage of 100 volts to 550 volts.

(iii) An IR value of less than 500 megohms for any individual insulated conductor tested at or corrected to a temperature of 23 °C is considered a failure. If the cable sample is more than 7.5 meters (25 feet) long, all failing conductors shall be retested and reported in 7.5 meter (25 foot) segments.

(iv) The pair count, gauge, footage, and number of insulation faults shall be recorded. This information shall be retained on a 6 month running basis for review by RUS when requested.

(v) A fault rate, in a continuous length in any one reel, in excess of one fault per 3,000 conductor meters (10,000 conductor feet) due to manufacturing defects is cause for rejection. A minimum of 6,000 conductor meters (20,000 conductor feet) is required to develop a noncompliance in a reel.

(9) Repairs to the conductor insulation during manufacturing are permissible. The method of repair shall be accepted by RUS prior to its use. The repaired insulation shall be capable of meeting the relevant electrical requirements of this section.

(10) All repaired sections of insulation shall be retested in the same manner as originally tested for compliance with paragraph (b)(7) of this section.

(11) The colored composite insulating material removed from or tested on the conductor, from a finished cable, shall be capable of meeting the following performance requirements:

Property	Composite insulation
Tensile Strength, Minimum Megapascals (MPa) (Pounds per square inch (psi))	16.5 (2400)
Ultimate Elongation Percent, Minimum	125
Cold Bend Failures, Maximum	0/10
Shrinkback, Maximum Millimeter (mm) (Inches (in.))	9.5 (3/8)
Adhesion, Maximum Newtons (N) (Pound-force (lbf))	13.3 (3)
Compression Minimum, N (lbf)	1780 (400)

(12) *Testing procedures.* The procedures for testing the composite insulation samples for compliance with paragraph (b)(11) of this section shall be as follows:

(i) *Tensile strength and ultimate elongation.* Samples of the insulation material, removed from the conductor, shall be tested in accordance with ASTM D 2633–82(1989), except that the speed of jaw separation shall be 50 millimeters/minute (50 mm/min) (2 inches/minute (2 in./min)).

NOTE: Quality assurance testing at a jaw separation speed of 500 mm/min (20 in./min) is permissible. Failures at this rate shall be retested at the 50 mm/min (2 in./min) rate to determine specification compliance.

(ii) *Cold bend.* Samples of the insulation material on the conductor shall be tested in accordance with ASTM D 4565–90a at a temperature of −40±1 °C with a mandrel diameter of 6 mm (0.25 in.). There shall be no cracks visible to normal or corrected-to-normal vision.

(iii) *Shrinkback.* Samples of insulation shall be tested for four hours at a temperature of 115±1 °C in accordance with ASTM D 4565–90a.

(iv) *Adhesion.* Samples of insulation material on the conductor shall be tested in accordance with ASTM D 4565–90a with a crosshead speed of 50 mm/min (2 in./min).

(v) *Compression.* Samples of the insulation material on the conductor shall be tested in accordance with ASTM D 4565–90a with a crosshead speed of 5 mm/min (0.2 in./min).

(13) Other methods of testing may be used if acceptable to RUS.

(c) *Identification of pairs and twisting of pairs.* (1) The PVC skin shall be colored to identify:

(i) The tip and ring conductor of each pair; and

(ii) Each pair in the completed cable.

(2) The colors used to provide identification of the tip and ring conductor of each pair shall be as shown in the following table:

Pair No.	Color	
	Tip	Ring
1	White	Blue
2	White	Orange
3	White	Green
4	White	Brown
5	White	Slate
6	Red	Blue
7	Red	Orange
8	Red	Green
9	Red	Brown
10	Red	Slate
11	Black	Blue
12	Black	Orange
13	Black	Green
14	Black	Brown
15	Black	Slate
16	Yellow	Blue
17	Yellow	Orange
18	Yellow	Green
19	Yellow	Brown
20	Yellow	Slate
21	Violet	Blue
22	Violet	Orange
23	Violet	Green
24	Violet	Brown
25	Violet	Slate

(3) *Standards of color.* The colors of the insulated conductors supplied in accordance with this section are specified in terms of the Munsell Color System (ASTM D 1535–89) and shall comply with the "Table of Wire and Cable Limit Chips" as defined in ANSI/EIA–359–A–84. (Visual color standards meeting these requirements may be obtained directly from the Munsell Color Company, Inc., 2441 North Calvert Street, Baltimore, Maryland 21218).

(4) Positive identification of the tip and ring conductors of each pair by marking each conductor of a pair with the color of its mate is permissible. The method of marking shall be accepted by RUS prior to its use.

(5) Other methods of providing positive identification of the tip and ring conductors of each pair may be employed if accepted by RUS prior to its use.

(6) The insulated conductors shall be twisted into pairs.

(7) In order to provide sufficiently high crosstalk isolation, the pair twists shall be designed to enable the cable to meet the capacitance unbalance and the crosstalk loss requirements of paragraphs (h)(2), (h)(3), and (h)(4) of this section.

(8) The average length of pair twists in any pair in the finished cable, when measured on any 3 meter (m) (10 foot (ft)) length, shall not exceed 152 mm (6 in.).

(d) *Forming of the cable core.* (1) Twisted pairs shall be assembled in such a way as to form a substantially cylindrical group.

(2) When desired for lay-up reasons, the basic group may be divided into two or more subgroups called units.

(3) Each group, or unit in a particular group, shall be enclosed in bindings of the colors indicated for its particular pair count. The pair count, indicated by the color of insulation, shall be consecutive as indicated in paragraph (d)(5) of this section through units in a group.

(4) Threads or tapes used as binders shall be nonhygroscopic and nonwicking. The threads shall consists of a suitable number of ends of each color arranged as color bands. When tapes are used as binders, they shall be colored. Binders shall be applied with a lay of not more than 100 mm (4 in.). The colored binders shall be readily

recognizable as the basic intended color and shall be distinguishable from all other colors.

(5) The colors of the bindings and their significance with respect to pair count shall be as shown in the following table:

Group No.	Color of bindings	Group pair count
1	White-Blue	1–25
2	White-Orange	26–50
3	White-Green	51–75
4	White-Brown	76–100
5	White-Slate	101–125
6	Red-Blue	126–150
7	Red-Orange	151–175
8	Red-Green	176–200
9	Red-Brown	201–225
10	Red-Slate	226–250
11	Black-Blue	251–275
12	Black-Orange	276–300
13	Black-Green	301–325
14	Black-Brown	326–350
15	Black-Slate	351–375
16	Yellow-Blue	376–400
17	Yellow-Orange	401–425
18	Yellow-Green	426–450
19	Yellow-Brown	451–475
20	Yellow-Slate	476–500
21	Violet-Blue	501–525
22	Violet-Orange	526–550
23	Violet-Green	551–575
24	Violet-Brown	576–600

(6) The use of the white unit binder in cables of 100 pair or less is optional.

(7) When desired for manufacturing reasons, two or more 25 pair groups may be bound together with nonhygroscopic and nonwicking threads or tapes into super-units. The group binders and the super-unit binders shall be colored such that the combination of the two binders shall positively identify each 25 pair group from every other 25 pair group in the cable.

(8) Super-unit binders shall be of the colors shown in the following table:

SUPER-UNIT BINDER COLORS

Pair No.	Binder color
1–600	White
601–1200	Red

(e) *Core wrap.* (1) The core shall be completely covered with a layer of nonhygroscopic and nonwicking dielectric material. The core wrap shall be applied with an overlap.

(2) The core wrap shall provide a sufficient heat barrier to prevent visible evidence of conductor insulation deformation or adhesion between conductors, caused by adverse heat transfer during the jacketing operation.

(3) Engineering Information: If required for manufacturing reasons, white or uncolored binders of nonhygroscopic and nonwicking material may be applied over the core and/or core wrap.

(f) *Shield.* (1) An aluminum shield, plastic coated on one side, shall be applied longitudinally over the core wrap.

(2) The shield may be applied over the core wrap with or without corrugations (smooth) and shall be bonded to the outer jacket.

(3) The shield overlap shall be a minimum of 3 mm (0.125 in.) for cables with core diameters of 15 mm (0.625 in.) or less and a minimum of 6 mm (0.25 in.) for cables with core diameters greater than 15 mm (0.625 in.). The core diameter is defined as the diameter under the core wrap and binding.

(4) General requirements for application of the shielding material shall be as follows:

(i) Successive lengths of shielding tapes may be joined during the manufacturing process by means of cold weld, electric weld, soldering with a nonacid flux, or other acceptable means;

(ii) The metal shield with the plastic coating shall have the coating removed prior to joining the metal ends together. After joining, the plastic coating shall be restored without voids using good manufacturing techniques;

(iii) The shields of each length of cable shall be tested for continuity. A one meter (3 ft) section of shield containing a factory joint shall exhibit not more than 110 percent of the resistance of a shield of equal length without a joint;

(iv) The breaking strength of any section of a shield tape containing a factory joint shall not be less than 80 percent of the breaking strength of an adjacent section of the shield of equal length without a joint;

(v) The reduction in thickness of the shielding material due to the corrugating or application process shall be kept to a minimum and shall not exceed 10 percent at any spot; and

(vi) The shielding material shall be applied in such a manner as to enable the cable to pass the bend test as specified in paragraph (i)(1) of this section.

(5) The dimensions of the uncoated aluminum tape shall be 0.2030±0.0254 mm (0.0080±0.0010 in.).

(6) The aluminum tape shall conform to either Alloy AA–1100–0, AA–1145–0, or AA–1235–0 as covered in the latest edition of Aluminum Standards and Data, issued by the Aluminum Association, except that requirements for tensile strength are waived.

(7) The single-sided plastic coated aluminum shield shall conform to the requirements of ASTM B 736–92a, Type I Coating, Class 1 or 2, or Type II Coating, Class 1. The minimum thickness of the Type I Coating shall be 0.038 mm (0.0015 in.). The minimum thickness of the Type II Coating shall be 0.008 mm (0.0003 in.).

(8) The plastic coated aluminum shield shall be tested for resistance to water migration by immersing a one meter (3 ft) length of tape under a one meter (3 ft) head of water containing a soluble dye plus 0.25 percent (%) wetting agent.

(i) After a minimum of 5 minutes, no dye shall appear between the interface of the shield tape and the plastic coating.

(ii) The actual test method shall be acceptable to RUS.

(9) The bond between the plastic coated shield and the jacket shall conform to the following requirements:

(i) Prepare test strips approximately 200 mm (8 in.) in length. Slit the jacket and shield longitudinally to produce 4 strips evenly spaced and centered in 4 quadrants on the jacket circumference. One of the strips shall be centered over the overlapped edge of the shielding tape. The strips shall be 13 mm (0.5 in.) wide. For cable diameters less than 19 mm (0.75 in.) make two strips evenly spaced.

(ii) Separate the shield and jacket for a sufficient distance to allow the shield and jacket to be fitted in the upper and lower jaws of a tensile machine. Record the maximum force required to separate the shield and jacket to the nearest newton (pound-force). Repeat this action for each test strip.

(iii) The force required to separate the jacket from the shield shall not be less than 9 N (2 lbf) for any individual strip when tested in accordance with paragraph (f)(9)(ii) of this section. The average force for all strips of any cable shall not be less than 18 N (4 lbf).

(g) *Cable jacket and extraneous material.* (1) The jacket shall provide the cable with a tough, flexible, protective covering which can withstand stresses reasonably expected in normal installation and service.

(2) The jacket shall be free from holes, splits, blisters, or other imperfections and shall be as smooth and concentric as is consistent with the best commercial practice.

(3) The raw material used for the cable jacket shall be one of the following four types:

(i) Type PVC–55554EOXO in accordance with ASTM D 2287–81(1988);

(ii) Type PVC–65554EOXO in accordance with ASTM D 2287–81(1988);

(iii) Type PVC–55556EOXO in accordance with ASTM D 2287–81(1988); or

(iv) Type PVC–66554EOXO in accordance with ASTM D 2287–81(1988).

(4) The jacketing material removed from or tested on the cable shall be capable of meeting the following performance requirements:

Property	Jacket performance
Tensile Strength-Unaged Minimum, MPa (psi)	13.8 (2000)
Ultimate Elongation-Unaged Minimum, Percent (%)	200
Tensile Strength-Aged Minimum, % of original value	80
Ultimate Elongation-Aged Minimum, % of original value	50
Impact Failures, Maximum	2/10

(5) *Testing procedures.* The procedures for testing the jacket samples for compliance with paragraph (g)(4) of this section shall be as follows:

(i) *Tensile strength and ultimate elongation-unaged.* The test shall be performed in accordance with ASTM D 2633–82(1989), using a jaw separation speed of 50 mm/min (2 in./min).

NOTE: Quality assurance testing at a jaw separation speed of 500 mm/min (20 in./min) is permissible. Failures at this rate shall be retested at the 50 mm/min (2 in./min) rate to determine specification compliance.

(ii) *Tensile strength and ultimate elongation-aged.* The test shall be performed in accordance with paragraph (g)(5)(i) of this section after being aged for 7 days at a temperature of 100±1 °C in a circulating air oven conforming to ASTM D 2436–85.

(iii) *Impact*. The test shall be performed in accordance with ASTM D 4565–90a using an impact force of 4 newton-meter (3 pound force-foot) at a temperature of −10±1 °C. The cylinder shall strike the sample at the shield overlap. A crack or split in the jacket constitutes failure.

(6) *Jacket thickness*. The nominal jacket thickness shall be as specified in the following table. The test method used shall be either the End Sample Method (paragraph (g)(6)(i) of this section) or the Continuous Uniformity Thickness Gauge Method (paragraph (g)(6)(ii) of this section):

No. of pairs	Nominal jacket thickness mm (in.)
25 or less	1.4 (0.055)
50	1.5 (0.060)
100	1.7 (0.065)
200	1.9 (0.075)
300	2.2 (0.085)
400	2.4 (0.095)
600	2.9 (0.115)
800 and over	3.3 (0.130)

(i) *End sample method*. The jacket shall be capable of meeting the following requirements:

Minimum Average Thickness—90% of nominal thickness

Minimum Thickness—70% of nominal thickness

(ii) *Continuous uniformity thickness gauge method*. (A) The jacket shall be capable of meeting the following requirements:

Minimum Average Thickness—90% of nominal thickness

Minimum (Min.) Thickness—70 % of nominal thickness

Maximum (Max.) Eccentricity—55%

Eccentricity = Max. Thickness—Min. Thickness (Average Thickness) × 100

(B) *Maximum and minimum thickness values*. The maximum and minimum thickness values shall be based on the average of each axial section.

(7) The color of the jacket shall be either black or dark grey in conformance with the Munsell Color System specified in ASTM D 1535–89.

(8) There shall be no water or other contaminants in the finished cable which would have a detrimental effect on its performance or its useful life.

(h) *Electrical requirements*—(1) *Mutual capacitance and conductance*. (i) The average mutual capacitance (corrected for length) of all pairs in any reel shall not exceed the following when tested in accordance with ASTM D 4566–90 at a frequency of 1.0±0.1 kilohertz (kHz) and a temperature of 23±3 °C:

Number of cable pairs	Mutual capacitance	
	Nanofarad/ kilometer	(Nanofarad/ mile)
12	52±4	(83±7)
Over 12	52±2	(83±4)

(ii) The root mean square (rms) deviation of the mutual capacitance of all pairs from the average mutual capacitance of that reel shall not exceed 3.0 % when calculated in accordance with ASTM D 4566–90.

(iii) The mutual conductance (corrected for length and gauge) of any pair shall not exceed 3.7 micromhos/kilometer (micromhos/km) (6.0 micromhos/mile) when tested in accordance with ASTM D 4566–90 at a frequency of 1.0±0.1 kHz and a temperature of 23±3 °C.

(2) *Pair-to-pair capacitance unbalance*. The capacitance unbalance as measured on the completed cable shall not exceed 45.3 picofarad/kilometer (pF/km) (25 picofarad/1000 ft (pF/1000 ft)) rms when tested in accordance with ASTM D 4566–90 at a frequency of 1.0±0.1 kHz and a temperature of 23±3 °C.

(3) *Pair-to-ground capacitance unbalance*. (i) The average capacitance unbalance as measured on the completed cable shall not exceed 574 pF/km (175 pF/1000 ft) when tested in accordance with ASTM D 4566–90 at a frequency of 1±0.1 kHz and a temperature of 23±3 °C.

(ii) When measuring pair-to-ground capacitance unbalance all pairs except the pair under test are grounded to the shield except when measuring cable containing super-units in which case all other pairs in the same super-unit shall be grounded to the shield.

(iii) Pair-to-ground capacitance unbalance may vary directly with the length of the cable.

(4) *Crosstalk loss*. (i) The rms output-to-output far-end crosstalk loss (FEXT) measured on the completed cable in accordance with ASTM D 4566–90 at a test frequency of 150 kHz shall

not be less than 68 decibel/kilometer (dB/km) (73 decibel/1000 ft (dB/1000 ft)). The rms calculation shall be based on the combined total of all adjacent and alternate pair combinations within the same layer and center to first layer pair combinations.

(ii) The FEXT crosstalk loss between any pair combination of a cable shall not be less than 58 dB/km (63 dB/1000 ft) at a frequency of 150 kHz. If the loss K_o at a frequency F_o for length L_o is known, then K_x can be determined for any other frequency F_x or length L_x by:

$$\text{FEXT loss}\,(K_x) = K_o - 20\log 10\frac{F_x}{F_o} - 10\log 10\frac{L_x}{L_o}$$

(iii) The near-end crosstalk loss (NEXT) as measured within and between units of a completed cable in accordance with ASTM D 4566–90 at a frequency of 772 kHz shall not be less than the following mean minus sigma (M-S) crosstalk requirement for any unit within the cable:

Unit size	M-S decibel (dB)
Within Unit:	
12 and 13 pairs	56
18 and 25 pairs	60
Between Unit:	
Adjacent 13 pairs	65
Adjacent 25 pairs	66
Nonadjacent (all)	81

Where M-S is the Mean near-end coupling loss based on the combined total of all pair combinations, less one Standard Deviation, Sigma, of the mean value.

(5) *Insulation resistance.* Each insulated conductor in each length of completed cable, when measured with all other insulated conductors and the shield grounded, shall have an insulation resistance of not less than 152 megohm-kilometer (500 megohm-mile) at 20±1 °C. The measurement shall be made in accordance with the procedures of ASTM D 4566–90.

(6) *High voltage test.* (i) In each length of completed cable, the dielectric strength of the insulation between conductors shall be tested in accordance with ASTM D 4566–90 and shall withstand, for 3 seconds, a direct current (dc) potential whose value is not less than:

(A) 3.6 kilovolts for 22-gauge conductors; or

(B) 3.0 kilovolts for 24-gauge conductors.

(ii) In each length of completed cable, the dielectric strength between the shield and all conductors in the core shall be tested in accordance with ASTM D 4566–90 and shall withstand, for 3 seconds, a dc potential whose value is not less than 10 kilovolts.

(7) *Conductor resistance.* The dc resistance of any conductor shall be measured in the completed cable in accordance with ASTM D 4566–90 and shall not exceed the following values when measured at or corrected to a temperature of 20±1 °C:

AWG	Maximum resistance	
	ohms/kilometer	(ohms/1000 ft)
22	60.7	(18.5)
24	95.1	(29.0)

(8) *Resistance unbalance.* (i) The difference in dc resistance between the two conductors of a pair in the completed cable shall not exceed the values listed in this paragraph when measured in accordance with the procedures of ASTM D 4566–90:

AWG	Resistance unbalance	Maximum for any reel
	Average percent	Individual pair percent
22	1.5	4.0
24	1.5	5.0

(ii) The resistance unbalance between tip and ring conductors shall be random with respect to the direction of unbalance. That is, the resistance of the tip conductors shall not be consistently higher with respect to the ring conductors and vice versa.

(9) *Electrical variations.* (i) Pairs in each length of cable having either a ground, cross, short, or open circuit condition shall not be permitted.

(ii) The maximum number of pairs in a cable which may vary as specified in paragraph (h)(9)(iii) of this section from the electrical parameters given in this section are listed in this paragraph. These pairs may be excluded from the arithmetic calculation:

Nominal pair count	Maximum No. of pairs with allowable electrical variation
12–100	1
101–300	2
301–400	3
401–600	4
601 and above	6

(iii) *Parameter variations*—(A) *Capacitance unbalance-to-ground.* If the cable fails either the maximum individual pair or average capacitance unbalance-to-ground requirement and all individual pairs are 3280 pF/km (1000 pF/1000 ft) or less the number of pairs specified in paragraph (h)(9)(ii) of this section may be eliminated from the average and maximum individual calculations.

(B) *Resistance unbalance.* Individual pair of not more than 7 percent for all gauges.

(C) *Far end crosstalk.* Individual pair combination of not less than 52 dB/km (57 dB/1000 ft).

NOTE: RUS recognizes that in large pair count cables (600 pair and above) a cross, short, or open circuit condition occasionally may develop in a pair which does not affect the performance of the other cable pairs. In these circumstances rejection of the entire cable may be economically unsound or repairs may be impractical. In such circumstances the manufacturer may desire to negotiate with the customer for acceptance of the cable. No more than 0.5 percent of the pairs may be involved.

(i) *Mechanical requirements*—(1) *Cable cold bend test.* The completed cable shall be capable of meeting the requirements of ASTM D 4565–90a after conditioning at −20 ±2 °C except the mandrel diameters shall be as specified below:

Cable outside diameter	Mandrel diameter
<40 mm (1.5 in.)	15x
≥40 mm (1.5 in.)	20x

(2) *Cable flame test.* The completed cable shall be capable of meeting a maximum flame height of 3.7 m (12.0 ft) when tested in accordance with Underwriters Laboratories (UL) 1666 dated January 22, 1991.

(3) *Cable listing.* All cables manufactured to the specification of this section at a minimum shall be listed as Communication Riser Cable (Type CMR) in accordance with Sections 800–50 and 800–51(b) of the 1993 National Electrical Code.

(j) *Sheath slitting cord (optional).* (1) Sheath slitting cords may be used in the cable structure at the option of the manufacturer.

(2) When a sheath slitting cord is used it shall be nonhygroscopic and nonwicking, continuous throughout a length of cable, and of sufficient strength to open the sheath without breaking the cord.

(3) Sheath slitting cords shall be capable of consistently slitting the jacket and/or shield for a continuous length of 0.6 m (2 ft) when tested in accordance with the procedure specified in appendix B of this section.

(k) *Identification marker and length marker.* (1) Each length of cable shall be permanently identified as to manufacturer and year of manufacture.

(2) The number of conductor pairs and their gauge size shall be marked on the jacket.

(3) The marking shall be printed on the jacket at regular intervals of not more than 1.5 m (5 ft).

(4) An alternative method of marking may be used if accepted by RUS prior to its use.

(5) The completed cable shall have sequentially numbered length markers in FEET OR METERS at regular intervals of not more than 1.5 m (5 ft) along the outside of the jacket.

(6) The method of length marking shall be such that for any single length of cable, continuous sequential numbering shall be employed.

(7) The numbers shall be dimensioned and spaced to produce good legibility and shall be approximately 3 mm (0.125

in.) in height. An occasional illegible marking is permissible if there is a legible marking located not more than 1.5 m (5 ft) from it.

(8) The method of marking shall be by means of suitable surface markings producing a clear, distinguishable, contrasting marking acceptable to RUS. Where direct or transverse printing is employed, the characters should be indented to produce greater durability of marking. Any other method of length marking shall be acceptable to RUS as producing a marker suitable for the field. Size, shape and spacing of numbers, durability, and overall legibility of the marker shall be considered in acceptance of the method.

(9) The accuracy of the length marking shall be such that the actual length of any cable section is never less than the length indicated by the marking and never more than one percent greater than the length indicated by the marking.

(10) The color of the initial marking for a black colored jacket shall be either white or silver. The color of the initial marking for a dark grey colored jacket shall be either red or black. If the initial marking of the black colored jacket fails to meet the requirements of the preceding paragraphs, it will be permissible to either remove the defective marking and re-mark with the white or silver color or leave the defective marking on the cable and re-mark with yellow. If the initial marking of the dark grey colored jacket fails to meet the requirements of the preceding paragraphs, it will be permissible to either remove the defective marking and re-mark with the red or black color or leave the defective marking on the cable and re-mark with yellow. No further re-marking is permitted. Any re-marking shall be on a different portion of the cable circumference than any existing marking when possible and have a numbering sequence differing from any other existing marking by at least 5,000.

(11) Any reel of cable which contains more than one set of sequential markings shall be labeled to indicate the color and sequence of marking to be used. The labeling shall be applied to the reel and also to the cable.

(l) *Preconnectorized cable* (optional). (1) At the option of the manufacturer and upon request by the purchaser, cables 100 pairs and larger may be factory terminated in 25 pair splicing modules.

(2) The splicing modules shall meet the requirements of RUS Bulletin 345–54, PE–52, RUS Specification for Telephone Cable Splicing Connectors (Incorporated by Reference at §1755.97), and be accepted by RUS prior to their use.

(m) *Acceptance testing and extent of testing.* (1) The tests described in appendix A of this section are intended for acceptance of cable designs and major modifications of accepted designs. RUS decides what constitutes a major modification. These tests are intended to show the inherent capability of the manufacturer to produce cable products having long life and stability.

(2) For initial acceptance, the manufacturer shall submit:

(i) An original signature certification that the product fully complies with each section of the specification;

(ii) Qualification Test Data, per appendix A of this section;

(iii) To periodic plant inspections;

(iv) A certification that the product does or does not comply with the domestic origin manufacturing provisions of the "Buy American" requirements of the Rural Electrification Act of 1938 (7 U.S.C. 901 *et seq.*);

(v) Written user testimonials concerning performance of the product; and

(vi) Other nonproprietary data deemed necessary by the Chief, Outside Plant Branch (Telephone).

(3) For requalification acceptance, the manufacturer shall submit an original signature certification that the product fully complies with each section of the specification, excluding the Qualification Section, and a certification that the product does or does not comply with the domestic origin manufacturing provisions of the "Buy American" requirements of the Rural Electrification Act of 1938 (7 U.S.C. 901 *et seq.*) for acceptance by June 30 every three years. The required data and certification shall have been gathered within 90 days of the submission.

(4) Initial and requalification acceptance requests should be addressed to:

Chairman, Technical Standards Committee "A" (Telephone), Telecommunications Standards Division, Rural Utilities Service, Washington, DC 20250–1500.

(5) *Tests on 100 percent of completed cable.* (i) The shield of each length of cable shall be tested for continuity using the procedures of ASTM D 4566–90.

(ii) Dielectric strength between all conductors and the shield shall be tested to determine freedom from grounds in accordance with paragraph (h)(6)(ii) of this section.

(iii) Each conductor in the completed cable shall be tested for continuity using the procedures of ASTM D 4566–90.

(iv) Dielectric strength between conductors shall be tested to ensure freedom from shorts and crosses in accordance with paragraph (h)(6)(i) of this section.

(v) Each conductor in the completed preconnectorized cable shall be tested for continuity.

(vi) Each length of completed preconnectorized cable shall be tested for split pairs.

(vii) The average mutual capacitance shall be measured on all cables. If the average mutual capacitance for the first 100 pairs tested from randomly selected groups is between 50 and 53 nF/km (80 to 85 nF/mile), the remainder of the pairs need not to be tested on the 100 percent basis. (See paragraph (h)(1) of this section).

(6) *Capability tests.* Tests on a quality assurance basis shall be made as frequently as is required for each manufacturer to determine and maintain compliance with:

(i) Performance requirements for conductor insulation and jacket material;

(ii) Bonding properties of coated or laminated shielding materials;

(iii) Sequential marking and lettering;

(iv) Capacitance unbalance and crosstalk;

(v) Insulation resistance;

(vi) Conductor resistance and resistance unbalance;

(vii) Cable cold bend and cable flame tests; and

(viii) Mutual conductance.

(n) *Summary of records of electrical and physical tests.* (1) Each manufacturer shall maintain a suitable summary of records for a period of at least 3 years for all electrical and physical tests required on completed cable by this section as set forth in paragraphs (m)(5) and (m)(6) of this section. The test data for a particular reel shall be in a form that it may be readily available to the purchaser or to RUS upon request.

(2) Measurements and computed values shall be rounded off to the number of places of figures specified for the requirement according to ASTM E 29–90.

(o) *Manufacturing irregularities.* (1) Repairs to the shield are not permitted in cable supplied to the end user under this section.

(2) No repairs or defects in the jacket are allowed.

(p) *Preparation for shipment.* (1) The cable shall be shipped on reels unless otherwise specified or agreed to by the purchaser. The diameter of the drum shall be large enough to prevent damage to the cable from reeling or unreeling. The reels shall be substantial and so constructed as to prevent damage to the cable during shipment and handling.

(2) A waterproof corrugated board or other means of protection acceptable to RUS shall be applied to the reel and shall be suitably secured in place to prevent damage to the cable during storage and shipment.

(3) The outer end of the cable shall be securely fastened to the reel head so as to prevent the cable from becoming loose in transit. The inner end of the cable shall be securely fastened in such a way as to make it readily available if required for electrical testing. Spikes, staples, or other fastening devices which penetrate the cable jacket shall not be used. The method of fastening the cable ends shall be accepted by RUS prior to it being used.

(4) Each length of cable shall be wound on a separate reel unless otherwise specified or agreed to by the purchaser.

(5) The arbor hole shall admit a spindle 63 mm (2.5 in.) in diameter without binding. Steel arbor hole liners may be used but shall be acceptable to RUS prior to their use.

(6) Each reel shall be plainly marked to indicate the direction in which it should be rolled to prevent loosening of the cable on the reel.

(7) Each reel shall be stenciled or labeled on either one or both sides with the name of the manufacturer, year of manufacture, actual shipping length, an inner and outer end sequential length marking, description of the cable, reel number and the RUS cable designation:

Cable Designation

CT
Cable Construction
Pair Count
Conductor Gauge

A = Coated Aluminum Shield

P = Preconnectorized Cable

Example: CTAP 100–22

Terminating Cable, Coated Aluminum Shield, Preconnectorized, 100 pairs, 22 AWG.

(8) When preconnectorized cable is shipped, the splicing modules shall be protected to prevent damage during shipment and handling. The protection method shall be acceptable to RUS prior to its use.

(The information collection and recordkeeping requirements of this section have been approved by the Office of Management and Budget (OMB) under control number 0572–0059)

APPENDIX A TO § 1755.870—QUALIFICATION TEST METHODS

(I) The test procedures described in this appendix are for qualification of initial designs and major modifications of accepted designs. Included in paragraph (V) of this appendix are suggested formats that may be used in submitting test results to RUS.

(II) *Sample Selection and Preparation*. (1) All testing shall be performed on lengths removed sequentially from the same 25 pair, 22 gauge jacketed cable. This cable shall not have been exposed to temperatures in excess of 38 °C since its initial cool down after sheathing. The lengths specified are minimum lengths and if desirable from a laboratory testing standpoint longer lengths may be used.

(a) Length A shall be 12 ±0.2 meters (40 ±0.5 feet) long. Prepare the test sample by removing the jacket, shield, and core wrap for a sufficient distance on both ends to allow the insulated conductors to be flared out. Remove sufficient conductor insulation so that appropriate electrical test connections can be made at both ends. Coil the sample with a diameter of 15 to 20 times its sheath diameter. Two lengths are required.

(b) Length B shall be 300 millimeters (1 foot) long. Three lengths are required.

(c) Length C shall be 3 meters (10 feet) long and shall be maintained at 23 ±3 °C for the duration of the test. Two lengths are required.

(2) *Data Reference Temperature*. Unless otherwise specified, all measurements shall be made at 23 ±3 °C.

(III) *Environmental Tests*—(1) *Heat Aging Test*—(a) *Test Samples*. Place one sample each of lengths A and B in an oven or environmental chamber. The ends of sample A shall exit from the chamber or oven for electrical tests. Securely seal the oven exit holes.

(b) *Sequence of Tests*. Sample B referenced in paragraph (III)(1)(a) of this appendix shall be subjected to the insulation compression test outlined in paragraph (III)(2) of this appendix.

(c) *Initial Measurements*. (i) For sample A, measure the open circuit capacitance and conductance for each odd pair at 1, 150, and 772 kilohertz after conditioning the sample at the data reference temperature for 24 hours. Calculate the average and standard deviation for the data of the 13 pairs on a per kilometer (per mile) basis.

(ii) Record on suggested formats in paragraph (V) of this appendix or on other easily readable formats.

(d) *Heat Conditioning*. (i) Immediately after completing the initial measurements, condition the sample for 14 days at a temperature of 65 ±2 °C.

(ii) At the end of this period. Measure and calculate the parameters given in paragraph (III)(1)(c) of this appendix. Record on suggested formats in paragraph (V) of this appendix or on other easily readable formats.

(e) *Overall Electrical Deviation*. (i) Calculate the percent change in all average parameters between the final parameters after conditioning with the initial parameters in paragraph (III)(1)(c) of this appendix.

(ii) The stability of the electrical parameters after completion of this test shall be within the following prescribed limits:

(A) *Capacitance*. The average mutual capacitance shall be within 10 percent of its original value;

(B) The change in average mutual capacitance shall be less than 10 percent over the frequency range of 1 to 150 kilohertz; and

(C) *Conductance*. The average mutual conductance shall not exceed 3.7 micromhos/kilometer (6 micromhos/mile) at a frequency of 1 kilohertz.

(2) *Insulation Compression Test*—(a) *Test Sample B*. Remove jacket, shield, and core wrap being careful not to damage the conductor insulation. Remove one pair from the core and carefully separate and straighten the insulated conductors. Retwist the two

insulated conductors together under sufficient tension to form 10 evenly spaced 360 degree twists in a length of 100 millimeters (4 inches).

(b) *Sample Testing*. Center the mid 50 millimeters (2 inches) of the twisted pair between two smooth rigid parallel metal plates measuring 50 millimeters (2 inches) in length or diameter. Apply a 1.5 volt direct current potential between the conductors, using a light or buzzer to indicate electrical contact between the conductors. Apply a constant load of 67 newtons (15 pound-force) on the sample for one minute and monitor for evidence of contact between the conductors. Record results on suggested formats in paragraph (V) of this appendix or on other easily readable formats.

(3) *Temperature Cycling*. (a) Repeat paragraphs (III)(1)(a) through (III)(1)(c)(ii) of this appendix for a separate set of samples A and B which have not been subjected to prior environmental conditioning.

(b) Immediately after completing the measurements, subject the test samples to 10 cycles of temperature between −40 °C and + 60 °C. The test samples shall be held at each temperature extreme for a minimum of 1.5 hours during each cycle of temperature. The air within the temperature cycling chamber shall be circulated throughout the duration of the cycling.

(c) Repeat paragraphs (III)(1)(d)(ii) through (III)(2)(b) of this appendix.

(IV) *Control Sample*—(1) *Test Samples*. One length of sample B shall have been maintained at 23 ±3 °C for at least 48 hours before the testing.

(2) Repeat paragraphs (III)(2) through (III)(2)(b) of this appendix.

(3) *Surge Test*. (a) One length of sample C shall be used to measure the breakdown between conductors while the other length of C shall be used to measure core to shield breakdown.

(b) The samples shall be capable of withstanding, without damage, a single surge voltage of 20 kilovolts peak between conductors, and 35 kilovolts peak between conductors and the shield as hereinafter described. The surge voltage shall be developed from a capacitor discharge through a forming resistor connected in parallel with the dielectric of the test sample. The surge generator constants shall be such as to produce a surge of 1.5 × 40 microseconds wave shape.

(c) The shape of the generated wave shall be determined at a reduced voltage by connecting an oscilloscope across the forming resistor with the cable sample connected in parallel with the forming resistor. The capacitor bank is charged to the test voltage and then discharged through the forming resistor and test sample. The test sample shall be considered to have passed the test if there is no distinct change in the wave shape obtained with the initial reduced voltage compared to that obtained after the application of the test voltage.

(V) The following suggested formats may be used in submitting the test results to RUS:

Environmental Conditioning ______________

FREQUENCY 1 KILOHERTZ

Pair No.	Capacitance nF/km (nF/mile)		Conductance micromhos/km (micromhos/mile)	
	Initial	Final	Initial	Final
1.				
3.				
5.				
7.				
9.				
11.				
13.				
15.				
17.				
19.				
21.				
23.				
25.				
Average x̄.				
Overall Percent Difference in Average x̄	....	....	....	....

Environmental Conditioning ______________

FREQUENCY 150 KILOHERTZ

Pair No.	Capacitance nF/km (nF/mile)		Conductance micromhos/km (micromhos/mile)	
	Initial	Final	Initial	Final
1.				
3.				
5.				
7.				
9.				
11.				
13.				
15.				
17.				
19.				
21.				
23.				
25.				
Average $\bar{x}$.				
Overall Percent Difference in Average $\bar{x}$	....	....	....	....

Environmental Conditioning ______________

FREQUENCY 772 KILOHERTZ

Pair No.	Capacitance nF/km (nF/mile)		Conductance micromhos/km (micromhos/mile)	
	Initial	Final	Initial	Final
1.				
3.				
5.				
7.				
9.				
11.				
13.				
15.				
17.				
19.				
21.				
23.				
25.				
Average $\bar{x}$.				
Overall Percent Difference in. Average $\bar{x}$	....	....	....	....

	Failures
Insulation Compression:	
Control.	
Heat Age.	
Temperature Cycling.	
Surge Test (kilovolts):	
Conductor-to-Conductor.	
Shield-to-Conductors	....

APPENDIX B TO § 1755.870—SHEATH SLITTING CORD QUALIFICATION

(I) This test procedure described in this appendix is for qualification of initial and subsequent changes in sheath slitting cords.

(II) *Sample selection.* All testing shall be performed on two 1.2 m (4 ft) lengths of cable removed sequentially from the same 25 pair, 22 gauge jacketed cable. This cable shall not have been exposed to temperatures in excess of 38 °C since its initial cool down after sheathing.

(III) *Test procedure.* (1) Using a suitable tool, expose enough of the sheath slitting cord to permit grasping with needle nose pliers.

(2) The prepared test specimens shall be maintained at a temperature of 23 ±1 °C for at least 4 hours immediately prior to and during the test.

(3) Wrap the sheath slitting cord around the plier jaws to ensure a good grip.

(4) Grasp and hold the cable in a convenient position while gently and firmly pulling the sheath slitting cord longitudinally in the direction away from the cable end. The angle of pull may vary to any convenient and functional degree. A small starting notch is permissible.

(5) The sheath slitting cord is considered acceptable if the cord can slit the jacket and/

or shield for a continuous length of 0.6 m (2 ft) without breaking the cord.

[59 FR 30507, June 14, 1994; 59 FR 34899, July 7, 1994, as amended at 60 FR 1711, Jan. 5, 1995; 69 FR 18803, Apr. 9, 2004]

§§ 1755.871–1755.889 [Reserved]

§ 1755.890 RUS specification for filled telephone cables with expanded insulation.

(a) *Scope.* (1) This section covers the requirements for filled telephone cables intended for direct burial installation either by trenching or by direct plowing, for underground application by placement in a duct, or for aerial installation by attachment to a support strand.

(i) The conductors are solid copper, individually insulated with an extruded cellular insulating compound which may be either totally expanded or expanded with a solid skin coating.

(ii) The insulated conductors are twisted into pairs which are then stranded or oscillated to form a cylindrical core.

(iii) For high frequency applications, the cable core may be separated into compartments with screening shields.

(iv) A moisture resistant filling compound is applied to the stranded conductors completely covering the insulated conductors and filling the interstices between pairs and units.

(v) The cable structure is completed by the application of suitable core wrapping material, a flooding compound, a shield or a shield/armor, and an overall plastic jacket.

(2) The number of pairs and gauge size of conductors which are used within the RUS program are provided in the following table:

AWG	19	22	24	26
Pairs	6	6	6	
	12	12	12	
	18	18	18	
	25	25	25	25
		50	50	50
		75	75	75
		100	100	100
		150	150	150
		200	200	200
		300	300	300
		400	400	400
		600	600	600
		900	900	900
		1000	1000	1000
			1200	1200
			1500	1500
			1800	1800
				2100
				2400
				2700

NOTE: Cables larger in pair sizes than those shown in this table must meet all requirements of this section.

(3) Screened cable, when specified, must meet all requirements of this section. The pair sizes of screened cables used within the RUS program are referenced in paragraph (e)(2)(i) of this section.

(4) All cables sold to RUS borrowers for projects involving RUS loan funds under this section must be accepted by RUS Technical Standards Committee "A" (Telephone). For cables manufactured to the specification of this section, all design changes to an accepted design must be submitted for acceptance. RUS will be the sole authority on what constitutes a design change.

(5) Materials, manufacturing techniques, or cable designs not specifically addressed by this section may be allowed if accepted by RUS. Justification for acceptance of modified materials, manufacturing techniques, or cable designs must be provided to substantiate product utility and long-term stability and endurance.

(6) The American National Standard Institute/Insulated Cable Engineers Association, Inc. (ANSI/ICEA) S-84-608-1988, Standard For Telecommunications Cable, Filled, Polyolefin Insulated, Copper Conductor Technical Requirements referenced throughout this section is incorporated by reference by RUS. This incorporation by reference was approved by the Director of the Federal Register in accordance with 5 U.S.C. 552(a) and 1 CFR part 51. Copies

of ANSI/ICEA S-84-608-1988 are available for inspection during normal business hours at RUS, room 2845, U.S. Department of Agriculture, Washington, DC 20250, or at the National Archives and Records Administration (NARA). For information on the availability of this material at NARA, call 202–741–6030, or go to: *http://www.archives.gov/federal_register/code_of_federal_regulations/ibr_locations.html.* Copies are available from ICEA, P. O. Box 440, South Yarmouth, MA 02664, telephone number (508) 394–4424.

(7) American Society for Testing and Materials specifications (ASTM) A 505-87, Standard Specification for Steel, Sheet and Strip, Alloy, Hot-Rolled and Cold-Rolled, General Requirements For; ASTM B 193-87, Standard Test Method for Resistivity of Electrical Conductor Materials; ASTM B 224-80, Standard Classification of Coppers; ASTM B 694-86, Standard Specification for Copper, Copper Alloy, and Copper-Clad Stainless Steel Sheet and Strip for Electrical Cable Shielding; ASTM D 4565-90a, Standard Test Methods for Physical and Environmental Performance Properties of Insulations and Jackets for Telecommunications Wire and Cable; and ASTM D 4566-90, Standard Test Methods for Electrical Performance Properties of Insulations and Jackets for Telecommunications Wire and Cable referenced in this section are incorporated by reference by RUS. These incorporations by references were approved by the Director of the Federal Register in accordance with 5 U.S.C. 552(a) and 1 CFR part 51. Copies of the ASTM standards are available for inspection during normal business hours at RUS, room 2845, U.S. Department of Agriculture, Washington, DC 20250, or at the National Archives and Records Administration (NARA). For information on the availability of this material at NARA, call 202–741–6030, or go to: *http://www.archives.gov/federal_register/code_of_federal_regulations/ibr_locations.html.* Copies are available from ASTM, 1916 Race Street, Philadelphia, PA 19103–1187, telephone number (215) 299–5585.

(b) *Conductors and conductor insulation.* (1) The gauge sizes of the copper conductors covered by this section must be 19, 22, 24, and 26 American Wire Gauge (AWG).

(2) Each conductor must comply with the requirements specified in ANSI/ICEA S-84-608-1988, paragraph 2.1.

(3) Factory joints made in conductors during the manufacturing process must comply with the requirements specified in ANSI/ICEA S-84-608-1988, paragraph 2.2.

(4) The raw materials used for conductor insulation must comply with the requirements specified in ANSI/ICEA S-84-608-1988, paragraphs 3.1 through 3.1.3.

(5) The finished conductor insulation must comply with the requirements specified in ANSI/ICEA S-84-608-1988, paragraphs 3.2.2, 3.2.3, and 3.3.

(6) Insulated conductor must not have an overall diameter greater than 2 millimeters (mm) (0.081 inch (in.)).

(7) A permissible overall performance level of faults in conductor insulation must average not greater than one fault per 12,000 conductor meters (40,000 conductor feet) for each gauge of conductor.

(i) All insulated conductors must be continuously tested for insulation faults during the twinning operation with a method of testing acceptable to RUS. The length count and number of faults must be recorded. The information must be retained for a period of 6 months and be available for review by RUS when requested.

(ii) The voltages for determining compliance with the requirements of this section are as follows:

AWG	Direct Current Voltages (kilovolts)
19	4.5
22	3.6
24	3.0
26	2.4

(8) Repairs to the conductor insulation during manufacture are permissible. The method of repair must be accepted by RUS prior to its use. The repaired insulation must be capable of meeting the relevant electrical requirements of this section.

(9) All repaired sections of insulation must be retested in the same manner as originally tested for compliance with paragraph (b)(7) of this section.

(10) The colored insulating material removed from or tested on the conductor, from a finished cable, must meet the performance requirements specified in ANSI/ICEA S-84-608-1988, paragraphs 3.4.1 through 3.4.6.

(c) *Identification of pairs and twisting of pairs.* (1) The insulation must be colored to identify:

(i) The tip and ring conductor of each pair; and

(ii) Each pair in the completed cable.

(2) The colors to be used in the pairs in the 25 pair group, together with the pair numbers must be in accordance with the table specified in ANSI/ICEA S-84-608-1988, paragraph 3.5.

(3) Positive identification of the tip and ring conductors of each pair by marking each conductor of a pair with the color of its mate is permissible. The method of marking must be accepted by RUS prior to its use.

(4) Other methods of providing positive identification of the tip and ring conductors of each pair may be employed if accepted by RUS prior to its use.

(5) The insulated conductors must be twisted into pairs.

(6) In order to provide sufficiently high crosstalk isolation, the pair twists must be designed to enable the cable to meet the capacitance unbalance and crosstalk loss requirements of paragraphs (k)(5), (k)(6), and (k)(8) this section.

(7) The average length of pair twists in any pair in the finished cable, when measured on any 3 meter (10 foot) length, must not exceed the requirement specified in ANSI/ICEA S-84-608-1988, paragraph 3.5.

(d) *Forming of the cable core.* (1) Twisted pairs must be assembled in such a way as to form a substantially cylindrical group.

(2) When desired for lay-up reasons, the basic group may be divided into two or more subgroups called units.

(3) Each group, or unit in a particular group, must be enclosed in bindings of the colors indicated for its particular pair count. The pair count, indicated by the colors of insulation, must be consecutive as indicated in paragraph (d)(6) of this section through units in a group.

(4) The filling compound must be applied to the cable core in such a way as to provide as near a completely filled core as is commercially practical.

(5) Threads and tapes used as binders must comply with the requirements specified in ANSI/ICEA S-84-608-1988, paragraphs 4.2 and 4.2.1.

(6) The colors of the bindings and their significance with respect to pair count must be as follows:

Group No.	Color of Bindings	Group Pair Count
1	White-Blue	1–25
2	White-Orange	26–50
3	White-Green	51–75
4	White-Brown	76–100
5	White-Slate	101–125
6	Red-Blue	126–150
7	Red-Orange	151–175
8	Red-Green	176–200
9	Red-Brown	201–225
10	Red-Slate	226–250
11	Black-Blue	251–275
12	Black-Orange	276–300
13	Black-Green	301–325
14	Black-Brown	326–350
15	Black-Slate	351–375
16	Yellow-Blue	376–400
17	Yellow-Orange	401–425
18	Yellow-Green	426–450
19	Yellow-Brown	451–475
20	Yellow-Slate	476–500
21	Violet-Blue	501–525
22	Violet-Orange	526–550
23	Violet-Green	551–575
24	Violet-Brown	576–600

(7) The use of the white unit binder in cables of 100 pairs or less is optional.

(8) When desired for manufacturing reasons, two or more 25 pair groups may be bound together with nonhygroscopic and nonwicking threads or tapes into a super-unit. Threads or tapes must meet the requirements specified in paragraph (d)(5) of this section. The group binders and the super-unit binders must be color coded such that the combination of the two binders must positively identify each 25 pair group from every other 25 pair group in the cable. Super-unit binders must be of the color shown in the following table:

SUPER-UNIT BINDER COLORS

Pair Numbers	Binder Color
1–600	White
601–1200	Red
1201–1800	Black
1801–2400	Yellow
2401–3000	Violet
3001–3600	Blue
3601–4200	Orange
4201–4800	Green
4801–5400	Brown
5401–6000	Slate

(9) Color binders must not be missing for more than 90 meters (300 feet) from any 25 pair group or from any subgroup used as part of a super-unit. At any cable cross-section, no adjacent 25 pair groups and no more than one subgroup of any super-unit may have missing binders. In no case must the total number of missing binders exceed three. Missing super-unit binders must not be permitted for any distance.

(10) Any reel of cable which contains missing binders must be labeled indicating the colors and location of the binders involved. The labeling must be applied to the reel and also to the cable.

(e) *Screened cable.* (1) Screened cable must be constructed such that a metallic, internal screen(s) must be provided to separate and provide sufficient isolation between the compartments to meet the requirements of this section.

(2) At the option of the user or manufacturer, identified service pairs providing for voice order and fault location may be placed in screened cables.

(i) The number of service pairs provided must be one per twenty-five operating pairs plus two for a cable size up to and including 400 pairs, subject to a minimum of four service pairs. The pair counts for screened cables are as follows:

SCREENED CABLE PAIR COUNTS

Carrier Pair Count	Service Pairs	Total Pair Count
24	4	28
50	4	54
100	6	106
150	8	158
200	10	210
300	14	314
400	18	418

(ii) The service pairs must be equally divided among the compartments. The color sequence must be repeated in each compartment.

(iii) The electrical and physical characteristics of each service pair must meet all the requirements set forth in this section.

(iv) The colors used for the service pairs must be in accordance with the requirements of paragraph (b)(5) of this section. The color code used for the service pairs together with the service pair number are shown in the following table:

COLOR CODE FOR SERVICE PAIRS

Service Pair No.	Color	
	Tip	Ring
1	White	Red
2	"	Black
3	"	Yellow
4	"	Violet
5	Red	Black
6	"	Yellow
7	"	Violet
8	Black	Yellow
9	"	Violet

(3) The screen tape must comply with the requirements specified in ANSI/ICEA S-84-608-1988, paragraphs 5.1 through 5.4.

(4) The screen tape must be tested for dielectric strength by completely removing the protective coating from one end to be used for grounding purposes.

(i) Using an electrode, over a 30 centimeter (1 foot) length, apply a direct current (dc) voltage at the rate of rise of 500 volts/second until failure.

(ii) No breakdown should occur below 8 kilovolts.

(f) *Filling compound.* (1) After or during the stranding operation and prior to application of the core wrap, filling compound must be applied to the cable core. The compound must be as nearly colorless as is commercially feasible and consistent with the end product requirements and pair identification.

(2) The filling compound must comply with the requirements specified in ANSI/ICEA S-84-608-1988, paragraphs 4.4 through 4.4.4.

(3) The individual cable manufacturer must satisfy RUS that the filling compound selected for use is suitable for its intended application. The filling compound must be applied to the cable

in such a manner that the cable components will not be degraded.

(g) *Core wrap.* (1) The core wrap must comply with the requirements specified in ANSI/ICEA-S-84-608-1988, paragraph 4.3.

(2) If required for manufacturing reasons, white or colored binders of nonhygroscopic and nonwicking material may be applied over the core and/or wrap. When used, binders must meet the requirements specified in paragraph (d)(5) of this section.

(3) Sufficient filling compound must have been applied to the core wrap so that voids or air spaces existing between the core and the inner side of the core wrap are minimized.

(h) *Flooding compound.* (1) Sufficient flooding compound must be applied on all sheath interfaces so that voids and air spaces in these areas are minimized. When the optional armored design is used, the flooding compound must be applied between the core wrap and shield, between the shield and armor, and between the armor and the jacket so that voids and air spaces in these areas are minimized. The use of floodant over the outer metallic substrate is not required if uniform bonding, per paragraph (i)(7) of this section, is achieved between the plastic-clad metal and the jacket.

(2) The flooding compound must comply with the requirements specified in ANSI/ICEA S-84-608-1988, paragraph 4.5 and the jacket slip test requirements of appendix A, paragraph (III)(5) of this section.

(3) The individual cable manufacturer must satisfy RUS that the flooding compound selected for use is acceptable for the application.

(i) *Shield and optional armor.* (1) A single corrugated shield must be applied longitudinally over the core wrap.

(2) For unarmored cable the shield overlap must comply with the requirements specified in ANSI/ICEA S-84-608-1988, paragraph 6.3.2. Core diameter is defined as the diameter under the core wrap and binding.

(3) For cables containing the coated aluminum shield/coated steel armor (CACSP) sheath design, the coated aluminum shield must be applied in accordance with the requirements specified in ANSI/ICEA S-84-608-1988, paragraph 6.3.2, Dual Tape Shielding System.

(4) General requirements for application of the shielding material are as follows:

(i) Successive lengths of shielding tapes may be joined during the manufacturing process by means of cold weld, electric weld, soldering with a nonacid flux or other acceptable means.

(ii) Shield splices must comply with the requirements specified in ANSI/ICEA S-84-608-1988, paragraph 6.3.3.

(iii) The corrugations and the application process of the coated aluminum and copper bearing shields must comply with the requirements specified in ANSI/ICEA S-84-608-1988, paragraph 6.3.1.

(iv) The shielding material must be applied in such a manner as to enable the cable to pass the cold bend test specified in paragraph (l)(3) of this section.

(5) The following is a list of acceptable materials for use as cable shielding. Other types of shielding materials may also be used provided they are accepted by RUS prior to their use.

Standard Cable	Gopher Resistant Cable
8-mil Coated Aluminum [1] 5-mil Copper	10-mil Copper 6-mil Copper-Clad Stainless Steel 5 mil Copper-Clad Stainless Steel 5 mil Copper-Clad Alloy Steel 7-mil Alloy 194 6-mil Alloy 194 8-mil Coated Aluminum [1] and 6-mil Coated Steel [1]

[1] Dimensions of uncoated metal.

(i) The 8-mil aluminum tape must be plastic coated on both sides and must comply with the requirements of ANSI/ICEA S-84-608-1988, paragraph 6.2.2.

(ii) The 5-mil copper tape must comply with the requirements specified in ANSI/ICEA S-84-608-1988, paragraph 6.2.3.

(iii) The 10-mil copper tape must comply with the requirements specified in ANSI/ICEA S-84-608-1988, paragraph 6.2.4.

(iv) The 6-mil copper clad stainless steel tape must comply with the requirements specified in ANSI/ICEA S-84-608-1988, paragraph 6.2.5.

(v) The 5-mil copper clad stainless steel tape must be in the fully annealed condition and must conform to the requirements of American Society for Testing and Materials (ASTM) B 694–86, with a cladding ratio of 16/68/16.

(A) The electrical conductivity of the clad tape must be a minimum of 28 percent of the International Annealed Copper Standard (IACS) when measured per ASTM B 193–87.

(B) The tape must be nominally 0.13 millimeter (0.005 inch) thick with a minimum thickness of 0.11 millimeter (0.0045 inch).

(vi) The 5-mil copper clad alloy steel tape must be in the fully annealed condition and the copper component must conform to the requirements of ASTM B 224–80 and the alloy steel component must conform to the requirements of ASTM A 505–87, with a cladding ratio of 16/68/16.

(A) The electrical conductivity of the copper clad alloy steel tape must comply with the requirement specified in (5)(v)(A) of this section.

(B) The thickness of the copper clad alloy steel tape must comply with the requirements specified in (5)(v)(B) of this section.

(vii) The 6-mil and 7-mil 194 copper alloy tapes must comply with the requirements specified in ANSI/ICEA S-84–608–1988, paragraph 6.2.6.

(6) The corrugation extensibility of the coated aluminum shield must comply with the requirements specified in ANSI/ICEA S-84–608–1988, paragraph 6.4.

(7) When the jacket is bonded to the plastic coated aluminum shield, the bond between the jacket and shield must comply with the requirements specified in ANSI/ICEA S-84–608–1988, paragraph 7.2.6.

(8) A single plastic coated steel corrugated armor must be applied longitudinally directly over the coated aluminum shield listed in paragraph (i)(5) of this section with an overlap complying with the requirements specified in ANSI/ICEA S-84–608–1988, paragraph 6.3.2, Outer Steel Tape.

(9) Successive lengths of steel armoring tapes may be joined during the manufacturing process by means of cold weld, electric weld, soldering with a nonacid flux or other acceptable means. Armor splices must comply with the breaking strength and resistance requirements specified in ANSI/ICEA S-84–608–1988, paragraph 6.3.3.

(10) The corrugations and the application process of the coated steel armor must comply with the requirements specified in ANSI/ICEA S-84–608–1988, paragraph 6.3.1.

(i) The corrugations of the armor tape must coincide with the corrugations of the coated aluminum shield.

(ii) Overlapped portions of the armor tape must be in register (corrugations must coincide at overlap) and in contact at the outer edge.

(11) The armoring material must be so applied to enable the cable to pass the cold bend test specified in paragraph (l)(3) of this section.

(12) The 6-mil steel tape must be electrolytic chrome coated steel (ECCS) plastic coated on both sides and must comply with the requirements specified in ANSI/ICEA S-84–608–1988, paragraph 6.2.8.

(13) When the jacket is bonded to the plastic coated steel armor, the bond between the jacket and armor must comply with the requirement specified in ANSI/ICEA-S-84–608–1988, paragraph 7.2.6.

(j) *Cable jacket.* (1) The jacket must comply with the requirements specified in ANSI/ICEA S-84–608–1988, paragraph 7.2.

(2) The raw materials used for the cable jacket must comply with the requirements specified in ANSI/ICEA S-84–608–1988, paragraph 7.2.1.

(3) Jacketing material removed from or tested on the cable must meet the performance requirements specified in ANSI/ICEA S-84–608–1988, paragraphs 7.2.3 and 7.2.4.

(4) The thickness of the jacket must comply with the requirements specified in ANSI/ICEA S-84–608–1988, paragraph 7.2.2.

(k) *Electrical requirements*—(1) *Conductor resistance.* The direct current resistance of any conductor in a completed cable and the average resistance of all conductors in a Quality Control Lot must comply with the requirements specified in ANSI/ICEA S-84–608–1988, paragraph 8.1.

(2) *Resistance unbalance.* (i) The direct current resistance unbalance between the two conductors of any pair in a

completed cable and the average resistance unbalance of all pairs in a completed cable must comply with the requirements specified in ANSI/ICEA S-84–608–1988, paragraph 8.2.

(ii) The resistance unbalance between tip and ring conductors shall be random with respect to the direction of unbalance. That is, the resistance of the tip conductors shall not be consistently higher with respect to the ring conductors and vice versa.

(3) *Mutual capacitance.* The average mutual capacitance of all pairs in a completed cable and the individual mutual capacitance of any pair in a completed cable must comply with the requirements specified in ANSI/ICEA S-84–608–1988, paragraph 8.3.

(4) *Capacitance difference.* (i) The capacitance difference for completed cables having 75 pairs or greater must comply with the requirement specified in ANSI/ICEA S-84–608–1988, paragraph 8.4.

(ii) When measuring screened cable, the inner and outer pairs must be selected from both sides of the screen.

(5) *Pair-to-pair capacitance unbalance*—(i) *Pair-to-pair.* The capacitance unbalance as measured on the completed cable must comply with the requirements specified in ANSI/ICEA S-84–608–1988, paragraph 8.5.

(ii) *Screened cable.* In cables with 25 pairs or less and within each group of multigroup cables, the pair-to-pair capacitance unbalance between any two pairs in an individual compartment must comply with the requirements specified in ANSI/ICEA S-84–608–1988, paragraph 8.5. The pair-to-pair capacitance unbalances to be considered must be:

(A) Between pairs adjacent in a layer in an individual compartment;

(B) Between pairs in centers of 4 pairs or less in an individual compartment; and

(C) Between pairs in adjacent layers in an individual compartment when the number of pairs in the inner (smaller) layer is 6 or less. The center is counted as a layer.

(iii) In cables with 25 pairs or less, the root-mean-square (rms) value is to include all the pair-to-pair unbalances measured for each compartment separately.

(iv) In cables containing more than 25 pairs, the rms value must include the pair-to-pair unbalances in the separate compartments.

(6) *Pair-to-ground capacitance unbalance*—(i) *Pair-to-ground.* The capacitance unbalance as measured on the completed cable must comply with the requirements specified in ANSI/ICEA S-84–608–1988, paragraph 8.6.

(ii) When measuring pair-to-ground capacitance unbalance all pairs except the pair under test are grounded to the shield and/or shield/armor except when measuring cables containing super units in which case all other pairs in the same super unit must be grounded to the shield.

(iii) The screen tape must be left floating during the test.

(iv) Pair-to-ground capacitance unbalance may vary directly with the length of the cable.

(7) *Attenuation.* (i) For nonscreened and screened cables, the average attenuation of all pairs on any reel when measured at 150 and 772 kilohertz must comply with the requirements specified in ANSI/ICEA S-84–608–1988, paragraph 8.7, Foam and/or Foam-Skin Column.

(ii) For T1C type cables over 12 pairs, the maximum average attenuation of all pairs on any reel must not exceed the values listed below when measured at a frequency of 1576 kilohertz at or corrected to a temperature of 20 ±1 °C. The test must be conducted in accordance with ASTM D 4566–90.

AWG	Maximum Average Attenuation decibel/kilometer (dB/km) (decibel/mile)
19	14.9 (24.0)
22	21.6 (34.8)
24	27.2 (43.8)

(8) *Crosstalk loss.* (i) The equal level far-end power sum crosstalk loss (FEXT) as measured on the completed cable must comply with the requirements specified in ANSI/ICEA S-84–608–1988, paragraph 8.8, FEXT Table.

(ii) The near-end power sum crosstalk loss (NEXT) as measured on completed cable must comply with the requirements specified in ANSI/ICEA S-84–608–1988, paragraph 8.8, NEXT Table.

(iii) *Screened cable.* (A) For screened cables the NEXT as measured on the completed cable must comply with the requirements specified in ANSI/ICEA S-84-608-1988, paragraphs 8.9 and 8.9.1.

(B) For T1C screened cable the NEXT as measured on the completed cable must comply with the requirements specified in ANSI/ICEA S-84-608-1988, paragraphs 8.9 and 8.9.2.

(9) *Insulation resistance.* The insulation resistance of each insulated conductor in a completed cable must comply with the requirement specified in ANSI/ICEA S-84-608-1988, paragraph 8.11.

(10) *High voltage test.* (i) In each length of completed cable, the insulation between conductors must comply with the requirements specified in ANSI/ICEA S-84-608-1988, paragraph 8.12, Foam and/or Foam-Skin Column.

(ii) In each length of completed cable, the dielectric between the shield and/or armor and conductors in the core must comply with the requirements specified in ANSI/ICEA S-84-608-1988, paragraph 8.13, Single Jacketed, Foam and/or Foam-Skin Column. In screened cable the screen tape must be left floating.

(iii) *Screened cable.* (A) In each length of completed screened cable, the dielectric between the screen tape and the conductors in the core must comply with the requirement specified in ANSI/ICEA S-84-608-1988, paragraph 8.14.

(B) In this test, the cable shield and/or armor must be left floating.

(11) *Electrical variations.* (i) Pairs in each length of cable having either a ground, cross, short, or open circuit condition will not be permitted.

(ii) The maximum number of pairs in a cable which may vary as specified in paragraph (k)(11)(iii) of this section from the electrical parameters given in this section are listed below. These pairs may be excluded from the arithmetic calculation.

Nominal Pair Count	Maximum Number of Pairs With Allowable Electrical Variation
6–100	1
101–300	2
301–400	3
401–600	4
601 and above	6

(iii) *Parameter variations.* (A) *Capacitance unbalance-to-ground.* If the cable fails either the maximum individual pair or average capacitance unbalance-to-ground requirement and all individual pairs are 3937 picofarad/kilometer (1200 picofarad/1000 feet) or less, the number of pairs specified in paragraph (k)(11)(ii) of this section may be eliminated from the average and maximum individual calculations.

(B) *Resistance unbalance.* Individual pair of 7 percent for all gauges.

(C) *Conductor resistance, maximum.* The following table shows maximum conductor resistance:

AWG	ohms/kilometer	(ohms/1000 feet)
19	29.9	(9.1)
22	60.0	(18.3)
24	94.5	(28.8)
26	151.6	(46.2)

NOTE: RUS recognizes that in large pair count cable (600 pair and above) a cross, short, or open circuit condition occasionally may develop in a pair which does not affect the performance of the other cable pairs. In these circumstances rejection of the entire cable may be economically unsound or repairs may be impractical. In such circumstances the manufacturer may desire to negotiate with the customer for acceptance of the cable. No more than 0.5 percent of the pairs may be involved.

(l) *Mechanical requirements*—(1) *Compound flow test.* All cables manufactured in accordance with the requirements of this section must be capable of meeting the compound flow test specified in ANSI/ICEA S-84-608-1988, paragraph 9.1 using a test temperature of 80 ±1 °C.

(2) *Water penetration test.* All cables manufactured in accordance with the requirements of this section must be capable of meeting the water penetration test specified in ANSI/ICEA S-84-608-1988, paragraph 9.2.

(3) *Cable cold bend test.* All cables manufactured in accordance with the requirements of this section must be capable of meeting the cable cold bend test specified in ANSI/ICEA S-84-608-1988, paragraph 9.3.

(4) *Cable impact test.* All cables manufactured in accordance with the requirements of this section must be capable of meeting the cable impact test specified in ANSI/ICEA S-84-608-1988, paragraph 9.4.

(5) *Jacket notch test (CACSP sheath only).* All cables utilizing the coated aluminum/coated steel sheath (CACSP) design manufactured in accordance with the requirements of this section must be capable of meeting the jacket notch test specified in ANSI/ICEA S-84-608-1988, paragraph 9.5.

(6) *Cable torsion test (CACSP sheath only).* All cables utilizing the coated aluminum/coated steel sheath (CACSP) design manufactured in accordance with the requirements of this section must be capable of meeting the cable torsion test specified in ANSI/ICEA S-84-608-1988, paragraph 9.6.

(m) *Sheath slitting cord (optional).* (1) Sheath slitting cord may be used in the cable structure at the option of the manufacturer unless specified by the end user.

(2) When a sheath slitting cord is used it must be nonhygroscopic and nonwicking, continuous throughout a length of cable and of sufficient strength to open the sheath without breaking the cord.

(n) *Identification marker and length marker.* (1) Each length of cable must be identified in accordance with ANSI/ICEA S-84-608-1988, paragraphs 10.1 through 10.1.4. The color of the ink used for the initial outer jacket marking must be either white or silver.

(2) The markings must be printed on the jacket at regular intervals of not more than 0.6 meter (2 feet).

(3) The completed cable must have sequentially numbered length markers in accordance with ANSI/ICEA S-84-608-1988, paragraph 10.1.5. The color of the ink used for the initial outer jacket marking must be either white or silver.

(o) *Preconnectorized cable (optional).* (1) At the option of the manufacturer and upon request by the purchaser, cables 100 pairs and larger may be factory terminated in 25 pair splicing modules.

(2) The splicing modules must meet the requirements of RUS Bulletin 345-54, PE-52, RUS Specification for Telephone Cable Splicing Connectors (Incorporated by Reference at §1755.97), and be accepted by RUS prior to their use.

(p) *Acceptance testing and extent of testing.* (1) The tests described in appendix A of this section are intended for acceptance of cable designs and major modifications of accepted designs. What constitutes a major modification is at the discretion of RUS. These tests are intended to show the inherent capability of the manufacturer to produce cable products having long life and stability.

(2) For initial acceptance, the manufacturer must submit:

(i) An original signature certification that the product fully complies with each section of the specification;

(ii) Qualification Test Data, per appendix A of this section;

(iii) To periodic plant inspections;

(iv) A certification that the product does or does not comply with the domestic origin manufacturing provisions of the "Buy American" requirements of the Rural Electrification Act of 1938 (7 U.S.C. 901 *et seq.*);

(v) Written user testimonials concerning field performance of the product; and

(vi) Other nonproprietary data deemed necessary by the Chief, Outside Plant Branch (Telephone).

(3) For requalification acceptance, the manufacturer must submit an original signature certification that the product fully complies with each section of the specification, excluding the Qualification Section, and a certification that the product does or does not comply with the domestic origin manufacturing provisions of the "Buy American" requirements of the Rural Electrification Act of 1938 (7 U.S.C. 901 *et seq.*), for acceptance by August 30 of each year. The required data must have been gathered within 90 days of the submission. If the initial acceptance of a product to this specification was within 180 days of August 30, then requalification for that product will not be required for that year.

(4) Initial and requalification acceptance requests should be addressed to:

Chairman, Technical Standards Committee "A" (Telephone), Telecommunications Standard Division, Rural Utilities Service, Washington, DC 20250-1500.

(5) *Tests on 100 percent of completed cable.* (i) The shield and/or armor of each length of cable must be tested for continuity in accordance with ANSI/ICEA S-84-608-1988, paragraph 8.16.

(ii) The screen tape of each length of screened cable must be tested for continuity in accordance with ANSI/ICEA S-84-608-1988, paragraph 8.16.

(iii) Dielectric strength between conductors and shield and/or armor must be tested to determine freedom from grounds in accordance with paragraph (k)(10)(ii) of this section.

(iv) Dielectric strength between conductors and screen tape must be tested to determine freedom from grounds in accordance with paragraph (k)(10)(iii) of this section.

(v) Each conductor in the completed cable must be tested for continuity in accordance with ANSI/ICEA S-84-608-1988, paragraph 8.16.

(vi) Dielectric strength between conductors, in each length of completed cable, must be tested to insure freedom from shorts and crosses in each length of completed cable in accordance with paragraph (k)(10)(i) of this section.

(vii) Each conductor in the completed preconnectorized cable must be tested for continuity.

(viii) Each length of completed preconnectorized cable must be tested for split pairs.

(ix) The average mutual capacitance must be measured on all cables. If the average mutual capacitance for the first 100 pairs tested from randomly selected groups is between 50 and 53 nanofarads/kilometer (nF/km) (80 and 85 nanofarad/mile), the remainder of the pairs need not be tested on the 100 percent basis (See paragraph (k)(3) of this section).

(6) *Capability tests.* Tests on a quality assurance basis must be made as frequently as is required for each manufacturer to determine and maintain compliance with:

(i) Performance requirements for conductor insulation, jacketing material, and filling and flooding compounds;

(ii) Bonding properties of coated or laminated shielding and armoring materials and performance requirements for screen tape;

(iii) Sequential marking and lettering;

(iv) Capacitance difference, capacitance unbalance, crosstalk, and attenuation;

(v) Insulation resistance, conductor resistance, and resistance unbalance;

(vi) Cable cold bend and cable impact tests;

(vii) Water penetration and compound flow tests; and

(viii) Jacket notch and cable torsion tests.

(q) *Summary of records of electrical and physical tests.* (1) Each manufacturer must maintain suitable summary records for a period of at least 3 years of all electrical and physical tests required on completed cable by this section as set forth in paragraphs (p)(5) and (p)(6) of this section. The test data for a particular reel must be in a form that it may be readily available to the purchaser or to RUS upon request.

(2) Measurements and computed values must be rounded off to the number of places or figures specified for the requirement according to ANSI/ICEA S-84-608-1988, paragraph 1.3.

(r) *Manufacturing irregularities.* (1) Repairs to the shield and/or armor are not permitted in cable supplied to end users under this section.

(2) Minor defects in jackets (defects having a dimension of 3 millimeters (0.125 inch.) or less in any direction) may be repaired by means of heat fusing in accordance with good commercial practices utilizing sheath grade compounds.

(s) *Preparation for shipment.* (1) The cable must be shipped on reels. The diameter of the drum must be large enough to prevent damage to the cable from reeling or unreeling. The reels must be substantial and so constructed as to prevent damage to the cable during shipment and handling.

(2) The thermal wrap must comply with the requirements of ANSI/ICEA S-84-608-1988, paragraph 10.3. When a thermal reel wrap is supplied, the wrap must be applied to the reel and must be suitably secured in place to minimize thermal exposure to the cable during storage and shipment. The use of the thermal reel wrap as a means of reel protection will be at the option of the

manufacturer unless specified by the end user.

(3) The outer end of the cable must be securely fastened to the reel head so as to prevent the cable from becoming loose in transit. The inner end of the cable must be securely fastened in such a way as to make it readily available if required for electrical testing. Spikes, staples, or other fastening devices which penetrate the cable jacket must not be used. The method of fastening the cable ends must be acceptable to RUS and accepted prior to its use.

(4) Each length of cable must be wound on a separate reel unless otherwise specified or agreed to by the purchaser.

(5) The arbor hole must admit a spindle 63 millimeters (2.5 inches) in diameter without binding. Steel arbor hole liners may be used but must be accepted by RUS prior to their use.

(6) Each reel must be plainly marked to indicate the direction in which it should be rolled to prevent loosening of the cable on the reel.

(7) Each reel must be stenciled or labeled on either one or both sides with the information specified in ANSI/ICEA S-84-608-1988, paragraph 10.4 and the RUS cable designation:

Cable Designation

BFCE

Cable Construction

Pair Count

Conductor Gauge

E = Expanded Insulation
A = Coated Aluminum Shield
C = Copper Shield
Y = Gopher Resistant Shield
X = Armored, Separate Shield
H = T1 Screened Cable
H1C = T1C Screened Cable
P = Preconnectorized

Example: BFCEXH100-22

Buried Filled Cable, Expanded Insulation, Armored (w/separate shield), T1 Screened Cable, 100 pair, 22 AWG.

(8) When cable manufactured to the requirements of this specification is shipped, both ends must be equipped with end caps acceptable to RUS.

(9) When preconnectorized cables are shipped, the splicing modules must be protected to prevent damage during shipment and handling. The protection method must be acceptable to RUS and accepted prior to its use.

(10) All cables ordered for use in underground duct applications must be equipped with a factory-installed pulling-eye on the outer end in accordance with ANSI/ICEA S-84-608-1988, paragraph 10.5.2.

(The information and recordkeeping requirements of this section have been approved by the Office of Management and Budget (OMB) under the control number 0572-0059)

APPENDIX A TO § 1755.890—QUALIFICATION TEST METHODS

(I) The test procedures described in this appendix are for qualification of initial cable designs and major modifications of accepted designs. Included in (V) of this appendix are suggested formats that may to be used in submitting test results to RUS.

(II) *Sample selection and preparation.* (1) All testing must be performed on lengths removed sequentially from the same 25 pair, 22 gauge jacketed cable. This cable must not have been exposed to temperatures in excess of 38 °C since its initial cool down after sheathing. The lengths specified are minimum lengths and if desirable from a laboratory testing standpoint longer lengths may be used.

(a) Length A must be 10 ±0.2 meters (33 ±0.5 feet) long and must be maintained at 23 ±3 °C. One length is required.

(b) Length B must be 12 ±0.2 meters (40 ±0.5 feet) long. Prepare the test sample by removing the jacket, shield or shield/armor, and core wrap for a sufficient distance on both ends to allow the insulated conductors to be flared out. Remove sufficient conductor insulation so that appropriate electrical test connections can be made at both ends. Coil the sample with a diameter of 15 to 20 times its sheath diameter. Three lengths are required.

(c) Length C must be one meter (3 feet) long. Four lengths are required.

(d) Length D must be 300 millimeters (1 foot) long. Four lengths are required.

(e) Length E must be 600 millimeters (2 feet) long. Four lengths are required.

(f) Length F must be 3 meters (10 feet) long and must be maintained at 23 ±3 °C for the duration of the test. Two lengths are required.

(2) *Data reference temperature.* Unless otherwise specified, all measurements must be made at 23 ±3 °C.

(III) *Environmental tests*—(1) *Heat aging test*—(a) *Test samples.* Place one sample each of lengths B, C, D, and E in an oven or environmental chamber. The ends of Sample B must exit from the chamber or oven for electrical tests. Securely seal the oven exit holes.

(b) *Sequence of tests.* The samples are to be subjected to the following tests after conditioning:

(i) Water Immersion Test outlined in (III)(2) of this appendix;

(ii) Water Penetration Test outlined in (III)(3) of this appendix;

(iii) Insulation Compression Test outlined in (III)(4) of this appendix; and

(iv) Jacket Slip Strength Test outlined in (III)(5) of this appendix.

(c) *Initial Measurements.* (i) For Sample B measure the open circuit capacitance for each odd numbered pair at 1, 150, and 772 kilohertz, and the attenuation at 150 and 772 kilohertz after conditioning the sample at the data reference temperature for 24 hours. Calculate the average and standard deviation for the data of the 13 pairs on a per kilometer or (on a per mile) basis.

(ii) The attenuation at 150 and 772 kilohertz may be calculated from open circuit admittance (Yoc) and short circuit impedance (Zsc) or may be obtained by direct measurement of attenuation.

(iii) Record on suggested formats in (V) of this appendix or on other easily readable formats.

(d) *Heat conditioning.* (i) Immediately after completing the initial measurements, condition the sample for 14 days at a temperature of 65 ±2 °C.

(ii) At the end of this period note any exudation of cable filler. Measure and calculate the parameters given in (III)(1)(c) of this appendix. Record on suggested formats in (V) of this appendix or other easily readable formats.

(iii) Cut away and discard a one meter (3 foot) section from each end of length B.

(e) *Overall electrical deviation.* (i) Calculate the percent change in all average parameters between the final parameters after conditioning and the initial parameters in (III)(1)(c) of this appendix.

(ii) The stability of the electrical parameters after completion of this test must be within the following prescribed limits:

(A) *Capacitance.* The average mutual capacitance must be within 5 percent of its original value;

(B) The change in average mutual capacitance must be less than 5 percent over frequency 1 to 150 kilohertz; and

(C) *Attenuation.* The 150 and 772 kilohertz attenuation must not have increased by more than 5 percent over their original values.

(2) *Water immersion electrical test*—(a) *Test sample selection.* The 10 meter (33 foot) section of length B must be tested.

(b) *Test sample preparation.* Prepare the sample by removing the jacket, shield or shield/armor, and core wrap for sufficient distance to allow one end to be accessed for test connections. Cut out a series of 6 millimeter (0.25 inch.) diameter holes along the test sample, at 30 centimeters (1 foot) intervals progressing successively 90 degrees around the circumference of the cable. Assure that the cable core is exposed at each hole by slitting the core wrapper. Place the prepared sample in a dry vessel which when filled will maintain a one meter (3 foot) head of water over 6 meters (20 feet) of uncoiled cable. Extend and fasten the ends of the cable so they will be above the water line and the pairs are rigidly held for the duration of the test.

(c) *Capacitance testing.* Measure the initial values of mutual capacitance of all odd pairs in each cable at a frequency of 1 kilohertz before filling the vessel with water. Be sure the cable shield or shield/armor is grounded to the test equipment. Fill the vessels until there is a one meter (3 foot) head of water on the cables.

(i) Remeasure the mutual capacitance after the cables have been submerged for 24 hours and again after 30 days.

(ii) Record each sample separately on suggested formats attached or on other easily readable formats.

(d) *Overall electrical deviation.* (i) Calculate the percent change in all average parameters between the final parameters after conditioning with the initial parameters in (III)(2)(c) of this appendix.

(ii) The average mutual capacitance must be within 5 percent of its original value.

(3) *Water penetration testing.* (a) A watertight closure must be placed over the jacket of length C. The closure must not be placed over the jacket so tightly that the flow of water through pre-existing voids of air spaces is restricted. The other end of the sample must remain open.

(b) Test per Option A or Option B—(i) *Option A.* Weigh the sample and closure prior to testing. Fill the closure with water and place under a continuous pressure of 10 ±0.7 kilopascals (1.5 ±0.1 pounds per square inch gauge) for one hour. Collect the water leakage from the end of the test sample during the test and weigh to the nearest 0.1 gram. Immediately after the one hour test, seal the ends of the cable with a thin layer of grease and remove all visible water from the closure, being careful not to remove water that penetrated into the core during the test. Reweigh the sample and determine the weight of water that penetrated into the core. The weight of water that penetrated into the core must not exceed 6 grams.

(ii) *Option B.* Fill the closure with a 0.2 gram sodium fluorscein per liter water solution and apply a continuous pressure 10 ±0.7 kilopascals (1.5 ±0.1 pounds per square inch gauge) for one hour. Catch and weigh any water that leaks from the end of the cable during the one hour period. If no water leaks from the sample, carefully remove the water from the closure. Then carefully remove the jacket, shield or shield/ armor, and core wrap one at a time, examining with an ultraviolet light source for water penetration. After removal of the core wrap, carefully dissect the core and examine for water penetration within the core. Where water penetration is observed, measure the penetration distance. The distance of water penetration into the core must not exceed 127 millimeters (5.0 inches).

(4) *Insulation compression test*—(a) *Test sample D.* Remove jacket, shield or shield/armor, and core wrap being careful not to damage the conductor insulation. Remove one pair from the core and carefully separate, wipe off core filler and straighten the insulated conductors. Retwist the two insulated conductors together under sufficient tension to form 10 evenly spaced 360 degree twists in a length of 10 centimeters (4 inches).

(b) *Sample testing.* Center the mid 50 millimeters (2 inches) of the twisted pair between 2 smooth rigid parallel metal plates that are 50 millimeters × 50 millimeters (2 inches × 2 inches). Apply a 1.5 volt direct current potential between the conductors, using a light or buzzer to indicate electrical contact between the conductors. Apply a constant load of 67 newtons (15 pound-force) on the sample for one minute and monitor for evidence of contact between the conductors. Record results on suggested formats in (V) of this appendix or on other easily readable formats.

(5) *Jacket slip strength test*—(a) *Sample selection.* Test Sample E from (III)(1)(a) of this appendix.

(b) *Sample preparation.* Prepare test sample in accordance with the procedures specified in ASTM D 4565–90a.

(c) *Sample conditioning and testing.* Remove the sample from the tensile tester prior to testing and condition for one hour at 50 ±2 °C. Test immediately in accordance with the procedures specified in ASTM D 4565–90a. A minimum jacket slip strength of 67 newtons (15 pound-force) is required. Record the highest load attained.

(6) *Humidity exposure.* (a) Repeat steps (III)(1)(a) through (III)(1)(c)(iii) of this appendix for separate set of samples B, C, D, and E which have not been subjected to prior environmental conditioning.

(b) Immediately after completing the measurements, expose the test sample to 100 temperature cyclings. Relative humidity within the chamber must be maintained at 90 ±2 percent. One cycle consists of beginning at a stabilized chamber and test sample temperature of 52 ±1 °C, increasing the temperature to 57 ±1 °C, allowing the chamber and test samples to stabilize at this level, then dropping the temperature back to 52 ±1 °C.

(c) Repeat steps (III)(1)(d)(ii) through (III)(5)(c) of this appendix.

(7) *Temperature cycling.* (a) Repeat steps (III)(1)(a) through (III)(1)(c)(iii) of this appendix for separate set of samples B, C, D, and E which have not been subjected to prior environmental conditioning.

(b) Immediately after completing the measurements, subject the test sample to the 10 cycles of temperature between a minimum of −40 °C and + 60 °C. The test sample must be held at each temperature extreme for a minimum of 1½ hours during each cycle of temperature. The air within the temperature cycling chamber must be circulated throughout the duration of the cycling.

(c) Repeat steps (III)(1)(d)(ii) through (III)(5)(c) of this appendix.

(IV) *Control sample*—(1) *Test samples.* A separate set of lengths A, C, D, E, and F must have been maintained at 23 ±3 °C for at least 48 hours before the testing.

(2) Repeat steps (III)(2) through (III)(5)(c) of this appendix except use length A instead of length B.

(3) *Surge test.* (a) One length of sample F must be used to measure the breakdown between conductors while the other length of F must be used to measure the core to shield breakdown.

(b) The samples must be capable of withstanding without damage, a single surge voltage of 15 kilovolts peak between conductors, and a 25 kilovolts peak surge voltage between conductors and the shield or shield/armor as hereinafter described. The surge voltage must be developed from a capacitor discharged through a forming resistor connected in parallel with the dielectric of the test sample. The surge generator constants must be such as to produce a surge of 1.5 × 40 microsecond wave shape.

(c) The shape of the generated wave must be determined at a reduced voltage by connecting an oscilloscope across the forming resistor with the cable sample connected in parallel with the forming resistor. The capacitor bank is charged to the test voltage and then discharged through the forming resistor and test sample. The test sample will be considered to have passed the test if there is no distinct change in the wave shape obtained with the initial reduced voltage compared to that obtained after the application of the test voltage.

(V) The following suggested formats may be used in submitting the test results to RUS:

ENVIRONMENTAL CONDITIONING ______________

FREQUENCY 1 KILOHERTZ

Pair Number	Capacitance	
	nF/km (nanofarad/mile)	
	Initial	Final
1	______	______
3	______	______
5	______	______
7	______	______
9	______	______
11	______	______
13	______	______
15	______	______
17	______	______
19	______	______
21	______	______
23	______	______
25	______	______
Average x̄	______	______

Overall Percent Difference in Average x̄ ______

ENVIRONMENTAL CONDITIONING ______________

FREQUENCY 772 KILOHERTZ

Pair Number	Capacitance		Attenuation	
	nF/km (nanofarad/mile)		dB/km (decibel/mile)	
	Initial	Final	Initial	Final
1	______	______	______	______
3	______	______	______	______
5	______	______	______	______
7	______	______	______	______
9	______	______	______	______
11	______	______	______	______
13	______	______	______	______
15	______	______	______	______
17	______	______	______	______
19	______	______	______	______
21	______	______	______	______
23	______	______	______	______
25	______	______	______	______
Average x̄	______	______	______	______

Overall Percent Difference in Average x̄ Capacitance: ______ Conductance: ______

ENVIRONMENTAL CONDITIONING ______________

FREQUENCY 150 KILOHERTZ

Pair Number	Capacitance		Attenuation	
	nF/km (nanofarad/mile)		dB/km (decibel/mile)	
	Initial	Final	Initial	Final
1	______	______	______	______
3	______	______	______	______
5	______	______	______	______
7	______	______	______	______
9	______	______	______	______
11	______	______	______	______
13	______	______	______	______
15	______	______	______	______
17	______	______	______	______
19	______	______	______	______
21	______	______	______	______
23	______	______	______	______
25	______	______	______	______
Average x̄	______	______	______	______

Overall Percent Difference in Average x̄ Capacitance: ______ Conductance: ______

ENVIRONMENTAL CONDITIONING ______________

WATER IMMERSION TEST (1 KILOHERTZ)

Pair Number	Capacitance		
	nF/km (nanofarad/mile)		
	Initial	24 Hours	Final
1	______	______	______
3	______	______	______
5	______	______	______
7	______	______	______
9	______	______	______
11	______	______	______
13	______	______	______
15	______	______	______
17	______	______	______
19	______	______	______
21	______	______	______
23	______	______	______
25	______	______	______
Average x̄	______	______	______

Overall Percent Difference in Average x̄ ______

WATER PENETRATION TEST

	Option A		Option B	
	End Leakage grams	Weight Gain grams	End Leakage grams	Penetration mm (in.)
Control.				
Heat Age.				
Humidity Exposure.				
Temperature Cycling.				

INSULATION COMPRESSION

	Failures
Control	______
Heat Age	______
Humidity Exposure	______
Temperature Cycling	______

JACKET SLIP STRENGTH @ 50 °C

	Load in newtons (pound-force)
Control	______
Heat Age	______
Humidity Exposure	______

JACKET SLIP STRENGTH @ 50 °C—Continued

	Load in newtons (pound-force)
Temperature Cycling	____________

FILLER EXUDATION (GRAMS)

Heat Age	____________
Humidity Exposure	____________
Temperature Cycling	____________

SURGE TEST (KILOVOLTS)

Conductor to Conductor	____________
Shield to Conductors	____________

[58 FR 29328, May 20, 1993, as amended at 60 FR 1711, Jan. 5, 1995; 69 FR 18803, Apr. 9, 2004]

§ 1755.900 Abbreviations and Definitions.

The following abbreviations and definitions apply to §§ 1755.901 and 1755.902:

(a) *Abbreviations.* (1) ADSS All dielectric self-supporting;

(2) ASTM American Society for Testing and Materials;

(3) °C Centigrade temperature scale;

(4) dB Decibel;

(5) CSM Central strength member;

(6) dB/km Decibels per 1 kilometer;

(7) ECCS Electrolytic chrome coated steel;

(8) EIA Electronic Industries Alliance;

(9) EIA/TIA Electronic Industries Alliance/Telecommunications Industry Association;

(10) FTTH Fiber-to-the-Home;

(11) Gbps Gigabit per second or Gbit/s;

(12) GE General Electric;

(13) HDPE High density polyethylene;

(14) ICEA Insulated Cable Engineers Association, Inc.;

(15) Km kilometer(s;)

(16) LDPE Low density polyethylene;

(17) m meter(s;)

(18) Max. Maximum;

(19) Mbit Megabits;

(20) MDPE Medium density polyethylene;

(21) MHz-km Megahertz-kilometer;

(22) Min. Minimum;

(23) MFD Mode-Field Diameter;

(24) nm Nanometer(s;)

(25) N Newton(s;)

(26) NA Numerical aperture;

(27) NESC National Electrical Safety Code;

(28) OC Optical cable;

(29) O.D. Outside Diameter;

(30) OF Optical fiber;

(31) OSHA Occupational Safety and Health Administration;

(32) OTDR Optical Time Domain Reflectometer;

(33) % Percent;

(34) ps/(nm · km) Picosecond per nanometer times kilometer;

(35) ps/(nm^2 · km) Picosecond per nanometer squared times kilometer;

(36) PMD Polarization Mode Dispersion;

(37) RUS Rural Utilities Service;

(38) s Second(s);

(39) SI International System (of Units) (From the French *Système international d'unités*); and

(40) µm Micrometer.

(b) *Definitions*—(1) *Accept; Acceptance* means Agency action of providing the manufacturer of a product with a letter by mail or facsimile that the Agency has determined that the manufacturer's product meets its requirements. For information on how to obtain Agency product acceptance, refer to the procedures listed at *http://www.usda.gov/rus/telecom/listing_procedures/index_listing_procedures.htm,* as well as additional information in RUS Bulletin 345-3, *Acceptance of Standards, Specifications, Equipment Contract Forms, Manual Sections, Drawings, Materials and Equipment for the Telephone Program,* available for download at *http://www.usda.gov/rus/telecom/publications/bulletins.htm.*

(2) *Agency* means the Rural Utilities Service, an Agency which delivers the United States Department of Agriculture's Rural Development Utilities Programs.

(3) *Armor* means a metal tape installed under the outer jacket of the cable intended to provide mechanical protection during cable installation and environmental protection against rodents, termites, etc.

(4) *Attenuation* means the loss of power as the light travels in the fiber usually expressed in dB/km.

(5) *Bandwidth* means the range of signal frequencies that can be transmitted

by a communications channel with defined maximum loss or distortion. Bandwidth indicates the information-carrying capacity of a channel.

(6) *Birefringence* means the decomposition of a pulse of light entering the fiber into "two polarized pulses" traveling at different velocities due to the different refractive indexes in the polarization axes in which the electric fields oscillate. Different refractive indexes in the fiber may be caused by an asymmetric fiber core, internal manufacturing stresses, or through external stresses from cabling and installation of the fiber optic cable, such as bending and twisting.

(7) *Cable cutoff wavelength* means the shortest wavelength at which only one mode light can be transmitted in any of the single mode fibers of an optical fiber cable.

(8) *Chromatic dispersion* means the broadening of a light pulse as it travels down the length of an optical fiber, resulting in different spectral components of the light pulse traveling at different speeds, due to the fact that the index of refraction of the fiber core is different for different wavelengths.

(9) *Cladding* means the outer layer of an optical fiber made of glass or other transparent material that is fused to the fiber core. The cladding concentrically surrounds the fiber core. It has a lower refractive index than the core, so light travelling in the fiber is maintained in the core by internal reflection at the core-cladding interface.

(10) *Core* means the central region of an optical waveguide or fiber which has a higher refractive index than the cladding through which light is transmitted.

(11) *Cutoff wavelength* means, in single mode fiber, the shortest wavelength at which only the fundamental mode of an optical wavelength can propagate.

(12) *Dielectric cable* means a cable which has neither metallic members nor other electrically conductive materials or elements.

(13) *Differential group delay* means the arrival time differential of the two polarized light components of a light pulse traveling through the optical fiber due to birefringence.

(14) *Graded Refractive Index Profile* means the refractive index profile of an optical fiber that varies smoothly with radius from the center of the fiber to the outer boundary of the cladding.

(15) *List of Acceptable Materials* means the latest edition of RUS Informational Publication 344–2, "*List of Materials Acceptable for Use on Telecommunications Systems of RUS Borrowers.*" This document contains a convenient listing of products which have been determined to be acceptable by the Agency. The List of Acceptable Materials is available on the Internet at *http://www.usda.gov/rus/telecom/materials/lstomat.htm.*

(16) *Loose tube buffer* means the protective tube that loosely contains the optical fibers within the fiber optic cable, often filled with suitable water blocking material.

(17) *Matched cable* means fiber optic cable manufactured to meet the requirement of this section for which the calculated splice loss using the formula below is ≤0.06 dB for any two cabled fibers to be spliced.

$$\text{LOSS (dB)} = -10\ \text{LOG}_{10}\ [4/(\text{MFD}_1/\text{MFD}_2 + \text{MFD}_2/\text{MFD}_1)^2],$$

where subscripts 1 and 2 refer to any two cabled fibers to be spliced.

(18) *Mil* means a measurement unit of length indicating one thousandth of an inch.

(19) *Minimum bending diameter* means the smallest diameter that must be maintained while bending a fiber optic cable to avoid degrading cable performance indicated as a multiple of the cable diameter (Bending Diameter/Cable Diameter).

(20) *Mode-field diameter* means the diameter of the cross-sectional area of an optical fiber which includes the core and portion of the cladding where the majority of the light travels in a single mode fiber.

(21) *Multimode fiber* means an optical fiber in which light travels in more than one bound mode. A multimode fiber may either have a graded index or step index refractive index profile.

(22) *Numerical Aperture (NA)* means an optical fiber parameter that indicates the angle of acceptance of light into a fiber.

(23) *Optical fiber* means any fiber made of dielectric material that guides light.

(24) *Optical point discontinuities* means the localized deviations of the optical fiber loss characteristic which location and magnitude may be determined by appropriate OTDR measurements of the fiber.

(25) *Optical waveguide* means any structure capable of guiding optical power. In optical communications, the term generally refers to a fiber designed to transmit optical signals.

(26) *Polarization mode dispersion* means, for a particular length of fiber, the average of the differential group delays of the two polarized components of light pulses traveling in the fiber, when the light pulses are generated from a sufficient narrow band source. The differential group delay varies randomly with time and wavelength. The term PMD is used in the industry in the general sense to indicate the phenomenon of birefringence (polarized light having different group velocities), and used specifically to refer to the value of time delay expected in a specific length of fiber.

(27) PMD_Q means the statistical upper bound for the PMD coefficient of a fiber optic cable link composed of M number of randomly chosen concatenated fiber optic cable sections of the same length. The upper bound is defined in terms of a probability level Q, which is the probability that a concatenated PMD coefficient value exceeds PMD_Q, ITU G recommendations for fiber optic cables call for M = 20 and Q = 0.01%. This PMD_Q value is the one used in the design of fiber optic links.

(28) *Ribbon* means a planar array of parallel optical fibers.

(29) *Shield* means a conductive metal tape placed under the cable jacket to provide lightning protection, bonding, grounding, and electrical shielding.

(30) *Single mode fiber* means an optical fiber in which only one bound mode of light can propagate at the wavelength of interest.

(31) *Step Refractive Index Profile* means an index profile characterized by a uniform refractive index within the core, a sharp decrease in refractive index at the core-cladding interface, and a uniform refractive index within the cladding.

(32) *Tight tube buffer* means one or more layers of buffer material tightly surrounding a fiber that is in contact with the coating of the fiber.

[74 FR 20561, May 5, 2009]

§ 1755.901 Incorporation by Reference.

(a) *Incorporation by reference.* The materials listed here are incorporated by reference where noted. These incorporations by reference were approved by the Director of the Federal Register in accordance with 5 U.S.C. 552(a) and 1 CFR part 51. These materials are incorporated as they exist on the date of the approval, and notice of any change in these materials will be published in the FEDERAL REGISTER. The materials are available for purchase at the corresponding addresses noted below. All are available for inspection at the Rural Development Utilities Programs, during normal business hours at room 2849–S, U.S. Department of Agriculture, Washington, DC 20250. Telephone (202) 720–0699, and e-mail *norberto.esteves@wdc.usda.gov*. The materials are also available for inspection at the National Archives and Records Administration (NARA). For information on the availability of these materials at NARA, call (202) 741–6030, or go to: *http://www.archives.gov/federal_register/code_of_federal_regulations/ibr_locations.html.*

(b) The American National Standards Institute/Institute of Electrical and Electronics Engineers, Inc. ANSI/IEEE C2–2007, *The National Electrical Safety Code,* 2007 edition, approved April 20, 2006, (“ANSI/IEEE C2–2007”), incorporation by reference approved for § 1755.902(a), § 1755.902(p), § 1755.903(a), § 1755.903(k) and § 1755.903(n). ANSI/IEEE C2–2007 is available for purchase from IEEE Service Center, 445 Hoes Lane, Piscataway, NJ 08854, telephone 1–800–678–4333 or online at *http://standards.ieee.org/nesc/index.html.*

(c) The following Insulated Cable Engineers Association standards are available for purchase from the Insulated Cable Engineers, Inc, (ICEA), P.O. Box 1568, Carrollton, GA 30112 or from Global Engineering Documents, 15 Iverness Way East, Englewood, CO 80112, telephone 1–800–854–7179 (USA and Canada) or 303–792–2181 (International), or online at *http://global.ihs.com*:

(1) ICEA S–110–717–2003, *Standard for Optical Drop Cable,* 1st edition, September 2003 (''ICEA S–110–717''), incorporation by reference approved for §1755.903(a), §1755.903(b), §1755.903(c), §1755.903(d), §1755.903(e), §1755.903(f), §1755.903(g), §1755.903(l), §1755.903(n), §1755.903(p), §1755.903(u); and

(2) ANSI/ICEA S–87–640–2006, *Standard for Optical Fiber Outside Plant Communications Cable,* 4th edition, December 2006 (''ANSI/ICEA S–87–640''), incorporation by reference approved for §1755.902(a), §1755.902(b), §1755.902(c), §1755.902(d), §1755.902(e), §1755.902(i), §1755.902(l), §1755.902(m), §1755.902(n), §1755.902(p), §1755.902(q), §1755.902(r), §1755.902(u), §1755.903(b), §1755.903(g), §1755.903(l), §1755.903(o), §1755.903(p), and §1755.903(s).

(d) The following American Society for Testing and Materials (ASTM) standards are available for purchase from ASTM International, 100 Barr Harbor Drive, P.O. Box C700, West Conshohocken, PA 19428–2959. Telephone (610) 832–9585, Fax (610) 832–9555, by e-mail at *service@astm.org,* or online at *http://www.astm.org* or from ANSI, 1916 Race Street, Philadelphia, PA 19103, telephone (215) 299–5585, or online at *http://webstore.ansi.org/ansidocstore/default.asp:*

(1) ASTM A 640–97, (Reapproved 2002)^{e1}, *Standard Specification for Zinc-Coated Steel Strand for Messenger Support of Figure 8 Cable,* approved September 2002 (''ASTM A 640''), incorporation by reference approved for §1755.902(n);

(2) ASTM B 736–00, *Standard Specification for Aluminum, Aluminum Alloy and Aluminum-Clad Steel Cable Shielding Stock,* approved May 10, 2000 (''ASTM B 736''), incorporation by reference approved for §1755.902(m) and §1755.903(j);

(3) ASTM D 4565–99, *Standard Test Methods for Physical and Environmental Performance Properties of Insulations and Jackets for Telecommunications Wire and Cable,* approved March 10, 1999 (''ASTM D 4565''), incorporation by reference approved for §1755.902(c), §1755.902(m), §1755.903(c) and §1755.903(j);

(4) ASTM D 4566–98, *Standard Test Methods for Electrical Performance Properties of Insulations and Jackets for Telecommunications Wire and Cable,* approved December 10, 1998 (''ASTM D 4566''), incorporation by reference approved for §1755.902(f), §1755.902(t) and §1755.903(t); and

(5) ASTM D 4568–99, *Standard Test Methods for Evaluating Compatibility Between Cable Filling and Flooding Compounds and Polyolefin Wire and Cable Materials,* approved April 10, 1999 (''ASTM D 4568''), incorporation by reference approved for §1755.902(h).

(e) The following Telecommunications Industry Association/Electronics Industries Association (TIA/EIA) standards are available from Electronic Industries Association, Engineering Department, 1722 Eye Street, NW., Washington, DC 20006; or from Global Engineering Documents, 15 Iverness Way East, Englewood, CO 80112, telephone 1–800–854–7179 (USA and Canada) or (303) 792–2181 (International), or online at *http://global.ihs.com;* or from TIA, 2500 Wilson Blvd, Suite 300, Arlington, VA 22201, telephone 1–800–854–7179 or online *http://www.tiaonline.org/standards/catalog:*

(1) TIA/EIA Standard 455–3A, *FOTP–3, Procedure to Measure Temperature Cycling on Optical Fibers, Optical Cable, and Other Passive Fiber Optic Components,* approved May 1989, (''TIA/EIA Standard 455–3A''), incorporation by reference approved for §1755.902(r).

(2) [Reserved]

(f) The following International Telecommunication Union (ITU) recommendations may be obtained from ITU, Place des Nations, 1211 Geneva 20, Switzerland, telephone + 41 22 730 6141 or online at *http://www.itu.int/ITU-T/publications/recs.html:*

(1) ITU–T Recommendation G.652, *Series G: Transmission Systems and Media, Digital Systems and Networks, Transmission media characteristics—Optical fibre cables, Characteristics of a single-mode optical fibre and cable,* approved June 2005 (''ITU–T Recommendation G.652''), incorporation by reference approved for §1755.902(b), §1755.902(q), §1755.903(b) and §1755.903(o);

(2) ITU–T Recommendation G.655, *Series G: Transmission Systems and Media, Digital Systems and Networks, Transmission media characteristics—Optical fibre cables, Characteristics of a non-zero dispersion-shifted single-mode optical fibre and cable,* approved March 2006

("ITU–T Recommendation G.655"), incorporation by reference approved for § 1755.902(b) and § 1755.902(q);

(3) ITU–T Recommendation G.656, *Series G: Transmission Systems and Media, Digital Systems and Networks, Transmission media characteristics—Optical fibre cables, Characteristics of a fibre and cable with non-zero dispersion for wideband optical transport*, approved December 2006 ("ITU–T Recommendation G.656"), incorporation by reference approved for § 1755.902(b) and § 1755.902(q);

(4) ITU–T Recommendation G.657, *Series G: Transmission Systems and Media, Digital Systems and Networks, Transmission media characteristics—Optical fibre cables, Characteristics of a bending loss insensitive single mode optical fibre and cable for the access network*, approved December 2006 ("ITU–T Recommendation G.657"), incorporation by reference approved for § 1755.902(b) and § 1755.902(q); and

(5) ITU–T Recommendation L.58, *Series L: Construction, Installation and Protection of Cables and Other Elements of Outside Plant, Optical fibre cables: Special Needs for Access Network*, approved March 2004 ("ITU–T Recommendation L.58"), incorporation by reference approved for § 1755.902(a).

[74 FR 20561, May 5, 2009]

§ 1755.902 Minimum performance Specification for fiber optic cables.

(a) *Scope*. This section is intended for cable manufacturers, Agency borrowers, and consulting engineers. It covers the requirements for fiber optic cables intended for aerial installation either by attachment to a support strand or by an integrated self-supporting arrangement, for underground application by placement in a duct, or for buried installations by trenching, direct plowing, and directional or pneumatic boring.

(1) *General*. (i) Specification requirements are given in SI units which are the controlling units in this part. Approximate English equivalent of units are given for information purposes only.

(ii) The optical waveguides are glass fibers having directly-applied protective coatings, and are called "fibers," herein. These fibers may be assembled in either loose fiber bundles with a protective core tube, encased in several protective buffer tubes, in tight buffer tubes, or ribbon bundles with a protective core tube.

(iii) Fillers, strength members, core wraps, and bedding tapes may complete the cable core.

(iv) The core or buffer tubes containing the fibers and the interstices between the buffer tubes, fillers, and strength members in the core structure are filled with a suitable material or water swellable elements to exclude water.

(v) The cable structure is completed by an extruded overall plastic jacket. A shield or armor or combination thereof may be included under the jacket. The jacket may have strength members embedded in it, in some designs.

(vi) Buried installation requires armor under the outer jacket.

(vii) For self-supporting cable, the outer jacket may be extruded over the support messenger and cable core.

(viii) Cables for mid-span applications for network access must be designed for easy mid-span access to the fibers. The manufacturer may use reversing oscillating stranding (SZ) described in section 6.4 of ITU–T Recommendation L.58, *Construction, Installation and Protection of Cables and Other Elements of Outside Plant*, 2004 (incorporated by reference at § 1755.901(f)). The cable end user is cautioned that installed cable must be properly terminated. This includes properly securing rigid strength members (*i.e.*, central strength member) and clamping the cable and jacket. It is important that cable components be secured to prevent movement of the cable or components over the operating conditions. Central strength member (CSM) clamps must prevent movement of the CSM; positive stop CSM clamps are recommended. The CSM must be routed as straight and as short as practical to prevent bowing or breaking of the CSM. The cable and jacket retention must be sufficient to prevent jacket slippage over the operating temperature range.

(2) The normal temperature ranges for cables must meet paragraph 1.1.3 of ANSI/ICEA S–87–640, *Standard for Optical Fiber Outside Plant Communications Cable* (incorporated by reference at § 1755.901(c)).

(3) *Tensile rating*. The standard installation tensile rating for cables is 2670 N (600 lbf), unless installation involves micro type cables that utilize less stress related methods of installation, *i.e.*, blown micro-fiber cable or All-Dielectric Self-Supporting (ADSS) cables (see paragraph (c)(4) of this section).

(4) *ADSS and other self-supporting cables*. Based on the storm loading districts referenced in Section 25, Loading of Grades B and C, of ANSI/IEEE C2–2007, *National Electrical Safety Code*, 2007 (incorporated by reference at §1755.901(b)) and the maximum span and location of cable installation provided by the end user, the manufacturer must provide a cable design with sag and tension tables showing the maximum span and sag information for that particular installation. The information included must be for Rule B, *Ice and Wind Loading*, and when applicable, information on Rule 250C, *Extreme Wind Loading*. Additionally, to ensure the proper ground clearance, typically a minimum of 4.7 m (15.5 feet), the end user should factor in the maximum sag under loaded conditions, as well as, height of attachment for each application.

(5) *Minimum bend diameter*. For cable under loaded and unloaded conditions, the cable must have the minimum bend diameters indicated in paragraph 1.1.5, *Minimum Bend Diameter*, of the ANSI/ICEA S–87–640 (incorporated by reference at §1755.901(c)). For very small cables, manufacturers may specify fixed cable minimum bend diameters that are independent of the outside diameter. For cables having a non-circular cross-section, the bend diameter is to be determined using the thickness of the cable associated with the preferential bending axis.

(6) The cable is fully color coded so that each fiber is distinguishable from every other fiber. A basic color scheme of twelve colors allows individual fiber identification. Colored tubes, binders, threads, strippings, or markings provide fiber group identification.

(7) Cables must demonstrate compliance with the qualification testing requirements of this section to ensure satisfactory end-use performance characteristics for the intended applications.

(8) Optical cable designs not specifically addressed by this section may be allowed if accepted by the Agency. Justification for acceptance of a modified design must be provided to substantiate product utility and long term stability and endurance. For information on how to obtain Agency product acceptance, refer to the procedures listed at *http://www.usda.gov/rus/telecom/listing_procedures/index_listing_procedures.htm*, as well as additional information in RUS Bulletin 345–3, *Acceptance of Standards, Specifications, Equipment Contract Forms, Manual Sections, Drawings, Materials and Equipment for the Telephone Program* (hereinafter "RUS Bulletin 345–3"), available for download at *http://www.usda.gov/rus/telecom/publications/bulletins.htm*.

(9) All cables sold to RUS telecommunications borrowers for projects involving RUS loan funds must be accepted by the Agency's Technical Standards Committee "A" (Telecommunications). Any design change to existing acceptable designs must be submitted to the Agency for acceptance. As stated in paragraph 8 above, refer to the procedures listed at *http://www.usda.gov/rus/telecom/listing_procedures/index_listing_procedures.htm* as well as RUS Bulletin 345–3.

(10) The Agency intends that the optical fibers contained in the cables meeting the requirements of this section have characteristics that will allow signals having a range of wavelengths to be carried simultaneously.

(b) *Optical fibers*. (1) The solid glass optical fibers must consist of a cylindrical core and cladding covered by either an ultraviolet-cured acrylate or other suitable coating. Each fiber must be continuous throughout its length.

(2) *Zero-dispersion*. Optical fibers must meet the fiber attributes of Table 2, *G.652.B attributes*, found in ITU–T Recommendation G.652 (incorporated by reference at §1755.901(f)). However, when the end user stipulates a low water peak fiber, the optical fibers must meet the fiber attributes of Table 4, *G.652.D attributes*, found in ITU–T Recommendation G.652; or when the end user stipulates a low bending loss fiber, the optical fibers must meet the fiber attributes of Table 7–1, *G.657 class*

A attributes, found in the ITU–T Recommendation G.657 (incorporated by reference at § 1755.901(f)).

(3) *Non-zero-dispersion*. Optical fibers must meet the fiber attributes of Table 1, *G.656 attributes*, found in ITU–T Recommendation G.656 (incorporated by reference at § 1755.901(f)). However, when the end user specifies Recommendation A, B, C, D, or E of ITU–T Recommendation G.655 (incorporated by reference at § 1755.901(f)), the optical fibers must meet the fiber attributes of ITU–T Recommendation G.655.

(4) *Multimode fibers*. Optical fibers must meet the requirements of paragraphs 2.1 and 2.3.1 of ANSI/ICEA S–87–640 (incorporated by reference at § 1755.901(c)).

(5) *Matched cable*. Unless otherwise specified by the buyer, all single mode fiber cables delivered to a RUS-financed project must be manufactured to the same MFD specification. However, notwithstanding the requirements of paragraphs (d)(2) and (d)(3) of this section, the maximum MFD tolerance allowed for cable meeting the requirements of this section must be of a magnitude meeting the definition of "matched cable," as defined in paragraph (b) of § 1755.900. With the use of cables meeting this definition the user can reasonably expect that the average bi-directional loss of a fusion splice to be ≤0.1 dB.

(6) Buyers will normally specify the MFD for the fibers in the cable. When a buyer does not specify the MFD at 1310 nm, the fibers must be manufactured to an MFD of 9.2 μm with a maximum tolerance range of ±0.5 μm (362 ±20 microinch), unless the end user agrees to accept cable with fibers specified to a different MFD. When the end user does specify a MFD and tolerance conflicting with the MFD maximum tolerance allowed by paragraph (d)(5) of this section, the requirements of paragraph (d)(5) must prevail.

(7) Factory splices are not allowed.

(8) *Coating*. The optical fiber must be coated with a suitable material to preserve the intrinsic strength of the glass having an outside diameter of 250 ±15 micrometers (10 ±0.6 mils). Dimensions must be measured per the methods of paragraph 7.13 of ANSI/ICEA S–87–640 (incorporated by reference at § 1755.901(c)). The protective coverings must be free from holes, splits, blisters, and other imperfections and must be as smooth and concentric as is consistent with the best commercial practice. The diameter of the fiber, as the fiber is used in the cable, includes any coloring thickness or the uncolored coating, as the case may be. The strip force required to remove 30 ±3 millimeters (1.2 ±0.1 inch) of protective fiber coating must be between 1.0 N (0.2 pound-force) and 9.0 N (2 pound-force).

(9) All optical fibers in any single length of cable must be of the same type, unless otherwise specified by end user.

(10) Optical fiber dimensions and data reporting must be as required by paragraph 7.13.1.1 of ANSI/ICEA S–87–640 (incorporated by reference at § 1755.901(c)).

(c) *Buffers*. (1) The optical fibers contained in a tube buffer (loose tube), an inner jacket (unit core), a channel, or otherwise loosely packaged must have a clearance between the fibers and the inside of the container sufficient to allow for thermal expansions of the tube buffer without constraining the fibers. The protective container must be manufactured from a material having a coefficient of friction sufficiently low to allow the fibers free movement. The loose tube must contain a suitable water blocking material. Loose tubes must be removable without damage to the fiber when following the manufacturer's recommended procedures.

(2) The tubes for single mode loose tube cables must be designed to allow a maximum mid-span buffer tube exposure of 6.096 meters (20 feet). The buyer should be aware that certain housing hardware may require cable designed for 6.096 meters of buffer tube storage.

(3) Optical fibers covered in near contact with an extrusion (tight tube) must have an intermediate soft buffer to allow for thermal expansions and minor pressures. The buffer tube dimension must be established by the manufacturer to meet the requirement of this section. Tight buffer tubes must be removable without damage to the fiber when following the manufacturer's recommended procedures. The tight buffered fiber must be strippable per paragraph 7.20 of ANSI/ICEA S–87–

640 (incorporated by reference at §1755.901(c)).

(4) Both loose tube and tight tube coverings of each color and other fiber package types removed from the finished cable must meet the following shrinkback and cold bend performance requirements. The fibers may be left in the tube.

(i) *Shrinkback.* Testing must be conducted per paragraph 14.1 of ASTM D 4565 (incorporated by reference at §1755.901(d)), using a talc bed at a temperature of 95 °C (203 °F). Shrinkback must not exceed 5 percent of the original 150 millimeter (6 inches) length of the specimen. The total shrinkage of the specimen must be measured. (Buffer tube material meeting this test may not meet the mid-span test in paragraph (t)(15) of this section).

(ii) *Cold bend.* Testing must be conducted on at least one tube from each color in the cable. Stabilize the specimen to −30 ±1 °C (−22 ±2 °F) for a minimum of four hours. While holding the specimen and mandrel at the test temperature, wrap the tube in a tight helix ten times around a mandrel with a diameter to be greater than five times the tube diameter or 50 mm (2 inches). The tube must show no evidence of cracking when observed with normal or corrected-to-normal vision.

NOTE TO PARAGRAPH (c)(4)(ii): Channel cores and similar slotted single component core designs do not need to be tested for cold bend.

(d) *Fiber identification.* (1) Each fiber within a unit and each unit within the cable must be identifiable per paragraphs 4.2.1 and 4.3.1 of ANSI/ICEA S–87–640 (incorporated by reference at §1755.901(c)).

(2) For the following items the colors designated for identification within the cable must comply with paragraphs 4.2.2 and 4.3.2 of ANSI/ICEA S–87–640 (incorporated by reference at §1755.901(c)): loose buffer tubes, tight tube buffer fibers, individual fibers in multi-fiber tubes, slots, bundles or units of fibers, and the units in cables with more than one unit.

(e) *Optical fiber ribbon.* (1) Each ribbon must be identified per paragraphs 3.4.1 and 3.4.2 of ANSI/ICEA S–87–640 (incorporated by reference at §1755.901(c)).

(2) Ribbon fiber count must be specified by the end user, *i.e.*, 2, 4, 6, 12, etc.

(3) Ribbon dimensions must be as agreed by the end user and manufacturer per paragraph 3.4.4.1 of ANSI/ICEA S–87–640 (incorporated by reference at §1755.901(c)).

(4) Ribbons must meet each of the following tests. These tests are included in the paragraphs of ANSI/ICEA S–87–640 (incorporated by reference at §1755.901(c)), indicated in parenthesis below.

(i) Ribbon Dimensions (ANSI/ICEA S–87–640 paragraphs 7.14 through 7.14.2)—measures ribbon dimension.

(ii) Ribbon Twist Test (ANSI/ICEA S–87–640 paragraphs 7.15 through 7.15.2)—evaluates the ability of the ribbon to resist splitting or other damage while undergoing dynamic cyclically twisting the ribbon under load.

(iii) Ribbon Residual Twist Test (ANSI/ICEA S–87–640 paragraphs 7.16 through 7.16.2)—evaluates the degree of permanent twist in a cabled optical ribbon.

(iv) Ribbon Separability Test (ANSI/ICEA S–87–640 paragraphs 7.17 through 7.17.2)—evaluates the ability to separate fibers.

(5) Ribbons must meet paragraph 3.4.4.6 of ANSI/ICEA S–87–640 (incorporated by reference at §1755.901(c)), Ribbon Strippability.

(f) *Strength members.* (1) Strength members may be an integral part of the cable construction, but are not considered part of the support messenger for self-supporting optical cable.

(2) The strength members may be metallic or nonmetallic.

(3) The combined strength of all the strength members must be sufficient to support the stress of installation and to protect the cable in service.

(4) Strength members may be incorporated into the core as a central support member or filler, as fillers between the fiber packages, as an annular serving over the core, as an annular serving over the intermediate jacket, embedded in the outer jacket, or as a combination of any of these methods.

(5) The central support member or filler must contain no more than one splice per kilometer of cable. Individual fillers placed between the fiber packages and placed as annular

servings over the core must contain no more than one splice per kilometer of cable. Cable sections having central member or filler splices must meet the same physical requirements as unspliced cable sections.

(6) In each length of completed cable having a metallic central member, the dielectric strength between the shield or armor, when present, and the metallic center member must withstand at least 15 kilovolts when tested per ASTM D 4566 (incorporated by reference at § 1755.901(d)). The voltage must be applied for 3 seconds minimum; no failures are allowed.

(g) *Cable core.* (1) Protected fibers may be assembled with the optional central support member, fillers and strength members in such a way as to form a cylindrical group.

(2) The standard cylindrical group or core designs commonly consist of 4, 6, 12, 18, or 24 fibers. Cylindrical groups or core designs larger than the sizes shown above must meet all the applicable requirements of this section.

(3) When threads or tapes are used in cables using water blocking elements as core binders, they must be a non-hygroscopic and non-wicking dielectric material or be rendered by the gel or water blocking material produced by the ingress of water.

(4) When threads or tapes are used as unit binders to define optical fiber units in loose tube, tight tube, slotted, or bundled cored designs, they must be non-hygroscopic and non-wicking dielectric material or be rendered by the filling compound or water blocking material contained in the binder. The colors of the binders must be per paragraphs (f)(2) and (f)(3) of this section.

(h) *Core water blocking.* (1) To prevent the ingress of water into the core and water migration, a suitable filling compound or water blocking elements must be applied into the interior of the loose fiber tubes and into the interstices of the core. When a core wrap is used, the filling compound or water blocking elements, as the case may be, must also be applied to the core wrap, over the core wrap and between the core wrap and inner jacket when required.

(2) The materials or elements must be homogeneous and uniformly mixed; free from dirt, metallic particles and other foreign matter; easily removed; nontoxic and present no dermal hazards. The filling compound and water blocking elements must contain a suitable antioxidant or be of such composition as to provide long term stability.

(3) The individual cable manufacturer must satisfy the Agency that the filling compound or water blocking elements selected for use is suitable for its intended application by submitting test data showing compliance with ASTM D 4568 (incorporated by reference at § 1755.901(d)). The filling compound and water blocking elements must be compatible with the cable components when tested per ASTM D 4568 at a temperature of 80 °C (176 °F). The jacket must retain a minimum of 85% of its un-aged tensile and elongation values.

(i) *Water blocking material.* (1) Sufficient flooding compound or water blocking elements must be applied between the inner jacket and armor and between the armor and outer jacket so that voids and air spaces in these areas are minimized. The use of flooding compound or water blocking elements between the armor and outer jacket is not required when uniform bonding, paragraph (o)(9) of this section, is achieved between the plastic-clad armor and the outer jacket.

(2) The flooding compound or water blocking elements must be compatible with the jacket when tested per paragraphs 7.19 and 7.19.1 of ANSI/ICEA S–87–640 (incorporated by reference at § 1755.901(c)). The aged jacket must retain a minimum of 85% of its un-aged tensile strength and elongation values when tested per paragraph 7.19.2.3. The flooding compound must exhibit adhesive properties sufficient to prevent jacket slip when tested per paragraph 7.30.1 of ANSI/ICEA S–87–640 and meets paragraph 7.30.2 of ANSI/ICEA S–87–640 for minimum sheath adherence of 14 N/mm for armored cables.

(3) The individual cable manufacturer must satisfy the Agency by submitting test data showing compliance with the appropriate cable performance testing requirements of this section that the flooding compound or water blocking elements selected for use is acceptable for the application.

(j) *Core wrap*. (1) At the option of the manufacturer, one or more layers of dielectric material may be applied over the core.

(2) The core wrap(s) can be used to provide a heat barrier to prevent deformation or adhesion between the fiber tubes or can be used to contain the core.

(k) *Inner jackets*. (1) For designs with more than one jacket, the inner jackets must be applied directly over the core or over the strength members when required by the end user. The jacket must be free from holes, splits, blisters, or other imperfections and must be as smooth and concentric as is consistent with the best commercial practice. The inner jacket must not adhere to other cable components such as fibers, buffer tubes, etc.

(2) For armored and unarmored cable, an inner jacket is optional. The inner jacket may absorb stresses in the cable core that may be introduced by armor application or by armored cable installation.

(3) The inner jacket material and test requirements must be the same as the outer jacket material, except that either black or natural polyethylene may be used and the thickness requirements are included in paragraph (m)(4) of this section. In the case of natural polyethylene, the requirements for absorption coefficient and the inclusion of furnace black are waived.

(4) The inner jacket thickness must be determined by the manufacturer, but must be no less than a nominal jacket thickness of 0.5 mm (0.02 inch) with a minimum jacket thickness of 0.35 mm (0.01 inch).

(l) *Outer jacket*. (1) The outer jacket must provide the cable with a tough, flexible, protective covering which can withstand exposure to sunlight, to atmosphere temperatures, and to stresses reasonably expected in normal installation and service.

(2) The jacket must be free from holes, splits, blisters, or other imperfections and must be as smooth and concentric as is consistent with the best commercial practice.

(3) The jacket must contain an antioxidant to provide long term stabilization and must contain a minimum of 2.35 percent concentration of furnace black to provide ultraviolet shielding measures as required by paragraph 5.4.2 of ANSI/ICEA S–87–640 (incorporated by reference at § 1755.901(c)), except that the concentration of furnace black does not necessarily need to be initially contained in the raw material and may be added later during the jacket making process.

(4) The raw material used for the outer jacket must be one of the types listed below.

(i) *Type L1*. Low density, polyethylene (LDPE) must conform to the requirements of paragraph 5.4.2 of ANSI/ICEA S–87–640 (incorporated by reference at § 1755.901(c)).

(ii) *Type L2*. Linear low density, polyethylene (LLDPE) must conform to the requirements of paragraph 5.4.2 of ANSI/ICEA S–87–640 (incorporated by reference at § 1755.901(c)).

(iii) *Type M*. Medium density polyethylene (MDPE) must conform to the requirements of paragraph 5.4.2 of ANSI/ICEA S–87–640 (incorporated by reference at § 1755.901(c)).

(iv) *Type H*. High density polyethylene (HDPE) must conform to the requirements of paragraph 5.4.2 of ANSI/ICEA S–87–640 (incorporated by reference at § 1755.901(c)).

(5) Particle size of the carbon selected for use must not average greater than 20 nm.

(6) The outer jacketing material removed from or tested on the cable must be capable of meeting the performance requirements of Table 5.1 found in ANSI/ICEA S–87–640 (incorporated by reference at § 1755.901(c)).

(7) *Testing Procedures*. The procedures for testing the jacket specimens for compliance with paragraph (n)(5) of this section must be as follows:

(i) *Jacket material density measurement*. Test per paragraphs 7.7.1 and 7.7.2 of ANSI/ICEA S–87–640 (incorporated by reference at § 1755.901(c)).

(ii) *Tensile strength, yield strength, and ultimate elongation*. Test per paragraphs 7.8.1 and 7.8.2 of ANSI/ICEA S–87–640 (incorporated by reference at § 1755.901(c)).

(iii) *Jacket material absorption coefficient test*. Test per paragraphs 7.9.1 and 7.9.2 of ANSI/ICEA S–87–640 (incorporated by reference at § 1755.901(c)).

(iv) *Environmental stress crack resistance test.* For large cables (outside diameter ≥ 30 mm (1.2 inch)), test per paragraphs 7.10.1 through 7.10.1.2 of ANSI/ICEA S–87–640 (incorporated by reference at § 1755.901(c)). For small cables (Diameter < 30 mm (1.2 inch)), test per paragraphs 7.10.2 through and 7.10.2.2 of ANSI/ICEA S–87–640. A crack or split in the jacket constitutes failure.

(v) *Jacket shrinkage test.* Test per paragraphs 7.11.1 and 7.11.2 of ANSI/ ICEA S–87–640 (incorporated by reference at § 1755.901(c)).

(8) *Jacket thickness.* The outer jacket must meet the requirements of paragraphs 5.4.5.1 and 5.4.5.2 of ANSI/ICEA S–87–640 (incorporated by reference at § 1755.901(c)).

(9) *Jacket repairs.* Repairs are allowed per paragraph 5.5 of ANSI/ICEA S–87–640 (incorporated by reference at § 1755.901(c)).

(m) *Armor.* (1) A steel armor, plastic coated on both sides, is required for direct buried cable manufactured under this section. Armor is optional for duct and aerial cable, as required by the end user. The plastic coated steel armor must be applied longitudinally directly over the core wrap or the intermediate jacket and have a minimum overlap of 3.0 millimeters (118 mils), except for small diameter cables with diameters of less than 10 mm (394 mils) for which the minimum overlap must be 2 mm (79 mils). When a cable has a shield, the armor should normally be applied over the shielding tape.

(2) The uncoated steel tape must be electrolytic chrome coated steel (ECCS) and must meet the requirements of paragraph B.2.4 of ANSI/ICEA S–87–640 (incorporated by reference at § 1755.901(c)).

(3) The reduction in thickness of the armoring material due to the corrugating or application process must be kept to a minimum and must not exceed 10 percent at any spot.

(4) The armor of each length of cable must be electrically continuous with no more than one joint or splice allowed in any length of one kilometer of cable. This requirement does not apply to a joint or splice made in the raw material by the raw material manufacturer.

(5) The breaking strength of any section of an armor tape, containing a factory splice joint, must not be less than 80 percent of the breaking strength of an adjacent section of the armor of equal length without a joint.

(6) For cables containing no flooding compound over the armor, the overlap portions of the armor tape must be bonded in cables having a flat, non-corrugated armor to meet the mechanical requirements of paragraphs (t)(1) through (t)(16)(ii) of this section. If the tape is corrugated, the overlap portions of the armor must be sufficiently bonded and the corrugations must be sufficiently in register to meet the requirements of paragraphs (t)(1) through (t)(16)(ii) of this section.

(7) The armor tape must be so applied as to enable the cable to pass the Cable Low (−30 °C (−22 °F)) and High (60 °C (140 °F)) Temperatures Bend Test, as required by paragraph (t)(3) of this section.

(8) The protective coating on the steel armor must meet the Bonding-to-Metal, Heat Sealability, Lap-Shear and Moisture Resistance requirements of Type I, Class 2 coated metals per ASTM B 736 (incorporated by reference in § 1755.901(d)).

(9) When the jacket is bonded to the plastic coated armor, the bond between the plastic coated armor and the outer jacket must not be less than 525 Newtons per meter (36 pound-force) over at least 90 percent of the cable circumference when tested per ASTM D 4565 (incorporated by reference at § 1755.901(d)). For cables with strength members embedded in the jacket, and residing directly over the armor, the area of the armor directly under the strength member is excluded from the 90 percent calculation.

(n) *Figure 8 aerial cables.* (1) When self-supporting aerial cable containing an integrated support messenger is supplied, the support messenger must comply with the requirements specified in paragraphs D.2.1 through D.2.4 of ANSI/ICEA S–87–640 (incorporated by reference at § 1755.901(c)), with exceptions and additional provisions as follows:

(i) Any section of a completed strand containing a joint must have minimum tensile strength and elongation of

29,500 Newtons (6,632 pound-force) and 3.5 percent, respectively, when tested per the procedures specified in ASTM A 640 (incorporated by reference in § 1755.901(d)).

(ii) The individual wires from a completed strand which contains joints must not fracture when tested per the "Ductility of Steel" procedures specified in ASTM A 640 (incorporated by reference at § 1755.901(d)), except that the mandrel diameter must be equal to 5 times the nominal diameter of the individual wires.

(iii) The support strand must be completely covered with a flooding compound that offers corrosion protection. The flooding compound must be homogeneous and uniformly mixed.

(iv) The flooding compound must be nontoxic and present no dermal hazard.

(v) The flooding compound must be free from dirt, metallic particles, and other foreign matter that may interfere with the performance of the cable.

(2) Other methods of providing self-supporting cable specifically not addressed in this section may be allowed if accepted. Justification for acceptance of a modified design must be provided to substantiate product utility and long term stability and endurance. To obtain the Agency's acceptance of a modified design, refer to the product acceptance procedures available at *http://www.usda.gov/rus/telecom/listing_procedures/index_listing_procedures.htm*, as well as RUS Bulletin 345–3.

(3) *Jacket thickness requirements*. Jackets applied over an integral messenger must meet the following requirements:

(i) The minimum jacket thickness at any point over the support messenger must meet the requirements of paragraph D.3 of ANSI/ICEA S–87–640 (incorporated by reference at § 1755.901(c)).

(ii) The web dimension for self-supporting aerial cable must meet the requirements of paragraph D.3 of ANSI/ICEA S–87–640 (incorporated by reference at § 1755.901(c)).

(o) *Sheath slitting cord*. (1) A sheath slitting cord or ripcord is optional.

(2) When a sheath slitting cord is used it must be capable of slitting the jacket or jacket and armor, at least one meter (3.3 feet) length without breaking the cord at a temperature of 23 ±5 °C (73 ±9 °F).

(3) The sheath slitting cord must meet the sheath slitting cord test described in paragraph (t)(1) of this section.

(p) *Identification markers*. (1) Each length of cable must be permanently identified. The method of marking must be by means of suitable surface markings producing a clear distinguishable contrasting marking meeting paragraph 6.1.1 of ANSI/ICEA S–87–640 (incorporated by reference at § 1755.901(c)), and must meet the durability requirements of paragraphs 7.5.2 through 7.5.2.2 of ANSI/ICEA S–87–640.

(2) The color of the initial marking must be white or silver. If the initial marking fails to meet the requirements of the preceding paragraphs, it will be permissible to either remove the defective marking and re-mark with the white or silver color or leave the defective marking on the cable and re-mark with yellow. No further re-marking is permitted. Any re-marking must be done on a different portion of the cable's circumference where the existing marking is found and have a numbering sequence differing from any other marking by at least 3,000. Any reel of cable that contains more than one set of sequential markings must be labeled to indicate the color and sequence of marking to be used. The labeling must be applied to the reel and also to the cable.

(3) Each length of cable must be permanently labeled OPTICAL CABLE, OC, OPTICAL FIBER CABLE, or OF on the outer jacket and identified as to manufacturer and year of manufacture.

(4) Each length of cable intended for direct burial installation must be marked with a telephone handset in compliance with requirements of the Rule 350G of the ANSI/IEEE C2–2007 (incorporated by reference at § 1755.901(b)).

(5) Each length of cable must be identified as to the manufacturer and year of manufacturing. The manufacturer and year of manufacturing may also be indicated by other means as indicated in paragraphs 6.1.2 through 6.1.4 of ANSI/ICEA S–87–640 (incorporated by reference at § 1755.901(c)).

(6) The number of fibers on the jacket must be marked on the jacket.

(7) The completed cable must have sequentially numbered length markers in METERS or FEET at regular intervals of not more than 2 feet or not more than 1 meter along the outside of the jacket. Continuous sequential numbering must be employed in a single length of cable. The numbers must be dimensioned and spaced to produce good legibility and must be approximately 3 millimeters (118 mils) in height. An occasional illegible marking is permissible when it is located within 2 meters of a legible making for cables marked in meters or 4 feet for cables marked in feet.

(8) Agreement between the actual length of the cable and the length marking on the cable jacket must be within the limits of + 1 percent and − 0 percent.

(9) *Jacket print test.* Cables must meet the Jacket Print Test described in paragraphs 7.5.2.1 and 7.5.2.2 of ANSI/ICEA S–87–640 (incorporated by reference at § 1755.901(c)).

(q) *Performance of a finished cable*—(1) *Zero dispersion optical fiber cable.* Unless otherwise specified by the end user, the optical performance of a finished cable must comply with the attributes of Table 2, *G.652.B attributes,* found in ITU Recommendation G.652 (incorporated by reference at § 1755.901(f)). However, when the end user stipulates a low water peak fiber the finished cable must meet the attributes of Table 4, *G.652.D attributes,* found in ITU–T Recommendation G.652; or when the end user stipulates a low bending loss fiber, the finished cable must meet the attributes of Table 7–1, *G.657 class A attributes,* found in ITU–T Recommendation G.657 (incorporated by reference at § 1755.901(f)).

(i) The attenuation methods must be per Table 8.4, *Optical attenuation measurement methods,* of ANSI/ICEA S–87–640 (incorporated by reference at § 1755.901(c)).

(ii) The cable must have a maximum attenuation of 0.1 dB at a point of discontinuity (a localized deviation of the optical fiber loss). Per paragraphs 8.4 and 8.4.1 of ANSI/ICEA S–87–640 (incorporated by reference at § 1755.901(c)), measurements must be conducted at 1310 and 1550 nm, and at 1625 nm when specified by the end user.

(iii) The cable cutoff wavelength (γ_{cc}) must be reported per paragraph 8.5.1 of ANSI/ICEA S–87–640 (incorporated by reference in § 1755.901(c)).

(2) *Nonzero dispersion optical fiber cable.* Unless otherwise specified by the end user, the optical performance of the finished cable must comply with the attributes of Table 1, *G.656 attributes,* found in ITU–T Recommendation G.656 (incorporated by reference at § 1755.901(f)). When the buyer specifies Recommendation A, B, C, D or E of ITU–T Recommendation G.655 (incorporated by reference at § 1755.901(f)), the finished cable must comply with the attributes of ITU–T Recommendation G.655.

(i) The attenuation methods must be per Table 8.4, *Optical attenuation measurement methods* of ANSI/ICEA S–87–640 (incorporated by reference at § 1755.901(c)).

(ii) The cable must have a maximum attenuation of 0.1 dB at a point of discontinuity (a localized deviation of the optical fiber loss). Per paragraphs 8.4 and 8.4.1 of ANSI/ICEA S–87–640 (incorporated by reference at § 1755.901(c)), measurements must be conducted at 1310 and 1550 nm, and at 1625 nm when specified by the end user.

(iii) The cable cutoff wavelength (γ_{cc}) must be reported per paragraph 8.5.1 of ANSI/ICEA S–87–640 (incorporated by reference at § 1755.901(c)).

(3) *Multimode optical fiber cable.* Unless otherwise specified by the end user, the optical performance of the fibers in a finished cable must comply with Table 8.1, *Attenuation coefficient performance requirement (dB/k),* Table 8.2, *Multimode bandwidth coefficient performance requirements (MHz-km)* and Table 8.3, *Points discontinuity acceptance criteria (dB),* of ANSI/ICEA S–87–640 (incorporated by reference at § 1755.901(c)).

(4) Because the accuracy of attenuation measurements for single mode fibers becomes questionable when measured on short cable lengths, attenuation measurements are to be made utilizing characterization cable lengths. Master Cable reels must be tested and the attenuation values

measured will be used for shorter ship lengths of cable.

(5) Because the accuracy of attenuation measurements for multimode fibers becomes questionable when measured on short cable lengths, attenuation measurements are to be made utilizing characterization cable lengths. If the ship length of cable is less than one kilometer, the attenuation values measured on longer lengths of cable (characterization length of cable) before cutting to the ship lengths of cable may be applied to the ship lengths.

(6) Attenuation must be measured per Table 8.4, *Optical Attenuation Measurement Methods,* of ANSI/ICEA S–87–640 (incorporated by reference at § 1755.901(c)).

(7) The bandwidth of multimode fibers in a finished cable must be no less than the values specified in ANSI/ICEA S–87–640 (incorporated by reference at § 1755.901(c)), Table 8.2 per paragraphs 8.3.1 and 8.3.2.

(r) *Mechanical requirements.* Fiber optic cables manufactured under the requirements of this section must be tested by the manufacturer to determine compliance with such requirements. Unless otherwise specified, testing must be performed at the standard conditions defined in paragraph 7.3.1 of ANSI/ICEA S–87–640 (incorporated by reference at § 1755.901(c)). The standard optical test wavelengths to be used are 1550 nm single mode and 1300 nm multimode, unless otherwise specified in the individual test.

(1) *Sheath slitting cord test.* All cables manufactured under the requirements of this section must meet the Ripcord Functional Test described in paragraphs 7.18.1 and 7.18.2 of ANSI/ICEA S–87–640 (incorporated by reference at § 1755.901(c)).

(2) *Material compatibility and cable aging test.* All cables manufactured under the requirements of this section must meet the Material Compatibility and Cable Aging Test described in paragraphs 7.19 through 7.19.2.4 of ANSI/ICEA S–87–640 (incorporated by reference at § 1755.901(c)).

(3) *Cable low and high bend test.* Cables manufactured under the requirements of this section must meet the Cable Low (−30 °C (−22 °F)) and High (60 °C (140 °F)) Temperatures Bend Test per paragraphs 7.21 and 7.21.2 of ANSI/ICEA S–87–640 (incorporated by reference at § 1755.901(c)).

(4) *Compound flow test.* All cables manufactured under the requirements of this section must meet the test described in paragraphs 7.23, 7.23.1, and 7.23.2 of ANSI/ICEA S–87–640 (incorporated by reference at § 1755.901(c)).

(5) *Cyclic flexing test.* All cables manufactured under the requirements of this section must meet the Flex Test described in paragraphs 7.27 through 7.27.2 of the ANSI/ICEA S–87–640 (incorporated by reference at § 1755.901(c)).

(6) *Water penetration test.* All cables manufactured under the requirements of this section must meet paragraphs 7.28 through 7.28.2 of ANSI/ICEA S–87–640 (incorporated by reference at § 1755.901(c)).

(7) *Cable impact test.* All cables manufactured under the requirements of this section must meet the Cable Impact Test described in paragraphs 7.29.1 and 7.29.2 of ANSI/ICEA S–87–640 (incorporated by reference at § 1755.901(c)).

(8) *Cable tensile loading and fiber strain test.* Cables manufactured under the requirements of this section must meet the Cable Loading and Fiber Strain Test described in paragraphs 7.30 through 7.30.2 of ANSI/ICEA S–87–640 (incorporated by reference at § 1755.901(c)). This test does not apply to aerial self-supporting cables.

(9) *Cable compression test.* All cables manufactured under requirements of this section must meet the Cable Compressive Loading Test described in paragraphs 7.31 through 7.31.2 of ANSI/ICEA S–87–640 (incorporated by reference at § 1755.901(c)).

(10) *Cable twist test.* All cables manufactured under the requirements of this section must meet the Cable Twist Test described in paragraphs 7.32 through 7.32.2 of ANSI/ICEA S–87–640 (incorporated by reference at § 1755.901(c)).

(11) *Cable Lighting damage susceptibility test.* Cables manufactured under the requirements of this section must meet the Cable Lighting Damage Susceptibility Test described in paragraphs 7.33 and 7.33.1 of ANSI/ICEA S–87–640 (incorporated by reference at § 1755.901(c)).

(12) *Cable external freezing test.* All cables manufactured under the requirements of this section must meet the Cable External Freezing Test described in paragraphs 7.22 and 7.22.1 of ANSI/ICEA S–87–640 (incorporated by reference at § 1755.901(c)).

(13) *Cable temperature cycling test.* All cables manufactured under the requirements of this section must meet the Cable Temperature Cycling Test described in paragraph 7.24.1 of ANSI/ICEA S–87–640 (incorporated by reference at § 1755.901(c)).

(14) *Cable sheath adherence test.* All cables manufactured under the requirements of this section must meet the Cable Sheath Adherence Test described in paragraphs 7.26.1 and 7.26.2 of ANSI/ICEA S–87–640 (incorporated by reference at § 1755.901(c)).

(15) *Mid-span test.* This test is applicable only to cables of a loose tube design specified for mid-span applications with tube storage. Cable of specialty design may be exempted from this requirement when this requirement is not applicable to such design. All buried and underground loose tube single mode cables manufactured per the requirements in this section and intended for mid-span applications with tube storage must meet the following mid-span test without exhibiting an increase in fiber attenuation greater than 0.1 dB and a maximum average increase over all fibers of 0.05 dB.

(i) The specimen must be installed in a commercially available pedestal or closure or in a device that mimics their performance, as follows: A length of cable sheath, equal to the mid-span length, must be removed from the middle of the test specimen so as to allow access to the buffer tubes. All binders, tapes, strength members, etc. must be removed. The buffer tubes must be left intact. The cable ends defining the ends of the mid-span length must be properly secured in the closure to the more stringent of the cable or hardware manufacturer's recommendations. Strength members must be secured with an end stop type clamp and the outer jacket must be clamped to prevent slippage. A minimum of 6.096 meters (20 feet) of cable must extend from the entry and exit ports of the closure for the purpose of making optical measurements. If a device that mimics the performance of pedestals or closures is used, the buffer tubes must be wound in a coil with a minimum width of 3 inches and minimum length of 12 inches.

(ii) The expressed buffer tubes must be loosely constrained during the test.

(iii) The enclosure, with installed cable, must be placed in an environmental chamber for temperature cycling. It is acceptable for some or all of the two 20 feet (6.096 meters) cable segments to extend outside the environmental chamber.

(iv) Lids, pedestal enclosures, or closure covers must be removed if possible to allow for temperature equilibrium of the buffer tubes. If this is not possible, the manufacturer must demonstrate that the buffer tubes are at temperature equilibrium prior to beginning the soak time.

(v) Measure the attenuation of single mode fibers at 1550 ±10 nm. The supplier must certify the performance of lower specified wavelengths comply with the mid-span performance requirements.

(vi) After measuring the attenuation of the optical fibers, test the cable sample per TIA/EIA Standard 455–3A (incorporated by reference at § 1755.901(e)). Temperature cycling, measurements, and data reporting must conform to TIA/EIA Standard 455–3A. The test must be conducted for at least five complete cycles. The following detailed test conditions must apply:

(A) TIA/EIA Standard 455–3A (incorporated by reference at § 1755.901(e)), Section 4.1—Loose tube single mode optical cable sample must be tested.

(B) TIA/EIA Standard 455–3A (incorporated by reference at § 1755.901(e)), Section 4.2—An Agency accepted 8 to 12 inch diameter optical buried distribution pedestal or a device that mimics their performance must be tested.

(C) Mid-span opening for installation of loose tube single mode optical cable in pedestal must be 6.096 meters (20 feet).

(D) TIA/EIA Standard 455–3A (incorporated by reference at § 1755.901(e)), Section 5.1—3 hours soak time.

(E) TIA/EIA Standard 455–3A (incorporated by reference at § 1755.901(e)),

Section 5.2—Test Condition C–2, minimum −40 °C (−40 °F) and maximum 70 °Celsius (158 °F).

(F) TIA/EIA Standard 455–3A (incorporated by reference at § 1755.901(e)), Section 5.7.2—A statistically representative amount of transmitting fibers in all express buffer tubes passing through the pedestal and stored must be measured.

(G) The buffer tubes in the closure or pedestal must not be handled or moved during temperature cycling or attenuation measurements.

(vii) Fiber cable attenuation measured through the express buffer tubes during the last cycle at −40 °C (−40 °F) and + 70 °C (158 °F) must not exceed a maximum increase of 0.1 dB and must not exceed a 0.05 dB average across all tested fibers from the initial baseline measurements. At the conclusion of the temperature cycling, the maximum attenuation increase at 23 °C from the initial baseline measurement must not exceed 0.05 dB which allows for measurement noise that may be encountered during the test. The cable must also be inspected at room temperature at the conclusion of all measurements; the cable must not show visible evidence of fracture of the buffer tubes nor show any degradation of all exposed cable assemblies.

(16) *Aerial self-supporting cables.* The following tests apply to aerial cables only:

(i) *Static tensile testing of aerial self-supporting cables.* Aerial self-supporting cable must meet the test described in paragraphs D.4.1.1 through D.4.1.5 of ANSI/ICEA S–87–640 (incorporated by reference at § 1755.901(c)).

(ii) *Cable galloping test.* Aerial self-supporting cable made to the requirements of this section must meet the test described in paragraphs D.4.2 through D.4.2.3 of ANSI/ICEA S–87–640 (incorporated by reference at § 1755.901(c)).

(s) *Pre-connectorized cable.* (1) At the option of the manufacturer and upon request by the end user, the cable may be factory terminated with connectors.

(2) All connectors must be accepted by the Agency prior to their use. To obtain the Agency's acceptance of connectors, refer to product acceptance procedures available at *http://www.usda.gov/rus/telecom/listing_procedures/index_listing_procedures.htm* as well as RUS Bulletin 345–3.

(t) *Acceptance testing.* (1) The tests described in the Appendix to this section are intended for acceptance of cable designs and major modifications of accepted designs. What constitutes a major modification is at the discretion of the Agency. These tests are intended to show the inherent capability of the manufacturer to produce cable products that have satisfactory performance characteristics, long life, and long-term optical stability but are not intended as field tests. After initial Rural Development product acceptance is granted, the manufacturer will need to apply for continued product acceptance in January of the third year after the year of initial acceptance. For information on Agency acceptance, refer to the product acceptance procedures available at *http://www.usda.gov/rus/telecom/listing_procedures/index_listing_procedures.htm,* as well as RUS Bulletin 345–3.

(2) *Acceptance.* For initial acceptance, the manufacturer must submit:

(i) An original signature certification that the product fully complies with each paragraph of this section;

(ii) Qualification Test Data, per the Appendix to this section;

(iii) A set of instructions for handling the cable;

(iv) OSHA Material Safety Data Sheets for all components;

(v) Agree to periodic plant inspections;

(vi) A certification stating whether the cable, as sold to RUS Telecommunications borrowers, complies with the following two provisions:

(A) Final assembly or manufacture of the product, as the product would be used by an RUS Telecommunications borrower, is completed in the United States or eligible countries (currently, Mexico, Canada and Israel); and

(B) The cost of United States and eligible countries' components (in any combination) within the product is more than 50 percent of the total cost of all components utilized in the product. The cost of non-domestic components (components not manufactured within the United States or eligible countries) which are included in the finished product must include all duties, taxes, and delivery charges to the point of assembly or manufacture;

(vii) Written user testimonials concerning performance of the product; and

(viii) Other nonproprietary data deemed necessary.

(3) *Re-qualification acceptance.* For submission of a request for continued product acceptance after the initial acceptance, follow paragraph (v)(1) of this section and then, in January every three years, the manufacturer must submit an original signature certification stating that the product fully complies with each paragraph of this section, excluding the Qualification Section, and a certification that the products sold to RUS Telecommunications borrowers comply with paragraphs (v)(2)(vi) through (v)(2)(vi)(B) of this section. The tests of the Appendix to this section must be conducted and records kept for at least three years and the data must be made available to the Agency on request. The required data must have been gathered within 90 days of the submission. A certification must be submitted to the Agency stating that the cable manufactured to the requirements of this section has been tested per the Appendix of this section and that the cable meets the test requirements.

(4) *Initial and re-qualification acceptance requests should be addressed to:* Chairman, Technical Standards Committee "A" (Telecommunications), STOP 1550, Advanced Services Division, Rural Development Telecommunications Program, Washington, DC 20250–1500.

(5) *Tests on 100 Percent of Completed Cable.* (i) The armor for each length of cable must be tested for continuity using the procedures of ASTM D 4566 (incorporated by reference at §1755.901(d)).

(ii) Attenuation for each optical fiber in the cable must be measured.

(iii) Optical discontinuities greater than 0.1 dB must be isolated and their location and amplitude recorded.

(6) *Capability tests.* The manufacturer must establish a quality assurance system. Tests on a quality assurance basis must be made as frequently as is required for each manufacturer to determine and maintain compliance with all the mechanical requirements and the fiber and cable attributes required by this section, including:

(i) Numerical aperture and bandwidth of multimode fibers;

(ii) Cut off wavelength of single mode fibers;

(iii) Dispersion of single mode fibers;

(iv) Shrinkback and cold testing of loose tube and tight tube buffers, and mid-span testing of cables of a loose tube design with tube storage;

(v) Adhesion properties of the protective fiber coating;

(vi) Dielectric strength between the armor and the metallic central member;

(vii) Performance requirements for the fibers.

(viii) Performance requirements for the inner and outer jacketing materials;

(ix) Performance requirements for the filling and flooding compounds;

(x) Bonding properties of the coated armoring material;

(xi) Sequential marking and lettering; and

(xii) Mechanical tests described in paragraphs (t)(1) through (t)(16)(ii) of this section.

(u) *Records tests.* (1) Each manufacturer must maintain suitable summary records for a period of at least 3 years of all optical and physical tests required on completed cable by section as set forth in paragraphs (v)(5) and (v)(6) of this section. The test data for a particular reel must be in a form that it may be readily available to the Agency upon request. The optical data must be furnished to the end user on a suitable and easily readable form.

(2) Measurements and computed values must be rounded off to the number of places or figures specified for the requirement per paragraph 1.3 of ANSI/ICEA S–87–640 (incorporated by reference at §1755.901(c)).

(v) *Manufacturing irregularities.* (1) Under this section, repairs to the armor, when present, are not permitted in cable supplied to the end user.

(2) Minor defects in the inner and outer jacket (defects having a dimension of 3 millimeter or less in any direction) may be repaired by means of

heat fusing per good commercial practices utilizing sheath grade compounds.

(w) *Packaging and preparation for shipment.* (1) The cable must be shipped on reels containing one continuous length of cable. The diameter of the drum must be large enough to prevent damage to the cable from reeling and unreeling. The diameter must be at least equal to the minimum bending diameter of the cable. The reels must be substantial and so constructed as to prevent damage during shipment and handling.

(2) A circumferential thermal wrap or other means of protection must be secured between the outer edges of the reel flange to protect the cable against damage during storage and shipment. The thermal wrap must meet the requirements included in the *Thermal Reel Wrap Test,* described below. This test procedure is for qualification of initial and subsequent changes in thermal reel wraps.

(i) *Sample selection.* All testing must be performed on two 450 millimeter (18 inches) lengths of cable removed sequentially from the same fiber jacketed cable. This cable must not have been exposed to temperatures in excess of 38 °C (100 °F) since its initial cool down after sheathing.

(ii) *Test procedure.* (A) Place the two samples on an insulating material such as wood.

(B) Tape thermocouples to the jackets of each sample to measure the jacket temperature.

(C) Cover one sample with the thermal reel wrap.

(D) Expose the samples to a radiant heat source capable of heating the uncovered sample to a minimum of 71 °C (160 °F). A GE 600 watt photoflood lamp or an equivalent lamp having the light spectrum approximately that of the sun must be used.

(E) The height of the lamp above the jacket must be 380 millimeters (15 inches) or an equivalent height that produces the 71 °C (160 °F) jacket temperature on the unwrapped sample must be used.

(F) After the samples have stabilized at the temperature, the jacket temperatures of the samples must be recorded after one hour of exposure to the heat source.

(G) Compute the temperature difference between jackets.

(H) The temperature difference between the jacket with the thermal reel wrap and the jacket without the reel wrap must be greater than or equal to 17 °C (63 °F).

(3) Cables must be sealed at the ends to prevent entrance of moisture.

(4) The end-of-pull (outer end) of the cable must be securely fastened to prevent the cable from coming loose during transit. The start-of-pull (inner end) of the cable must project through a slot in the flange of the reel, around an inner riser, or into a recess on the flange near the drum and fastened in such a way to prevent the cable from becoming loose during installation.

(5) Spikes, staples or other fastening devices must be used in a manner which will not result in penetration of the cable.

(6) The arbor hole must admit a spindle 63.5 millimeters (2.5 inches) in diameter without binding.

(7) Each reel must be plainly marked to indicate the direction in which it should be rolled to prevent loosening of the cable on the reel.

(8) Each reel must be stenciled or lettered with the name of the manufacturer.

(9) The following information must be either stenciled on the reel or on a tag firmly attached to the reel: Optical Cable, Type and Number of Fibers, Armored or Non-armored, Year of Manufacture, Name of Cable Manufacturer, Length of Cable, Reel Number, 7 CFR 1755.902, Minimum Bending Diameter for both Residual and Loaded Condition during installation.

Example: Optical Cable, G.657 class A, 4 fibers, Armored, XYZ Company, 1050 meters, Reel Number 3, 7 CFR 1755.902. Minimum Bending Diameter: Residual (Installed): 20 times Cable O.D., Loaded Condition: 40 times Cable O.D.

APPENDIX TO § 1755.902

FIBER OPTIC CABLES

BULLETIN 1753F–601(PE–90) QUALIFICATIONS TEST DATA

[Initial qualification and three year re-qualification test data required for TELECOMMUNICATIONS PROGRAM product acceptance. Please note that some tests may apply only to a particular cable design.]

Paragraph	Test	Initial qualification	3 Year re-qualification
(e)(4)(i)	Shrinkback	X	
(e)(4)(ii)	Cold Bend	X	
(t)(1)	Sheath Slitting Cord	X	
(t)(2)	Material Compatibility	X	
(t)(3)	Cable Low & High Bend	X	X
(t)(4)	Compound Flow	X	
(t)(5)	Cyclic Flexing	X	X
(t)(6)	Water Penetration	X	X
(t)(7)	Cable Impact	X	X
(t)(8)	Cable Tensile Loading & Fiber Strain	X	X
(t)(9)	Cable Compression	X	
(t)(10)	Cable Twist	X	X
(t)(11)	Cable Lighting Damage Susceptibility	X	
(t)(12)	Cable External Freezing	X	
(t)(13)	Cable Temperature Cycling	X	X
(t)(14)	Cable Sheath Adherence	X	
(t)(15)	Mid-Span	X	X
(t)(16)(i)	Static Tensile Testing of Aerial Self-Supporting Cables	X	X
(t)(16)(ii)	Cable Galloping	X	
(y)(2)(i)	Thermal Reel Wrap test	X	

[74 FR 20561, May 5, 2009]

§ 1755.903 Fiber optic service entrance cables.

(a) *Scope.* This section covers Agency requirements for fiber optic service entrance cables intended for aerial installation either by attachment to a support strand or by an integrated self-supporting arrangement, for underground application by placement in a duct, or for buried installations by trenching, direct plowing, directional or pneumatic boring. Cable meeting this section is recommended for fiber optic service entrances having 12 or fewer fibers with distances less than 100 meters (300 feet).

(1) *General.* (i) Specification requirements are given in SI units which are the controlling units in this part. Approximate English equivalent of units are given for information purposes only.

(ii) The optical waveguides are glass fibers having directly-applied protective coatings, and are called "fibers," herein. These fibers may be assembled in either loose fiber bundles with a protective core tube, encased in several protective buffer tubes, in tight buffer tubes, or ribbon bundles with a protective core tube.

(iii) Fillers, strength members, core wraps, and bedding tapes may complete the cable core.

(iv) The core or buffer tubes containing the fibers and the interstices between the buffer tubes, fillers, and strength members in the core structure are filled with a suitable material or water swellable elements to exclude water.

(v) The cable structure is completed by an extruded overall plastic jacket. A shield or armor or combination thereof may be included under the jacket. This jacket may have strength members embedded in it, in some designs.

(vi) For rodent resistance or for additional protection with direct buried installations, it is recommended the use of armor under the outer jacket.

(vii) For self-supporting cable the outer jacket may be extruded over the support messenger and cable core.

(viii) For detection purposes, the cable may have toning elements embedded or extruded with the outer jacket.

(2) The cable is fully color coded so that each fiber is distinguishable from every other fiber. A basic color scheme of twelve colors allows individual fiber identification. Colored tubes, binders, threads, striping, or markings provide fiber group identification.

(3) Cables manufactured to the requirements of this section must demonstrate compliance with the qualification testing requirements to ensure satisfactory end-use performance characteristics for the intended applications.

(4) Optical cable designs not specifically addressed by this section may be allowed. Justification for acceptance of a modified design must be provided to substantiate product utility and long term stability and endurance. For information on how to obtain Agency's acceptance of such a modified design, refer to the product acceptance procedures available at *http://www.usda.gov/rus/telecom/listing_procedures/index_listing_procedures.htm* as well as RUS Bulletin 345–3.

(5) The cable must be designed for the temperatures ranges of Table 1–1, *Cable Normal Temperature Ranges*, of ICEA S–110–717 (incorporated by reference at §1755.901(c)).

(6) *Tensile rating*. The cable must have ratings that are no less than the tensile ratings indicated in paragraph 1.1.4, *Tensile Rating*, of Part 1 of the ICEA S–110–717 (incorporated by reference at §1755.901(c)).

(7) *Self-supporting cables*. Based on the storm loading districts referenced in Section 25, Loading of Grades B and C, of ANSI/IEEE C2–2007 (incorporated by reference at §1755.901(b)), and the maximum span and location of cable installation provided by the end user, the manufacturer must provide a cable design with sag and tension tables showing the maximum span and sag information for that particular installation. The information included must be for Rule B, *Ice and Wind Loading*, and when applicable, information on Rule 250C, *Extreme Wind Loading*. Additionally, to ensure the proper ground clearance, typically a minimum of 4.7 m (15.5 feet), the end user should factor in the maximum sag under loaded conditions as well as height of attachment for each application.

(8) *Minimum bend diameter*. For cable under loaded and unloaded conditions, the cable must have the minimum bend diameters indicated in paragraph 1.1.5, *Minimum Bend Diameter*, of Part 1 of ICEA S–110–717 (incorporated by reference at §1755.901(c)). For very small cables, manufacturers may specify fixed cable minimum bend diameters that are independent of the outside diameter.

(9) All cables sold to RUS Telecommunications borrowers must be accepted by the Agency's Technical Standards Committee ''A'' for projects involving RUS loan funds. All design changes to Agency acceptable designs must be submitted to the Agency for acceptance. Optical cable designs not specifically addressed by this section may be allowed, if accepted by the Agency. Justification for acceptance of a modified design must be provided to substantiate product utility and long term stability and endurance. For information on how to obtain the Agency's acceptance of cables, refer to the product acceptance procedures available at *http://www.usda.gov/rus/telecom/listing_procedures/index_listing_procedures.htm* as well as RUS Bulletin 345–3.

(10) The Agency intends that the optical fibers contained in the cables meeting the requirement of this section have characteristics that will allow signals, having a range of wavelengths, to be carried simultaneously.

(11) The manufacturer is responsible to establish a quality assurance system meeting industry standards described in paragraph 1.8 of ICEA S–110–717 (incorporated by reference at §1755.901(c)).

(12) The cable made must meet paragraph 1.10 of ICEA S–110–717 (incorporated by reference at §1755.901(c)).

(b) *Optical fibers*. (1) The solid glass optical fibers must consist of a cylindrical core and cladding covered by either an ultraviolet-cured acrylate or other suitable coating. Each fiber must be continuous throughout its length.

(2) Optical fibers must meet the fiber attributes of Table 2, *G.652.B attributes*, of ITU–T Recommendation G.652 (incorporated by reference at §1755.901(f)), unless the end user specifically asks for another type of fiber. However, when the end user stipulates a low water peak fiber, the optical fibers must meet the fiber attributes of Table 4, *G.652.D attributes*, of ITU–T Recommendation G.652; or when the end user stipulates a low bending loss fiber, the optical fibers must meet the fiber

attributes of Table 7–1, *G.657 class A attributes*, of ITU–T Recommendation G.657 (incorporated by reference at §1755.901(f)).

(i) Additionally, optical ribbon fibers must meet paragraph 3.3, *Optical Fiber Ribbons*, of Part 3 of ICEA S–110–717 (incorporated by reference at §1755.901(c)).

(ii) [Reserved]

(3) *Multimode fibers.* Optical fibers must meet the requirements of paragraphs 2.1 and 2.3.1 of ANSI/ICEA S–87–640 (incorporated by reference at §1755.901(c)).

(4) *Matched cable.* Unless otherwise specified by the buyer, all single mode fiber cables delivered to an Agency-financed project must be manufactured to the same MFD specification. However, notwithstanding the requirements indicated in paragraphs (d)(2) and (d)(3) of this section, the maximum MFD tolerance allowed for cables meeting the requirements of this section must be of a magnitude meeting the definition of "matched cable," as defined in paragraph (b) of §1755.900. With the use of cables meeting this definition the user can reasonably expect that the average bi-directional loss of a fusion splice to be ≤0.1 dB.

(5) Buyers will normally specify the MFD for the fibers in the cable. When a buyer does not specify the MFD at 1310 nm, the fibers must be manufactured to an MFD of 9.2 μm with a maximum tolerance range of ±0.5 μm (362 ±20 microinch), unless the buyer agrees to accept cable with fibers specified to a different MFD. When the buyer does specify a MFD and tolerance conflicting with the MFD maximum tolerance allowed by paragraph (d)(4) of this section, the requirements of paragraph (d)(4) must prevail.

(6) Factory splices are not allowed.

(7) All optical fibers in any single length of cable must be of the same type unless otherwise specified by end user.

(8) Optical fiber dimensions and data reporting must be as required by paragraph 7.13.1.1 of ANSI/ICEA S–87–640 (incorporated by reference at §1755.901(c)).

(c) *Buffers/coating.* (1) The optical fibers contained in a buffer tube (loose tube) loosely packaged must have a clearance between the fibers and the inside of the container sufficient to allow for thermal expansions without constraining the fibers. The protective container must be manufactured from a material having a coefficient of friction sufficiently low to allow the fibers free movement. The design may contain more than one tube. Loose buffer tubes must meet the requirements of Paragraph 3.2.1, *Loose Buffer Tube Dimensions*, of Part 3 of ICEA S–110–717 (incorporated by reference at §1755.901(c)).

(2) The loose tube coverings of each color and other fiber package types removed from the finished cable must meet the following shrinkback and cold bend performance requirements. The fibers may be left in the tube.

(i) *Shrinkback.* Testing must be conducted per ASTM D 4565 (incorporated by reference at §1755.901(d)), paragraph 14.1, using a talc bed at a temperature of 95 °C. Shrinkback must not exceed 5 percent of the original 150 millimeter length of the specimen. The total shrinkage of the specimen must be measured.

(ii) *Cold bend.* Testing must be conducted on at least one tube from each color in the cable. Stabilize the specimen to −20 ±1 °C for a minimum of four hours. While holding the specimen and mandrel at the test temperature, wrap the tube in a tight helix ten times around a mandrel with a diameter the greater of five times the tube diameter or 50 mm. The tube must show no evidence of cracking when observed with normal or corrected-to-normal vision.

(3) Optical fiber coating must meet the requirements of paragraph 2.4, *Optical Fiber Coatings and Requirements*, of Part 2 of ICEA S–110–717 (incorporated by reference at §1755.901(c)).

(i) All protective coverings in any single length of cable must be continuous and be of the same material except at splice locations.

(ii) The protective coverings must be free from holes, splits, blisters, and other imperfections and must be as smooth and concentric as is consistent with the best commercial practice.

(iii) Repairs to the fiber coatings are not allowed.

(d) *Fiber and buffer tube identification.* Fibers within a unit and the units within a cable must be identified as indicated in paragraphs 4.2 and 4.3 of Part 4 of ICEA S–110–717 (incorporated by reference at §1755.901(c)), respectively.

(e) *Strength members.* (1) Combined strength of all the strength members must be sufficient to support the stress of installation and to protect the cable in service. Strength members must meet paragraph 4.4, *Strength Members*, of ICEA S–110–717 (incorporated by reference at §1755.901(c)). Self supporting aerial cables using the strength members as an integral part of the cable strength must comply with paragraph C.4, *Static Tensile Testing of Aerial Self-Supporting Cables*, of ANNEX C of ICEA S–110–717.

(2) Strength members may be incorporated into the core as a central support member or filler, as fillers between the fiber packages, as an annular serving over the core, as an annular serving over the intermediate jacket, embedded in the outer jacket or as a combination of any of these methods.

(3) The central support member or filler must contain no more than one splice per kilometer of cable. Individual fillers placed between the fiber packages and placed as annular servings over the core must contain no more than one splice per kilometer of cable. Cable sections having central member or filler splices must meet the same physical requirements as unspliced cable sections.

(4) Notwithstanding what has been indicated in other parts of this document, in each length of completed cable having a metallic central member, the dielectric strength between the optional armor and the metallic center member must withstand at least 15 kilovolts direct current for 3 seconds.

(f) *Forming the cable core.* (1) Protected fibers must be assembled with the optional central support member and strength members in such a way as to form a cylindrical group or other acceptable core constructions and must meet Section 4.5, Assembly of Cables, of Part 4 of ICEA S–110–717 (incorporated by reference at 1755.901(c)). Other acceptable cable cores include round, figure 8, flat or oval designs.

(2) The standard cylindrical group or core designs must consist of 12 fibers or less.

(3) When threads or tapes are used as core binders, they must be colored either white or natural and must be a non-hygroscopic and non-wicking dielectric material. Water swell-able threads and tapes are permitted.

(g) *Filling/flooding compounds and water blocking elements.* (1) To prevent the ingress and migration of water through the cable and core, filling/flooding compounds or water blocking elements must be used.

(i) Filling compounds must be applied into the interior of the loose fiber tubes and into the interstices of the core. When a core wrap is used, the filling compound must also be applied to the core wrap, over the core wrap and between the core wrap and inner jacket when required.

(ii) Flooding compounds must be sufficiently applied between the optional inner jacket and armor and between the armor and outer jacket so that voids and air spaces in these areas are minimized. The use of floodant between the armor and outer jacket is not required when uniform bonding, per paragraph l(9) of this section, is achieved between the plastic-clad armor and the outer jacket. Floodant must exhibit adhesive properties sufficient to prevent jacket slip when tested per the requirements of paragraphs 7.26 through 7.26.2 of Part 7, *Testing, Test Methods, and Requirements*, of ANSI/ICEA S–87–640 (incorporated by reference at 1755.901(c)).

(iii) Water blocking elements must achieve equal or better performance in preventing the ingress and migration of water as compared to filling and flooding compounds. In lieu of a flooding compound, water blocking elements may be applied between the optional inner jacket and armor and between the armor and outer jacket to prevent water migration. The use of the water blocking elements between the armor and outer jacket is not required when uniform bonding, per paragraph (l)(10) of this section, is achieved between the plastic-clad armor and the outer jacket.

(2) The materials must be homogeneous and uniformly mixed; free

from dirt, metallic particles and other foreign matter; easily removed; nontoxic and present no dermal hazards.

(3) The individual cable manufacturer must satisfy the Agency that the filling compound or water blocking elements selected for use is suitable for its intended application.

(i) Filling/Flooding compound materials must be compatible with the cable components when tested per paragraph 7.16, *Material Compatibility and Cable Aging Test*, of Part 7 of ICEA S–110–717 (incorporated by reference at §1755.901(c)).

(ii) Water blocking elements must be compatible with the cable components when tested per paragraph 7.16, *Material Compatibility and Cable Aging Test*, of Part 7 of ICEA S–110–717 (incorporated by reference at §1755.901(c)).

(h) *Core wrap (optional).* (1) At the option of the manufacturer, one or more layers of non-hygroscopic and non-wicking dielectric material may be applied with an overlap over the core.

(2) The core wrap(s) can be used to provide a heat barrier to prevent deformation or adhesion between the fiber tubes or can be used to contain the core.

(3) When core wraps are used, sufficient filling compound must be applied to the core wraps so that voids or air spaces existing between the core wraps and between the core and the inner side of the core wrap are minimized.

(i) *Inner jacket (optional).* (1) Inner jackets may be applied directly over the core or over the strength members. Inner jackets are optional.

(2) The inner jacket material and test requirements must be the same as for the outer jacket material per paragraph (n) of this section, except that either black or natural polyethylene may be used. In the case of natural polyethylene, the requirements for absorption coefficient and the inclusion of furnace black are waived.

(j) *Armor (optional).* (1) A steel armor, plastic coated on both sides, is recommended for direct buried service entrance cable in gopher areas. Armor is also optional for duct and aerial cable as required by the end user. The plastic coated steel armor must be applied longitudinally directly over the core wrap or the intermediate jacket and must have an overlapping edge.

(2) The uncoated steel tape must be electrolytic chrome coated steel (ECCS) with a thickness of 0.155 ±0.015 millimeters.

(3) The reduction in thickness of the armoring material due to the corrugating or application process must be kept to a minimum and must not exceed 10 percent at any spot.

(4) The armor of each length of cable must be electrically continuous with no more than one joint or splice allowed per kilometer of cable. This requirement does not apply to a joint or splice made in the raw material by the raw material manufacturer.

(5) The breaking strength of any section of an armor tape, containing a factory splice joint, must not be less than 80 percent of the breaking strength of an adjacent section of the armor of equal length without a joint.

(6) For cables containing no floodant over the armor, the overlap portions of the armor tape must be bonded in cables having a flat, non-corrugated armor to meet the requirements of paragraphs (r)(1) and (r)(2) of this section. If the tape is corrugated, the overlap portions of the armor must be sufficiently bonded and the corrugations must be sufficiently in register to meet the requirements of paragraphs (r)(1) and (r)(2) of this section.

(7) The armor tape must be so applied as to enable the cable to meet the testing requirements of paragraphs (r)(1) and (r)(2) of this section.

(8) The protective coating on the steel armor must meet the Bonding-to-Metal, Heat Sealability, Lap-Shear and Moisture Resistance requirements of Type I, Class 2 coated metals per ASTM B 736 (incorporated by reference at §1755.901(d)).

(9) When the jacket is bonded to the plastic coated armor, the bond between the plastic coated armor and the outer jacket must not be less than 525 Newtons per meter over at least 90 percent of the cable circumference when tested per ASTM D 4565 (incorporated by reference at §1755.901(d)). For cables with strength members embedded in the jacket, and residing directly over the armor, the area of the armor directly

under the strength member is excluded from the 90 percent calculation.

(k) *Optional support messenger (aerial cable).* (1) Integrated messenger(s) for self-supporting cable must provide adequate strength to operate under the appropriate weather loading conditions over the maximum specified span.

(2) Based on the storm loading districts referenced in Section 25, Loading of Grades B and C, of ANSI/IEEE C2–2007 (incorporated by reference at §1755.901(b)), and the maximum span and location of cable installation provided by the end user, the manufacturer must provide a cable design with sag and tension tables showing the maximum span and sag information for that particular installation. The information included must be for Rule B, *Ice and Wind Loading*, and when applicable, information on Rule 250C, *Extreme Wind Loading*. Additionally, to ensure the proper ground clearance, typically a minimum of 4.7 m (15.5 feet) the end user should factor in the maximum sag under loaded conditions as well as height of attachment for each application.

(l) *Outer jacket.* (1) The outer jacket must provide the cable with a tough, flexible, protective covering which can withstand exposure to sunlight, to atmosphere temperatures and to stresses reasonably expected in normal installation and service.

(2) The jacket must be free from holes, splits, blisters, or other imperfections, and must be as smooth and concentric as is consistent with the best commercial practice.

(3) Jacket materials must meet the stipulations of paragraph 5.4 of ANSI/ICEA S–87–640 (incorporated by reference at §1755.901(c)), except that the concentration of furnace black does not necessarily need to be initially contained in the raw material and may be added later during the jacket making process. Jacket thickness must have a 0.50 mm minimum thickness over the core or over any radial strength member used as the primary strength element(s), 0.20 mm when not used as the primary strength member, and 0.30 mm over any optional toning elements.

(4) Jacket Repairs must meet the stipulations of paragraph 5.5, *Jacket Repairs*, of ICEA S–110–717 (incorporated by reference at §1755.901(c)).

(5) *Jacket Testing:* The jacket must be tested to determine compliance with requirements of this section. The specific tests for the jacket are described in paragraphs 7.6 through 7.11.2 of Part 7, *Testing, Test Methods, and Requirements*, of ANSI/ICEA S–87–640 (incorporated by reference at §1755.901(c)).

(m) *Sheath slitting cord (optional).* (1) A sheath slitting cord is optional.

(2) When a sheath slitting cord is used it must be non-hygroscopic and non-wicking, or be rendered such by the filling or flooding compound, continuous throughout a length of cable and of sufficient strength to open the sheath over at least a one meter length without breaking the cord at a temperature of 23 ±5 °C.

(n) *Identification and length markers.* (1) Each length of cable must be permanently labeled OPTICAL CABLE, OC, OPTICAL FIBER CABLE, or OF on the outer jacket and identified as to manufacturer and year of manufacture.

(2) Each length of cable intended for direct burial installation must be marked with a telephone handset in compliance with the requirements of the Rule 350G of ANSI/IEEE C2–2007 (incorporated by reference at §1755.901(b)).

(3) Mark the number of fibers on the jacket.

(4) The identification and date marking must conform to paragraph 6.1, Identification and Date Marking, of ICEA S–110–717 (incorporated by reference at §1755.901(c)).

(5) The length marking must conform to paragraph 6.3, Length Marking, of ICEA S–110–717 (incorporated by reference at §1755.901(c)).

(o) *Optical performance of a finished cable.* (1) Unless otherwise specified by the end user, the optical performance of a finished cable must comply with the attributes of Table 2, *G.652.B attributes*, found in ITU Recommendation G.652 (incorporated by reference at §1755.901(f)). However, when the end user stipulates a low water peak fiber the finished cable must meet the attributes of Table 4, *G.652.D attributes*, found in ITU–T Recommendation G.652; or when the end user stipulates a low bending loss fiber, the finished cable must meet the attributes of Table 7–1, *class A attributes,* of ITU–T Recommendation G.657 (incorporated by reference at § 1755.901(f)).

(i) The attenuation methods must be per Table 8.4, *Optical attenuation measurement methods,* of ANSI/ICEA S–87–640 (incorporated by reference at § 1755.901(c)).

(ii) The cable must have a maximum attenuation of 0.1 dB at a point of discontinuity (a localized deviation of the optical fiber loss). Per paragraphs 8.4 and 8.4.1 of ANSI/ICEA S–87–640 (incorporated by reference at § 1755.901(c)), measurements must be conducted at 1310 and 1550 nm, and at 1625 nm when specified by the end user.

(iii) The cable cutoff wavelength (γ_{cc}) must be reported per paragraph 8.5.1 of ANSI/ICEA S–87–640 (incorporated by reference at § 1755.901(c)).

(2) *Multimode optical fiber cable.* Unless otherwise specified by the end user, the optical performance of the fibers in a finished cable must comply with Table 8.1, *Attenuation coefficient performance requirement (dB/km),* Table 8.2, *Multimode bandwidth coefficient performance requirements (MHz-km),* and Table 8.3, *Points discontinuity acceptance criteria (d),* of ANSI/ICEA S–87–640 (incorporated by reference at § 1755.901(c)).

(3) Because the accuracy of attenuation measurements for single mode fibers becomes questionable when measured on short cable lengths, attenuation measurements are to be made utilizing characterization cable lengths. Master Cable reels must be tested and the attenuation values measured will be used for shorter ship lengths of cable.

(4) Because the accuracy of attenuation measurements for multimode fibers becomes questionable when measured on short cable lengths, attenuation measurements are to be made utilizing characterization cable lengths. If the ship length of cable is less than one kilometer, the attenuation values measured on longer lengths of cable (characterization length of cable) before cutting to the ship lengths of cable may be applied to the ship lengths.

(5) Attenuation must be measured per Table 8.4, *Optical Attenuation Measurement Methods,* ANSI/ICEA S–87–640 (incorporated by reference at § 1755.901(c)).

(6) The bandwidth of multimode fibers in a finished cable must be no less than the values specified in Table 8.2 per paragraph 8.3.1 of ANSI/ICEA S–87–640 (incorporated by reference at § 1755.901(c)).

(p) *Mechanical requirements.* (1) *Cable Testing:* Cable designs must meet the requirements of Part 7, Testing and Test Methods, of ICEA S–110–717 (incorporated by reference at § 1755.901(c)), except for paragraph 7.15 applicable to tight tube fibers.

(2) *Bend test.* All cables manufactured must meet the "Cable Low and High Temperature Bend Test" described in Section 7.21 (paragraphs 7.21, 7.21.1, and 7.21.2) of ANSI/ICEA S–87–640 (incorporated by reference at § 1755.901(c)).

(q) *Pre-connectorized cable (optional).* (1) At the option of the manufacturer and upon request by the end user, the cable may be factory terminated with connectors.

(2) All connectors must be accepted by the Agency prior to their use. For information on how to obtain the Agency's acceptance, refer to the product acceptance procedures available at *http://www.usda.gov/rus/telecom/listing_procedures/index_listing_procedures.htm* as well as RUS Bulletin 345–3.

(r) *Acceptance testing and extent of testing.* (1) The tests described in this section are intended for acceptance of cable designs and major modifications of accepted designs. What constitutes a major modification is at the discretion of the Agency. These tests are intended to show the inherent capability of the manufacturer to produce cable products that have satisfactory performance characteristics, long life, and long-term optical stability, but are not intended as field tests. For information on how to obtain the Agency's acceptance, refer to the product acceptance procedures available at *http://www.usda.gov/rus/telecom/listing_procedures/index_listing_procedures.htm* as well as RUS Bulletin 345–3.

(2) For initial acceptance, the manufacturer must submit:

(i) An original signature certification that the product fully complies with each paragraph of this section;

(ii) Qualification Test Data for demonstrating that the cable meets the requirements of this section;

(iii) A set of instructions for handling the cable;

(iv) OSHA Material Safety Data Sheets for all components;

(v) Agree to periodic plant inspections;

(vi) Agency's "Buy American" Requirements. For each cable for which the Agency acceptance is requested, the manufacturer must include a certification stating whether the cable complies with the following two domestic origin manufacturing provisions:

(A) Final assembly or manufacture of the product, as the product would be used by an Agency's borrower, is completed in the United States or eligible countries. For a list of eligible countries, see *http://www.usda.gov/rus/telecom/publications/eligible.htm;* and

(B) The cost of United States and eligible countries' components (in any combination) within the product is more than 50 percent of the total cost of all components utilized in the product. The cost of non-domestic components (components not manufactured within the United States or eligible countries) which are included in the finished product must include all duties, taxes, and delivery charges to the point of assembly or manufacture;

(vii) Written user testimonials concerning performance of the product; and

(viii) Other nonproprietary data deemed necessary by the Chief, Technical Support Branch (Telecommunications).

(3) For continued Agency product acceptance, the manufacturer must submit an original signature certification that the product fully complies with each paragraph of this section and a certification stating whether the cable meets the two domestic provisions of paragraph (t)(2)(vi) above for acceptance by January every three years. The certification must be based on test data showing compliance with the requirements of this section. The test data must have been gathered within 90 days of the submission and must be kept on files per paragraph (u)(1).

(4) Initial and re-qualification acceptance requests should be addressed to: Chairman, Technical Standards Committee "A" (Telecommunications), STOP 1550, Advanced Services Division, Rural Development Utilities Program, Washington, DC 20250–1550.

(s) *Records of optical and physical tests.* (1) Each manufacturer must maintain suitable summary records for a period of at least 3 years of all optical and physical tests required on completed cable manufactured under the requirement of this section. The test data for a particular reel must be in a form that it may be readily available to the Agency upon request. The optical data must be furnished to the end user on a suitable and easily readable form.

(2) Measurements and computed values must be rounded off to the number of places or figures specified for the requirement per paragraph 1.3 of ANSI/ICEA S–87–640 (incorporated by reference at §1755.901(c)).

(t) *Manufacturing irregularities.* (1) Repairs to the armor, when present, are not permitted in cable supplied to the end user under the requirement of this section. The armor for each length of cable must be tested for continuity using the procedures of ASTM D 4566 (incorporated by reference at §1755.901(d)).

(2) Minor defects in the inner and outer jacket (defects having a dimension of 3 millimeter or less in any direction) may be repaired by means of heat fusing per good commercial practices utilizing sheath grade compounds.

(3) Buffer tube repair is permitted only in conjunction with fiber splicing.

(u) *Packaging and preparation for shipment.* (1) All cables must comply with paragraph 6.5, *Packaging and Marking,* of ICEA S–110–717 (incorporated by reference at §1755.901(c)).

(2) For cables shipped on reels a circumferential thermal wrap or other means of protection complying with section (w)(3) of this section must be secured between the outer edges of the reel flange to protect the cable against damage during storage and shipment. This requirement applies to reels weighing more than 75 lbs. The thermal

wrap is optional for reels weighing 75 lbs or less.

(3) The thermal wrap must meet the requirements included in the *Thermal Reel Wrap Test,* described below in paragraphs (w)(3)(i) and (w)(3)(ii) of this section. This test procedure is for qualification of initial and subsequent changes in thermal reel wraps.

(i) *Sample selection.* All testing must be performed on two 450 millimeter (18 inches) lengths of cable removed sequentially from the same fiber jacketed cable. This cable must not have been exposed to temperatures in excess of 38 °C (100 °F) since its initial cool down after sheathing.

(ii) *Test procedure.* (A) Place the two samples on an insulating material such as wood.

(B) Tape thermocouples to the jackets of each sample to measure the jacket temperature.

(C) Cover one sample with the thermal reel wrap.

(D) Expose the samples to a radiant heat source capable of heating the uncovered sample to a minimum of 71 °C (160 °F). A GE 600 watt photoflood lamp or an equivalent lamp having the light spectrum approximately that of the sun must be used.

(E) The height of the lamp above the jacket must be 380 millimeters (15 inches) or an equivalent height that produces the 71 °C (160 °F) jacket temperature on the unwrapped sample must be used.

(F) After the samples have stabilized at the temperature, the jacket temperatures of the samples must be recorded after one hour of exposure to the heat source.

(G) Compute the temperature difference between jackets.

(H) The temperature difference between the jacket with the thermal reel wrap and the jacket without the reel wrap must be greater than or equal to 17 °C (63 °F).

(4) Cable must be sealed at the ends to prevent entrance of moisture.

(5) The end-of-pull (outer end) of the cable must be securely fastened to prevent the cable from coming loose during transit. The start-of-pull (inner end) of the cable must project through a slot in the flange of the reel, around an inner riser, or into a recess on the flange near the drum and fastened in such a way to prevent the cable from becoming loose during installation.

(6) Spikes, staples or other fastening devices must be used in a manner which will not result in penetration of the cable.

(7) The minimum size arbor hole must be 44.5 mm (1.75 inch) and must admit a spindle without binding.

(8) Each reel must be plainly marked to indicate the direction in which it should be rolled to prevent loosening of the cable on the reel.

(9) Each reel must be stenciled or lettered with the name of the manufacturer.

(10) The following information must be either stenciled on the reel or on a tag firmly attached to the reel: Optical Cable, Type and Number of Fibers, Armored or Nonarmored, Year of Manufacture, Name of Cable Manufacturer, Length of Cable, Reel Number, REA 7 CFR 1755.903.

Example: Optical Cable, G.657 class A, 4 fibers, Armored. XYZ Company, 1050 meters, Reel Number 3, REA 7 CFR 1755.903.

(11) When pre-connectorized cable is shipped, the splicing modules must be protected to prevent damage during shipment and handling.

[74 FR 20561, May 5, 2009]

§ 1755.910 RUS specification for outside plant housings and serving area interface systems.

(a) *Scope.* (1) The purpose of this specification is to inform manufacturers and users of outside plant housings and serving area interface (SAI) systems of the engineering and technical requirements that are considered necessary for satisfactory performance in outside plant environments. Included are the mechanical, electrical, and environmental requirements, desired design features, and test methods for evaluation of the product.

(2) The housing and terminal requirements reflect the best engineering judgment available at the present time and may be subject to change due to advances in technology, economic conditions, or other factors.

(3) The test procedures described in this section are required by RUS to demonstrate the functional reliability

of the product. However, other standard or unique test procedures may serve the same function. In such cases, RUS shall evaluate the test procedures and results on an individual basis.

(4) The test procedures specified herein satisfy the requirements of housings as well as the requirements of terminals that may be installed within housings. Some of the requirements are interrelated to several tests designed to determine the performance aspects of terminals and are directly affected by testing required for housings. Therefore, the manufacturer should carefully review all the test requirements in order to develop a testing schedule that is comprehensive, efficient in terms of the number of test specimens required and can be accomplished in an orderly and logical sequence.

(5) The specified tests may require special facilities to comply with Federal, State, or local regulatory requirements. Some test procedures are potentially hazardous to personnel because of the high voltages and mechanical forces involved. Safety precautions are necessary to prevent injury.

(6) Underwriters Laboratories, Inc. (UL) 94, Tests for Flammability of Plastic Materials for Parts in Devices and Appliances, fourth edition, dated June 18, 1991, referenced in this section is incorporated by reference by RUS. This incorporation by reference was approved by the Director of the Federal Register in accordance with 5 U.S.C. 552(a) and 1 CFR part 51. A copy of the UL standard is available for inspection during normal business hours at RUS, room 2845–S, U.S. Department of Agriculture, Washington, DC 20250–1500, or at the National Archives and Records Administration (NARA). For information on the availability of this material at NARA, call 202–741–6030, or go to: *http://www.archives.gov/federal_register/code_of_federal_regulations/ibr_locations.html.* Copies are available from UL Inc., 333 Pfingsten Road, Northbrook, Illinois 60062–2096, telephone number (708) 272–8800.

(7) The American Society for Testing and Materials Specifications (ASTM) A 109–91, Standard Specification for Steel, Strip, Carbon, Cold-Rolled; ASTM A 153–82 (Reapproved 1987), Standard Specification for Zinc Coating (Hot-Dip) on Iron and Steel Hardware; ASTM A 366/A 366M–91, Standard Specification for Steel, Sheet, Carbon, Cold-Rolled, Commercial Quality; ASTM A 525–91b, Standard Specification for General Requirements for Steel Sheet, Zinc-Coated (Galvanized) by the Hot-Dip Process; ASTM A 526/A 526M–90, Standard Specification for Steel Sheet, Zinc-Coated (Galvanized) by the Hot-Dip Process, Commercial Quality; ASTM A 569/A 569M–91a, Standard Specification for Steel, Carbon (0.15 Maximum, Percent), Hot-Rolled Sheet and Strip Commercial Quality; ASTM A 621/A 621M–92, Standard Specification for Steel, Sheet and Strip, Carbon, Hot-Rolled, Drawing Quality; ASTM B 117–90, Standard Test Method of Salt Spray (Fog) Testing; ASTM B 539–90, Standard Test Methods for Measuring Contact Resistance of Electrical Connections (Static Contacts); ASTM B 633–85, Standard Specification for Electrodeposited Coatings of Zinc on Iron and Steel; ASTM D 523–89, Standard Test Method for Specular Gloss; ASTM D 610–85 (Reapproved 1989), Standard Test Method for Evaluating Degree of Rusting on Painted Steel Surfaces; ASTM D 822–89, Standard Practice for Conducting Tests on Paint and Related Coatings and Materials using Filtered Open-Flame Carbon-Arc Light and Water Exposure Apparatus; ASTM D 1535–89, Standard Test Method for Specifying Color by the Munsell System; ASTM D 1654–92, Standard Test Method for Evaluation of Painted or Coated Specimens Subjected to Corrosive Environments; ASTM D 1693–70 (Reapproved 1988), Standard Test Method for Environmental Stress-Cracking of Ethylene Plastics; ASTM D 2197–86 (Reapproved 1991), Standard Test Method for Adhesion of Organic Coatings by Scrape Adhesion; ASTM D 2247–92, Standard Practice for Testing Water Resistance of Coatings in 100% Relative Humidity; ASTM D 2565–92, Standard Practice for Operating Xenon Arc-Type Light-Exposure Apparatus With and Without Water for Exposure of Plastics; ASTM D 2794–92, Standard Test Method for Resistance of Organic Coatings to the Effects of Rapid Deformation (Impact); ASTM D 3928–89, Standard Test Method

for Evaluation of Gloss or Sheen Uniformity; ASTM D 4568–86, Standard Test Methods for Evaluating Compatibility Between Cable Filling and Flooding Compounds and Polyolefin Cable Materials; ASTM G 21–90, Standard Practice for Determining Resistance of Synthetic Polymeric Materials to Fungi; and ASTM G 23–90, Standard Practice for Operating Light-Exposure Apparatus (Carbon-Arc Type) With and Without Water for Exposure of Nonmetallic Materials, referenced in this section are incorporated by reference by RUS. These incorporations by references were approved by the Director of the Federal Register in accordance with 5 U.S.C. 552(a) and 7 CFR part 51. Copies of the ASTM standards are available for inspection during normal business hours at RUS, room 2845–S, U.S. Department of Agriculture, Washington, DC 20250–1500, or at the National Archives and Records Administration (NARA). For information on the availability of this material at NARA, call 202–741–6030, or go to: *http://www.archives.gov/federal_register/code_of_federal_regulations/ibr_locations.html*. Copies are available from ASTM, 1916 Race Street, Philadelphia, Pennsylvania 19103–1187, telephone number (215) 299–5585.

(b) *General information*. (1) Outside plant housings are fabricated of either metallic or nonmetallic materials in different sizes and configurations to suit a variety of applications. The purpose of a housing is to protect its contents from environmental elements, rodents, insects, or vandalism and unauthorized access. Housings are designed with internal brackets for accommodating splicing, bonding and grounding connections, cable terminals, cross-connect facilities, load coils, and optical and electronic equipment.

(2) Pedestals are housings primarily intended to house, organize, and protect cable terminations incorporating terminal blocks, splice connectors and modules, ground lugs and load coils. Activities typically performed in a pedestal are cable splicing, shield bonding and grounding, inductive loading, and connection of subscriber drops.

(3) Serving area interface (SAI) cabinets are housings intended to perform some of the same functions as pedestals but are primarily intended to serve as the connecting terminal between feeder cable and distribution cables.

(4) Outside plant housings shall be manufactured in accordance with National Electrical Code (NEC) requirements, Underwriters' Laboratories (UL) requirements, Department of Labor, Occupational Safety and Health Administration Standards (OSHA), and all other applicable Federal, State, and local requirements including, but not limited to, statutes, rules, regulations, orders, or ordinances otherwise imposed by law.

(c) *General documentation requirements*—(1) *Installation and maintenance instructions*. (i) Each product shall have available a set of instructions designed to provide sufficient information for the successful installation of the housing, cables, auxiliary equipment, and the associated splice preparation. The instructions shall be of sufficient size to be easily read and shall be printed using waterproof ink. Pedestal instruction sheets shall include a list of miscellaneous replacement parts that may be purchased locally. SAI systems shall be supplied with complete instructions for installation and use.

(ii) When requested by RUS, or an RUS borrower, the manufacturer shall prepare a training package for the purpose of training technicians in the use and installation of the product and its auxiliary equipment.

(iii) The manufacturer shall provide ordering information for repair parts. Repair parts shall be obtainable through a local distributor or shall be easily obtainable. Information describing equivalent parts and their sources should be provided for those parts that may also be obtained from other sources.

(2) *Quality assurance*. The manufacturer shall demonstrate the existence of an ongoing quality assurance program that includes controls, procedures, and standards used for vendor certification, source inspection, incoming inspection, manufacture, in process testing, calibration and maintenance of tools and test equipment, final product inspection and testing, periodic qualification testing and control of nonconforming materials and products. The

manufacturer shall maintain quality assurance records for five years.

(3) *RUS acceptance applications.* (i) The tests described in this specification are required for acceptance of product designs and major modifications of accepted designs. All modifications shall be considered major unless otherwise declared by RUS. The tests are intended to show the inherent capability of the manufacturer to produce products which have an expected service life of 30 years.

(ii) For initial acceptance the manufacturer shall:

(A) Submit an original signature certification that the product complies with each section of the specification;

(B) Provide qualification test data;

(C) Provide OSHA Material Safety Data Sheets for the product;

(D) Provide a detailed explanation concerning the intended use and capacity of the product;

(E) Provide a complete set of instructions, recommendations for equipment organization and splicing;

(F) Agree to periodic plant inspections;

(G) Provide a certification that the product does or does not comply with the domestic origin manufacturing provisions of the "Buy American" requirements of the Rural Electrification Act of 1938 (52 Stat. 818);

(H) Provide user testimonials concerning field performance of the product;

(I) Provide product samples if requested by RUS; and

(J) Provide any other data required by the Chief, Outside Plant Branch (Telephone).

(iii) Each requirement of this section must be addressed in submissions for acceptance. The designation N/A may be entered when the requirements do not apply.

(iv) Acceptance requests should be addressed to: Chairman, Technical Standards, Committee "A" (Telephone), Telecommunications Standards Division, Rural Utilities Service, Washington, DC 20250–1500.

(d) *Functional design criteria for housings*—(1) *General requirements.* (i) The functional requirements for housings concern materials, finishes, environmental factors, and design features that are applicable to most above ground housings used in the outside plant.

(ii) Housings shall be of sufficient size to permit easily managed installation, operational, testing, and maintenance operations. The general shape of outside plant housings is usually comparable to that of a rectangular column or cylinder, with the shape of any particular housing being left to the manufacturer's discretion. Each design is subject to acceptance by RUS.

(2) *Housing types and capacities.* (i) Housings used in outside plant are either the smaller housings generally known as pedestals or larger housings known as equipment or splice cabinets. Both categories may have designs intended for stake mounting, pole mounting, or pad mounting.

(ii) The classifications of pedestals are the general purpose channel Type (H) and the dome Type (M). The Type H pedestal has either front only access or back and front access while the Type M pedestal has top only access. Pedestals are further designated as follows:

Stake mounted	Type	Pole mounted	Pole mounted (extra high)
BD3	H	BD3A	
BD4	H	BD4A	
BD5	H	BD5A	
BD7	H	BD7A	
BD14	M	BD14A	BD14AG
BD15	M	BD15A	BD15AG
BD16	M	BD16A	BD16AG

(iii) The minimum volume associated with the pedestal designations shall be as shown in the following table:

Pedestal [1] housing designation	Minimum volume	
	Cubic centimeters cm^3	(Cubic Inches) ($in.^3$)
BD3, BD3A [2]	9,000	(550)
BD4, BD4A [2]	15,000	(900)
BD5, BD5A [2]	35,000	(2,100)
BD7([2])	72,000	(4,400)
BD14, BD14A, BD14AG [3]	9,000	(550)
BD15, BD15A, BD15AG [3]	27,000	(1,600)
BD16, BD16A, BD16AG [3]	38,000	(2,300)

NOTE 1: Housings designed for unique purposes will be evaluated on a case-by-case basis.

NOTE 2: For Type H pedestals, the minimum volume is that space as measured 5 centimeters (cm) (2 inches (in.)) below the top of the housing to a point 40 cm (16 in.) above the bottom of the lower cover plate.

NOTE 3: The minimum volume of the Type M pedestals shall be the space within the dome measured from the lower edge of the dome to a point 5 cm (2 in.) from the top.

(iv) Equipment cabinets intended for use as SAI housings shall be assigned size designations according to their maximum pair termination capacities. The capacity will vary depending on the type of terminating equipment used. SAI cabinets shall be suffix designated with an "A" for pole mounting, "X" for pad mounting, and "S" for stake mounting.

(v) Large pair count splice cabinets are classified according to their splice capacity. Approximately 48 cm^3 (3.0 $in.^3$) of splice area per pair straight spliced shall be permitted.

(vi) The minimum volume associated with large pair count splice cabinets shall be as shown in the following table:

Splice cabinet[1] designation	Minimum volume		Maximum splice capacity (pairs)
	($cm.^3$)	($in.^3$)	
BD6000	295,000	(18,000)	6,000
BD8000	393,000	(24,000)	8,000
BD10000	491,000	(30,000)	10,000

NOTE 1: Additional sizes of splice cabinets shall be considered by RUS on a case-by-case basis.

(3) *Design and fabrication requirements for housings.* (i) Type H pedestal housings may consist of an enclosed channel incorporating an integrally mounted stake that serves as a backplate, or they may be designed for universal mounting on stakes or poles. The body of the housing shall have two major components; an upper cover and a base cover. The upper cover shall have a top, front and back plate with the front cover removable to permit entry and provide increased work space. The base cover shall consist of a front plate and back plate. The base cover back plate may be an extension of the upper back plate cover.

(ii) Type M pedestal housings shall consist of a one piece upper sleeve designed to fit over the base cover trapping air to prohibit water from entering the splice area when installed in locations prone to temporary flooding. Pedestals designed to be mounted extra high on poles for locations susceptible to deep snow shall have a bottom close-off option available to prohibit the ingress of birds, rodents and insects.

(iii) The external housing components on all outside plant housings shall provide reasonable protection against accidental removal or vandalism. Housings shall be equipped with a cover plate retaining bolt and cup washer that may be opened only with an industry accepted socket type can wrench. Housings may be equipped with provisions to allow the purchaser to install a padlock.

(iv) Installed housings shall resist the disassembling force of frost heaving applied to the bottom of ground line cover plates. The base cover must remain stationary to stabilize the contents of the housing cavity.

(v) In an effort to provide protection against dust penetration, blowing snow, rain, and ultraviolet light degradation of internal components, all mechanical gaps shall be restricted. The use of seals, overlaps, gaskets, and/or dovetailing is required to assure satisfactory protection of housed equipment.

(vi) Knockouts, cutouts, or notches designed to accommodate aerial service drops shall not be permitted. A design option for housings intended to accommodate service drops shall include a separate channel or equivalent in the base cover to allow future additions of service drops without the removal of gravel or the moisture barrier in the base of the housing. Service wire channels must be designed to prevent the entry of birds, reptiles, rodents and insects.

(vii) Minimal venting of SAI housings may be necessary to relieve internal pressure and condensation.

(viii) There shall be no aluminum housing components that will become buried in the soil when the housing is properly installed.

(ix) Housing components may be assembled using rivets, welds, glue, bolts and nuts, or other techniques suitable for the materials involved.

(x) Housings and their components that require field assembly must be capable of being assembled with tools normally available to outside plant technicians.

(xi) Hinged doors on SAI housings and large pair count splice housings shall be equipped with a device that restrains the doors in the open position.

(xii) Outside plant housings shall be free of sharp edges, burrs, etc., that could present a safety hazard to personnel involved in installation and use of the product or to the general public. Surfaces inside housings must not allow pinching of conductors during installation of cover plates or the opening and closing of doors.

(xiii) A ground line mark shall be provided, approximately 15 cm (6 in.) below the top edge of the housing base cover plate on housings intended for ground level mounting. Base cover plates shall have a minimum height of 31 cm (12 in.).

(xiv) Any housing, which weighs in excess of 91 kilograms (kg) (200 pounds (lb)), including its contents, shall be equipped with lifting brackets for attaching hoisting cables or chains.

(xv) Housing stakes shall be a minimum of 107 cm (42 in.) in length. If fabricated from steel, they shall have a minimum thickness of No. 13 gauge as measured according to American Society for Testing and Materials (ASTM) A 525–91b. Stakes shall be formed into a "U" channel with a minimum depth of 2 cm (0.75 in.). The stake shall be a single part of suitable design strength for driving 91 cm (36 in.) into the soil with hand tools without damage such as bending or warping. The stake shall have adequate mounting holes having a minimum separation of 15 cm (6 in.) for mounting the housing baseplate. The stake material must resist corrosion and deterioration when exposed to soil and atmospheric conditions.

(xvi) The housing design must permit a logical progression of installation steps that would normally be encountered in typical field installations.

(xvii) Provisions for attaching housings to stakes, poles, walls, other housings, or pads shall be provided for each design intended for those purposes. Locations of holes for mounting attachments may be provided by knockouts on above ground components. Mounting hole locations for below ground components may be predrilled.

(xviii) Pole mounting hardware shall provide at least 1.3 cm (0.5 in.) clearance from the pole to the housing. Pole mounting brackets shall accommodate the wide range of pole sizes used in the telephone industry.

(xix) Pad-mounted housings shall have hardware available for anchoring the housing base to the pad. A template may be provided to assist in the location of mounting attachment details for pad preparation.

(xx) Housings equipped with stub cables shall have strain relief devices to permit shipping and handling of the housing without damage to the housing or stub cables. Only RUS accepted cable shall be used for stub cables. The cable manufacturer's recommendations concerning minimum bend radius shall be observed. The minimum bend radius for most copper cables is 10 times the cable diameter.

(xxi) Cable supports shall be provided near the top of the ground line cover and other appropriate locations within the housing to provide cable stability consistent with the intended use and capacity of the housing. Cable supports shall be capable of holding a minimum load of 23 kg (50 lb).

(xxii) An adequate supply of nonmetallic retainer clips or tie wraps capable of supporting a minimum load of 23 kg (50 lb) shall be provided with the housing. Adequate spaces for installation of the clips or tie wraps must be provided on the housing backplate and cable supports.

(xxiii) Housing chambers designed for splicing operations shall be equipped with insulated supporting straps or rods suitable for supporting splice bundles. The insulation on the straps or rods shall extend for the entire length of the device and shall have a dielectric strength of 15 kilovolts (kv) direct current (dc) minimum. Housings having an "H" frame design where both front and rear covers may be removed may incorporate insulated tie bars to be used as cable supports.

(xxiv) Housings designed to contain equipment in addition to splices shall be equipped with a device for physically separating the splice area from the service area of the housing.

(xxv) A dielectric shield rated at 15 kv dc shall be provided to enclose the cable splice area. The shield shall extend from the lower cable supports to within 2.5 cm (1 in.) of the top of the housing. The shield shall be equipped with Velcro or equivalent fastening devices designed to hold the shield in both the open or closed positions. The fastening devices shall extend along the entire vertical edge of the dielectric shield.

(xxvi) Mounting arrangements for a variety of terminal blocks and other equipment shall be provided by means of good housekeeping panels or other devices that may enhance the service aspect of the housing.

(xxvii) Housings designed for SAI cabinets may be shipped with terminal blocks installed and stub cables attached. If this option is exercised, the stub cables and terminal blocks must be RUS accepted. In all cases, SAI cabinets must be equipped with appropriate mounting devices for installing the peripheral equipment required for a serving area interface.

(xxviii) SAI cabinets shall be designed to provide physical separation between the splicing area and the area provided for running cross-connect jumpers.

(xxix) SAI cabinets and large splice housings must have an external feature for attaching a padlock to prevent unauthorized entry.

(xxx) Each housing shall have a tinned or zinc electroplated copper alloy or equivalent connector plate or bar to be used for terminating ground and cable shield bond connections. The device shall be equipped with captive studs and nuts with captive lock washers designed for attaching 6 American Wire Gauge (AWG) copper bonding harness wire or braid and a 6 AWG copper ground wire. Connector plates shall be equipped with enough studs and nuts to provide individual connections equivalent to the maximum number of cable sheaths recommended for the housing. Housings shall incorporate design features that enable the field installation of at least one additional connector plate for service conditions that require numerous connections. A bonding and grounding system capable of providing support and strain relief for service wires shall be provided for housings intended for use as distribution points. The bonding system shall be designed to provide sheath continuity as cable and service wires are installed, and prior to any other operation being performed. The bonding arrangement shall provide electrical continuity between all bonds and the ground connector plate. The bonding and grounding arrangement shall permit the lifting of individual cable ground connections for testing and cable locating activities without jeopardizing the grounding potential of other cables that may enter the housing. The bonding and grounding system shall be capable of conducting a current of 1000 amperes for at least 20 seconds.

(4) *Warning sign.* (i) A buried cable warning sign shall be securely attached to the outside of each housing. The lettering information on the sign shall be permanent.

(ii) For pedestals, the sign shall be centered horizontally on the front cover and the top of the sign shall be not more than 10 cm (4 in.) from the top of the housing.

(iii) For SAI cabinets, the sign shall be centered horizontally and vertically on the door. If there are two doors, the sign shall be mounted on the left door.

(iv) Deviations from warning sign location requirements are permitted only for housing design constraints. Alternate sign locations will be considered by RUS.

(v) The RUS standard sign design is shown in Figure 1.

(5) *Housing materials.* (i) Materials used in housings shall present no environmental or safety hazard as defined by industry standards or Federal, State, or local laws and regulations. Figure 1 is as follows:

FIGURE 1

WARNING SIGN

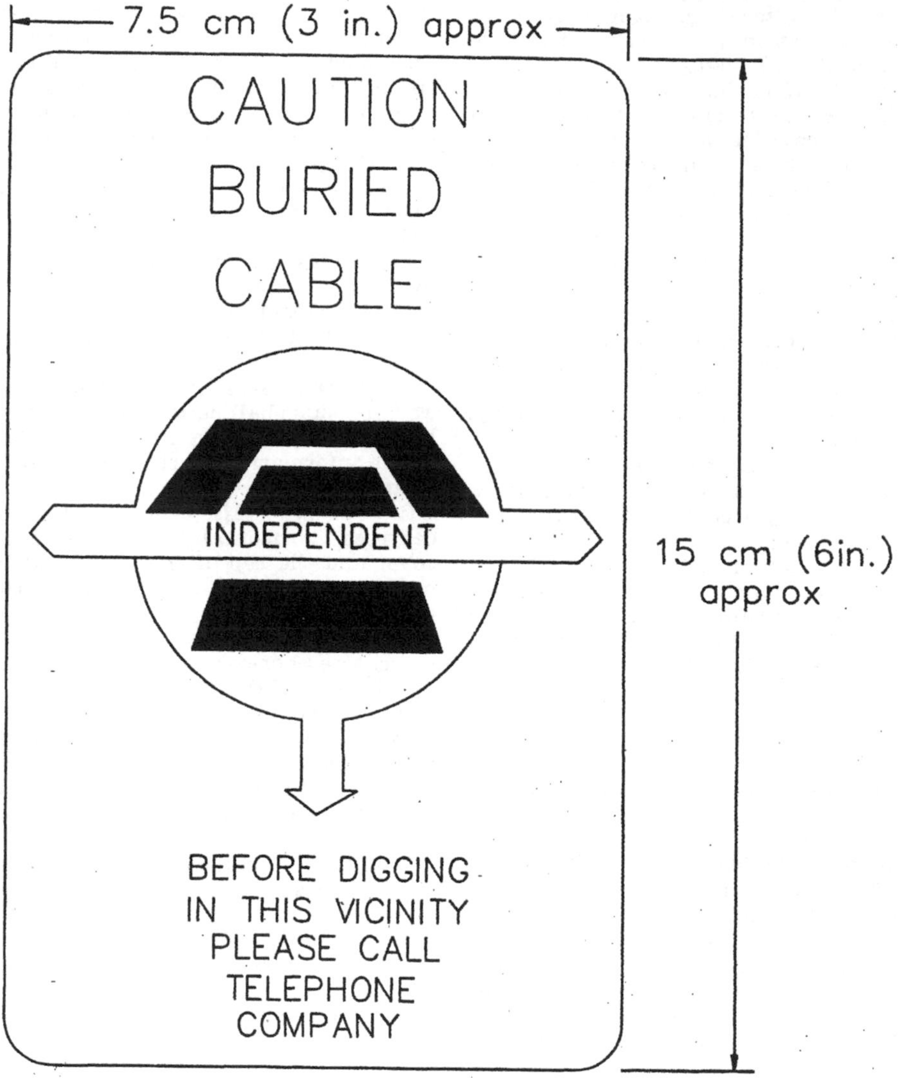

(ii) All materials are required to have fire resistance ratings consistent with recognized industry standards. External materials must be flame resistant.

(iii) All materials used in the manufacture of housings or component parts must achieve the required strength properties, resist deterioration when exposed to outdoor conditions, and be acceptable to RUS for the specific application. New materials or materials not familiar to the RUS staff shall be supported by test and performance data which demonstrates their suitability for the intended use.

(iv) Nonmetallic housing materials shall have a fungus growth rating no greater than one according to ASTM G 21–90.

(v) Metallic components shall be either corrosion resistant or protected against corrosion and must not produce galvanic corrosion in wet or humid conditions on other metals that may be present in the housing environment.

(vi) Mill galvanized steel used in the manufacture of housings shall comply with the appropriate requirements of one of the following standards:

(A) ASTM A 109–91;

(B) ASTM A 366/A 366M–91;

(C) ASTM A 525–91b; or

(D) ASTM A 526/A 526M–90.

(vii) Hot rolled steel shall comply with the appropriate requirements of one of the following standards:

(A) ASTM A 569/A 569M–91a; or

(B) ASTM A 621/A 621M–92.

(viii) Cold rolled steel shall comply with the appropriate requirements of one of the following standards:

(A) ASTM A 109–91; or

(B) ASTM A 366/A 366M–91.

(ix) Steel parts used for internal housing brackets shall be hexavalent chromate coated or zinc plated in accordance with ASTM B 633–85.

(x) Hardware items used for assembling or fastening housing components shall be 300 series or passivated 400 series stainless steel or hot dip galvanized in accordance with ASTM A 153–82 (1987). Other materials will be considered by RUS on an individual basis.

(xi) Aluminum components shall be fabricated from alloy types 5052 or 6061 or other types that have been recognized as having acceptable corrosion resistance and formability and weldability features.

(xii) Nonmetallic parts must be resistant to solvents and stress cracking and shall be compatible with metals and other materials such as conductor insulations and filling compounds used in the manufacture of cable. Plastic materials must be noncorrosive to metals and resist deterioration when exposed to industrial chemical pollutants, ultra-violet rays, road salts, cleaning agents, insecticides, fertilizers, or other detrimental elements normally encountered in the outdoor environment.

(xiii) Housing door seals and gaskets may be manufactured from rubber or synthetic rubber-like elastomer materials. Seals and gaskets shall exhibit a high degree of weatherability with an effective life of at least 30 years in the outdoor environment. The material shall be tear resistant and have a low compression set.

(6) *Housing finish requirements.* (i) All interior and exterior surfaces of housings shall be free from blisters, wrinkles, cracks, scratches, dents, heat marks, and other defects.

(ii) There shall be inherent design provisions to prevent objectionable deterioration of the housing such as rusting, exposure of fiber or delamination. Secondary protection, such as galvanizing over steel per ASTM A 526/A 526M–90 or anodizing over aluminum, shall be provided to ensure reliability over the projected 30 year design life of the housing.

(iii) Painted metal housings shall have a minimum gloss of 60 (60 °specular) in accordance with ASTM D 523–89.

(iv) All painted surfaces shall have a uniform color and texture in accordance with ASTM D 3928–89. Nonmetallic housings shall meet recognized industry standards concerning optical appearance for gloss and haze as applicable for the material.

(v) The colors of housings that RUS will consider for acceptance shall be as follows:

Color	Standard
Gray-Green	Munsell 6.5 GY 6.03/1.6 Munsell 4.4 GY 6.74/1.5
Green	Munsell 8.8 G 2.65/5.3
Orange	Federal Standard 595A Color Number I2246 Munsell 0.15YR 5.26/13.15
Chocolate	Munsell 5.27YR 2.40/2.60 Color Number 835

(7) *Installation requirements.* (i) The design of the housing must provide for a logical and normal installation sequence, i.e., excavation, installation of

a foundation or base and anchoring devices, addition of hardware, installation and bonding of cables, splicing, addition of service, and final closing.

(ii) No special tools or equipment other than that usually carried by outside plant technicians and construction crews must be required for installation of the housing. Security devices are the exception to this requirement.

(iii) Installation hardware shall maintain housings in an erect and stable position when subjected to normal storm loads. Pad-mounted designs must accommodate precast or cast-in-place reinforced concrete or other suitable prefabricated material. Brackets, inserts for fastening, conduit openings, or other items necessary for a pad-mounted installation must be provided. The manufacturer shall provide detailed drawings or a template for locating inserts, conduit openings, or slots for cast-in-place pad construction.

(e) *Performance criteria and test procedures for housings*—(1) *General information.* (i) The housing manufacturer shall perform adequate inspections and tests to demonstrate that housings and housing components comply with RUS requirements.

(ii) Testing shall be performed at a room temperature of 24±3 °C (75±5 °C). Temperatures for testing performed at other than room temperature shall be determined as near the center of the product under test as practical.

(2) *Description of test housing.* (i) Each distinctly designed and configured family of housings intended to perform a particular function shall be tested.

(ii) The typical test sample shall consist of the exterior housing components such as covers, backplates, good housekeeping panels, cap assembly, anchor posts, decals, etc. Interior components must include the bonding and grounding hardware for cables and service wires and the dielectric shield. The housing may include terminal blocks or cross-connect modules, cable splices, or the typical outside plant equipment the housing is designed to contain and protect.

(3) *Environmental requirement for housings*—(i) *Thermal shock.* The test housing shall be placed in a test chamber and exposed to the temperature cycle of Figure 2 for five complete cycles. The step function nature of the temperature changes may be achieved by insertion and removal of the test housing from the chamber. The soak time at each temperature shall be four hours. The housing shall be removed from the test chamber at the conclusion of the five-cycle period. After the test housing temperature has stabilized to room temperature, the housing must be inspected for deterioration of materials and satisfactory operation of mechanical functions. Figure 2 is as follows:

FIGURE 2

THERMAL SHOCK TEMPERATURE CYCLE

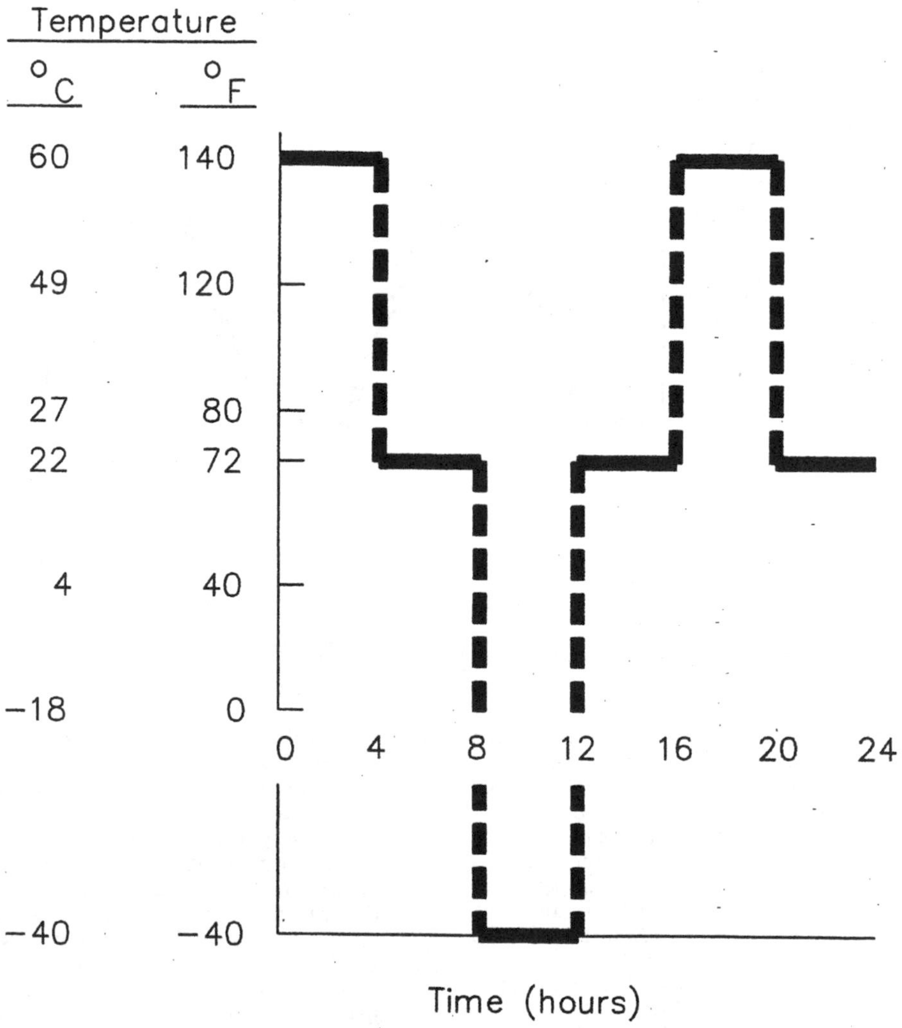

(ii) *Thermal shock and humidity*. The test housing shall be placed in an environmental test chamber at 95 ±3 percent (%) relative humidity (RH) and temperature cycled per Figure 3 for a period of 30 days. At the end of the test there shall be no rust or corrosion of any closure components. Minor corrosion due to surface scratches, nicks, etc. is permitted. If the closure is made of a nonmetallic material, there shall be no signs of degradation. Figure 3 is as follows:

FIGURE 3

ENVIRONMENTAL TEST CHAMBER TEMPERATURE CYCLE

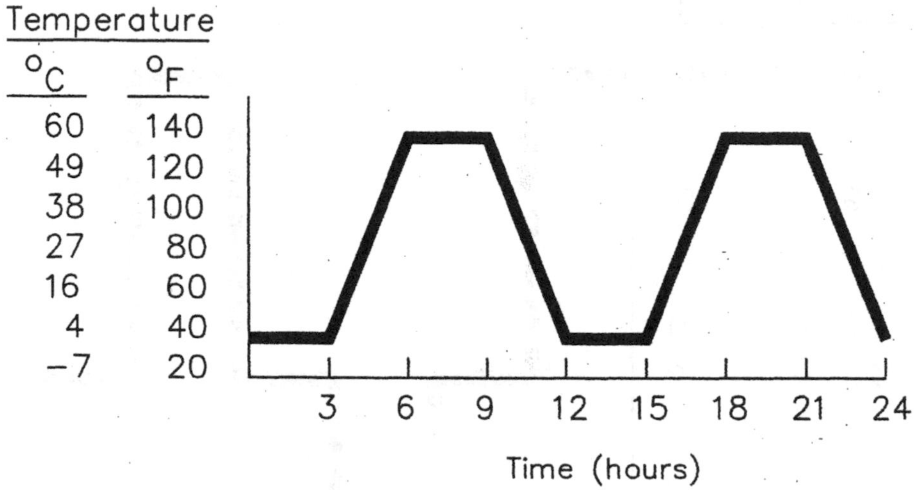

Note: Relative Humidity = 95% ± 3%

(iii) *Humidity and condensation.* Test panels shall be placed in an environmental chamber and subjected to 1,008 hours (42 cycles) of exposure per ASTM D 2247–92. One cycle consists of 24 hours of 100% humidity (with condensation on the panels) at a cabinet temperature of 38±1 °C (100±2 °F) and an ambient temperature of 25±1 °C (77±2 °F) without heat input. Upon completion of cycling, the test panels shall be subjected to an 11 newton-meter (N-m) (100 pound-inches (lb-in.)) impact test using the Gardner-Impact Tester or equivalent. Test panels shall show no substrate or coating cracking or loss of coating adhesion on either side.

(iv) *Weatherability.* Three test panels shall be tested for weatherability in accordance with the appropriate procedures of either ASTM D 822–89 or ASTM G 23–90. Total exposure time shall be a minimum of 800 hours. Failure is defined as fading, cracking, blistering, or delamination on any of the three test panels.

(v) *Low temperature durability.* Low temperature durability shall be proven by exposing the three test panels from (e)(3)(iv) of this section to at least 25 continuous cycles of the following test sequence:

(A) To insure complete saturation of the three test panels, soak them for 96 hours in a container of distilled water 22±2 °C (71.6±4 °F);

(B) Lower the temperature of the water and the immersed test panels to −28±2 °C (−18.4±4 °F) and stabilize for 24 hours;

(C) Thaw the water with the samples to 22±2 °C (71.6±4 °F) and stabilize for 24 hours;

(D) Repeat the procedure 24 times. Any cracking, crazing, deforming, or delaminating on any of the three test panels shall be considered a failure; and

(E) Remove the samples from the water and impact test the three panels by delivering a force of 11.3 N-m (100 lb-in.) using a Gardner-Impact Tester to each specimen at 71, 22, and −28±2 °C (159.8, 71.6, and −18.4±4 °F), after stabilizing them at those temperatures for at least two hours. Visual inspection shall reveal no deformation or perforations on any of the test panels.

(vi) *Corrosion resistance.* Corrosivity shall be tested in accordance with the requirements of ASTM B 117–90. Both scribed and unscribed panels shall be evaluated following the procedures of ASTM D 1654–92. Scribed panels shall have a rating of at least six, following 500 hours of exposure to salt fog, and the unscribed panels shall have a rating no lower than 10, after 1,000 hours exposure. Visual rust inspection shall confirm no more than 0.03% rusting (rust grade 9) of the surface area of the test sample when evaluated in accordance with ASTM D 610–85(1989). The unscribed samples shall be impacted with an 11.3 N-m (100 lb-in.) force, using a Gardner-Impact Tester or equivalent. Visual inspection of the impacted samples shall reveal no loss of adhesion between the base material and the coating or cracking at the finish on the test panels.

(vii) *Fungi resistance.* Fungi resistance of nonmetallic housing materials shall be tested according to the procedures of ASTM G 21–90. Any rating greater than one shall be considered a failure.

(viii) *Stress crack resistance.* The stress cracking characteristics of nonmetallic housing components shall be tested in accordance with ASTM D 1693–70 (Reapproved 1988). The tests shall be performed at 49±2½ C (120±4½ F) for 14 days and exposed to the following materials:

(A) Industry recognized filling compounds;

(B) Isopar M;

(C) Industry recognized solvents;

(D) Industry recognized encapsulants; and

(E) Commonly used insect, pest, and weed control products and agricultural fertilizers.

(ix) *Chemical resistance.* (A) Chemical resistance shall be determined by immersing representative nonmetallic material samples in each of the following solutions for 72 hours at 22±2 °C (71.6±4 °F):

(*1*) 3% sulfuric acid;

(*2*) 100 parts per million (ppm) trichloroethane in water;

(*3*) 0.2 N sodium hydroxide; and

(*4*) Unleaded high octane gasoline.

(B) There shall be no swelling, deformation, or softening of the material samples or any discoloration of the solution.

(x) *Ultraviolet resistance.* Test panels of metallic and nonmetallic outer housing materials shall be subjected to 700 hours exposure per ASTM D 2565–92 using the type BH apparatus. The panels shall not exhibit fading, blistering, checking, or delamination.

(xi) *Weathertightness.* The housing shall be mounted in its typical field installation position and sprayed with water. The temperature of the water shall be adjusted to be equal to or warmer than the temperature of the cabinet interior to avoid the possibility of condensation. A water spray head shall be used to direct water at the housing so that the water stream will strike the assembly at a downward angle of 45 degrees. The flow of the water shall be 3.8 liters per minute (one gallon per minute), with 276 kilopascals (40 pounds per square inch) head of pressure. The spray head shall be held 1.8 meters (m) (6 feet (ft)) from the test cabinet. The spray head shall be adjusted so that water impinges uniformly over the housing surface. The duration of the test shall be five minutes. All vertical cabinet surfaces shall be tested by this procedure. The exterior of the cabinet shall be thoroughly dried with towels (no heat drying) prior to examination of the housing interior. The interior of the housing shall be checked for presence of water. Wetting of over-lapping surfaces is permitted. There shall be no presence of water inside the housing.

(xii) *Wind Resistance.* (A)(*1*) Stub pole or wall mounted SAI and large pair count splice housings shall be subjected to a load (F) as shown in Figure 4 and the following table to simulate the turning moment equivalent to a uniform wind load of 161 kilometers per hour (km/h) (100 miles per hour (mi/h)) perpendicular to the largest surface area.

Maximum area of largest surface square centimeters cm^2 (Square inches) ($in.^2$)	Load	
	kg	(lb)
5,200 (800) or less	18	(40)
5,201 to 9,100 (801 to 1,400)	32	(70)
9,101 to 13,000 (1,401 to 2,000)	45	(100)
13,001 to 16,200 (2,001 to 2,500)	57	(125)

NOTE: The procedures for housings with larger surface area will be evaluated by RUS on a case-by-case basis.

(*2*) The housing shall remain in its original mounting position throughout the test and exhibit no mechanical deformation.

(*3*) Figure 4 is as follows:

FIGURE 4

TURNING MOMENT – POLE MOUNTED HOUSINGS

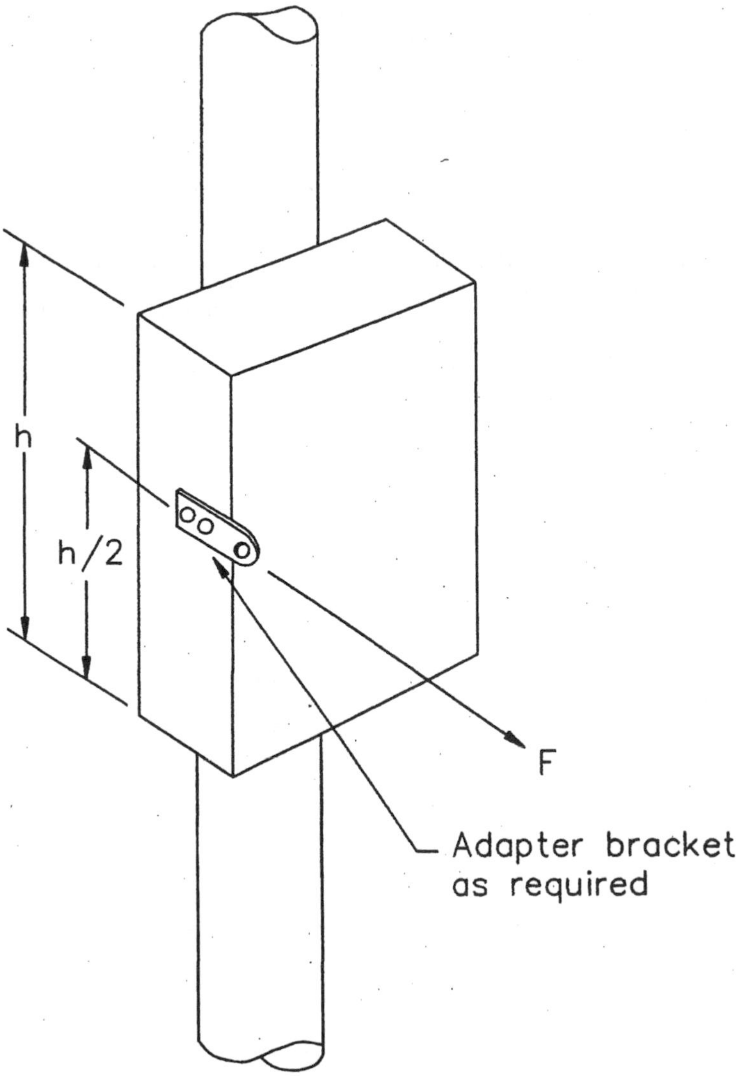

(B)(*1*) Pad or ground mounted SAI or splice housings shall be subjected to a load (F) as shown in Figure 5 and the following table to simulate the overturning moment equivalent to a uniform wind load of 161 km/h (100 mi/h) perpendicular to the largest surface area.

Height cm (in.)	Maximum area of largest surface cm^2 ($in.^2$)	Load	
		kg	(lb)
122 (48) or less	11,000 (1,700) or less	91	(200)
	11,001–13,000 (1,701–2,000)	104	(230)
	13,001–14,900 (2,001–2,300)	118	(260)
123–152 (49–60)	11,700 (1,800) or less	91	(200)
	11,701–14,300 (1,801–2,200)	109	(240)
	14,301–16,200 (2,201–2,500)	127	(280)
	16,201–18,800 (2,501–2,900)	145	(320)
	18,801–20,800 (2,901–3,200)	163	(360)
	20,801–23,400 (3,201–3,600)	181	(400)
153–183 (61–72)	14,300 (2,200) or less	109	(240)
	14,301–16,900 (2,201–2,600)	127	(280)
	16,901–19,500 (2,601–3,000)	150	(330)
	19,501–22,700 (3,001–3,500)	172	(380)
	22,701–25,300 (3,501–3,900)	190	(420)
	25,301–27,900 (3,901–4,300)	213	(470)

NOTE: The procedures for housings with larger surface areas will be evaluated by RUS on a case-by-case basis

(*2*) The housing shall remain in its original mounting position throughout the test and exhibit no mechanical deformation.

(*3*) Figure 5 is as follows:

FIGURE 5

OVERTURNING MOMENT, PAD MOUNTED HOUSINGS

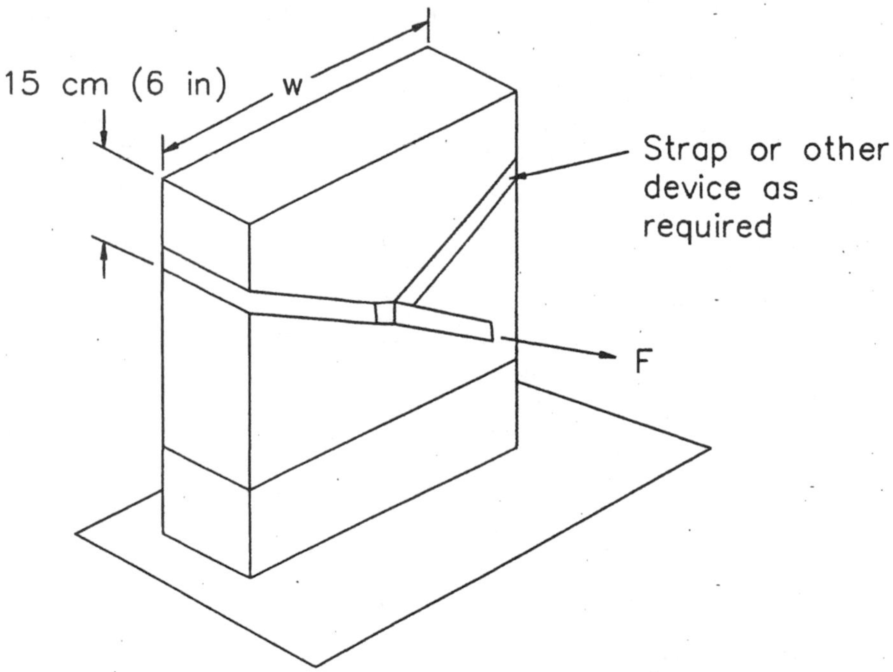

Notes:

1. The load "F" shall be applied perpendicular to the cabinet width where the width is greater than the depth.
2. If a foundation is used, the load "F" shall be applied toward the edge nearest to the cabinet.

(xiii) *Fire resistance.* (A) The test housing shall be installed in a manner typical of field installation. U.S. No. 1 wheat straw shall be placed on the ground around the housing base in an one meter (3 ft) radius at an approximate depth of 10 cm (4 in.). The straw shall be ignited and permitted to burn fully. After the housing has cooled, its contents shall be inspected for evidence of ignition, melting, burning, or structural damage. Damage sufficient to impair service constitutes failure.

(B) Polymeric materials shall be tested in accordance with the Underwriters Laboratories Publication (UL) 94, dated June 18, 1991. Materials used in housing components shall have a rating of 94V–0 or 94V–1 and shall not sustain combustion when an open flame source is removed.

(4) *Mechanical requirements for housings*—(i) *Impact resistance.* The test housing shall be subjected to the following impacts according to its minimum volume or minimum width and depth as shown in the following table:

Minimum volume cm^3 ($in.^3$)	Minimum width or depth cm (in.)	Impact force	
		N-m	(lb-ft)
Less than 35,000 (2,100).	Less than 13 (5)	68	(50)
35,000 (2,100) or greater.	13 (5) or greater	136	(100)

(A) The impact force shall be delivered to the front, back, and top surfaces. Circular housings shall be impacted on side surfaces 180 °apart and on the top. The device used to deliver the force shall be spherical and approximately 25 to 31 cm (10 to 12 in.) in diameter. A typical test procedure may include the use of a hard rubber bowling ball, weighing 6 to 7 kg (13 to 16 lb), enclosed in a mesh bag, attached to a rope with a metal ring. The load shall be dropped vertically on the top surface and applied to the sides with a pendulum motion using the appropriate height and extension arm to achieve the required impact force. The housing must be impacted at the approximate mid-point of the surface area.

(B) Housings shall be conditioned for a minimum of eight hours at −40 °C (−40 °F) in an environmental chamber prior to testing. If the chamber is insufficient in size to conduct tests within the chamber, the housing may be removed and shall be tested within 10 minutes after removal.

(C) After impact testing, the housing shall not exhibit fractured or ruptured surfaces sufficient to allow the ingress of moisture or dust. The housing shall not exhibit mechanical damage that would impair the functioning of hinges, latches, locks, etc.

(ii) *Load deflection.* Free standing buried plant housings shall be tested for load deflection in accordance with Figure 6. The assembled housing shall be rigidly held in place by a mechanical means to simulate a normal field installation. A length of wire or cable, or other suitable material, shall be placed around the top section of the housing and deadended. The wire or cable shall be initially tensioned to 23 kg (50 lb). A measurement shall then be taken of the deflection of the housing at the top as shown in Figure 6. The deflection shall be recorded at incremental loads of 23 kg (50 lb) until destruction of the housing occurs. The average load for the three directions shall not be less than 136 kg (300 lb) and the minimum load in any direction shall be 113 kg (250 lb). Failure is defined as housing component fracture or crazing of the housing's surface finish. Figure 6 is as follows:

FIGURE 6

MEASURING LOAD DEFLECTION

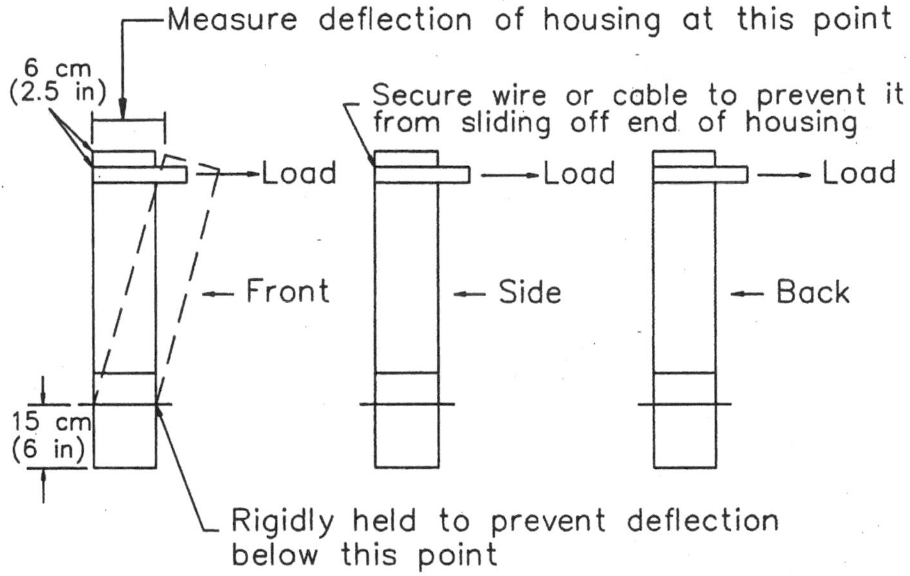

Notes:

1. One pedestal-mounted housing of each BD classification shall be tested to failure in each of the directions shown above.
2. A total of three pedestal-mounted housings of each BD classification shall be subjected to the required loads in each direction.
3. The average load for the three directions shall not be less than 136 kilograms (300 pounds). The minimum load shall be 113 kilograms (250 pounds).
4. Pole mounted housings shall be subjected to the same loading criteria.

(iii) *Vibration requirements*. The test housing and its contents shall be subjected to acceleration at a sine wave frequency sweep rate as shown in Figure 7 for a housing packaged for shipment and Figure 8 for an unpackaged housing. The frequency sweep may be performed continually or sequentially. The test shall be conducted once along each of three mutually perpendicular axes of the housing. There shall be no mechanical or electrical degradation of the housing or its contents. Noticeable damage to the housing constitutes failure. Figure 7 and Figure 8 are as follows:

FIGURE 7

VIBRATION TEST FOR PACKAGED HOUSINGS

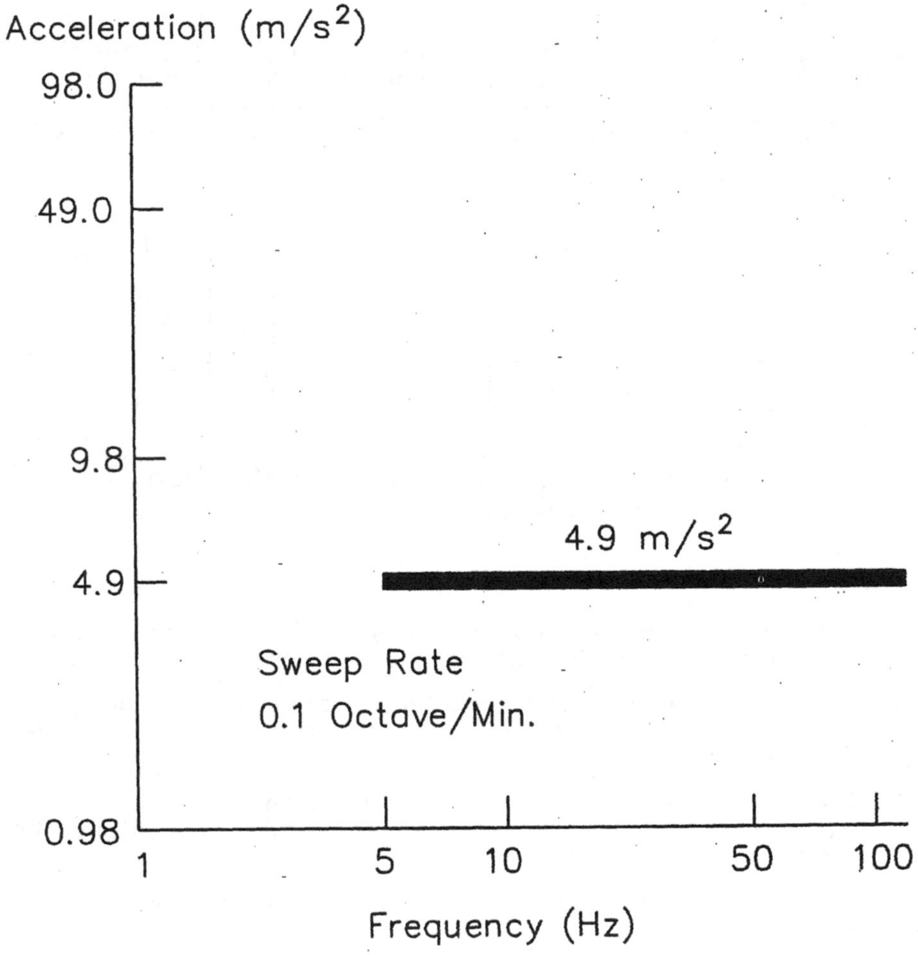

FIGURE 8

VIBRATION TEST FOR UNPACKAGED HOUSINGS

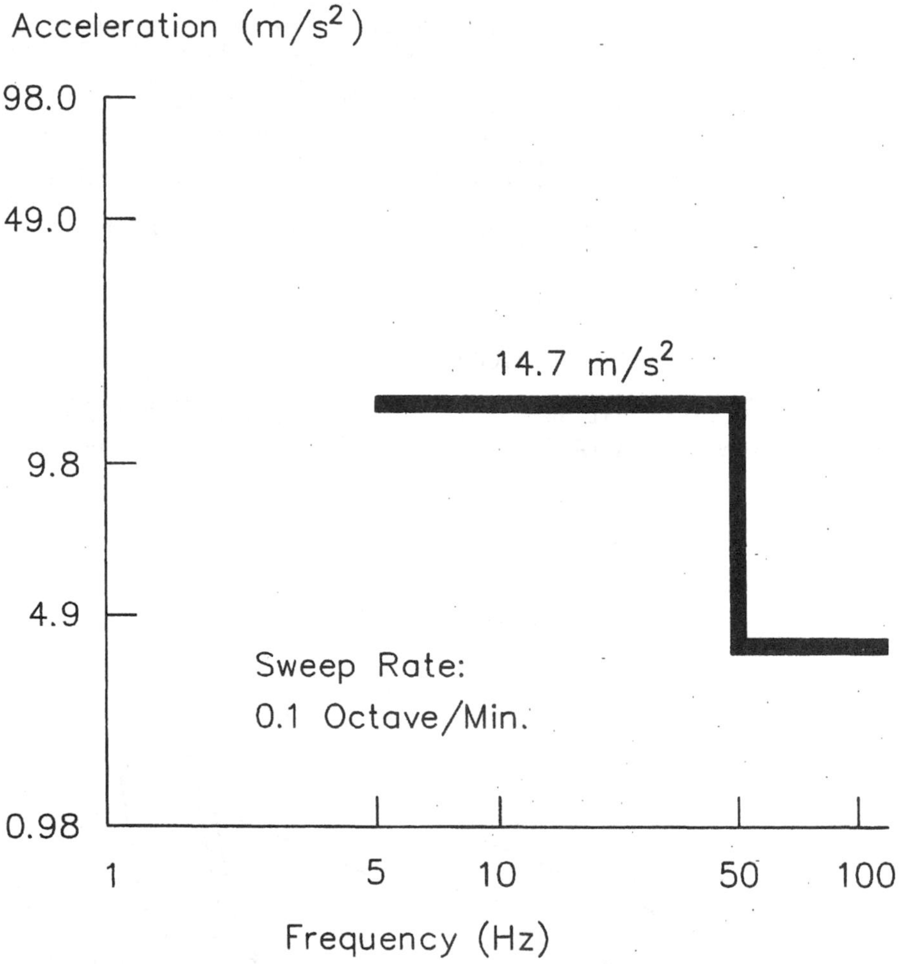

(iv) *Drop test requirements.* Housings shall be subjected to appropriate drop tests according to their weight. The drop tests shall be performed on housings and their contents as normally packaged as well as on unpackaged housings. The tests shall be conducted on a smooth level concrete floor or similar unyielding surface. For corner drops, the packaged housing and its contents shall be oriented at impact such that a straight line drawn through the struck corner and package geometric center is approximately perpendicular to the impact surface.

(A) Packaged housings and their contents weighing 91 kg (200 lb) or less shall be capable of enduring a single drop on each face or corner without damage from a height specified as follows:

Packaged housing including contents weight kg (lb)	Drop height cm (in.)
0 to 9 (0 to 20)	76 (30)
10 to 23 (21 to 50)	61 (24)
24 to 45 (51 to 100)	53 (21)
46 to 91 (101 to 200)	46 (18)

(B) Packaged housings and their contents weighing more than 91 kg (200 lb) shall be capable of enduring a single drop on each of two diagonally opposite corners of the package without significant damage from a height specified as follows:

Packaged housing including contents weight kg (lb)	Drop height cm (in.)
92 to 453 (201 to 1000)	30 (12)
Over to 453 (1000)	15 (6)

(*1*) The packaged housing and contents shall be placed on its normal shipping base with one corner supported 15 cm (6 in.) above the floor and the other corner of the same end supported 30 cm (12 in.) above the floor as shown in Figure 9. The unsupported end of the package shall be raised so that the lowest corner reaches the height listed above and then allowed to fall freely. Figure 9 is as follows:

FIGURE 9

CORNER DROP TESTS FOR PACKAGED HOUSINGS WEIGHING MORE THAN 91 KILOGRAMS (200 POUNDS)

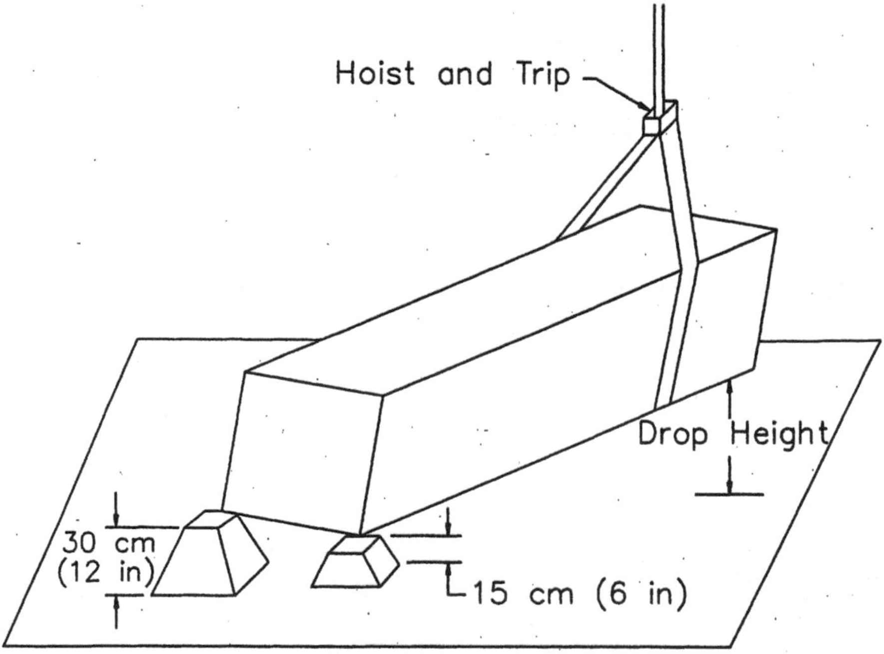

(*2*) The procedure of paragraph (e)(4)(iv)(B)(*1*) of this section shall be repeated for the diagonally opposite corner.

(*3*) The packaged housing and contents shall be capable of enduring a single drop on each edge of the base of its normal shipping position from the required height without damage and shall remain operational without function impairment. The packaged housing and contents shall be placed on its base with one edge supported on a sill 15 cm (6 in.) high and the unsupported edge raised to the required height as shown in Figure 10 and allowed to fall freely. Figure 10 is as follows:

FIGURE 10

EDGE DROP TEST FOR PACKAGED HOUSINGS WEIGHING MORE THAN 91 KILOGRAMS (200 POUNDS)

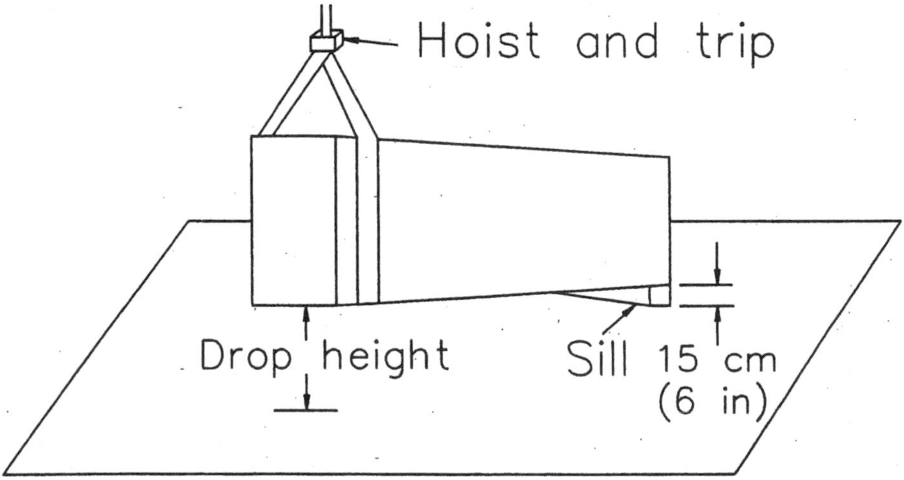

(*4*) The procedure of (e)(4)(iv)(B)(*3*) of this section shall be repeated for all edges of the base.

(C) Unpackaged housings and their contents weighing 23 kg (50 lb) or less shall be capable of enduring a single drop on each face and adjacent corners without significant damage from a height specified as follows:

Packaged housing including contents weight kg (lb)	Drop height cm (in.)
0 to 9 (0 to 20)	10 (4)
10 to 23 (21 to 50)	8 (3)

(D)(*1*) Unpackaged housings and their contents weighing more than 23 kg (50 lb) shall be capable of enduring a single drop without significant damage when lifted by its normal hoisting supports as shown in Figure 11 and with its lowest point at a height specified as follows:

Packaged housing including contents weight kg (lb)	Drop height cm (in.)
23 to 45 (51 to 100)	5 (2)

(*2*) Figure 11 is as follows:

FIGURE 11

DROP TEST FOR UNPACKAGED HOUSINGS WEIGHING MORE THAN 23 KILOGRAMS (50 POUNDS)

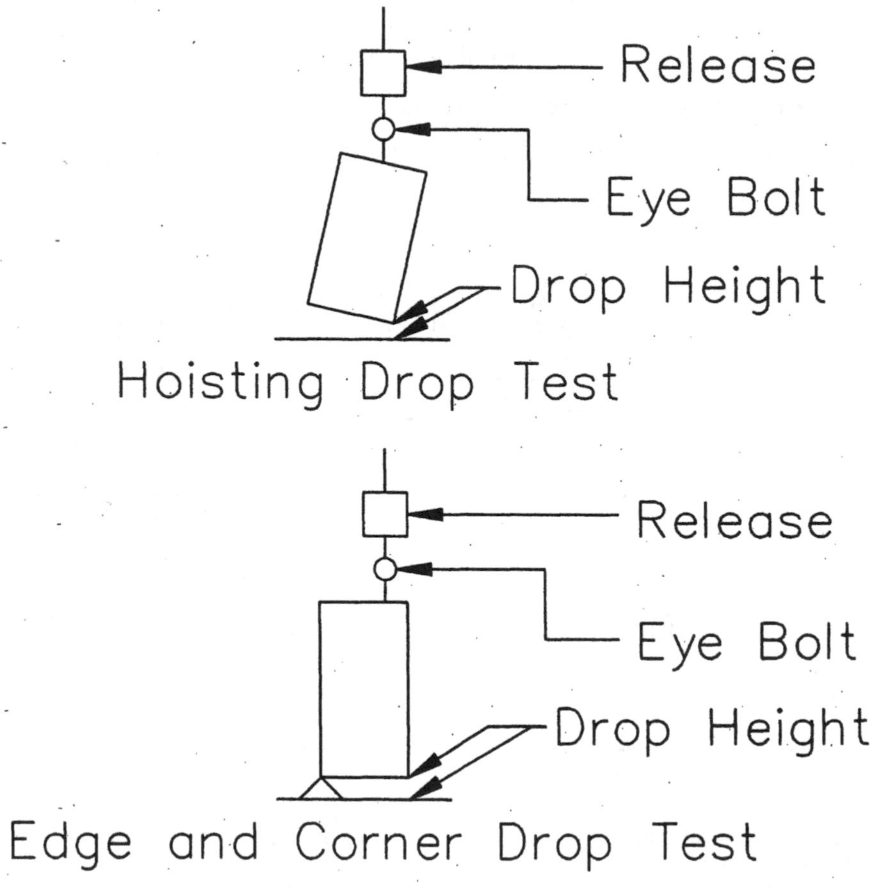

(v) *Firearms resistance.* All housings shall be tested for resistance to penetration by direct impact from a 12 gauge shotgun equipped with a modified choke and the use of a 3¾ dram equivalent powder charge and 35 grams #6 lead shot fired from a distance of 15 m (50 ft). The 12 gauge shotgun shall be fired from a normal standing position at the front side of the housing. Penetration through the housing wall by the lead shot shall constitute failure.

(vi) *Lifting hardware requirements.* The lifting hardware on housings and their contents that weigh more than 91 kg (200 lb) shall be tested. The housing shall be fastened to a restraining device such as a concrete slab and subjected to loading through the lifting attachments to simulate the lifting load. For the first test a lifting line equipped with a dynamometer shall be attached to the housing lifting hardware and a load applied equal to three times the weight of a fully equipped housing. Deformation or damage to the housing or lifting hardware constitutes failure. A second test shall be conducted with the same arrangements as for the first except that a load shall be applied equal to six times the weight of a fully equipped housing. There shall be no catastrophic failure of the lifting hardware or housing.

(vii) *Stub cable strain relief tests.* Housings equipped with cable stubs and cable shipping retainer shall be tested by lifting a test housing, with the maximum length and weight of cable orderable, in a manner causing the full weight of the cable to be supported by the cabinet. Examination of the cable sheath after lifting shall reveal no tearing, rupturing, or other damage. The cable conductors and shield shall be tested for shorts and opens. Electrical defects to the stub cable or damage to the housing constitutes failure.

(viii) *Door restrainer evaluation.* (A) The housing shall be positioned with the door held in the open position by the door restraining device. A load, determined in accordance with the following table, shall be applied to the center of the door, perpendicular to the door and in each of the opening and closing directions.

Maximum area of door surface cm² (in².)	Load kg (lb)
5,200 (800) or less	72 (160)
5,201 to 9,100 (801 to 1,400)	127 (280)
9,101 to 13,000 (1,401 to 2,000)	181 (400)

NOTE: Test procedures for housings with larger doors will be evaluated by RUS on a case-by-case basis.

(B) There shall be no functional failure of the restraining device nor mechanical damage to the housing.

(ix) *Security evaluation.* The security locking device shall be capable of withstanding a maximum torque of 2.8 N-m (25 lb-in.) without incurring physical damage to the closure, thereby resulting in a condition where the closure cannot be either accessed or locked.

(5) *Electrical requirements for housings.* Each bonding stud and nut location shall be evaluated by attaching one lead from a dc or alternating current (ac) power source to a bonding stud with the nut torqued as specified by its manufacturer and the other power source lead connected to the closure grounding conductor connector. The current path thus established must be capable of sustaining a current of 1,000 amperes root-mean-square for at least 20 seconds without fusing or causing any damage to the closure or its contents.

(6) *Finish requirements*—(i) *Impact resistance.* The finish on painted metal surfaces shall not exhibit radial cracking on the impact surface (intrusion) when indented at 18 N-m (160 lb-in.) with a 1.6 cm (0.6 in.) diameter spherical indentor. This test shall be performed in accordance with ASTM D 2794–92 with the exception that the test panel shall be of the same material, thickness, and finish as the pedestal housing being evaluated.

(ii) *Finish adhesion.* Painted finishes shall be tested for adhesion of finish in accordance with ASTM D 2197–86 (Reapproved 1991), Method A. There shall be no gouging in the top coat when tested with an 8 kg (17.7 lb) load. Gouging is defined as removal or separation of paint particles or breaking of the finish by the scraping loop to the extent of exposing base metal.

(iii) *Color evaluation.* The color of the housing finish should be compared against the Munsell system of color notation, as described in ASTM D 1535–89 to determine color consistency with that desired.

(iv) *Gloss evaluation.* The finish on painted housings shall be tested on two approximately 20 cm × 20 cm (8 in. × 8 in.) samples for each color used in accordance with the procedures of ASTM D 523–89. The finish shall have a minimum gloss of 60 (60 °Specular).

(v) *Secondary finish evaluation.* Evidence of secondary protection shall be required for RUS acceptance. Typical secondary protection is galvanizing per ASTM A 526/A 526M–90 for steel surfaces.

(f) *Functional design criteria for binding post terminal blocks used in SAI cabinets*—(1) *General description.* A conventional binding post terminal consists of a metallic element or post, one end of which is configured for the permanent connection of 22, 24, or 26 AWG solid copper conductors and the opposite end is configured for recurring connections and disconnections of solid copper cross-connect wire using a threaded screw or stud and nut combination for gripping the wire. The terminal is usually housed in a SAI cabinet. However, the terminal may receive limited use in smaller pedestal-type housings and pole mounted cabinets in the outside plant environment.

(2) *Design and fabrication requirements.* (i) Terminal blocks used in outside plant housings are expected to perform satisfactorily for a nominal design life of 30 years.

(ii) All individual terminals or terminal fields must be enclosed and the terminal enclosure must be totally filled with an encapsulating grease or gel which prevents connection degradation caused by moisture and corrosion. The encapsulant must provide complete encapsulation of terminal metallic connections and surfaces and totally fill all voids and cavities within individual terminal enclosures or terminal field enclosures to prevent ingress of moisture. The encapsulant must not restrict access to the terminal or restrict craft personnel from making connections. The encapsulant must be compatible with the standard materials used in cross-connect hardware and wiring.

(iii) Binding post terminals shall not be susceptible to damage under normal use of standard tools used by outside plant technicians such as screwdrivers and test set clips. In addition, use of other tools such as scissors, diagonal cutters and long nose pliers for tightening and loosening screws shall not result in damage to the terminal.

(iv) Terminals shall be designed so that a typical technician using customary tools shall be able to terminate cross-connect wire on a pair of terminals, or to remove it, without causing an electrical short between any two terminals or any other adjacent terminals.

(v) The terminal count sequence shall be indicated using numerals of at least 0.25 cm (0.10 in.) in height.

(vi) A means shall be provided to distinguish feeder terminals from distribution terminals.

(vii) A means shall be provided to identify tip terminals and ring terminals in a terminal field. The identification convention shall indicate tip on the left with ring on the right for horizontal spacing and tip on the top with ring on the bottom for vertical spacing.

(viii) The preferred height of the highest terminal in the connector field in a ground mounted SAI unit shall be 168 cm (66 in.) or less as measured from the top surface of the mounting pad. The bottom or lowest terminals in the connector field shall be at least 46 cm (18 in.) from the top surface of the pad.

(ix) Pole mounted aerial units shall be 84 cm (33 in.) or less in width. The maximum allowable height of the highest terminals in a pole mounted aerial unit is 168 cm (66 in.) as measured from the top surface of the standard balcony seat used with the interface. For computation purposes, 15 cm (6 in.) shall be allowed for the distance between the bottom of the interface and the top of the balcony seat.

(3) *Auxiliary features.* (i) SAI cabinets with terminal designs which do not permit direct attachment of common test instrument clips to terminal pairs without the occurrence of shorts shall be equipped with single pair auxiliary test contacts. The auxiliary test contacts shall attach to a terminal pair and provide a set of secondary terminals which will accept typical test instrument clips without the occurrence of shorts. Wire used to connect the auxiliary test contacts to the secondary terminals shall be 20 gauge minimum stranded conductor copper wire with a minimum dielectric strength between conductors of 15 kv. The test connector shall be functional on all terminal pairs.

(ii) A 25 or 50 pair test connector shall be available which can be used to make reliable electrical contact to terminals associated with discrete 25 pair binder groups. The multi-pair test connector shall be provided with a minimum of 1.8 m (6 ft) of suitable cabling terminated to a connector, for interfacing with test sets common to the industry. The multi-pair test connector shall be functional on all terminal groups.

(iii) A special service marker shall be available which must attach to a binding post terminal to identify special circuits and insulate exposed metal parts from accidental shorts from tools and wires. A supply of 25 special service markers shall be provided with each SAI cabinet. The color of special service markers shall be red.

(iv)(A) A supply of twisted pair cross-connect wire shall be supplied with housings that are equipped with cross-connect terminals or that have provisions for mounting cross-connect terminals. The minimum length of cross-connect wire supplied is dependent on the SAI cabinet terminal capacity as follows:

Cabinet termination capacity (pairs)	Wire length
1 to 600	60 m (200 ft)
601 to 1200	120 m (400 ft)
Over 1200	180 m (600 ft)

(B) The cabinet shall be equipped to store the length of wire in a manner designed for convenient dispensing. The cross-connect wire supply shall be easily replaceable.

(g) *Performance criteria and test procedures for binding post terminal blocks used in SAI cabinets*—(1) *General.* Many of the tests described in this section require that the terminal block be installed in an appropriate housing in its typical field configuration.

(2) *Environmental requirements*—(i) *Insulation resistance/high humidity and salt fog exposure.* A test specimen shall consist of a standard ground or pole mounted housing equipped with a full complement of binding post terminals equipped with 25 special service markers. The minimum number of terminals to be tested shall be 100 pair (100 tips and 100 associated rings). The test terminals shall be selected to form a terminal array of approximate square dimensions. A 1 cm (36 in.) length of cross-connect wire shall be installed on each test terminal. All tips shall be joined together and all rings shall be joined together with a 48 volt dc potential applied as shown in Figure 12 during the high humidity/salt fog and simulated rain exposures. The 48 volt dc may be temporarily removed from the test samples during the measurement process and the ring terminal being measured shall be isolated from the remaining ring terminals. The terminal insulation resistance shall be measured at a potential of 100 volts dc using suitable instrumentation with a minimum measurement range of 10^4 to 10^{12} ohms. Figure 12 is as follows:

FIGURE 12

BINDING POST ARRANGEMENT FOR INSULATION RESISTANCE TESTING

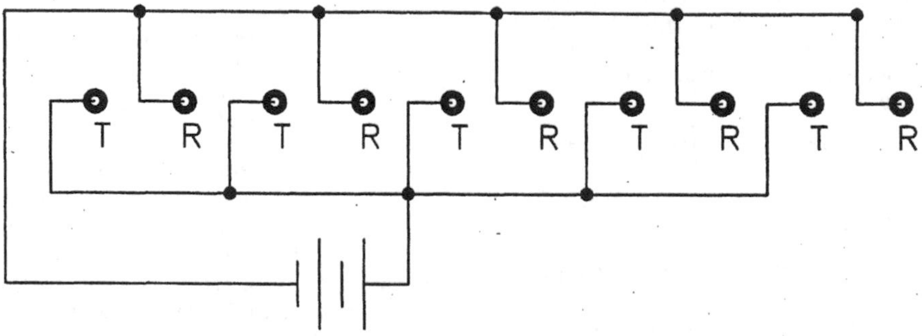

Note: 5 pair specimen arrangement shown is typical for entire (100 pair specimen minimum) test population.

(A) *High humidity*. The test housing shall be placed in an environmental test chamber at 95±3% RH and the temperature cycled as shown in Figure 3 in paragraph (e)(3)(ii) of this section for a period of 30 days. The cabinet doors shall remain in the fully open position. The insulation resistance between the ring terminal of each sample and all the common tip terminals shall be measured each 24 hours when the temperature is between 38 and 57 °C (100 and 135 °F) and increasing. The minimum insulation resistance when measured in accordance with paragraph (g)(2)(i) of this section shall not be less than 1×10^6 ohms.

(B) *Salt fog*. A test housing with its doors closed shall be placed in a salt fog 35 °C (95 °F) test chamber and exposed to a salt fog spray per ASTM B 117–90 for a period of 30 days. The insulation resistance should be measured every 24 hours as indicated in paragraph (g)(2)(i) of the section and shall not be less than 1×10^6 ohms. The special service markers shall exhibit no sign of fading, corrosion, swelling, warping, running color, or other signs of deterioration.

(ii) *Insulation resistance/simulated rain exposure*. (A) A test housing as described in paragraph (g)(2)(i) of this section shall be tested for water infiltration. The test shall be conducted using the method described in paragraph (e)(3)(xi) of this section. The cabinet doors shall remain closed for the duration of the test. The insulation resistance between the ring terminals and the common tip terminals shall be measured during and immediately following the spray application as indicated in paragraph (g)(2)(i) of this section and shall not be less than 1×10^6 ohms.

(B) With the cabinet doors open, a spray of tap water at a rate of 3.8 liters per minute (1 gallon per minute) at 276 kilo-pascals (40 pounds per square inch) shall be directed on the terminal array for a period of 1 minute saturating all of the terminals. Following the spray application the doors shall be closed. The cabinet shall be maintained in a temperature environment of 26 to 28 °C (78 to 82 °F) at 95±3% RH for 6 hours. The insulation resistance shall then be measured as specified in paragraph (g)(2)(i) of this section. The minimum insulation resistance shall not be less than 1×10^6 ohms.

(iii) *Contact resistance.* A minimum of 100 terminals equipped with cross-connect wire that has been installed in a manner typical of that used in the industry shall be temperature cycled.

(A) The test shall consist of eight-hour temperature cycles with one-hour dwells at extreme temperatures of −40 °C to + 60 °C (−40 °F to + 140 °F), and temperature changes at an average rate of 16 °C (60 °F) per hour between the extremes. The relative humidity shall be maintained at 95±3%. The eight-hour test shall be conducted for 512 cycles. Millivolt drop measurements shall be made initially and after 2, 8, 16, 32, 64, 256, and 512 cycles with the samples at room temperature. The resistance measurement technique must conform to ASTM B 539–90. The measurement method must have an accuracy of at least ±30 microohms for resistances less than 50 milliohms. The change in contact resistance shall not exceed 2 milliohms.

(B) A minimum of 100 terminals equipped with cross-connect wire installed in a manner typical of the industry shall be maintained at 118 °C (245 °F) during the test period, except during disturbance measurement periods where each wire connection to the terminals shall have a 0.23 kg (0.5 lb) force momentarily applied in a manner to stress the connection. Initial millivolt measurements shall be made without disturbing the joints in accordance with paragraph (g)(2)(iii)(A) of this section with the samples at room temperature. After initial measurement each sample shall be disturbed followed by a millivolt drop measurement after 1, 2, 4, 8, 16, and 33 days. The change in contact resistance should be less than 2 milliohms when compared to the initial measurement.

(iv) *Fire resistance.* A fully equipped cabinet including a full complement of cross-connect jumpers shall be installed in the standard field arrangement and tested for fire resistance in accordance with paragraphs (e)(3)(xiii) introductory text through (e)(3)(xiii)(B) of this section. After cooling, the cabinet, terminals, and associated wiring shall be inspected for signs of ignition, melting, burning, or structural damage of sufficient consequences such that the results are service affecting.

(v) *Encapsulant material compatibility.* The terminal connection encapsulant compound must be compatible with the standard materials used in cross-connect hardware and wiring when aged in accordance with ASTM D 4568–86 at a temperature of 80±1 °C (176 ±2 °F). The conductor insulation shall retain a minimum of 85% of its unaged tensile strength and elongation values. The cross-connect hardware shall exhibit no visible material degradation.

(vi) *Encapsulant flow test.* Terminal connection encapsulant must remain stable at 80±1 °C (176±2 °F) when tested in an environmental chamber. Test specimens shall be suspended in a preheated oven over a glass dish or other drip-catching medium for a period of 24 hours. At the end of the test period, the glass dish shall be examined for evidence of flowing or dripping of encapsulant from the cross-connect terminal. More than 0.5 gram of encapsulant in the dish at the end of the test constitutes failure.

(3) *Mechanical requirements*—(i) *Vibration.* A test housing equipped with a full complement of cross-connect terminals and jumper wiring shall be subjected to vibration testing in accordance with paragraph (e)(4)(iii) of this section.

(ii) *Torsional capacity of binding posts.* The test specimens shall consist of the complete binding post terminal consisting of the screw or nut, washers if required, and threaded post or stud respectively.

(A) Test specimens shall include the terminals along the matrix edge at mid-span locations as well as centrally located terminals. Tests shall be conducted using a torque indicating screwdriver, or wrench, with an accuracy of

±0.17 N-m (±1.5 lb-in.) or better. The torque indicating device shall be used to tighten a screw or nut until failure of the screw or nut is achieved. Tests shall be conducted while the test specimen is stabilized at temperatures of −40 °C, 20 °C, and 71 °C (−40 °F, + 68 °F, and at + 160 °F). Record the torques at terminal failure. At least 10 test specimens shall be tested at each temperature. The failure torque shall not be less than 2.8 N-m (25.0 lb-in.) for each temperature.

(B) The post or stud of the binding post terminal shall not fail before the screw or nut when increasing torque. The faceplate or receptacle restraining the post or stud shall not fail before the screw or nut when increasing torque.

(iii) *Lateral loading capacity of binding posts.* A minimum of three sets of 25 terminals shall be tested with the test specimens stabilized at temperatures of −40 °C, 20 °C and 71 °C (−40 °F, + 68 °F, and 100 °F). The test arrangement shall include the terminals along the matrix edge at mid-span locations as well as centrally located terminals. A force measuring device, such as a dynamometer, shall be attached to the end of a binding post terminal and a 16 kg (35 lb) force applied orthogonally to the terminal axis in 4 perpendicular directions as shown in Figure 13. Permanent deformation in excess of 0.08 cm (0.03 in.) or any structural damage in either the terminal or faceplate constitutes a failure. Figure 13 is as follows:

FIGURE 13

LATERAL LOADING OF BINDING POST TERMINALS

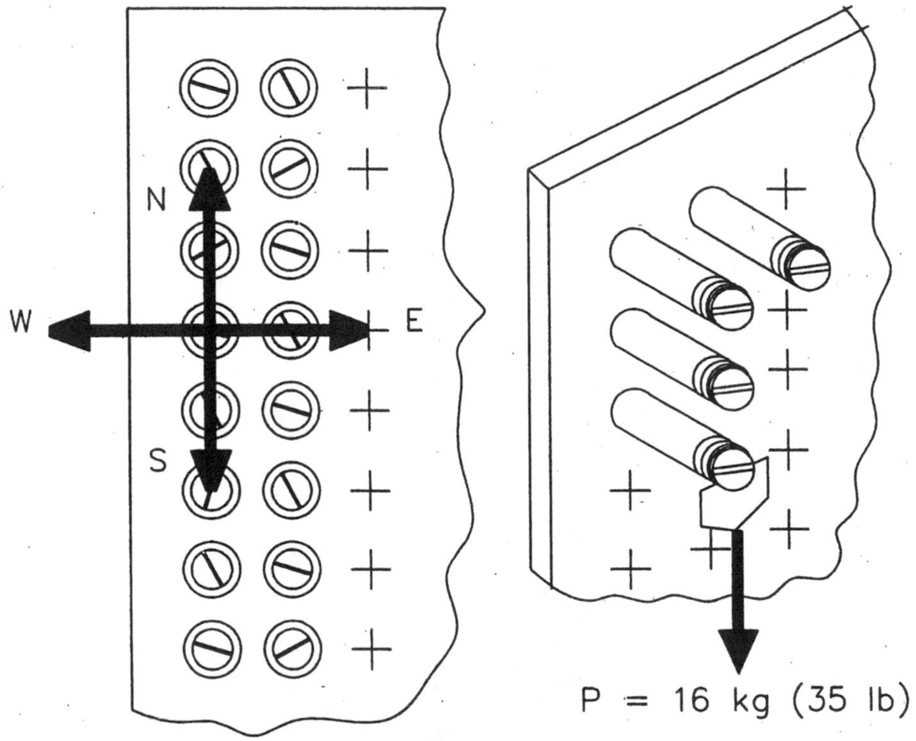

Note: Apply load P in N, E, S, and W directions

(iv) *Axial pullout resistance.* A minimum of three sets of 25 terminals shall be tested with the test specimens stabilized at temperatures of −40 °C, 20 °C, and 71 °C (−40 °F, + 68 °F, and 100 °F). The test arrangement shall include the terminals along the matrix edge at mid-span locations as well as centrally located terminals. A force measuring device, such as a dynamometer, shall be attached to a terminal and a force of 16 kg (35 lb) applied on axis as shown in Figure 14. There shall be no permanent deformation in excess of 0.08 cm (0.03 in.), any structural damage, or terminal pull-out in either the terminal or the faceplate. Figure 14 is as follows:

FIGURE 14

AXIAL PULLOUT OF BINDING POST TERMINALS

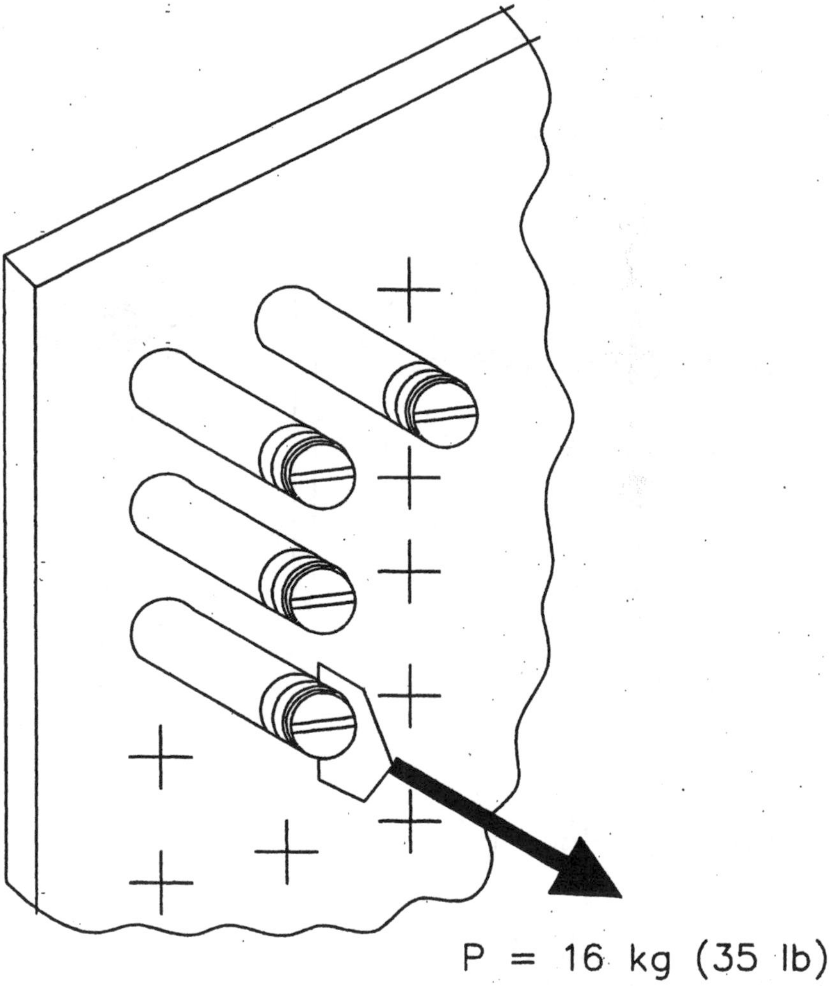

(v) *Test connector reliability*. (A) A single pair connector shall be capable of making a minimum of 100 successive connections to binding post terminals without the occurrence of an open circuit. The test shall include terminals along the matrix edge, center, top, and bottom.

(B) A multi-pair test connector shall be attached to the binding post terminal field and tests for opens between the binding post terminals and the test connector shall be conducted. All circuits must prove good. The test shall be repeated along the terminal matrix edges, center, top, and bottom.

(vi) *Service cycle reliability*. A torque indicating device or wrench with an accuracy of ±0.17 N-m (±1.5 lb-in.) or better shall be used to tighten the terminal screw or nut as appropriate to 1.7 N-m (15.0 lb-in.). The terminal nut or screw is then loosened and retightened to 1.7 N-m (15 lb-in.). After 50 repeated connections and disconnections, the terminal shall be placed in an environmental chamber at 95% RH where the temperature shall be cycled as indicated in Figure 3 in paragraph (e)(3)(ii) of this section for a duration of 72 hours. The terminal shall then be momentarily removed from the chamber and the test procedure repeated. After a total of 250 loosening and retightening cycles have accumulated, the terminal must be capable of withstanding a torque of 1.7 N-m (15 lb-in.).

(4) *Dielectric strength*. All housing components in the vicinity of unsheathed field cable conductors, unsheathed housing stub cable or harness conductors, terminals, or cross-connect wire paths shall have a minimum dielectric strength of 500 volts ac to the cabinet grounding and bonding bracket. Dielectric strength is tested by connecting one lead from a 500-volt ac at 0.5 ampere source to the cabinet ground connector and the other lead is passed along the surfaces of all cabinet components in the vicinity of unsheathed cable or harness conductors, cross-connect wire paths, and in the splice area where unsheathed field cable conductors may be located. Sparkover constitutes failure.

(5) *Operational requirements*—(i) *Durability*. In order to verify the durability requirements while minimizing the number of test housings required to complete the test program, the binding posts selected for tests shall be separately identified and then checked to establish compliance after the various tests have been conducted.

(ii) Twenty-five jumper connections shall be made on each of two binding post connectors chosen at random from a representative sample in an assembled interface unit. After exposure to this test, these and adjacent connectors shall be inspected for damage such as cracks or chips in metal or plastic parts. Failure consists of structural damage, open circuits through the connector, or inability to pass the torsional, lateral loading, or axial pullout tests described in paragraphs (g)(3)(ii) through (g)(3)(iv) of this section.

(iii) Select six binding posts at random in a representative interface. On each connector, attach any test cord included with the unit and then remove the test cord as follows. On binding post sample 1, remove the cord normally ten times. On binding post sample 2, remove the cord ten times by jerking the test leads straight out. In these and the remaining tests, do this without releasing any manual attachment mechanisms. On sample 3, remove ten times by jerking downward at 45 °from horizontal; sample 4, upward at 45 °ten times; sample 5, left 45 °ten times; sample 6, right 45 °ten times. Check for opens and damage in the test cord, clips, and connectors. Failure consists of structural damage, open circuits through the connector, or inability of the terminal blocks to pass the torsional, lateral loading, axial pullout, test connector reliability, or dielectric strength tests described in paragraphs (g)(3)(ii) through (g)(3)(v)(B), and paragraph (g)(4) of this section.

(iv) Use craft tools such as scissors, diagonal cutters, and long nose pliers to loosen and tighten screws where the binding post design does not prohibit the possibility. Failure consists of severe structural damage.

(h) *Functional design criteria for insulation displacement type cross-connect modules used in SAI cabinets*—(1) *General description*. Cross-connect modules normally consist of multiple metallic contact elements that are retained by nonmetallic fixtures. The contact elements are spliced with permanent wire leads compatible for splicing to 22, 24, or 26 gauge cable on one side and configured for the acceptance of recurring connections and disconnections of plastic insulated cross-connect wire on the other side. Cross-connect modules are usually housed in a SAI cabinet. However, modules may receive limited usage in smaller pedestal-type housings and cabinets in the outside plant environment.

(2) *Design and fabrication requirements.* (i) All individual terminals or terminal fields must be enclosed and the terminal enclosures must be totally filled with an encapsulating grease or gel which prevents connection degradation caused by moisture and corrosion. The encapsulant must provide complete encapsulation of terminal metallic connections and surfaces and totally fill all voids and cavities within individual terminal enclosures or terminal field enclosures to prevent ingress of moisture. The encapsulant must not restrict access to the terminal or restrict craft personnel from making connections. The encapsulant must be compatible with the standard materials used in cross-connect hardware and wiring.

(ii) The cross-connect module manufacturer shall make available any nonstandard tools and test apparatus which are required for splicing, placing of jumpers, and the performance of maintenance operations.

(iii) The module shall be designed so that a typical outside plant technician using tools shall be able to terminate cross-connect wire on terminals, or to remove them without causing electrical shorts between any other terminals.

(iv) The pair count sequence terminated on a module shall be easily visible and shall have numerals of at least 0.25 cm (0.10 in.) in height.

(v) Feeder terminations shall be easily distinguished from distribution terminations.

(vi) Tip and ring terminations shall be easily visible and shall be identifiable as described in paragraph (f)(2)(vi) of this section.

(vii) The preferred locations for cross-connect modules to be mounted inside a housing is the same as those for terminals and are described in paragraphs (f)(2)(vii) and (f)(2)(viii) of this section.

(3) *Auxiliary features.* (i) Housings equipped with cross-connect modules shall be equipped with auxiliary test contacts as described in paragraphs (f)(3)(i) and (f)(3)(ii) of this section.

(ii) Special service markers shall be available for cross-connect modules as described in paragraph (f)(3)(iii) of this section.

(iii) Housings equipped with, or designed for, cross-connect modules shall contain a supply of cross-connect wire as described in paragraph (f)(3)(iv) of this section.

(i) *Performance criteria and test procedures for insulation displacement type cross-connect modules*—(1) *General.* Many of the tests described in this section require that the cross-connect module be installed in an appropriate housing in its typical field configuration for testing. Resistance measurements should be made with an electrical device which measures changes in resistance for each test parameter measured. The tests specified provide an indication of the stability of the electrical connections under the test conditions encountered.

(2) *Environmental requirements.* (i) A fully equipped arrangement of cross-connect modules having approximately 25 special service markers shall successfully complete environmental testing in accordance with paragraphs (e)(3) introductory text through (e)(3)(xiii)(B) of this section.

(ii) *Insulation resistance/high humidity and salt fog exposure.* Insulation resistance measurements shall not be less than 1×10^6 ohms when cross-connect modules are tested by a procedure similar to that described in paragraphs (g)(2)(i) introductory text through (g)(2)(i)(B) of this section.

(iii) *Insulation resistance/simulated rain exposure.* Insulation resistance measurements shall not be less than 1×10^6 ohms when cross-connect modules are tested by a procedure similar to that described in and paragraphs (g)(2)(ii) introductory text through (g)(2)(ii)(B) of this section.

(iv) *Contact resistance.* The change in contact resistance should not exceed 2 milliohms when cross-connect modules are tested by a procedure similar to that described in paragraphs (g)(2)(iii) introductory text through (g)(2)(iii)(B) of this section.

(v) *Fire resistance.* A housing fully equipped with cross-connect modules and jumper wiring shall be tested for fire resistance by a procedure similar to that described in paragraph (g)(2)(iv) of this section.

(vi) *Encapsulant material compatibility.* Cross-connect wire insulation and cross-connect hardware shall exhibit no visible material degradation when tested by the procedure described in paragraph (g)(2)(v) of this section.

(vii) *Encapsulant flow test.* The cross-connect contact encapsulant shall drip no more than 0.5 gram when tested by the procedure described in paragraph (g)(2)(vi) of this section.

(3) *Mechanical requirements*—(i) *Vibration.* A housing fully equipped with cross-connect modules shall be vibration tested in accordance with paragraph (g)(3)(i) of this section.

(ii) *Test connector reliability.* The test connectors supplied with housings intended for cross-connect modules shall successfully complete 100 successive connections as described in paragraphs (g)(3)(v) introductory text through (g)(3)(v)(B) of this section.

(iii) *Service cycle reliability.* A combination of multiple insertions of jumper wires, vibration, and temperature cycling shall be performed on cross-connect modules. The multiple insertions on approximately 100 connections shall be accomplished by 300 operations consisting of insertion, removal and reinsertion of new jumper wire. Contact resistance shall be measured and the final insertion of jumper wire shall not be removed from the connectors but must be subjected to vibration testing in accordance with paragraph (g)(3)(i) of this section and temperature cycled as indicated in Figure 3 in paragraph (e)(3)(ii) of this section for a duration of 72 hours. After vibration and temperature cycling, the average change in contact resistance shall be no greater than 2 milliohms.

(iv) *Jumper wire pull-out resistance.* Test modules that have received no prior conditioning shall be equipped with 100 38 cm (15 in.) jumper connections of the gauges recommended for use with the module using the insertion tool recommended by the cross-connect module manufacturer. With the test samples suitably supported, wires from each sample shall be pulled, one at a time, by a tensile machine at a cross-head speed of 6 centimeters per minute (cm/min) (2.4 inches per minute (in./min)). Wires shall be pulled both perpendicular and parallel to the plane of the cross-connect field and shall withstand a load of at least 1.1 kg (2.5 lb) before pulling out.

(v) *Cable conductor pull-out resistance.* Test modules that have received no prior conditioning shall be equipped with 100 26, 24, and 22 AWG 38 cm (15 in.) cable conductors using the insertion tool recommended by the cross-connect module manufacturer. With the test samples suitably supported, conductors from each sample shall be pulled, one at a time, by a tensile machine at a cross-head speed of 6 cm/min (2.4 in./min). Wires shall be pulled both perpendicular and parallel to the plane of the face of the splice module and shall withstand a load of at least 1.1 kg (2.5 lb) before pulling out.

(4) *Electrical requirements*—(i) *Dielectric strength.* A housing fully equipped with cross-connect modules shall be tested for dielectric strength in accordance with (g)(4) of this section.

(ii) The dielectric strength of a contact within the cross-connect module to contacts on either side shall be tested. The module shall be tested in a dry environment with an ac power source capable of supplying 8 kv at a rate of increase of 500 volts per second, a circuit breaker to open at breakdown, and a voltmeter to record the breakdown potential. Cross-connect modules shall be prepared in accordance with industry accepted splicing techniques with leads trimmed to approximately 38 cm (15 in.). The dielectric strength of each contact to the contacts on either side shall have an average dielectric strength of approximately 5.0 kv.

(5) *Operational requirements*—(i) *Durability.* In order to verify the durability requirements while minimizing the number of test housings required to complete the test program, the contacts selected for tests shall be separately identified and then checked to establish compliance after the various tests have been conducted.

(ii) Twenty-five jumper connections shall be made on each of two contacts chosen at random from a representative sample in an assembled interface unit. After this test, these and surrounding contacts shall be inspected for damage such as cracks or chips in metal or plastic parts. Failure consists of structural damage; open circuits through the connector, or inability to pass the jumper wire pullout tests described in paragraph (i)(3)(iv) of this section.

(iii) Select six contacts at random in a representative interface. On each of these contacts attach any test cord included with the unit as specified under normal use of that cord and then remove the test cord as follows. On sample 1, remove the cord normally ten times. On sample 2, remove the clip ten times by jerking the test leads straight out. In these and the remaining tests, do this without releasing any manual attachment mechanisms. On sample 3, remove ten times by jerking downward at 45 °from horizontal; sample 4, upward 45 °ten times; sample 5, left 45 °ten times; sample 6, right 45 °ten times. Check for opens and damage in the test cord, clips, and cross-connect modules. Failure consists of structural damage, open circuits through the connector, or inability of module to pass the test connector reliability, jumper wire pullout, and dielectric strength tests described in paragraphs (i)(3)(ii), (i)(3)(iv), and (i)(4)(ii) of this section.

(j) *Packaging and identification requirements*—(1) *Product identification.* (i) Each housing, terminal block, or cross-connect module shall be permanently marked with the manufacturer's name or trade mark.

(ii) The date of manufacture, model number, serial number and RUS assigned designations shall be placed on a decal inside housings. The product identification nomenclature must correspond with the nomenclature used in the manufacturer's quality assurance program.

(2) *Packaging requirements.* (i) Buried plant housings shall be packaged securely in an environmentally safe container to prevent either deterioration or physical damage to the unit during shipment, handling and storage.

(ii) The product with all the necessary parts shall be shipped in one container unless significant advantages to the user can be obtained otherwise. Packaging of parts in the carton shall be such that the parts become available in the order in which they are needed. The package should be clearly marked as to which end to open. Packages shall be clearly labeled, and correspond to the names given in the instructions.

(iii) Products packed in shipping containers shall be cushioned, blocked, braced, and anchored to prevent movement and damage.

(iv) All products shall be secured to pallets with non-metallic strapping. The strapping and the manner employed shall be of sufficient quantity, width, and thickness to preclude failure during transit and handling.

(v) The use of shrink or stretch film to secure the load to the pallet is permitted. However, such film must be applied over the required strapping.

(vi) Containers that are too large or heavy to be palletized, such as crates, shall be shipped in their own containers. When practical, these containers shall be provided with skids to facilitate fork-lift handling.

(vii) When packaged, the outer cartons shall meet the requirements of the Uniform Freight Classification and the National Motor Freight Classification.

(3) *Container marking requirements.* (i) The package shall be readily identifiable as to the manufacturer, model number, date of manufacture, and serial number.

(ii) The RUS assigned housing designation shall be stamped or marked on the outside of the package container with letter and number sizes large enough for easy identification.

(iii) Each package shall be marked with its approximate gross weight.

(iv) All containers carrying delicate or fragile items shall be marked to clearly identify this condition.

(v) All marking shall be clear, legible, and as large as space permits.

(The information and recordkeeping requirements of this section have been approved by the Office of Management and Budget under control number 0572–0059)

[59 FR 53044, Oct. 21, 1994, as amended at 69 FR 18803, Apr. 9, 2004]

PART 1757—TELEPHONE SYSTEMS OPERATIONS AND MAINTENANCE [RESERVED]

PARTS 1758–1759 [RESERVED]

FINDING AIDS

A list of CFR titles, subtitles, chapters, subchapters and parts and an alphabetical list of agencies publishing in the CFR are included in the CFR Index and Finding Aids volume to the Code of Federal Regulations which is published separately and revised annually.

Table of CFR Titles and Chapters

(Revised as of January 1, 2022)

Title 1—General Provisions

Title 2—Grants and Agreements

Title 2—Grants and Agreements—Continued

Title 3—The President

Title 4—Accounts

Title 5—Administrative Personnel

Title 6—Domestic Security

Title 7—Agriculture

Chap.	
IX	Agricultural Marketing Service (Marketing Agreements and Orders; Fruits, Vegetables, Nuts), Department of Agriculture (Parts 900—999)
X	Agricultural Marketing Service (Marketing Agreements and Orders; Milk), Department of Agriculture (Parts 1000—1199)
XI	Agricultural Marketing Service (Marketing Agreements and Orders; Miscellaneous Commodities), Department of Agriculture (Parts 1200—1299)
XIV	Commodity Credit Corporation, Department of Agriculture (Parts 1400—1499)
XV	Foreign Agricultural Service, Department of Agriculture (Parts 1500—1599)
XVI	[Reserved]
XVII	Rural Utilities Service, Department of Agriculture (Parts 1700—1799)
XVIII	Rural Housing Service, Rural Business-Cooperative Service, Rural Utilities Service, and Farm Service Agency, Department of Agriculture (Parts 1800—2099)
XX	[Reserved]
XXV	Office of Advocacy and Outreach, Department of Agriculture (Parts 2500—2599)
XXVI	Office of Inspector General, Department of Agriculture (Parts 2600—2699)
XXVII	Office of Information Resources Management, Department of Agriculture (Parts 2700—2799)
XXVIII	Office of Operations, Department of Agriculture (Parts 2800—2899)
XXIX	Office of Energy Policy and New Uses, Department of Agriculture (Parts 2900—2999)
XXX	Office of the Chief Financial Officer, Department of Agriculture (Parts 3000—3099)
XXXI	Office of Environmental Quality, Department of Agriculture (Parts 3100—3199)
XXXII	Office of Procurement and Property Management, Department of Agriculture (Parts 3200—3299)
XXXIII	Office of Transportation, Department of Agriculture (Parts 3300—3399)
XXXIV	National Institute of Food and Agriculture (Parts 3400—3499)
XXXV	Rural Housing Service, Department of Agriculture (Parts 3500—3599)
XXXVI	National Agricultural Statistics Service, Department of Agriculture (Parts 3600—3699)
XXXVII	Economic Research Service, Department of Agriculture (Parts 3700—3799)
XXXVIII	World Agricultural Outlook Board, Department of Agriculture (Parts 3800—3899)
XLI	[Reserved]
XLII	Rural Business-Cooperative Service and Rural Utilities Service, Department of Agriculture (Parts 4200—4299)

Title 7—Agriculture—Continued

Title 8—Aliens and Nationality

Title 9—Animals and Animal Products

Title 10—Energy

Title 11—Federal Elections

Title 12—Banks and Banking

Title 12—Banks and Banking—Continued

Title 13—Business Credit and Assistance

Title 14—Aeronautics and Space

Title 15—Commerce and Foreign Trade

Title 15—Commerce and Foreign Trade—Continued

Title 16—Commercial Practices

Title 17—Commodity and Securities Exchanges

Title 18—Conservation of Power and Water Resources

Title 19—Customs Duties

Title 20—Employees' Benefits

Chap.

Chap.	
I	Office of Workers' Compensation Programs, Department of Labor (Parts 1—199)
II	Railroad Retirement Board (Parts 200—399)
III	Social Security Administration (Parts 400—499)
IV	Employees' Compensation Appeals Board, Department of Labor (Parts 500—599)
V	Employment and Training Administration, Department of Labor (Parts 600—699)
VI	Office of Workers' Compensation Programs, Department of Labor (Parts 700—799)
VII	Benefits Review Board, Department of Labor (Parts 800—899)
VIII	Joint Board for the Enrollment of Actuaries (Parts 900—999)
IX	Office of the Assistant Secretary for Veterans' Employment and Training Service, Department of Labor (Parts 1000—1099)

Title 21—Food and Drugs

I	Food and Drug Administration, Department of Health and Human Services (Parts 1—1299)
II	Drug Enforcement Administration, Department of Justice (Parts 1300—1399)
III	Office of National Drug Control Policy (Parts 1400—1499)

Title 22—Foreign Relations

I	Department of State (Parts 1—199)
II	Agency for International Development (Parts 200—299)
III	Peace Corps (Parts 300—399)
IV	International Joint Commission, United States and Canada (Parts 400—499)
V	United States Agency for Global Media (Parts 500—599)
VII	U.S. International Development Finance Corporation (Parts 700—799)
IX	Foreign Service Grievance Board (Parts 900—999)
X	Inter-American Foundation (Parts 1000—1099)
XI	International Boundary and Water Commission, United States and Mexico, United States Section (Parts 1100—1199)
XII	United States International Development Cooperation Agency (Parts 1200—1299)
XIII	Millennium Challenge Corporation (Parts 1300—1399)
XIV	Foreign Service Labor Relations Board; Federal Labor Relations Authority; General Counsel of the Federal Labor Relations Authority; and the Foreign Service Impasse Disputes Panel (Parts 1400—1499)
XV	African Development Foundation (Parts 1500—1599)
XVI	Japan-United States Friendship Commission (Parts 1600—1699)
XVII	United States Institute of Peace (Parts 1700—1799)

Title 23—Highways

Title 24—Housing and Urban Development

Title 24—Housing and Urban Development—Continued

Chap.

XX Office of Assistant Secretary for Housing—Federal Housing Commissioner, Department of Housing and Urban Development (Parts 3200—3899)

XXIV Board of Directors of the HOPE for Homeowners Program (Parts 4000—4099) [Reserved]

XXV Neighborhood Reinvestment Corporation (Parts 4100—4199)

Title 25—Indians

I Bureau of Indian Affairs, Department of the Interior (Parts 1—299)

II Indian Arts and Crafts Board, Department of the Interior (Parts 300—399)

III National Indian Gaming Commission, Department of the Interior (Parts 500—599)

IV Office of Navajo and Hopi Indian Relocation (Parts 700—899)

V Bureau of Indian Affairs, Department of the Interior, and Indian Health Service, Department of Health and Human Services (Part 900—999)

VI Office of the Assistant Secretary, Indian Affairs, Department of the Interior (Parts 1000—1199)

VII Office of the Special Trustee for American Indians, Department of the Interior (Parts 1200—1299)

Title 26—Internal Revenue

I Internal Revenue Service, Department of the Treasury (Parts 1—End)

Title 27—Alcohol, Tobacco Products and Firearms

I Alcohol and Tobacco Tax and Trade Bureau, Department of the Treasury (Parts 1—399)

II Bureau of Alcohol, Tobacco, Firearms, and Explosives, Department of Justice (Parts 400—799)

Title 28—Judicial Administration

I Department of Justice (Parts 0—299)

III Federal Prison Industries, Inc., Department of Justice (Parts 300—399)

V Bureau of Prisons, Department of Justice (Parts 500—599)

VI Offices of Independent Counsel, Department of Justice (Parts 600—699)

VII Office of Independent Counsel (Parts 700—799)

VIII Court Services and Offender Supervision Agency for the District of Columbia (Parts 800—899)

IX National Crime Prevention and Privacy Compact Council (Parts 900—999)

Title 29—Labor

Title 30—Mineral Resources

Title 31—Money and Finance: Treasury

Title 32—National Defense

Title 33—Navigation and Navigable Waters

Title 34—Education

Chap.

SUBTITLE A—OFFICE OF THE SECRETARY, DEPARTMENT OF EDUCATION (PARTS 1—99)

SUBTITLE B—REGULATIONS OF THE OFFICES OF THE DEPARTMENT OF EDUCATION

I Office for Civil Rights, Department of Education (Parts 100—199)

II Office of Elementary and Secondary Education, Department of Education (Parts 200—299)

III Office of Special Education and Rehabilitative Services, Department of Education (Parts 300—399)

IV Office of Career, Technical, and Adult Education, Department of Education (Parts 400—499)

V Office of Bilingual Education and Minority Languages Affairs, Department of Education (Parts 500—599) [Reserved]

VI Office of Postsecondary Education, Department of Education (Parts 600—699)

VII Office of Educational Research and Improvement, Department of Education (Parts 700—799) [Reserved]

SUBTITLE C—REGULATIONS RELATING TO EDUCATION

XI [Reserved]

XII National Council on Disability (Parts 1200—1299)

Title 35 [Reserved]

Title 36—Parks, Forests, and Public Property

I National Park Service, Department of the Interior (Parts 1—199)

II Forest Service, Department of Agriculture (Parts 200—299)

III Corps of Engineers, Department of the Army (Parts 300—399)

IV American Battle Monuments Commission (Parts 400—499)

V Smithsonian Institution (Parts 500—599)

VI [Reserved]

VII Library of Congress (Parts 700—799)

VIII Advisory Council on Historic Preservation (Parts 800—899)

IX Pennsylvania Avenue Development Corporation (Parts 900—999)

X Presidio Trust (Parts 1000—1099)

XI Architectural and Transportation Barriers Compliance Board (Parts 1100—1199)

XII National Archives and Records Administration (Parts 1200—1299)

XV Oklahoma City National Memorial Trust (Parts 1500—1599)

XVI Morris K. Udall Scholarship and Excellence in National Environmental Policy Foundation (Parts 1600—1699)

Title 37—Patents, Trademarks, and Copyrights

I United States Patent and Trademark Office, Department of Commerce (Parts 1—199)

II U.S. Copyright Office, Library of Congress (Parts 200—299)

Title 37—Patents, Trademarks, and Copyrights—Continued

Title 38—Pensions, Bonuses, and Veterans' Relief

Title 39—Postal Service

Title 40—Protection of Environment

Title 41—Public Contracts and Property Management

Title 41—Public Contracts and Property Management—Continued

Title 42—Public Health

Title 43—Public Lands: Interior

Title 44—Emergency Management and Assistance

Title 45—Public Welfare

Title 46—Shipping

Title 47—Telecommunication

Chap.

Chap.	
I	Federal Communications Commission (Parts 0—199)
II	Office of Science and Technology Policy and National Security Council (Parts 200—299)
III	National Telecommunications and Information Administration, Department of Commerce (Parts 300—399)
IV	National Telecommunications and Information Administration, Department of Commerce, and National Highway Traffic Safety Administration, Department of Transportation (Parts 400—499)
V	The First Responder Network Authority (Parts 500—599)

Title 48—Federal Acquisition Regulations System

1	Federal Acquisition Regulation (Parts 1—99)
2	Defense Acquisition Regulations System, Department of Defense (Parts 200—299)
3	Department of Health and Human Services (Parts 300—399)
4	Department of Agriculture (Parts 400—499)
5	General Services Administration (Parts 500—599)
6	Department of State (Parts 600—699)
7	Agency for International Development (Parts 700—799)
8	Department of Veterans Affairs (Parts 800—899)
9	Department of Energy (Parts 900—999)
10	Department of the Treasury (Parts 1000—1099)
12	Department of Transportation (Parts 1200—1299)
13	Department of Commerce (Parts 1300—1399)
14	Department of the Interior (Parts 1400—1499)
15	Environmental Protection Agency (Parts 1500—1599)
16	Office of Personnel Management Federal Employees Health Benefits Acquisition Regulation (Parts 1600—1699)
17	Office of Personnel Management (Parts 1700—1799)
18	National Aeronautics and Space Administration (Parts 1800—1899)
19	Broadcasting Board of Governors (Parts 1900—1999)
20	Nuclear Regulatory Commission (Parts 2000—2099)
21	Office of Personnel Management, Federal Employees Group Life Insurance Federal Acquisition Regulation (Parts 2100—2199)
23	Social Security Administration (Parts 2300—2399)
24	Department of Housing and Urban Development (Parts 2400—2499)
25	National Science Foundation (Parts 2500—2599)
28	Department of Justice (Parts 2800—2899)
29	Department of Labor (Parts 2900—2999)
30	Department of Homeland Security, Homeland Security Acquisition Regulation (HSAR) (Parts 3000—3099)
34	Department of Education Acquisition Regulation (Parts 3400—3499)

Title 49—Transportation

Title 50—Wildlife and Fisheries

Alphabetical List of Agencies Appearing in the CFR

(Revised as of January 1, 2022)

Agency	CFR Title, Subtitle or Chapter
Administrative Conference of the United States	1, III
Advisory Council on Historic Preservation	36, VIII
Advocacy and Outreach, Office of	7, XXV
Afghanistan Reconstruction, Special Inspector General for	5, LXXXIII
African Development Foundation	22, XV
Federal Acquisition Regulation	48, 57
Agency for International Development	2, VII; 22, II
Federal Acquisition Regulation	48, 7
Agricultural Marketing Service	7, I, VIII, IX, X, XI; 9, II
Agricultural Research Service	7, V
Agriculture, Department of	2, IV; 5, LXXIII
Advocacy and Outreach, Office of	7, XXV
Agricultural Marketing Service	7, I, VIII, IX, X, XI; 9, II
Agricultural Research Service	7, V
Animal and Plant Health Inspection Service	7, III; 9, I
Chief Financial Officer, Office of	7, XXX
Commodity Credit Corporation	7, XIV
Economic Research Service	7, XXXVII
Energy Policy and New Uses, Office of	2, IX; 7, XXIX
Environmental Quality, Office of	7, XXXI
Farm Service Agency	7, VII, XVIII
Federal Acquisition Regulation	48, 4
Federal Crop Insurance Corporation	7, IV
Food and Nutrition Service	7, II
Food Safety and Inspection Service	9, III
Foreign Agricultural Service	7, XV
Forest Service	36, II
Information Resources Management, Office of	7, XXVII
Inspector General, Office of	7, XXVI
National Agricultural Library	7, XLI
National Agricultural Statistics Service	7, XXXVI
National Institute of Food and Agriculture	7, XXXIV
Natural Resources Conservation Service	7, VI
Operations, Office of	7, XXVIII
Procurement and Property Management, Office of	7, XXXII
Rural Business-Cooperative Service	7, XVIII, XLII
Rural Development Administration	7, XLII
Rural Housing Service	7, XVIII, XXXV
Rural Utilities Service	7, XVII, XVIII, XLII
Secretary of Agriculture, Office of	7, Subtitle A
Transportation, Office of	7, XXXIII
World Agricultural Outlook Board	7, XXXVIII
Air Force, Department of	32, VII
Federal Acquisition Regulation Supplement	48, 53
Air Transportation Stabilization Board	14, VI
Alcohol and Tobacco Tax and Trade Bureau	27, I
Alcohol, Tobacco, Firearms, and Explosives, Bureau of	27, II
AMTRAK	49, VII
American Battle Monuments Commission	36, IV
American Indians, Office of the Special Trustee	25, VII
Animal and Plant Health Inspection Service	7, III; 9, I
Appalachian Regional Commission	5, IX
Architectural and Transportation Barriers Compliance Board	36, XI

Agency	CFR Title, Subtitle or Chapter
Arctic Research Commission	45, XXIII
Armed Forces Retirement Home	5, XI; 38, II
Army, Department of	32, V
Engineers, Corps of	33, II; 36, III
Federal Acquisition Regulation	48, 51
Benefits Review Board	20, VII
Bilingual Education and Minority Languages Affairs, Office of	34, V
Blind or Severely Disabled, Committee for Purchase from People Who Are	41, 51
Federal Acquisition Regulation	48, 19
Career, Technical, and Adult Education, Office of	34, IV
Census Bureau	15, I
Centers for Medicare & Medicaid Services	42, IV
Central Intelligence Agency	32, XIX
Chemical Safety and Hazard Investigation Board	40, VI
Chief Financial Officer, Office of	7, XXX
Child Support Enforcement, Office of	45, III
Children and Families, Administration for	45, II, III, IV, X, XIII
Civil Rights, Commission on	5, LXVIII; 45, VII
Civil Rights, Office for	34, I
Coast Guard	33, I; 46, I; 49, IV
Coast Guard (Great Lakes Pilotage)	46, III
Commerce, Department of	2, XIII; 44, IV; 50, VI
Census Bureau	15, I
Economic Affairs, Office of the Under-Secretary for	15, XV
Economic Analysis, Bureau of	15, VIII
Economic Development Administration	13, III
Emergency Management and Assistance	44, IV
Federal Acquisition Regulation	48, 13
Foreign-Trade Zones Board	15, IV
Industry and Security, Bureau of	15, VII
International Trade Administration	15, III; 19, III
National Institute of Standards and Technology	15, II; 37, IV
National Marine Fisheries Service	50, II, IV
National Oceanic and Atmospheric Administration	15, IX; 50, II, III, IV, VI
National Technical Information Service	15, XI
National Telecommunications and Information Administration	15, XXIII; 47, III, IV
National Weather Service	15, IX
Patent and Trademark Office, United States	37, I
Secretary of Commerce, Office of	15, Subtitle A
Commercial Space Transportation	14, III
Commodity Credit Corporation	7, XIV
Commodity Futures Trading Commission	5, XLI; 17, I
Community Planning and Development, Office of Assistant Secretary for	24, V, VI
Community Services, Office of	45, X
Comptroller of the Currency	12, I
Construction Industry Collective Bargaining Commission	29, IX
Consumer Financial Protection Bureau	5, LXXXIV; 12, X
Consumer Product Safety Commission	5, LXXI; 16, II
Copyright Royalty Board	37, III
Corporation for National and Community Service	2, XXII; 45, XII, XXV
Cost Accounting Standards Board	48, 99
Council on Environmental Quality	40, V
Council of the Inspectors General on Integrity and Efficiency	5, XCVIII
Court Services and Offender Supervision Agency for the District of Columbia	5, LXX; 28, VIII
Customs and Border Protection	19, I
Defense, Department of	2, XI; 5, XXVI; 32, Subtitle A; 40, VII
Advanced Research Projects Agency	32, I
Air Force Department	32, VII
Army Department	32, V; 33, II; 36, III; 48, 51
Defense Acquisition Regulations System	48, 2
Defense Intelligence Agency	32, I

Agency	CFR Title, Subtitle or Chapter
Defense Logistics Agency	32, I, XII; 48, 54
Engineers, Corps of	33, II; 36, III
National Imagery and Mapping Agency	32, I
Navy, Department of	32, VI; 48, 52
Secretary of Defense, Office of	2, XI; 32, I
Defense Contract Audit Agency	32, I
Defense Intelligence Agency	32, I
Defense Logistics Agency	32, XII; 48, 54
Defense Nuclear Facilities Safety Board	10, XVII
Delaware River Basin Commission	18, III
Denali Commission	45, IX
Disability, National Council on	5, C; 34, XII
District of Columbia, Court Services and Offender Supervision Agency for the	5, LXX; 28, VIII
Drug Enforcement Administration	21, II
East-West Foreign Trade Board	15, XIII
Economic Affairs, Office of the Under-Secretary for	15, XV
Economic Analysis, Bureau of	15, VIII
Economic Development Administration	13, III
Economic Research Service	7, XXXVII
Education, Department of	2, XXXIV; 5, LIII
Bilingual Education and Minority Languages Affairs, Office of	34, V
Career, Technical, and Adult Education, Office of	34, IV
Civil Rights, Office for	34, I
Educational Research and Improvement, Office of	34, VII
Elementary and Secondary Education, Office of	34, II
Federal Acquisition Regulation	48, 34
Postsecondary Education, Office of	34, VI
Secretary of Education, Office of	34, Subtitle A
Special Education and Rehabilitative Services, Office of	34, III
Educational Research and Improvement, Office of	34, VII
Election Assistance Commission	2, LVIII; 11, II
Elementary and Secondary Education, Office of	34, II
Emergency Oil and Gas Guaranteed Loan Board	13, V
Emergency Steel Guarantee Loan Board	13, IV
Employee Benefits Security Administration	29, XXV
Employees' Compensation Appeals Board	20, IV
Employees Loyalty Board	5, V
Employment and Training Administration	20, V
Employment Policy, National Commission for	1, IV
Employment Standards Administration	20, VI
Endangered Species Committee	50, IV
Energy, Department of	2, IX; 5, XXIII; 10, II, III, X
Federal Acquisition Regulation	48, 9
Federal Energy Regulatory Commission	5, XXIV; 18, I
Property Management Regulations	41, 109
Energy, Office of	7, XXIX
Engineers, Corps of	33, II; 36, III
Engraving and Printing, Bureau of	31, VI
Environmental Protection Agency	2, XV; 5, LIV; 40, I, IV, VII
Federal Acquisition Regulation	48, 15
Property Management Regulations	41, 115
Environmental Quality, Office of	7, XXXI
Equal Employment Opportunity Commission	5, LXII; 29, XIV
Equal Opportunity, Office of Assistant Secretary for	24, I
Executive Office of the President	3, I
Environmental Quality, Council on	40, V
Management and Budget, Office of	2, Subtitle A; 5, III, LXXVII; 14, VI; 48, 99
National Drug Control Policy, Office of	2, XXXVI; 21, III
National Security Council	32, XXI; 47, II
Presidential Documents	3
Science and Technology Policy, Office of	32, XXIV; 47, II
Trade Representative, Office of the United States	15, XX

Agency	CFR Title, Subtitle or Chapter
Federal Property Management Regulations	41, 101
Federal Travel Regulation System	41, Subtitle F
General	41, 300
Payment From a Non-Federal Source for Travel Expenses	41, 304
Payment of Expenses Connected With the Death of Certain Employees	41, 303
Relocation Allowances	41, 302
Temporary Duty (TDY) Travel Allowances	41, 301
Geological Survey	30, IV
Government Accountability Office	4, I
Government Ethics, Office of	5, XVI
Government National Mortgage Association	24, III
Grain Inspection, Packers and Stockyards Administration	7, VIII; 9, II
Great Lakes St. Lawrence Seaway Development Corporation	33, IV
Gulf Coast Ecosystem Restoration Council	2, LIX; 40, VIII
Harry S. Truman Scholarship Foundation	45, XVIII
Health and Human Services, Department of	2, III; 5, XLV; 45, Subtitle A
Centers for Medicare & Medicaid Services	42, IV
Child Support Enforcement, Office of	45, III
Children and Families, Administration for	45, II, III, IV, X, XIII
Community Services, Office of	45, X
Family Assistance, Office of	45, II
Federal Acquisition Regulation	48, 3
Food and Drug Administration	21, I
Indian Health Service	25, V
Inspector General (Health Care), Office of	42, V
Public Health Service	42, I
Refugee Resettlement, Office of	45, IV
Homeland Security, Department of	2, XXX; 5, XXXVI; 6, I; 8, I
Coast Guard	33, I; 46, I; 49, IV
Coast Guard (Great Lakes Pilotage)	46, III
Customs and Border Protection	19, I
Federal Emergency Management Agency	44, I
Human Resources Management and Labor Relations Systems	5, XCVII
Immigration and Customs Enforcement Bureau	19, IV
Transportation Security Administration	49, XII
HOPE for Homeowners Program, Board of Directors of	24, XXIV
Housing, Office of, and Multifamily Housing Assistance Restructuring, Office of	24, IV
Housing and Urban Development, Department of	2, XXIV; 5, LXV; 24, Subtitle B
Community Planning and Development, Office of Assistant Secretary for	24, V, VI
Equal Opportunity, Office of Assistant Secretary for	24, I
Federal Acquisition Regulation	48, 24
Federal Housing Enterprise Oversight, Office of	12, XVII
Government National Mortgage Association	24, III
Housing—Federal Housing Commissioner, Office of Assistant Secretary for	24, II, VIII, X, XX
Housing, Office of, and Multifamily Housing Assistance Restructuring, Office of	24, IV
Inspector General, Office of	24, XII
Public and Indian Housing, Office of Assistant Secretary for	24, IX
Secretary, Office of	24, Subtitle A, VII
Housing—Federal Housing Commissioner, Office of Assistant Secretary for	24, II, VIII, X, XX
Housing, Office of, and Multifamily Housing Assistance Restructuring, Office of	24, IV
Immigration and Customs Enforcement Bureau	19, IV
Immigration Review, Executive Office for	8, V
Independent Counsel, Office of	28, VII
Independent Counsel, Offices of	28, VI
Indian Affairs, Bureau of	25, I, V
Indian Affairs, Office of the Assistant Secretary	25, VI

Agency	CFR Title, Subtitle or Chapter
Indian Arts and Crafts Board	25, II
Indian Health Service	25, V
Industry and Security, Bureau of	15, VII
Information Resources Management, Office of	7, XXVII
Information Security Oversight Office, National Archives and Records Administration	32, XX
Inspector General	
Agriculture Department	7, XXVI
Health and Human Services Department	42, V
Housing and Urban Development Department	24, XII, XV
Institute of Peace, United States	22, XVII
Inter-American Foundation	5, LXIII; 22, X
Interior, Department of	2, XIV
American Indians, Office of the Special Trustee	25, VII
Endangered Species Committee	50, IV
Federal Acquisition Regulation	48, 14
Federal Property Management Regulations System	41, 114
Fish and Wildlife Service, United States	50, I, IV
Geological Survey	30, IV
Indian Affairs, Bureau of	25, I, V
Indian Affairs, Office of the Assistant Secretary	25, VI
Indian Arts and Crafts Board	25, II
Land Management, Bureau of	43, II
National Indian Gaming Commission	25, III
National Park Service	36, I
Natural Resource Revenue, Office of	30, XII
Ocean Energy Management, Bureau of	30, V
Reclamation, Bureau of	43, I
Safety and Environmental Enforcement, Bureau of	30, II
Secretary of the Interior, Office of	2, XIV; 43, Subtitle A
Surface Mining Reclamation and Enforcement, Office of	30, VII
Internal Revenue Service	26, I
International Boundary and Water Commission, United States and Mexico, United States Section	22, XI
International Development, United States Agency for	22, II
Federal Acquisition Regulation	48, 7
International Development Cooperation Agency, United States	22, XII
International Development Finance Corporation, U.S.	5, XXXIII; 22, VII
International Joint Commission, United States and Canada	22, IV
International Organizations Employees Loyalty Board	5, V
International Trade Administration	15, III; 19, III
International Trade Commission, United States	19, II
Interstate Commerce Commission	5, XL
Investment Security, Office of	31, VIII
James Madison Memorial Fellowship Foundation	45, XXIV
Japan–United States Friendship Commission	22, XVI
Joint Board for the Enrollment of Actuaries	20, VIII
Justice, Department of	2, XXVIII; 5, XXVIII; 28, I, XI; 40, IV
Alcohol, Tobacco, Firearms, and Explosives, Bureau of	27, II
Drug Enforcement Administration	21, II
Federal Acquisition Regulation	48, 28
Federal Claims Collection Standards	31, IX
Federal Prison Industries, Inc.	28, III
Foreign Claims Settlement Commission of the United States	45, V
Immigration Review, Executive Office for	8, V
Independent Counsel, Offices of	28, VI
Prisons, Bureau of	28, V
Property Management Regulations	41, 128
Labor, Department of	2, XXIX; 5, XLII
Benefits Review Board	20, VII
Employee Benefits Security Administration	29, XXV
Employees' Compensation Appeals Board	20, IV
Employment and Training Administration	20, V
Federal Acquisition Regulation	48, 29

Agency	CFR Title, Subtitle or Chapter
National Technical Information Service	15, XI
National Telecommunications and Information Administration	15, XXIII; 47, III, IV, V
National Transportation Safety Board	49, VIII
Natural Resource Revenue, Office of	30, XII
Natural Resources Conservation Service	7, VI
Navajo and Hopi Indian Relocation, Office of	25, IV
Navy, Department of	32, VI
Federal Acquisition Regulation	48, 52
Neighborhood Reinvestment Corporation	24, XXV
Northeast Interstate Low-Level Radioactive Waste Commission	10, XVIII
Nuclear Regulatory Commission	2, XX; 5, XLVIII; 10, I
Federal Acquisition Regulation	48, 20
Occupational Safety and Health Administration	29, XVII
Occupational Safety and Health Review Commission	29, XX
Ocean Energy Management, Bureau of	30, V
Oklahoma City National Memorial Trust	36, XV
Operations Office	7, XXVIII
Patent and Trademark Office, United States	37, I
Payment From a Non-Federal Source for Travel Expenses	41, 304
Payment of Expenses Connected With the Death of Certain Employees	41, 303
Peace Corps	2, XXXVII; 22, III
Pennsylvania Avenue Development Corporation	36, IX
Pension Benefit Guaranty Corporation	29, XL
Personnel Management, Office of	5, I, IV, XXXV; 45, VIII
Federal Acquisition Regulation	48, 17
Federal Employees Group Life Insurance Federal Acquisition Regulation	48, 21
Federal Employees Health Benefits Acquisition Regulation	48, 16
Human Resources Management and Labor Relations Systems, Department of Homeland Security	5, XCVII
Pipeline and Hazardous Materials Safety Administration	49, I
Postal Regulatory Commission	5, XLVI; 39, III
Postal Service, United States	5, LX; 39, I
Postsecondary Education, Office of	34, VI
President's Commission on White House Fellowships	1, IV
Presidential Documents	3
Presidio Trust	36, X
Prisons, Bureau of	28, V
Privacy and Civil Liberties Oversight Board	6, X
Procurement and Property Management, Office of	7, XXXII
Public and Indian Housing, Office of Assistant Secretary for	24, IX
Public Contracts, Department of Labor	41, 50
Public Health Service	42, I
Railroad Retirement Board	20, II
Reclamation, Bureau of	43, I
Refugee Resettlement, Office of	45, IV
Relocation Allowances	41, 302
Research and Innovative Technology Administration	49, XI
Rural Business-Cooperative Service	7, XVIII, XLII
Rural Development Administration	7, XLII
Rural Housing Service	7, XVIII, XXXV
Rural Utilities Service	7, XVII, XVIII, XLII
Safety and Environmental Enforcement, Bureau of	30, II
Science and Technology Policy, Office of	32, XXIV; 47, II
Secret Service	31, IV
Securities and Exchange Commission	5, XXXIV; 17, II
Selective Service System	32, XVI
Small Business Administration	2, XXVII; 13, I
Smithsonian Institution	36, V
Social Security Administration	2, XXIII; 20, III; 48, 23
Soldiers' and Airmen's Home, United States	5, XI
Special Counsel, Office of	5, VIII
Special Education and Rehabilitative Services, Office of	34, III
State, Department of	2, VI; 22, I; 28, XI

List of CFR Sections Affected

All changes in this volume of the Code of Federal Regulations (CFR) that were made by documents published in the FEDERAL REGISTER since January 1, 2017 are enumerated in the following list. Entries indicate the nature of the changes effected. Page numbers refer to FEDERAL REGISTER pages. The user should consult the entries for chapters, parts and subparts as well as sections for revisions.

For changes to this volume of the CFR prior to this listing, consult the annual edition of the monthly List of CFR Sections Affected (LSA). The LSA is available at *www.govinfo.gov.* For changes to this volume of the CFR prior to 2001, see the "List of CFR Sections Affected, 1949–1963, 1964–1972, 1973–1985, and 1986–2000" published in 11 separate volumes. The "List of CFR Sections Affected 1986–2000" is available at *www.govinfo.gov.*

2017

7 CFR

82 FR Page

Chapter XVII

1703 Authority citation revised.....55925
1703.100—1703.119 (Subpart D) Removed.....55925
1703.120—1703.129 (Subpart E) Removed.....55925
1703.130—1703.139 (Subpart F) Removed.....55925
1703.140—1703.147 (Subpart G) Removed.....55925
1703.300 Revised.....55925
1734 Added.....55925
1735.30 (d)(1)(v) revised.....55939

2018

7 CFR

83 FR Page

Chapter XVII

1709.3 Amended.....45032
1709.114 Revised.....45032
1709.121 (d) revised.....45033
1728.97 (a)(24) revised.....55467
1728.98 (a)(24) revised.....55467
1734 Regulation at 82 FR 55925 confirmed.....10357
1739.1 (a) amended.....45033
1739.2 Revised.....45033
1739.3 Amended.....45033
1739.15 Introductory text revised.....45033

7 CFR—Continued

83 FR Page

Chapter XVII—Continued

1739.16 (a) amended.....45033
1739.17 (d)(7) revised.....45033

2019

7 CFR

84 FR Page

Chapter XVI

Chapter XVI Removed; eff. 1-6-2020.....59920

Chapter XVII

1700.1 (b) revised; (c) removed; eff. 1-6-2020.....59920
1700.25 Amended; eff. 1-6-2020.....59920
1700.29 Introductory text revised; eff. 1-6-2020.....59920
1710.2 (a) amended.....32610
1710.2 Regulation at 84 FR 32610 eff. date revised.....33163
1710.106 (c)(4) removed.....32610
1710.106 Regulation at 84 FR 32610 eff. date revised.....33163
1710.112 (c) removed.....32610
1710.112 Regulation at 84 FR 32610 eff. date revised.....33163
1710.154 Added.....32610
1710.154 Regulation at 84 FR 32610 eff. date revised.....33163
1710.202 Revised.....32610
1710.202 Regulation at 84 FR 32610 eff. date revised.....33163
1710.203 Revised.....32610

7 CFR—Continued

84 FR Page

7 CFR—Continued

84 FR Page

7 CFR—Continued

84 FR Page

Chapter XVII—Continued

1724.54 (c)(1) and (d)(1)(i) introductory text revised....32617
1724.54 Regulation at 84 FR 32617 eff. date revised....33163
1724.55 (a)(7) revised....32617
1724.55 Regulation at 84 FR 32617 eff. date revised....33163
1726.35 (c)(5) and (e)(3)(vi)(A) removed; (c)(6), (7), and (e)(3)(vi)(B) through (E) redesignated as (c)(5), (6), and (e)(3)(vi)(A) through (D); (d) and (e)(4)(iii)(E)(*1*) revised....32617
1726.35 Regulation at 84 FR 32617 eff. date revised....33163
1726.403 (d)(2)(ii) revised....32618
1726.403 Regulation at 84 FR 32618 eff. date revised....33163
1728.97 (g)(1) redesignated as (g)(3); (d)(1) and (2) redesignated as new (g)(1) and (2); (d) removed; (e) through (i) redesignated as new (d) through (h); introductory text, (a) introductory text, (21), (b), (c), new (d) introductory text, (e), (f) introductory text, (g), and (h) revised....28190
1728.98 (a)(21) revised....28191
1728.201 Revised....28191
1728.202 Revised....28198
1728.203 Added....28200
1730.20—1730.59 (Subpart B) Appendix A amended....32618
1730.20—1730.59 (Subpart B) Regulation at 84 FR 32618 eff. date revised....33163
1735.2 Amended; eff. 1-6-2020....59920
1735.10 (b) and (e) revised; eff. 1-6-2020....59921
1735.17 (b) revised; eff. 1-6-2020....59921
1735.30 (b)(2), (f), and (g) revised; eff. 1-6-2020....59921
1735.31 Heading, (a) introductory text, (b), (c) introductory text, (d), (e), and (f) revised; eff. 1-6-2020....59921
1735.32 (a) revised; eff. 1-6-2020....59921
1735.43 (b) through (e) revised; (f) removed; eff. 1-6-2020....59921
1735.44 Revised; eff. 1-6-2020....59922
1735.46 (f) revised; eff. 1-6-2020....59922
1735.47 (a)(2) revised; eff. 1-6-2020....59922
1737.2 Amended; eff. 1-6-2020....59922
1737.11 (i) revised; eff. 1-6-2020....59922

7 CFR—Continued

84 FR Page

Chapter XVII—Continued

1737.60 (a) introductory text amended; (a)(1) and (2) revised; eff. 1-6-2020....59922
1737.71 (a) amended; eff. 1-6-2020....59922
1740 Removed; eff. 1-6-2020....59922
1744.21 Amended; eff. 1-6-2020....59922
1744.67 (a)(2) introductory text revised; eff. 1-6-2020....59922
1744.68 (a) introductory text and (1) revised; (d) amended; eff. 1-6-2020....59922
1744.200 (b) removed; eff. 1-6-2020....59923
1744.201 Amended; eff. 1-6-2020....59923
1751.100 Amended; eff. 1-6-2020....59923
1751.103 (a) introductory text amended; eff. 1-6-2020....59923
1755.97 Revised....28201
1755.98 Table revised....28201

2020

7 CFR

85 FR Page

Chapter XVII

1710.101 (a)(2) revised; (a)(3) added....18418
1710.106 (g) added....57081
1719 Added....18418
1719 Regulation at 85 FR 18418 confirmed....67427
1738 Revised; interim....14398
1739.3 Amended; interim....14408
1739.8 Added; interim....14408
1739.15 (l) redesignated as (m); (d) introductory text and new (m)(8) amended; new (l) added; interim....14409
1739.19 (f) added; interim....14409
1752 Added....10558

2021

7 CFR

86 FR Page

Chapter XVII

1710.109 (c) revised....36196
1710.251 (b) and (c)(11) amended; (c)(7) and (12) revised; (c)(13) added....36196
1710.252 (b) amended....36196
1710.501 (a)(3) introductory text revised....36196
1714.56 (a) and (b) revised....36196
1717.154 (c)(1) amended....36197
1717.604 (b) and (c) revised....36197

7 CFR—Continued

86 FR Page

7 CFR—Continued

86 FR Page

○

Lightning Source UK Ltd.
Milton Keynes UK
UKHW021300260123
416011UK00033B/387